Duxbury Titles of Related Interest

Carver, *Doing Data Analysis with MINITAB 12*

Elliott, *Learning SAS in the Computer Lab,* 2nd ed.

Johnson, *Applied Multivariate Methods for Data Analysis*

Kleinbaum/Kupper/Muller/Nizam, *Applied Regression Analysis and Multivariable Methods,* 3rd ed.

Kuehl, *Design of Experiments: Statistical Principles of Research Design and Analysis,* 2nd ed.

Lehmann/Zeitz, *Statistical Explorations with Microsoft Excel*

Lohr, *Sampling: Design and Analysis*

MathSoft Inc., *S-Plus 4.5 for Windows, Student Edition*

SAS Institute Inc., *JMP-In: Statistical Discovery Software*

Selvin, *Practical Biostatistical Methods*

INDEX OF APPLICATIONS (continued)

CARDIOLOGY

CARDIOVASCULAR DISEASE

CEREBROVASCULAR DISEASE

DEMOGRAPHY

(continued in the back of the book)

Fifth Edition

Fundamentals of Biostatistics

Fifth Edition

Fundamentals of Biostatistics

Bernard Rosner

Harvard University

Australia • Canada • Mexico • Singapore • Spain • United Kingdom • United States

Sponsoring Editor: *Carolyn Crockett*
Marketing Team: *Samantha Cabaluna, Beth Kronke, and Tom Ziolkowski*
Editorial Assistants: *Ann Day and Kimberly Raburn*
Production Editor: *Mary Vezilich*
Manuscript Editor: *Linda Purrington*
Permissions Editor: *Lillian Campobasso*
Production Service: *Scratchgravel Publishing Services*
Cover Design: *Laurie Albrecht*
Cover Photos: *PhotoDisc*
Interior Design: *Devenish Design*
Art Editor: *Lisa Torri*
Art Illustration: *Suffolk Graphics*
Print Buyer: *Vena Dyer*
Printing and Binding: *R. R. Donnelley/Crawfordsville*

For more information, contact:

DUXBURY
511 Forest Lodge Road
Pacific Grove, CA 93950 USA
www.duxbury.com

Printed in the United States of America

10 9 8 7 6 5 4 3

Library of Congress Cataloging-in-Publication Data
Rosner, Bernard (Bernard A.)
Fundamentals of biostatistics / Bernard Rosner.—5th ed.
p. cm.
Includes bibliographical references and indexes.
ISBN 0-534-37068-3 (alk. paper)
1. Biometry. 2. Medical statistics. I. Title.

QH323.5 .R674 2000
570'.1'5195—dc21 99-050394

This book is dedicated to my wife, Cynthia, and my children, Sarah, David, and Laura

Preface

This introductory-level biostatistics text is designed for upper-level undergraduate or graduate students interested in medicine or other health-related areas. It requires no previous background in statistics, and its mathematical level assumes only a knowledge of algebra.

Fundamentals of Biostatistics evolved from notes that I have used in a biostatistics course taught to Harvard University undergraduates and Harvard Medical School students over the past twenty-five years. I wrote this book to help motivate students to master the statistical methods that are most often used in the medical literature. From the student's viewpoint, it is important that the example material used to develop these methods is representative of what actually exists in the literature. Therefore, most of the examples and exercises in this book are based either on actual articles from the medical literature or on actual medical research problems I have encountered during my consulting experience at the Harvard Medical School.

The Approach

Most introductory statistics texts either use a completely nonmathematical, cookbook approach or develop the material in a rigorous, sophisticated mathematical framework. In this book, however, I follow an intermediate course, minimizing the amount of mathematical formulation but giving complete explanations of all the important concepts. Every new concept in this book is developed systematically through completely worked-out examples from current medical research problems. In addition, I introduce computer output where appropriate to illustrate these concepts.

I initially wrote this text for the introductory biostatistics course. However, the field has changed rapidly over the past several years; because of the increased power of newer statistical packages, we can now perform more sophisticated data analyses than ever before. Therefore, a second goal of this text is to present these new techniques *at an introductory level* so that students can become familiar with them without having to wade through specialized (and, usually, more advanced) statistical texts.

To differentiate these two goals more clearly, I included most of the content for the introductory course in the first 12 chapters. I then added a new chapter (Chapter 13), "Design and Analysis Techniques for Epidemiologic Studies." This new chapter, together with Chapter 14, "Hypothesis Testing: Person-Time Data," covers more advanced statistical techniques used in recent epidemiologic studies.

Changes in the Fifth Edition

For this edition, I have added 11 new sections and substantially revised eight other sections. Features new to this edition include the following:

- An expanded set of computer exercises based on real data sets. This edition contains 23 data sets, which are contained on the disk bound in the back of the book. For the first time you will find each data set available in the Excel-readable format, in addition to the MINITAB® and ASCII formats found in the previous edition.
- An additional case study—concerning the effects of cigarette smoking on bone density—used in several chapters throughout the book.
- New or expanded coverage of the following topics:
  - ROC curves (Section 3.7.1)
  - Use of electronic tables based on Microsoft® Excel (Section 4.8.2)
  - Covariance (Sections 5.6.1 and 11.7)
  - Expected value and variance of linear combinations of dependent random variables (Section 5.6.1)
  - Use of Excel to perform hypothesis tests and obtain confidence intervals (Chapters 6–14)
  - Sample size estimation for longitudinal studies (Section 8.11)
  - The delta method (Section 13.3)
  - Equivalence studies (Section 13.9)
  - Meta-analysis (Section 13.8)
  - Variance-covariance matrix (Section 12.5)
  - Sample size estimation for correlation coefficients (Section 11.8.4)
  - Clustered binary data (Section 13.11)
  - Measurement error methods (Section 13.12)
  - Power and sample size estimation for person-time data (Sections 14.4 and 14.6)
  - Sample size estimation for survival analysis (Section 14.12)
  - One-sample inference for incidence rate data (Section 14.2)
  - Confidence intervals for incidence rates (Section 14.2)

The new sections and the expanded sections for this edition have been indicated by an asterisk in the table of contents.

Exercises

This edition contains 1166 exercises (compared with 893 in the previous edition). All data-set-based problems are included. Problems marked by an asterisk (*) at the end of each chapter have corresponding brief solutions in the answer section at the back of the book. Based on requests from students for more completely solved problems, approximately 600 additional problems are presented in the *Study Guide to Accompany Fundamentals of Biostatistics,* 5th edition (ISBN 0–534–37120–5). The study guide presents a complete solution for each of the problems it contains. In addition, approximately 100 of these problems are included in a Miscellaneous Problems section and are randomly ordered so that they are not tied to a specific chapter in the book. This gives the student additional practice in determining what method to use in what situation. Finally, the appendix to the *Study Guide* has a brief description of the statistical commands in Excel used in this textbook.

Computation Method

The method of handling computations is similar to the fourth edition. All intermediate results are carried to full precision (10+ significant digits), even though they are presented with fewer significant digits (usually 2 or 3) in the text. Thus, intermediate results may seem inconsistent with final results in some instances; this, however, is not the case.

Organization

Fundamentals of Biostatistics, 5th edition, is organized as follows.

Chapter 1 is an *introductory* chapter that contains an outline of the development of an actual medical study with which I was involved. It provides a unique sense of the role of biostatistics in medical research.

Chapter 2 concerns *descriptive statistics* and presents all the major numeric and graphic tools used for displaying medical data. This chapter is especially important for both consumers and producers of medical literature because much actual communication of information is accomplished via descriptive material.

Chapters 3 through 5 discuss *probability*. The basic principles of probability are developed, and the most common probability distributions—such as the binomial and normal distributions—are introduced. These distributions are used extensively in later chapters of the book.

Chapters 6 through 10 cover some of the basic methods of *statistical inference.*

Chapter 6 introduces the concept of drawing random samples from populations. The difficult notion of a sampling distribution is developed and includes an introduction to the most common sampling distributions, such as the t and chi-square distributions. The basic methods of *estimation,* including an extensive discussion of confidence intervals, are also presented.

Chapters 7 and 8 contain the basic principles of *hypothesis testing.* The most elementary hypothesis tests for normally distributed data, such as the t test, are also fully discussed for one- and two-sample problems.

Chapter 9 covers the basic principles of *nonparametric statistics.* The assumptions of normality are relaxed, and distribution-free analogues are developed for the tests in Chapters 7 and 8.

Chapter 10 contains the basic concepts of *hypothesis testing* as applied to categorical data, including some of the most widely used statistical procedures, such as the chi-square test and Fisher's exact test.

Chapter 11 develops the principles of *regression analysis.* The case of simple linear regression is thoroughly covered, and extensions are provided for the multiple regression case. Important sections on goodness-of-fit of regression models are also included. Finally, rank correlation is introduced.

Chapter 12 introduces the basic principles of the *analysis of variance* (ANOVA). The one-way analysis of variance fixed and random effects models are discussed. In addition, two-way ANOVA and the analysis of covariance are covered. Finally, we discuss nonparametric approaches to one-way ANOVA.

Chapter 13 discusses methods of design and analysis for *epidemiologic studies.* The most important study designs, including the prospective study, the case–control study, the cross-sectional study, and the cross-over design are introduced. The concept of a confounding variable—that is, a variable related to both the disease and the exposure variable—is introduced, and methods for controlling for confounding, which include the Mantel-Haenszel test and multiple-logistic regression, are discussed in detail. This discussion is followed by the exploration of topics of current interest in epidemiologic data analysis, including: meta-analysis (the combination of results from more than one study); correlated binary data techniques (techniques that can be applied when replicate measures, such as data from multiple teeth from the same person, are available for an individual); measurement error methods (useful when there is substantial measurement error in the exposure data collected); and equivalence studies (whose objective it is to establish bioequivalence between two treatment modalities rather than that one treatment is superior to the other).

Chapter 14 introduces methods of analysis for person-time data. The methods covered in this chapter include those for incidence-rate data, as well as several methods of survival analysis: the Kaplan-Meier survival curve estimator, the log rank test, and the Cox proportional hazards model.

Throughout the text—particularly in Chapter 13—I discuss the elements of study designs, including the concepts of matching; cohort studies; case–control studies; retrospective studies; prospective studies; and the sensitivity, specificity, and predictive value of screening tests. These designs are presented in the context of actual samples. In addition, Chapters 7, 8, 10, 11, 13, and 14 contain specific sections on sample-size estimation for different statistical situations.

A flowchart of appropriate methods of statistical inference (see pages 776–780) is a handy reference guide to the methods developed in this book. At the end of each of Chapters 7 through 14, I refer students to this flowchart in order to give them some perspective on how the methods discussed in a given chapter fit in with all the other statistical methods introduced in this book.

In addition, I have provided an index of applications, grouped by *medical specialty,* that summarizes all the examples and problems that this book covers.

Acknowledgments

I am indebted to Debra Sheldon, the late Marie Sheehan, and Harry Taplin for their invaluable help typing the manuscript, and to Marion McPhee for helping to prepare the disk. I am also indebted to the manuscript reviewers, among them: John E. Alcaraz, San Diego State University; Stewart J. Anderson, the University of Pittsburgh; Christiana Drake, the University of California—Davis; P. D. M. Macdonald, McMaster University; Craig D. Turnbull, the University of North Carolina at Chapel Hill; Mark J. van der Laan, the University of California—Berkeley; and Dennis Wallace, the University of Kansas School of Medicine. I would also like to thank my colleagues Nancy Cook, Robert Glynn, Cathy Berkey, and Donna Spiegelman for their helpful reviews of new material in this edition.

In addition, I wish to thank Carolyn Crockett, Mary Vezilich, Anne Draus, Greg Draus, and Linda Purrington, who were instrumental in providing editorial advice and in preparing the manuscript.

I am also indebted to my colleagues at the Channing Laboratory—most notably, the late Edward Kass, Frank Speizer, Charles Hennekens, the late Frank Polk, Ira Tager, Jerome Klein, James Taylor, Stephen Zinner, Scott Weiss, Frank Sacks, Walter Willett, Alvaro Munoz, Graham Colditz, and Susan Hankinson—and to my other colleagues at the Harvard Medical School—most notably, Frederick Mosteller, Eliot Berson, Robert Ackerman, Mark Abelson, Arthur Garvey, Leo Chylack, Eugene Braunwald, and Arthur Dempster, who inspired me to write this book. I also wish to acknowledge John Hopper and Philip Landrigan for providing the data for our case studies.

Finally, I would like to acknowledge Leslie Miller, Andrea Wagner, Loren Fishman, and Frank Santopietro, without whose clinical help the current edition of this book would not have been possible.

Bernard Rosner

About the Author

Bernard Rosner is Professor of Medicine (Biostatistics) at Harvard Medical School and Professor of Biostatistics in the Harvard School of Public Health. He received a B.A. in Mathematics from Columbia University in 1967, an M.S. in Statistics from Stanford University in 1968, and a Ph.D. in Statistics from Harvard University in 1971.

He has more than 25 years of biostatistical consulting experience with other investigators at the Harvard Medical School. Special areas of interest include: cardiovascular disease, hypertension, breast cancer, and ophthalmology. Many of the examples and exercises used in the text reflect data collected from actual studies in conjunction with his consulting experience. In addition, he has developed new biostatistical methods, mainly in the areas of longitudinal data analysis, analysis of clustered data (such as data collected in families or from paired organ systems in the same person), measurement error methods, and outlier detection methods. You will see some of these methods introduced in this book at an elementary level. He was married in 1972 to his wife, Cynthia, and has three children, Sarah, David, and Laura, each of whom has contributed examples for this book.

Contents

CHAPTER 1

General Overview / 1

CHAPTER 2

Descriptive Statistics / 7

CHAPTER 3

Probability / 45

*The new sections and the expanded sections for this edition are indicated by an asterisk.

CHAPTER 4

Discrete Probability Distributions / 79

CHAPTER 5

Continuous Probability Distributions / 117

CHAPTER 6

Estimation / 157

CHAPTER 7

Hypothesis Testing: One-Sample Inference / 211

CHAPTER 8

Hypothesis Testing: Two-Sample Inference / 273

CHAPTER 9

Nonparametric Methods / 331

CHAPTER 10

Hypothesis Testing: Categorical Data / 355

CHAPTER 11

Regression and Correlation Methods / 425

CHAPTER 12

Multisample Inference / 511

CHAPTER 13

Design and Analysis Techniques for Epidemiologic Studies / 577

CHAPTER 14

Hypothesis Testing: Person-Time Data / 677

1

General Overview

Statistics is the science whereby inferences are made about specific random phenomena on the basis of relatively limited sample material. The field of statistics can be subdivided into two main areas: mathematical statistics and applied statistics. **Mathematical statistics** concerns the development of new methods of statistical inference and requires detailed knowledge of abstract mathematics for its implementation. **Applied statistics** concerns the application of the methods of mathematical statistics to specific subject areas, such as economics, psychology, and public health. **Biostatistics** is the branch of applied statistics that concerns the application of statistical methods to medical and biological problems. There is of course some overlap among these areas of statistics. For example, in some instances, given a certain biostatistical application, standard methods do not apply and must be modified. Biostatisticians are involved in the development of new methods in this circumstance.

A good way to learn about biostatistics and its role in the research process is to follow the flow of a research study from its inception at the planning stage to its completion, which usually occurs when a manuscript reporting the results of the study is published. As an example, I will describe one such study in which I previously participated.

A friend called one morning and in the course of our conversation mentioned that he had recently used a new, automated blood-pressure device of the type seen in many banks, hotels, and department stores. The machine had read his average diastolic blood pressure on several occasions as 115 mm Hg; the highest reading was 130 mm Hg. I was horrified to hear of his experience, because if these readings were true my friend might be in imminent danger of having a stroke or developing some other serious cardiovascular disease. I referred him to a clinical colleague of mine who, using a standard blood-pressure cuff, measured my friend's diastolic blood pressure as 90 mm Hg. The contrast in the readings aroused my interest, and I began to jot down the readings on the digital display every time I passed the machine at my local bank. I got the distinct impression that a large percentage of the reported readings were in the hypertensive range. Although one would expect that hypertensives would be more likely to use such a machine, I still believed that blood-pressure readings obtained with the machine might not be comparable with those obtained using standard methods of blood-pressure measurement. I spoke to Dr. B. Frank Polk about my suspicion and succeeded in interesting him in a small-scale evaluation of such machines. We decided to send a human observer who was well trained in blood-pressure measurement techniques to several of these machines. He would offer to pay subjects 50¢ for the cost of using the machine if they would agree to fill out a short questionnaire and have their blood pressure measured by both a human observer and the machine.

At this stage we had to make several important decisions, each of which would prove vital to the success of the study. The decisions were based on the following questions:

(1) How many machines should we test?

(2) How many people should we test at each machine?

(3) In what order should the measurements be taken—should the human observer or the machine be used first? Ideally, we would have preferred to avoid this problem by taking both the human and machine readings simultaneously, but this procedure was logistically impossible.

(4) What other data should we collect on the questionnaire that might influence the comparison between methods?

(5) How should the data be recorded to facilitate their computerization at a later date?

(6) How should the accuracy of the computerized data be checked?

We resolved these problems as follows:

(1) and (2) We decided to test more than one machine (four to be exact), because we were not sure if the machines were comparable in quality. However, we wanted to sample enough subjects from each machine so that we would have an accurate comparison of the standard and automated methods for each machine. We tried to predict how large a discrepancy there might be between the two methods. Using the methods of sample-size determination discussed in this book, we calculated that we would need 100 subjects at each site to have an accurate comparison.

(3) We then had to decide in what order the measurements should be taken for each person. According to some reports, one problem that occurs with repeated blood-pressure measurements is that people tense up at the initial measurement, yielding higher blood pressure than at subsequent repeated measurements. Thus, we would not always want to use the automated or manual method first, because the effect of the method would get confused with the order-of-measurement effect. A conventional technique that we used here was to **randomize** the order in which the measurements were taken, so that for any person it was equally likely that the machine or the human observer would take the first measurement. This random pattern could be implemented by flipping a coin or, more likely, by using a table of **random numbers** similar to Table 4 of the Appendix.

(4) We felt that the major extraneous factor that might influence the results would be body size, because we might have more difficulty getting accurate readings from people with fatter arms than from those with leaner arms. We also

wanted to get some idea of the type of people who use these machines. Thus, we asked questions about age, sex, and previous hypertensive history.

(5) To record the data, we developed a coding form that could be filled out on site and from which data could be easily entered on a computer terminal for subsequent analysis. Each person in the study was assigned an identification (ID) number by which the computer could uniquely identify that person. The data on the coding forms were then keyed and verified. That is, the same form was entered twice, and a comparison was made between the two records to make sure they were the same. If the records were not the same, the form was reentered.

(6) After data entry we ran some editing programs to ensure that the data were accurate. Checking each item on each form was impossible because of the large amount of data. Alternatively, we checked that the values for individual variables were within specified ranges and printed out aberrant values for manual checking. For example, we checked that all blood-pressure readings were at least 50 and no more than 300 and printed out all readings that fell outside this range.

After completing the data-collection, data-entry, and data-editing phases, we were ready to look at the results of the study. The first step in this process is to get a general feel for the data by summarizing the information in the form of several descriptive statistics. This descriptive material can be numeric or graphic. If numeric, it can be in the form of a few summary statistics, which can be presented in tabular form or, alternatively, in the form of a **frequency distribution**, which lists each value in the data and how frequently it occurs. If graphic, the data are summarized pictorially and can be presented in one or more figures. The appropriate type of descriptive material will vary with the type of distribution considered. If the distribution is **continuous**—that is, if there are essentially an infinite number of possible values, as would be the case for blood pressure—then means and standard deviations might be the appropriate descriptive statistics. However, if the distribution is **discrete**—that is, if there are only a few possible values, as would be the case for sex—then percentages of people taking on each value would be the appropriate descriptive measure. In some cases both types of descriptive statistics are used for continuous distributions by condensing the range of possible values into a few groups and giving the percentage of people that fall into each group (e.g., the percentages of people who have blood pressures between 120 and 129 mm Hg, between 130 and 139 mm Hg, etc.).

In this study we decided first to look at mean blood pressure for each method at each of the four sites. Table 1.1 summarizes this information [1].

You might notice from this table that we did not obtain meaningful data from all the 100 people interviewed at each site, because we could not obtain valid readings from the machine for many of the people. This type of missing-data problem is very common in biostatistics and should be anticipated at the planning stage when deciding on sample sizes (which was not done in this study).

Our next step in the study was to determine whether the apparent differences in blood pressure between machine and human measurements at two of the locations (C, D) were "real" in some sense or were "due to chance." This type of question falls into the area of **inferential statistics.** We realized that although there was

TABLE 1.1 **Mean blood pressures and differences between machine and human readings at four locations**

		Systolic blood pressure (mm Hg)					
		Machine		Human		Difference	
Location	*Number of people*	*Mean*	*Standard deviation*	*Mean*	*Standard deviation*	*Mean*	*Standard deviation*
A	98	142.5	21.0	142.0	18.1	0.5	11.2
B	84	134.1	22.5	133.6	23.2	0.5	12.1
C	98	147.9	20.3	133.9	18.3	14.0	11.7
D	62	135.4	16.7	128.5	19.0	6.9	13.6

Source: By permission of the American Heart Association, Inc.

a 14-mm Hg difference in mean systolic blood pressure between the two methods for the 98 people we interviewed at location C, this difference might not hold up if we interviewed 98 other people at a different time, and we wanted to have some idea as to the **error in the estimate** of 14 mm Hg. In statistical jargon this group of 98 people represents a **sample** from the **population** of all people who might use that machine. We were interested in the population and we wished to use the sample to help us learn something about the population. In particular, we wanted to know how different the **estimated** mean difference of 14 mm Hg in our sample was likely to be from the **true** mean difference in the population of all people who might use this machine. More specifically, we wanted to know if it was still possible that there was no underlying difference between the two methods and that our results were due to chance. The 14-mm Hg difference in our group of 98 people is referred to as an **estimate** of the true mean difference (d) in the population. The problem of inferring characteristics of a population from a sample is the central concern of statistical inference and is a major topic in this text. To accomplish this aim, we needed to develop a **probability model,** which would tell us how likely it is that we would obtain a 14-mm Hg difference between the two methods in a sample of 98 people if there were no real difference between the two methods over the entire population of users of the machine. If this probability were sufficiently small, then we would begin to believe that a real difference existed between the two methods. In this particular case, using a probability model based on the t distribution, we were able to conclude that this probability was less than 1 in 1000 for each of machines C and D. This probability was sufficiently small for us to conclude that there was a real difference between the automatic and manual methods of taking blood pressure for two of the four machines tested.

We used a statistical package to perform the preceding data analyses. A package is a collection of statistical programs that describe data and perform various statistical tests on the data. Currently the most widely used statistical packages include SAS, SPSS, BMDP, MINITAB, Stata, and Excel.

The final step in this study, after completing the data analysis, was to compile the results in the form of a publishable manuscript. Inevitably, because of space considerations, much of the material developed during the data-analysis phase was weeded out and only the essential items were presented for publication.

The review of this study should give you some idea of what medical research is about and what the role of biostatistics is in this process. The material in this text parallels the description of the data-analysis phase of the study described. Chapter 2 summarizes different types of descriptive statistics. Chapters 3 through 5 present some basic principles of probability and various probability models for use in later discussions of inferential statistics. Chapters 6 through 14 discuss the major topics of inferential statistics as used in biomedical practice. Issues of study design or data collection are brought up only as they relate to other topics discussed in the text.

REFERENCE

[1] Polk, B. F., Rosner, B., Feudo, R., & Vandenburgh, M. (1980). An evaluation of the Vita-Stat automatic blood pressure measuring device. *Hypertension, 2*(2), 221–227.

Descriptive Statistics

SECTION 2.1 Introduction

The first step in looking at data is to describe the data at hand in some concise way. In smaller studies this step can be accomplished by listing each data point. In general, however, this procedure is tedious or impossible and, even if it were possible, would not give an overall picture of what the data look like.

Example 2.1 **Cancer, Nutrition** Some investigators have proposed that consumption of vitamin A prevents cancer. To test this theory, a dietary questionnaire to collect data on vitamin-A consumption among 200 hospitalized cancer cases and 200 controls might be used. The controls would be matched on age and sex to the cancer cases and would be in the hospital at the same time for an unrelated disease. What should be done with these data after they are collected?

Before any formal attempt to answer this question can be made, the vitamin-A consumption among cases and controls must be described. Consider Figure 2.1. The **bar graphs** show visually that the controls have a higher vitamin-A consumption than the cases do, particularly at consumption levels exceeding the Recommended Daily Allowance (RDA).

Example 2.2 **Pulmonary Disease** Medical researchers have often suspected that passive smokers—people who themselves do not smoke but who live or work in an environment where others smoke—might have impaired pulmonary function as a result. In 1980 a research group in San Diego published results indicating that passive smokers did indeed have significantly lower pulmonary function than comparable nonsmokers who did not work in smoky environments [1]. As supporting evidence, the authors measured the carbon-monoxide (CO) concentrations in the working environments of passive smokers and of nonsmokers (where no smoking was permitted in the workplace) to see if the relative CO concentration changed over the course of the day. These results are displayed in the form of a **scatter plot** in Figure 2.2.

Figure 2.2 clearly shows that the CO concentrations in the two working environments are about the same early in the day but diverge widely in the middle of the day and then converge again after the working day is over at 7 P.M.

FIGURE 2.1 **Daily vitamin-A consumption among cancer cases and controls**

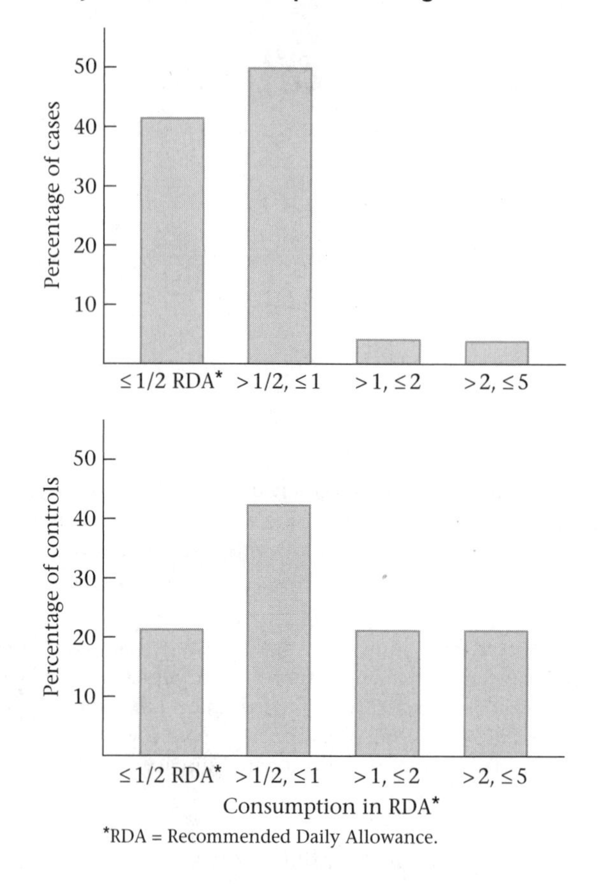

Graphic displays illustrate the important role of descriptive statistics, which is to quickly display data to give the researcher a clue as to the principal trends in the data and suggest hints as to where a more detailed look at the data, using the methods of inferential statistics, might be worthwhile. Descriptive statistics are also crucially important in conveying the final results of studies in written publications. Unless it is one of their primary interests, most readers will not have time to critically evaluate the work of others but will be influenced mainly by the descriptive statistics presented.

What makes a good graphic or numeric display? The principal guideline is that the material should be as self-contained as possible and should be understandable without reading the text. These attributes require clear labeling. The captions, units, and axes on graphs should be clearly labeled, and the statistical terms used in tables and figures should be well defined. The quantity of material presented is equally important. If bar graphs are constructed, then care must be taken that neither too many nor too few groups be displayed. The same is true of tabular material.

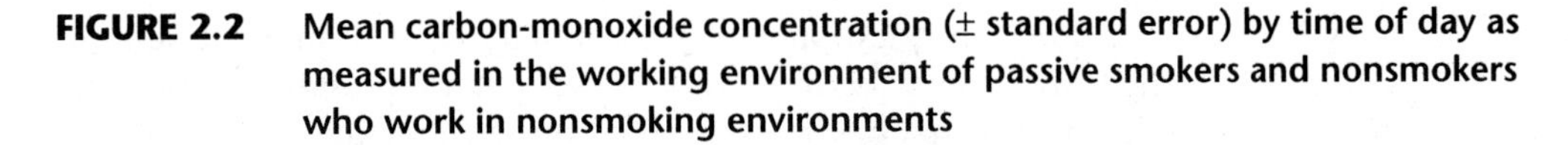

FIGURE 2.2 **Mean carbon-monoxide concentration (± standard error) by time of day as measured in the working environment of passive smokers and nonsmokers who work in nonsmoking environments**

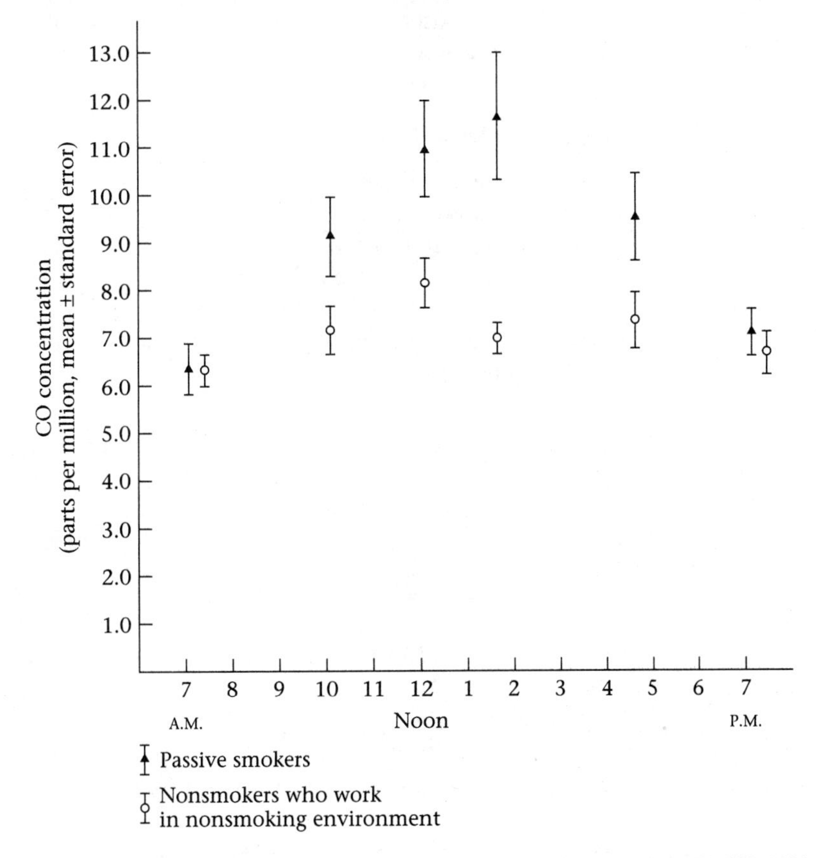

Source: Reproduced with permission of *The New England Journal of Medicine, 302*, 720–723, 1980.

Many methods are available for summarizing data in both numeric and graphic form. In this chapter the methods are summarized and their strengths and weaknesses given.

SECTION 2.2 Measures of Location

The basic problem of statistics can be stated as follows: Consider a sample of data $x_1, \ldots, x_n$, where x_1 corresponds to the first sample point and x_n corresponds to the nth sample point. Presuming that the sample is drawn from some population P, what inferences or conclusions can be made about P from the sample?

Before this question can be answered, the data must be summarized as succinctly as possible, because the number of sample points is frequently large and it is easy to lose track of the overall picture by looking at all the data at once. One type of measure useful for summarizing data defines the center, or middle, of the sample. This type of measure is a **measure of location.**

2.2.1 The Arithmetic Mean

How to define the middle of a sample may seem obvious, but the more you think about it, the less obvious it becomes. Suppose the sample consists of birthweights of all live-born infants born at a private hospital in San Diego, California, during a 1-week period. This sample is shown in Table 2.1.

One measure of location for this sample is the arithmetic mean (colloquially referred to as the average). The arithmetic mean (or mean or sample mean) is usually denoted by $\bar{x}$.

DEFINITION 2.1 The **arithmetic mean** is the sum of all the observations divided by the number of observations. It is written in statistical terms as

$$\bar{x} = \frac{1}{n}\sum_{i=1}^{n} x_i$$

The sign Σ (sigma) in Definition 2.1 is a summation sign. The expression

$$\sum_{i=1}^{n} x_i$$

is simply a short way of writing the quantity $(x_1 + x_2 + \cdots + x_n)$.

If a and b are integers, where $a \le b$, then

$$\sum_{i=a}^{b} x_i$$

means $x_a + x_{a+1} + \cdots + x_b$.

TABLE 2.1 Sample of birthweights (g) of live-born infants born at a private hospital in San Diego, California, during a 1-week period

i	x_i	i	x_i	i	x_i	i	x_i
1	3265	6	3323	11	2581	16	2759
2	3260	7	3649	12	2841	17	3248
3	3245	8	3200	13	3609	18	3314
4	3484	9	3031	14	2838	19	3101
5	4146	10	2069	15	3541	20	2834

If $a = b$, then $\sum_{i=a}^{b} x_i = x_a$. One fundamental property of summation signs is that if each term in the summation is a multiple of the same constant c, then c can be factored out from the summation; that is,

$$\sum_{i=1}^{n} cx_i = c\left(\sum_{i=1}^{n} x_i\right)$$

Example 2.3 If $x_1 = 2 \quad x_2 = 5 \quad x_3 = -4$

find $\sum_{i=1}^{3} x_i \quad \sum_{i=2}^{3} x_i \quad \sum_{i=1}^{3} x_i^2 \quad \sum_{i=1}^{3} 2x_i$

Solution

$$\sum_{i=1}^{3} x_i = 2 + 5 - 4 = 3 \qquad \sum_{i=2}^{3} x_i = 5 - 4 = 1$$

$$\sum_{i=1}^{3} x_i^2 = 4 + 25 + 16 = 45 \qquad \sum_{i=1}^{3} 2x_i = 2\sum_{i=1}^{3} x_i = 6$$

It is important to become familiar with summation signs, because they are used extensively in the remainder of the text.

Example 2.4 What is the arithmetic mean for the sample of birthweights in Table 2.1?

$$\bar{x} = (3265 + 3260 + \cdots + 2834)/20 = 3166.9 \text{ g}$$

The arithmetic mean is, in general, a very natural measure of location. One of its principal limitations, however, is that it is overly sensitive to extreme values. In this instance it may not be representative of the location of the great majority of the sample points. For example, if the first infant in Table 2.1 happened to be a premature infant weighing 500 g rather than 3265 g, then the arithmetic mean of the sample would be reduced to 3028.7 g. In this instance, 7 of the birthweights would be lower than the arithmetic mean, and 13 would be higher than the arithmetic mean. It is possible in extreme cases for all but one of the sample points to be on one side of the arithmetic mean. The arithmetic mean is a poor measure of central location in these types of samples, because it does not reflect the center of the sample. Nevertheless, the arithmetic mean is by far the most widely used measure of central location.

2.2.2 The Median

An alternative measure of location, perhaps second in popularity to the arithmetic mean, is the **median** or, more precisely, the **sample median.**

Suppose there are n observations in a sample. If these observations are ordered from smallest to largest, then the median is defined as follows:

DEFINITION 2.2 The **sample median** is

(1) The $\left(\frac{n+1}{2}\right)$th largest observation if n is odd

(2) The average of the $\left(\frac{n}{2}\right)$th and $\left(\frac{n}{2}+1\right)$th largest observations if n is even

The rationale for these definitions is to ensure an equal number of sample points on both sides of the sample median. The median is defined differently when n is even and odd because it is impossible to achieve this goal with one uniform definition. For samples with an odd sample size, there is a unique central point; for example, for samples of size 7, the fourth largest point is the central point in the sense that 3 points are smaller and 3 points are larger than it. For samples with an even sample size, there is no unique central point and the middle two values must be averaged. Thus, for samples of size 8, the fourth and fifth largest points would be averaged to obtain the median, since neither is the central point.

Example 2.5 Compute the sample median for the sample in Table 2.1.

Solution First, arrange the sample in ascending order:

2069, 2581, 2759, 2834, 2838, 2841, 3031, 3101, 3200, 3245, 3248, 3260, 3265, 3314, 3323, 3484, 3541, 3609, 3649, 4146

Because n is even,

$$\begin{aligned}\text{Sample median} &= \text{average of the 10th and 11th largest observations}\\ &= (3245 + 3248)/2 = 3246.5 \text{ g}\end{aligned}$$

Example 2.6 **Infectious Disease** Consider the data set in Table 2.2, which consists of white-blood counts taken on admission of all patients entering a small hospital in Allentown, Pennsylvania, on a given day. Compute the median white-blood count.

TABLE 2.2 **Sample of admission white-blood counts (× 1000) for all patients entering a hospital in Allentown, PA, on a given day**

i	x_i	i	x_i
1	7	6	3
2	35	7	10
3	5	8	12
4	9	9	8
5	8		

Solution First, order the sample as follows: 3, 5, 7, 8, 8, 9, 10, 12, 35. Because n is odd, the sample median is given by the fifth largest point, which equals 8.

The principal strength of the sample median is that it is insensitive to very large or very small values. In particular, if the second patient in Table 2.2 had a white count of 65,000 rather than 35,000, the sample median would remain unchanged, because the fifth largest value is still 8000. Conversely, the arithmetic mean would increase dramatically from 10,778 in the original sample to 14,111 in the new sample. The principal weakness of the sample median is that it is determined mainly by the middle points in a sample and is less sensitive to the actual numeric values of the remaining data points.

2.2.3 Comparison of the Arithmetic Mean and the Median

If a distribution is **symmetric**, then the relative position of the points on each side of the sample median will be the same. Examples of distributions expected to be roughly symmetric include the distribution of birthweights in Table 2.1 and the distribution of systolic blood-pressure measurements taken on all 30- to 39-year-old factory workers in a given workplace (Figure 2.3a).

If a distribution is **positively skewed** (skewed to the right), then points above the median will tend to be farther from the median in absolute value than points

FIGURE 2.3 Graphic displays of (a) symmetric, (b) positively skewed, and (c) negatively skewed distributions

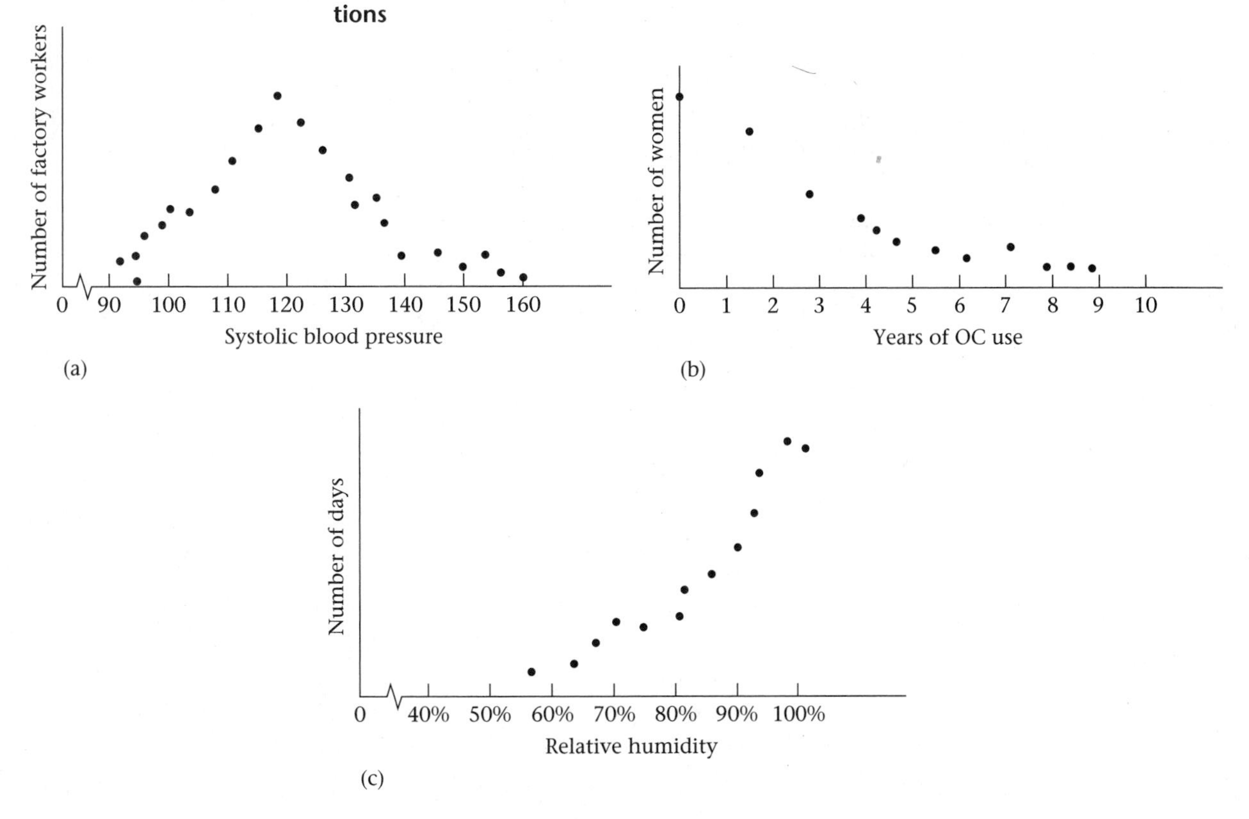

below the median. An example of such a distribution would be the distribution of the number of years of oral contraceptive (OC) use among a group of women ages 20–29 years (Figure 2.3b). Similarly, if a distribution is **negatively skewed** (skewed to the left), then points below the median will tend to be farther from the median in absolute value than points above the median. An example of such a distribution would be the distribution of relative humidities observed in a humid climate at the same time of day over a number of days. In this case, most humidities will be at or close to 100%, with a few very low humidities on dry days (Figure 2.3c).

In many samples, the relationship between the arithmetic mean and the sample median can be used to assess the symmetry of a distribution. In particular, for symmetric distributions, the arithmetic mean will be approximately the same as the median. For positively skewed distributions, the arithmetic mean will tend to be larger than the median; for negatively skewed distributions, the arithmetic mean will tend to be smaller than the median.

2.2.4 The Mode

Another widely used measure of location is the mode.

DEFINITION 2.3 The **mode** is the most frequently occurring value among all the observations in a sample.

Example 2.7 **Gynecology** Consider the sample of time intervals between successive menstrual periods for a group of 500 college women aged 18–21, as shown in Table 2.3. The frequency column gives the number of women who reported each of the respective durations. The mode is 28 because it is the most frequently occurring value.

TABLE 2.3 **Sample of time intervals between successive menstrual periods (days) of college-age women**

Value	Frequency	Value	Frequency	Value	Frequency
24	5	29	96	34	7
25	10	30	63	35	3
26	28	31	24	36	2
27	64	32	9	37	1
28	185	33	2	38	1

Example 2.8 Compute the mode of the distribution in Table 2.2.

Solution The mode is 8000, because it occurs more frequently than any other white count.

Some distributions have more than one mode. In fact, one useful method of classifying distributions is by the number of modes present. A distribution with one mode is referred to as **unimodal**; two modes, **bimodal**; three modes, **trimodal**; and so forth.

Example 2.9 Compute the mode of the distribution in Table 2.1.

Solution There is no mode, because all the values occur exactly once.

Example 2.9 illustrates a common problem with the mode: It is not a useful measure of location if there are a large number of possible values, each of which occurs infrequently. In such cases the mode will be either far from the center of the sample or, in extreme cases, will not exist, as in Example 2.9. The mode is not used in this text because its mathematical properties are, in general, rather intractable, and in most common situations it is inferior to the arithmetic mean.

2.2.5 The Geometric Mean

Many types of laboratory data, specifically data in the form of concentrations of one substance in another, as assessed by serial dilution techniques, can be expressed either as multiples of 2 or as a constant multiplied by a power of 2; that is, outcomes can only be of the form $2^k c$, $k = 0, 1, \ldots$, for some constant c. For example, the data in Table 2.4 represent the minimal inhibitory concentration (MIC) of penicillin G in the urine for *N. gonorrhoeae* in 74 patients [2]. The arithmetic mean would not be appropriate as a measure of location in this situation, because the distribution is very skewed.

TABLE 2.4 **Distribution of minimal inhibitory concentration (MIC) of penicillin G for *N. gonorrhoeae***

(μg/mL) Concentration	Frequency	(μg/mL) Concentration	Frequency
$0.03125 = 2^0(0.03125)$	21	$0.250 = 2^3(0.03125)$	19
$0.0625 = 2^1(0.03125)$	6	$0.50 = 2^4(0.03125)$	17
$0.125 = 2^2(0.03125)$	8	$1.0 = 2^5(0.03125)$	3

Source: Reproduced with permission from *JAMA, 220,* 205–208, 1972. Copyright 1972, American Medical Association.

However, the data do have a certain pattern, because the only possible values are of the form $2^k(0.03125)$ for $k = 0, 1, 2, \ldots$. One solution is to work with the distribution of the logs of the concentrations. The log concentrations have the property that successive possible concentrations differ by a constant; that is, $\log(2^{k+1}c) - \log(2^k c) = \log(2^{k+1}) + \log c - \log(2^k) - \log c = (k+1)\log 2 - k \log 2 = \log 2$. Thus, the log concentrations are equally spaced from each other, and the resulting distribution is now not as skewed as the concentrations themselves. The arithmetic mean could then be computed in the log scale; that is,

$$\overline{\log x} = \frac{1}{n}\sum_{i=1}^{n} \log x_i$$

and used as a measure of location. However, it is usually preferable to work in the original scale by taking the antilogarithm of $\overline{\log x}$ to form the geometric mean, which leads to the following definition:

DEFINITION 2.4 The **geometric mean** is the antilogarithm of $\overline{\log x}$, where

$$\overline{\log x} = \frac{1}{n}\sum_{i=1}^{n} \log x_i$$

Any base can be used to compute logarithms for the geometric mean. The geometric mean will be the same regardless of which base is used. The only requirement is that the logs and antilogs in Definition 2.4 should be in the same base. Bases often used in practice are base 10 and base e; logs and antilogs using these bases can be computed using many pocket calculators.

Example 2.10 **Infectious Disease** Compute the geometric mean for the sample in Table 2.4.

Solution (1) For convenience, use base 10 to compute the logs and antilogs in this example.

(2) Compute

$$\overline{\log x} = \left[\begin{array}{l} 21\log(0.03125) + 6\log(0.0625) + 8\log(0.125) \\ + 19\log(0.250) + 17\log(0.50) + 3\log(1.0) \end{array}\right] \Big/ 74 = -0.846$$

(3) The geometric mean = the antilogarithm of $-0.846 = 0.143$.

SECTION 2.3 Some Properties of the Arithmetic Mean

Consider a sample $x_1, \ldots, x_n$, which will be referred to as the original sample. To create a **translated sample** $x_1 + c, \ldots, x_n + c$, add a constant c to each data point. Let $y_i = x_i + c$, $i = 1, \ldots, n$. Suppose we want to compute the arithmetic mean of the translated sample. We can show that the following relationship holds:

EQUATION 2.1

If $\quad y_i = x_i + c, \quad i = 1, \ldots, n$

then $\quad \bar{y} = \bar{x} + c$

Therefore, to find the arithmetic mean of the y's, compute the arithmetic mean of the x's and add the constant c.

This principle is useful because it is sometimes convenient to change the "origin" of the sample data, that is, to compute the arithmetic mean after the translation and transform back to the original origin.

Example 2.11 In Table 2.3 it is more convenient to work with numbers that are near zero than with numbers near 28 to compute the arithmetic mean of the time interval between menstrual periods. Thus, a translated sample might first be created by subtracting 28 days from each outcome in Table 2.3. The arithmetic mean of the translated sample could then be found and 28 added to get the actual arithmetic mean. The calculations are shown in Table 2.5.

TABLE 2.5 **Translated sample for duration between successive menstrual periods in college-aged women**

Value	Frequency	Value	Frequency	Value	Frequency
–4	5	1	96	6	7
–3	10	2	63	7	3
–2	28	3	24	8	2
–1	64	4	9	9	1
0	185	5	2	10	1

Note: $\bar{y} = [(-4)(5) + (-3)(10) + \cdots + (10)(1)]/500 = 0.54$
$\bar{x} = \bar{y} + 28 = 0.54 + 28 = 28.54$ days

Similarly, systolic blood-pressure scores are usually between 100 and 200. It is easier to subtract 100 from each blood-pressure score, find the mean of the translated sample, and add 100 to obtain the mean of the original sample.

What happens to the arithmetic mean if the units or scale being worked with are changed? A **rescaled sample** can be created:

$$y_i = cx_i, \quad i = 1, \ldots, n$$

The following result holds:

EQUATION 2.2

If $y_i = cx_i,\ i = 1, \ldots, n$
then $\bar{y} = c\bar{x}$

Therefore, to find the arithmetic mean of the y's, compute the arithmetic mean of the x's and multiply it by the constant c.

Example 2.12 Express the mean birthweight for the data in Table 2.1 in ounces rather than grams.

Solution We know that 1 oz = 28.35 g and that $\bar{x}$ = 3166.9 g. Thus, if the data were expressed in terms of ounces,

$$c = \frac{1}{28.35} \quad \text{and} \quad \bar{y} = \frac{1}{28.35}(3166.9) = 111.71 \text{ oz}$$

Sometimes we want to change both the origin and the scale of the data at the same time. To do this, apply Equations 2.1 and 2.2 as follows:

EQUATION 2.3

Let $x_1, \ldots, x_n$ be the original sample of data and let $y_i = c_1x_i + c_2, i = 1, \ldots, n$, represent a transformed sample obtained by multiplying each original sample point by a factor c_1 and then shifting over by a constant c_2.

If $y_i = c_1x_i + c_2, \quad i = 1, \ldots, n$
then $\bar{y} = c_1\bar{x} + c_2$

Example 2.13 If we have a sample of temperatures in °C with an arithmetic mean of 11.75°C, then what is the arithmetic mean in °F?

Solution Let y_i denote the °F temperature that corresponds to a °C temperature of x_i. The required transformation to convert the data to °F would be

$$y_i = \frac{9}{5}x_i + 32, \quad i = 1, \ldots, n$$

so the arithmetic mean would be

$$\bar{y} = \frac{9}{5}(11.75) + 32 = 53.15°\text{F}$$

SECTION 2.4 Measures of Spread

Consider the two samples shown in Figure 2.4. They represent two samples of cholesterol measurements, each on the same person, but using different measurement techniques. They appear to have about the same center, and whatever measure of central location is used will probably be about the same in the two samples. In fact, the arithmetic means are both 200 mg/dL. However, the two samples visually appear to be radically different. This difference lies in the greater **variability,** or **spread,** of the Autoanalyzer method relative to the Microenzymatic method. In this section, the notion of variability will be quantified. Many samples can be well described by the combination of a measure of location and a measure of spread.

2.4.1 The Range

Several different measures can be used to describe the variability of a sample. Perhaps the simplest measure is the range.

FIGURE 2.4 **Two samples of cholesterol measurements on a given person using an Autoanalyzer and a Microenzymatic measurement technique**

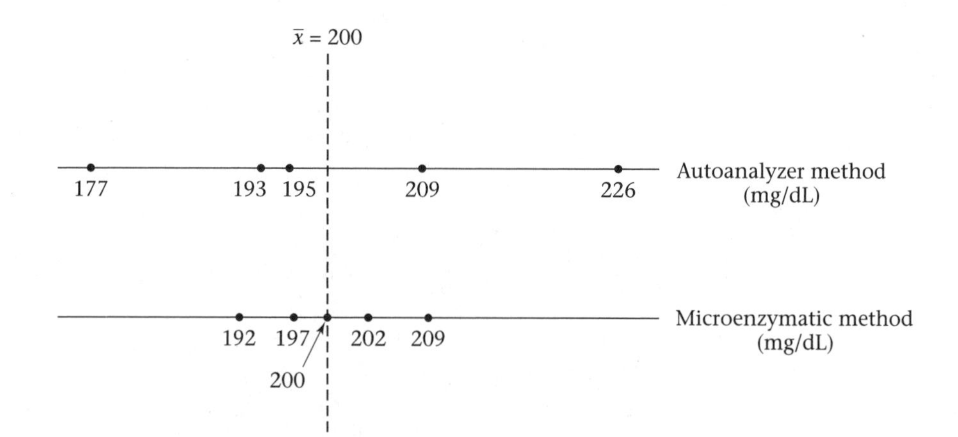

DEFINITION 2.5 The **range** is the difference between the largest and smallest observations in a sample.

Example 2.14 The range in the sample of birthweights in Table 2.1 is

$$4146 - 2069 = 2077 \text{ g}$$

Example 2.15 Compute the ranges for the Autoanalyzer- and Microenzymatic-method data in Figure 2.4, and compare the variability of the two methods.

Solution The range for the Autoanalyzer method = 226 – 177 = 49 mg/dL. The range for the Microenzymatic method = 209 – 192 = 17 mg/dL. The Autoanalyzer method clearly seems more variable.

One advantage of the range is that it is very easy to compute once the sample points are ordered. One striking disadvantage is that it is very sensitive to extreme observations. Hence, if the lightest infant in Table 2.1 weighed 500 g rather than 2069 g, then the range would increase dramatically to 4146 – 500 = 3646 g. Another disadvantage of the range is that it depends on the sample size (n). That is, the larger n is, the larger the range tends to be. This complication makes it difficult to compare ranges from different-size data sets.

2.4.2 Quantiles

Another approach that addresses some of the shortcomings of the range in quantifying the spread in a data set is the use of **quantiles** or **percentiles.** Intuitively, the pth percentile is the value V_p such that p percent of the sample points are less than or equal to V_p. The median, being the 50th percentile, is a special case of a quantile. As was the case for the median, a different definition is needed for the pth percentile, depending on whether $np/100$ is an integer or not.

DEFINITION 2.6 The pth **percentile** is defined by

(1) The $(k + 1)$th largest sample point if $np/100$ is not an integer (where k is the largest integer less than $np/100$)

(2) The average of the $(np/100)$th and $(np/100 + 1)$th largest observations if $np/100$ is an integer.

Percentiles are also sometimes referred to as **quantiles.**

The spread of a distribution can be characterized by specifying several percentiles. For example, the 10th and 90th percentiles are often used to characterize spread. Percentiles have the advantage over the range of being less sensitive to outliers and of not being greatly affected by the sample size (n).

Example 2.16 Compute the 10th and 90th percentiles for the birthweight data in Table 2.1.

Solution Because $20 \times .1 = 2$ and $20 \times .9 = 18$ are integers, the 10th and 90th percentiles are defined by

10th percentile: average of the 2nd and 3rd largest values
= (2581 + 2759)/2 = 2670 g

90th percentile: average of the 18th and 19th largest values
= (3609 + 3649)/2 = 3629 g

We would estimate that 80% of birthweights will fall between 2670 g and 3629 g, which gives us an overall impression of the spread of the distribution.

Example 2.17 Compute the 20th percentile for the white-count data in Table 2.2.

Solution Because $np/100 = 9 \times .2 = 1.8$ is not an integer, the 20th percentile is defined by the (1 + 1)th largest value = 2nd largest value = 5000.

To compute percentiles, the sample points must be ordered. This can be difficult if n is even moderately large. An easy way to accomplish this is to use a stem-and-leaf plot (see Section 2.8.3), or a computer program.

There is no limit to the number of percentiles that can be computed. The most useful percentiles are often determined by the sample size and by subject-matter considerations. Frequently used percentiles are quartiles (25th, 50th, and 75th percentiles), quintiles (20th, 40th, 60th, and 80th percentiles), and deciles (10th, 20th, . . . , 90th percentiles). It is almost always instructive to look at some of the quantiles to get an overall impression of the spread and the general shape of a distribution.

2.4.3 The Variance and Standard Deviation

The principal difference between the Autoanalyzer- and Microenzymatic-method data in Figure 2.4 is that the Microenzymatic-method values are in some sense closer to the center of the sample than the Autoanalyzer-method values are. If the center of the sample is defined as the arithmetic mean, then a measure that can summarize the difference (or deviations) between the individual sample points and the arithmetic mean is needed; that is,

$$x_1 - \bar{x}, x_2 - \bar{x}, \ldots, x_n - \bar{x}$$

One simple measure that would seem to accomplish this goal is

$$d = \frac{\sum_{i=1}^{n} (x_i - \bar{x})}{n}$$

Unfortunately, this measure will not work, because of the following principle:

EQUATION 2.4

> The sum of the deviations of the individual observations of a sample about the sample mean is always 0.

Example 2.18 Compute the sum of the deviations about the mean for the Autoanalyzer- and Microenzymatic-method data in Figure 2.4.

Solution For the Autoanalyzer-method data,

$$d = (177 - 200) + (193 - 200) + (195 - 200) + (209 - 200) + (226 - 200)$$
$$= -23 - 7 - 5 + 9 + 26 = 0$$

For the Microenzymatic-method data,

$$d = (192 - 200) + (197 - 200) + (200 - 200) + (202 - 200) + (209 - 200)$$
$$= -8 - 3 + 0 + 2 + 9 = 0$$

Thus, d does not help distinguish the difference in spreads between the two methods.

A second possible measure is

$$\sum_{i=1}^{n} | x_i - \bar{x} | / n$$

which is referred to as the mean deviation. The mean deviation is a reasonable measure of spread, but does not characterize the spread as well as the standard deviation (see Definition 2.8) if the underlying distribution is bell-shaped.

A third idea is to use the average of the squares of the deviations from the sample mean rather than the deviations themselves. The resulting measure of spread, denoted by s^2, is

$$s^2 = \frac{\sum_{i=1}^{n} (x_i - \bar{x})^2}{n}$$

The more usual form for this measure is with $n - 1$ in the denominator rather than with n. The resulting measure is called the sample variance (or variance).

DEFINITION 2.7 The **sample variance, or variance**, is defined as follows:

$$s^2 = \frac{\sum_{i=1}^{n} (x_i - \bar{x})^2}{n - 1}$$

A rationale for using $n - 1$ in the denominator rather than n is presented in the discussion of estimation in Chapter 6.

Another commonly used measure of spread is the sample standard deviation.

DEFINITION 2.8 The **sample standard deviation**, or **standard deviation**, is defined as follows:

$$s = \sqrt{\frac{\sum_{i=1}^{n} (x_i - \bar{x})^2}{n - 1}} = \sqrt{\text{sample variance}}$$

Example 2.19 Compute the variance and standard deviation for the Autoanalyzer- and Microenzymatic-method data in Figure 2.4.

Solution

Autoanalyzer Method

$$s^2 = \left[(177-200)^2 + (193-200)^2 + (195-200)^2 + (209-200)^2 + (226-200)^2\right]/4$$
$$= (529 + 49 + 25 + 81 + 676)/4 = 1360/4 = 340$$
$$s = \sqrt{340} = 18.4$$

Microenzymatic Method

$$s^2 = \left[(192-200)^2 + (197-200)^2 + (200-200)^2 + (202-200)^2 + (209-200)^2\right]/4$$
$$= (64 + 9 + 0 + 4 + 81)/4 = 158/4 = 39.5$$
$$s = \sqrt{39.5} = 6.3$$

Thus, the Autoanalyzer method has a standard deviation roughly three times as large as that of the Microenzymatic method.

SECTION 2.5 Some Properties of the Variance and Standard Deviation

The same questions can be asked of the variance and standard deviation as of the arithmetic mean: namely, how are the variance and standard deviation affected by a change in origin or a change in the units being worked with? Suppose there is a sample $x_1, \ldots, x_n$ and all data points in the sample are shifted by a constant c; that is, a new sample $y_1, \ldots, y_n$ is created such that $y_i = x_i + c$, $i = 1, \ldots, n$.

In Figure 2.5, we would clearly expect that the variance and standard deviation would remain the same, because the relationship of the points in the sample relative to one another remains the same. This property is stated as follows:

EQUATION 2.5

Suppose there are two samples

$$x_1, \ldots, x_n \quad \text{and} \quad y_1, \ldots, y_n$$

where $y_i = x_i + c, \quad i = 1, \ldots, n$

If the respective sample variances of the two samples are denoted by

$$s_x^2 \quad \text{and} \quad s_y^2$$

then $s_y^2 = s_x^2$

FIGURE 2.5 **Comparison of the variances of two samples, where one sample has an origin shifted relative to the other**

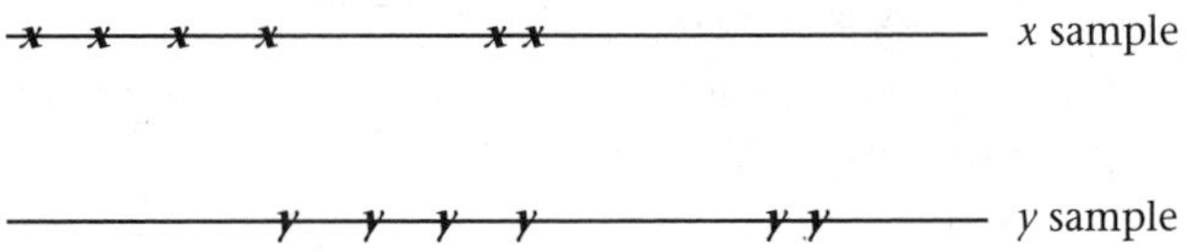

x sample

y sample

Example 2.20 Compare the variances and standard deviations for the menstrual-period data in Tables 2.3 and 2.5.

Solution The variance and standard deviation of the two samples are the same, because the second sample was obtained from the first by subtracting 28 days from each data value; that is,

$$y_i = x_i - 28$$

Suppose the units are now changed so that a new sample $y_1, \ldots, y_n$ is created such that $y_i = cx_i$, $i = 1, \ldots, n$. The following relationship holds between the variances of the two samples.

EQUATION 2.6

Suppose there are two samples

$$x_1, \ldots, x_n \quad \text{and} \quad y_1, \ldots, y_n$$

where $y_i = cx_i, \quad i = 1, \ldots, n, \quad c > 0$

Then $s_y^2 = c^2 s_x^2 \quad s_y = cs_x$

This can be shown by noting that

$$s_y^2 = \frac{\sum_{i=1}^{n}(y_i - \bar{y})^2}{n-1} = \frac{\sum_{i=1}^{n}(cx_i - c\bar{x})^2}{n-1}$$

$$= \frac{\sum_{i=1}^{n}[c(x_i - \bar{x})]^2}{n-1} = \frac{\sum_{i=1}^{n}c^2(x_i - \bar{x})^2}{n-1}$$

$$= \frac{c^2\sum_{i=1}^{n}(x_i - \bar{x})^2}{n-1} = c^2 s_x^2$$

$$s_y = \sqrt{c^2 s_x^2} = cs_x$$

Example 2.21 Compute the variance and standard deviation of the birthweight data in Table 2.1 in both grams and ounces.

Solution The original data are given in grams, so first compute the variance and standard deviation in these units.

$$s^2 = \frac{(3265 - 3166.9)^2 + \cdots + (2834 - 3166.9)^2}{19}$$

$$= 3{,}768{,}147.8/19 = 198{,}323.6 \text{ g}^2$$

$$s = 445.3 \text{ g}$$

To compute the variance and standard deviation in ounces, note that

$$1 \text{ oz} = 28.35 \text{ g} \quad \text{or} \quad y_i = \frac{1}{28.35}x_i$$

Thus $$s^2(\text{oz}) = \frac{1}{28.35^2}s^2(\text{g}) = 246.8 \text{ oz}^2$$

$$s(\text{oz}) = \frac{1}{28.35}s(\text{g}) = 15.7 \text{ oz}$$

Thus, if the sample points change in scale by a factor of c, the variance changes by a factor of c^2 and the standard deviation changes by a factor of c. This relationship is the main reason why the standard deviation is more often used than the variance as a measure of spread, because the standard deviation and the arithmetic mean are in the same units, whereas the variance and the arithmetic mean are not. Thus, as illustrated in Examples 2.12 and 2.21, both the mean and the standard deviation change by a factor of 1/28.35 in the birthweight data of Table 2.1 when the units are expressed in terms of ounces rather than grams.

The mean and standard deviation are the most widely used measures of location and spread in the literature. One of the principal reasons for this is that the normal (or bell-shaped) distribution is defined explicitly in terms of these two parameters, and this distribution has wide applicability in many biological and medical settings. The normal distribution is discussed extensively in Chapter 5.

SECTION 2.6 The Coefficient of Variation

It is useful to relate the arithmetic mean and the standard deviation to each other, because, for example, a standard deviation of 10 would mean something different conceptually if the arithmetic mean were 10 than if it were 1000. A special measure, the coefficient of variation, is often used for this purpose.

DEFINITION 2.9 The **coefficient of variation** (CV) is defined by

$$100\% \times (s/\bar{x})$$

This measure remains the same regardless of what units are used, because if the units are changed by a factor c, both the mean and standard deviation change by the factor c; the CV, which is the ratio between them, remains unchanged.

Example 2.22 Compute the coefficient of variation for the data in Table 2.1 when the birthweights are expressed in either grams or ounces.

Solution

$$CV = 100\% \times (s/\bar{x}) = 100\% \times (445.3 \text{ g}/3166.9 \text{ g}) = 14.1\%$$

If the data were expressed in ounces, then

$$CV = 100\% \times (15.7 \text{ oz}/111.71 \text{ oz}) = 14.1\%$$

The coefficient of variation is most useful in comparing the variability of several different samples, each with different arithmetic means. This is because a higher variability is usually expected when the mean increases, and the CV is a measure that accounts for this variability. Thus, if we are conducting a study where air pollution is measured at several sites and we wish to compare day-to-day variability at the different sites, we might expect a higher variability for the more highly polluted sites. A more accurate comparison could be made by comparing the CV's at different sites than by comparing the standard deviations.

TABLE 2.6 **Reproducibility of cardiovascular risk factors in children, Bogalusa Heart Study, 1978–1979**

	n	Mean	sd	*CV* (%)
Height (cm)	364	142.6	0.31	0.2
Weight (kg)	365	39.5	0.77	1.9
Triceps skin fold (mm)	362	15.2	0.51	3.4
Systolic blood pressure (mm Hg)	337	104.0	4.97	4.8
Diastolic blood pressure (mm Hg)	337	64.0	4.57	7.1
Total cholesterol (mg/dL)	395	160.4	3.44	2.1
HDL cholesterol (mg/dL)	349	56.9	5.89	10.4

The coefficient of variation is also useful for comparing the reproducibility of different variables. Consider, for example, data from the Bogalusa Heart Study, a large study of cardiovascular risk factors in children [3].

Children in the study were seen at approximately 3-year intervals. Every 3 years, a subset of the children had replicate measurements a short time apart of cardiovascular risk factors. In Table 2.6 we present reproducibility data on a selected subset of cardiovascular risk factors. We note that the coefficient of variation ranges from 0.2% for height to 10.4% for HDL cholesterol. The standard deviations reported here are within-subject standard deviations based on repeated assessments on the same child. Details on how within- and between-subject variation are computed will be covered in Chapter 12 when we discuss the random-effects analysis-of-variance model.

SECTION 2.7 Grouped Data

Sometimes the sample size is prohibitively large to display all the raw data. Also, data are frequently collected in grouped form, because the required degree of accuracy to specify a measured quantity exactly is often lacking, because of either measurement error or imprecise patient recall. For example, systolic blood-pressure measurements taken with a standard cuff are usually specified to the nearest 2 mm Hg, because assessing them with any more precision is difficult using this instrument. Thus, a stated measurement of 120 mm Hg may actually imply that the reading is some number ≥119 mm Hg and <121 mm Hg. Similarly, because dietary recall is generally not very accurate, the most precise estimate of fish consumption might take the following form: 2–3 servings per day, 1 serving per day, 5–6 servings per week, 2–4 servings per week, 1 serving per week, <1 serving per week and ≥1 serving per month, never.

Consider the data set in Table 2.7, which represents the birthweights from 100 consecutive deliveries at a Boston hospital. Suppose we wish to display these data for publication purposes. How can we do this? If the data are on a computer, then the simplest way to display the data would be to generate a frequency distribution using one of the common statistical packages.

TABLE 2.7 **Sample of birthweights (oz) from 100 consecutive deliveries**

58	118	92	108	132	32	140	138	96	161
120	86	115	118	95	83	112	128	127	124
123	134	94	67	124	155	105	100	112	141
104	132	98	146	132	93	85	94	116	113
121	68	107	122	126	88	89	108	115	85
111	121	124	104	125	102	122	137	110	101
91	122	138	99	115	104	98	89	119	109
104	115	138	105	144	87	88	103	108	109
128	106	125	108	98	133	104	122	124	110
133	115	127	135	89	121	112	135	115	64

DEFINITION 2.10

A **frequency distribution** is an ordered display of each value in a data set together with its **frequency**; that is, the number of times that value occurs in the data set. In addition, the percentage of sample points that take on a particular value is also typically given.

A frequency distribution of the sample of 100 birthweights in Table 2.7 was generated using the MINITAB package and is displayed in Table 2.8.

The MINITAB Tally program provides the frequency (COUNT), relative frequency (PERCENT), cumulative frequency (CUMCNT), and cumulative percent (CUMPCT) for each birthweight present in the sample. For any particular birthweight b, the cumulative frequency, or CUMCNT, is the number of birthweights in the sample that are less than or equal to b. The PERCENT = $100 \times$ COUNT/n, while the cumulative percent (CUMPCT) = $100 \times$ CUMCNT/n = the percentage of birthweights less than or equal to b.

If the number of unique sample values is large, then a frequency distribution may still be too detailed a summary for publication purposes. Instead, the data could be grouped into broader categories. Some general instructions for categorizing the data are provided in the following guidelines:

(1) Subdivide the data into k intervals, starting at some lower bound y_1 and ending at some upper bound y_{k+1}.

(2) The first interval is from y_1 inclusive to y_2 exclusive; the second interval is from y_2 inclusive to y_3 exclusive; . . . ; the kth and last interval is from y_k inclusive to y_{k+1} exclusive. The rationale for this representation is to make certain that the group intervals include all possible values *and* do not overlap. These errors are common in the presentation of grouped data.

(3) The group intervals are generally chosen to be equal, although the appropriateness of equal group sizes should be dictated more by subject-matter considerations. Thus, equal intervals might be appropriate for the blood-pressure or birthweight data but not for the dietary-recall data, where the nature of the data dictates unequal group sizes corresponding to how most people remember what they eat.

TABLE 2.8 Frequency distribution of birthweight data in Table 2.7 using the MINITAB program

birthwgt	COUNT	PERCENT	CUMCNT	CUMPCT
32	1	1.00	1	1.00
58	1	1.00	2	2.00
64	1	1.00	3	3.00
67	1	1.00	4	4.00
68	1	1.00	5	5.00
83	1	1.00	6	6.00
85	2	2.00	8	8.00
86	1	1.00	9	9.00
87	1	1.00	10	10.00
88	2	2.00	12	12.00
89	3	3.00	15	15.00
91	1	1.00	16	16.00
92	1	1.00	17	17.00
93	1	1.00	18	18.00
94	2	2.00	20	20.00
95	1	1.00	21	21.00
96	1	1.00	22	22.00
98	3	3.00	25	25.00
99	1	1.00	26	26.00
100	1	1.00	27	27.00
101	1	1.00	28	28.00
102	1	1.00	29	29.00
103	1	1.00	30	30.00
104	5	5.00	35	35.00
105	2	2.00	37	37.00
106	1	1.00	38	38.00
107	1	1.00	39	39.00
108	4	4.00	43	43.00
109	2	2.00	45	45.00
110	2	2.00	47	47.00
111	1	1.00	48	48.00
112	3	3.00	51	51.00
113	1	1.00	52	52.00
115	6	6.00	58	58.00
116	1	1.00	59	59.00
118	2	2.00	61	61.00
119	1	1.00	62	62.00
120	1	1.00	63	63.00
121	3	3.00	66	66.00
122	4	4.00	70	70.00
123	1	1.00	71	71.00
124	4	4.00	75	75.00
125	2	2.00	77	77.00
126	1	1.00	78	78.00
127	2	2.00	80	80.00
128	2	2.00	82	82.00
132	3	3.00	85	85.00
133	2	2.00	87	87.00
134	1	1.00	88	88.00
135	2	2.00	90	90.00
137	1	1.00	91	91.00
138	3	3.00	94	94.00
140	1	1.00	95	95.00
141	1	1.00	96	96.00
144	1	1.00	97	97.00
146	1	1.00	98	98.00
155	1	1.00	99	99.00
161	1	1.00	100	100.00
N =	100			

TABLE 2.9 **General layout of grouped data**

Group interval	Frequency
$y_1 \le x < y_2$	f_1
$y_2 \le x < y_3$	f_2
.	.
.	.
.	.
$y_i \le x < y_{i+1}$	f_i
.	.
.	.
.	.
$y_k \le x < y_{k+1}$	f_k

(4) A count is made of the number of units that fall in each interval, which is denoted by the frequency within that interval.

(5) Finally, the group intervals and their frequencies, f_i, are then displayed concisely in a table such as Table 2.9.

For example, the raw data in Table 2.7 might be displayed in grouped form as shown in Table 2.10.

TABLE 2.10 **Grouped frequency distribution of birthweight (oz) from 100 consecutive deliveries**

Group interval	Frequency
$29.5 \le x < 69.5$	5
$69.5 \le x < 89.5$	10
$89.5 \le x < 99.5$	11
$99.5 \le x < 109.5$	19
$109.5 \le x < 119.5$	17
$119.5 \le x < 129.5$	20
$129.5 \le x < 139.5$	12
$139.5 \le x < 169.5$	6
	100

SECTION 2.8 Graphic Methods

In Sections 2.1–2.7 we concentrated on methods for describing data in numeric and tabular form. In this section, these techniques are supplemented by presenting certain commonly used graphic methods for displaying data. The purpose of using graphic displays is to give a quick overall impression of the data, which is sometimes difficult to obtain with numeric measures.

2.8.1 Bar Graphs

One of the most widely used methods for displaying grouped data is the bar graph.

A **bar graph** can be constructed as follows:

(1) The data are divided in a number of groups using the guidelines provided in Section 2.7.

(2) For each group a rectangle is constructed with a base of a constant width and a height proportional to the frequency within that group.

(3) The rectangles are generally not contiguous and are equally spaced from each other.

A bar graph of daily vitamin-A consumption among 200 cancer cases and 200 age- and sex-matched controls was presented in Figure 2.1.

2.8.2 Histograms

The bar graph tends to work well with grouped data when the groups are characterized by nonnumeric attributes, such as {current smoker/ex-smoker/never smoker} or {patient gets worse/patient gets better/patient stays the same}. If the groups are characterized by a numeric attribute, such as systolic blood pressure or birthweight, then a histogram is preferable. For a histogram, the position of the rectangle will correspond to the location of the group interval along the x-axis, and the size (area) of the rectangle will correspond to the frequency within the group.

A **histogram** is constructed as follows:

(1) The data are divided into groups as described in Section 2.7.

(2) A rectangle is constructed for each group. The location of the base of the rectangle corresponds to the position of the ends of the group interval along the x-axis, and the **area** of the rectangle is proportional to the frequency within the group.

(3) The scale used along either axis should allow all the rectangles to fit into the space allotted for the graph.

Note that the area, rather than the height, is proportional to the frequency. If the length of each group interval is the same, then the area and the height are in the same proportions and the height will be proportional to the frequency as well. However, if one group interval is 5 times as long as another and the two group intervals have the same frequency, then the first group interval should have a height $\frac{1}{5}$ as high as the second group interval so that the areas will be the same. A common mistake in the literature is to construct histograms with group intervals of different lengths but with the height proportional to the frequency. This representation gives a misleading impression of the data. A histogram for the birthweight data in Table 2.10 is given in Figure 2.6.

FIGURE 2.6 Histogram for the birthweight data in Table 2.10

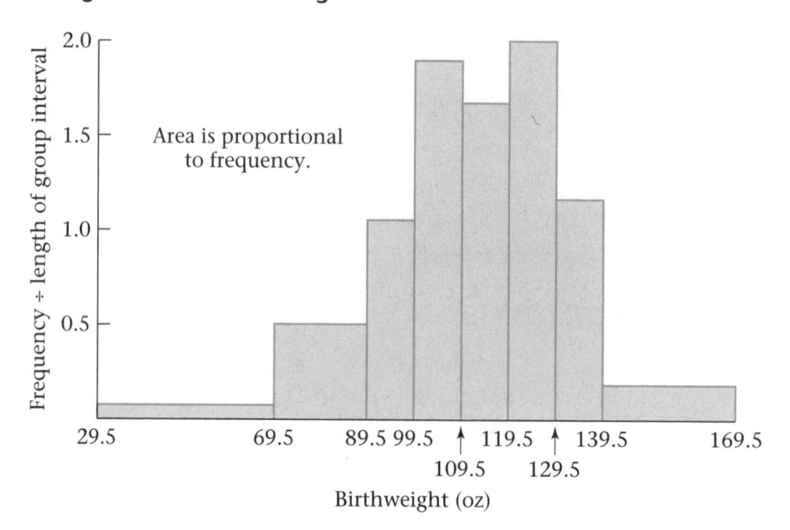

2.8.3 Stem-and-Leaf Plots

Two problems with histograms are that (1) they are somewhat difficult to construct and (2) the sense of what the actual sample points are within the respective groups is lost. One type of graphic display that overcomes these problems is the stem-and-leaf plot.

A **stem-and-leaf** plot can be constructed as follows:

(1) Separate each data point into a stem component and a leaf component, respectively, where the stem component consists of the number formed by all but the rightmost digit of the number, and the leaf component consists of the rightmost digit. Thus the stem of the number 483 is 48, and the leaf is 3.

(2) Write the smallest stem in the data set in the upper left-hand corner of the plot.

(3) Write the second stem, which equals the first stem + 1, below the first stem.

(4) Continue with step 3 until you reach the largest stem in the data set.

(5) Draw a vertical bar to the right of the column of stems.

(6) For each number in the data set, find the appropriate stem and write the leaf to the right of the vertical bar.

The collection of leaves thus formed will take on the general shape of the distribution of the sample points. Furthermore, the actual sample values are preserved and yet there is a grouped display for the data, which is a distinct advantage over a histogram. Finally, it is also easy to compute the median and other quantiles from a stem-and-leaf plot. A stem-and-leaf plot is given in Figure 2.7 using MINITAB for

FIGURE 2.7 Stem-and-leaf plot for the birthweight data (oz) in Table 2.7

```
Stem-and-leaf of birthwgt N = 100
Leaf Unit = 1.0
  1     3 | 2
  1     4 |
  2     5 | 8
  5     6 | 478
  5     7 |
 15     8 | 3556788999
 26     9 | 12344568889
 45    10 | 0123444445567888899
(17)   11 | 00122235555556889
 38    12 | 01112222344445567788
 18    13 | 222334557888
  6    14 | 0146
  2    15 | 5
  1    16 | 1
```

the birthweight data in Table 2.7. Thus, the point 5 | 8 represents 58, 11 | 8 represents 118, and so forth. Notice how this plot gives an overall feel for the distribution without losing the individual values. Also, the cumulative frequency count from either the lowest or the highest value is given in the first column. For the 11 stem, the absolute count is given in parentheses (17) instead of the cumulative total, because the highest or lowest value would exceed 50% (50).

There are variations of stem-and-leaf plots where the leaf can consist of more than one digit. This variation might be appropriate for the birthweight data in Table 2.1, because the number of three-digit stems required would be very large relative to the number of data points. In this case, the leaf would consist of the rightmost two digits and the stem the leftmost two digits, and the pairs of digits to the right of the vertical bar would be underlined to distinguish between two different leaves. The stem-and-leaf display for the data in Table 2.1 is presented in Figure 2.8.

Another common variation on the ordinary stem-and-leaf plot if the number of leaves is large is to allow more than one line for each stem. Similarly, one can position the largest stem at the top of the plot and the smallest stem at the bottom of the plot. In Figure 2.9 some graphic displays using the SAS UNIVARIATE procedure are given to illustrate this technique.

Notice that each stem is allowed two lines, with the leaves from 5 to 9 on the upper line and the leaves from 0 to 4 on the lower line. Furthermore, the leaves are ordered on each line, and a count of the number of leaves on each line is provided under the # column to allow easy computation of the median and other quantiles. Thus, the number 7 in the # column on the upper line for stem 12 indicates that there are 7 birthweights from 125 to 129 oz in the sample, whereas the number 13

FIGURE 2.8 Stem-and-leaf plot for the birthweight data (g) in Table 2.1

Stem	Leaf
20	69
21	
22	
23	
24	
25	81
26	
27	59
28	41 38 34
29	
30	31
31	01
32	65 60 45 00 48
33	23 14
34	84
35	41
36	49 09
37	
38	
39	
40	
41	46

indicates that there are 13 birthweights from 120 to 124 oz. Finally, a multiplication factor is given in the bottom of the display to allow for the representation of decimal numbers in stem-and-leaf form. In particular, if no multiplication factor (m) is present, then it is assumed that all numbers have actual value stem.leaf; whereas if m is present, then the actual value of the number is assumed to be stem.leaf $\times 10^m$. Thus, for example, as the multiplication factor is 10^1, the value 6 4 on the stem-and-leaf plot represents the number $6.4 \times 10^1 = 64$ oz.

2.8.4 Box Plots

In Section 2.2.3 the comparison of the arithmetic mean and the median was discussed as a method for looking at the skewness of a distribution. This goal can also be accomplished by a graphic technique known as the **box plot**. A box plot uses the relationships among the median, upper quartile, and lower quartile to describe the skewness of a distribution.

FIGURE 2.9 **Stem-and-leaf and box plots for the birthweight data (oz) in Table 2.7 as generated by the SAS UNIVARIATE procedure**

```
Stem | Leaf                 #        Boxplot
  16 | 1                    1           |
  15 | 5                    1           |
  15 |                                  |
  14 | 6                    1           |
  14 | 014                  3           |
  13 | 557888               6           |
  13 | 222334               6           |
  12 | 5567788              7           |
  12 | 0111222234444       13        +-----+
  11 | 5555556889          10        |     |
  11 | 0012223              7        *--+--*
  10 | 5567888899          10        |     |
  10 | 012344444            9        |     |
   9 | 568889               6        +-----+
   9 | 12344                5           |
   8 | 556788999            9           |
   8 | 3                    1           |
   7 |                                  |
   7 |                                  |
   6 | 78                   2           |
   6 | 4                    1           |
   5 | 8                    1           0
   5 |
   4 |
   4 |
   3 |
   3 | 2                    1           0
       ----+----+----+----+
Multiply Stem.Leaf by 10**+1
```

The upper and lower quartiles can be thought of conceptually as the approximate 75th and 25th percentiles of the sample, that is, the points $\frac{3}{4}$ and $\frac{1}{4}$ along the way in the ordered sample.

How can the median, upper quartile, and lower quartile be used to judge the symmetry of a distribution?

(1) If the distribution is symmetric, then the upper and lower quartiles should be approximately equally spaced from the median.

(2) If the upper quartile is farther from the median than the lower quartile, then the distribution is positively skewed.

(3) If the lower quartile is farther from the median than the upper quartile, then the distribution is negatively skewed.

These relationships are illustrated graphically in a box plot. In Figure 2.9 the top of the box corresponds to the upper quartile, whereas the bottom of the box corresponds to the lower quartile. A horizontal line is also drawn at the median value. Furthermore, in the SAS implementation of the box plot, the sample mean is indicated by a + sign distinct from the edges of the box.

Example 2.23 What can be learned about the symmetry properties of the distribution of birthweights from the box plot in Figure 2.9?

Solution In Figure 2.9, because the lower quartile is farther from the median than the upper quartile, the distribution is slightly negatively skewed. This pattern is true of many birthweight distributions.

In addition to displaying the symmetry properties of a sample, a box plot can also be used to give a feel for the spread of a sample and can help identify possible outlying values, that is, values that seem inconsistent with the rest of the points in the sample. In the context of box plots, outlying values are defined as follows:

DEFINITION 2.11 An **outlying value** is a value x such that either

(1) $x >$ upper quartile + 1.5 × (upper quartile − lower quartile) or

(2) $x <$ lower quartile − 1.5 × (upper quartile − lower quartile)

DEFINITION 2.12 An **extreme outlying value** is a value x such that either

(1) $x >$ upper quartile + 3.0 × (upper quartile − lower quartile) or

(2) $x <$ lower quartile − 3.0 × (upper quartile − lower quartile)

The box plot is then completed by

(1) Drawing a vertical bar from the upper quartile to the largest nonoutlying value in the sample

(2) Drawing a vertical bar from the lower quartile to the smallest nonoutlying value in the sample

(3) Individually identifying the outlying and extreme outlying values in the sample by 0's and *'s, respectively

Example 2.24 Using the box plot in Figure 2.9, comment on the spread of the sample in Table 2.7 and the presence of outlying values.

Solution It can be shown from Definition 2.6 that the upper and lower quartiles are 124.5 and 98.5 oz, respectively. Hence, an outlying value x must satisfy the following relations:

$$x > 124.5 + 1.5 \times (124.5 - 98.5) = 124.5 + 39.0 = 163.5$$

or $$x < 98.5 - 1.5 \times (124.5 - 98.5) = 98.5 - 39.0 = 59.5$$

Similarly, an extreme outlying value x must satisfy the following relations:

$x > 124.5 + 3.0 \times (124.5 - 98.5) = 124.5 + 78.0 = 202.5$
or $x < 98.5 - 3.0 \times (124.5 - 98.5) = 98.5 - 78.0 = 20.5$

Thus, the values 32 and 58 oz are outlying values but not extreme outlying values. These values are identified by 0's on the box plot. A vertical bar extends from 64 oz (the smallest nonoutlying value) to the lower quartile and from 161 oz (the largest nonoutlying value = the largest value in the sample) to the upper quartile. The accuracy of the two identified outlying values should probably be checked.

The methods used to identify outlying values in Definitions 2.11 and 2.12 are descriptive and unfortunately are sensitive to sample size, with more outliers detected for large sample sizes. Alternative methods for identifying outliers based on a hypothesis-testing framework are given in Chapter 8.

Many more details on stem-and-leaf plots, box plots, and other exploratory data methods are given in Tukey [4].

SECTION 2.9 Case Study 1: Effects of Lead Exposure on Neurological and Psychological Function in Children

A study of the effects of exposure to lead on the psychological and neurological well-being of children was performed [5]. The complete raw data for this study are provided in Data Set LEAD.DAT, and the documentation for this file is given in Data Set LEAD.DOC. This Data Set was provided by Dr. Philip Landrigan. All Data Sets are on the data disk.

In summary, a group of children who lived near a lead smelter in El Paso, Texas, were identified and their blood levels of lead were measured. An exposed group of 46 children were identified who had blood-lead levels $\geq 40\ \mu g/mL$ in 1972 (or for a few children in 1973). This group is defined by the variable GROUP = 2. A control group of 78 children were also identified who had blood-lead levels $< 40\ \mu g/ml$ in both 1972 and 1973. This group is defined by the variable GROUP = 1. All children lived close to the lead smelter.

Two important outcome variables that were studied were (1) the number of finger–wrist taps in the dominant hand (a measure of neurological function) and (2) the Wechsler full-scale IQ score. To explore the relationship of lead exposure to the outcome variables, we used MINITAB to obtain box plots for these two variables for children in the exposed and control groups, respectively. These are given in Figures 2.10 and 2.11, respectively. For this purpose, because the dominant hand was not identified in the data base, we used the larger of the finger–wrist tapping scores for the right and left hand as a proxy for the number of finger–wrist taps in the dominant hand.

We note that although there is considerable spread within each group, both finger–wrist tapping scores (MAXFWT) and full-scale IQ scores (IQF) seem to be slightly lower in the exposed group than the control group. We will be analyzing these data in more detail in subsequent chapters, using *t* tests, analysis of variance, and regression methods.

FIGURE 2.10 Number of finger–wrist taps in the dominant hand for the exposed and control groups, El Paso Lead Study

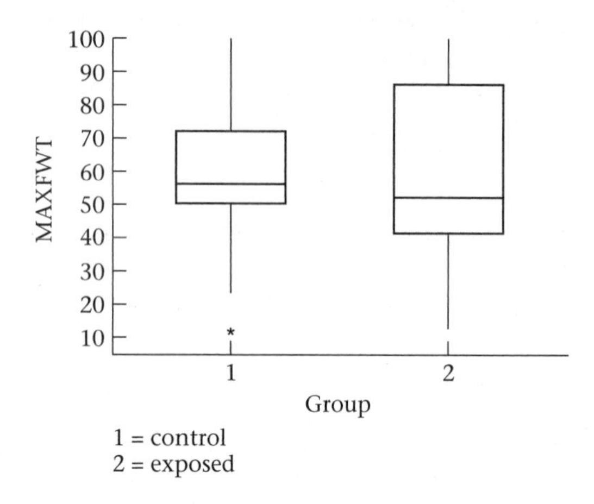

FIGURE 2.11 Wechsler full-scale IQ scores for the exposed and control groups, El Paso Lead Study

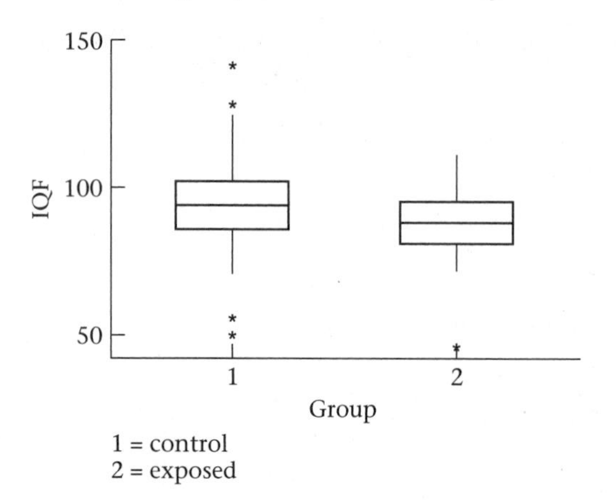

SECTION 2.10 Case Study 2: Effects of Tobacco Use on Bone-Mineral Density in Middle-Aged Women

A twin study was performed concerning the relationship between bone density and cigarette consumption [6]. Forty-one pairs of middle-aged female twins who were discordant for tobacco consumption (i.e., had different smoking histories) were enrolled in a study in Australia and invited to visit a hospital in Victoria, Australia, for a measurement of bone density. Additional information was also obtained from the participants by questionnaire, including details concerning tobacco use; alcohol, coffee, and tea consumption; intake of calcium from dairy products; menopausal, reproductive, and fracture history; use of oral contraceptives or estrogen replacement therapy; and assessment of physical activity. Dr. John Hopper provided the data set for this study, which is available on the data disk under the file name BONEDEN.DAT with documentation in BONEDEN.DOC.

Tobacco consumption was expressed in terms of *pack-years*. One pack-year is defined as 1 pack of cigarettes per day (usually about 20 cigarettes per pack) consumed for 1 year. One advantage of using twins is that genetic influences on bone density are inherently controlled for. To analyze the data, the investigators first identified the heavier- and lighter-smoking twins in terms of pack-years. The lighter-smoking twin usually had 0 pack-years (indicating that she had never smoked) or occasionally either smoked very few cigarettes per day and/or smoked for only a short time. The researchers then looked at the difference in bone-mineral density (BMD) (BMD heavier-smoking twin minus BMD lighter-smoking twin) (expressed as a percentage of the average bone density of the 2 twins) as a function of the difference in tobacco use (pack-years for the heavier-smoking twin minus pack-years for the lighter-smoking twin). BMD was assessed separately at 3 sites, the lumbar spine (lower back), the femoral neck (hip) and the femoral shaft (hip). A *scatter plot* showing the relationship between the difference in BMD versus the difference in tobacco use is given in Figure 2.12.

Note that for the lumbar spine an inverse relationship appears between the difference in BMD and the difference in tobacco use (i.e., a downward trend). Especially for twins with a large difference in tobacco use (i.e., ≥ 30 pack-years), virtually all the differences in BMD are below 0, indicating that the heavier-smoking twin had a lower BMD than the lighter-smoking twin. A similar relationship holds for the femoral neck. Results are less clear for the femoral shaft.

This is a classic example of a *matched-pair study,* which we discuss in detail beginning in Chapter 8. For such a study, the exposed (heavier-smoking twin) and control (lighter-smoking twin) are matched on other characteristics related to the outcome (BMD). In this case, the matching is based on having similar genes. We analyze this data set in more detail in subsequent chapters using methods based on the binomial distribution, *t* tests, and regression methods.

SECTION 2.11 Summary

This chapter presented several **numeric and graphic methods for describing data.** These techniques are used to

(1) quickly summarize a data set and/or

(2) present results to others.

In general, a data set can be described numerically in terms of a **measure of location** and a **measure of spread.** Several alternatives were introduced, including the **arithmetic mean, median, mode,** and **geometric mean,** as possible choices for measures of location, and the **standard deviation, quantiles,** and **range** as possible choices for measures of spread. Criteria were discussed for choosing the appropriate measures in particular circumstances. Several graphic techniques for summarizing data, including traditional methods, such as the **bar graph** and **histogram,**

FIGURE 2.12 **Within-pair differences in bone density at the lumbar spine, femoral neck, and femoral shaft as a function of within-pair differences in pack-years of tobacco use in 41 pairs of female twins. Monozygotic (identical) twins are represented by solid circles, and dizygotic (fraternal) twins by open circles. The difference in bone density between the members of a pair is expressed as the percentage of the mean bone density for the pair.**

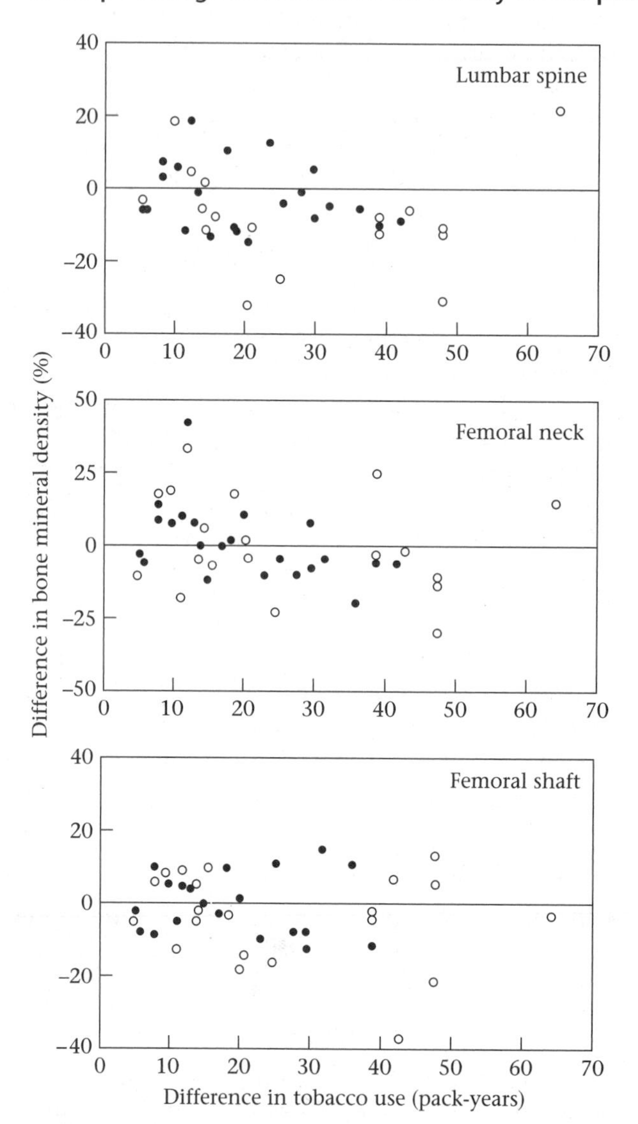

and some more modern methods characteristic of exploratory data analysis (EDA), such as the **stem-and-leaf plot** and **box plot**, were introduced.

How do the descriptive methods in this chapter fit in with the methods of statistical inference discussed later in this book? Specifically, if, based on some prespecified hypotheses, some interesting trends using descriptive methods can be found, then we need some method to judge how "significant" these trends are. For this purpose, several commonly used **probability models** are introduced in Chapters 3 through 5 and approaches for testing the validity of these models using the methods of **statistical inference** are explored in Chapters 6 through 14.

PROBLEMS

Infectious Disease

The data in Table 2.11 are a sample from a larger data set collected on persons discharged from a selected Pennsylvania hospital as part of a retrospective chart review of antibiotic usage in hospitals [7]. The data are also given in Data Set HOSPITAL.DAT with documentation in HOSPITAL.DOC on the data disk.

TABLE 2.11 Hospital-stay data

ID no.	Duration of hospital stay	Age	Sex 1 = M 2 = F	First temp. following admission	First WBC ($\times 10^3$) following admission	Received anti-biotic 1 = yes 2 = no	Received bacterial culture 1 = yes 2 = no	Service 1 = med. 2 = surg.
1	5	30	2	99.0	8	2	2	1
2	10	73	2	98.0	5	2	1	1
3	6	40	2	99.0	12	2	2	2
4	11	47	2	98.2	4	2	2	2
5	5	25	2	98.5	11	2	2	2
6	14	82	1	96.8	6	1	2	2
7	30	60	1	99.5	8	1	1	1
8	11	56	2	98.6	7	2	2	1
9	17	43	2	98.0	7	2	2	1
10	3	50	1	98.0	12	2	1	2
11	9	59	2	97.6	7	2	1	1
12	3	4	1	97.8	3	2	2	2
13	8	22	2	99.5	11	1	2	2
14	8	33	2	98.4	14	1	1	2
15	5	20	2	98.4	11	2	1	2
16	5	32	1	99.0	9	2	2	2
17	7	36	1	99.2	6	1	2	2
18	4	69	1	98.0	6	2	2	2
19	3	47	1	97.0	5	1	2	1
20	7	22	1	98.2	6	2	2	2
21	9	11	1	98.2	10	2	2	2
22	11	19	1	98.6	14	1	2	2
23	11	67	2	97.6	4	2	2	1
24	9	43	2	98.6	5	2	2	2
25	4	41	2	98.0	5	2	2	1

2.1 Compute the mean and median for the duration of hospitalization for the 25 patients.

2.2 Compute the standard deviation and range for the duration of hospitalization for the 25 patients.

2.3 It is of clinical interest to know if the duration of hospitalization is affected by whether or not a patient has received antibiotics. Answer this question descriptively using either numeric or graphic methods.

Suppose the scale for a data set is changed by multiplying each observation by a positive constant.

***2.4** What is the effect on the median?

***2.5** What is the effect on the mode?

***2.6** What is the effect on the geometric mean?

***2.7** What is the effect on the range?

Ophthalmology

Table 2.12 comes from a paper giving the distribution of astigmatism in 1033 young men, aged 18–22, who were accepted for military service in Great Britain [8]. Assume that astigmatism is rounded to the nearest 10th of a diopter and that each subject in a group has the average astigmatism within that group (e.g., for the group 0.2–0.3 diopters, the actual range is from 0.15 to 0.35 diopters, and assume that each man in the group has an astigmatism of (0.15 + 0.35)/2 = 0.25 diopters).

TABLE 2.12 Distribution of astigmatism in 1033 young men aged 18–22

Degree of astigmatism (diopters)	Frequency
0.0 or less than 0.2	458
0.2–0.3	268
0.4–0.5	151
0.6–1.0	79
1.1–2.0	44
2.1–3.0	19
3.1–4.0	9
4.1–5.0	3
5.1–6.0	2
	1033

Source: Reprinted with permission of the Editor, the authors and the Journal from the *British Medical Journal,* May 7, 1394–1398, 1960.

2.8 Compute the arithmetic mean.

2.9 Compute the standard deviation.

2.10 Plot a histogram to properly illustrate these data.

Cardiovascular Disease

The data in Table 2.13 are a sample of cholesterol levels taken from 24 hospital employees who were on a standard American diet and who agreed to adopt a vegetarian diet for 1 month. Serum-cholesterol measurements were made before adopting the diet and 1 month after.

TABLE 2.13 Serum-cholesterol levels (mg/dL) before and after adopting a vegetarian diet

Subject	Before	After	Before – after
1	195	146	49
2	145	155	–10
3	205	178	27
4	159	146	13
5	244	208	36
6	166	147	19
7	250	202	48
8	236	215	21
9	192	184	8
10	224	208	16
11	238	206	32
12	197	169	28
13	169	182	–13
14	158	127	31
15	151	149	2
16	197	178	19
17	180	161	19
18	222	187	35
19	168	176	–8
20	168	145	23
21	167	154	13
22	161	153	8
23	178	137	41
24	137	125	12

***2.11** Compute the mean change in cholesterol.

***2.12** Compute the standard deviation of the change in cholesterol levels.

2.13 Construct a stem-and-leaf plot of the cholesterol changes.

***2.14** Compute the median change in cholesterol.

2.15 Construct a box plot of the cholesterol changes to the right of the stem-and-leaf plot.

*Asterisk indicates that the answer to the problem is given in the Answer Section at the back of the book.

2.16 Comment on the symmetry of the distribution of change scores based on your answers to Problems 2.11 through 2.15.

2.17 Some investigators feel that the effects of diet on cholesterol are more evident in people with high rather than low cholesterol levels. If you split the data in Table 2.13 according to whether baseline cholesterol is above or below the median, can you comment descriptively on this issue?

Hypertension

An experiment was performed to look at the effect of position on level of blood pressure [9]. In the experiment 32 subjects had their blood pressures measured while lying down with their arms at their sides and again standing with their arms supported at heart level. The data are given in Table 2.14.

2.18 Compute the arithmetic mean and median for the difference in systolic and diastolic blood pressure, respectively, taken in different positions (i.e., recumbent minus standing).

2.19 Construct stem-and-leaf and box plots for the difference scores for each type of blood pressure.

2.20 Based on your answers to Problems 2.18 and 2.19, comment on the effect of position on the levels of systolic and diastolic blood pressure.

Pulmonary Disease

FEV (forced expiratory volume) is an index of pulmonary function that measures the volume of air expelled after 1 second of constant effort. The Data Set FEV.DAT (on the data disk) contains determinations of FEV in 1980 on 654 children ages 3–19 who were seen in the Childhood Respiratory Disease Study (CRD Study) in East Boston, Massachusetts. These data are part of a longitudinal study to follow the change in pulmonary function over time in children [10].

The data in Table 2.15 are available for each child.

TABLE 2.15 Format for FEV.DAT

Column	Variable	Format or code
1–5	ID number	
7–8	Age (years)	
10–15	FEV (liters)	X.XXX
17–20	Height (inches)	XX.X
22	Sex	0 = female/1 = male
24	Smoking status	0 = noncurrent smoker/ 1 = current smoker

TABLE 2.14 Effect of position on blood pressure

	Blood pressure (mm Hg)			
Subject	Recumbent, arm at side		Standing, arm at heart level	
B. R. A.	99[a]	71[b]	105[a]	79[b]
J. A. B.	126	74	124	76
F. L. B.	108	72	102	68
V. P. B.	122	68	114	72
M. F. B.	104	64	96	62
E. H. B.	108	60	96	56
G. C.	116	70	106	70
M. M. C.	106	74	106	76
T. J. F.	118	82	120	90
R. R. F.	92	58	88	60
C. R. F.	110	78	102	80
E. W. G.	138	80	124	76
T. F. H.	120	70	118	84
E. J. H.	142	88	136	90
H. B. H.	118	58	92	58
R. T. K.	134	76	126	68
W. E. L.	118	72	108	68
R. L. L.	126	78	114	76
H. S. M.	108	78	94	70
V. J. M.	136	86	144	88
R. H. P.	110	78	100	64
R. C. R.	120	74	106	70
J. A. R.	108	74	94	74
A. K. R.	132	92	128	88
T. H. S.	102	68	96	64
O. E. S.	118	70	102	68
R. E. S.	116	76	88	60
E. C. T.	118	80	100	84
J. H. T.	110	74	96	70
F. P. V.	122	72	118	78
P. F. W.	106	62	94	56
W. J. W.	146	90	138	94

[a]Systolic blood pressure
[b]Diastolic blood pressure
Source: Reprinted with permission of the *American Journal of Medicine.*

2.21 For each variable (other than ID), obtain appropriate descriptive statistics (both numeric and graphic).

2.22 Use both numeric and graphic measures to assess the relationship of FEV to age, height, and smoking status. (Do this separately for boys and girls.)

2.23 Compare the pattern of growth of FEV by age for boys and girls. Are there any similarities? Any differences?

scatterplots

Nutrition

The food-frequency questionnaire (FFQ) is an instrument that is often used in dietary epidemiology to assess consumption of specific foods. A person is asked to write down the number of servings per day typically eaten in the past year of over 100 individual food items. A food-composition table is then used to compute nutrient intakes (protein, fat, etc.), based on aggregating responses for individual foods. The FFQ is inexpensive to administer but is considered less accurate than the diet record (DR) (the gold standard of dietary epidemiology). For the diet record, a participant writes down the amount of each specific food eaten over the past week in a food diary and a nutritionist using a special computer program computes nutrient intakes from the food diaries. This is a much more expensive method of dietary recording. To validate the FFQ, 173 nurses participating in the Nurses' Health Study completed 4 weeks of diet recording about equally spaced over a 12-month period and an FFQ at the end of diet recording [11]. Data are presented in the Data Set VALID.DAT (on the data disk) for saturated fat, total fat, total alcohol consumption, and total caloric intake for both the DR and FFQ. For the DR, average nutrient intakes were computed over the 4 weeks of diet recording. The format of this file is shown in Table 2.16.

TABLE 2.16 Format for VALID.DAT

Column	Variable	Format or code
1–6	ID number	XXXXX.XX
8–15	Saturated fat–DR (g)	XXXXX.XX
17–24	Saturated fat–FFQ (g)	XXXXX.XX
26–33	Total fat–DR (g)	XXXXX.XX
35–42	Total fat–FFQ (g)	XXXXX.XX
44–51	Alcohol consumption–DR (oz)	XXXXX.XX
53–60	Alcohol consumption–FFQ (oz)	XXXXX.XX
62–70	Total calories–DR	XXXXXX.XX
72–80	Total calories–FFQ	XXXXXX.XX

2.24 Compute appropriate descriptive statistics for each nutrient for both DR and FFQ using both numeric and graphic measures.

2.25 Use descriptive statistics to relate nutrient intake for the DR and FFQ. Do you think that the FFQ is a reasonably accurate approximation to the DR? Why or why not?

2.26 A frequently used method for quantifying dietary intake is in the form of quintiles. Compute quintiles for each nutrient and each method of recording and relate the nutrient composition for DR and FFQ using the quintile scale. (That is, how does the quintile category based on DR relate to the quintile category based on FFQ for the same individual?) Do you get the same impression about the concordance between DR and FFQ using quintiles as in Problem 2.25, where raw (ungrouped) nutrient intake is considered?

In nutritional epidemiology, it is customary to assess nutrient intake in relationship to total caloric intake. One measure used to accomplish this is *nutrient density*, which is defined as 100% × (caloric intake of a nutrient/total caloric intake). For fat consumption, 1 gram of fat is equivalent to 9 calories.

2.27 Compute the nutrient density for total fat for DR and FFQ, and obtain appropriate descriptive statistics for this variable. How do they compare?

2.28 Relate the nutrient density for total fat for the DR versus the FFQ using the quintile approach in Problem 2.26. Is the concordance between total fat for DR and FFQ stronger, weaker, or the same when total fat is expressed in terms of nutrient density as opposed to raw nutrient?

Environmental Health, Pediatrics

In Section 2.9, we described the Data Set LEAD.DAT (on the data disk) concerning the effect of lead exposure on neurological and psychological function in children.

2.29 Compare the exposed and control groups on age and gender, using appropriate numeric and graphic descriptive measures.

2.30 Compare the exposed and control groups on verbal and performance IQ, using appropriate numeric and graphic descriptive measures.

Cardiovascular Disease

Activated-Protein-C (APC) resistance is a serum marker that has been associated with thrombosis (the formation of blood clots often leading to heart attacks) among adults. A study is performed to assess this risk factor among adolescents. To assess the reproducibility of the assay, a split-sample technique was used where each of 10 blood samples was split into two aliquots (subsamples) and each aliquot was assessed separately. The results are given in Table 2.17.

TABLE 2.17 APC resistance split-samples data

Sample number	*A*	*B*	*A* – *B*
1	2.22	1.88	0.34
2	3.42	3.59	−0.17
3	3.68	3.01	0.67
4	2.64	2.37	0.27
5	2.68	2.26	0.42
6	3.29	3.04	0.25
7	3.85	3.57	0.28
8	2.24	2.29	−0.05
9	3.25	3.39	−0.14
10	3.30	3.16	0.14

2.31 What measure can be used to assess the reproducibility of the assay? Please compute the measure for this sample and interpret what it means. (*Hint:* Consider using a standard deviation.)

2.32 Suppose the variation between split samples is thought to be a function of the mean level, where more variation is expected as the mean level increases. What measure of reproducibility can be used under these circumstances?

2.33 Compare the two measures of reproducibility in Problems 2.31 and 2.32. Which do you feel is more appropriate? (Explain.)

Microbiology

The purpose of this study was to demonstrate that soy beans inoculated with nitrogen-fixing bacteria yield more and grow adequately without the use of expensive environmentally deleterious synthesized fertilizers. The trial was conducted under controlled conditions with uniform amounts of soil. The initial hypothesis was that inoculated plants would outperform their uninoculated counterparts. This assumption is based on the fact that plants need nitrogen to manufacture vital proteins and amino acids, and that nitrogen-fixing bacteria would make more of this substance available to plants, increasing their size and yield. There were 8 inoculated plants (*I*) and 8 uninoculated plants (*U*). The plant yield as measured by pod weight for each plant is given in Table 2.18.

2.34 Compute appropriate descriptive statistics for *I* and *U* plants.

TABLE 2.18 Pod weight (gm) from inoculated (*I*) and uninoculated (U) plants

Sample number	*I*	*U*
1	1.76	0.49
2	1.45	0.85
3	1.03	1.00
4	1.53	1.54
5	2.34	1.01
6	1.96	0.75
7	1.79	2.11
8	1.21	0.92

Note: The data for this problem were supplied by David Rosner.

2.35 Use graphic methods to compare the two groups.

2.36 What is your overall impression concerning the pod weight in the two groups?

Endocrinology

In Section 2.10, we described the Data Set BONEDEN.DAT (on the data disk) concerning the effect of tobacco use on bone-mineral density (BMD).

2.37 For each pair of twins, for the lumbar spine, compute

A = BMD for the heavier-smoking twin − BMD for the lighter-smoking twin = $x_1 - x_2$

B = mean BMD for the twinship = $(x_1 + x_2)/2$

C = 100% × (A/B)

Derive appropriate descriptive statistics for *C* over the entire study population.

2.38 Suppose we group the twin pairs according to the difference in tobacco use expressed in 10 pack-year groups (0–9.9 pack-years/10–19.9 pack-years/20–29.9 pack-years/30–39.9 pack-years/40+ pack-years). Compute appropriate descriptive statistics and provide a scatter plot for *C* grouped by the difference in tobacco use in pack-years.

2.39 Do you have any impression of the relationship between BMD and tobacco use based on Problem 2.38?

2.40–2.42 Answer Problems 2.37–2.39 for BMD for the femoral neck.

2.43–2.45 Answer Problems 2.37–2.39 for BMD for the femoral shaft.

REFERENCES

[1] White, J. R., & Froeb, H. E. (1980). Small-airways dysfunction in nonsmokers chronically exposed to tobacco smoke. *New England Journal of Medicine, 302*(33), 720–723.

[2] Pedersen, A., Wiesner, P., Holmes, K., Johnson, C., & Turck, M. (1972). Spectinomycin and Penicillin G in the treatment of gonorrhea. *JAMA, 220*(2), 205–208.

[3] Foster, T. A., & Berenson, G. (1987). Measurement error and reliability in four pediatric cross-sectional surveys of cardiovascular disease risk factor variables—the Bogalusa Heart Study. *Journal of Chronic Disease, 40*(1), 13–21.

[4] Tukey, J. (1977). *Exploratory data analysis.* Reading, MA: Addison-Wesley.

[5] Landrigan, P. J., Whitworth, R. H., Baloh, R. W., Staehling, N. W., Barthel, W. F., & Rosenblum, B. F. (1975, March 29). Neuropsychological dysfunction in children with chronic low-level lead absorption. *Lancet, 1,* 708–715.

[6] Hopper, J. H., & Seeman, E. (1994). The bone density of female twins discordant for tobacco use. *New England Journal of Medicine, 330,* 387–392.

[7] Townsend, T. R., Shapiro, M., Rosner, B., & Kass, E. H. (1979). Use of antimicrobial drugs in general hospitals I. Description of population and definition of methods. *Journal of Infectious Diseases, 139*(6), 688–697.

[8] Sorsby, A., Sheridan, M., Leary, G. A., & Benjamin, B. (1960, May 7). Vision, visual acuity and ocular refraction of young men in a sample of 1033 subjects. *British Medical Journal,* 1394–1398.

[9] Kossmann, C. E. (1946). Relative importance of certain variables in clinical determination of blood pressure. *American Journal of Medicine, 1,* 464–467.

[10] Tager, I. B., Weiss, S. T., Rosner, B., & Speizer, F. E. (1979). Effect of parental cigarette smoking on pulmonary function in children. *American Journal of Epidemiology, 110,* 15–26.

[11] Willett, W. C., Sampson, L., Stampfer, M. J., Rosner, B., Bain, C., Witschi, J., Hennekens, C. H., & Speizer, F. E. (1985). Reproducibility and validity of a semi-quantitative food frequency questionnaire. *American Journal of Epidemiology, 122,* 51–65.

3 Probability

SECTION 3.1 Introduction

Chapter 2 outlined various techniques for concisely describing data. But we usually want to do more with data than just describe them. In particular, we might want to test certain specific inferences about the behavior of the data.

Example 3.1 **Cancer** One theory concerning the etiology of breast cancer states that women in a given age group who give birth to their first child relatively late in life (after 30) are at greater risk for eventually developing breast cancer over some time period t than are women who give birth to their first child early in life (before 20). Because women in the upper social classes tend to have children later, this theory has been used to explain why these women have a higher risk of developing breast cancer than women in the lower social classes. To test this hypothesis, we might identify 2000 women from a particular census tract who are currently ages 45–54 and have never had breast cancer, of whom 1000 had their first child before the age of 20 (call this group A) and 1000 after the age of 30 (group B). These 2000 women might be followed for 5 years and asked if they developed breast cancer during this period. Suppose that there are 4 new cases of breast cancer in group A and 5 new cases in group B.

Is this enough evidence to confirm a difference in risk between the two groups? Most people would feel uneasy about concluding this on the basis of such a limited amount of data.

Suppose we had a more ambitious plan and sampled 10,000 women from groups A and B, respectively, and found 40 new cases in group A and 50 new cases in group B and asked the same question. Although we might be more comfortable with the conclusion because of the larger sample size, we would still have to admit that this apparent difference in the rates could be due to chance.

The problem is that we need a conceptual framework to make these decisions but have not explicitly stated what the framework is. This framework is provided by the underlying concept of **probability.** In this chapter probability is defined and some rules for working with probabilities are introduced. Understanding probability is essential in calculating and interpreting p-values in the statistical tests of subsequent chapters. It also permits a discussion of sensitivity, specificity, and predictive values of screening tests, which are discussed in Section 3.7.

SECTION 3.2 Definition of Probability

Example 3.2 **Obstetrics** Suppose we are interested in the probability of a male live childbirth (or livebirth) among all livebirths in the United States. Conventional wisdom tells us that this probability should be close to .5. We can explore this subject by looking at some vital-statistics data, as presented in Table 3.1 [1]. The probability of a male livebirth based on 1965 data is .51247, based on 1965–1969 data .51248, and based on 1965–1974 data .51268. These are **empirical** probabilities based on a finite amount of data. In principle, the sample size could be expanded indefinitely and an increasingly more precise estimate of the probability obtained.

TABLE 3.1 **Probability of a male livebirth during the period 1965–1974**

Time period	Number of male livebirths (a)	Total number of livebirths (b)	Empirical probability of a male livebirth (a/b)
1965	1,927,054	3,760,358	.51247
1965–1969	9,219,202	17,989,361	.51248
1965–1974	17,857,857	34,832,051	.51268

This principle leads to the following definition of probability:

DEFINITION 3.1 The **sample space** is the set of all possible outcomes. In referring to probabilities of events, an **event** is any set of outcomes of interest. The **probability** of an event is the relative frequency (see p. 26) of this set of outcomes over an indefinitely large (or infinite) number of trials.

Example 3.3 **Pulmonary Disease** The tuberculin skin test is a routine screening test used to detect tuberculosis. The results of this test can be categorized as either positive, negative, or uncertain. If the probability of a positive test is .1, it means that if a large number of such tests were performed, about 10% of them would be positive. The actual percentage of positive tests will be increasingly close to .1 the more tests are performed.

Example 3.4 **Cancer** The probability of developing a new case of breast cancer in 30 years in 40-year-old women who have never had breast cancer is approximately 1/11. This probability means that over a large sample of 40-year-old women who have never had breast cancer, approximately 1 in 11 will develop the disease by age 70, with this proportion becoming increasingly close to 1 in 11 as the number of women sampled increases.

In real life, experiments cannot be performed an infinite number of times. Instead, probabilities of events are estimated from the empirical probabilities obtained from large samples (as was done in Examples 3.2–3.4). In other instances, theoretical probability models are constructed from which probabilities of many different kinds of events can be computed. An important issue in statistical inference is to compare empirical probabilities with theoretical probabilities, that is, to assess the goodness-of-fit of probability models. This topic is covered in Section 10.7.

Example 3.5 **Cancer** The probability of developing a new case of stomach cancer over a 1-year period for 45- to 49-year-old women based on Connecticut Tumor Registry data from 1963 to 1965 is 14 per 100,000 [2]. Suppose we have studied cancer rates in a small group of Connecticut nurses over this period and wish to compare how close the rates from this limited sample are to the tumor-registry figures. The value 14 per 100,000 would be the best estimate of the probability before collecting any data, and we would then see how closely our new sample data conformed with this probability.

From Definition 3.1 and from the preceding examples, we can deduce that probabilities have the following basic properties:

EQUATION 3.1

(1) The probability of an event E, denoted by $Pr(E)$, always satisfies $0 \le Pr(E) \le 1$.

(2) If outcomes A and B are two events that cannot both happen at the same time, then $Pr(A \text{ or } B \text{ occurs}) = Pr(A) + Pr(B)$.

Example 3.6 **Hypertension** Let A be the event that a person has normotensive diastolic blood-pressure (DBP) readings (DBP < 90), and let B be the event that a person has borderline DBP readings ($90 \le \text{DBP} < 95$). Suppose that $Pr(A) = .7$, $Pr(B) = .1$. Let C be the event that a person has DBP < 95. Then

$$Pr(C) = Pr(A) + Pr(B) = .8$$

because the events A and B cannot occur at the same time.

DEFINITION 3.2 Two events A and B are **mutually exclusive** if they cannot both happen at the same time.

1 event's outcome doesn't affect another event's outcome

Thus, the events A and B in Example 3.6 are mutually exclusive.

Example 3.7 **Hypertension** Let X be DBP, C be the event that $X \ge 90$, and D be the event that $75 \le X \le 100$. The events C and D are *not* mutually exclusive because they both occur when $90 \le X \le 100$.

SECTION 3.3 Some Useful Probabilistic Notation

DEFINITION 3.3 The symbol { } is used as shorthand for the phrase "the event."

DEFINITION 3.4 $A \cup B$ is the event that either A or B occurs, or they both occur.

FIGURE 3.1 **Diagrammatic representation of $A \cup B$: (a) A, B mutually exclusive; (b) A, B not mutually exclusive**

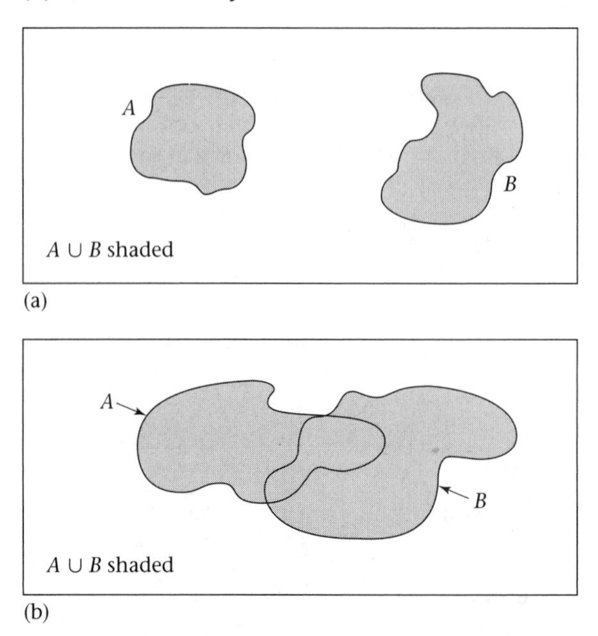

Figure 3.1 diagrammatically depicts $A \cup B$ both for the case where A and B are and are not mutually exclusive.

Example 3.8 **Hypertension** Let the events A and B be defined as in Example 3.6: $A = \{X < 90\}$, $B = \{90 \le X < 95\}$, where X = DBP. Then $A \cup B = \{X < 95\}$.

Example 3.9 **Hypertension** Let the events C and D be defined as in Example 3.7:

$$C = \{X \ge 90\} \qquad D = \{75 \le X \le 100\}$$

Then $\qquad C \cup D = \{X \ge 75\}$

DEFINITION 3.5 $A \cap B$ is the event that both A and B occur simultaneously. $A \cap B$ is depicted diagrammatically in Figure 3.2.

Example 3.10 **Hypertension** Let the events C and D be defined as in Example 3.7; that is,

$$C = \{X \ge 90\} \qquad D = \{75 \le X \le 100\}$$

Then, $\qquad C \cap D = \{90 \le X \le 100\}$

Notice that $A \cap B$ is not well defined for the events A and B in Example 3.6, because both A and B cannot occur simultaneously. This situation is true for any mutually exclusive events.

FIGURE 3.2 **Diagrammatic representation of $A \cap B$**

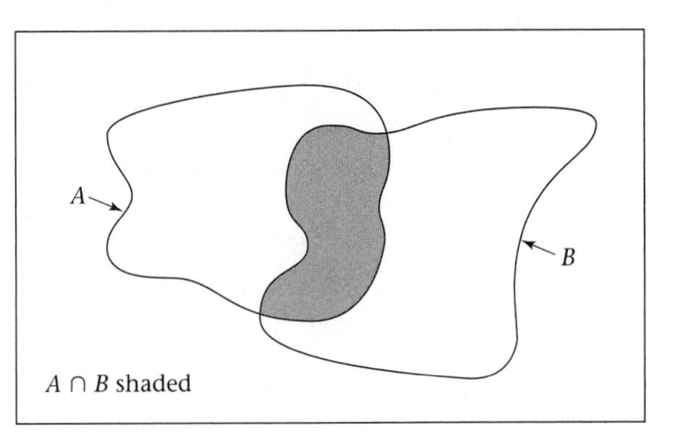

DEFINITION 3.6

$\overline{A}$ is the event that A does not occur. It is referred to as the **complement** of A. Notice that $Pr(\overline{A}) = 1 - Pr(A)$, because $\overline{A}$ occurs only when A does not occur. The event $\overline{A}$ is depicted diagrammatically in Figure 3.3.

FIGURE 3.3 **Diagrammatic representation of $\overline{A}$**

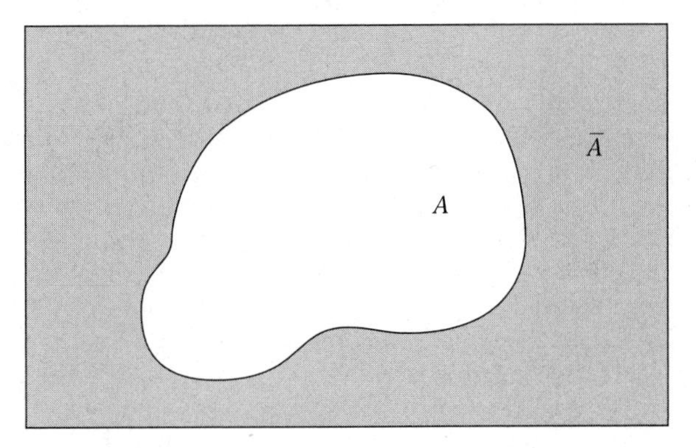

Example 3.11 **Hypertension** Let the events A and C be defined as in Examples 3.6 and 3.7; that is,

$$A = \{X < 90\} \qquad C = \{X \geq 90\}$$

Then, $C = \overline{A}$, because C can only occur when A does not occur. Notice that

$$Pr(C) = Pr(\overline{A}) = 1 - .7 = .3$$

Thus, if 70% of people have DBP < 90, then 30% of people must have DBP ≥ 90.

SECTION 3.4 The Multiplication Law of Probability

In the preceding section, events in general were described. In this section, certain specific types of events are discussed.

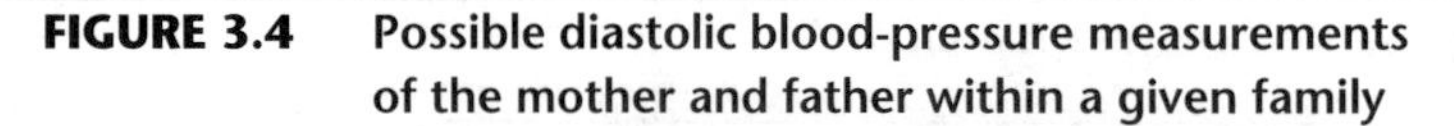

FIGURE 3.4 **Possible diastolic blood-pressure measurements of the mother and father within a given family**

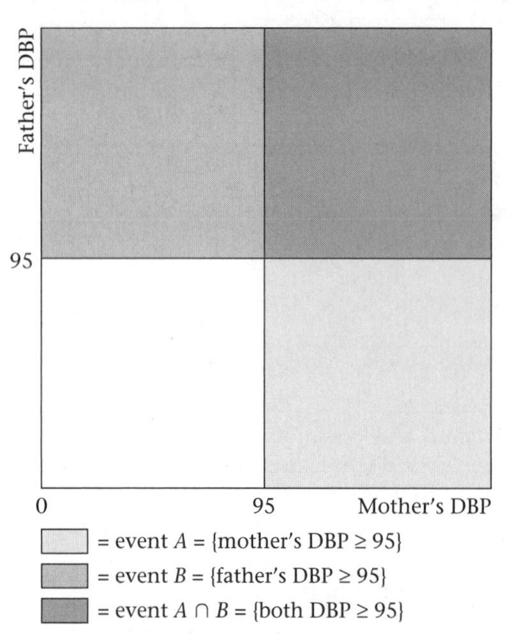

Example 3.12 **Hypertension, Genetics** Suppose we are conducting a hypertension-screening program in the home. Consider all possible pairs of DBP measurements of the mother and father within a given family, assuming that the mother and father are not genetically related. This sample space consists of all pairs of numbers of the form (X, Y) where $X > 0$, $Y > 0$. Certain specific events might be of interest in this context. In particular, we might be interested in whether the mother or father is hypertensive, which is described, respectively, by the events A = {mother's DBP ≥ 95}, B = {father's DBP ≥ 95}. These events are depicted graphically in Figure 3.4.

Suppose we know that $Pr(A) = .1$, $Pr(B) = .2$. What can we say about $Pr(A \cap B) =$ Pr(mother's DBP ≥ 95 and father's DBP ≥ 95) = Pr(both mother and father are hypertensive)? We can say nothing unless we are willing to make certain assumptions.

DEFINITION 3.7 Two events A and B are referred to as **independent events** if

$$Pr(A \cap B) = Pr(A) \times Pr(B)$$

Example 3.13 **Hypertension, Genetics** Compute the probability that both the mother and father are hypertensive if the events in Example 3.12 are independent.

Solution If A and B are independent events, then

$$Pr(A \cap B) = Pr(A) \times Pr(B) = .1(.2) = .02$$

One way to interpret this example is to assume that the hypertensive status of the mother does not depend at all on the hypertensive status of the father. Thus, if these events are independent, then in 10% of all households where the father is hypertensive the mother is also hypertensive, and in 10% of all households where the father is *not* hypertensive the mother is hypertensive. We would expect these two events to be independent if the primary determinants of elevated blood pressure were genetic. However, if the primary determinants of elevated blood pressure were, to some extent, environmental, then we would expect that the mother would be more likely to have elevated blood pressure (A true) if the father had elevated blood pressure (B true) than if the father did not have elevated blood pressure (B not true). In this latter case the events would not be independent. The implications of this lack of independence are discussed later in this chapter.

If two events are not independent, then they are said to be dependent.

DEFINITION 3.8 Two events A, B are **dependent** if

$$Pr(A \cap B) \neq Pr(A) \times Pr(B)$$

Example 3.14 is a classic example of dependent events.

Example 3.14 **Hypertension, Genetics** Consider all possible diastolic blood-pressure measurements from a mother and her first-born child. Let

$$A = \{\text{mother's DBP} \geq 95\} \qquad B = \{\text{first-born child's DBP} \geq 80\}$$

Suppose $Pr(A \cap B) = .05 \quad Pr(A) = .1 \quad Pr(B) = .2$

Then $Pr(A \cap B) = .05 > Pr(A) \times Pr(B) = .02$

and the events A, B would be dependent.

This outcome would be expected, because the mother and first-born child both share the same environment and are genetically related. In other words, the first-born child is more likely to have elevated blood pressure in households where the mother is hypertensive than in households where the mother is not hypertensive.

Example 3.15 **Sexually Transmitted Disease** Suppose two doctors, A and B, test all patients coming into a VD clinic for syphilis. Let the events $A^+ = \{\text{doctor } A \text{ makes a positive diagnosis}\}$, $B^+ = \{\text{doctor } B \text{ makes a positive diagnosis}\}$. Suppose that doctor A diagnoses 10% of all patients as positive, doctor B diagnoses 17% of all patients as positive, and both doctors diagnose 8% of all patients as positive. Are the events A^+, B^+ independent?

Solution We are given that

$$Pr(A^+) = .1 \quad Pr(B^+) = .17 \quad Pr(A^+ \cap B^+) = .08$$

Thus, $Pr(A^+ \cap B^+) = .08 > Pr(A^+) \times Pr(B^+) = .1(.17) = .017$

and the events are dependent. This result would be expected, because there should be a similarity between how two doctors diagnose patients for syphilis.

Definition 3.7 can be generalized to the case of $k(>2)$ independent events. This is often referred to as the *multiplication law of probability.*

EQUATION 3.2

Multiplication Law of Probability

If $A_1, \ldots, A_k$ are mutually independent events, then

$$Pr(A_1 \cap A_2 \cap \cdots \cap A_k) = Pr(A_1) \times Pr(A_2) \times \cdots \times Pr(A_k)$$

SECTION 3.5 The Addition Law of Probability

We have seen from the definition of probability that if A and B are mutually exclusive events, then $Pr(A \cup B) = Pr(A) + Pr(B)$. A more general formula for $Pr(A \cup B)$ can be developed when the events A and B are not necessarily mutually exclusive. This formula is referred to as the *addition law of probability* and is stated as follows:

EQUATION 3.3

Addition Law of Probability

If A and B are any events, then

$$Pr(A \cup B) = Pr(A) + Pr(B) - Pr(A \cap B)$$

This principle is depicted diagrammatically in Figure 3.5. Thus, to compute $Pr(A \cup B)$, add the probabilities of A and B separately and then subtract the overlap, which is $Pr(A \cap B)$.

FIGURE 3.5 **Diagrammatic representation of the addition law of probability**

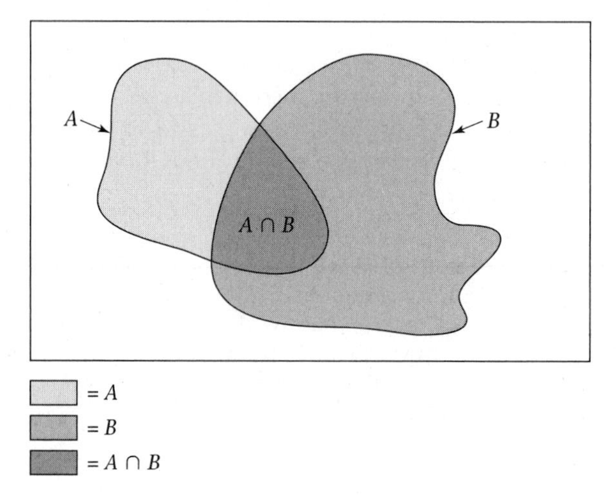

Example 3.16 **Sexually Transmitted Disease** Consider the data given in Example 3.15. Suppose a patient is referred for further lab tests if either doctor A or B makes a positive diagnosis. What is the probability that a patient will be referred for further lab tests?

Solution The event that either doctor makes a positive diagnosis can be represented by $A^+ \cup B^+$. We know that

$$Pr(A^+) = .1 \qquad Pr(B^+) = .17 \qquad Pr(A^+ \cap B^+) = .08$$

Therefore, from the addition law of probability,

$$Pr(A^+ \cup B^+) = Pr(A^+) + Pr(B^+) - Pr(A^+ \cap B^+) = .1 + .17 - .08 = .19$$

Thus, 19% of all patients will be referred for further lab tests.

There are special cases of the addition law that are of interest. First, if the events A and B are *mutually exclusive,* then $Pr(A \cap B) = 0$ and the addition law reduces to $Pr(A \cup B) = Pr(A) + Pr(B)$. This property is given in Equation 3.1 for probabilities over any two mutually exclusive events. Second, if the events A and B are *independent,* then by definition $Pr(A \cap B) = Pr(A) \times Pr(B)$ and $Pr(A \cup B)$ can be rewritten as $Pr(A) + Pr(B) - Pr(A) \times Pr(B)$. This leads to the following important special case of the addition law.

EQUATION 3.4

Addition Law of Probability for Independent Events

If two events A and B are independent, then

$$Pr(A \cup B) = Pr(A) + Pr(B) \times [1 - Pr(A)]$$

This special case of the addition law can be interpreted as follows: The event $A \cup B$ can be separated into two mutually exclusive events: {A occurs} and {B occurs and A does not occur}. Furthermore, because of the independence of A and B, the probability of the latter event can be written as $Pr(B) \times [1 - Pr(A)]$. This probability is depicted diagrammatically in Figure 3.6.

FIGURE 3.6 Diagrammatic representation of the addition law of probability for independent events

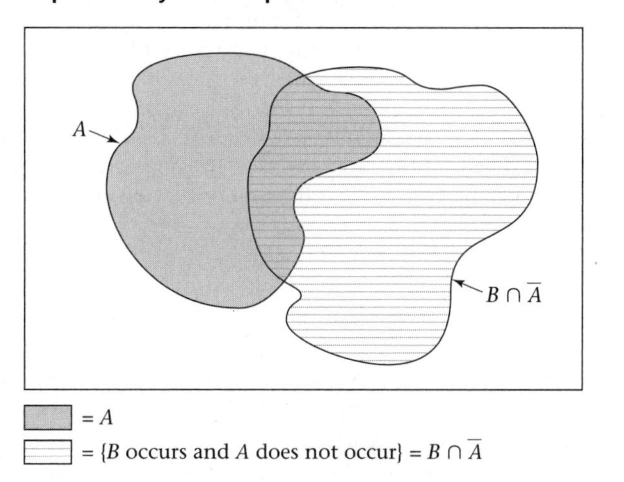

Example 3.17 **Hypertension** Refer to Example 3.12, where

$$A = \{\text{mother's DBP} \geq 95\} \quad \text{and} \quad B = \{\text{father's DBP} \geq 95\}$$

$Pr(A) = .1$, $Pr(B) = .2$, and assume that A and B are independent events. Suppose a "hypertensive household" is defined as one in which either the mother or the father is hypertensive, and hypertension is defined for the mother and father, respectively, in terms of the events A and B. What is the probability of a hypertensive household?

Solution Pr(hypertensive household) is

$$Pr(A \cup B) = Pr(A) + Pr(B) \times [1 - Pr(A)] = .1 + .2(.9) = .28$$

Thus, 28% of all households will be hypertensive.

It is possible to extend the addition law to more than two events. In particular, if there are three events A, B, and C, then

$$Pr(A \cup B \cup C) = Pr(A) + Pr(B) + Pr(C) - Pr(A \cap B) - Pr(A \cap C) - Pr(B \cap C) + Pr(A \cap B \cap C)$$

This result can be generalized to an arbitrary number of events, although this is beyond the scope of this text (see [3]).

SECTION 3.6 Conditional Probability

Suppose we want to compute the probability of several events occurring simultaneously. If the events are independent, then the multiplication law of probability can be used to accomplish this. If some of the events are dependent, then some quantitative measure of dependence is needed in order to extend the multiplication law to the case of dependent events. Consider the following example:

Example 3.18 **Cancer** It is recommended that all women over age 50 be screened for breast cancer. The definitive test for identifying breast tumors is a breast biopsy. However, this procedure is too expensive and invasive to recommend for *all* women over the age of 50. Instead, women in this age group are encouraged to have a mammogram every 1 to 2 years. Women with positive mammograms are then tested further with a biopsy. The ideal situation would be if the probability of breast cancer among women who are mammogram positive is 1, while the probability of breast cancer among women who are mammogram negative is 0. The two events {mammogram positive} and {breast cancer} would then be completely dependent; the results of the screening test would automatically determine the disease state. The opposite extreme is achieved when the events {mammogram positive} and {breast cancer} are completely independent. In this case, the probability of breast cancer is the same whether or not the mammogram is positive, and the mammogram would not be useful in screening for breast cancer and should not be used.

These concepts can be quantified in the following way. Let A = {mammogram$^+$}, B = {breast cancer} and suppose that we are interested in the probability of breast cancer (B) given that the mammogram is positive (A). This probability can be written as $Pr(A \cap B)/Pr(A)$.

DEFINITION 3.9

The quantity $Pr(A \cap B)/Pr(A)$ is defined as the **conditional probability of B given A,** which is written as $Pr(B|A)$.

However, from Section 3.4 we know that, by definition, if two events are independent, then $Pr(A \cap B) = Pr(A) \times Pr(B)$. If both sides are divided by $Pr(A)$, then $Pr(B) = Pr(A \cap B)/Pr(A) = Pr(B|A)$. Similarly, we can show that if A and B are independent events, then $Pr(B|\overline{A}) = Pr(B|A) = Pr(B)$. This relationship leads to the following alternative interpretation of independence in terms of conditional probabilities.

EQUATION 3.5

(1) If A and B are independent events, then $Pr(B|A) = Pr(B) = Pr(B|\overline{A})$.
(2) If two events A, B are dependent, then $Pr(B|A) \neq Pr(B) \neq Pr(B|\overline{A})$ and $Pr(A \cap B) \neq Pr(A) \times Pr(B)$.

DEFINITION 3.10

The **relative risk (RR)** of B given A is

$$Pr(B|A)/Pr(B|\overline{A})$$

Notice that if two events A, B are independent, then the relative risk will be 1. If two events A, B are dependent, then the relative risk will be different from 1. Heuristically, the more the dependence between events increases, the further the relative risk is from 1.

Example 3.19 **Cancer** Suppose that among 100,000 women with negative mammograms 20 will have breast cancer diagnosed within 2 years, or $Pr(B|\overline{A}) = 20/10^5 = .0002$, whereas 1 woman in 10 with positive mammograms will have breast cancer diagnosed within 2 years, or $Pr(B|A) = .1$. The two events A and B would be highly dependent, because

$$RR = Pr(B|A)/Pr(B|\overline{A}) = .1/.0002 = 500$$

In other words, women with positive mammograms are 500 times as likely to develop breast cancer over the next 2 years than women with negative mammograms. This is the rationale for using the mammogram as a screening test for breast cancer. If the events A and B were independent, then the relative risk would be 1; women with positive or negative mammograms would be equally likely to have breast cancer and the mammogram would not be useful as a screening test for breast cancer.

Example 3.20 **Sexually Transmitted Disease** Using the data in Example 3.15, find the conditional probability that doctor B makes a positive diagnosis of syphilis given that doctor A makes a positive diagnosis. What is the conditional probability that doctor B makes a positive diagnosis of syphilis given that doctor A makes a negative diagnosis? What is the relative risk of B^+ given A^+?

Solution

$$Pr(B^+|A^+) = Pr(B^+ \cap A^+)/Pr(A^+) = .08/.1 = .8$$

Thus doctor B will confirm doctor A's positive diagnoses 80% of the time. Similarly,

$$Pr(B^+|A^-) = Pr(B^+ \cap A^-)/Pr(A^-) = Pr(B^+ \cap A^-)/.9$$

We must compute $Pr(B^+ \cap A^-)$. We know that if doctor B diagnoses a patient as positive, then doctor A either does or does not confirm the diagnosis. Thus

$$Pr(B^+) = Pr(B^+ \cap A^+) + Pr(B^+ \cap A^-)$$

because the events $B^+ \cap A^+$ and $B^+ \cap A^-$ are mutually exclusive. If we subtract $Pr(B^+ \cap A^+)$ from both sides of the equation, then

$$Pr(B^+ \cap A^-) = Pr(B^+) - Pr(B^+ \cap A^+) = .17 - .08 = .09$$

Therefore, $Pr(B^+|A^-) = .09/.9 = .1$

Thus, when doctor A diagnoses a patient as negative, doctor B will contradict the diagnosis 10% of the time. The relative risk of the event B^+ given A^+ is

$$Pr(B^+|A^+)/Pr(B^+|A^-) = .8/.1 = 8$$

This indicates that doctor B is 8 times as likely to diagnose a patient as positive when doctor A diagnoses the patient as positive than when doctor A diagnoses the patient as negative. These results quantify the dependence between the two doctors' diagnoses.

The conditional $(Pr(B|A), Pr(B|\overline{A}))$ and unconditional $(Pr(B))$ probabilities mentioned previously are related in the following way:

EQUATION 3.6

For any events A and B,

$$Pr(B) = Pr(B|A) \times Pr(A) + Pr(B|\overline{A}) \times Pr(\overline{A})$$

This formula tells us that the unconditional probability of B is the sum of the conditional probability of B given A *times* the unconditional probability of A *plus* the conditional probability of B given A *not* occurring *times* the unconditional probability of A *not* occurring.

To derive this, we note that if the event B occurs, it must occur either with A or without A. Therefore,

$$Pr(B) = Pr(B \cap A) + Pr(B \cap \overline{A})$$

From the definition of conditional probability, we see that

$$Pr(B \cap A) = Pr(A) \times Pr(B|A)$$

and

$$Pr(B \cap \overline{A}) = Pr(\overline{A}) \times Pr(B|\overline{A})$$

By substitution, it follows that

$$Pr(B) = Pr(B|A)Pr(A) + Pr(B|\overline{A})Pr(\overline{A})$$

Example 3.21 **Cancer** Let A and B be defined as in Example 3.19 and suppose that 7% of the general population of women will have a positive mammogram. What is the probability of developing breast cancer over the next 2 years among women in the general population?

Solution

$$\begin{aligned} Pr(B) &= Pr(\text{breast cancer}) \\ &= Pr(\text{breast cancer}|\text{mammogram}^+) \times Pr(\text{mammogram}^+) \\ &\quad + Pr(\text{breast cancer}|\text{mammogram}^-) \times Pr(\text{mammogram}^-) \\ &= .1(.07) + .0002(.93) = .00719 = 719/10^5 \end{aligned}$$

Thus, the unconditional probability of developing breast cancer over the next 2 years in the general population ($719/10^5$) is a weighted average of the conditional probability of developing breast cancer over the next 2 years among women with a positive mammogram (.1) and the conditional probability of developing breast cancer over the next 2 years among women with a negative mammogram ($20/10^5$).

In Equation 3.6 the probability of the event B is expressed in terms of two mutually exclusive events A and $\bar{A}$. In many instances the probability of an event B will need to be expressed in terms of more than two mutually exclusive events, denoted by $A_1, A_2, \ldots, A_k$.

DEFINITION 3.11 A set of events $A_1, \ldots, A_k$ is **exhaustive** if at least one of the events must occur.

Assume that the events $A_1, \ldots, A_k$ are mutually exclusive and exhaustive; that is, at least one of the events $A_1, \ldots, A_k$ must occur and no two events can occur simultaneously. Thus, exactly one of the events $A_1, \ldots, A_k$ must occur.

EQUATION 3.7 **Total Probability Rule**

Let $A_1, \ldots, A_k$ be mutually exclusive and exhaustive events. The unconditional probability of B ($Pr(B)$) can then be written as a weighted average of the conditional probabilities of B given $A_i\left(Pr(B|A_i)\right)$ as follows:

$$Pr(B) = \sum_{i=1}^{k} Pr(B|A_i) \times Pr(A_i)$$

To show this, we note that if B occurs, then it must occur together with one and only one of the events, $A_1, \ldots, A_k$. Therefore,

$$Pr(B) = \sum_{i=1}^{k} Pr(B \cap A_i)$$

Also, from the definition of conditional probability,

$$Pr(B \cap A_i) = Pr(A_i) \times Pr(B|A_i)$$

By substitution, we obtain Equation 3.7.

An application of the total probability rule is given in the following example:

Example 3.22 **Ophthalmology** We are planning a 5-year study of cataract in a population of 5000 people 60 years of age and older. We know from census data that 45% of this population are ages 60–64, 28% are ages 65–69, 20% are ages 70–74, and 7% are age 75 or older. We also know from the Framingham Eye Study that 2.4%, 4.6%, 8.8%, and 15.3% of the people in these respective age groups will develop cataract over the next 5 years [4]. What percentage of our population will develop cataract over the next 5 years, and how many people with cataract does this percentage represent?

Solution Let A_1 = {ages 60–64}, A_2 = {ages 65–69}, A_3 = {ages 70–74}, A_4 = {age 75+}. These events are mutually exclusive and exhaustive, because exactly one event must occur for each person in our population. Furthermore, from the conditions of the problem we know that $Pr(A_1) = .45$, $Pr(A_2) = .28$, $Pr(A_3) = .20$, $Pr(A_4) = .07$, $Pr(B|A_1) = .024$, $Pr(B|A_2) = .046$, $Pr(B|A_3) = .088$, and $Pr(B|A_4) = .153$, where B = {develop cataract in the next 5 years}. Finally, using the total probability rule,

$$\begin{aligned} Pr(B) &= Pr(B|A_1) \times Pr(A_1) + Pr(B|A_2) \times Pr(A_2) \\ &\quad + Pr(B|A_3) \times Pr(A_3) + Pr(B|A_4) \times Pr(A_4) \\ &= .024(.45) + .046(.28) + .088(.20) + .153(.07) = .052 \end{aligned}$$

Thus 5.2% of our population will develop cataract over the next 5 years, which represents a total of $5000 \times .052 = 260$ people with cataract.

The definition of conditional probability allows the multiplication law of probability to be extended to the case of dependent events.

EQUATION 3.8

Generalized Multiplication Law of Probability

If $A_1, \ldots, A_k$ are an arbitrary set of events, then

$$\begin{aligned} &Pr(A_1 \cap A_2 \cap \cdots \cap A_k) \\ &\quad = Pr(A_1) \times Pr(A_2|A_1) \times Pr(A_3|A_2 \cap A_1) \times \cdots \times Pr(A_k|A_{k-1} \cap \cdots \cap A_2 \cap A_1) \end{aligned}$$

If the events are independent, then the conditional probabilities on the right-hand side of Equation 3.8 reduce to unconditional probabilities, and the generalized multiplication law reduces to the multiplication law for independent events given in Equation 3.2. Equation 3.8 also generalizes the relationship $Pr(A \cap B) = Pr(A) \times Pr(B|A)$ given in Definition 3.9 for two events to the case of more than two events.

SECTION 3.7 Bayes' Rule and Screening Tests

The mammography test given in Example 3.18 illustrates the general concept of the predictive value of a screening test, which can be defined as follows:

DEFINITION 3.12 The **predictive value positive** (PV^+) of a screening test is the probability that a person has a disease given that the test is positive

$$Pr(\text{disease}|\text{test}^+)$$

The **predictive value negative** (PV^-) of a screening test is the probability that a person does *not* have a disease given that the test is negative

$Pr(\text{no disease}|\text{test}^-)$

Example 3.23 **Cancer** Find the predictive value positive and the predictive value negative for the mammogram given the data in Example 3.19.

Solution We see that $PV^+ = Pr(\text{breast cancer}|\text{mammogram}^+) = .1$

whereas
$$\begin{aligned} PV^- &= Pr(\text{no breast cancer}|\text{mammogram}^-) \\ &= 1 - Pr(\text{breast cancer}|\text{mammogram}^-) = 1 - .0002 = .9998 \end{aligned}$$

Thus, if the mammogram is negative, the woman is virtually certain not to develop breast cancer over the next 2 years ($PV^- \approx 1$); whereas if the mammogram is positive, the woman has only a 10% chance of developing breast cancer ($PV^+ = .10$).

A symptom or a set of symptoms can also be regarded as a screening test for disease. The higher the predictive value of the screening test or symptoms, the more valuable the test. Ideally, we would like to find a set of symptoms such that both PV^+ and PV^- are 1. Then we would be able to accurately diagnose disease for each patient.

Clinicians often cannot directly measure the predictive value of a set of symptoms. However, they can measure how often specific symptoms occur in diseased and normal people. These measures are defined as follows:

DEFINITION 3.13 The **sensitivity** of a symptom (or set of symptoms or screening test) is the probability that the symptom is present given that the person has a disease.

DEFINITION 3.14 The **specificity** of a symptom (or set of symptoms or screening test) is the probability that the symptom is *not* present given that the person does *not* have a disease.

DEFINITION 3.15 A **false negative** is defined as a person who tests out as negative but who is actually positive. A **false positive** is defined as a person who tests out as positive but who is actually negative.

For a symptom to be effective in predicting disease, it is important that both the sensitivity and specificity be high.

Example 3.24 **Cancer** Suppose that the disease is lung cancer and the symptom is cigarette smoking. If we assume that 90% of people with lung cancer and 30% of people without lung cancer (essentially the entire general population) are smokers, then the sensitivity and specificity are .9 and .7, respectively. Obviously, cigarette smoking cannot be used by itself as a diagnostic tool for predicting lung cancer, because there will be too many false positives (normal people who are smokers).

Example 3.25 **Cancer** Suppose that the disease is breast cancer in women and the symptom is having a family history of breast cancer (either a mother or a sister with breast cancer). If we assume that 5% of women with breast cancer have a family history of breast cancer whereas only 2% of women without breast cancer have such a history, then the sensitivity is .05 and the specificity is .98 = (1 – .02). A family history of breast cancer cannot be used by itself to diagnose breast cancer because there will be too many false negatives (women without a family history but who have the disease).

How can the sensitivity and specificity of a symptom (or set of symptoms), which are quantities a physician can estimate, be used to compute predictive values, which are quantities a physician needs to make appropriate diagnoses?

Let A = symptom and B = disease. From Definitions 3.12, 3.13, and 3.14, we have

$$\text{Predictive value positive} = PV^+ = Pr(B|A)$$

$$\text{Predictive value negative} = PV^- = Pr(\overline{B}|\overline{A})$$

$$\text{Sensitivity} = Pr(A|B)$$

$$\text{Specificity} = Pr(\overline{A}|\overline{B})$$

Let $Pr(B)$ = probability of disease in the reference population. We wish to compute $Pr(B|A)$ and $Pr(\overline{B}|\overline{A})$ in terms of the other quantities. This relationship is known as Bayes' rule.

EQUATION 3.9

Bayes' Rule

Let A = symptom and B = disease.

$$PV^+ = Pr(B|A) = \frac{Pr(A|B) \times Pr(B)}{Pr(A|B) \times Pr(B) + Pr(A|\overline{B}) \times Pr(\overline{B})}$$

In words, this can be written as

$$PV^+ = \frac{\text{sensitivity} \times x}{\text{sensitivity} \times x + (1 - \text{specificity}) \times (1 - x)}$$

where $x = Pr(B)$ = prevalence of disease in the reference population. Similarly,

$$PV^- = \frac{\text{specificity} \times (1 - x)}{\text{specificity} \times (1 - x) + (1 - \text{sensitivity}) \times x}$$

To derive this, we have, from the definition of conditional probability,

$$PV^+ = Pr(B|A) = \frac{Pr(B \cap A)}{Pr(A)}$$

Also, from the definition of conditional probability,

$$Pr(B \cap A) = Pr(A|B) \times Pr(B)$$

Finally, from the total probability rule,

$$Pr(A) = Pr(A|B) \times Pr(B) + Pr(A|\overline{B}) \times Pr(\overline{B})$$

If the expressions for $Pr(B \cap A)$ and $Pr(A)$ are substituted into the equation for PV^+, we obtain

$$PV^+ = Pr(B|A) = \frac{Pr(A|B) \times Pr(B)}{Pr(A|B) \times Pr(B) + Pr(A|\overline{B}) \times Pr(\overline{B})}$$

That is, PV^+ can be expressed as a function of sensitivity, specificity, and probability of disease in the reference population. A similar derivation can be used to obtain PV^-.

Example 3.26 **Hypertension** Suppose that 84% of hypertensives and 23% of normotensives are classified as hypertensive by an automated blood-pressure machine. What are the predictive value positive and predictive value negative of the machine, assuming that 20% of the adult population is hypertensive?

Solution The sensitivity = .84 and specificity = 1 − .23 = .77. Thus from Bayes' rule it follows that

$$\begin{aligned} PV^+ &= (.84)(.2)/[(.84)(.2) + (.23)(.8)] \\ &= .168/.352 = .48 \end{aligned}$$

Similarly, $$\begin{aligned} PV^- &= (.77)(.8)/[(.77)(.8) + (.16)(.2)] \\ &= .616/.648 = .95 \end{aligned}$$

Thus a negative result from the machine is very predictive, because we are 95% sure that a person with a negative result based on the machine is normotensive. However, a positive result is not very predictive, because we are only 48% sure that a person with a positive result based on the machine is hypertensive.

Example 3.26 considered only two possible disease states: hypertensive and normotensive. In clinical medicine there are often more than two possible disease states. We would like to be able to predict the most likely disease state given a specific symptom (or set of symptoms). We will assume that the probability of having these symptoms among people in a given disease state is known from clinical experience, as is the probability of each of the disease states in the reference population. This leads us to the generalized Bayes' rule:

EQUATION 3.10

Generalized Bayes' Rule

Let $B_1, B_2, \ldots, B_k$ be a set of mutually exclusive and exhaustive disease states; that is, at least one disease state must occur and no two disease states can occur at the same time. Let A represent the presence of a symptom or set of symptoms. Then

$$Pr(B_i|A) = Pr(A|B_i) \times Pr(B_i) \Big/ \left[\sum_{j=1}^{k} Pr(A|B_j) \times Pr(B_j)\right]$$

This result is obtained in a similar manner to that of Bayes' rule for two disease states in Equation 3.9. Specifically, from the definition of conditional probability, note that

$$Pr(B_i|A) = \frac{Pr(B_i \cap A)}{Pr(A)}$$

Also, from the definition of conditional probability,

$$Pr(B_i \cap A) = Pr(A|B_i) \times Pr(B_i)$$

From the total probability rule,

$$Pr(A) = Pr(A|B_1) \times Pr(B_1) + \cdots + Pr(A|B_k) \times Pr(B_k)$$

If the expressions for $Pr(B_i \cap A)$ and $Pr(A)$ are substituted, we obtain

$$Pr(B_i|A) = \frac{Pr(A|B_i) \times Pr(B_i)}{\sum_{j=1}^{k} Pr(A|B_j) \times Pr(B_j)}$$

Example 3.27 **Pulmonary Disease** Suppose that a 60-year-old man who has never smoked cigarettes presents to a physician with symptoms consisting of a chronic cough and occasional breathlessness. The physician becomes concerned and orders the patient admitted to the hospital for a lung biopsy. Suppose that the results of the lung biopsy are consistent either with lung cancer or with sarcoidosis, a fairly common, nonfatal lung disease. In this case

Symptoms A = {chronic cough, results of lung biopsy}
Disease state B_1 = normal
B_2 = lung cancer
B_3 = sarcoidosis

Suppose that $Pr(A|B_1) = .001 \quad Pr(A|B_2) = .9 \quad Pr(A|B_3) = .9$
and that in 60-year-old, never-smoking men

$$Pr(B_1) = .99 \quad Pr(B_2) = .001 \quad Pr(B_3) = .009$$

The first set of probabilities $Pr(A|B_i)$ could be obtained from clinical experience with the previous diseases, whereas the latter set of probabilities $Pr(B_i)$ would have to be obtained from age-sex-smoking specific prevalence rates for the diseases in question. The interesting question now is what are the probabilities $Pr(B_i|A)$ of the three disease states given the previous symptoms?

Solution Bayes' rule can be used to answer this question. Specifically,

$$\begin{aligned} Pr(B_1|A) &= Pr(A|B_1) \times Pr(B_1) \Big/ \left[\sum_{j=1}^{3} Pr(A|B_j) \times Pr(B_j)\right] \\ &= .001(.99)/[.001(.99) + .9(.001) + .9(.009)] \\ &= .00099/.00999 = .099 \end{aligned}$$

$$Pr(B_2|A) = .9(.001)/[.001(.99) + .9(.001) + .9(.009)]$$
$$= .00090/.00999 = .090$$
$$Pr(B_3|A) = .9(.009)/[.001(.99) + .9(.001) + .9(.009)]$$
$$= .00810/.00999 = .811$$

Thus, although the unconditional probability of sarcoidosis is very low (.009), the conditional probability of the disease given these symptoms and this age/sex/smoking group is .811. Also, although the symptoms are consistent with both lung cancer and sarcoidosis, the latter is much more likely among patients in this age/sex/smoking group.

Example 3.28 **Pulmonary Disease** Now, suppose that the patient in Example 3.27 was a smoker of two packs of cigarettes per day for 40 years. Then, assume that $Pr(B_1) = .98$, $Pr(B_2) = .015$, $Pr(B_3) = .005$ in this type of person. What are the probabilities of the three disease states given these symptoms for this type of patient?

Solution

$$Pr(B_1|A) = .001(.98)/[.001(.98) + .9(.015) + .9(.005)]$$
$$= .00098/.01898 = .052$$
$$Pr(B_2|A) = .9(.015)/.01898 = .01350/.01898 = .711$$
$$Pr(B_3|A) = .9(.005)/.01898 = .237$$

Thus, in this type of patient, lung cancer is the most likely diagnosis.

3.7.1 ROC Curves

In some instances, a test provides several categories of response rather than simply test positive or test negative. In other instances, the results of the test are reported as a continuous variable. In either case, the designation of a cut-off point for distinguishing test positive versus test negative is arbitrary.

Example 3.29 **Radiology** The following data, provided by Hanley and McNeil [5], are ratings of computed tomographic (CT) images by a single radiologist in a sample of 109 subjects with possible neurological problems. The true disease status is also known for each of these subjects. The data are presented in Table 3.2. How can we quantify the diagnostic accuracy of the test?

TABLE 3.2 **Ratings of 109 CT images by a single radiologist**

	CT rating by radiologist					
True disease status	Definitely normal (1)	Probably normal (2)	Questionable (3)	Probably abnormal (4)	Definitely abnormal (5)	**Total**
Normal	33	6	6	11	2	58
Abnormal	3	2	2	11	33	51
Total	36	8	8	22	35	109

TABLE 3.3 **Sensitivity and specificity of the radiologist ratings according to different criteria for test positive based on the data in Table 3.2**

Test positive criteria	Sensitivity	Specificity
1+	1.0	0
2+	.94	.57
3+	.90	.67
4+	.86	.78
5+	.65	.97
6+	0	1.0

Unlike the previous examples, there is no obvious cut-off point to use for designating a subject as positive based on the CT scan. For example, if we designate a subject as test positive if he or she is either probably abnormal or definitely abnormal (i.e., a rating of 4 or 5, or 4+), then the sensitivity of the test is (11 + 33)/51 = 44/51 = .86, whereas the specificity is (33 + 6 + 6)/58 = 45/58 = .78. In Table 3.3, we compute the sensitivity and specificity of the radiologist ratings according to different criteria for test positive.

To display these data, we construct a receiver operating characteristic (ROC) curve.

DEFINITION 3.16 A **receiver operating characteristic (ROC) curve** is a plot of the sensitivity versus (1 – specificity) of a screening test, where the different points on the curve correspond to different cut-off points used to designate test positive.

Example 3.30 **Radiology** Construct an ROC curve based on the data in Table 3.3.

Solution We plot sensitivity versus (1 – specificity) using the data in Table 3.3. The plot is shown in Figure 3.7.

FIGURE 3.7 **ROC curve for the data in Table 3.3**

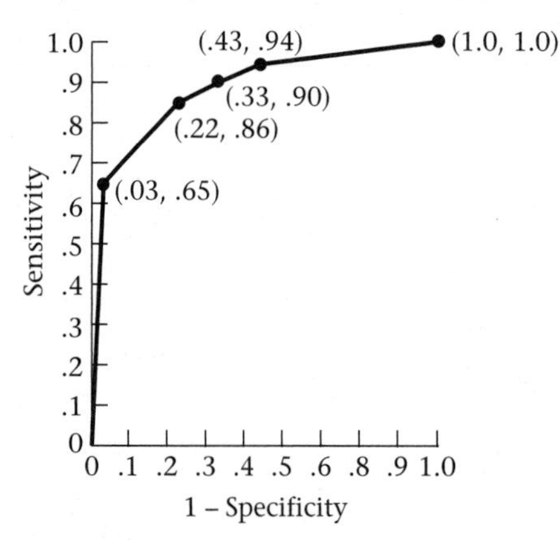

The area under the ROC curve is a reasonable summary of the overall diagnostic accuracy of the test. It can be shown [5] that this area, when calculated by the trapezoidal rule, corresponds to the probability that for a randomly selected pair of normal and abnormal subjects, the radiologist will correctly identify the normal subject given the CT ratings. It is assumed that for untied ratings the radiologist designates the subject with the lower test score as normal, and the subject with the higher test score as abnormal. For tied ratings, it is assumed that the radiologist randomly chooses one patient as normal and the other as abnormal.

EXAMPLE 3.31 **Radiology** Calculate the area under the ROC curve in Figure 3.7, and interpret what it means.

Solution The area under the ROC curve, when evaluated by the trapezoidal rule, is given by

$$.5(.94 + 1.0)(.57) + .5\,(.90 + .94)(.10) + .5(.86 + .90)(.11) + .5(.65 + .86)(.19) + .5(0 + .65)(.03) = .89$$

This means that the radiologist has an 89% probability of correctly distinguishing a normal from an abnormal subject based on the relative ordering of their CT ratings. For normal and abnormal subjects with the same ratings, it is assumed that the radiologist selects one of the two subjects at random.

In general, for two screening tests for the same disease, the test with the higher area under its ROC curve is considered the better test, unless some particular level of sensitivity or specificity is especially important in comparing the two tests.

SECTION 3.8 Prevalence and Incidence

In clinical medicine, the terms *prevalence* and *incidence* are used to denote probabilities in a special context and are used frequently in this text.

DEFINITION 3.17 The **prevalence** of a disease is the probability of currently having the disease regardless of the duration of time one has had the disease. It is obtained by dividing the number of people who currently have the disease by the number of people in the study population.

Example 3.32 **Hypertension** The prevalence of hypertension in 1974 among all people 17 years of age and older was reported to be 15.7%, as assessed by a government study [6]. It was computed by dividing the number of people who had elevated blood pressure and were 17 years of age and older (22,626) by the total number of people 17 years of age and older in the study population (144,380).

DEFINITION 3.18 The **cumulative incidence** of a disease is the probability that a person with no prior disease will develop a new case of the disease over some specified time period.

In Chapter 14 we will distinguish between *cumulative incidence,* which is defined over a long period of time, and *incidence density,* which is defined over a very short (or instantaneous) period of time. For simplicity, before Chapter 14 we will use the abbreviated term *incidence* to denote *cumulative incidence.*

Example 3.33 **Cancer** The cumulative incidence rate of breast cancer in 40- to 44-year-old Connecticut women over the time period January 1, 1970, through December 31, 1970, was approximately 1 per 1000 [2]. This rate means that about 1 woman in 1000 of the 40- to 44-year-old women who had never had breast cancer on January 1, 1970, would have developed a new case of breast cancer by December 31, 1970.

SECTION 3.9 Summary

In this chapter, probabilities and how to work with them using the addition and multiplication laws were discussed. An important distinction was made between independent events, which are unrelated to each other, and dependent events, which are related to each other. The general concepts of conditional probability and relative risk were introduced to quantify the dependence between two events. These ideas were then applied to the special area of screening populations for disease. In particular, the notions of sensitivity, specificity, and predictive value, which are used to define the accuracy of screening tests, were developed as applications of conditional probability. We also used an ROC curve to extend the concepts of sensitivity and specificity when the designation of the cut-off point for test positive versus test negative is arbitrary.

On some occasions, only sensitivities and specificities are available and we wish to compute the predictive value of screening tests. This task can be accomplished using Bayes' rule. Indeed, Bayes' rule can be used generally to change the direction of conditional probabilities. Finally, prevalence and incidence, which are probabilistic parameters that are often used to describe the magnitude of disease in a population, were defined.

In the next two chapters, these general principles of probability are applied to derive some of the important probabilistic models often used in biomedical research, including the binomial, Poisson, and normal models. These models will be used eventually to test hypotheses about data.

PROBLEMS

Consider a family with a mother, father, and two children. Let A_1 = {mother has influenza}, A_2 = {father has influenza}, A_3 = {first child has influenza}, A_4= {second child has influenza}, B = {at least one child has influenza}, C = {at least one parent has influenza}, D = {at least one person in the family has influenza}.

***3.1** What does $A_1 \cup A_2$ mean?

***3.2** What does $A_1 \cap A_2$ mean?

***3.3** Are A_3 and A_4 mutually exclusive?

***3.4** What does $A_3 \cup B$ mean?

***3.5** What does $A_3 \cap B$ mean?

***3.6** Express C in terms of A_1, A_2, A_3, A_4.

***3.7** Express D in terms of B and C.

***3.8** What does $\overline{A_1}$ mean?

***3.9** What does $\overline{A_2}$ mean?

***3.10** Represent $\overline{C}$ in terms of A_1, A_2, A_3, A_4.

***3.11** Represent $\overline{D}$ in terms of B and C.

Suppose that an influenza epidemic strikes a city. In 10% of families the mother has influenza; in 10% of families the father has influenza; and in 2% of families both the mother and father have influenza.

3.12 Are the events A_1, A_2 independent?

Suppose there is a 20% chance that each child will get influenza, whereas in 10% of two-child families, both children get the disease.

3.13 What is the probability that at least one child will get influenza?

Hypertension

Multiple drugs are often used in treating hypertension. Suppose that 10% of patients taking antihypertensive agent A experience gastrointestinal (GI) side effects, whereas 20% of patients taking antihypertensive agent B experience such side effects.

3.14 If the side effects of the two agents are assumed to be independent events, then what is the probability that a patient taking the two agents simultaneously will experience GI side effects from at least one of the two agents?

Refer to Problem 3.12.

3.15 What is the conditional probability that the father has influenza given that the mother has influenza?

3.16 What is the conditional probability that the father has influenza given that the mother does not have influenza?

Mental Health

Estimates of the prevalence of Alzheimer's disease have recently been provided by Pfeffer et al. [7]. The estimates are given in Table 3.4.

TABLE 3.4 Prevalence of Alzheimer's disease (cases per 100 population)

Age group	Males	Females
65–69	1.6	0.0
70–74	0.0	2.2
75–79	4.9	2.3
80–84	8.6	7.8
85+	35.0	27.9

Suppose an unrelated 77-year-old man, 76-year-old woman, and 82-year-old woman are selected from a community.

3.17 What is the probability that all three of these individuals have Alzheimer's disease?

3.18 What is the probability that at least one of the women has Alzheimer's disease?

3.19 What is the probability that at least one of the three individuals has Alzheimer's disease?

3.20 What is the probability that exactly one of the three individuals has Alzheimer's disease?

3.21 Suppose we know that one of the three individuals has Alzheimer's disease, but we don't know which one. What is the conditional probability that the affected individual is a woman?

3.22 Suppose we know that two of the three individuals have Alzheimer's disease. What is the conditional probability that they are both women?

3.23 Suppose we know that two of the three individuals have Alzheimer's disease. What is the conditional probability that they are both less than 80 years old?

Suppose the probability that both members of a married couple will have the disease, where each member is 75–79 years old, is .0015.

3.24 What is the conditional probability that the man will be affected given that the woman is affected? How does this value compare to the prevalence in Table 3.4? Why should it be the same (or different)?

3.25 What is the conditional probability that the woman will be affected given that the man is affected? How does this value compare to the prevalence in Table 3.4? Why should it be the same (or different)?

3.26 What is the probability that at least one member of the couple is affected?

Suppose a study of Alzheimer's disease is proposed in a retirement community with persons 65+ years of age, where the age–sex distribution is as shown in Table 3.5.

TABLE 3.5 Age–sex distribution of retirement community

	Male (%)[a]	Female (%)
65–69	5	10
70–74	9	17
75–79	11	18
80–84	8	12
85+	4	6

[a]Percentage of total population.

3.27 What is the expected overall prevalence of Alzheimer's disease in the community if the prevalence estimates in Table 3.4 for specific age–sex groups holds?

3.28 If there are 1000 people 65+ years of age in the community, then what is the expected number of cases of Alzheimer's disease in the community?

Occupational Health

A study is conducted among men 50–69 years old working in a chemical plant. We are interested in comparing the mortality experience of the workers in the plant with national mortality rates. Suppose that of the 500 workers in this age group in the plant, 35% are 50–54, 30% are 55–59, 20% are 60–64, and 15% are 65–69.

***3.29** If the annual national mortality rates are 0.9% in 50- to 54-year-old men, 1.4% in 55- to 59-year-old men, 2.2% in 60- to 64-year-old men, and 3.3% in 65- to 69-year-old men, then what is the projected annual mortality rate in the plant as a whole?

The SMR (standardized mortality ratio) is often used in occupational studies as a measure of risk. It is defined as 100% *times* the observed number of events in the exposed group *divided by* the expected number of events in the exposed group (based on some reference population).

***3.30** If 15 deaths are observed over 1 year among the 500 workers, then what is the SMR?

Genetics

Suppose that a disease is inherited via a **dominant** mode of inheritance and that one of two parents is affected with the disease whereas one is not. The implications of this mode of inheritance are that the probability is $\frac{1}{2}$ that any particular offspring will get the disease.

3.31 What is the probability that in a family with two children, both siblings are affected?

3.32 What is the probability that exactly one sibling is affected?

3.33 What is the probability that neither sibling will be affected?

3.34 Suppose that the older child is affected. What is the probability that the younger child will be affected?

3.35 If *A*, *B* are two events such that *A* = {older child is affected}, *B* = {younger child is affected}, then are the events *A*, *B* independent?

Suppose that a disease is inherited via an **autosomal recessive** mode of inheritance. The implications of this mode of inheritance are that the children in a family each have a probability of $\frac{1}{4}$ of inheriting the disease.

3.36 What is the probability that in a family with two children, both siblings are affected?

3.37 What is the probability that exactly one sibling is affected?

3.38 What is the probability that neither sibling is affected?

Suppose that a disease is inherited via a **sex-linked** mode of inheritance. The implications of this mode of inheritance are that each male offspring has a 50% chance of inheriting the disease, whereas the female offspring have no chance of getting the disease.

3.39 In a family with one male and one female offspring, what is the probability that both siblings are affected?

3.40 What is the probability that exactly one sibling is affected?

3.41 What is the probability that neither sibling is affected?

3.42 Answer Problem 3.39 for families with two male siblings.

3.43 Answer Problem 3.40 for families with two male siblings.

3.44 Answer Problem 3.41 for families with two male siblings.

Suppose that in a family with two male siblings, both siblings are affected with a genetically inherited disease. Suppose also that, although the genetic history of the family is unknown, only a dominant, recessive, or sex-linked mode of inheritance is possible.

3.45 Assume that the dominant, recessive, and sex-linked modes of inheritance follow the probability laws given in Problems 3.31, 3.36, and 3.39 and that, without prior knowledge about the family in question, each is equally likely to occur. What is the probability of each mode of inheritance in this family?

3.46 Answer Problem 3.45 for a family with two male siblings where only one sibling is affected.

3.47 Answer Problem 3.45 for a family with one male and one female sibling where both siblings are affected.

3.48 Answer Problem 3.47 where only the male sibling is affected.

Obstetrics

The following data are derived from the 1973 Final Natality Statistics Report issued by the National Center for Health Statistics [8]. These data are pertinent to livebirths only.

Suppose that infants are classified as low birthweight if they have a birthweight $\leq$2500 g and as normal birthweight if they have a birthweight $\geq$2501 g. Suppose that infants are also classified by length of gestation in the following four categories: $<$20 weeks, 20–27 weeks, 28–36 weeks, $>$36 weeks. Assume that the probabilities of the different periods of gestation are as given in Table 3.6.

TABLE 3.6 Distribution of length of gestation

Length of gestation	Probability
<20 weeks	.0004
20–27 weeks	.0059
28–36 weeks	.0855
>36 weeks	.9082

Also assume that the probability of low birthweight given that length of gestation is <20 weeks is .540, the probability of low birthweight given that length of gestation is 20–27 weeks is .813, the probability of low birthweight given that length of gestation is 28–36 weeks is .379, and the probability of low birthweight given that length of gestation is >36 weeks is .035.

***3.49** What is the probability of having a low birthweight infant?

3.50 Show that the events {length of gestation ≤27 weeks} and {low birthweight} are not independent.

***3.51** What is the probability of having a length of gestation ≤36 weeks given that a child is low birthweight?

Pulmonary Disease

A 1974 paper by Colley et al. looked at the relationship between parental smoking and the incidence of pneumonia and/or bronchitis in children in the first year of life [9]. One important finding of the paper was that 7.8% of children with nonsmoking parents had episodes of pneumonia and/or bronchitis in the first year of life, whereas, respectively, 11.4% of children with one smoking parent and 17.6% of children with two smoking parents had such an episode. Suppose that in the general population both parents are smokers in 40% of households, one parent smokes in 25% of households, and neither parent smokes in 35% of households.

3.52 What percentage of children in the general population will have pneumonia and/or bronchitis in the first year of life?

A group of families in which both parents smoke at the time of the first prenatal visit decide, after counseling by the nurse practitioner, to give up smoking. Suppose that in 10% of these families both parents resume smoking and in 30% of these families one parent resumes smoking. In the remainder of the families both parents have not resumed smoking at the time of birth of the child. Assume also that the smoking status of the parents at the time of the birth is maintained during the first year of life of the child.

3.53 What is the probability of pneumonia and/or bronchitis in children from families in this group?

3.54 Among families where both parents smoke, what percentage of cases of pneumonia and/or bronchitis have been prevented by this type of counseling in comparison to the number of cases that would have occurred if no counseling was provided?

Pulmonary Disease

The familial aggregation of respiratory disease is a well-established clinical phenomenon. However, whether this aggregation is due to genetic or environmental factors or both is somewhat controversial. An investigator wishes to study a particular environmental factor, namely, the relationship of cigarette-smoking habits in the parents to the presence or absence of asthma in their oldest child living in the household in the 5- to 9-year-old age range (referred to below as their offspring). Suppose that the investigator finds that (1) if both the mother and father are current smokers, then the probability of their offspring having asthma is .15; (2) if the mother is a current smoker and the father is not, then the probability of their offspring having asthma is .13; (3) if the father is a current smoker and the mother is not, then the probability of their offspring having asthma is .05; (4) if neither parent is a current smoker, then the probability of their offspring having asthma is .04.

***3.55** Suppose the smoking habits of the parents are independent and the probability that the mother is a current smoker is .4, whereas the probability that the father is a current smoker is .5. What is the probability that both the father and mother are current smokers?

***3.56** Consider the subgroup of families where the mother is not a current smoker. What is the probability that the father is a current smoker among such families? How does this probability differ from that calculated in Problem 3.55?

Suppose, alternatively, that if the father is a current smoker, then the probability that the mother is a current smoker is .6; whereas if the father is not a current smoker, then the probability that the mother is a current smoker is .2. Also assume that statements 1, 2, 3, and 4, above hold.

***3.57** If the probability that the father is a current smoker is .5, what is the probability that the father is a current smoker *and* that the mother is not a current smoker?

***3.58** Are the current smoking habits of the father and the mother independent? Why or why not?

***3.59** Find the unconditional probability that the offspring will have asthma under the assumptions in Problems 3.57 and 3.58.

***3.60** Suppose a child has asthma. What is the probability that the father is a current smoker?

***3.61** What is the probability that the mother is a current smoker if the child has asthma?

***3.62** Answer Problem 3.60 if the child does not have asthma.

***3.63** Answer Problem 3.61 if the child does not have asthma.

***3.64** Are the child's asthma status and the father's smoking status independent? Why or why not?

***3.65** Are the child's asthma status and the mother's smoking status independent? Why or why not?

Pulmonary Disease

Smoking cessation is an important dimension in public health programs aimed at preventing cancer and heart and lung diseases. For this purpose, data were accumulated starting in 1962 on a group of currently smoking men as part of the Normative Aging Study, a longitudinal study of the Veterans Administration in Boston. No interventions were attempted on this group of men, but the data in Table 3.7 were obtained as to annual quitting rates among initially healthy men who remained healthy during the entire period [10]:

TABLE 3.7 Annual quitting rates of men who smoked, from the Normative Aging Study, 1962–1975

Time period	Light smokers (≤ one pack per day) average annual quitting rate per 100 persons	Heavy smokers (> one pack per day) average annual quitting rate per 100 persons
1962–1966	3.1	2.0
1967–1970	7.1	5.0
1971–1975	4.7	4.1

Note that the quitting rates increased from 1967 to 1970, which was around the time of the first Surgeon General's report on cigarette smoking.

3.66 Suppose a man was a light smoker on January 1, 1962. What is the probability that he quit smoking by the end of 1975? (Assume he remained a light smoker until just prior to quitting and remained a quitter until 1975.)

3.67 Answer Problem 3.66 for a heavy smoker on January 1, 1962 (assuming he remained a heavy smoker until just prior to quitting and remained a quitter until 1975).

Pulmonary Disease

Research into cigarette-smoking habits, smoking prevention, and cessation programs necessitates accurate measurement of smoking behavior. However, decreasing social acceptability of smoking appears to cause significant underreporting. Chemical markers for cigarette use can provide objective indicators of smoking behavior. One widely used noninvasive marker is the level of saliva thiocyanate (SCN). In a Minneapolis school district, 1332 students in eighth grade (ages 12–14) participated in a study [11] whereby they

(1) Viewed a film illustrating how recent cigarette use could be readily detected from small samples of saliva

(2) Provided a personal sample of saliva thiocyanate

(3) Provided a self-report of the number of cigarettes smoked per week

The results are given in Table 3.8.

TABLE 3.8 Relationship between saliva thiocyanate levels (SCN) and self-reported cigarettes smoked per week

Self-reported cigarettes smoked in last week	Number of students	Percent with SCN ≥ 100 μg/mL
None	1163	3.3
1–4	70	4.3
5–14	30	6.7
15–24	27	29.6
25–44	19	36.8
45+	23	65.2

Source: Reprinted with permission from the *American Journal of Public Health, 71*(12), 1320, 1981.

Suppose the self-reports are completely accurate and are representative of the amount eighth-grade students smoke in the general community. We are considering using an SCN level ≥100 μg/mL as a test criterion for identifying cigarette smokers. Regard a student as positive if he or she smokes one or more cigarettes per week.

***3.68** What is the sensitivity of the test for light-smoking students (students who smoke ≤14 cigarettes per week)?

***3.69** What is the sensitivity of the test for moderate-smoking students (students who smoke 15–44 cigarettes per week)?

***3.70** What is the sensitivity of the test for heavy-smoking students (students who smoke ≥45 cigarettes per week)?

***3.71** What is the specificity of the test?

***3.72** What is the predictive value positive of the test?

***3.73** What is the predictive value negative of the test?

Suppose we regard the self-reports of all students who report some cigarette consumption as valid but estimate that 20% of students who report no cigarette consumption actually smoke 1–4 cigarettes per week and an additional 10% smoke 5–14 cigarettes per week.

***3.74** If we assume that the percentage of students with SCN ≥100 μg/mL in these two subgroups is the same as in those who truly report 1–4 and 5–14 cigarettes per week, then compute the specificity under these assumptions.

***3.75** Compute the predictive value negative (PV^-)under these altered assumptions. How does the true PV^- using a screening criteria of SCN ≥100 μg/mL for identifying smokers compare with the PV^- based on self-reports obtained in Problem 3.73?

Hypertension

Laboratory measures of cardiovascular reactivity are receiving increasing attention. Much of the expanded interest is based on the belief that these measures, obtained under challenge from physical and psychological stressors, may yield a more biologically meaningful index of cardiovascular function than more traditional static measures. Typically, measurement of cardiovascular reactivity involves the use of an automated blood-pressure monitor to examine the changes in blood pressure before and after a stimulating experience (such as playing a video game). For this purpose, BP measurements were made with the Vita-Stat BP machine both before and after playing a video game. Similar measurements were obtained using manual methods for measuring blood pressure. A person was classified as a "reactor" if his or her diastolic blood pressure (DBP) increased by 10 mm Hg or more after playing the game and as a nonreactor otherwise. The results are given in Table 3.9.

TABLE 3.9 Classification of cardiovascular reactivity using an automated and a manual sphygmomanometer

	ΔDBP, manual	
ΔDBP, automated	<10	≥10
<10	51	7
≥10	15	6

3.76 If the manual measurements are regarded as the "true" measure of reactivity, then what is the sensitivity of automated DBP measurements?

3.77 What is the specificity of automated DBP measurements?

3.78 If the population tested is representative of the general population, then what are the predictive values positive and negative using this test?

Otolaryngology

The data set in Table 3.10 is based on 214 children with acute otitis media (OME) who participated in a randomized clinical trial [12]. Each child had OME at the beginning of the study in either one (unilateral cases) or both (bilateral cases) ears. Each child was randomly assigned to receive a 14-day course of one of two antibiotics, either cefaclor (CEF) or amoxicillin (AMO). The data here concerns the 203 children whose middle-ear status was determined at a 14-day follow-up visit. The data in Table 3.10 are presented in Data Set EAR.DAT (on the data disk).

3.79 Does there seem to be any difference in the effect of the antibiotics on clearance of otitis media? Try to express your results in terms of relative risk. Consider separate analyses for unilateral and bilateral cases. Also consider an analysis combining the two types of cases.

3.80 The investigators recorded the age of the children because they felt this might be an important factor in determining outcome. Were they right? Try to express your results in terms of relative risk.

3.81 While controlling for age, propose an analysis comparing the effectiveness of the two antibiotics. Express your results in terms of relative risk.

3.82 Another issue in this trial is the possible dependence between ears for the bilateral cases. Can you comment on this issue based on the data collected?

The concept of a **randomized clinical trial** is discussed more completely in Chapter 6. The analysis of **contingency-table data** is studied in Chapters 10 and 13, where many of the formal methods for analyzing this type of data are discussed.

TABLE 3.10 Format for EAR.DAT

Column	Variable	Format or code
1–3	ID	
5	Clearance by 14 days	1 = yes/0 = no
7	Antibiotic	1 = CEF/2 = AMO
9	Age	1 = < 2 yrs/2 = 2–5 yrs/3 = 6+ yrs
11	Ear	1 = 1st ear/2 = 2nd ear

Cardiovascular Disease

An experiment was set up by a group from the University of Utah to use Bayes' rule to help make clinical diagnoses [13]. In particular, a detailed medical history questionnaire and electrocardiogram were administered to each patient referred to a cardiovascular laboratory and suspected of having congenital heart disease. From the experience of this laboratory and from estimates based on other published data, two sets of probabilities were generated:

(1) The unconditional probability of each of several disease states (see prevalence column in Table 3.11).

(2) The conditional probability of specific symptoms given specific disease states (see the rest of Table 3.11).

Thus, the probability that a person has chest pain given that he or she is normal is .05. Similarly, the proportion of persons with isolated pulmonary hypertension is .020. A subset of the data is given in Table 3.11.

Assume that these diagnoses are the only ones possible and that a patient can have one and only one diagnosis.

3.83 What is the probability of having a symptom of chest pain given that you have isolated pulmonary hypertension?

3.84 What is the probability of being more than 20 years old in this clinic?

3.85 Suppose we assume that the probability of any set of symptoms are independent given a specific diagnosis (e.g., the probability of being >20 years old and having both chest pain and mild cyanosis given that one is normal is $.50 \times .05 \times .01 = .00025$). What is the probability of being diagnosed as normal given that you have the following symptoms: (1) age 1–20 years, (2) repeated respiratory infections, and (3) easy fatigue?

3.86 What is the *most likely* diagnosis given that you have all the following symptoms: (1) mild cyanosis, (2) age >20 years, (3) EKG axis more than 110°? What is the second most likely diagnosis?

Suppose the symptom of an EKG axis of more than 110° is used as a screening criterion for diagnosing atrial septal defect without pulmonary stenosis or pulmonary hypertension.

TABLE 3.11 Prevalence of symptoms and diagnoses for patients suspected of having congenital heart disease

		Symptoms						
Diagnosis	**Prevalence**	X_1	X_2	X_3	X_4	X_5	X_6	X_7
Y_1	.155	.49	.50	.01	.10	.05	.05	.01
Y_2	.126	.50	.50	.02	.50	.02	.40	.70
Y_3	.084	.55	.05	.25	.90	.05	.10	.95
Y_4	.020	.45	.45	.01	.95	.10	.10	.95
Y_5	.098	.10	.00	.20	.70	.01	.05	.40
Y_6	.391	.70	.15	.01	.30	.01	.15	.30
Y_7	.126	.60	.10	.30	.70	.10	.20	.70

Y_1 = normal
Y_2 = atrial septal defect without pulmonary stenosis or pulmonary hypertension[a]
Y_3 = ventricular septal defect with valvular pulmonary stenosis
Y_4 = isolated pulmonary hypertension[a]
Y_5 = transposed great vessels
Y_6 = ventricular septal defect without pulmonary hypertension[a]
Y_7 = ventricular septal defect with pulmonary hypertension[a]
X_1 = age 1–20 years old
X_2 = age >20 years old
X_3 = mild cyanosis
X_4 = easy fatigue
X_5 = chest pain
X_6 = repeated respiratory infections
X_7 = EKG axis more than 110°

[a]Pulmonary hypertension is defined as pulmonary artery pressure ≥ systematic arterial pressure.
Source: Reprinted with permission of the American Medical Association from *JAMA, 177*(3), 177–183, 1961. Copyright 1961, American Medical Association.

3.87 What is the sensitivity of this test?

3.88 What is the specificity of this test?

3.89 Suppose we want to use another symptom in addition to EKG axis more than 110° to diagnose atrial septal defect without pulmonary stenosis or pulmonary hypertension. If we use the symptoms of age 1–20 years old and EKG axis more that 110°, then what is the predictive value positive?

3.90 Suppose we want to use two symptoms to diagnose atrial septal defect without pulmonary stenosis or pulmonary hypertension (not necessarily including EKG axis more than 110°). Which two symptoms can be used to maximize the predictive value positive? (Use a computer to answer this question.)

3.91 Answer Problem 3.90 using three symptoms rather than two. (Use a computer to answer this question.)

3.92 Write a computer program to compute the probability of each of the disease states given the presence or absence of any combination of the 7 symptoms. Note that some of the symptoms are mutually exclusive and thus cannot occur simultaneously; for example, symptom 1 = age 1 month to 1 year and symptom 2 = age 1–20 years.

3.93 Test your program using your answers to Problems 3.90 and 3.91.

Gynecology

A drug company is developing a new pregnancy-test kit for use on an outpatient basis. The company uses the pregnancy test on 100 women who are known to be pregnant, of whom 95 are positive using the test. The company uses the pregnancy test on 100 other women who are known to *not* be pregnant, of whom 99 are negative using the test.

***3.94** What is the sensitivity of the test?

***3.95** What is the specificity of the test?

The company anticipates that of the women who will use the pregnancy-test kit, 10% will actually be pregnant.

***3.96** What is the predictive value positive of the test?

***3.97** Suppose the "cost" of a false negative ($2c$) is twice that of a false positive (c) (because for a false negative prenatal care would be delayed during the first trimester of pregnancy). If the standard home pregnancy-test kit (made by another drug company) has a sensitivity of .98 and a specificity of .98, then which test (the new or standard) will cost the least per woman using it in the general population and by how much?

Mental Health

The Chinese Mini-Mental Status Test (CMMS) is a test consisting of 114 items intended to identify people with Alzheimer's disease and senile dementia among people in China [14]. An extensive clinical evaluation was performed of this instrument, whereby participants were interviewed by psychiatrists and nurses and a definitive diagnosis of dementia was made. Table 3.12 shows the results obtained on the subgroup of people with at least some formal education.

TABLE 3.12 Relationship of clinical dementia to outcome on the Chinese Mini-Mental Status Test

CMMS score	Nondemented	Demented
0–5	0	2
6–10	0	1
11–15	3	4
16–20	9	5
21–25	16	3
26–30	18	1
	46	16

Suppose a cutoff value of ≤20 on the test is used to identify people with dementia.

3.98 What is the sensitivity of the test?

3.99 What is the specificity of the test?

3.100 The cutoff value of 20 on the CMMS used to identify people with dementia is arbitrary. Suppose we consider changing the cutoff. What are the sensitivity and specificity if cutoffs of 5, 10, 15, 20, 25, or 30 are used, respectively? Make a table of your results.

3.101 Construct an ROC curve based on the table constructed in Problem 3.100.

3.102 Suppose we want both the sensitivity and specificity to be at least 70%. Use the ROC curve to identify the best value(s) to use as the cutoff for identifying people with dementia.

3.103 Calculate the area under the ROC curve. Interpret what this area means in words in the context of this problem.

Demography

A study based on data collected from the Medical Birth Registry of Norway looked at fertility rates according to survival outcomes of previous births [15]. The data are presented in Table 3.13.

3.104 What is the probability of having a livebirth (L) at a second birth given that the outcome of the first pregnancy was a stillbirth (D), that is, death?

3.105 Answer Problem 3.104 if the outcome of the first pregnancy was a livebirth.

TABLE 3.13 Relationship of fertility rates to survival outcome of previous births in Norway

Perinatal outcome	First birth *n*	Continuing to second birth *n*	Second birth outcome	*n*	Continuing to third birth *n*	Third birth outcome	*n*
D	7,022	5,924	D	368	277	D	39
						L	238
			L	5,556	3,916	D	115
						L	3,801
L	350,693	265,701	D	3,188	2,444	D	140
						L	2,304
			L	262,513	79,450	D	1,005
						L	78,445

Note: D = dead, L = alive at birth and for at least one week.

3.106 What is the probability of 0, 1, and 2+ additional pregnancies if the first birth was a stillbirth?

3.107 Answer Problem 3.106 if the first birth was a live-birth.

Mental Health

The ε4 allelle of the gene encoding apolipoprotein E (APOE) is strongly associated with Alzheimer's disease, but its value in making the diagnosis remains uncertain. A study was conducted among 2188 patients who were evaluated at autopsy for Alzheimer's disease by previously established pathological criteria [16]. Patients were also evaluated clinically for the presence of Alzheimer's disease. The data in Table 3.14 were presented

Suppose the pathological diagnosis is considered to be the gold standard for Alzheimer's disease.

TABLE 3.14 Relationship between clinical and pathological diagnoses of Alzheimer's disease

	Pathological diagnosis	
Clinical diagnosis	Alzheimer's disease	Other causes of dementia
Alzheimer's disease	1643	190
Other causes of dementia	127	228

3.108 If the clinical diagnosis is considered as a screening test for Alzheimer's disease, then what is the sensitivity of this test?

3.109 What is the specificity of this test?

To possibly improve on the diagnostic accuracy of the clinical diagnosis for Alzheimer's disease, information on both the APOE genotype as well as the clinical diagnosis were considered. The data are presented in Table 3.15.

Suppose we consider the combination of both a clinical diagnosis for Alzheimer's disease *and* the presence of ≥1 ε4 allele as a screening test for Alzheimer's disease.

3.110 What is the sensitivity of this test?

3.111 What is the specificity of this test?

Cardiovascular Disease

One of the fascinating subjects of current interest is the "Hispanic paradox": Census data "show" that Hispanic persons tend to have a lower prevalence of coronary heart disease (CHD) than non-Hispanic Whites (NHW) based on health interviews of representative samples of persons from different ethnic groups from the U.S. population, although their risk factor profile is generally worse (more hypertension, diabetes, obesity for Hispanics than for NHW). To study this further, researchers took a group of 1000 Hispanic men ages 50–64 who were free of CHD in 1990 from several counties in Texas and followed them for 5 years. It was found that 100 of the men had developed CHD (either fatal cases, or nonfatal cases who survived a heart attack).

TABLE 3.15 Influence of the APOE genotype in diagnosing Alzheimer's disease

APOE genotype	Both clinical and pathological criteria for AD[a]	Only clinical criteria for AD	Only pathological criteria for AD	Neither clinical nor pathological criteria for AD
≥1 ε4 allele	1076	66	66	67
No ε4 allele	567	124	61	161
Total	1643	190	127	228

[a]AD = Alzheimer's disease.

3.112 Is the proportion 100 out of 1000 a prevalence rate, an incidence rate, or neither?

3.113 Given other surveys over the same time period among non-Hispanic Whites in these counties, it was expected that the comparable rate of CHD for non-Hispanic Whites would be 8%. Is there a significant excess of CHD among Hispanics based on these data?

Another important parameter in the epidemiology of CHD is the *case fatality rate* (i.e., the proportion of people who die among those who have a heart attack). Among the 100 CHD cases ascertained above, 50 were fatal cases.

3.114 What is the expected proportion of Hispanic men who will be identified as having a previous heart attack in the past 5 years by health surveys (who are by definition survivors) if we assume that the proportion of men with more than one nonfatal heart attack is negligible? What is the comparable proportion for non-Hispanic White men if the expected case fatality rate is 20% among NHW men with CHD?

3.115 Are these proportions prevalence rates, incidence rates, or neither? Do the results in this problem give insight into why the Hispanic paradox occurs (do Hispanic men truly have lower risk of CHD as government surveys would indicate)? Why or why not?

Genetics

A dominantly inherited genetic disease is identified over several generations of a large family. However, about half of families have dominant disease with *complete penetrance,* whereby if a parent is affected, there is a 50% probability that any one offspring will be affected. Similarly, about half of families have dominant disease with *reduced penetrance,* whereby if a parent is affected, there is a 25% probability that any one offspring will be affected.

Suppose in a particular family one parent and two out of 2 offspring are affected.

3.116 What is the probability that exactly two out of two offspring will be affected in a family with dominant disease with complete penetrance?

3.117 What is the probability that exactly two out of two offspring will be affected in a family with dominant disease with reduced penetrance?

3.118 What is the probability that the mode of transmission for this particular family is dominant with complete penetrance?

3.119 Suppose you are a genetic counselor and are asked by the parents what the probability is, if they have another (i.e., a 3rd) child, that it will be affected by the disease. What is the answer?

Hypertension

A study was performed in Bergen, Norway [17]. In 1963, 70,000 people in the general population had their blood pressure measured; two readings were obtained and the second reading was used in the analysis. The people were followed for mortality outcome over a 10-year period after the blood-pressure measurement using death files in the Norwegian Central Bureau of Statistics. The results shown in Table 3.16 were obtained from the subgroup of 5034 men ages 50–59 in 1963:

TABLE 3.16 Relationship of 10-year mortality to diastolic blood pressure (DBP) at baseline in Norway

	10-year mortality outcome		
DBP (mm Hg)	Dead	Alive	Total
100+	124	295	419
≤99	764	3851	4615
Total	888	4146	5034

3.120 If we regard a diastolic blood pressure of ≥100 mm Hg as a screening test for predicting mortality over the next 10 years, then what is the sensitivity of the test?

3.121 What is the specificity of the test?

3.122 If the subjects in the study sample are considered representative of the general population, then what is the predictive value positive and negative of the test?

3.123 Suppose the threshold for positivity were changed from 100+ to 95+. Would the sensitivity and specificity increase, decrease, or remain the same?

Infectious Disease

Suppose a standard antibiotic is known to eliminate a type of bacteria 80% of the time. A new antibiotic is produced that is reputed to have better efficacy than the standard antibiotic. It is proposed to try the new antibiotic on 100 patients infected with the bacteria. Using principles of hypothesis testing (to be covered in Chapter 7), researchers will deem the new antibiotic "significantly better" than the old if it kills the bacteria for at least 88 out of the 100 infected patients.

3.124 Suppose there is a true probability (true efficacy) of 85% that the new antibiotic will work *for an individual patient.* Perform a "simulation study" on the computer based on random number generation (using, for example, MINITAB or Excel) for a group of 100 randomly simulated patients. Repeat this exercise 20 times with separate columns for each simulated sample of 100 patients. For what percentage of the 20 samples is the new antibiotic considered "significantly better" than the standard antibiotic? (This percentage is referred to as the *statistical power* of the experiment.)

3.125 Repeat the procedure in Problem 3.124 assuming the true efficacy of the new antibiotic is (a), 80%, (b) 90%, (c) 95%, respectively, and compute the statistical power for each of (a), (b), and (c).

3.126 Plot the statistical power versus the true efficacy. Do you think that 100 patients is a sufficiently large sample to discover if the new drug is "significantly better" if the true efficacy of the drug is 90%?

REFERENCES

[1] National Center for Health Statistics. (1976, February 13). *Monthly vital statistics report, advance report, final natality statistics (1974), 24*(11) (Suppl. 2).

[2] Doll, R., Muir, C., & Waterhouse, J. (Eds.). (1970). *Cancer incidence in five continents II.* Berlin: Springer-Verlag.

[3] Feller, W. (1960). *An introduction to probability theory and its applications* (Vol. I). New York: Wiley.

[4] Podgor, M. J., Leske, M. C., & Ederer, F. (1983). Incidence estimates for lens changes, macular changes, open-angle glaucoma, and diabetic retinopathy. *American Journal of Epidemiology, 118*(2), 206–212.

[5] Hanley, J. A., & McNeil, B. J. (1982). The meaning and use of the area under a Receiver Operating Characteristic (ROC) curve. *Diagnostic Radiology, 143,* 29–36.

[6] National Center for Health Statistics. (1976, November 8). *Advance data from vital and health statistics, 2.*

[7] Pfeffer, R. I., Afifi, A. A., & Chance, J. M. (1987). Prevalence of Alzheimer's disease in a retirement community. *American Journal of Epidemiology, 125*(3), 420–436.

[8] National Center for Health Statistics. (1975, January 30). *Monthly vital statistics report, final natality statistics (1973), 23*(11) (Suppl.).

[9] Colley, J. R. T., Holland, W. W., & Corkhill, R. T. (1974). Influence of passive smoking and parental phlegm on pneumonia and bronchitis in early childhood. *Lancet, II,* 1031.

[10] Garvey, A. J., Bossé, R., Glynn, R. J., & Rosner, B. (1983). Smoking cessation in a prospective study of healthy adult males: Effects of age, time period, and amount smoked. *American Journal of Public Health, 73*(4), 446–450.

[11] Luepker, R. V., Pechacek, T. F., Murray, D. M., Johnson, C. A., Hund, F., & Jacobs, D. R. (1981). Saliva thiocyanate: A chemical indicator of cigarette smoking in adolescents. *American Journal of Public Health, 71*(12), 1320.

[12] Mandel, E., Bluestone, C. D., Rockette, H. E., Blatter, M. M., Reisinger, K. S., Wucher, F. P., & Harper, J. (1982). Duration of effusion after antibiotic treatment for acute otitis media: Comparison of cefaclor and amoxicillin. *Pediatric Infectious Diseases, 1,* 310–316.

[13] Warner, H., Toronto, A., Veasey, L. G., & Stephenson, R. (1961). A mathematical approach to medical diagnosis. *JAMA, 177*(3), 177–183.

[14] Katzman, R., Zhang, M. Y., Ouang-Ya-Qu, Wang, Z. Y., Liu, W. T., Yu, E., Wong, S. C., Salmon, D. P., &

Grant, I. (1988). A Chinese version of the Mini-Mental State Examination; impact of illiteracy in a Shanghai dementia survey. *Journal of Clinical Epidemiology, 41*(10), 971–978.

[15] Skjaerven, R., Wilcox, A. J., Lie,, R. T., & Irgens, L. M. (1988). Selective fertility and the distortion of perinatal mortality. *American Journal of Epidemiology, 128*(6), 1352–1363.

[16] Mayeux, R., Saunders, A. M., Shea, S., Mirra, S., Evans, D., Roses, A. D., Hyman, B. T., Crain, B., Tang, M. X., & Phelps, C. H. (1998). Utility of the apolipoprotein E genotype in the diagnosis of Alzheimer's disease. Alzheimer's Disease Centers Consortium on Apolipoprotein E and Alzheimer's Disease. *New England Journal of Medicine, 338*(8), 506–511.

[17] Waaler, H. T. (1980). Specificity and sensitivity of blood pressure measurements. *Journal of Epidemiology and Community Health, 34*(1), 52–58.

4 Discrete Probability Distributions

SECTION 4.1 Introduction

Chapter 3 defined probability and introduced some basic tools used in working with probabilities. We now look at problems that can be put in a probabilistic framework. That is, by assessing the probabilities of certain events from actual past data, we can consider specific probability models that fit our problems.

Example 4.1 **Ophthalmology** Retinitis pigmentosa is a progressive ocular disease that in some cases eventually results in blindness. The three main genetic types of the disease are the dominant, the recessive, and the sex-linked. Each genetic type has a different rate of progression, the dominant mode being the slowest to progress and the sex-linked mode the fastest. Suppose the prior history of disease in a family is unknown. However, 1 of 2 male children is affected, whereas 0 of 1 female children are affected. Can this information help identify the genetic type?

The **binomial distribution** can be applied to calculate the probability of this event occurring (1 out of 2 males affected, 0 out of 1 females affected) under each of the genetic types mentioned, and these results can then be used to infer the most likely genetic type. In fact, this distribution can be used to make an inference for any family where we know that k_1 out of n_1 male children are affected and k_2 out of n_2 female children are affected.

Example 4.2 **Cancer** A second example of a commonly used probability model concerns a cancer scare in Woburn, Massachusetts. A news story reported an "excessive" number of cancer deaths in young children in this town and speculated whether or not this high rate was due to the dumping of industrial wastes in the northeastern portion of town [1]. Suppose that 12 cases of leukemia were reported in a town where 6 would normally be expected. Is this difference sufficient evidence for concluding that there is an excessive number of leukemia cases in the town?

The **Poisson distribution** can be used to calculate the probability of 12 or more cases if typical national rates for leukemia were present in this town. If this probability were sufficiently small, then we would conclude that there was an excessive number; otherwise, we would conclude that a longer surveillance of the town was necessary before arriving at a conclusion.

This chapter introduces the general concept of a discrete random variable and describes the binomial and Poisson distributions in depth. This forms the basis for the discussion (in Chapters 7 and 10) of hypothesis tests based on the binomial and Poisson distributions.

SECTION 4.2 Random Variables

In Chapter 3 we dealt with very specific events, such as the outcome of a tuberculin skin test or blood-pressure measurements taken on different members of a family. We now want to introduce ideas that will let us refer, in general terms, to different types of events having the *same probabilistic structure.* For this purpose the concept of a random variable is introduced.

DEFINITION 4.1 A **random variable** is a numeric function that assigns probabilities to different events in a sample space.

Two types of random variables are discussed in this text: discrete and continuous.

DEFINITION 4.2 A random variable for which there exists a discrete set of values with specified probabilities is a **discrete random variable**.

Example 4.3 **Otolaryngology** Otitis media is a disease of the middle ear and is one of the most frequent reasons for visiting a doctor in the first 2 years of life other than a routine well-baby visit. Let X be the random variable that represents the number of episodes of otitis media in the first 2 years of life. Then X is a discrete random variable, which takes on the values 0, 1, 2, . . .

Example 4.4 **Hypertension** Many new drugs have been introduced in the last several decades to bring hypertension under control, that is, to reduce high blood pressure to normotensive levels. Suppose a physician agrees to use a new antihypertensive drug on a trial basis on the first 4 untreated hypertensives she encounters in her practice before deciding whether to adopt the drug for routine use. Let X = the number of patients out of 4 who are brought under control. Then X is a discrete random variable, which takes on the values 0, 1, 2, 3, 4.

DEFINITION 4.3 A random variable whose possible values cannot be enumerated is a **continuous random variable**.

Example 4.5 **Environmental Health** The possible health effects on workers exposed to low levels of radiation over long periods of time are an issue of public health interest. One problem in assessing this issue is how to measure the cumulative exposure of a worker. A study was performed at the Portsmouth Naval Shipyard, whereby each exposed worker wore a badge, or dosimeter, which measured annual radiation exposure in rem [2]. The cumulative exposure over a worker's lifetime could then be obtained by summing the yearly exposures. The cumulative lifetime exposure is a good example of a continuous random variable, because it varied in this study from 0.000 rem to 91.414 rem, which would be regarded as taking on an essentially infinite number of values, which cannot be enumerated.

SECTION 4.3 The Probability Mass Function for a Discrete Random Variable

The values taken by a discrete random variable and its associated probabilities can be expressed by a rule or relationship called a *probability mass function.*

DEFINITION 4.4 A **probability mass function** is a mathematical relationship, or rule, that assigns to any possible value r of a discrete random variable X the probability $Pr(X = r)$. This assignment is made for all values r that have positive probability. The probability mass function is sometimes also referred to as a **probability distribution**.

The probability mass function can be displayed as a table giving the values and their associated probabilities or it can be expressed as a mathematical formula giving the probabilities of all possible values.

Example 4.6 **Hypertension** Consider Example 4.4. Suppose from previous experience with the drug, the drug company expects that for any clinical practice the probability that 0 patients out of 4 will be brought under control is .008, 1 patient out of 4 is .076, 2 patients out of 4 is .265, 3 patients out of 4 is .411, and all 4 patients is .240. This probability mass function, or probability distribution, is displayed in Table 4.1.

TABLE 4.1 **Probability mass function for the hypertension-control example**

$Pr(X = r)$	.008	.076	.265	.411	.240
r	0	1	2	3	4

Notice that for any probability mass function, the probability of any particular value must be between 0 and 1 and the sum of the probabilities of all values must exactly equal 1. Thus, $0 < Pr(X = r) \le 1, \sum Pr(X = r) = 1$, where the summation is taken over all possible values that have positive probability.

Example 4.7 **Hypertension** In Table 4.1, for any clinical practice, the probability that between 0 and 4 hypertensives are brought under control = 1; that is,

$$.008 + .076 + .265 + .411 + .240 = 1$$

4.3.1 Relationship of Probability Distributions to Sample Distributions

In Chapters 1 and 2 the concept of a **frequency distribution** in the context of a sample was discussed. It was described as a list of each value in the data set and a corresponding count of how frequently the value occurs. If each count is divided by the total number of points in the sample, then the frequency distribution can be considered as a sample analogue to a probability distribution. In particular, a probability distribution can be thought of as a model based on an infinitely large sample, giving the fraction of data points in a sample that *should* be allocated to each specific value. Because the frequency distribution gives the actual proportion of points in a sample that correspond to specific values, the appropriateness of the model can be assessed by comparing the observed sample frequency distribution to the probability distribution. The formal statistical procedure for making this comparison is called a **goodness-of-fit test**, and is discussed in Chapter 10.

Example 4.8 **Hypertension** How can the probability mass function in Table 4.1 be used to judge if the drug behaves with the same efficacy in actual practice as predicted by the drug company? The drug company might distribute the drug to 100 physicians and ask each of them to treat their first 4 untreated hypertensives with it. Each physician would then report his or her results to the drug company, and the combined results could be compared with the expected results in Table 4.1. For example, suppose that out of 100 physicians who agree to participate, 19 are able to bring all their first 4 untreated hypertensives under control, 48 are able to bring 3 of the 4 hypertensives under control, 24 are able to bring 2 out of 4 under control, the remaining 9 are able to bring only 1 of 4 under control. The sample-frequency distribution can be compared with the probability distribution given in Table 4.1, as shown in Table 4.2 and Figure 4.1.

TABLE 4.2 **Comparison of the sample-frequency distribution and the theoretical-probability distribution for the hypertension-control example**

Number of hypertensives under control = r	Probability distribution $Pr(X = r)$	Frequency distribution
0	.008	.000 = 0/100
1	.076	.090 = 9/100
2	.265	.240 = 24/100
3	.411	.480 = 48/100
4	.240	.190 = 19/100

The distributions look reasonably similar. The role of statistical inference is to compare the two distributions to judge if the differences between the two can be attributed to chance or whether real differences exist between the drug's performance in actual clinical practice and expectations from previous drug-company experience.

Students often ask, Where does a probability mass function come from? In some instances previous data can be obtained on the same type of random variable being studied and the probability mass function can be computed from these data. In other instances, previous data may not be available, but the probability mass

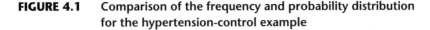

FIGURE 4.1 **Comparison of the frequency and probability distribution for the hypertension-control example**

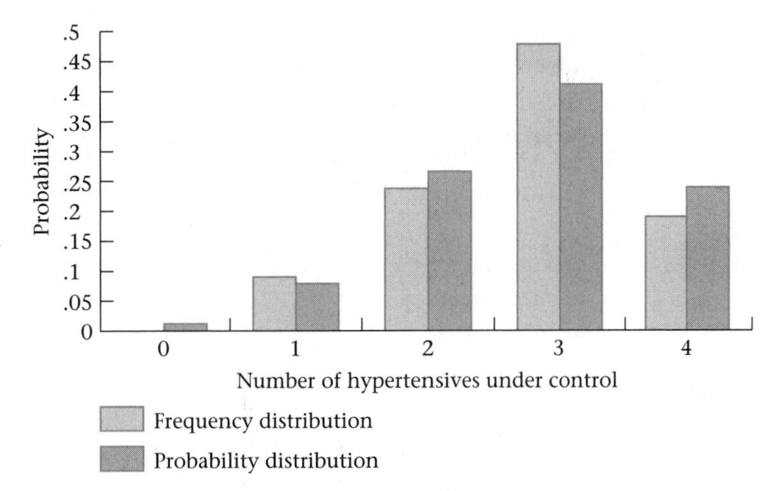

function from some well-known distribution may be used to see how well it fits actual sample data. In fact, this approach was used in Table 4.2, where the probability mass function was derived from the binomial distribution and then compared with the frequency distribution from the sample of 100 physician practices.

SECTION 4.4 The Expected Value of a Discrete Random Variable

If a random variable has a large number of values with positive probability, then the probability mass function is not a useful summary measure. Indeed, we are faced with the same problem as in trying to summarize a sample by enumerating each data value.

Measures of location and spread can be developed for a random variable in much the same way as they were developed for samples. The analogue to the arithmetic mean $\bar{x}$ is referred to as the expected value of a random variable, or population mean, and is denoted by $E(X)$ or μ. The expected value represents the "average" value of the random variable. It is obtained by multiplying each possible value by its respective probability and summing these products over all the values that have positive (that is, nonzero) probability.

DEFINITION 4.5 The **expected value of a discrete random variable** is defined as

$$E(X) \equiv \mu = \sum_{i=1}^{R} x_i Pr(X = x_i)$$

where the x_i's are the values the random variable assumes with positive probability.

Note that the sum in the definition of μ is over R possible values. R may be either finite or infinite. In either case, the individual values must be distinct from each other.

Example 4.9 **Hypertension** Find the expected value for the random variable shown in Table 4.1.

Solution

$$E(X) = 0(.008) + 1(.076) + 2(.265) + 3(.411) + 4(.240) = 2.80$$

Thus, on average about 2.8 hypertensives would be expected to be brought under control for every 4 that are treated.

Example 4.10 **Otolaryngology** Consider the random variable mentioned in Example 4.3 representing the number of episodes of otitis media in the first 2 years of life. Suppose this random variable has a probability mass function as given in Table 4.3.

TABLE 4.3 **Probability mass function for the number of episodes of otitis media in the first 2 years of life**

r	0	1	2	3	4	5	6
$Pr(X = r)$	.129	.264	.271	.185	.095	.039	.017

What is the expected number of episodes of otitis media in the first 2 years of life?

Solution

$$E(X) = 0(.129) + 1(.264) + 2(.271) + 3(.185) + 4(.095) + 5(.039) + 6(.017) = 2.038$$

Thus, on average a child would be expected to have approximately 2 episodes of otitis media in the first two years of life.

In Example 4.8 the probability mass function for the random variable representing the number of previously untreated hypertensives brought under control was compared with the actual number of hypertensives brought under control in 100 clinical practices. In much the same way, the expected value of a random variable can be compared with the actual sample mean in a data set $(\bar{x})$.

Example 4.11 **Hypertension** Compare the average number of hypertensives brought under control in the 100 clinical practices $(\bar{x})$ with the expected number of hypertensives brought under control (μ) per 4 patient practice.

Solution From Table 4.2 we have

$$\bar{x} = \left[0(0) + 1(9) + 2(24) + 3(48) + 4(19)\right]/100 = 2.77$$

hypertensives controlled per clinical practice while $\mu = 2.80$. This agreement is rather good. The specific methods for comparing the observed average value and expected value of a random variable ($\bar{x}$ and μ) will be covered in the material on statistical inference in Chapter 7. Notice that $\bar{x}$ could be written in the form

$$\bar{x} = 0(0/100) + 1(9/100) + 2(24/100) + 3(48/100) + 4(19/100)$$

that is, a weighted average of the number of hypertensives brought under control, where the weights are the observed probabilities. The expected value, in comparison, can be written as a similar weighted average, where the weights are the theoretical probabilities:

$$\mu = 0(.008) + 1(.076) + 2(.265) + 3(.411) + 4(.240)$$

Thus, the two quantities are actually obtained in the same way, one with weights given by the "observed" probabilities and the other with weights given by the "theoretical" probabilities.

SECTION 4.5 The Variance of a Discrete Random Variable

The analogue to the sample variance (s^2) for a random variable is called the *variance of the random variable,* or *population variance,* and is denoted by *Var*(*X*). The variance represents the spread, relative to the expected value, of all values that have positive probability. In particular, the variance is obtained by multiplying the squared distance of each possible value from the expected value by its respective probability and summing over all the values that have positive probability.

DEFINITION 4.6

The **variance of a discrete random variable**, denoted by $Var(X)$, is defined by

$$Var(X) = \sigma^2 = \sum_{i=1}^{R} (x_i - \mu)^2 Pr(X = x_i)$$

where the x_i's are the values for which the random variable takes on positive probability. The **standard deviation of a random variable** X, denoted by *sd*(*X*) or σ, is defined by the square root of its variance.

The population variance can also be expressed in a different ("short") form as follows:

EQUATION 4.1

A **short form for the population variance** is given by

$$\sigma^2 = E(X - \mu)^2 = \left[\sum_{i=1}^{R} x_i^2 Pr(X = x_i)\right] - \mu^2$$

Example 4.12 **Otolaryngology** Compute the variance and standard deviation for the random variable depicted in Table 4.3.

Solution We know from Example 4.10 that $\mu = 2.038$. Furthermore

$$\begin{aligned}\sum_{i=1}^{R} x_i^2 Pr(X = x_i) &= 0^2(.129) + 1^2(.264) + 2^2(.271) + 3^2(.185) \\ &\quad + 4^2(.095) + 5^2(.039) + 6^2(.017) \\ &= 0(.129) + 1(.264) + 4(.271) + 9(.185) \\ &\quad + 16(.095) + 25(.039) + 36(.017) \\ &= 6.12\end{aligned}$$

Thus $Var(X) = \sigma^2 = 6.12 - (2.038)^2 = 1.967$. The standard deviation of X is $\sigma = \sqrt{1.967} = 1.402$.

How can we get a feel for what the standard deviation of a random variable means? The following often-used principle is true for many, but not all, random variables:

EQUATION 4.2

> Approximately 95% of the probability mass falls within two standard deviations (2σ) of the mean of a random variable.

If 1.96σ is substituted for 2σ in Equation 4.2, this statement holds exactly for normally distributed random variables and approximately for certain other random variables. Normally distributed random variables are discussed in detail in Chapter 5.

Example 4.13 **Otolaryngology** Find a, b such that approximately 95% of infants will have between a and b episodes of otitis media in the first 2 years of life.

Solution The random variable depicted in Table 4.3 has mean $(\mu) = 2.038$ and standard deviation $(\sigma) = 1.402$. The interval $\mu \pm 2\sigma$ is given by

$$2.038 \pm 2(1.402) = 2.038 \pm 2.805$$

or from −0.77 to 4.84. Because only positive integer values are possible for this random variable, the valid range is from $a = 0$ to $b = 4$ episodes. Table 4.3 gives the probability of having ≤4 episodes as

$$.129 + .264 + .271 + .185 + .095 = .944$$

The rule allows us to quickly summarize the range of values that have most of the probability mass for a random variable without specifying each individual value. Chapter 6 will discuss the type of random variable for which Equation 4.2 applies.

SECTION 4.6 The Cumulative-Distribution Function of a Discrete Random Variable

Many random variables are displayed in tables or figures in terms of a cumulative-distribution function rather than a distribution of probabilities of individual values as in Table 4.1. The basic concept is to assign to each individual value the sum of probabilities of all values that are no larger than the value being considered. This function is defined as follows:

DEFINITION 4.7

The **cumulative-distribution function (cdf)** of a random variable X is denoted by $F(X)$ and, for a specific value x of X, is defined by $Pr(X \le x)$ and denoted by $F(x)$.

Example 4.14 **Otolaryngology** Compute the cumulative-distribution function for the otitis-media random variable in Table 4.3 and display it graphically.

Solution The cumulative-distribution function is given by

$$F(x) = 0 \quad \text{if} \quad x < 0$$
$$F(x) = .129 \quad \text{if} \quad 0 \le x < 1$$
$$F(x) = .393 \quad \text{if} \quad 1 \le x < 2$$

FIGURE 4.2 **Cumulative-distribution function for the number of episodes of otitis media in the first 2 years of life**

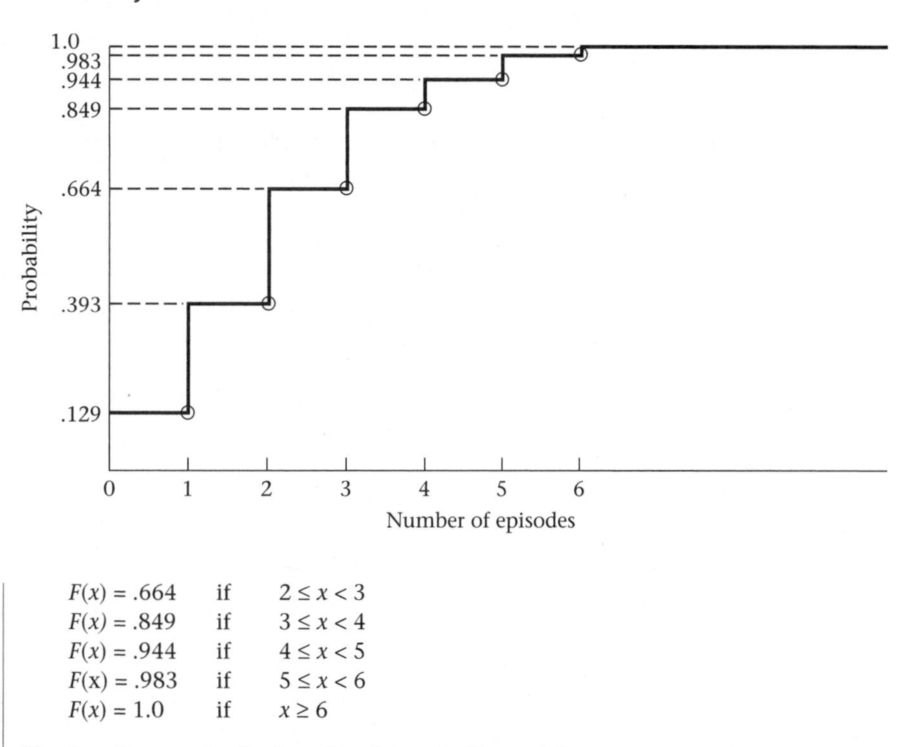

$F(x) = .664$ if $2 \le x < 3$
$F(x) = .849$ if $3 \le x < 4$
$F(x) = .944$ if $4 \le x < 5$
$F(x) = .983$ if $5 \le x < 6$
$F(x) = 1.0$ if $x \ge 6$

The function can be displayed as shown in Figure 4.2.

Another way to distinguish between a discrete and continuous random variable is by each variable's cdf. For a discrete random variable, the cdf looks like a series of steps and is sometimes referred to as a *step function.* For a continuous random variable, the cdf is a smooth curve. As the number of values increases, the cdf for a discrete random variable approaches that of a smooth curve. In Chapter 5, we will discuss the cdf for continuous random variables in more detail.

SECTION 4.7 Permutations and Combinations

Sections 4.2 through 4.6 introduced the concept of a discrete random variable in very general terms. The remainder of this chapter focuses on some specific discrete random variables that occur frequently in medical and biological work. Consider the following example.

Example 4.15 **Infectious Disease** One of the most common laboratory tests performed on any routine medical examination is a blood count. The two main aspects to a blood count are (1) counting the number of white blood cells (referred to as the "white count") and (2) differentiating white blood cells that do exist into five categories—namely, neutrophils, lymphocytes, monocytes, eosinophils, and basophils (referred to as the "differential"). Both the white count and the differential are extensively used in making clinical diagnoses. We will concentrate here on the differential, particularly on the distribution of the number of neutrophils k out of 100 white blood cells (which is the typical number counted). We will see that the number of neutrophils follows a binomial distribution.

To study the binomial distribution, **permutations** and **combinations**—important topics in probability—must first be understood.

Example 4.16 **Mental Health** Suppose we identify 5 male subjects ages 50–59 with schizophrenia in a community, and wish to match these subjects with normal controls of the same sex and age living in the same community. Suppose we wish to employ a **matched-pair design**, where each case is matched with a normal control of the same sex and age. Five psychologists are employed by the study, with each psychologist interviewing a single case and his matched control. If there are 10 eligible 50- to 59-year-old male controls in the community (labeled $A, B, \ldots, J$), then how many ways are there of choosing controls for the study if a control can never be used more than once?

Solution The first control can be any of $A, \ldots, J$ and thus can be chosen in 10 ways. Once the first control is chosen, he can no longer be selected as the second control; therefore, the second control can be chosen in 9 ways. Thus, the first two controls can be chosen in any one of $10 \times 9 = 90$ ways. Similarly, the third control can be chosen in any one of 8 ways, the fourth control in 7 ways, and the fifth control in 6 ways. In total, there are $10 \times 9 \times 8 \times 7 \times 6 = 30{,}240$ ways of choosing the 5 controls. For example, one possible selection is *ACDFE*. This means that control A is matched to the first case, control C to the second case, and so on. The order of selection of the controls is important, because different psychologists may be assigned to interview each matched pair. Thus, the selection *ABCDE* is different from *CBAED,* even though the same group of controls is selected.

We can now ask the general question, How many ways can k objects be selected out of n where the order of selection matters? Note that the first object can be selected in any one of $n = (n + 1) - 1$ ways. Given that the first object has been selected, the second object can be selected in any one of $n - 1 = (n + 1) - 2$ ways, . . . ; the kth object can be selected in any one of $n - k + 1 = (n + 1) - k$ ways.

DEFINITION 4.8 The number of **permutations** of n things taken k at a time is

$$_nP_k = n(n-1) \times \cdots \times (n-k+1)$$

It represents the number of ways of selecting k items out of n, where the order of selection is important.

Example 4.17 **Mental Health** Suppose 3 schizophrenic women aged 50–59 and 6 eligible controls live in the same community. How many ways are there of selecting three controls?

Solution To answer this question, consider the number of permutations of 6 things taken 3 at a time.

$$_6P_3 = 6 \times 5 \times 4 = 120$$

Thus there are 120 ways of choosing the controls. For example, one way would be to match control *A* to case 1, control *B* to case 2, and control *C* to case 3 (*ABC*). Another way would be to match control *F* to case 1, control *C* to case 2, and control *D* to case 3 (*FCD*). The order of selection is important because, for example, the selection *ABC* is different from the selection *BCA*.

In some instances we are interested in a special type of permutation: selecting n objects out of n, where the order of selection matters (ordering n objects). By the preceding principle,

$$_nP_n = n(n-1)\times\cdots\times[(n-n+1)] = n(n-1)\times\cdots\times 2\times 1$$

The special symbol generally used for this quantity is $n!$, which is called n factorial and is defined as follows:

DEFINITION 4.9 $n! =$ ***n* factorial** is defined as $n(n-1)\times\cdots\times 2\times 1$

Example 4.18 Evaluate 5 factorial.

$$5! = 5\times 4\times 3\times 2\times 1 = 120$$

The quantity 0! has no intuitive meaning, but for consistency it will be defined as 1.

Another way of writing $_nP_k$ is in terms of factorials. Specifically, from Definition 4.8, we can re-express $_nP_k$ in the form

$$\begin{aligned} _nP_k &= n(n-1)\times\cdots\times(n-k+1) \\ &= \frac{n(n-1)\times\cdots\times(n-k+1)\times(n-k)\times(n-k-1)\times\cdots\times 1}{(n-k)\times(n-k-1)\times\cdots\times 1} \\ &= n!/(n-k)! \end{aligned}$$

EQUATION 4.3

Alternative Form for Permutations An alternative formula expressing permutations in terms of factorials is given by

$$_nP_k = n!/(n-k)!$$

Example 4.19 **Mental Health** Suppose 4 schizophrenic women and 7 eligible controls live in the same community. How many ways are there of selecting 4 controls?

Solution The number of ways $= {}_7P_4 = 7(6)(5)(4) = 840$.
Alternatively, $_7P_4 = 7!/3! = 5040/6 = 840$.

Example 4.20 **Mental Health** Consider a somewhat different study design for the situation described in Example 4.16. Suppose an **unmatched study design**, where *all* cases and controls will be interviewed by the same psychologist, is used. If there are 10 eligible controls, then how many ways are there of choosing 5 controls for the study?

Solution In this case, because the same psychologist interviews all patients, what is important is which controls are selected, not the order of selection. Thus the question is, How many ways can 5 out of 10 eligible controls be selected, where order is not important? Note that for each set of 5 controls (say *A, B, C, D, E*), there are $5 \times 4 \times 3 \times 2 \times 1 = 5!$ ways of ordering the controls among themselves (e.g., *ACBED* and *DBCAE* are two possible orders). Thus, the number of ways of selecting 5 out of 10 controls for the study without respect to order = (the number of ways of selecting 5 controls out of 10 where order is important)/5! = ${}_{10}P_5/5! = (10 \times 9 \times 8 \times 7 \times 6)/120 = 30{,}240/120 = 252$ ways. Thus, *ABCDE* and *CDFIJ* are two possible selections. Also, *ABCDE* and *BCADE* are *not* counted twice.

The number of ways of selecting 5 objects out of 10 without respect to order is referred to as the number of **combinations** of 10 things taken 5 at a time and is denoted by ${}_{10}C_5$ or $\binom{10}{5} = 252$.

This discussion can be generalized to evaluate the number of combinations of n things taken k at a time. Note that for every selection of k distinct items out of n, there are $k(k-1) \times \cdots \times 2 \times 1 = k!$ ways of ordering the items among themselves. Thus, we have the following definition:

DEFINITION 4.10 The number of **combinations** of n things taken k at a time is

$$ {}_nC_k = \binom{n}{k} = \frac{n(n-1) \times \cdots \times (n-k+1)}{k!} $$

Alternatively, if we express permutations in terms of factorials as in Equation 4.3, we obtain

$$ {}_nC_k = \binom{n}{k} = {}_nP_k/k! $$
$$ = n!/[(n-k)!\ k!] $$

Thus, we have the following definition of combinations:

DEFINITION 4.11 The number of **combinations** of n things taken k at a time is

$$ {}_nC_k = \binom{n}{k} = \frac{n!}{k!\,(n-k)!} $$

It represents the number of ways of selecting k objects out of n where the order of selection does not matter.

Example 4.21 Evaluate ${}_7C_3$.

$$ {}_7C_3 = \frac{7 \times 6 \times 5}{3 \times 2 \times 1} = 7 \times 5 = 35 $$

Henceforth, for consistency we will always use the more common notation $\binom{n}{k}$ for combinations. In words, this is expressed as "n choose k."

A special situation arises upon evaluating $\binom{n}{0}$. By definition, $\binom{n}{0} = n!/(0!\,n!)$, and 0! was defined as 1. Hence, $\binom{n}{0} = 1$ for any n.

Frequently, $\binom{n}{k}$ will need to be computed for $k = 0, 1, \ldots, n$. The combinatorials have the following symmetry property, which makes this calculation easier than it appears at first.

EQUATION 4.4

For any nonnegative integers n, k, where $n \ge k$,

$$\binom{n}{k} = \binom{n}{n-k}$$

To see this, note from Definition 4.11 that

$$\binom{n}{k} = \frac{n!}{k!(n-k)!}$$

If $n - k$ is substituted for k in this expression, then we obtain

$$\binom{n}{n-k} = \frac{n!}{(n-k)![n-(n-k)]!} = \frac{n!}{(n-k)!\,k!} = \binom{n}{k}$$

Intuitively, this result makes sense, because $\binom{n}{k}$ represents the number of ways of selecting k objects out of n without regard to order. However, for every selection of k objects, we have also, in a sense, identified the other $n - k$ objects that were not selected. Thus, the number of ways of selecting k objects out of n without regard to order should be the same as the number of ways of selecting $n - k$ objects out of n without regard to order.

Hence, we need only evaluate combinatorials $\binom{n}{k}$ for the integers $k \le n/2$. If $k > n/2$, then the relationship $\binom{n}{n-k} = \binom{n}{k}$ can be used.

Example 4.22 Evaluate

$$\binom{7}{0}, \binom{7}{1}, \ldots, \binom{7}{7}$$

Solution

$$\binom{7}{0} = 1 \quad \binom{7}{1} = 7 \quad \binom{7}{2} = \frac{7 \times 6}{2 \times 1} = 21 \quad \binom{7}{3} = \frac{7 \times 6 \times 5}{3 \times 2 \times 1} = 35$$

$$\binom{7}{4} = \binom{7}{3} = 35 \quad \binom{7}{5} = \binom{7}{2} = 21 \quad \binom{7}{6} = \binom{7}{1} = 7 \quad \binom{7}{7} = \binom{7}{0} = 1$$

SECTION 4.8 The Binomial Distribution

All examples involving the binomial distribution have a common structure: a sample of n independent trials, each of which can have only two possible outcomes, which are denoted as "success" and "failure." Furthermore, the probability of a success at each trial is assumed to be some constant p, and hence the probability of a failure at each trial is $1 - p = q$. The term "success" is used in a general way, without any specific contextual meaning.

For Example 4.15, $n = 100$ and a "success" occurs when a cell is a neutrophil.

Example 4.23 **Infectious Disease** Reconsider Example 4.15 with 5 cells rather than 100 and ask the more limited question, What is the probability that the second and fifth cells considered will be neutrophils and the remaining cells nonneutrophils, given the probability that any one cell is a neutrophil is .6?

Solution If a neutrophil is denoted by an x and a nonneutrophil by an o, then the question being asked is, What is the probability of the outcome $oxoox = Pr(oxoox)$? Because the probabilities of success and failure are given respectively by .6 and .4, and the outcomes for different cells are presumed to be independent, then the probability is

$$q \times p \times q \times q \times p = p^2q^3 = (.6)^2(.4)^3$$

Example 4.24 **Infectious Disease** Now consider the more general question, What is the probability that any 2 cells out of 5 will be neutrophils?

Solution The arrangement $oxoox$ is only one of 10 possible orderings that result in 2 neutrophils. The 10 possible orderings are given in Table 4.4.

TABLE 4.4 **Possible orderings for 2 neutrophils out of 5 cells**

xxooo	oxxoo	ooxox
xoxoo	oxoxo	oooxx
xooxo	oxoox	
xooox	ooxxo	

In terms of combinations, the number of orderings = the number of ways of selecting 2 cells to be neutrophils out of 5 cells = $\binom{5}{2} = (5 \times 4)/(2 \times 1) = 10$.

The probability of any of the orderings in Table 4.4 is the same as that for the ordering $oxoox$, namely, $(.6)^2(.4)^3$. Thus, the probability of obtaining 2 neutrophils in 5 cells is $\binom{5}{2}(.6)^2(.4)^3 = 10(.6)^2(.4)^3 = .230$.

Suppose the neutrophil problem is now considered more generally, with n trials rather than 5 trials, and the question is asked, What is the probability of k successes (rather than 2 successes) in these n trials? The probability that the k successes will occur at k **specified** trials within the n trials and that the remaining trials will be failures is given by $p^k(1 - p)^{n-k}$. To compute the probability of k successes in any of the n trials, this probability must be multiplied by the number of ways in which k trials for the successes and $n - k$ trials for the failures can be selected = $\binom{n}{k}$, as was done in Table 4.4. Thus, the probability of k successes in n trials, or k neutrophils in n cells, is

$$\binom{n}{k}p^k(1-p)^{n-k} = \binom{n}{k}p^kq^{n-k}$$

EQUATION 4.5

The distribution of the number of successes in n statistically independent trials, where the probability of success on each trial is p, is known as the **binomial distribution** and has a probability mass function given by

$$Pr(X=k)=\binom{n}{k}p^k q^{n-k}, \qquad k=0,1,\ldots,n$$

Example 4.25 What is the probability of obtaining 2 boys out of 5 children if the probability of a boy is .51 at each birth and the sexes of successive children are considered independent random variables?

Solution Use a binomial distribution with $n = 5$, $p = .51$, $k = 2$. Compute

$$Pr(X=2)=\binom{5}{2}(.51)^2(.49)^3=\frac{5\times 4}{2\times 1}(.51)^2(.49)^3$$
$$=10(.51)^2(.49)^3=.306$$

4.8.1 Using Binomial Tables

Frequently, a number of binomial probabilities will need to be evaluated for the same n and p, which would be tedious if each probability had to be calculated from Equation 4.5. Instead, for small n ($n \le 20$) and selected values of p, refer to Table 1 in the Appendix, where individual binomial probabilities are calculated. In this table, the number of trials (n) is provided in the first column, the number of successes (k) out of the n trials is given in the second column, and the probability of success for an individual trial (p) is given in the first row. Binomial probabilities are provided for $n = 2, 3, \ldots, 20$; $p = .05, .10, \ldots, .50$.

Example 4.26 **Infectious Disease** Evaluate the probability of 2 lymphocytes out of 10 white blood cells if the probability that any one cell is a lymphocyte is .2.

Solution Refer to Table 1 with $n = 10$, $k = 2$, $p = .20$. The appropriate probability, given in the $k = 2$ row and $p = .20$ column under $n = 10$, is .3020.

Example 4.27 **Pulmonary Disease** An investigator notices that children develop chronic bronchitis in the first year of life in 3 out of 20 households where both parents are chronic bronchitics, as compared with the national incidence rate of chronic bronchitis, which is 5% in the first year of life. Is this difference "real" or can it be attributed to chance? Specifically, how likely are infants in at least 3 out of 20 households to develop chronic bronchitis if the probability of developing disease in any one household is .05?

Solution Suppose the underlying rate of disease in the offspring is .05. Under this assumption, the number of households where the infants develop chronic bronchitis will follow a binomial distribution with parameters $n = 20$, $p = .05$. Thus, the probability of observing k cases out of 20 with disease is given by

$$\binom{20}{k}(.05)^k(.95)^{20-k}, \quad k=0,1,\ldots,20$$

The question is, What is the probability of observing at least 3 cases? The answer is

$$Pr(X\ge 3)=\sum_{k=3}^{20}\binom{20}{k}(.05)^k(.95)^{20-k}=1-\sum_{k=0}^{2}\binom{20}{k}(.05)^k(.95)^{20-k}$$

These 3 probabilities in the sum can be evaluated using the binomial table (Table 1). Refer to $n = 20$, $p = .05$ and note that $Pr(X = 0) = .3585$, $Pr(X = 1) = .3774$, $Pr(X = 2) = .1887$.

Thus

$$Pr(X \geq 3) = 1 - (.3585 + .3774 + .1887) = .0754$$

Thus, $X \geq 3$ is an unusual event, but not very unusual. Usually .05 or less is the range of probabilities used to identify unusual events. This criterion is discussed in more detail in our work on p-values in Chapter 7. If 3 infants out of 20 were to develop the disease, it would be difficult to judge whether the familial aggregation was real until a larger sample was available.

One question that is sometimes asked is, Why did we use a criterion of $Pr(X \geq 3 \text{ cases})$ to define unusualness in Example 4.27, rather than $Pr(X = 3 \text{ cases})$? The latter is what we actually observed. An intuitive answer is that if the number of households studied where both parents were chronic bronchitics was very large (for example, $n = 1500$), then the probability of any specific occurrence would be small. For example, suppose 75 cases occurred among 1500 households where both parents are chronic bronchitics. If the incidence of chronic bronchitis was .05 in such families, then the probability of 75 cases among 1500 households would be

$$\binom{1500}{75}(.05)^{75}(.95)^{1425} = .047$$

This result is exactly consistent with the national incidence rate (5% of households with cases in the first year of life) and yet yields a small probability. This doesn't make intuitive sense. The alternative approach is to calculate the probability of obtaining a result at least as extreme as the one obtained (a probability of at least 75 cases out of 1500 households) if the incidence rate of .05 were applicable to families where both parents were chronic bronchitics. This would yield a probability of approximately .50 in the preceding example and would indicate that nothing very unusual at all is occurring in such families, which is clearly the correct conclusion. If this probability were small enough, then this would cast doubt on the assumption that the true incidence rate was .05 for such families. This was the approach used in Example 4.27 and is developed in more detail in our work on hypothesis testing in Chapter 7. Alternative approaches to the analysis of these data also exist, based on *Bayesian inference*, but are beyond the scope of this text.

One question that arises is how to use the binomial tables if the probability of success on an individual trial (p) is greater than .5. Recall that

$$\binom{n}{k} = \binom{n}{n-k}$$

Let X be a binomial random variable with parameters n and p, and Y be a binomial random variable with parameters n and $q = 1 - p$. Then Equation 4.5 can be rewritten as

EQUATION 4.6

$$Pr(X = k) = \binom{n}{k} p^k q^{n-k} = \binom{n}{n-k} q^{n-k} p^k = Pr(Y = n - k)$$

In words, the probability of obtaining k successes for a binomial random variable X with parameters n and p is the same as the probability of obtaining $n - k$ successes for a binomial random variable Y with parameters n and q. Clearly, if $p > .5$, then $q = 1 - p < .5$, and Table 1 can be used with sample size n, referring to the $n - k$ row and the q column to obtain the appropriate probability.

Example 4.28 **Infectious Disease** Evaluate the probabilities of obtaining k neutrophils out of 5 cells for $k = 0, 1, 2, 3, 4, 5$, where the probability that any one cell is a neutrophil is .6.

Solution Because $p > .5$, refer to the random variable Y with parameters $n = 5$, $p = 1 - .6 = .4$.

$$Pr(X=0)=\binom{5}{0}(.6)^0(.4)^5=\binom{5}{5}(.4)^5(.6)^0=Pr(Y=5)=.0102$$

on referring to the $k = 5$ row and $p = .40$ column under $n = 5$. Similarly,

$Pr(X = 1) = Pr(Y = 4) = .0768$ on referring to the 4 row and .40 column under $n = 5$
$Pr(X = 2) = Pr(Y = 3) = .2304$ on referring to the 3 row and .40 column under $n = 5$
$Pr(X = 3) = Pr(Y = 2) = .3456$ on referring to the 2 row and .40 column under $n = 5$
$Pr(X = 4) = Pr(Y = 1) = .2592$ on referring to the 1 row and .40 column under $n = 5$
$Pr(X = 5) = Pr(Y = 0) = .0778$ on referring to the 0 row and .40 column under $n = 5$

4.8.2 Using "Electronic" Tables

In many instances we will want to evaluate binomial probabilities for $n > 20$ and/or for values of p not given in Table 1 of the Appendix. For sufficiently large n, the normal distribution can be used to approximate the binomial distribution, and tables of the normal distribution can be used to evaluate binomial probabilities. This procedure is usually less tedious than evaluating binomial probabilities directly using Equation 4.5 and is studied in detail in Chapter 5. Alternatively, if the sample size is not large enough to use the normal approximation and if the value of p is not in Table 1, then an electronic table can be used to evaluate binomial probabilities.

One example of an electronic table is provided by Microsoft Excel. A menu of statistical functions is available to the user, including calculation of probabilities for many probability distributions including, but not limited to, those discussed in this text. For example, one element in the statistical-functions menu is the binomial distribution function. Using this function, we can calculate the probability mass function and cdf for virtually any binomial distribution.

Example 4.29 **Pulmonary Disease** Compute the probability of obtaining exactly 75 cases of chronic bronchitis and the probability of obtaining at least 75 cases of chronic bronchitis in the first year of life among 1500 families, where both parents are chronic bronchitics, if the underlying incidence rate of chronic bronchitis in the first year of life is .05.

Solution We use the binomial distribution function of Excel 97 to solve this problem. Table 4.5 gives the results. First we compute $Pr(X = 75)$, which is .047, which is unusual. We then use the cdf option to compute $Pr(X \le 74)$, which equals .483. Finally, we compute the probability of obtaining at least 75 cases by

$$Pr(X \ge 75) = 1 - Pr(X \le 74) = .517$$

Hence, obtaining 75 cases out of 1500 children is clearly not unusual.

TABLE 4.5 **Calculation of binomial probabilities using Excel 97**

n	1500
k	75
p	0.05
Pr(X = 75)	0.047210
Pr(X <= 74)	0.483458
Pr(X >= 75)	0.516542

Example 4.30 **Infectious Disease** Suppose that a group of 100 men aged 60–64 received a new flu vaccine in 1986 and that 5 of them died within the next year. Is this event unusual, or can this death rate be expected for people of this age–sex group? Specifically, how likely are at least 5 out of 100 60- to 64-year-old men who receive a flu vaccine to die in the next year?

Solution We first find the expected annual death rate in 60- to 64-year-old males. From a 1986 U.S. life table, we find that 60- to 64-year-old men have an approximate probability of death in the next year of .020 [3]. Thus, from the binomial distribution the probability that k out of 100 men will die during the next year is given by $\binom{100}{k}(.02)^k(.98)^{100-k}$. We want to know if 5 deaths in a sample of 100 men is an "unusual" event. One criterion for this evaluation might be to find the probability of finding at least 5 deaths in this group = $Pr(X \geq 5)$ given that the probability of death for an individual man is .02. This probability can be expressed as

$$\sum_{k=5}^{100}\binom{100}{k}(.02)^k(.98)^{100-k}$$

Because this sum of 96 probabilities is tedious to compute, we instead compute

$$Pr(X < 5) = \sum_{k=0}^{4}\binom{100}{k}(.02)^k(.98)^{100-k}$$

and then evaluate $Pr(X \geq 5) = 1 - Pr(X < 5)$. The binomial tables cannot be used because $n > 20$. Therefore, the sum of 5 binomial probabilities is evaluated using Excel, as shown in Table 4.6.

TABLE 4.6 **Calculation of probability of at least 5 deaths among 100 men 60–64 years old in 1986**

n	100
p	0.02
Pr(X <= 4)	0.94917
Pr(X >= 5)	0.05083

We see that

$$Pr(X \leq 4) = .949$$

and $Pr(X \geq 5) = 1 - Pr(X \leq 4) = .051$

Thus, at least 5 deaths in 100 is a slightly unusual, but not a very unusual, event. If there were 10 deaths rather than 5, then using the same approach,

$$Pr(X \geq 10) = 1 - Pr(X < 10) < .001$$

which is very unlikely and would probably be grounds for considering halting use of the vaccine in the absence of any other evidence.

SECTION 4.9 Expected Value and Variance of the Binomial Distribution

The expected value and variance of the binomial distribution are important both in terms of our general knowledge about the binomial distribution and for our later work on estimation and hypothesis testing. From Definition 4.5 we know that the general formula for the expected value of a discrete random variable is

$$E(X) = \sum_{i=1}^{R} x_i Pr(X = x_i)$$

In the special case of a binomial distribution, the only values that take on positive probability are 0, 1, 2, . . . , n, and these values occur with probabilities

$$\binom{n}{0} p^0 q^n, \binom{n}{1} p^1 q^{n-1}, \ldots$$

Thus $E(X) = \sum_{k=0}^{n} k \binom{n}{k} p^k q^{n-k}$

It can be shown that this summation reduces to the simple expression np. Similarly, using Definition 4.6, we can show that

$$Var(X) = \sum_{k=0}^{n} (k - np)^2 \binom{n}{k} p^k q^{n-k} = npq$$

which leads directly to the following result:

EQUATION 4.7

The **expected value** and the **variance of a binomial distribution** are np and npq, respectively.

These results make good sense, because the expected number of successes in n trials is simply the probability of success on one trial multiplied by n, which equals np. Furthermore, for a given number of trials n, the binomial distribution has the highest variance when $p = \frac{1}{2}$, as shown in Figure 4.3. The variance of the distribution decreases as p moves away from $\frac{1}{2}$ in either direction, becoming 0 when $p = 0$ or 1. This result makes sense, because when $p = 0$ there must be 0 successes in n trials, and when $p = 1$ there must be n successes in n trials, and there is no variability

FIGURE 4.3 Plot of pq versus p

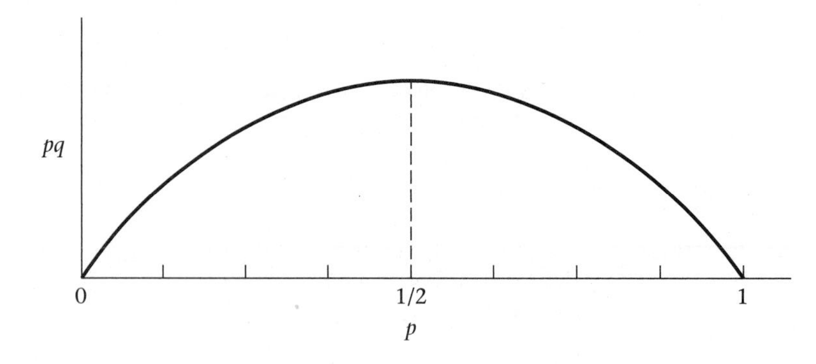

in either instance. Furthermore, when p is near 0 or near 1, the distribution of the number of successes is clustered near 0 and n, respectively, and there is comparatively little variability as compared with the situation when $p = \frac{1}{2}$. This point is depicted in Figure 4.4.

SECTION 4.10 The Poisson Distribution

The Poisson distribution is perhaps the second most frequently used discrete distribution after the binomial distribution. This distribution is usually associated with rare events.

Example 4.31 **Infectious Disease** Consider the distribution of the number of deaths attributed to typhoid fever over a long period of time, for example, 1 year. Assuming that the probability of a new death from typhoid fever in any one day is very small and that the number of cases reported in any two distinct periods of time are independent random variables, then the number of deaths over a 1-year period will follow a Poisson distribution.

Example 4.32 **Bacteriology** The preceding example concerns a rare event occurring over time. Rare events can also be considered not only over time but also on a surface area, such as the distribution of the number of bacterial colonies growing on an agar plate. Suppose we have a 100-cm^2 agar plate and that the probability of finding any bacterial colonies at any 1 point a (or more precisely in a small area around a) is very small and that the events of finding bacterial colonies at any 2 points a_1, a_2 are independent. The number of bacterial colonies over the entire agar plate will follow a Poisson distribution.

Consider Example 4.31. Ask the question, What is the distribution of the number of deaths due to typhoid fever from time 0 to time t (where t is some long period of time, such as 1 year or 20 years)?

Three assumptions must be made about the incidence of the disease. Consider any general *small* subinterval of the time period t, denoted by Δt.

FIGURE 4.4 **The binomial distribution for various values of p when $n = 10$**

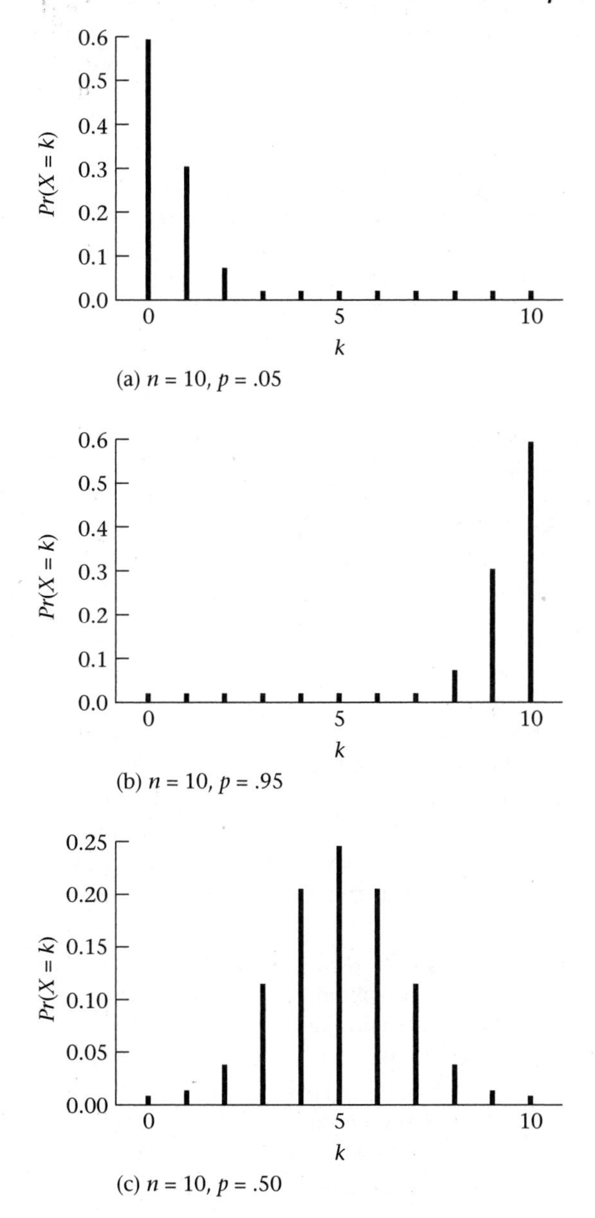

ASSUMPTION 4.1 Assume that

(1) The probability of observing 1 death is directly proportional to the length of the time interval Δt. That is, $Pr(1 \text{ death}) \approx \lambda \Delta t$ for some constant λ.

(2) The probability of observing 0 deaths over Δt is approximately $1 - \lambda \Delta t$.

(3) The probability of observing more than 1 death over this time interval is essentially 0.

ASSUMPTION 4.2 **Stationarity** Assume that the number of deaths per unit time is the same throughout the entire time interval t. Thus, an increase in the incidence of the disease as time goes on within the time period t would violate this assumption. Note that t should not be overly long, because this assumption is less likely to hold as t increases.

ASSUMPTION 4.3 **Independence** If a death occurs within one time subinterval, it has no bearing on the probability of death in the next time subinterval. This assumption would be violated in an epidemic situation, because if a new case of disease occurs, then subsequent deaths are likely to build up over a short period of time until after the epidemic subsides.

Given these assumptions, the Poisson probability distribution can be derived:

EQUATION 4.8 The probability of k events occurring in a time period t for a Poisson random variable with parameter λ is

$$Pr(X = k) = e^{-\mu}\mu^k/k!, \qquad k = 0, 1, 2, \ldots$$

where $\mu = \lambda t$ and e is approximately 2.71828.

Thus, the Poisson distribution depends on one parameter $\mu = \lambda t$. Note that the parameter λ represents the *expected number of events per unit time,* whereas the parameter μ represents the *expected number of events over the time period t.* One important difference between the Poisson distribution and the binomial distribution concerns the numbers of trials and events. For a binomial distribution there are a finite number of trials n, and the number of events can be no larger than n. For a Poisson distribution the number of trials is essentially infinite and the number of events (or number of deaths) can be indefinitely large, although the probability of k events will get very small as k gets large.

Example 4.33 **Infectious Disease** Consider the typhoid-fever example. Suppose the number of deaths attributable to typhoid fever over a 1-year period is Poisson with parameter μ = 4.6. What is the probability distribution of the number of deaths over a 6-month period? A 3-month period?

Solution Let X = the number of deaths in 6 months. Because $\mu = 4.6$, $t = 1$ year, it follows that λ = 4.6 deaths per year. For a 6-month period we have λ = 4.6 deaths per year, t = .5 year. Thus, $\mu = \lambda t = 2.3$. Therefore,

$$Pr(X = 0) = e^{-2.3} = .100$$

$$Pr(X = 1) = \frac{2.3}{1!}e^{-2.3} = .231$$

$$Pr(X = 2) = \frac{2.3^2}{2!}e^{-2.3} = .265$$

$$Pr(X = 3) = \frac{2.3^3}{3!}e^{-2.3} = .203$$

$$Pr(X = 4) = \frac{2.3^4}{4!}e^{-2.3} = .117$$

$$Pr(X = 5) = \frac{2.3^5}{5!}e^{-2.3} = .054$$

$$Pr(X \geq 6) = 1 - (.100 + .231 + .265 + .203 + .117 + .054) = .030$$

Let Y = the number of deaths in 3 months. For a 3-month period, we have $\lambda = 4.6$ deaths per year, $t = .25$ year, $\mu = \lambda t = 1.15$. Therefore,

$$Pr(Y = 0) = e^{-1.15} = .317$$

$$Pr(Y = 1) = \frac{1.15}{1!}e^{-1.15} = .364$$

$$Pr(Y = 2) = \frac{1.15^2}{2!}e^{-1.15} = .209$$

$$Pr(Y = 3) = \frac{1.15^3}{3!}e^{-1.15} = .080$$

$$Pr(Y \geq 4) = 1 - (.317 + .364 + .209 + .080) = .030$$

These distributions are plotted in Figure 4.5. Note that the distribution tends to become more symmetric as the time interval increases or, more specifically, as μ increases.

FIGURE 4.5 **Distribution of the number of deaths attributable to typhoid fever over various time intervals**

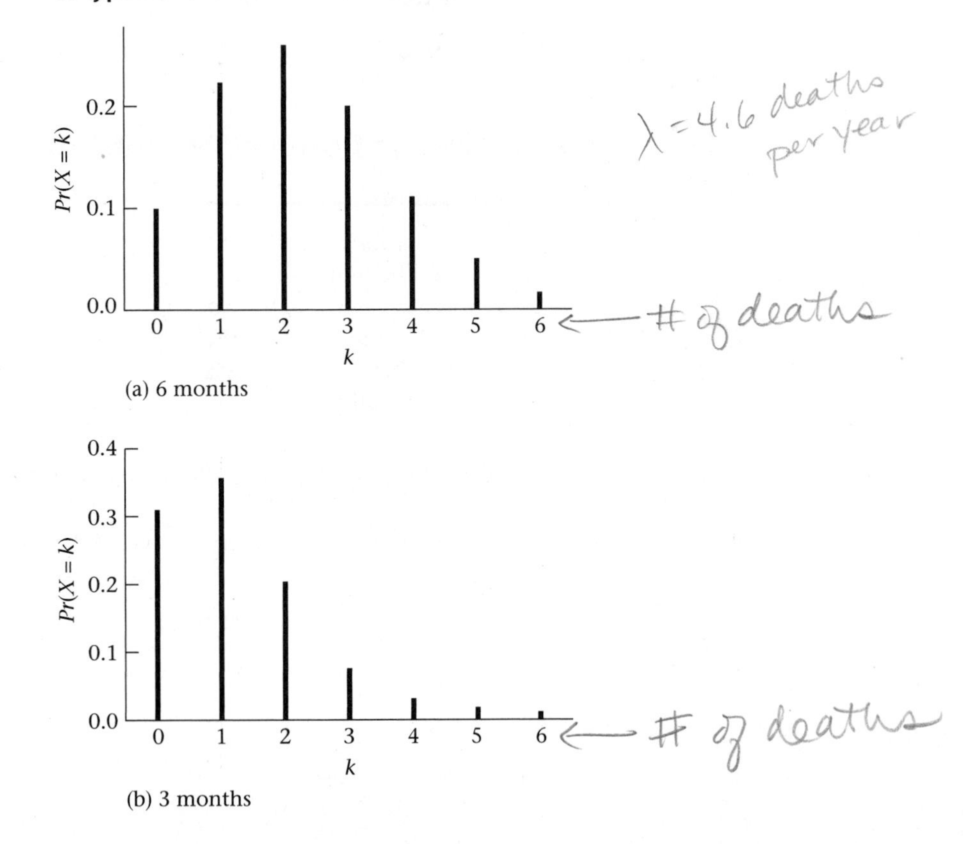

The Poisson distribution can also be applied to Example 4.32, where the distribution of the number of bacterial colonies in an agar plate of area A is discussed. Assuming that the probability of finding 1 colony in an area the size of ΔA at any point on the plate is $\lambda \Delta A$ for some λ and that the number of bacterial colonies found at 2 different points of the plate are independent random variables, then the probability of finding k bacterial colonies in an area of size A is given by $e^{-\mu}\mu^k/k!$, where $\mu = \lambda A$.

Example 4.34 **Bacteriology** If $A = 100$ cm^2 and $\lambda = .02$ colonies per cm^2, calculate the probability distribution of the number of bacterial colonies.

Solution We have that $\mu = \lambda A = 100(.02) = 2$. Let X = the number of colonies.

$$Pr(X = 0) = e^{-2} = .135$$

$$Pr(X = 1) = e^{-2}2^1/1! = 2e^{-2} = .271$$

$$Pr(X = 2) = e^{-2}2^2/2! = 2e^{-2} = .271$$

$$Pr(X = 3) = e^{-2}2^3/3! = \frac{4}{3}e^{-2} = .180$$

$$Pr(X = 4) = e^{-2}2^4/4! = \frac{2}{3}e^{-2} = .090$$

$$Pr(X \geq 5) = 1 - (.135 + .271 + .271 + .180 + .090) = .053$$

Clearly, the larger λ is, the more bacterial colonies we would expect to find.

SECTION 4.11 Computation of Poisson Probabilities

4.11.1 Using Poisson Tables

A number of Poisson probabilities for the same parameter μ often need to be evaluated. This task would be tedious if Equation 4.8 had to be applied repeatedly. Instead, for $\mu \leq 20$ refer to Table 2 in the Appendix, in which individual Poisson probabilities are specifically calculated. In this table the Poisson parameter μ is given in the first row, the number of events (k) is given in the first column, and the corresponding Poisson probability is given in the k row and μ column.

Example 4.35 Compute the probability of obtaining a least 5 events for a Poisson distribution with parameter $\mu = 3$.

Solution Refer to Table 2 under the 3.0 column. Let X = the number of events.

$$Pr(X = 0) = .0498$$

$$Pr(X = 1) = .1494$$

$$Pr(X = 2) = .2240$$

$$Pr(X = 3) = .2240$$

$$Pr(X = 4) = .1680$$

Thus $Pr(X \geq 5) = 1 - Pr(X \leq 4)$
$= 1 - (.0498 + .1494 + .2240 + .2240 + .1680)$
$= 1 - .8152 = .1848$

4.11.2 Electronic Tables for the Poisson Distribution

In many instances we will want to evaluate a collection of Poisson probabilities for the same μ, but μ will not be given in Table 2 of the Appendix. For large μ ($\mu \geq 10$), a normal approximation, as given in Chapter 5, can be used. Otherwise, an electronic table similar to that presented for the binomial distribution can be used.

Example 4.36 **Infectious Disease** Calculate the probability distribution of deaths caused by typhoid fever over a 1-year period using the information given in Example 4.33.

In this case, we model the number of deaths caused by typhoid fever by a Poisson distribution with $\mu = 4.6$. We will use the Poisson distribution program of Excel 97. The results are given in Table 4.7. We see that 9 or more deaths caused by typhoid fever would be unusual over a 1-year period.

TABLE 4.7 **Calculation of the probability distribution of the number of deaths caused by typhoid fever over a 1-year period**

Number of deaths	Probability
0	0.010
1	0.046
2	0.106
3	0.163
4	0.188
5	0.173
6	0.132
7	0.087
8	0.050
<=8	0.955
>=9	0.045

SECTION 4.12 Expected Value and Variance of the Poisson Distribution

In many instances we cannot predict whether the assumptions for the Poisson distribution given in Section 4.10 are satisfied. Fortunately, the relationship between the expected value and variance of the Poisson distribution provides an important guideline that helps identify random variables that follow this distribution. This relationship can be stated as follows:

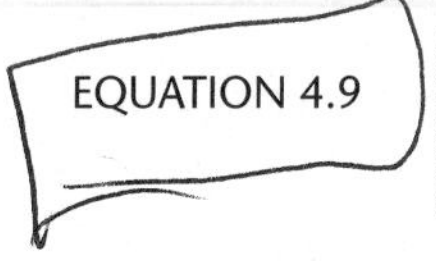

For a Poisson distribution with parameter μ, the mean and variance are both equal to μ.

This fact is useful, because if we have a data set from a discrete distribution where the *mean and variance are about the same,* then we can preliminarily identify it as a Poisson distribution and use various tests to confirm this hypothesis.

Example 4.37 **Infectious Disease** The number of deaths attributable to polio during the years 1968–1976 is given in Table 4.8 [4, 5]. Comment on the applicability of the Poisson distribution to this data set.

Solution The sample mean and variance of the annual number of deaths caused by polio during the period 1968–1976 are 11.3 and 51.5, respectively. The Poisson distribution clearly will not fit well here, because the variance is 4.5 times as large as the mean. The larger variance is probably caused by the clustering of polio deaths at certain times and geographical locations, which leads to a violation of both the independence assumption and the assumption of constant incidence over time.

TABLE 4.8 **Number of deaths attributable to polio during the years 1968–1976**

Year	1968	1969	1970	1971	1972	1973	1974	1975	1976
Number of deaths	24	13	7	18	2	10	3	9	16

Suppose we are studying a rare phenomenon and wish to apply the Poisson distribution. A question that often arises is how to estimate the parameter μ of the Poisson distribution in this context. Because the expected value of the Poisson distribution is μ, μ can be estimated by the observed mean number of events over time t (e.g., 1 year), if such data are available. If the data are not available, other data sources can be used to estimate μ.

Example 4.38 **Occupational Health** A public health issue arose concerning the possible carcinogenic potential of food ingredients containing ethylene dibromide (EDB). In some instances foods were removed from public consumption if they were shown to have excessive quantities of EDB. A study was previously performed looking at the mortality experience of 161 White male employees of two plants in Texas and Michigan who were exposed to EDB over the time period 1940–1975 [6]. Seven deaths due to cancer were observed among these employees. For this time period, 5.8 cancer deaths were expected as calculated from overall mortality rates for U.S. White men. Assess if the observed number of cancer deaths was excessive in this group.

Solution Estimate the parameter μ from the expected number of cancer deaths from U.S. White male mortality rates; that is, $\mu = 5.8$. Then calculate $Pr(X \geq 7)$, where X is a Poisson random variable with parameter $\mu = 5.8$. Use the relationship

$$Pr(X \geq 7) = 1 - Pr(X \leq 6)$$

where $Pr(X = k) = e^{-5.8}(5.8)^k/k!$

Because $\mu = 5.8$ is not in Table 2 of the Appendix we will use Excel 97 to perform the calculations. The results are given in Table 4.9.

TABLE 4.9 **Calculation of the probability distribution of the number of cancer deaths in the EDB example**

Mean Number of Deaths	Probability
0	0.003
1	0.018
2	0.051
3	0.098
4	0.143
5	0.166
6	0.160
<=6	0.638
>=7	0.362

Thus, $Pr(X \geq 7) = 1 - Pr(X \leq 6)$

$$= 1 - .638 = .362$$

Clearly, the observed number of cancer deaths is not excessive in this group.

SECTION 4.13 Poisson Approximation to the Binomial Distribution

As noted in the preceding section, the Poisson distribution appears to fit well in some applications. Another important use for the Poisson distribution is as an approximation to the binomial distribution. Consider the binomial distribution for large n and small p. The mean of this distribution is given by np and the variance by npq. Note that $q \approx$ (is approximately equal to) 1 for small p, and thus $npq \approx np$. Therefore, the mean and variance of the binomial distribution are almost equal in this case, which suggests the following rule:

EQUATION 4.10

> **Poisson Approximation to the Binomial Distribution** The binomial distribution with large n and small p can be accurately approximated by a Poisson distribution with parameter $\mu = np$.

The rationale for using this approximation is that the Poisson distribution is easier to work with than the binomial distribution. The binomial distribution involves expressions such as $\binom{n}{k}$ and $(1 - p)^{n-k}$, which are cumbersome for large n.

Example 4.39 **Cancer, Genetics** Suppose we are interested in the genetic susceptibility to breast cancer. We find that 4 out of 1000 women aged 40–49 whose mothers have had breast cancer develop breast cancer over the next year of life. We would expect from large population studies that 1 in 1000 women of this age group will develop a new case of the disease over this period of time. How unusual is this event?

Solution The exact binomial probability could be computed by letting $n = 1000$, $p = 1/1000$. Hence,

$$Pr(X \geq 4) = 1 - Pr(X \leq 3)$$
$$= 1 - \begin{bmatrix} \binom{1000}{0}(.001)^0(.999)^{1000} + \binom{1000}{1}(.001)^1(.999)^{999} \\ + \binom{1000}{2}(.001)^2(.999)^{998} + \binom{1000}{3}(.001)^3(.999)^{997} \end{bmatrix}$$

Instead, use the Poisson approximation with $\mu = 1000(.001) = 1$, which is obtained as follows:

$$Pr(X \geq 4) = 1 - [Pr(X = 0) + Pr(X = 1) + Pr(X = 2) + Pr(X = 3)]$$

Using Table 2 of the Appendix under the $\mu = 1.0$ column, we find that

$$Pr(X = 0) = .3679$$
$$Pr(X = 1) = .3679$$
$$Pr(X = 2) = .1839$$
$$Pr(X = 3) = .0613$$

Thus, $Pr(X \geq 4) = 1 - (.3679 + .3679 + .1839 + .0613)$
$= 1 - .9810 = .0190$

This event is indeed unusual and suggests a genetic susceptibility to breast cancer among female offspring of women who have had breast cancer. For comparative purposes, we have computed the exact binomial probabilities of obtaining 0, 1, 2, and 3 events, which are given by .3677, .3681, .1840, and .0613, respectively. The corresponding exact binomial probability of obtaining 4 or more breast-cancer cases is .0189, which agrees almost exactly with the Poisson approximation of .0190 just given.

How large should n be or how small should p be before the approximation is "adequate"? A conservative rule is to use the approximation when $n \geq 100$ and $p \leq .01$. As an example, we give the exact binomial probability and the Poisson approximation for $n = 100$, $p = .01$, $k = 0, 1, 2, 3, 4, 5$ in Table 4.10. The two probability distributions agree to within .002 in all instances.

TABLE 4.10 **An example of the Poisson approximation to the binomial distribution for $n = 100$, $p = .01$, $k = 0, 1, \ldots, 5$**

k	Exact binomial probability	Poisson approximation	k	Exact binomial probability	Poisson approximation
0	.366	.368	3	.061	.061
1	.370	.368	4	.015	.015
2	.185	.184	5	.003	.003

Example 4.40 **Infectious Disease** An outbreak of poliomyelitis occurred in Finland in 1984 after 20 years without a single case being reported in the country. As a result, an intensive immunization campaign was conducted within 5 weeks between February 9 and March 15, 1985; it covered 94% of the population and was highly successful. During and after the campaign, several patients with Guillain-Barré syndrome (GBS), a rare neurologic disease often resulting in paralysis, were admitted to the neurologic units of hospitals in Finland [7].

The authors provided data on monthly incidence of GBS from April 1984 to October 1985. These data are given in Table 4.11.

TABLE 4.11 **Monthly incidence of GBS in Finland from April 1984 to October 1985**

Month	Number of GBS cases	Month	Number of GBS cases	Month	Number of GBS cases
April 1984	3	October 1984	2	April 1985	7
May 1984	7	November 1984	2	May 1985	2
June 1984	0	December 1984	3	June 1985	2
July 1984	3	January 1985	3	July 1985	6
August 1984	4	February 1985	8	August 1985	2
September 1984	4	March 1985	14	September 1985	2
				October 1985	6

Determine if the number of cases in March 1985 is excessive compared to the experience in the other 18 months based on the data in Table 4.11.

Solution If there are n people in Finland who could get GBS and the monthly incidence of GBS (p) is low, then we could model the number of GBS cases in 1 month (X) by a binomial distribution with parameters n and p. Since n is large and p is small, it is reasonable to approximate the distribution of the number of GBS cases in 1 month (X) by a Poisson distribution with parameter $\mu = np$. To estimate μ, we use the average monthly number of GBS cases during the 18-month period from April 1984 to October 1985, excluding the vaccine month of March 1985. The mean number of cases per month = $(3 + 7 + \cdots + 6)/18 = 3.67$. We now assess if the number of cases in March 1985 (14) is excessive, by computing $Pr(X \geq 14|\mu = 3.67)$. We use Excel 97 to perform this computation, as shown in the table.

Probability of observing 14 or more cases of GBS in Finland during March 1985

Mean	3.67
Pr(X <= 13)	0.999969
Pr(X >= 14)	3.09E-05

The results indicate that $Pr(X \geq 14|\mu = 3.67) = 3.09 \times 10^{-5}$. Thus, 14 cases in one month is very unusual given the 18-month experience in nonvaccine months and possibly indicates that the large number of cases seen in March 1985 are attributable to the vaccination campaign.

SECTION 4.14 Summary

In this chapter, random variables were discussed and a distinction between discrete and continuous random variables was made. Specific attributes of random variables, including the notions of probability mass function (or probability distribution), cumulative-distribution function, expected value, and variance were introduced. These notions were shown to be related to similar concepts for finite samples, which were discussed in Chapter 2. In particular, the sample frequency distribution is a sample realization of a probability distribution, whereas the sample mean ($\bar{x}$) and variance (s^2) are sample analogues of the expected value and variance, respectively, of a random variable. The relationship between attributes of probability models and finite samples is explored in more detail in Chapter 6.

Finally, some specific probability models were introduced, focusing on the binomial and Poisson distributions. The binomial distribution was shown to be applicable to binary outcomes, that is, if only two outcomes are possible, where outcomes on different trials are independent. These two outcomes are labeled as "success" and "failure," where the probability of success is the same for each trial. The Poisson distribution is a classic model used to describe the distribution of rare events.

The study of probability models continues in Chapter 5, where the focus is on continuous random variables.

PROBLEMS

Let X be the random variable representing the number of hypertensive adults in Example 3.12.

***4.1** Derive the probability mass function for X.

***4.2** What is its expected value?

***4.3** What is its variance?

***4.4** What is its cumulative-distribution function?

Suppose we wish to check the accuracy of self-reported diagnoses of angina by getting further medical records on a subset of the cases.

4.5 If we have 50 reported cases of angina and we wish to select 5 for further review, then how many ways can we select these cases if the order of selection matters?

4.6 Answer Problem 4.5 if the order of selection does not matter.

4.7 Evaluate $\binom{10}{0}, \binom{10}{1}, \ldots, \binom{10}{10}$.

***4.8** Evaluate 9!.

4.9 Suppose that 6 out of 15 students in a grade-school class develop influenza, whereas 20% of grade-school students nationwide develop influenza. Is there evidence of an excessive number of cases in the class? That is, what is the probability of obtaining at least 6 cases in this class if the nationwide rate holds true?

4.10 What is the expected number of students in the class who will develop influenza?

***4.11** What is the probability of obtaining exactly 6 events for a Poisson distribution with parameter $\mu = 4.0$?

***4.12** What is the probability of obtaining at least 6 events for a Poisson distribution with parameter $\mu = 4.0$?

***4.13** What is the expected value and variance for a Poisson distribution with parameter $\mu = 4.0$?

Infectious Disease

Newborns were screened for human immunodeficiency virus (HIV or AIDS virus) in five Massachusetts hospitals. The data obtained [8] are shown in Table 4.12.

4.14 If 500 newborns are screened at the inner-city hospital, then what is the exact binomial probability of precisely 5 HIV-positive test results?

4.15 If 500 newborns are screened at the inner-city hospital, then what is the exact binomial probability of at least 5 HIV-positive test results?

TABLE 4.12 Seroprevalence of HIV antibody in newborns' blood samples, according to hospital category

Hospital	Type	Number tested	Number positive	Number positive (per 1000)
A	Inner city	3,741	30	8.0
B	Urban/suburban	11,864	31	2.6
C	Urban/suburban	5,006	11	2.2
D	Suburban/rural	3,596	1	0.3
E	Suburban/rural	6,501	8	1.2

4.16 Answer Problems 4.14 and 4.15 using an approximation rather than an exact probability.

4.17 Answer Problem 4.14 for a mixed urban/suburban hospital (hospital C).

4.18 Answer Problem 4.15 for a mixed urban/suburban hospital (hospital C).

4.19 Answer Problem 4.16 for a mixed urban/suburban hospital (hospital C).

4.20 Answer Problem 4.14 for a mixed suburban/rural hospital (hospital E).

4.21 Answer Problem 4.15 for a mixed suburban/rural hospital (hospital E).

4.22 Answer Problem 4.16 for a mixed suburban/rural hospital (hospital E).

Occupational Health

Many investigators have suspected that workers in the tire industry have an unusual incidence of cancer.

***4.23** Suppose the expected number of deaths due to bladder cancer for all workers in a tire plant on January 1, 1964, over the next 20 years (1/1/64–12/31/83) based on U.S. mortality rates is 1.8. If the Poisson distribution is assumed to hold and there are 6 reported deaths caused by bladder cancer among the tire workers, then how unusual is this event?

***4.24** Suppose a similar analysis is done for stomach cancer. In this plant, 4 deaths caused by stomach cancer are observed for the workers, whereas 2.5 are expected based on U.S. mortality rates. How unusual is this event?

Infectious Disease

One hypothesis is that gonorrhea tends to cluster in central cities.

4.25 Suppose that 10 gonorrhea cases are reported over a 3-month period among 10,000 people living in an urban county. The statewide incidence of gonorrhea is 50 per 100,000 over a 3-month period. Is the number of gonorrhea cases in this county unusual for this time period?

Otolaryngology

Assume that the number of episodes per year of otitis media, a common disease of the middle ear in early childhood, follows a Poisson distribution with parameter $\lambda = 1.6$ episodes per year.

***4.26** Find the probability of getting 3 or more episodes of otitis media in the first 2 years of life.

***4.27** Find the probability of not getting any episodes of otitis media in the first year of life.

An interesting question in pediatrics is whether the tendency for children to have many episodes of otitis media is inherited in a family.

***4.28** What is the probability that 2 siblings will both have 3 or more episodes of otitis media in the first 2 years of life?

***4.29** What is the probability that exactly 1 of the siblings will have 3 or more episodes in the first 2 years of life?

***4.30** What is the probability that neither sibling will have 3 or more episodes in the first 2 years of life?

***4.31** What is the expected number of siblings in a 2-sibling family that will have 3 or more episodes in the first 2 years of life?

Hypertension

Hypertension has often been claimed to have a "familial aggregation." That is, if 1 person in a family is hypertensive, then his or her siblings are more likely to be hypertensive. Suppose that the prevalence of hypertension among 50- to 59-year-olds in the general population is 18%. Suppose we identify sibships of size 3 in a community where all members of the sibship are 50- to 59 years old.

4.32 What is the probability that 0, 1, 2, or 3 hypertensives will be identified in such sibships if the hypertensive status of 2 siblings in the same family are independent events?

4.33 Suppose that among 25 sibships of this type, 5 have at least 2 affected siblings. Are these data consistent with the independence assumption in Problem 4.32?

Environmental Health, Obstetrics

Suppose that the rate of major congenital malformations in the general population is 2.5 malformations per 100 deliveries. A study is set up to investigate if the offspring of Vietnam-veteran fathers are at special risk of having congenital malformations.

***4.34** If 100 infants are identified in a birth registry as being offspring of Vietnam-veteran fathers and 4 have a major congenital malformation, then is there an excess risk of malformations in this group?

Using the same birth-registry data, let us look at the effect of maternal use of marijuana on the rate of major congenital malformations.

***4.35** Of 75 offspring of mothers who used marijuana, 8 are found to have a major congenital malformation. Is there an excess risk of malformations in this group?

Hypertension

A national study found that treating people appropriately for high blood pressure reduced their overall mortality by 20%. Treating people adequately for hypertension has been difficult, because it is estimated that 50% of hypertensives do not know they have high blood pressure; 50% of those who do know are inadequately treated by their physicians; and 50% who are appropriately treated fail to comply with this treatment by taking the appropriate number of pills.

4.36 What is the probability that among 10 true hypertensives at least 50% are being treated appropriately and are complying with this treatment?

4.37 What is the probability that at least 7 of the 10 hypertensives know they have high blood pressure?

4.38 If the preceding 50% rates were each reduced to 40% by a massive education program, then what effect would this change have on the overall mortality rate among true hypertensives; that is, would the mortality rate decrease, and if so, what percent of deaths among hypertensives could be prevented by the education program?

Renal Disease

The presence of bacteria in a urine sample (bacteriuria) is sometimes associated with symptoms of kidney disease in women. Suppose that a determination of bacteriuria has been made over a large population of women at one point in time and that 5% of those sampled are positive for bacteriuria.

***4.39** If a sample of size 5 is selected from this population, what would be the probability that 1 or more women would be positive for bacteriuria?

***4.40** Suppose 100 women from this population are sampled. What is the probability that 3 or more women would be positive for bacteriuria?

One interesting phenomenon of bacteriuria is that there is a "turnover"; that is, if bacteriuria is measured on the same woman at 2 different points in time, the results are not necessarily the same. Assume that 20% of all women who are bacteriuric at time 0 are again bacteriuric at time 1 (1 year later), whereas only 4.2% of women who were not bacteriuric at time 0 *are* bacteriuric at time 1. Let X be the random variable representing the number of bacteriuric events over the 2 time periods for 1 woman and still assume that the probability that a woman will be positive for bacteriuria at any one exam is 5%.

***4.41** What is the probability distribution of X?

***4.42** What is the mean of X?

***4.43** What is the variance of X?

Demography

In Table 4.13 we provide life-table data for the United States in 1986 [3]. This table can be used to estimate the probability of survival between any 2 ages for persons of a given race or sex. For example, for White males, to calculate the probability of survival from age 60 to age 62, we refer to the age 60 and 62 lines under the White male column and obtain a probability of 79,669/82,435 = .966. Refer to the 11 males among the 25 people described in Table 2.11. (The race of the subjects is not known, so use the "All races" section of Table 4.13.)

4.44 What is the expected number of deaths among the 11 males over the next year based on the life-table data?

4.45 Answer Problem 4.44 for a 2-year period.

4.46 Answer Problem 4.44 for a 3-year period.

Use a computer, if necessary, to answer Problems 4.47–4.52.

4.47 What is the probability of exactly 2 deaths among the 11 males over the next year?

4.48 Answer Problem 4.47 for a 2-year period.

4.49 Answer Problem 4.47 for a 3-year period.

4.50 What is the probability of at least 4 deaths among the 11 males over the next year?

4.51 Answer Problem 4.50 for a 2-year period.

4.52 Answer Problem 4.50 for a 3-year period.

TABLE 4.13 Number of survivors at single years of age, out of 100,000 born alive, by race and sex: United States, 1986

Age	All races			White			All other					
	Both sexes	Male	Female	Both sexes	Male	Female	Total			Black		
							Both sexes	Male	Female	Both sexes	Male	Female
0	100,000	100,000	100,000	100,000	100,000	100,000	100,000	100,000	100,000	100,000	100,000	100,000
1	98,964	98,845	99,090	99,106	98,998	99,220	98,426	98,262	98,597	98,190	97,996	98,391
2	98,892	98,764	99,028	99,040	98,923	99,164	98,332	98,158	98,513	98,085	97,882	98,296
3	98,838	98,704	98,980	98,991	98,869	99,121	98,256	98,075	98,444	98,000	97,790	98,218
4	98,796	98,658	98,942	98,953	98,828	99,087	98,194	98,007	98,388	97,932	97,716	98,155
5	98,762	98,620	98,912	98,923	98,794	99,060	98,143	97,951	98,342	97,876	97,655	98,104
6	98,733	98,587	98,887	98,897	98,765	99,038	98,100	97,904	98,304	97,830	97,605	98,062
7	98,707	98,557	98,866	98,874	98,738	99,019	98,064	97,863	98,272	97,792	97,563	98,028
8	98,684	98,530	98,847	98,853	98,712	99,002	98,033	97,828	98,246	97,760	97,526	97,999
9	98,663	98,505	98,830	98,834	98,689	98,987	98,006	97,797	98,223	97,732	97,494	97,975
10	98,645	98,484	98,815	98,817	98,669	98,973	97,982	97,769	98,203	97,706	97,464	97,954
11	98,628	98,464	98,801	98,802	98,651	98,960	97,959	97,742	98,184	97,681	97,435	97,934
12	98,610	98,443	98,786	98,785	98,632	98,946	97,935	97,713	98,166	97,655	97,404	97,914
13	98,587	98,415	98,768	98,763	98,606	98,929	97,907	97,677	98,146	97,625	97,365	97,893
14	98,554	98,372	98,745	98,730	98,564	98,906	97,871	97,629	98,123	97,587	97,314	97,869
15	98,507	98,309	98,716	98,683	98,501	98,876	97,825	97,565	98,095	97,538	97,247	97,840
16	98,445	98,223	98,678	98,620	98,414	98,838	97,767	97,484	98,061	97,477	97,161	97,805
17	98,369	98,116	98,633	98,542	98,306	98,792	97,696	97,384	98,021	97,403	97,056	97,763
18	98,280	97,991	98,583	98,452	98,180	98,741	97,612	97,264	97,975	97,315	96,930	97,715
19	98,183	97,852	98,530	98,355	98,041	98,688	97,515	97,123	97,923	97,214	96,781	97,662
20	98,081	97,702	98,477	98,254	97,894	98,635	97,405	96,959	97,867	97,098	96,607	97,603
21	97,974	97,542	98,424	98,150	97,740	98,583	97,281	96,771	97,805	96,967	96,407	97,538
22	97,861	97,372	98,370	98,043	97,578	98,532	97,143	96,560	97,737	96,821	96,181	97,467
23	97,745	97,196	98,316	97,935	97,412	98,482	96,993	96,331	97,664	96,661	95,933	97,390
24	97,628	97,018	98,261	97,827	97,247	98,432	96,834	96,089	97,586	96,490	95,670	97,306
25	97,512	96,843	98,204	97,720	97,086	98,381	96,669	95,839	97,502	96,311	95,395	97,216
26	97,397	96,672	98,147	97,616	96,931	98,330	96,499	95,582	97,413	96,124	95,110	97,119
27	97,283	96,504	98,088	97,514	96,780	98,278	96,322	95,317	97,318	95,927	94,814	97,015
28	97,168	96,336	98,027	97,412	96,632	98,225	96,137	95,043	97,217	95,719	94,503	96,902
29	97,050	96,164	97,964	97,309	96,481	98,171	95,942	94,756	97,107	95,497	94,174	96,778
30	96,927	95,986	97,897	97,202	96,325	98,115	95,735	94,455	96,988	95,258	93,823	96,642
31	96,798	95,800	97,826	97,090	96,162	98,056	95,515	94,139	96,858	95,001	93,450	96,492
32	96,663	95,606	97,751	96,973	95,993	97,994	95,283	93,807	96,718	94,727	93,054	96,329
33	96,522	95,405	97,672	96,852	95,818	97,928	95,038	93,459	96,568	94,436	92,636	96,154
34	96,377	95,199	97,588	96,727	95,639	97,859	94,781	93,093	96,411	94,131	92,196	95,970
35	96,227	94,988	97,500	96,598	95,457	97,786	94,512	92,708	96,247	93,812	91,735	95,779
36	96,072	94,771	97,407	96,465	95,271	97,708	94,230	92,303	96,076	93,480	91,252	95,581
37	95,910	94,546	97,308	96,326	95,079	97,625	93,934	91,877	95,896	93,132	90,745	95,373
38	95,739	94,311	97,202	96,180	94,878	97,534	93,622	91,429	95,705	92,765	90,212	95,152
39	95,557	94,065	97,086	96,024	94,667	97,434	93,291	90,958	95,498	92,373	89,649	94,911
40	95,363	93,805	96,958	95,856	94,443	97,323	92,938	90,462	95,271	91,952	89,052	94,646
41	95,153	93,528	96,816	95,674	94,203	97,199	92,560	89,939	95,022	91,500	88,419	94,353
42	94,927	93,232	96,659	95,476	93,945	97,061	92,156	89,386	94,751	91,015	87,749	94,033
43	94,683	92,916	96,487	95,261	93,668	96,909	91,727	88,806	94,458	90,501	87,046	93,687
44	94,421	92,579	96,300	95,029	93,371	96,743	91,276	88,201	94,145	89,962	86,315	93,318
45	94,139	92,218	96,097	94,778	93,051	96,563	90,803	87,572	93,812	89,400	85,559	92,928
46	93,836	91,831	95,878	94,507	92,705	96,367	90,307	86,919	93,458	88,816	84,781	92,518
47	93,508	91,413	95,639	94,212	92,329	96,153	89,784	86,237	93,081	88,206	83,974	92,082
48	93,151	90,960	95,377	93,888	91,919	95,917	89,226	85,514	92,673	87,559	83,125	91,614
49	92,758	90,464	95,087	93,531	91,469	95,656	88,620	84,733	92,226	86,862	82,217	91,104
50	92,325	89,919	94,766	93,137	90,973	95,365	87,957	83,881	91,733	86,104	81,237	90,545
51	91,848	89,320	94,409	92,701	90,427	95,042	87,230	82,951	91,189	85,280	80,178	89,931
52	91,324	88,665	94,016	92,221	89,827	94,684	86,440	81,945	90,594	84,389	79,042	89,261
53	90,752	87,949	93,586	91,694	89,167	94,291	85,591	80,869	89,951	83,437	77,836	88,540
54	90,130	87,169	93,122	91,117	88,441	93,864	84,693	79,735	89,268	82,433	76,575	87,774
55	89,458	86,322	92,623	90,488	87,644	93,401	83,749	78,549	88,548	81,382	75,266	86,968
56	88,733	85,405	92,088	89,803	86,773	92,901	82,763	77,316	87,794	80,287	73,915	86,124
57	87,951	84,413	91,513	89,058	85,823	92,360	81,729	76,030	87,000	79,142	72,514	85,236
58	87,103	83,339	90,888	88,247	84,788	91,770	80,630	74,674	86,148	77,930	71,046	84,284
59	86,180	82,173	90,203	87,361	83,661	91,121	79,444	73,227	85,215	76,626	69,487	83,244
60	85,173	80,908	89,449	86,393	82,435	90,406	78,156	71,675	84,185	75,215	67,822	82,097
61	84,077	79,539	88,619	85,338	81,105	89,619	76,757	70,011	83,047	73,687	66,042	80,834
62	82,891	78,065	87,712	84,193	79,669	88,758	75,254	68,243	81,808	72,051	64,157	79,461
63	81,618	76,492	86,731	82,961	78,131	87,823	73,665	66,388	80,485	70,328	62,189	77,999
64	80,264	74,827	85,681	81,645	76,497	86,817	72,016	64,473	79,104	68,548	60,167	76,479
65	78,833	73,076	84,565	80,246	74,770	85,740	70,325	62,516	77,684	66,732	58,113	74,922
66	77,327	71,244	83,381	78,766	72,955	84,590	68,600	60,526	76,232	64,890	56,039	73,337
67	75,740	69,325	82,122	77,199	71,046	83,360	66,834	58,499	74,737	63,015	53,941	71,715
68	74,059	67,305	80,776	75,532	69,027	82,040	65,011	56,423	73,180	61,092	51,805	70,035
69	72,267	65,161	79,330	73,750	66,877	80,618	63,109	54,279	71,534	59,098	49,612	68,270
70	70,353	62,881	77,772	71,841	64,581	79,082	61,111	52,055	69,778	57,018	47,350	66,401
71	68,312	60,462	76,096	69,801	62,139	77,427	59,013	49,749	67,903	54,848	45,018	64,419
72	66,149	57,916	74,299	67,634	59,561	75,649	56,823	47,373	65,916	52,598	42,630	62,332
73	63,872	55,256	72,383	65,347	56,860	73,748	54,557	44,945	63,830	50,284	40,203	60,154
74	61,491	52,503	70,350	62,949	54,057	71,724	52,236	42,489	61,666	47,926	37,763	57,904
75	59,016	49,675	68,200	60,450	51,169	69,577	49,877	40,023	59,437	45,540	35,329	55,597
76	56,452	46,785	65,931	57,853	48,210	67,303	47,486	37,558	57,145	43,133	32,914	53,236
77	53,800	43,842	63,538	55,162	45,191	64,896	45,061	35,097	54,783	40,706	30,522	50,815
78	51,061	40,855	61,012	52,376	42,122	62,350	42,594	32,639	52,339	38,254	28,154	48,324
79	48,235	37,833	58,347	49,498	39,013	59,657	40,078	30,181	49,799	35,773	25,811	45,749
80	45,324	34,789	55,535	46,530	35,879	56,812	37,506	27,723	47,151	33,260	23,495	43,081
81	42,333	31,739	52,570	43,478	32,737	53,809	34,876	25,270	44,387	30,716	21,215	40,314
82	39,269	28,705	49,450	40,350	29,611	50,643	32,192	22,831	41,506	28,147	18,982	37,447
83	36,144	25,712	46,172	37,158	26,528	47,313	29,461	20,419	38,509	25,564	16,812	34,485
84	32,972	22,791	42,736	33,916	23,520	43,817	26,695	18,053	35,405	22,982	14,728	31,435
85	29,771	19,977	39,143	30,642	20,625	40,155	23,912	15,755	32,206	20,419	12,755	28,312

Pediatrics, Otolaryngology

Otitis media is a disease that occurs frequently in the first few years of life and is one of the most common reasons for physician visits after the routine check-up. A study was conducted to assess the frequency of otitis media in the general population in the first year of life. Table 4.14 gives the number of infants out of 2500 infants who were first seen at birth who remained disease-free by the end of the ith month of life, $i = 0, 1, \ldots, 12$. (Assume that no infants have been lost to follow-up.)

***4.53** What is the probability that an infant will have 1 or more episodes of otitis media by the end of the sixth month of life? The first year of life?

TABLE 4.14 Number of infants (out of 2500) who remain disease-free at the end of each month during the first year of life

month i	Disease-free infants at the end of month i
0	2500
1	2425
2	2375
3	2300
4	2180
5	2000
6	1875
7	1700
8	1500
9	1300
10	1250
11	1225
12	1200

***4.54** What is the probability that an infant will have 1 or more episodes of otitis media by the end of the ninth month of life given that no episodes have been observed by the end of the third month of life?

***4.55** Suppose an "otitis-prone family" is defined as one where at least 3 siblings out of 5 develop otitis media in the first 6 months of life. What proportion of 5-sibling families are otitis prone if we assume that the disease occurs independently for different siblings in a family?

***4.56** What is the expected number of otitis-prone families out of 100 5-sibling families?

Cancer, Epidemiology

An experiment is designed to test the potency of a drug on 20 rats. Previous animal studies have shown that a 10-mg dose of the drug is lethal 5% of the time within the first 4 hours; of the animals alive at 4 hours, 10% will die in the next 4 hours.

4.57 What is the probability that 3 or more rats will die in the first 4 hours?

4.58 Suppose 2 rats die in the first 4 hours. What is the probability that 2 or fewer rats will die in the next 4 hours?

4.59 What is the probability that 0 rats will die in the 8-hour period?

4.60 What is the probability that 1 rat will die in the 8-hour period?

4.61 What is the probability that 2 rats will die in the 8-hour period?

4.62 Can you write a general formula for the probability that x rats will die in the 8-hour period? Evaluate this formula for $x = 0, 1, \ldots, 10$.

Environmental Health

One of the important issues in assessing nuclear energy is whether there are excess disease risks in the communities surrounding nuclear-power plants. A study was undertaken in the community surrounding Hanford, Washington, looking at the prevalence of selected congenital malformations in the counties surrounding the nuclear-test facility [9].

***4.63** Suppose that 27 cases of Down's syndrome are found and only 19 are expected based on Birth Defects Monitoring Program prevalence estimates conducted in the states of Washington, Idaho, and Oregon. Is there a significant excess number of cases in the area surrounding the nuclear-power plant?

Suppose that 12 cases of cleft palate are observed, while only 7 are expected based on Birth Defects Monitoring Program estimates.

***4.64** What is the probability of observing exactly 12 cases of cleft palate if there is no excess risk of cleft palate in the study area?

***4.65** Do you feel there is a meaningful excess number of cases of cleft palate in the area surrounding the nuclear-power plant?

Health Promotion

A study was conducted among 234 people who had expressed a desire to stop smoking but who had not yet stopped. On the day they quit smoking, their carbon-monoxide level (CO) was measured and the time was noted from the time they smoked their last cigarette to the time of the CO measurement. The CO level provides an "objective" indicator of the number of cigarettes smoked per day during the time immediately prior to the quit attempt. However, it

TABLE 4.15 Format of SMOKE.DAT

Variable	Columns	Code
ID number	1–3	
Age	4–5	
Gender	6	1 = male, 2 = female
Cigarettes/day	7–8	
Carbon monoxide (CO) (× 10)	9–11	
Minutes elapsed since the last cigarette smoked	12–15	
LogCOAdj[a] (× 1000)	16–19	
Days abstinent[b]	20–22	

[a]This variable represents adjusted carbon-monoxide (CO) values. CO values were adjusted for minutes elapsed since the last cigarette smoked using the formula, $\log_{10}\text{CO (adjusted)} = \log_{10}\text{CO} - (-0.000638) \times (\text{min} - 80)$, where min is the number of minutes elapsed since the last cigarette smoked.
[b]Those abstinent less than 1 day were given a value of 0.

is known to also be influenced by the time since the last cigarette was smoked. Thus, this time is provided as well as a "corrected CO level," which is adjusted for the time since last smoked. Information is also provided on the age and sex of the subjects as well as the subject's self-report of the number of cigarettes per day. The subjects were followed for one year for the purpose of determining the number of days they remained abstinent. The number of days abstinent ranges from 0 days for those who quit for less than 1 day to 365 days for those who were abstinent for the full year. Assume that all people were followed for the entire year.

The data were provided by Dr. Arthur J. Garvey, Boston, Massachusetts, and are given in Data Set SMOKE.DAT, on the data disk. The format of this file is given in Table 4.15.

4.66 Develop a life table similar to Table 4.14, giving the number of people who remained abstinent at 1, 2, . . . , 12 months of life (assume for simplicity that there are 30 days in each of the first 11 months after quitting and 35 days in the 12th month). Plot these data either by hand or on the computer. Compute the probability that a person will remain abstinent at 1, 3, 6, and 12 months after quitting.

4.67 Develop life tables for subsets of the data based on age, sex, number of cigarettes per day, and carbon-monoxide level (one variable at a time). Given these data, do you feel that age, sex, number of cigarettes per day, or CO level are related to success in quitting? (Methods of analysis for life-table data are discussed in more detail in Chapter 14.)

Genetics

4.68 A topic of some interest in the genetic literature over at least the last 30 years has been the study of sex-ratio data. In particular, one hypothesis suggested is that there are enough families with a preponderance of males (females) that the sexes of successive childbirths are not independent random variables but are related to each other. This hypothesis has been extended beyond just successive births so that some authors also consider relationships between offspring two birth orders apart (first and third offspring, second and fourth offspring, etc.). Sex-ratio data from the first 5 births in 51,868 families are given in Data Set SEXRAT.DAT (on the data disk). The format of this file is given in Table 4.16 [10]. What are your conclusions concerning the preceding hypothesis based on your analysis of these data?

TABLE 4.16 Format of SEXRAT.DAT

Variable	Column
Number of children[a]	1
Sex of children[b]	3–7
Number of families	9–12

[a]For families with 5+ children, the sexes of the first 5 children are listed. The number of children is given as 5 for such families.
[b]The sex of successive births is given. Thus, MMMF means that the first 3 children were males and the fourth child was a female. There were 484 such families.

Infectious Disease

A study was conducted of risk factors for HIV infection among intravenous drug users [11]. It was found that 40% of users who had ≤ 100 injections per month (light users) and 55% of users who had > 100 injections per month (heavy users) were HIV positive.

4.69 What is the probability that exactly 3 of 5 light users are HIV positive?

4.70 What is the probability that at least 3 of 5 light users are HIV positive?

4.71 Suppose we have a group of 10 light users and 10 heavy users. What is the probability that exactly 3 of the 20 users will be HIV positive?

4.72 What is the probability that at least 4 of the 20 users will be HIV positive?

Ophthalmology, Diabetes

In a recent study [12] of incidence rates of blindness among insulin-dependent diabetics, it was reported that the annual incidence rate of blindness per year was 0.67% among 30- to 39-year-old male insulin-dependent diabetics (male IDDM) and 0.74% among 30- to 39-year-old female insulin-dependent diabetics (female IDDM).

4.73 If a group of 200 IDDM 30- to 39-year-old men is followed, what is the probability that exactly 2 will become blind over a 1-year period?

4.74 If a group of 200 IDDM 30- to 39-year-old women is followed, what is the probability that at least 2 will become blind over a 1-year period?

4.75 What is the probability that a 30-year-old male IDDM patient will become blind over the next 10 years?

4.76 After how many years of follow-up would we expect the cumulative incidence of blindness to be 10% among 30-year-old IDDM women if the incidence rate remains constant over time?

4.77 What does cumulative incidence mean, in words, in the context of this problem?

Cardiovascular Disease

An article was recently published [13] concerning the evidence of cardiac death attributable to the earthquake in Los Angeles County on January 17, 1994. In the week before the earthquake there were an average of 15.6 cardiac deaths per day in Los Angeles County. On the day of the earthquake, there were 51 cardiac deaths.

4.78 What is the exact probability of 51 deaths occurring on one day if the cardiac death rate in the previous week continued to hold on the day of the earthquake?

4.79 Is the occurrence of 51 deaths an unusual occurrence? (*Hint:* Use the same methodology as in Example 4.30.)

4.80 What is the maximum number of cardiac deaths that could have occurred on the day of the earthquake that would have been consistent with the rate of cardiac deaths in the past week? (*Hint:* Use a cutoff of .05 to determine the maximum number.)

Environmental Health

Some previous studies have shown that there is a relationship between emergency-room admissions per day and level of pollution on a given day. A small local hospital finds that the number of admissions to the emergency ward on a single day ordinarily (unless there is unusually high pollution) follows a Poisson distribution with mean = 2.0 admissions per day. Suppose that each admitted person to the emergency ward stays there for exactly 1 day and is then discharged.

4.81 The hospital is planning a new emergency-room facility. It wants to have enough beds in the emergency ward so that for at least 95% of normal-pollution days it will have enough beds without turning anyone away. What is the smallest number of beds it should have, to satisfy this criterion?

4.82 The hospital also finds that on high-pollution days the number of admissions is Poisson-distributed with mean = 4.0 admissions per day. Answer Problem 4.81 for high-pollution days.

4.83 On a random day during the year, what is the probability that there will be 4 admissions to the emergency ward, assuming that there are 345 normal-pollution days and 20 high-pollution days?

Women's Health

The number of legal induced abortions per year per 1000 U.S. women ages 15–44 [14] are given as follows:

TABLE 4.17 Annual incidence of legal induced abortions by time period

Year	Legal induced abortions per year per 1000 women ages 15–44
1975–1979	21
1980–1984	25
1985–1989	24
1990–1994	24
1995–2004	20

For example, of 1000 women ages 15–44 in 1980, 25 had a legal induced abortion during 1980.

4.84 If we assume that (1) no woman has more than 1 abortion and (2) the probability of having an abortion is independent across different years, what is the probability that a 15-year-old woman in 1975 will have an abortion over her 30 years of reproductive life (ages 15–44, or 1975–2004)?

Studies have been undertaken to assess the relationship between abortion and the development of breast cancer. In one study among nurses (the Nurses' Health Study II), there were 16,359 abortions among 2,169,321 person-years of follow-up for women of reproductive age. (*Note:* 1 person-year = 1 woman followed for 1 year.)

4.85 What is the expected number of abortions among nurses over this time period if the incidence of abortion is 25 per 1000 women per year and no woman has more than 1 abortion?

4.86 Does the abortion rate among nurses differ significantly from the national experience? Why or why not? (*Hint:* Use the Poisson distribution.) A yes/no answer is not acceptable.

Endocrinology

4.87 Consider the Data Set BONEDEN.DAT on the data disk. Calculate the difference in bone density of the lumbar spine (g/cm^2) between the heavier-smoking twin and the lighter-smoking twin (bone density for the heavier-smoking twin minus bone density for the lighter-smoking twin) for each of the 41 twin pairs. Suppose smoking has no relationship to bone density. What would be the expected number of twin pairs with negative difference scores? What is the actual number of twin pairs with negative difference scores? Do you feel smoking is related to bone density of the lumbar spine, given the observed results? Why or why not? A yes/no answer is not acceptable. (*Hint:* Use the binomial distribution.)

4.88 Sort the differences in smoking between members of a twin pair (expressed in pack-years). Identify the subgroup of 20 twin pairs with the largest differences in smoking. Answer Problem 4.87 based on this subgroup of 20 twin pairs.

4.89 Answer Problem 4.87 for bone density of the femoral neck.

4.90 Answer Problem 4.88 for bone density of the femoral neck.

4.91 Answer Problem 4.87 for bone density of the femoral shaft.

4.92 Answer Problem 4.88 for bone density of the femoral shaft.

Simulation

One of the attractive features of modern statistical packages such as MINITAB or Excel is the ability to use the computer to simulate random variables on the computer and compare the characteristics of the observed samples with the theoretical properties of the random variables.

4.93 Draw 100 random samples from a binomial distribution, each based on 10 trials with probability of success = .05 on each trial. Obtain a frequency distribution of the number of successes over the 100 random samples, and plot the distribution. How does it compare with Figure 4.4(a)?

4.94 Answer Problem 4.93 for a binomial distribution with parameters $n = 10$ and $p = .95$. Compare your results with Figure 4.4(b).

4.95 Answer Problem 4.93 for a binomial distribution with parameters $n = 10$ and $p = .5$. Compare your results with Figure 4.4(c).

Simulation

4.96 Draw 200 random samples from a binomial distribution, each based on 100 trials with a probability of success = .01. Obtain a frequency distribution of the number of successes over the 200 random samples, and plot the distribution.

4.97 Draw 200 random samples from a Poisson distribution, each with mean = 1. Obtain a frequency distribution of the number of successes over the 200 random samples, and plot the distribution.

4.98 Compare your results in Problems 4.96 and 4.97. Do you feel that the Poisson approximation to the binomial distribution is adequate in this situation?

REFERENCES

[1] *Boston Globe,* October 7, 1980.

[2] Rinsky, R. A., Zumwalde, R. O., Waxweiler, R. J., Murray, W. E., Bierbaum, P. J., Landrigan, P. J., Terpilak, M., & Cox, C. (1981, January 31). Cancer mortality at a naval nuclear shipyard. *Lancet,* 231–235.

[3] U.S. Department of Health and Human Services. (1986). *Vital statistics of the United States, 1986.*

[4] National Center for Health Statistics. (1974, June 27). *Monthly vital statistics report, annual summary for the United States (1973), 22*(13).

[5] National Center for Health Statistics. (1978, December 7). *Monthly vital statistics report, annual summary for the United States (1977), 26*(13).

[6] Ott, M. G., Scharnweber, H. C., & Langner, R. (1980). Mortality experience of 161 employees exposed to ethylene dibromide in two production units. *British Journal of Industrial Medicine, 37,* 163–168.

[7] Kinnunen, E., Junttila, O., Haukka, J., & Hovi, T. (1998). Nationwide oral poliovirus vaccination campaign and the incidence of Guillain-Barré syndrome. *American Journal of Epidemiology, 147*(1), 69–73.

[8] Hoff, R., Berardi, V. P., Weiblen, B. J., Mahoney-Trout, L., Mitchell, M. L., & Grady, G. F. (1988). Seroprevalence of human immunodeficiency virus among childbearing women. *New England Journal of Medicine, 318*(9), 525–530.

[9] Sever, L. E., Hessol, N. A., Gilbert, E. S., & McIntyre, J. M. (1988). The prevalence at birth of congenital malformations in communities near the Hanford site. *American Journal of Epidemiology, 127*(2), 243–254.

[10] Renkonen, K. O., Mäkelä, O., & Lehtovaara, R. (1961). Factors affecting the human sex ratio. *Annales Medicinae Experimentalis et Biologiae Fenniae, 39,* 173–184.

[11] Schoenbaum, E. E., Hartel, D., Selwyn, P. A., Klein, R. S., Davenny, K., Rogers, M., Feiner, C., & Friedland, G. (1989). Risk factors for human immunodeficiency virus infection in intravenous drug users. *New England Journal of Medicine, 321*(13), 874–879.

[12] Sjolie, A. K., & Green, A. (1987). Blindness in insulin-treated diabetic patients with age at onset less than 30 years. *Journal of Chronic Disease, 40*(3), 215–220.

[13] Leor, J., Poole, W. K., & Kloner, R. A. (1996). Sudden cardiac death triggered by an earthquake. *New England Journal of Medicine, 334*(7), 413–419.

[14] National Center for Health Statistics. (1997, December 5). *Morbidity and mortality weekly report* (1980), *46*(48).

5 Continuous Probability Distributions

SECTION 5.1 Introduction

In this chapter continuous probability distributions are discussed. In particular, the normal distribution, which is the most widely used distribution in statistical work, is explored in depth.

The normal, or Gaussian or "bell-shaped" distribution is the cornerstone of most of the methods of estimation and hypothesis testing that are developed in the rest of this text. Many random variables, such as the distribution of birthweights or blood pressures in the general population, tend to approximately follow a normal distribution. In addition, many random variables that are not themselves normal are closely approximated by a normal distribution when summed many times. In such cases, using the normal distribution is desirable, because the normal distribution is easy to use and tables for the normal distribution are more widely available than those for many other distributions.

Example 5.1 **Infectious Disease** The number of neutrophils in a sample of 2 white blood cells is not normally distributed, but the number in a sample of 100 white blood cells is very close to being normally distributed.

SECTION 5.2 General Concepts

We want to develop an analogue for a continuous random variable to the concept of a probability mass function, as was developed for a discrete random variable in Section 4.3. Thus, we would like to know which values are more probable than others and how probable they are.

Example 5.2 **Hypertension** Consider the distribution of diastolic blood-pressure (DBP) measurements in 35- to 44-year-old men. In actual practice this distribution is discrete because only a finite number of blood-pressure values are possible, because the measurement is only accurate to within 2 mm Hg, or in some cases 5 mm Hg. However, assume that there is no measurement error and hence the random variable can take on a continuum of possible values. One consequence of this assumption is that the probabilities of specific blood-pressure measurement values such as 117.3 are 0, and thus the concept of a probability mass function cannot be used. The proof of this statement is beyond the scope of this text. Instead, we speak in terms of the probability that blood pressure falls within a range of values. Thus, the probabilities of blood pressures (denoted by X) falling in the ranges of $90 \le X < 100$, $100 \le X < 110$, and $X \ge 110$ might be 15%, 5%, and 1%, respectively. People whose blood pressures fall in these ranges might be denoted as mild hypertensive, moderate hypertensive, and severe hypertensive, respectively.

Although the probability of exactly obtaining any value is 0, we still have the intuitive notion that certain ranges of values occur more frequently than others. This notion can be quantified using the concept of a probability density function.

DEFINITION 5.1

The **probability density function** (or pdf) of the random variable X is a function such that the area under the curve corresponding to the function between any two points a and b is equal to the probability that the random variable X falls between a and b. Thus, the total area under the curve over the entire range of possible values for the random variable is 1.

The probability density function takes on high values in regions of high probability and low values in regions of low probability.

Example 5.3

Hypertension The probability density function for diastolic blood pressure (DBP) in 35- to 44-year-old men is given in Figure 5.1. Areas A, B, and C correspond to the probabilities of being mildly hypertensive, moderately hypertensive, and severely hypertensive, respectively. Furthermore, the most likely range of values for diastolic blood pressure occurs around 80 mm Hg, with the values becoming increasingly less likely as we move further away from 80.

Example 5.4

Cardiovascular Disease Serum triglycerides is an asymmetric, positively skewed, continuous random variable whose probability density function appears in Figure 5.2.

FIGURE 5.1 **Probability density function of diastolic blood pressure in 35- to 44-year-old men**

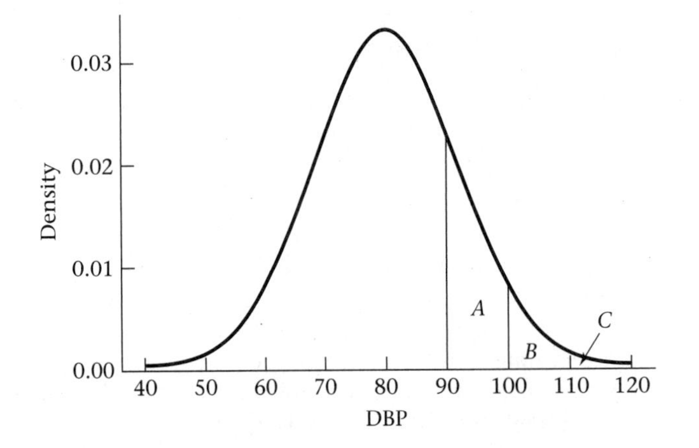

FIGURE 5.2 **Probability density function for serum triglycerides**

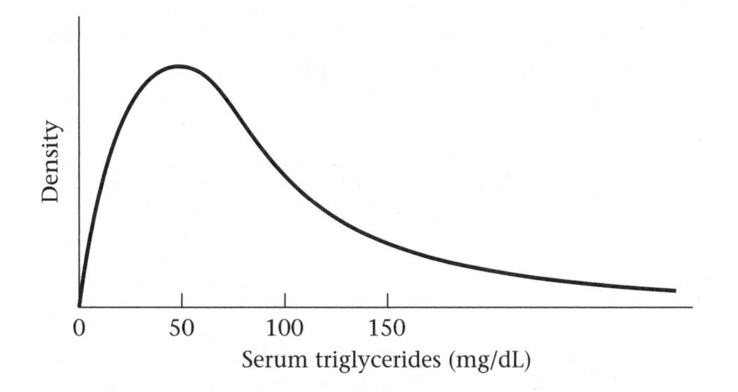

The cumulative-distribution function (or cdf) is defined similarly to that for a discrete random variable (see Section 4.6).

DEFINITION 5.2 The **cumulative-distribution function (cdf)** for the random variable X evaluated at the point a is defined as the probability that X will take on values $\leq a$. It is represented by the area under the probability density function to the left of a.

Example 5.5 **Obstetrics** The probability density function for the random variable representing the distribution of birthweights (oz) in the general population is given in Figure 5.3. The cumulative-distribution function evaluated at 88 oz = $Pr(X \leq 88)$ is represented by the area under this curve to the left of 88 oz. The region $X \leq 88$ oz has a special meaning in obstetrics, because 88 oz is the cutoff point usually used by obstetricians for identifying low-birthweight infants. Such infants are generally at higher risk for various unfavorable outcomes, such as mortality in the first year of life.

Generally, a distinction will not be made between the probabilities $Pr(X < x)$ and $Pr(X \leq x)$ when X is a continuous random variable. The reason is that they

FIGURE 5.3 **Probability density function for birthweight**

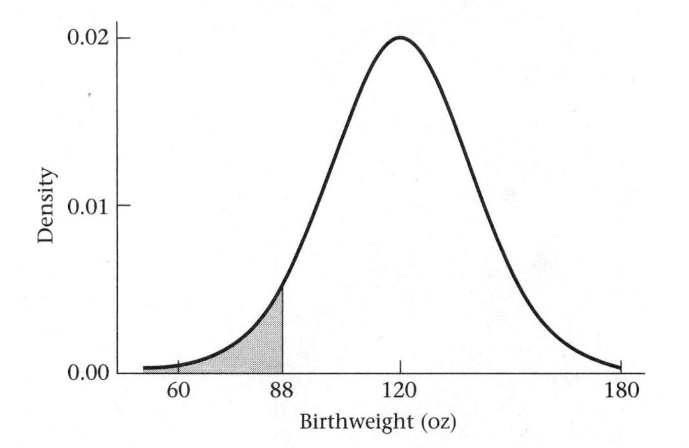

represent the same quantity, because the probability of individual values is 0; that is, $Pr(X = x) = 0$.

The expected value and variance for continuous random variables have the same meaning as for discrete random variables (Sections 4.4 and 4.5). However, the mathematical definition of these terms is beyond the scope of this book.

DEFINITION 5.3 The **expected value** of a continuous random variable X, denoted by $E(X)$, or μ, is the average value taken on by the random variable.

DEFINITION 5.4 The **variance** of a continuous random variable X, denoted by $Var(X)$ or σ^2, is the average squared distance of each value of the random variable from its expected value, which is given by $E(X - \mu)^2$ and can be re-expressed in short form as $E(X^2) - \mu^2$. The standard deviation, or σ, is the square root of the variance, that is, $\sigma = \sqrt{Var(X)}$.

Example 5.6 **Hypertension** The expected value and standard deviation of the distribution of diastolic blood pressures in 35- to 44-year-old men are 80 mm Hg and 12 mm Hg, respectively.

SECTION 5.3 The Normal Distribution

The normal distribution is the most widely used continuous distribution. It is also frequently referred to as the Gaussian distribution, after the well-known mathematician Karl Friedrich Gauss (Figure 5.4.)

FIGURE 5.4 Karl Friedrich Gauss (1777–1855)

Example 5.7 **Hypertension** The distribution of body weights or of diastolic blood pressures for a group of 35- to 44-year-old men will follow a normal distribution.

Many other distributions that are not themselves normal can be made normal by transforming the data onto a different scale.

Example 5.8 **Cardiovascular Disease** The distribution of serum-triglyceride concentrations from this same group of 35- to 44-year-old men is likely to be positively skewed. However, the log transformation of these measurements will usually follow a normal distribution.

Generally speaking, any random variable that can be expressed as a sum of many other random variables can be well approximated by a normal distribution. For example, many physiologic measures are determined in part by a combination of several genetic and environmental risk factors and can often be well approximated by a normal distribution.

Example 5.9 **Infectious Disease** The distribution of the number of lymphocytes in a differential of 100 white blood cells (refer to Example 4.15 for the definition of a differential) will tend to be normally distributed, because this random variable is a sum of 100 random variables, each representing whether or not an individual cell is a lymphocyte.

Thus, because of its omnipresence, the normal distribution is vital to statistical work, and most estimation procedures and hypothesis tests that we will study assume that the random variable being considered has an underlying normal distribution.

Another important area of application of the normal distribution is as an approximating distribution to other distributions. The normal distribution is generally more convenient to work with than any other distribution, particularly in hypothesis testing. Thus, if an accurate normal approximation to some other distribution can be found, we will often use it.

DEFINITION 5.5

The **normal distribution** is defined by its **probability density function** (or pdf), which is given as

$$f(x) = \frac{1}{\sqrt{2\pi}\sigma} \exp\left[-\frac{1}{2\sigma^2}(x-\mu)^2\right], \qquad -\infty < x < \infty$$

for some parameters μ, σ, where $\sigma > 0$.

The exp function merely implies that the quantity to the right in brackets is the power to which "e" (≈ 2.71828) is raised. A plot of this probability density function is given in Figure 5.5 for a normal distribution with $\mu = 50$ and $\sigma^2 = 100$.

The density function follows a bell-shaped curve, with the mode at μ and most frequently occurring values around μ. The curve is symmetric about μ, with points of inflection on each side of μ at $\mu - \sigma$ and $\mu + \sigma$, respectively. A point of inflection is a point where the slope of the curve changes direction. In Figure 5.5, the slope of

FIGURE 5.5 **Probability density function for a normal distribution with mean μ (50) and variance σ² (100)**

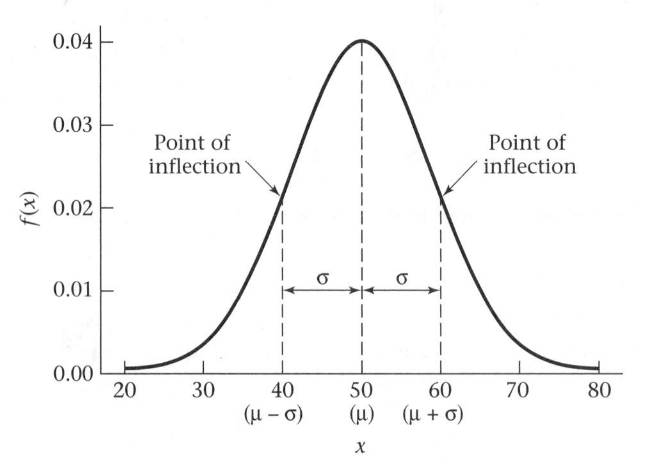

the curve increases to the left of $\mu - \sigma$ and then starts to decrease to the right of $\mu - \sigma$ and continues to decrease until reaching $\mu + \sigma$, after which it starts increasing again. Thus, the distances from μ to the points of inflection provide a good visual sense of the magnitude of the parameter σ.

You may wonder why the parameters μ and σ^2 have been used to define the normal distribution when the expected value and variance of an arbitrary distribution were previously defined as μ and σ^2. Indeed, from the definition of the normal distribution it can be shown, using calculus methods, that μ and σ^2 are, respectively, the expected value and variance of this distribution.

Example 5.10 For diastolic blood pressure the parameters might be $\mu = 80$ mm Hg, $\sigma = 12$ mm Hg; for birthweight they might be $\mu = 120$ oz, $\sigma = 20$ oz.

Interestingly, the entire shape of the normal distribution is determined by the two parameters μ and σ^2. If the two normal distributions with the same variance σ^2 and different means μ_1, μ_2, where $\mu_2 > \mu_1$, are compared, then their density functions will appear as in Figure 5.6, where $\mu_1 = 50$, $\mu_2 = 62$, and $\sigma = 7$.

Similarly, two normal distributions with the same mean but different variances $\left(\sigma_2^2 > \sigma_1^2\right)$ can be compared, as shown in Figure 5.7, with $\mu = 50$, $\sigma_1 = 5$, and $\sigma_2 = 10$. Note that the area under any normal density function must be 1. Thus, the two normal distributions shown in Figure 5.7 must cross, because otherwise one curve would remain completely above the other and the areas under both curves could not simultaneously be 1.

DEFINITION 5.6 A **normal distribution with mean** μ **and variance** σ^2 will generally be referred to as an $N(\mu, \sigma^2)$ distribution.

FIGURE 5.6 **Comparison of two normal distributions with the same variance and different means**

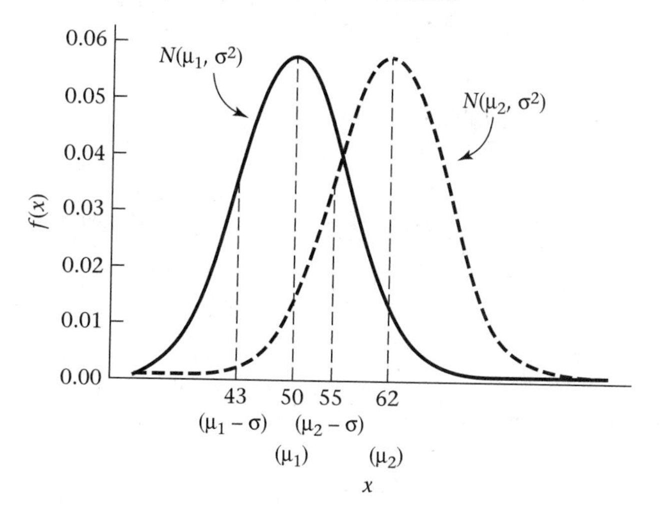

FIGURE 5.7 **Comparison of two normal distributions with the same means and different variances**

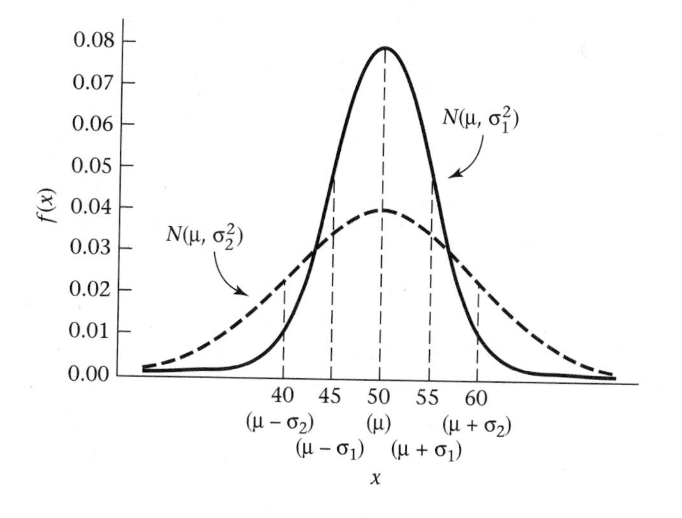

Note that the second parameter is always the variance σ^2 and not the standard deviation σ.

Another property of the normal distribution is that the height $= 1/\left(\sqrt{2\pi}\sigma\right)$. Thus, the height is inversely proportional to σ. This helps us to visualize σ, because the height of the $N\left(\mu, \sigma_1^2\right)$ distribution in Figure 5.7 is greater than the height of the $N\left(\mu, \sigma_2^2\right)$ distribution.

DEFINITION 5.7

A normal distribution with mean 0 and variance 1 will be referred to as a **standard**, or **unit**, normal distribution. This distribution is denoted by $N(0, 1)$.

We will see that any information concerning an $N(\mu, \sigma^2)$ distribution can be obtained from appropriate manipulations of an $N(0, 1)$ distribution.

SECTION 5.4 Properties of the Standard Normal Distribution

To become familiar with the $N(0, 1)$ distribution, some of its properties will be reviewed. First, the probability density function in this case reduces to

EQUATION 5.1

$$f(x) = \frac{1}{\sqrt{2\pi}} e^{(-1/2)x^2}, \quad -\infty < x < +\infty$$

This distribution is symmetrical about 0, because $f(x) = f(-x)$, as depicted in Figure 5.8.

EQUATION 5.2

It can be shown that about 68% of the area under the standard normal density lies between +1 and −1, about 95% of the area lies between +2 and −2, and about 99% lies between +2.5 and −2.5.

These relationships can be expressed more precisely by saying that

$$Pr(-1 < X < 1) = .6827 \qquad Pr(-1.96 < X < 1.96) = .95$$
$$Pr(-2.576 < X < 2.576) = .99$$

Thus, the standard normal distribution slopes off very rapidly, and absolute values greater than 3 are unlikely. These relationships are shown in Figure 5.9.

FIGURE 5.8 **Probability density function for a standard normal distribution**

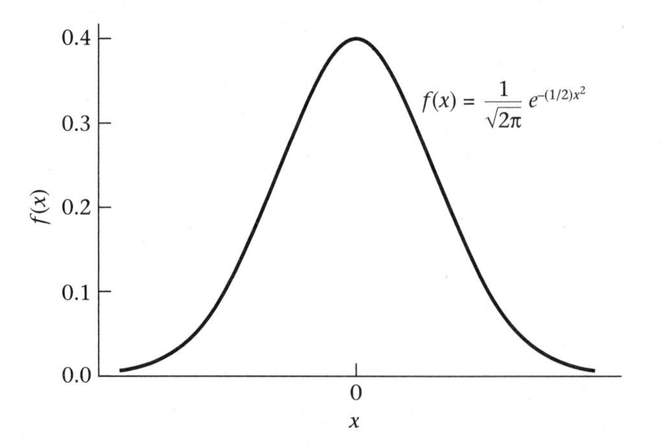

FIGURE 5.9 Empirical properties of the standard normal distribution

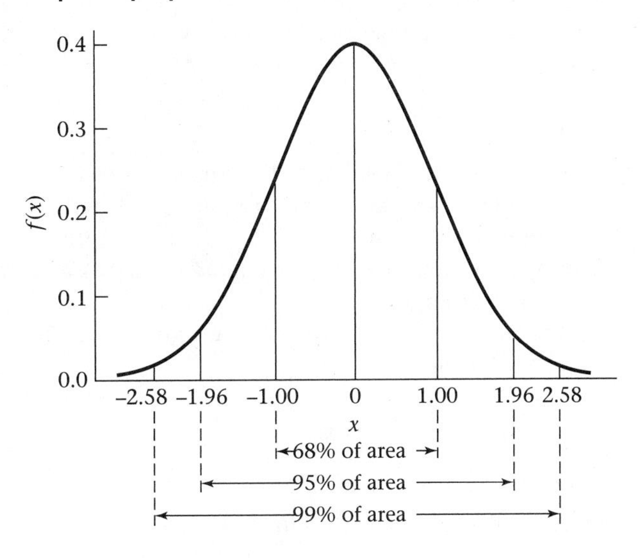

Tables of the area under the normal density function, or so-called normal tables, take advantage of the symmetry properties of the normal distribution and generally are concerned with areas for positive values of x.

DEFINITION 5.8 The **cumulative-distribution function (cdf) for a standard normal distribution** is denoted by

$$\Phi(x) = Pr(X \leq x)$$

where X follows an $N(0, 1)$ distribution. This function is depicted in Figure 5.10.

FIGURE 5.10 Cumulative-distribution function [$\Phi(x)$] for a standard normal distribution

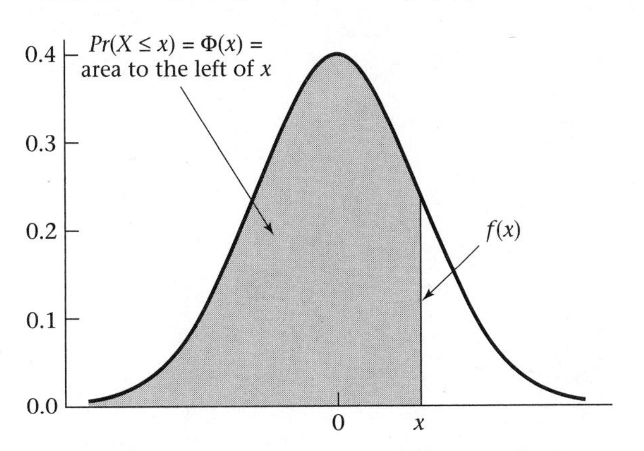

DEFINITION 5.9

The symbol ~ is used as shorthand for the phrase "**is distributed as.**" Thus, $X \sim N(0, 1)$ means that the random variable X is distributed as an $N(0, 1)$ distribution.

5.4.1 Using Normal Tables

Under column A in Table 3 of the Appendix, $\Phi(x)$ for various positive values of x for a standard normal distribution are presented. This cumulative distribution function is depicted in Figure 5.11. Notice that the area to the left of 0 is .5. Furthermore, the area to the left of x approaches 0 as x becomes small and approaches 1 as x becomes large.

The right-hand tail of the standard normal distribution = $Pr(X \geq x)$ is provided in column B of Table 3.

Example 5.11 If $X \sim N(0, 1)$

then find $Pr(X \leq 1.96)$ and $Pr(X \leq 1)$

Solution From the Appendix, Table 3, column A,

$$\Phi(1.96) = .975 \quad \text{and} \quad \Phi(1) = .8413$$

EQUATION 5.3

Symmetry Properties of the Standard Normal Distribution From the symmetry properties of the standard normal distribution,

$$\Phi(-x) = Pr(X \leq -x) = Pr(X \geq x) = 1 - Pr(X \leq x) = 1 - \Phi(x)$$

This symmetry property is depicted in Figure 5.12 for $x = 1$.

Example 5.12 Calculate $Pr(X \leq -1.96)$ if $X \sim N(0,1)$.

Solution $Pr(X \leq -1.96) = Pr(X \geq 1.96) = .0250$ from column B of Table 3.

FIGURE 5.11 **Cumulative-distribution function for a standard normal distribution [$\Phi(x)$]**

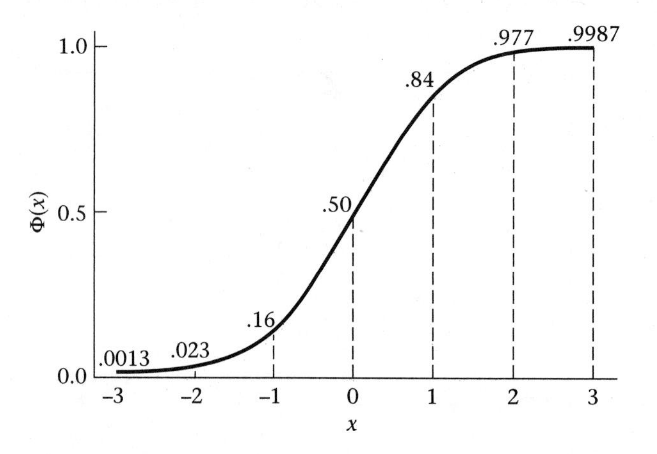

FIGURE 5.12 **Illustration of the symmetry properties of the normal distribution**

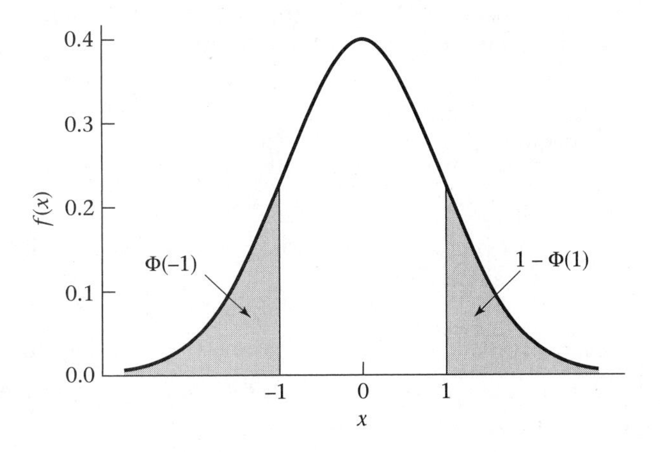

Furthermore, for any numbers a, b, we have $Pr(a \le X \le b) = Pr(X \le b) - Pr(X \le a)$ and thus we can evaluate $Pr(a \le X \le b)$ for any a, b from Table 3.

Example 5.13 Compute $Pr(-1 \le X \le 1.5)$ if $X \sim N(0, 1)$.

Solution

$$\begin{aligned} Pr(-1 \le X \le 1.5) &= Pr(X \le 1.5) - Pr(X \le -1) \\ &= Pr(X \le 1.5) - Pr(X \ge 1) = .9332 - .1587 \\ &= .7745 \end{aligned}$$

Example 5.14 **Pulmonary Disease** Forced vital capacity (FVC) is a standard measure of pulmonary function and represents the volume of air a person can expel in 6 seconds. A topic of current research interest is to look at potential risk factors, such as cigarette smoking, air pollution, or the type of stove used in the home, that may affect FVC in grade-school children. One problem is that pulmonary function is affected by age, sex, and height, and these variables must be corrected for before looking at other risk factors. One way to make these adjustments for a particular child is to find the mean μ and standard deviation σ for children of the same age (in 1-year age groups), sex, and height (in 2-in. height groups) from large national surveys and compute a **standardized FVC**, which is defined as $(X - \mu)/\sigma$, where X is the original FVC. The standardized FVC would then approximately follow an $N(0, 1)$ distribution. Suppose that a child is considered in poor pulmonary health if his or her standardized FVC < -1.5. What percentage of children are in poor pulmonary health?

Solution

$$Pr(X < -1.5) = Pr(X > 1.5) = .0668$$

Thus about 7% of children are in poor pulmonary health.

In many instances we will be concerned with tail areas on either side of 0 for a standard normal distribution. For example, the *normal range* for a biological quantity is often defined by a range within x standard deviations of the mean for some specified value of x. The probability of a value falling in this range is given by $Pr(-x \le X \le x)$ for a standard normal distribution. This quantity is tabulated in column D of Table 3 for various values of x.

Example 5.15 **Pulmonary Disease** Suppose a child is considered to have normal lung growth if his or her standardized FVC is within 1.5 standard deviations of the mean. What proportion of children are within the normal range?

Solution Compute $Pr(-1.5 \le X \le 1.5)$. Under 1.50 in Table 3, column D, this quantity is given as .8664. Thus, about 87% of children are considered to have normal lung growth, using this definition.

Finally, in column C of Table 3, the area under the standard normal density from 0 to x is provided, because these areas will occasionally prove useful in work on statistical inference.

Example 5.16 Find the area under the standard normal density from 0 to 1.45.

Solution Refer to column C of Table 3 under 1.45. The appropriate area is given by .4265.

Of course, the areas given in columns A, B, C, and D are redundant in that *all* computations concerning the standard normal distribution could be performed using any one of these columns. In particular, we have seen that $B(x) = 1 - A(x)$. Also, from the symmetry of the normal distribution, we can easily show that $C(x) = A(x) - .5$, $D(x) = 2 \times C(x) = 2 \times A(x) - 1.0$. However, this redundancy is deliberate, because for some applications one or the other of these columns will be more convenient to use.

It is also possible to use "electronic tables" to compute areas under a standard normal distribution. For example, in Excel 97 the function NORMSDIST(x) will provide the cdf for a standard normal distribution for any value of x.

Example 5.17 Using an electronic table, find the area under the standard normal density to the left of 2.824.

Solution We use the Excel 97 function NORMSDIST evaluated at 2.824 [NORMSDIST(2.824)], with the result given as follows:

```
     x              2.824
NORMSDIST(x)     0.997629
```

The area is .9976.

The percentiles of a normal distribution are often referred to in statistical inference. For example, we might be interested in the upper and lower fifth percentiles of the distribution of FVC in children so as to define a normal range of values. For this purpose, the definition of the percentiles of a standard normal distribution is introduced:

DEFINITION 5.10 The **$(100 \times u)$th percentile** of a standard normal distribution is denoted by z_u. It is defined by the relationship

$$Pr(X < z_u) = u, \quad \text{where } X \sim N(0, 1)$$

Figure 5.13 depicts z_u graphically.

FIGURE 5.13 **Graphic display of the (100 × *u*)th percentile of a standard normal distribution (z_u)**

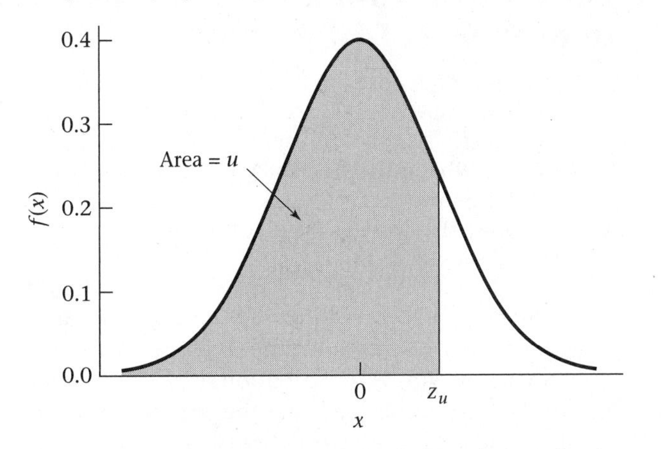

The function z_u is sometimes referred to as the *inverse normal function.* In previous uses of the normal table, we were given a value x and have used the normal tables to evaluate the area to the left of x—i.e., $\Phi(x)$—for a standard normal distribution.

To obtain z_u, we perform this operation in reverse. Thus, to evaluate z_u, we must find the area u in column A of Table 3 and then find the value z_u that corresponds to this area. If $u < .5$, then we use the symmetry properties of the normal distribution to obtain $z_u = -z_{1-u}$, where z_{1-u} can be obtained from Table 3.

Example 5.18 Compute $z_{.975}$, $z_{.95}$, $z_{.5}$, and $z_{.025}$.

Solution From Table 3 we have that

$$\Phi(1.96) = .975$$
$$\Phi(1.645) = .95$$
$$\Phi(0) = .5$$
$$\Phi(-1.96) = 1 - \Phi(1.96) = 1 - .975 = .025$$

Thus,
$$z_{.975} = 1.96$$
$$z_{.95} = 1.645$$
$$z_{.5} = 0$$
$$z_{.025} = -1.96$$

where for $z_{.95}$ we interpolate between 1.64 and 1.65 to obtain 1.645.

Example 5.19 Compute the value x such that the area to the left of $x = .85$.

Solution We use the Excel 97 function NORMSINV evaluated at .85 [NORMSINV(.85)] with the result given as follows:

```
x                  0.85
NORMSINV(x)    1.036433
```

Thus the area to the left of 1.036 is .85.

The percentile z_u will be frequently used in our work on estimation in Chapter 6 and hypothesis testing in Chapters 7–14.

SECTION 5.5 Conversion from an $N(\mu, \sigma^2)$ Distribution to an $N(0, 1)$ Distribution

Example 5.20 **Hypertension** Suppose a mild hypertensive is defined as a person whose diastolic blood pressure is between 90 and 100 mm Hg inclusive, and the subjects are 35- to 44-year-old males whose blood pressures are normally distributed with mean 80 and variance 144. What is the probability that a randomly selected person from this population will be a mild hypertensive? This question can be stated more precisely: If $X \sim N(80, 144)$, then what is $Pr(90 < X < 100)$?

(The solution is given below.)

More generally, the following question can be asked: If $X \sim N(\mu, \sigma^2)$, then what is $Pr(a < X < b)$ for any *a, b*? To solve this, we convert the probability statement about an $N(\mu, \sigma^2)$ distribution to an equivalent probability statement about an $N(0, 1)$ distribution. Consider the random variable $Z = (X - \mu)/\sigma$. We can show that the following relationship holds:

EQUATION 5.4

If $X \sim N(\mu, \sigma^2)$ and $Z = (X - \mu)/\sigma$, then $Z \sim N(0, 1)$.

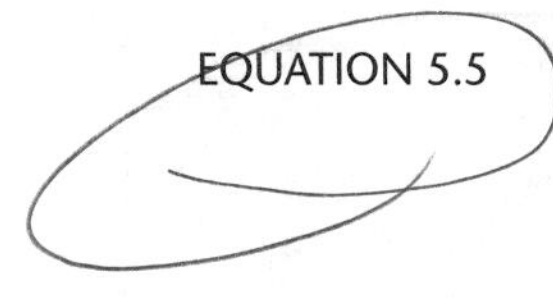

EQUATION 5.5

Evaluation of Probabilities for Any Normal Distribution via Standardization

If $X \sim N(\mu, \sigma^2)$ and $Z = (X - \mu)/\sigma$

then $$Pr(a < X < b) = Pr\left(\frac{a-\mu}{\sigma} < Z < \frac{b-\mu}{\sigma}\right) = \Phi[(b-\mu)/\sigma] - \Phi[(a-\mu)/\sigma]$$

Because the Φ function, which is the cumulative distribution function for a standard normal distribution, is given in column A of Table 3 of the Appendix, probabilities for *any* normal distribution can now be evaluated using the tables in this text. This procedure is depicted in Figure 5.14 for $\mu = 80, \sigma = 12, a = 90, b = 100$, where the areas in Figure 5.14a and 5.14b are the same.

The procedure in Equation 5.5 is known as **standardization of a normal variable.**

EQUATION 5.6

The general principle is that for any probability statement concerning normal random variables of the form $Pr(a < X < b)$, the population mean μ is subtracted from each boundary point and divided by the standard deviation σ to obtain an equivalent probability statement for the standard normal random variable Z,

$$Pr[(a-\mu)/\sigma < Z < (b-\mu)/\sigma]$$

The standard normal tables are then used to evaluate this latter probability.

Solution to Example 5.20 The probability of being a mild hypertensive among the group of 35- to 44-year-old men can now be calculated.

$$Pr(90 < X < 100) = Pr\left(\frac{90-80}{12} < Z < \frac{100-80}{12}\right)$$
$$= Pr(0.83 < Z < 1.67) = \Phi(1.67) - \Phi(0.83)$$
$$= .9522 - .7977 = .155$$

Thus, approximately 15.5% of this population will have mild hypertension.

FIGURE 5.14 **Evaluation of probabilities for any normal distribution using standardization**

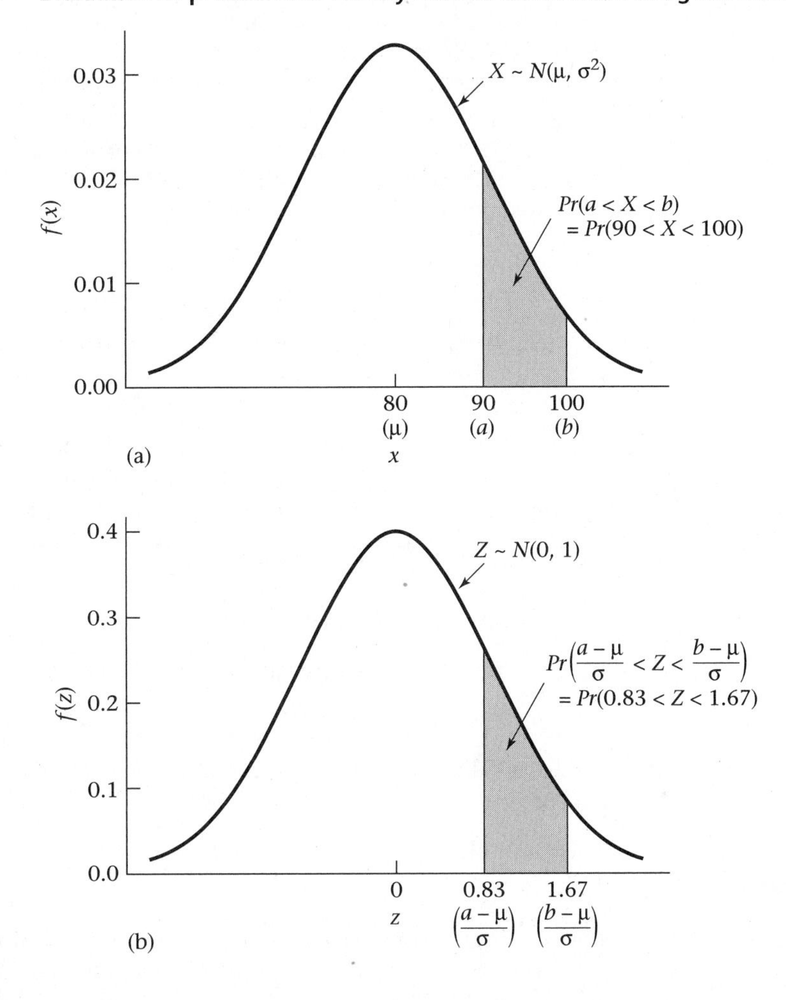

Example 5.21 **Botany** Suppose that tree diameters of a certain species of tree from some defined forest area are assumed to be normally distributed with mean 8 in. and standard deviation 2 in. Find the probability of a tree having an unusually large diameter, which is defined as > 12 in.

Solution We have $X \sim N(8, 4)$ and require

$$Pr(X > 12) = 1 - Pr(X < 12) = 1 - Pr\left(Z < \frac{12-8}{2}\right)$$
$$= 1 - Pr(Z < 2.0) = 1 - .977 = .023$$

Thus, 2.3% of trees from this area have an unusually large diameter.

Example 5.22 **Cerebrovascular Disease** Diagnosing stroke strictly on the basis of clinical symptoms is difficult. A standard diagnostic test used in clinical medicine to detect stroke in patients is the angiogram. This test has some risks for the patient, and several noninvasive techniques have been developed that are hoped to be as effective as the angiogram. One such method uses the measurement of cerebral blood flow (CBF) in the brain, because stroke patients tend to have lower levels of CBF than normal. Assume that in the general population, CBF is normally distributed with mean 75 and standard deviation 17. A patient is classified as being at risk for stroke if his or her CBF is less than 40. What proportion of normal patients will be mistakenly classified as being at risk for stroke?

Solution Let X be the random variable representing CBF. Then $X \sim N(75, 17^2) = N(75, 289)$. We want to find $Pr(X < 40)$. We standardize the limit of 40 so as to use the standard normal distribution. The standardized limit is $(40 - 75)/17 = -2.06$. Thus, if Z represents the standardized normal random variable $= (X - \mu)/\sigma$, then

$$\begin{aligned} Pr(X < 40) &= Pr(Z < -2.06) \\ &= \Phi(-2.06) = 1 - \Phi(2.06) = 1 - .9803 \approx .020 \end{aligned}$$

Thus, about 2.0% of normal patients will be incorrectly classified as being at risk for stroke.

If we use electronic tables, then the pdf, cdf, and inverse normal distribution can be obtained for any normal distribution, and standardization is unnecessary. For example, using Excel 97, the two functions NORMDIST and NORMINV are available for this purpose. To find the probability p that a $N(\mu, \sigma^2)$ distribution is $\le x$, we use the function

$$p = \text{NORMDIST}(x, \mu, \sigma, \text{TRUE})$$

To find the probability density f at x, we use the function

$$f = \text{NORMDIST}(x, \mu, \sigma, \text{FALSE})$$

To find the value x such that the cdf for an $N(\mu, \sigma^2)$ distribution is equal to p, we use the function

$$x = \text{NORMINV}(p, \mu, \sigma)$$

EQUATION 5.7

The pth percentile of a general normal distribution (x) can also be written in terms of the percentiles of a standard normal distribution as follows:

$$x = \mu + z_p\sigma$$

Example 5.23 **Ophthalmology** Glaucoma is an eye disease that is manifested by high intraocular pressure. The distribution of intraocular pressure in the general population is approximately normal with mean 16 mm Hg and standard deviation 3 mm Hg. If the normal range for intraocular pressure is considered to be between 12 mm Hg and 20 mm Hg, then what percentage of the general population would fall within this range?

Solution We wish to calculate $Pr(12 \le X \le 20)$, where $X \sim N(16, 9)$, as shown in Figure 5.15.

We use the NORMDIST function of Excel 97 to perform these computations. First, we compute $p1 = Pr[X \le 20 | X \sim N(16,9)]$ given by NORMDIST(20, 16, 3, TRUE). Second, we compute $p2 = Pr[X \le 12 | X \sim N(16,9)]$ given by NORMDIST(12, 16, 3, TRUE). Thus, $Pr(12 \le X \le 20) = p1 - p2 = .818$. The computations are shown in the following spreadsheet.

FIGURE 5.15 Calculation of the proportion of people with intraocular pressure in the normal range

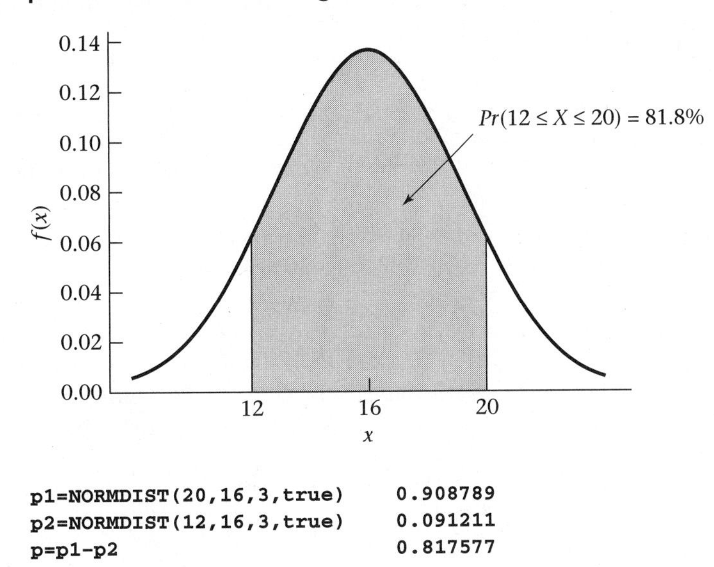


```
p1=NORMDIST(20,16,3,true)     0.908789
p2=NORMDIST(12,16,3,true)     0.091211
p=p1-p2                       0.817577
```

Thus, 81.8% of the population have intraocular pressures in the normal range.

Example 5.24 **Hypertension** Suppose the distribution of diastolic blood pressure in 35- to 44-year-old men is normally distributed with mean = 80 and variance = 144. Find the upper and lower fifth percentiles of this distribution.

Solution We could do this either using Table 3 (Appendix) or using Excel. If we use Table 3 and we denote the upper and lower 5th percentiles by $x_{.05}$ and $x_{.95}$, respectively, then from Equation 5.7 we have

$$\begin{aligned} x_{.05} &= 80 + z_{.05}(12) \\ &= 80 - 1.645(12) = 60.3 \text{ mm Hg} \\ x_{.95} &= 80 + z_{.95}(12) \\ &= 80 + 1.645(12) = 99.7 \text{ mm Hg} \end{aligned}$$

If we use Excel, then we have

$$x_{.05} = \text{NORMINV}(.05,\ 80,\ 12)$$
$$x_{.95} = \text{NORMINV}(.95,\ 80,\ 12)$$

The results denoted by $x(.05)$ and $x(.95)$ on the spreadsheet are shown as follows:

```
x(.05)    NORMINV(.05,80,12)    60.3
x(.95)    NORMINV(.95,80,12)    99.7
```

SECTION 5.6 Linear Combinations of Random Variables

In work on statistical inference, sums or differences or more complicated linear functions of random variables (either continuous or discrete) are often used. For this reason, the properties of linear combinations of random variables are important to consider.

DEFINITION 5.11

A **linear combination** L of the random variables $X_1, \ldots, X_n$ is defined as any function of the form $L = c_1X_1 + \cdots + c_nX_n$. A linear combination is sometimes also referred to as a **linear contrast.**

Example 5.25 **Renal Disease** Let X_1, X_2 be random variables representing serum-creatinine levels for white and black individuals with end-stage renal disease. Represent the sum, difference, and average of the random variables X_1, X_2 as linear combinations of the random variables X_1, X_2.

Solution The sum is $X_1 + X_2$, where $c_1 = 1$, $c_2 = 1$. The difference is $X_1 - X_2$, where $c_1 = 1$, $c_2 = -1$. The average is $(X_1 + X_2)/2$, where $c_1 = 0.5$, $c_2 = 0.5$.

It will often be necessary to compute the expected value and variance of linear combinations of random variables. To find the expected value of L, the principle that the expected value of the sum of n random variables is the sum of the n respective expected values is used. Applying this principle,

$$\begin{aligned} E(L) &= E(c_1X_1 + \cdots + c_nX_n) \\ &= E(c_1X_1) + \cdots + E(c_nX_n) = c_1E(X_1) + \cdots + c_nE(X_n) \end{aligned}$$

EQUATION 5.8

Expected Value of Linear Combinations of Random Variables The expected value of the linear combination $L = \sum_{i=1}^{n} c_iX_i$ is $E(L) = \sum_{i=1}^{n} c_iE(X_i)$.

Example 5.26 **Renal Disease** Suppose the expected values of serum creatinine for the white and black individuals in Example 5.25 are 1.3 and 1.5, respectively. What is the expected value of the average serum-creatinine level of a single white and a single black individual?

Solution The expected value of the average serum-creatinine level = $E(0.5X_1 + 0.5X_2) = 0.5E(X_1) + 0.5E(X_2) = 0.65 + 0.75 = 1.4$.

To compute the variance of linear combinations of random variables, we first assume that the random variables are independent. Under this assumption, it can be shown that the variance of the sum of n random variables is the sum of the respective variances. Applying this principle,

$$\begin{aligned} Var(L) &= Var(c_1X_1 + \cdots + c_nX_n) \\ &= Var(c_1X_1) + \cdots + Var(c_nX_n) = c_1^2Var(X_1) + \cdots + c_n^2Var(X_n) \end{aligned}$$

because

$$Var(c_iX_i) = c_i^2Var(X_i)$$

EQUATION 5.9

Variance of Linear Combinations of Independent Random Variables The variance of the linear combination $L = \sum_{i=1}^{n} c_iX_i$, where $X_1, \ldots, X_n$ are independent is $Var(L) = \sum_{i=1}^{n} c_i^2Var(X_i)$.

Example 5.27 **Renal Disease** Suppose X_1, X_2 are defined as in Example 5.26. If we know that $Var(X_1) = Var(X_2) = 0.25$, then what is the variance of the average serum-creatinine level over a single white and a single black individual?

Solution We wish to compute $Var(0.5X_1 + 0.5X_2)$. Applying Equation 5.9,

$$Var(0.5X_1 + 0.5X_2) = (0.5)^2 Var(X_1) + (0.5)^2 Var(X_2)$$
$$= 0.25(0.25) + 0.25(0.25) = 0.125$$

The results for the expected value and variance of linear combinations in Equations 5.8 and 5.9 *do not* depend on the assumption of normality. However, linear combinations of normal random variables are often of specific concern. It can be shown that any linear combination of normal random variables is itself normally distributed. This leads to the following important result:

EQUATION 5.10

If $X_1, \ldots, X_n$ are independent normal random variables with expected values $\mu_1, \ldots, \mu_n$ and variances $\sigma_1^2, \ldots, \sigma_n^2$, and L is any linear combination $= \sum_{i=1}^{n} c_i X_i$, then L is normally distributed with

$$\text{Expected value} = E(L) = \sum_{i=1}^{n} c_i \mu_i \quad \text{and} \quad \text{variance} = Var(L) = \sum_{i=1}^{n} c_i^2 \sigma_i^2$$

Example 5.28 **Renal Disease** If X_1, X_2 are defined as in Examples 5.25–5.27 and are each normally distributed, then what is the distribution of the average $= 0.5X_1 + 0.5X_2$?

Solution Based on the solutions to Examples 5.26 and 5.27, we know that $E(L) = 1.4$, $Var(L) = 0.125$. Therefore, $(X_1 + X_2)/2 \sim N(1.4, 0.125)$.

5.6.1 Dependent Random Variables

In some instances, we will be interested in studying linear contrasts of random variables that are not independent.

Example 5.29 **Renal Disease** A hypothesis exists that a diet high in protein may aggravate the course of kidney disease among diabetic patients. To test the feasibility of administering a low-protein diet to such patients, a small "pilot" study is set up where 20 diabetic patients are followed for 1 year on the diet. Serum creatinine is a parameter that is often used to monitor kidney function. Let X_1 be the serum creatinine at baseline and X_2 the serum creatinine after 1 year. We wish to compute the expected value and variance of the change in serum creatinine represented by the random variable $D = X_1 - X_2$ under the assumption that $E(X_1) = E(X_2) = 1.5$ and $Var(X_1) = Var(X_2) = .25$.

Solution The random variables X_1 and X_2 are not independent, because they represent serum-creatinine values on the same subject. However, we can still compute the expected value of D using Equation 5.8, because the formula for the expected value of a linear contrast is valid whether the random variables involved are independent or dependent. Therefore, $E(D) = E(X_1) - E(X_2) = 0$. However, the formula in Equation 5.9 for the variance of a linear contrast is not valid for dependent random variables.

The *covariance* is a measure used to quantify the relationship between 2 random variables.

DEFINITION 5.12 The **covariance** between 2 random variables X and Y is denoted by $Cov(X, Y)$ and is defined by

$$Cov(X,Y) = E[(X - \mu_x)(Y - \mu_y)]$$

which can also be written as $E(XY) - \mu_x\mu_y$, where μ_x is the expected value of X, μ_y is the expected value of Y, and $E(XY)$ = average value of the product of X and Y.

It can be shown that if the random variables X and Y are independent, then the covariance between them is 0. If large values of X and Y tend to occur among the same subjects (as well as small values of X and Y), then the covariance will be positive. If large values of X and small values of Y (or conversely, small values of X and large values of Y) tend to occur among the same subjects, then the covariance will be negative.

One issue is that the covariance between 2 random variables X and Y is in the units of X multiplied by the units of Y. Thus, it is difficult to interpret the strength of association between 2 variables from the magnitude of the covariance. To obtain a measure of relatedness that is independent of the units of X and Y, we consider the *correlation coefficient.*

DEFINITION 5.13 The **correlation coefficient** between 2 random variables X and Y is denoted by $Corr(X, Y)$ or ρ and is defined by

$$\rho = Corr(X,Y) = Cov(X,Y)/(\sigma_x\sigma_y)$$

where σ_x and σ_y are the standard deviations of X and Y, respectively.

Unlike the covariance, the correlation coefficient is a dimensionless quantity that is independent of the units of X and Y and ranges between –1 and 1. For random variables that are approximately linearly related, a correlation coefficient of 0 implies independence. A correlation coefficient close to 1 implies nearly perfect positive dependence with large values of X corresponding to large values of Y and small values of X corresponding to small values of Y. An example of a strong positive correlation is between forced expiratory volume (FEV), a measure of pulmonary function and height (see Figure 5.16a). A somewhat weaker positive correlation exists between serum cholesterol and dietary intake of cholesterol (see Figure 5.16b). A correlation coefficient close to –1 implies ≈ perfect negative dependence, with large values of X corresponding to small values of Y and vice versa, as is evidenced by the relationship between resting pulse rate and age in children under the age of 10 (see Figure 5.16c). A somewhat weaker negative correlation exists between FEV and number of cigarettes smoked per day in children (see Figure 5.16d).

For variables that are not linearly related, it is difficult to infer independence or dependence from a correlation coefficient.

Example 5.30 Let X be the random variable height z-score for 7-year-old children, where height z-score = (height – μ)/σ.

μ = mean height for 7-year-old children
σ = standard deviation of height for 7-year-old children

FIGURE 5.16 **Interpretation of various degrees of correlation**

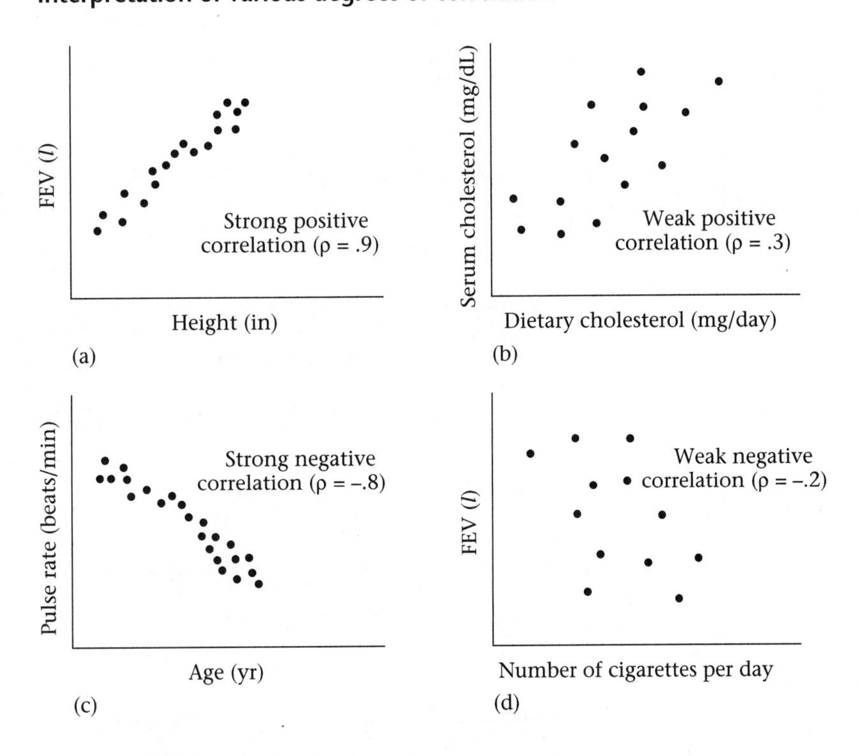

Let Y = height z-score^2 = X^2. Compute the correlation coefficient between X and Y, assuming that X is normally distributed.

Solution Since X is symmetric about 0 and Y is the same for positive and negative values of X with the same absolute value, it is easy to show that $Corr(X, Y) = 0$. However, X and Y are totally dependent because if we know X, then Y is totally determined. For example, if $X = 2$, then $Y = 4$. Thus, it would be a mistake to assume that the random variables X and Y are independent if the correlation coefficient between them is 0. This relationship can only be inferred for linearly related variables. We will discuss linear and nonlinear relationships between variables in more detail in Chapter 11.

To compute the variance of a linear contrast involving 2 dependent random variables, X_1 and X_2, we can generalize Equation 5.9 as follows:

EQUATION 5.11

$$\begin{aligned} Var(c_1X_1 + c_2X_2) &= c_1^2 Var(X_1) + c_2^2 Var(X_2) \\ &\quad + 2c_1c_2 Cov(X_1, X_2) \\ &= c_1^2 Var(X_1) + c_2^2 Var(X_2) \\ &\quad + 2c_1c_2\sigma_x\sigma_y Corr(X_1, X_2) \end{aligned}$$

Example 5.31 **Renal Disease** Suppose that the correlation coefficient between 2 determinations of serum creatinine 1 year apart is .5. Compute the variance of the change in serum creatinine over 1 year using the parameters given in Example 5.29.

Solution We have that $Var(X_1) = Var(X_2) = .25$ and $Corr(X_1, X_2) = .5$. Also, for the linear contrast $D = X_1 - X_2$, we have $c_1 = 1$ and $c_2 = -1$. Therefore, from Equation 5.11, we have that

$$\begin{aligned} Var(D) &= (1)^2(.25) + (-1)^2(.25) + 2(1)(-1)(.5)(.5)(.5) \\ &= .25 + .25 - .25 = .25 \end{aligned}$$

Notice that this variance is much smaller than the variance of the difference in serum creatinine between 2 different subjects (at the same or different times). In this case, because values for 2 different subjects are independent, $Corr(X_1, X_2) = 0$, and it follows that

$$Var(D) = (1)^2(.25) + (-1)^2(.25) + 0 = .50$$

This is the rationale for using each person as his or her own control, because it greatly reduces variability. In Chapter 8 we will discuss in more detail the paired sample and independent sample experimental designs for comparing two groups (e.g., a treated and a control group).

To compute the variance of a linear contrast involving n (possibly) dependent random variables, it can be shown that

EQUATION 5.12

Variance of Linear Combination of Random Variables (General Case)

The variance of the linear contrast $L = \sum_{i=1}^{n} c_i X_i$ is

$$\begin{aligned} Var(L) &= \sum_{i=1}^{n} c_i^2 Var(X_i) + 2\sum_{\substack{i=1 \\ i<j}}^{n} \sum_{j=1}^{n} c_i c_j Cov(X_i X_j) \\ &= \sum_{i=1}^{n} c_i^2 Var(X_i) + 2\sum_{\substack{i=1 \\ i<j}}^{n} \sum_{j=1}^{n} c_i c_j \sigma_i \sigma_j Corr(X_i, X_j) \end{aligned}$$

We will discuss covariance and correlation in more detail in Chapter 11.

Finally, we can generalize Equation 5.10 by stating that

EQUATION 5.13

If $X_1, \ldots, X_n$ are normal random variables with expected values $\mu_1, \ldots, \mu_n$ and variances $\sigma_1^2, \ldots, \sigma_n^2$ and L is any linear contrast $= \sum_{i=1}^{n} c_i X_i$, then L is normally distributed with expected value $= \sum_{i=1}^{n} c_i \mu_i$ and variance equals

$$Var(L) = \sum_{i=1}^{n} c_i^2 \sigma_i^2 + 2\sum_{\substack{i=1 \\ i<j}}^{n} \sum_{j=1}^{n} c_i c_j \sigma_i \sigma_j \rho_{ij}$$

where ρ_{ij} = correlation between X_i and X_j, $i \neq j$.

SECTION 5.7 Normal Approximation to the Binomial Distribution

In Chapter 4 the binomial distribution was introduced to assess the probability of k successes in n independent trials, where the probability of success (p) is the same for each trial. If n is large, the binomial distribution is very cumbersome to work with and an approximation is easier to use rather than the exact binomial distribution. The normal distribution is often used to approximate the binomial, because it is very easy to work with. The key question is, When will the normal distribution provide an accurate approximation to the binomial?

Suppose a binomial distribution has parameters n and p. If n is large and p is either near 0 or near 1, then the binomial distribution will be very positively or negatively skewed, respectively. See Figure 5.17a and 5.17b. Similarly, when n is small, for any p, the distribution will tend to be skewed. See Figure 5.17c. However, if n is moderately large and p is not too extreme, then the binomial distribution will tend to be symmetric and will be well approximated by a normal distribution. See Figure 5.17d.

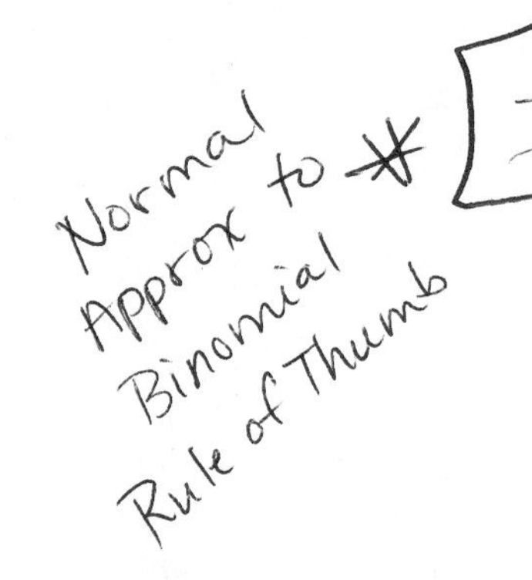

We know from Chapter 4 that the mean and variance of a binomial distribution are np and npq, respectively. A natural approximation to use is a normal distribution with the *same* mean and variance, that is, $N(np, npq)$. Suppose we want to compute $Pr(a \le X \le b)$ for some integers a, b, where X is binomially distributed with parameters n and p. This probability might be approximated by the area under the normal curve from a to b. However, we can show empirically that a better approximation to this probability is given by the area under the normal curve from $a - \frac{1}{2}$

FIGURE 5.17 **Symmetry properties of the binomial distribution**

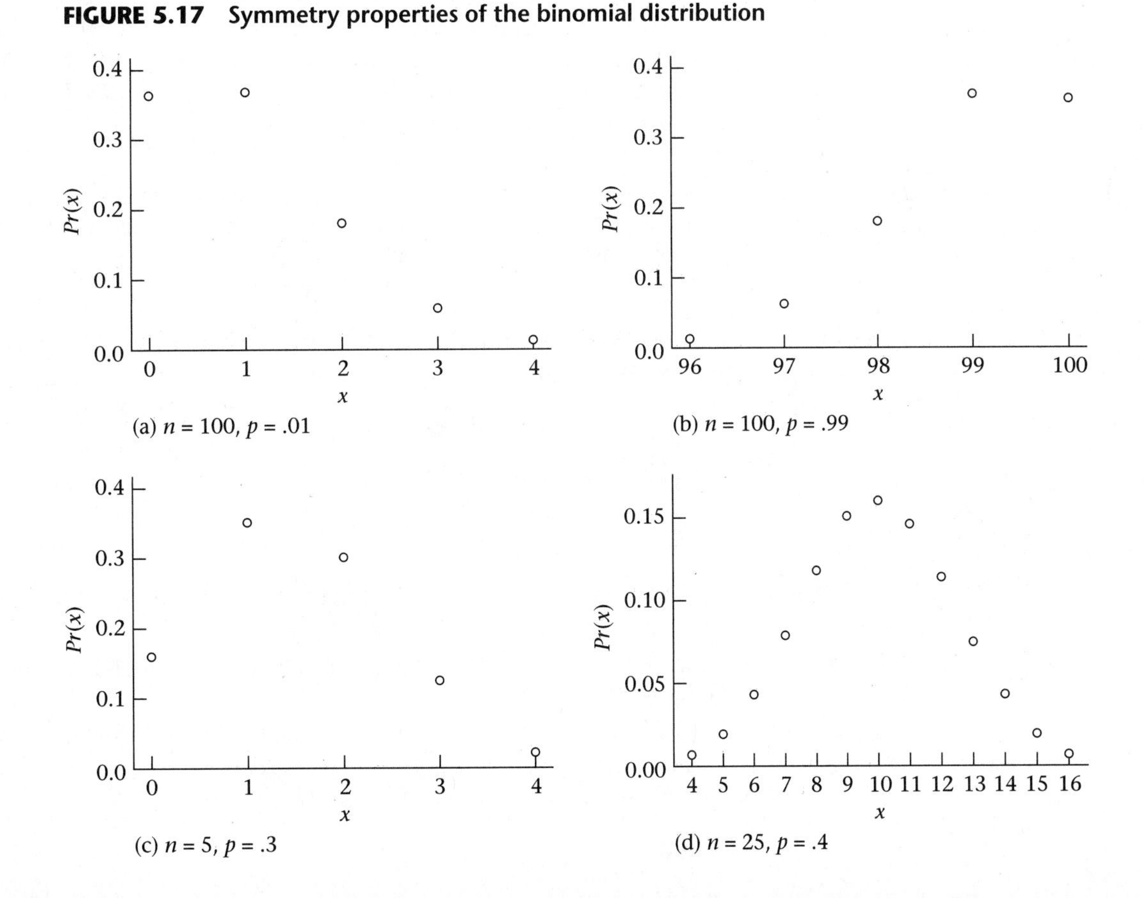

(a) $n = 100$, $p = .01$

(b) $n = 100$, $p = .99$

(c) $n = 5$, $p = .3$

(d) $n = 25$, $p = .4$

to $b+\frac{1}{2}$. This will generally be the case when any discrete distribution is approximated by the normal distribution. Thus the following rule applies:

EQUATION 5.14

Normal Approximation to the Binomial Distribution If X is a binomial random variable with parameters n and p, then $Pr(a \le X \le b)$ is approximated by the area under an $N(np, npq)$ curve from $\left(a-\frac{1}{2}\right)$ to $\left(b+\frac{1}{2}\right)$. This rule implies that for the special case $a = b$, the binomial probability $Pr(X = a)$ is approximated by the area under the normal curve from $\left(a-\frac{1}{2}\right)$ to $\left(a+\frac{1}{2}\right)$. The only exception to this rule is that $Pr(X = 0)$ and $Pr(X = n)$ are approximated by the area under the normal curve to the left of $\frac{1}{2}$ and to the right of $n-\frac{1}{2}$, respectively.

We saw in Equation 5.10 that if $X_1, \ldots, X_n$ are independent normal random variables, then any linear combination of these variables $L = \sum_{i=1}^{n} c_i X_i$ will be normally distributed. In particular, if $c_1 = \cdots = c_n = 1$, then a sum of normal random variables $L = \sum_{i=1}^{n} X_i$ will be normally distributed.

The normal approximation to the binomial distribution is a special case of a very important statistical principle, the central-limit theorem, which is a generalization of Equation 5.10. Under this principle, for large n, a sum of n random variables will be approximately normally distributed *even if the individual random variables being summed are not themselves normal.*

DEFINITION 5.14

Let X_i be a random variable that takes on the value 1 with probability p and the value 0 with probability $q = 1 - p$. This type of random variable is defined as a **Bernoulli trial.** This is a special case of a binomial random variable with $n = 1$.

We know from the definition of an expected value that $E(X_i) = 1(p) + 0(q) = p$ and that $E(X_i^2) = 1^2(p) + 0^2(q) = p$. Therefore,

$$Var(X_i) = E(X_i^2) - [E(X_i)]^2 = p - p^2 = p(1-p) = pq$$

Now consider the random variable

$$X = \sum_{i=1}^{n} X_i$$

This random variable simply represents the number of successes among n trials.

Example 5.32 Interpret $X_1, \ldots, X_n$ and X in the case of the number of neutrophils among 100 white blood cells (see Example 4.15).

Solution In this case, $n = 100$ and $X_i = 1$ if the ith white blood cell is a neutrophil and $X_i = 0$ if the ith white blood cell is not a neutrophil, where $i = 1, \ldots, 100$. X represents the number of neutrophils among $n = 100$ white blood cells.

Given Equations 5.8 and 5.9, we know that

$$E(X) = E\left(\sum_{i=1}^{n} X_i\right) = p + p + \cdots + p = np$$

and

$$Var(X) = Var\left(\sum_{i=1}^{n} X_i\right) = \sum_{i=1}^{n} Var(X_i) = pq + pq + \cdots + pq = npq$$

Based on the normal approximation to the binomial distribution, we approximate the distribution of X by a normal distribution with mean = np and variance = npq. We discuss the central-limit theorem in more detail in Section 6.5.3.

Example 5.33 Suppose a binomial distribution has parameters $n = 25$, $p = .4$. How can $Pr(7 \le X \le 12)$ be approximated?

Solution We have $np = 25(.4) = 10$, $npq = 25(.4)(.6) = 6.0$. Thus, this distribution is approximated by a normal random variable Y with mean 10 and variance 6. We specifically want to compute the area under this normal curve from 6.5 to 12.5. We have

$$\begin{aligned} Pr(6.5 \le Y \le 12.5) &= \Phi\left(\frac{12.5 - 10}{\sqrt{6}}\right) - \Phi\left(\frac{6.5 - 10}{\sqrt{6}}\right) \\ &= \Phi(1.021) - \Phi(-1.429) = \Phi(1.021) - [1 - \Phi(1.429)] \\ &= \Phi(1.021) + \Phi(1.429) - 1 = .8463 + .9235 - 1 = .770 \end{aligned}$$

This approximation is depicted in Figure 5.18. For comparative purposes, we have also computed $Pr(7 \le X \le 12)$ based on exact binomial probabilities using Excel and obtain .773, which compares well with the normal approximation of .770.

Example 5.34 **Infectious Disease** Suppose we want to compute the probability that between 50 and 75 of 100 white blood cells will be neutrophils, where the probability that any one cell

FIGURE 5.18 **The approximation of the binomial random variable X with parameters $n = 25$, $p = .4$ by the normal random variable Y with mean = 10 and variance = 6**

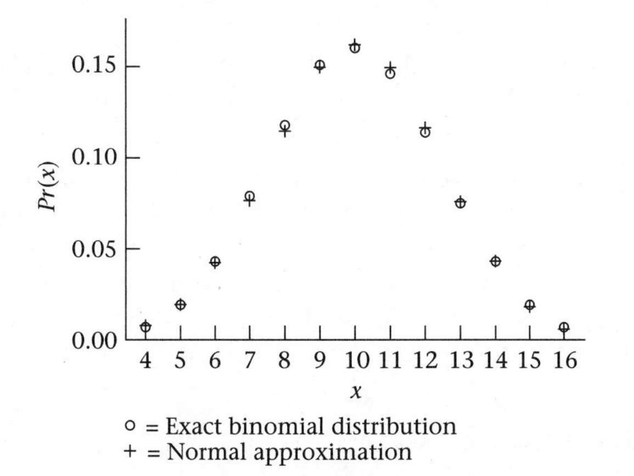

is a neutrophil is .6. These values are chosen as proposed limits to the range of neutrophils in normal people, and we wish to predict what proportion of people will be in the normal range according to this definition.

Solution The exact probability is given by

$$\sum_{k=50}^{75} \binom{100}{k} (.6)^k (.4)^{100-k}$$

The normal approximation is used to approximate the exact probability. The mean of the binomial distribution in this case is 100(.6) = 60, and the variance is 100(.6)(.4) = 24. Thus, we find the area between 49.5 and 75.5 for an $N(60, 24)$ distribution. This area is

$$\begin{aligned}\Phi\left(\frac{75.5-60}{\sqrt{24}}\right) - \Phi\left(\frac{49.5-60}{\sqrt{24}}\right) &= \Phi(3.164) - \Phi(-2.143) \\ &= \Phi(3.164) + \Phi(2.143) - 1 \\ &= .9992 + .9840 - 1 = .983\end{aligned}$$

Thus, 98.3% of the people will be normal.

Example 5.35 **Infectious Disease** Suppose a neutrophil count is defined as abnormally high if the number of neutrophils is ≥ 76, and abnormally low if the number of neutrophils is ≤ 49. Calculate the proportion of people whose neutrophil counts will be abnormally high or low.

Solution The probability of being abnormally high is given by $Pr(X \geq 76) \approx Pr(Y \geq 75.5)$, where X is a binomial random variable with parameters $n = 100$, $p = .6$, and $Y \sim N(60, 24)$. This latter probability is

$$1 - \Phi\left(\frac{75.5-60}{\sqrt{24}}\right) = 1 - \Phi(3.164) = .001$$

Similarly, the probability of being abnormally low is

$$\begin{aligned}Pr(X \leq 49) \approx Pr(Y \leq 49.5) &= \Phi\left(\frac{49.5-60}{\sqrt{24}}\right) \\ &= \Phi(-2.143) = 1 - \Phi(2.143) \\ &= 1 - .9840 = .016\end{aligned}$$

Thus, 0.1% of people will have abnormally high neutrophil counts and 1.6% will have abnormally low neutrophil counts. These probabilities are depicted in Figure 5.19.

For comparative purposes, we have also computed the proportion of people who are in the normal range, abnormally high, and abnormally low based on exact binomial probabilities using Excel. We obtain $Pr(50 \leq X \leq 75) = .983$, $Pr(X \geq 76) = .0006$, and $Pr(X \leq 49) = .017$, which corresponds almost exactly with the normal approximations used in Examples 5.34 and 5.35.

Under what conditions should this approximation be used?

EQUATION 5.15

Rule of Thumb →

> The normal distribution with mean np and variance npq can be used to approximate a binomial distribution with parameters n and p when $npq \geq 5$.

FIGURE 5.19 **Normal approximation to the distribution of neutrophils**

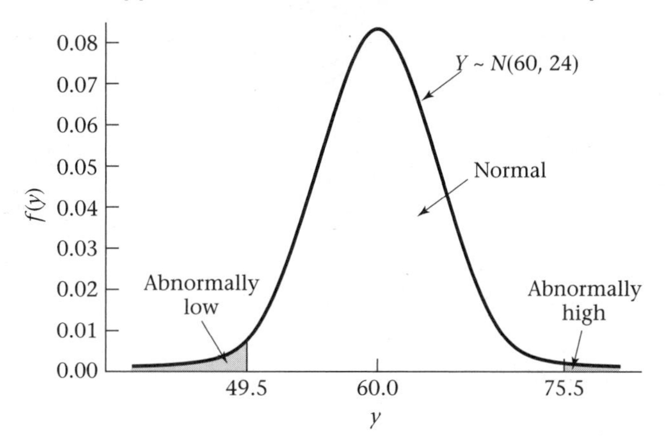

This condition will be satisfied if n is moderately large and p is not too small. To illustrate this condition, the binomial probability distributions for $p = .1$, $n = 10$, 20, 50, and 100 are plotted in Figure 5.20a through d and $p = .2$, $n = 10$, 20, 50, and 100 are plotted in Figure 5.21a through d using MINITAB.

FIGURE 5.20 **MINITAB plot of binomial distribution, $n = 10, 20, 50, 100$, $p = .1$**

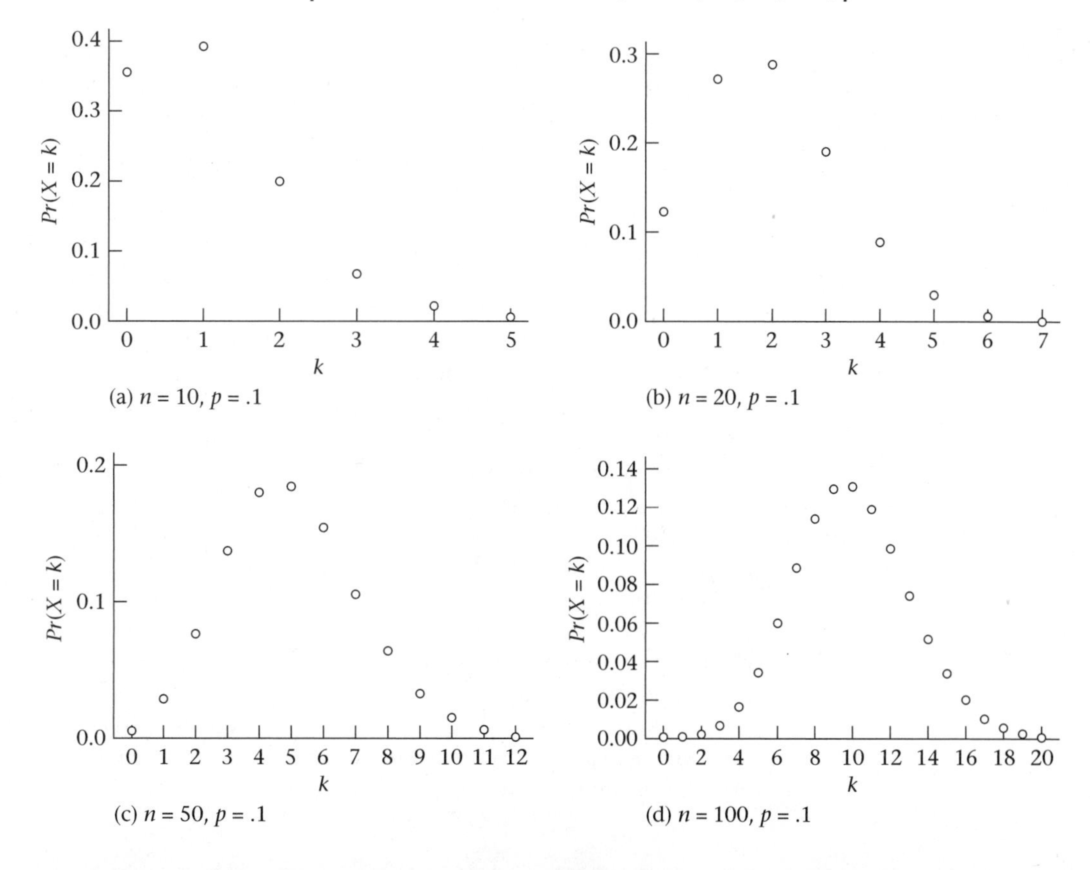

FIGURE 5.21 MINITAB plot of binomial distribution, $n = 10, 20, 50, 100$, $p = .2$

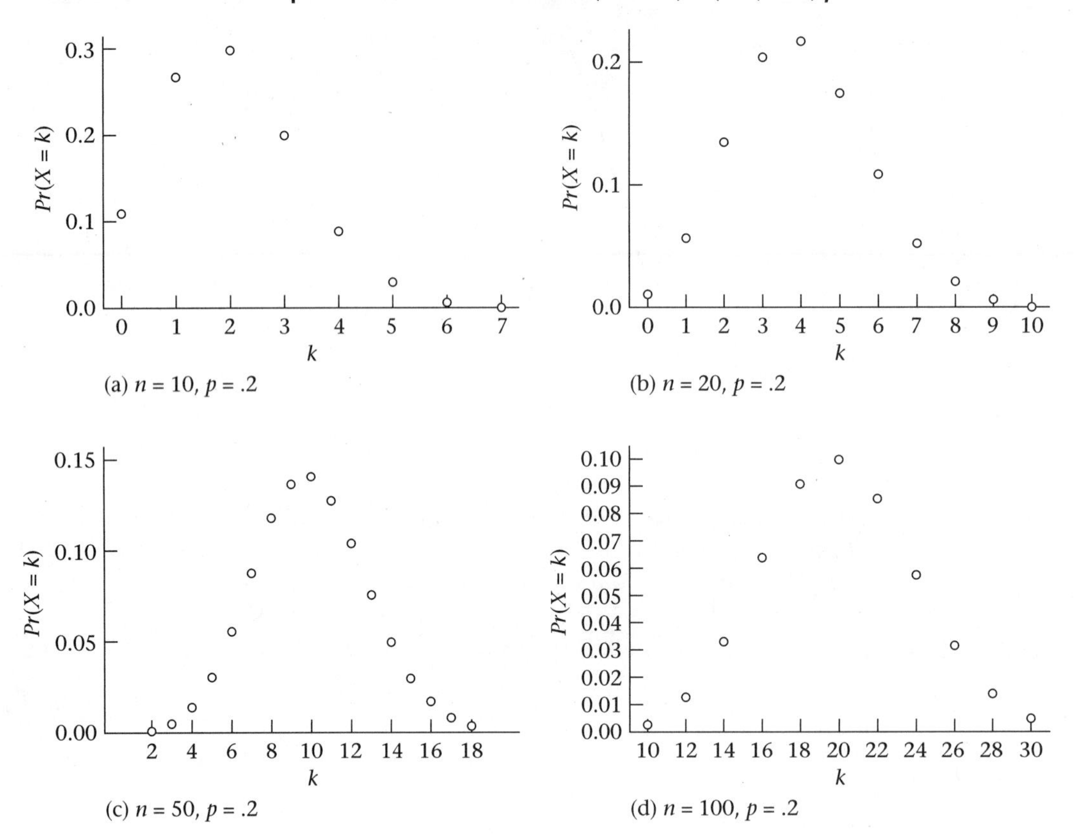

Notice that the normal approximation to the binomial distribution does not fit well in Figure 5.20a, $n = 10$, $p = .1$ ($npq = 0.9$), or Figure 5.20b, $n = 20$, $p = .1$ ($npq = 1.8$). The approximation is marginally adequate in Figure 5.20c, $n = 50$, $p = .1$ ($npq = 4.5$), where the right-hand tail is only slightly longer than the left-hand tail. The approximation is quite good in Figure 5.20d, $n = 100$, $p = .1$ ($npq = 9.0$), where the distribution appears to be quite symmetric. Similarly, for $p = .2$, although the normal approximation is not good for $n = 10$ (Figure 5.21a, $npq = 1.6$), it becomes marginally adequate for $n = 20$ (Figure 5.21b, $npq = 3.2$) and quite good for $n = 50$ (Figure 5.21c, $npq = 8.0$) and $n = 100$ (Figure 5.21d, $npq = 16.0$).

Note that the conditions under which the normal approximation to the binomial distribution works well (namely, $npq \geq 5$), which corresponds to n moderate and p not too large or too small, are generally *not* the same as the conditions for which the Poisson approximation to the binomial distribution works well [n large (≥ 100) and p very small ($p \leq .01$)]. However, occasionally both of these criteria will be met. In such cases (for example, when $n = 1000$, $p = .01$), the two approximations will yield about the same results. The normal approximation is preferable because it is easier to apply.

SECTION 5.8 Normal Approximation to the Poisson Distribution

The normal distribution can also be used to approximate discrete distributions other than the binomial distribution, particularly the Poisson distribution. The motivation for this is that the Poisson distribution is cumbersome to use for large values of μ.

The same technique is used as for the binomial distribution; that is, the mean and variance of the Poisson distribution and the approximating normal distribution are equated.

EQUATION 5.16

Normal Approximation to the Poisson Distribution A Poisson distribution with parameter μ is approximated by a normal distribution with mean and variance both equal to μ. $Pr(X = x)$ is approximated by the area under an $N(\mu, \mu)$ density from $x - \frac{1}{2}$ to $x + \frac{1}{2}$ for $x > 0$ or by the area to the left of $\frac{1}{2}$ for $x = 0$. This approximation is used when $\mu \geq 10$.

The Poisson distributions for $\mu = 2, 5, 10$, and 20 are plotted using MINITAB in Figure 5.22a through d, respectively. The normal approximation is clearly

FIGURE 5.22 MINITAB plot of Poisson distribution, μ = 2, 5, 10, 20

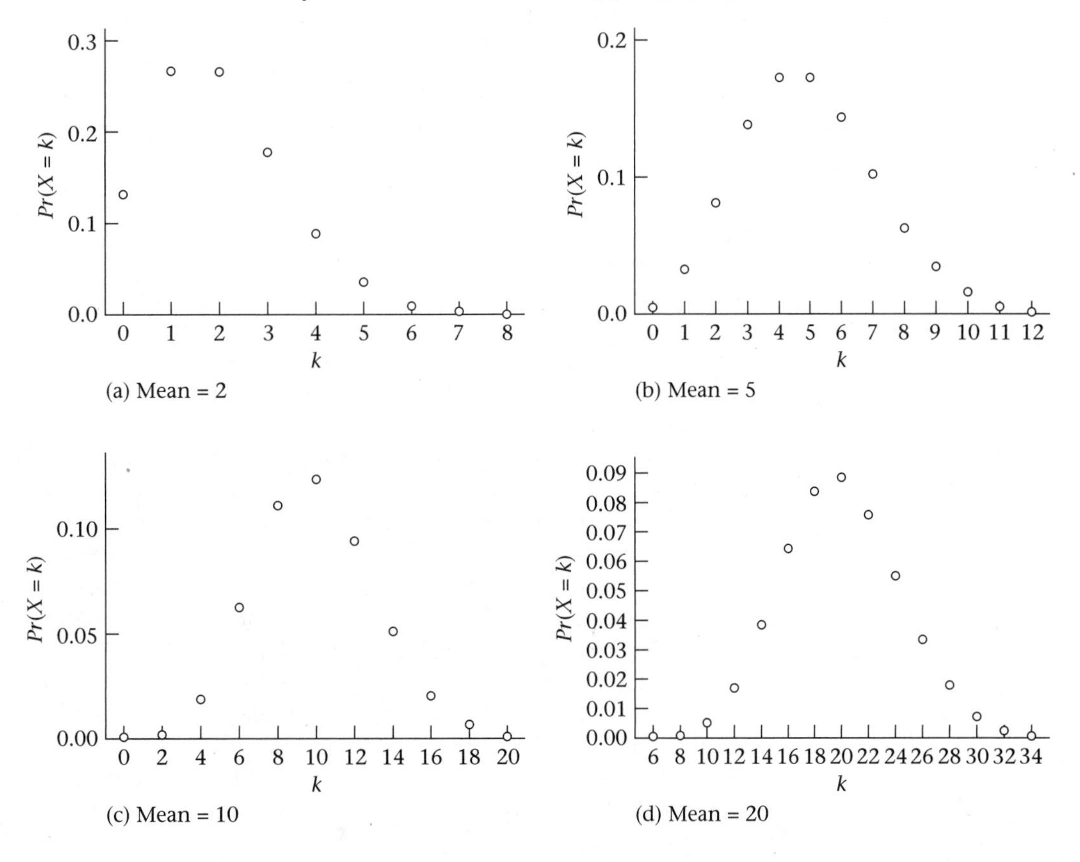

inadequate for $\mu = 2$ (Figure 5.22a), marginally adequate for $\mu = 5$ (Figure 5.22b), and adequate for $\mu = 10$ (Figure 5.22c) and $\mu = 20$ (Figure 5.22d).

Example 5.36 **Bacteriology** Consider again the distribution of the number of bacteria in a petri plate of area A. Assume that the probability of observing x bacteria is given exactly by a Poisson distribution with parameter $\mu = \lambda A$, where $\lambda = 0.1$ bacteria/cm^2 and $A = 100$ cm^2. Suppose 20 bacteria are observed in this area. How unusual is this event?

Solution The exact distribution of the number of bacteria observed in 100 cm^2 is Poisson with parameter $\mu = 10$. We approximate this distribution by a normal distribution with mean = 10 and variance = 10. Therefore, we compute

$$Pr(X \geq 20) \approx Pr(Y \geq 19.5)$$

where $Y \sim N(\lambda A, \lambda A) = N(10, 10)$

We have

$$\begin{aligned} Pr(Y \geq 19.5) &= 1 - Pr(Y \leq 19.5) = 1 - \Phi\left(\frac{19.5 - 10}{\sqrt{10}}\right) \\ &= 1 - \Phi\left(\frac{9.5}{\sqrt{10}}\right) = 1 - \Phi(3.004) \\ &= 1 - .9987 = .0013 \end{aligned}$$

Thus, 20 or more colonies in 100 cm^2 would be expected only 1.3 times in 1000 plates, a rare event indeed. For comparison, we have also computed the exact Poisson probability of obtaining 20 or more bacteria using Excel and obtain $Pr(X \geq 20 | \mu = 10) = .0035$. Thus, the normal approximation is only fair in this case, but does result in the same conclusion that obtaining 20 or more bacteria in 100 cm^2 is a rare event.

SECTION 5.9 Summary

In this chapter continuous random variables were discussed. The concept of a probability density function, which is the analogue to a probability mass function for discrete random variables, was introduced. In addition, generalizations of the concepts of expected value, variance, and cumulative distribution were presented for continuous random variables.

The normal distribution, the most important continuous distribution, was then studied in detail. The normal distribution is often used in statistical work, because many random phenomena follow this probability distribution, particularly those that can be expressed as a sum of many random variables. It was shown that the normal distribution is indexed by two parameters, the mean μ and the variance σ^2. Fortunately, all computations concerning any normal random variable can be accomplished using the standard, or unit, normal probability law, which has mean 0 and variance 1. Normal tables were introduced to use when working with the standard normal distribution. Alternatively, electronic tables can be used to evaluate areas and/or percentiles for *any* normal distribution. Also, because the normal distribution is easy to use, it is often employed to approximate other distributions. In particular, we studied the normal approximations to the binomial and Poisson distributions. This is a special case of the central-limit theorem, which is covered in more detail in Chapter 6. Also, to facilitate applications of the central-limit theorem, the properties of linear combinations of random variables were discussed, both for independent and dependent random variables.

In the next three chapters, the normal distribution is used extensively as a foundation for work on statistical inference.

PROBLEMS

Cardiovascular Disease

Because serum cholesterol is related to age and sex, some investigators prefer to express it in terms of z-scores. If X = raw serum cholesterol, then $z = \frac{X - \mu}{\sigma}$, where μ is the mean and σ is the standard deviation of serum cholesterol for a given age–sex group. Suppose z is regarded as a standard normal distribution.

***5.1** What is $Pr(z < 0.5)$?

***5.2** What is $Pr(z > 0.5)$?

***5.3** What is $Pr(-1.0 < z < 1.5)$?

Suppose a person is regarded as having high cholesterol if $z > 2.0$ and borderline cholesterol if $1.5 < z < 2.0$.

***5.4** What proportion of people have high cholesterol?

***5.5** What proportion of people have borderline cholesterol?

Nutrition

Suppose that total carbohydrate intake in 12- to 14-year-old boys is normally distributed with mean 124 g/1000 cal and standard deviation 20 g/1000 cal.

5.6 What percentage of boys in this age range have carbohydrate intake above 140 g/1000 cal?

5.7 What percentage of boys in this age range have carbohydrate intake below 90 g/1000 cal?

Suppose boys in this age range who live below the poverty level have a mean carbohydrate intake of 121 g/1000 cal with a standard deviation of 19 g/1000 cal.

5.8 Answer Problem 5.6 for boys in this age range and economic environment.

5.9 Answer Problem 5.7 for boys in this age range and economic environment.

Diabetes

A number of clinical characteristics were ascertained in a large group of subjects with insulin-dependent diabetes mellitus (IDDM). Suppose the distribution of percentage of ideal body weight in this group of patients is normal with mean 110 and standard deviation of 13.

5.10 What percentage of subjects with IDDM are above their ideal body weight, i.e., above 100% ideal body weight?

5.11 What percentage of subjects with IDDM are overweight (defined as 10% or more above ideal body weight)?

5.12 What percentage of subjects with IDDM are obese (defined as 20% or more above ideal body weight)?

5.13 What percentage of subjects with IDDM are underweight (defined as 10% or more below ideal body weight)?

5.14 What percentage of subjects with IDDM have normal body weight (within 10% of ideal body weight)?

Pulmonary Disease

Many investigators have studied the relationship between asbestos exposure and death due to chronic obstructive pulmonary disease (COPD).

5.15 Suppose that among workers exposed to asbestos in a shipyard in 1980, 33 died over a 10-year period from COPD, whereas only 24 such deaths could be expected based on statewide mortality rates. Is the number of deaths due to COPD in this group excessive?

5.16 Twelve cases of leukemia are reported in people living in a particular census tract over a 5-year period. Is this number of cases abnormal if only 6.7 cases would be expected based on national cancer-incidence rates?

Cardiovascular Disease, Pulmonary Disease

The duration of cigarette smoking has been linked to many diseases, including lung cancer and various forms of heart disease. Suppose we know that among men aged 30–34 who have ever smoked, the mean number of years they smoked is 12.8 with a standard deviation of 5.1 years. For women in this age group, the mean number of years they smoked is 9.3 with a standard deviation of 3.2.

***5.17** Assuming that the duration of smoking is normally distributed, what proportion of men in this age group have smoked for more than 20 years?

***5.18** Answer Problem 5.17 for women.

Cancer

Previous census data have indicated that approximately 0.2% of women aged 45–54 will have had cervical cancer at some point in their lives. However, the general feeling is that the rate of cervical cancer has decreased.

5.19 If a new study by mail questionnaire is performed and it is found that 100 out of 100,000 women have had cervical cancer, then is this proportion consistent with the census rate?

Cardiovascular Disease

Serum cholesterol is an important risk factor for coronary disease. We can show that serum cholesterol is approximately normally distributed with mean 219 mg/dL and standard deviation 50 mg/dL.

***5.20** If the clinically desirable range for cholesterol is < 200 mg/dL, then what proportion of people have clinically desirable levels of cholesterol?

***5.21** Some investigators feel that only cholesterol levels of over 250 mg/dL indicate a high-enough risk for heart disease to warrant treatment. What proportion of the population does this group represent?

***5.22** What proportion of the general population have borderline high-cholesterol levels—that is, > 200, but < 250 mg/dL?

Nutrition, Cancer

Beta-carotene is a substance that is hypothesized to prevent cancer. A dietary survey was undertaken for the purpose of measuring the level of beta-carotene intake in the typical American diet. Assume that the distribution of ln beta-carotene intake is normal with mean 8.34 and standard deviation 1.00. (Units are in ln IU.)

5.23 What percentage of people have dietary beta-carotene intake below 2000 IU? (*Note:* ln 2000 = 7.60.)

5.24 What percentage of people have dietary beta-carotene intake below 1000 IU? (*Note:* ln 1000 = 6.91.)

5.25 Some studies suggest that beta-carotene intake over 10,000 IU may protect against cancer. What percentage of people have a dietary intake of at least 10,000 IU?

Suppose that each person took a beta-carotene supplement pill of dosage 5000 IU in addition to his or her normal diet. Assume that the resulting distribution of ln beta-carotene intake is normally distributed with mean 9.12 and standard deviation 1.00.

5.26 What percentage of people would have an intake from diet and supplements of at least 10,000 IU?

Hypertension

People are classified as hypertensive if their systolic blood pressure is higher than a specified level for their age group, according to the scheme in Table 5.1.

TABLE 5.1 Mean and standard deviation of systolic blood pressure (mm Hg) in specific age groups

Age group	Mean	Standard deviation	Specified hypertension level
1–14	105.0	5.0	115.0
15–44	125.0	10.0	140.0

Assume that systolic blood pressure is normally distributed with mean and standard deviation given in Table 5.1 for the age groups 1–14 and 15–44, respectively. Define a *family* as a group of 2 people in the age group 1–14 and 2 people in the age group 15–44. A family is classified as hypertensive if *any one* family member is hypertensive.

***5.27** What proportion of 1- to 14-year-olds are hypertensive?

***5.28** What proportion of 15- to 44-year-olds are hypertensive?

***5.29** What proportion of families are hypertensive? (Assume that the hypertensive status of different members of a family are independent random variables.)

***5.30** Suppose an apartment building has 200 families living in it. What is the probability that between 10 and 25 families are hypertensive?

Pulmonary Disease

Forced expiratory volume (FEV) is an index of pulmonary function that measures the volume of air expelled after 1 second of constant effort. FEV is known to be influenced by age, sex, and cigarette smoking. Assume that in 45- to 54-year-old nonsmoking men FEV is normally distributed with mean 4.0 liters and standard deviation 0.5 liter.

In comparably aged currently smoking men FEV is normally distributed with mean 3.5 liters and standard deviation 0.6 liter.

5.31 If an FEV of less than 2.5 liters is regarded as showing some functional impairment (occasional breathlessness, inability to climb stairs, etc.), then what is the probability that a currently smoking man has functional impairment?

5.32 Answer Problem 5.31 for a nonsmoking man.

Some people are not functionally impaired now but their pulmonary function usually declines with age and they eventually will be functionally impaired. Assume that the *decline* in FEV over n years is normally distributed with mean $0.03n$ liters and standard deviation $0.02n$ liters.

5.33 What is the probability that a 45-year-old man with an FEV of 4.0 liters will be functionally impaired by the age of 75?

5.34 Answer Problem 5.33 for a 25-year-old man with an FEV of 4.0 liters.

Infectious Disease

The differential is a standard measurement made during a blood test. It consists of classifying white blood cells into the following 5 categories: (1) basophils, (2) eosinophils, (3) monocytes, (4) lymphocytes, and (5) neutrophils. The usual practice is to look at 100 randomly selected cells under a microscope and count the number of cells within each of the 5 categories. Assume that a normal adult will have the following proportions of cells in each category: basophils, 0.5%; eosinophils, 1.5%; monocytes, 4%; lymphocytes, 34%; and neutrophils, 60%.

***5.35** An excess of eosinophils is sometime consistent with a violent allergic reaction. What is the exact probability that a normal adult will have 5 or more eosinophils?

***5.36** An excess of lymphocytes is consistent with various forms of viral infection, such as hepatitis. What is the probability that a normal adult will have 40 or more lymphocytes?

***5.37** What is the probability that a normal adult will have 50 or more lymphocytes?

***5.38** How many lymphocytes would have to appear in the differential before you would feel that the "normal" pattern was violated?

***5.39** An excess of neutrophils is consistent with several types of bacterial infection. Suppose an adult has x neutrophils. How large would x have to be in order for the probability of a normal adult having x or more neutrophils to be $\leq 5\%$?

***5.40** How large would x have to be for the probability of a normal adult having x or more neutrophils to be $\leq 1\%$.

Blood Chemistry

In pharmacologic research a variety of clinical chemistry measurements are routinely monitored closely for evidence of side effects of the medication under study. Suppose typical blood-glucose levels are normally distributed with mean 90 mg/dL and standard deviation 38 mg/dL.

5.41 If the normal range is 65–120 mg/dL, then what percentage of values will fall in the normal range?

5.42 In some studies only values that are at least 1.5 times as high as the upper limit of normal are identified as abnormal. What percentage of values would fall in this range?

5.43 Answer Problem 5.42 for 2.0 times the upper limit of normal.

5.44 Frequently, tests that yield abnormal results are repeated for confirmation. What is the probability that for a normal person a test will be at least 1.5 times as high as the upper limit of normal on two separate occasions?

5.45 Suppose that in a pharmacologic study involving 6000 patients, 75 patients have blood-glucose levels at least 1.5 times the upper limit of normal on one occasion. What is the probability that this result could be due to chance?

Cancer

A treatment trial is proposed to test the efficacy of vitamin E as a preventive agent for cancer. One problem with such a study is how to assess compliance among study participants. A small pilot study is undertaken to establish criteria for compliance with the proposed study agents. In this regard, 10 patients are given 400 IU/day of vitamin E and 10 patients are given similar-sized tablets of placebo over a 3-month period. Their serum vitamin-E levels are measured before and after the 3-month period and the change (3-month – baseline) is shown in Table 5.2.

TABLE 5.2 Change in serum vitamin E (mg/dL) in pilot study

Group	Mean	sd	n
Vitamin E	0.80	0.48	10
Placebo	0.05	0.16	10

***5.46** Suppose a change of 0.30 mg/dL in serum levels is proposed as a test criterion for compliance; that is, a patient who shows a change of ≥ 0.30 mg/dL is considered a compliant vitamin-E taker. If normality is assumed, what percentage of the vitamin-E group would be expected to show a change of at least 0.30 mg/dL?

***5.47** Is the measure in Problem 5.46 a measure of sensitivity, specificity, or predictive value?

***5.48** What percentage of the placebo group would be expected to show a change of not more than 0.30 mg/dL?

***5.49** Is the measure in Problem 5.48 a measure of sensitivity, specificity, or predictive value?

***5.50** Suppose a new threshold of change, Δ mg/dL, is proposed for establishing compliance. We wish to use a level of Δ such that the compliance measures in Problems 5.46 and 5.48 for the patients in the vitamin-E and placebo groups are the same. What should Δ be? What would be the compliance in the vitamin-E and placebo groups using this threshold level?

Mental Health

5.51 Refer to Tables 3.4 and 3.5 (p. 67). Suppose a study of Alzheimer's disease is planned in more than one retirement community. How many retired people need to be studied to have a 90% chance of detecting at least 100 people with Alzheimer's disease, assuming that the age-sex-specific prevalence rates and age–sex distribution in Tables 3.4 and 3.5 hold?

5.52 Answer Problem 5.51 for 50 rather than 100 people.

Pulmonary Disease

Refer to the pulmonary function data in the Data Set FEV.DAT on the data disk (see Problem 2.21, p. 42). We are interested in whether there is a relationship between smoking status and level of pulmonary function. However, FEV is affected by age and sex; also, smoking children tend to be older than nonsmoking children. For these reasons, FEV should be standardized for age and sex. To accomplish this, use the z-score approach outlined above in Problem 5.1, where the z-scores here are defined by age–sex groups.

5.53 Plot the distribution of z-scores for smokers and nonsmokers separately. Do these distributions look normal? Does there appear to be any relationship between smoking and pulmonary function in these data?

5.54 Repeat the analyses in Problem 5.53 for the subgroup of children 10+ years of age (because smoking is very rare before this age). Do you reach similar conclusions?

5.55 Repeat the analyses in Problem 5.54 separately for boys and girls. Are your conclusions the same in the two groups?

Note: Formal methods for comparing mean FEVs between smokers and nonsmokers are discussed in the material on statistical inference in Chapter 8.

Cardiovascular Disease

A clinical trial was conducted to test the efficacy of nifedipine, a new drug for reducing chest pain in patients with angina severe enough to require hospitalization. The duration of the study was 14 days in the hospital unless the patient was withdrawn prematurely from therapy, was discharged from the hospital, or died prior to this time. Patients were randomly assigned to either nifedipine or propranolol and were given the same dosage of each drug in identical capsules at level 1 of therapy. If pain did not cease at this level of therapy, or if pain recurred after a period of pain cessation, then the patient progressed to level 2, whereby the dosage of each drug was increased according to a prespecified schedule. Similarly, if pain continued or recurred at level 2, then the patient progressed to level 3, whereby the dosage of the anginal drug was increased again. Patients randomized to either group were allowed to receive nitrates in any amount that was deemed clinically appropriate to help control pain.

The main objective of the study was to compare the degree of pain relief with nifedipine and propranolol. A secondary objective was to better understand the effects of these agents on other physiologic parameters including heart rate and blood pressure. Data on these latter parameters are given in the Data Set NIFED.DAT (on the data disk); the format of this file is given in Table 5.3.

5.56 Describe the effect of each treatment regimen on changes in heart rate and blood pressure. Do the distribution of changes in these parameters look normal or not?

TABLE 5.3 Format of NIFED.DAT

Column	Variable	Code
1–2	ID	
4	Treatment group	N = nifedipine/ P = propanolol
6–8	Baseline heart rate[a]	beats/min
10–12	Level 1 heart rate[b]	beats/min
14–16	Level 2 heart rate	beats/min
18–20	Level 3 heart rate	beats/min
22–24	Baseline SBP[a]	mm Hg
26–28	Level 1 SBP[b]	mm Hg
30–32	Level 2 SBP	mm Hg
34–36	Level 3 SBP	mm Hg

[a]Heart rate and systolic blood pressure immediately prior to randomization.
[b]Highest heart rate and systolic blood pressure (SBP) at each level of therapy.
Note: Missing values indicate that either
(1) the patient withdrew from the study prior to entering this level of therapy;
(2) the patient achieved pain relief before reaching this level of therapy; or,
(3) the patient encountered this level of therapy, but this particular piece of data was missing.

5.57 Compare graphically the effects of the treatment regimens on heart rate and blood pressure. Do you notice any difference between treatments?

(*Note:* Formal tests for comparing changes in heart rate and blood pressure in the two treatment groups are covered in Chapter 8.)

Hypertension

It is well known that there are racial differences in blood pressure between white and black adults. These differences generally do not exist between white and black children. Because aldosterone levels have been related to blood-pressure levels in adults in previous research, an investigation was performed to look at aldosterone levels among black children and white children [1].

***5.58** If the mean plasma-aldosterone level in black children was 230 pmol/L with $sd = 203$ pmol/L, then what percentage of black children have levels ≤ 300 pmol/L if normality is assumed?

***5.59** If the mean plasma-aldosterone level in white children is 400 pmol/L with $sd = 218$ pmol/L, then what percentage of white children have levels ≤ 300 pmol/L if normality is assumed?

***5.60** The distribution of plasma-aldosterone concentration in 53 white and 46 black children is shown in Figure 5.23. Does the assumption of normality seem reasonable? Why or why not? (*Hint:* Qualitatively compare the observed number of children who have levels below 300 pmol/L with the expected number in each group under the assumption of normality.)

Hepatic Disease

Suppose we observe 84 alcoholics with cirrhosis of the liver, of whom 29 have hepatomas—that is, liver-cell carcinoma. Suppose we know, based on a large sample, that the risk of hepatoma among alcoholics without cirrhosis of the liver is 24%.

5.61 What is the probability that we observe exactly 29 alcoholics with cirrhosis of the liver who have hepatomas if the true rate of hepatoma among alcoholics (with or without cirrhosis of the liver) is .24?

5.62 What is the probability of observing at least 29 hepatomas among the 84 alcoholics with cirrhosis of the liver under the assumptions in Problem 5.61?

5.63 What is the smallest number of hepatomas that would have to be observed among the group of alcoholics with cirrhosis of the liver in order for the hepatoma experience in this group to be different from the hepatoma experience among alcoholics without cirrhosis of the liver? (*Hint:* Use a 5% probability of getting a result at least as extreme to denote differences between the hepatoma experiences of the two groups.)

FIGURE 5.23 Plasma-aldosterone concentrations in 53 white and 46 black children. Values within the shaded area were undetectable (< 50 pmol per liter). The solid horizontal lines indicate the mean values, and the broken horizontal lines the mean ± *se*. The concept of standard error (*se*) is discussed in Chapter 6.

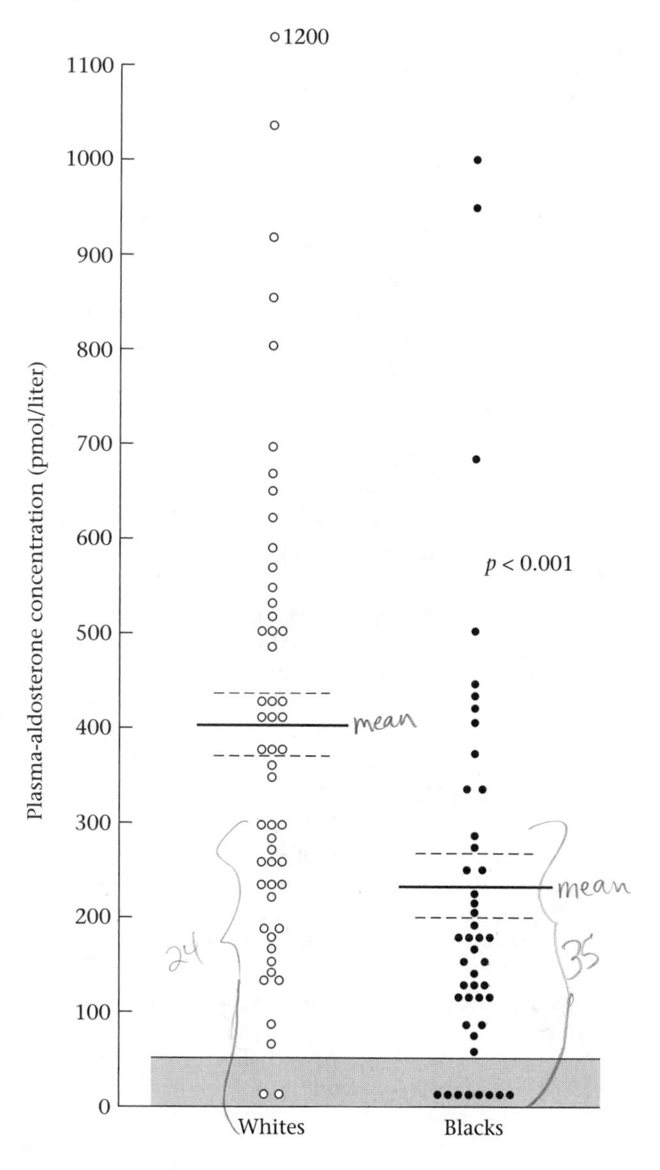

Hypertension

The Second Task Force Report on Blood Pressure Control in Children [2] reports blood-pressure norms for children by age and sex group. The mean ± standard deviation for 17-year-old boys for diastolic blood pressure is 63.7 ± 11.4 mm Hg, based on a large sample.

5.64 One approach for defining elevated blood pressure is to use 90 mm Hg—the standard for elevated adult diastolic blood pressure—as the cutoff. What percentage of 17-year-old boys would be found to have elevated blood pressure, using this approach?

5.65 Suppose there are 2000 17-year-old boys in the 11th grade, of whom 25 have elevated blood pressure by the criteria in Problem 5.64. Is this an unusually high number of boys with elevated blood pressure? Why or why not?

Environmental Health

5.66 A study was conducted relating particulate air pollution and daily mortality in Steubenville, Ohio [3]. On average over the last 10 years there have been 3 deaths per day. Suppose that on 90 high-pollution days—days where the total suspended particulates are in the highest quartile among all days—the death rate is 3.2 deaths per day, or 288 deaths observed over the 90 high-pollution days. Are there an unusual number of deaths on high-pollution days?

Refer to the Data Set VALID.DAT (on the data disk) described in Table 2.16 (p. 42).

5.67 Consider the nutrients saturated fat, total fat, and total calories. Plot the distribution of each nutrient for both the diet record (DR) and the food-frequency questionnaire (FFQ). Do you think a normal distribution is appropriate for these nutrients?

Hint: Compute the observed proportion of women who fall within 1.0, 1.5, 2.0, and 2.5 standard deviations of the mean. Compare the observed proportions with the expected proportions based on the assumption of normality.

5.68 Answer Problem 5.67 using the ln(nutrient) transformation for each nutrient value. Is the normality assumption more appropriate for log-transformed or untransformed values, or neither?

5.69 A special problem arises for the nutrient alcohol consumption. There are often a large number of nondrinkers (alcohol consumption = 0) and another large group of drinkers (alcohol consumption > 0). The overall distribution of alcohol consumption appears bimodal. Plot the distribution of alcohol consumption for both the DR and the FFQ. Do the distributions appear unimodal or bimodal? Do you think that the normality assumption is appropriate for this nutrient?

Occupational Health

Table 5.4 is obtained from the 1975 *Statistical Abstract of the United States* published by the Census Bureau with the primary data obtained from the National Center for Health Statistics [4]. The right-hand side of the table provides age-race-sex-specific 1-year mortality rates for the United States in 1973. Please note that the entries on the right side of the table are the number of deaths per 1000 individuals; they are not percentages. Suppose we are investigating workers in a nuclear-power plant and wish to ascertain whether the mortality of workers in this plant is higher or lower than expected. On January 1, 1973, we have the following age distribution in the plant as given in Table 5.5.

5.70 Suppose we follow this group of men over a 5-year period from January 1, 1973, to December 31, 1977, and find that 20 of the men have died over this period. Is this an unusual number of deaths? Justify your answer. Please assume that the mortality rate of a 45-year-old, for example, remains constant over the 5-year period. (*Hint:* Consider using an approximation to solve this problem.)

TABLE 5.5 Age distribution in plant

	Age	*n*
	45[a]	30
	50	80
	55	70
	60	20
Total		200

[a] We assume for simplicity that 30 of the workers are exactly 45 years old, 80 are exactly 50 years old, 70 are exactly 55 years old, and 20 are exactly 60 years old. We also assume that 80% of the workers are white and 20% are black within each age group.

Cancer, Neurology

A study was performed concerning the risk of cancer among patients with cystic fibrosis [5]. Based on registries of patients with cystic fibrosis in the United States and Canada, cancer incidence among cystic-fibrosis patients between January 1, 1985, and December 31, 1992, was compared with expected cancer-incidence rates based on the SEER program from the National Cancer Institute from 1984 to 1988.

5.71 Among cystic-fibrosis patients, 37 cancers were observed, whereas 45.6 cancers were expected. What distribution can be used to model the distribution of the number of cancers among cystic-fibrosis patients?

5.72 Are there an unusually low number of cancers among cystic-fibrosis patients?

TABLE 5.4 **Expectation of life and mortality rates, by age, race, and sex, 1973**

Age (years)	Expectation of life in years					Mortality rate per 1000 living at specified age				
		White		Black and other			White		Black and other	
	Total	*Male*	*Female*	*Male*	*Female*	Total	*Male*	*Female*	*Male*	*Female*
40	34.9	32.2	38.5	28.7	34.4	2.95	3.20	1.82	8.12	4.45
41	34.0	31.3	37.6	27.9	33.6	3.20	3.50	1.99	8.53	4.79
42	33.1	30.4	36.7	27.1	32.7	3.50	3.87	2.18	9.07	5.19
43	32.2	29.5	35.7	26.4	31.9	3.85	4.31	2.42	9.79	5.65
44	31.3	28.6	34.8	25.6	31.1	4.25	4.81	2.68	10.66	6.17
45	30.5	27.8	33.9	24.9	30.3	4.70	5.39	2.97	11.63	6.74
46	29.6	26.9	33.0	24.2	29.5	5.17	6.00	3.27	12.62	7.32
47	28.8	26.1	32.1	23.5	28.7	5.64	6.61	3.57	13.56	7.88
48	27.9	25.3	31.2	22.8	27.9	6.08	7.19	3.85	14.42	8.39
49	27.1	24.4	30.3	22.1	27.1	6.53	7.77	4.12	15.23	8.87
50	26.3	23.6	29.5	21.5	26.4	6.99	8.38	4.41	16.03	9.35
51	25.5	22.8	28.6	20.8	25.6	7.51	9.09	4.74	16.94	9.90
52	24.6	22.0	27.7	20.2	24.9	8.16	9.97	5.14	18.05	10.57
53	23.8	21.2	26.9	19.5	24.1	8.97	11.07	5.64	19.47	11.42
54	23.1	20.5	26.0	18.9	23.4	9.90	12.36	6.22	21.13	12.40
55	22.3	19.7	25.2	18.3	22.7	10.92	13.76	6.86	22.93	13.46
56	21.5	19.0	24.4	17.7	22.0	11.97	15.22	7.52	24.75	14.53
57	20.8	18.3	23.5	17.2	21.3	13.06	16.76	8.19	26.51	15.58
58	20.0	17.6	22.7	16.6	20.6	14.18	18.36	8.84	28.12	16.58
59	19.3	16.9	21.9	16.1	20.0	15.32	20.04	9.49	29.64	17.54
60	18.6	16.2	21.1	15.6	19.3	16.56	21.82	10.21	31.25	18.64
61	17.9	15.6	20.3	15.0	18.7	17.89	23.73	11.02	32.98	19.85
62	17.2	15.0	19.6	14.5	18.0	19.27	25.75	11.88	34.66	20.93
63	16.6	14.3	18.8	14.0	17.4	20.68	27.88	12.81	36.24	21.82
64	15.9	13.7	18.0	13.5	16.8	22.17	30.15	13.82	37.81	22.63
65	15.3	13.2	17.3	13.1	16.2	23.73	32.56	14.94	39.20	23.18
70	12.2	10.4	13.7	10.7	13.2	35.38	47.85	23.63	57.22	40.74
75	9.5	8.1	10.4	9.2	11.3	55.13	72.33	41.66	78.02	55.91
80	7.3	6.3	7.9	7.9	9.4	82.71	107.02	68.61	93.95	65.28
85 and over	5.4	4.7	5.7	6.3	7.3	1000.00	1000.00	1000.00	1000.00	1000.00

Source: U.S. National Center for Health Statistics, *Vital Statistics of the United States,* annual.

5.73 In the same study 13 cancers of the digestive tract were observed, whereas only 2 cancers were expected. Are there an unusually high number of digestive cancers among cystic-fibrosis patients?

Hypertension

A doctor diagnoses a patient as hypertensive and prescribes an antihypertensive medication. To assess the clinical status of the patient, she takes n replicate blood-pressure measurements before the patient starts the drug (baseline) and n replicate blood-pressure measurements 4 weeks after starting the drug (follow-up). She will use the average of the n replicates at baseline minus the average of the n replicates at follow-up to assess the clinical status of the patient. She knows based on previous clinical experience with the drug that the mean diastolic blood-pressure change over a 4-week period over a large number of patients after starting the drug is 5.0 mm Hg with variance $33/n$ where n is the number of replicate measures obtained at both baseline and follow-up.

5.74 If we assume that the change in mean diastolic blood pressure is normally distributed, then what is the probability that a subject will decline by at least 5 mm Hg if 1 replicate measure is obtained at baseline and follow-up?

5.75 The physician also knows that if a patient is untreated (or does not take the prescribed medication), then the mean diastolic blood pressure over 4 weeks will decline by 2 mm Hg with variance $33/n$. What is the probability that an untreated subject will decline by at least 5 mm Hg if 1 replicate measure is obtained at both baseline and follow-up?

5.76 Suppose the physician is not sure whether the patient is actually taking the prescribed medication. She wants to take enough replicate measurements at baseline and follow-up so that the probability in Problem 5.74 is at least 5 times the probability in Problem 5.75. How many replicate measurements should she take?

Endrocrinology

A study was performed comparing different treatments for preventing bone loss among postmenopausal women under 60 years of age [6]. The mean change in bone-mineral density of the lumbar spine over a 2-year period for women in the placebo group was –1.8 percent (a mean decrease) with a standard deviation of 4.3%. Assume that the change in bone-mineral density is normally distributed.

5.77 If a decline of 2% in bone-mineral density is considered clinically significant, then what percentage of women in the placebo group can be expected to have a decline at least this large?

The change in bone-mineral density of the lumbar spine over a 2-year period among women in the alendronate 5-mg group was +3.5% (a mean increase), with a standard deviation of 4.2%.

5.78 What percentage of women in the alendronate 5-mg group can be expected to have a clinically significant decline as defined in Problem 5.77?

5.79 Suppose that 10% of the women assigned to the alendronate 5-mg group are actually not taking their pills (noncompliers). If noncompliers are assumed to have a similar response as women in the placebo group, then what percentage of women who are complying with the alendronate 5-mg treatment would be expected to have a clinically significant decline? (*Hint:* Use the total-probability rule.)

Cardiovascular Disease

Obesity is an important determinant of cardiovascular disease because it directly affects several established cardiovascular risk factors, including hypertension and diabetes. It is estimated that the average weight for an 18-year-old woman is 123 lbs, and increases to 142 lbs at 50 years of age. Also, let us assume that the average systolic blood pressure (SBP) for a 50-year-old woman is 125 mm Hg, with a standard deviation = 15 mm Hg and that systolic blood pressure is normally distributed.

5.80 What proportion of 50-year-old women are hypertensive, if hypertension is defined as SBP $\geq$ 140 mm Hg?

From previous clinical trials, it is estimated that for every 10 lbs of weight loss, on average there is a corresponding reduction in mean SBP of 3 mm Hg.

5.81 Suppose that an average woman did not gain any weight from age 18 to 50. What would be the expected average SBP for 50-year-old women under these assumptions?

5.82 If the standard deviation of SBP under the assumption in Problem 5.81 remained the same (15 mm Hg), and the distribution of SBP remained normal, then what would be the expected proportion of hypertensive women under the assumption in Problem 5.81?

5.83 What percentage of hypertension at age 50 is attributable to the weight gain from age 18 to 50?

Simulation

5.84 Draw 100 random samples from a binomial distribution with parameters n = 10 and p = .4. Consider an approximation to this distribution by a normal distribution with mean = np = 4 and variance = npq = 2.4. Draw 100 random samples from the normal approximation. Plot the 2 frequency distributions on the same graph, and compare results. Do you think that the normal approximation is adequate here?

5.85 Answer the question in Problem 5.84 for a binomial distribution with parameters n = 20 and p = .4 and the corresponding normal approximation.

5.86 Answer the question in Problem 5.84 for a binomial distribution with parameters n = 50 and p = .4 and the corresponding normal approximation.

Simulation

An apparatus exists whereby a collection of balls is displaced to the top of a stack by suction. At the top level (Level 1) each ball is shifted 1 unit to the left or 1 unit to the right at random with equal probability (see Figure 5.24). The ball then drops down to Level 2. At Level 2, each ball is again shifted 1 unit to the left or 1 unit to the right at random. The process continues for 15 levels and the balls are collected at the bottom for a short time until being collected by suction to the top. (*Note:* A similar apparatus is located in the Museum of Science, Boston, Massachusetts, and is displayed in Figure 5.25.)

FIGURE 5.24 Apparatus for random displacement of balls

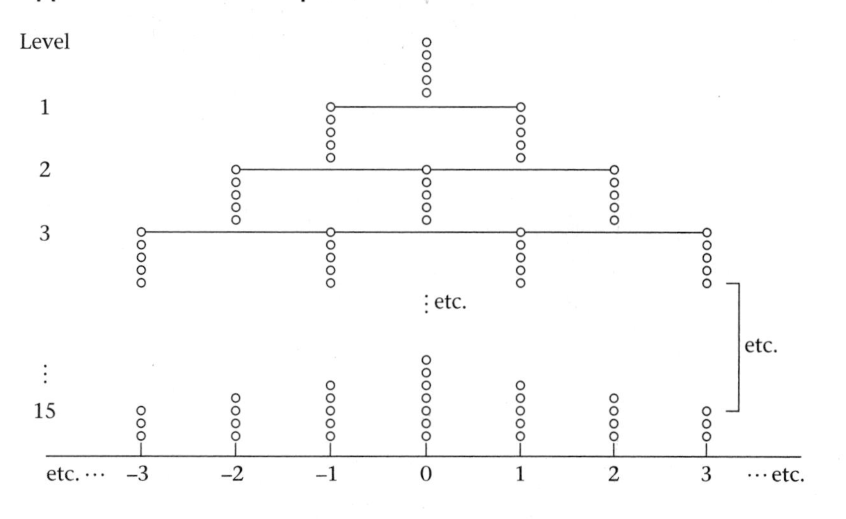

FIGURE 5.25 Probability apparatus at the Museum of Science, Boston

Photo taken by David Rosner.

5.87 What is the exact probability distribution of the position of the balls at the bottom with respect to the entry position (which is arbitrarily denoted by 0)?

5.88 Can you think of an approximation to the distribution derived in Problem 5.87?

5.89 Perform a simulation of this process (e.g., using MINITAB or Excel) with 100 balls, and plot the frequency distribution of the position of the balls at the bottom with respect to the entry position. Does the distribution appear to conform to the distributions derived in Problems 5.87 and 5.88?

REFERENCES

[1] Pratt, J. H., Jones, J. J., Miller, J. Z., Wagner, M. A., & Fineberg, N. S. (1989, October). Racial differences in aldosterone excretion and plasma aldosterone concentrations in children. *New England Journal of Medicine, 321*(17), 1152–1157.

[2] *Report of the Second Task Force on Blood Pressure Control in Children—1987*. (1987, January). Bethesda, MD. National Heart, Lung and Blood Institute. *Pediatrics, 79*(1), 1–25.

[3] Schwartz, J., & Dockery, D. W. (1992, January). Particulate air pollution and daily mortality in Steubenville, Ohio. *American Journal of Epidemiology, 135*(1), 12–19.

[4] U.S. National Center for Health Statistics. (1975). *Vital statistics of the United States*. Washington, DC: U.S. Government Printing Office.

[5] Neglia, J. F., Fitzsimmons, S. C., Maisonneauve, P., Schoni, M. H., Schoni-Affolter, F., Corey, M., Lowenfels, A. B., & the Cystic Fibrosis and Cancer Study Group. (1995). The risk of cancer among patients with cystic fibrosis. *New England Journal of Medicine, 332*, 494–400.

[6] Hosking, D., Chilvers, C. E. D., Christiansen, C., Ravn, P., Wasnich, R., Ross, P., McClung, M., Belske, A., Thompson, D., Daley, M. T., & Yates, A. J. (1998). Prevention of bone loss with alendronate in postmenopausal women under 60 years of age. *New England Journal of Medicine, 338*, 485–492.

Estimation

SECTION 6.1 Introduction

Chapters 3 through 5 explored the properties of different probability models. In doing this, we always assumed that the specific probability distributions were known.

Example 6.1 **Infectious Disease** We assumed that the number of neutrophils in a sample of 100 white blood cells was binomially distributed with parameter $p = .6$.

Example 6.2 **Bacteriology** We assumed that the number of bacterial colonies on a 100-cm^2 agar plate was Poisson-distributed with parameter $\mu = 2$.

Example 6.3 **Hypertension** We assumed that the distribution of diastolic blood-pressure measurements in 35- to 44-year-old men was normal with mean $\mu = 80$ mm Hg and $\sigma = 12$ mm Hg.

In general, we have been assuming that the properties of the underlying distributions from which our data are drawn are known and that the only question that remains is what can be predicted about the behavior of the data given a knowledge of these properties.

Example 6.4 **Hypertension** Using the data in Example 6.3, we could predict that about 95% of all diastolic blood pressures from 35- to 44-year-old men should fall between 56 mm Hg and 104 mm Hg.

The problem addressed in the remainder of this text, and the more basic statistical problem, is that we have a data set and we want to **infer** the properties of the underlying distribution from this data set. This inference usually involves **inductive** rather than **deductive** reasoning; that is, in principle, a variety of different probability models must at least be explored to see which model best "fits" the data.

Statistical inference can be further subdivided into the two main areas of estimation and hypothesis testing. **Estimation** is concerned with estimating the values of specific population parameters; **hypothesis testing** is concerned with testing whether the value of a population parameter is equal to some specific value. Problems of estimation are covered in this chapter, while problems of hypothesis testing are discussed in Chapters 7 through 10.

Some typical problems that involve estimation follow.

Example 6.5 **Hypertension** Suppose we measure the systolic blood pressures of a group of Samoan villagers and we believe the underlying distribution is normal. How can the parameters of this distribution (μ, σ^2) be estimated?

Example 6.6 **Infectious Disease** Suppose we look at people living within a low-income census tract in an urban area and we wish to estimate the prevalence of HIV-positive individuals in the community. We assume that the number of cases among n people sampled will be binomially distributed with some parameter p. How is the parameter p estimated?

In Examples 6.5 and 6.6, we were interested in obtaining specific numbers as estimates of our parameters. These numbers are often referred to as **point estimates.** Sometimes we want to specify a range within which the parameter values are likely to fall. If this range is narrow, then we may feel that our point estimate is a good one. This type of problem involves **interval estimation.**

Example 6.7 **Ophthalmology** A study is proposed to screen a group of 1000 people ages 65 or older to identify those with visual impairment, that is, a visual acuity of 20–50 or worse in both eyes, even with the aid of glasses. Suppose we assume that the number of such people ascertained in this manner is binomially distributed with parameters $n = 1000$ and unknown p. We would like to obtain a point estimate of p and to provide an interval about this point estimate to see how precise our point estimate is. For example, we would feel better about a point estimate of 5% if this interval were .04–.06 than if it were .01–.10.

SECTION 6.2 The Relationship Between Population and Sample

Example 6.8 **Obstetrics** Suppose we wish to characterize the distribution of birthweights of all liveborn infants who were born in the United States in 1998. Assume that the underlying distribution of birthweight has an expected value (or mean) μ and variance σ^2. Ideally, we wish to estimate μ and σ^2 exactly, based on the entire population of U.S. liveborn infants in 1998. But this task is difficult with such a large group. Instead, we decide to select a random sample of n infants who are *representative of* this large group and use the birthweights $x_1, \ldots, x_n$ from this sample to help us estimate μ and σ^2. What is a random sample?

DEFINITION 6.1 A **random sample** is a selection of some members of the population such that each member is independently chosen and has a known nonzero probability of being selected.

DEFINITION 6.2 A **simple random sample** is a random sample in which each group member has the same probability of being selected.

DEFINITION 6.3 The **reference**, **target**, or **study** population is the group we wish to study. The random sample is selected from the study population.

For ease of discussion, the abbreviated term "random sample" will be used to denote a simple random sample.

Although many samples in practice are random samples, this is not the only type of sample used in practice. A popular alternative design is that of **cluster sampling.**

Example 6.9 **Cardiovascular Disease** The Minnesota Heart Study seeks to accurately assess the prevalence and incidence of different types of cardiovascular morbidity (such as heart attack and stroke) in the state of Minnesota, as well as trends in these rates over time. It is impossible to survey every individual in the state. It is also impractical to survey, in person, a random sample of individuals in the state, because it would require a large number of interviewers to be dispersed throughout the state. Instead, the state of Minnesota is divided into geographically compact regions or clusters. A random sample of clusters is then chosen for study, and several interviewers are sent to each cluster selected. The goal is first to enumerate all households in a cluster, and then to survey all members in these households. If some cardiovascular morbidity is identified by interviewers, then the relevant individuals are invited to be examined in more detail at a centrally located health site within the cluster. The total sample of all interviewed subjects over the entire state is referred to as a *cluster sample*. Similar strategies are also used in many national health surveys. Cluster samples require statistical methods that are beyond the scope of this book. See Cochran [1] for more discussion of cluster sampling.

In this book, we will assume that all samples are random samples from a reference population.

Example 6.10 **Epidemiology** The Nurses' Health Study is a large epidemiologic study involving over 100,000 female nurses residing in 11 large states in the United States. The nurses were first contacted by mail in 1976 and since then have been followed every 2 years by mail. Suppose we want to select a test sample of 100 nurses to test a new procedure for obtaining serum samples by mail. One way of selecting the sample is to assign each nurse an ID number and then select the nurses with the lowest 100 ID numbers. This is definitely *not* a random sample because each nurse is not equally likely to be chosen. Indeed, because the first two digits of the ID number are assigned according to state, the 100 nurses with the lowest ID numbers would all come from the same state. An alternative method of selecting the sample is to have a computer generate a set of 100 **random numbers** (from among the numbers 1 to over 100,000), one to be assigned to each nurse. By doing this, each member is equally likely to be included in the sample. This would be a truly random sample. (More details on random numbers are given in Section 6.3.)

In practice, there is rarely an opportunity to enumerate each member of the reference population so as to select a random sample, and the assumption that the sample selected has all the properties of a random sample without formally being a random sample must be made.

In Example 6.8 the reference population is finite and well defined and can be enumerated. In many instances, the reference population is effectively infinite and is not well defined.

Example 6.11 **Cancer** Suppose we wish to estimate the 5-year survival rate of women who are initially diagnosed as having breast cancer at the ages of 45–54 and who undergo radical mastectomy at this time. Our reference population is all women who have ever had a first diagnosis of breast cancer in the past when they were 45–54 years old or who ever will have such a diagnosis in the future when they are 45–54 years old and who receive radical mastectomies.

This population is effectively infinite. The population cannot be formally enumerated and thus a truly random sample cannot be selected from it. However, we will again assume that the sample we have selected behaves as if it were a random sample.

In this text we will assume that all reference populations discussed are **effectively infinite**, although, as in Examples 6.8 and 6.10, many are actually very large but finite. Sampling theory is the special branch of statistics that treats statistical inference for finite populations; it is beyond the scope of this text. See reference [1] for a good treatment of this subject.

SECTION 6.3 Random-Number Tables

In this section practical methods for selecting random samples are discussed.

Example 6.12 **Hypertension** Suppose we wish to study how effective a hypertension treatment program is in controlling the blood pressure of its participants. We are given a roster of all 1000 participants in the program but due to limited resources only 20 people can be surveyed. We would like the 20 people chosen to be a random sample from the population of all participants in the program. How should this random sample be selected?

A computer-generated list of random numbers would probably be used to select this sample.

DEFINITION 6.4 A **random number** (or **random digit**) is a random variable X that takes on the values 0, 1, 2, . . . , 9 with equal probability. Thus,

$$Pr(X = 0) = Pr(X = 1) = \cdots = Pr(X = 9) = \frac{1}{10}$$

DEFINITION 6.5 Computer-generated **random numbers** are collections of digits that satisfy the following two properties:

(1) Each digit 0, 1, 2, . . . , 9 is equally likely to occur.

(2) The value of any particular digit is independent of the value of any other digit selected.

Table 4 in the Appendix lists 1000 random digits generated on the computer.

Example 6.13 Suppose that 5 is a particular random digit selected. Does this mean that 5's are more likely to occur in the next few digits selected?

Solution No. Each digit either after or before the 5 is still equally likely to be any of the digits, 0, 1, 2, . . . , 9.

Computer programs generate large sequences of random digits that approximately satisfy the conditions in Definition 6.5. Thus, such numbers are sometimes referred to as **pseudorandom numbers**, because they are simulated to satisfy the properties in Definition 6.5.

Example 6.14 **Hypertension** How can the random digits in Table 4 be used to select 20 random participants in the hypertension treatment program in Example 6.12?

Solution A roster of the 1000 participants must be compiled and each participant must then be assigned a number from 000 to 999. Perhaps an alphabetical list of the participants already exists, which would make this task easy. Twenty groups of three digits would then be selected, starting at any position in the random-number table. For example, if we start at the first row of Table 4, we have the numbers listed in Table 6.1.

TABLE 6.1 **20 random participants chosen from 1000 participants in the hypertension treatment program**

First 3 rows of random-number table				Actual random numbers chosen				
32924	22324	18125	09077	329	242	232	418	125
54632	90374	94143	49295	090	775	463	290	374
88720	43035	97081	83373	941	434	929	588	720
				430	359	708	183	373

Therefore, our random sample would consist of the people numbered 329, 242, . . . , 373 in the alphabetical list. In this particular case there were no repeats in the 20 three-digit numbers selected. If there had been repeats, then more three-digit numbers would have been selected until 20 different numbers were selected. This process is referred to as **random selection.**

Example 6.15 **Diabetes** Suppose we wish to conduct a clinical trial of an oral hypoglycemic agent for diabetes and compare the oral hypoglycemic agent with standard insulin therapy. A small study of this type will be conducted on 10 patients: 5 patients will be randomly assigned to the oral agent and 5 to insulin therapy. How can the table of random numbers be used to make the assignments?

Solution The prospective patients are numbered from 0 to 9, and five unique random digits are selected from some arbitrary position in the random-number table (e.g., from the 28th row). The first five unique digits are 6, 9, 4, 3, 7. Thus, the patients numbered 3, 4, 6, 7, 9 will be assigned to the oral hypoglycemic agent and the remaining patients (numbered 0, 1, 2, 5, 8) to standard insulin therapy. In some studies the prospective patients are not known in advance and are recruited over time. In this case, if 00 is identified with the first patient recruited, 01 with the second patient recruited, . . . , and 09 with the tenth patient recruited, then the oral hypoglycemic agent would be assigned to the fourth (3 + 1), fifth (4 + 1), seventh (6 + 1), eighth (7 + 1), and tenth (9 + 1) patients recruited and the standard therapy to the first (0 + 1), second (1 + 1), third (2 + 1), sixth (5 + 1), and ninth (8 + 1) patients recruited.

This process is referred to as **random assignment.** It differs from random selection (Example 6.14) in that, typically, the number, in this case, of patients to be assigned to each type of treatment (5), is fixed in advance. The random-number table helps select the 5 patients who are to receive one of the two treatments (oral hypoglycemic agent). By default, the patients not selected for the oral agent are assigned to the alternative treatment (standard insulin therapy). No additional random numbers need be chosen for the second group of 5 patients. If random selection were used instead, then one approach might be to draw a random digit for each patient. If the random digit is from 0 to 4, then the patient is assigned to the oral agent; if the random digit is from 5 to 9, then the patient is assigned to insulin therapy. One problem with this approach is that in a finite sample, equal numbers of patients will not necessarily be assigned to each therapy, which is usually the most efficient design. Indeed, referring to the first ten digits in the 28th row of the random-number table (69644 37198), we see that 4 patients would be assigned to oral therapy (patients 4, 5, 6, and 8) and 6 patients would be assigned to insulin therapy (patients 1, 2, 3, 7, 9, 10) if the method of random selection were used. Random assignment is preferable in this instance, because it ensures an equal number of patients assigned to each treatment group.

Example 6.16 **Obstetrics** The birthweights from 1000 consecutive deliveries at Boston City Hospital (serving a low-income population) are enumerated in Table 6.2. For the purpose of this example, consider this population as effectively infinite. Suppose we wish to draw 5 random samples of size 10 from this population using the computer. How can these samples be selected?

Solution MINITAB has a function that allows sampling from columns. The user must specify the number of rows to be sampled (the size of the random sample to be selected). Thus, if the 1000 birthweights are stored in a single column (e.g., C1), and we specify 10 rows to be sampled, then we will obtain a random sample of size 10 from this population. This random sample of size 10 can be stored in a different column (e.g., C2). This process can be repeated 5 times and results stored in 5 separate columns. It is also possible to calculate the mean and *sd* for each random sample. The results are shown in

TABLE 6.2 Sample of birthweights (oz) obtained from 1000 consecutive deliveries at Boston City Hospital

ID Numbers	0	1	2	3	4	5	6	7	8	9	10	11	12	13	14	15	16	17	18	19
000–019	116	124	119	100	127	103	140	82	107	132	100	92	76	129	138	128	115	133	70	121
020–039	114	114	121	107	120	123	83	96	116	110	71	86	136	118	120	110	107	157	89	71
040–059	98	105	106	52	123	101	111	130	129	94	124	127	128	112	83	95	118	115	86	120
060–079	106	115	100	107	131	114	121	110	115	93	116	76	138	126	143	93	121	135	81	135
080–099	108	152	127	118	110	115	109	133	116	129	118	126	137	110	32	139	132	110	140	119
100–119	109	108	103	88	87	144	105	138	115	104	129	108	92	100	145	93	115	85	124	123
120–139	141	96	146	115	124	113	98	110	153	165	140	132	79	101	127	137	129	144	126	155
140–159	120	128	119	108	113	93	144	124	89	126	87	120	99	60	115	86	143	97	106	148
160–179	113	135	117	129	120	117	92	118	80	132	121	119	57	126	126	77	135	130	102	107
180–199	115	135	112	121	89	135	127	115	133	64	91	126	78	85	106	94	122	111	109	89
200–219	99	118	104	102	94	113	124	118	104	124	133	80	117	112	112	112	102	118	107	104
220–239	90	113	132	122	89	111	118	108	148	103	112	128	86	111	140	126	143	120	124	110
240–259	142	92	132	128	97	132	99	131	120	106	115	101	130	120	130	89	107	152	90	116
260–279	106	111	120	198	123	152	135	83	107	55	131	108	100	104	112	121	102	114	102	101
280–299	118	114	112	133	139	113	77	109	142	144	114	117	97	96	93	120	149	107	107	117
300–319	93	103	121	118	110	89	127	100	156	106	122	105	92	128	124	125	118	113	110	149
320–339	98	98	141	131	92	141	110	134	90	88	111	137	67	95	102	75	108	118	99	79
340–359	110	124	122	104	133	98	108	125	106	128	132	95	114	67	134	136	138	122	103	113
360–379	142	121	125	111	97	127	117	122	120	80	114	126	103	98	108	100	106	98	116	109
380–399	98	97	129	114	102	128	107	119	84	117	119	128	121	113	128	111	112	120	122	91
400–419	117	100	108	101	144	104	110	146	117	107	126	120	104	129	147	111	106	138	97	90
420–439	120	117	94	116	119	108	109	106	134	121	125	105	177	109	109	109	79	118	92	103
440–459	110	95	111	144	130	83	93	81	116	115	131	135	116	97	108	103	134	140	72	112
460–479	101	111	129	128	108	90	113	99	103	41	129	104	144	124	70	106	118	99	85	93
480–499	100	105	104	113	106	88	102	125	132	123	160	100	128	131	49	102	110	106	96	116
500–519	128	102	124	110	129	102	101	119	101	119	141	112	100	105	155	124	67	94	134	123
520–539	92	56	17	135	141	105	133	118	117	112	87	92	104	104	132	121	118	126	114	90
540–559	109	78	117	165	127	122	108	109	119	98	120	101	96	76	143	83	100	128	124	137
560–579	90	129	89	125	131	118	72	121	91	113	91	137	110	137	111	135	105	88	112	104
580–599	102	122	144	114	120	136	144	98	108	130	119	97	142	115	129	125	109	103	114	106
600–619	109	119	89	98	104	115	99	138	122	91	161	96	138	140	32	132	108	92	118	58
620–639	158	127	121	75	112	121	140	80	125	73	115	120	85	104	95	106	100	87	99	113
640–659	95	146	126	58	64	137	69	90	104	124	120	62	83	96	126	155	133	115	97	105
660–679	117	78	105	99	123	86	126	121	109	97	131	133	121	125	120	97	101	92	111	119
680–699	117	80	145	128	140	97	126	109	113	125	157	97	119	103	102	128	116	96	109	112
700–719	67	121	116	126	106	116	77	119	119	122	109	117	127	114	102	75	88	117	99	136
720–739	127	136	103	97	130	129	128	119	22	109	145	129	96	128	122	115	102	127	109	120
740–759	111	114	115	112	146	100	106	137	48	110	97	103	104	107	123	87	140	89	112	123
760–779	130	123	125	124	135	119	78	125	103	55	69	83	106	130	98	81	92	110	112	104
780–799	118	107	117	123	138	130	100	78	146	137	114	61	132	109	133	132	120	116	133	133
800–819	86	116	101	124	126	94	93	132	126	107	98	102	135	59	137	120	119	106	125	122
820–839	101	119	97	86	105	140	89	139	74	131	118	91	98	121	102	115	115	135	100	90
840–859	110	113	136	140	129	117	117	129	143	88	105	110	123	87	97	99	128	128	110	132
860–879	78	128	126	93	148	121	95	121	127	80	109	105	136	141	103	95	140	115	118	117
880–899	114	109	144	119	127	116	103	144	117	131	74	109	117	100	103	123	93	107	113	144
900–919	99	170	97	135	115	89	120	106	141	137	107	132	132	58	113	102	120	98	104	108
920–939	85	115	108	89	88	126	122	107	68	121	113	116	94	85	93	132	146	98	132	104
940–959	102	116	108	107	121	132	105	114	107	121	101	110	137	122	102	125	104	124	121	111
960–979	101	93	93	88	72	142	118	157	121	58	92	114	104	119	91	52	110	116	100	147
980–999	114	99	123	97	79	81	146	92	126	122	72	153	97	89	100	104	124	83	81	129

Table 6.3. One issue in obtaining random samples on the computer is whether the samples are obtained with or without replacement. The default option is sampling without replacement, whereby the same data point from the population cannot be selected more than once in a specific sample. Under sampling with replacement, repetitions are permissible. Table 6.3 uses sampling without replacement.

TABLE 6.3 **5 random samples of size 10 from the population of infants whose birthweights (oz) appear in Table 6.2**

	Sample				
Individual	1	2	3	4	5
1	97	177	97	101	137
2	117	198	125	114	118
3	140	107	62	79	78
4	78	99	120	120	129
5	99	104	132	115	87
6	148	121	135	117	110
7	108	148	118	106	106
8	135	133	137	86	116
9	126	126	126	110	140
10	121	115	118	119	98
$\bar{x}$	116.90	132.80	117.00	106.70	111.90
s	21.70	32.62	22.44	14.13	20.46

SECTION 6.4 Randomized Clinical Trials

An important advance in clinical research design is the use of randomization and the randomized clinical trial (RCT).

DEFINITION 6.6 A **randomized clinical trial (RCT)** is a type of research design for comparing different treatments, in which the assignment of treatments to patients is by some random mechanism. The process of assignment of treatments to patients is called **randomization.** Randomization means that the types of patients assigned to different treatment modalities will be similar if the sample sizes are large. However, if the sample sizes are small, then patient characteristics of treatment groups may not be comparable. Thus, it is customary to present a table of characteristics of different treatment groups in RCTs, to check that the randomization process is working well.

Example 6.17 **Hypertension** The SHEP (Systolic Hypertension in the Elderly Program) trial is a study designed to assess the ability of antihypertensive drug treatment to reduce the risk of stroke among people 60 years of age or older with isolated systolic hypertension. Isolated systolic hypertension is defined as elevated systolic blood pressure level (≥160 mm Hg), but normal diastolic blood pressure level (<90 mm Hg) [2]. Of the 4736 people studied, 2365 were randomly assigned to active drug treatment and 2371 were randomly assigned to placebo. The baseline characteristics of the participants were compared by treatment group to check that the randomization achieved its goal of providing comparable groups of patients in the two treatment groups (see Table 6.4). We see that the patient characteristics of the two treatment groups are generally very comparable.

TABLE 6.4 Baseline characteristics of randomized SHEP participants by treatment group[a]

Characteristic	Active-treatment group	Placebo group	Total
Number randomized	2365	2371	4736
Age, *y*			
Average[b]	71.6 (6.7)	71.5 (6.7)	71.6 (6.7)
%			
60–69	41.1	41.8	41.5
70–79	44.9	44.7	44.8
≥80	14.0	13.4	13.7
Race–sex, %[c]			
Black men	4.9	4.3	4.6
Black women	8.9	9.7	9.3
White men	38.8	38.4	38.6
White women	47.4	47.7	47.5
Education, *y*[b]	11.7 (3.5)	11.7 (3.4)	11.7 (3.5)
Blood pressure, mm Hg[b]			
Systolic	170.5 (9.5)	170.1 (9.2)	170.3 (9.4)
Diastolic	76.7 (9.6)	76.4 (9.8)	76.6 (9.7)
Antihypertensive medication at initial contact, %	33.0	33.5	33.3
Smoking, %			
Current smokers	12.6	12.9	12.7
Past smokers	36.6	37.6	37.1
Never smokers	50.8	49.6	50.2
Alcohol use, %			
Never	21.5	21.7	21.6
Formerly	9.6	10.4	10.0
Occasionally	55.2	53.9	54.5
Daily or nearly daily	13.7	14.0	13.8
History of myocardial infarction, %	4.9	4.9	4.9
History of stroke, %	1.5	1.3	1.4
History of diabetes, %	10.0	10.2	10.1
Carotid bruits, %	6.4	7.9	7.1
Pulse rate, beats/min[b d]	70.3 (10.5)	71.3 (10.5)	70.8 (10.5)
Body-mass index, kg/m^2 [b]	27.5 (4.9)	27.5 (5.1)	27.5 (5.0)
Serum cholesterol, mmol/L[b]			
Total	6.1 (1.2)	6.1 (1.1)	6.1 (1.1)
High-density lipoprotein	1.4 (0.4)	1.4 (0.4)	1.4 (0.4)
Depressive symptoms, %[e]	11.1	11.0	11.1
Evidence of cognitive impairment, %[f]	0.3	0.5	0.4
No limitation of activities of daily living, %[d]	95.4	93.8	94.6
Baseline electrocardiographic abnormalities, %[g]	61.3	60.7	61.0

[a]SHEP indicates the Systolic Hypertension in the Elderly Program.
[b]Values are mean (*sd*).
[c]Included among the whites were 204 Asians (5% of whites), 84 Hispanics (2% of whites), and 41 classified as "other" (1% of whites).
[d]$P < .05$ for the active-treatment group compared with the placebo group.
[e]Depressive symptom scale score of 7 or greater.
[f]Cognitive impairment scale score of 4 or greater.
[g]One or more of the following Minnesota codes: 1. 1 to 1.3 (Q/QS), 3.1 to 3.4 (high R waves), 4.1 to 4.4 (ST depression), 5.1 to 5.4 (T wave changes), 6.1 to 6.8 (AV conduction defects), 7.1 to 7.8 (ventricular conduction defects), 8.1 to 8.6 (arrhythmias), and 9.1 to 9.3 and 9.5 (miscellaneous items).

The importance of randomization in modern clinical research cannot be overestimated. Prior to randomization, comparison of different treatments were often based on selected samples, which are often not comparable.

Example 6.18 **Infectious Disease** Aminoglycosides are a type of antibiotic that are effective against certain types of Gram-negative organisms. They are often given to critically ill patients (such as cancer patients, to prevent secondary infections that are caused by the treatment received). However, there are also side effects of aminoglycosides including nephrotoxicity (damage to the kidney) and ototoxicity (temporary hearing loss). For several decades, studies have been comparing the efficacy and safety of different aminoglycosides. Many studies have compared the most common aminoglycoside, gentamicin, with other antibiotics in this class (such as tobramycin). The earliest studies were nonrandomized studies. Typically, physicians would compare outcomes for all patients treated with gentamicin in an infectious disease service over a defined period of time with outcomes for all patients treated with another aminoglycoside. No random mechanism was used to assign treatments to patients. The problem is that patients prescribed tobramycin might be sicker than patients prescribed gentamicin, especially if tobramycin is perceived as a more effective antibiotic and is "the drug of choice" for the sickest patient. Ironically, in a nonrandomized study, the more effective antibiotic might actually perform worse, because this antibiotic is prescribed more often for the sickest patients. Recent clinical studies are virtually all randomized studies. Patients assigned to different antibiotics will tend to be similar in randomized studies, and comparison of different types of antibiotics can be performed based on comparable patient populations.

6.4.1 Design Features of Randomized Clinical Trials

The actual method of randomization differs widely in different studies. Random selection, random assignment, or some other random process may be used as the method of randomization. In clinical trials, random assignment is sometimes referred to as **block randomization.**

DEFINITION 6.7 **Block randomization** is defined as follows in clinical trials comparing two treatments (referred to as treatment A and treatment B). A block size of $2n$ is determined in advance, where for every $2n$ patients entering the study, n patients are randomly assigned to treatment A and the remaining n patients are assigned to treatment B. A similar approach can be used in clinical trials with more than 2 treatment groups. For example, if there are k treatment groups, then the block size might be kn, where for every kn patients, n patients are randomly assigned to the first treatment, n patients are randomly assigned to the second treatment, and so on, n patients are randomly assigned to the kth treatment.

Thus, if there are 2 treatment groups, then under block randomization, for every $2n$ patients there will be an equal number assigned to each treatment. The advantage is that treatment groups will be of equal size in both the short and the long run. Because the eligibility criteria, types of patients entering a trial, or other procedures in a clinical trial sometimes change as a study progresses, this ensures comparability of treatment groups over short periods of time as the study procedures evolve. One disadvantage of blocking is that it may become evident what the randomization scheme is after a while, and physicians may defer entering patients into the study until the treatment they perceive as better is more likely to be selected. To avert this problem, a variable block size is sometimes used. For example, the block size might be 8 for the first block, 6 for the second block, 10 for the third block, and so on.

Another technique that is sometimes used in the randomization process is **stratification.**

DEFINITION 6.8 In some clinical studies, patients are subdivided into subgroups, or strata, according to characteristics that are thought to be important for patient outcome. Separate randomization lists are maintained for each stratum to ensure that there are comparable patient populations within each stratum. This procedure is called **stratification.** Either random selection (ordinary randomization) or random assignment (block randomization) might be used for each stratum. Typical characteristics used to define strata are age, sex, or overall clinical condition of the patient.

Another important advance in modern clinical research is the use of **blinding.**

DEFINITION 6.9 A clinical trial is referred to as **double blind** if neither the physician nor the patient knows what treatment he or she is getting. A clinical trial is referred to as **single blind** if the patient is blinded as to treatment assignment but the physician is not. A clinical trial is **unblinded** if both the physician and patient are aware of the treatment assignment.

Currently, the gold standard of clinical research is the randomized double-blind study, in which patients are assigned to treatments at random and neither the patient nor the physician is aware of the treatment assignment.

Example 6.19 **Hypertension** The SHEP study referred to in Example 6.17 was a double-blind study. Neither the patient nor the physician knew whether the antihypertensive medication was an active drug or a placebo. Blinding is always preferable to prevent biased reporting of outcome by the patient and/or the physician. However, it is not always feasible in all research settings.

Example 6.20 **Cerebrovascular Disease** Atrial fibrillation (AF) is a common symptom in the elderly, characterized by a specific type of abnormal heart rhythm. For example, former President George Bush had this condition while in office. It is well known that the risk of stroke is much higher among people with AF than for other people of comparable age and sex, particularly among the elderly. Warfarin is a drug considered effective in preventing stroke among people with AF. However, warfarin can cause bleeding complications and it is important to determine the optimal dose for a patient so as to maximize the benefit of stroke prevention while minimizing the risk of bleeding. Unfortunately, to monitor the dose requires periodic assessments of the prothrombin time (a measure of the clot-forming capacity of blood), with blood tests every few weeks, when the dose may be increased, decreased, or kept the same, depending on the prothrombin time. Because it is usually felt to be impractical to subject control patients to regular sham blood tests, the dilemma arises of selecting a good control treatment to compare with warfarin, in a clinical trial setting. In most clinical trials involving warfarin, patients are assigned at random to either warfarin or control treatment, where control is simply nontreatment. However, it is important in this setting that people making the sometimes subjective determination of whether or not a stroke has occurred be blinded as to treatment assignment of individual patients.

Another issue with blinding is that patients may be initially blinded as to treatment assignment, but the nature of side effects may strongly indicate the actual treatment received.

Example 6.21 **Cardiovascular Disease** In the Physicians' Health Study, a randomized study was performed comparing aspirin with aspirin placebo in the prevention of cardiovascular disease. One side effect of regular intake of aspirin is gastrointestinal bleeding. The presence of this side effect strongly indicates that the type of treatment received was aspirin.

SECTION 6.5 Estimation of the Mean of a Distribution

Now that you understand the meaning of a random sample from a population and have explored some practical methods for selecting such samples using computer-generated random numbers, we will move on to estimation. The question remains, How is a specific random sample $x_1, \ldots, x_n$ used to estimate μ and σ^2, the mean and variance of the underlying distribution? Estimating the mean is the focus of this section, and estimating the variance is covered in Section 6.7.

6.5.1 Point Estimation

A natural estimator to use for estimating the population mean μ is the sample mean

$$\overline{X} = \sum_{i=1}^{n} X_i/n$$

What properties of $\overline{X}$ make it a desirable estimator of μ? Forget about our particular sample for the moment, and consider the set of all possible samples of size n that could have been selected from the population. The values of $\overline{X}$ in each of these samples will, in general, be different. These values will be denoted by $\bar{x}_1$, $\bar{x}_2$, and so forth. The key conceptual point in this instance is to forget about our sample as a unique entity and to consider it instead as representative of all possible samples of size n that could have been drawn from the population. Stated another way, $\bar{x}$ is regarded as a single realization of a random variable $\overline{X}$ over all possible samples of size n that could have been selected from the population. In the remainder of this text, the symbol X will be used to denote a random variable, while x will be used to denote a specific realization of the random variable X in a sample.

DEFINITION 6.10 The **sampling distribution** of $\overline{X}$ is the distribution of values of $\bar{x}$ over all possible samples of size n that could have been selected from the reference population.

FIGURE 6.1 **Sampling distribution of $\bar{x}$ over 200 samples of size 10 selected from the population of 1000 birthweights given in Table 6.2 (100 = 100.0–100.9, etc.)**

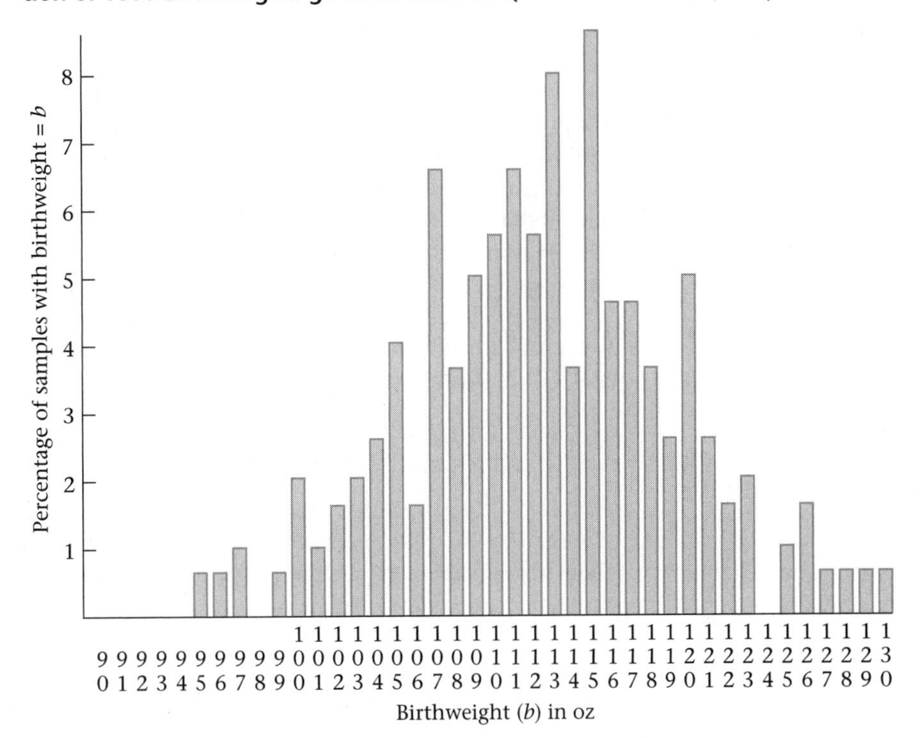

In Figure 6.1, an example of such a sampling distribution is provided. This example consists of a frequency distribution of the sample mean from 200 randomly selected samples of size 10 drawn from the distribution of 1000 birthweights given in Table 6.2, as generated by the Statistical Analysis System (SAS) procedure PROC CHART.

We can show that the average of these sample means ($\bar{x}$'s) when taken over a large number of random samples of size n will approximate μ as the number of samples selected becomes large. In other words, the expected value of $\bar{X}$ over its sampling distribution is equal to μ. This result is summarized as follows:

EQUATION 6.1

Let $X_1, \ldots, X_n$ be a random sample drawn from some population with mean μ. Then for the sample mean $\bar{X}$, $E(\bar{X}) = \mu$.

Note that Equation 6.1 holds for any population regardless of its underlying distribution. In words, we refer to $\bar{X}$ as an unbiased estimator of μ.

DEFINITION 6.11

We will refer to an estimator of a parameter θ as $\hat{\theta}$. An **estimator** $\hat{\theta}$ of a parameter θ is **unbiased** if $E(\hat{\theta}) = \theta$. This means that the average value of $\hat{\theta}$ over a large number of repeated samples of size n will be θ.

The unbiasedness of $\overline{X}$ is not a sufficient reason to use it as an estimator of μ. Many unbiased estimators of μ exist, including the sample median and the average value of the largest and smallest data points in a sample. Why is $\overline{X}$ chosen rather than any of the other unbiased estimators? The reason is that if the underlying distribution of the population is normal, then it can be shown that the unbiased estimator with the smallest variance is given by $\overline{X}$. Thus, $\overline{X}$ is referred to as the **minimum variance unbiased estimator** of μ.

This concept is illustrated in Figure 6.2a–c, where for 200 random samples of size 10 drawn from the population of 1000 birthweights in Table 6.2, the sampling distribution of the sample mean ($\bar{x}$) is plotted in Figure 6.2a, the sample median in Figure 6.2b, and the average of the smallest and largest observations in the sample in Figure 6.2c. Note that the variability of the distribution of sample means is slightly smaller than that of the sample median and considerably smaller than that of the average of the smallest and largest observations.

6.5.2 Standard Error of the Mean

From Equation 6.1 we see that $\overline{X}$ will be an unbiased estimator of μ for any sample size n. Why then is it preferable to estimate parameters from large samples rather than from small ones? The intuitive reason is that the larger the sample size, the more precise an estimator $\overline{X}$ will be.

Example 6.22 **Obstetrics** Consider Table 6.3 (p. 164). Notice that the 50 individual birthweights range from 62 to 198 oz and have a sample standard deviation of 23.79 oz. The 5 sample means range from 106.7 to 132.8 oz and have a sample standard deviation of 9.77 oz. Thus, the sample means based on 10 observations are less variable from sample to sample than are the individual observations, which can be considered as sample means from samples of size 1.

Indeed, we would expect that the sample means from repeated samples of size 100 would be less variable than those from samples of size 10. We can show that this is true. Using the properties of linear combinations of independent random variables given in Equation 5.9,

$$
\begin{aligned}
Var(\overline{X}) &= \left(\frac{1}{n^2}\right) Var\left(\sum_{i=1}^{n} X_i\right) \\
&= \left(\frac{1}{n^2}\right) \sum_{i=1}^{n} Var(X_i)
\end{aligned}
$$

However, by definition $Var(X_i) = \sigma^2$. Therefore,

$$Var(\overline{X}) = (1/n^2)(\sigma^2 + \sigma^2 + \cdots + \sigma^2) = (1/n^2)(n\sigma^2) = \sigma^2/n$$

The standard deviation $(sd) = \sqrt{\text{variance}}$; thus, $sd(\overline{X}) = \sigma/\sqrt{n}$. We have the following summary:

FIGURE 6.2 **Sampling distributions of several estimators of μ for 200 random samples of size 10 selected from the population of 1000 birthweights given in Table 6.2 (100 = 100.0–101.9, etc.)**

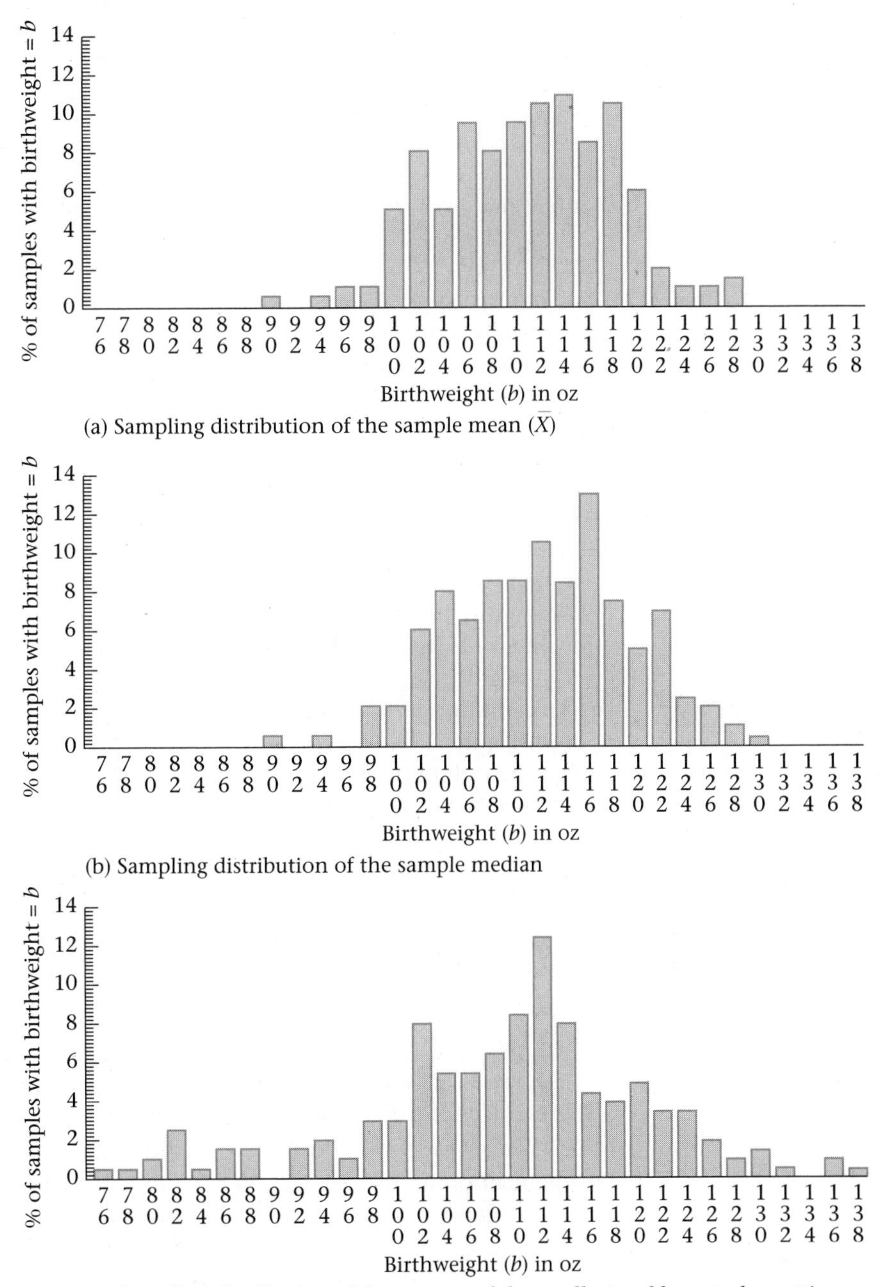

(a) Sampling distribution of the sample mean ($\bar{X}$)

(b) Sampling distribution of the sample median

(c) Sampling distribution of the average of the smallest and largest observations

EQUATION 6.2

Let $X_1, \ldots, X_n$ be a random sample from a population with underlying mean μ and variance σ^2. The set of sample means in repeated random samples of size n from this population has variance σ^2/n. The standard deviation of this set of sample means is thus $\sigma/\sqrt{n}$ and is referred to as the *standard error of the mean (sem)* or the *standard error.*

In practice, the population variance σ^2 is rarely known. We will see later in Section 6.7 that a reasonable estimator for the population variance σ^2 is the sample variance s^2, which leads to the following definition:

DEFINITION 6.12

The **standard error of the mean (sem)**, or the **standard error**, is given by $\sigma/\sqrt{n}$ and is estimated by $s/\sqrt{n}$. The standard error represents the estimated standard deviation obtained from a set of sample means from repeated samples of size n from a population with underlying variance σ^2.

Note that the standard error is *not* the standard deviation of an individual observation X_i but rather of the sample mean $\overline{X}$. The standard error of the mean is illustrated in Figure 6.3a–c. In Figure 6.3a, the frequency of distribution of the sample mean is plotted for 200 samples of size 1 drawn from the collection of birthweights in Table 6.2. Similar frequency distributions are plotted for 200 sample means from samples of size 10 in Figure 6.3b and from samples of size 30 in Figure 6.3c. Notice that the spread of the frequency distribution in Figure 6.3a, corresponding to $n = 1$, is much larger than the spread of the frequency distribution in Figure 6.3b, corresponding to $n = 10$. Furthermore, the spread of the frequency distribution in Figure 6.3b, corresponding to $n = 10$, is much larger than the spread of the frequency distribution in Figure 6.3c, corresponding to $n = 30$.

Example 6.23 **Obstetrics** Compute the standard error of the mean for the third sample of birthweights in Table 6.3 (p. 164).

Solution The standard error of the mean is given by

$$s/\sqrt{n} = 22.44/\sqrt{10} = 7.09$$

The standard error is a quantitative measure of the variability of sample means obtained from repeated random samples of size n drawn from the same population. Notice that the standard error is directly proportional to both $1/\sqrt{n}$ and to the population standard deviation σ of individual observations. It justifies the concern with sample size in assessing the precision of our estimate $\overline{x}$ of the unknown population mean μ. The reason it is preferable to estimate μ from a sample of size 400 rather than from one of size 100 is that the standard error from the first sample will be $\frac{1}{2}$ as large as in the second sample. Thus, the larger sample should provide a more precise estimate of μ. Notice that the precision of our estimate is also af-

FIGURE 6.3 **Illustration of the standard error of the mean (100 = 100.0–103.9, etc.)**

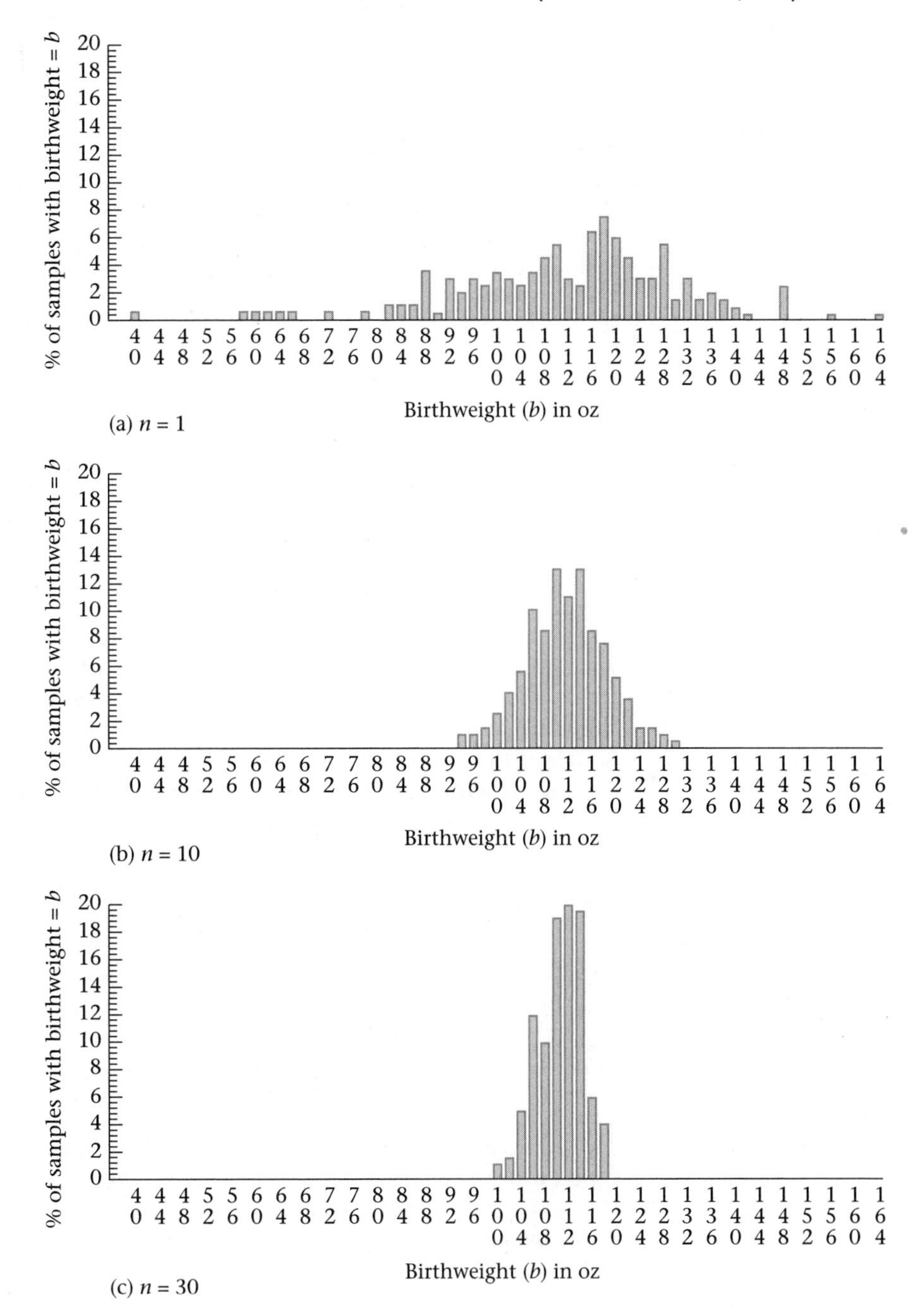

fected by the underlying variance σ^2 from the population of individual observations, a quantity which is unrelated to the sample size n. However, σ^2 can sometimes be affected by experimental technique. For example, in measuring blood pressure, σ^2 can be reduced by better standardization of blood-pressure observers and/or by using additional replicates for individual subjects (for example, using an average of two blood-pressure readings rather than a single reading).

Example 6.24 **Gynecology** Suppose a woman wishes to estimate her exact day of ovulation for contraceptive purposes. A theory exists that at the time of ovulation the body temperature rises by an amount from 0.5°F to 1.0°F. Thus, changes in body temperature can be used to guess the day of ovulation.

To use this method, we need a good estimate of basal body temperature during a period when ovulation is definitely not occurring. Suppose that for this purpose a woman measures her body temperature on awakening on the first 10 days after menstruation and obtains the following data: 97.2°, 96.8°, 97.4°, 97.4°, 97.3°, 97.0°, 97.1°, 97.3°, 97.2°, 97.3°. What is the best estimate of her underlying basal body temperature (μ)? How precise is this estimate?

Solution The best estimate of her underlying body temperature during the nonovulation period (μ) is given by

$$\bar{x} = (97.2 + 96.8 + \cdots + 97.3)/10 = 97.20°$$

The standard error of this estimate is given by

$$s/\sqrt{10} = 0.189/\sqrt{10} = 0.06°$$

In our work on confidence intervals in Section 6.5.5 we will show that for many underlying distributions, we can be fairly certain that the true mean μ is approximately within two standard errors of $\bar{x}$. In this case, true mean basal body temperature (μ) is within $97.20° \pm 2(0.06)° \approx (97.1°\text{–}97.3°)$. Thus, if the temperature is elevated by at least 0.5° above this range on a given day, then it might indicate that the woman was ovulating, and for contraceptive purposes, intercourse should not be attempted on that day.

6.5.3 Central-Limit Theorem

If the underlying distribution is normal, then it can be shown that the sample mean will itself be normally distributed with mean μ and variance σ^2/n (see Section 5.6). In other words, $\bar{X} \sim N(\mu, \sigma^2/n)$. If the underlying distribution is *not* normal, we would still like to make some statement about the sampling distribution of the sample mean. This statement is given by the following theorem:

EQUATION 6.3

Central-Limit Theorem Let $X_1, \ldots, X_n$ be a random sample from some population with mean μ and variance σ^2. Then for large n, $\bar{X} \dot{\sim} N(\mu, \sigma^2/n)$ even if the underlying distribution of individual observations in the population is not normal. (The symbol $\dot{\sim}$ is used to represent "approximately distributed.")

This theorem is very important because many of the distributions encountered in practice are not normal. In such cases the central-limit theorem can often be applied; this will allow us to perform statistical inference based on the approximate normality of the sample mean, despite the nonnormality of the distribution of individual observations.

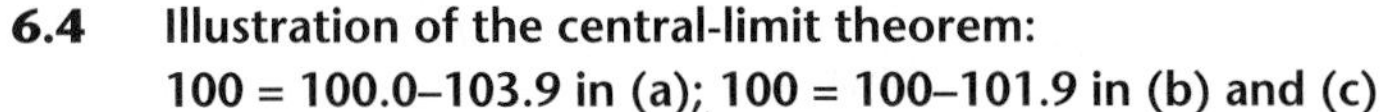

FIGURE 6.4 **Illustration of the central-limit theorem: 100 = 100.0–103.9 in (a); 100 = 100–101.9 in (b) and (c)**

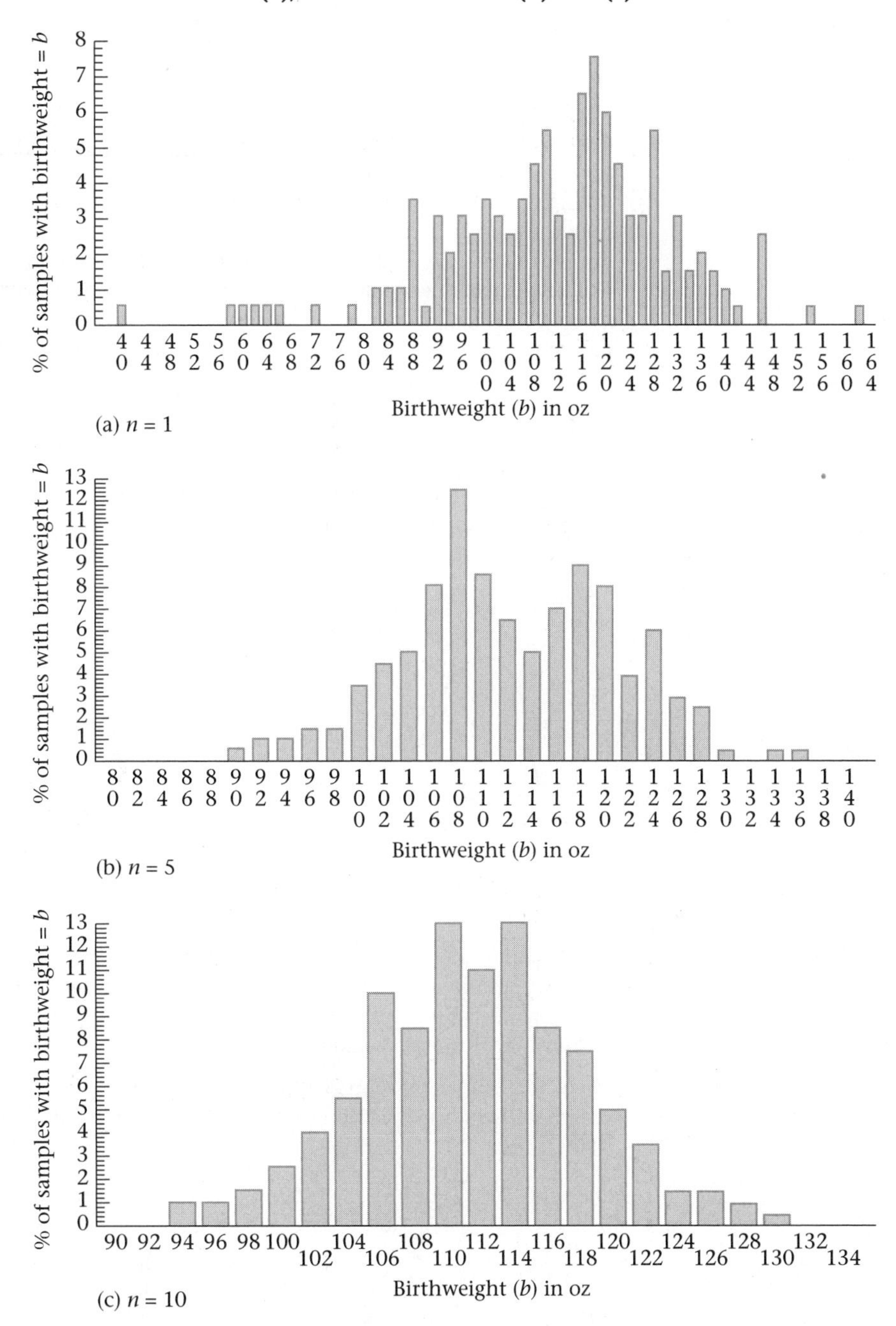

Example 6.25 **Obstetrics** The central-limit theorem is illustrated by plotting, in Figure 6.4a, the sampling distribution of mean birthweights obtained by drawing 200 random samples of size 1 from the collection of birthweights in Table 6.2. Similar sampling distributions

of sample means are plotted from samples of size 5, in Figure 6.4b, and samples of size 10, in Figure 6.4c. Notice that the distribution of individual birthweights (i.e., sample means from samples of size 1) is slightly skewed to the left. However, the distribution of sample means becomes increasingly closer to bell-shaped as the sample size increases to 5, in Figure 6.4b, and 10, in Figure 6.4c,

Example 6.26 **Cardiovascular Disease** Serum triglycerides are an important risk factor for certain types of coronary disease. Their distribution tends to be positively skewed or skewed to the right, with a few people with very high values, as is shown in Figure 6.5. However, hypothesis tests can be performed based on mean serum triglycerides over moderate samples of people, because from the central-limit theorem the distribution of means will be approximately normal, even if the underlying distribution of individual measurements is not. To further ensure normality, the data can also be transformed onto a different scale. For example, if a log transformation is used, then the skewness of the distribution will be reduced and the central-limit theorem will be applicable for smaller sample sizes than if the data are kept in the original scale.

FIGURE 6.5 **Distribution of single serum-triglyceride measurements and of means of such measurements over samples of size *n***

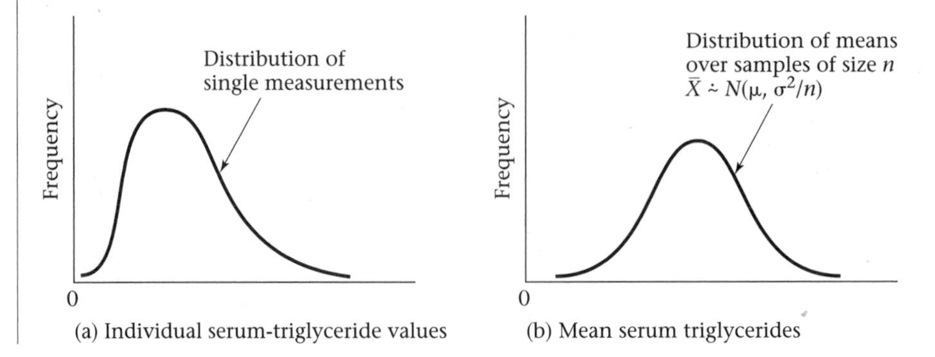

Example 6.27 **Obstetrics** Compute the probability that the mean birthweight from a sample of 10 infants from the Boston City Hospital population in Table 6.2 will fall between 98.0 and 126.0 oz (i.e., $98 \le \overline{X} < 126$) if the mean birthweight for the 1000 birthweights from the Boston City Hospital population is 112.0 oz with a standard deviation of 20.6 oz.

Solution The central-limit theorem is applied and it is assumed that $\overline{X}$ follows a normal distribution with mean $\mu = 112.0$ oz and standard deviation $\sigma/\sqrt{n} = 20.6/\sqrt{10} = 6.51$ oz. It follows that

$$\begin{aligned} Pr(98.0 \le \overline{X} < 126.0) &= \Phi\left(\frac{126.0-112.0}{6.51}\right) - \Phi\left(\frac{98.0-112.0}{6.51}\right) \\ &= \Phi(2.15) - \Phi(-2.15) \\ &= \Phi(2.15) - [1-\Phi(2.15)] = 2\Phi(2.15) - 1 \end{aligned}$$

Refer to Table 3 in the Appendix and obtain

$$Pr(98.0 \le \overline{X} < 126.0) = 2(.9842) - 1.0 = .968$$

Thus, 96.8% of the samples of size 10 would be expected to have mean birthweights between 98 and 126 oz if the central-limit theorem holds. This value can be checked by referring to Figure 6.2a. Note that the 90 column corresponds to the birthweight interval 90.0–91.9, the 92 column to 92.0–93.9, and so forth. Note that 0.5% of the birthweights are in the 90 column, 0.5% in the 94 column, 1% in the 96 column, 1% in the 126 column, and 1.5% in the 128 column. Thus 2% of the distribution is less than 98.0 oz, and 2.5% of the distribution is greater than or equal to 126.0 oz. It follows that 100% – 4.5% = 95.5% of the distribution is actually between 98 and 126 oz. This value corresponds well to the 96.8% predicted by the central-limit theorem, confirming that the central-limit theorem holds approximately for averages from samples of size 10 drawn from this population.

6.5.4 Interval Estimation

We have been discussing the rationale for using $\bar{x}$ to estimate the mean of a distribution and have given a measure of variability of this estimate, namely, the standard error. These statements hold for any underlying distribution. However, we frequently wish to obtain an interval of plausible estimates of the mean as well as a best estimate of its precise value. Our interval estimates will hold exactly only if the underlying distribution is normal and only approximately if the underlying distribution is not normal, as stated in the central-limit theorem.

Example 6.28 **Obstetrics** Suppose the first sample of 10 birthweights given in Table 6.3 has been drawn. Our best estimate of the population mean μ would be the sample mean $\bar{x}$ = 116.9 oz. Although 116.9 oz is our best estimate of μ, we still are not certain the μ is 116.9 oz. Indeed, if the second sample of 10 birthweights had been drawn, a point estimate of 132.8 oz would have been obtained. Our point estimate would certainly have a different meaning if it was highly likely that μ was within 1 oz of 116.9 rather than within 1 lb (16 oz).

We have assumed previously that the distribution of birthweights in Table 6.2 was normal with mean μ and variance σ^2. It follows from our previous discussion of the properties of the sample mean that $\bar{X} \sim N(\mu, \sigma^2/n)$. Thus, if μ and σ^2 were known, then the behavior of the set of sample means over a large number of samples of size n would be precisely known. In particular, 95% of all such sample means will fall within the interval $(\mu - 1.96\sigma/\sqrt{n}, \mu + 1.96\sigma/\sqrt{n})$.

EQUATION 6.4

Alternatively, if we re-express $\bar{X}$ in standardized form by

$$Z = \frac{\bar{X} - \mu}{\sigma/\sqrt{n}}$$

then Z should follow a standard normal distribution. Hence, 95% of the Z values from repeated samples of size n will fall between –1.96 and +1.96 because these values correspond to the 2.5th and 97.5th percentiles from a standard normal distribution. However, the assumption that σ is known is somewhat artificial, because σ is rarely known in practice.

6.5.5 *t* Distribution

Because σ is unknown, it is reasonable to estimate σ by the sample standard deviation s and to try to construct confidence intervals using the quantity $(\overline{X}-\mu)/(s/\sqrt{n})$. The problem is that this quantity is no longer normally distributed.

This problem was first solved in 1908 by a statistician named William Gossett. For his entire professional life, Gossett worked for the Guinness Brewery in Ireland. He chose to identify himself by the pseudonym "Student," and thus the distribution of $(\overline{X}-\mu)/(s/\sqrt{n})$ is usually referred to as **Student's *t* distribution.** Gossett found that the shape of the distribution depends on the sample size n. Thus, the t distribution is not a unique distribution but is instead a family of distributions indexed by a parameter referred to as the **degrees of freedom (*df*)** of the distribution.

EQUATION 6.5

If $X_1,\ldots,X_n \sim N(\mu,\sigma^2)$ and are independent, then $(\overline{X}-\mu)/(S/\sqrt{n})$ is distributed as a t distribution with $(n-1)$ degrees of freedom (df).

Once again, Student's t distribution is not a unique distribution but is a family of distributions indexed by the degrees of freedom d. The t distribution with d degrees of freedom is sometimes referred to as the t_d distribution.

DEFINITION 6.13

The **100 × *u*th percentile of a *t* distribution with *d* degrees of freedom** is denoted by $t_{d,u}$, that is,

$$Pr\left(t_d < t_{d,u}\right) \equiv u$$

Example 6.29 What does $t_{20,\ .95}$ mean?

Solution $t_{20,\ .95}$ is the 95th percentile or the upper 5th percentile of a t distribution with 20 degrees of freedom.

It is interesting to compare a t distribution with d degrees of freedom to an $N(0, 1)$ distribution. The density functions corresponding to these distributions are depicted in Figure 6.6 for the special case where $d = 5$.

Notice that the t distribution is symmetric about 0 but is more spread out than the $N(0, 1)$ distribution. It can be shown that for any α, where $\alpha > .5, t_{d,1-\alpha}$ is always larger than the corresponding percentile for an $N(0, 1)$ distribution ($z_{1-\alpha}$). This relationship is depicted in Figure 6.6. However, as d becomes large, the t distribution converges to an $N(0, 1)$ distribution. An explanation for this principle is that for finite samples the sample variance (s^2) is an approximation to the population variance (σ^2). This approximation gives the statistic $(\overline{X}-\mu)/(S/\sqrt{n})$ more variability than the corresponding statistic $(\overline{X}-\mu)/(\sigma/\sqrt{n})$. As n becomes large, this approximation gets better and s^2 will converge to σ^2. The two distributions thus get more and more alike as n becomes large. The upper 2.5th percentile of the t distribution for various degrees of freedom and the corresponding percentile for the normal distribution are given in Table 6.5.

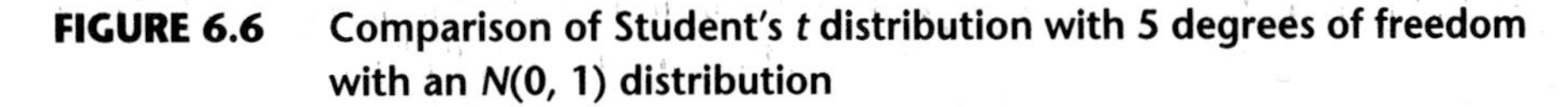

FIGURE 6.6 **Comparison of Student's *t* distribution with 5 degrees of freedom with an *N*(0, 1) distribution**

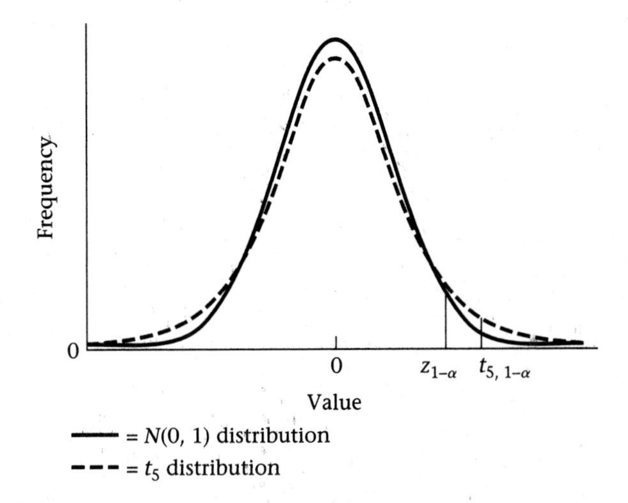

TABLE 6.5 **Comparison of the 97.5th percentile of the *t* distribution and the normal distribution**

d	$t_{d,.975}$	$z_{.975}$	d	$t_{d,.975}$	$z_{.975}$
4	2.776	1.960	60	2.000	1.960
9	2.262	1.960	∞	1.960	1.960
29	2.045	1.960			

The difference between the t distribution and the normal distribution is greatest for small values of n ($n < 30$). Table 5 in the Appendix gives the percentage points of the t distribution for various degrees of freedom. The degrees of freedom are given in the first column of the table, and the percentiles are given across the first row. The uth percentile of a t distribution with d degrees of freedom is found by reading across the row marked d and reading down the column marked u.

Example 6.30 Find the upper 5th percentile of a t distribution with 23 *df*.

Solution Find $t_{23,\,.95}$, which is given in row 23 and column .95 of Table 5 and is 1.714.

Statistical packages such as MINITAB, Excel, or SAS will also compute exact probabilities associated with the t distribution. This is particularly useful for values of the degrees of freedom (d) that are not given in Table 5.

If σ is unknown, we can replace σ by S in Equation 6.4 and correspondingly replace the z statistic by a t statistic given by

$$t = \frac{\overline{X} - \mu}{S/\sqrt{n}}$$

The t statistic should follow a t distribution with $n - 1$ *df*. Hence, 95% of the t statistics in repeated samples of size n should fall between the 2.5th and 97.5th percentiles of a t_{n-1} distribution, or

$$Pr\left(t_{n-1,.025} < t < t_{n-1,.975}\right) = 95\%$$

More generally, 100% × (1 – α) of the t statistics should fall between the lower and upper α/2 percentile of a t_{n-1} distribution in repeated samples of size n or

$$Pr\left(t_{n-1,\alpha/2} < t < t_{n-1,1-\alpha/2}\right) = 1 - \alpha$$

This inequality can be written in the form of two inequalities:

$$t_{n-1,\alpha/2} < \frac{\overline{X} - \mu}{S/\sqrt{n}} \quad \text{and} \quad \frac{\overline{X} - \mu}{S/\sqrt{n}} < t_{n-1,1-\alpha/2}$$

If we multiply both sides of each inequality by $S/\sqrt{n}$ and add μ to both sides, we obtain

$$\mu + t_{n-1,\alpha/2}\, S/\sqrt{n} < \overline{X} \quad \text{and} \quad \overline{X} < t_{n-1,1-\alpha/2}\, S/\sqrt{n} + \mu$$

Finally, if we subtract $t_{n-1,\alpha/2}\, S/\sqrt{n}$ from both sides of the first inequality and $t_{n-1,1-\alpha/2}\, S/\sqrt{n}$ from both sides of the second inequality, we get:

$$\mu < \overline{X} - t_{n-1,\alpha/2}\, S/\sqrt{n} \quad \text{and} \quad \overline{X} - t_{n-1,1-\alpha/2}\, S/\sqrt{n} < \mu$$

Expressed as one inequality, this is

$$\overline{X} - t_{n-1,1-\alpha/2}\, S/\sqrt{n} < \mu < \overline{X} - t_{n-1,\alpha/2}\, S/\sqrt{n}$$

From the symmetry of the t distribution, $t_{n-1,\alpha/2} = -t_{n-1,1-\alpha/2}$, and thus this inequality can be rewritten as

$$\overline{X} - t_{n-1,1-\alpha/2}\, S/\sqrt{n} < \mu < \overline{X} + t_{n-1,1-\alpha/2}\, S/\sqrt{n}$$

and we can say that

$$Pr\left(\overline{X} - t_{n-1,1-\alpha/2}\, S/\sqrt{n} < \mu < \overline{X} + t_{n-1,1-\alpha/2}\, S/\sqrt{n}\right) = 1 - \alpha$$

The interval $\left(\overline{x} - t_{n-1,1-\alpha/2}\, s/\sqrt{n}, \overline{x} + t_{n-1,1-\alpha/2}\, s/\sqrt{n}\right)$ is referred to as a 100% × (1 – α) confidence interval (CI) for μ. This can be summarized as follows:

EQUATION 6.6

Confidence Interval for the Mean of a Normal Distribution A 100% × (1 – α) **confidence interval (CI) for the mean μ of a normal distribution with unknown variance** is given by

$$\left(\overline{x} - t_{n-1,1-\alpha/2}\, s/\sqrt{n}, \overline{x} + t_{n-1,1-\alpha/2}\, s/\sqrt{n}\right)$$

A shorthand notation for the confidence interval is

$$\overline{x} \pm t_{n-1,1-\alpha/2}\, s/\sqrt{n}$$

Example 6.31 Compute a 95% CI for the mean birthweight based on the first sample of size 10 in Table 6.3.

Solution

We have that $n = 10$, $\bar{x} = 116.90$, $s = 21.70$. Because we want a 95% CI, $\alpha = .05$. Therefore, from Equation 6.6 the 95% CI is

$$\left[116.9 - t_{9,.975}(21.70)/\sqrt{10}, 116.9 + t_{9,.975}(21.70)/\sqrt{10}\right]$$

From Table 5, $t_{9,.975} = 2.262$. Therefore, the 95% CI is

$$\left[116.9 - 2.262(21.70)/\sqrt{10}, 116.9 + 2.262(21.70)/\sqrt{10}\right]$$
$$= (116.9 - 15.5, 116.9 + 15.5)$$
$$= (101.4, 132.4)$$

Note that if the sample size is large (say >200), then the percentiles of a t distribution are virtually the same as for a normal distribution. In this case, a reasonable approximate $100\% \times (1 - \alpha)$ confidence interval for μ is given as follows:

EQUATION 6.7

Confidence Interval for the Mean of a Normal Distribution (Large-Sample Case) An approximate $100\% \times (1 - \alpha)$ CI for the mean μ of a normal distribution with unknown variance is given by:

$$\left(\bar{x} - z_{1-\alpha/2}\, s/\sqrt{n}, \bar{x} + z_{1-\alpha/2}\, s/\sqrt{n}\right)$$

This interval should only be used if $n > 200$. In addition, Equation 6.7 can also be used for $n \le 200$ if the standard deviation (σ) is known, by replacing s by σ.

same formula

You may be puzzled at this point as to what the confidence interval means. The parameter μ is a fixed unknown constant. How can we state that the probability that it lies within some specific interval is, for example, 95%? The important point to understand is that the boundaries of the interval depend on the sample mean and sample variance and will vary from sample to sample. Furthermore, 95% of such intervals that could be constructed from repeated random samples of size n will contain the parameter μ.

Example 6.32

Obstetrics Consider the 5 samples of size 10 from the population of birthweights as shown in Table 6.3 (p. 164). Since $t_{9,.975} = 2.262$, the 95% CI is given by

$$\left(\bar{x} - t_{9,.975}\, s/\sqrt{n}, \bar{x} + t_{9,.975}\, s/\sqrt{n}\right) = \left(\bar{x} - \frac{2.262s}{\sqrt{10}}, \bar{x} + \frac{2.262s}{\sqrt{10}}\right)$$
$$= (\bar{x} - 0.715s, \bar{x} + 0.715s)$$

The interval will be different for each sample and is given in Figure 6.7. A dashed line has been added to represent an imaginary value for μ. The idea is that over a large number of hypothetical samples of size 10, 95% of such intervals will contain the parameter μ. Any one interval from a particular sample *may* or *may not* contain the parameter μ. In Figure 6.7, by chance all 5 intervals contain the parameter μ. However with additional random samples, this need not be the case.

Therefore, we cannot say that there is a 95% chance that the parameter μ will fall within a particular 95% CI. However, we can say the following:

FIGURE 6.7 A collection of 95% confidence intervals for the mean μ as computed from repeated samples of size 10 (see Table 6.3) from the population of birthweights given in Table 6.2

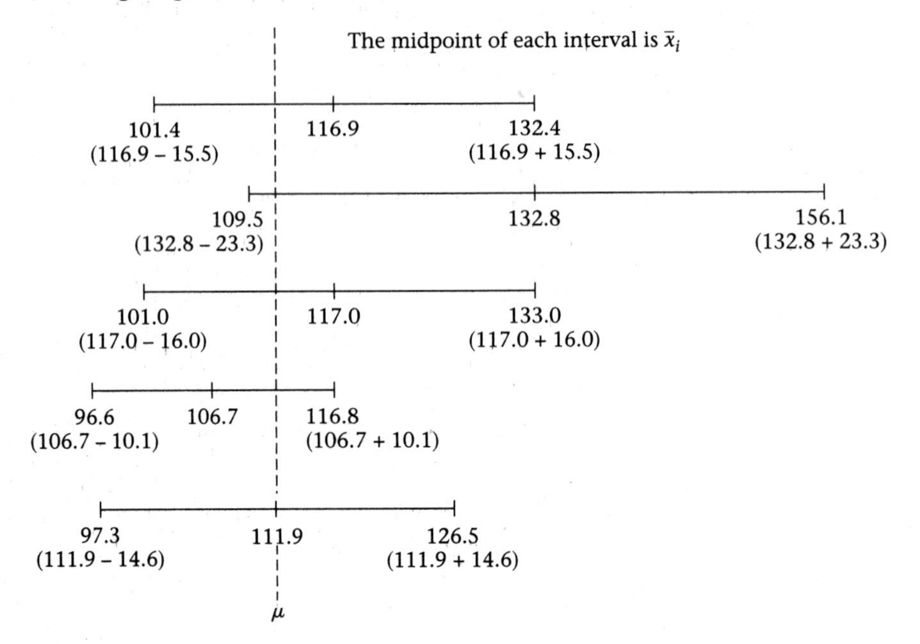

EQUATION 6.8

> Over the collection of all 95% confidence intervals that could be constructed from repeated random samples of size n, 95% will contain the parameter μ.

The length of the confidence interval gives some idea of the precision of the point estimate $\bar{x}$. In this particular case, the length of each confidence interval ranges from 20 oz to 47 oz, which makes the precision of the point estimate $\bar{x}$ doubtful and implies that a larger sample size is needed to get a more precise estimate of μ.

Example 6.33 **Gynecology** Compute a 95% CI for the underlying mean basal body temperature using the data in Example 6.24 (p. 174).

Solution The 95% CI is given by

$$\bar{x} \pm t_{9,.975}\, s/\sqrt{n} = 97.2° \pm 2.262(0.189)/\sqrt{10} = 97.2° \pm 0.13°$$
$$= (97.07°, 97.33°)$$

We can also consider confidence intervals with a level of confidence other than 95%.

Example 6.34 Suppose the first sample in Table 6.3 has been drawn. Compute a 99% CI for the underlying mean birthweight.

Solution The 99% CI is given by

$$\left(116.9 - t_{9,.995}(21.70)/\sqrt{10}, 116.9 + t_{9,.995}(21.70)/\sqrt{10}\right)$$

From Table 5 of the Appendix we see that $t_{9,.995} = 3.250$, and therefore the 99% CI is

$$\left(116.9 - 3.250(21.70)/\sqrt{10}, 116.9 + 3.250(21.70)/\sqrt{10}\right) = (94.6, 139.2)$$

Notice that the 99% confidence interval (94.6, 139.2) computed in Example 6.34 is wider than the corresponding 95% confidence interval (101.4, 132.4) computed in Example 6.31. The rationale for this difference is that the higher the level of confidence desired that μ lies within an interval, the wider the confidence interval must be. Indeed, for 95% confidence intervals the length was $2(2.262)s/\sqrt{n}$; for 99% confidence intervals the length was $2(3.250)s/\sqrt{n}$. In general, the length of the $100\% \times (1-\alpha)$ confidence interval is given by

$$2t_{n-1,1-\alpha/2}\, s/\sqrt{n}$$

Therefore, we can see that the length of a confidence interval is governed by three variables: n, s, and α.

EQUATION 6.9

Factors Affecting the Length of a Confidence Interval The length of a $100\% \times (1-\alpha)$ confidence interval for μ equals $2t_{n-1,1-\alpha/2}\, s/\sqrt{n}$ and is determined by n, s, and α.

- n As the sample size (n) increases, the length of the confidence interval decreases.
- s As the standard deviation (s), which reflects the variability of the distribution of individual observations, increases, the length of the confidence interval increases.
- α As the confidence desired increases (α decreases), the length of the confidence interval increases.

Example 6.35 **Gynecology** Compute a 95% CI for the underlying mean basal body temperature using the data in Example 6.24, assuming that the number of days sampled is 100 rather than 10.

Solution The 95% CI is given by

$$97.2° \pm t_{99,.975}(0.189)/\sqrt{100} = 97.2° \pm 1.984(0.189)/10 = 97.2° \pm 0.04° = (97.16°,\ 97.24°)$$

where we use the TINV function of Excel to estimate $t_{99,.975}$ by 1.984. Notice that this interval is much narrower than the corresponding interval (97.07°, 97.33°) based on a sample of 10 days given in Example 6.33.

Example 6.36 Compute a 95% CI for the underlying mean basal temperature using the data in Example 6.24, assuming that the standard deviation of basal body temperature is 0.4° rather than 0.189° with a sample size of 10.

Solution The 95% CI is given by

$$97.2° \pm 2.262(0.4)/\sqrt{10} = 97.2° \pm 0.29° = (96.91°, 97.49°)$$

Notice that this interval is much wider than the corresponding interval (97.07°, 97.33°) based on a standard deviation of 0.189° with a sample size of 10.

Usually only n and α can be controlled. s is a function of the type of variable being studied, although s itself can sometimes be decreased, if changes in technique can reduce the amount of measurement error, day-to-day variability, and so forth. An important way in which s can be reduced is by obtaining replicate measurements for each individual and using the average of several replicates for an individual, rather than a single measurement.

Up to this point, confidence intervals have been used mainly as descriptive tools for characterizing the precision with which the parameters of a distribution can be estimated. Another use for confidence intervals is in making decisions on the basis of data.

Example 6.37 **Cardiovascular Disease, Pediatrics** Suppose we know from large studies that the mean cholesterol level in children ages 2–14 is 175 mg/dL. We wish to see if there is a familial aggregation of cholesterol levels. Specifically, we identify a group of fathers who have had a heart attack and have elevated cholesterol levels (≥ 250 mg/dL) and measure the cholesterol levels of their offspring within the 2–14 age range.

Suppose we find that the mean cholesterol level in a group of 100 such children is 207.3 mg/dL with standard deviation = 30 mg/dL. Is this value sufficiently far from 175 mg/dL for us to believe that the underlying mean cholesterol level in the population of all children selected in this way is different from 175 mg/dL?

Solution One approach would be to construct a 95% confidence interval for μ on the basis of our sample data. We then could use the following decision rule: If the interval contains 175 mg/dL, then we cannot say that the underlying mean for this group is any different from the mean for all children (175), because 175 is among the plausible values for μ provided by the 95% confidence interval. We would decide that there is no demonstrated familial aggregation of cholesterol levels. If the confidence interval does not contain 175, then we would conclude that the true underlying mean for this group is different from 175. If the lower bound of the CI is above 175, then there is a demonstrated familial aggregation of cholesterol levels. The basis for this decision rule is discussed in the chapters on hypothesis testing.

The confidence interval in this case is given by

$$207.3 \pm t_{99,.975}(30)/\sqrt{100} = 207.3 \pm 6.0 = (201.3, 213.3)$$

Clearly, 175 is far from the lower bound of the interval, and we thus conclude that there is familial aggregation of cholesterol.

SECTION 6.6 Case Study: Effects of Tobacco Use on Bone-Mineral Density in Middle-Aged Women

There were 41 twin pairs in this study. We wish to assess whether there is a relationship between bone mineral density (BMD) and cigarette smoking. One way to approach this problem is to calculate the difference in bone mineral density between the heavier-smoking twin and the lighter-smoking twin for each twin pair and then calculate the average of these differences over the 41 twin pairs. In this study, there was a mean difference in BMD of -0.036 ± 0.014 g/cm^2 (mean ± *se*) for the 41 twin pairs. We can use the confidence-interval methodology to address this question. Specifically, the 95% CI for the true mean difference (μ_d) in BMD between the heavier- and lighter-smoking twins is

$$-0.036 \pm t_{40,.975}\left(s/\sqrt{41}\right)$$

However, because $se = s/\sqrt{41}$, another way to express this formula is

$$\begin{aligned}-0.036 \pm t_{40,.975}(se) &= -0.036 \pm 2.021(0.014)\\ &= -0.036 \pm 0.028 = (-0.064, -0.008)\end{aligned}$$

Because the upper bound of the 95% CI is less than 0, we can be fairly confident that the true mean difference is less than 0. Stated another way, we can be fairly confident that the true mean BMD for the heavier-smoking twins is lower than for the lighter-smoking twins. In statistical terms, we say that there is a significant association between BMD and cigarette smoking. We will discuss assessment of statistical significance in more detail in Chapter 7.

SECTION 6.7 Estimation of the Variance of a Distribution

6.7.1 Point Estimation

In Chapter 2, the sample variance was defined as

$$s^2 = \frac{1}{n-1}\sum_{i=1}^{n}\left(x_i - \bar{x}\right)^2$$

This definition is somewhat counterintuitive, because the denominator would be expected to be n rather than $n - 1$. A more formal justification for this definition is now given. If our sample $x_1, \ldots, x_n$ is considered as coming from some population with mean μ and variance σ^2, then how can the unknown population variance σ^2 be estimated from our sample? The following principle is useful in this regard:

EQUATION 6.10

> Let $X_1, \ldots, X_n$ be a random sample from some population with mean μ and variance σ^2. The **sample variance S^2 is an unbiased estimator** of σ^2 over all possible random samples of size n that could have been drawn from this population; that is, $E(S^2) = \sigma^2$.

Therefore, if repeated random samples of size n are selected from the population, as was done in Table 6.3, and the sample variance s^2 is computed from each sample, then the average of these sample variances over a large number of such samples of size n will be the population variance σ^2. This statement holds for any underlying distribution.

Example 6.38 **Gynecology** Estimate the variance of the distribution of basal body temperature using the data in Example 6.24.

Solution We have

$$s^2 = \frac{1}{9}\sum_{i=1}^{n}\left(x_i - \overline{x}\right)^2 = 0.189^2 = 0.0356$$

which is an unbiased estimate of σ^2.

Note that the intuitive estimator for σ^2 with n in the denominator rather than $n-1$, that is,

$$\frac{1}{n}\sum_{i=1}^{n}\left(x_i - \overline{x}\right)^2$$

will tend to underestimate the underlying variance σ^2 by a factor of $(n-1)/n$. This factor is considerable for small samples but tends to be negligible for large samples. A more complete discussion of the relative merits of different estimators for σ^2 is given in [3].

6.7.2 The Chi-Square Distribution

The problem of interval estimation of the mean of a normal distribution was discussed in Sections 6.5.4 and 6.5.5. We often want to obtain interval estimates of the variance as well. Once again, as was the case for the mean, the interval estimates will hold exactly only if the underlying distribution is normal. The interval estimates will perform much more poorly for the variance than for the mean if the underlying distribution is not normal, and they should be used with caution in this case.

Example 6.39 **Hypertension** A new machine has been produced, called an *arteriosonde machine,* that "prints" blood-pressure readings on a tape so that the measurement can be read rather than heard. A major argument for using such a machine is that the variability of measurements obtained by different observers on the same person will be lower than with a standard blood-pressure cuff.

Suppose we have the data presented in Table 6.6, consisting of systolic blood-pressure measurements obtained on 10 people and read by 2 observers. We will use the difference d_i between the first and second observer to assess interobserver variability. In particular, if we assume that the underlying distribution of these differences is normal with mean μ and variance σ^2, then it is of primary interest to estimate σ^2. The higher σ^2 is, the higher the interobserver variability.

We have seen previously that an unbiased estimator of the variance σ^2 is given by the sample variance S^2. In this case,

$$\text{The mean difference} = (-6+3+\cdots-2)/10 = -0.2 = \overline{d}$$

$$\text{The sample variance} = s^2 = \sum_{i=1}^{n}(d_i - \overline{d})^2/9$$

$$= \left[(-6+0.2)^2 + \cdots + (-2+0.2)^2\right]/9 = 8.178$$

How can an interval estimate for σ^2 be obtained?

TABLE 6.6 **Systolic blood-pressure measurements (mm Hg) from an arteriosonde machine obtained from 10 people and read by 2 observers**

	Observer		
Person (i)	1	2	Difference (d_i)
1	194	200	−6
2	126	123	+3
3	130	128	+2
4	98	101	−3
5	136	135	+1
6	145	145	0
7	110	111	−1
8	108	107	+1
9	102	99	+3
10	126	128	−2

To obtain an interval estimate for σ^2, a new family of distributions, called chi-square (χ^2) distributions, must be introduced to enable us to find the sampling distribution of S^2 from sample to sample.

DEFINITION 6.14 If $G = \sum_{i=1}^{n} X_i^2$

where $X_1, \ldots, X_n \sim N(0,1)$

and are independent, then G is said to follow a **chi-square distribution with n degrees of freedom (df)**. The distribution is often denoted by χ_n^2.

The chi-square distribution is actually a family of distributions indexed by the parameter n referred to, again, as the degrees of freedom, as was the case for the t distribution. Unlike the t distribution, which is always symmetric about 0 for any degrees of freedom, the chi-square distribution only takes on positive values and is always skewed to the right. The general shape of these distributions is indicated in Figure 6.8.

For $n = 1, 2$, the distribution has a mode at 0 [3]. For $n \geq 3$, the distribution has a mode greater than 0 and is skewed to the right. The skewness diminishes as n increases. It can be shown that the expected value of a χ_n^2 distribution is n and the variance is $2n$.

DEFINITION 6.15 The **uth percentile of a χ_n^2 distribution** for a chi-square distribution with n df is denoted by $\chi_{n,u}^2$, where $Pr(\chi_n^2 < \chi_{n,u}^2) \equiv u$. These percentiles are depicted in Figure 6.9 for a chi-square distribution with 5 df and appear in Table 6 in the Appendix.

Table 6 is constructed similarly to the t table (Table 5), with the degrees of freedom (d) indexed in the first column and the percentile (u) indexed in the first row.

FIGURE 6.8 General shape of various χ^2 distributions with *n df*

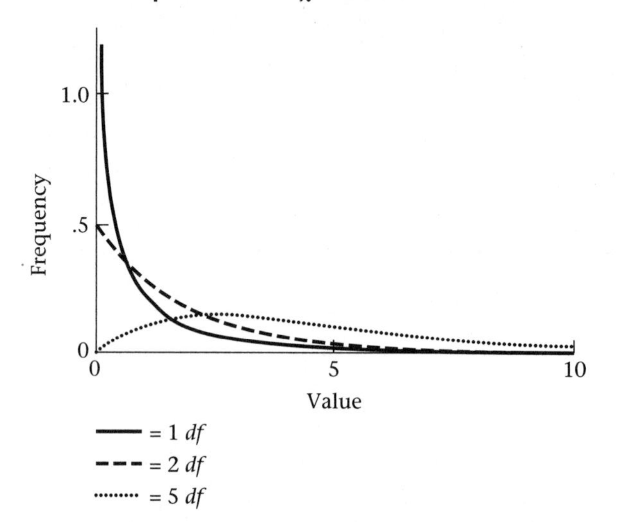

FIGURE 6.9 Graphic display of the percentiles of a χ_5^2 distribution

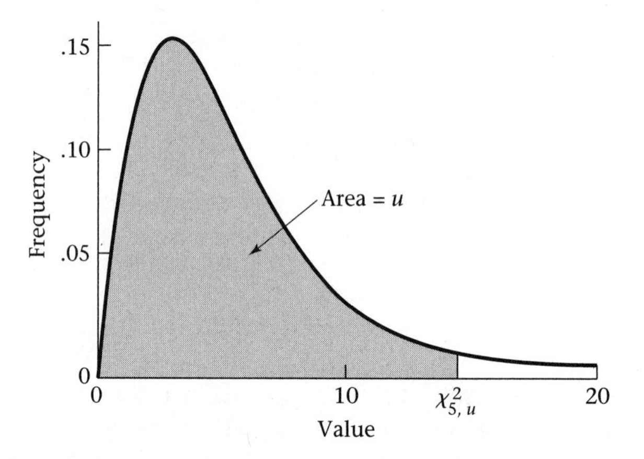

The principal difference between the two tables is that both *lower* ($u \le 0.5$) and *upper* ($u > 0.5$) percentiles are given for the chi-square distribution, whereas only upper percentiles are given for the *t* distribution. The *t* distribution is symmetric about 0, and therefore any lower percentile can be obtained as the negative of the corresponding upper percentile. Because the chi-square distribution is, in general, a skewed distribution, there is no simple relationship between the upper and lower percentiles.

Example 6.40 Find the upper and lower 2.5th percentiles of a chi-square distribution with 10 *df*.

Solution According to Table 6, the upper and lower percentiles are given by

$$\chi^2_{10,.975} = 20.48 \qquad \text{and} \qquad \chi^2_{10,.025} = 3.25 \qquad \text{respectively}$$

For values of d not given in Table 6, a computer program, such as MINITAB or Excel, can be used to obtain percentiles.

6.7.3 Interval Estimation

To obtain an interval estimate of σ^2, we need to find the sampling distribution of S^2. Suppose we assume that $X_1, \ldots, X_n \sim N(\mu, \sigma^2)$. Then it can be shown that

EQUATION 6.11

$$S^2 \sim \frac{\sigma^2 \chi^2_{n-1}}{n-1}$$

To see this, we recall from Section 5.5 that if $X \sim N(\mu, \sigma^2)$, then if we standardize X (that is, we subtract μ and divide by σ), thus creating a new random variable $Z = (X - \mu)/\sigma$, then Z will be normally distributed with mean 0 and variance 1. Thus, from Definition 6.14 we see that

EQUATION 6.12

$$\sum_{i=1}^{n} Z_i^2 = \sum_{i=1}^{n} (X_i - \mu)^2 / \sigma^2 \sim \chi^2_n = \text{chi-square distribution with } n \ df$$

Because we usually don't know μ, we estimate μ by $\bar{x}$. However, it can be shown that if we substitute $\bar{X}$ for μ in Equation 6.12, then we lose 1 df [3], resulting in the relationship

EQUATION 6.13

$$\sum_{i=1}^{n} (X_i - \bar{X})^2 / \sigma^2 \sim \chi^2_{n-1}$$

However, we recall from the definition of a sample variance that $S^2 = \sum_{i=1}^{n} (X_i - \bar{X})^2 / (n-1)$. Thus, multiplying both sides by $(n - 1)$ yields the relationship

$$(n-1)S^2 = \sum_{i=1}^{n} (X_i - \bar{X})^2$$

Substituting into Equation 6.13, we obtain

EQUATION 6.14

$$\frac{(n-1)S^2}{\sigma^2} \sim \chi^2_{n-1}$$

If we multiply both sides of Equation 6.14 by $\sigma^2/(n-1)$, we obtain Equation 6.11,

$$S^2 \sim \frac{\sigma^2}{n-1} \chi^2_{n-1}$$

Thus, from Equation 6.11 we see that S^2 follows a chi-square distribution with $n - 1$ *df* multiplied by the constant $\sigma^2/(n-1)$. Manipulations similar to those given in Section 6.5.5 can now be used to obtain a $100\% \times (1-\alpha)$ confidence interval for σ^2.

In particular, from Equation 6.11 it follows that

$$Pr\left(\frac{\sigma^2\chi^2_{n-1,\alpha/2}}{n-1} < S^2 < \frac{\sigma^2\chi^2_{n-1,1-\alpha/2}}{n-1}\right) = 1-\alpha$$

This inequality can be represented as two separate inequalities:

$$\frac{\sigma^2\chi^2_{n-1,\alpha/2}}{n-1} < S^2 \quad \text{and} \quad S^2 < \frac{\sigma^2\chi^2_{n-1,1-\alpha/2}}{n-1}$$

If both sides of the first inequality are multiplied by $(n-1)/\chi^2_{n-1,\alpha/2}$, and both sides of the second inequality are multiplied by $(n-1)/\chi^2_{n-1,1-\alpha/2}$, then

$$\sigma^2 < \frac{(n-1)S^2}{\chi^2_{n-1,\alpha/2}} \quad \text{and} \quad \frac{(n-1)S^2}{\chi^2_{n-1,1-\alpha/2}} < \sigma^2$$

or, upon combining these two inequalities,

$$\frac{(n-1)S^2}{\chi^2_{n-1,1-\alpha/2}} < \sigma^2 < \frac{(n-1)S^2}{\chi^2_{n-1,\alpha/2}}$$

It follows that

$$Pr\left[\frac{(n-1)S^2}{\chi^2_{n-1,1-\alpha/2}} < \sigma^2 < \frac{(n-1)S^2}{\chi^2_{n-1,\alpha/2}}\right] = 1-\alpha$$

Thus, the interval $\left[(n-1)s^2/\chi^2_{n-1,1-\alpha/2}, (n-1)s^2/\chi^2_{n-1,\alpha/2}\right]$ is a $100\% \times (1-\alpha)$ confidence interval for σ^2.

EQUATION 6.15

A 100% × (1 – α) confidence interval for σ^2 is given by

$$\left[(n-1)s^2/\chi^2_{n-1,1-\alpha/2}, (n-1)s^2/\chi^2_{n-1,\alpha/2}\right]$$

Example 6.41 **Hypertension** We now return to the specific data set in Example 6.39. Suppose we wish to construct a 95% confidence interval for the interobserver variability as defined by σ^2.

Solution Because there are 10 people and $s^2 = 8.178$, the required interval is given by

$$\left(9s^2/\chi^2_{9,.975}, 9s^2/\chi^2_{9,.025}\right) = \left[9(8.178)/19.02, 9(8.178)/2.70\right] = (3.87, 27.26)$$

Similarly, a 95% confidence interval for σ is given by $\left(\sqrt{3.87}, \sqrt{27.26}\right) = (1.97, 5.22)$. Notice that the confidence interval for σ^2 is *not* symmetric about $s^2 = 8.178$, in contrast to the confidence interval for μ, which *was* symmetric about $\bar{x}$. This characteristic is common in confidence intervals for the variance.

We could use the confidence interval for σ^2 to make decisions concerning the variability of the arteriosonde machine if we had a good estimate of the interobserver variability of blood-pressure readings from a standard cuff. For example, suppose we know from previous work that if two people are listening to blood-pressure recordings from a standard cuff, then the interobserver variability as measured by the variance of the set of differences between the readings of two observers is 35. This value is outside the range of the 95% confidence interval for $\sigma^2 (3.87, 27.26)$, and we thus conclude that the interobserver variability is reduced by using an arteriosonde machine. Alternatively, if this prior variance were 15, then we could not say that the variances obtained from using the two methods are different.

Note that the CI for σ^2 in Equation 6.15 is only valid for normally distributed samples. If the underlying distribution is not normal, then the level of confidence for this interval may not be $1 - \alpha$ even if the sample size is large. This is different from the CI for μ given in Equation 6.6, which will be valid for large n based on the central-limit theorem, even if the underlying distribution is not normal.

SECTION 6.8 Estimation for the Binomial Distribution

6.8.1 Point Estimation

Point estimation for the parameter p of a binomial distribution is discussed in this section.

Example 6.42 **Cancer** Consider the problem of estimating the prevalence of malignant melanoma in 45- to 54-year-old women in the United States. Suppose that a random sample of 5000 women is selected from this age group and that 28 are found to have the disease. Let the random variable X_i represent the disease status for the ith woman, where $X_i = 1$ if the ith woman has the disease and 0 if she does not; $i = 1, \ldots, 5000$. The random variable X_i was also defined as a Bernoulli trial in Definition 5.14. Suppose that the prevalence of the disease in this age group $= p$. How can p be estimated?

We let $X = \sum_{i=1}^{n} X_i$ = the number of women with malignant melanoma among the n women. Based on Equation 5.8 and Example 5.32, we have that $E(X) = np$ and $Var(X) = npq$. Note that X can also be looked at as a binomial random variable with parameters n and p, because it represents the number of events in n independent trials.

Finally, consider the random variable $\hat{p}$ = sample proportion of events. In our example, $\hat{p}$ = proportion of women with malignant melanoma. Thus,

$$\hat{p} = \frac{1}{n}\sum_{i=1}^{n} X_i = X/n$$

Because $\hat{p}$ is a sample mean, the results of Equation 6.1 apply and we see that $E(\hat{p}) = E(X_i) \equiv \mu = p$. Furthermore, from Equation 6.2 it follows that

$$Var(\hat{p}) = \sigma^2/n = pq/n \quad \text{and} \quad se(\hat{p}) = \sqrt{pq/n}$$

Thus, for any sample of size n, the sample proportion $\hat{p}$ is an unbiased estimator of the population proportion p. The standard error of this proportion is given exactly by $\sqrt{pq/n}$ and is estimated by $\sqrt{\hat{p}\hat{q}/n}$. These principles can be summarized as follows:

EQUATION 6.16

Point Estimation of the Binomial Parameter *p* Let X be a binomial random variable with parameters n and p. An unbiased estimator of p is given by the sample proportion of events $\hat{p}$. Its standard error is given exactly by $\sqrt{pq/n}$ and is estimated by $\sqrt{\hat{p}\hat{q}/n}$.

Example 6.43 Estimate the prevalence of malignant melanoma in Example 6.42 and provide its standard error.

Solution Our best estimate of the prevalence rate of malignant melanoma among 45- to 54-year-old women is 28/5000 = .0056. Its estimated standard error is

$$\sqrt{.0056(.9944)/5000} = .0011$$

6.8.2 Interval Estimation—Normal-Theory Methods

Point estimation of the parameter p of a binomial distribution was covered in Section 6.8.1. How can an **interval estimate** of the parameter p be obtained?

Example 6.44 **Cancer** Suppose we are interested in estimating the prevalence rate of breast cancer among 50- to 54-year-old women whose mothers have had breast cancer. Suppose that in a random sample of 10,000 such women, 400 are found to have had breast cancer at some point in their lives. We have shown that the best point estimate of the prevalence rate p is given by the sample proportion $\hat{p} = 400/10,000 = .040$. How can an interval estimate of the parameter p be obtained? (See the solution in Example 6.45.)

We will assume that the normal approximation to the binomial distribution is valid—whereby from Equation 5.14 the number of events X observed out of n women will be approximately normally distributed with mean np and variance npq or, correspondingly, the proportion of women with events $= \hat{p} = X/n$ is normally distributed with mean p and variance pq/n.

The normal approximation can actually be justified on the basis of the central-limit theorem. Indeed, in the previous section we showed that $\hat{p}$ could be represented as an average of n Bernoulli trials, each of which has mean p and variance pq. Thus, for large n, from the central-limit theorem, we can see that $\hat{p} = \overline{X}$ is normally distributed with mean $\mu = p$ and variance $\sigma^2/n = pq/n$, or

EQUATION 6.17

$$\hat{p} \dot{\sim} N(p, pq/n)$$

Alternatively, since the number of successes in n Bernoulli trials $= X = n\hat{p}$ (which is the same as a binomial random variable with parameters n and p), if Equation 6.17 is multiplied by n,

EQUATION 6.18

$$X \dot{\sim} N(np, npq)$$

This formulation is indeed the same as that for the normal approximation to the binomial distribution, which was given in Equation 5.14. How large should n be before this approximation can be used? In Chapter 5 we said that the normal approximation to the binomial distribution is valid if $npq \geq 5$. However, in Chapter 5 we assumed that p was known, whereas here we assume that it is unknown. Thus, we shall estimate p by $\hat{p}$ and q by $\hat{q} = 1 - \hat{p}$ and will apply the normal approximation to the binomial if $n\hat{p}\hat{q} \geq 5$. Therefore, the results of this section should only be used if $n\hat{p}\hat{q} \geq 5$. An approximate $100\% \times (1-\alpha)$ confidence interval for p can now be derived from Equation 6.17 using methods similar to those given in Section 6.5.5. In particular, from Equation 6.17, we see that

$$Pr\left(p - z_{1-\alpha/2}\sqrt{pq/n} < \hat{p} < p + z_{1-\alpha/2}\sqrt{pq/n}\right) = 1 - \alpha$$

This inequality can be written in the form of two inequalities:

$$p - z_{1-\alpha/2}\sqrt{pq/n} < \hat{p} \qquad \text{and} \qquad \hat{p} < p + z_{1-\alpha/2}\sqrt{pq/n}$$

To explicitly derive a confidence interval based on these inequalities requires solving a quadratic equation for p in terms of $\hat{p}$. To avoid this complexity, it is customary to approximate $\sqrt{pq/n}$ by $\sqrt{\hat{p}\hat{q}/n}$ and rewrite the inequalities in the form

$$p - z_{1-\alpha/2}\sqrt{\hat{p}\hat{q}/n} < \hat{p} \qquad \text{and} \qquad \hat{p} < p + z_{1-\alpha/2}\sqrt{\hat{p}\hat{q}/n}$$

We now add $z_{1-\alpha/2}\sqrt{\hat{p}\hat{q}/n}$ to both sides of the first inequality and subtract this quantity from both sides of the second inequality, obtaining

$$p < \hat{p} + z_{1-\alpha/2}\sqrt{\hat{p}\hat{q}/n} \qquad \text{and} \qquad \hat{p} - z_{1-\alpha/2}\sqrt{\hat{p}\hat{q}/n} < p$$

Combining these two inequalities, we get

$$\hat{p} - z_{1-\alpha/2}\sqrt{\hat{p}\hat{q}/n} < p < \hat{p} + z_{1-\alpha/2}\sqrt{\hat{p}\hat{q}/n}$$

or $$Pr\left(\hat{p} - z_{1-\alpha/2}\sqrt{\hat{p}\hat{q}/n} < p < \hat{p} + z_{1-\alpha/2}\sqrt{\hat{p}\hat{q}/n}\right) = 1 - \alpha$$

The approximate $100\% \times (1-\alpha)$ confidence interval for p is given by

$$\left(\hat{p} - z_{1-\alpha/2}\sqrt{\hat{p}\hat{q}/n}, \hat{p} + z_{1-\alpha/2}\sqrt{\hat{p}\hat{q}/n}\right)$$

EQUATION 6.19

Normal-Theory Method for Obtaining a Confidence Interval for the Binomial Parameter *p* An approximate $100\% \times (1-\alpha)$ confidence interval for the binomial parameter p based on the normal approximation to the binomial distribution is given by

$$\hat{p} \pm z_{1-\alpha/2}\sqrt{\hat{p}\hat{q}/n}$$

This method of interval estimation should only be used if $n\hat{p}\hat{q} \geq 5$.

Example 6.45 **Cancer** Using the data in Example 6.44, derive a 95% confidence interval for the prevalence rate of breast cancer among 50- to 54-year-old women whose mothers have had breast cancer.

Solution $\hat{p} = .040 \quad \alpha = .05 \quad z_{1-\alpha/2} = 1.96 \quad n = 10{,}000$

Therefore, an approximate 95% confidence interval is given by

$$\left[.040 - 1.96\sqrt{.04(.96)/10{,}000}, .040 + 1.96\sqrt{.04(.96)/10{,}000}\right]$$
$$= (.040 - .004, .040 + .004) = (.036, .044)$$

Suppose we know that the prevalence rate of breast cancer among all 50- to 54-year-old American women is 2%. Because 2% is less than .036 (the lower confidence limit), we can be quite confident that the underlying rate for the group of women whose mothers have had breast cancer is higher than the rate in the general population.

6.8.3 Interval Estimation—Exact Methods

The question remains, How is a confidence interval for the binomial parameter p obtained when either the normal approximation to the binomial distribution is not valid or a more exact confidence interval is desired?

Example 6.46 **Cancer, Nutrition** Suppose we want to estimate the rate of bladder cancer in rats that have been fed a diet high in saccharin. We feed this diet to 20 rats and find that 2 develop bladder cancer. In this case our best point estimate of p is $\hat{p} = \frac{2}{20} = .1$. However, because

$$n\hat{p}\hat{q} = 20(2/20)(18/20) = 1.8 < 5$$

the normal approximation to the binomial distribution cannot be used and thus normal-theory methods for obtaining confidence intervals are not valid. How can an interval estimate be obtained in this case?

A small-sample method for obtaining confidence limits will be presented.

EQUATION 6.20

Exact Method for Obtaining a Confidence Interval for the Binomial Parameter *p* An exact $100\% \times (1-\alpha)$ confidence interval for the binomial parameter p that is always valid is given by (p_1, p_2), where p_1, p_2 satisfy the equations

$$Pr(X \geq x | p = p_1) = \frac{\alpha}{2} = \sum_{k=x}^{n} \binom{n}{k} p_1^k (1-p_1)^{n-k}$$
$$Pr(X \leq x | p = p_2) = \frac{\alpha}{2} = \sum_{k=0}^{x} \binom{n}{k} p_2^k (1-p_2)^{n-k}$$

A rationale for this confidence interval will be given in our discussion of hypothesis testing for the binomial distribution in Section 7.9.2.

The main problem with using this method is the difficulty in computing expressions such as

$$\sum_{k=0}^{x} \binom{n}{k} p^k (1-p)^{n-k}$$

Fortunately, special tables exist for the evaluation of such expressions, one of which is given in Table 7 in the Appendix. This table can be used as follows:

EQUATION 6.21

Exact Confidence Limits for Binomial Proportions

(1) The sample size (n) is given along each curve. Two curves should correspond to a given sample size. One curve is used to obtain the lower confidence limit and the other to obtain the upper confidence limit.

(2) If $0 \le \hat{p} \le .5$, then

- (a) Refer to the lower horizontal axis and find the point corresponding to $\hat{p}$.
- (b) Draw a line perpendicular to the horizontal axis and find the two points where this line intersects the two curves identified in (1).
- (c) Read across to the left vertical axis; the smaller value corresponds to the lower confidence limit and the larger value to the upper confidence limit.

(3) If $.5 < \hat{p} \le 1.0$, then

- (a) Refer to the upper horizontal axis and find the point corresponding to $\hat{p}$.
- (b) Draw a line perpendicular to the horizontal axis and find the two points where this line intersects the two curves identified in (1).
- (c) Read across to the right vertical axis; the smaller value corresponds to the lower confidence limit and the larger value to the upper confidence limit.

Example 6.47 **Cancer** Derive an exact 95% confidence interval for the probability of developing bladder cancer using the data given in Example 6.46.

Solution We refer to Table 7a in the Appendix ($\alpha = .05$), and identify the two curves with $n = 20$. Because $\hat{p} = .1 \le .5$, we refer to the lower horizontal axis and draw a vertical line at .10 until it intersects the two curves marked $n = 20$. We then read across to the left vertical axis and find the confidence limits of .01 and .32. Thus, the exact 95% confidence interval = (.01, .32). Notice that this confidence interval is *not* symmetric about $\hat{p} = .10$.

Example 6.48 **Health Promotion** Suppose that as part of a program for counseling patients with many risk factors for heart disease, 100 smokers are identified. Of this group, 10 give up smoking for at least one month. After a 1-year follow-up, 6 of the 10 patients are found to have taken up smoking again. The proportion of ex-smokers who start smoking again is referred to as the *recidivism rate*. Derive a 99% confidence interval for the recidivism rate.

Solution Exact binomial confidence limits must be used, because

$$n\hat{p}\hat{q} = 10(.6)(.4) = 2.4 < 5$$

We refer to the upper horizontal axis of the chart marked $\alpha = .01$ in Table 7b and note the point $\hat{p} = .60$. We then draw a vertical line at .60 until it intersects the two curves marked $n = 10$. We then read across to the right vertical axis and find the confidence limits of .19 and .92. Thus, the exact 99% confidence interval = (.19, .92).

More extensive and precise exact binomial confidence limits are available in Geigy Scientific Tables [4]. Also, calculation of exact binomial confidence limits are available directly in some statistical packages, including STATA, or using the BINOMDIST function of Excel.

SECTION 6.9 Estimation for the Poisson Distribution

6.9.1 Point Estimation

In this section, we discuss point estimation for the parameter λ of a Poisson distribution.

Example 6.49 **Cancer, Environmental Health** A study was performed in Woburn, Massachusetts, in the 1970s to look at possible excess cancer risk in children, with a particular focus on leukemia, which was later portrayed in the book and movie titled *A Civil Action*. An important environmental issue in the investigation concerned the possible contamination of the town's water supply. Specifically, 12 cases of childhood leukemia (<19 years old) were diagnosed in Woburn during the 1970s (January 1, 1970, to December 31, 1979). A key statistical issue is whether this represents an excessive number of leukemia cases, assuming that Woburn has had a constant 12,000 child residents (≤age 19) during this period and that the incidence rate of leukemia in children nationally is 5 cases per 100,000 person-years. Can we estimate the incidence rate of childhood leukemia in Woburn during the 1970s and provide a confidence interval about this estimate?

We let X = the number of children who develop leukemia in Woburn during the 1970s. Because X represents a rare event, we will assume that X follows a Poisson distribution with parameter $\mu = \lambda T$. We know from Chapter 4 that for a Poisson distribution, $E(X) = \lambda T$, where T = time and λ = number of events per unit time.

DEFINITION 6.16 A **person-year** is a unit of time defined as 1 person being followed for 1 year.

This unit of follow-up time is commonly used in longitudinal studies, that is, studies where the same individual is followed over time.

Example 6.50 **Cancer, Environmental Health** How many person-years were accumulated in the Woburn study in Example 6.49?

Solution In the Woburn study, there were 12,000 children who were each followed for 10 years. Thus, a total of 120,000 person-years were accumulated. This is actually an approximation, because the children who developed leukemia over the 10-year period would only be followed up to the time they developed the disease. It is also common to curtail follow-up for other reasons such as death and development of other types of cancer. However, the number of children for whom follow-up is curtailed for these reasons is probably small and the approximation is likely to be accurate.

Finally, although children have moved in and out of Woburn over the 10-year period, we assume that there is no net migration in and out of the area during the 1970s.

We now wish to assess how to estimate λ based on an observed number of events X over T person-years.

EQUATION 6.22

Point Estimation for the Poisson Distribution Suppose we assume that the number of events X over T person-years is Poisson-distributed with parameter $\mu = \lambda T$. An unbiased estimator of λ is given by $\hat{\lambda} = X/T$, where X is the observed number of events over T person-years.

If λ is the incidence rate per person-year, T = number of person-years of follow-up, and we assume a Poisson distribution for the number of events X over T person-years, then the expected value of X is given by $E(X) = \lambda T$. Therefore,

$$\begin{aligned} E(\hat{\lambda}) &= E(X)/T \\ &= \lambda T/T = \lambda \end{aligned}$$

Thus, $\hat{\lambda}$ is an unbiased estimator of λ.

Example 6.51 **Cancer, Environmental Health** Estimate the incidence rate of childhood leukemia in Woburn during the 1970s based on the data provided in Example 6.49.

Solution There were 12 events over 120,000 person-years, so the estimated incidence rate = 12/120,000 = 1/10,000 = 0.0001 events per person-year. Because cancer incidence rates per person-year are usually very low, it is customary to express such rates per 100,000 (or 10^5) person-years; that is, to change the unit of time to 10^5 person-years. Thus, if the unit of time = 10^5 person-years, then $T = 1.2$ and $\lambda = 0.0001(10^5) = 10$ events per 100,000 person-years.

6.9.2 Interval Estimation

The question remains as to how to obtain an interval estimate for λ. We use a similar approach as was used to obtain exact confidence limits for the binomial proportion p in Equation 6.20. For this purpose, it will be easier to first obtain a confidence interval for μ = expected number of events over time T of the form (μ_1, μ_2) and then obtain the corresponding confidence interval for λ from $(\mu_1/T, \mu_2/T)$. The approach is given as follows:

EQUATION 6.23

Exact Method for Obtaining a Confidence Interval for the Poisson Parameter λ An exact $100\% \times (1-\alpha)$ confidence interval for the Poisson parameter λ is given by $(\mu_1/T, \mu_2/T)$, where μ_1, μ_2 satisfy the equations

$$\begin{aligned} Pr(X \ge x \mid \mu = \mu_1) &= \frac{\alpha}{2} = \sum_{k=x}^{\infty} e^{-\mu_1} \mu_1^k / k! \\ &= 1 - \sum_{k=0}^{x-1} e^{-\mu_1} \mu_1^k / k! \\ Pr(X \le x \mid \mu = \mu_2) &= \frac{\alpha}{2} = \sum_{k=0}^{x} e^{-\mu_2} \mu_2^k / k! \end{aligned}$$

and x = observed number of events, T = number of person-years of follow-up.

As was the case in obtaining exact confidence limits for the binomial parameter p, it is difficult to compute μ_1, μ_2 exactly so as to satisfy Equation 6.23. Table 8 in the Appendix provides the solution to these equations. This table can be used to obtain 90%, 95%, 98%, 99%, or 99.8% confidence intervals for μ if the observed number of events (x) is ≤ 50. The observed number of events (x) is listed in the first column, and the level of confidence is given in the first row. The confidence interval is obtained by cross-referencing the x row and the $1-\alpha$ column.

Example 6.52 Suppose we observe 8 events and assume that the number of events is Poisson-distributed with parameter μ. Find a 95% confidence interval for μ.

Solution We refer to Table 8 under the 8 row and the 0.95 column to obtain the 95% CI for μ = (3.45, 15.76).

We see that this confidence interval is *not* symmetric about x (8), because $15.76 - 8 = 7.76 > 8 - 3.45 = 4.55$. This is true for all exact CI's based on the Poisson distribution unless x is very large.

Example 6.53 **Cancer, Environmental Health** Compute a 95% confidence interval for both the expected number of childhood leukemias (μ) and the incidence rate of childhood leukemia per 10^5 person-years (λ) in Woburn based on the data provided in Example 6.49.

Solution We observed 12 cases of childhood leukemia over 10 years. Thus, from Table 8, referring to $x = 12$ and level of confidence 95%, we find that the 95% CI for $\mu = (6.20, 20.96)$. Because there were 120,000 person-years = T, a 95% CI for the incidence rate = $\left(\frac{6.20}{120{,}000}, \frac{20.96}{120{,}000}\right)$ events per person-year or $\left(\frac{6.20}{120{,}000} \times 10^5, \frac{20.96}{120{,}000} \times 10^5\right)$ events per 10^5 person-years = (5.2, 17.5) events per 10^5 person-years = 95% CI for λ.

Example 6.54 **Cancer, Environmental Health** Interpret the results in Example 6.53. Specifically, do you feel there was an excess childhood leukemia risk in Woburn, Massachusetts, relative to expected U.S. incidence rates?

Solution Referring to Example 6.49, we note that the incidence rate of childhood leukemia in the United States during the 1970s was 5 events per 10^5 person-years. We will denote this rate by λ_0. Referring to Example 6.53, we see that the 95% CI for λ in Woburn = (5.2, 17.5) events per 10^5 person-years. The lower bound of the 95% CI exceeds $\lambda_0 \,(= 5)$, so we can conclude that there was a significant excess of childhood leukemia in Woburn during the 1970s. Another way to express these results is in terms of the standardized morbidity ratio (or SMR) defined by

$$\text{SMR} = \frac{\text{incidence rate in Woburn for childhood leukemia}}{\text{U.S. incidence rate for childhood leukemia}} = \frac{10/10^5}{5/10^5} = 2$$

If the U.S. incidence rate is assumed to be known, then a 95% CI for SMR is given by $\left(\frac{5.2}{5}, \frac{17.5}{5}\right) = (1.04, 3.50)$. Because the lower bound of the CI for SMR is > 1, we conclude there is a significant excess risk in Woburn. We pursue a different approach in Chapter 7, addressing this issue in terms of hypothesis testing and p-values.

In some instances, a random variable representing a rare event over time is assumed to follow a Poisson distribution but the actual amount of person-time is either unknown or is not reported in an article from the literature. In this instance, it is still possible to use Table 8 to obtain a confidence interval for μ, although it is impossible to obtain a confidence interval for λ.

Example 6.55 **Occupational Health** In Example 4.38, a study was described concerning the possible excess cancer risk among employees with high exposure to ethylene dibromide (EDB) in two plants in Texas and Michigan. Seven deaths due to cancer were reported over the period 1940–1975, whereas only 5.8 cancer deaths were expected based on mortality rates for U.S. white men. Find a 95% CI for the expected number of deaths among the exposed workers and assess whether their risk differs from the general population.

Solution In this case, the actual number of person-years used in computing the expected number of deaths was not reported in the original article. Indeed, the computation of the expected number of deaths is complex because

(1) Each worker is of a different age at the start of follow-up.

(2) The age of a worker changes over time.

(3) Mortality rates for men of the same age change over time.

However, we can use Table 8 to obtain a 95% CI for μ. Because $x = 7$ events, we have a 95% CI for $\mu = (2.81, 14.42)$. The expected number of deaths based on U.S. mortality rates for white males = 5.8, which falls within the preceding interval. Thus, we conclude that the risk among exposed workers does not differ from the general population.

Table 8 can also be used for applications of the Poisson distribution other than those based specifically on rare events over time.

Example 6.56 **Bacteriology** Suppose we observe 15 bacteria in a petri dish and assume that the number of bacteria is Poisson-distributed with parameter μ. Find a 90% confidence interval for μ.

Solution We refer to the 15 row and the 0.90 column in Table 8 to obtain the 90% confidence interval (9.25, 23.10).

SECTION 6.10 One-Sided Confidence Intervals

In the previous discussion of interval estimation, what are known as *two-sided confidence intervals* have been described. Frequently, the following type of problem occurs.

Example 6.57 **Cancer** A standard treatment exists for a certain type of cancer, and the patients receiving the treatment have a 5-year survival rate of 30%. A new treatment is proposed that has some unknown survival rate p. We would only be interested in using the new treatment if it were better than the standard treatment. Suppose that 40 out of 100 patients who receive the new treatment survive for 5 years. Can we say that the new treatment is better than the standard treatment?

One way to assess these data is to construct a one-sided confidence interval, where we are interested in only *one* bound of the interval, in this case the lower bound. If 30% is below the lower bound, then it is an unlikely estimate of the 5-year survival rate for patients getting the new treatment. We could reasonably conclude from this that the new treatment is better than the standard treatment in this case.

EQUATION 6.24

Upper One-Sided Confidence Interval for the Binomial Parameter p—Normal-Theory Method An upper one-sided $100\% \times (1-\alpha)$ confidence interval is of the form $p > p_1$ such that

$$Pr(p > p_1) = 1 - \alpha$$

If we assume that the normal approximation to the binomial holds true, then we can show that this confidence interval is given approximately by

$$p > \hat{p} - z_{1-\alpha}\sqrt{\hat{p}\hat{q}/n}$$

This interval estimator should only be used if $n\hat{p}\hat{q} \geq 5$.

To see this, note that if the normal approximation to the binomial distribution holds, then $\hat{p} \sim N(p, pq/n)$. Therefore, by definition

$$Pr\left(\hat{p} < p + z_{1-\alpha}\sqrt{pq/n}\right) = 1 - \alpha$$

We approximate $\sqrt{pq/n}$ by $\sqrt{\hat{p}\hat{q}/n}$ and subtract $z_{1-\alpha}\sqrt{\hat{p}\hat{q}/n}$ from both sides of the equation, yielding

$$\hat{p} - z_{1-\alpha}\sqrt{\hat{p}\hat{q}/n} < p$$

or $\quad p > \hat{p} - z_{1-\alpha}\sqrt{\hat{p}\hat{q}/n} \quad$ and $\quad Pr\left(p > \hat{p} - z_{1-\alpha}\sqrt{\hat{p}\hat{q}/n}\right) = 1 - \alpha$

Therefore, if the normal approximation to the binomial distribution holds, then $p > \hat{p} - z_{1-\alpha}\sqrt{\hat{p}\hat{q}/n}$ is an approximate upper $100\% \times (1-\alpha)$ one-sided confidence interval for p.

Notice that $z_{1-\alpha}$ is used in constructing one-sided intervals, whereas $z_{1-\alpha/2}$ was used in constructing two-sided intervals.

Example 6.58 Suppose a 95% confidence interval for a binomial parameter p is desired. What percentile of the normal distribution should be used for a one-sided interval? A two-sided interval?

Solution For $\alpha = .05$, we use $z_{1-.05} = z_{.95} = 1.645$ for a one-sided interval and $z_{1-.05/2} = z_{.975}$ $= 1.96$ for a two-sided interval.

Example 6.59 **Cancer** Construct an upper one-sided 95% confidence interval for the survival rate based on the cancer-treatment data in Example 6.57.

Solution First check that $n\hat{p}\hat{q} = 100(.4)(.6) = 24 \geq 5$. The confidence interval is then given by

$$Pr\left[p > .40 - z_{.95}\sqrt{.4(.6)/100}\right] = .95$$
$$Pr\left[p > .40 - 1.645(.049)\right] = .95$$
$$Pr(p > .319) = .95$$

Because .30 is not within the given interval [i.e., (.319, 1.0)], we would conclude that the new treatment is better than the standard treatment.

If we were interested in 5-year death rates rather than survival rates, then a one-sided interval of the form $Pr(p < p_2) = 1 - \alpha$ would be appropriate, because we would only be interested in the new treatment if its death rate were lower than that of the standard treatment.

EQUATION 6.25

Lower One-Sided Confidence Interval for the Binomial Parameter *p*—Normal-Theory Method The interval $p < p_2$ such that

$$Pr(p < p_2) = 1 - \alpha$$

is referred to as a **lower one-sided 100% × (1 – α) confidence interval** and is given approximately by

$$p < \hat{p} + z_{1-\alpha}\sqrt{\hat{p}\hat{q}/n}$$

This expression can be derived in the same manner as in Equation 6.24 by starting with the relationship

$$Pr\left(\hat{p} > p - z_{1-\alpha}\sqrt{pq/n}\right) = 1 - \alpha$$

If we approximate $\sqrt{pq/n}$ by $\sqrt{\hat{p}\hat{q}/n}$ and add $z_{1-\alpha}\sqrt{\hat{p}\hat{q}/n}$ to both sides of the equation, we get

$$Pr\left(p < \hat{p} + z_{1-\alpha}\sqrt{\hat{p}\hat{q}/n}\right) = 1 - \alpha$$

Example 6.60 **Cancer** Compute a lower one-sided 95% confidence interval for the death rate using the cancer-treatment data in Example 6.57.

Solution We have that $\hat{p} = .6$. Thus, the 95% confidence interval is given by

$$Pr\left[p < .6 + 1.645\sqrt{.6(.4)/100}\right] = .95$$
$$Pr\left[p < .6 + 1.645(.049)\right] = .95$$
$$Pr(p < .681) = .95$$

Because 70% is not within this interval [i.e., (0, .681)], we can conclude that the new treatment has a lower death rate than the old treatment does.

Similar methods can be used to obtain one-sided confidence intervals for the mean and variance of a normal distribution, for the binomial parameter *p*, and for the Poisson expectation μ using exact methods.

SECTION 6.11 Summary

In this chapter, the concept of a sampling distribution was introduced. This concept is crucial to understanding the principles of statistical inference. The fundamental idea is to forget about our sample as a unique entity; instead, regard it as a random sample from all possible samples of size n that could have been drawn from the population under study. Using this concept, $\overline{X}$ was shown to be an unbiased estimator of the population mean μ; that is, the average of all sample means over all possible random samples of size n that could have been drawn will equal the population mean. Furthermore, if our population follows a normal distribution, then $\overline{X}$ has minimum variance among all possible unbiased estimators and is thus referred to as a *minimum-variance unbiased estimator of* μ. Finally, if our population follows a normal distribution, then $\overline{X}$ will also follow a normal distribution. However, even if our population is not normal, the sample mean will still approximately follow a normal distribution for a sufficiently large sample size. This very important idea, which justifies many of the hypothesis tests we study in the remainder of this book, is called the *central-limit theorem.*

The idea of an interval estimate (or confidence interval) was then introduced. Specifically, a 95% confidence interval is defined as an interval that will contain the true parameter for 95% of all random samples that could have been obtained from the reference population. The preceding principles of point and interval estimation were applied to

(1) estimating the mean μ of a normal distribution

(2) estimating the variance σ^2 of a normal distribution

(3) estimating the parameter p of a binomial distribution

(4) estimating the parameter λ of a Poisson distribution

(5) estimating the expected value μ of a Poisson distribution

The t and chi-square distributions were introduced to obtain interval estimates for (1) and (2), respectively.

In Chapters 7 through 14, the discussion of statistical inference continues, focusing primarily on testing hypotheses rather than on parameter estimation. In this regard, some parallels between inference from the points of view of hypothesis testing and confidence intervals are discussed.

PROBLEMS

Gastroenterology

Suppose we wish to construct a list of treatment assignments for patients entering a study comparing different treatments for duodenal ulcer.

6.1 Anticipating that 20 patients will be entered in the study and 2 treatments will be used, construct a list of random-treatment assignments starting in the 28th row of the random-number table (Table 4 in the Appendix).

6.2 Count the number of people assigned to each treatment group. How does this number compare with the expected number in each group?

6.3 Suppose we change our minds and decide to enroll 40 patients and use four treatment groups. Use a computer program (such as MINITAB or Excel) to construct the list of random-treatment assignments referred to in Problem 6.1.

6.4 Answer Problem 6.2 for the list of treatment assignments derived in Problem 6.3.

Pulmonary Disease

The data in Table 6.7 concern the mean triceps skin-fold thickness in a group of normal men and a group of men with chronic airflow limitation [5].

TABLE 6.7 Triceps skin-fold thickness in normal men and men with chronic airflow limitation

Group	Mean	*sd*	n
Normal	1.35	0.5	40
Chronic airflow limitation	0.92	0.4	32

Source: Reprinted with permission of *Chest, 85*(6), 58S–59S, 1984.

***6.5** What is the standard error of the mean for each group?

6.6 Assume that the central-limit theorem is applicable. What does it mean in this context?

Cardiology

The data in Table 6.8 on left ventricular ejection fraction (LVEF) were collected from a group of 27 patients with acute dilated cardiomyopathy [6].

6.7 Calculate the standard deviation of LVEF for these patients.

6.8 Calculate the standard error of the mean for LVEF.

6.9 Using the computer, draw 50 subsamples of size 10 from the sample of 27 subjects and calculate the sample mean for each subsample. Do you think that the distribution of sample means is normally distributed? Is the central-limit theorem applicable to samples of size 10?

6.10 Find the upper 1st percentile of a *t* distribution with 16 *df.*

6.11 Find the lower 10th percentile of a *t* distribution with 28 *df.*

6.12 Find the upper 2.5th percentile of a *t* distribution with 7 *df.*

6.13 Compute a 95% confidence interval for the mean duration of hospitalization using the data in Table 2.11.

6.14 Compute a 95% confidence interval for the mean white blood count following admission using the data in Table 2.11.

6.15 Answer Problem 6.14 for a 90% confidence interval.

TABLE 6.8 Left ventricular ejection fraction (LVEF) for 27 patients with acute dilated cardiomyopathy

Patient	LVEF	Patient	LVEF
1	0.19	15	0.24
2	0.24	16	0.18
3	0.17	17	0.22
4	0.40	18	0.23
5	0.40	19	0.14
6	0.23	20	0.14
7	0.20	21	0.30
8	0.20	22	0.07
9	0.30	23	0.12
10	0.19	24	0.13
11	0.24	25	0.17
12	0.32	26	0.24
13	0.32	27	0.19
14	0.28		

Source: Reprinted with permission of the *New England Journal of Medicine, 312*(14), 885–890, 1985.

6.16 What is the relationship between your answers to Problems 6.14 and 6.15?

6.17 What are the upper and lower 2.5th percentiles for a chi-square distribution with 2 *df*? What notation is used to denote these percentiles?

Refer to the data in Table 2.11. Regard this hospital as typical of Pennsylvania hospitals.

***6.18** What is the best point estimate of the percentage of males among patients discharged from Pennsylvania hospitals?

***6.19** What is the standard error of the estimate obtained in Problem 6.18?

***6.20** Provide a 95% confidence interval for the percentage of males among patients discharged from Pennsylvania hospitals.

6.21 What is the best point estimate of the percentage of discharged patients, exclusive of women of childbearing age (ages 18–45), who received a bacterial culture while in the hospital?

6.22 Provide a 95% confidence interval corresponding to the estimate in Problem 6.21.

6.23 Answer Problem 6.22 for a 99% confidence interval.

Microbiology

A nine-laboratory cooperative study was performed to evaluate quality control for susceptibility tests with 30 μg

TABLE 6.9 Mean zone diameters with 30 μg netilmicin disks tested in nine separate laboratories

	Type of control strain					
	E. coli		*S. aureus*		*P. aeruginosa*	
Laboratory	Different media	Common medium	Different media	Common medium	Different media	Common medium
A	27.5	23.8	25.4	23.9	20.1	16.7
B	24.6	21.1	24.8	24.2	18.4	17.0
C	25.3	25.4	24.6	25.0	16.8	17.1
D	28.7	25.4	29.8	26.7	21.7	18.2
E	23.0	24.8	27.5	25.3	20.1	16.7
F	26.8	25.7	28.1	25.2	20.3	19.2
G	24.7	26.8	31.2	27.1	22.8	18.8
H	24.3	26.2	24.3	26.5	19.9	18.1
I	24.9	26.3	25.4	25.1	19.3	19.2

netilmicin disks [7]. Each laboratory tested 3 standard control strains on a different lot of Mueller-Hinton agar, with 150 tests performed per laboratory. For protocol control, each laboratory also performed 15 additional tests on each of the control strains using the *same* lot of Mueller-Hinton agar across laboratories. The mean zone diameters for each of the nine laboratories are given in Table 6.9.

***6.24** Provide a point and interval estimate (95% confidence interval) for the mean zone diameter across laboratories for each type of control strain, if each laboratory uses different media to perform the susceptibility tests.

***6.25** Answer Problem 6.24 if each laboratory uses a common medium to perform the susceptibility tests.

***6.26** Provide a point and interval estimate (95% confidence interval) for the interlaboratory standard deviation of mean zone diameters for each type of control strain, if each laboratory uses different media to perform the susceptibility tests.

***6.27** Answer Problem 6.26 if each laboratory uses a common medium to perform the susceptibility tests.

6.28 Are there any advantages to using a common medium versus using different media for performing the susceptibility tests with regard to standardization of results across laboratories?

Renal Disease

A study of psychological and physiological changes in a cohort of dialysis patients with end-stage renal disease was conducted [8]. 102 patients were initially ascertained at baseline; 69 of the 102 patients were reascertained at an 18-month follow-up visit. The data in Table 6.10 were reported.

6.29 Provide a point and interval estimate (95% confidence interval) for the mean of each of the parameters at baseline and follow-up.

TABLE 6.10 Psychological and physiological parameters in patients with end-stage renal disease

	Baseline (n = 102)		18-month follow-up (n = 69)	
Variable	Mean	*sd*	Mean	*sd*
Serum creatinine (mmol/L)	0.97	0.22	1.00	0.19
Serum potassium (mmol/L)	4.43	0.64	4.49	0.71
Serum phosphate (mmol/L)	1.68	0.47	1.57	0.40
Psychological Adjustment to Illness (PAIS) scale	36.50	16.08	23.27	13.79

6.30 Do you have any opinion on the physiological and psychological changes in this group of patients?

Hypertension

In an effort to detect hypertension in young children, blood-pressure measurements were taken on 30 children aged 5–6 years living in a specific community. For these children the mean diastolic blood pressure was found to be 56.2 mm Hg with standard deviation 7.9 mm Hg. From a nationwide study, we know that the mean diastolic blood pressure is 64.2 mm Hg for 5- to 6-year-old children.

6.31 Is there evidence that the mean diastolic blood pressure for children in the community is different from the nationwide average of children of the same age group?

6.32 Provide a 95% confidence interval for the standard deviation of the diastolic blood pressure of 5- to 6-year-old children in this community based on the observed 30 children.

Ophthalmology, Hypertension

A special study is conducted to test the hypothesis that people with glaucoma have higher blood pressure than average. In the study 200 people with glaucoma are recruited with a mean systolic blood pressure of 140 mm Hg and a standard deviation of 25 mm Hg.

6.33 Construct a 95% confidence interval for the true mean systolic blood pressure among people with glaucoma.

6.34 If the average systolic blood pressure for people of comparable age is 130 mm Hg, then is there an association between glaucoma and blood pressure?

Sexually Transmitted Disease

Suppose a clinical trial is conducted to test the efficacy of a new drug, spectinomycin, in the treatment of gonorrhea for females. Forty-six patients are given a 4-g daily dose of the drug and are seen one week later, at which time six of the patients still have gonorrhea.

***6.35** What is the best point estimate for *p*, the probability of a failure with the drug?

***6.36** What is a 95% confidence interval for *p*?

***6.37** Suppose we know that penicillin G at a daily dose of 4.8 mega units has a 10% failure rate. What can be said in comparing the two drugs?

Hepatic Disease

Suppose we are experimenting with a group of guinea pigs and inoculate them with a fixed dose of a particular toxin causing liver enlargement. We find that out of 40 guinea pigs, 15 actually have enlarged livers.

6.38 What is the best point estimate *p* of the probability of a guinea pig having an enlarged liver?

6.39 What is a two-sided 95% confidence interval for *p* assuming that the normal approximation is valid?

6.40 Answer Problem 6.39 if we do *not* assume that the normal approximation is valid.

Pharmacology

Suppose we wish to estimate the concentration (μg/mL) of a specific dose of ampicillin in the urine after various periods of time. We recruit 25 volunteers who have received ampicillin and find that they have a mean concentration of 7.0 μg/mL with a standard deviation of 2.0 μg/mL. Assume that the underlying population distribution of concentrations is normally distributed.

***6.41** Find a 95% confidence interval for the population mean concentration.

***6.42** Find a 99% confidence interval for the population variance of the concentrations.

***6.43** How large a sample would be needed to ensure that the length of the confidence interval in Problem 6.41 is 0.5 μg/mL if we assume that the sample standard deviation remains at 2.0 μg/mL?

Environmental Health

Much discussion has taken place concerning possible health hazards from exposure to anesthetic gases. In one study a group of 525 Michigan nurse anesthetists was surveyed by mail questionnaires and telephone interviews in 1972 to determine the incidence rate of cancer [9]. Of this group, 7 women reported having a new malignancy other than skin cancer during 1971.

6.44 What is the best estimate of the 1971 incidence rate from these data?

6.45 Provide a 95% confidence interval for the true incidence rate.

A comparison was made between the Michigan report and the 1969 cancer-incidence rates from the Connecticut tumor registry, where the expected incidence rate was determined to be 402.8 per 100,000 person-years.

6.46 Comment on the comparison between the observed incidence rate and the Connecticut tumor-registry data.

Obstetrics, Serology

A new assay is developed to obtain the concentration of *M. hominis* mycoplasma in the serum of pregnant women. The developers of this assay wish to make a statement as to the variability of their laboratory technique. For this purpose, 10 subsamples of 1 ml each are drawn from a large serum sample for *one* woman, and the assay is performed on each subsample. The concentrations are given as follows: 2^4, 2^3, 2^5, 2^4, 2^5, 2^4, 2^3, 2^4, 2^4, 2^5.

***6.47** If the distribution of concentrations in the log scale to the base 2 is assumed to be normal, then obtain the best estimate of the variance of the concentrations from these data.

***6.48** Compute a 95% confidence interval for the variance of the concentrations.

***6.49** Assuming that the point estimate in Problem 6.47 is the true population parameter, what is the probability that a particular assay, when expressed in the log scale to the base 2, is no more than 1.5 log units off from its true value for a particular woman?

***6.50** Answer Problem 6.49 for 2.5 log units.

Hypertension

Suppose 100 hypertensive people are given an antihypertensive drug and the drug is *effective* in 20 of the people. By *effective*, we mean that their diastolic blood pressure is lowered by at least 10 mm Hg as judged from a repeat measurement 1 month after taking the drug.

6.51 What is the best point estimate of the probability p of the drug being effective?

6.52 Suppose we know that 10% of all hypertensive patients who are given a placebo will have their diastolic blood pressure lowered by 10 mm Hg after 1 month. Can we carry out some procedure to be sure that we are not simply observing the placebo effect?

6.53 What assumptions have you made to carry out the procedure in Problem 6.52?

Suppose we decide that a better measure of the effectiveness of the drug is the mean decrease in blood pressure rather than the measure of effectiveness used previously. Let $d_i = x_i - y_i$, $i = 1, \ldots, 100$, where x_i = diastolic blood pressure for the ith person before taking the drug and y_i = diastolic blood pressure for the ith person 1 month after taking the drug. Suppose that the sample mean of the d_i is +5.3 and the sample variance is 144.0.

6.54 What is the standard error of $\bar{d}$?

6.55 What is a 95% confidence interval for the population mean of d?

6.56 Can we make a statement about the effectiveness of the drug?

6.57 What does a 95% confidence interval mean, in words, in this case?

Draw 6 random samples of size 5 from the data in Table 6.2.

6.58 Compute the mean birthweight for each of the 6 samples.

6.59 Compute the standard deviation based on the sample of 6 means. What is another name for this quantity?

6.60 Select the third point from each of the 6 samples, and compute the sample standard deviation from the collection of 6 third points.

6.61 What theoretical relationship should there be between the standard deviation in Problem 6.59 and the standard deviation in Problem 6.60?

6.62 How do the actual sample results in Problems 6.59 and 6.60 compare?

Obstetrics

In Figure 6.4b a plot of the sampling distribution of the sample mean from 200 samples of size 5 from the population of 1000 birthweights given in Table 6.2 was provided. The mean of the 1000 birthweights in Table 6.2 is 112.0 oz with standard deviation 20.6 oz.

***6.63** If the central-limit theorem holds, then what proportion of the sample means should fall within 0.5 lb of the population mean (112.0 oz)?

***6.64** Answer Problem 6.63 for 1 lb rather than 0.5 lb.

***6.65** Compare your results in Problems 6.63 and 6.64 with the actual proportion of sample means that fall in these ranges.

***6.66** Do you feel the central-limit theorem is applicable for samples of size 5 from this population?

Hypertension, Pediatrics

The etiology of high blood pressure remains a subject of active investigation. One widely accepted hypothesis is that excessive sodium intake adversely affects blood-pressure outcomes. To explore this hypothesis, an experiment was set up to measure responsiveness to the taste of salt and to relate the responsiveness to blood-pressure level. The protocol used involved testing 3-day-old infants in the newborn nursery by giving them a drop of various solutions and thus eliciting the sucking response and noting the vigor with which they sucked—denoted by MSB = mean number of sucks per burst of sucking. The content of the solution was changed over 10 consecutive periods: (1) water, (2) water, (3) 0.1 molar salt + water, (4) 0.1 molar salt + water, (5) water, (6) water, (7) 0.3 molar salt + water, (8) 0.3 molar salt + water, (9) water, (10) water. In addition, as a control, the response of the baby to the taste of sugar was also measured after the salt-taste protocol was completed. In this experiment, the sucking response was measured over 5 different periods with the following stimuli: (1) nonnutritive sucking, that is, a pure sucking response was elucidated without using any external substance; (2) water; (3) 5% sucrose + water; (4) 15% sucrose + water; (5) nonnutritive sucking.

The data for the first 100 infants in the study are given in Data Set INFANTBP.DAT, on the data disk. The format of the data is given in Data Set INFANTBP.DOC, on the data disk.

Construct a variable measuring the response to salt. For example, one possibility is to compute the average MSB for trials 3 and 4 – average MSB for trials 1 and 2 = average MSB when the solution was 0.1 molar salt + water – average MSB when the solution was water. A similar index could be computed comparing trials 7 and 8 to trials 5 and 6.

6.67 Obtain descriptive statistics and graphic displays for these salt-taste indices. Do the indices appear to be normally distributed? Why or why not? Compute the sample mean for this index, and obtain 95% confidence limits about the point estimate.

6.68 Construct indices measuring the responsiveness to the sugar taste and provide descriptive statistics and graphical displays for these indices. Do the indices appear to be normally distributed? Why or why not? Compute the sample mean and associated 95% confidence limits for these indices.

6.69 We wish to relate the indices to blood-pressure level. Provide a scatter plot relating mean systolic blood pressure (SBP) and mean diastolic blood pressure (DBP), respectively, to each of the salt-taste and sugar-taste indices. Does there appear to be a relation between the indices and blood-pressure level? We will discuss this in more detail in our work on regression analysis in Chapter 11.

Genetics

In Data Set SEXRAT.DAT, on the data disk, the sexes of children born in over 50,000 families with more than one child are listed.

6.70 Use interval-estimation methods to determine if the sex of successive births is predictable from the sex of previous births.

Nutrition

In Data Set VALID.DAT, on the data disk, estimated daily consumption of total fat, saturated fat, and alcohol as well as total caloric intake using two different methods of dietary assessment are provided for 173 subjects.

6.71 Use a computer to draw repeated random samples of size 5 from this population. Does the central-limit theorem seem to hold for these dietary attributes based on samples of size 5?

6.72 Answer Problem 6.71 for random samples of size 10.

6.73 Answer Problem 6.71 for random samples of size 20.

6.74 How do the sampling distributions compare based on samples of size 5, 10, and 20? Use graphic and numeric methods to answer this question.

Infectious Disease

A cohort of hemophiliacs is followed to elicit information on the distribution of time to onset of AIDS following seroconversion (referred to as *latency time*). All patients who seroconvert become symptomatic within 10 years, according to the following distribution:

Latency time (years)	Number of patients
0	2
1	6
2	9
3	33
4	49
5	66
6	52
7	37
8	18
9	11
10	4

6.75 Assuming an underlying normal distribution, compute 95% confidence intervals for the mean and variance of the latency times.

6.76 Still assuming normality, estimate the probability p that a patient's latency time will be at least 8 years.

6.77 Now suppose we are unwilling to assume a normal distribution for latency time. Re-estimate the probability p that a patient's latency time will be at least 8 years and provide a 95% confidence interval for p.

Environmental Health

We have previously described the Data Set LEAD.DAT (which is on the data disk). Children were classified according to blood-lead level in 1972 and 1973 by the variable GROUP, where 1 = blood-lead level $< 40\ \mu g/100$ mL in both 1972 and 1973, 2 = blood-lead level $\geq 40\ \mu g/100$ mL in 1973, 3 = blood-lead level $\geq 40\ \mu g/100$ mL in 1972, but $< 40\ \mu g/100$ mL in 1973.

6.78 Compute the mean, standard deviation, standard error, and 95% CI for the mean verbal IQ for children with specific values of the variable GROUP. Provide a box plot comparing the distribution of verbal IQ for subjects with GROUP = 1, 2, and 3. Summarize your findings concisely.

6.79 Answer Problem 6.78 for performance IQ.

6.80 Answer Problem 6.78 for full-scale IQ.

Cardiology

The Data Set NIFED.DAT (on the data disk) was described earlier. We wish to look at the effect of each treatment separately on heart rate and systolic blood pressure.

6.81 Provide a point estimate and a 95% CI for the changes in heart rate and systolic blood pressure (level 1 to baseline), separately for the subjects randomized to nifedipine and propranolol, respectively. Also provide box plots of the change scores in the two treatment groups.

6.82 Answer Problem 6.81 for level 2 to baseline.

6.83 Answer Problem 6.81 for level 3 to baseline.

6.84 Answer Problem 6.81 for the last available level to baseline.

6.85 Answer Problem 6.81 for the average heart rate (or blood pressure) over all available levels to baseline.

Occupational Health

***6.86** Refer to Problem 4.23. Provide a 95% CI for the expected number of deaths due to bladder cancer over 20 years among tire workers. Is there an excess number of cases of bladder cancer in this group?

***6.87** Refer to Problem 4.24. Provide a 95% CI for the expected number of deaths due to stomach cancer over 20 years among tire workers. Is there an excess number of cases of stomach cancer in this group?

Cancer

The value of mammography as a screening test for breast cancer has been controversial, particularly among young women. A study was recently performed looking at the rate of false positives among repeated screening mammograms among approximately 10,000 women who were members of Harvard Pilgrim Health Care, a large HMO (health-maintenance organization) in New England [10].

It was reported that a total of 1996 tests were given to women aged 40–49, of which 156 were false positives.

6.88 What does a false-positive test mean, in words, in this context?

6.89 Some physicians feel that a mammogram is not cost-effective unless one can be reasonably certain (e.g., 95% certain) that the false-positive rate is less than 10%. Can you address this issue based on the preceding data? (*Hint*: Use a confidence-interval (CI) approach.)

6.90 Suppose a woman is given a mammogram every 2 years starting at age 40. What is the probability that she will have at least one false-positive test among 5 screening tests during her forties? (Assume that the repeated screening tests are independent.)

6.91 Provide a two-sided 95% CI for the probability estimate in Problem 6.90.

Nutrition

On the computer, we draw 500 random samples of size 5 from the distribution of 173 values of ln(alcohol DR [diet record] + 1) in the Data Set VALID.DAT, where alcohol DR is the amount of alcohol consumed as reported by diet record by a group of 173 American nurses who recorded each food eaten on a real-time basis, over four 1-week periods spaced approximately 3 months apart over a 1-year period. For each sample of size 5, we compute the sample mean $\bar{x}$, the sample standard deviation s, and the test statistic t given by

$$t = \frac{\bar{x} - \mu_0}{s/\sqrt{n}}$$

where $n = 5$ and μ_0 = overall mean of ln(alcohol DR + 1) over the 173 nurses = 1.7973.

6.92 What distribution should the t-values follow if the central-limit theorem holds? Assume that μ_0 is the population mean for ln(alcohol DR + 1).

6.93 If the central-limit theorem holds, then what percentage of t-values should exceed 2.776 in absolute value?

6.94 The actual number of t-values that exceed 2.776 in absolute value is 38. Do you feel that the central-limit theorem is applicable to these data for samples of size 5?

Cardiovascular Disease

A study was performed to investigate the variability of cholesterol and other lipid measures in children. The reported within-subject *sd* for cholesterol in children was 7.8 mg/dL [11].

6.95 Suppose 2 total cholesterol determinations are obtained from 1 child, yielding an average value of 200 mg/dL. What is a *two-sided* 90% confidence interval for the true total cholesterol for that child? (*Hint:* Assume the sample standard deviation of cholesterol for that child is known to be 7.8 mg/dL.)

6.96 Suppose an average of 2 total cholesterol determinations are to be used as a screening tool to identify children with high cholesterol. The investigators wish to find a value c, such that all children whose mean cholesterol values over 2 determinations are $\ge c$ will be called back for further screening, whereas children whose mean cholesterol values are $< c$ will not be followed any further. To determine c, the investigators want to choose a value c such that the lower *one-sided* 90% confidence interval for μ if the observed average cholesterol over 2 determinations = c would exclude 250 mg/dL. What is the largest value of c that satisfies this requirement?

Endocrinology

Refer to the Data Set BONEDEN.DAT on the data disk.

6.97 Assess whether there is a relationship between bone density at the femoral neck and cigarette smoking using confidence-interval methodology. (*Hint:* Refer to Section 6.6.)

6.98 Assess whether there is a relationship between bone density at the femoral shaft and cigarette smoking using confidence-interval methodology. (*Hint:* Refer to Section 6.6.)

Simulation

6.99 Using the computer, generate 200 random samples from a binomial distribution with $n = 10$ and $p = .6$. Derive a 90% confidence interval for p based on each sample.

6.100 What percent of the confidence intervals include the parameter p?

6.101 Do you think that the large-sample binomial confidence-limit formula is adequate for this distribution?

6.102 Answer the same question as in Problem 6.99 for a binomial distribution with $n = 20$ and $p = .6$.

6.103 Answer the same question as in Problem 6.100 for a binomial distribution with $n = 20$ and $p = .6$.

6.104 Answer the same question as in Problem 6.101 for a binomial distribution with $n = 20$ and $p = .6$.

6.105 Answer the same question as in Problem 6.99 for a binomial distribution with $n = 50$ and $p = .6$.

6.106 Answer the same question as in Problem 6.100 for a binomial distribution with $n = 50$ and $p = .6$.

6.107 Answer the same question as in Problem 6.101 for a binomial distribution with $n = 50$ and $p = .6$.

REFERENCES

[1] Cochran, W. G. (1963). *Sampling techniques* (2nd ed.). New York: Wiley.

[2] SHEP Cooperative Research Group. (1991). Prevention of stroke by antihypertensive drug treatment in older persons with isolated systolic hypertension: Final results of the Systolic Hypertension in the Elderly Program (SHEP). *JAMA, 265*(24): 3255–3264.

[3] Mood, A., & Graybill, F. (1973). *Introduction to the theory of statistics* (3rd ed.). New York: McGraw-Hill.

[4] *Documenta Geigy scientific tables*, vol. 2 (8th ed.). (1982). Basel: Ciba-Geigy.

[5] Arora, N. S., & Rochester, D. F. (1984). Effect of chronic airflow limitation (CAL) on sternocleidomastoid muscle thickness. *Chest, 85*(6), 58S–59S.

[6] Dec, G. W., Jr., Palacios, I. F., Fallon, J. T., Aretz, H. T., Mills, J., Lee, D. C. S., & Johnson, R. A. (1985). Active myocarditis in the spectrum of acute dilated cardiomyopathies. *New England Journal of Medicine, 312*(14), 885–890.

[7] Barry, A. L., Gavan, T. L., & Jones, R. N. (1983). Quality control parameters for susceptibility data with 30 μg netilmicin disks. *Journal of Clinical Microbiology, 18*(5), 1051–1054.

[8] Oldenburg, B., Macdonald, G. J., & Perkins, R. J. (1988). Prediction of quality of life in a cohort of end-stage renal disease patients. *Journal of Clinical Epidemiology, 41*(6), 555–564.

[9] Corbett, T. H., Cornell, R. G., Leiding, K., & Endres, J. L. (1973). Incidence of cancer among Michigan nurse-anesthetists. *Anesthesiology, 38*(3), 260–263.

[10] Elmore, J. G., Barton, M. B., Moceri, V. M., Polk, S., Arena, P. J., & Fletcher, S. W. (1998). Ten year risk of false positive screening mammograms and clinical breast examinations. *New England Journal of Medicine, 338*, 1089–1096.

[11] Elveback, L. R., Weidman, W. H., & Ellefson, R. D. (1980). Day to day variability and analytic error in determination of lipids in children. *Mayo Clinic Proceedings, 55*, 267–269.

7

Hypothesis Testing: One-Sample Inference

SECTION 7.1 Introduction

In Chapter 6, methods of point and interval estimation for parameters of various distributions were discussed. However, researchers often have preconceived ideas about what these parameters might be and wish to test whether the data conform with these ideas.

Example 7.1 **Cardiovascular Disease, Pediatrics** A current area of research interest is the familial aggregation of cardiovascular risk factors in general and lipid levels in particular. Suppose it is known that the "average" cholesterol level in children is 175 mg/dL. A group of men who have died from heart disease within the past year are identified, and the cholesterol levels of their offspring are measured. Two hypotheses will be considered:

(1) The average cholesterol level of these children is 175 mg/dL.

(2) The average cholesterol level of these children is greater than 175 mg/dL.

This type of question is formulated in a hypothesis-testing framework by specifying two hypotheses—a null and an alternative hypothesis. We wish to compare the relative probabilities of obtaining the sample data under each of these hypotheses. In Example 7.1, the null hypothesis is that the average cholesterol level of the children is 175 mg/dL and the alternative hypothesis is that the average cholesterol level of the children is greater than 175 mg/dL.

Why is hypothesis testing so important? Hypothesis testing provides an objective framework for making decisions using probabilistic methods, rather than relying on subjective impressions. People can form different opinions by looking at data, but a hypothesis test provides a uniform decision-making criterion that is consistent for all people.

In this chapter some of the basic concepts of hypothesis testing are developed and applied to one-sample problems of statistical inference. In a **one-sample problem**, hypotheses are specified about a single distribution; in a **two-sample problem**, two different distributions are compared.

SECTION 7.2 General Concepts

Example 7.2 **Obstetrics** Suppose we want to test the hypothesis that mothers with low socioeconomic status (SES) deliver babies whose birthweights are lower than "normal." To test this hypothesis, a list is obtained of birthweights from 100 consecutive, full-term, liveborn deliveries from the maternity ward of a hospital in a low-SES area. The mean birthweight ($\bar{x}$) is found to be 115 oz with a sample standard deviation (s) of 24 oz. Suppose we know from nationwide surveys based on millions of deliveries that the mean birthweight in the United States is 120 oz. Can we actually say that the underlying mean birthweight from this hospital is lower than the national average?

Assume that the 100 birthweights from this hospital come from an underlying normal distribution with unknown mean μ. The methods in Section 6.10 could be used to construct a 95% lower one-sided confidence interval for μ based on the sample data, that is, an interval of the form $\mu < c$. If this interval contains 120 oz (if $c \geq 120$), then the hypothesis would be accepted that these birthweights are similar to the national average. If it does not contain 120 oz ($c < 120$), then the hypothesis would be accepted that these birthweights tend to be lower than the national average.

Another way of looking at this problem is in terms of hypothesis testing. In particular, the hypotheses being considered can be formulated in terms of null and alternative hypotheses, which can be defined as follows:

DEFINITION 7.1 The **null hypothesis**, denoted by H_0, is the hypothesis that is to be tested. The **alternative hypothesis**, denoted by H_1, is the hypothesis that in some sense contradicts the null hypothesis.

Example 7.3 **Obstetrics** In Example 7.2, the null hypothesis (H_0) is the hypothesis that the mean birthweight in the low-SES-area hospital (μ) is equal to the mean birthweight in the United States (μ_0). This is the hypothesis we want to test. The alternative hypothesis (H_1) is the hypothesis that the mean birthweight in this hospital (μ) is less than the mean birthweight in the United States (μ_0). We want to compare the relative probabilities of obtaining the sample data under each of these two hypotheses.

We also assume that the underlying distribution is normal under either hypothesis. These hypotheses can be written more succinctly in the following form:

EQUATION 7.1

$$H_0: \mu = \mu_0 \quad \text{vs.} \quad H_1: \mu < \mu_0$$

Suppose the only possible decisions are whether H_0 is true or H_1 is true. Actually, for ease of notation, all outcomes in a hypothesis-testing situation generally refer to the null hypothesis. Hence, if we decide that H_0 is true, then we say that we accept H_0. If we decide that H_1 is true, then we state that H_0 is not true or, equivalently, that we reject H_0. Thus four possible outcomes can occur:

(1) We accept H_0, and H_0 is in fact true.

(2) We accept H_0, and H_1 is in fact true.

(3) We reject H_0, and H_0 is in fact true.

(4) We reject H_0, and H_1 is in fact true.

These four possibilities are depicted in Table 7.1.

TABLE 7.1 Four possible outcomes in hypothesis testing

Decision	Truth: H_0	Truth: H_1
Accept H_0.	H_0 is true and H_0 is accepted.	H_1 is true and H_0 is accepted.
Reject H_0.	H_0 is true and H_0 is rejected.	H_1 is true and H_0 is rejected.

In actual practice, it is impossible, using hypothesis-testing methods, to *prove that the null hypothesis is true*. Thus, in particular, if we *accept* H_0, then we have actually failed to reject H_0.

If H_0 is true and H_0 is accepted, or if H_1 is true and H_0 is rejected, then the correct decision has been made. If H_0 is true and H_0 is rejected, or if H_1 is true and H_0 is accepted, then an *error* has been made. The two types of errors are generally treated differently.

DEFINITION 7.2 The probability of a **type I error** is the probability of rejecting the null hypothesis when H_0 is true.

DEFINITION 7.3 The probability of a **type II error** is the probability of accepting the null hypothesis when H_1 is true. This probability is a function of μ as well as other factors.

Example 7.4 **Obstetrics** In the context of the birthweight data in Example 7.2, a type I error would be the probability of deciding that the mean birthweight in the hospital was less than 120 oz when in fact it was 120 oz. A type II error would be the probability of deciding that the mean birthweight was 120 oz when in fact it was less than 120 oz.

Example 7.5 **Cardiovascular Disease, Pediatrics** What are the type I and type II errors for the cholesterol data in Example 7.1?

Solution The type I error is the probability of deciding that the offspring of men who have died from heart disease have an average cholesterol greater than 175 mg/dL when in fact their average cholesterol level is 175 mg/dL. The type II error is the probability of deciding that the offspring have normal cholesterol levels when in fact their cholesterol levels are above average.

Type I and type II errors often result in monetary and nonmonetary costs.

Example 7.6 **Obstetrics** The birthweight data in Example 7.2 might be used to decide if a special-care nursery for low-birthweight babies is needed in this hospital. If H_1 were true—that is, if the birthweights in this hospital did tend to be lower than the national average—then the hospital might be justified in having its own special-care nursery. If H_0 were true and the mean birthweight was no different from the U.S. average, then the hospital probably does not need such a nursery. If a type I error is made, then a special-care nursery will be recommended, with all the extra costs involved, when in fact it is not needed. If a type II error is made, a special-care nursery will not be funded, when in fact it is needed. The nonmonetary cost of this decision is that some low-birthweight babies may not survive without the unique equipment in a special-care nursery.

DEFINITION 7.4 The probability of a **type I error** is usually denoted by α and is commonly referred to as the **significance level** of a test.

DEFINITION 7.5 The probability of a **type II error** is usually denoted by β.

DEFINITION 7.6 The **power** of a test is defined as

$$1 - \beta = 1 - \text{probability of a type II error} = Pr(\text{rejecting } H_0 | H_1 \text{ true})$$

Example 7.7 **Rheumatology** Suppose a new drug is to be tested for the purpose of pain relief among patients with osteoarthritis (OA). The measure of pain relief will be the percent change in level of pain as reported by the patient after taking the medication for 1 month. Fifty OA patients will participate in the study. What are the hypotheses to be tested? What do type I error, type II error, and power mean in this situation?

Solution The hypotheses to be tested are H_0: $\mu = 0$ versus H_1: $\mu > 0$, where μ = mean % change in level of pain over a 1-month period. It is assumed that a positive value for μ indicates improvement, whereas a negative value indicates decline.

A type I error is the probability of deciding that the drug is an effective pain reliever based on data from 50 patients, given that the true state of nature is that the drug has no effect on pain relief. The true state of nature here means the effect of the drug when tested on a large (infinite) number of patients.

An type II error is the probability of deciding that the drug has no effect on pain relief based on data from 50 patients given that the true state of nature is that the drug is an effective pain reliever.

The power of the test is the probability of deciding that the drug is effective as a pain reliever based on data from 50 patients when the true state of nature is that it is effective. It is important to note that the power is not a single number, but will depend on the true degree of pain relief offered by the drug as measured by the true mean change in pain-relief score (δ). The higher δ is, the higher the power will be. In Section 7.5, we will study formulas for calculating power in more detail.

The general aim in hypothesis testing is to use statistical tests that make α and β as small as possible. This goal requires compromise, because making α small involves rejecting the null hypothesis less often, whereas making β small involves accepting the null hypothesis less often. These actions are contradictory; that is, as

α increases, β will decrease, and as α decreases, β will increase. Our general strategy will be to fix α at some specific level (for example, .10, .05, .01, . . .) and to use the test that minimizes β or, equivalently, maximizes the power.

SECTION 7.3 One-Sample Test for the Mean of a Normal Distribution: One-Sided Alternatives

The appropriate hypothesis test for the birthweight data in Example 7.2 will now be developed. The statistical model in this case is that the birthweights come from a normal distribution with mean μ and unknown variance σ^2. We wish to test the null hypothesis, H_0, that $\mu = 120$ oz versus the alternative hypothesis, H_1, that $\mu <$ 120 oz. Suppose a more specific alternative, namely, H_1: $\mu = \mu_1 = 110$ oz, is selected. We will show that the nature of the best test does not depend on the value chosen for μ_1 provided that μ_1 is less than 120 oz. We will also fix the α level at .05 for concreteness.

Example 7.8 A very simple test could be used by referring to the table of random digits in Table 4 in the Appendix. Suppose two digits are selected from this table and the null hypothesis is rejected if these two digits are between 00 and 04 inclusive and is accepted if these two digits are between 05 and 99. Clearly, from the properties of the random-number table, the type I error of this test = α = Pr(rejecting the null hypothesis | H_0 true) = Pr(drawing two random digits between 00 and 04) = $\frac{5}{100}$ = .05. Thus, the proposed test satisfies the α-level criterion given previously. The problem with this test is that it has very low power. Indeed, the power of the test = Pr(rejecting the null hypothesis | H_1 true) = Pr(drawing two random digits between 00 and 04) = $\frac{5}{100}$ = .05.

Note that the outcome of the test has nothing to do with the sample birthweights drawn. H_0 will be rejected just as often when the sample mean birthweight ($\bar{x}$) is 110 oz as when it is 120 oz. Thus, this test must be very poor, because we would expect to reject H_0 with near certainty if $\bar{x}$ is small enough and would expect never to reject H_0 if $\bar{x}$ is large enough.

It can be shown that the best (most powerful) test in this situation is based on the sample mean ($\bar{x}$). If $\bar{x}$ is sufficiently smaller than μ_0, then H_0 is rejected; otherwise, H_0 is accepted. This test is reasonable, because if H_0 is true, then the most likely values of $\bar{x}$ will tend to cluster around μ_0, whereas if H_1 is true, the most likely values of $\bar{x}$ will tend to cluster around μ_1. By "most powerful," we mean that the test based on the sample mean has the highest power among all tests with a given type I error of α.

DEFINITION 7.7 The **acceptance region** is the range of values of $\bar{x}$ for which H_0 is accepted.

DEFINITION 7.8 The **rejection region** is the range of values of $\bar{x}$ for which H_0 is rejected.

For the birthweight data in Example 7.2, the rejection region consists of small values of $\bar{x}$ because the underlying mean under the alternative hypothesis (μ_1) is less than the underlying mean under the null hypothesis. This type of test is called a *one-tailed test.*

DEFINITION 7.9

A **one-tailed test** is a test in which the values of the parameter being studied (in this case μ) under the alternative hypothesis are allowed to be either greater than or less than the values of the parameter under the null hypothesis (μ_0), *but not both.*

Example 7.9

Cardiovascular Disease, Pediatrics The hypotheses for the cholesterol data in Example 7.1 are H_0: $\mu = \mu_0$ versus H_1: $\mu > \mu_0$, where μ is the true mean cholesterol level for children of men who have died from heart disease. This test is one-tailed, because the alternative mean is only allowed to be greater than the null mean.

How small should $\bar{x}$ be for H_0 to be rejected? This issue can be settled by recalling that the significance level of the test is set at α. Suppose that H_0 is rejected for all values of $\bar{x} < c$ and accepted otherwise. The value c should be selected such that the type I error = α.

It will be more convenient to define test criteria in terms of standardized values rather than in terms of $\bar{x}$. Specifically, if we subtract μ_0 and divide by $S/\sqrt{n}$, we obtain the random variable $t = (\bar{X} - \mu_0)/(S/\sqrt{n})$, which, based on Equation 6.5, follows a t_{n-1} distribution under H_0. We note that under H_0, based on the definition of the percentiles of a t distribution, $Pr(t < t_{n-1,\alpha}) = \alpha$. This leads us to the following test procedure.

EQUATION 7.2

One-Sample *t* Test for the Mean of a Normal Distribution with Unknown Variance (Alternative Mean < Null Mean) To test the hypothesis

$$H_0\text{: } \mu = \mu_0, \sigma \text{ unknown} \quad \text{vs.} \quad H_1\text{: } \mu < \mu_0, \sigma \text{ unknown}$$

with a significance level of α, we compute

$$t = \frac{\bar{x} - \mu_0}{s/\sqrt{n}}$$

If $t < t_{n-1,\alpha}$, then we reject H_0.
If $t \ge t_{n-1,\alpha}$, then we accept H_0.

DEFINITION 7.10

The value t in Equation 7.2 is called a **test statistic**, because the test procedure is based on this statistic.

DEFINITION 7.11

The value $t_{n-1,\alpha}$ in Equation 7.2 is called a **critical value** because the outcome of the test depends on whether the test statistic $t < t_{n-1,\alpha}$ = critical value, whereby we reject H_0 or $t \ge t_{n-1,\alpha}$, whereby we accept H_0.

DEFINITION 7.12 The general approach where we compute a test statistic and determine the outcome of a test by comparing the test statistic to a critical value determined by the type I error is called the **critical-value method** of hypothesis testing.

Example 7.10 **Obstetrics** Use the one-sample t test to test the hypothesis H_0: $\mu = 120$ versus H_1: $\mu < 120$ based on the birthweight data given in Example 7.2 using a significance level of .05.

Solution We compute the test statistic

$$t = \frac{\bar{x} - \mu_0}{s/\sqrt{n}}$$

$$= \frac{115 - 120}{24/\sqrt{100}}$$

$$= \frac{-5}{2.4} = -2.08$$

Using Excel, we see that the critical value $= t_{99,\ .05} = -1.66$. Because $t = -2.08 < -1.66$, it follows that we can reject H_0 at a significance level of .05.

Example 7.11 **Obstetrics** Use the one-sample t test to test the hypothesis given in Example 7.10 using a significance level of .01.

Solution Using Excel, the critical value is $t_{99,\ .01} = -2.36$. Because $t = -2.08 > -2.36$, it follows that we accept H_0 at significance level = .01.

If we use the critical-value method, how do we know what level of α to use? The actual α level used should depend on the relative importance of type I and type II errors, because for a fixed sample size (n), the smaller α is made, the larger β becomes. Most people feel uncomfortable with α levels much greater than .05. Traditionally, an α level of exactly .05 is used most frequently.

In general, a number of significance tests could be performed at different α levels, as was done in Examples 7.10 and 7.11, and whether H_0 would be accepted or rejected in each instance could be noted. This can be somewhat tedious and is unnecessary because, instead, significance tests can be effectively performed *at all* α *levels* by obtaining the p-value for the test.

DEFINITION 7.13 The ***p*-value** for any hypothesis test is the α level at which we would be indifferent between accepting or rejecting H_0 given the sample data at hand. That is, the p-value is the α level at which the given value of the test statistic (such as t) would be on the borderline between the acceptance and rejection regions.

According to the test criterion in Equation 7.2, if a significance level of p is used, then H_0 would be rejected if $t < t_{n-1,\ p}$ and accepted if $t \geq t_{n-1,\ p}$. We would be indifferent to the choice between accepting or rejecting H_0 if $t = t_{n-1,\ p}$. We can solve for p as a function of t by

EQUATION 7.3

$$p = Pr(t_{n-1} \le t)$$

Thus, p is the area to the left of t under a t_{n-1} distribution.

The p-value can be depicted graphically, as shown in Figure 7.1.

FIGURE 7.1 **Graphic display of a *p*-value**

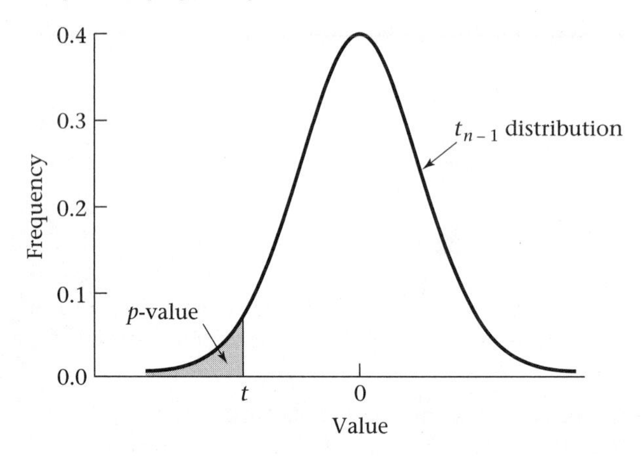

Example 7.12 **Obstetrics** Compute the p-value for the birthweight data in Example 7.2.

Solution From Equation 7.3, we have that the p-value is given by

$$Pr(t_{99} \le -2.08)$$

Using the TDIST function of Excel, we find that this probability is given by .020, which is the p-value.

An alternative definition of a p-value that will be useful in other hypothesis-testing problems is given as follows:

DEFINITION 7.14 The ***p*-value** can also be thought of as the probability of obtaining a test statistic as extreme as or more extreme than the actual test statistic obtained given that the null hypothesis is true.

We know that under the null hypothesis, the t statistic follows a t_{n-1} distribution. Hence, the probability of obtaining a t statistic that is no larger than t under the null hypothesis is $Pr(t_{n-1} \le t) = p$-value, as shown in Figure 7.1.

Example 7.13 **Cardiology** A topic of recent clinical interest is the possibility of using drugs to reduce infarct size in patients who have had a myocardial infarction within the past 24 hours. Suppose we know that in untreated patients the mean infarct size is 25 ($ck - g - EQ/m^2$). Furthermore, in 8 patients treated with a drug, the mean infarct size is 16 with a standard deviation of 10. Is the drug effective in reducing infarct size?

Solution The hypotheses are H_0: $\mu = 25$ versus H_1: $\mu < 25$. The *p*-value is computed using Equation 7.3. First, we compute the *t* statistic given by

$$t = \frac{16 - 25}{10/\sqrt{8}} = -2.55$$

The *p*-value is then given by $p = Pr(t_7 < -2.55)$. Referring to Table 5 in the Appendix, we see that $t_{7,\ .975} = 2.365$, and $t_{7,\ .99} = 2.998$. Because $2.365 < 2.55 < 2.998$, it follows that $1 - .99 < p < 1 - .975$ or $.01 < p < .025$. Thus, H_0 is rejected and we conclude that the drug reduces infarct size (all other things being equal).

This can also be interpreted as the probability that mean infarct size among a random sample of 8 patients will be no larger than 16, given that the null hypothesis is true. In this example, the null hypothesis is that the drug is ineffective, or in other words, that true mean infarct size for the population of all patients with myocardial infarction who are treated with drug = true mean infarct size for untreated patients = 25.

The importance of the *p*-value is that it tells us *exactly* how significant our results are without performing repeated significance tests at different α levels. A question typically asked is, How small should the *p*-value be for results to be considered statistically significant? Although this question has no definite answer, some commonly used criteria follow:

EQUATION 7.4

Guidelines for Judging the Significance of a *p*-Value

If $.01 \le p < .05$, then the results are *significant.*
If $.001 \le p < .01$, then the results are *highly significant.*
If $p < .001$, then the results are *very highly significant.*
If $p > .05$, then the results are considered *not statistically significant* (sometimes denoted by NS).
However, if $.05 \le p < .10$, then a trend toward statistical significance is sometimes noted.

Authors frequently do not specify the exact *p*-value beyond giving ranges of the type shown here, because whether the *p*-value is .024 or .016 is thought to be unimportant. Other authors give an exact *p*-value even for results that are not statistically significant so that the reader can appreciate how close to statistical significance the results have come. With the advent of statistical packages such as Excel or MINITAB, exact *p*-values are easy to obtain. These different approaches lead to the following general principle:

EQUATION 7.5

Determination of Statistical Significance for Results from Hypothesis Tests
Either of the following methods can be used to establish whether results from hypothesis tests are statistically significant:

(1) The test statistic *t* can be computed and compared with the critical value $t_{n-1,\alpha}$ at an α level of .05. Specifically, if H_0: $\mu = \mu_0$ versus H_1: $\mu < \mu_0$ is being tested and $t < t_{n-1,.05}$, then H_0 is rejected and the results are declared *statistically significant* ($p < .05$). Otherwise, H_0 is accepted and the results are declared *not statistically significant* ($p \ge .05$). We have referred to this approach as the **critical-value method** (see Definition 7.12).

(2) The exact p-value can be computed, and if $p < .05$, then H_0 is rejected and the results are declared *statistically significant.* Otherwise, if $p \geq .05$, then H_0 is accepted and the results are declared *not statistically significant.* We will refer to this approach as the ***p*-value method.**

These two approaches are equivalent regarding the determination of statistical significance (whether $p < .05$ or $p \geq .05$). The p-value method is somewhat more precise in that it yields an exact p-value. The two approaches in Equation 7.5 can also be used to determine statistical significance in other hypothesis-testing problems.

Example 7.14 **Obstetrics** Assess the statistical significance of the birthweight data in Example 7.12.

Solution Because the p-value is .020, the results would be considered statistically significant and we would conclude that the true mean birthweight is significantly lower in this hospital than in the general population.

Example 7.15 **Cardiology** Assess the significance of the infarct-size data in Example 7.13.

Solution The p-value $= Pr(t_7 < -2.55)$. Using the TDIST function of Excel, we find that $p = .019$. Thus, the results are highly significant.

In writing up the results of a study, a distinction between scientific and statistical significance should be made, because the two terms do not necessarily coincide. The results of a study can be statistically significant but still not be scientifically important. This situation would occur if a small difference was found to be statistically significant because of a large sample size. Conversely, some statistically nonsignificant results can be scientifically important, encouraging researchers to perform larger studies to confirm the direction of the findings and possibly reject H_0 with a larger sample size. This statement is true not only for the one-sample t test, but for virtually any hypothesis test.

Example 7.16 **Obstetrics** Suppose the mean birthweight in Example 7.2 was 119 oz, based on a sample of size 10,000. Assess the results of the study.

Solution The test statistic would be given by

$$t = \frac{119 - 120}{24/\sqrt{10{,}000}} = -4.17$$

Thus the p-value is given by $Pr(t_{9999} < -4.17)$. Because a t distribution with 9999 df is virtually the same as an $N(0,1)$ distribution, we can approximate the p-value by $\Phi(-4.17) < .001$. The results are thus very highly significant but are clearly not very important because of the small difference in mean birthweight (1 oz) between this hospital and the national average.

Example 7.17 **Obstetrics** Suppose that the mean birthweight in Example 7.2 was 110 oz, based on a sample size of 10. Assess the results of the study.

Solution The test statistic would be given by

$$t = \frac{110 - 120}{24/\sqrt{10}} = -1.32$$

The p-value is given by $Pr(t_9 < -1.32)$. From Table 5, because $t_{9,\,.85} = 1.100$ and $t_{9,\,.90} = 1.383$ and $1.100 < 1.32 < 1.383$, it follows that $1 - .90 < p < 1 - .85$ or $.10 < p < .15$. These results are not statistically significant but could be important if the same trends were also apparent in a larger study.

The test criterion in Equation 7.2 was based on an alternative hypothesis that $\mu < \mu_0$. In many situations we wish to use an alternative hypothesis that $\mu > \mu_0$. In this case H_0 would be rejected if $\bar{x}$, or correspondingly our test statistic t, were large ($>c$) and accepted if t were small ($\le c$). To ensure a type I error of α, find c such that

$$\begin{aligned}\alpha &= Pr\left(t > c \middle| H_0\right) = Pr\left(t > c \middle| \mu = \mu_0\right)\\ &= 1 - Pr\left(t \le c \middle| \mu = \mu_0\right)\end{aligned}$$

Because t follows a t_{n-1} distribution under H_0, we have that

$$\alpha = 1 - Pr\left(t_{n-1} \le c\right) \quad \text{or} \quad 1 - \alpha = Pr\left(t_{n-1} \le c\right)$$

Because $Pr(t_{n-1} < t_{n-1,1-\alpha}) = 1 - \alpha$, we have $c = t_{n-1,1-\alpha}$. Thus, at level α, H_0 is rejected if $t > t_{n-1,1-\alpha}$ and accepted otherwise. The p-value is the probability of observing a test statistic at least as large as t under the null hypothesis. Thus, because t follows a t_{n-1} distribution under H_0, we have

$$p = 1 - Pr(t_{n-1} \le t) = Pr(t_{n-1} \ge t)$$

The test procedure is summarized as follows:

EQUATION 7.6

One-Sample *t* Test for the Mean of a Normal Distribution with Unknown Variance (Alternative Mean > Null Mean) To test the hypothesis

$$H_0\colon \mu = \mu_0 \quad \text{vs.} \quad H_1\colon \mu > \mu_0$$

with a significance level of α, the best test is based on t, where

$$t = \frac{\bar{x} - \mu_0}{s/\sqrt{n}}$$

If $t > t_{n-1,1-\alpha}$ then H_0 is rejected

If $t \le t_{n-1,1-\alpha}$ then H_0 is accepted

The p-value for this test is given by

$$p = Pr\left(t_{n-1} > t\right)$$

The p-value for this test is depicted in Figure 7.2.

Example 7.18 **Cardiovascular Disease, Pediatrics** Suppose the mean cholesterol level of 10 children whose fathers died from heart disease in Example 7.1 is 200 mg/dL and the sample

FIGURE 7.2 ***p*-value for the one-sample *t* test when the alternative mean (μ_1) > null mean (μ_0)**

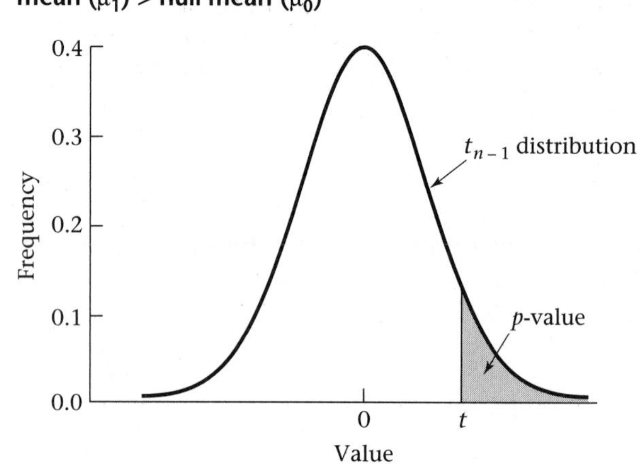

standard deviation is 50 mg/dL. Test the hypothesis that the mean cholesterol level is higher in this group than in the general population.

Solution The hypothesis

$$H_0: \mu = 175 \quad \text{vs.} \quad H_1: \mu > 175$$

is tested using an α level of .05. H_0 is rejected if

$$t > t_{n-1,1-\alpha} = t_{9,.95}$$

In this case,

$$t = \frac{200 - 175}{50/\sqrt{10}} = \frac{25}{15.81} = 1.58$$

From Table 5, we see that $t_{9,.95} = 1.833$. Because $1.833 > 1.58$, it follows that we accept H_0 at the 5% level of significance.

If we use the *p*-value method, the exact *p*-value is given by

$$p = Pr(t_9 > 1.58)$$

Using Table 5, we find $t_{9,.90} = 1.383$ and $t_{9,.95} = 1.833$. Thus, because $1.383 < 1.58 < 1.833$, it follows that $.05 < p < .10$. Alternatively, using the TDIST function of Excel, we can get the exact *p*-value = $Pr(t_9 > 1.58) = .074$. Because $p > .05$, we conclude that our results are not statistically significant, and the null hypothesis is accepted. Thus, the mean cholesterol level of these children is not significantly different from that of an average child.

SECTION 7.4 One-Sample Test for the Mean of a Normal Distribution: Two-Sided Alternatives

In the previous section the alternative hypothesis was assumed to be in a *specific direction* relative to the null hypothesis.

Example 7.19 **Obstetrics** In Example 7.2, it was assumed that the mean birthweight of infants from a low-SES-area hospital was either the same or lower than average. In Example 7.1, it was assumed that the mean cholesterol level of children of men who died from heart disease was either the same or higher than average.

In most instances this *prior knowledge* is unavailable. If the null hypothesis is not true, then we have no idea in which direction the alternative mean will fall.

Example 7.20 **Cardiovascular Disease** Suppose we want to compare fasting serum-cholesterol levels among recent Asian immigrants to the United States with typical levels found in the general population in the United States. Suppose we assume that cholesterol levels in women aged 21–40 in the United States are approximately normally distributed with mean 190 mg/dL. It is unknown whether cholesterol levels among recent Asian immigrants are higher or lower than those in the general U.S. population. We will assume that levels among recent female Asian immigrants are normally distributed with unknown mean μ. Hence, we wish to test the null hypothesis H_0: $\mu = \mu_0 = 190$ versus the alternative hypothesis H_1: $\mu \neq \mu_0$. Blood tests are performed on 100 female Asian immigrants aged 21–40 and the mean level ($\bar{x}$) is found to be 181.52 mg/dL with standard deviation = 40 mg/dL. What can be concluded on the basis of this evidence?

The type of alternative given in Example 7.20 is known as a *two-sided* alternative, because the alternative mean can be either less than or greater than the null mean.

DEFINITION 7.15 A **two-tailed test** is a test in which the values of the parameter being studied (in this case μ) under the alternative hypothesis are allowed to be either *greater than or less than* the values of the parameter under the null hypothesis (μ_0).

The best test here depends on the sample mean $\bar{x}$ or, equivalently, on the test statistic *t*, as it did in the one-sided situation developed in Section 7.3. We showed in Equation 7.2 that to test the hypothesis H_0: $\mu = \mu_0$ versus H_1: $\mu < \mu_0$, the best test was of the form: reject H_0 if $t < t_{n-1,\alpha}$ and accept H_0 if $t \geq t_{n-1,\alpha}$. This test is clearly only appropriate for alternatives on one side of the null mean, namely, $\mu < \mu_0$. We also showed in Equation 7.6 that to test the hypothesis

$$H_0\text{: } \mu = \mu_0 \quad \text{vs.} \quad H_1\text{: } \mu > \mu_0$$

the best test was correspondingly of the form: reject H_0 if $t > t_{n-1,1-\alpha}$ and accept H_0 if $t \leq t_{n-1,1-\alpha}$.

EQUATION 7.7

A reasonable decision rule to test for alternatives on *either* side of the null mean is to *reject H_0 if t is either too small or too large*. Another way of stating the rule is that H_0 will be rejected if t is either $<c_1$ or $>c_2$ for some constants c_1, c_2 and H_0 will be accepted if $c_1 \le t \le c_2$.

The question remains, What are appropriate values for c_1 and c_2? These values are again determined by the type I error (α). The constants c_1, c_2 should be chosen such that

EQUATION 7.8

$$\begin{aligned}Pr(\text{reject } H_0|H_0 \text{ true}) &= Pr(t < c_1 \text{ or } t > c_2|H_0 \text{ true})\\ &= Pr(t < c_1|H_0 \text{ true}) + Pr(t > c_2|H_0 \text{ true}) = \alpha\end{aligned}$$

Half of the type I error is arbitrarily assigned to each of the probabilities on the left side of the second line of Equation 7.8. Thus, we wish to find c_1, c_2 such that

EQUATION 7.9

$$Pr(t < c_1|H_0 \text{ true}) = Pr(t > c_2|H_0 \text{ true}) = \alpha/2$$

We know that t follows a t_{n-1} distribution under H_0. Because $t_{n-1,\alpha/2}$ and $t_{n-1,1-\alpha/2}$ are the lower and upper $100\% \times \alpha/2$ percentiles of a t_{n-1} distribution, it follows that

$$Pr(t < t_{n-1,\alpha/2}) = Pr(t > t_{n-1,1-\alpha/2}) = \alpha/2$$

Therefore,

$$c_1 = t_{n-1,\alpha/2} \quad \text{and} \quad c_2 = t_{n-1,1-\alpha/2}$$

This test procedure can be summarized as follows:

EQUATION 7.10

One-Sample t Test for the Mean of a Normal Distribution (Two-Sided Alternative) To test the hypothesis H_0: $\mu = \mu_0$ versus H_1: $\mu \ne \mu_0$, with a significance level of α, the best test is based on $t = (\bar{x} - \mu_0)/(s/\sqrt{n})$. If

$$|t| > t_{n-1,1-\alpha/2}$$

then H_0 is rejected. If

$$|t| \le t_{n-1,1-\alpha/2}$$

then H_0 is accepted.

The acceptance and rejection regions for this test are depicted in Figure 7.3.

Example 7.21 **Cardiovascular Disease** Test the hypothesis that the mean cholesterol level of recent Asian immigrants is different from the mean in the general U.S. population, using the data in Example 7.20.

FIGURE 7.3 **One-sample t test for the mean of a normal distribution (two-sided alternative)**

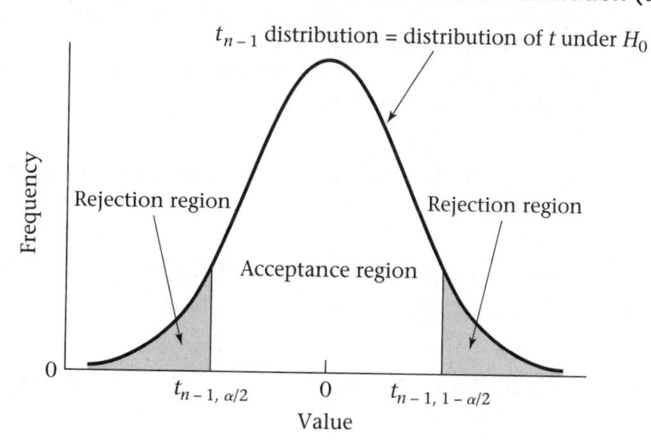

Solution We compute the test statistic

$$t = \frac{\bar{x} - \mu_0}{s/\sqrt{n}} = \frac{181.52 - 190}{40/\sqrt{100}} = \frac{-8.48}{4} = -2.12$$

For a two-sided test with $\alpha = .05$, the critical values are $c_1 = t_{99,.025}$, $c_2 = t_{99,.975}$.

From Table 5 in the Appendix, because $t_{99,.975} < t_{60,.975} = 2.000$, it follows that $c_2 < 2.000$. Also, because $c_1 = -c_2$ it follows that $c_1 > -2.000$. Because $t = -2.12 < -2.000 < c_1$, it follows that we can reject H_0 at the 5% level of significance. We conclude that the mean cholesterol level of recent Asian immigrants is significantly lower than the mean for the general U.S. population.

Alternatively, we might want to compute a p-value as we did in the one-sided case. The p-value is computed in two different ways, depending on whether t is less than or greater than 0.

EQUATION 7.11

p-Value for the One-Sample t Test for the Mean of a Normal Distribution (Two-Sided Alternative)

Let $t = \dfrac{\bar{x} - \mu_0}{s/\sqrt{n}}$

$$p = \begin{cases} 2 \times Pr(t_{n-1} \le t), & \text{if } t \le 0 \\ 2 \times \left[1 - Pr(t_{n-1} \le t)\right], & \text{if } t > 0 \end{cases}$$

Thus, in words, if $t \le 0$, then p = 2 times the area under a t_{n-1} distribution to the left of t; if $t > 0$, then p = 2 times the area under a t_{n-1} distribution to the right of t. One way to interpret the p-value is as follows:

EQUATION 7.12

> The **p-value** is the probability under the null hypothesis of obtaining a test statistic as extreme as or more extreme than the observed test statistic, where, because a two-sided alternative hypothesis is being used, extremeness is measured by the **absolute value** of the test statistic.

Hence if $t > 0$, the p-value is the area to the right of t plus the area to the left of $-t$ under a t_{n-1} distribution.

However, this area simply amounts to twice the right-hand tail area, because the t distribution is symmetric about 0. A similar interpretation holds if $t < 0$.

These areas are illustrated in Figure 7.4.

FIGURE 7.4 Illustration of the *p*-value for a one-sample *t* test for the mean of a normal distribution (two-sided alternative)

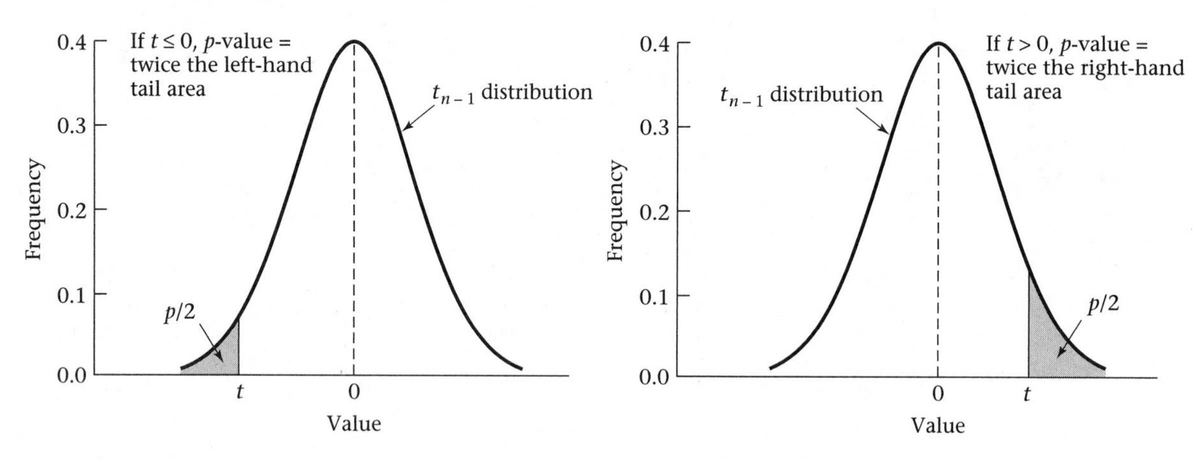

Example 7.22 **Cardiovascular Disease** Compute the p-value for the hypothesis test in Example 7.20.

Solution Because $t = -2.12$, the p-value for the test would be twice the left-hand tail area, or

$$p = 2 \times Pr(t_{99} < -2.12) = .037$$

Hence, the results are statistically significant with a p-value of .037.

Finally, if n is large (say, >200), then the percentiles of the t distribution $(t_{n-1,\,1-\alpha/2})$ used in determining the critical values in Equation 7.10 can be approximated by the corresponding percentiles of an $N(0, 1)$ distribution $(z_{1-\alpha/2})$. Similarly, in computing p-values in Equation 7.11, if $n > 200$, then $Pr(t_{n-1} \le t)$ can be approximated by $Pr[N(0, 1) < t] = \Phi(t)$. We used similar approximations in our work on confidence intervals for the mean of a normal distribution with unknown variance in Section 6.5.5.

When is a one-sided test more appropriate than a two-sided test? Generally, the sample mean falls in the expected direction from μ_0 and it is *easier* to reject H_0 using a one-sided test than using a two-sided test. However, this is not necessarily always the case. Suppose we guess from a previous review of the literature that the cholesterol level of Asian immigrants is likely to be lower than that of the general U.S. population due to better dietary habits. In this case, we would use a one-sided test of the form: H_0: $\mu = 190$ versus H_1: $\mu < 190$. From Equation 7.3, the one-sided p-value $= Pr(t_{99} < -2.12) = .018 = \frac{1}{2}$(two-sided p-value). Alternatively, suppose we guess from a previous literature review that the cholesterol level of Asian immigrants is likely to be higher than that of the general U.S. population due to more stressful living conditions. In this case, we would use a one-sided test of the form: H_0: $\mu = 190$ versus H_1: $\mu > 190$. From Equation 7.6, the p-value $= Pr\left(t_{99} > -2.12\right) =$ .982. Thus, we would accept H_0 if we use a one-sided test and the sample mean is on the opposite side of the null mean from the alternative hypothesis. Generally, a two-sided test is always appropriate, because there can be no question about the conclusions. Also, as illustrated above, a two-sided test can be more conservative, since one does not have to guess the appropriate side of the null hypothesis for the alternative hypothesis. However, in certain situations only alternatives on one side of the null mean are of interest or are possible, and in this case a one-sided test is better because it has more power (i.e., it will be easier to reject H_0 based on a finite sample if H_1 is actually true) than its two-sided counterpart. In all instances, it is important to decide whether to use a one-sided or a two-sided test *before* data analysis (or preferably before data collection) begins so as not to bias conclusions based on results of hypothesis testing. In particular, you should not change from a two-sided to a one-sided test *after* looking at the data.

Example 7.23 **Hypertension** Suppose we are testing the efficacy of a drug to reduce blood pressure. We will assume that the change in blood pressure (baseline blood pressure minus follow-up blood pressure) is normally distributed with mean μ and variance σ^2. An appropriate hypothesis test might be H_0: $\mu = 0$ versus H_1: $\mu > 0$, because the drug is of interest only if it reduces the level of blood pressure, not if it raises it.

In the remainder of this text, we focus primarily on two-sided tests, due to their much wider use in the literature.

In this section and in Section 7.3, we have presented the **one-sample t test,** which is used for testing hypotheses concerning the mean of a normal distribution when the variance is unknown. This test is featured in the flowchart in Figure 7.18 (p. 262) where we display techniques for determining appropriate methods of statistical inference. Beginning at the Start box, we arrive at the one-sample t test box by answering yes to each of the following: (1) one variable of interest? (2) one-sample problem? (3) underlying distribution normal or can central-limit theorem be assumed to hold? and (4) inference concerning μ? and no to (5) σ known?

7.4.1 One-Sample *z*-Test

In Equations 7.10 and 7.11, the critical values and p-values for the one-sample t test have been specified in terms of percentiles of the t distribution, assuming that the underlying variance is unknown. In some applications, the variance may be assumed known from prior studies. In this case, the test statistic t can be replaced by the test statistic $z = (\bar{x} - \mu_0)/(\sigma/\sqrt{n})$. Also, the critical values based on the t distribution can be replaced by the corresponding critical values of a standard normal distribution. This leads to the following test procedure:

EQUATION 7.13

One-Sample z Test for the Mean of a Normal Distribution with Known Variance (Two-Sided Alternative) To test the hypothesis H_0: $\mu = \mu_0$ vs. H_1: $\mu \neq \mu_0$ with a significance level of α, where the underlying standard deviation σ is known, the best test is based on $z = (\bar{x} - \mu_0)/(\sigma/\sqrt{n})$. If

$$z < z_{\alpha/2} \quad \text{or} \quad z > z_{1-\alpha/2}$$

then H_0 is rejected. If

$$z_{\alpha/2} \le z \le z_{1-\alpha/2}$$

then H_0 is accepted.

To compute a two-sided p-value, we have

$$\begin{aligned} p &= 2\Phi(z) \quad &\text{if } z \le 0 \\ &= 2[1 - \Phi(z)] \quad &\text{if } z > 0 \end{aligned}$$

Example 7.24 **Cardiovascular Disease** Consider the cholesterol data in Example 7.21. Assume that the standard deviation is known to be 40 and the sample size is 200 instead of 100. Assess the significance of the results.

Solution The test statistic is

$$\begin{aligned} z &= \frac{181.52 - 190}{40/\sqrt{200}} \\ &= \frac{-8.48}{2.828} = -3.00 \end{aligned}$$

We first use the critical-value method with $\alpha = 0.05$. Based on Equation 7.13, the critical values are -1.96 and 1.96. Because $z = -3.00 < -1.96$ we can reject H_0 at a 5% level of significance. The two-sided p-value is given by $2 \times \Phi(-3.00) = .003$.

Similarly, we can consider the one-sample z test for a one-sided alternative as follows:

EQUATION 7.14

One-Sample z Test for the Mean of a Normal Distribution with Known Variance (One-Sided Alternative) ($\mu_1 < \mu_0$) To test the hypothesis H_0: $\mu = \mu_0$ versus H_1: $\mu < \mu_0$ with a significance level of α, where the underlying standard deviation σ is known, the best test is based on

$$z = (\bar{x} - \mu_0)/(\sigma/\sqrt{n})$$

If $z < z_\alpha$, then H_0 is rejected; if $z \ge z_\alpha$, then H_0 is accepted. The p-value is given by $p = \Phi(z)$.

EQUATION 7.15

One-Sample z Test for the Mean of a Normal Distribution with Known Variance (One-Sided Alternative) ($\mu_1 > \mu_0$) To test the hypothesis H_0: $\mu = \mu_0$ versus H_1: $\mu > \mu_0$ with a significance level of α, where the underlying standard deviation σ is known:

If $z > z_{1-\alpha}$, then H_0 is rejected. If $z \le z_{1-\alpha}$, then H_0 is accepted. The p-value is given by $p = 1 - \Phi(z)$.

In Section 7.4.1 we presented the **one-sample z test**, which is used for testing hypotheses concerning the mean of a normal distribution when the variance is known. Beginning at the Start box of the flowchart (Figure 7.18, p. 262), we arrive at the one-sample z test box by answering yes to each of the following: (1) one variable of interest? (2) one-sample problem? (3) underlying distribution normal, or can central-limit theorem be assumed to hold? (4) inference concerning μ? and (5) σ known?

SECTION 7.5 The Power of a Test

The calculation of power is used to plan a study, usually before any data have been obtained, except possibly from a small preliminary study called a pilot study. Also, we usually make a projection concerning the standard deviation without actually having any data to estimate it. Therefore, we will assume that the standard deviation is known, and base power calculations on the one-sample z test as given in Equations 7.13 and 7.14.

7.5.1 One-Sided Alternatives

Example 7.25 **Ophthalmology** A new drug is proposed for people with high intraocular pressure (IOP), to prevent the development of glaucoma. A pilot study is conducted with the drug among 10 patients and their mean IOP decreased by 5 mm Hg with a *sd* of 10 mm Hg after 1 month of using the drug. The investigators propose to study 100 subjects in the main study. Is this a sufficient sample size for the study?

Solution To determine whether 100 subjects are enough, we need to do a power calculation. The power of the study is the probability that we will be able to declare a significant difference with a sample of size 100 if the true mean decline in IOP is 5 mm Hg with a *sd* of 10 mm Hg. Usually we want a power of at least 80% to perform a study. In this section we will learn about formulas for computing power and addressing the question just asked.

In Section 7.4 (Equation 7.14) the appropriate hypothesis test was derived to test

$$H_0\text{: } \mu = \mu_0 \text{ vs. } H_1\text{: } \mu = \mu_1 < \mu_0$$

where the underlying distribution was assumed to be normal and the population variance was assumed to be known. The best test was based on the test statistic z. In particular, from Equation 7.14 for a type I error of α, H_0 is rejected if $z < z_\alpha$ and H_0 is accepted if $z \ge z_\alpha$. The form of the best test *does not depend on the alternative mean chosen* (μ_1) as long as this mean is less than the null mean μ_0.

Hence, in Example 7.2, where $\mu_0 = 120$ oz, if we were interested in an alternative mean of $\mu_1 = 115$ oz rather than $\mu_1 = 110$ oz, then the same test procedure would still be used. However, what differs for the two alternative means is the power of the test = 1 − *Pr*(type II error). Recall from Definition 7.6 that

$$\begin{aligned}\text{Power } &= Pr\left(\text{reject } H_0 \middle| H_0 \text{ false}\right) = Pr\left(Z < z_\alpha \middle| \mu = \mu_1\right)\\ &= Pr\left[\frac{\overline{X}-\mu_0}{\sigma/\sqrt{n}} < z_\alpha \middle| \mu = \mu_1\right]\\ &= Pr\left[\overline{X} < \mu_0 + z_\alpha\,\sigma/\sqrt{n} \middle| \mu = \mu_1\right]\end{aligned}$$

We know that under $H_1, \overline{X} \sim N\left(\mu_1, \sigma^2/n\right)$. Hence, upon standardization of limits,

$$\text{Power } = \Phi\left[\left(\mu_0 + z_\alpha\,\sigma/\sqrt{n} - \mu_1\right)/\left(\sigma/\sqrt{n}\right)\right] = \Phi\left[z_\alpha + \frac{\left(\mu_0-\mu_1\right)}{\sigma}\sqrt{n}\right]$$

This power is depicted graphically in Figure 7.5.

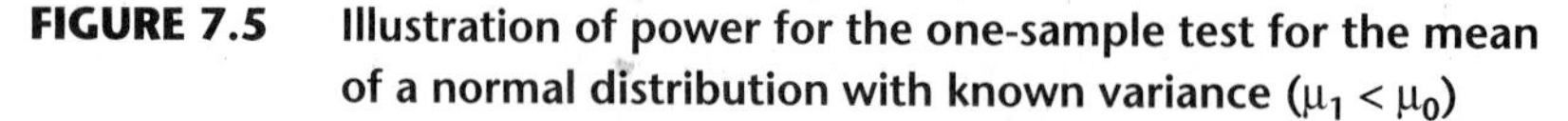

FIGURE 7.5 **Illustration of power for the one-sample test for the mean of a normal distribution with known variance ($\mu_1 < \mu_0$)**

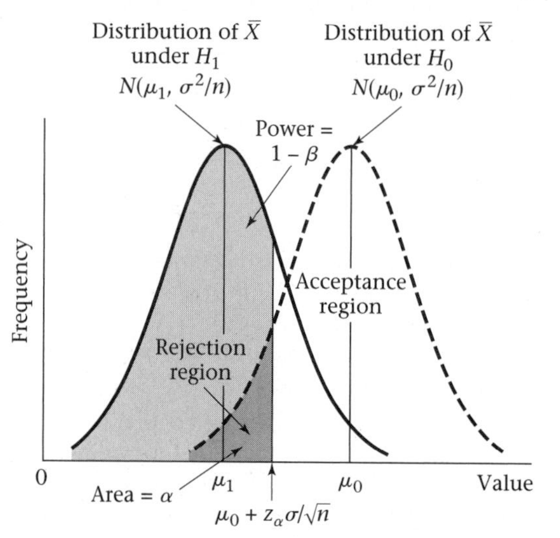

Note that the area to the left of $\mu_0 + z_\alpha\,\sigma/\sqrt{n}$ under the H_0 distribution is the significance level α, whereas the area to the left of $\mu_0 + z_\alpha\,\sigma/\sqrt{n}$ under the H_1 distribution is the power = $1 - \beta$.

Why should power concern us? The power of a test tells us how likely it is that a significant difference will be found given that the alternative hypothesis is true, that is, given that the true mean μ is different from the mean under the null hypothesis (μ_0). If the power is too low, then there is little chance of finding a significant difference and nonsignificant results are likely even if real differences exist between the true mean μ of the group being studied and the null mean μ_0. An inadequate sample size is usually the cause of low power to detect a scientifically meaningful difference.

Example 7.26 **Obstetrics** Compute the power of the test for the birthweight data in Example 7.2 (p. 212) with an alternative mean of 115 oz and $\alpha = .05$, assuming that the true standard deviation = 24 oz.

Solution We have $\mu_0 = 120$ oz, $\mu_1 = 115$ oz, $\alpha = .05$, $\sigma = 24$, $n = 100$. Thus,

$$\text{Power} = \Phi\left[z_{.05} + (120 - 115)\sqrt{100}/24\right] = \Phi\left[-1.645 + 5(10)/24\right] = \Phi(.438) = .669$$

Therefore, there is about a 67% chance of detecting a significant difference using a 5% significance level with this sample size.

We have focused on the situation where $\mu_1 < \mu_0$. We are also interested in power when testing the hypothesis

$$H_0\colon \mu = \mu_0 \quad \text{vs.} \quad H_1\colon \mu = \mu_1 > \mu_0$$

as was the case with the cholesterol data in Example 7.1. The best test for this situation was presented in Equation 7.15, where H_0 is rejected if $z > z_{1-\alpha}$ and accepted if $z \le z_{1-\alpha}$. Notice that if $z > z_{1-\alpha}$, then

EQUATION 7.16

$$\frac{\bar{x} - \mu_0}{\sigma/\sqrt{n}} > z_{1-\alpha}$$

If we multiply both sides of Equation 7.16 by $\sigma/\sqrt{n}$, and add μ_0, we can re-express the rejection criteria in terms of $\bar{x}$, as follows:

EQUATION 7.17

$$\bar{x} > \mu_0 + z_{1-\alpha}\,\sigma/\sqrt{n}$$

Similarly, the acceptance criteria, $z \le z_{1-\alpha}$, can also be expressed as

EQUATION 7.18

$$\bar{x} \le \mu_0 + z_{1-\alpha}\,\sigma/\sqrt{n}$$

Hence the power of the test is given by

$$\begin{aligned}\text{Power} &= Pr\left(\bar{x} > \mu_0 + z_{1-\alpha}\,\sigma/\sqrt{n} \mid \mu = \mu_1\right) = 1 - Pr\left(\bar{x} < \mu_0 + z_{1-\alpha}\,\sigma/\sqrt{n} \mid \mu = \mu_1\right) \\ &= 1 - \Phi\left(\frac{\mu_0 + z_{1-\alpha}\,\sigma/\sqrt{n} - \mu_1}{\sigma/\sqrt{n}}\right) = 1 - \Phi\left[z_{1-\alpha} + \frac{(\mu_0 - \mu_1)\sqrt{n}}{\sigma}\right]\end{aligned}$$

Using the relationships $\Phi(-x) = 1 - \Phi(x)$ and $z_\alpha = -z_{1-\alpha}$, this expression can be rewritten as

$$\Phi\left[-z_{1-\alpha} + \frac{(\mu_1 - \mu_0)\sqrt{n}}{\sigma}\right] = \Phi\left[z_\alpha + \frac{(\mu_1 - \mu_0)\sqrt{n}}{\sigma}\right] \quad \text{if} \quad \mu_1 > \mu_0$$

This power is depicted graphically in Figure 7.6.

FIGURE 7.6 **Illustration of power for the one-sample test for the mean of a normal distribution with known variance ($\mu_1 > \mu_0$)**

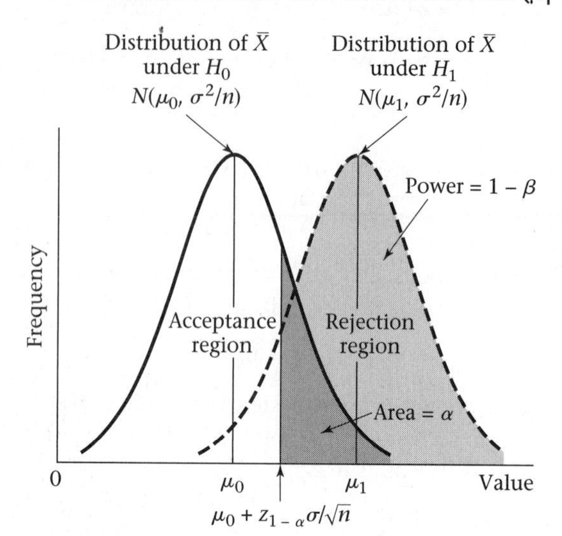

Example 7.27 **Cardiovascular Disease, Pediatrics** Using a 5% level of significance and a sample of size 10, compute the power of the test for the cholesterol data in Example 7.18, with an alternative mean of 190 mg/dL, a null mean of 175 mg/dL and a standard deviation (σ) of 50 mg/dL.

Solution We have $\mu_0 = 175,\ \mu_1 = 190,\ \alpha = .05,\ \sigma = 50,\ n = 10$. Thus,

$$\begin{aligned}\text{Power} &= \Phi\left[-1.645 + (190 - 175)\sqrt{10}/50\right]\\ &= \Phi\left[-1.645 + 15\sqrt{10}/50\right] = \Phi(-0.70)\\ &= 1 - \Phi(0.70) = 1 - .757 = .243\end{aligned}$$

Therefore, the chance of finding a significant difference in this case is only 24%. Thus, it is not surprising that a significant difference was not found in Example 7.18, because the sample size was too small.

The power formulas presented in this section can be summarized as follows:

EQUATION 7.19

Power for the One-Sample *z* Test for the Mean of a Normal Distribution with Known Variance (One-Sided Alternative) The power of the test for the hypothesis

$$H_0: \mu = \mu_0 \quad \text{vs.} \quad H_1: \mu = \mu_1$$

where the underlying distribution is normal and the population variance (σ^2) is assumed known is given by

$$\Phi\left(z_\alpha + |\mu_0 - \mu_1|\sqrt{n}/\sigma\right) = \Phi\left(-z_{1-\alpha} + |\mu_0 - \mu_1|\sqrt{n}/\sigma\right)$$

Notice from Equation 7.19 that the power depends on four factors: α, $|\mu_0 - \mu_1|$, n, and σ.

EQUATION 7.20

Factors Affecting the Power

(1) If the significance level is made smaller (α decreases), z_α decreases and hence the power decreases.

(2) If the alternative mean is shifted further away from the null mean ($|\mu_0 - \mu_1|$ increases), then the power increases.

(3) If the standard deviation of the distribution of individual observations increases (σ increases), then the power decreases.

(4) If the sample size increases (n increases), then the power increases.

Example 7.28 **Cardiovascular Disease, Pediatrics** Compute the power of the test for the cholesterol data in Example 7.27 with a significance level of .01 versus an alternative mean of 190 mg/dL.

Solution If $\alpha = .01$, then the power is given by

$$\Phi\left[z_{.01} + (190 - 175)\sqrt{10}/50\right] = \Phi\left(-2.326 + 15\sqrt{10}/50\right)$$
$$= \Phi(-1.38) = 1 - \Phi(1.38) = 1 - .9158 \approx 8\%$$

which is lower than the power of 24% for $\alpha = .05$, as computed in Example 7.27. What does this mean? It means that if the α level is lowered from .05 to .01, the β error will be higher or, equivalently, the power, which decreases from .24 to .08, will be lower.

Example 7.29 **Obstetrics** Compute the power of the test for the birthweight data in Example 7.26 with $\mu_1 = 110$ oz rather than 115 oz.

Solution If $\mu_1 = 110$ oz, then the power is given by

$$\Phi\left[-1.645 + (120 - 110)10/24\right] = \Phi(2.522) = .994 \approx 99\%$$

which is higher than the power of 67%, as computed in Example 7.26 for $\mu_1 = 115$ oz. What does this mean? It means that if the alternative mean changes from 115 oz to 110 oz, then the chance of finding a significant difference increases from 67% to 99%.

Example 7.30 **Cardiology** Compute the power of the test for the infarct-size data in Example 7.13 with $\sigma = 10$ and $\sigma = 15$ using an alternative mean of 20 ($ck - g - EQ/m^2$) and $\alpha = .05$.

Solution In Example 7.13, $\mu_0 = 25$ and $n = 8$. Thus, if $\sigma = 10$, then

$$\text{Power} = \Phi\left[-1.645 + (25 - 20)\sqrt{8}/10\right] = \Phi(-0.23)$$
$$= 1 - \Phi(0.23) = 1 - .591 = .409 \approx 41\%$$

whereas if $\sigma = 15$, then

$$\text{Power} = \Phi\left[-1.645 + (25 - 20)\sqrt{8}/15\right] = \Phi(-0.70)$$
$$= 1 - .759 = .241 \approx 24\%$$

What does this mean? It means that the chance of finding a significant difference declines from 41% to 24% if σ increases from 10 to 15.

Example 7.31 **Obstetrics** Assuming a sample size of 10 rather than 100, compute the power for the birthweight data in Example 7.26 with an alternative mean of 115 oz and $\alpha = .05$.

Solution We have $\mu_0 = 120$ oz, $\mu_1 = 115$ oz, $\alpha = .05$, $\sigma = 24$, and $n = 10$. Thus,

$$\text{Power} = \Phi\left[z_{.05} + (120 - 115)\sqrt{10}/24\right] = \Phi\left(-1.645 + 5\sqrt{10}/24\right)$$
$$= \Phi(-0.99) = 1 - .838 = .162$$

What does this mean? It means there is only a 16% chance of finding a significant difference with a sample size of 10, whereas there was a 67% chance with a sample size of 100 (see Example 7.26). These results imply that if 10 infants were sampled, we would have virtually no chance of finding a significant difference and would almost surely report a false-negative result.

For given levels of α (.05), σ (24 oz), n (100), and μ_0 (120 oz), a **power curve** can be drawn for the power of the test for various alternatives μ_1. Such a power curve is shown in Figure 7.7 for the birthweight data in Example 7.2. The power ranges from 99% for $\mu = 110$ oz to about 20% when $\mu = 118$ oz.

FIGURE 7.7 **Power curve for the birthweight data in Example 7.2**

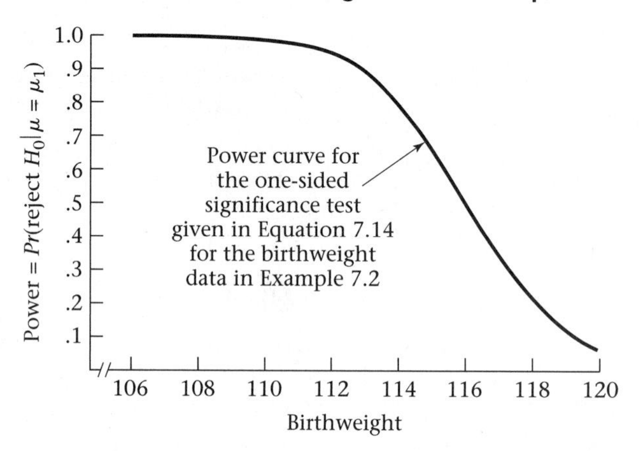

7.5.2 Two-Sided Alternatives

The power formula given in Equation 7.19 is appropriate for a one-sided significance test at level α for the mean of a normal distribution with known variance. Using a two-sided test with hypotheses H_0: $\mu = \mu_0$ versus H_1: $\mu \neq \mu_0$, the following power formula is used:

EQUATION 7.21

Power for the One-Sample z Test for the Mean of a Normal Distribution (Two-Sided Alternative) The power of the two-sided test H_0: $\mu = \mu_0$ versus H_1: $\mu \neq \mu_0$ for the specific alternative $\mu = \mu_1$, where the underlying distribution is normal and the population variance (σ^2) is assumed known, is given exactly by

$$\Phi\left[-z_{1-\alpha/2} + \frac{(\mu_0 - \mu_1)\sqrt{n}}{\sigma}\right] + \Phi\left[-z_{1-\alpha/2} + \frac{(\mu_1 - \mu_0)\sqrt{n}}{\sigma}\right]$$

and approximately by

$$\Phi\left[-z_{1-\alpha/2}+\frac{|\mu_0-\mu_1|\sqrt{n}}{\sigma}\right]$$

To see this, note that from Equation 7.13 we reject H_0 if

$$z=\frac{\bar{x}-\mu_0}{\sigma/\sqrt{n}}<z_{\alpha/2} \quad \text{or} \quad z=\frac{\bar{x}-\mu_0}{\sigma/\sqrt{n}}>z_{1-\alpha/2}$$

If we multiply each inequality by $\sigma/\sqrt{n}$ and add μ_0, we can re-express the rejection criteria in terms of $\bar{x}$, as follows:

EQUATION 7.22

$$\bar{x}<\mu_0+z_{\alpha/2}\,\sigma/\sqrt{n} \quad \text{or} \quad \bar{x}>\mu_0+z_{1-\alpha/2}\,\sigma/\sqrt{n}$$

The power of the test versus the specific alternative $\mu=\mu_1$ is given by

EQUATION 7.23

$$\begin{aligned}\text{Power} &= Pr\left(\bar{X}<\mu_0+z_{\alpha/2}\,\sigma/\sqrt{n}\,|\mu=\mu_1\right)+Pr\left(\bar{X}>\mu_0+z_{1-\alpha/2}\,\sigma/\sqrt{n}\,|\mu=\mu_1\right)\\ &=\Phi\left(\frac{\mu_0+z_{\alpha/2}\,\sigma/\sqrt{n}-\mu_1}{\sigma/\sqrt{n}}\right)+1-\Phi\left(\frac{\mu_0+z_{1-\alpha/2}\,\sigma/\sqrt{n}-\mu_1}{\sigma/\sqrt{n}}\right)\\ &=\Phi\left[z_{\alpha/2}+\frac{(\mu_0-\mu_1)\sqrt{n}}{\sigma}\right]+1-\Phi\left[z_{1-\alpha/2}+\frac{(\mu_0-\mu_1)\sqrt{n}}{\sigma}\right]\end{aligned}$$

Using the relationship $1-\Phi(x)=\Phi(-x)$, the last two terms can be combined as follows:

EQUATION 7.24

$$\text{Power}=\Phi\left[z_{\alpha/2}+\frac{(\mu_0-\mu_1)\sqrt{n}}{\sigma}\right]+\Phi\left[-z_{1-\alpha/2}+\frac{(\mu_1-\mu_0)\sqrt{n}}{\sigma}\right]$$

Finally, recalling the relationship $z_{\alpha/2}=-z_{1-\alpha/2}$, we have

EQUATION 7.25

$$\text{Power}=\Phi\left[-z_{1-\alpha/2}+\frac{(\mu_0-\mu_1)\sqrt{n}}{\sigma}\right]+\Phi\left[-z_{1-\alpha/2}+\frac{(\mu_1-\mu_0)\sqrt{n}}{\sigma}\right]$$

Equation 7.25 is more tedious to use than is usually necessary. Specifically, if $\mu_1<\mu_0$, then the second term is usually negligible relative to the first term. On the other hand, if $\mu_1>\mu_0$, then the first term is usually negligible relative to the second term. Therefore, the approximate power formula in Equation 7.21 is usually used for a two-sided test, since it represents the first term in Equation 7.25 if $\mu_0>\mu_1$ and the second term in Equation 7.25 if $\mu_1>\mu_0$. The power is depicted

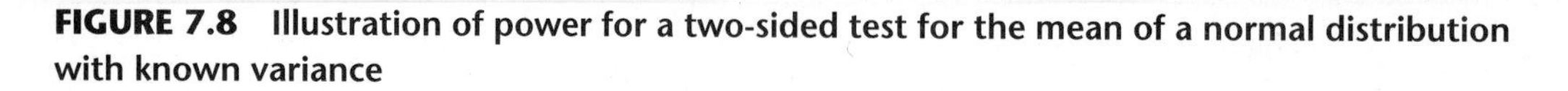

FIGURE 7.8 Illustration of power for a two-sided test for the mean of a normal distribution with known variance

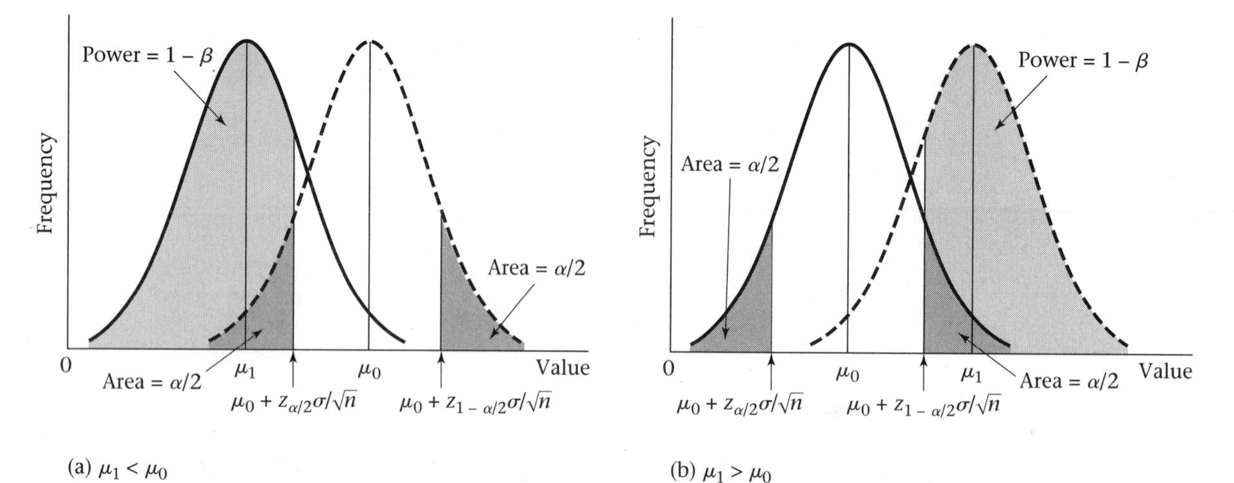

graphically in Figure 7.8. Note that the approximate power formula for the two-sided test in Equation 7.21 is the same as the formula for the one-sided test in Equation 7.19, with α replaced by α/2.

Example 7.32 **Cardiology** A new drug in the class of calcium channel blockers is to be tested for the treatment of patients with unstable angina, a severe type of angina. The effect this drug will have on heart rate is unknown. Suppose that 20 patients are to be studied and the change in heart rate after 48 hours is known to have a standard deviation of 10 beats per minute. What power would such a study have of detecting a significant difference in heart rate over 48 hours if it is hypothesized that the true mean change in heart rate from baseline to 48 hours could be either a mean increase or a decrease of 5 beats per minute?

Solution Use Equation 7.21 with $\sigma = 10$, $|\mu_0 - \mu_1| = 5$, $\alpha = .05$, $n = 20$. We have

$$\text{Power} = \Phi\left(-z_{1-.05/2} + 5\sqrt{20}/10\right) = \Phi(-1.96 + 2.236) = \Phi(0.276) = .609 \approx .61$$

Thus, the study would have a 61% chance of detecting a significant difference.

SECTION 7.6 Sample-Size Determination

7.6.1 One-Sided Alternatives

Frequently, for planning purposes, we need to have some idea of an appropriate sample size for investigation before a study actually begins. One possible result of making these calculations is finding out that the appropriate sample size is far beyond the financial means of the investigator(s) and thus abandoning the proposed investigation. Obviously, reaching this conclusion before a study starts is far better than after it is in progress.

What does "an appropriate sample size for investigation" actually mean? Consider the birthweight data in Example 7.2. We are testing the null hypothesis H_0: $\mu = \mu_0$ versus the alternative hypothesis H_1: $\mu = \mu_1$, assuming that the distribution of birthweights is normal in both cases and that the standard deviation σ is known. We are presumably going to conduct a test with significance level α and have some idea of what the magnitude of the alternative mean μ_1 is likely to be. If the test procedure in Equation 7.14 is used, then H_0 would be rejected if $z < z_\alpha$ or equivalently if $\bar{x} < \mu_0 + z_\alpha \sigma/\sqrt{n}$ and accepted if $z \ge z_\alpha$ or equivalently if $\bar{x} \ge \mu_0 + z_\alpha \sigma/\sqrt{n}$. Suppose the alternative hypothesis is actually true. The investigator should have some idea as to what he or she would like the probability of rejecting H_0 to be in this instance. This probability is, of course, nothing other than the power, or $1 - \beta$. Typical values for the desired power are 80%, 90%, . . . , and so forth. The problem of determining **sample size** can be summarized as follows: Given that a significance test will be conducted at level α and that the true alternative mean is expected to be μ_1, what sample size is needed to be able to detect a significant difference with probability $1 - \beta$? The situation is depicted in Figure 7.9.

FIGURE 7.9 **Requirements for appropriate sample size**

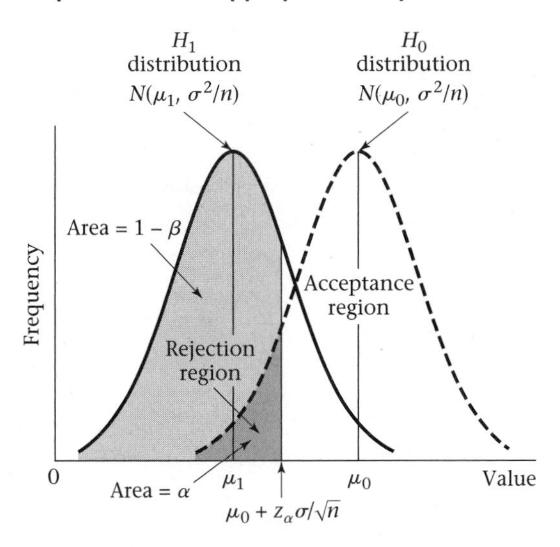

In Figure 7.9, the underlying sampling distribution of $\bar{X}$ is shown under the null and alternative hypotheses, respectively, and the point $\mu_0 + z_\alpha \sigma/\sqrt{n}$ has been identified. H_0 will be rejected if $\bar{x} < \mu_0 + z_\alpha \sigma/\sqrt{n}$. Hence, the area to the left of $\mu_0 + z_\alpha \sigma/\sqrt{n}$ under the rightmost curve is α. However, we also want the area to the left of $\mu_0 + z_\alpha \sigma/\sqrt{n}$ under the leftmost curve, which represents the power, to be $1 - \beta$. These requirements will be met if n is made sufficiently large, because the variance of each curve (σ^2/n) will decrease as n increases and thus the curves will separate. From the power formula in Equation 7.19,

$$\text{Power} = \Phi\left(z_\alpha + |\mu_0 - \mu_1|\sqrt{n}/\sigma\right) = 1 - \beta$$

We wish to solve for n in terms of α, β, $|\mu_0 - \mu_1|$, and σ. To accomplish this, recall that $\Phi(z_{1-\beta}) = 1 - \beta$ and, therefore,

$$z_\alpha + |\mu_0 - \mu_1|\sqrt{n}/\sigma = z_{1-\beta}$$

Subtract z_α from both sides of the equation and multiply by $\sigma/|\mu_0 - \mu_1|$ to obtain

$$\sqrt{n} = \frac{(-z_\alpha + z_{1-\beta})\sigma}{|\mu_0 - \mu_1|}$$

Replace $-z_\alpha$ by $z_{1-\alpha}$ and square both sides of the equation to obtain

$$n = \frac{(z_{1-\alpha} + z_{1-\beta})^2 \sigma^2}{(\mu_0 - \mu_1)^2}$$

Similarly, if we were to test the hypothesis

$$H_0: \mu = \mu_0 \quad \text{vs.} \quad H_1: \mu = \mu_1 > \mu_0$$

as was the case with the cholesterol data in Example 7.1, using a significance level of α and a power of $1 - \beta$, then, from Equation 7.19, the same sample-size formula would hold. This procedure can be summarized as follows:

EQUATION 7.26

Sample-Size Estimation When Testing for the Mean of a Normal Distribution (One-Sided Alternative) Suppose we wish to test

$$H_0: \mu = \mu_0 \quad \text{vs.} \quad H_1: \mu = \mu_1$$

where the data are normally distributed with mean μ and known variance σ^2. The **sample size** needed to conduct a one-sided test with significance level α and probability of detecting a significant difference = $1 - \beta$ is

$$n = \frac{\sigma^2 (z_{1-\beta} + z_{1-\alpha})^2}{(\mu_0 - \mu_1)^2}$$

Example 7.33 **Obstetrics** Consider the birthweight data in Example 7.2. Suppose that $\mu_0 = 120$ oz, $\mu_1 = 115$ oz, $\sigma^2 = 576$, $\alpha = .05$, $1 - \beta = .80$ and we will use a one-sided test. Compute the appropriate sample size needed to conduct the test.

Solution

$$n = \frac{576(z_{.8} + z_{.95})^2}{25} = 23.04(0.84 + 1.645)^2 = 23.04(6.175) = 142.3$$

The sample size is always rounded up so that we can be sure to achieve at least the required level of power (in this case, 80%). Thus, a sample size of 143 is needed to have an 80% chance of detecting a significant difference at the 5% level if the alternative mean is 115 oz and a one-sided test is used.

Notice that the sample size is very sensitive to the alternative mean chosen. We see from Equation 7.26 that the sample size is inversely proportional to $(\mu_0 - \mu_1)^2$. Thus, if the absolute value of the distance between the null and alternative means is halved, then the sample size needed is 4 times as large. Similarly, if the distance between the null and alternative means is doubled, then the sample size needed is $\frac{1}{4}$ as large.

Example 7.34 **Obstetrics** Compute the sample size for the birthweight data in Example 7.2 if μ_1 = 110 oz rather than 115 oz.

Solution The required sample size would be $\frac{1}{4}$ as large, because $(\mu_0 - \mu_1)^2 = 100$ rather than 25. Thus, $n = 35.6$, or 36 people would be needed.

Example 7.35 **Cardiovascular Disease, Pediatrics** Consider the cholesterol data in Example 7.1. Suppose that the null mean is 175 mg/dL, the alternative mean is 190 mg/dL, the standard deviation is 50, and we wish to conduct a one-sided significance test at the 5% level with a power of 90%. How large should the sample size be?

Solution

$$n = \frac{\sigma^2(z_{1-\beta} + z_{1-\alpha})^2}{(\mu_0 - \mu_1)^2} = \frac{50^2(z_{.9} + z_{.95})^2}{(190-175)^2}$$

$$= \frac{2500(1.28 + 1.645)^2}{15^2} = \frac{2500(8.556)}{225} = 95.1$$

Thus, 96 people are needed to achieve a power of 90% using a 5% significance level. We should not be surprised that we did not find a significant difference with a sample size of 10 in Example 7.18.

Clearly, the required sample size is related to the following four quantities:

EQUATION 7.27

Factors Affecting the Sample Size

(1) The sample size increases as σ^2 increases.

(2) The sample size increases as the significance level is made smaller (α decreases).

(3) The sample size increases as the required power increases ($1 - \beta$ increases).

(4) The sample size decreases as the absolute value of the distance between the null and alternative means ($|\mu_0 - \mu_1|$) increases.

Example 7.36 **Obstetrics** What would happen to the sample-size estimate in Example 7.33 if σ were increased to 30? If α were reduced to .01? If the required power were increased to 90%? If the alternative mean were changed to 110 oz (keeping all other parameters the same in each instance)?

Solution From Example 7.33 we see that 143 infants need to be studied to achieve a power of 80% using a 5% significance level with a null mean of 120 oz, an alternative mean of 115 oz, and a standard deviation of 24 oz.

If σ increases to 30, then we need

$$n = 30^2(z_{.8} + z_{.95})^2 / (120-115)^2 = 900(0.84 + 1.645)^2 / 25 = 222.3, \text{ or } 223 \text{ infants}$$

If α were reduced to .01, then we need

$$n = 24^2(z_{.8} + z_{.99})^2 / (120-115)^2 = 576(0.84 + 2.326)^2 / 25 = 230.9, \text{ or } 231 \text{ infants}$$

If $1 - \beta$ were increased to .9, then we need

$$n = 24^2(z_{.9} + z_{.95})^2 / (120-115)^2 = 576(1.28 + 1.645)^2 / 25 = 197.1, \text{ or } 198 \text{ infants}$$

If μ_1 is decreased to 110 or, equivalently, if $|\mu_0 - \mu_1|$ is increased from 5 to 10, then we need

$$n = 24^2 \left(z_{.8} + z_{.95}\right)^2 / (120 - 110)^2 = 576(0.84 + 1.645)^2 / 100 = 35.6, \text{ or } 36 \text{ infants}$$

Thus the required sample size increases if σ increases, α decreases, or $1 - \beta$ increases, respectively. The required sample size decreases if the absolute value of the distance between the null and alternative means increases.

One question that arises is how to estimate the parameters necessary to compute sample size. It usually is easy to specify the magnitude of the null mean (μ_0). Similarly, by convention the type I error (α) is usually set at .05. What the level of the power should be is somewhat less clear, although most investigators seem to feel uncomfortable with a power of less than .80. The appropriate values for μ_1 and σ^2 are usually unknown. The parameters μ_1, σ^2 may be obtained from previous work, similar experiments, or prior knowledge of the underlying distribution. In the absence of such information, the parameter μ_1 is sometimes estimated by assessing what a *scientifically important difference* $|\mu_0 - \mu_1|$ would be in the context of the problem being studied. Conducting a small *pilot study* is sometimes valuable. Such a study is generally inexpensive, and one of its principal aims is to obtain estimates of μ_1 and σ^2 for the purpose of estimating the sample size needed to conduct the major investigation.

Keep in mind that most sample-size estimates are "ballpark estimates" because of the inaccuracy in estimating μ_1 and σ^2. These estimates are often used merely to check that the proposed sample size of a study is close to what is actually needed rather than to identify a precise sample size.

7.6.2 Sample-Size Determination (Two-Sided Alternatives)

The sample-size formula given in Equation 7.26 was appropriate for a one-sided significance test at level α for the mean of a normal distribution with known variance. If it is not known whether the alternative mean (μ_1) is greater or less than the null mean (μ_0), then a two-sided test is appropriate, and the corresponding sample size needed to conduct a study with power $1 - \beta$ is given by

$$n = \frac{\sigma^2 \left(z_{1-\beta} + z_{1-\alpha/2}\right)^2}{(\mu_0 - \mu_1)^2}$$

To see this, use the approximate power formula in Equation 7.21 and solve for n in terms of the other parameters, whereby

$$\Phi\left(-z_{1-\alpha/2} + \frac{|\mu_0 - \mu_1|\sqrt{n}}{\sigma}\right) = 1 - \beta$$

or $$-z_{1-\alpha/2} + \frac{|\mu_0 - \mu_1|\sqrt{n}}{\sigma} = z_{1-\beta}$$

If $z_{1-\alpha/2}$ is added to both sides of the equation and the result is multiplied by $\sigma / |\mu_0 - \mu_1|$, we get

$$\sqrt{n} = \frac{\left(z_{1-\beta} + z_{1-\alpha/2}\right)\sigma}{|\mu_0 - \mu_1|}$$

If both sides of the equation are squared, we get

$$n = \frac{\left(z_{1-\beta} + z_{1-\alpha/2}\right)^2 \sigma^2}{\left(\mu_0 - \mu_1\right)^2}$$

This procedure can be summarized as follows:

EQUATION 7.28

Sample-Size Estimation When Testing for the Mean of a Normal Distribution (Two-Sided Alternative) Suppose we wish to test H_0: $\mu = \mu_0$ versus H_1: $\mu = \mu_1$, where the data are normally distributed with mean μ and known variance σ^2. The **sample size** needed to conduct a two-sided test with significance level α and power $1 - \beta$ is

$$n = \frac{\sigma^2\left(z_{1-\beta} + z_{1-\alpha/2}\right)^2}{\left(\mu_0 - \mu_1\right)^2}$$

Note that this sample size is always larger than the corresponding sample size for a one-sided test, given in Equation 7.26, because $z_{1-\alpha/2}$ is larger than $z_{1-\alpha}$.

Example 7.37 **Cardiology** Consider a study of the effect of a calcium channel blocking agent on heart rate for patients with unstable angina, as described in Example 7.32. Suppose we want to have at least 80% power for detecting a significant difference if the effect of the drug is to change mean heart rate by 5 beats per minute over 48 hours in either direction and $\sigma = 10$ beats per minute. How many patients should be enrolled in such a study?

Solution We assume that $\alpha = .05$ and $\sigma = 10$ beats per minute, as in Example 7.32. We intend to use a two-sided test, since we are not sure in what direction the heart rate will change after using the drug. Therefore, the sample size is estimated using the two-sided formulation in Equation 7.28.

$$\begin{aligned} n &= \frac{\sigma^2\left(z_{1-\beta} + z_{1-\alpha/2}\right)^2}{\left(\mu_0 - \mu_1\right)^2} \\ &= \frac{10^2\left(z_{.8} + z_{.975}\right)^2}{5^2} = \frac{100(0.84 + 1.96)^2}{25} \\ &= 4(7.84) = 31.36, \text{ or } 32 \text{ patients} \end{aligned}$$

Thus, 32 patients need to be studied to have at least an 80% chance of finding a significant difference using a two-sided test with $\alpha = .05$ if the true mean change in heart rate from using the drug is 5 beats per minute. Note that in Example 7.32 the investigators proposed a study with 20 patients, which would have provided only 61% power for testing the preceding hypothesis, which would have been inadequate.

If the direction of effect of the drug on heart rate were well known, then a one-sided test might be justified. In this case, the appropriate sample size could be obtained from the one-sided formulation in Equation 7.26, whereby

$$n = \frac{\sigma^2(z_{1-\beta} + z_{1-\alpha})^2}{(\mu_0 - \mu_1)^2} = \frac{10^2(z_{.8} + z_{.95})^2}{5^2}$$

$$= \frac{100(0.84 + 1.645)^2}{25} = 4(6.175) = 24.7, \text{ or } 25 \text{ patients}$$

Thus, only 25 patients would need to be studied for a one-sided test instead of the 32 patients needed for a two-sided test.

7.6.3 Sample-Size Estimation Based on Confidence-Interval Width

In some instances, it is well known that the treatment will have a significant effect on some physiologic parameter. Interest focuses instead on estimating the effect with a given degree of precision.

Example 7.38 **Cardiology** Suppose it is well known that propranolol will lower heart rate over 48 hours when given to patients with angina at standard dosage levels. A new study is proposed using a higher dose of propranolol than the standard one. Investigators are interested in estimating the drop in heart rate with high precision. How can this be done?

Suppose we quantify the precision of estimation by the width of the two-sided $100\% \times (1-\alpha)$ CI. Based on Equation 6.6, the $100\% \times (1-\alpha)$ CI for μ = true decline in heart rate is $\bar{x} \pm t_{n-1,1-\alpha/2}\, s/\sqrt{n}$. The width of this confidence interval is $2t_{n-1,1-\alpha/2}\, s/\sqrt{n}$. If we wish this interval to be no wider than L, then

$$2t_{n-1,1-\alpha/2}\, s/\sqrt{n} = L$$

We multiply both sides of the equation by $\sqrt{n}/L$ and obtain

$$2t_{n-1,1-\alpha/2}\, s/L = \sqrt{n}$$

or, on squaring both sides,

$$n = 4t^2_{n-1,1-\alpha/2}\, s^2/L^2$$

We usually approximate $t_{n-1,1-\alpha/2}$ by $z_{1-\alpha/2}$ and obtain the following result:

EQUATION 7.29

Sample-Size Estimation Based on Confidence-Interval Width Suppose we wish to estimate the mean of a normal distribution with sample variance s^2 and require that the two-sided $100\% \times (1-\alpha)$ CI for μ be no wider than L. The number of subjects needed is approximately

$$n = 4z^2_{1-\alpha/2}\, s^2/L^2$$

Example 7.39 **Cardiology** Find the minimum sample size needed to estimate the change in heart rate (μ) in Example 7.38 if we require that the two-sided 95% CI for μ be no wider than 5 beats per minute and the sample standard deviation for change in heart rate equals 10 beats per minute.

Solution We have $\alpha = .05,\ s = 10,\ L = 5$. Therefore, from Equation 7.29:

$$n = 4(z_{.975})^2 (10)^2/5^2$$
$$= 4(1.96)^2 (100)/25 = 61.5$$

Thus, 62 patients need to be studied.

SECTION 7.7 The Relationship Between Hypothesis Testing and Confidence Intervals

A test procedure was presented in Equation 7.10 for testing the hypothesis H_0: $\mu = \mu_0$ versus H_1: $\mu \neq \mu_0$. Similarly, in Section 6.5.5 a method for obtaining a two-sided confidence interval for the parameter μ of a normal distribution when the variance is unknown was discussed. The relationship between these two procedures can be stated as follows:

EQUATION 7.30

The Relationship Between Hypothesis Testing and Confidence Intervals (Two-Sided Case) Suppose we are testing H_0: $\mu = \mu_0$ versus H_1: $\mu \neq \mu_0$. H_0 is rejected with a two-sided level α test if and only if the two-sided $100\% \times (1-\alpha)$ confidence interval for μ *does not* contain μ_0. H_0 is accepted with a two-sided level α test if and only if the two-sided $100\% \times (1-\alpha)$ confidence interval for μ *does* contain μ_0.

Recall that the two-sided $100\% \times (1-\alpha)$ CI for $\mu = (c_1, c_2) = \bar{x} \pm t_{n-1,1-\alpha/2}\, s/\sqrt{n}$. Suppose we reject H_0 at level α. Then either $t < -t_{n-1,1-\alpha/2}$ or $t > t_{n-1,1-\alpha/2}$. Suppose that

$$t = (\bar{x} - \mu_0)/(s/\sqrt{n}) < -t_{n-1,1-\alpha/2}$$

We multiply both sides by $s/\sqrt{n}$ and obtain

$$\bar{x} - \mu_0 < -t_{n-1,1-\alpha/2}\, s/\sqrt{n}$$

If we add μ_0 to both sides, then

$$\bar{x} < \mu_0 - t_{n-1,1-\alpha/2}\, s/\sqrt{n}$$

or

$$\mu_0 > \bar{x} + t_{n-1,1-\alpha/2}\, s/\sqrt{n} = c_2$$

Similarly, if $t > t_{n-1,1-\alpha/2}$, then

$$\bar{x} - \mu_0 > t_{n-1,1-\alpha/2}\, s/\sqrt{n}$$

or

$$\mu_0 < \bar{x} - t_{n-1,1-\alpha/2}\, s/\sqrt{n} = c_1$$

Thus, if we reject H_0 at level α using a two-sided test, then either $\mu_0 < c_1$ or $\mu_0 > c_2$; i.e., μ_0 must fall outside the two-sided $100\% \times (1-\alpha)$ CI for μ. Similarly, it can be shown that if we accept H_0 at level α using a two-sided test, then μ_0 must fall within the two-sided $100\% \times (1-\alpha)$ CI for μ (or, $c_1 \le \mu_0 \le c_2$).

Hence, this relationship is the rationale for using confidence intervals in Chapter 6 to decide on the reasonableness of specific values for the parameter μ. If any specific proposed value μ_0 did not fall in the two-sided $100\% \times (1-\alpha)$ confidence interval for μ, then we said that is was an unlikely value for the parameter μ. Equivalently, we could have tested the hypothesis H_0: $\mu = \mu_0$ versus H_1: $\mu \ne \mu_0$ and rejected H_0 at significance level α.

Another way of expressing this relationship is as follows:

EQUATION 7.31

The two-sided $100\% \times (1-\alpha)$ confidence interval for μ contains all values μ_0 such that we accept H_0 using a two-sided test with significance level α, where the hypotheses are H_0: $\mu = \mu_0$ versus H_1: $\mu \ne \mu_0$. Conversely, the $100\% \times (1-\alpha)$ confidence interval *does not* contain any value μ_0 for which we can reject H_0, using a two-sided test with significance level α, where H_0: $\mu = \mu_0$ and H_1: $\mu \ne \mu_0$.

Example 7.40 **Cardiovascular Disease** Consider the cholesterol data in Example 7.20. We have $\bar{x}$ = 181.52 mg/dL, s = 40 mg/dL, and n = 100. The two-sided 95% confidence interval for μ is given by

$$\left(\bar{x} - t_{99,.975}\, s/\sqrt{n}, \bar{x} + t_{99,.975}\, s/\sqrt{n}\right)$$
$$= \left[181.52 - \frac{1.984(40)}{10}, 181.52 + \frac{1.984(40)}{10}\right]$$
$$= (181.52 - 7.94, 181.52 + 7.94) = (173.58, 189.46)$$

This confidence interval contains all values for μ_0 for which we accept H_0: $\mu = \mu_0$ and does not contain any value μ_0 for which we could reject H_0 at the 5% level. Specifically, the 95% confidence interval (173.58, 189.46) does not contain μ_0 = 190, which corresponds to the decision rule in Example 7.21, where we were able to reject H_0: μ = 190 at the 5% level of significance.

Another way of stating this is that the p-value computed in Example 7.22 for μ_0 = 190 = .037, which is less than .05.

Example 7.41 **Cardiovascular Disease** Suppose the sample mean for cholesterol was 185 mg/dL for the cholesterol data in Example 7.20. The 95% confidence interval would be

$$(185 - 7.94, 185 + 7.94) = (177.06, 192.94)$$

which contains the null mean (190). The p-value for the hypothesis test would be

$$p = 2 \times Pr\left[t_{99} < (185 - 190)/4\right] = 2 \times Pr(t_{99} < -1.25)$$
$$= 2(.1071) = .214 > .05$$

using the TDIST function of Excel. Thus, we are able to accept H_0 using α = .05, if μ_0 = 190, which is consistent with the statement that 190 falls within the above 95% confidence interval. Thus, the conclusions based on the confidence-interval and hypothesis-testing approaches are also the same here.

A similar relationship exists between the one-sided hypothesis test developed in Section 7.3 and the one-sided confidence interval for the parameter μ developed in Section 6.10. Equivalent confidence-interval statements can also be made about most of the other one-sided or two-sided hypothesis tests covered in this text.

Because the hypothesis-testing and confidence-interval approaches yield the same conclusions, is there any advantage to using one method over the other? The *p*-value from a hypothesis test tells us precisely how statistically significant the results are. However, often results that are statistically significant are not very important in the context of the subject matter, because the actual difference between $\bar{x}$ and μ_0 may not be very large, but the results are statistically significant because of a large sample size. A 95% confidence interval for μ would give additional information, because it would give a range of values within which μ is likely to fall. Conversely, the 95% confidence interval does not contain all the information contained in a *p*-value. It does not tell us precisely how significant the results are but merely tells us whether or not they are significant at the 5% level. Hence, it is good practice to compute both a *p*-value and a 95% confidence interval for μ.

Unfortunately, some researchers have become polarized on this issue, with some statisticians favoring only the hypothesis-testing approach, and some epidemiologists favoring only the confidence-interval approach. These issues have correspondingly influenced editorial policy, with some journals *requiring* that results be presented in one format or the other. The crux of the issue is that, traditionally, results need to be statistically significant (at the 5% level) in order to demonstrate the validity of a particular finding. One advantage of this approach is that a uniform statistical standard is provided (the 5% level) for *all* researchers in order to demonstrate evidence of an association. This protects the research community against scientific claims not based on any statistical or empirical criteria whatsoever (such as solely on the basis of clinical case reports). Advocates of the confidence-interval approach contend that the width of the confidence interval provides information on the likely magnitude of the differences between groups, regardless of the level of significance. My opinion is that both significance levels and confidence limits provide complementary information and both should be reported, where possible.

Example 7.42 **Cardiovascular Disease** Consider the cholesterol data in Examples 7.22 and 7.40. The *p*-value of .037, computed in Example 7.22, tells us precisely how significant the results are. The 95% confidence interval for μ = (173.58, 189.46) computed in Example 7.40 gives a range of likely values that μ might assume. The two types of information are complementary.

SECTION 7.8 One-Sample χ^2 Test for the Variance of a Normal Distribution

Example 7.43 **Hypertension** Consider Example 6.39, concerning the variability of blood-pressure measurements taken on an arteriosonde machine. We were concerned with the difference between measurements taken by two observers on the same person = $d_i = x_{1i} - x_{2i}$, where x_{1i} = the measurement on the ith person by the first observer and x_{2i} = the measurement on the ith person by the second observer. We will assume that this difference is a good measure of interobserver variability, and we wish to compare this variability with the variability using a standard blood-pressure cuff. We have reason to believe that the variability of the arteriosonde machine might be different from that of a standard cuff. Intuitively, we think the variability of the new method should be lower. However, because the new method is not as widely used, the observers are probably less experienced at using the method, and the variability of the new method could possibly be higher than that of the old method. Thus, a two-sided test will be used to study this question. Suppose we know from previously published work that $\sigma^2 = 35$ for d_i obtained from the standard cuff. We wish to test the hypothesis H_0: $\sigma^2 = \sigma_0^2 = 35$ versus H_1: $\sigma^2 \neq \sigma_0^2$. How should this test be performed?

If $x_1, \ldots, x_n$ are a random sample, then we can reasonably base the test on s^2, because it is an unbiased estimator of σ^2. We know from Equation 6.14 that if $x_1, \ldots, x_n$ are a random sample from an $N(\mu, \sigma^2)$ distribution, then under H_0,

$$X^2 = \frac{(n-1)S^2}{\sigma^2} \sim \chi^2_{n-1}$$

Therefore,

$$Pr\left(X^2 < \chi^2_{n-1,\alpha/2}\right) = \alpha/2 = Pr\left(X^2 > \chi^2_{n-1,1-\alpha/2}\right)$$

Hence, the test procedure is given as follows:

EQUATION 7.32

One-Sample χ^2 Test for the Variance of a Normal Distribution (Two-Sided Alternative) We compute the test statistic $X^2 = (n-1)s^2/\sigma_0^2$.

If $X^2 < \chi^2_{n-1,\alpha/2}$ or $X^2 > \chi^2_{n-1,1-\alpha/2}$, then H_0 is rejected.

If $\chi^2_{n-1,\alpha/2} \le X^2 \le \chi^2_{n-1,1-\alpha/2}$, then H_0 is accepted.

The acceptance and rejection regions for this test are depicted in Figure 7.10.

FIGURE 7.10 **Acceptance and rejection regions for the one-sample χ^2 test for the variance of a normal distribution (two-sided alternative)**

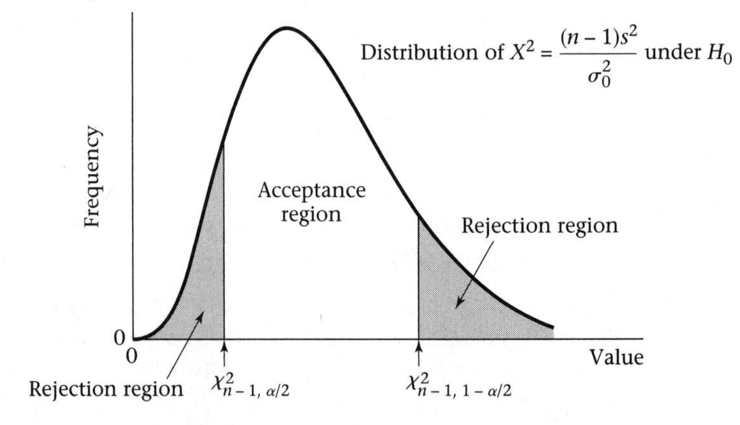

Alternatively, we may wish to compute a p-value for our experiment. The computation of the p-value will depend on whether $s^2 \le \sigma_0^2$ or $s^2 > \sigma_0^2$. The rule is given as follows:

EQUATION 7.33

p-Value for a One-Sample χ^2 Test for the Variance of a Normal Distribution (Two-Sided Alternative)

Let the test statistic $X^2 = \dfrac{(n-1)s^2}{\sigma_0^2}$.

If $s^2 \le \sigma_0^2$, then p-value $= 2 \times$ (area to the left of X^2 under a χ_{n-1}^2 distribution).

If $s^2 > \sigma_0^2$, then p-value $= 2 \times$ (area to the right of X^2 under a χ_{n-1}^2 distribution).

The p-values are illustrated in Figure 7.11.

FIGURE 7.11 **Illustration of the p-value for a one-sample χ^2 test for the variance of a normal distribution (two-sided alternative)**

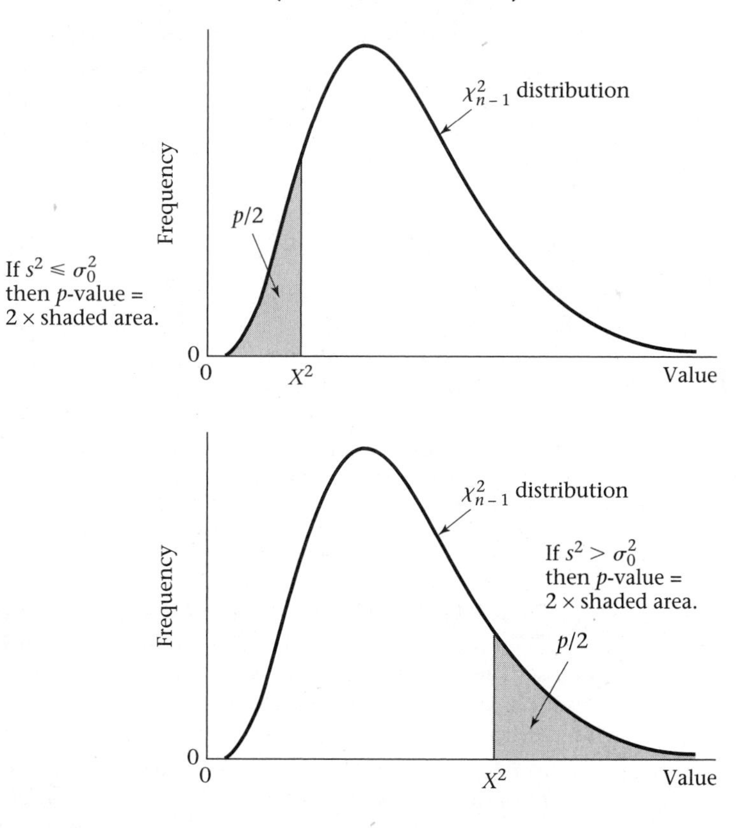

Example 7.44 **Hypertension** Assess the statistical significance of the arteriosonde-machine data in Example 7.43.

Solution We know from Example 6.39 that $s^2 = 8.178$, $n = 10$. From Equation 7.32, we compute the test statistic X^2 given by

$$X^2 = \frac{(n-1)s^2}{\sigma_0^2} = \frac{9(8.178)}{35} = 2.103$$

Under H_0, X^2 follows a χ^2 distribution with nine degrees of freedom. Thus, the critical values are $\chi^2_{9,.025} = 2.70$ and $\chi^2_{9,.975} = 19.02$. Because $X^2 = 2.103 < 2.70$, it follows that H_0 is rejected using a two-sided test with $\alpha = .05$. To obtain the p-value, refer to Equation 7.33. Because $s^2 = 8.178 < 35 = \sigma_0^2$, the p-value is computed as follows:

$$p = 2 \times Pr(\chi_9^2 < 2.103)$$

From Table 6 of the Appendix we see that

$$\chi^2_{9,.025} = 2.70,\ \chi^2_{9,.01} = 2.09$$

Thus, because $2.09 < 2.103 < 2.70$, we have $.01 < p/2 < .025$ or $.02 < p < .05$.

To obtain the exact p-value, use Microsoft Excel 97 to evaluate areas under the χ^2 distribution. The CHIDIST function computes the right-hand tail area = .9897. Thus, subtract from 1 and multiply by 2 to obtain the exact two-sided p-value = .021. The details are given in Table 7.2.

TABLE 7.2 **Computation of the exact p-value for the arteriosonde-machine data in Example 7.44 using a one-sample χ^2 test with Microsoft Excel 97**

Microsoft Excel 97	
x	2.103
df	9
cdf=chisq(2.103,9)	0.989732
p-value(1-tail)	0.010268
p-value(2-tail)	0.020536

Therefore, the results are statistically significant, and we conclude that the interobserver variance using the arteriosonde machine is significantly different from the interobserver variance using the standard cuff. To quantify how different the two variances are, a two-sided 95% confidence interval for σ^2 could be obtained, as was done in Example 6.41. This interval was (3.87, 27.26). Of course, it does not contain 35 because the p-value is less than .05.

In general, the assumption of normality is particularly important for hypothesis testing and confidence-interval estimation for variances. If this assumption is not satisfied, then the critical regions and p-values in Equations 7.32 and 7.33 and the confidence limits in Equation 6.15 will not be valid.

In this section, we have presented the **one-sample chi-square test for variances**, which is used for testing hypotheses concerning the variance of a normal distribution. Beginning at the Start box of the flowchart (Figure 7.18, p. 262), we arrive at the one-sample chi-square test for variances by answering yes to each of the following: (1) one variable of interest? (2) one-sample problem? and (3) underlying distribution normal or can central-limit theorem be assumed to hold? and by answering no to (4) inference concerning μ? and yes to (5) inference concerning σ?

SECTION 7.9 One-Sample Test for a Binomial Proportion

7.9.1 Normal-Theory Methods

Example 7.45 **Cancer** Consider the breast-cancer data in Example 6.44. In that example we were interested in the effect of having a family history of breast cancer on the incidence of breast cancer. Suppose that out of 10,000 50- to 54-year-old women sampled whose mothers had breast cancer, 400 had breast cancer at some time in their lives. Based on large studies, assume that the prevalence rate of breast cancer for U.S. women in this age group is about 2%. The question is, How compatible is the sample rate of 4% with a population rate of 2%?

Another way of asking this question is to restate it in terms of hypothesis testing: If p = prevalence rate of breast cancer in 50- to 54-year-old women whose mothers have had breast cancer, then we wish to test the hypothesis H_0: $p = .02 = p_0$ versus H_1: $p \neq .02$. How can this be done?

The significance test will be based on the sample proportion of cases $\hat{p}$. We will assume that the normal approximation to the binomial distribution is valid. This assumption is reasonable when $np_0q_0 \geq 5$. Therefore, from Equation 6.17 we know that under H_0

$$\hat{p} \sim N\left(p_0, \frac{p_0q_0}{n}\right)$$

It will be more convenient to standardize $\hat{p}$. For this purpose, we subtract the expected value of $\hat{p}$ under $H_0 = p_0$ and divide by the standard error of $\hat{p}$ under $H_0 = \sqrt{p_0q_0/n}$, thus creating the test statistic z given by

$$z = \frac{\hat{p} - p_0}{\sqrt{p_0q_0/n}}$$

It follows that under $H_0, z \sim N(0,1)$. Thus,

$$Pr(z < z_{\alpha/2}) = Pr(z > z_{1-\alpha/2}) = \alpha/2$$

Thus the test takes the following form:

EQUATION 7.34

One-Sample Test for a Binomial Proportion—Normal-Theory Method (Two-Sided Alternative) Let the test statistic $z = (\hat{p} - p_0)/\sqrt{p_0q_0/n}$.

If $z < z_{\alpha/2}$ or $z > z_{1-\alpha/2}$, then H_0 is rejected. If $z_{\alpha/2} \leq z \leq z_{1-\alpha/2}$, then H_0 is accepted.

The acceptance and rejection regions are shown in Figure 7.12.

Alternatively, a p-value could be computed. The computation of the p-value will depend on whether $\hat{p} \leq p_0$ or $\hat{p} > p_0$. If $\hat{p} \leq p_0$, then

p-value = 2 × area to the left of z under an $N(0, 1)$ curve

If $\hat{p} > p_0$, then

p-value = 2 × area to the right of z under an $N(0, 1)$ curve

FIGURE 7.12 **Acceptance and rejection regions for the one-sample binomial test—normal-theory method (two-sided alternative)**

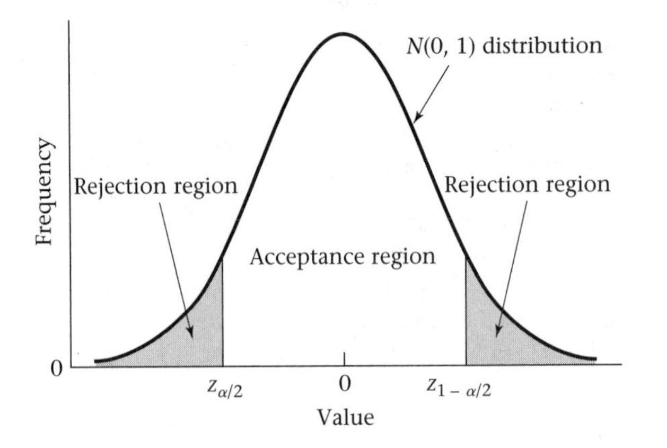

This is summarized as follows:

EQUATION 7.35

Computation of the *p*-Value for the One-Sample Binomial Test—Normal-Theory Method (Two-Sided Alternative)

Let the test statistic $z = (\hat{p} - p_0) / \sqrt{p_0 q_0 / n}$.

If $\hat{p} < p_0$, then *p*-value = $2 \times \Phi(z)$ = twice the area to the left of z under an $N(0, 1)$ curve. If $\hat{p} \geq p_0$, then *p*-value = $2 \times [1 - \Phi(z)]$ = twice the area to the right of z under an $N(0, 1)$ curve. The calculation of the *p*-value is illustrated in Figure 7.13.

These definitions of a *p*-value are again compatible with the idea of a *p*-value as the probability of obtaining results as extreme as or more extreme than the results in our particular sample.

Example 7.46 **Cancer** Assess the statistical significance of the data in Example 7.45.

Solution Using the critical-value method, we compute the test statistic

$$z = \frac{\hat{p} - p_0}{\sqrt{p_0 q_0 / n}}$$

$$= \frac{.04 - .02}{\sqrt{.02(.98)/10{,}000}} = \frac{.02}{.0014} = 14.3$$

Because $z_{1-\alpha/2} = z_{.975} = 1.96$, it follows that H_0 can be rejected using a two-sided test with $\alpha = .05$. To compute the *p*-value, because $p = .04 > p_0 = .02$, it follows that

$$p\text{-value} = 2 \times [1 - \Phi(z)]$$

$$= 2 \times [1 - \Phi(14.3)] < .001$$

Thus, the results are very highly significant.

FIGURE 7.13 **Illustration of the p-value for a one-sample binomial test—normal-theory method (two-sided alternative)**

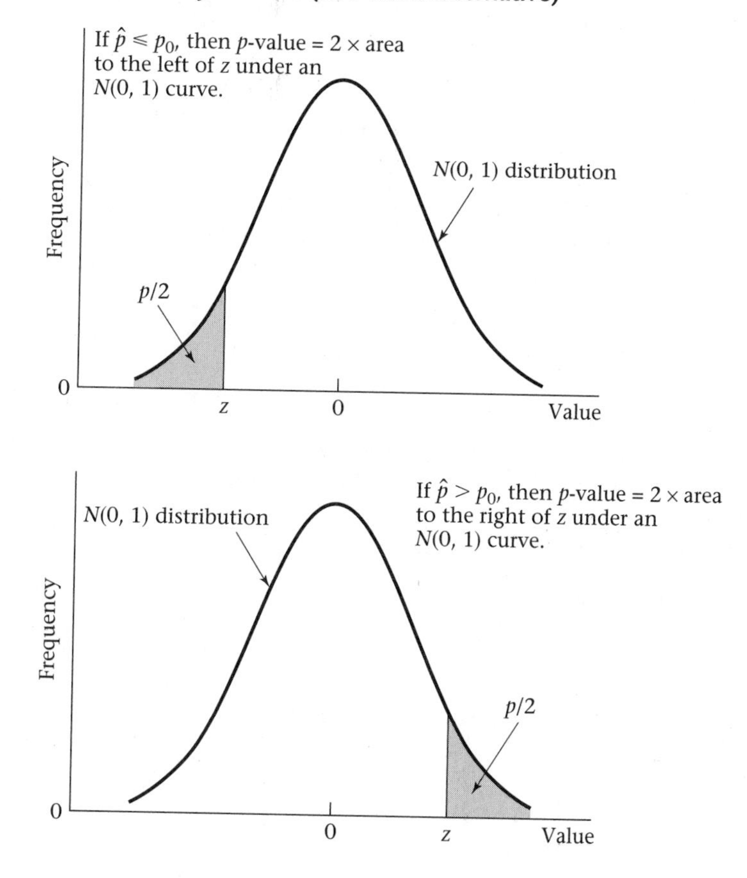
If $\hat{p} \le p_0$, then p-value = 2 × area
to the left of z under an
$N(0, 1)$ curve.
$N(0, 1)$ distribution
Frequency
$p/2$
0
z
0
Value
$N(0, 1)$ distribution
If $\hat{p} > p_0$, then p-value = 2 × area
to the right of z under an
$N(0, 1)$ curve.
Frequency
$p/2$
0
0
z
Value

7.9.2 Exact Methods

The test procedure presented in Equation 7.34 to test the hypothesis H_0: $p = p_0$ versus H_1: $p \ne p_0$ depends on the assumption that the normal approximation to the binomial distribution is valid. This assumption will only be true if $np_0q_0 \ge 5$. How can the preceding hypothesis be tested if this criterion is not satisfied?

Our approach will be to base our test on *exact* binomial probabilities. In particular, let X be a binomial random variable with parameters n and p_0 and let $\hat{p} = x/n$, where x is the observed number of events. The computation of the p-value depends on whether $\hat{p} \le p_0$ or $\hat{p} > p_0$. If $\hat{p} \le p_0$, then

$$\begin{aligned} p/2 &= Pr(\le x \text{ successes in } n \text{ trials} | H_0) \\ &= \sum_{k=0}^{x} \binom{n}{k} p_0^k (1-p_0)^{n-k} \end{aligned}$$

If $\hat{p} > p_0$, then

$$p/2 = Pr(\geq x \text{ successes in } n \text{ trials}|H_0)$$
$$= \sum_{k=x}^{n} \binom{n}{k} p_0^k (1-p_0)^{n-k}$$

This is summarized as follows:

EQUATION 7.36

Computation of the *p*-Value for the One-Sample Binomial Test—Exact Method (Two-Sided Alternative)

$$\text{If } \hat{p} \leq p_0, p = 2 \times Pr(X \leq x) = \min\left[2\sum_{k=0}^{x} \binom{n}{k} p_0^k (1-p_0)^{n-k}, 1\right]$$

$$\text{If } \hat{p} > p_0, p = 2 \times Pr(X \geq x) = \min\left[2\sum_{k=x}^{n} \binom{n}{k} p_0^k (1-p_0)^{n-k}, 1\right]$$

The computation of the *p*-value is depicted in Figure 7.14, in the case of $n = 30$, $p_0 = .5$ and $x = 10$ and 20, respectively.

FIGURE 7.14 Illustration of the *p*-value for a one-sample binomial test—exact method (two-sided alternative)

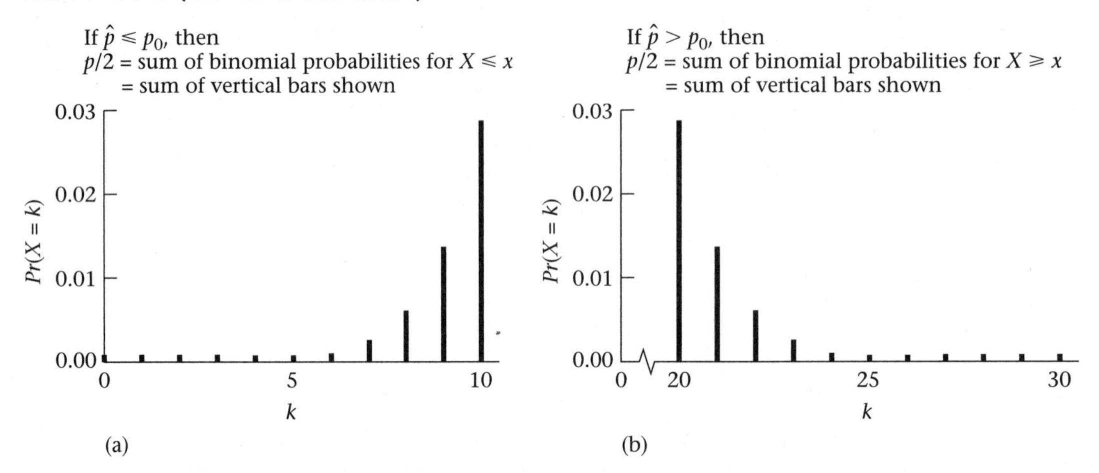
If $\hat{p} \leq p_0$, then
$p/2$ = sum of binomial probabilities for $X \leq x$
= sum of vertical bars shown
0.03
0.02
0.01
0.00
$Pr(X = k)$
0
5
10
k
(a)
If $\hat{p} > p_0$, then
$p/2$ = sum of binomial probabilities for $X \geq x$
= sum of vertical bars shown
0.03
0.02
0.01
0.00
$Pr(X = k)$
0
20
25
30
k
(b)

In either case the *p*-value corresponds to the sum of the probabilities of all events that are as extreme as or more extreme than the sample result obtained.

Example 7.47 **Occupational Medicine, Cancer** The safety of those who work at or live close to nuclear power plants has been the subject of widely publicized debate in recent years. One possible health hazard due to radiation exposure is an excess of cancer deaths among those exposed. One problem with studying this question is that the number of deaths attributable to either cancer in general or specific types of cancer is small, and reaching statistically significant conclusions is difficult, except after long periods of follow-up. An alternative approach is to perform a *proportional-mortality study*, whereby the proportion of deaths attributed to a specific cause in an exposed group is compared with the corresponding proportion in a large population. Suppose, for example, that 13 deaths have occurred among 55- to 64-year-old male workers in a nuclear power plant and that in 5 of them the cause of death was cancer. Assume, based on vital-statistics reports, that approximately 20% of all deaths can be attributed to some form of cancer. Is this result significant?

Solution We wish to test the hypothesis H_0: $p = .20$ versus H_1: $p \neq .20$, where p = probability that the cause of death was cancer in nuclear-power workers. The normal approximation to the binomial cannot be used, because

$$np_0q_0 = 13(.2)(.8) = 2.1 < 5$$

However, the exact procedure in Equation 7.36 can be used:

$$\hat{p} = \frac{5}{13} = .38 > .20$$

$$\text{Therefore,} \quad p = 2\sum_{k=5}^{13}\binom{13}{k}(.2)^k(.8)^{13-k} = 2 \times \left[1 - \sum_{k=0}^{4}\binom{13}{k}(.2)^k(.8)^{13-k}\right]$$

From Table 1 in the Appendix, with $n = 13$, $p = .2$ we have

$$Pr(0) = .0550$$
$$Pr(1) = .1787$$
$$Pr(2) = .2680$$
$$Pr(3) = .2457$$
$$Pr(4) = .1535$$

$$\text{Therefore,} \quad p = 2 \times [1 - (.0550 + .1787 + .2680 + .2457 + .1535)]$$
$$= 2 \times (1 - .9009) = .198$$

In summary, the proportion of deaths due to cancer is not significantly different for nuclear power plant workers than for comparably aged men in the general population.

7.9.3 Power and Sample-Size Estimation

The power of the one-sample binomial test can also be considered using the large-sample test procedure given in Section 7.9.1. Suppose we are conducting a two-tailed test at level α, where $p = p_0$ under the null hypothesis. Under the alternative hypothesis of $p = p_1$, the power is given by the following formula:

EQUATION 7.37

Power for the One-Sample Binomial Test (Two-Sided Alternative) The **power** of the one-sample binomial test for the hypothesis

$$H_0: p = p_0 \quad \text{vs.} \quad H_1: p \neq p_0$$

for the specific alternative $p = p_1$ is given by

$$\Phi\left[\sqrt{\frac{p_0q_0}{p_1q_1}}\left(z_{\alpha/2} + \frac{|p_0 - p_1|\sqrt{n}}{\sqrt{p_0q_0}}\right)\right]$$

To use this formula, we assume that $np_0q_0 \geq 5$ so that the normal-theory methods in Section 7.9.1 are valid.

Example 7.48 **Cancer** Suppose we wish to test the hypothesis that women with a sister history of breast cancer are at higher risk of developing breast cancer themselves. Suppose we assume, as in Example 7.45, that the prevalence rate of breast cancer is 2% among 50- to 54-year-old U.S. women, whereas it is 5% among women with a sister history. We propose to interview 500 50- to 54-year-old women with a sister history of the disease. What is the power of such a study assuming that we conduct a two-sided test with $\alpha = .05$?

Solution We have $\alpha = .05,\ p_0 = .02,\ p_1 = .05,\ n = 500$. The power, as given by Equation 7.37, is

$$\text{Power} = \Phi\left[\sqrt{\frac{.02(.98)}{.05(.95)}}\left(z_{.025} + \frac{.03\sqrt{500}}{\sqrt{.02(.98)}}\right)\right]$$
$$= \Phi[.642(-1.96 + 4.792)] = \Phi(1.819) = .966$$

Thus, there should be a 96.6% chance of finding a significant difference based on a sample size of 500, if the true rate of breast cancer among women with a sister history is 2.5 times as high as that of typical 50- to 54-year-old women.

Similarly, we can consider the issue of appropriate sample size if the one-sample binomial test for a given α, p_0, p_1, and power is being used. The sample size is given by the following formula:

EQUATION 7.38

Sample-Size Estimation for the One-Sample Binomial Test (Two-Sided Alternative) Suppose we wish to test H_0: $p = p_0$ versus H_1: $p \neq p_0$. The sample size needed to conduct a two-sided test with significance level α and power $1 - \beta$ versus the specific alternative hypothesis $p = p_1$ is

$$n = \frac{p_0 q_0\left(z_{1-\alpha/2} + z_{1-\beta}\sqrt{\frac{p_1 q_1}{p_0 q_0}}\right)^2}{(p_1 - p_0)^2}$$

Example 7.49 **Cancer** How many women should be interviewed in the study proposed in Example 7.48 to achieve 90% power if a two-sided significance test with $\alpha = .05$ is being used?

Solution We have $\alpha = .05,\ 1-\beta = .90,\ p_0 = .02,\ p_1 = .05$. The sample size is given by Equation 7.38:

$$n = \frac{.02(.98)\left[z_{.975} + z_{.90}\sqrt{\frac{.05(.95)}{.02(.98)}}\right]^2}{(.03)^2}$$
$$= \frac{.0196[1.96 + 1.28(1.557)]^2}{.0009} = \frac{.0196(15.623)}{.0009} = 340.2, \text{ or } 341 \text{ women}$$

Thus, 341 women need to be interviewed to have a 90% chance of detecting a significant difference using a two-sided test with $\alpha = .05$ if the true rate of breast cancer among women with a sister history is 2.5 times as high as that of a typical 50- to 54-year-old woman.

Note that if we wish to perform a one-sided test rather than a two-sided test at level α, then α is substituted for $\alpha/2$ in the power formula in Equation 7.37 and the sample-size formula in Equation 7.38.

In this section, we have presented the **one-sample binomial test,** which is used for testing hypotheses concerning the parameter p of a binomial distribution. Beginning at the Start box of the flowchart (Figure 7.18, p. 262), we arrive at the one-sample binomial test by answering yes to (1) one variable of interest? and to (2) one-sample problem? no to (3) underlying distribution normal or can central-limit theorem be assumed to hold? and yes to (4) underlying distribution is binomial?

SECTION 7.10 One-Sample Inference for the Poisson Distribution

Example 7.50 **Occupational Health** Many studies have looked at possible health hazards faced by rubber workers. In one such study, a group of 8418 white male workers ages 40–84 (either active or retired) on January 1, 1964, were followed for 10 years for various mortality outcomes [1]. Their mortality rates were then compared with U.S. white male mortality rates in 1968. In one of the reported findings, 4 deaths due to Hodgkin's disease were observed compared with 3.3 deaths expected from U.S. mortality rates. Is this difference significant?

One problem with this type of study is that workers of different ages in 1964 have very different mortality risks over time. Thus the test procedures in Equations 7.34 and 7.36, which assume a constant p for all persons in the sample, are not applicable. However, these procedures can be generalized to take account of the different mortality risks of different individuals. Let

X = total observed number of deaths for members of the study population

p_i = probability of death for the ith individual

Under the null hypothesis that the death rates for the study population are the same as those for the general U.S. population, the expected number of events μ_0 is given by

$$\mu_0 = \sum_{i=1}^{n} p_i$$

If the disease under study is rare, then the observed number of events may be considered approximately Poisson-distributed with unknown expected value = μ. We wish to test the hypothesis H_0: $\mu = \mu_0$ versus H_1: $\mu \neq \mu_0$.

One approach for significance testing is to use the critical-value method. We know from Section 7.7 that the two-sided $100\% \times (1-\alpha)$ confidence interval for μ given by (c_1, c_2) contains all values μ_0 for which we would accept H_0 for the preceding hypothesis test. Thus if $c_1 \leq \mu_0 \leq c_2$, then we accept H_0, while if either $\mu_0 < c_1$ or $\mu_0 > c_2$, then we reject H_0. Table 8 in the Appendix contains exact confidence limits for the Poisson expectation μ, and this leads us to the following simple approach for hypothesis testing.

EQUATION 7.39

One-Sample Inference for the Poisson Distribution (Small-Sample Test—Critical-Value Method) Let X be a Poisson random variable with expected value $= \mu$. To test the hypothesis H_0: $\mu = \mu_0$ versus H_1: $\mu \neq \mu_0$ using a two-sided test with significance level α,

(1) Obtain the two-sided $100\% \times (1-\alpha)$ confidence interval for μ based on the observed value x of X. Denote this confidence interval (c_1, c_2).

(2) If $\mu_0 < c_1$ or $\mu_0 > c_2$, then reject H_0.
If $c_1 \le \mu_0 \le c_2$, then accept H_0.

Example 7.51 **Occupational Health** Test for the significance of the findings in Example 7.50 using the critical-value method with a two-sided significance level of .05.

Solution We wish to test the hypothesis H_0: $\mu = 3.3$ versus H_1: $\mu \neq 3.3$. We observed 4 events $= x$. Hence, referring to Table 8, the two-sided 95% confidence interval for μ based on $x = 4$ is (1.09, 10.24). From Equation 7.39, because $1.09 \le 3.3 \le 10.24$, we accept H_0 at the 5% significance level.

Another approach to use for significance testing is the p-value method. We wish to reject H_0 if x is either much larger or much smaller than μ_0. This leads to the following test procedure.

EQUATION 7.40

One-Sample Inference for the Poisson Distribution (Small-Sample Test—*p*-Value Method) Let $\mu =$ expected value of a Poisson distribution. To test the hypothesis H_0: $\mu = \mu_0$ versus H_1: $\mu \neq \mu_0$

(1) Compute

$x =$ observed number of deaths in the study population

(2) Under H_0, the random variable X will follow a Poisson distribution with parameter μ_0. Thus, the exact two-sided p-value is given by

$$\min\left(2 \times \sum_{k=0}^{x} \frac{e^{-\mu_0}\mu_0^k}{k!}, 1\right) \qquad \text{if } x < \mu_0$$

$$\min\left[2 \times \left(1 - \sum_{k=0}^{x-1} \frac{e^{-\mu_0}\mu_0^k}{k!}\right), 1\right] \qquad \text{if } x \ge \mu_0$$

These computations are depicted in Figure 7.15 for the case of $\mu_0 = 5$, with $x = 3$ and 8, respectively.

Example 7.52 **Occupational Health** Test for the significance of the findings in Example 7.50 using the p-value method.

Solution We refer to Equation 7.40. Because $x = 4 > \mu_0 = 3.3$, the p-value is given by

$$p = 2 \times \left[1 - \sum_{k=0}^{3} \frac{e^{-3.3}(3.3)^k}{k!}\right]$$

FIGURE 7.15 Computation of the exact *p*-value for the one-sample Poisson test

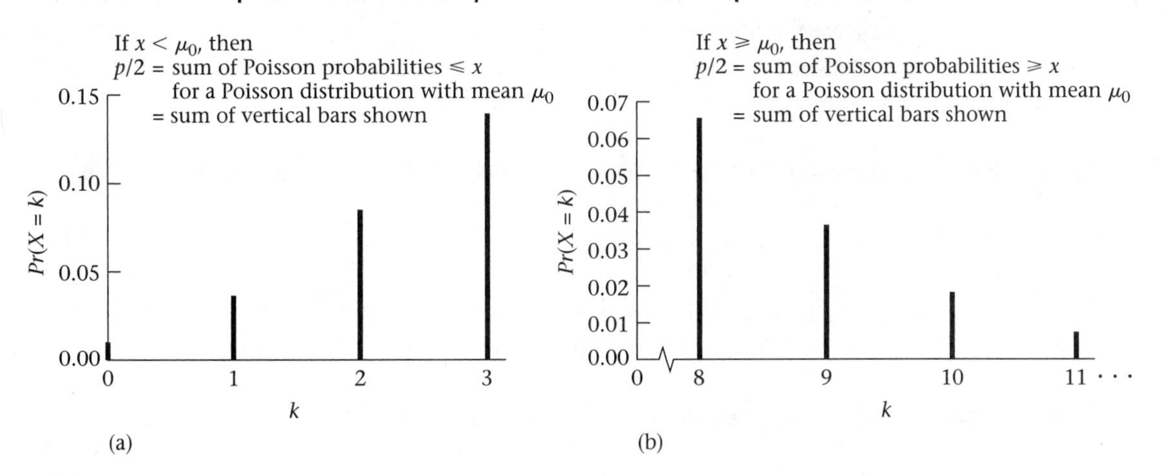

From the Poisson distribution, we have

$$Pr(0) = e^{-3.3} = .0369$$

$$Pr(1) = \frac{e^{-3.3}(3.3)}{1!} = .1217$$

$$Pr(2) = \frac{e^{-3.3}(3.3)^2}{2!} = .2008$$

$$Pr(3) = \frac{e^{-3.3}(3.3)^3}{3!} = .2209$$

Thus $\quad p = 2 \times [1 - (.0369 + .1217 + .2008 + .2209)]$
$\quad = 2 \times (1 - .5803) = .839$

Thus, there is no significant excess or deficit of Hodgkin's disease in this population.

An index that is frequently used to quantify risk in a study population relative to the general population is the standardized mortality ratio (or SMR).

DEFINITION 7.16 The **standardized mortality ratio (SMR)** is defined by 100% × O/E = 100% × the observed number of deaths in the study population divided by the expected number of deaths in the study population under the assumption that the mortality rates for the study population are the same as those for the general population. For nonfatal conditions the standardized mortality ratio is sometimes known as the **standardized morbidity ratio.**

Thus,

- If SMR > 100%, there is an excess risk in the study population relative to the general population.

- If SMR < 100%, there is a reduced risk in the study population relative to the general population.
- If SMR = 100%, there is neither an excess nor a deficit of risk in the study population relative to the general population.

Example 7.53 **Occupational Health** What is the standardized mortality ratio for Hodgkin's disease using the data in Example 7.50?

Solution SMR = 100% × 4/3.3 = 121%

Another way to interpret the test procedures in Equations 7.39 and 7.40 are as tests of whether the SMR is significantly different from 100%.

Example 7.54 **Occupational Health** In the rubber-worker data described in Example 7.50, there were 21 observed cases of bladder cancer and an expected number of events from general-population cancer mortality rates of 18.1. Evaluate the statistical significance of the results.

Solution We refer to the 21 row and .95 column in Table 8 and find the 95% CI for μ = (13.00, 32.10). Because μ_0 = expected number of deaths = 18.1 is within the 95% CI, it follows that we can accept H_0 at the 5% level of significance. To obtain an exact p-value, we refer to Equation 7.40 and compute

$$p = 2 \times \left(1 - \sum_{k=0}^{20} e^{-18.1} (18.1)^k / k!\right)$$

This is a tedious calculation, so we have used Excel as shown in Table 7.3. From Excel 97, we see that $Pr(X \le 20 | \mu = 18.1) = .7227$. Therefore, the p-value = 2 × (1 − .7227) = .55. Thus, there is no significant excess or deficit of bladder cancer in the rubber-workers population. The SMR for bladder cancer = 100% × 21/18.1 = 116%. Another interpretation of the significance tests in Equations 7.39 and 7.40 is that the underlying SMR in the reference population is not significantly different from 100%.

TABLE 7.3 **Computation of the exact p-value for the bladder-cancer data in Example 7.54**

Microsoft Excel 97	
mean	18.1
k	20
Pr(X<=k)=Poisson(20,18.1,true)	0.722696

The test procedures in Equations 7.39 and 7.40 are exact methods. If the expected number of events is large, then the following approximate method can be used.

EQUATION 7.41

One-Sample Inference for the Poisson Distribution (Large-Sample Test) Let μ = expected value of a Poisson random variable. To test the hypothesis H_0: $\mu = \mu_0$ versus H_1: $\mu \ne \mu_0$,

(1) Compute x = observed number of events in the study population.

(2) Compute the test statistic

$$X^2 = \frac{(x-\mu_0)^2}{\mu_0} = \mu_0\left(\frac{\text{SMR}}{100}-1\right)^2 \sim \chi_1^2 \text{ under } H_0$$

(3) For a two-sided test at level α, H_0 is rejected if

$$X^2 > \chi^2_{1,1-\alpha}$$

and H_0 is accepted if $\quad X^2 \le \chi^2_{1,1-\alpha}$

(4) The exact p-value is given by $Pr(\chi_1^2 > X^2)$.

(5) This test should only be used if $\mu_0 \ge 10$.

The acceptance and rejection regions for this test are depicted in Figure 7.16. The computation of the exact p-value is given in Figure 7.17.

FIGURE 7.16 **Acceptance and rejection regions for the one-sample Poisson test (large-sample test)**

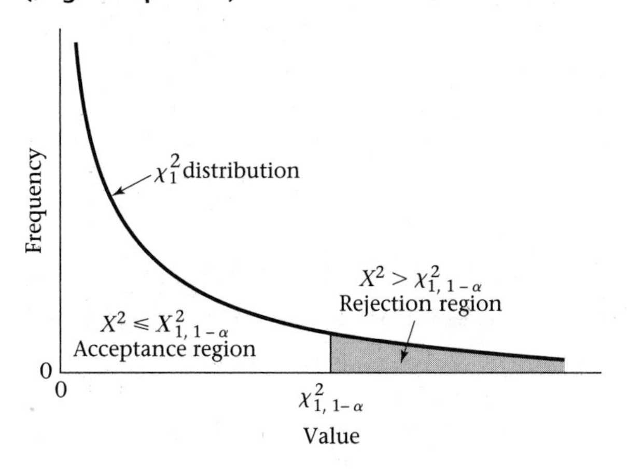

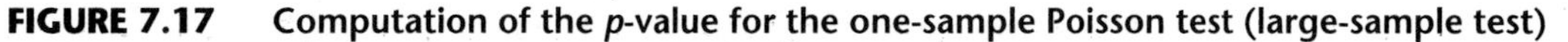

FIGURE 7.17 **Computation of the *p*-value for the one-sample Poisson test (large-sample test)**

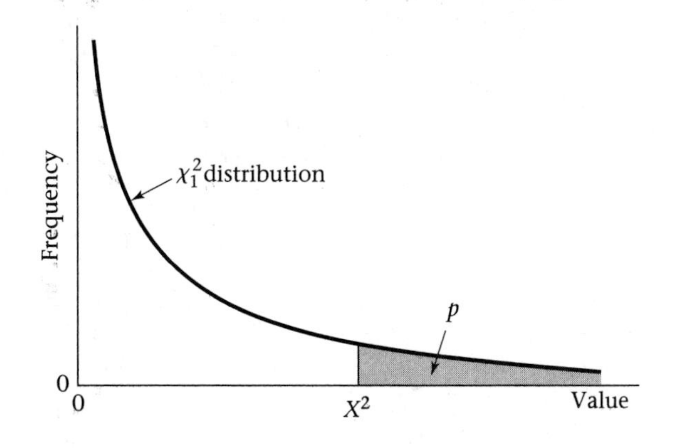

Example 7.55 **Occupational Health** Assess the statistical significance of the bladder-cancer data in Example 7.54 using the large-sample test.

Solution We wish to test the hypothesis H_0: $\mu = 18.1$ versus H_1: $\mu \neq 18.1$. In this case, $x = 21$ and SMR = 100% × 21/18.1 = 116%. Hence, we have the test statistic

$$X^2 = \frac{(21-18.1)^2}{18.1} \quad \text{or} \quad 18.1 \times (1.16-1)^2$$
$$= \frac{8.41}{18.1} = 0.46 \sim \chi_1^2 \text{ under } H_0$$

Because $\chi^2_{1,.95} = 3.84 > X^2$, $p > .05$ and H_0 is accepted. Furthermore, from Table 6 in the Appendix we note that $\chi^2_{1,.50} = 0.45$, $\chi^2_{1,.75} = 1.32$, and $0.45 < X^2 < 1.32$. Thus, $1 - .75 < p < 1 - .50$, or $.25 < p < .50$. Therefore, the rubber workers in this plant do not have a significantly increased or decreased risk of bladder cancer relative to the general population. Using MINITAB to compute the p-value for the large-sample test yields a p-value = $Pr(\chi_1^2 > 0.46) = .50$ to two decimal places. From Example 7.54, the exact p-value based on the Poisson distribution = .55. In general, exact methods are strongly preferred for inference concerning the Poisson distribution.

In this section, we have presented the **one-sample Poisson test,** which is used for testing hypotheses concerning the parameter μ of a Poisson distribution. Beginning at the Start box of the flowchart (Figure 7.18, p. 262), we arrive at the one-sample Poisson test by answering yes to (1) one variable of interest? and to (2) one-sample problem? no to (3) underlying distribution normal or can central-limit theorem be assumed to hold? no to (4) underlying distribution is binomial? and yes to (5) underlying distribution is Poisson?

SECTION 7.11 Case Study: Effects of Tobacco Use on Bone Mineral Density in Middle-Aged Women

In Chapter 6, we compared the bone mineral density of the lumbar spine between the heavier- and lighter-smoking twins using confidence-interval methodology. We now want to consider a similar issue based on hypothesis-testing methods.

Example 7.56 **Endocrinology** The mean difference in bone mineral density (BMD) between the heavier- and lighter-smoking twins when expressed as a percentage of the twin pair mean was –5.0% ± 2.0% (mean ± *se*) based on 41 twin pairs. Assess the statistical significance of the results.

Solution We will use the one-sample t test to test the hypothesis H_0: $\mu = 0$ versus H_1: $\mu \neq 0$, where μ = underlying mean difference in BMD between the heavier- and lighter-smoking twins. Based on Equation 7.10, we have the test statistic

$$t = \frac{\overline{x} - \mu_0}{s/\sqrt{n}}$$

Because $\mu_0 = 0$ and $s/\sqrt{n} = se$, it follows that

$$t = \frac{\overline{x}}{se} = \frac{-5.0}{2.0} = -2.5 \sim t_{40} \text{ under } H_0$$

Using Table 5 in the Appendix, we see that $t_{40,.99} = 2.423$, $t_{40,.995} = 2.704$. Because $2.423 < 2.5 < 2.704$, it follows that $1 - .995 < p/2 < 1 - .99$ or $.005 < p/2 < .01$ or $.01 < p < .02$. Hence, there is a significant difference in mean BMD between the heavier- and lighter-smoking twins with the heavier -smoking twins having lower mean BMD.

SECTION 7.12 Summary

In this chapter some of the fundamental ideas of hypothesis testing were introduced: (1) specification of the null (H_0) and alternative (H_1) hypotheses; (2) type I error (α), type II error (β), and power ($1 - \beta$) of a hypothesis test; (3) the p-value of a hypothesis test; and (4) the distinction between one-sided and two-sided tests. Methods for estimating the appropriate sample size for a proposed study as determined by the prespecified null and alternative hypotheses and type I and type II errors were also discussed.

These general concepts were applied to several one-sample hypothesis-testing situations:

(1) The mean of a normal distribution with unknown variance (one-sample t test)

(2) The mean of a normal distribution with known variance (one-sample z test)

(3) The variance of a normal distribution (one-sample chi-square test)

(4) The parameter p of a binomial distribution (one-sample binomial test)

(5) The expected value μ of a Poisson distribution (one-sample Poisson test)

Each of the hypothesis tests can be conducted in one of two ways:

(1) Specify critical values to determine the acceptance and rejection regions (critical-value method) based on a specified type I error α.

(2) Compute p-values (p-value method).

These methods were shown to be equivalent in the sense that they yield the same inferences regarding acceptance and rejection of the null hypothesis.

Finally, the relationship between the hypothesis-testing methods in this chapter and the confidence-interval methods in Chapter 6 was explored. We showed that the inferences that can be drawn from using these methods are usually the same.

Many hypothesis tests are covered in this book. A flowchart is provided at the back of the book to help clarify the decision process in selecting the appropriate test. The flowchart can be used to choose the proper test by answering a series of yes/no questions. The specific hypothesis tests covered in this chapter have been shaded in an excerpt from the flowchart shown in Figure 7.18 and have been referred to in several places in this chapter. For example, if we are interested in performing hypothesis tests concerning the mean of a normal distribution with known variance, then, beginning at the Start box of the flowchart, we would answer *yes* to each of the following questions: (1) only one variable of interest? (2) one-sample problem? (3) underlying distribution normal, or can central limit theorem be assumed to hold? (4) inference concerning μ? (5) σ known? The flowchart leads us to the box on the lower left of the figure, indicating that the one-sample z test should be used. The boxes marked Go to 1 and Go to 4 refer to other parts of the flowchart in the back of the book.

FIGURE 7.18 Flowchart for appropriate methods of statistical inference

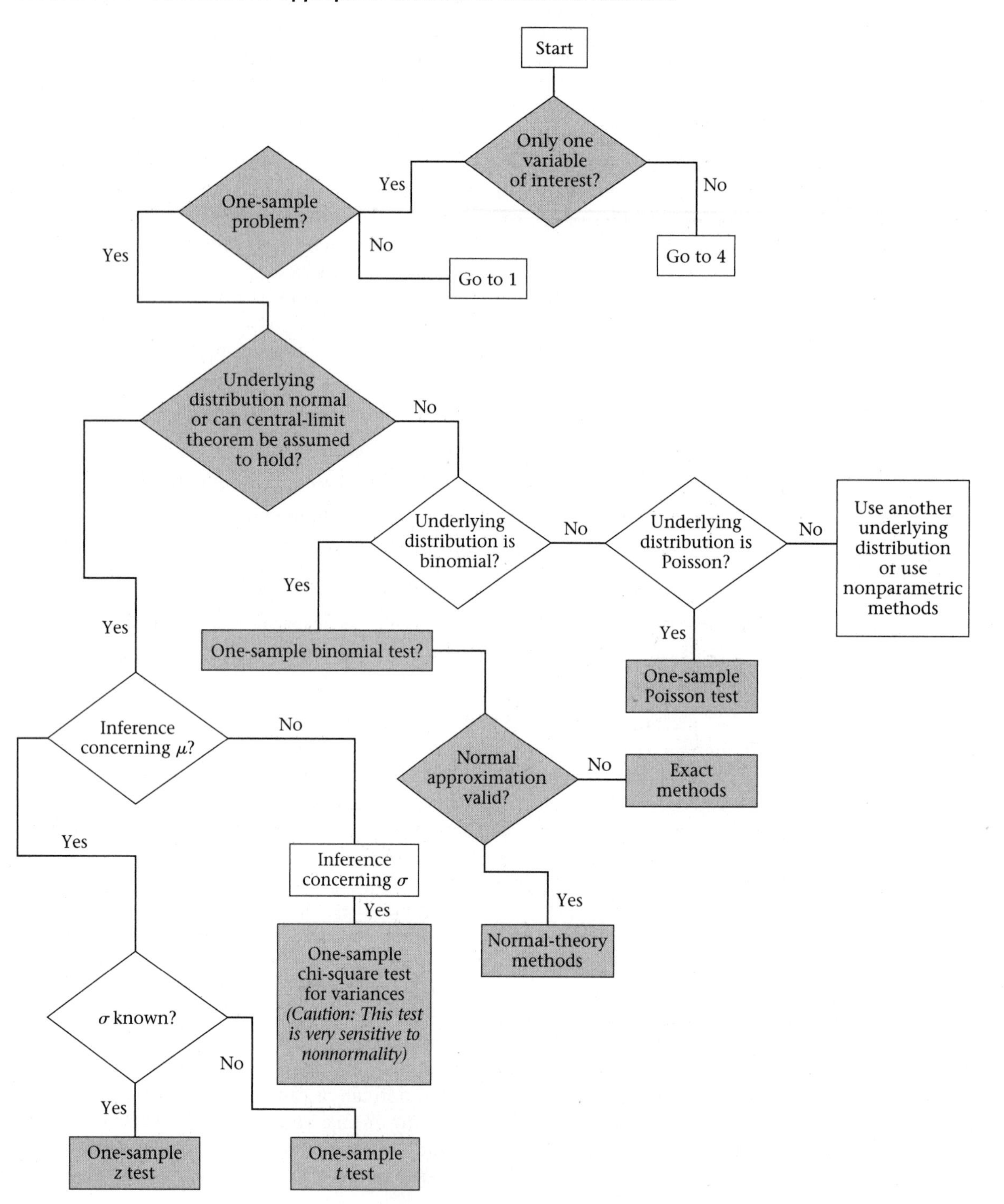

The study of hypothesis testing is extended in Chapter 8 to situations in which two different samples are compared. This topic corresponds to the answer *yes* to the question (1) only one variable of interest? and the answer *no* to (2) one-sample problem?

PROBLEMS

Renal Disease

The mean serum-creatinine level measured in 12 patients 24 hours after they received a newly proposed antibiotic was 1.2 mg/dL.

***7.1** If the mean and standard deviation of serum creatinine in the general population are 1.0 and 0.4 mg/dL, respectively, then, using a significance level of .05, test if the mean serum-creatinine level in this group is different from that of the general population.

***7.2** What is the *p*-value for the test?

7.3 Suppose $\frac{\bar{x}-\mu_0}{s/\sqrt{n}} = -1.52$ and a one-sample *t* test is performed based on 7 subjects. What is the two-tailed *p*-value?

***7.4** Suppose the sample standard deviation of serum creatinine in Problem 7.1 is 0.6 mg/dL. Assume that the standard deviation of serum creatinine is not known, and perform the hypothesis test in Problem 7.1. Report a *p*-value.

Diabetes

Plasma-glucose levels are used to determine the presence of diabetes. Suppose the mean ln (plasma-glucose) concentration (mg/dL) in 35- to 44-year-olds is 4.86 with standard deviation 0.54. A study of 100 sedentary persons in this age group is planned to test if they have a higher or lower level of plasma glucose than the general population.

7.5 If the expected difference is 0.10 ln units, then what is the power of such a study if a two-sided test is to be used with $\alpha = .05$?

7.6 Answer Problem 7.5 if the expected difference is 0.20 ln units.

7.7 How many people would need to be studied to have 80% power under the assumptions in Problem 7.5?

***7.8** Compute a two-sided 95% confidence interval for the true mean serum-creatinine level in Problem 7.4.

***7.9** How does your answer to Problem 7.8 relate to your answer to Problem 7.4?

Cardiovascular Disease

Suppose the incidence rate of MI (myocardial infarction) per year was 5 per 1000 among 45- to 54-year-old men in 1970. To look at changes in incidence over time, 5000 45- to 54-year-old men were followed for 1 year starting in 1980. Fifteen new cases of MI were found.

7.10 Using the critical-value method with $\alpha = .05$, test the hypothesis that incidence rates of MI changed from 1970 to 1980.

7.11 Report a *p*-value to correspond to your answer to Problem 7.10.

Suppose that 25% of MI cases in 1970 died within 24 hours. This proportion is called the 24-hour case-fatality rate.

7.12 Of the 15 new MI cases in the preceding study, 5 died within 24 hours. Test if the 24-hour case-fatality rate changed from 1970 to 1980.

7.13 Suppose we eventually plan to accumulate 50 MI cases during the period 1980–1985. Assume that the 24-hour case-fatality rate is truly 20% during this period. How much power would such a study have in distinguishing between case-fatality rates in 1970 and 1980–1985 if a two-sided test with significance level .05 is planned?

7.14 How large a sample is needed in Problem 7.13 to achieve 90% power?

Pulmonary Disease

Suppose the annual incidence of asthma in the general population among children 0–4 years of age is 1.4% for boys and 1% for girls.

***7.15** If 10 cases are observed over one year among 500 boys 0–4 years of age with smoking mothers, then test if there is a significant difference in asthma incidence between this group and the general population using the critical-value method with a two-sided test.

***7.16** Report a *p*-value corresponding to your answer to Problem 7.15.

***7.17** Suppose that 4 cases are observed over one year among 300 girls 0–4 years of age with smoking mothers. Answer Problem 7.15 based on these data.

***7.18** Report a *p*-value corresponding to your answer to Problem 7.17.

Genetics

Ribosomal 5S RNA can be represented as a sequence of 120 nucleotides. Each nucleotide can be represented by one of four characters: A (adenine), G (guanine), C (cytosine), or U (uracil). The characters occur with different probabilities for each position. We wish to test if a new sequence is the same as ribosomal 5S RNA. For this purpose, we replicate the new sequence 100 times and find that there are 60 A's in the 20th position.

7.19 If the probability of an A in the 20th position is .79 in ribosomal 5S RNA, then test the hypothesis that the new sequence is the same as ribosomal 5S RNA using the critical-value method.

7.20 Report a *p*-value corresponding to your results in Problem 7.19.

Cancer

7.21 Suppose we identify fifty 50- to 54-year-old women who have both a mother *and* a sister with a history of breast cancer. Five of these women themselves have developed breast cancer at some time in their lives. If we assume that the expected prevalence rate of breast cancer in women whose mothers have had breast cancer is 4%, then does having a sister with the disease add to the risk?

Obstetrics

7.22 The probability of having twins in the United States is approximately 1 in 90. This proportion is thought to be affected by a number of factors, including age, race, and parity. To study the effect of age, hospital records are abstracted. Of 538 deliveries for women under 20, 2 were found to have resulted in twins. What can be said about the effect of maternal age on having twins?

Obstetrics

Erythromycin is a drug that has been proposed to possibly lower the risk of premature delivery. A related area of interest is its association with the incidence of side effects during pregnancy. Assume that 30% of all pregnant women complain of nausea between the 24th and 28th week of pregnancy. Furthermore, suppose that of 200 women who are taking erythromycin regularly during the period, 110 complain of nausea.

***7.23** Test the hypothesis that the incidence rate of nausea for the erythromycin group is the same as that for a typical pregnant woman.

Epidemiology

One hundred volunteers agree to participate in a clinical trial involving a dietary intervention. The investigators want to check how representative this sample is of the general population. One interesting finding is that 10 of the volunteers are current cigarette smokers.

7.24 Assuming that 30% of the general population of adults are current smokers, state the hypotheses needed to test whether the volunteer group is representative of the general population regarding cigarette smoking.

7.25 Carry out the test described in Problem 7.24 and report a *p*-value.

Cardiovascular Disease, Nutrition

Much discussion has appeared in the medical literature in recent years on the role of diet in the development of heart disease. The serum-cholesterol levels of a group of people who eat a primarily macrobiotic diet are measured. Among 24 of them, aged 20–39, the mean cholesterol level was found to be 175 mg/dL with a standard deviation of 35 mg/dL.

7.26 If the mean cholesterol level in the general population in this age group is 230 mg/dL and the distribution is assumed to be normal, then test the hypothesis that the group of people on a macrobiotic diet have cholesterol levels different from those of the general population.

7.27 Compute a 95% confidence interval for the true mean cholesterol level in this group.

7.28 What type of complementary information is provided by the hypothesis test and confidence interval in this case?

Cardiovascular Disease

High cholesterol levels have for some time been suspected as predictors of future heart attacks. The problem is that other factors such as age, smoking, body weight, family history, and so forth enter into the picture, and the different factors are difficult to separate out. Suppose we isolate a group of 100 men who have "high" cholesterol levels and can predict, on the basis of other factors, that 10% of these men will have a heart attack in the next 5 years.

7.29 Suppose that the incidence rate of heart attacks in the next 5 years in this group is 13%. Is this finding indicative of anything about cholesterol?

7.30 Suppose we had 1000 men instead of 100 men and also found an incidence rate of 13%. Is this finding indicative of anything about cholesterol?

7.31 How large a sample would be needed to have an 80% chance of finding a significant difference if the true rate of heart disease in this group is 13%?

Hypertension

A pilot study of a new antihypertensive agent is performed for the purpose of planning a larger study. Five patients who have a mean diastolic blood pressure of at least 95 mm Hg are recruited for the study and are kept on the agent for 1 month. After 1 month the observed mean decline in diastolic blood pressure in these 5 patients is 4.8 mm Hg with a standard deviation of 9 mm Hg.

***7.32** If μ_d = true mean difference in diastolic blood pressure between baseline and 1 month, then how many patients would be needed to have a 90% chance of detecting a significant difference using a one-tailed test with a significance level of 5%? Assume that the true mean and standard deviation of the blood-pressure difference was the same as that observed in the pilot study.

***7.33** Suppose we conduct a study of the preceding hypothesis based on 20 subjects. What is the probability that we will be able to reject H_0 using a one-sided test at the 5% level if the true mean and standard deviation of the blood-pressure difference is the same as that in the pilot study?

Occupational Health

The proportion of deaths due to lung cancer in males aged 15–64 in England and Wales during the period 1970–1972 was 12%. Suppose that of 20 deaths that occur among male workers in this age group who have worked for at least 1 year in a chemical plant, 5 are due to lung cancer. We wish to determine if there is a difference between the proportion deaths due to lung cancer in this plant and the proportion in the general population.

7.34 State the hypotheses to be used in answering this question.

7.35 Is a one-sided or two-sided test appropriate here?

7.36 Perform the hypothesis test and report a *p*-value.

After reviewing the results from 1 plant, the company decides to expand its study to include results from 3 additional plants. They find that of 90 deaths occurring among 15- to 64-year-old male workers who have worked for a least 1 year in these 4 plants, 19 are due to lung cancer.

7.37 Answer Problem 7.36 using the data from 4 plants and report a *p*-value.

One criticism of studies of this type is that they are biased because of the "healthy worker" effect. That is, workers in general are healthier than the general population, particularly regarding cardiovascular endpoints, which makes the proportion of deaths due to noncardiovascular causes seem abnormally high.

7.38 If the proportion of deaths due to ischemic heart disease (IHD) is 40% for all 15- to 64-year-old men in England and Wales, whereas 18 of the preceding 90 deaths are attributed to IHD, then answer Problem 7.37 if deaths due to IHD are *excluded* from the total.

Nutrition

Iron-deficiency anemia is an important nutritional health problem in the United States. A dietary assessment was performed on fifty-one 9- to 11-year-old male children whose families were below the poverty level. The mean daily iron intake among these children was found to be 12.50 mg with standard deviation 4.75 mg. Suppose that the mean daily iron intake among a large population of 9- to 11-year-old boys from all income strata is 14.44 mg. We wish to test if the mean iron intake among the low-income group is different from that of the general population.

***7.39** State the hypotheses that can be used to consider this question.

***7.40** Carry out the hypothesis test in Problem 7.39 using the critical-value method with an α level of .05 and summarize your findings.

***7.41** What is the *p*-value for the test conducted in Problem 7.40?

The standard deviation of daily iron intake in the larger population of 9- to 11-year-old boys was 5.56 mg. We wish to test if the standard deviation from the low-income group is comparable to that of the general population.

***7.42** State the hypotheses that can be used to answer this question.

***7.43** Carry out the test in Problem 7.42 using the critical-value method with an α level of .05 and summarize your findings.

***7.44** What is the *p*-value for the test conducted in Problem 7.43?

***7.45** Compute a 95% confidence interval for the underlying variance of daily iron intake in the low-income group. What can you infer from the confidence interval?

7.46 Compare the inferences you made from the procedures in Problems 7.43, 7.44, and 7.45.

Gynecology

A survey of contraceptive methods conducted by the National Center for Health Statistics in 1965 indicated that among 30- to 39-year-old, married, nonpregnant women who practiced contraception, 20% used some form of permanent contraception (either tubal ligation for the women or vasectomy for their spouses). The frequency of use of permanent contraception was suspected to have changed in the next decade. To test this hypothesis, records were selected of 50 women subscribing to a prepaid health plan in 1975 who satisfied the preceding demographic criteria. From the records, 35 of the women were found to have practiced some form of contraception, of whom 10 used the method of permanent contraception. Suppose we wish to test the hypothesis that a change has occurred in the percentage of contraceptors who use permanent contraception, without specifying the direction of the change.

7.47 Specify the hypotheses needed to perform this test.

7.48 Use the critical-value method to conduct the hypothesis test with an α level of .01.

7.49 What is the exact p-value for this test?

The investigators are encouraged by the results of the small study and wish to enlarge the study.

7.50 How large a sample would be needed to test the preceding hypotheses if the following assumptions hold:
(a) The true rate of permanent contraception among contraceptors is 30%.
(b) Seventy percent of the women use some form of contraception.
(c) We wish to have a 90% chance of finding a significant difference using a two-sided test with an α level of .05.

Demography

A study is performed using census data to look at various health parameters for a group of 10,000 Americans of Chinese descent living in the Chinatown areas of New York and San Francisco in 1970. The comparison group for this study is the total U.S. population in 1970. Suppose it is found that 100 of the Chinese have died over a 1-year period and that this represents a 15% decline from the expected mortality rate based on 1970 U.S. age- and sex-specific mortality rates.

7.51 Test if the total mortality experience in this group differs significantly from that in the total U.S. population. Report an exact p-value.

Suppose that 8 deaths due to tuberculosis are found in this group, which is twice the rate expected based on the total U.S. population in 1970.

7.52 Test if the mortality experience due to tuberculosis in this group differs significantly from that in the total U.S. population. Report an exact p-value.

Occupational Health

The mortality experience of 8146 male employees of a research, engineering, and metal-fabrication plant in Tonawanda, New York, was studied from 1946 to 1981 [2]. Potential workplace exposures included welding fumes, cutting oils, asbestos, organic solvents, and environmental ionizing radiation, as a result of waste disposal during the Manhattan Project of World War II. Comparisons were made for specific causes of death between mortality rates in workers and U.S. white male mortality rates from 1950 to 1978.

Suppose that among workers who were hired prior to 1946 and who had worked in the plant for 10 or more years that 17 deaths due to cirrhosis of the liver were observed, while 6.3 were expected based on U.S. white male mortality rates.

7.53 What is the standardized mortality ratio for this group?

7.54 Perform a significance test to assess whether there is an association between long duration of employment and mortality due to cirrhosis of the liver in the group hired prior to 1946. Report a p-value.

7.55 A similar analysis was performed among workers who were hired after 1945 and who were employed for 10 or more years. It found 4 deaths due to cirrhosis of the liver, whereas only 3.4 were expected. What is the SMR for this group?

7.56 Perform a significance test to assess whether there is an association between mortality due to cirrhosis of the liver and duration of employment in the group hired after 1945. Report a p-value.

Ophthalmology

It has been reported that the incidence rate of cataract may be elevated among people with excessive exposure to sunlight. To confirm this, a pilot study is conducted among 200 people age 65–69 who report excessive tendency to burn on exposure to sunlight. It is found that 4 of the 200 people develop cataract over a 1-year period. Suppose that the expected rate of cataract among 65- to 69-year-olds is 1% over a 1-year period.

***7.57** What test procedure can be used to compare the 1-year rate of cataract in this population with that in the general population?

***7.58** Implement the test procedure in Problem 7.57, and report a *p*-value (two-sided).

It is decided to extend the study to a 5-year period, and it is found that 20 of the 200 people develop cataract over a 5-year period. Suppose the expected incidence of cataract among 65- to 69-year-olds in the general population is 5% over a 5-year period.

***7.59** Test the hypothesis that the 5-year incidence rate of cataract is different in the excessive sunlight exposure group compared to the general population and report a *p*-value (two-sided).

***7.60** Construct a 95% confidence interval for the 5-year true rate of cataract among the excessive sunlight exposure group.

Cancer

A study was performed in Basel, Switzerland, on the relationship between the concentration of plasma antioxidant vitamins and cancer risk [3]. Table 7.4 shows data for plasma vitamin-A concentration for stomach-cancer cases and controls.

TABLE 7.4 Plasma vitamin-A concentration (μmol/L) for stomach-cancer cases and controls

	Mean	*se*	*n*
Stomach-cancer cases	2.65	0.11	20
Controls	2.88		2421

7.61 If we assume that the mean plasma vitamin-A concentration among controls is known without error, then what procedure can be used to test if the mean concentration is the same for stomach-cancer cases and controls?

7.62 Perform the test in Problem 7.61 and report a *p*-value (two-sided).

7.63 How many stomach-cancer cases are needed to achieve 80% power if the mean plasma vitamin-A concentration among controls is known without error, the true difference in mean concentration is 0.20 μmol/L, and a two-sided test is used with $\alpha = .05$?

Nutrition, Cardiovascular Disease

Previous studies have shown that supplementation of the diet with oat bran may lower serum-cholesterol levels. However, it is not known whether the cholesterol is reduced by a direct effect of oat bran or by the replacement of fatty foods in the diet. To address this question, a study was performed to compare the effect of dietary supplementation with high-fiber oat bran (87 g per day) to dietary supplementation with a low-fiber refined wheat product on the serum cholesterol of 20 healthy subjects ages 23–49 years old [4]. Each subject had a cholesterol level measured at baseline and then was randomly assigned to receive either a high-fiber or a low-fiber diet for 6 weeks. A 2-week period followed during which no supplements were taken. Subjects then took the alternate supplement for a 6-week period. The results are shown in Table 7.5.

7.64 Test the hypothesis that the high-fiber diet has an effect on cholesterol levels as compared with baseline (report your results as $p < .05$ or $p > .05$).

7.65 Test the hypothesis that the low-fiber diet has an effect on cholesterol levels as compared with baseline (report your results as $p < .05$ or $p > .05$).

TABLE 7.5 Serum-cholesterol levels before and during high-fiber–low-fiber supplemention[a]

	n	Baseline	High fiber	Low fiber	Difference (high fiber – low fiber	Difference (high fiber – baseline)	Difference (low fiber – baseline)
Total cholesterol (mg/dL)	20	186 ± 31	172 ± 28	172 ± 25	–1(–8, +7)	–14(–21, –7)	–13(–20, –6)

[a]Plus–minus values are mean ± *sd*. Values in parentheses are 95% confidence limits.

7.66 Test the hypothesis that the high-fiber diet has a differential effect on cholesterol levels compared to a low-fiber diet (report your results as $p < .05$ or $p > .05$).

7.67 What is the approximate standard error of the mean for the high-fiber compared to the low-fiber diet (that is, the mean difference in cholesterol level between high- and low-fiber diets)?

7.68 How many subjects would be needed to have a 90% chance of finding a significant difference in cholesterol lowering between the high- and low-fiber diets if the high-fiber diet lowers mean cholesterol by 5 mg/dL more than the low-fiber diet and a two-sided test is used with significance level = .05?

Nutrition

Refer to Data Set VALID.DAT, on the data disk.

7.69 Assess if reported nutrient consumption (saturated fat, total fat, alcohol consumption, total caloric intake) is comparable for the diet record and the food-frequency questionnaire. Use either hypothesis-testing and/or confidence-interval methodology.

7.70 Answer Problem 7.69 for the percentage of calories due to fat (separately for total fat and saturated fat) as reported on the diet record and the food-frequency questionnaire. Assume there are 9 calories due to fat for every gram of fat consumed.

Demography

Refer to Data Set SEXRAT.DAT, on the data disk.

7.71 Apply hypothesis-testing methods to answer the questions posed in Problem 4.68.

Cardiology

Refer to Data Set NIFED.DAT, on the data disk.

7.72 Use hypothesis-testing methods to assess if either treatment has an effect on blood pressure or heart rate in patients with severe angina.

Cancer

The combination of photochemotherapy with oral methoxsalen (psoralen) and ultraviolet A radiation (called PUVA treatment) is an effective treatment for psoriasis. However, PUVA is mutagenic, increases the risk of squamous-cell skin cancer, and can cause irregular pigmented skin lesions. Stern et al. [5] performed a study to assess the incidence of melanoma among patients treated with PUVA. The study identified 1380 patients with psoriasis who were first treated with PUVA in 1975 or 1976. Patients were subdivided according to the total number of treatments received (<250 or ≥250 from 1975 to 1996). Within each group, the observed number of melanomas was determined from 1975 to 1996 and compared with the expected number of melanomas as determined by published U.S. age- and sex-specific melanoma incidence rates. The results were as follows:

	Observed	Expected
<250 treatments	5	3.7
≥250 treatments	6	1.1

7.73 Suppose we wish to compare the observed and expected number of events among the group with <250 treatments. Perform an appropriate significance test, and report a two-tailed *p*-value.

7.74 Provide a 95% CI for the expected number of events in the group with ≥250 treatments.

7.75 Interpret the results for Problems 7.73 and 7.74.

Cerebrovascular Disease

A study is to be performed comparing treatment with warfarin versus treatment with aspirin among patients who have had a recent stroke. The primary endpoint is the prevention of a second stroke over an 18-month follow-up period. A substudy of the main study is to assess other risk factors for a second stroke among patients randomized to treatment with aspirin. One potential risk factor is the F_{12} level, a hemostatic factor in blood.

In a previous study, in a group of 63 patients treated with placebo, the mean F_{12} level was 1.57, with a standard deviation 0.794.

7.76 What was the standard error of the mean in the previous study?

7.77 What is the difference in interpretation between the standard deviation and the standard error in this case?

7.78 In the new substudy, one goal is to compare the mean level of F_{12} at baseline between patients with cryptogenic (group C) versus noncryptogenic stroke (group D) at baseline. Suppose (1) the mean level of F_{12} for patients in group C = the mean level in the placebo group in the previous study; (2) the mean level for F_{12} for patients in group D is reduced by 30% relative to group C; (3) the standard deviation of both groups is the same as the placebo group in the previous study; (4) we wish to conduct a two-sided test with type I error = .05; (5) we require a type II error = .20; (6) there are an equal number of subjects in groups C and D in the substudy. How many subjects need to be enrolled in the substudy?

7.79 Suppose that the actual number of patients enrolled in the substudy is 40 patients in group C and 30 patients in group D. How much power will the substudy have if assumptions (1)–(4) in Problem 7.78 hold?

Cancer

Breast cancer is a disease known to be strongly influenced by a woman's reproductive history. In particular, the longer the length of time from the age at menarche (the age when periods begin) to the age at first childbirth, the greater the risk.

A projection was made based on a mathematical model that the 30-year risk of a woman (from age 40 to age 70) of developing breast cancer for women in the general U.S. population is 7%. Suppose a special subgroup of five hundred 40-year-old women without breast cancer was studied whose age at menarche was 17 (compared with an average age at menarche of 13 in the general population) and age at first birth was 20 (compared with an average age at first birth of 25 in the general population). These women were followed for the development of breast cancer from age 40 to age 70. It is found that 18 of them develop breast cancer from age 40 to age 70.

7.80 Test the hypothesis that the underlying rate of breast cancer is the same or different in this group compared with the general population.

7.81 Provide a 95% CI for the true incidence rate of breast cancer over the period from age 40 to 70 in this special subgroup.

7.82 Suppose 100 million women in the U.S. population have not developed breast cancer by the age of 40. What is our best estimate of the number of cases of breast cancer that would be prevented from age 40 to 70 if all women in the U.S. population had an age at menarche of 17 and an age at first birth of 20? Provide a 95% CI for the number of breast-cancer cases prevented.

Ophthalmology

A study is being planned to assess whether a topical anti-allergic eye drop is effective in preventing the signs and symptoms of allergic conjunctivitis. In a pilot study, at an initial visit, subjects are given an allergen challenge; that is, they are subjected to an allergic substance (e.g., cat dander) and their redness score is noted 10 minutes after the allergen challenge (visit 1 score). At a follow-up visit, the same procedure is followed, except that subjects are given an active eye drop in one eye and the placebo in the fellow eye 3 hours before the challenge and a visit 2 score is obtained 10 minutes after the challenge. The following data were collected.

	Active eye mean ± *sd*	Placebo eye mean ± *sd*	Active – placebo eye mean ± *sd*
Change in average redness score[a] (visit 2 – visit 1 score)	–0.61 ± 0.70	–0.04 ± 0.68	–0.57 ± 0.86

[a]The redness score ranges from 0 to 4 in increments of 0.5 where 0 is no redness at all and 4 is severe redness.

7.83 Suppose we wish to estimate the number of subjects needed in the main study so that there will be a 90% chance of finding a significant difference using a two-sided test with a significance level of .05 and we expect the active eyes to have a mean redness score 0.5 unit less than that of the placebo eyes. How many subjects are needed in the main study?

7.84 Suppose 60 subjects are enrolled in the study. How much power would the study have to detect a 0.5-unit mean difference if a two-sided test is used with a significance level of .05?

Ophthalmology

An investigator wants to test a new eye drop that is supposed to prevent ocular itching during allergy season. To study the drug she uses a *contralateral design* whereby for each subject one eye is randomized (using a random-number table) to get active drug (A) while the other eye gets placebo (P). The subjects use the eye drops 3 times a day for a 1-week period and then report their degree of itching in each eye on a 4-point scale (1 = none, 2 = mild, 3 = moderate, 4 = severe) after 1 week (without knowing which eye drop is used in which eye). Ten subjects are randomized into the study.

7.85 What is the principal advantage of the contralateral design?

Suppose the randomization assignment is as given in Table 7.6.

TABLE 7.6 Randomization assignment

	Eye[a]			Eye	
Subject	L	R	Subject	L	R
1	A	P	6	A	P
2	P	A	7	A	P
3	A	P	8	P	A
4	A	P	9	A	P
5	P	A	10	A	P

[a]A = active drug, P = placebo.

7.86 There seem to be more left eyes assigned to A than to P, and the investigator wonders if the assignments are really random. Perform a significance test to assess how well the randomization is working. (*Hint*: Use the binomial tables.)

The itching scores reported by the subjects are given in Table 7.7.

TABLE 7.7 Itching scores reported by subjects

	Eye		
Subject	L	R	Difference[a]
1	1	2	−1
2	3	3	0
3	4	3	1
4	2	4	−2
5	4	1	3
6	2	3	−1
7	2	4	−2
8	3	2	1
9	4	4	0
10	1	2	−1
Mean	2.60	2.80	−0.20
sd	1.17	1.03	1.55
N	10	10	10

[a]Itching score left eye – itching score right eye

7.87 What test can be used to test the hypothesis that the mean degree of itching is the same for active versus placebo eyes?

7.88 Implement the test in Problem 7.87 using a two-sided test (please report a *p*-value).

Endocrinology

Refer to the Data Set BONEDEN.DAT on the data disk.

7.89 Perform a hypothesis test to assess whether there are significant differences in mean bone mineral density (BMD) for the femoral neck between the heavier- and lighter-smoking twins.

7.90 Answer Problem 7.89 for mean BMD at the femoral shaft.

Simulation

Consider the birthweight data in Example 7.2.

7.91 Suppose that the true mean birthweight for the low-SES babies is 120 oz, the true standard deviation is 24 oz, and the distribution of birthweights is normal. Generate 100 random samples of size 100 each from this distribution. Perform the appropriate t test for each sample to test the hypothesis stated in Example 7.2, and compute the proportion of samples where we declare a significant difference using a 5% level of significance with a one-tailed test.

7.92 What should this proportion be for a large number of simulated samples? How do the results in Problem 7.91 compare with this?

7.93 Now assume that the true mean birthweight is 115 oz and repeat the exercise in Problem 7.91 assuming that the other conditions stated in Problem 7.91 are still correct.

7.94 For what proportion of samples do you declare a significant difference? How does this proportion relate to the results in Figure 7.7 (p. 234)?

REFERENCES

[1] Andjelkovic, D., Taulbee, J., & Symons, M. (1976). Mortality experience of a cohort of rubber workers, 1964–1973. *Journal of Occupational Medicine, 18*(6), 387–394.

[2] Teta, M. J., & Ott, M. G. (1988). A mortality study of a research, engineering and metal fabrication facility in western New York State. *American Journal of Epidemiology, 127*(3), 540–551.

[3] Stähelin, H. B., Gey, K. F., Eichholzer, M., Ludin, E., Bernasconi, F., Thurneysen, J., & Brubacher, G. (1991). Plasma antioxidant vitamins and subsequent cancer mortality in the 12-year follow-up of the prospective Basel Study. *American Journal of Epidemiology, 133*(8), 766–775.

[4] Swain, J. F., Rouse, I. L., Curley, C. B., & Sacks, F. M. (1990). Comparison of the effects of oat bran and low-fiber wheat on serum lipoprotein levels and blood pressure. *New England Journal of Medicine, 322*(3), 147–152.

[5] Stern, R. S., Nichols, K. J., & Vakeva, L. H. (1997). Malignant melanoma in patients treated for psoriasis with methoxsalen (Psoralen) and ultraviolet A radiation (PUVA). The PUVA follow-up study. *New England Journal of Medicine, 336*, 1041–1045.

Hypothesis Testing: Two-Sample Inference

SECTION 8.1 Introduction

All the tests introduced in Chapter 7 were one-sample tests. The underlying parameters of the population from which the sample was drawn were compared with comparable values from other generally large populations *whose parameters were assumed to be known.*

Example 8.1 **Obstetrics** In the birthweight data in Example 7.2, the underlying mean birthweight in one hospital was compared with the underlying mean birthweight in the United States, *whose value was assumed known.*

A more frequently encountered situation is the two-sample hypothesis-testing problem.

DEFINITION 8.1 In a **two-sample** hypothesis-testing problem, the underlying parameters of two different populations, *neither of whose values is assumed known,* are compared.

Example 8.2 **Cardiovascular Disease, Hypertension** We might be interested in the relationship between the use of oral contraceptives (OC) and the level of blood pressure (BP) in women.

Two different experimental designs can be used to assess this relationship. One method would involve the following design:

EQUATION 8.1

Longitudinal Study

(1) Identify a group of nonpregnant, premenopausal women of childbearing age (16–49) who are not currently OC users, and measure their blood pressure (BP), which will be referred to as *baseline blood pressure.*

(2) Rescreen these women 1 year later to ascertain a subgroup who have remained nonpregnant throughout the year and have become OC users. This subgroup will be the study population.

(3) Measure the BP of the study population at the follow-up visit.

(4) Compare the baseline and follow-up BP of the women in the study population to determine the difference between the BP of women when they *were* using the pill at follow-up and when they *were not* using the pill at baseline

Another method would involve the following design:

EQUATION 8.2

Cross-Sectional Study

(1) Identify both a group of OC users and a group of non-OC users among non-pregnant, premenopausal women of childbearing age (16–49), and measure their BP.

(2) Compare the BP of the OC users and nonusers.

DEFINITION 8.2 The first type of study is called a **longitudinal** or **follow-up study**, because the same group of women are followed *over time*.

DEFINITION 8.3 The second type of study is called a **cross-sectional study**, because the women are seen at only one point in time.

There is another important difference between these two designs. The first study represents a *paired-sample* design, because each woman is used as her own control. The second study represents an *independent-sample* design, because two completely different groups of women are being compared.

DEFINITION 8.4 Two samples are said to be **paired** when each data point of the first sample is matched and is related to a unique data point of the second sample.

Example 8.3 The paired samples may represent two sets of measurements on the same people. In this case each person is serving as his or her own control, as is the case in Equation 8.1. The paired samples may also represent measurements on different people who are chosen on an individual basis using matching criteria, such as age and sex, to be very similar to each other.

DEFINITION 8.5 Two samples are said to be **independent** when the data points in one sample are unrelated to the data points in the second sample.

Example 8.4 The samples in Equation 8.2 are completely independent, because the data are obtained from unrelated groups of women.

Which type of study is better in this case? The first type of study is probably more definitive, because most other factors that influence a woman's blood pressure at the first screening (referred to as confounders) will also be present at the second screening and will not influence the comparison of blood-pressure levels at the first and second screenings. The second type of study, by itself, can only be considered suggestive, because other confounding factors may influence blood pressure in the two samples and cause an apparent difference to be found where none is actually present.

For example, OC users are known to weigh less than non-OC users. Low weight tends to be associated with low BP, so OC users' blood-pressure levels as a group would appear lower than the levels of non-OC users.

However, a follow-up study is more expensive than a cross-sectional study. Therefore, a cross-sectional study may be the only financially feasible way of doing the study.

In this chapter the appropriate methods of hypothesis testing for both the paired-sample and independent-sample situations are studied.

SECTION 8.2 The Paired *t* Test

Suppose the paired-sample study design in Equation 8.1 is adopted and the sample data in Table 8.1 are obtained. The systolic blood-pressure (SBP) level of the ith woman is denoted at baseline by x_{i1} and at follow-up by x_{i2}.

TABLE 8.1 Systolic blood-pressure levels (mm Hg) in 10 women while not using (baseline) and while using (follow-up) oral contraceptives

i	SBP level while not using OC's (x_{i1})	SBP level while using OC's (x_{i2})	d_i*
1	115	128	13
2	112	115	3
3	107	106	–1
4	119	128	9
5	115	122	7
6	138	145	7
7	126	132	6
8	105	109	4
9	104	102	–2
10	115	117	2

*$d_i = x_{i2} - x_{i1}$

EQUATION 8.3

> Assume that the SBP of the ith woman is normally distributed at baseline with mean μ_i and variance σ^2 and at follow-up with mean $\mu_i + \Delta$ and variance σ^2.

We are thus assuming that the underlying mean difference in SBP between follow-up and baseline is Δ. If $\Delta = 0$, then there is no difference between mean baseline and follow-up SBP. If $\Delta > 0$, then the use of OC pills is associated with an increase in mean SBP. If $\Delta < 0$, then the use of OC pills is associated with a decrease in mean SBP.

We wish to test the hypothesis H_0: $\Delta = 0$ versus H_1: $\Delta \neq 0$. How should this be done? The problem is that μ_i is unknown, and we are assuming, in general, that it is different for each woman. However, consider the difference $d_i = x_{i2} - x_{i1}$. From Equation 8.3 we know that d_i is normally distributed with mean Δ and a variance that shall be denoted by σ_d^2. Thus, although BP levels μ_i are different for each woman, the differences in BP between baseline and follow-up have the same underlying mean (Δ) and variance (σ_d^2) over the entire population of women. The hypothesis-testing problem can thus be considered a *one-sample t test based on the differences* (d_i). From our work on the one-sample t test in Section 7.4, we know that the best test of the hypothesis H_0: $\Delta = 0$ versus H_1: $\Delta \neq 0$, when the variance is unknown, is based on the mean difference

$$\bar{d} = \left(d_1 + d_2 + \cdots + d_n\right)/n$$

Specifically, from Equation 7.10 for a two-sided level α test, we have the following test procedure, which is referred to as the paired t test:

EQUATION 8.4

Paired t Test Denote the test statistic $\bar{d}/(s_d/\sqrt{n})$ by t, where s_d is the sample standard deviation of the observed differences:

$$s_d = \sqrt{\left[\sum_{i=1}^{n} d_i^2 - \left(\sum_{i=1}^{n} d_i\right)^2 \Big/ n\right] \Big/ (n-1)}$$

n = number of matched pairs

If $t > t_{n-1,1-\alpha/2}$ or $t < -t_{n-1,1-\alpha/2}$

then H_0 is rejected. If

$$-t_{n-1,1-\alpha/2} \le t \le t_{n-1,1-\alpha/2}$$

then H_0 is accepted. The acceptance and rejection regions for this test are depicted in Figure 8.1.

FIGURE 8.1 **Acceptance and rejection regions for the paired *t* test**

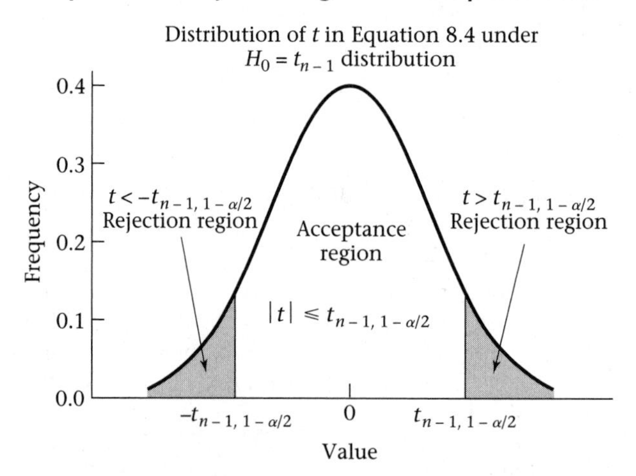

Similarly, from Equation 7.11, a *p*-value for the test can be computed as follows:

EQUATION 8.5

Computation of the *p*-Value for the Paired *t* Test

If $t < 0$,

$$p = 2 \times [\text{the area to the left of } t = \bar{d}/(s_d/\sqrt{n}) \text{ under a } t_{n-1} \text{ distribution}]$$

If $t \geq 0$,

$$p = 2 \times [\text{the area to the right of } t \text{ under a } t_{n-1} \text{ distribution}]$$

The computation of the *p*-value is illustrated in Figure 8.2.

FIGURE 8.2 **Computation of the *p*-value for the paired *t* test**

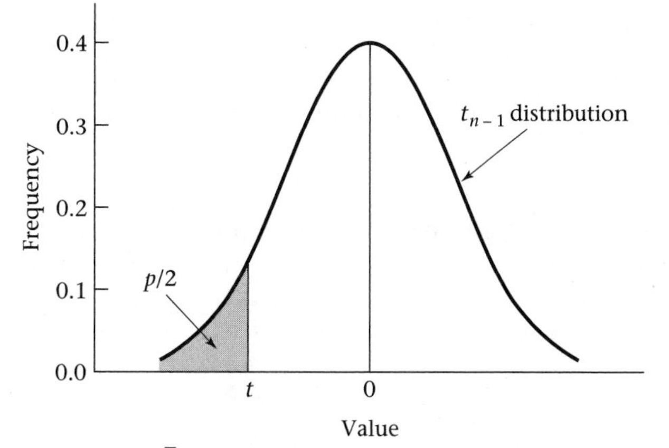

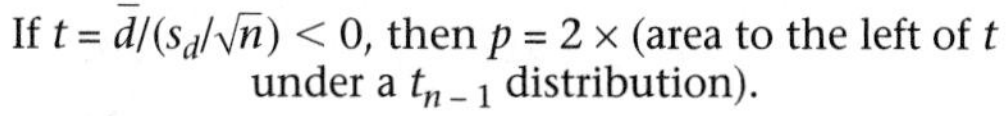
If $t = \bar{d}/(s_d/\sqrt{n}) < 0$, then $p = 2 \times$ (area to the left of t under a t_{n-1} distribution).

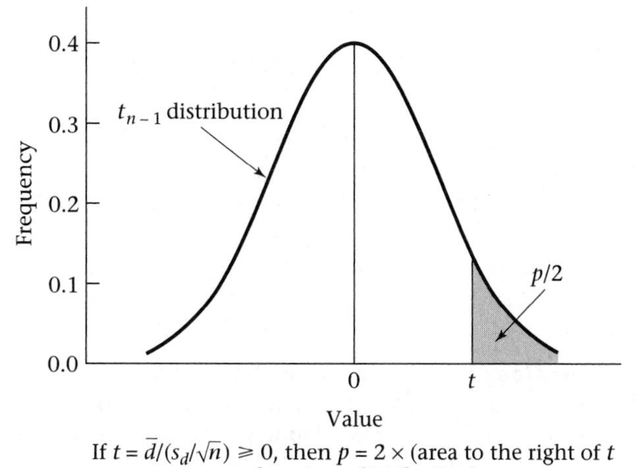

If $t = \bar{d}/(s_d/\sqrt{n}) \geq 0$, then $p = 2 \times$ (area to the right of t under a t_{n-1} distribution).

Example 8.5 **Cardiovascular Disease, Hypertension** Assess the statistical significance of the OC–BP data in Table 8.1.

Solution

$$\bar{d} = (13 + 3 + \cdots + 2)/10 = 4.80$$
$$s_d^2 = \left[(13 - 4.8)^2 + \cdots + (2 - 4.8)^2\right]/9 = 20.844$$
$$s_d = \sqrt{20.844} = 4.566$$
$$t = 4.80/\left(4.566/\sqrt{10}\right) = 4.80/1.444 = 3.32$$

The critical-value method is first used to perform the significance test. There are 10 – 1 = 9 degrees of freedom, and from Table 5 in the Appendix we see that $t_{9,\,.975} = 2.262$. Because $t = 3.32 > 2.262$, it follows from Equation 8.4 that H_0 can be rejected using a two-sided significance test with $\alpha = .05$. To compute an approximate p-value, refer to Table 5 and note that $t_{9,\,.9995} = 4.781$, $t_{9,\,.995} = 3.250$. Thus, because $3.25 < 3.32 < 4.781$, it follows that $.0005 < p/2 < .005$ or $.001 < p < .01$. To compute a more exact p-value, a computer program must be used. The results in Table 8.2 were obtained using the Microsoft Excel 97 T-TEST program.

To use the program, the user specifies the arrays being compared on the spreadsheet (B3:B12 and C3:C12), the number of tails for the p-value (2), and the type of t test (paired t test, type 1).

Note from Table 8.2 that the exact two-sided p-value = .009. Therefore, we can conclude that starting oral-contraceptive use is associated with a significant increase in blood pressure.

TABLE 8.2 **Use of the Microsoft Excel T-TEST program to analyze the blood-pressure data in Table 8.1**

SBP while not using OC's	SBP while using OC's	Paired t test p-value*
115	128	0.008874337
112	115	
107	106	*TTEST(B3:B12,C3:C12,2,1)
119	128	
115	122	
138	145	
126	132	
105	109	
104	102	
115	117	

Example 8.5 is a classic example of a paired study, because each woman is used as her own control. In many other paired studies, different people are used for the two groups, but they are matched individually on the basis of specific matching characteristics.

Example 8.6 **Gynecology** A topic of recent clinical interest is the effect of different contraceptive methods on fertility. In particular, suppose we wish to compare how long it takes users of oral contraceptives and diaphragms, respectively, to become pregnant after stopping contraception. A study group of 20 oral-contraceptive users is formed and diaphragm users who match each OC user on age (within 5 years), race, parity (number of previous pregnancies), and socioeconomic status (SES) are found. The differences in time to fertility between previous OC and diaphragm users are computed and it is found that the mean difference $\bar{d}$ (OC minus diaphragm) in time to fertility is 4 months with a standard deviation (s_d) of 8 months. What can we conclude from these data?

Solution Perform the paired t test. We have

$$t = \bar{d}\Big/\left(s_d/\sqrt{n}\right) = 4\Big/\left(8/\sqrt{20}\right) = 4/1.789 = 2.24 \sim t_{19}$$

under H_0. Referring to Table 5 in the Appendix, we find that

$$t_{19,\,.975} = 2.093 \qquad \text{and} \qquad t_{19,\,.99} = 2.539$$

Then, because $2.093 < 2.24 < 2.539$, it follows that $.01 < p/2 < .025$ or $.02 < p < .05$. Therefore, previous OC users take a significantly longer time to become pregnant than do previous diaphragm users.

In this section, we have introduced the paired t test, which is used to compare the mean level of a normally distributed random variable (or a random variable with samples large enough so that the central-limit theorem can be assumed to hold) between two paired samples. If we refer to the flowchart (Figure 8.13, p. 313), starting from position 1, we would answer yes to (1) two-sample problem? (2) underlying distribution normal or can central-limit theorem be assumed to hold? and (3) inferences concerning means? We would answer no to (4), are samples independent? This leads us to the box entitled "Use paired t test."

SECTION 8.3 Interval Estimation for the Comparison of Means from Two Paired Samples

In the previous section, methods of hypothesis testing for comparing means from two paired samples were discussed. It is also useful to construct confidence limits for the true mean difference (Δ). The observed difference scores = d_i are normally distributed with mean Δ and variance σ_d^2. Thus, the sample mean difference = $\bar{d}$ is normally distributed with mean Δ and variance σ_d^2/n, where σ_d^2 is unknown. The methods of confidence-interval estimation in Equation 6.6 can be used to derive a $100\% \times (1-\alpha)$ confidence interval for Δ, which is given by

$$\left(\bar{d} - t_{n-1,1-\alpha/2}\, s_d/\sqrt{n},\ \bar{d} + t_{n-1,1-\alpha/2}\, s_d/\sqrt{n}\right)$$

EQUATION 8.6

Confidence Interval for the True Difference (Δ) Between the Underlying Means of Two Paired Samples (Two-Sided) A two-sided $100\% \times (1-\alpha)$ confidence interval for the true mean difference (Δ) between two paired samples is given by

$$\left(\bar{d} - t_{n-1,1-\alpha/2}\, s_d/\sqrt{n},\ \bar{d} + t_{n-1,1-\alpha/2}\, s_d/\sqrt{n}\right)$$

Example 8.7 **Cardiovascular Disease, Hypertension** Using the data in Table 8.1, compute a 95% confidence interval for the true increase in mean systolic blood pressure after starting oral contraceptives.

Solution From Example 8.5 we have $\bar{d} = 4.80$ mm Hg, $s_d = 4.566$ mm Hg, $n = 10$. Thus, from Equation 8.6, a 95% confidence interval for the true mean blood-pressure change is given by

$$\bar{d} \pm t_{n-1,.975}\, s_d/\sqrt{n} = 4.80 \pm t_{9,.975}(1.444)$$
$$= 4.80 \pm 2.262(1.444) = 4.80 \pm 3.27 = (1.53, 8.07) \text{ mm Hg}$$

Thus, the true change in mean BP is most likely between 1.5 and 8.1 mm Hg.

Example 8.8 **Gynecology** Using the data in Example 8.6, compute a 95% confidence interval for the true mean difference between OC users and diaphragm users in time to fertility.

Solution From Example 8.6, we have that $\bar{d} = 4$ months, $s_d = 8$ months, $n = 20$. Thus, the 95% confidence interval for μ_d is given by

$$\bar{d} \pm \frac{t_{n-1,.975}s_d}{\sqrt{n}} = 4 \pm \frac{t_{19,.975}(8)}{\sqrt{20}}$$
$$= 4 \pm \frac{2.093(8)}{\sqrt{20}} = 4 \pm 3.74 = (0.26, 7.74) \text{ months}$$

Thus, the true lag in time to fertility can be anywhere from about 0.25 month to nearly 8 months. A much larger study is needed to narrow the width of this confidence interval.

SECTION 8.4 Two-Sample *t* Test for Independent Samples with Equal Variances

The question posed in Example 8.2 will now be discussed, assuming that the cross-sectional study defined in Equation 8.2 rather than the longitudinal study defined in Equation 8.1 is being used.

Example 8.9 **Cardiovascular Disease, Hypertension** Suppose a sample of eight 35- to 39-year-old nonpregnant, premenopausal OC users who work in a company are identified who have mean systolic blood pressure of 132.86 mm Hg and sample standard deviation of 15.34 mm Hg. A sample of twenty-one 35- to 39-year-old nonpregnant, premenopausal non-OC users are similarly identified who have mean systolic blood pressure of 127.44 mm Hg and sample standard deviation of 18.23 mm Hg. What can be said about the underlying mean difference in blood pressure between the two groups?

Assume that systolic blood pressure (SBP) is normally distributed in the first group with mean μ_1 and variance σ_1^2 and in the second group with mean μ_2 and variance σ_2^2. We wish to test the hypothesis H_0: $\mu_1 = \mu_2$ versus H_1: $\mu_1 \neq \mu_2$. Assume in this section that the underlying variances in the two groups are the same

$(\sigma_1^2 = \sigma_2^2 = \sigma^2)$. The means and variances in the two samples are denoted by $\bar{x}_1, \bar{x}_2, s_1^2, s_2^2$, respectively.

It seems reasonable to base the significance test on the difference between the two sample means, $\bar{x}_1 - \bar{x}_2$. If this difference is far from 0, then H_0 will be rejected; otherwise, it will be accepted. Thus, we wish to study the behavior of $\bar{x}_1 - \bar{x}_2$ under H_0. We know that $\bar{X}_1$ is normally distributed with mean μ_1 and variance σ^2/n_1 and that $\bar{X}_2$ is normally distributed with mean μ_2 and variance σ^2/n_2. Hence, from Equation 5.10, since the two samples are independent, $\bar{X}_1 - \bar{X}_2$ is normally distributed with mean $\mu_1 - \mu_2$ and variance $\sigma^2(1/n_1 + 1/n_2)$. In symbols,

EQUATION 8.7

$$(\bar{X}_1 - \bar{X}_2) \sim N\left[\mu_1 - \mu_2, \sigma^2\left(\frac{1}{n_1} + \frac{1}{n_2}\right)\right]$$

Under H_0, we know that $\mu_1 = \mu_2$. Thus, Equation 8.7 reduces to

EQUATION 8.8

$$(\bar{X}_1 - \bar{X}_2) \sim N\left[0, \sigma^2\left(\frac{1}{n_1} + \frac{1}{n_2}\right)\right]$$

If σ^2 were known, then $\bar{X}_1 - \bar{X}_2$ could be divided by $\sigma\sqrt{1/n_1 + 1/n_2}$. From Equation 8.8,

EQUATION 8.9

$$\frac{\bar{X}_1 - \bar{X}_2}{\sigma\sqrt{\frac{1}{n_1} + \frac{1}{n_2}}} \sim N(0,1)$$

and the test statistic in Equation 8.9 could be used as a basis for the hypothesis test. Unfortunately, σ^2 in general is unknown and must be estimated from the data. How can σ^2 be best estimated in this situation?

From the first and second sample, the sample variances are s_1^2, s_2^2, respectively, each of which could be used to estimate σ^2. The average of s_1^2 and s_2^2 could simply be used as the estimate of σ^2. However, this average will weight the sample variances equally even if the sample sizes are very different from each other. The sample variances should not be weighted equally, because the sample variance from the larger sample is probably more precise and should be weighted more heavily. The best estimate of the population variance σ^2, which is denoted by s^2, is given by a weighted average of the two sample variances, where the weights are the number of degrees of freedom in each sample.

EQUATION 8.10

The **pooled estimate of the variance** from two independent samples is given by

$$s^2 = \frac{(n_1 - 1)s_1^2 + (n_2 - 1)s_2^2}{n_1 + n_2 - 2}$$

In particular, s^2 will then have $n_1 - 1$ *df* from the first sample and $n_2 - 1$ *df* from the second sample, or

$$(n_1 - 1) + (n_2 - 1) = n_1 + n_2 - 2 \; df$$

overall. Then s can be substituted for σ in Equation 8.9, and the resulting test statistic can then be shown to follow a t distribution with $n_1 + n_2 - 2$ *df* rather than an $N(0, 1)$ distribution, because σ^2 is unknown. Thus the following test procedure is used:

EQUATION 8.11

Two-Sample *t* Test for Independent Samples with Equal Variances Suppose we wish to test the hypothesis H_0: $\mu_1 = \mu_2$ versus H_1: $\mu_1 \neq \mu_2$ with a significance level of α for two normally distributed populations, where σ^2 is assumed to be the same for each population.

Compute the test statistic:

$$t = \frac{\bar{x}_1 - \bar{x}_2}{s\sqrt{\frac{1}{n_1} + \frac{1}{n_2}}}$$

where $s = \sqrt{\left[(n_1 - 1)s_1^2 + (n_2 - 1)s_2^2\right] / (n_1 + n_2 - 2)}$

If $t > t_{n_1+n_2-2,1-\alpha/2}$ or $t < -t_{n_1+n_2-2,1-\alpha/2}$

then H_0 is rejected. If

$$-t_{n_1+n_2-2,1-\alpha/2} \le t \le t_{n_1+n_2-2,1-\alpha/2}$$

then H_0 is accepted.

The acceptance and rejection regions for this test are depicted in Figure 8.3.

FIGURE 8.3 **Acceptance and rejection regions for the two-sample *t* test for independent samples with equal variances**

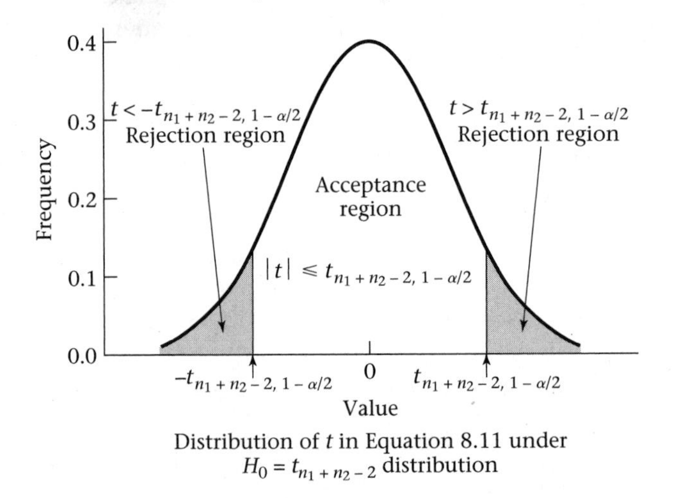

Similarly, a p-value can be computed for the test. The computation of the p-value will depend on whether $\bar{x}_1 \le \bar{x}_2$ ($t \le 0$) or $\bar{x}_1 > \bar{x}_2$ ($t > 0$). In each case, the p-value corresponds to the probability of obtaining a test statistic at least as extreme as the observed value t. This is given in Equation 8.12.

EQUATION 8.12

Computation of the p-Value for the Two-Sample t Test for Independent Samples with Equal Variances

Compute the test statistic:

$$t = \frac{\bar{x}_1 - \bar{x}_2}{s\sqrt{\frac{1}{n_1} + \frac{1}{n_2}}}$$

where $s = \sqrt{\left[(n_1 - 1)s_1^2 + (n_2 - 1)s_2^2\right]/(n_1 + n_2 - 2)}$

If $t \le 0$, $p = 2 \times$ (area to the left of t under a $t_{n_1+n_2-2}$ distribution)

If $t > 0$, $p = 2 \times$ (area to the right of t under a $t_{n_1+n_2-2}$ distribution)

The computation of the p-value is illustrated in Figure 8.4.

Example 8.10 **Cardiovascular Disease, Hypertension** Assess the statistical significance of the data in Example 8.9.

Solution The common variance is first estimated:

$$s^2 = \frac{7(15.34)^2 + 20(18.23)^2}{27} = \frac{8293.9}{27} = 307.18$$

or $s = 17.527$. The following test statistic is then computed:

$$t = \frac{132.86 - 127.44}{17.527\sqrt{1/8 + 1/21}} = \frac{5.42}{17.527 \times 0.415} = \frac{5.42}{7.282} = 0.74$$

If the critical-value method is used, then note that under H_0, t comes from a t_{27} distribution. Referring to Table 5 in the Appendix, we see that $t_{27,.975} = 2.052$. Because $-2.052 \le 0.74 \le 2.052$, it follows that H_0 is accepted using a two-sided test at the 5% level, and we conclude that the mean blood pressures of the two groups of OC users and non-OC users are not significantly different from each other. In a sense, this result shows the superiority of the longitudinal design in Example 8.5. Despite the similarity in the magnitudes of the mean blood-pressure differences between users and nonusers in the two studies, significant differences could be detected in Example 8.5, in contrast to the nonsignificant results that were obtained using the preceding cross-sectional design. The longitudinal design is usually more efficient because it uses people as their own controls.

To compute an approximate p-value, note from Table 5 that $t_{27,.75} = 0.684$, $t_{27,.80} = 0.855$. Because $0.684 < 0.74 < 0.855$, it follows that $.2 < p/2 < .25$ or $.4 < p < .5$. The exact p-value obtained from MINITAB is $p = 2 \times P(t_{27} > 0.74) = .46$.

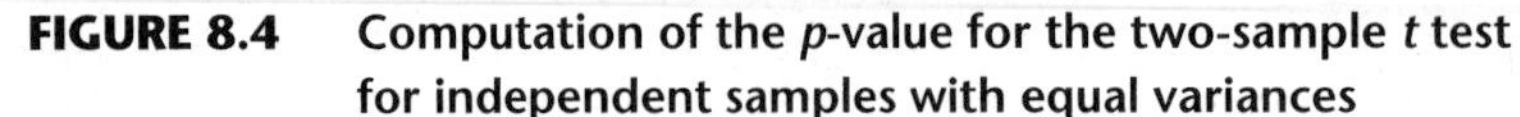

FIGURE 8.4 **Computation of the *p*-value for the two-sample *t* test for independent samples with equal variances**

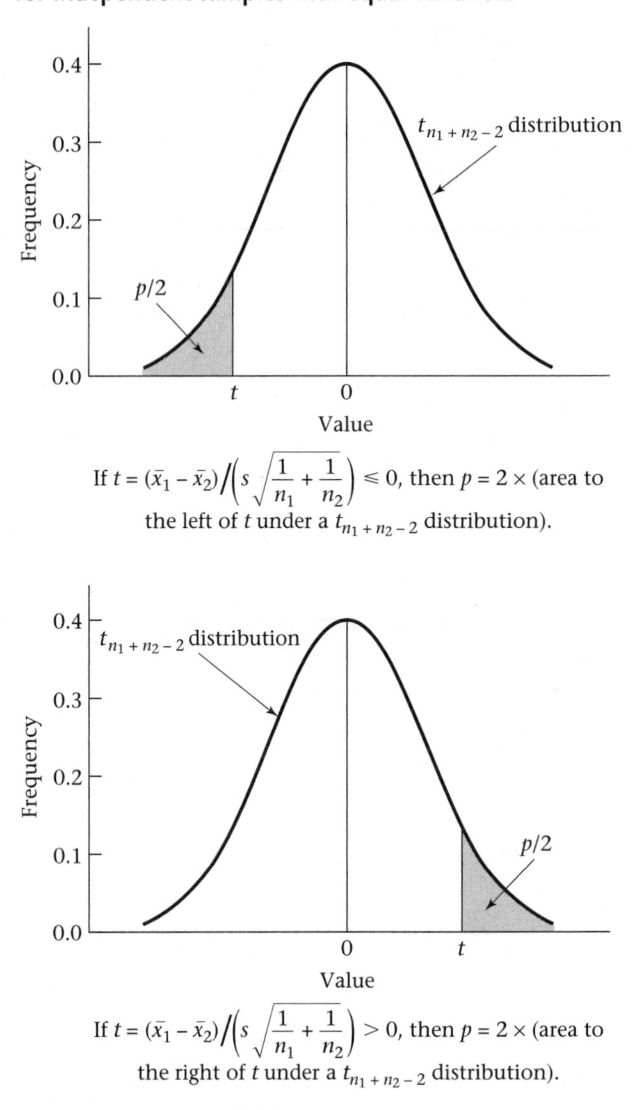
0.4
0.3
0.2
0.1
0.0
Frequency
$t_{n_1+n_2-2}$ distribution
$p/2$
t
0
Value
If $t = (\bar{x}_1 - \bar{x}_2) \Big/ \left(s\sqrt{\frac{1}{n_1}+\frac{1}{n_2}}\right) \le 0$, then $p = 2 \times$ (area to the left of t under a $t_{n_1+n_2-2}$ distribution).
0.4
0.3
0.2
0.1
0.0
Frequency
$t_{n_1+n_2-2}$ distribution
$p/2$
0
t
Value
If $t = (\bar{x}_1 - \bar{x}_2) \Big/ \left(s\sqrt{\frac{1}{n_1}+\frac{1}{n_2}}\right) > 0$, then $p = 2 \times$ (area to the right of t under a $t_{n_1+n_2-2}$ distribution).

SECTION 8.5 Interval Estimation for the Comparison of Means from Two Independent Samples (Equal Variance Case)

In the previous section, methods of hypothesis testing for the comparison of means from two independent samples were discussed. It is also useful to compute $100\% \times (1-\alpha)$ confidence limits for the true mean difference between the two groups = $\mu_1 - \mu_2$. From Equation 8.7, if σ is known, then $(\overline{X}_1 - \overline{X}_2) \sim N[\mu_1 - \mu_2, \sigma^2(1/n_1 + 1/n_2)]$ or, equivalently,

$$\frac{\left(\bar{X}_1 - \bar{X}_2\right) - \left(\mu_1 - \mu_2\right)}{\sigma\sqrt{\dfrac{1}{n_1} + \dfrac{1}{n_2}}} \sim N(0,1)$$

If σ is unknown, then σ is estimated by s from Equation 8.10 and

$$\frac{\left(\bar{x}_1 - \bar{x}_2\right) - \left(\mu_1 - \mu_2\right)}{s\sqrt{\dfrac{1}{n_1} + \dfrac{1}{n_2}}} \sim t_{n_1+n_2-2}$$

To construct a two-sided $100\% \times (1-\alpha)$ confidence interval, note that

$$Pr\left[-t_{n_1+n_2-2,1-\alpha/2} \le \frac{\left(\bar{x}_1 - \bar{x}_2\right) - \left(\mu_1 - \mu_2\right)}{s\sqrt{\dfrac{1}{n_1} + \dfrac{1}{n_2}}} \le t_{n_1+n_2-2,1-\alpha/2}\right] = 1-\alpha$$

This can be written in the form of two inequalities:

$$-t_{n_1+n_2-2,1-\alpha/2} \le \frac{\left(\bar{x}_1 - \bar{x}_2\right) - \left(\mu_1 - \mu_2\right)}{s\sqrt{\dfrac{1}{n_1} + \dfrac{1}{n_2}}}$$

and
$$\frac{\left(\bar{x}_1 - \bar{x}_2\right) - \left(\mu_1 - \mu_2\right)}{s\sqrt{\dfrac{1}{n_1} + \dfrac{1}{n_2}}} \le t_{n_1+n_2-2,1-\alpha/2}$$

Each inequality is multiplied by $s\sqrt{\dfrac{1}{n_1} + \dfrac{1}{n_2}}$ and $\mu_1 - \mu_2$ is added to both sides to obtain

$$\left(\mu_1 - \mu_2\right) - t_{n_1+n_2-2,1-\alpha/2}\, s\sqrt{\frac{1}{n_1} + \frac{1}{n_2}} \le \bar{x}_1 - \bar{x}_2$$

and $\bar{x}_1 - \bar{x}_2 \le \left(\mu_1 - \mu_2\right) + t_{n_1+n_2-2,1-\alpha/2}\, s\sqrt{\dfrac{1}{n_1} + \dfrac{1}{n_2}}$

Finally, $t_{n_1+n_2-2,1-\alpha/2}\, s\sqrt{\dfrac{1}{n_1} + \dfrac{1}{n_2}}$ is added to both sides of the first inequality and subtracted from both sides of the second inequality to obtain

$$\mu_1 - \mu_2 \le \left(\bar{x}_1 - \bar{x}_2\right) + t_{n_1+n_2-2,1-\alpha/2}\, s\sqrt{\frac{1}{n_1} + \frac{1}{n_2}}$$

$$\left(\bar{x}_1 - \bar{x}_2\right) - t_{n_1+n_2-2,1-\alpha/2}\, s\sqrt{\frac{1}{n_1} + \frac{1}{n_2}} \le \mu_1 - \mu_2$$

If these two inequalities are combined, the required confidence interval is obtained.

$$\left[(\bar{x}_1-\bar{x}_2)-t_{n_1+n_2-2,1-\alpha/2}s\sqrt{\frac{1}{n_1}+\frac{1}{n_2}},\ (\bar{x}_1-\bar{x}_2)+t_{n_1+n_2-2,1-\alpha/2}s\sqrt{\frac{1}{n_1}+\frac{1}{n_2}}\right]$$

This is summarized as follows:

EQUATION 8.13

Confidence Interval for the Underlying Mean Difference ($\mu_1 - \mu_2$) Between Two Groups (Two-Sided) ($\sigma_1^2 = \sigma_2^2$) A two-sided $100\% \times (1 - \alpha)$ confidence interval for the true mean difference $\mu_1 - \mu_2$ based on two independent samples is given by

$$\left(\bar{x}_1-\bar{x}_2-t_{n_1+n_2-2,1-\alpha/2}s\sqrt{\frac{1}{n_1}+\frac{1}{n_2}},\ \bar{x}_1-\bar{x}_2+t_{n_1+n_2-2,1-\alpha/2}s\sqrt{\frac{1}{n_1}+\frac{1}{n_2}}\right)$$

Example 8.11 **Cardiovascular Disease, Hypertension** Using the data in Examples 8.9 and 8.10, compute a 95% confidence interval for the true mean difference in blood pressure between the two OC-use groups.

Solution A 95% confidence interval for the underlying mean difference in systolic blood pressure between the population of 35- to 39-year-old OC users and non-OC users is given by

$$\begin{aligned}&\left[5.42-t_{27,.975}(7.282),5.42+t_{27,.975}(7.282)\right]\\&=\left[5.42-2.052(7.282),5.42+2.052(7.282)\right]=(-9.52,20.36)\end{aligned}$$

This interval is rather wide and indicates that a much larger sample is needed to accurately assess the true mean difference.

In this section, we have introduced the two-sample t test for independent samples with equal variances. This test is used to compare the mean level of a normally distributed random variable (or a random variable with samples large enough so that the central-limit theorem can be assumed to hold) between two independent samples with similar variances. If we refer to the flowchart (Figure 8.13, p. 313), starting from position 1 we would answer yes to (1) two-sample problem? (2) underlying distribution normal or can central-limit theorem be assumed to hold? (3) inferences concerning means? (4) are samples independent? and no to (5) are variances of two samples significantly different? (discussed in Section 8.6). This leads us to the box entitled "Use the two-sample t test with equal variances."

SECTION 8.6 Testing for the Equality of Two Variances

In Section 8.4, when a two-sample t test for independent samples was conducted, it was assumed that the underlying variances of the two samples were the same. The common variance was then estimated using a weighted average of the individual sample variances. In this section a significance test to validate this assumption is developed. In particular, we want to test the hypothesis H_0: $\sigma_1^2 = \sigma_2^2$ versus H_1: $\sigma_1^2 \neq \sigma_2^2$, where the two samples are assumed to be independent random samples from an $N(\mu_1, \sigma_1^2)$ and $N(\mu_2, \sigma_2^2)$ distribution, respectively.

Example 8.12 **Cardiovascular Disease, Pediatrics** Consider a problem that was discussed earlier, namely the familial aggregation of cholesterol levels. In particular, suppose cholesterol levels are assessed in one hundred 2- to 14-year-old children of men who have died from heart disease and it is found that the mean cholesterol level in the group $(\bar{x}_1)$ is 207.3 mg/dL. Suppose the sample standard deviation in this group (s_1) is 35.6 mg/dL. Previously, the cholesterol levels in this group of children were compared with 175 mg/dL, which was assumed to be the underlying mean level in children in this age group based on previous large studies.

A better experimental design would be to select a group of control children whose fathers are alive and do not have heart disease and who are from the same census tract as the case children and then compare their cholesterol levels with those of the case children. If the case fathers are identified by a search of death records from the census tract, then control children who live in the same census tract as the case families but whose fathers have no history of heart disease can be selected. The case and control children come from the same census tract but are *not* individually matched. Thus, they are considered as two independent samples rather than as two paired samples. The cholesterol levels in these children can then be measured. Suppose this procedure is done and it is found that among 74 control children, the mean cholesterol level $(\bar{x}_2)$ is 193.4 mg/dL with a sample standard deviation (s_2) of 17.3 mg/dL. We would like to compare the means of these two groups using the two-sample t test for independent samples given in Equation 8.11, but we are hesitant to assume equal variances because the sample variance of the case group is about 4 times as large as that of the control group:

$$35.6^2/17.3^2 = 4.23$$

What should be done?

What is needed is a significance test to determine if the underlying variances are in fact equal; that is, we wish to test the hypothesis H_0: $\sigma_1^2 = \sigma_2^2$ versus H_1: $\sigma_1^2 \neq \sigma_2^2$. It seems reasonable to base the significance test on the relative magnitudes of the sample variances (s_1^2, s_2^2). The best test in this case is based on the ratio of the sample variances (s_1^2/s_2^2) rather than on the difference between the sample variances $(s_1^2 - s_2^2)$. Thus, H_0 would be rejected if the variance ratio is either too large or too small and accepted otherwise. To implement this test, the sampling distribution of s_1^2/s_2^2 under the null hypothesis $\sigma_1^2 = \sigma_2^2$ must be determined.

8.6.1 The *F* Distribution

The distribution of the variance ratio (s_1^2/s_2^2) was studied by the statisticians R. A. Fisher and G. Snedecor. It can be shown that the variance ratio follows an ***F* distribution** under the null hypothesis that $\sigma_1^2 = \sigma_2^2$. There is not a unique F distribution but instead a family of F distributions. This family is indexed by two parameters termed the *numerator* and *denominator degrees of freedom* (*df*), respectively. Specifically, if the sample sizes of the first and second samples are n_1 and n_2, respectively,

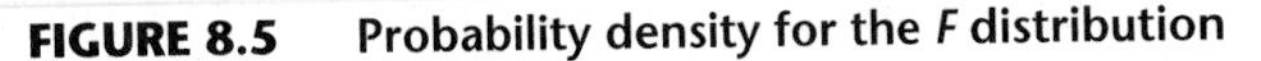

FIGURE 8.5 **Probability density for the *F* distribution**

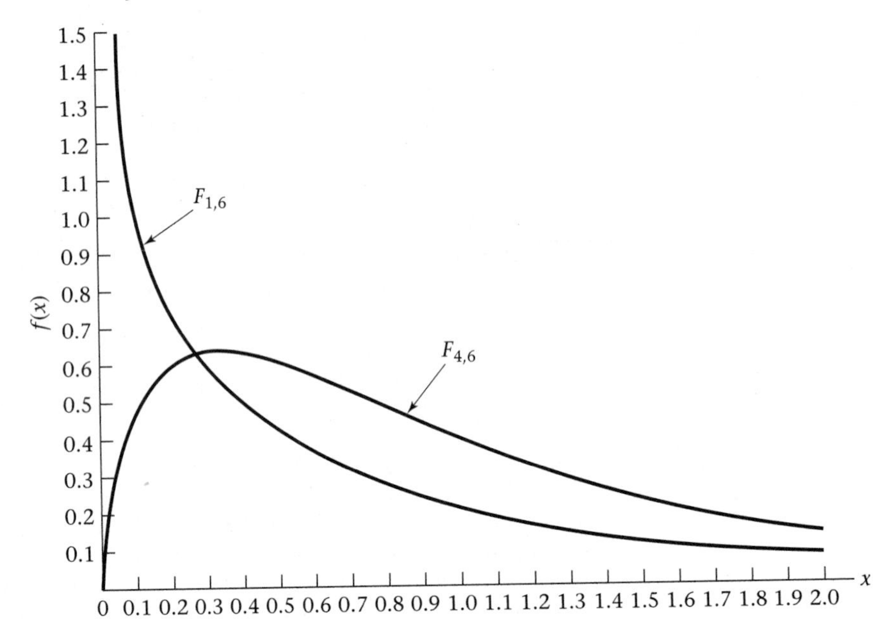

then the variance ratio follows an F distribution with $n_1 - 1$ (numerator df) and $n_2 - 1$ (denominator df), which is denoted by F_{n_1-1, n_2-1}.

The F distribution is generally positively skewed, with the skewness dependent on the relative magnitudes of the two degrees of freedom. If the numerator degrees of freedom is 1 or 2, then the distribution has a mode at 0; otherwise it has a mode at some point greater than 0. The distribution is illustrated in Figure 8.5. Table 9 in the Appendix gives the percentiles of the F distribution.

DEFINITION 8.6 The **100 × *p*th percentile of an *F* distribution** with d_1 and d_2 degrees of freedom is denoted by $F_{d_1, d_2, p}$. Thus

$$Pr\left(F_{d_1, d_2} \le F_{d_1, d_2, p}\right) = p$$

The F table is organized such that the different numerator df (d_1) is shown in the first row, the different denominator df (d_2) is shown in the first column, and the various percentiles (p) are shown in the second column.

Example 8.13 Find the upper 1st percentile of an F distribution with 5 and 9 degrees of freedom.

Solution $F_{5, 9, .99}$ must be found. Look in the 5 column, the 9 row, and the sub row marked .99 to obtain

$$F_{5, 9, .99} = 6.06$$

Generally, F distribution tables give only upper percentage points because the symmetry properties of the F distribution make it possible to derive the lower percentage points of any F distribution from the corresponding upper percentage points of an F distribution with the appropriate degrees of freedom. Specifically, note that under H_0, s_2^2/s_1^2 follows an F_{d_2,d_1} distribution. Therefore,

$$Pr\left(s_2^2/s_1^2 \ge F_{d_2,d_1,1-p}\right) = p$$

By taking the inverse of each side and reversing the direction of the inequality, we get

$$Pr\left(\frac{s_1^2}{s_2^2} \le \frac{1}{F_{d_2,d_1,1-p}}\right) = p$$

Under H_0, however, s_1^2/s_2^2 follows an F_{d_1,d_2} distribution. Therefore,

$$Pr\left(\frac{s_1^2}{s_2^2} \le F_{d_1,d_2,p}\right) = p$$

It follows from the last two inequalities that

$$F_{d_1,d_2,p} = \frac{1}{F_{d_2,d_1,1-p}}$$

This principle is summarized as follows:

EQUATION 8.14

Computation of the Lower Percentiles of an *F* Distribution The lower pth **percentile** of an F distribution with d_1 and d_2 *df* is the reciprocal of the **upper pth percentile** of an F distribution with d_2 and d_1 *df*. In symbols,

$$F_{d_1,d_2,p} = 1/F_{d_2,d_1,1-p}$$

Thus, from Equation 8.14 we see that the lower pth percentile of an F distribution is the same as the inverse of the upper pth percentile of an F distribution with the degrees of freedom reversed.

Example 8.14 Estimate $F_{6,8,.05}$.

Solution From Equation 8.14, $F_{6,8,.05} = 1/F_{8,6,.95} = 1/4.15 = 0.241$

8.6.2 The *F* Test

We now return to the significance test for the equality of two variances. We wish to test the hypothesis H_0: $\sigma_1^2 = \sigma_2^2$ versus H_1: $\sigma_1^2 \ne \sigma_2^2$. We stated that the test would be based on the variance ratio s_1^2/s_2^2, which under H_0 follows an F distribution with $n_1 - 1$ and $n_2 - 1$ *df*. This is a two-sided test, so we wish to reject H_0 for both small and large values of s_1^2/s_2^2. This procedure can be made more specific, as follows:

EQUATION 8.15

***F* Test for the Equality of Two Variances** Suppose we wish to conduct a test of the hypothesis H_0: $\sigma_1^2 = \sigma_2^2$ versus H_1: $\sigma_1^2 \neq \sigma_2^2$ with significance level α.

Compute the test statistic $F = s_1^2/s_2^2$. If

$$F > F_{n_1-1,n_2-1,1-\alpha/2} \quad \text{or} \quad F < F_{n_1-1,n_2-1,\alpha/2}$$

then H_0 is rejected. If

$$F_{n_1-1,n_2-1,\alpha/2} \leq F \leq F_{n_1-1,n_2-1,1-\alpha/2}$$

then H_0 is accepted. The acceptance and rejection regions for this test are shown in Figure 8.6.

FIGURE 8.6 Acceptance and rejection regions for the *F* test for the equality of two variances

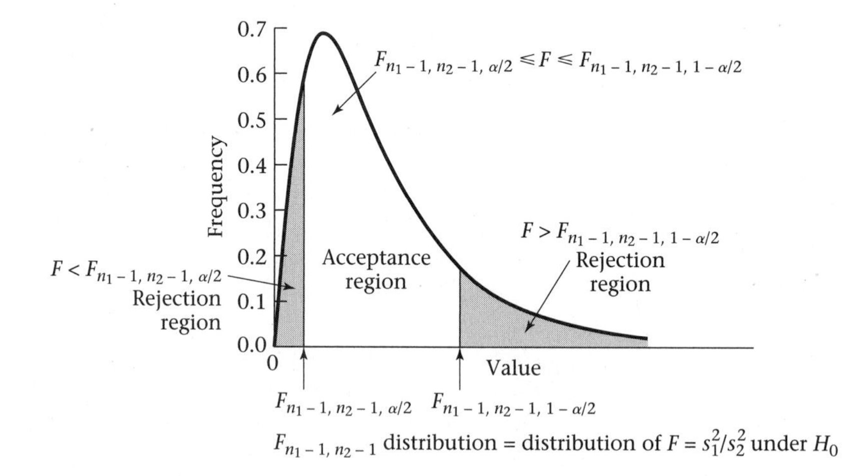

Alternatively, the exact *p*-value is given by

EQUATION 8.16

Computation of the *p*-Value for the *F* Test for the Equality of Two Variances
Compute the test statistic $F = s_1^2/s_2^2$.

If $F \geq 1$, then $\quad p = 2 \times Pr\left(F_{n_1-1,n_2-1} > F\right)$

If $F < 1$, then $\quad p = 2 \times Pr\left(F_{n_1-1,n_2-1} < F\right)$

This computation is illustrated in Figure 8.7.

Example 8.15 **Cardiovascular Disease, Pediatrics** Test for the equality of the two variances given in Example 8.12.

Solution $F = s_1^2/s_2^2 = 35.6^2/17.3^2 = 4.23$

Because the two samples have 100 and 74 people, respectively, we know from Equation 8.15 that under H_0, $F \sim F_{99,73}$. Thus, H_0 is rejected if

$$F > F_{99,73,.975} \quad \text{or} \quad F < F_{99,73,.025}$$

FIGURE 8.7 Computation of the p-value for the F test for the equality of two variances

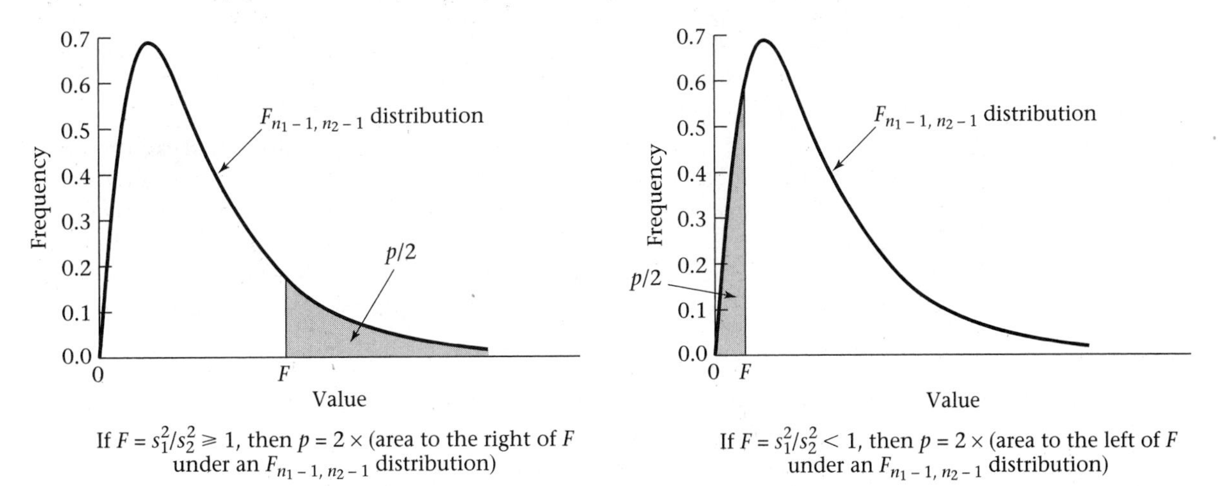

0.7
0.6
0.5
0.4
0.3
0.2
0.1
0.0
Frequency
$F_{n_1-1,\, n_2-1}$ distribution
$p/2$
0
F
Value
If $F = s_1^2/s_2^2 \geq 1$, then $p = 2 \times$ (area to the right of F under an $F_{n_1-1,\, n_2-1}$ distribution)
$p/2$
0 F
Value
If $F = s_1^2/s_2^2 < 1$, then $p = 2 \times$ (area to the left of F under an $F_{n_1-1,\, n_2-1}$ distribution)

Note that neither 99 *df* nor 73 *df* appear in Table 9 in the Appendix. One approach is to obtain the percentiles using a computer program. In this example, we wish to find the value $c_1 = F_{99,73,.025}$ and $c_2 = F_{99,73,.975}$ such that

$$Pr\left(F_{99,73} \leq c_1\right) = .025 \quad \text{and} \quad Pr\left(F_{99,73} \geq c_2\right) = .975$$

The result is shown in Table 8.3 using the FINV function of Microsoft Excel 97, where the first argument of FINV is the desired right-hand tail area and the next two arguments are the numerator and denominator *df*, respectively.

TABLE 8.3 Computation of critical values for the cholesterol data in Example 8.15 using Excel 97

Numerator df	99	
Denominator df	73	
Percentile		
0.025	0.6547598	FINV(.975, 99, 73)
0.975	1.54907909	FINV(.025, 99, 73)

Thus, $c_1 = 0.6548$ and $c_2 = 1.5491$. Because $F = 4.23 > c_2$, it follows that $p < .05$. Alternatively, we could compute the exact p-value. This is given by $p = 2 \times Pr(F_{99,\,73} \geq 4.23)$.

TABLE 8.4 Computation of the exact p-value in Example 8.15 using Excel 97

Numerator df	99		
Denominator df	73		
x	4.23		
one-tailed p-value		4.41976E-10	FDIST(4.23, 99, 73)
two-tailed p-value		8.83951E-10	2*FDIST(4.23, 99, 73)

Using the Excel 97 FDIST function, which calculates the right-hand tail area, we see from Table 8.4 that to four decimal places, the *p*-value = $2 \times Pr(F_{99,73} \geq 4.23) \leq .0001$. Thus, the two sample variances are significantly different. The two-sample *t* test with equal variances, as given in Section 8.4, cannot be used, because this test depends on the assumption that the variances are equal.

A question often asked about the *F* test is whether or not it makes a difference which sample is selected as the numerator sample and which as the denominator sample. The answer is that, for a two-sided test, it does *not* make a difference, because of the rules for calculating lower percentiles given in Equation 8.14. A variance ratio > 1 is usually more convenient, so there is no need to use Equation 8.14. Thus, the larger variance is usually put in the numerator and the smaller variance in the denominator.

Example 8.16 **Cardiovascular Disease, Hypertension** Using the data in Example 8.9, test whether or not the variance of blood pressure is significantly different between OC users and non-OC users.

Solution The sample standard deviation of blood pressure for the 8 OC users was 15.34 and for the 21 non-OC users was 18.23. Hence the variance ratio is

$$F = (18.23/15.34)^2 = 1.41$$

Under H_0, *F* follows an *F* distribution with 20 and 7 *df*, whose percentiles do not appear in Table 9. However, the percentiles of $F_{24,7}$ are provided in Table 9. Also, it can be shown that for a specified upper percentile (e.g, the 97.5th percentile), as either the numerator or denominator *df* increases, the corresponding percentile decreases. Therefore,

$$F_{20,7,.975} \geq F_{24,7,.975} = 4.42 > 1.41$$

It follows that $p > 2(.025) = .05$, and the underlying variances of the two samples are not significantly different from each other. Thus, it was correct to use the two-sample *t* test for independent samples with *equal variances* for these data, where the variances were assumed to be the same.

To compute an exact *p*-value, a computer program must be used to evaluate the area under the *F* distribution. The exact *p*-value for Example 8.16 has been evaluated using Excel 97, with the results given in Table 8.5. The program evaluates the right-hand tail area $= Pr(F_{20,7} \geq 1.41) = .334$. The two-tailed *p*-value $= 2 \times Pr(F_{20,7} \geq 1.41) = 2 \times .334 = .669$.

TABLE 8.5 **Computation of the exact *p*-value for the blood-pressure data in Example 8.16 using the *F* test for the equality of two variances with the Excel 97 FDIST program**

Numerator df	20	
Denominator df	7	
x	1.412285883	
one-tailed p-value	0.334279505	FDIST(1.41,20,7)
two-tailed p-value	0.66855901	2*FDIST(1.41,20,7)

If the numerator and denominator samples are reversed, then the F statistic = $1/1.41 = 0.71 \sim F_{7,20}$ under H_0. We use the FDIST program of Excel 97 to calculate $Pr(F_{7,20} \geq 0.71)$. This is given by FDIST(0.71, 7, 20) = .666. Because $F < 1$, we have $p\text{-value} = 2 \times Pr(F_{7,20} \leq 0.71) = 2 \times (1 - .666) = .669$, which is the same as the p-value just given. Thus, it was correct to use the two-sample t test for independent samples with *equal variances* for these data, where the variances were assumed to be the same.

In this section, we have introduced the F test for the equality of two variances. This test is used to compare variance estimates from two normally distributed samples. If we refer to the flowchart (Figure 8.13, p. 313), then starting from position 1 we would answer yes to (1) two-sample problem? and (2) underlying distribution normal or can central-limit theorem be assumed to hold? We would answer no to (3) inference concerning means? and yes to (4) inference concerning variances? This leads us to the box entitled "Two-sample F test to compare variances." We should be cautious about using this test with nonnormally distributed samples.

SECTION 8.7 Two-Sample *t* Test for Independent Samples with Unequal Variances

The F test for the equality of two variances from two independent, normally distributed samples was presented in Equation 8.15. If the two variances *are not* significantly different, then the two-sample t test for independent samples with *equal variances* outlined in Section 8.4 can be used. If the two variances *are* significantly different, then a two-sample t test for independent samples with *unequal variances*, which is presented in this section, must be used.

Specifically, assume that there are two normally distributed samples, where the first sample is a random sample of size n_1 from an $N(\mu_1, \sigma_1^2)$ distribution, the second sample is a random sample from an $N(\mu_2, \sigma_2^2)$ distribution, and $\sigma_1^2 \neq \sigma_2^2$. We again wish to test the hypothesis H_0: $\mu_1 = \mu_2$ versus H_1:. $\mu_1 \neq \mu_2$. Statisticians refer to this problem as the **Behrens-Fisher problem.**

It still makes sense to base the significance test on the difference between the sample means $\bar{x}_1 - \bar{x}_2$. Under either hypothesis, $\overline{X}_1$ is normally distributed with mean μ_1 and variance σ_1^2/n_1, and $\overline{X}_2$ is normally distributed with mean μ_2 and variance σ_2^2/n_2. Hence it follows that

EQUATION 8.17

$$\left(\overline{X}_1 - \overline{X}_2\right) \sim N\left(\mu_1 - \mu_2, \frac{\sigma_1^2}{n_1} + \frac{\sigma_2^2}{n_2}\right)$$

Under H_0, $\mu_1 - \mu_2 = 0$. Thus, from Equation 8.17,

EQUATION 8.18

$$\left(\overline{X}_1 - \overline{X}_2\right) \sim N\left(0, \frac{\sigma_1^2}{n_1} + \frac{\sigma_2^2}{n_2}\right)$$

If σ_1^2 and σ_2^2 were known, then the test statistic

EQUATION 8.19

$$z = (\bar{x}_1 - \bar{x}_2) \Big/ \sqrt{\frac{\sigma_1^2}{n_1} + \frac{\sigma_2^2}{n_2}}$$

could be used for the significance test, which under H_0 would be distributed as an $N(0,1)$ distribution. However, σ_1^2 and σ_2^2 are usually unknown and are estimated by s_1^2 and s_2^2, respectively (the sample variances in the two samples). Notice that a pooled estimate of the variance was not computed as in Equation 8.10, because the variances (σ_1^2, σ_2^2) are assumed to be different. If s_1^2 is substituted for σ_1^2 and s_2^2 for σ_2^2 in Equation 8.19, then the following test statistic is obtained:

EQUATION 8.20

$$t = (\bar{x}_1 - \bar{x}_2) \Big/ \sqrt{s_1^2/n_1 + s_2^2/n_2}$$

The exact distribution of t under H_0 is difficult to derive. However, several approximate solutions have been proposed that have appropriate type I error. The Satterthwaite approximation is presented here. Its advantage is its easy implementation using the ordinary t tables [1].

EQUATION 8.21

Two-Sample *t* Test for Independent Samples with Unequal Variances (Satterthwaite's Method)

(1) Compute the test statistic

$$t = \frac{\bar{x}_1 - \bar{x}_2}{\sqrt{\frac{s_1^2}{n_1} + \frac{s_2^2}{n_2}}}$$

(2) Compute the approximate degrees of freedom d', where

$$d' = \frac{\left(s_1^2/n_1 + s_2^2/n_2\right)^2}{\left(s_1^2/n_1\right)^2/(n_1 - 1) + \left(s_2^2/n_2\right)^2/(n_2 - 1)}$$

(3) Round d' down to the nearest integer d''.

If $\quad t > t_{d'',1-\alpha/2} \quad$ or $\quad t < -t_{d'',1-\alpha/2}$

then reject H_0.

If $\quad -t_{d'',1-\alpha/2} \le t \le t_{d'',1-\alpha/2}$

then accept H_0.

The acceptance and rejection regions for this test are illustrated in Figure 8.8.

Similarly, the approximate p-value for the hypothesis test can be computed as follows:

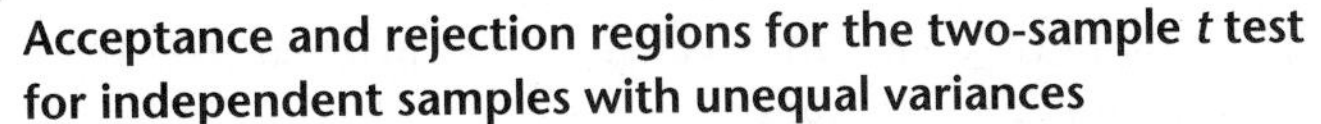

FIGURE 8.8 **Acceptance and rejection regions for the two-sample t test for independent samples with unequal variances**

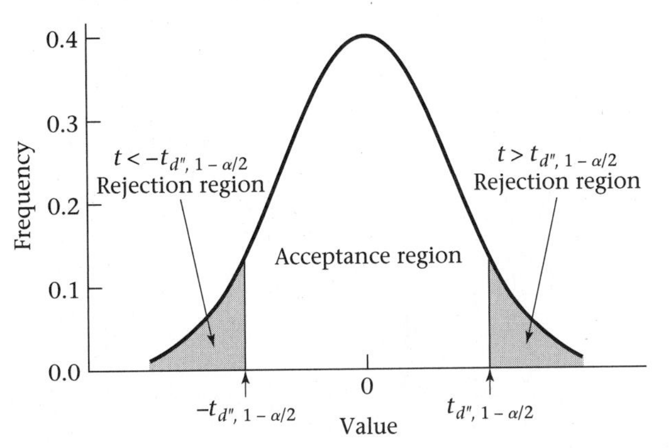

$t_{d''}$ distribution = approximate distribution of t in Equation 8.21 under H_0

EQUATION 8.22

Computation of the p-Value for the Two-Sample Test for Independent Samples with Unequal Variances (Satterthwaite Approximation)

Compute the test statistic

$$t = \frac{\bar{x}_1 - \bar{x}_2}{\sqrt{\frac{s_1^2}{n_1} + \frac{s_2^2}{n_2}}}$$

If $t \le 0$, then $p = 2 \times$ (area to the left of t under a $t_{d''}$ distribution)

If $t > 0$, then $p = 2 \times$ (area to the right of t under a $t_{d''}$ distribution)

where d'' is given in Equation 8.21.

The computation of the p-value is illustrated in Figure 8.9.

Example 8.17 **Cardiovascular Disease, Pediatrics** Consider the cholesterol data in Example 8.12. Test for the equality of the mean cholesterol levels of the children whose fathers have died from heart disease versus the children whose fathers do not have a history of heart disease.

Solution We have already tested for the equality of the two variances in Example 8.15 and found them to be significantly different. Thus the t test for unequal variances in Equation 8.21 must be used. The test statistic is

$$t = \frac{207.3 - 193.4}{\sqrt{35.6^2/100 + 17.3^2/74}} = \frac{13.9}{4.089} = 3.40$$

The approximate degrees of freedom are now computed:

FIGURE 8.9 **Computation of the p-value for the two-sample t test for independent samples with unequal variances**

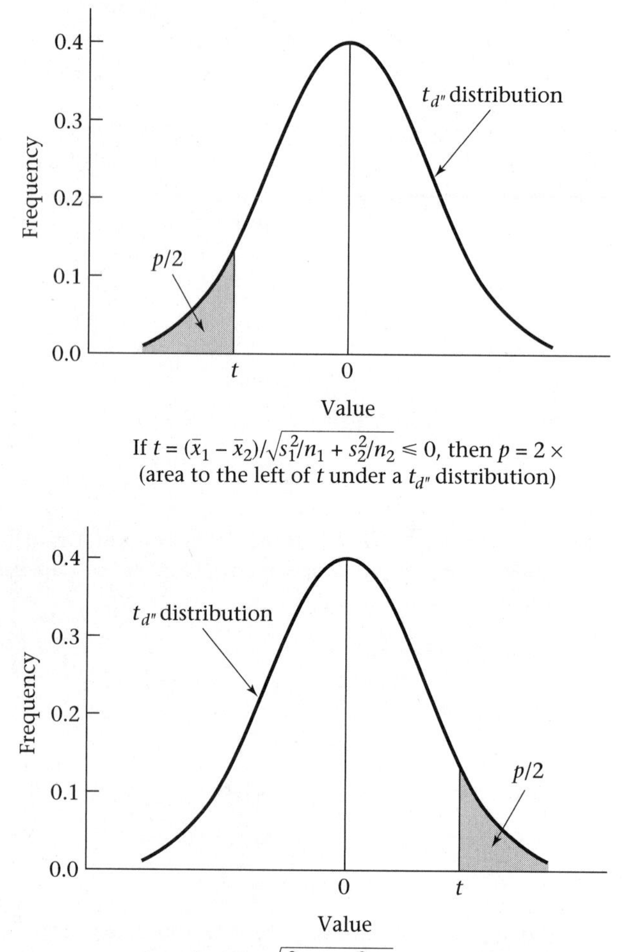

If $t = (\bar{x}_1 - \bar{x}_2)/\sqrt{s_1^2/n_1 + s_2^2/n_2} \le 0$, then $p = 2 \times$ (area to the left of t under a $t_{d''}$ distribution)

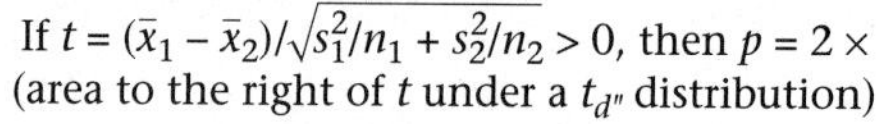
If $t = (\bar{x}_1 - \bar{x}_2)/\sqrt{s_1^2/n_1 + s_2^2/n_2} > 0$, then $p = 2 \times$ (area to the right of t under a $t_{d''}$ distribution)

$$d' = \frac{\left(s_1^2/n_1 + s_2^2/n_2\right)^2}{\left(s_1^2/n_1\right)^2/(n_1 - 1) + \left(s_2^2/n_2\right)^2/(n_2 - 1)}$$

$$= \frac{\left(35.6^2/100 + 17.3^2/74\right)^2}{\left(35.6^2/100\right)^2/99 + \left(17.3^2/74\right)^2/73} = \frac{16.718^2}{1.8465} = 151.4$$

Therefore, the approximate degrees of freedom $= d'' = 151$. If the critical-value method is used, note that $t = 3.40 > t_{120,.975} = 1.980 > t_{151,.975}$. Therefore H_0 can be rejected using a two-sided test with $\alpha = .05$. Furthermore, $t = 3.40 > t_{120,.9995} = 3.373 > t_{151,.9995}$, which implies that the p-value $< 2 \times (1.0 - .9995) = .001$. To compute the exact p-value, we use Excel 97, as shown in Table 8.6.

TABLE 8.6 **Computation of the exact p-value for Example 8.17**

t	3.4	
df	151	
two-tailed p-value	0.000862	TDIST(3.4,151,2)

We see from Table 8.6 that the p-value $= 2 \times [1 - Pr(t_{151} \le 3.40)] = .0009$. We conclude that the mean cholesterol levels in children whose fathers have died from heart disease are significantly greater than the mean cholesterol levels in children of fathers without heart disease. It would be of great interest to identify the source of this difference; that is, whether it is due to genetic factors, environmental factors such as diet, or both.

In this chapter, two procedures for comparing two means from independent, normally distributed samples have been presented. The first step in this process is to test for the equality of the two variances using the F test in Equation 8.15. If this test is not significant, then the t test with equal variances is used; otherwise, the t test with unequal variances is used. This overall strategy is illustrated in Figure 8.10.

FIGURE 8.10 **Strategy for testing for the equality of means in two independent, normally distributed samples**

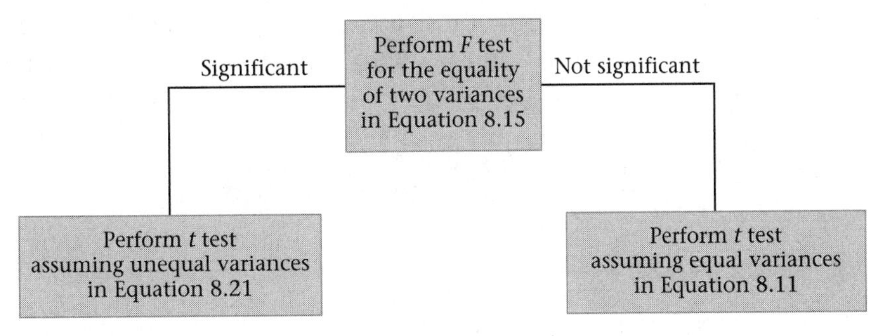

Example 8.18 **Infectious Disease** Using the data in Table 2.11, compare the mean duration of hospitalization between antibiotic users and nonantibiotic users.

Solution Refer to Table 8.7, where the PC-SAS T-TEST program (PROC TTEST) was used to analyze these data. Among the 7 antibiotic users (ANTIB = yes), the mean duration of hospitalization was 11.57 days with standard deviation 8.81 days; among the 18 nonantibiotic users (ANTIB = no), the mean duration of hospitalization was 7.44 days with standard deviation 3.70 days. Both the F test and the t test with equal and unequal variances are displayed in this program. Using Figure 8.10, note that the first step in comparing the two means is to perform the F test for the equality of two variances in order to decide whether to use the t test with equal or unequal variances. The F statistic is denoted in Table 8.7 by $F' = 5.68$, with p-value (labeled Prob > F') = .004. Thus the variances are significantly different, and a two-sample t test with unequal variances should be used. Therefore, refer to the Unequal Variance row, where the t statistic (as given in Equation 8.21) is found to be 1.20 with degrees of freedom $d'(df) = 6.8$. The corresponding two-tailed p-value (labeled Prob > $|T|$) = .271. Thus there is no significant difference between the mean duration of hospitalization in these two groups.

TABLE 8.7 Use of the PC-SAS T-TEST program (PROC TTEST) to analyze the association between antibiotic use and duration of hospitalization (raw data presented in Table 2.11)

```
                              The SAS System
                             TTEST PROCEDURE
Variable: DUR

ANTIB    N         Mean      Std Dev    Std Error   Variances        T     DF   Prob>|T|

 yes     7  11.57142857   8.81016729   3.32993024   Unequal     1.1990    6.8     0.2705
  no    18   7.44444444   3.69772939   0.87156317   Equal       1.6816   23.0     0.1062

For H0: Variances are equal, F' = 5.68   DF = (6, 17)   Prob>F' = 0.0043
```

If the results of the F test had revealed a nonsignificant difference between the variances of the two samples, then the t test with equal variances would have been used, which is provided in the Equal Variance row of the SAS output. In this example, considerable differences are present in both the test statistics (1.68 versus 1.20), and the two-tailed p-values (.106 versus .271) resulting from using these two procedures.

Using similar methods to those developed in Section 8.5, we can show that a two-sided $100\% \times (1-\alpha)$ confidence interval for the underlying mean difference $\mu_1 - \mu_2$ in the case of unequal variances is given as follows:

EQUATION 8.23

Two-Sided 100% × (1 − α) Confidence Interval for $\mu_1 - \mu_2$ $(\sigma_1^2 \neq \sigma_2^2)$

$$\left(\bar{x}_1 - \bar{x}_2 - t_{d'',1-\alpha/2}\sqrt{s_1^2/n_1 + s_2^2/n_2},\ \bar{x}_1 - \bar{x}_2 + t_{d'',1-\alpha/2}\sqrt{s_1^2/n_1 + s_2^2/n_2}\right)$$

where d'' is given in Equation 8.21.

Example 8.19 **Infectious Disease** Using the data in Table 8.7, compute a 95% confidence interval for the mean difference in duration of hospital stay between patients who do and patients who do not receive antibiotics.

Solution Using Table 8.7, the 95% confidence interval is given by

$$\left[(11.571 - 7.444) - t_{6,.975}\sqrt{8.810^2/7 + 3.698^2/18},\right.$$
$$\left.(11.571 - 7.444) + t_{6,.975}\sqrt{8.810^2/7 + 3.698^2/18}\right]$$
$$= \left[4.127 - 2.447(3.442), 4.127 + 2.447(3.442)\right]$$
$$= (4.127 - 8.423,\ 4.127 + 8.423) = (-4.30, 12.55)$$

In this section, we have introduced the two-sample t test for independent samples with unequal variances. This test is used to compare the mean level of a normally distributed random variable (or a random variable with samples that are large enough so that the central-limit theorem can be assumed to hold) between two independent samples with unequal variances. If we refer to the flowchart (Figure 8.13, p. 313), then starting from position 1 we would answer yes to (1) two-sample problem? (2) underlying distribution normal or can central-limit theorem be assumed to hold? (3) inference concerning means? and (4) are samples independent? We would answer yes to (5) are variances of two samples significantly different? This leads us to the box entitled "Use two-sample t test with unequal variances."

SECTION 8.8 Case Study: Effects of Lead Exposure on Neurological and Psychological Function in Children

Example 8.20 **Environmental Health, Pediatrics** In Section 2.9, we described a study performed in El Paso, Texas, to look at the association between lead exposure and developmental features in children [2]. There are different ways to quantify lead exposure. One method consists of defining a control group of children whose blood-lead levels were < 40 μg/100 mL in both 1972 and 1973 ($n = 78$) and an exposed group of children who had blood-lead levels ≥ 40 μg/100 mL in either 1972 or 1973 ($n = 46$). Two important outcome variables in the study were the number of finger-wrist taps per 10 seconds in the dominant hand (a measure of neurological function) as well as the Wechsler full-scale IQ score (a measure of intellectual development). Because only children ≥ 5 years old were given the neurological tests, we actually have 35 exposed and 64 control children who have finger-wrist tapping scores. The distributions of these variables by group were displayed in a box plot in Figures 2.10 and 2.11, respectively. The distributions appeared to be reasonably symmetric, particularly in the exposed group, although there is a hint that there may be a few outliers present. We will discuss the detection of outliers more formally in Section 8.9. We also note from these figures that the exposed group seems to have lower levels than the control group for both of these variables. How can we confirm whether this impression is correct?

One approach is to use a two-sample t test to compare the mean level of the exposed group to the mean level of the control group on these variables. We have used the PC-SAS TTEST procedure for this purpose, as shown in Tables 8.8 and 8.9. The program actually performs three different significance tests each time the t test procedure is specified. Following the flowchart in Figure 8.10, we first perform the F test for the equality of two variances. In Table 8.8, the F statistic (labeled as F') = 1.19 with 34 and 63 df. The p-value (labeled Prob > F') equals 0.5408, which implies that we can accept H_0 that the variances are *not* significantly different. Therefore, following Figure 8.10, we should perform the two-sample t test with equal variances (Equation 8.11). The t statistic is found in the T column and the Equal row to be 2.6772 with 97 df. The two-tailed p-value, found in the column headed Prob > $|T|$ and the Equal row is 0.0087, which implies that there *is* a significant difference in mean finger-wrist tapping scores between the exposed and the control group, with the exposed group having lower mean scores. If there had been a

TABLE 8.8 Comparison of mean finger-wrist tapping scores for the exposed versus control group, using the SAS t test procedure

The SAS System
TTEST PROCEDURE

Variable: MAXFWT

CSCN2	N	Mean	Std Dev	Std Error	Variances	T	DF	Prob>|T|
control	64	54.43750000	12.05657958	1.50707245	Unequal	2.6091	65.0	0.0113
case	35	47.42857143	13.15582115	2.22373964	Equal	2.6772	97.0	0.0087

For H0: Variances are equal, F' = 1.19 DF = (34,63) Prob>F' = 0.5408

TABLE 8.9 Comparison of mean full-scale IQ scores for the exposed versus control group, using the SAS t test procedure

The SAS System
TTEST PROCEDURE

Variable: IQF

CSCN2	N	Mean	Std Dev	Std Error	Variances	T	DF	Prob>|T|
control	78	92.88461538	15.34451192	1.73742384	Unequal	1.9439	111.4	0.0544
case	46	88.02173913	12.20653583	1.79975552	Equal	1.8334	122.0	0.0692

For H0: Variances are equal, F' = 1.58 DF = (77,45) Prob>F' = 0.0982

significant difference between the variances from the F test (that is, if (Prob > F') < 0.05), then we would use the two-sample t test with unequal variances. The program automatically performs both t tests and lets the user decide which one to use. If a two-sample t test with unequal variances were used, then referring to the Unequal row, the t statistic equals 2.6091 (as given in Equation 8.21) with 65 df (d' in Equation 8.21) with a two-sided p-value equal to 0.0113. The program also provides the mean, standard deviation (Std Dev), and standard error (Std Error) for each group. Referring to Table 8.9, for the analysis of the full-scale IQ scores, we see that the p-value for the F test is 0.0982, which is not statistically significant. Therefore we again use the equal variance t test. The t statistic is 1.8334 with 122 df, with two-tailed p-value equal to 0.0692. Thus there is *not* a significant difference between the mean full-scale IQ scores for the two groups.

SECTION 8.9 The Treatment of Outliers

We saw in the case study in Section 8.8 that there was a suggestion that there might be some outliers in the finger-wrist tapping and IQ scores. Outliers can potentially have an important impact on the conclusions of a study. It is important to definitely identify outliers and either exclude them outright, or at least perform alternative analyses with and without the outliers present. Therefore, in this section we study some decision rules for outlier detection.

We refer to Figures 8.11 and 8.12, which provide stem-and-leaf and box plots from SAS of the finger-wrist tapping scores and the full-scale IQ scores for the control group and the exposed group, respectively. According to the box plots in Figure 8.11, there are potential outlying finger-wrist tapping scores (denoted by zeros in the plot) of 13, 23, 26, and 84 taps per 10 seconds for the control group and 13, 14, and 83 taps per 10 seconds for the exposed group. According to the box plots in Figure 8.12, there are potential outlying full-scale IQ scores of 50, 56, 125, 128, and 141 for the control group and 46 for the exposed group. All the potentially outlying values are far from the mean in absolute value. Therefore, a useful way to quantify an extreme value is by the number of standard deviations that a value is from the mean. This statistic applied to the most extreme value in a sample is referred to as the "Extreme Studentized Deviate" (or ESD) and is defined as follows:

DEFINITION 8.7 The **Extreme Studentized Deviate** (or **ESD statistic**) = $\max_{i=1,\ldots,n} |x_i - \bar{x}|/s$.

Example 8.21 Compute the ESD statistic for the finger-wrist tapping scores for the control group.

Solution From Table 8.8, we see that $\bar{x} = 54.4, s = 12.1$. From Figure 8.11a we note that the distance from the mean for the smallest and largest values are $|13 - 54.4| = 41.4$ and $|84 - 54.4| = 29.6$, respectively. Therefore, since $41.4 > 29.6$, it follows that ESD = $41.4/12.1 = 3.44$.

How large must the ESD statistic be for us to conclude that the most extreme value is an outlier? Remember that in a sample of size n without outliers, we would expect the largest value to correspond approximately to the $100\% \times \left(\frac{n}{n+1}\right)$th percentile. Thus, for a sample of size 64 from a normal distribution this would correspond to approximately the $100 \times 64/65$th percentile $\approx$ 98.5th percentile = 2.17. If an outlier is present, then the ESD statistic will be larger than 2.17. The appropriate critical values will depend on the sampling distribution of the ESD statistic for samples of size n from a normal distribution. Critical values from Rosner [3] based on an approximation provided by Quesenberry and David [4] are presented in Table 10 in the Appendix. The critical values depend on the sample size n and the percentile p. The pth percentile for the ESD statistic based on a sample of size n is denoted by $\text{ESD}_{n,p}$.

Example 8.22 Find the upper 5th percentile for the ESD statistic based on a sample of size 50.

Solution The appropriate percentile = $\text{ESD}_{50,\,.95}$ is found by referring to the 50 row and the .95 column and is 3.13.

For values of n that are not in the table, we can sometimes assess significance by using the principle that for a given level of significance, the critical values increase as the sample size increases. This leads to the following procedure for the detection of a single outlier in normally distributed samples:

FIGURE 8.11 Stem-and-leaf and box plots of finger-wrist tapping score by group, El Paso Lead Study

(a) Control

```
Stem | Leaf                  #     Boxplot
  8  | 4                     1        0
  7  | 69                    2        |
  7  | 224                   3        |
  6  | 55558                 5        |
  6  | 01122344              8     +-----+
  5  | 566677778999         12     |     |
  5  | 000000011222334      15     *--+--*
  4  | 566666888999         12     +-----+
  4  | 02                    2        |
  3  | 8                     1        |
  3  |
  2  | 6                     1        0
  2  | 3                     1        0
  1  |
  1  | 3                     1        0
       ----+----+----+----+
Multiply Stem.Leaf by 10**+1
```

(b) Exposed

```
Stem | Leaf                  #     Boxplot
  8  | 3                     1        0
  7  |
  7  | 0                     1        |
  6  |                                |
  6  | 2                     1        |
  5  | 567789                6     +-----+
  5  | 0122244               7     |     |
  4  | 56889                 5     *--+--*
  4  | 0012244               7     +-----+
  3  | 5788                  4        |
  3  | 4                     1        |
  2  |
  2  |
  1  |
  1  | 34                    2        0
       ----+----+----+----+
Multiply Stem.Leaf by 10**+1
```

FIGURE 8.12 Stem-and-leaf and box plots of full-scale IQ by group, El Paso Lead Study

(a) Control

```
Stem | Leaf                  #     Boxplot
 14  | 1                     1        0
 13  |
 13  |
 12  | 58                    2        0
 12  | 0                     1        |
 11  | 558                   3        |
 11  | 1                     1        |
 10  | 55677778              8        |
 10  | 011244                6     +-----+
  9  | 566666667789999      15     |     |
  9  | 123444                6     *--+--*
  8  | 5555666677888999     16     +-----+
  8  | 0004                  4        |
  7  | 566666778             9        |
  7  | 0234                  4        |
  6  |
  6  |
  5  | 6                     1        0
  5  | 0                     1        0
       ----+----+----+----+
Multiply Stem.Leaf by 10**+1
```

(b) Exposed

```
Stem | Leaf                  #     Boxplot
     |                                |
 11  | 124                   3        |
 10  |                                |
 10  | 01144                 5        |
  9  | 678                   3     +-----+
  9  | 0111222334           10     *--+--*
  8  | 55568889              8     |     |
  8  | 0002233               7     +-----+
  7  | 5567899               7        |
  7  | 12                    2        |
  6  |
  6  |
  5  |
  5  |
  4  | 6                     1        0
       ----+----+----+----+
Multiply Stem.Leaf by 10**+1
```

EQUATION 8.24

ESD Single-Outlier Procedure Suppose we have a sample $x_1, \ldots, x_n \sim N(\mu, \sigma^2)$, but feel that there may be some outliers present. To test the hypothesis, H_0: that no outliers are present versus H_1: that a single outlier is present, with a type I error of α,

(1) We compute the Extreme Studentized Deviate test statistic = ESD = $\max_{i=1,\ldots,n} \frac{|x_i - \bar{x}|}{s}$. The sample value x_i, such that ESD = $\frac{|x_i - \bar{x}|}{s}$ is referred to as $x^{(n)}$.

(2) We refer to Table 10 in the Appendix to obtain the critical value = $\text{ESD}_{n,1-\alpha}$.

(3) If $\text{ESD} > \text{ESD}_{n,1-\alpha}$, then we reject H_0 and declare that $x^{(n)}$ is an outlier. If $\text{ESD} \le \text{ESD}_{n,1-\alpha}$, then we declare that no outliers are present.

Example 8.23 Evaluate whether outliers are present for the finger-wrist tapping scores in the control group.

Solution Following Equation 8.24, we compute the ESD test statistic. From Example 8.21, we have ESD = 3.44 with 13 being the most extreme value. To assess statistical significance with $\alpha = .05$, we refer to Table 10. From Table 10, $\text{ESD}_{70,.95} = 3.26$. Because ESD = $3.44 > \text{ESD}_{70,.95} = 3.26 > \text{ESD}_{64,.95}$ it follows that $p < .05$. Therefore, we infer that the finger-wrist tapping score of 13 taps per 10 seconds is an outlier.

In some instances, when multiple outliers are present, it is difficult to identify specific data points as outliers using the single-outlier detection procedure. This is because the standard deviation can get inflated in the presence of multiple outliers, thus reducing the magnitude of the ESD test statistic in Equation 8.24 and making it difficult to identify specific data points as outliers.

Example 8.24 Evaluate whether any outliers are present for the finger-wrist tapping scores in the exposed group.

Solution Referring to Table 8.8, we see that $\bar{x} = 47.4$, $s = 13.2$, and $n = 35$ in the exposed group. Furthermore, the minimum and maximum values are 13 and 83, respectively. Because $|83 - 47.4| = 35.6 > |13 - 47.4| = 34.4$, it follows that the ESD statistic is 35.6/13.2 = 2.70. From Table 10, we see that $\text{ESD}_{35,.95} = 2.98 > \text{ESD} = 2.70$. Therefore $p > .05$, and we accept the null hypothesis that no outliers are present.

The solution to Example 8.24 is unsettling, because it is inconsistent with Figure 8.11b. It appears that the values 13, 14, and 83 are outliers, yet no outliers are identified by the single-outlier procedure in Equation 8.24. The problem is that the multiple outliers have artificially inflated the standard deviation. This is referred to as the *masking problem*, because multiple outliers have made it difficult to identify the single most extreme sample point as an outlier. This is particularly true if the multiple outliers are roughly equally distant from the sample mean as is the case in Figure 8.11b. To overcome this problem, we must employ a flexible procedure that can accurately identify either single or multiple outliers and will be less susceptible to the masking problem. For this purpose, we first must decide what is a reasonable upper bound for the number of outliers present in a data set. In my experience, a reasonable upper bound for the number of possible outliers is min([n/10], 5), where [n/10] is the largest integer $\le n/10$. If there are more than five outliers in a data set, then we most likely have an underlying nonnormal distribution, unless the sample size is very large. The following multiple-outlier procedure [3] achieves this goal.

EQUATION 8.25

ESD Many-Outlier Procedure Suppose $x_1, \ldots, x_n \sim N(\mu, \sigma^2)$ for a large majority of the sample points, but that we suspect that we may have as many as k outliers, where $k = \min([n/10], 5)$, where [n/10] is the largest integer $\le n/10$. We wish to have a type I error of α to test the hypothesis H_0: there are no outliers versus H_1: there are between 1 and k outliers, and would like to use a decision rule that can specifically identify the outliers. For this purpose,

(1) We compute the ESD statistic based on the full sample $= \max_{i=1,\ldots,n} |x_i - \bar{x}|/s$. We denote this statistic by $\text{ESD}^{(n)}$ and the most outlying data point by $x^{(n)}$.

(2) We remove $x^{(n)}$ from the sample and compute the mean, *sd*, and ESD statistic from the remaining $n - 1$ data points. We denote the ESD statistic from the reduced sample by $\text{ESD}^{(n-1)}$.

(3) We continue to remove the most outlying sample points and recompute the ESD statistic until we have computed k ESD statistics denoted by $\text{ESD}^{(n)}$, $\text{ESD}^{(n-1)}, \ldots, \text{ESD}^{(n-k+1)}$ based on the original sample size of n, and successively reduced samples of size $n - 1, \ldots, n - k + 1$. The most outlying values identified at each of the k steps are denoted by $x^{(n)}, x^{(n-1)}, \ldots, x^{(n-k+1)}$.

(4) The critical values corresponding to the ESD statistics are $\text{ESD}_{n,1-\alpha}$, $\text{ESD}_{n-1,1-\alpha,\ldots,}\text{ESD}_{n-k+1,1-\alpha}$.

(5) We then use the following decision rule for outlier detection:

If $\text{ESD}^{(n-k+1)} > \text{ESD}_{n-k+1,1-\alpha}$, then we declare the k values $x^{(n)}, \ldots, x^{(n-k+1)}$ as outliers

else If $\text{ESD}^{(n-k+2)} > \text{ESD}_{n-k+2,1-\alpha}$, then we declare the $k - 1$ values $x^{(n)}, \ldots, x^{(n-k+2)}$ as outliers

⋮

else If $\text{ESD}^{(n)} > \text{ESD}_{n,1-\alpha}$, then we declare one outlier, $x^{(n)}$

else If $\text{ESD}^{(n)} \le \text{ESD}_{n,1-\alpha}$, then we declare no outliers present

Thus, we have the ability to declare either 0, 1, . . . , or k sample points as outliers.

(6) We should only use Table 10 in the Appendix to implement this procedure if $n \ge 20$.

Note that we must compute all k outlier test statistics $\text{ESD}^{(n)}, \text{ESD}^{(n-1)}, \ldots, \text{ESD}^{(n-k+1)}$ regardless of whether any specific test statistic (e.g., $\text{ESD}^{(n)}$) is significant or not. This procedure has good power to either declare no outliers, or detect from 1 up to k outliers with little susceptibility to masking effects unless the true number of outliers is larger than k.

Example 8.25 Reanalyze the finger-wrist tapping scores for the exposed group in Figure 8.11b, using the multiple-outlier procedure in Equation 8.25.

Solution We will set the maximum number of outliers to be detected to be [35/10] = 3. From Example 8.24, we see that $ESD^{(35)} = 2.70$ and the most outlying value = $x^{(35)} = 83$. We remove 83 from the sample and recompute the sample mean (46.4) and standard deviation (11.8) from the reduced sample of size 34. Because $|13 - 46.4| = 33.4 > |70 - 46.4| = 23.6$, 13 is the most extreme value and $ESD^{(34)} = 33.4/11.8 = 2.83$. We then remove 13 from the sample and recompute the sample mean (47.4) and standard deviation (10.4) from the reduced sample of size 33. Since $|14 - 47.4| = 33.4 > |70 - 47.4| = 22.6$, it follows that $ESD^{(33)} = 33.4/10.4 = 3.22$.

To assess statistical significance, we first compare 3.22 with the critical value $ESD_{33,.95}$. From Table 10 in the Appendix, we see that $ESD^{(33)} = 3.22 > ESD_{35,.95} = 2.98 > ESD_{33,.95}$. Therefore, $p < .05$, and we declare the three most extreme values (83, 13, and 14) as outliers. Note that although significance was achieved by an analysis of the third most extreme value (14), once it is identified as an outlier, then the more extreme points (13, 83) are also designated as outliers. Also, note that the results are consistent with Figure 8.11b and are different from the single-outlier procedure, where no outliers were declared.

Example 8.26 Assess whether any outliers are present for the finger-wrist tapping scores for controls.

Solution Because $n = 64$, min([64/10], 5) = min(6, 5) = 5. Therefore, we set the maximum number of outliers to be detected to 5 and organize the appropriate test statistics and critical values in a table (Table 8.10).

TABLE 8.10 **Test statistics and critical values for Example 8.26**

n	$\bar{x}$	s	$x^{(n)}$	$ESD^{(n)}$	$ESD_{n,.95}$	p-value
64	54.4	12.1	13	3.44	$ESD_{64,.95}$[a]	<.05
63	55.1	10.9	23	2.94	$ESD_{63,.95}$[b]	NS
62	55.6	10.2	26	2.90	$ESD_{62,.95}$[b]	NS
61	56.1	9.6	84	2.92	$ESD_{61,.95}$[b]	NS
60	55.6	8.9	79	2.62	3.20	NS

[a] $ESD_{64,.95} < ESD_{70,.95} = 3.26$
[b] $ESD_{63,.95}, \ldots, ESD_{61,.95}$ are all $> ESD_{60,.95} = 3.20$

From Table 8.10 we see that 79, 84, 26, and 23 are *not* identified as outliers, whereas 13 *is* identified as an outlier. Thus, we declare one outlier is present. This decision is consistent with the single-outlier test in Example 8.23.

In general, it is suggested that the multiple-outlier test in Equation 8.25 be used rather than the single-outlier test in Equation 8.24, unless we are very confident that there is at most one outlier.

The issue remains—what should we do now that we have identified one outlier among the controls and three outliers among the exposed? We have elected to reanalyze the data using a two-sample t test after deleting the outlying observations.

Example 8.27 Reanalyze the finger-wrist tapping score data given in Table 8.8 after exclusion of the outliers identified in Examples 8.25 and 8.26.

Solution The *t* test results after exclusion of the outliers is given in Table 8.11.

TABLE 8.11 Comparison of mean finger-wrist tapping scores for the exposed versus control groups after excluding outliers using the SAS *t* test procedure

The SAS System
TTEST PROCEDURE

Variable: MAXFWT

CSCN2	N	Mean	Std Dev	Std Error	Variances	T	DF	Prob>|T|
control	63	55.09523810	10.93487213	1.37766439	Unequal	3.2485	77.0	0.0017
case	32	48.43750000	8.58332028	1.51733099	Equal	3.0035	93.0	0.0034

For H0: Variances are equal, F' = 1.62 DF = (62,31) Prob>F' = 0.1424

We see that there remains a significant difference between the mean finger-wrist tapping scores for the exposed (case) and control groups ($p = .003$). Indeed, the results are more significant than previously, because the standard deviations are lower after the exclusion of outliers, particularly for the exposed group.

There are several approaches to the treatment of outliers in performing data analyses. One approach is to use efficient methods of outlier detection and either exclude outliers from further data analyses or perform data analyses with and without outliers present and compare results. Another possibility is to not exclude the outliers, but use a method of analysis that minimizes their effect on the overall results. One method for accomplishing this is to convert continuous variables such as finger-wrist tapping score to categorical variables (for example, high = above the median versus low = below the median) and analyze the data using categorical-data methods. We discuss this approach in Chapter 10. Another possibility is to use nonparametric methods to analyze the data. These methods make much weaker assumptions about the underlying distributions than do the normal-theory methods such as the *t* test. We discuss this approach in Chapter 9. Another approach is to use "robust" estimators of important population parameters (such as μ). These estimators give less weight to extreme values in the sample, but do not entirely exclude them. The subject of robust estimation is beyond the scope of this book. Using each of these methods may result in a loss of power relative to using ordinary *t* tests if no outliers exist, but offer the advantage of a gain in power if some outliers are present. In general, there is no one correct way to analyze data; the conclusions from a study are strengthened if they are consistently found using more than one analytic technique.

SECTION 8.10 Estimation of Sample Size and Power for Comparing Two Means

8.10.1 Estimation of Sample Size

Methods of sample-size estimation for the one-sample z test for the mean of a normal distribution with known variance were presented in Section 7.6. Estimates of sample size that are useful in planning studies in which *two* samples are to be compared are covered in this section.

Example 8.28 **Cardiovascular Disease, Hypertension** Consider the blood-pressure data for OC users and non-OC users in Example 8.9 (p. 280) as a pilot study conducted to obtain parameter estimates to plan for a larger study. Suppose we assume that the true blood-pressure distribution of 35- to 39-year-old OC users is normal with mean μ_1 and variance σ_1^2. Similarly, for non-OC users we assume that the distribution is normal with mean μ_2 and variance σ_2^2. We wish to test the hypothesis H_0: $\mu_1 = \mu_2$ versus H_1: $\mu_1 \neq \mu_2$. How can the sample size needed for the larger study be estimated?

Suppose we assume that σ_1^2 and σ_2^2 are known and we anticipate equal sample sizes in the two groups. To conduct a two-sided test with significance level α and power of $1 - \beta$, the appropriate sample size for *each* group is as follows:

EQUATION 8.26

Sample Size Needed for Comparing the Means of Two Normally Distributed Samples of Equal Size Using a Two-Sided Test with Significance Level α and Power $1 - \beta$

$$n = \frac{\left(\sigma_1^2 + \sigma_2^2\right)\left(z_{1-\alpha/2} + z_{1-\beta}\right)^2}{\Delta^2} = \text{sample size for each group}$$

where $\Delta = |\mu_2 - \mu_1|$. The means and variances of the two respective groups are (μ_1, σ_1^2) and (μ_2, σ_2^2).

In words, n is the appropriate sample size in each group to have a probability of $1 - \beta$ of finding a significant difference based on a two-sided level α significance test, if the absolute value of the true difference in means between the two groups is $\Delta = |\mu_2 - \mu_1|$.

Example 8.29 **Cardiovascular Disease, Hypertension** Determine the appropriate sample size for the large study proposed in Example 8.28 using a two-sided test with a significance level of .05 and a power of .80.

Solution In the small study, $\bar{x}_1 = 132.86$, $s_1 = 15.34$, $\bar{x}_2 = 127.44$, and $s_2 = 18.23$.

If the sample data $(\bar{x}_1, s_1^2, \bar{x}_2, s_2^2)$ are used as estimates of the population parameters $(\mu_1, \sigma_1^2, \mu_2, \sigma_2^2)$, then ensuring an 80% chance of finding a significant difference using a two-sided significance test with $\alpha = .05$ would require a sample size of

$$n = \left(15.34^2 + 18.23^2\right)\left(1.96 + 0.84\right)^2 / \left(132.86 - 127.44\right)^2 = 151.5$$

or 152 people in *each* group. It is not surprising that a significant difference was not found with sample sizes of 8 and 21 in the two groups, respectively.

In many instances an imbalance between the groups can be anticipated and it can be predicted in advance that the number of people in one group will be k times the number in the other group for some number $k \neq 1$. In this case, where $n_2 = kn_1$, the appropriate sample size in the two groups for achieving a power of $1 - \beta$ using a two-sided level α significance test is given by the following formulas:

EQUATION 8.27

Sample Size Needed for Comparing the Means of Two Normally Distributed Samples of Unequal Size Using a Two-Sided Test with Significance Level α and Power $1 - \beta$

$$n_1 = \frac{\left(\sigma_1^2 + \sigma_2^2/k\right)\left(z_{1-\alpha/2} + z_{1-\beta}\right)^2}{\Delta^2} = \text{sample size of first group}$$

$$n_2 = \frac{\left(k\sigma_1^2 + \sigma_2^2\right)\left(z_{1-\alpha/2} + z_{1-\beta}\right)^2}{\Delta^2} = \text{sample size of second group}$$

where $\Delta = |\mu_2 - \mu_1|$; (μ_1, σ_1^2), (μ_2, σ_2^2), are the means and variances of the two respective groups and $k = n_2/n_1$ = the projected ratio of the two sample sizes.

Note that if $k = 1$, then the sample-size estimates given in Equation 8.27 are the same as those in Equation 8.26.

Example 8.30 **Cardiovascular Disease** Suppose we anticipate twice as many non-OC users as OC users entering the study proposed in Example 8.28. Project the required sample size if a two-sided test is used with a 5% significance level and an 80% power is desired.

Solution If Equation 8.27 is used with $\mu_1 = 132.86$, $\sigma_1 = 15.34$, $\mu_2 = 127.44$, $\sigma_2 = 18.23$, $k = 2$, $\alpha = .05$, and $1 - \beta = .8$, then to achieve an 80% power in the study using a two-sided significance test with $\alpha = .05$, we need to enroll

$$n_1 = \frac{\left(15.34^2 + 18.23^2/2\right)\left(1.96 + 0.84\right)^2}{\left(132.86 - 127.44\right)^2} = 107.1, \text{ or } 108 \text{ OC users}$$

and $n_2 = 2(108) = 216$ non-OC users

If the variances in the two groups are the same, then for a given α, β, the smallest total sample size needed is achieved by the *equal sample size allocation rule* in Equation 8.26. Thus in the case of equal variances, the sample sizes in the two groups should be as nearly equal as possible.

Finally, to perform a one-sided rather than a two-sided test, we substitute α for $\alpha/2$ in Equations 8.26 and 8.27.

8.10.2 Estimation of Power

In many situations, a predetermined sample size is available for study and how much power the study will have for detecting specific alternatives needs to be determined.

Example 8.31 **Cardiovascular Disease** Suppose 100 OC users and 100 non-OC users are available for study and a true difference in mean systolic blood pressure (SBP) of 5 mm Hg is anticipated, with OC users having the higher mean SBP. How much power would such a study have assuming that the variance estimates in the pilot study in Example 8.9 are correct?

Assuming that σ_1^2 and σ_2^2 are known, to conduct a two-sided test with significance level α, the power is given by:

EQUATION 8.28

Power for Comparing the Means of Two Normally Distributed Samples Using a Two-Sided Test with Significance Level α To test the hypothesis H_0: $\mu_1 = \mu_2$ versus H_1: $\mu_1 \neq \mu_2$ for the specific alternative $|\mu_1 - \mu_2| = \Delta$, with significance level α,

$$\text{Power} = \Phi\left(-z_{1-\alpha/2} + \frac{\sqrt{n_1}\,\Delta}{\sqrt{\sigma_1^2 + \sigma_2^2/k}}\right)$$

where (μ_1, σ_1^2), (μ_2, σ_2^2) are the means and variances of the two respective groups and $k = n_2/n_1$ = the projected ratio of the two sample sizes.

Example 8.32 **Cardiovascular Disease** Estimate the power available for the study proposed in Example 8.31 using a two-sided test with significance level = .05.

Solution From Example 8.31, $n_1 = n_2 = 100$, $k = n_2/n_1 = 1$, $\Delta = 5$, $\sigma_1 = 15.34$, $\sigma_2 = 18.23$, and $\alpha = .05$. Therefore, from Equation 8.28,

$$\text{Power} = \Phi\left(-z_{.975} + \frac{\sqrt{100}(5)}{\sqrt{15.34^2 + 18.23^2/1}}\right) = \Phi\left[-1.96 + \frac{10(5)}{23.83}\right]$$
$$= \Phi(-1.96 + 2.099) = \Phi(0.139) = .555$$

Thus, there is a 55.5% chance of detecting a significant difference using a two-sided test with significance level = .05.

To calculate power for a one-sided rather than a two-sided test, simply substitute α for $\alpha/2$ in Equation 8.28.

SECTION 8.11 Sample-Size Estimation for Longitudinal Studies

Example 8.33 **Hypertension** Suppose we are planning a longitudinal study to compare the mean change in systolic blood pressure (SBP) between a treated and a control group. It is projected, based on previous data, that the standard deviation of SBP at both baseline and follow-up is 15 mm Hg, and that the correlation coefficient between repeated SBP values 1 year apart is approximately .70. How many subjects do we need to study to have 80% power to detect a significant difference between groups using a two-sided test with $\alpha = .05$ if the true mean change in SBP over 1 year is 8 mm Hg for the treated group and 3 mm Hg for the control group?

To answer the question posed in Example 8.33, we would like to apply the sample-size formula given in Equation 8.26. However, using Equation 8.26 requires knowledge of the variances of change in SBP for each of the treated and control groups. Considering the control group first, let

x_{1i} = SBP for the ith subject in the control group at baseline

x_{2i} = SBP for the ith subject in the control group at 1 year

Therefore,

$d_i = x_{2i} - x_{1i}$ = change in SBP for the ith subject in the control group over 1 year

If x_{1i} and x_{2i} were independent, then from Equation 5.9 it would follow that

$Var(d_i) = \sigma_2^2 + \sigma_1^2$, where

σ_1^2 = variance of baseline SBP in the control group
σ_2^2 = variance of 1-year SBP in the control group

However, repeated SBP measures on the same person are usually not independent. The correlation between them will, in general, depend on the time interval between the baseline and follow-up measures. Let us assume that the correlation coefficient between measures 1 year apart is ρ. We have defined a correlation coefficient in Chapter 5, and in Chapter 11 we will discuss how to estimate correlation coefficients from sample data. Then from Equation 5.11, we have

EQUATION 8.29

$$Var(x_{2i} - x_{1i}) = \sigma_2^2 + \sigma_1^2 - 2\rho\sigma_1\sigma_2 = \sigma_d^2$$

where σ_d^2 = variance of change in SBP.

For simplicity, we will assume that σ_1^2, σ_2^2, ρ, and σ_d^2 are the same in the treated and control groups. We wish to test the hypothesis H_0: $\mu_1 = \mu_2$ versus H_1: $\mu_1 \neq \mu_2$, where

μ_1 = true mean change in the control group
μ_2 = true mean change in the active group

Based on Equations 8.26 and 8.29, we obtain the following sample-size estimate:

EQUATION 8.30

Sample Size Needed for Longitudinal Studies Comparing Mean Change in Two Normally Distributed Samples with Two Time Points Suppose we are planning a longitudinal study with an equal number of subjects (n) in each of two groups. We wish to test the hypothesis H_0: $\mu_1 = \mu_2$ versus H_1: $\mu_1 \neq \mu_2$, where

μ_1 = underlying mean change over time t in group 1
μ_2 = underlying mean change over time t in group 2

We will conduct a two-sided test at level α and wish to have a power of $1 - \beta$ of detecting a significant difference if $|\mu_1 - \mu_2| = \delta$ under H_1. The required sample size per group is

$$n = \frac{2\sigma_d^2\left(z_{1-\alpha/2} + z_{1-\beta}\right)^2}{\delta^2}$$

where

$\sigma_d^2 = \sigma_1^2 + \sigma_2^2 - 2\rho\sigma_1\sigma_2$
σ_1^2 = variance of baseline values within a treatment group
σ_2^2 = variance of follow-up values within a treatment group
ρ = correlation coefficient between baseline and follow-up values within a treatment group

Solution to Example 8.33 We have that $\sigma_1^2 = \sigma_2^2 = 15^2 = 225$, $\rho = .70$, and $\delta = 8 - 3 = 5$ mm Hg. Therefore,

$$\sigma_d^2 = 225 + 225 - 2(.70)(15)(15) = 135$$

Also, $z_{1-\alpha/2} = z_{.975} = 1.96$, $z_{1-\beta} = z_{.80} = 0.84$. Thus,

$$n = \frac{2(135)(1.96 + 0.84)^2}{5^2}$$
$$= \frac{2116.8}{25} = 84.7, \text{ or 85 subjects in each group}$$

Similar to Equation 8.30, we can also consider the power of a longitudinal study given α, σ_1^2, σ_2^2, ρ, and a specified sample size per group (n).

EQUATION 8.31

Power of a Longitudinal Study Comparing Mean Change Between Two Normally Distributed Samples with Two Time Points To test the hypothesis H_0: $\mu_1 = \mu_2$ versus H_1: $\mu_1 \neq \mu_2$, where

μ_1 = underlying mean change over time t in treatment group 1
μ_2 = underlying mean change over time t in treatment group 2

for the specific alternative $|\mu_1 - \mu_2| = \delta$, with two-sided significance level α, and sample of size n in each group, the power is given by

$$\text{Power} = \Phi\left(-z_{1-\alpha/2} + \frac{\sqrt{n}\,\delta}{\sigma_d\sqrt{2}}\right)$$

where

$\sigma_d^2 = \sigma_1^2 + \sigma_2^2 - 2\rho\sigma_1\sigma_2$
σ_1^2 = variance of baseline values within a treatment group
σ_2^2 = variance of follow-up values within a treatment group
ρ = correlation between baseline and follow-up values over time t within a treatment group

Example 8.34 **Hypertension** Suppose that 75 subjects per group are recruited for the study described in Example 8.33. How much power will the study have under the same assumptions as in Example 8.33?

Solution We have $n = 75$, $\alpha = .05$, $\delta = 5$ mm Hg, $\sigma_d^2 = 135$ (from the solution to Example 8.33). Thus,

$$\begin{aligned}\text{Power} &= \Phi\left(-z_{.975} + \frac{\sqrt{75}(5)}{\sqrt{135(2)}}\right) \\ &= \Phi\left(-1.96 + \frac{43.30}{16.43}\right) \\ &= \Phi(0.675) = .750\end{aligned}$$

Thus, the study will have 75% power to detect this difference.

Note that based on Equations 8.30 and 8.31, as the correlation coefficient between repeated measures decreases, the variance of change scores (σ_d^2) will increase, resulting in an increase in the required sample size for a given level of power (Equation 8.30), and a decrease in power for a given sample size (Equation 8.31). Thus measures that are less reproducible over time will require a larger sample size for hypothesis-testing purposes.

Also, as the length of follow-up (t) increases, the correlation between repeated measures usually decreases. Therefore, studies with longer follow-up (say, 2 years) will require a larger sample size than studies with a shorter follow-up (say, 1 year) to detect a difference of the same magnitude (δ). However, in some instances the expected difference between groups (δ) may increase as t increases. Thus the overall impact of length of follow-up on sample size is uncertain.

Finally, if data already exist on change scores over the time period t (either from a pilot study or from the literature), then the variance of change (σ_d^2) can be estimated directly from the sample variance of change (in the pilot study or the literature), and it is unnecessary to use Equation 8.29 to compute σ_d^2. However, a common mistake is to run a pilot study based on repeated measures a short time apart (e.g., 1 week) and base the estimate of σ_d^2 on difference scores from this pilot study, even when the length of the main investigation is much longer (e.g., 1 year). This usually will result in an underestimate of σ_d^2 (or correspondingly, an overestimate of ρ) and will result in an underestimate (sometimes sizable) of required sample size for a given level of power, or an overestimate of power for a given sample size [5].

The methods described in this section are for studies with a single follow-up visit. For studies with more follow-up visits, more complicated methods of sample size and power estimation are needed [5].

SECTION 8.12 Summary

In this chapter, methods of hypothesis testing for comparing the means and variances of two samples that are assumed to be normally distributed were studied. The basic strategy is outlined in the shaded boxes of the flowchart in Figure 8.13, which

FIGURE 8.13 **Flowchart summarizing two-sample statistical inference—normal-theory methods. Material covered in Chapter 8 is shaded.**

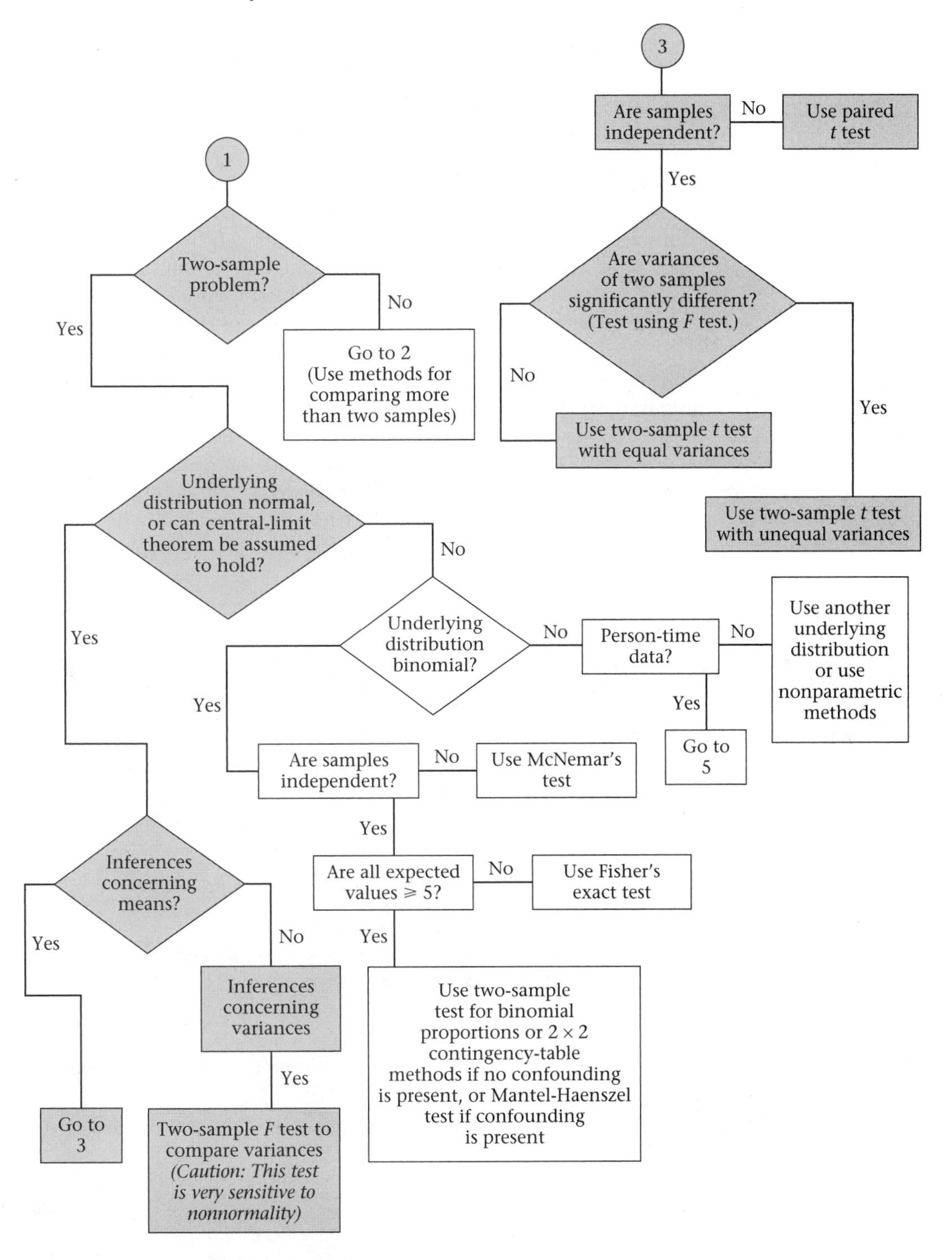

is an extract from the larger flowchart in the back of this book (pp. 776–780). Referring to 1 in the upper left, first note that we are dealing with the case of a two-sample problem in which either the underlying distributions are normal or the central-limit theorem can be assumed to hold. If we are interested in comparing the means of the two samples, then we refer to box 3. If our two samples are paired—that is, if each person is used as his or her own control or if the samples consist of different people who are matched on a one-to-one basis—then the paired t test is appropriate. If the samples are independent, then the F test for the equality of two variances is used to decide whether or not the variances are significantly different. If the variances are not significantly different, then the two-sample t test with equal variances is used; if the variances are significantly different, then the two-sample t test with unequal variances is used. If we are only comparing the variances of the two samples, then only the F test for comparing variances is used, as indicated in the lower left of Figure 8.13.

The chapter concluded by providing methods for the detection of outliers and presenting the appropriate sample size and power formulas for planning investigations in which the goal is to compare the means from two independent samples. We considered samples size and power formulas for both cross-sectional and longitudinal studies. In Chapter 9, we extend our work on the comparison of two samples to the case where there are two groups to be compared but the assumption of normality is questionable. We will introduce nonparametric methods to solve this problem to complement the parametric methods discussed in Chapters 7 and 8.

PROBLEMS

Cardiovascular Disease

Twenty volunteers adopt a low-cholesterol diet for 3 months. The mean ±1 *sd* of changes (baseline – 3 months) in serum cholesterol over the 3-month period was 20.0 ± 35.0 (mg/dL).

8.1 Test for significant changes in mean cholesterol over 3 months.

An important component of cholesterol is HDL cholesterol, which is widely believed to have a beneficial effect on heart disease. The mean ±1 *sd* of changes (baseline – 3 months) in HDL cholesterol over the 3-month period was 3.0 ± 12.0 (mg/dL).

8.2 Test for significant changes in mean HDL cholesterol over 3 months.

The mean ±1 *sd* of weight loss over the 3-month period was 5.2 ± 8.0 (lb).

8.3 Test for significant changes in mean weight over 3 months.

8.4 Find the lower 2.5th percentile of an *F* distribution with 14 and 7 *df*.

Nutrition

The mean ±1 *sd* of ln [calcium intake (mg)] among twenty-five 12- to 14-year-old females below the poverty level is 6.56 ± 0.64. Similarly, the mean ±1 *sd* of ln [calcium intake (mg)] among forty 12- to 14-year-old females above the poverty level is 6.80 ± 0.76.

8.5 Test for a significant difference between the variances of the two groups.

8.6 What is the appropriate procedure to test for a significant difference in means between the two groups?

8.7 Implement the procedure in Problem 8.6 using the critical-value method.

8.8 What is the *p*-value corresponding to your answer to Problem 8.7?

8.9 Compute a 95% confidence interval for the difference in means between the two groups.

Refer to the data in Table 2.11.

8.10 Test for a significant difference in the variances of the initial white blood count (WBC) between patients who did and patients who did not receive a bacterial culture.

8.11 What is the appropriate test procedure to test for significant differences in mean WBC between people who do and people who do not receive a bacterial culture?

8.12 Perform the procedure in Problem 8.11 using the critical-value method.

8.13 What is the *p*-value corresponding to your answer to Problem 8.12?

8.14 Compute a 95% confidence interval for the true difference in mean WBC between the two groups.

Refer to Problem 8.5.

***8.15** Suppose an equal number of 12- to 14-year-old girls below and above the poverty level are recruited to study differences in calcium intake. How many girls should be recruited to have an 80% chance of detecting a significant difference using a two-sided test with $\alpha = .05$?

***8.16** Answer Problem 8.15 if a one-sided rather than a two-sided test is used.

***8.17** Using a two-sided test with $\alpha = .05$, answer Problem 8.15 anticipating that 2 girls above the poverty level will be recruited for every girl below the poverty level.

***8.18** Suppose 50 girls above the poverty level and 50 girls below the poverty level are recruited for the study. How much power will the study have of finding a significant difference using a two-sided test with $\alpha = .05$ assuming that the population parameters are the same as the sample estimates in Problem 8.5?

***8.19** Answer Problem 8.18 if a one-sided rather than a two-sided test is used.

***8.20** Suppose that 50 girls above the poverty level and 25 girls below the poverty level are recruited for the study. How much power will the study have if a two-sided test is used with $\alpha = .05$?

***8.21** Answer Problem 8.20 if a one-sided test is used with $\alpha = .05$?

Gynecology

A study was conducted to compare the age at menarche (the age at the first menstrual period) of girls entering the first-year class of a small U.S. private college in the year 1975 with that of girls entering the first-year class of the same college in 1985. This study is in response to reports of differences over time in other countries. Suppose that 30 girls in the 1975 class have a mean age at menarche of 12.78 years with a standard deviation of 0.43 years and 40 girls in the class of 1985 have a mean age at menarche of 12.42 years with a standard deviation of 0.67 years.

8.22 What are the appropriate null and alternative hypotheses to test whether or not the mean ages at menarche are comparable in the two groups? Justify your choice of the appropriate hypotheses.

8.23 Perform the significance test indicated in Problem 8.22 and state your conclusions.

8.24 What is the advantage of comparing two different classes of the same school as opposed to comparing the 1975 entering class of one school with the 1985 entering class of another school?

Ophthalmology

Diflunisal is a drug used to treat mild to moderate pain, osteoarthritis, or rheumatoid arthritis. Diflunisal's ocular effects had not been studied until a study on its effect on intraocular pressure in glaucoma patients who were already receiving maximum therapy for glaucoma was conducted [6].

***8.25** Suppose the change (mean ± *sd*) in ocular pressure after administration of diflunisal (follow-up—baseline) among 10 patients whose standard therapy was methazolamide and topical glaucoma medications was -1.6 ± 1.5 mm Hg. Assess the statistical significance of the results.

***8.26** The change in ocular pressure after administration of diflunisal among 30 patients whose standard therapy was topical drugs only was -0.7 ± 2.1 mm Hg. Assess the statistical significance of these results.

***8.27** Compute 95% confidence limits for the mean change in pressure in each of the two groups identified in Problems 8.25 and 8.26.

***8.28** Compare the mean change in ocular pressure in the two groups identified in Problems 8.25 and 8.26 using hypothesis-testing methods.

Cardiovascular Disease, Pediatrics

A study in Pittsburgh looked at various cardiovascular risk factors in children, as measured at birth and during their first five years of life [7]. In particular, heart rate was assessed at birth, 5 months, 15 months, 24 months, and annually thereafter until 5 years of age. Heart rate was related to age, sex, race, and socioeconomic status. The data in Table 8.12 were presented relating heart rate to race among newborns.

TABLE 8.12 Relationship of heart rate to race among newborns

Race	Mean (beats per minute)	*sd*	*n*
White	125	11	218
Black	133	12	156

Source: Reprinted with permission of the *American Journal of Epidemiology, 119*(4), 554–563.

8.29 Test for a significant difference in mean heart rate between white and black newborns.

8.30 Report a p-value for the test performed in Problem 8.29.

Pharmacology

One method for assessing the effectiveness of a drug is to note its concentration in blood and/or urine samples at certain periods of time after giving the drug. Suppose we wish to compare the concentrations of two types of aspirin (types A and B) in urine specimens taken from the same person, 1 hour after he or she has taken the drug. Hence, a specific dosage of either type A or type B aspirin is given at one time and the 1-hour urine concentration is measured. One week later, after the first aspirin has presumably been cleared from the system, the same dosage of the other aspirin is given to the same person and the 1-hour urine concentration is noted. Because the order of giving the drugs may affect the results, a table of random numbers is used to decide which of the two types of aspirin to give first. This experiment is performed on 10 people; the results are given in Table 8.13.

TABLE 8.13 Concentration of aspirin in urine samples

Person	Aspirin A 1-hour concentration (mg%)	Aspirin B 1-hour concentration (mg%)
1	15	13
2	26	20
3	13	10
4	28	21
5	17	17
6	20	22
7	7	5
8	36	30
9	12	7
10	18	11
Mean	19.20	15.60
sd	8.63	7.78

Suppose we wish to test the hypothesis that the concentrations of the two drugs are the same in urine specimens.

***8.31** What are the appropriate hypotheses?

***8.32** What are the appropriate procedures to test these hypotheses?

***8.33** Conduct the tests mentioned in Problem 8.32.

***8.34** What is the best point estimate of the mean difference in concentrations between the two drugs?

***8.35** What is a 95% confidence interval for the mean difference?

8.36 Suppose an α level of .05 is used for the test in Problem 8.33. What is the relationship between the decision reached with the test procedure in Problem 8.33 and the nature of the confidence interval in Problem 8.35?

Nutrition

A hypothesis of ongoing clinical interest is that vitamin C prevents the common cold. A study is organized to test this hypothesis using 20 prisoners as participants. In the study, 10 are randomly allocated to receive vitamin C capsules and 10 are randomly allocated to receive placebo capsules. The number of colds over a 12-month period for each participant is given in Table 8.14. We wish to test the hypothesis that vitamin C prevents the common cold.

TABLE 8.14 Number of colds over a 12-month period for people taking vitamin C and placebo capsules

	Vitamin C	Placebo	
i	x_{i1}	x_{i2}	$d_i(x_{i1} - x_{i2})$
1	4	7	−3
2	0	8	−8
3	3	4	−1
4	4	6	−2
5	4	6	−2
6	3	4	−1
7	4	6	−2
8	3	4	−1
9	2	6	−4
10	6	6	0

8.37 Is a one-sample or two-sample test needed here?

8.38 Is a one-sided or two-sided test needed here?

8.39 Which of the following test procedures should be used to test this hypothesis? (More than one may be necessary.) (Refer to the flowchart in Figure 8.13.)

(a) Paired t test

(b) Two-sample t test with equal variances

(c) Two-sample t test with unequal variances

(d) F test for the equality of two variances

(e) One-sample t test

8.40 Carry out the test procedure(s) in Problem 8.39 and report a p-value.

8.41 Derive a lower one-sided 95% confidence interval for the mean difference (vitamin C – placebo) in number of colds per year between the two groups.

8.42 What is the relationship between your answers to Problems 8.40 and 8.41?

Pulmonary Disease

A 1980 study was conducted whose purpose was to compare the indoor air quality in offices where smoking was permitted with that in offices where smoking was not permitted [8]. Measurements were made of carbon monoxide (CO) at 1:20 P.M. in 40 work areas where smoking was permitted and 40 work areas where smoking was not permitted. Where smoking was permitted, the mean CO = 11.6 parts per million (ppm) and the standard deviation CO = 7.3 ppm. Where smoking was not permitted, the mean CO = 6.9 ppm and the standard deviation CO = 2.7 ppm.

8.43 Test for whether or not the standard deviation of CO is significantly different in the two types of working environments.

8.44 Test for whether or not the mean CO is significantly different in the two types of working environments.

8.45 Provide a 95% confidence interval for the difference in mean CO between the smoking and nonsmoking working environments.

Ophthalmology

A camera has been developed to detect the presence of cataract more accurately. Using this camera, the gray level of each point (or pixel) in the lens of a human eye can be characterized into 256 gradations, where a gray level of 1 represents black and a gray level of 256 represents white. To test the camera, photographs were taken of 6 randomly selected normal eyes and 6 randomly selected cataractous eyes (the two groups consist of different people). The median gray level of each eye was computed over the 10,000 + pixels in the lens. The data are given in Table 8.15.

TABLE 8.15 Median gray level for cataractous and normal eyes

Patient number	Cataractous median gray level	Normal median gray level
1	161	158
2	140	182
3	136	185
4	171	145
5	106	167
6	149	177
$\bar{x}$	143.8	169.0
s	22.7	15.4

8.46 What statistical procedure can be used to test if there is a significant difference in the median gray levels between cataractous and normal eyes?

8.47 Carry out the test procedure mentioned in Problem 8.46, and report a *p*-value.

8.48 Provide a 99% confidence interval for the mean difference in gray levels between cataractous and normal eyes.

Obstetrics

A clinical trial is conducted at the gynecology unit of a major hospital to determine the effectiveness of drug A in preventing premature birth. In the trial, 30 pregnant women are to be studied, 15 assigned to a treatment group that will receive drug A and 15 assigned to a control group that will receive a placebo. The patients are to take a fixed dose of each drug on a one-time-only basis between the 24th and 28th weeks of pregnancy. The patients are assigned to groups using computer-generated random numbers, where for every 2 patients eligible for the study, one is assigned randomly to the treatment group and the other to the control group.

8.49 Suppose *you* are conducting the study. What would be a reasonable way of allocating women to the treatment and control groups?

Suppose the weights of the babies are those given in Table 8.16.

TABLE 8.16 Birthweights in a clinical trial to test a drug for preventing low-birthweight deliveries

Patient number	Treatment group baby weight (lb)	Control group baby weight (lb)
1	6.9	6.4
2	7.6	6.7
3	7.3	5.4
4	7.6	8.2
5	6.8	5.3
6	7.2	6.6
7	8.0	5.8
8	5.5	5.7
9	5.8	6.2
10	7.3	7.1
11	8.2	7.0
12	6.9	6.9
13	6.8	5.6
14	5.7	4.2
15	8.6	6.8

8.50 How would you assess the effects of drug A in light of your answer to Problem 8.49? Specifically, would you use a paired or an unpaired analysis, and of what type?

8.51 Perform both a paired and an unpaired analysis of the data. Does the type of analysis affect the assessment of the results?

8.52 Suppose patient 3 in the control group subsequently moves to another city before giving birth and her child's weight is unknown. Does this event affect the analyses in Problem 8.51? If so, how?

Pharmacology

In a pediatric clinic a study is carried out to see how effective aspirin is in reducing temperature. Twelve 5-year-old children suffering from influenza had their temperatures taken immediately before and 1 hour after administration of aspirin. The results are given in Table 8.17. Suppose we assume normality and want to test the hypothesis that aspirin is reducing the temperature.

TABLE 8.17 Body temperature (°F) before and after taking aspirin

Patient	Before	After
1	102.4	99.6
2	103.2	100.1
3	101.9	100.2
4	103.0	101.1
5	101.2	99.8
6	100.7	100.2
7	102.5	101.0
8	103.1	100.1
9	102.8	100.7
10	102.3	101.1
11	101.9	101.3
12	101.4	100.2

8.53 What are the null and alternative hypotheses in this situation?

8.54 In words, what is meant by a type I error in this situation?

8.55 Suppose the alternative that aspirin reduces the mean temperature by 1 degree is considered. What is meant by the power of the test against this specific alternative?

8.56 How would the power in Problem 8.55 change if the alternative were a mean temperature reduction of 2 degrees?

8.57 Compute the power for the alternatives mentioned in Problems 8.55 and 8.56.

8.58 Perform a significance test for the hypotheses in Problem 8.53. Are there any possible alternative explanations for the results obtained?

Pulmonary Disease

A possible important environmental determinant of lung function in children is the amount of cigarette smoking in the home. Suppose this question is studied by selecting two groups: Group 1 consists of 23 nonsmoking children 5–9 years of age, *both* of whose parents smoke, who have a mean FEV of 2.1 L and *sd* of 0.7 L; group 2 consists of 20 nonsmoking children of comparable age, *neither* of whose parents smoke, who have a mean FEV of 2.3 L and *sd* of 0.4 L.

***8.59** What are the appropriate null and alternative hypotheses in this situation?

***8.60** What is the appropriate test procedure for the hypotheses in Problem 8.59?

***8.61** Carry out the test in Problem 8.60 using the critical-value method.

***8.62** Provide a 95% confidence interval for the true mean difference in FEV between 5- to 9-year-old children whose parents smoke and comparable children whose parents do not smoke.

***8.63** If this is regarded as a pilot study, then how many children are needed in each group (assuming equal numbers in each group) to have a 95% chance of detecting a significant difference using a two-sided test with $\alpha = .05$?

***8.64** Answer the question in Problem 8.63 if the investigators intend to use a one-sided rather than a two-sided test.

Suppose 40 children, both of whose parents smoke, and 50 children, neither of whose parents smoke, are recruited for the study.

***8.65** How much power would such a study have using a two-sided test with significance level = .05, assuming that the estimates of the population parameters in the pilot study are correct?

***8.66** Answer Problem 8.65 if a one-sided rather than a two-sided test is to be used.

Infectious Disease

The degree of clinical agreement among different physicians on the presence or absence of generalized lymphadenopathy was assessed in 32 randomly selected participants from a prospective study of male sexual contacts of men with AIDS or an AIDS-related condition (ARC) [9]. The total number of palpable lymph nodes was assessed by each of three physicians. Results from two of the three physicians are presented in Table 8.18.

TABLE 8.18 Reproducibility of assessment of number of palpable lymph nodes among sexual contacts of AIDS or ARC patients

	Number of palpable lymph nodes		
Patient	Doctor A	Doctor B	Difference
1	4	1	3
2	17	9	8
3	3	2	1
4	11	13	−2
5	12	9	3
6	5	2	3
7	5	6	−1
8	6	3	3
9	3	0	3
10	5	0	5
11	9	6	3
12	1	1	0
13	5	4	1
14	8	4	4
15	7	7	0
16	8	6	2
17	4	1	3
18	12	9	3
19	10	7	3
20	9	11	−2
21	5	0	5
22	3	0	3
23	12	12	0
24	5	1	4
25	13	9	4
26	12	6	6
27	6	9	−3
28	19	9	10
29	8	4	4
30	15	9	6
31	6	1	5
32	5	4	1
Mean	7.91	5.16	2.75
sd	4.35	3.93	2.83
n	32	32	32

8.67 What is the appropriate test procedure to test if there is a systematic difference between the assessment of Doctor A versus Doctor B?

8.68 Should a one-sided or a two-sided test be performed? Why?

8.69 Perform the test in Problem 8.67 and report a *p*-value.

8.70 Compute a 95% confidence interval for the true mean difference between observers. How does it relate to your answer to Problem 8.69?

8.71 Suppose the results of Problem 8.69 shows no significant difference. Does this mean that this type of assessment is highly reproducible? Why or why not?

Renal Disease

Ten patients with advanced diabetic nephropathy (kidney complications of diabetes) were treated with captopril over an 8-week period [10]. Urinary protein was measured before and after drug therapy, with results listed in Table 8.19 in both the raw and ln scale.

TABLE 8.19 Changes in urinary protein after treatment with captopril

	Raw scale Urinary protein (g/24 hr)		ln scale Urinary protein (g/24 hr)	
Patient	Before	After	Before	After
1	25.6	10.1	3.24	2.31
2	17.0	5.7	2.83	1.74
3	16.0	5.6	2.77	1.72
4	10.4	3.4	2.34	1.22
5	8.2	6.5	2.10	1.87
6	7.9	0.7	2.07	−0.36
7	5.8	6.1	1.76	1.81
8	5.4	4.7	1.69	1.55
9	5.1	2.0	1.63	0.69
10	4.7	2.9	1.55	1.06

***8.72** What is the appropriate statistical procedure to test if mean urinary protein has changed over the 8-week period?

***8.73** Perform the test in Problem 8.72 using both the raw and ln scale, and report a *p*-value. Are there any advantages in using the raw or the ln scale?

***8.74** What is your best estimate of the percentage change in urinary protein based on the data in Table 8.19?

***8.75** Provide a 95% confidence interval associated with your estimate in Problem 8.74.

Nutrition

An important hypothesis in hypertension research is that sodium restriction may lower blood pressure. However, it is difficult to achieve sodium restriction over the long term and dietary counseling in a group setting is sometimes used to achieve this goal. The data on urinary sodium in Table 8.20 were obtained on 8 individuals enrolled in a sodium-restriction group. Data were collected at baseline and after 1 week of dietary counseling.

TABLE 8.20 Overnight Na excretion (mEq/8hr) before and after dietary counseling

Person	Week 0 (baseline)	Week 1	Difference
1	7.85	9.59	−1.74
2	12.03	34.50	−22.47
3	21.84	4.55	17.29
4	13.94	20.78	−6.84
5	16.68	11.69	4.99
6	41.78	32.51	9.27
7	14.97	5.46	9.51
8	12.07	12.95	−0.88
$\bar{x}$	17.65	16.50	1.14
s	10.56	11.63	12.22

8.76 What are appropriate hypotheses to test if dietary counseling is effective in reducing sodium intake over a 1-week period (as measured by overnight urinary sodium excretion)?

8.77 Conduct the test mentioned in Problem 8.76, and report a *p*-value.

8.78 Provide a 95% confidence interval for the true mean change in overnight Na excretion over a 1-week period.

8.79 How many subjects would be needed to have a 90% chance of detecting a significant change in mean urinary sodium excretion if a one-sided test is used with $\alpha = .05$ and the estimates from the data in Table 8.20 are used as the true population parameters?

Refer to Data Set NIFED.DAT on the data disk. See p. 150 for a complete description of the data set.

8.80 Assess if there is any difference between the nifedipine and propranolol groups regarding their effects on blood pressure and heart rate. Refer to the indices of change defined in Problems 6.81–6.85.

Genetics

A study was conducted of genetic and environmental influences on cholesterol levels. The data set used for the study were obtained from a twin registry in Sweden [11]. Specifically, four populations of adult twins were studied: (1) monozygotic (MZ) twins reared apart, (2) MZ twins reared together, (3) dizygotic (DZ) twins reared apart, and (4) DZ twins reared together. One issue is whether it is necessary to correct for sex before performing more complex genetic analyses. The data in Table 8.21 were presented for total cholesterol levels for MZ twins reared apart by sex.

TABLE 8.21 Comparison of mean total cholesterol for MZ twins reared apart, by sex

	Men	Women
Mean	253.3	271.0
sd	44.1	44.1
n	44	48

Note: n = number of people (e.g., for males, 22 pairs of twins = 44 people)

***8.81** If we assume that (a) serum cholesterol is normally distributed, (b) the samples are independent, and (c) the standard deviations for men and women are the same, then what is the name of the statistical procedure that can be used to compare the two groups?

***8.82** Suppose we wish to use the procedure in Problem 8.81 using a two-sided test. State the hypotheses being tested and implement the method. Please report a *p*-value.

***8.83** Suppose we wish to use the procedure in Problem 8.81 using a one-sided test where the alternative hypothesis is that men have higher cholesterol levels than women. State the hypotheses being tested and implement the method in Problem 8.81. Please report a *p*-value.

***8.84** Are the assumptions in Problem 8.81 likely to hold for these samples? Why or why not?

Pulmonary Disease

A study was performed looking at the effect of mean ozone exposure on change in pulmonary function. Fifty hikers were recruited into a study; 25 of the hikers hiked on days with low-ozone exposure, and 25 hiked on days with high-ozone exposure. The change in pulmonary function after a 4-hour hike was recorded for each subject. The results are given in Table 8.22.

TABLE 8.22 Comparison of change in FEV on high-ozone versus low-ozone days

	Mean change In FEV[a]	*sd*	*n*
High-ozone days	0.101	0.253	25
Low-ozone days	0.042	0.106	25

[a]Change in FEV, forced expiratory volume, in 1 second (liters) (baseline – follow-up)

8.85 What test can be used to determine if the mean change in FEV differs between the high-ozone and low-ozone days?

8.86 Implement the test in Problem 8.85, and report a *p*-value (two-tail).

8.87 Suppose we determine a 95% confidence interval for the true mean change in pulmonary function on high-ozone days. Is this CI narrower, wider, or the same width as a 90% confidence interval? (*Do not* actually derive the confidence interval.)

Rheumatology

A study was conducted [12] comparing muscle function between patients with rheumatoid arthritis (RA) and osteoarthritis (OA). A 10-point scale was used to assess balance and coordination where a high score indicates better coordination. The results were as shown in Table 8.23 for 36 RA subjects and 30 OA subjects.

TABLE 8.23 Comparison of balance scores for rheumatoid arthritis versus osteoarthritis patients

	Mean balance score	*sd*	*n*
RA	3.4	3.0	36
OA	2.5	2.8	30

***8.88** What test can be used to test if the mean balance score is the same for RA and OA subjects? What are some of the assumptions of this test?

***8.89** Perform the test mentioned in Problem 8.88, and report a *p*-value.

***8.90** What is your best estimate of the proportion of RA and OA patients, with impaired balance, where impaired balance is defined as a balance score less than or equal to 2 and normality is assumed?

***8.91** Suppose a larger study is planned. How many subjects would be needed to detect a difference of 1 unit in mean balance score with 80% power if the number of subjects in each group is intended to be the same and a two-sided test is used with $\alpha = .05$?

Ophthalmology

A group of eight unrelated patients with retinitis pigmentosa (RP) were identified with a distinct point mutation in the gene coding for rhodopsin. The mutation corresponds to a single amino-acid substitution in rhodopsin protein. Their ocular findings were compared with 140 unrelated RP patients without this mutation. The results are given in Table 8.24 for ln (visual field area).

TABLE 8.24 Comparison of ln (visual field area) between RP patients with and without the rhodopsin mutation

	Mean	*sd*	*n*
Mutation present	7.11	1.21	8
Mutation absent	7.99	1.32	140

8.92 What test procedure can be used to test if the variances of the two groups are the same or not?

8.93 Implement the test in Problem 8.92 and report a *p*-value (two-sided).

8.94 Assuming that the underlying variances in the two groups are the same, what test procedure can be used to test if the means of the two groups are the same or not?

8.95 Implement the test procedure mentioned in Problem 8.94 and report a *p*-value (two-sided).

8.96 Provide a 95% CI for the true mean difference between ln (visual field area) and the percent difference in visual field area between the two groups.

Cardiology

A clinical trial was undertaken comparing percutaneous transluminal coronary angioplasty (PTCA) with medical therapy in the treatment of single-vessel coronary-artery disease [13]. One hundred and seven patients were randomly assigned to medical therapy and 105 to PTCA. Patients were given exercise tests at baseline and after 6 months of follow-up. Exercise tests were performed up to maximal effort until symptoms (such as angina) were present. The results shown in Table 8.25 were obtained for change in total duration of exercise (min)(6 months – baseline).

TABLE 8.25 Change in total duration of exercise for patients with coronary-artery disease randomized to medical therapy versus PTCA

	Mean change (min)	*sd*	*n*
Medical Therapy	0.5	2.2	100
PTCA	2.1	3.1	99

***8.97** What test can be performed to see if there has been a change in mean total duration of exercise for a *specific* treatment group?

***8.98** Perform the test in Problem 8.97 for the medical therapy group, and report a *p*-value.

***8.99** What test can be performed to compare the mean change in duration of exercise *between* the two treatment groups?

***8.100** Perform the test mentioned in Problem 8.99, and report a *p*-value.

Hypertension

A case-control study was performed among participants in the Kaiser Permanente Health Plan to compare body-fat distribution among "hypertensive" people who were normotensive at entry into the plan and became hypertensive over time as compared with "normotensives" who remained normotensive throughout their participation in the plan. A match was found for each "hypertensive" person of the same sex, race, year of birth, and year of entry into the plan; 609 matched pairs were created. The data shown in Table 8.26 were presented regarding body-mass index at baseline in the two groups [14].

8.101 What are the advantages of using a matched design for this study?

8.102 What test procedure can be used to test for differences in mean body-mass index between the two groups?

8.103 Implement the test procedure in Problem 8.102, and report a *p*-value.

8.104 Compute a 90% confidence interval for the mean difference in body-mass index between the two groups.

Hepatic Disease

An experiment was conducted to examine the influence of avian pancreatic polypeptide (aPP), cholecystokinin (CCK), vasoactive intestinal peptide (VIP), and secretin on pancreatic and biliary secretions in laying hens. In particular, researchers were concerned with the extent to which these hormones increase or decrease biliary and pancreatic flows and their pH values.

White leghorn hens, 14–29 weeks old, were surgically fitted with cannulas for collecting pancreatic and biliary secretions and a jugular cannula for continuous infusion of aPP, CCK, VIP, or secretin. One trial per day was conducted on a hen, as long as her implanted cannulas remained functional. Thus, there were varying numbers of trials per hen.

Each trial began with infusion of physiological saline for 20 minutes. At the end of this period, pancreatic and biliary secretions were collected and the cannulas were attached to new vials. The biliary and pancreatic flow rates (in microliters per minute) and pH values (if possible) were measured. Infusion of a hormone was then begun, and continued for 40 minutes. Measurements were then repeated.

Data Set HORMONE.DAT (on the data disk) contains data for the four hormones and saline, where saline indicates trials where physiological saline was infused in place of an active hormone during the second period. Each trial is one record in the file. There are 11 variables associated with each trial, as shown in Table 8.27.

8.105 Assess if there are significant changes in secretion rates or pH levels with any of the hormones or with saline.

8.106 Compare the changes in secretion rate or pH levels for each active hormone versus the placebo (saline) group. Use methods of hypothesis testing and/or confidence intervals to express these comparisons statistically.

8.107 For each active hormone group, categorize dosage by high dose (above the median) versus low dose (at or below the median) and assess if there is any dose–response relationship (any differences in mean changes in secretion rates or pH between the high- and low-dose groups).

Refer to Data Set FEV.DAT, on the data disk.

8.108 Compare the level of mean FEV between males and females separately in three distinct age groups (5–9, 10–14, 15–19).

TABLE 8.26 Comparison of body-mass index for people who developed hypertension versus people who remained normotensive

	Cases		Controls		Mean difference	Test statistic
	Mean	*sd*	Mean	*sd*		
Body-mass index (kg/m^2)	25.39	3.75	24.10	3.42	1.29	6.66

TABLE 8.27 Format of HORMONE.DAT

Column	Record number	Format of HORMONE.DAT	Code
1–8	1	Unique identification number for each chicken	xx.x
10–17	1	Biliary secretion rate (pre)	xx.x
19–26	1	Biliary pH (pre)	xx.x
28–35	1	Pancreatic secretion rate (pre)	xx.x
37–44	1	Pancreatic pH (pre)	x.x
46–53	1	Dosage of hormone	xx.x
55–62	1	Biliary secretion rate (post)	xx.x
64–71	1	Biliary pH (post)	x.x
1–8	2	Pancreatic secretion rate (post)	xx.x
10–17	2	Pancreatic pH (post)	x.x
19–26	2	Hormone (1 = saline; 2 = aPP, 3 = CCK; 4 = secretin; 5 = VIP)	xx.x
		Zero values for pH indicate missing values. The units for dosages are nanograms per mL of plasma for aPP, and μg per kg per hour for CCK, VIP, and secretin.	

8.109 Compare the level of mean FEV between smokers and nonsmokers separately for 10- to 14-year-old boys, 10- to 14-year-old girls, 15- to 19-year-old boys, and 15- to 19-year-old girls.

Hypertension, Pediatrics

Refer to Data Set INFANTBP.DAT and INFANTBP.DOC, both on the data disk.

Consider again the salt-taste indices and sugar-taste indices constructed in Problems 6.67–6.68.

8.110 Obtain a frequency distribution and subdivide infants as high or low according to whether the children are above or below the median value for the indices. Use hypothesis-testing and confidence-interval methodology to compare mean blood-pressure levels for children in the high and low groups.

8.111 Answer Problem 8.110 in a different way by subdividing the salt- and sugar-taste indices more finely (such as quintiles or deciles). Compare mean blood-pressure level for children at the extremes (such as children at the highest quintile versus the lowest quintile). Do you get the impression that the indices are related to blood-pressure level? Why or why not? We will discuss this from a different point of view in our work on regression analysis in Chapter 11 and the analysis of variance in Chapter 12.

Sports Medicine

Tennis elbow is a painful condition that afflicts many tennis players at some time. A number of different treatments are used for this condition, including rest, heat, and anti-inflammatory medications. A clinical trial was conducted among 87 subjects comparing the effectiveness of Motrin (generic name, *ibuprofen*), a widely used anti-inflammatory agent, to placebo. Subjects received both drugs, but the order of administration of the two drugs was determined by randomization. Specifically, approximately half the subjects (group A) received an initial 3-week course of Motrin, while the other subjects (group B) received an initial 3-week course of placebo. After the 3-week period, subjects were given a 2-week **washout period** during which they received no study medication. The purpose of the washout period was to eliminate any residual biological effect of the first-period medication. After the washout period, a second period of active drug administration began with group A subjects receiving a 3-week course of placebo, and group B subjects receiving a 3-week course of Motrin. At the end of each active drug period as well as at the end of the washout period, participants were asked to rate their degree of pain compared with baseline (before the beginning of the first active drug period). The goal of the study was to compare the degree of pain while on Motrin versus the degree of pain while on a placebo. This type of study is called a **cross-over design**, which we discuss in more detail in Chapter 13. The degree of pain versus baseline was measured on a 1–6 scale with 1 being "worse than baseline" and 6 being "completely improved." The comparison was made in four different ways: (1) during maximum activity, (2) 12 hours following maximum activity, (3) during the average day, and (4) by overall impression of drug efficacy. The data are given in Data Set TENNIS2.DAT with documentation in TENNIS2.DOC, both on the data disk.

8.112 Compare the degree of pain while on Motrin with the degree of pain while on placebo during maximal activity.

8.113 Answer Problem 8.112 for the degree of pain 12 hours following maximum activity.

8.114 Answer Problem 8.112 for the degree of pain during the average day.

8.115 Answer Problem 8.112 for the overall impression of drug efficacy.

Environmental Health, Pediatrics

Refer to Figure 8.12 and Table 8.9.

8.116 Assess if there are any outliers for Full-Scale IQ in the control group.

8.117 Assess if there are any outliers for Full-Scale IQ in the exposed group.

8.118 Based on your answers to Problems 8.116 and 8.117, compare the mean Full-Scale IQ between the exposed and the control group, after exclusion of outliers.

Pulmonary Disease

8.119 Refer to Data Set FEV.DAT on the data disk. Assess if there are any outliers in FEV for the following groups: 5- to 9-year-old boys; 5- to 9-year-old girls; 10- to 14-year-old boys; 10- to 14-year-old girls; 15- to 19-year-old boys; 15- to 19-year-old girls.

Ophthalmology

A study was performed comparing the mean ERG (electroretinogram) amplitude of patients with different genetic types of retinitis pigmentosa (RP), a genetic eye disease that often results in blindness. The following results were obtained for ln (ERG amplitude) among patients 18–29 years of age.

Genetic type	mean ± *sd*	*N*
Dominant	0.85 ± 0.18	62
Recessive	0.38 ± 0.21	35
X-linked	−0.09 ± 0.21	28

8.120 What is the standard error of ln (ERG amplitude) among patients with dominant RP? How does it differ from the standard deviation in the table?

8.121 What test can be used to compare the variance of ln (ERG amplitude) between patients with dominant versus recessive RP?

8.122 Implement the test in Problem 8.121 and report a *p*-value (two-tail). (*Hint*: $F_{34, 61, .975} = 1.778$.)

8.123 What test can be used to compare the mean ln (ERG amplitude) between patients with dominant versus recessive RP?

8.124 Implement the test in Problem 8.123 and report a *p*-value (two-tail).

Diabetes

A study was performed to compare the serum levels of oxidation products and antioxidant enzymes between diabetic and normal subjects and also between individuals with type I diabetes (IDDM) versus individuals with type II diabetes (NIDDM)—insulin-dependent and non-insulin-dependent diabetes mellitus, respectively [15]. The mean levels of lipid peroxidase (LP) among IDDM patients and normal controls is given in the following table.

Mean ± *sd* of LP (mmol/L) by group

	Mean	*sd*	*N*
IDDM	2.5	1.22	27
Normal	2.2	0.95	39

8.125 What test can be used to compare the variances between the two groups?

8.126 Implement the test in Problem 8.125, and assess whether there is a significant difference between the variances using an alpha level of .05. (*Please note:* $F_{26, 38, .95} = 1.79$, $F_{26, 38, .975} = 2.00$.)

8.127 What test can be used to compare the mean LP level between the two groups?

8.128 Implement the test in Problem 8.127, and report a two-sided *p*-value.

8.129 How much power did the preceding study have to detect a true mean difference between the groups of 1.0 mmol/L using a two-sided test with $\alpha = .05$ if the sample standard deviations in the preceding study are assumed to be the true standard deviations?

Hypertension

A study was performed comparing different nonpharmacologic treatments for people with high normal diastolic blood pressure (DBP) (80–89 mm Hg). One of the modes of treatment studied was stress management. People were randomly assigned to a stress management intervention (SMI) and a control group. Subjects randomized to SMI were given instruction in a group setting concerning different techniques for stress management and met periodically over a 1-month period. Subjects randomized to control were advised to pursue their normal lifestyles and were told that their blood pressure would be closely monitored and that their physician would be notified of any consistent elevation. The results for the SMI group ($n = 242$) at the end of the study (18 months) were as follows:

Mean (change) = –5.53 mm Hg (follow-up minus baseline), *sd* (change) = 6.48 mm Hg

8.130 What test can be used to assess if mean blood pressure has changed significantly in the SMI group?

8.131 Implement the test in Problem 8.130 and report a *p*-value.

The results for the control group ($n = 320$) at the end of the study were as follows:

Mean (change) = –4.77 mm Hg, *sd* (change) = 6.09 mm Hg.

8.132 What test can be used to compare mean blood-pressure change between the two groups?

8.133 Implement the test in Problem 8.132, and report a *p*-value. (*Hint*: For reference, $F_{241, 319, .90} = 1.166$, $F_{241, 319, .95} = 1.218$, $F_{241, 319, .975} = 1.265$.)

8.134 How much power did the study have for detecting a significant difference between groups (using a two-sided test with a 5% level of significance) if the true effect of the SMI intervention is to reduce mean DBP by 2mm Hg more than the control group and the standard deviation of change within a group is 6 mm Hg?

Cardiovascular Disease, Nutrition

A recent randomized trial examined the impact of fat-restricted diets on cholesterol-lowering in hypercholesterolemic men [16]. In 71 men randomized to a diet containing 22% calories from fat, the mean reduction over 1 year in low-density lipoprotein cholesterol was 8.4% (*sd* = 11.2%). In 59 men randomized to a diet containing 18% calories from fat, the mean reduction over 1 year was 13.0% (*sd* = 15.7%).

8.135 Assume that these percent reductions follow normal distributions. Test the null hypothesis that the mean reduction in the men on the diet containing 22% calories from fat is significantly different from 0.

8.136 Give a 95% confidence interval for this mean reduction. *Note:* $t_{70, .975} = 1.994$.

8.137 Now test the hypothesis that the mean reductions differ between the two groups, that is, that those assigned to a diet containing 18% calories from fat have a different mean reduction from those assigned to the diet containing 22% calories from fat.

8.138 Provide a 95% confidence interval for the mean difference between groups.

Endocrinology

A study was performed of the effect of introducing a low-fat diet on hormone levels of 73 postmenopausal women not using exogenous hormones [17]. The following data were presented for plasma estradiol in picograms/milligram.

	Estradiol (pg/ml)[a]
Preintervention	0.71 (0.26)
Postintervention	0.63 (0.26)
Difference (postintervention – preintervention)	–0.08 (0.20)

[a]Values are mean and *sd* (in parentheses) for log (base 10) of preintervention and postintervention measurements and for their difference.

8.139 What test can be performed to assess the effects of adopting a low-fat diet on mean plasma estradiol levels?

8.140 Implement the test in Problem 8.139, and report a *p*-value.

8.141 Provide a 95% CI for the change in mean $\log_{10}$ (plasma estradiol). (*Hint:* The 95th percentile of a *t* distribution with 72 *df* = 1.6663; the 97.5th percentile of a *t* distribution with 72 *df* = 1.9935.)

8.142 Suppose a similar study is planned among women who use exogenous hormones. How many subjects need to be enrolled if the mean change in $\log_{10}$ (plasma estradiol) is –0.08, the *sd* of change is 0.20, and we wish to conduct a two-sided test with an α level of .05 and a power of .80?

Cardiology

A study was performed concerning risk factors for carotid-artery stenoisis (narrowing) among 464 men born in 1914 and residing in the city of Malmö, Sweden [18]. The following data were reported for blood-glucose level:

	No stenosis ($n = 356$)		Stenosis ($n = 108$)	
	mean	*sd*	mean	*sd*
Blood glucose(mmol/L)	5.3	1.4	5.1	0.8

8.143 What test can be performed to assess if there is a significant difference in mean blood-glucose level between men with and without stenosis? (*Hint*: $F_{355, 107, .95} = 1.307$; $F_{355, 107, .975} = 1.377$.)

8.144 Implement the test mentioned in Problem 8.143, and report a *p*-value (two-tail). (*Hint:* $t_{282, .975} = 1.968$.)

Ophthalmology

A study is being planned to assess whether a topical anti-allergic eye drop is effective in preventing the signs and symptoms of allergic conjunctivitis. In a pilot study, at an initial visit, subjects are given an allergen challenge, that is, they are subjected to an allergic substance (e.g., cat dander) and their redness score is noted 10 minutes after the allergen challenge (visit 1 score). At a follow-up visit, the same procedure is followed, except that subjects are given an active eye drop in one eye and the placebo in the fellow eye 3 hours before the challenge and a visit 2 score is obtained 10 minutes after the challenge. The following data were collected.

	Active eye	Placebo eye	Active-placebo eye
	Mean ± *sd*	Mean ± *sd*	Mean ± *sd*
Change in average redness score[a] (visit 2 – visit 1 score)	−0.61 ± 0.70	−0.04 ± 0.68	−0.57 ± 0.86

[a]The redness score ranges from 0 to 4 in increments of 0.5 where 0 is no redness at all and 4 is severe redness.

8.145 Suppose we wish to estimate the number of subjects needed in the main study so that there will be a 90% chance of finding a significant difference between active and placebo eyes using a two-sided test with a significance level of .05 and we expect the active eyes to have a mean redness score 0.5 unit less than that of the placebo eyes. How many subjects are needed in the main study?

8.146 Suppose 60 subjects are enrolled in the main study. How much power would the study have to detect a 0.5-unit mean difference if a two-sided test is used with a significance level of .05?

8.147 In a substudy, subjects will be subdivided into 2 equal groups according to the severity of previous allergy symptoms and the effectiveness of the eye drop (vs. placebo) will be compared between the 2 groups. If 60 subjects are enrolled in the main study (in the 2 groups combined), then how much power will the substudy have if there is a true mean difference in effectiveness of 0.25 [i.e., (active-placebo, subgroup 1) minus (active-placebo, subgroup 2) = 0.25] between the 2 groups and a two-sided test is used with a significance level of .05?

Microbiology

The purpose of this study was to demonstrate that soy beans inoculated with nitrogen-fixing bacteria yield more and grow adequately without the use of expensive environmentally deleterious synthesized fertilizers. The trial was conducted under controlled conditions with uniform amounts of soil. The initial hypothesis was that inoculated plants would outperform their uninoculated counterparts. This assumption is based on the fact that plants need nitrogen to manufacture vital proteins and amino acids, and that nitrogen-fixing bacteria would make more of this substance available to plants, increasing their size and yield. There were 8 inoculated plants (I) and 8 uninoculated plants (U). The plant yield as measured by pod weight for each plant is given in Table 8.28.

Table 8.28 Pod weight (gm) from inoculated (I) and uninoculated (U) plants[a]

	I	U
	1.76	0.49
	1.45	0.85
	1.03	1.00
	1.53	1.54
	2.34	1.01
	1.96	0.75
	1.79	2.11
	1.21	0.92
Mean	1.634	1.084
sd	0.420	0.510
n	8	8

[a]The data for this problem were supplied by David Rosner.

8.148 Provide a 95% CI for the mean pod weight in each group.

8.149 Suppose there is some overlap between the 95% CIs in Problem 8.148. Does this imply that there is no significant difference between the mean pod weights for the 2 groups? Why or why not?

8.150 What test can be used to compare the mean pod weight between the 2 groups?

8.151 Perform the test in Problem 8.150, and report a *p*-value (two-tail).

8.152 Provide a 95% CI for the difference in mean pod weight between the two groups.

Cardiovascular Disease

A study was performed to assess whether hyperinsulinemia is an independent risk factor for ischemic heart disease [19]. A group of 91 men who developed clinical manifestations of ischemic heart disease (IHD) over a 5-year period were compared with 105 control men who did not develop IHD over this period and were comparable to the case men with respect to age, obesity (body-mass index = wt/ht^2), cigarette smoking, and alcohol intake. The primary exposure variable of interest was level of fasting insulin at baseline. The following data were presented.

	Controls (n = 105)	Cases (n = 91)
Fasting insulin (pmol/L)	78.2 ± 28.8[a]	92.1 ± 27.5

[a]Plus–minus values are mean ± *sd*.

8.153 What test can be performed to compare the mean level of fasting insulin between case and control patients? (*Hint*: $F_{104, 90, .975} = 1.498$.)

8.154 Implement the test in Problem 8.153, and report a *p*-value (two-tail).

8.155 Provide a 95% CI for the mean difference in fasting insulin between the two groups. (*Note*: $t_{194, .975} = 1.972$.)

8.156 Suppose a 99% CI for the mean difference were also desired. Would this interval be of the same length, longer, or shorter than the 95% CI (do not actually compute the interval)?

Cerebrovascular Disease

A study is to be performed comparing treatment with warfarin versus treatment with aspirin among patients who have had a recent stroke. The primary endpoint is the prevention of a second stroke over an 18-month follow-up period. A substudy of the main study is to assess other risk factors for a second stroke among patients randomized to aspirin. One potential risk factor is the F_{12} level, a hemostatic factor in blood.

In a previous study, in a group of 63 patients treated with placebo, the mean F_{12} level was 1.57 with standard deviation 0.794.

8.157 What was the standard error of the mean in the previous study?

8.158 What is the difference in interpretation between the standard deviation and the standard error in this case?

8.159 In the new substudy one goal is to compare the mean level of F_{12} at baseline between patients with cryptogenic (group C) versus noncryptogenic stroke (group D) at baseline. Suppose (1) the mean level of F_{12} for patients in group C = the mean level in the placebo group in the previous study, (2) the mean level of F_{12} for patients in group D is reduced by 30% relative to group C; (3) the standard deviation of both groups is the same as in the placebo group in the previous study; (4) we wish to conduct a two-sided test with type I error = .05; (5) we require a type II error = .20; and (6) there are an equal number of subjects in group C and D in the substudy. How many subjects need to be enrolled in the substudy?

8.160 Suppose that the actual number of patients enrolled in the substudy is 40 patients in group C and 30 patients in group D. How much power will the substudy have if assumptions (1)–(4) in Problem 8.159 hold?

Renal Disease

The Swiss Analgesic Study was a study whose goal was to assess the effect of intake of phenacetin-containing analgesics on kidney function and other health parameters. A group of 624 women were identified from workplaces near Basel, Switzerland, with high intake of phenacetin-containing analgesics. The level of NAPAP (N-acetyl-P-aminophenyl) in urine was used as a marker of phenacetin intake. This constitutes the "study" group. In addition, a control group of 626 women were identified, from the same workplaces and with normal NAPAP levels, who were presumed to have low or no phenacetin intake. The study group was then subdivided into a high-NAPAP and a low-NAPAP subgroup according to the absolute NAPAP level. The women were examined at baseline during 1967/1968 and also in 1969, 1970, 1971, 1972, 1975, and 1978 and had their kidney function evaluated by several objective laboratory tests. The Data Set SWISS.DAT contains longitudinal data on serum-creatinine levels (an important index of kidney function) for women in both the study group and the control group. Documentation for this data set is given in SWISS.DOC.

8.161 One hypothesis is that analgesic abusers would have different serum-creatinine profiles at baseline. Using the data from the baseline visit, can you address this question?

8.162 A major hypothesis of the study is that women with high phenacetin intake would show a greater change in serum creatinine compared to women with low phenacetin intake. Can you assess this issue using the longitudinal data available in the data set? (*Hint*: A simple approach for accomplishing this is to look at the change in serum creatinine between the baseline visit and the last follow-up visit. More complex approaches using all the available data will be considered in our work on regression analysis in Chapter 11.)

Health Promotion

A study is planned of the effect of a new health-education program promoting smoking cessation among heavy-smoking teenagers (≥20 cigarettes—equal to 1 pack—per day). A randomized study is planned whereby 50 heavy-smoking teenagers in two schools (Schools A and B) will receive an active intervention with group meetings run by trained psychologists according to an American Cancer Society protocol; 50 other heavy-smoking teenagers in two different schools (Schools C and D) will receive pamphlets from the American Cancer Society promoting smoking cessation but will receive no active intervention by psychologists. Random numbers are used to select two of the four schools to receive the active intervention and the remaining 2 schools to receive the control intervention. The intervention is planned to last for 1 year, after which study participants in all schools will provide self-reports of the number of cigarettes smoked, which will be confirmed by biochemical tests of urinary continine levels. The principal outcome variable is the change in the number of cigarettes smoked per day. A subject who completely stops smoking is scored as smoking 0 cigarettes per day.

It is hypothesized that the effect of the intervention will be to reduce the mean number of cigarettes smoked by 5 cigarettes per day over 1 year for the active intervention group. It is also hypothesized that subjects in the control group will increase their cigarette consumption by an average of 2 cigarettes per day over 1 year. Let us assume that the distribution of the number of cigarettes smoked per day at baseline in both groups is normal with mean = 30 cigarettes per day and standard deviation = 5 cigarettes per day. Furthermore, it is expected, based on previous intervention studies, that the standard deviation of the number of cigarettes per day will increase to 7 cigarettes per day after 1 year. Finally, past data also suggest that the correlation coefficient between the number of cigarettes smoked by the same person at baseline and 1 year will be .80.

8.163 How much power will the proposed study have if a two-sided test is used with $\alpha = .05$?

8.164 Suppose the organizers of the study are reconsidering their sample-size estimate. How many subjects should be enrolled in each of the active and control intervention groups to achieve 80% power if a two-sided test is used with $\alpha = .05$?

Hypertension

A study was recently reported comparing the effects of different dietary patterns on blood pressure within an 8-week follow-up period [20]. Subjects were randomized to 3 groups: A, a control diet group, $N = 154$; B, a fruit-and-vegetables diet group, $N = 154$; C, a combination-diet group consisting of a diet rich in fruits, vegetables, and low-fat dairy products and with reduced saturated and total fat, $N = 151$. The following results were reported for systolic blood pressure.

Mean change in fruit-and-vegetables group	−2.8 mm Hg
Minus mean change in control group	
(97.5% CI)	(−4.7 to −0.9)

8.165 Suppose we wish to compute a p-value (two-sided) for this comparison. *Without doing any further calculation,* which statement(s) must be false?

(1) $p = .01$ **(2)** $p = .04$ **(3)** $p = .07$ **(4)** $p = .20$

(*Note:* The actual p-value may be different from all these values.)

8.166 Suppose we assume that the standard deviation of change in blood pressure is the same in each group and is known without error. Compute the exact p-value from the information provided.

8.167 Suppose we wish to compute a two-sided 95% CI for the true mean change in the fruit and vegetable group minus the true mean change in the control group, which we represent by (c_1, c_2). *Without doing any further calculations,* which of the following statement(s) must be *false.*

(1) The lower confidence limit $(c_1) = -5.0$.
(2) The upper confidence limit $(c_2) = -1.0$.
(3) The width of the confidence interval $(c_2 - c_1) = 3.0$.

(*Note:* The actual values of c_1, c_2, or $(c_2 - c_1)$ may differ from those given in (1), (2), and (3).)

8.168 If we make the same assumption as in Problem 8.166, then compute the 95% CI from the information provided.

REFERENCES

[1] Satterthwaite, F. W. (1946). An approximate distribution of estimates of variance components. *Biometrics Bulletin, 2,* 110–114.

[2] Landrigan, P. J., Whitworth, R. H., Baloh, R. W., Staehling, N. W., Barthel, W. F., & Rosenblum, B. F. (1975, March 29). Neuropsychological dysfunction in children with chronic low-level lead absorption. *The Lancet,* 708–715.

[3] Rosner, B. (1983). Percentage points for a generalized ESD many-outlier procedure. *Technometrics, 25*(2), 165–172.

[4] Quesenberry, C. P., & David, H. A. (1961). Some tests for outliers. *Biometrika, 48,* 379–399.

[5] Cook, N. R., & Rosner, B. A. (1997). Sample size estimation for clinical trials with longitudinal measures: Application to studies of blood pressure. *Journal of Epidemiology and Biostatistics, 2,* 65–74.

[6] Yablonski, M. E., Maren, T. H., Hayashi, M., Naveh, N., Potash, S. D., & Pessah, N. (1988). Enhancement of the ocular hypertensive effect of acetazolamide by diflunisal. *American Journal of Ophthalmology, 106,* 332–336.

[7] Schachter, J., Kuller, L. H., & Perfetti, C. (1984). Heart rate during the first five years of life: Relation to ethnic group (black or white) and to parental hypertension. *American Journal of Epidemiology, 119*(4), 554–563.

[8] White, J. R., & Froeb, H. E. (1980). Small airway dysfunction in nonsmokers chronically exposed to tobacco smoke. *New England Journal of Medicine, 302*(13), 720–723.

[9] Coates, R. A., Fanning, M. M., Johnson, J. K., & Calzavara, L. (1988). Assessment of generalized lymphadenopathy in AIDS research: The degree of clinical agreement. *Journal of Clinical Epidemiology, 41*(3), 267–273.

[10] Taguma, Y., Kitamoto, Y., Futaki, G., Ueda, H., Monma, H., Ishizaki, M., Takahashi, H., Sekino, H., & Sasaki, Y. (1985). Effect of captopril on heavy proteinuria in azotemic diabetics. *New England Journal of Medicine, 313*(26), 1617–1620.

[11] Heller, D. A., DeFaire, U., Pederson, N. L., Dahlen, G., & McClearn, G. E. (1993). Genetic and environmental influences on serum lipid levels in twins. *New England Journal of Medicine, 328*(16), 1150–1156.

[12] Ekdahl, C., Andersson, S. I., & Svensson, B. (1989). Muscle function of the lower extremities in rheumatoid arthritis and osteoarthrosis. A descriptive study of patients in a primary health care district. *Journal of Clinical Epidemiology, 42*(10), 947–954.

[13] Parisi, A. F., Folland, E. D., & Hartigan, P. (1992). A comparison of angioplasty with medical therapy in the treatment of single-vessel coronary artery disease. *New England Journal of Medicine, 326*(1), 10–16.

[14] Selby, J. V., Friedman, G. D., & Quesenberry, C. P., Jr. (1989). Precursors of essential hypertension: The role of body fat distribution pattern. *American Journal of Epidemiology, 129*(1), 43–53.

[15] Hartnett, M.E., Stratton, R. D., Browne, R. W., Rosner, B. A., Lanham, R. J., & Armstrong, D. (1999). Serum markers of oxidative stress and severity of diabetic retinopathy. *Diabetes Care,* in press.

[16] Knopp, R. H., Walden, C. E., Retzlaff, B. M., McCann, B. S., Dowdy, A. A., & Albers, J. J. (1997). Long-term cholesterol-lowering effects of 4 fat-restricted diets in hypercholesterolemic and combined hyperlipidemic men. The Dietary Alternatives Study. *Journal of the American Medical Association, 278,* 1509–1515.

[17] Prentice, R., Thompson, D., Clifford, C., Gorbach, S., Goldin, B., & Byar, D. (1990). Dietary fat reduction and plasma estradiol concentration in healthy postmenopausal women. The Women's Health Trial Study Group. *Journal of the National Cancer Institute, 82,* 129–134.

[18] Jungquist, G., Hanson, B. S., Isacsson, S. O., Janzon, L., Steen, B., & Lindell, S. E. (1991). Risk factors for carotid artery stenosis: An epidemiological study of men aged 69 years. *Journal of Clinical Epidemiology, 44*(4/5), 347–353.

[19] Despres, J. P., Lamarche, B., Mauriege, P., Cantin, B., Dagenais, G. R., Moorjani, S., & Lupien, P. J. (1996). Hyperinsulinemia as an independent risk factor for ischemic heart disease. *New England Journal of Medicine, 334,* 952–957.

[20] Appel, L. J., Moore, T. J., Oberzanek, E., Vollmer, W. M., Svetkey, L. P., et. al. (1997). A clinical trial of the effects of dietary patterns on blood pressure. *New England Journal of Medicine, 336,* 1117–1124.

9

Nonparametric Methods

SECTION 9.1 Introduction

In the previous work in this text, the data were assumed to come from some underlying distribution, such as the normal or binomial distribution, whose general form is assumed known. Methods of estimation and hypothesis testing were developed, based on these assumptions. These procedures are usually referred to as **parametric statistical methods** because the parametric form of the distribution is assumed to be known. If these assumptions about the shape of the distribution are not made, and if the central-limit theorem also seems inapplicable, particularly if the sample size is small, then **nonparametric statistical methods,** which make fewer assumptions about the shape of the distribution, must be used.

Another assumption previously made in this text is that it is meaningful to measure the distance between possible data values. This assumption is characteristic of cardinal data.

DEFINITION 9.1 **Cardinal data** are data that are on a scale where it is meaningful to measure the distance between possible data values.

Example 9.1 Body weight is a cardinal variable because a difference of 6 lbs is twice as large as a difference of 3 lbs.

There are actually two types of cardinal data: interval-scale data and ratio-scale data.

DEFINITION 9.2 For cardinal data, if the zero point is arbitrary, then the data are on an **interval scale;** if the zero point is fixed, then the data are on a **ratio scale.**

Example 9.2 Body temperature is on an interval scale because the zero point is arbitrary. For example, the zero point has a different meaning for Fahrenheit and Celsius temperatures.

Example 9.3 Blood pressure and body weight are on ratio scales because the zero point is well defined in both instances.

It is meaningful to measure ratios between specific data values for data on a ratio scale (e.g., person A's weight is 10% higher than person B's), but it is not meaningful for data on an interval scale (e.g., the ratio of specific temperatures will be different in degrees F from what it is in degrees C). It is meaningful to use means and standard deviations for cardinal data of either type.

Another type of data that occurs frequently in medical and biological work but does not satisfy Definition 9.1 is ordinal data.

DEFINITION 9.3 **Ordinal data** are data that can be ordered but do not have specific numeric values. Thus, common arithmetic *cannot* be performed on ordinal data in a meaningful way.

Example 9.4 **Ophthalmology** Visual acuity can be measured on an ordinal scale, because we know that 20–20 vision is better than 20–30, which is better than 20–40, . . . , and so on. However, a numeric value cannot easily be assigned to each level of visual acuity that all ophthalmologists would agree on.

Example 9.5 In some clinical studies the major outcome variable is the change in a patient's condition after treatment. This variable is often measured on the following five-point scale: 1 = much improved, 2 = slightly improved, 3 = stays the same, 4 = slightly worse, 5 = much worse. This variable is ordinal because the different outcomes, 1, 2, 3, 4, 5, are ordered in the sense that condition 1 is better than condition 2, which is better than condition 3, . . . , and so on. However, we cannot say that the difference between categories 1 and 2 (2 − 1) is the same as the difference between categories 2 and 3 (3 − 2), . . . , and so on. If these categories were on a cardinal scale, then the variable would have this property.

Because ordinal variables cannot be given a numeric scale that makes sense, computing means and standard deviations for such data are not meaningful. Therefore, methods of estimation and hypothesis testing based on normal distributions, as discussed in Chapters 6 through 8, cannot be used. However, we are still interested in making comparisons between groups for variables such as visual acuity and outcome of treatment, and nonparametric methods can be used for this purpose.

Another type of data scale, which has even less structure than an ordinal scale concerning relationships between data values, is a nominal scale.

DEFINITION 9.4 Data are on a **nominal scale** if different data values can be classified into categories but the categories have no specific ordering.

Example 9.6 **Renal Disease** In classifying cause of death among deceased patients with documented analgesic abuse, the following categories were used: (1) cardiovascular disease, (2) cancer, (3) renal or urogenital disease, and (4) all other causes of death. Cause of death is a good example of a nominal scale, because the values (the categories of death) have no specific ordering with respect to each other.

In this chapter the most commonly used nonparametric statistical tests are developed, assuming that the data are on either a cardinal or an ordinal scale. If they are on a cardinal scale, then the methods will be most useful if there is reason to question the normality of the underlying distribution of the data. For nominal (or categorical) data, discrete data methods, to be described in Chapter 10, are used.

SECTION 9.2 The Sign Test

As discussed in Section 9.1, for ordinal data we can measure the relative ordering of different categories of a variable. In this section, we consider data with even more restrictive assumptions; namely, for any two subjects A, B we can identify whether the score for subject A is greater than, less than, or equal to the score for subject B, but not the relative magnitude of the differences.

Example 9.7 **Dermatology** Suppose we wish to compare the effectiveness of two ointments (A, B) in reducing excessive redness in people who cannot otherwise be exposed to sunlight. Ointment A is randomly applied either to the left or right arm, and ointment B is applied to the corresponding area on the other arm. The person is then exposed to 1 hour of sunlight and the two arms are compared for degrees of redness. Suppose only the following qualitative assessments can be made:

(1) The A arm is not as red as the B arm.

(2) The B arm is not as red as the A arm.

(3) The arms are equally red.

Of 45 people tested with the condition, 22 are better off on the A arm, 18 are better off on the B arm, and 5 are equally well off on both arms. How can we decide if this evidence is sufficient to conclude that ointment A is better than ointment B?

9.2.1 Normal-Theory Method

In this section, we will consider a large-sample method for addressing the question posed in Example 9.7.

Suppose that the degree of redness could be measured on a quantitative scale, with a higher number indicating more redness. Let x_i = degree of redness on the A arm, y_i = degree of redness on the B arm for the ith person. We will focus on $d_i = x_i - y_i$ = difference in redness between the A and B arms and will test the hypothesis H_0: $\Delta = 0$ versus H_1: $\Delta \neq 0$, where Δ = the population median of the d_i or the 50th percentile of the underlying distribution of the d_i.

(1) If $\Delta = 0$, then the ointments are equally effective.

(2) If $\Delta < 0$, then ointment A is better, because arm A is less red than arm B.

(3) If $\Delta > 0$, then ointment B is better, because arm A is more red than arm B.

Notice that the actual d_i cannot be observed; we can only observe if $d_i > 0$, $d_i < 0$, or $d_i = 0$. The people for whom $d_i = 0$ will be excluded, because we cannot tell which ointment is better for them. The test will be based on the number of people C for whom $d_i > 0$ out of the total number of people n with nonzero d_i. This test makes sense, because if C is large, treatment B is preferred by most people over treatment A, whereas if C is small, treatment A is preferred over treatment B. We would expect under H_0 that $Pr(\text{nonzero } d_i > 0) = \frac{1}{2}$. We will assume that the normal approximation to the binomial is valid. This assumption will be true if

$$npq \ge 5 \quad \text{or} \quad n\left(\frac{1}{2}\right)\left(\frac{1}{2}\right) \ge 5$$

$$\text{or} \quad \frac{n}{4} \ge 5 \quad \text{or} \quad n \ge 20$$

where n = the number of nonzero d_i's.

The following test procedure for a two-sided level α test. referred to as the **sign test**, can then be used:

EQUATION 9.1

The Sign Test To test the hypothesis H_0: $\Delta = 0$ versus H_1: $\Delta \ne 0$ where the number of nonzero d_i's $= n \ge 20$ and C = the number of d_i's where $d_i > 0$, if

$$C > c_2 = \frac{n}{2} + \frac{1}{2} + z_{1-\alpha/2}\sqrt{n/4} \quad \text{or} \quad C < c_1 = \frac{n}{2} - \frac{1}{2} - z_{1-\alpha/2}\sqrt{n/4}$$

then H_0 is rejected. Otherwise, H_0 is accepted.

The acceptance and rejection regions for this test are depicted in Figure 9.1.

FIGURE 9.1 **Acceptance and rejection regions for the sign test**

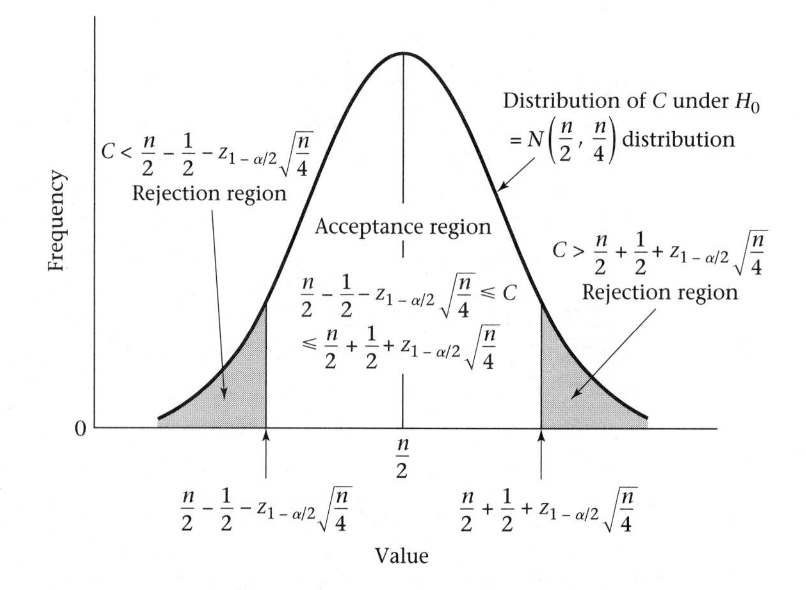

Similarly, the exact p-value of the procedure is given by the following formula:

EQUATION 9.2

Computation of the *p*-Value for the Sign Test (Normal-Theory Method)

$$p = 2 \times \left[1 - \Phi\left(\frac{C - \frac{n}{2} - .5}{\sqrt{n/4}}\right)\right] \quad \text{if} \quad C \geq \frac{n}{2}$$

$$p = 2 \times \Phi\left(\frac{C - \frac{n}{2} + .5}{\sqrt{n/4}}\right) \quad \text{if} \quad C < \frac{n}{2}$$

This computation is illustrated in Figure 9.2.

FIGURE 9.2 **Computation of the *p*-value for the sign test**

If $C < n/2$, then $p = 2 \times$ area to the left of $\left(C - \frac{n}{2} + \frac{1}{2}\right)\Big/\sqrt{\frac{n}{4}}$ under an $N(0, 1)$ distribution

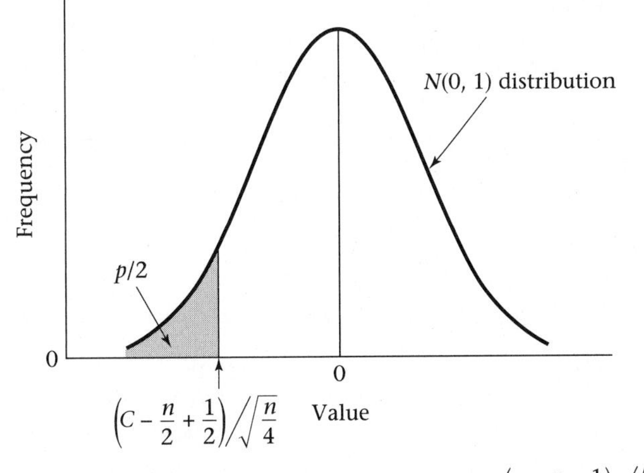

If $C \geq n/2$, then $p = 2 \times$ area to the right of $\left(C - \frac{n}{2} - \frac{1}{2}\right)\Big/\sqrt{\frac{n}{4}}$ under an $N(0, 1)$ distribution

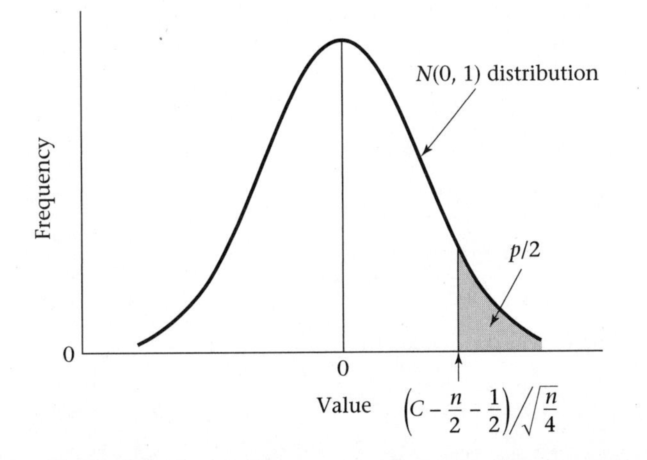

An alternative and equivalent formula for the p-value is given by

$$p = 2 \times \left[1 - \Phi\left(\frac{|C - D| - 1}{\sqrt{n}}\right)\right]$$

where C = the number of $d_i > 0$ and D = the number of $d_i < 0$.

This test is called the *sign test* because it depends only on the sign of the differences and not on their actual magnitude.

The sign test is actually a special case of the one-sample binomial test in Section 7.9, where the hypothesis H_0: $p = 1/2$ versus H_1: $p \neq 1/2$ is tested. In Equation 9.1 and Equation 9.2 a large-sample test is being used and we are assuming that the normal approximation to the binomial distribution is valid. Under H_0, $p = 1/2$ and $E(C) = np = n/2$, $Var(C) = npq = n/4$, and $C \,\dot{\sim}\, N(n/2, n/4)$. Furthermore, the .5 term in the computation of the critical region and p-value serves as a continuity correction and makes for a better approximation of the binomial distribution by the normal distribution.

Example 9.8 **Dermatology** Assess the statistical significance of the skin-ointment data in Example 9.7.

Solution In this case there are 40 untied pairs, and $C = 18 < n/2 = 20$. From Equation 9.1, the critical values are given by

$$\begin{aligned} c_2 &= n/2 + 1/2 + z_{1-\alpha/2}\sqrt{n/4} \\ &= 40/2 + 1/2 + z_{.975}\sqrt{40/4} = 20.5 + 1.96(3.162) = 26.7 \end{aligned}$$

and $c_1 = n/2 - 1/2 - z_{1-\alpha/2}\sqrt{n/4} = 19.5 - 1.96(3.162) = 13.3$

Because $13.3 \le C = 18 \le 26.7$, H_0 is accepted using a two-sided test with $\alpha = .05$ and we conclude that the ointments are not significantly different in effectiveness. From Equation 9.2, because $C = 18 < n/2 = 20$, the exact p-value is given by

$$p = 2 \times \Phi\left[\left(18 - 20 + \frac{1}{2}\right)\Big/\sqrt{40/4}\right] = 2 \times \Phi(-0.47) = 2 \times .3176 = .635$$

which is not statistically significant. Therefore, H_0, that the ointments are equally effective, is accepted.

Alternatively, we could compute the test statistic

$$z = \frac{|C - D| - 1}{\sqrt{n}}$$

where $C = 18$, $D = 22$, and $n = 40$, yielding

$$z = \frac{|18 - 22| - 1}{\sqrt{40}} = \frac{3}{\sqrt{40}} = 0.47$$

and obtain the p-value from

$$p = 2 \times [1 - \Phi(0.47)] = .635$$

9.2.2 Exact Method

If $n < 20$, then exact binomial probabilities rather than the normal approximation must be used to compute the p-value. H_0 should still be rejected if C is very large or very small. The expressions for the p-value based on exact binomial probabilities are given as follows:

EQUATION 9.3

Computation of the p-Value for the Sign Test (Exact Test)

If $C > n/2 \quad p = 2 \times \sum_{k=C}^{n} \binom{n}{k} \left(\frac{1}{2}\right)^n$

If $C < n/2, \quad p = 2 \times \sum_{k=0}^{C} \binom{n}{k} \left(\frac{1}{2}\right)^n$

If $C = n/2, \quad p = 1.0$

This computation is depicted in Figure 9.3.

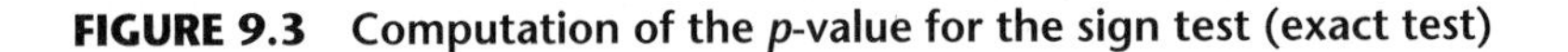

FIGURE 9.3 Computation of the p-value for the sign test (exact test)

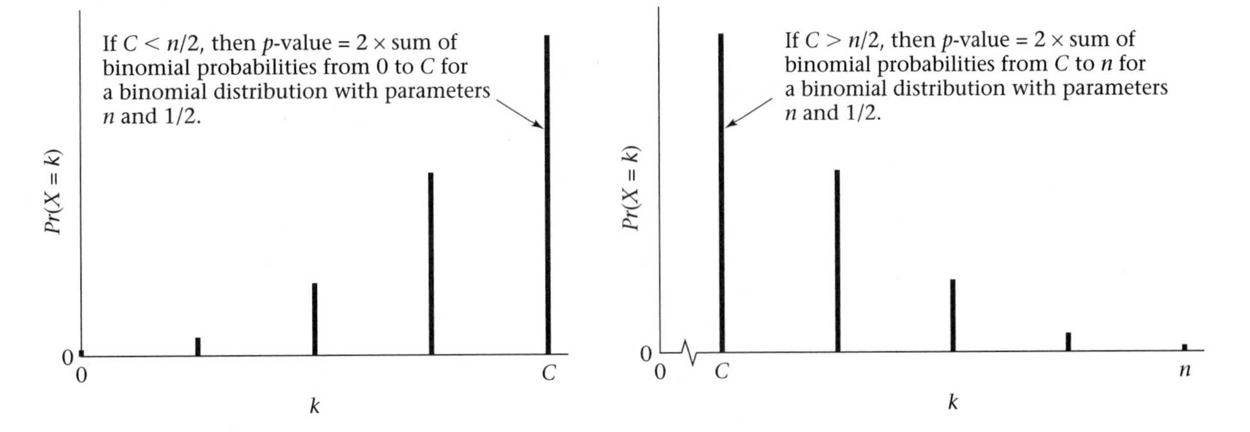

This test is a special case of the small-sample, one-sample binomial test described in Equation 7.36, where the hypothesis H_0: $p = \frac{1}{2}$ versus H_1: $p \neq \frac{1}{2}$ is tested.

Example 9.9 **Ophthalmology** Suppose we want to compare two different types of eye drops (A, B) that are intended to prevent redness in people with hay fever. Drug A is randomly given to one eye and drug B to the other eye. The redness is noted at baseline and after 10 minutes by an observer who is not aware of which drug is administered to which eye. We find that for 15 people with an equal amount of redness in each eye at baseline, after 10 minutes the drug A eye is less red than the drug B eye for 8 people; the drug B eye is less red than the drug A eye for 2 people; and the eyes are equally red for 5 people. Assess the statistical significance of the results.

Solution The test is based on the 10 people who had a differential response to the two types of eye drops. Because $n = 10 < 20$, the normal-theory method in Equation 9.2 cannot be used; the exact method in Equation 9.3 must be used instead. Because $C = 8 > \frac{10}{2} = 5$,

$$p = 2 \times \sum_{k=8}^{10} \binom{10}{k} \left(\frac{1}{2}\right)^{10}$$

Refer to the binomial tables (Table 1 in the Appendix) using $n = 10$, $p = .5$ and note that $Pr(X = 8) = .0439$, $Pr(X = 9) = .0098$, $Pr(X = 10) = .0010$. Thus $p = 2 \times Pr(X \geq 8) = 2(.0439 + .0098 + .0010) = 2 \times .0547 = .109$, which is not statistically significant. Thus, H_0, that the two types of eye drops are equally effective in reducing redness in people with hay fever, is accepted.

SECTION 9.3 The Wilcoxon Signed-Rank Test

Example 9.10 **Dermatology** Consider the data in Example 9.7 from a different perspective. We assumed that the only possible assessment was that the degree of sunburn with ointment A was either better or worse than that with ointment B. Suppose instead that the degree of burn can be quantified on a 10-point scale, with 10 being the worst burn and 1 being no burn at all. We can now compute $d_i = x_i - y_i$, where x_i = degree of burn for ointment A and y_i = degree of burn for ointment B. If d_i is positive, then ointment B is doing better than ointment A; if d_i is negative, then ointment A is doing better than ointment B. For example, if $d_i = +5$, then the degree of redness is 5 units greater on the ointment A arm than on the ointment B arm, whereas if $d_i = -3$, then the degree of redness is 3 units less on the ointment A arm than on the ointment B arm. How can this additional information be used to test if the ointments are equally effective?

Suppose the sample data in Table 9.1 are obtained. The f_i represent the frequency or the number of people with difference in redness d_i between the ointment A and ointment B arms.

Notice that there is only a slight excess of people with negative d_i (that is, who are better off with ointment A, 22) than with positive d_i (that is, who are better off with ointment B, 18). However, the extent to which the 22 people are better off appears to be far greater than that of the 18 people, because the negative d_i generally have a much greater absolute value than the positive d_i. This point is illustrated in Figure 9.4.

We wish to test the hypothesis H_0: $\Delta = 0$ versus H_1: $\Delta \neq 0$, where Δ = the median score difference between the ointment A and ointment B arms. If $\Delta < 0$, then ointment A is better; if $\Delta > 0$, then ointment B is better. We will assume that the d_i have an underlying continuous distribution.

Based on Figure 9.4, a seemingly reasonable test of this hypothesis would be to take account of both the magnitude and the sign of the differences d_i. A paired t test might be used here, but the problem is that the rating scale is ordinal. The measurement $d_i = -5$ does not mean the difference in degree of burn is 5 times as great as $d_i = -1$, but simply means there is a relative ranking of differences in degree of burn, with –8 being most favorable to ointment A, –7 the next most favorable, . . . , and so on. Thus a nonparametric test that is analogous to the paired t

TABLE 9.1 **Difference in degree of redness between ointment A and ointment B arms after 10 minutes of exposure to sunlight.**

	Negative		Positive		Number of people with same absolute value	Range of ranks	Average rank
$\lvert d_i \rvert$	d_i	f_i	d_i	f_i			
10	−10	0	10	0	0	—	—
9	−9	0	9	0	0	—	—
8	−8	1	8	0	1	40	40.0
7	−7	3	7	0	3	37–39	38.0
6	−6	2	6	0	2	35–36	35.5
5	−5	2	5	0	2	33–34	33.5
4	−4	1	4	0	1	32	32.0
3	−3	5	3	2	7	25–31	28.0
2	−2	4	2	6	10	15–24	19.5
1	−1	4	1	10	14	1–14	7.5
		22		18			
0	0	5					

FIGURE 9.4 **Bar graph of the differences in redness between the ointment A and ointment B arms for the data in Example 9.10**

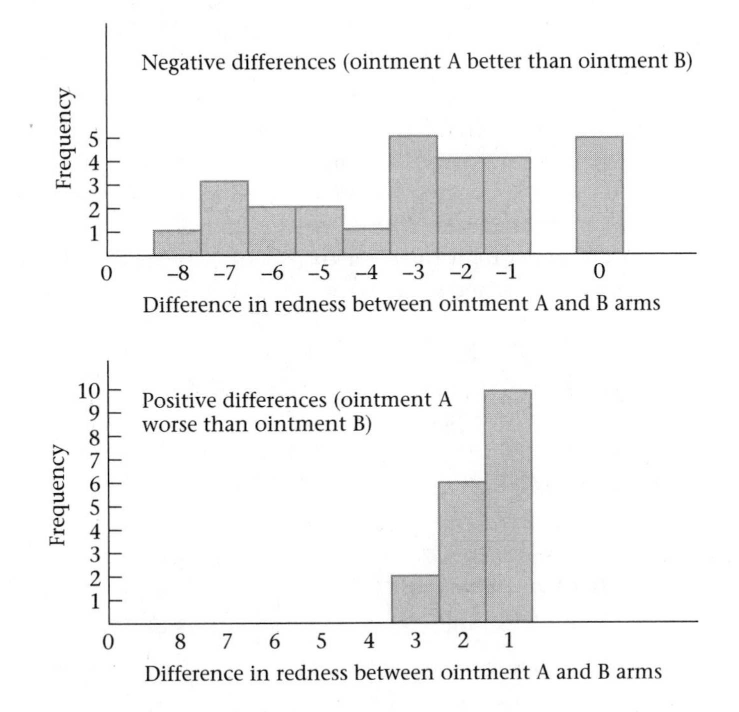

test is needed here. Such a test is the **Wilcoxon signed-rank test.** It is nonparametric because it is based on the ranks of the observations rather than on their actual values, as is the paired t test.

The first step in performing this test is to compute ranks for each of the observations, as follows:

EQUATION 9.4

Ranking Procedure for the Wilcoxon Signed-Rank Test

(1) Arrange the differences d_i in order of *absolute value* as has been done in Table 9.1.

(2) Count the number of differences with the same absolute value.

(3) Ignore the observations where $d_i = 0$ and rank the remaining observations from 1, for the observation with the lowest absolute value, up to n, for the observation with the highest absolute value.

(4) If there is a group of several observations with the same absolute value, then find the lowest rank in the range $= 1 + R$ and the highest rank in the range $= G + R$, where R = the highest rank used prior to considering this group and G = the number of differences in the *range of ranks* for the group. Assign the *average rank* = (lowest rank in the range + highest rank in the range)/2 as the rank for each difference in the group.

Example 9.11 **Dermatology** Compute the ranks for the skin-ointment data in Table 9.1.

Solution First collect the differences with the same absolute value. Fourteen people have absolute value 1; this group has a rank range from 1 to 14 and an average rank of $(1 + 14)/2 = 7.5$. The group of 10 people with absolute value 2 has a rank range from $(1 + 14)$ to $(10 + 14) = 15$ to 24 and an average rank $= (15 + 24)/2 = 19.5, \ldots,$ and so on.

The test is based on the sum of the ranks, or **rank sum** (R_1), for the group of people with positive d_i, that is, the rank sum for people for whom ointment A does worse than ointment B. A large rank sum indicates that differences in degree of burn in favor of treatment B tend to be larger than those for treatment A, whereas a small rank sum indicates that differences in degree of burn in favor of treatment A tend to be larger than those for treatment B. If the null hypothesis is true, then the expected value and variance of the rank sum (when there are no ties) are given by

$$E(R_1) = n(n+1)/4, \quad Var(R_1) = n(n+1)(2n+1)/24$$

where n is the number of nonzero differences.

If the number of nonzero d_i's is ≥ 16, then a normal approximation can be used for the test procedure. This test procedure, the Wilcoxon signed-rank test, is given as follows:

EQUATION 9.5

Wilcoxon Signed-Rank Test (Normal Approximation Method for Two-Sided Level α Test)

(1) Rank the differences as shown in Equation 9.4.

(2) Compute the rank sum R_1 of the positive differences.

(3) (a) Compute

$$T = \left[\left| R_1 - \frac{n(n+1)}{4} \right| - \frac{1}{2} \right] \Big/ \sqrt{n(n+1)(2n+1)/24}$$

if there *are no ties* (i.e., no groups of differences with the same absolute value).

(b) Compute

$$T = \left[\left| R_1 - \frac{n(n+1)}{4} \right| - \frac{1}{2} \right] \Big/ \sqrt{n(n+1)(2n+1)\Big/ 24 - \sum_{i=1}^{g} (t_i^3 - t_i)\Big/48}$$

if there *are ties*, where t_i refers to the number of differences with the same absolute value in the ith tied group and g is the number of tied groups.

(4) If

$$T > z_{1-\alpha/2}$$

then reject H_0. Otherwise, accept H_0.

(5) The p-value for the test is given by

$$p = 2 \times [1 - \Phi(T)]$$

(6) This test should be used only if the number of nonzero differences is ≥16 and if the difference scores have an underlying continuous symmetric distribution. The computation of the p-value is illustrated in Figure 9.5.

FIGURE 9.5 **Computation of the *p*-value for the Wilcoxon signed-rank test**

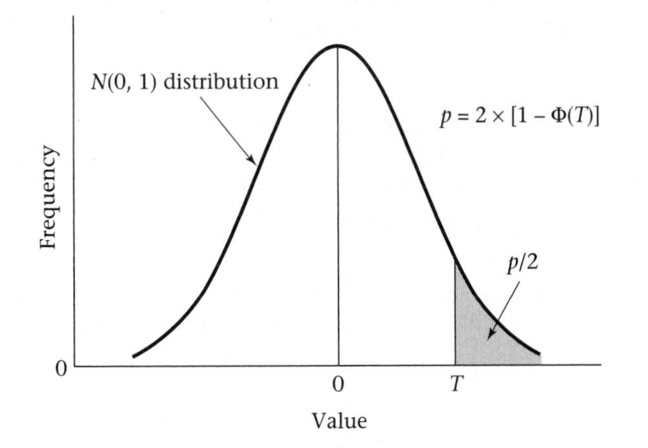

The term $\frac{1}{2}$ in the computation of T serves as a continuity correction in the same manner as for the sign test in Equations 9.1 and 9.2.

Example 9.12 **Dermatology** Perform the Wilcoxon signed-rank test for the data in Example 9.10.

Solution Since the number of nonzero differences $(22 + 18 = 40) \geq 16$, the normal approximation method in Equation 9.5 can be used. Compute the rank sum for the people with positive d_i, that is, where ointment B performs better than ointment A, as follows:

$$R_1 = 10(7.5) + 6(19.5) + 2(28.0) = 75 + 117 + 56 = 248$$

The expected rank sum is given by

$$E(R_1) = 40(41)/4 = 410$$

The variance of the rank sum corrected for ties is given by

$$\begin{aligned} Var(R_1) &= 40(41)(81)/24 - [(14^3 - 14) + (10^3 - 10) + (7^3 - 7) + (2^3 - 2) \\ &\quad + (2^3 - 2) + (3^3 - 3)]/48 \\ &= 5535 - (2730 + 990 + 336 + 6 + 6 + 24)/48 \\ &= 5535 - 4092/48 = 5449.75 \end{aligned}$$

Thus, $sd(R_1) = \sqrt{5449.75} = 73.82$. Therefore, the test statistic T is given by

$$T = \left(|248 - 410| - \frac{1}{2}\right)\Big/73.82 = 161.5/73.82 = 2.19$$

The p-value of the test is given by

$$p = 2[1 - \Phi(2.19)] = 2 \times (1 - .9857) = .029$$

We therefore can conclude that there is a significant difference between ointments, with ointment A doing better than ointment B, because the observed rank sum (248) is smaller than the expected rank sum (410). This conclusion is different from the conclusion based on the sign test in Example 9.8, where no significant difference between ointments was found. This result indicates that when the information is available, it is worthwhile to consider both the magnitude and the direction of the difference between treatments, as is done by the signed-rank test, rather than just the direction of the difference, as is done by the sign test.

In general, for a two-sided test, if the signed-rank test is based on negative differences rather than positive differences, the same test statistic and p-value will always result. Thus, the rank sum can be arbitrarily computed based on either positive or negative differences.

Example 9.13 **Dermatology** Perform the Wilcoxon signed-rank test for the data in Example 9.10 based on negative- rather than positive-difference scores.

Solution

$$\begin{aligned} R_2 &= \text{rank sum for negative differences} \\ &= 4(7.5) + 4(19.5) + 5(28.0) + 1(32.0) + 2(33.5) + 2(35.5) + 3(38.0) + 1(40.0) \\ &= 572 \end{aligned}$$

Thus,

$$\left|R_2 - \frac{n(n+1)}{4}\right| - .5 = |572 - 410| - .5 = 161.5 = \left|R_1 - \frac{n(n+1)}{4}\right| - .5$$

Because $Var(R_1) = Var(R_2)$, the same test statistic $T = 2.19$ and p-value = .029 are obtained as when positive-difference scores are used.

If the number of pairs with nonzero $d_i \le 15$, then the normal approximation is no longer valid, and special small-sample tables giving significance levels for this test must be used. Such a table is Table 11 in the Appendix, which gives upper and lower critical values for R_1 for a two-sided test with α levels of .10, .05, .02, and .01, respectively. In general, the results are statistically significant at a particular α level only if either $R_1 \le$ the lower critical value or $R_1 \ge$ the upper critical value for that α level.

Example 9.14 Suppose there are 9 untied pairs and a rank sum of 43. Evaluate the statistical significance of the results.

Solution Because $R_1 = 43 \ge 42$, it follows that $p < .02$. Because $R_1 = 43 < 44$, it follows that $p \ge .01$. Thus $.01 \le p < .02$, and the results are statistically significant.

An example of the signed-rank test with ordinal data has been presented. This test and the other nonparametric tests can be applied to cardinal data as well, particularly if the assumption of normality appears grossly violated. However, an assumption of the signed rank test is that one has a continuous, symmetric, although not necessarily normal distribution. If the actual distribution turns out to be normal, then the signed-rank test has less power than the paired t test, which is the penalty paid.

SECTION 9.4 The Wilcoxon Rank-Sum Test

In the previous section, a nonparametric analogue to the paired t test—namely, the Wilcoxon signed-rank test—was presented. In this section, a nonparametric analogue to the t test for two independent samples is described.

Example 9.15 **Ophthalmology** Different genetic types of the disease retinitis pigmentosa (RP) are thought to have different rates of progression, with the dominant form of the disease progressing the slowest, the recessive form of the disease the next slowest, and the sex-linked form of the disease the quickest. This hypothesis can be tested by comparing the visual acuity of people ages 10–19 who have different genetic types of RP. Suppose there are 25 people with dominant disease and 30 people with sex-linked disease. The best-corrected visual acuities (i.e., with appropriate glasses) in the better eye of these people are presented in Table 9.2. How can these data be used to test if the median visual acuity is different in the two groups?

We wish to test the hypothesis H_0: $\text{median}_D = \text{median}_{SL}$ versus H_1: $\text{median}_D \ne \text{median}_{SL}$, where median_D and median_{SL} are the median visual acuities in the dominant and sex-linked groups, respectively. The two-sample t test for independent samples, discussed in Sections 8.4 and 8.7, would ordinarily be used for this type of problem. However, visual acuity cannot be given a specific numeric value that all

TABLE 9.2 **Comparison of visual acuity in people age 10–19 with dominant and sex-linked retinitis pigmentosa**

Visual acuity	Dominant	Sex-linked	Combined sample	Range of ranks	Average rank
20–20	5	1	6	1–6	3.5
20–25	9	5	14	7–20	13.5
20–30	6	4	10	21–30	25.5
20–40	3	4	7	31–37	34.0
20–50	2	8	10	38–47	42.5
20–60	0	5	5	48–52	50.0
20–70	0	2	2	53–54	53.5
20–80	0	1	1	55	55.0
	25	30	55		

ophthalmologists would agree on. Thus, the t test is inapplicable, and a nonparametric analogue must be used. The nonparametric analogue to the independent-samples t test is the **Wilcoxon rank-sum test.** This test is nonparametric because it is based on the *ranks* of the individual observations rather than on their actual values, which would be used in the t test. The ranking procedure for this test is as follows:

EQUATION 9.6

Ranking Procedure for the Wilcoxon Rank-Sum Test

(1) Combine the data from the two groups and order the values from the lowest to highest, or in the case of visual acuity, from best visual acuity (20–20) to worst visual acuity (20–80).

(2) Assign ranks to the individual values, with the best visual acuity (20–20) having the lowest rank and the worst visual acuity (20–80) having the highest rank, or vice versa.

(3) If a group of observations has the same value, then compute the *range of ranks* for the group, as was done for the signed-rank test in Equation 9.4, and assign the *average rank* for each observation in the group.

Example 9.16 Compute the ranks for the visual-acuity data in Table 9.2.

Solution First collect all people with the same visual acuity over the two groups, as shown in Table 9.2. There are 6 people with visual acuity 20–20 who have a rank range of 1–6 and are assigned an average rank of $(1 + 6)/2 = 3.5$. There are 14 people for the two groups combined with visual acuity 20–25. The rank range for this group is from $(1 + 6)$ to $(14 + 6) = 7$ to 20. Thus, all people in this group are assigned the average rank $= (7 + 20)/2 = 13.5$, and similarly for the other groups.

The test statistic for this test is the sum of the ranks in the first sample (R_1). If this sum is large, then the dominant group has poorer visual acuity than the sex-linked group, whereas if it is small, the dominant group has better visual acuity. If the number of observations in the two groups are n_1 and n_2, respectively, then the average rank in the combined sample is $(1 + n_1 + n_2)/2$. Thus, under H_0 the expected rank sum in the first group $\equiv E(R_1) = n_1 \times$ average rank in the combined sample $= n_1(n_1 + n_2 + 1)/2$. It can be shown that the variance of R_1 under H_0 is given by $Var(R_1) = n_1 n_2(n_1 + n_2 + 1)/12$. Furthermore, we will assume that the smaller of the two groups is of size at least 10 and that the variable under study has an underlying continuous distribution. Under these assumptions, the distribution of the rank sum R_1 is approximately normal. Thus, the following test procedure is used:

EQUATION 9.7

Wilcoxon Rank-Sum Test (Normal Approximation Method for Two-Sided Level α Test)

(1) Rank the observations as shown in Equation 9.6.

(2) Compute the rank sum R_1 in the first sample (the choice of sample is arbitrary).

(3) (a) Compute

$$T = \left[\left|R_1 - \frac{n_1(n_1+n_2+1)}{2}\right| - \frac{1}{2}\right] \Big/ \sqrt{\left(\frac{n_1 n_2}{12}\right)(n_1+n_2+1)}$$

if there *are no ties*.

(b) Compute

$$T = \left[\left|R_1 - \frac{n_1(n_1+n_2+1)}{2}\right| - \frac{1}{2}\right] \Bigg/ \sqrt{\left(\frac{n_1 n_2}{12}\right)\left[n_1+n_2+1-\frac{\sum_{i=1}^{g} t_i(t_i^2-1)}{(n_1+n_2)(n_1+n_2-1)}\right]}$$

if there *are ties*, where t_i refers to the number of observations with the same value in the ith tied group, and g is the number of tied groups.

(4) If

$$T > z_{1-\alpha/2}$$

then reject H_0. Otherwise, accept H_0.

(5) Compute the exact p-value by

$$p = 2 \times [1 - \Phi(T)]$$

(6) This test should be used only if both n_1 and n_2 are at least 10, and if there is an underlying continuous distribution.

The computation of the p-value is illustrated in Figure 9.6.

Example 9.17 Perform the Wilcoxon rank-sum test for the data in Table 9.2.

Solution Because the minimum sample size in the two samples is $25 \geq 10$, the normal approximation can be used. The rank sum in the dominant group is given by

$$R_1 = 5(3.5) + 9(13.5) + 6(25.5) + 3(34) + 2(42.5)$$

$$= 17.5 + 121.5 + 153 + 102 + 85 = 479$$

Furthermore, $$E(R_1) = \frac{25(56)}{2} = 700$$

FIGURE 9.6 **Computation of the p-value for the Wilcoxon rank-sum test**

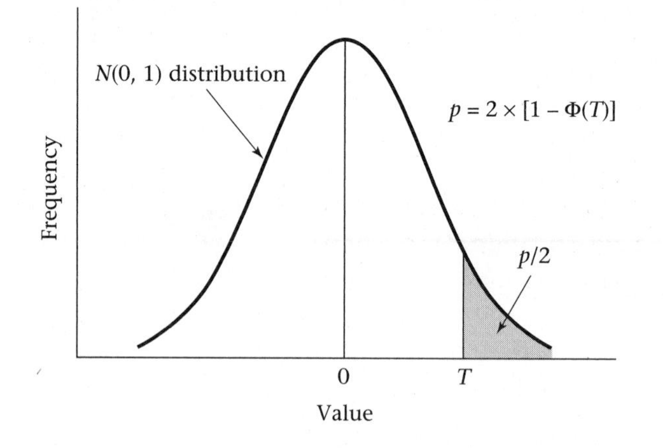

and $Var(R_1)$ corrected for ties is given by

$$[25(30)/12]\{56 - [6(6^2 - 1) + 14(14^2 - 1) + 10(10^2 - 1) + 7(7^2 - 1) + 10(10^2 - 1) + 5(5^2 - 1) + 2(2^2 - 1)]/[55(54)]\}$$
$$= 62.5(56 - 5382/2970) = 3386.74$$

Thus, the test statistic T is given by

$$T = \frac{\left(|479 - 700| - .5\right)}{\sqrt{3386.74}} = \frac{220.5}{58.2} = 3.79$$

which follows an $N(0, 1)$ distribution under H_0. The p-value of the test is

$$p = 2 \times [1 - \Phi(3.79)] < .001$$

We conclude that the visual acuities of the two groups are significantly different. Because the observed rank sum in the dominant group (479) is lower than the expected rank sum (700), the dominant group has better visual acuity than the sex-linked group.

If either sample size is less than 10, the normal approximation is not valid, and a small-sample table of exact significance levels must be used. Table 12 in the Appendix gives upper and lower critical values for the rank sum in the first of two samples (T) for a two-sided test with α levels of .10, .05, .02, and .01, respectively. In general, the results are statistically significant at a particular α level if either $T \le T_l$ = the lower critical value or $T \ge T_r$ = the upper critical value for that α level.

Example 9.18 Suppose there are two samples of size 8 and 15, with a rank sum of 73 in the sample size of 8. Evaluate the statistical significance of the results.

Solution Refer to $n_1 = 8$, $n_2 = 15$, $\alpha = .05$ and find that $T_l = 65$, $T_r = 127$. Because $T = 73 > 65$ and $T < 127$, the results are not statistically significant using a two-sided test at the 5% level.

The Wilcoxon rank-sum test is sometimes referred to in the literature as the **Mann-Whitney U test.** The test statistic for the Mann-Whitney U test is based on the number of pairs of observations (x_i, y_j), one from each sample, such that $x_i < y_j$; in addition, 0.5 is added to the test statistic for each (x_i, y_j) pair such that $x_i = y_j$.

The Mann-Whitney U test and the Wilcoxon rank-sum test are completely equivalent, because the same p-value is obtained by applying either test. Therefore, the choice of which test to use is a matter of convenience.

Because ranking all the observations in a large sample is tedious, a computer program is useful in performing this test. The MINITAB Mann-Whitney test procedure was used on the data set in Table 9.2; the results are given in Table 9.3.

TABLE 9.3 MINITAB Mann-Whitney test program used on the data in Table 9.2

```
Mann-Whitney Confidence Interval and Test

Dominant    N = 25     Median =        25.00
Sex-link    N = 30     Median =        50.00
Point estimate for ETA1-ETA2 is       -15.00
95.1 Percent C.I. for ETA1-ETA2 is (-25.00, -5.00)
W = 479.0
Test of ETA1 = ETA2  vs.  ETA1 ~ = ETA2 is significant at 0.0002
The test is significant at 0.0002 (adjusted for ties)
```

The median visual acuity in the dominant and sex-linked groups are listed first. The dominant group, with a lower median, tends to have better visual acuity than the sex-linked group. The Wilcoxon rank-sum (labeled W) = 479 is given in the output. The two-tailed p-value = .0002 is also given. Two p-values are given. The p-value without adjusting for ties is given in the next to the last row; the p-value adjusting for ties is given in the last row. In this case, they are the same (p = .0002). In addition, a point estimate and a 95% CI for the difference in median acuity between the two groups is also provided. The details concerning construction of these point and interval estimates is beyond the scope of this text.

Finally, a necessary condition for the strict validity of the rank-sum test is that the underlying distributions being compared must be continuous. However, McNeil has investigated the use of this test in comparing discrete distributions and has found only small losses in power when applying this test to grouped data from normal distributions, as compared with the actual ungrouped observations from such distributions [1]. He concludes that the rank-sum test is approximately valid in this case, with the appropriate provision for ties as given in Equation 9.7.

SECTION 9.5 Case Study: Effects of Lead Exposure on Neurological and Psychological Function in Children

In previous chapters, we have considered the effect of lead exposure on neurological and cognitive function in children as described in the Data Set LEAD.DAT on the data disk. Other effects of lead as described in the literature are behavioral in nature. One such variable is hyperactivity. In this study, the parents of the children were asked to rate the degree of hyperactivity of their children on a four-point scale from normal (0) to very hyperactive (3). The scale is ordinal in nature. Thus, to compare the degree of hyperactivity between the exposed and control group, nonparametric methods are appropriate. Because this question was only asked for the younger children, data are available for only 49 control children and 35 exposed children. The raw data are given in Table 9.4. The columns of the table correspond to the groups (1 = control, 2 = exposed). The rows of the table correspond to the degree of hyperactivity. Within each group, the percentage of children with a specific hyperactivity score (the column percentages) is also given. It appears that the exposed children are slightly more hyperactive than the control children.

TABLE 9.4 Raw data for case study

```
ROWS: HYPERACT          COLUMNS: GROUP
            1          2         ALL

0          24         15          39
        48.98      42.86       46.43

1          20         14          34
        40.82      40.00       40.48

2           3          5           8
         6.12      14.29        9.52

3           2          1           3
         4.08       2.86        3.57

ALL        49         35          84
       100.00     100.00      100.00

 CELL CONTENTS --
                 COUNT
                 % OF COL
```

We have used the Mann-Whitney *U* test to compare median hyperactivity level between the two groups. The results are given in Table 9.5. We see that the two-sided *p*-value (adjusted for ties) as given in the second-to-last row of the table is .46. Thus there is no significant difference in median hyperactivity level between the two groups.

TABLE 9.5 Results of Mann-Whitney *U* test

```
Mann-Whitney Confidence Interval and Test

hyper-1     N = 49      Median =       1.0000
hyper-2     N = 35      Median =       1.0000
Point estimate for ETA1-ETA2 is        0.0000
95.1 Percent C.I. for ETA1-ETA2 is (-0.0000, 0.0001)
W = 2008.5
Test of ETA1 = ETA2  vs.  ETA1 ~ = ETA2 is significant at 0.5049
The test is significant at 0.4649 (adjusted for ties)

Cannot reject at alpha = 0.05
```

SECTION 9.6 Summary

In this chapter some of the most widely used nonparametric statistical tests corresponding to the parametric procedures in Chapter 8 were presented. The main advantage of nonparametric methods is that the assumption of normality made in previous chapters can be relaxed when such assumptions are unreasonable. One drawback of nonparametric procedures is that some power is lost relative to using a parametric procedure (such as a t test) if the data truly follow a normal distribution or if the central-limit theorem is applicable. Also, the data typically have to be expressed in terms of ranks, a scale that some researchers find difficult to understand compared with maintaining the raw data in the original scale.

The specific procedures covered for the comparison of two samples include the sign test, the Wilcoxon signed-rank test, and the Wilcoxon rank-sum test. Both the sign test and the signed-rank test are nonparametric analogues to the paired t test. For the sign test it is only necessary to determine whether one member of a matched pair has a higher or lower score than the other member of the pair. For the signed-rank test the magnitude of the difference score (which is then ranked), as well as its direction, is used in performing the significance test. Furthermore, the Wilcoxon rank-sum test (also known as the Mann-Whitney U test) is an analogue to the two-sample t test for independent samples, in which the actual values are replaced by rank scores. Nonparametric procedures appropriate for regression, analysis of variance, and survival analysis are introduced in Chapters 11, 12, and 14.

The tests covered in this chapter are among the most basic of nonparametric tests. Hollander and Wolfe provide a more comprehensive treatment of nonparametric statistical methods [2].

PROBLEMS

Dentistry

In a study, 28 adults with mild periodontal disease are assessed before and 6 months after the implementation of a dental-education program intended to promote better oral hygiene. After 6 months, periodontal status improved in 15 patients, declined in 8, and remained the same in 5.

***9.1** Assess the impact of the program statistically (use a two-sided test).

Suppose that patients are graded on the degree of change in periodontal status on a 7-point scale, with +3 indicating the greatest improvement, 0 indicating no change, and –3 indicating the greatest decline. The data are given in Table 9.6.

9.2 What nonparametric test can be used to determine whether or not a significant change in periodontal status has occurred over time?

9.3 Implement the procedure in Problem 9.2, and report a p-value.

TABLE 9.6 Degree of change in periodontal status

Change score	Number of patients
+3	4
+2	5
+1	6
0	5
–1	4
–2	2
–3	2

9.4 Suppose there are two samples of size 6 and 7, with a rank sum of 58 in the sample of size 6. Using the Wilcoxon rank-sum test, evaluate the significance of the results.

9.5 Answer Problem 9.4 for two samples of size 7 and 10, with a rank sum of 47 in the sample of size 7.

9.6 Answer Problem 9.4 for two samples of size 12 and 15, with a rank sum of 220 in the sample of size 12 (assume that there are no ties).

Obstetrics

9.7 Reanalyze the data in Table 8.16 using nonparametric methods. Assume the samples are unpaired.

9.8 Would such methods be preferable to parametric methods in analyzing the data? Why or why not?

Health Services Administration

Suppose we wish to compare the length of stay in the hospital for patients with the same diagnosis at two different hospitals. The following results are found:

First hospital	21, 10, 32, 60, 8, 44, 29, 5, 13, 26, 33
Second hospital	86, 27, 10, 68, 87, 76, 125, 60, 35, 73, 96, 44, 238

***9.9** Why might a t test not be very useful in this case?

***9.10** Carry out a nonparametric procedure for testing the hypothesis that the lengths of stay are comparable in the two hospitals.

Infectious Disease

The distribution of white-blood count is typically positively skewed, and assumptions of normality are usually not valid.

9.11 To compare the white-blood counts of patients on the medical and surgical services in Table 2.11 when normality is not assumed, what test can be used?

9.12 Perform the test in Problem 9.11 and report a p-value.

Sports Medicine

Refer to the Data Set TENNIS2.DAT (on the data disk).

9.13 What test can be used to compare the degree of pain during maximal activity between people randomized to Motrin and placebo?

9.14 Perform the test in Problem 9.13 and report a p-value.

Otolaryngology, Pediatrics

A common symptom of otitis media in young children is the prolonged presence of fluid in the middle ear, known as *middle-ear effusion*. The presence of fluid may result in temporary hearing loss and interfere with normal learning skills in the first 2 years of life. One hypothesis is that babies who are breast-fed for at least 1 month build up some immunity against the effects of the disease and have less prolonged effusion than do bottle-fed babies. A small study of 24 pairs of babies is set up, where the babies are matched on a one-to-one basis according to age, sex, socioeconomic status, and type of medications taken. One member of the matched pair is a breast-fed baby whereas the other member is a bottle-fed baby. The outcome variable is the duration of middle-ear effusion after the first episode of otitis media. The results are given in Table 9.7.

TABLE 9.7 Duration of middle-ear effusion in breast-fed and bottle-fed babies

Pair number	Duration of effusion in breast-fed baby (days)	Duration of effusion in bottle-fed baby (days)
1	20	18
2	11	35
3	3	7
4	24	182
5	7	6
6	28	33
7	58	223
8	7	7
9	39	57
10	17	76
11	17	186
12	12	29
13	52	39
14	14	15
15	12	21
16	30	28
17	7	8
18	15	27
19	65	77
20	10	12
21	7	8
22	19	16
23	34	28
24	25	20

***9.15** What are the hypotheses being tested here?

***9.16** Why might a nonparametric test be useful in testing the hypotheses?

***9.17** Which nonparametric test should be used here?

***9.18** Test the hypothesis that the duration of effusion is less prolonged among breast-fed babies than among bottle-fed babies using a nonparametric test.

Hypertension

Polyunsaturated fatty acids in the diet favorably affect several risk factors for cardiovascular disease. The principal dietary polyunsaturated fat is linoleic acid. To test the effects of dietary supplements of linoleic acid on blood pressure, 17 adults consumed 23 g/day of safflower oil, high in linoleic acid, for 4 weeks. Blood-pressure measurements were taken at baseline (before ingestion of oil) and 1 month later, with the mean values over several readings at each visit given in Table 9.8.

TABLE 9.8 **Effect of linoleic acid on systolic blood pressure (SBP)**

Subject	Baseline SBP	1-month SBP	Baseline minus 1-month SBP
1	119.67	117.33	2.34
2	100.00	98.78	1.22
3	123.56	123.83	–0.27
4	109.89	107.67	2.22
5	96.22	95.67	0.55
6	133.33	128.89	4.44
7	115.78	113.22	2.56
8	126.39	121.56	4.83
9	122.78	126.33	–3.55
10	117.44	110.39	7.05
11	111.33	107.00	4.33
12	117.33	108.44	8.89
13	120.67	117.00	3.67
14	131.67	126.89	4.78
15	92.39	93.06	–0.67
16	134.44	126.67	7.77
17	108.67	108.67	0.00

9.19 What parametric test could be used to test for the effect of linoleic acid on blood pressure?

9.20 Perform the test in Problem 9.19 and report a *p*-value.

9.21 What nonparametric test could be used to test for the effect of linoleic acid on blood pressure?

9.22 Perform the test in Problem 9.21 and report a *p*-value.

9.23 Compare your results in Problems 9.20 and 9.22, and discuss which method you feel is appropriate here.

Hypertension

An instrument that is in fairly common use in blood-pressure epidemiology is the random-zero device, whereby the zero point of the machine is randomly set with each use and the observer is not aware of the actual level of blood pressure at the time of measurement. This instrument is intended to reduce observer bias. Before using such a machine, it is important to check that readings are, on average, comparable to those of a standard cuff. For this purpose, two measurements were made on 20 children with both the standard cuff and the random-zero machine. The mean systolic blood pressure (SBP) for the two readings for each machine are given in Table 9.9. Suppose observers are reluctant to assume that the distribution of blood pressure is normal.

TABLE 9.9 **Comparison of mean systolic blood pressure with the standard cuff and the random-zero machine (mm Hg)**

Person	Mean SBP (standard cuff)	Mean SBP (random zero)
1	79	84
2	112	99
3	103	92
4	104	103
5	94	94
6	106	106
7	103	97
8	97	108
9	88	77
10	113	94
11	98	97
12	103	103
13	105	107
14	117	120
15	94	94
16	88	87
17	101	97
18	98	93
19	91	87
20	105	104

***9.24** Which nonparametric test should be used to test the hypothesis that the mean SBP for the two machines are comparable?

***9.25** Conduct the test recommended in Problem 9.24.

Another aspect of the same study is to compare the variability of blood pressure with each method. This comparison is achieved by computing $|x_1 - x_2|$ for each method (i.e., the absolute difference between first and second readings) and comparing the absolute differences between machines. The data are given in Table 9.10. The observers are reluctant to assume that the distributions are normal.

***9.26** Which nonparametric test should be used to test the hypothesis that the variability of the two machines is comparable?

TABLE 9.10 Comparison of variability of systolic blood pressure with the standard cuff and the random-zero machine (mm Hg)

Person	Absolute difference, standard cuff (a_s)	Absolute difference, random zero (a_r)
1	2	12
2	4	6
3	6	0
4	4	2
5	8	4
6	4	4
7	2	6
8	2	8
9	4	2
10	2	4
11	0	6
12	2	6
13	6	6
14	2	4
15	8	8
16	0	2
17	6	6
18	4	6
19	2	14
20	2	4

***9.27** Conduct the test recommended in Problem 9.26.

Health Promotion

Refer to Data Set SMOKE.DAT, on the data disk.

9.28 Use nonparametric methods to test whether there is a difference between males and females regarding the number of days abstinent from smoking.

9.29 Divide the data set into age groups (above/below the median) and use nonparametric methods to test whether the number of days abstinent from smoking is related to age.

9.30 Use the same approach as in Problem 9.29 to test whether the amount previously smoked is related to the number of days abstinent from smoking.

9.31 Use the same approach as in Problem 9.29 to test whether the adjusted CO level is related to the number of days abstinent from smoking.

9.32 Why are nonparametric methods well suited to a study of risk factors for smoking cessation?

Refer to the urinary-sodium data in Table 8.20.

9.33 Use nonparametric methods to assess whether dietary counseling is effective in reducing sodium intake as judged by urinary-sodium excretion levels.

Refer to Data Set HORMONE.DAT, on the data disk.

9.34 Use nonparametric methods to answer Problem 8.105.

9.35 Use nonparametric methods to answer Problem 8.106.

9.36 Use nonparametric methods to answer Problem 8.107.

9.37 Compare your results in Problems 9.34–9.36 with the corresponding results using parametric methods in Problems 8.105–8.107.

Ophthalmology

Refer to the Data Set in Tables 7.6 and 7.7 (p. 270).

9.38 Answer the question in Problem 7.87 using nonparametric methods.

9.39 Implement the test suggested in Problem 9.38, and report a two-sided p-value.

9.40 Compare the results in Problem 9.39 with those obtained in Problem 7.88.

Endocrinology

Refer to the Data Set BONEDEN.DAT on the data disk.

9.41 Answer the question in Problem 7.89 using nonparametric methods, and compare your results with those obtained using parametric methods.

9.42 Answer the question in Problem 7.90 using nonparametric methods, and compare your results with those obtained using parametric methods.

Infectious Disease

9.43 Reanalyze the data in Table 8.18 using nonparametric methods, and compare your results with those obtained in Problem 8.69.

Microbiology

Refer to the data in Table 8.28 (p. 326).

9.44 What nonparametric test can be used to compare the median pod weight for inoculated versus noninoculated plants?

9.45 Implement the test in Problem 9.44, and report a p-value (two-tail).

9.46 Compare your results with those obtained in Problem 8.151.

REFERENCES

[1] McNeil, D. R. (1967). Efficiency loss due to grouping in distribution free tests. *Journal of the American Statistical Association, 62*, 954–965.

[2] Hollander, M., & Wolfe, D. (1973). *Nonparametric statistical methods*. New York: Wiley.

10 Hypothesis Testing: Categorical Data

SECTION 10.1 Introduction

In Chapters 7 and 8, the basic methods of hypothesis testing for continuous data were presented. For each test, the data consisted of one or two samples, which were assumed to come from an underlying normal distribution(s); appropriate procedures were developed based on this assumption. In Chapter 9, the assumption of normality was relaxed and a class of nonparametric methods were introduced. Using these methods, we assumed that the variable under study can be ordered without assuming any underlying distribution.

If the variable under study is not continuous but is instead classified into categories, which may or may not be ordered, then different methods of inference should be used. Consider the problems in Examples 10.1 through 10.3.

Example 10.1 **Cancer** Suppose we are interested in the association between the use of oral contraceptives (OC use) and the 1-year incidence of cervical cancer from January 1, 1988, to January 1, 1989. Women who are disease-free on January 1, 1988, are classified into two OC-use categories as of that date: ever users and never users. We are interested in whether or not the 1-year incidence of cervical cancer is different between ever users and never users. Hence, this is a two-sample problem comparing two binomial proportions, and the *t*-test methodology in Chapter 8 cannot be used because the outcome variable, the development of cervical cancer, is a discrete variable with two categories (yes/no), not a continuous variable.

Example 10.2 **Cancer** Suppose the OC users in Example 10.1 are subdivided into "heavy" users, who have used the pill for 5 years or more, and "light" users, who have used the pill for less than 5 years. We may be interested in comparing 1-year cervical-cancer incidence rates among heavy users, light users, and nonusers. In this problem, *three* binomial proportions are being compared, and we need to consider methods comparing more than two binomial proportions.

Example 10.3 **Infectious Disease** The fitting of a probability model based on the Poisson distribution to the random variable defined by the annual number of deaths due to polio in the United States during the period 1968–1976 has been discussed, as shown in Table 4.8. We want to develop a general procedure for testing the goodness of fit of this and other probability models on actual sample data.

In this chapter, methods of hypothesis testing for comparing two or more binomial proportions are developed. Methods for testing the goodness of fit of a previously specified probability model to actual data are also considered. We will also consider relationships between categorical and nonparametric approaches to data analysis.

SECTION 10.2 Two-Sample Test for Binomial Proportions

Example 10.4 **Cancer** A hypothesis has been proposed that breast cancer in women is caused in part by events that occur between the age at menarche (the age when menstruation begins) and the age at first childbirth. In particular, the hypothesis is that the risk of breast cancer increases as the length of this time interval increases. If this theory is correct, then an important risk factor for breast cancer is age at first birth. This theory would explain in part why breast-cancer incidence seems to be higher for women in the upper socioeconomic groups, because they tend to have their children relatively late.

An international study was set up to test this hypothesis [1]. Breast-cancer cases were identified among women in selected hospitals in the United States, Greece, Yugoslavia, Brazil, Taiwan, and Japan. Controls were chosen from women of comparable age who were in the hospital at the same time as the cases but who did *not* have breast cancer. All women were asked about their age at first birth.

The set of women with at least one birth was arbitrarily divided into two categories: (1) women whose age at first birth was ≤ 29, and (2) women whose age at first birth was ≥ 30. The following results were found among women with at least one birth: 683 out of 3220 (21.2%) women with breast cancer (case women) and 1498 out of 10,245 (14.6%) women without breast cancer (control women) had an age at first birth ≥ 30. How can we assess whether this difference is significant or simply due to chance?

Let p_1 = the probability that age at first birth is ≥ 30 in case women with at least one birth and p_2 = the probability that age at first birth is ≥ 30 in control women with at least one birth. The question is whether or not the underlying probability of having an age at first birth of ≥ 30 is different in the two groups. This problem is equivalent to testing the hypothesis H_0: $p_1 = p_2 = p$ versus H_1: $p_1 \neq p_2$ for some constant p.

Two approaches for testing the hypothesis will be presented. One approach uses normal-theory methods similar to those developed in Chapter 8 and is discussed in Section 10.2.1. A second approach uses contingency-table methods and is discussed in Section 10.2.2. These two approaches are *equivalent* in that they always yield the same p-values, so which one is used is a matter of convenience.

10.2.1 Normal-Theory Method

It is reasonable to base the significance test on the difference between the sample proportions $(\hat{p}_1 - \hat{p}_2)$. If this difference is very different from 0 (either positive or negative), then H_0 would be rejected; otherwise, H_0 would be accepted. The samples will be assumed large enough so that the *normal approximation to the binomial distribution is valid.* Then, under H_0, $\hat{p}_1$ is normally distributed with mean p and variance pq/n_1, and $\hat{p}_2$ is normally distributed with mean p and variance

pq/n_2. Therefore, from Equation 5.10, because the samples are independent, $\hat{p}_1 - \hat{p}_2$ is normally distributed with mean 0 and variance

$$\frac{pq}{n_1} + \frac{pq}{n_2} = pq\left(\frac{1}{n_1} + \frac{1}{n_2}\right)$$

If we divide $\hat{p}_1 - \hat{p}_2$ by its standard error,

$$\sqrt{pq\left(\frac{1}{n_1} + \frac{1}{n_2}\right)}$$

then under H_0,

EQUATION 10.1

$$z = \left(\hat{p}_1 - \hat{p}_2\right) \Big/ \sqrt{pq\left(1/n_1 + 1/n_2\right)} \sim N(0,1)$$

The problem is that p and q are unknown, and thus the denominator of z cannot be computed unless some estimate for p is found. The best estimator for p is based on a weighted average of the sample proportions $\hat{p}_1$, $\hat{p}_2$. This weighted average, referred to as $\hat{p}$, is given by

EQUATION 10.2

$$\hat{p} = \frac{n_1\hat{p}_1 + n_2\hat{p}_2}{n_1 + n_2} = \frac{x_1 + x_2}{n_1 + n_2}$$

where x_1 = the observed number of events in the first sample and x_2 = the observed number of events in the second sample. This estimate makes intuitive sense, since each of the sample proportions is weighted by the number of people in the sample. Thus, we substitute the estimate $\hat{p}$ in Equation 10.2 for p in Equation 10.1. Finally, to better accommodate the normal approximation to the binomial, a continuity correction is introduced in the numerator of Equation 10.1. If $\hat{p}_1 \geq \hat{p}_2$, then we subtract $\left(\frac{1}{2n_1} + \frac{1}{2n_2}\right)$; if $\hat{p}_1 < \hat{p}_2$, then we add $\left(\frac{1}{2n_1} + \frac{1}{2n_2}\right)$. Equivalently, we can rewrite the numerator in terms of $|\hat{p}_1 - \hat{p}_2| - \left(\frac{1}{2n_1} + \frac{1}{2n_2}\right)$ and reject H_0 only for large positive values of z. This suggests the following test procedure:

EQUATION 10.3

Two-Sample Test for Binomial Proportions (Normal-Theory Test) To test the hypothesis H_0: $p_1 = p_2$ versus H_1: $p_1 \neq p_2$, where the proportions are obtained from two independent samples, use the following procedure:

(1) Compute the test statistic

$$z = \frac{|\hat{p}_1 - \hat{p}_2| - \left(\frac{1}{2n_1} + \frac{1}{2n_2}\right)}{\sqrt{\hat{p}\hat{q}\left(\frac{1}{n_1} + \frac{1}{n_2}\right)}}$$

where $\hat{p} = \frac{n_1\hat{p}_1 + n_2\hat{p}_2}{n_1 + n_2} = \frac{x_1 + x_2}{n_1 + n_2}$, $\hat{q} = 1 - \hat{p}$

and x_1, x_2 are the number of events in the first and second samples, respectively.

(2) For a two-sided level α test, if

$$z > z_{1-\alpha/2}$$

then reject H_0; if $z \le z_{1-\alpha/2}$

then accept H_0.

(3) The approximate p-value for this test is given by

$$p = 2\left[1 - \Phi(z)\right]$$

(4) Use this test only when the normal approximation to the binomial distribution is valid for each of the two samples, that is, when $n_1\hat{p}\hat{q} \ge 5$ and $n_2\hat{p}\hat{q} \ge 5$.

The acceptance and rejection regions for this test are depicted in Figure 10.1. The computation of the exact p-value is illustrated in Figure 10.2.

Example 10.5 **Cancer** Assess the statistical significance of the results from the international study in Example 10.4.

Solution The sample proportion of case women whose age at first birth was ≥30 is 683/3220 = .212 = $\hat{p}_1$, and the sample proportion of control women whose age at first birth was ≥30 is 1498/10,245 = .146 = $\hat{p}_2$. To compute the test statistic z in Equation 10.3, the estimated common proportion $\hat{p}$ must be computed, which is given by

$$\hat{p} = (683 + 1498)/(3220 + 10{,}245) = 2181/13{,}465 = .162$$

$$\hat{q} = 1 - .162 = .838$$

Note that

$$n_1\hat{p}\hat{q} = 3220(.162)(.838) = 437.1 \ge 5$$

and $$n_2\hat{p}\hat{q} = 10{,}245(.162)(.838) = 1390.7 \ge 5$$

Thus, the test in Equation 10.3 can be used.

The test statistic is given by

$$z = \left\{|.212 - .146| - \left[\frac{1}{2(3220)} + \frac{1}{2(10{,}245)}\right]\right\} \Big/ \sqrt{.162(.838)\left(\frac{1}{3220} + \frac{1}{10{,}245}\right)}$$

$$= .0657/.00744$$
$$= 8.8$$

The p-value = $2 \times \left[1 - \Phi(8.8)\right] < .001$, and the results are extremely significant. Therefore, we can conclude that women with breast cancer are significantly more likely to have had their first child after the age of 30 than are comparable women without breast cancer.

Example 10.6 **Cardiovascular Disease** A study was conducted to look at the effects of oral contraceptives (OC) on heart disease in women 40 to 44 years of age. It is found that among 5000 current OC users at baseline, 13 women develop a myocardial infarction (MI) over a 3-year period, whereas among 10,000 non-OC users, 7 develop an MI over a 3-year period. Assess the statistical significance of the results.

FIGURE 10.1 **Acceptance and rejection regions for the two-sample test for binomial proportions (normal-theory test)**

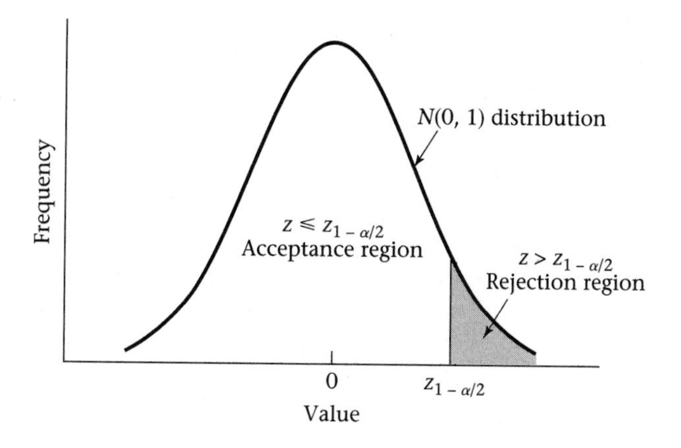

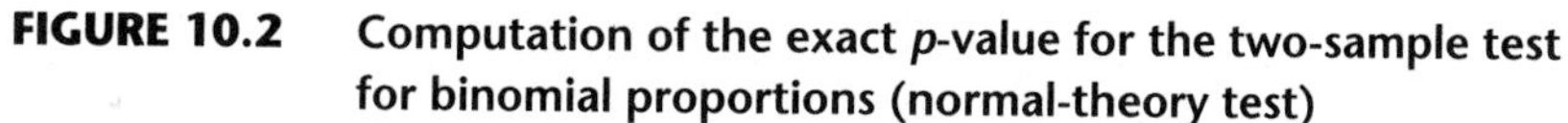

FIGURE 10.2 **Computation of the exact *p*-value for the two-sample test for binomial proportions (normal-theory test)**

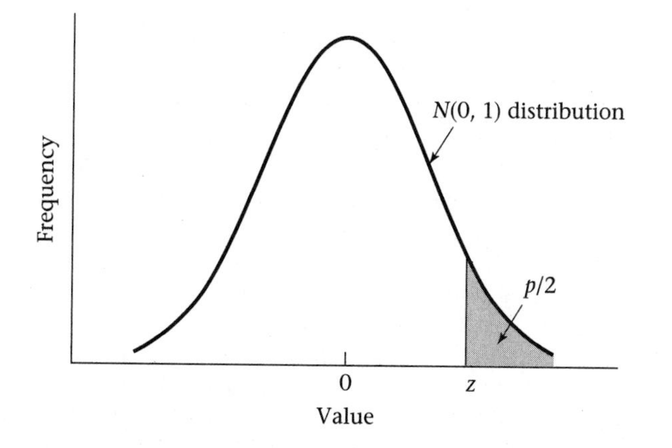

Solution Note that $n_1 = 5000$, $\hat{p}_1 = 13/5000 = .0026$, $n_2 = 10{,}000$, $\hat{p}_2 = 7/10{,}000 = .0007$. We wish to test the hypothesis H_0: $p_1 = p_2$ versus H_1: $p_1 \neq p_2$. The best estimate of the common proportion p is given by

$$\hat{p} = \frac{13+7}{15{,}000} = \frac{20}{15{,}000} = .00133$$

Because $n_1\hat{p}\hat{q} = 5000(.00133)(.99867) = 6.66$, $n_2\hat{p}\hat{q} = 10{,}000(.00133)(.99867) = 13.32$, the normal-theory test in Equation 10.3 can be used. The test statistic is given by

$$z = \frac{|.0026 - .0007| - \left[\frac{1}{2(5000)} + \frac{1}{2(10{,}000)}\right]}{\sqrt{.00133(.99867)(1/5000 + 1/10{,}000)}} = \frac{.00175}{.00063} = 2.77$$

The p-value is given by $2 \times [1 - \Phi(2.77)] = .006$. Thus, there is a highly significant difference between MI incidence rates for current OC users vs. non-OC users. In other words, OC use is significantly associated with MI incidence over a 3-year period.

10.2.2 Contingency-Table Method

The same test posed in Section 10.2.1 is now approached from a different perspective.

Example 10.7 **Cancer** Suppose all women with at least one birth in the international study in Example 10.4 are classified as either cases or controls and with age at first birth as either ≤ 29 or ≥ 30. The four possible combinations are displayed in Table 10.1.

TABLE 10.1 **Data for the international study in Example 10.4 comparing age at first birth in breast-cancer cases with comparable controls**

	Age at first birth		
Status	≥30	≤29	Total
Case	683	2537	3220
Control	1498	8747	10,245
Total	2181	11,284	13,465

Source: Reprinted with permission of *WHO Bulletin, 43,* 209–221, 1970.

The case–control status is displayed along the rows of the table and the age at first birth along the columns of the table. Hence, each woman falls into one of the four boxes, or *cells,* of the table. In particular, there are 683 women with breast cancer whose age at first birth is ≥30; 2537 women with breast cancer whose age at first birth is ≤29; 1498 control women whose age at first birth is ≥30; and 8747 control women whose age at first birth is ≤29. Furthermore, the number of women in each row and column can be totaled and displayed in the margins of the table. Thus, there are 3220 case women (683 + 2537); 10,245 control women (1498 + 8747); 2181 women with age at first birth ≥ 30 (683 + 1498); and 11,284 women with age at first birth ≤ 29 (2537 + 8747). These sums are referred to as row margins and column margins, respectively. Finally, the total number of units = 13,465 is given in the lower right-hand corner of the table, which can be obtained either by summing the four cells (683 + 2537 + 1498 + 8747) or by summing the row margins (3220 + 10,245) or the column margins (2181 + 11,284). This sum is sometimes referred to as the *grand total.*

Table 10.1 is called a 2 × 2 *contingency table,* because it has two categories for case–control status and two categories for age-at-first-birth status.

DEFINITION 10.1 A 2 × 2 **contingency table** is a table composed of two rows cross-classified by two columns. It is an appropriate way to display data that can be classified by two different variables, *each* of which has only two possible outcomes. One variable is arbitrarily assigned to the rows and the other to the columns. Each of the four cells represents the number of units (women, in the previous example), with a specific value for each of the two variables. The cells are sometimes referred to by number, with the (1, 1) cell being the cell in the first row and first column, the (1, 2) cell being the cell in the first row and second column, the (2, 1) cell being the cell in the second row and first column, and the (2, 2) cell being the cell in the second row and second column. The observed number of units in the four cells are likewise referred to as O_{11}, O_{12}, O_{21}, and O_{22}, respectively. Furthermore, it is customary to total

(1) The number of units in each row and display them in the right margins, which are called **row marginal totals or row margins.**

(2) The number of units in each column and display them in the bottom margins, which are called **column marginal totals or column margins.**

(3) The total number of units in the four cells, which is displayed in the lower right-hand corner of the table and is called the **grand total.**

Example 10.8 **Cardiovascular Disease** Display the myocardial-infarction data in Example 10.6 in the form of a 2×2 contingency table.

Solution Let the rows of the table represent the OC-use group, with the first row representing current OC users and the second row representing non-OC users. Let the columns of the table represent MI, with the first column representing "Yes" and the second column representing "No." We have studied 5000 current OC users, of whom 13 developed MI and 4987 did not. We have studied 10,000 non-OC users, of whom 7 developed MI and 9993 did not. Thus, the contingency table should look like Table 10.2.

TABLE 10.2 **2×2 contingency table for the OC–MI data in Example 10.6**

	MI status over 3 years		
OC-use group	Yes	No	Total
OC users	13	4987	5000
Non-OC users	7	9993	10,000
Total	20	14,980	15,000

Two different sampling designs lend themselves to a contingency-table framework. In the breast-cancer data in Example 10.4, there are two independent samples (i.e., case women and control women) and we want to compare the proportion of women in each group who have a first birth at a late age. Similarly, in the OC–MI data in Example 10.6, there are two independent samples of women with different contraceptive-use patterns and we wish to compare the proportion of women in each group who develop an MI. In both instances, we want to test whether or not the proportions are the same in the two independent samples. This test is referred to as a **test for homogeneity of binomial proportions.** In this situation, one set of margins is fixed (e.g., the rows) and the number of successes in each row is a random variable. For example, in Example 10.4, the total number of breast-cancer cases and controls is fixed, and the number of women with age at first birth ≥ 30 is a binomial random variable conditional on the fixed-row margins (i.e., 3220 cases and 10,245 controls).

Another possible situation from which contingency tables arise is in testing for the independence of two characteristics in the same sample when neither characteristic is particularly appropriate as a denominator. In this setting, both sets of margins are assumed to be fixed. The number of units in one particular cell of the table [e.g., the (1, 1) cell] is a random variable and all other cells can be determined from the fixed margins and the (1, 1) cell. An example of this design is given in Example 10.9.

Example 10.9 **Nutrition** The food-frequency questionnaire is widely used to measure dietary intake. A person specifies the number of servings consumed per week of each of many different food items. The total nutrient composition is then calculated from the specific dietary components of each food item. One way to judge how well a questionnaire measures dietary intake is by its reproducibility. To assess reproducibility the questionnaire is administered at two different times to 50 people and the reported nutrient intakes from the two questionnaires are compared. Suppose dietary cholesterol is quantified on each questionnaire as high if >300 mg/day and normal otherwise. The contingency table in Table 10.3 is a natural way to compare the results of the two surveys. Notice that there is no natural denominator in this example. We simply want to test whether there is some relationship between the two reported measures of dietary cholesterol for the same person. More specifically, we want to assess how unlikely it is that 15 women will report high dietary cholesterol intake on both questionnaires, given that 20 out of 50 women report high intake on the first questionnaire and 24 out of 50 women report high intake on the second questionnaire. This test is referred to as a **test of independence** or a **test of association** between the two characteristics.

TABLE 10.3 **A comparison of dietary cholesterol assessed by a food-frequency questionnaire at two different times**

	Second food-frequency questionnaire		
First food-frequency questionnaire	High	Normal	Total
High	15	5	20
Normal	9	21	30
Total	24	26	50

Fortunately, the same test procedure is used whether a test of homogeneity or a test of independence is performed, and we will no longer distinguish between these two tests in this section.

10.2.3 Significance Testing Using the Contingency-Table Approach

Table 10.1 is an **observed contingency table** or an **observed table.** In order to determine statistical significance, we need to develop an **expected table,** which is the contingency table that would be expected if there were no relationship between breast cancer and age at first birth, that is, if H_0: $p_1 = p_2 = p$ were true. In this example p_1 and p_2 are the probabilities (among women with at least one birth) of a breast-cancer case and a control, respectively, having a first birth at an age ≥ 30. For this purpose, a general observed table, if there were x_1 exposed out of n_1 women with breast cancer and x_2 exposed out of n_2 control women, is given in Table 10.4.

TABLE 10.4 **General contingency table for the international-study data in Example 10.4 if (1) of n_1 women in the case group, x_1 are exposed and (2) of n_2 women in the control group, x_2 are exposed (being exposed here means having an age at first birth ≥ 30)**

	Age at first birth		
Case–control status	≥ 30	≤ 29	Total
Case	x_1	$n_1 - x_1$	n_1
Control	x_2	$n_2 - x_2$	n_2
Total	$x_1 + x_2$	$n_1 + n_2 - (x_1 + x_2)$	$n_1 + n_2$

If H_0 were true, then the best estimate of the common proportion p is $\hat{p}$, which is given in Equation 10.2 as

$$(n_1\hat{p}_1 + n_2\hat{p}_2) / (n_1 + n_2)$$

or, alternatively, as $(x_1 + x_2) / (n_1 + n_2)$

where x_1 and x_2 are the number of exposed women in groups 1 and 2, respectively. Furthermore, under H_0 the expected number of units in the (1, 1) cell equals the expected number of women with age at first birth ≥ 30 among women with breast cancer, which is given by

$$n_1\hat{p} = n_1(x_1 + x_2) / (n_1 + n_2)$$

However, in Table 10.4 this number is simply the product of the first row margin (n_1) multiplied by the first column margin ($x_1 + x_2$), divided by the grand total ($n_1 + n_2$). Similarly, the expected number of units in the (2, 1) cell equals the expected number of control women with age at first birth ≥ 30:

$$n_2\hat{p} = n_2(x_1 + x_2) / (n_1 + n_2)$$

which is equal to the product of the second row margin multiplied by the first column margin, divided by the grand total. In general, the following rule can be applied:

EQUATION 10.4

Computation of Expected Values for 2 × 2 Contingency Tables The expected number of units in the ***(i, j)*** cell, which is usually denoted by E_{ij}, is the product of the ***i*th** row margin multiplied by the ***j*th** column margin, divided by the grand total.

Example 10.10 **Cancer** Compute the expected table for the breast-cancer data in Example 10.4.

Solution Refer to Table 10.1, which gives the observed table for these data. The row totals are 3220 and 10,245; the column totals are 2181 and 11,284; and the grand total is 13,465. Thus,

$$E_{11} = \text{expected number of units in the (1, 1) cell} = 3220(2181)/13{,}465 = 521.6$$

$$E_{12} = \text{expected number of units in the (1, 2) cell} = 3220(11{,}284)/13{,}465 = 2698.4$$

$$E_{21} = \text{expected number of units in the (2, 1) cell} = 10{,}245(2181)/13{,}465 = 1659.4$$

$$E_{22} = \text{expected number of units in the (2, 2) cell} = 10{,}245(11{,}284)/13{,}465 = 8585.6$$

These expected values are displayed in Table 10.5.

TABLE 10.5 **Expected table for the breast-cancer data in Example 10.4**

	Age at first birth		
Case–control status	≥30	≤29	Total
Case	521.6	2698.4	3220
Control	1659.4	8585.6	10,245
Total	2181	11,284	13,465

Example 10.11 **Cardiovascular Disease** Compute the expected table for the OC–MI data in Example 10.6.

Solution From Table 10.2, which gives the observed table for these data,

$$E_{11} = \frac{5000(20)}{15{,}000} = 6.7$$

$$E_{12} = \frac{5000(14{,}980)}{15{,}000} = 4993.3$$

$$E_{21} = \frac{10{,}000(20)}{15{,}000} = 13.3$$

$$E_{22} = \frac{10{,}000(14{,}980)}{15{,}000} = 9986.7$$

These expected values are displayed in Table 10.6.

TABLE 10.6 **Expected table for the OC–MI data in Example 10.6**

	MI status over 3 years		
OC-use group	Yes	No	Total
OC users	6.7	4993.3	5000
Non-OC users	13.3	9986.7	10,000
Total	20	14,980	15,000

We can show from Equation 10.4 that the *total* of the expected number of units in any row or column should be the same as the corresponding observed row or column total. This relationship provides a useful check that the expected values are computed correctly.

Example 10.12 Check that the expected values in Table 10.5 are computed correctly.

Solution The following information is given:

(1) The total of the expected values in the first row = $E_{11} + E_{12} = 521.6 + 2698.4 = 3220$ = first row total in the observed table.

(2) The total of the expected values in the second row = $E_{21} + E_{22} = 1659.4 + 8585.6 = 10{,}245$ = second row total in the observed table.

(3) The total of the expected values in the first column = $E_{11} + E_{21} = 521.6 + 1659.4 = 2181$ = first column total in the observed table.

(4) The total of the expected values in the second column = $E_{12} + E_{22} = 2698.4 + 8585.6 = 11{,}284$ = second column total in the observed table.

We now wish to compare the observed table in Table 10.1 with the expected table in Table 10.5. If the corresponding cells in these two tables are close, then H_0 will be accepted; if they are sufficiently different, then H_0 will be rejected. How should we decide how different the cells should be for us to reject H_0? It can be shown that the best way of comparing the cells in the two tables is to use the statistic $(O - E)^2/E$, where O and E are the observed and expected number of units, respectively, in a particular cell. In particular, under H_0 it can be shown that the sum of $(O - E)^2/E$ over the four cells in the table approximately follows a chi-square distribution with 1 *df*. H_0 is rejected only if this sum is large and accepted otherwise, because small values of this sum correspond to good agreement between the two tables, whereas large values correspond to poor agreement. This test procedure will be used only when the normal approximation to the binomial distribution is valid. In this setting the normal approximation can be shown to be approximately true if *no expected value in the table is less than 5*.

Furthermore, under certain circumstances a version of this test statistic with a *continuity correction* yields more accurate *p*-values than does the uncorrected version when approximated by a chi-square distribution. For the continuity corrected version, the statistic $\left(|O - E| - \frac{1}{2}\right)^2 / E$ rather than $(O - E)^2/E$ is computed for each cell and the preceding expression is summed over the four cells. This test procedure is referred to as the chi-square test using the Yates correction and is summarized as follows:

EQUATION 10.5

Yates-Corrected Chi-Square Test for a 2 × 2 Contingency Table Suppose we wish to test the hypothesis H_0: $p_1 = p_2$ versus H_1: $p_1 \neq p_2$ using a contingency-table approach, where O_{ij} represents the observed number of units in the (i, j) cell and E_{ij} represents the expected number of units in the (i, j) cell.

(1) Compute the test statistic

$$X^2 = \left(|O_{11} - E_{11}| - .5\right)^2 / E_{11} + \left(|O_{12} - E_{12}| - .5\right)^2 / E_{12} + \left(|O_{21} - E_{21}| - .5\right)^2 / E_{21} + \left(|O_{22} - E_{22}| - .5\right)^2 / E_{22}$$

which under H_0 approximately follows a χ_1^2 distribution.

(2) For a level α test, reject H_0 if $X^2 > \chi_{1,1-\alpha}^2$ and accept H_0 if $X^2 \leq \chi_{1,1-\alpha}^2$.

(3) The approximate p-value is given by the area to the right of X^2 under a χ_1^2 distribution.

(4) Use this test only if none of the four expected values is less than 5.

The acceptance and rejection regions for this test are depicted in Figure 10.3. The computation of the p-value is illustrated in Figure 10.4.

The Yates-corrected chi-square test is a *two-sided* test even though the critical region, based on the chi-square distribution, is one-sided. The rationale for this is that large values of $|O_{ij} - E_{ij}|$ and, correspondingly, of the test statistic X^2 will be obtained under H_1 regardless of whether $p_1 < p_2$ or $p_1 > p_2$. Small values of X^2 are evidence in favor of H_0.

Example 10.13 **Cancer** Assess the breast-cancer data in Example 10.4 for statistical significance using a contingency table approach.

Solution First compute the observed and expected tables as given in Tables 10.1 and 10.5, respectively. Check that all expected values in Table 10.5 are at least 5, which is clearly the case. Thus, following Equation 10.5,

$$\begin{aligned} X^2 &= \frac{\left(|683 - 521.6| - .5\right)^2}{521.6} + \frac{\left(|2537 - 2698.4| - .5\right)^2}{2698.4} \\ &+ \frac{\left(|1498 - 1659.4| - .5\right)^2}{1659.4} + \frac{\left(|8747 - 8585.6| - .5\right)^2}{8585.6} \\ &= \frac{160.9^2}{521.6} + \frac{160.9^2}{2698.4} + \frac{160.9^2}{1659.4} + \frac{160.9^2}{8585.6} \\ &= 49.661 + 9.599 + 15.608 + 3.017 = 77.89 \sim \chi_1^2 \text{ under } H_0 \end{aligned}$$

FIGURE 10.3 **Acceptance and rejection regions for the Yates-corrected chi-square test for a 2 × 2 contingency table**

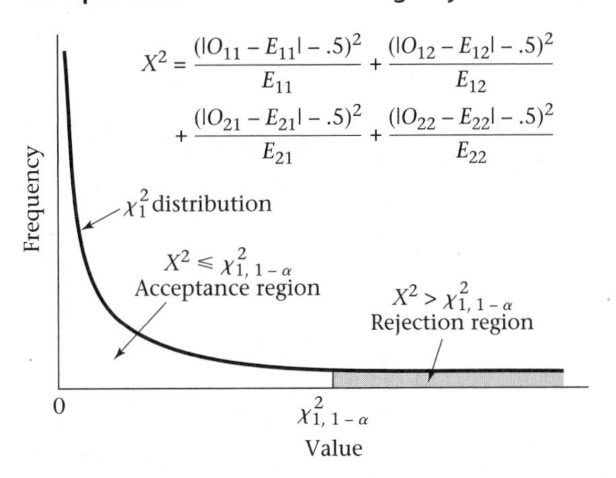

FIGURE 10.4 **Computation of the *p*-value for the Yates-corrected chi-square test for a 2 × 2 contingency table**

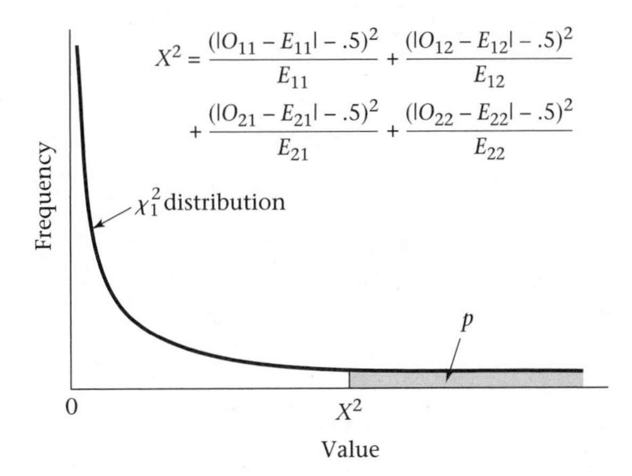

Because $\chi_{1,.999}^2 = 10.83 < 77.89 = X^2$

we have $p < 1 - .999 = .001$

and the results are extremely significant. Thus, breast cancer is significantly associated with having a first child after age 30.

Example 10.14 **Cardiovascular Disease** Assess the OC–MI data in Example 10.6 for statistical significance using a contingency table approach.

Solution First compute the observed and expected tables as given in Tables 10.2 and 10.6, respectively. Note that the minimum expected value in Table 10.6 is 6.7, which is ≥ 5. Thus the test procedure in Equation 10.5 can be used:

$$X^2 = \frac{(|13-6.7|-.5)^2}{6.7} + \frac{(|4987-4993.3|-.5)^2}{4993.3} + \frac{(|7-13.3|-.5)^2}{13.3} + \frac{(|9993-9986.7|-.5)^2}{9986.7}$$

$$= \frac{5.8^2}{6.7} + \frac{5.8^2}{4993.3} + \frac{5.8^2}{13.3} + \frac{5.8^2}{9986.7}$$

$$= 5.104 + 0.007 + 2.552 + 0.003 = 7.67 \sim \chi_1^2 \text{ under } H_0$$

Because $\chi^2_{1,.99} = 6.63, \chi^2_{1,.995} = 7.88$, and $6.63 < 7.67 < 7.88$, it follows that $1 - .995 < p < 1 - .99$, or $.005 < p < .01$, and the results are highly significant. Thus, there is a significant difference between incidence rates of MI for OC users versus non-OC users among 40- to 44-year-old women, with OC users having higher rates.

The test procedures in Equation 10.3 and Equation 10.5 are equivalent in the sense that they always give the same p-values and always result in the same decisions about accepting or rejecting H_0. Which test procedure is used is a matter of convenience. Most research workers find the contingency-table approach more understandable, and results are more frequently reported in this format in the scientific literature.

At this time statisticians disagree widely about whether or not a continuity correction is needed for the contingency-table test in Equation 10.5. Generally, p-values obtained using the continuity correction are slightly larger. Thus, results obtained are slightly less significant than comparable results obtained without using a continuity correction. However, the difference in results obtained using these two methods should be small for tables based on large sample sizes. I believe the Yates-corrected test statistic is slightly more widely used in the applied literature and therefore I use it in this section. Another possible approach for performing hypothesis tests based on 2×2 contingency tables is to use Fisher's exact test. This procedure will be discussed in Section 10.3.

10.2.4 Short Computational Form for the Yates-Corrected Chi-Square Test for 2 × 2 Contingency Tables

The test statistic X^2 in Equation 10.5 has another computational version that is more convenient to use with a hand calculator and does not require the computation of an expected table:

EQUATION 10.6

Short Computational Form for the Yates-Corrected Chi-Square Test for 2 × 2 Contingency Tables Suppose we have the 2×2 contingency table in Table 10.7. The X^2 test statistic in Equation 10.5 can be written in the form

$$X^2 = n\left(|ad - bc| - \frac{n}{2}\right)^2 \bigg/ \left[(a+b)(c+d)(a+c)(b+d)\right]$$

Thus, the test statistic X^2 depends only on (1) the grand total n, (2) the row and column margins $a + b, c + d, a + c, b + d$, and (3) the magnitude of the quantity $ad - bc$. To compute X^2, proceed as follows:

(1) Compute

$$\left(|ad - bc| - \frac{n}{2}\right)^2$$

Start with the first column margin and proceed counterclockwise.

(2) Divide by each of the two column margins.

(3) Multiply by the grand total.

(4) Divide by each of the two row margins.

This computation is particularly easy with a hand calculator, because previous products and quotients can be maintained in the display and used for further calculations.

TABLE 10.7 General contingency table

a	b	$a + b$
c	d	$c + d$
$a + c$	$b + d$	$n = a + b + c + d$

Example 10.15 **Nutrition** Compute the chi-square statistic for the nutrition data in Example 10.9 using the short computational form in Equation 10.6.

Solution From Table 10.3,

$$a = 15 \quad b = 5 \quad c = 9 \quad d = 21 \quad n = 50$$

Furthermore, the smallest expected value = $(24 \times 20)/50 = 9.6 \geq 5$. Thus, it is valid to use the chi-square test. Use the approach in Equation 10.6 as follows:

(1) Compute

$$\left(|ad - bc| - \frac{n}{2}\right)^2 = \left[|15 \times 21 - 5 \times 9| - \frac{50}{2}\right]^2$$
$$= (270 - 25)^2 = 245^2 = 60{,}025$$

(2) Divide the result in step 1 (60,025) by each of the two column margins (24 and 26), thus obtaining 96.194.

(3) Multiply the result in step 2 (96.194) by the grand total (50), thus obtaining 4809.70.

(4) Divide the result in step 3 (4809.70) by each of the two row margins (20 and 30), thus obtaining 8.02.

Since the critical value = $\chi^2_{1,.95} = 3.84$ and $X^2 = 8.02 > 3.84$, the results are statistically significant. To obtain a range for the p-value, note from the chi-square table that $\chi^2_{1,.995} = 7.88$, $\chi^2_{1,.999} = 10.83$, and thus, since $7.88 < 8.02 < 10.83$, $.001 < p < .005$.

These data have also been analyzed using the SPSS^X/PC CROSSTABS program, as shown in Table 10.8. The program prints out the cell counts, the row and column totals and percentages, and the grand total. Furthermore, it prints out the minimum expected frequency (min E.F. = 9.6), and it is noted that there are no expected frequencies <5. Finally, the significance test is performed using both the Yates-corrected chi-square test (chi-square = 8.02, $df = 1$, p-value = .0046) and the uncorrected chi-square test (chi-square = 9.74, $df = 1$, p-value = .0018).

TABLE 10.8 **Use of SPSS^X/PC CROSSTABS program to analyze the nutrition data in Table 10.3**

```
                            SPSSX/PC  Release 1.0
Crosstabulation:     CHOL1    1ST FOOD FREQUENCY QUESTIONNAIRE
                  By CHOL2    2ND FOOD FREQUENCY QUESTIONNAIRE

            Count   | HIGH     | NORMAL    |
 CHOL2->            |          |           |   Row
                    |     1.00 |      2.00 |  Total
CHOL1               |----------|-----------|
             1.00   |    15    |     5     |    20
 HIGH               |          |           |   40.0
                    |----------|-----------|
             2.00   |     9    |    21     |    30
 NORMAL             |          |           |   60.0
                    |----------|-----------|
           Column        24         26          50
            Total       48.0       52.0       100.0

Chi-Square     D.F.     Significance     Min E.F.     Cells with E.F.<5
----------     ----     ------------     --------     -----------------
   8.01616        1        0.0046           9.600       None
   9.73558        1        0.0018         (Before Yates Correction)

Number of Missing Observations =    0
```

The results show a highly significant association between dietary-cholesterol intake reported by the same person at two different points in time. This is to be expected, because, if the dietary instrument is at all meaningful, then there should be an association between the responses of the same person at two different points in time. We will discuss measures of reproducibility for categorical data in more detail in Section 10.8.

In this section, we have discussed the two-sample test for binomial proportions. This is the analogue to the two-sample t test for comparing means from two independent samples introduced in Chapter 8, except that here we are comparing proportions instead of means.

We refer to the master flowchart (p. 776). For all the methods in Chapter 10, we answer yes to (1) one variable of interest? no to (2) one-sample problem? yes to (3) two-sample problem? no to (4) is underlying distribution normal, or can central-limit theorem be assumed to hold? and yes to (5) is underlying distribution binomial?

We now refer to the flowchart at the end of this chapter (p. 412). We answer yes to (1) are samples independent? and (2) are all expected values ≥ 5? This leads us to the box entitled "Use the two-sample test for binomial proportions, or 2 × 2 contingency tables if no confounding is present, or the Mantel-Haenszel test if confounding is present." In brief, a confounder is another variable that is potentially related to both the row and column classification variables, and must be controlled for. We discuss methods for accomplishing this in Chapter 13. In this chapter, we assume no confounding is present. Thus, we use either the two-sample test for binomial proportions (Equation 10.3) or the equivalent chi-square test for 2×2 contingency tables (Equation 10.5).

SECTION 10.3 Fisher's Exact Test

In Section 10.2, we discussed methods for comparing two binomial proportions using either normal-theory or contingency-table methods. Both methods yield identical p-values. However, they require that the normal approximation to the binomial distribution be valid, which is not always the case, especially for small sample sizes.

Example 10.16 **Cardiovascular Disease, Nutrition** Suppose we wish to investigate the relationship between high salt intake and the occurrence of death from cardiovascular disease (CVD). Groups of high- and low-salt users could be identified and followed over a long period of time to compare the relative frequency of death from CVD in the two groups. In contrast, a much less expensive study would involve looking at death records, separating the CVD deaths from the non-CVD deaths, and then asking a close relative (such as a spouse) about the dietary habits of the deceased, and comparing salt intake of CVD deaths and non-CVD deaths.

The latter type of study, a retrospective study, may be impossible to perform for a number of reasons. But if it is possible, it will almost always be less expensive than the former type of study, a prospective study.

Example 10.17 **Cardiovascular Disease, Nutrition** Suppose a retrospective study is done among men aged 50–54 in a specific county who died over a 1-month period. The investigators attempt to include approximately an equal number of men who died from CVD (the cases) and men who died from other causes (the controls). It is found that of 35 people who died from CVD, 5 were on a high-salt diet before they died, whereas of 25 people who died from other causes, 2 were on such a diet. These data, presented in Table 10.9, are in the form of a 2×2 contingency table, and thus the methods of Section 10.2.2 may be applicable.

TABLE 10.9 **Data concerning the possible relationship between cause of death and high salt intake**

	Type of diet		
Cause of death	High salt	Low salt	Total
Non-CVD	2	23	25
CVD	5	30	35
Total	7	53	60

However, the expected values of this table are too small for such methods to be valid. Indeed,

$$E_{11} = 7(25)/60 = 2.92$$
$$E_{21} = 7(35)/60 = 4.08$$

and thus two of the four cells have expected values less than 5. How should the possible association between cause of death and type of diet be assessed?

In this case, **Fisher's exact test** can be used. This procedure gives exact results for any 2 × 2 table but is only necessary for tables with small expected values, where the standard chi-square test as given in Equation 10.5 is not applicable. For tables in which the use of the chi-square test is appropriate, the two tests give very similar results. Suppose the probability that a man was on a high-salt diet given that his cause of death was noncardiovascular (non-CVD) = p_1, and the probability that a man was on a high-salt diet given that his cause of death was cardiovascular (CVD) = p_2. We wish to test the hypothesis H_0: $p_1 = p_2 = p$ versus H_1: $p_1 \neq p_2$. Table 10.10 gives the general layout of the data.

TABLE 10.10 General layout of data for Fisher's exact test example

	Type of diet		
Cause of death	High salt	Low salt	Total
Non-CVD	a	b	$a + b$
CVD	c	d	$c + d$
Total	$a + c$	$b + d$	n

For mathematical convenience, we will assume that the margins of this table are *fixed*; that is, the numbers of non-CVD deaths and CVD deaths are fixed at $a + b$ and $c + d$, respectively, and the numbers of people on high- and low-salt diets are fixed at $a + c$ and $b + d$, respectively. Indeed, it is difficult to compute exact probabilities unless one makes the assumption of fixed margins. The *exact* probability of observing the table with cells a, b, c, d is given as follows:

EQUATION 10.7

Exact Probability of Observing a Table with Cells *a, b, c, d*

$$Pr(a,b,c,d) = \frac{(a+b)!(c+d)!(a+c)!(b+d)!}{n!\,a!\,b!\,c!\,d!}$$

The formula in Equation 10.7 is easy to remember, because the numerator is the product of the factorials of each of the row and column margins, and the denominator is the product of the factorial of the grand total and the factorials of the individual cells.

Example 10.18 Suppose we have the 2×2 table as shown in Table 10.11. Compute the exact probability of obtaining this table assuming that the margins are fixed.

Solution

$$Pr(2,5,3,1) = \frac{7!\,4!\,5!\,6!}{11!\,2!\,5!\,3!\,1!} = \frac{5040(24)(120)(720)}{39{,}916{,}800(2)(120)(6)} = \frac{1.0450944 \times 10^{10}}{5.7480192 \times 10^{10}} = .182$$

TABLE 10.11 **Hypothetical 2×2 contingency table in Example 10.18**

2	5	7
3	1	4
5	6	11

10.3.1 The Hypergeometric Distribution

Suppose we consider all possible tables with fixed row margins denoted by N_1 and N_2 and fixed column margins denoted by M_1 and M_2. We assume that the rows and columns have been rearranged so that $M_1 \le M_2$ and $N_1 \le N_2$. We will refer to each table by its (1, 1) cell, because all other cells are then determined from the fixed row and column margins. Let the random variable X denote the cell count in the (1, 1) cell. The probability distribution of X is given by

EQUATION 10.8

$$Pr(X = a) = \frac{N_1!\,N_2!\,M_1!\,M_2!}{N!\,a!(N_1 - a)!(M_1 - a)!(M_2 - N_1 + a)!}, a = 0,\ldots,\min(M_1, N_1)$$

and $N = N_1 + N_2 = M_1 + M_2$. This probability distribution is known as the **hypergeometric distribution.**

It will be useful for our subsequent work on combining evidence from more than one 2×2 table in Chapter 13 to refer to the expected value and variance of the hypergeometric distribution. These are given as follows:

EQUATION 10.9

Expected Value and Variance of the Hypergeometric Distribution Suppose we consider all possible tables with fixed row margins N_1, N_2 and fixed column margins M_1, M_2, where $N_1 \le N_2$ and $M_1 \le M_2$ and $N = N_1 + N_2 = M_1 + M_2$. Let the random variable X denote the cell count in the (1, 1) cell. The expected value and variance of X are given by

$$E(X) = \frac{M_1 N_1}{N}$$

$$Var(X) = \frac{M_1 M_2 N_1 N_2}{N^2(N-1)}$$

The basic strategy in testing the hypothesis H_0: $p_1 = p_2$ versus H_1: $p_1 \neq p_2$ will be to enumerate all possible tables with the same margins as the observed table and compute the exact probability for each such table based on the hypergeometric distribution. A method for accomplishing this task is given as follows:

EQUATION 10.10

Enumeration of All Possible Tables with the Same Margins as the Observed Table

(1) Rearrange the rows and columns of the observed table so that the smaller row total is in the first row and the smaller column total is in the first column.

Suppose that after the rearrangement, the cells in the observed table are a, b, c, d, as shown in Table 10.10.

(2) Start with the table with 0 in the (1, 1) cell. The other cells in this table are then determined from the row and column margins. Indeed, to maintain the same row and column margins as the observed table, the (1, 2) element must be $a + b$, the (2, 1) cell must be $a + c$, and the (2, 2) element must be $(c + d) - (a + c) = d - a$.

(3) Construct the next table by increasing the (1, 1) cell by 1 (i.e., from 0 to 1), decreasing the (1, 2) and (2, 1) cells by 1, and increasing the (2, 2) cell by 1.

(4) Continue increasing and decreasing the cells by 1, as in step 3, until one of the cells is 0, at which point all possible tables with the given row and column margins have been enumerated. Each table in the sequence of tables is referred to by its (1, 1) element. Thus, the first table is the 0 table, the next table is the 1 table, and so on.

Example 10.19 **Cardiovascular Disease, Nutrition** Enumerate all possible tables with the same row and column margins as the observed data in Table 10.9.

Solution The observed table has $a = 2$, $b = 23$, $c = 5$, $d = 30$. The rows or columns do not need to be rearranged, because the first row total is smaller than the second row total, and the first column total is smaller than the second column total. Start with the 0 table, which has 0 in the (1, 1) cell, 25 in the (1, 2) cell, 7 in the (2, 1) cell, and 30 – 2, or 28, in the (2, 2) cell. The 1 table then has 1 in the (1, 1) cell, 25 – 1 = 24 in the (1, 2) cell, 7 – 1 = 6 in the (2, 1) cell, and 28 + 1 = 29 in the (2, 2) cell. Continue in this fashion until the 7 table is reached, which has 0 in the (2, 1) cell, at which point all possible tables with the given row and column margins have been enumerated. The collection of tables and their associated probabilities based on the hypergeometric distribution in Equation 10.8 are given in Table 10.12.

The question now is what should be done with these probabilities to evaluate the significance of the results? The answer depends on whether a one-sided or a two-sided alternative is being used. In general, the following procedure can be used:

EQUATION 10.11

Fisher's Exact Test: General Procedure and Computation of *p*-Value To test the hypothesis H_0: $p_1 = p_2$ versus H_1: $p_1 \neq p_2$, where the expected value of at least one cell is <5 when the data are analyzed in the form of a 2 × 2 contingency table, use the following procedure:

TABLE 10.12 **Enumeration of all possible tables with fixed margins and their associated probabilities based on the hypergeometric distribution for Example 10.19**

0	25
7	28

.017

1	24
6	29

.105

2	23
5	30

.252

3	22
4	31

.312

4	21
3	32

.214

5	20
2	33

.082

6	19
1	34

.016

7	18
0	35

.001

(1) Enumerate all possible tables with the same row and column margins as the observed table, as shown in Equation 10.10.

(2) Compute the exact probability of each table enumerated in step 1, using either the computer or the method in Equation 10.7.

(3) Suppose that the observed table is the a table and that the last table enumerated is the k table.

(a) To test the hypothesis H_0: $p_1 = p_2$ versus H_1: $p_1 \neq p_2$, the p-value = $2 \times \min[Pr(0) + Pr(1) + \cdots + Pr(a), Pr(a) + Pr(a+1) + \cdots + Pr(k), .5]$.

(b) To test the hypothesis H_0: $p_1 = p_2$ versus H_1: $p_1 < p_2$, the p-value = $Pr(0) + Pr(1) + \cdots + Pr(a)$.

(c) To test the hypothesis H_0: $p_1 = p_2$ versus H_1: $p_1 > p_2$, the p-value = $Pr(a) + Pr(a+1) + \cdots + Pr(k)$.

For each of these three alternative hypotheses, the p-value can be interpreted as the probability of obtaining a table as extreme as or more extreme than the observed table.

Example 10.20 **Cardiovascular Disease, Nutrition** Evaluate the statistical significance of the data in Example 10.17.

Solution Suppose there is a two-sided alternative of the form H_0: $p_1 = p_2$ versus H_1: $p_1 \neq p_2$. Our table is the 2 table whose probability is .252 in Table 10.12. Thus, to compute the p-value, the smaller of the tail probabilities corresponding to the 2 table is computed and doubled. This strategy corresponds to the procedures for the various normal-theory tests studied in Chapters 7 and 8. First compute the left-hand tail area,

$$Pr(0) + Pr(1) + Pr(2) = .017 + .105 + .252 = .375$$

and the right-hand tail area,

$$Pr(2) + Pr(3) + \cdots + Pr(7) = .252 + .312 + .214 + .082 + .016 + .001 = .878$$

Then $p = 2 \times \min(.375, .878, .5) = 2(.375) = .749$

If a one-sided alternative of the form H_0: $p_1 = p_2$ versus H_1: $p_1 < p_2$ is used, then the p-value equals

$$Pr(0) + Pr(1) + Pr(2) = .017 + .105 + .252 = .375$$

Thus the two proportions in this example are *not* significantly different with either a one-sided or two-sided alternative, and we *cannot* say, on the basis of this limited amount of data, that there is a significant association between salt intake and cause of death.

In most instances, computer programs are used to implement Fisher's exact test using statistical packages such as SAS. There are other possible approaches to significance testing in the two-sided case. For example, the approach used by SAS is to compute

$$p\text{-value (two-tail)} = \sum_{\{i:\, Pr(i) \le Pr(a)\}} Pr(i)$$

In other words, the two-tail p-value using SAS is the sum of the probabilities of all tables whose probabilities are ≤ the probability of the observed table. Using this approach, the two-tail p-value would be

$$\begin{aligned} p\text{-value (two-tail)} &= Pr(0) + Pr(1) + Pr(2) + Pr(4) + Pr(5) + Pr(6) + Pr(7) \\ &= .017 + .105 + .252 + .214 + .082 + .016 + .001 = .688 \end{aligned}$$

In this section, we learned about Fisher's exact test, which is used for comparing binomial proportions from two independent samples in 2 × 2 tables with small expected counts (<5). This is the two-sample analogue to the exact one-sample binomial test given in Equation 7.36. If we refer to the flowchart at the end of this chapter (p. 412), we would answer yes to (1) Are samples independent? and no to (2) Are all expected values ≥5? This leads us to the box entitled "Use Fisher's exact test."

SECTION 10.4 Two-Sample Test for Binomial Proportions for Matched-Pair Data (McNemar's Test)

Example 10.21 **Cancer** Suppose we want to compare two different chemotherapy regimens for breast cancer after mastectomy. The two treatment groups should be as comparable as possible on other prognostic factors. To accomplish this goal, a matched study is set up such that a random member of each matched pair gets treatment A (chemotherapy) perioperatively (i.e., within 1 week after mastectomy) and for an additional 6 months, whereas the other member gets treatment B (chemotherapy only perioperatively). The patients are assigned to pairs matched on age (within 5 years) and clinical condition. The patients are followed for 5 years, with survival as the outcome variable. The data are displayed in a 2 × 2 table, as shown in Table 10.13. Notice the small difference in survival between the two treatment groups; the 5-year survival rate for treatment A = 526/621 = .847 and for treatment B = 515/621 = .829. Indeed, the Yates-corrected chi-square statistic as given in Equation 10.5 is 0.59 with 1 *df*, which is not significant. However, *the use of this test is valid only if the two samples are independent.* From the manner in which the samples were selected it is obvious that they are *not* independent, because members of each matched pair are similar in age and clinical condition. Thus, the Yates-corrected chi-square test *cannot* be used in this situation, since the p-value will not be correct. How then can the two treatments be compared using a hypothesis test?

TABLE 10.13 **A 2 × 2 contingency table comparing treatments A and B for breast cancer based on 1242 patients**

	Outcome		
Treatment	Survive for 5 years	Die within 5 years	Total
A	526	95	621
B	515	106	621
Total	1041	201	1242

Suppose a different kind of 2 × 2 table is constructed to display these data. In Table 10.13 the *person* was the unit of analysis, and the sample size was 1242 people. In Table 10.14 the *matched pair* is the unit of analysis and *pairs* are classified according to whether or not the members of that pair survived for 5 years. Notice that Table 10.14 has 621 units rather than the 1242 units in Table 10.13. Furthermore, there are 90 pairs in which both patients died within 5 years, 510 pairs in which both patients survived for 5 years, 16 pairs in which the treatment A patient survived and the treatment B patient died, and 5 pairs in which the treatment B patient survived and the treatment A patient died. The dependence of the two samples can be illustrated by noting that the probability that the treatment B member of the pair survived given that the treatment A member of the pair survived = 510/526 = .970, while the probability that the treatment B member of the pair survived given that the treatment A member of the pair died = 5/95 = .053. If the samples were independent, then these two probabilities should be about the same. Thus, we conclude that the samples are highly dependent and that the chi-square test cannot be used.

TABLE 10.14 **A 2 × 2 contingency table with the matched pair as the sampling unit based on 621 matched pairs**

	Outcome of treatment B patient		
Outcome of treatment A patient	Survive for 5 years	Die within 5 years	Total
Survive for 5 years	510	16	526
Die within 5 years	5	90	95
Total	515	106	621

In Table 10.14, for 600 pairs (90 + 510), the outcomes of the two treatments are the same, whereas for 21 pairs (16 + 5), the outcomes of the two treatments are different. The following special names are given to each of these types of pairs:

DEFINITION 10.2 A **concordant pair** is a matched pair in which the outcome is the same for each member of the pair.

DEFINITION 10.3 A **discordant pair** is a matched pair in which the outcomes are different for the members of the pair.

Example 10.22 There are 600 concordant pairs and 21 discordant pairs for the data in Table 10.14.

The concordant pairs provide no information about *differences between treatments* and will not be used in the assessment. Instead, we will focus on the discordant pairs, which can be divided into two types:

DEFINITION 10.4 A **type A discordant pair** is a discordant pair in which the treatment A member of the pair has the event and the treatment B member does not. Similarly, a **type B discordant pair** is a discordant pair in which the treatment B member of the pair has the event and the treatment A member does not.

Example 10.23 If we define having an event as dying within 5 years, then there are 5 type A discordant pairs and 16 type B discordant pairs from the data in Table 10.14.

Let p = the probability that a discordant pair is of type A. If the treatments are equally effective, then about an equal number of type A and type B discordant pairs would be expected, and p should $= \frac{1}{2}$. If treatment A is more effective than treatment B, then less type A than type B discordant pairs would be expected, and p should be $< \frac{1}{2}$. Finally, if treatment B is more effective than treatment A, then more type A than type B discordant pairs would be expected, and p should be $> \frac{1}{2}$.

Thus, we wish to test the hypothesis H_0: $p = \frac{1}{2}$ versus H_1: $p \neq \frac{1}{2}$.

10.4.1 Normal-Theory Test

Suppose that of n_D discordant pairs, n_A are type A. Then under H_0, $E(n_A) = n_D/2$ and $Var(n_A) = n_D/4$, from the mean and variance of a binomial distribution, respectively. We will assume that the normal approximation to the binomial distribution holds, but will use a continuity correction for a better approximation. This approximation will be valid if $npq = n_D/4 \geq 5$ or $n_D \geq 20$. The following test procedure, referred to as McNemar's test, can then be used:

EQUATION 10.12

McNemar's Test for Correlated Proportions—Normal-Theory Test

(1) Form a 2×2 table of matched pairs, where the outcomes for the treatment A members of the matched pairs are listed along the rows and the outcomes for the treatment B members are listed along the columns.

(2) Count the total number of discordant pairs (n_D) and the number of type A discordant pairs (n_A).

(3) Compute the test statistic

$$X^2 = \left(\left|n_A - \frac{n_D}{2}\right| - \frac{1}{2}\right)^2 \bigg/ \left(\frac{n_D}{4}\right)$$

An equivalent version of the test statistic is also given by

$$X^2 = \left(|n_A - n_B| - 1\right)^2 / (n_A + n_B)$$

(4) For a two-sided level α test, if

$$X^2 > \chi^2_{1,1-\alpha}$$

then reject H_0; if $X^2 \le \chi^2_{1,1-\alpha}$

then accept H_0.

(5) The exact p-value is given by p-value $= Pr(\chi^2_1 \ge X^2)$.

(6) Use this test only if $n_D \ge 20$.

The acceptance and rejection regions for this test are depicted in Figure 10.5. The computation of the p-value for McNemar's test is depicted in Figure 10.6.

This is a two-sided test despite the one-sided nature of the critical region in Figure 10.5. The rationale for this is that if either $p < \frac{1}{2}$ or $p > \frac{1}{2}$, $|n_A - n_D/2|$ will be

FIGURE 10.5 **Acceptance and rejection regions for McNemar's test—normal-theory method**

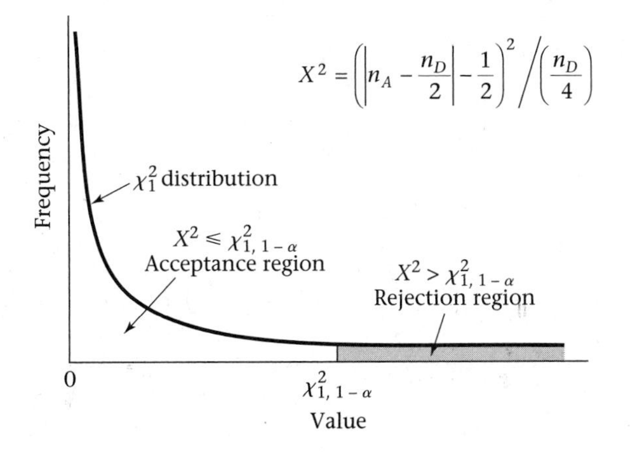

FIGURE 10.6 **Computation of the *p*-value for McNemar's test—normal-theory method**

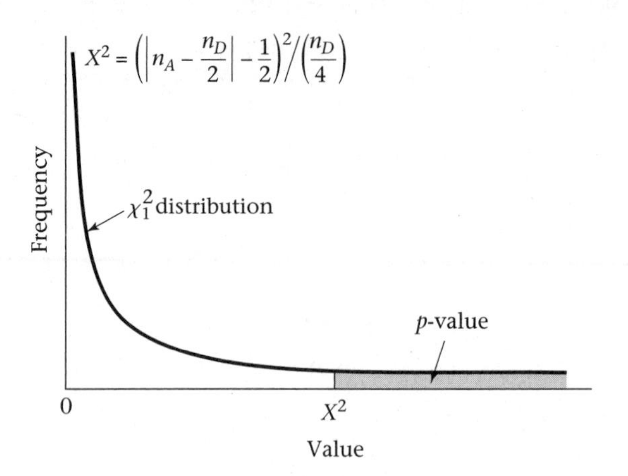

large and, correspondingly, X^2 will be large. Thus, for alternatives on either side of the null hypothesis $\left(p = \frac{1}{2}\right)$, H_0 is rejected if X^2 is large and accepted if X^2 is small.

Example 10.24 **Cancer** Assess the statistical significance of the data in Table 10.14.

Solution Note that $n_D = 21$. Since $n_D\left(\frac{1}{2}\right)\left(\frac{1}{2}\right) = 5.25 \ge 5$, the normal approximation to the binomial distribution is valid and the test in Equation 10.12 can be used. We have

$$X^2 = \frac{\left(|5 - 10.5| - \frac{1}{2}\right)^2}{21/4} = \frac{\left(5.5 - \frac{1}{2}\right)^2}{5.25} = \frac{5^2}{5.25} = \frac{25}{5.25} = 4.76$$

Equivalently, we could also compute the test statistic from

$$X^2 = \frac{(|5 - 16| - 1)^2}{5 + 16} = \frac{10^2}{21} = 4.76$$

From Table 6 in the Appendix, note that

$$\chi^2_{1,.95} = 3.84$$
$$\chi^2_{1,.975} = 5.02$$

Thus, because $3.84 < 4.76 < 5.02$, it follows that $.025 < p < .05$, and the results are statistically significant.

We conclude that *if the treatments give different results from each other* for the members of a matched pair, then the treatment A member of the pair is significantly more likely to survive for 5 years than the treatment B member. Thus, all other things being equal (such as toxicity, cost, etc.), treatment A would be the treatment of choice.

10.4.2 Exact Test

If $n_D/4 < 5$—that is, if $n_D < 20$—then the normal approximation to the binomial distribution cannot be used, and a test based on exact binomial probabilities is required. The details of the test procedure are similar to the one-sample binomial test in Equation 7.36 and are summarized as follows:

EQUATION 10.13

McNemar's Test for Correlated Proportions—Exact Test

(1) Follow the procedure in step 1 in Equation 10.12.

(2) Follow the procedure in step 2 in Equation 10.12.

(3) $$p = 2 \times \sum_{k=0}^{n_A} \binom{n_D}{k} \left(\frac{1}{2}\right)^{n_D} \quad \text{if } n_A < n_D/2$$

$$p = 2 \times \sum_{k=n_A}^{n_D} \binom{n_D}{k} \left(\frac{1}{2}\right)^{n_D} \quad \text{if } n_A > n_D/2$$

$$p = 1 \text{ if } n_A = n_D/2$$

(4) This test is valid for any number of discordant pairs (n_D) but is particularly useful for $n_D < 20$, when the normal-theory test in Equation 10.12 cannot be used.

The computation of the p-value for this test is depicted in Figure 10.7.

Example 10.25 **Hypertension** A recent phenomenon in the recording of blood pressure is the development of the automated blood-pressure machine, where for a small fee a person can sit in a booth and have his or her blood pressure measured by a computer device. A study is conducted to compare the computer device with standard methods of measuring blood pressure. Twenty patients are recruited, and their hypertensive status is assessed by both the computer device and a trained observer. Hypertensive status is defined as either hypertensive (+), if either systolic blood pressure ≥ 160 or diastolic blood pressure ≥ 95, or normotensive (–) otherwise. The data are given in Table 10.15. Assess the statistical significance of these findings.

Solution An ordinary Yates-corrected chi-square test cannot be used for these data, because each person is being used as his or her own control and there are *not* two independent samples. Instead, a 2×2 table of matched pairs is formed, as shown in Table 10.16. Note that 3 people are measured as hypertensive by both the computer device and the trained observer, 9 people are normotensive by both methods, 7 people are hypertensive by the computer device and normotensive by the trained observer, and 1 person is normotensive by the computer device and hypertensive by the trained observer. Therefore, there are 12 (9 + 3) concordant pairs and 8 (7 + 1) discordant pairs (n_D). Because $n_D < 20$, the exact method must be used. We see that $n_A = 7$, $n_D = 8$. Therefore, because $n_A > n_D/2 = 4$, it follows from Equation 10.13 that

$$p = 2 \times \sum_{k=7}^{8} \binom{8}{k} \left(\frac{1}{2}\right)^{8}$$

FIGURE 10.7 Computation of the p-value for McNemar's test—exact method

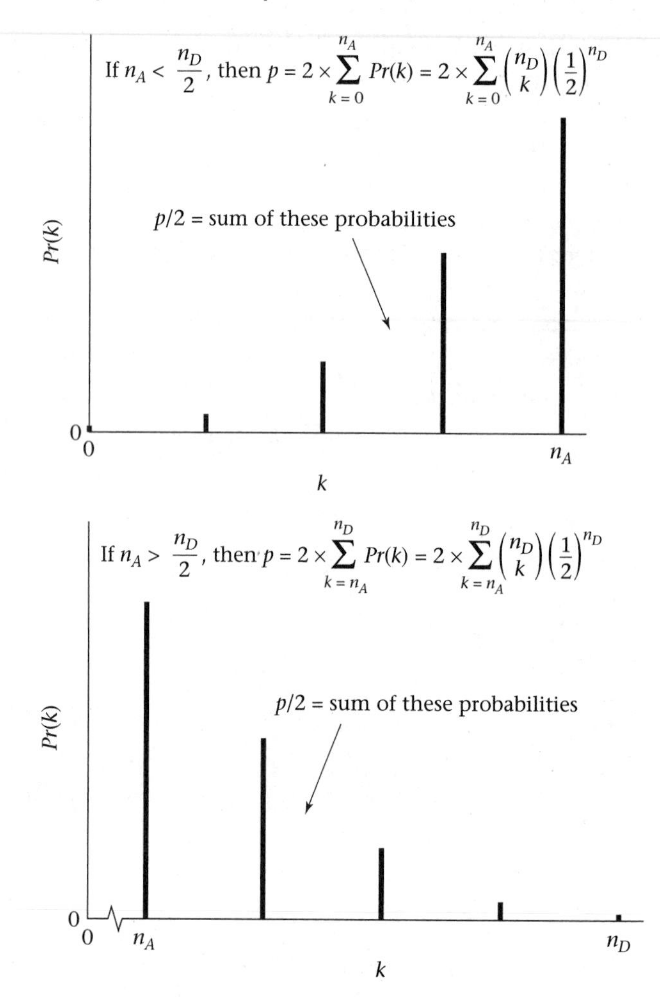

If $n_A < \frac{n_D}{2}$, then $p = 2 \times \sum_{k=0}^{n_A} Pr(k) = 2 \times \sum_{k=0}^{n_A} \binom{n_D}{k}\left(\frac{1}{2}\right)^{n_D}$
$p/2$ = sum of these probabilities
Pr(k)
0
0
n_A
k
If $n_A > \frac{n_D}{2}$, then $p = 2 \times \sum_{k=n_A}^{n_D} Pr(k) = 2 \times \sum_{k=n_A}^{n_D} \binom{n_D}{k}\left(\frac{1}{2}\right)^{n_D}$
$p/2$ = sum of these probabilities
Pr(k)
0
0
n_A
n_D
k

TABLE 10.15 Hypertensive status of 20 patients as judged by a computer device and a trained observer

	Hypertensive status			Hypertensive status	
Person	Computer device	Trained observer	Person	Computer device	Trained observer
1	–	–	11	+	–
2	–	–	12	+	–
3	+	–	13	–	–
4	+	+	14	+	–
5	–	–	15	–	+
6	+	–	16	+	–
7	–	–	17	+	–
8	+	+	18	–	–
9	+	+	19	–	–
10	–	–	20	–	–

TABLE 10.16 **Comparison of hypertensive status as judged by a computer device and a trained observer**

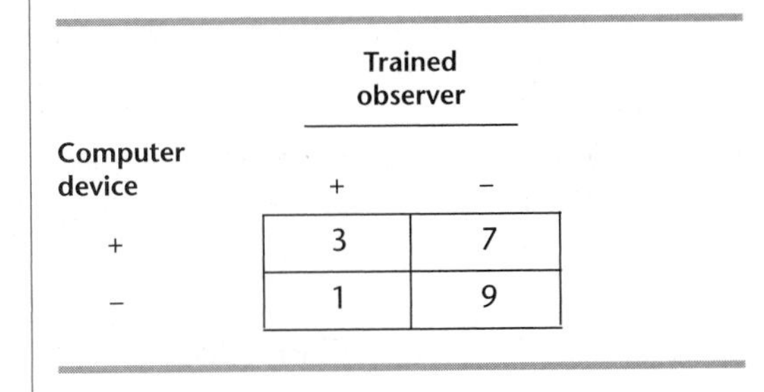

Computer device	Trained observer +	Trained observer −
+	3	7
−	1	9

This expression can be evaluated using Table 1 in the Appendix by referring to $n = 8$, $p = .5$ and noting that $Pr(X \geq 7|p = .5) = .0313 + .0039 = .0352$. Thus, the two-tailed p-value = $2 \times .0352 = .070$.

Alternatively, a computer program could be used to perform the computations, as shown in Table 10.17. Note that the first and second columns have been interchanged so that the discordant pairs appear in the diagonal elements (and are easier to identify). In summary, the results are not statistically significant, and we cannot conclude that there is a significant difference between the two methods, although a *trend* toward the computer device identifying more hypertensives than the trained observer can be detected.

TABLE 10.17 **Use of SPSS^X/PC McNemar's test program to evaluate the significance of the data in Table 10.16**

```
                         SPSSX/PC  Release 1.0
-----McNemar Test
     COMP       COMPUTER DEVICE
with OBS        TRAINED OBSERVER

                           OBS
                    2.00        1.00          Cases              20
                 +----------+----------+
           1.00  |    7     |    3     |
COMP             +----------+----------+    (Binomial)
           2.00  |    9     |    1     |    2-tailed P     0.0703
                 +----------+----------+
```

Note that for a two-sided one-sample binomial test, the hypothesis H_0: $p = p_0$ versus H_1: $p \neq p_0$ is tested. In the special case where $p_0 = 1/2$, the same test procedure as for McNemar's test is also followed.

Finally, if we interchange the designation of which of 2 outcomes is an event, then the p-values will be the same in Equations 10.12 and 10.13. For example, if we define an event as surviving for 5+ years, rather than dying within 5 years in Table 10.14, then $n_A = 16$, $n_B = 5$ (rather than $n_A = 5$, $n_B = 16$ in Example 10.23). However, the test statistic X^2 and the p-value are the same because $|n_A - n_B|$ remains the same in Equation 10.12. Similarly, the p-value remains the same in Equation 10.13 due to the symmetry of the binomial distribution when $p = 1/2$ (under H_0).

In this section, we have studied McNemar's test for correlated proportions, which is used to compare two binomial proportions from matched samples. We studied both a large-sample test when the normal approximation to the binomial distribution is valid (i.e., when the number of discordant pairs, n_D, ≥ 20) and a small-sample test when $n_D < 20$. Referring to the flowchart at the end of this chapter (p. 412), we would answer no to (1) are samples independent? which would lead us to the box entitled "Use McNemar's test."

SECTION 10.5 Estimation of Sample Size and Power for Comparing Two Binomial Proportions

In Section 8.10, methods for estimating the sample size needed to compare means from two normally distributed populations were presented. In this section, similar methods for estimating the sample size required to compare two proportions are developed.

10.5.1 Independent Samples

Example 10.26 **Cancer, Nutrition** Suppose we know from Connecticut tumor-registry data that the incidence rate of breast cancer over a 1-year period for initially disease-free women ages 45–49 is 150 cases per 100,000 [2]. We wish to study whether or not the ingestion of large doses of vitamin A in tablet form will prevent breast cancer. The study is set up with (1) a control group of 45- to 49-year-old women who are given placebo pills by mail and are anticipated to have the same disease rate as indicated in the Connecticut tumor-registry data and (2) a study group of similarly aged women who are given vitamin A pills by mail and are anticipated to have a 20% reduction in risk. How large a sample is needed if a two-sided test with a significance level of .05 is used and a power of 80% is desired?

We wish to test the hypothesis H_0: $p_1 = p_2$ versus H_1: $p_1 \neq p_2$. Suppose that we wish to conduct a test with significance level α and power $1 - \beta$ and anticipate that there will be k times as many people in group 2 as in group 1; that is, $n_2 = kn_1$. The sample size required in each of the two groups to achieve these objectives is given as follows:

EQUATION 10.14

Sample Size Needed to Compare Two Binomial Proportions Using a Two-Sided Test with Significance Level α and Power $1 - \beta$, Where One Sample (n_2) Is k Times as Large as the Other Sample (n_1) (Independent-Sample Case) To test the hypothesis H_0: $p_1 = p_2$ versus H_1: $p_1 \neq p_2$ for the specific alternative $|p_1 - p_2| = \Delta$, with a significance level α and power $1 - \beta$, the following sample size is required

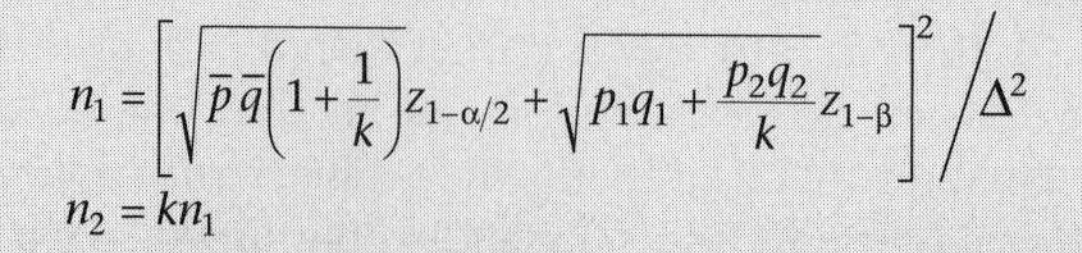

$$n_1 = \left[\sqrt{\bar{p}\bar{q}\left(1+\frac{1}{k}\right)}z_{1-\alpha/2} + \sqrt{p_1 q_1 + \frac{p_2 q_2}{k}}z_{1-\beta}\right]^2 \Big/ \Delta^2$$

$$n_2 = kn_1$$

where p_1, p_2 = projected true probabilities of success in the two groups

$$q_1, q_2 = 1 - p_1, 1 - p_2$$
$$\Delta = |p_2 - p_1|$$
$$\bar{p} = \frac{p_1 + kp_2}{1+k}$$
$$\bar{q} = 1 - \bar{p}$$

Example 10.27 **Cancer, Nutrition** Estimate the sample size required for the study proposed in Example 10.26 if an equal sample size is anticipated in each group.

Solution

$$p_1 = 150 \text{ per } 100{,}000 \text{ or } 150/10^5 = .00150$$
$$q_1 = 1 - .00150 = .99850$$

If we wish to detect a 20% reduction in risk, then $p_2 = 0.8p_1$ or

$$p_2 = (150 \times .8)/10^5 = 120/10^5 = .00120$$
$$q_2 = 1 - .00120 = .99880$$
$$\alpha = .05$$
$$1 - \beta = .8$$
$$k = 1 \; (\text{since } n_1 = n_2)$$
$$\bar{p} = \frac{.00150 + .00120}{2} = .00135$$
$$\bar{q} = 1 - .00135 = .99865$$
$$z_{1-\alpha/2} = z_{.975} = 1.96$$
$$z_{1-\beta} = z_{.80} = 0.84$$

Thus, referring to Equation 10.14,

$$n_1 = \frac{\left[\sqrt{.00135(.99865)(1+1)}\,(1.96) + \sqrt{.00150(.99850) + .00120(.99880)}\,(0.84)\right]^2}{(.00150 - .00120)^2}$$

$$= \frac{[.05193(1.96) + .05193(0.84)]^2}{.00030^2} = \frac{.14539^2}{.00030^2} = 234{,}881 = n_2$$

or about 235,000 women in each group.

To perform a one-tailed rather than a two-tailed test, simply substitute α for $\alpha/2$ in the sample-size formula in Equation 10.14.

Clearly, from the results in Example 10.27, it would not be feasible to conduct such a large study over a 1-year period. The sample size needed would be reduced considerably if the period of study was lengthened beyond 1 year, since the expected number of events would increase in a multiyear study.

In many instances, the sample size available for investigation is fixed by practical constraints, and what is desired is an estimate of statistical power with the anticipated available sample size. In other instances, after a study is completed, we want to calculate the power using the sample sizes that were actually used in the study. For these purposes the following estimate of power is provided to test the hypothesis H_0: $p_1 = p_2$ versus H_1: $p_1 \neq p_2$, with significance level α and sample sizes of n_1 and n_2 in the two groups.

EQUATION 10.15

Power Achieved in Comparing Two Binomial Proportions Using a Two-Sided Test with Significance Level α and Samples of Size n_1 and n_2 (Independent-Sample Case) To test the hypothesis H_0: $p_1 = p_2$ versus H_1: $p_1 \neq p_2$ for the specific alternative $|p_1 - p_2| = \Delta$, compute

$$\text{Power} = \Phi\left[\frac{\Delta}{\sqrt{p_1q_1/n_1 + p_2q_2/n_2}} - z_{1-\alpha/2}\frac{\sqrt{\bar{p}\bar{q}(1/n_1 + 1/n_2)}}{\sqrt{p_1q_1/n_1 + p_2q_2/n_2}}\right]$$

where

$p_1\, p_2$ = projected true probabilities of success in groups 1 and 2, respectively

$$q_1, q_2 = 1 - p_1, 1 - p_2$$

$$\Delta = |p_2 - p_1|$$

$$\bar{p} = \frac{n_1p_1 + n_2p_2}{n_1 + n_2}$$

$$\bar{q} = 1 - \bar{p}$$

Example 10.28 **Otolaryngology** Suppose a study comparing a medical and a surgical treatment for children who have an excessive number of episodes of otitis media (OTM) during the first 3 years of life is planned. Success rates of 50% and 70% are assumed in the medical and surgical groups, respectively, and the recruitment of 100 patients for each group is realistically anticipated. Success is define as ≤1 episode of OTM in the first 12 months after treatment. How much power does such a study have of detecting a significant difference if a two-sided test with an α level of .05 is to be used?

Solution Note that $p_1 = .5$, $p_2 = .7$, $q_1 = .5$, $q_2 = .3$, $n_1 = n_2 = 100$, $\Delta = .2$, $\bar{p} = (.5 + .7)/2 = .6$, $\bar{q} = .4$, $\alpha = .05$, $z_{1-\alpha/2} = z_{.975} = 1.96$. Thus, from Equation 10.15 the power can be computed as follows:

$$\text{Power} = \Phi\left[\frac{.2}{\sqrt{[.5(.5)+.7(.3)]/100}} - \frac{1.96\sqrt{.6(.4)(1/100+1/100)}}{\sqrt{[.5(.5)+.7(.3)]/100}}\right]$$

$$= \Phi\left[\frac{.2}{.0678} - 1.96\frac{(.0693)}{.0678}\right] = \Phi(2.949 - 2.002) = \Phi(0.947) = .83$$

Thus, there is an 83% chance of finding a significant difference using the anticipated sample sizes.

If a one-sided test is used, then Equation 10.15 can be used after replacing $z_{1-\alpha/2}$ by $z_{1-\alpha}$.

10.5.2 Paired Samples

In Section 10.4, McNemar's test for comparing binomial proportions in paired samples was introduced. As noted there, this test is a special case of the one-sample binomial test. Therefore, to estimate sample size and power, the more general formulas for the one-sample binomial test given in Section 7.9.3 can be used. Specifically, referring to Equation 7.38 to test the hypothesis H_0: $p = p_0$ versus H_1: $p \neq p_0$ using a two-sided test with significance level α and power $1 - \beta$ for the specific alternative $p = p_1$, a sample size of

$$n = \frac{p_0 q_0 \left[z_{1-\alpha/2} + z_{1-\beta} \sqrt{p_1 q_1 / (p_0 q_0)} \right]^2}{(p_1 - p_0)^2}$$

is needed. To use this formula in the case of McNemar's test, set $p_0 = q_0 = \frac{1}{2}$, $p_1 = p_A$ = the proportion of discordant pairs that are of type A, and $n = n_D$ = the number of discordant pairs. Upon substitution,

$$n_D = \frac{\left(z_{1-\alpha/2} + 2 z_{1-\beta} \sqrt{p_A q_A} \right)^2}{4(p_A - .5)^2}$$

However, the number of discordant pairs (n_D) = the total number of pairs (n) × the probability that a matched pair is discordant. If the latter probability is denoted by p_D, then $n_D = np_D$, or $n = n_D/p_D$. Thus, the following sample-size formula can be used:

EQUATION 10.16

Sample Size Needed to Compare Two Binomial Proportions Using a Two-Sided Test with Significance Level α and Power $1 - \beta$ (Paired-Sample Case) If McNemar's test for correlated proportions is used to test the hypothesis H_0: $p = \frac{1}{2}$ versus H_1: $p \neq \frac{1}{2}$, for the specific alternative $p = p_A$, where p = the probability that a discordant pair is of type A, then use

$$n = \frac{\left(z_{1-\alpha/2} + 2 z_{1-\beta} \sqrt{p_A q_A} \right)^2}{4(p_A - .5)^2 p_D} \text{ matched pairs}$$

$$\text{or} \quad 2n = \frac{\left(z_{1-\alpha/2} + 2 z_{1-\beta} \sqrt{p_A q_A} \right)^2}{2(p_A - .5)^2 p_D} \text{ individuals}$$

where p_D = projected proportion of discordant pairs among all pairs

p_A = projected proportion of discordant pairs of type A among discordant pairs

Example 10.29 **Cancer** Suppose we wish to compare two different regimens of chemotherapy (A, B) for the treatment of breast cancer where the outcome measure is recurrence of breast cancer over a 5-year period. A matched-pair design is used, where patients are matched on age and clinical stage of disease, and one patient in a matched pair is assigned to treatment A, while the other is assigned to treatment B. Based on previous work, it is estimated that patients in a matched pair will respond similarly to the treatments in 85% of matched pairs (i.e., both will either have a recurrence or both will not have a recurrence over 5 years). Furthermore, for the matched pairs where there is a difference in response, it is estimated that in $\frac{2}{3}$ of the pairs the treatment A patient will have a recurrence, and the treatment B patient will not; in $\frac{1}{3}$ of the pairs the treatment B patient will have a recurrence, and the treatment A patient will not. How many subjects (or matched pairs) need to be enrolled in the study to have a 90% chance of finding a significant difference using a two-sided test with type I error = .05?

Solution We have that $\alpha = .05$, $\beta = .10$, $p_D = 1 - .85 = .15$, $p_A = \frac{2}{3}$, $q_A = \frac{1}{3}$. Therefore, from Equation 10.16,

$$n \text{ (pairs)} = \frac{\left[z_{.975} + 2z_{.90}\sqrt{(2/3)(1/3)}\right]^2}{4(2/3 - 1/2)^2(.15)}$$

$$= \frac{[1.96 + 2(1.28)(.4714)]^2}{4(1/6)^2(.15)} = \frac{3.1668^2}{.0167} = 602 \text{ matched pairs}$$

$$2n = 2 \times 602 = 1204 \text{ individuals}$$

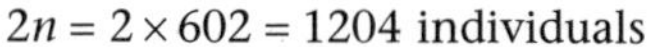

Therefore, 1204 individuals in 602 matched pairs need to be enrolled. This will yield approximately $.15 \times 602 = 90$ discordant pairs.

In some instances, the sample size is fixed and we want to determine what power a study has (or had) to detect specific alternatives. For a two-sided one-sample binomial test with significance level α, to test the hypothesis H_0: $p = p_0$ versus H_1: $p \neq p_0$ for the specific alternative $p = p_1$, the power is given by (see Equation 7.37)

$$\text{Power} = \Phi\left[\sqrt{p_0q_0/(p_1q_1)}\left(z_{\alpha/2} + \frac{|p_1 - p_0|\sqrt{n}}{\sqrt{p_0q_0}}\right)\right]$$

For McNemar's test, set $p_0 = q_0 = \frac{1}{2}$, $p_1 = p_A$, and $n = n_D$, yielding

$$\text{Power} = \Phi\left[\frac{1}{2\sqrt{p_Aq_A}}\left(z_{\alpha/2} + 2|p_A - .5|\sqrt{n_D}\right)\right]$$

On substituting $n_D = np_D$, the following power formula is obtained:

EQUATION 10.17

Power Achieved in Comparing Two Binomial Proportions Using a Two-Sided Test with Significance Level α (Paired-Sample Case) If McNemar's test for correlated proportions is used to test the hypothesis H_0: $p = \frac{1}{2}$ versus H_1: $p \neq \frac{1}{2}$, for the specific alternative $p = p_A$, where p = the probability that a discordant pair is of type A,

$$\text{Power} = \Phi\left[\frac{1}{2\sqrt{p_A q_A}}\left(z_{\alpha/2} + 2|p_A - .5|\sqrt{np_D}\right)\right]$$

where

n = number of matched pairs

p_D = projected proportion of discordant pairs among all pairs

p_A = projected proportion of discordant pairs of type A among discordant pairs

Example 10.30 **Cancer** Consider the study in Example 10.29. If 400 matched pairs are enrolled, how much power would such a study have?

Solution We have that $\alpha = .05$, $p_D = .15$, $p_A = \frac{2}{3}$, $n = 400$. Therefore, from Equation 10.17,

$$\text{Power} = \Phi\left\{\frac{1}{2\sqrt{(2/3)(1/3)}}\left[z_{.025} + 2|2/3 - .5|\sqrt{400(.15)}\right]\right\}$$

$$= \Phi\{1.0607[-1.96 + 2(1/6)(7.7460)]\}$$

$$= \Phi[1.0607(0.6220)] = \Phi(0.660) = .745$$

Therefore, the study would have 74.5% power, or a 74.5% chance of detecting a statistically significant difference.

To compute sample size and power for a one-sided alternative, substitute α for $\alpha/2$ in the formulas in Equations 10.16 and 10.17.

Note that a crucial element in calculating sample size and power for matched-pair studies based on binomial proportions using Equations 10.16 and 10.17 is knowledge of the probability of discordance between outcome for members of a matched pair (p_D). This probability will depend on the strictness of the matching criteria and on how strongly related the matching criteria are to the outcome variable.

10.5.3 Sample Size and Power in a Clinical Trial Setting

In Examples 10.27 and 10.28, we have estimated sample size and power in proposed clinical trials under the assumption that compliance with treatment regimens is perfect. To be more realistic, we should examine how these estimates will change if compliance is not perfect.

Suppose we are planning a clinical trial comparing an active treatment versus placebo. There are potentially two types of noncompliance to consider.

DEFINITION 10.5 The **dropout rate** is defined as the proportion of subjects in the active treatment group who don't comply; that is, who fail to actually receive the active treatment.

DEFINITION 10.6 The **drop-in rate** is defined as the proportion of subjects in the placebo group who actually receive the active treatment outside of the protocol of the study.

Example 10.31 **Cardiovascular Disease** The Physicians' Health Study was a randomized clinical trial, one of whose goals was to assess the effect of aspirin in preventing myocardial infarction (MI). Subjects were 22,000 male physicians ages 40–84 and free of cardiovascular disease in 1982. The physicians were randomized to either active aspirin (1 white pill containing 325 mg of aspirin to be taken every other day) or aspirin placebo (1 white aspirin placebo pill to be taken every other day). As the study progressed, it was estimated from self-report that 10% of the subjects in the aspirin group were not complying (i.e., were not taking their study [aspirin] capsules). Thus the dropout rate was 10%. Also, it was estimated from self-report that 5% of the subjects in the placebo group were taking aspirin regularly on their own outside of the study protocol. Thus, the drop-in rate was 5%. The issue is, How does this lack of compliance affect the sample size and power estimates for the study?

Let λ_1 = dropout rate, λ_2 = drop-in rate, p_1 = incidence of MI over a 5-year period among physicians who actually take aspirin, and p_2 = incidence of MI over a 5-year period among physicians who don't actually take aspirin under the assumption of perfect compliance. Finally, let p_1^*, p_2^* = observed risk of MI over a 5-year period in the aspirin and placebo groups, respectively (i.e., assuming that compliance is not perfect). We can estimate p_1^*, p_2^* using the total-probability rule. Specifically,

EQUATION 10.18

$$\begin{aligned} p_1^* &= Pr(\text{MI}|\text{assigned to aspirin group}) \\ &= Pr(\text{MI}|\text{aspirin-group complier}) \times Pr(\text{compliance in the aspirin group}) \\ &\quad + Pr(\text{MI}|\text{aspirin-group noncomplier}) \times Pr(\text{noncompliance in the aspirin group}) \\ &= p_1(1-\lambda_1) + p_2\lambda_1 \end{aligned}$$

where we have assumed that the observed risk for a noncompliant subject in the aspirin group = p_2. Similarly,

EQUATION 10.19

$$\begin{aligned} p_2^* &= Pr(\text{MI}|\text{assigned to placebo group}) \\ &= Pr(\text{MI}|\text{placebo-group complier}) \times Pr(\text{compliance in the placebo group}) \\ &\quad + Pr(\text{MI}|\text{placebo-group noncomplier}) \times Pr(\text{noncompliance in the placebo group}) \\ &= p_2(1-\lambda_2) + p_1\lambda_2 \end{aligned}$$

where we have assumed that noncompliance in the placebo group means that the subject takes aspirin on his or her own and that such a subject has risk = p_1 = risk for aspirin-group compliers. Placebo-group subjects who don't take their study capsules and refrain from taking aspirin outside the study are considered compliers from the point of view of the preceding discussion; that is, their risk is the same as that for placebo-group compliers = p_2.

If we subtract Equation 10.19 from Equation 10.18, we obtain

EQUATION 10.20

$$p_1^* - p_2^* = p_1(1-\lambda_1-\lambda_2) - p_2(1-\lambda_1-\lambda_2) = (p_1-p_2)(1-\lambda_1-\lambda_2)$$
$$= \text{compliance-adjusted risk difference}$$

In the presence of noncompliance, sample size and power estimates should be based on the compliance-adjusted risks (p_1^*, p_2^*) rather than on the perfect compliance risks (p_1, p_2). These results are summarized as follows:

EQUATION 10.21

Sample-Size Estimation to Compare Two Binomial Proportions in a Clinical Trial Setting (Independent-Sample Case) Suppose we wish to test the hypothesis H_0: $p_1 = p_2$ versus H_1: $p_1 \neq p_2$ for the specific alternative $|p_1 - p_2| = \Delta$ with a significance level α and a power $1-\beta$ in a randomized clinical trial, where group 1 receives active treatment and group 2 receives placebo and an equal number of subjects are allocated to each group. We assume that p_1, p_2 are the risks of disease in treatment groups 1 and 2 under the assumption of perfect compliance. We also assume that

λ_1 = dropout rate = proportion of subjects in the active-treatment group who fail to comply

λ_2 = drop-in rate = proportion of subjects in the placebo group who receive the active treatment outside of the study protocol

(1) The appropriate sample size in each group is

$$n_1 = n_2 = \left(\sqrt{2\bar{p}^*\bar{q}^*}\, z_{1-\alpha/2} + \sqrt{p_1^*q_1^* + p_2^*q_2^*}\, z_{1-\beta}\right)^2 \Big/ \Delta^{*2}$$

where

$$p_1^* = (1-\lambda_1)p_1 + \lambda_1 p_2$$
$$p_2^* = (1-\lambda_2)p_2 + \lambda_2 p_1$$
$$\bar{p}^* = (p_1^* + p_2^*)/2,\ \bar{q}^* = 1-\bar{p}^*,\ \Delta^* = |p_1^* - p_2^*| = (1-\lambda_1-\lambda_2)|p_1-p_2|$$
$$= (1-\lambda_1-\lambda_2)\Delta$$

(2) If noncompliance rates are low (λ_1, λ_2 each $\le .10$), then an approximate sample-size estimate is given by

$$n_{1,\text{approx}} = n_{2,\text{approx}} = \frac{\left(\sqrt{2\bar{p}\bar{q}}\, z_{1-\alpha/2} + \sqrt{p_1q_1 + p_2q_2}\, z_{1-\beta}\right)^2}{\Delta^2} \times \frac{1}{(1-\lambda_1-\lambda_2)^2}$$
$$= n_{\text{perfect compliance}} \Big/ (1-\lambda_1-\lambda_2)^2$$

where $n_{\text{perfect compliance}}$ is the sample size in each group under the assumption of perfect compliance, as computed in Equation 10.14 with $p_1^* = p_1$, $p_2^* = p_2$ and $k = n_2/n_1 = 1$.

Example 10.32 **Cardiovascular Disease** Refer to Example 10.31. Suppose we assume that the incidence of MI is .005 per year among subjects who actually take placebo and that aspirin prevents 20% of MI's (i.e., relative risk = $p_1/p_2 = 0.8$). We also assume that the duration of the study is 5 years and that the dropout rate in the aspirin group = 10% while the drop-in rate in the placebo group = 5%. How many subjects need to be enrolled in each group to achieve 80% power using a two-sided test with significance level = .05?

Solution This is a 5-year study, so the 5-year incidence of MI among subjects who actually take placebo = 5(.005) = .025 = p_2. Since the risk ratio = 0.8, we have $p_1/p_2 = 0.8$ or $p_1 = .020$ = 5-year incidence of MI among subjects who actually take aspirin. To estimate the true incidence rates to be expected in the study, we must factor in the expected rates of noncompliance. Based on Equation 10.21, the compliance-adjusted rates p_1^* and p_2^* are given by

$$\begin{aligned} p_1^* &= (1-\lambda_1)p_1 + \lambda_1 p_2 \\ &= .9(.020) + .1(.025) = .0205 \\ p_2^* &= (1-\lambda_2)p_2 + \lambda_2 p_1 \\ &= .95(.025) + .05(.020) = .02475 \end{aligned}$$

Also, $\Delta^* = |p_1^* - p_2^*| = .00425$

$$\bar{p}^* = \frac{p_1^* + p_2^*}{2} = \frac{.0205 + .02475}{2} = .02263, \quad \bar{q}^* = 1 - \bar{p}^* = .97737$$

Finally, $z_{1-\beta} = z_{.80} = 0.84$, $z_{1-\alpha/2} = z_{.975} = 1.96$. Therefore, from Equation 10.21, the required sample size in each group is

$$\begin{aligned} n_1 = n_2 &= \frac{\left[\sqrt{2(.02263)(.97737)}(1.96) + \sqrt{.0205(.9795) + .02475(.97525)}(0.84)\right]^2}{.00425^2} \\ &= \left[\frac{.2103(1.96) + .2103(0.84)}{.00425}\right]^2 = 19{,}196 \text{ per group} \end{aligned}$$

The total sample size needed = 38,392.

If we don't factor compliance into our sample-size estimates, then based on Equation 10.14, we would need

$$\begin{aligned} n_1 = n_2 &= \frac{\left(\sqrt{2\bar{p}\bar{q}}\,z_{1-\alpha/2} + \sqrt{p_1q_1 + p_2q_2}\,z_{1-\beta}\right)^2}{\Delta^2} \\ &= \frac{\left[\sqrt{2(.0225)(.9775)}\,1.96 + \sqrt{.02(.98) + .025(.975)}\,0.84\right]^2}{|.02 - .025|^2} \\ &= 13{,}794 \text{ per group} \end{aligned}$$

or a total sample size = 2(13,794) = 27,588.

The approximate sample-size formula in step 2 of Equation 10.21 would yield

$$n_{1,\text{approx}} = n_{2,\text{approx}} = \frac{n_{\text{perfect compliance}}}{(1-.10-.05)^2}$$
$$= \frac{13,794}{.85^2} = 19,093$$

or a total sample size of 2(19,093) = 38,186 subjects.

Thus, the effect of noncompliance is to narrow the observed difference in risk between the aspirin and placebo groups and as a result to increase the required sample size by approximately 100% × $(1/.85^2 - 1)$ = 38% or more exactly 100% × (38,392 – 27,588)/27,588 = 39%.

The Physicians' Health Study actually enrolled 22,000 subjects, thus implying that the power of the study with 5 years of follow-up would be somewhat lower than 80%. In addition, the physicians were much healthier than expected and the risk of MI in the placebo group was much lower than expected. However, aspirin proved to be much more effective than anticipated, preventing 40% of MI's (*RR* = 0.6) rather than the 20% that had been anticipated. This led to a highly significant treatment benefit for aspirin after 5 years of follow-up and an eventual change in the FDA-approved indications for aspirin to include labeling as a preventive agent for cardiovascular disease for men over the age of 50. A similar study is currently underway among female nurses to investigate whether the treatment benefits are generalizable to women.

The power formula for the comparison of binomial proportions in Equation 10.15 also assumes perfect compliance. To correct these estimates for noncompliance in a clinical trial setting, replace p_1, p_2, Δ, $\bar{p}$, and $\bar{q}$ in Equation 10.15 with p_1^*, p_2^*, Δ^*, $\bar{p}^*$, $\bar{q}^*$ as given in Equation 10.21. The resulting power is a compliance-adjusted power estimate.

SECTION 10.6 $R \times C$ Contingency Tables

10.6.1 Tests for Association for $R \times C$ Contingency Tables

In the previous sections of this chapter, methods of analyzing data that can be organized in the form of a 2 × 2 contingency table—that is, where each of the variables under study has only two categories—were studied. Frequently, one or both variables under study have more than two categories.

DEFINITION 10.7 An **$R \times C$ contingency table** is a table with R rows and C columns. It displays the relationship between two variables, where the variable depicted in the rows has R categories and the variable depicted in the columns has C categories.

Example 10.33 **Cancer** Suppose we wish to study further the relationship between age at first birth and the development of breast cancer, as given in Example 10.4. In particular, we would like to know if the effect of age at first birth follows a consistent trend, that is, (1) more protection for women whose age at first birth is <20 than for women whose age at first birth is 25–29 and (2) higher risk for women whose age at first birth is ≥35 than for women whose age at first birth is 30–34. The data are presented in Table 10.18, where case–control status is indicated along the rows and age at first birth is indicated along the columns. The data are arranged in the form of a 2×5 contingency table, because case–control status has two categories and age at first birth has five categories. We wish to test for a relationship between age at first birth and case–control status. How should this be done?

TABLE 10.18 **Data from the international study in Example 10.4 investigating the possible association between age at first birth and case–control status**

Case–control status	Age at first birth <20	20–24	25–29	30–34	≥35	Total
Case	320	1206	1011	463	220	3220
Control	1422	4432	2893	1092	406	10,245
Total	1742	5638	3904	1555	626	13,465
% cases	.184	.214	.259	.298	.351	.239

Source: Reprinted with permission of *WHO Bulletin, 43,* 209–221, 1970.

Generalizing our experience from the 2×2 situation, the expected table for an $R \times C$ table can be formed in the same way as for a 2×2 table.

EQUATION 10.22

Computation of the Expected Table for an $R \times C$ Contingency Table The expected number of units in the (i, j) cell $= E_{ij} =$ the product of the number of units in the ith row multiplied by the number of units in the jth column, divided by the total number of units in the table.

Example 10.34 **Cancer** Compute the expected table for the data in Table 10.18.

Solution

$$\text{Expected value of the }(1,1)\text{ cell} = \frac{\text{first row total} \times \text{first column total}}{\text{grand total}} = \frac{3220(1742)}{13,465} = 416.6$$

$$\text{Expected value of the }(1,2)\text{ cell} = \frac{\text{first row total} \times \text{second column total}}{\text{grand total}} = \frac{3220(5638)}{13,465} = 1348.3$$

$$\vdots$$

$$\text{Expected value of the }(2,5)\text{ cell} = \frac{\text{second row total} \times \text{fifth column total}}{\text{grand total}} = \frac{10,245(626)}{13,465} = 476.3$$

All ten expected values are given in Table 10.19.

The sum of the expected values across any row or column must equal the corresponding row or column total, as was the case for 2×2 tables. This fact provides a good check that the expected values are computed correctly. The expected values in Table 10.19 fulfill this criterion except for roundoff error.

TABLE 10.19 **Expected table for the international study data in Table 10.18**

Case–control status	Age at first birth					
	<20	20–24	25–29	30–34	≥35	Total
Case	416.6	1348.3	933.6	371.9	149.7	3220
Control	1325.4	4289.7	2970.4	1183.1	476.3	10,245
Total	1742	5638	3904	1555	626	13,465

We again want to compare the observed table with the expected table. The more similar these tables are, the more willing we will be to accept the null hypothesis that there is no relationship between the two variables. The more different the tables are, the more willing we will be to reject H_0. Again the criterion $(O - E)^2/E$ is used to compare the observed and expected counts for a particular cell. Furthermore, $(O - E)^2/E$ is summed over all the cells in the table to get an overall measure of agreement for the observed and expected tables. Under H_0, for an $R \times C$ contingency table, the sum of $(O - E)^2/E$ over the RC cells in the table will approximately follow a chi-square distribution with $(R - 1) \times (C - 1)$ *df*. H_0 will be rejected for large values of this sum and will be accepted for small values.

Generally speaking, the continuity correction is not used for contingency tables larger than 2×2, because it has been found empirically that the correction does not aid in the approximation of the test statistic by the chi-square distribution. As was the case for 2×2 tables, this test should not be used if the expected values of the cells are too small. Cochran has studied the validity of the approximation in this case and recommends its use under the following conditions [3]:

(1) No more than $\frac{1}{5}$ of the cells have expected values <5.

(2) No cell has expected value <1.

The test procedure can be summarized as follows:

EQUATION 10.23

Chi-Square Test for an $R \times C$ Contingency Table To test for the relationship between two discrete variables, where one variable has R categories and the other has C categories, use the following procedure:

(1) Analyze the data in the form of an $R \times C$ contingency table, where O_{ij} represents the observed number of units in the (i, j) cell.

(2) Compute the expected table as shown in Equation 10.22, where E_{ij} represents the expected number of units in the (i, j) cell.

(3) Compute the test statistic

$$X^2 = \left(O_{11} - E_{11}\right)^2 / E_{11} + \left(O_{12} - E_{12}\right)^2 / E_{12} + \cdots + \left(O_{RC} - E_{RC}\right)^2 / E_{RC}$$

which under H_0 approximately follows a chi-square distribution with $(R - 1) \times (C - 1)$ *df*.

(4) For a level α test, if

$$X^2 > \chi^2_{(R-1)\times(C-1),1-\alpha}$$

reject H_0. If $X^2 \le \chi^2_{(R-1)\times(C-1),1-\alpha}$ accept H_0.

(5) The approximate p-value is given by the area to the right of X^2 under a $\chi^2_{(R-1)\times(C-1)}$ distribution.

(6) Use this test only if the following two conditions are satisfied:

(a) No more than $\frac{1}{5}$ of the cells should have expected values < 5.

(b) No cell should have expected value < 1.

The acceptance and rejection regions for this test are depicted in Figure 10.8. The computation of the p-value for this test is illustrated in Figure 10.9.

FIGURE 10.8 **Acceptance and rejection regions for the chi-square test for an $R \times C$ contingency table**

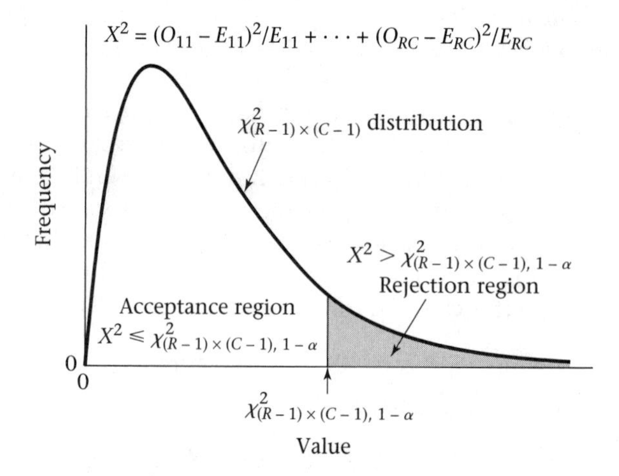

FIGURE 10.9 **Computation of the p-value for the chi-square test for an $R \times C$ contingency table**

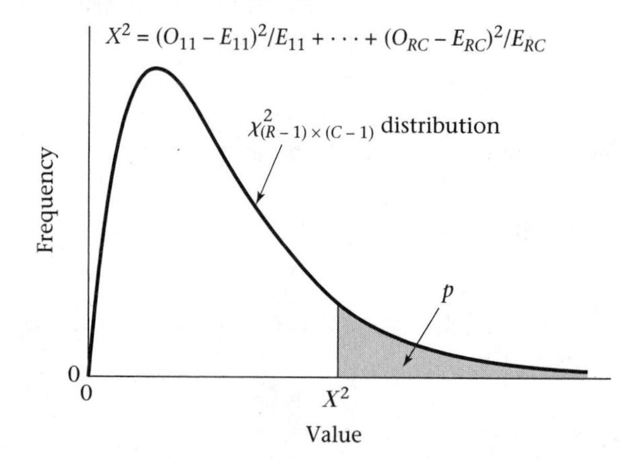

Example 10.35 **Cancer** Assess the statistical significance of the data in Example 10.33.

Solution From Table 10.19 we see that all expected values are ≥ 5; so the test procedure in Equation 10.23 can be used. From Tables 10.18 and 10.19,

$$X^2 = \frac{(320-416.6)^2}{416.6} + \frac{(1206-1348.3)^2}{1348.3} + \cdots + \frac{(406-476.3)^2}{476.3} = 130.3$$

Under H_0, X^2 follows a chi-square distribution with $(2-1) \times (5-1)$, or 4, *df*. Because

$$\chi^2_{4,.999} = 18.47 < 130.3 = X^2$$

if follows that $p < 1 - .999 = .001$

Therefore, the results are very highly significant, and we can conclude that there is a significant relationship between age at first birth and the development of breast cancer.

10.6.2 Chi-Square Test for Trend in Binomial Proportions

Refer again to the international study data in Table 10.18. In Example 10.35 the test procedure in Equation 10.23 was used to analyze the data. For the special case of a $2 \times k$ table, this test procedure enables us to test the hypothesis H_0: $p_1 = p_2 = \cdots = p_k$ versus H_1: at least two of the p_i are unequal, where p_i = probability of success for the ith group = probability that an observation from the ith column falls in the first row. When this test procedure was employed in Example 10.35, a chi-square statistic of 130.3 with 4 *df* was found, which was highly significant ($p < .001$). As a result, H_0 was rejected and we concluded that the proportion of breast-cancer cases in at least 2 of the 5 age-at-first-birth groups were different. However, although this result shows that some relationship exists between breast cancer and age at first birth, it does not tell us specifically about the nature of the relationship. In particular, from Table 10.18 we notice an increasing *trend* in the proportion of women with breast cancer in each succeeding column. We would like to employ a specific test to detect such trends. For this purpose a **score variable** S_i is introduced to correspond to the ith group. The score variable can represent some particular numeric attribute of the group. In other instances, for simplicity, 1 is assigned to the first group, 2 to the second group, . . . , k to the kth (last) group.

Example 10.36 **Cancer** Construct a score variable for the international study data in Table 10.18.

Solution It is natural to use the average age at first birth within a group as the score variable for that group. This rule presents no problem for the second, third, and fourth groups, in which the average age is estimated as 22.5 [(20 + 25)/2], 27.5, and 32.5 years, respectively. However, a similar calculation cannot be performed for the first and fifth groups, because they are defined as < 20 and ≥ 35, respectively. By symmetry, a score of 17.5 years could be assigned to the first group and 37.5 years to the fifth group. However, since the scores are equally spaced, our purposes will be equally well served by assigning scores of 1, 2, 3, 4, and 5 to the five groups. For simplicity, this scoring method will be adopted.

We wish to relate the proportion of breast-cancer cases in a group to the score variable for that group. In other words, we wish to test whether the proportion of breast-cancer cases increases or decreases as age at first birth increases. For this purpose the following test procedure is introduced:

EQUATION 10.24

Chi-Square Test for Trend in Binomial Proportions (Two-Sided Test) Suppose there are k groups and we wish to test if there is an increasing (or decreasing) trend in the proportion of "successes" p_i (i.e., the proportion of units in the first row of the ith group) as i increases.

(1) Set up the data in the form of a $2 \times k$ contingency table, where success or failure is listed along the rows and the k groups are listed along the columns.

(2) Denote the number of successes in the ith group by x_i, the total number of units in the ith group by n_i, and the proportion of successes in the ith group by $\hat{p}_i = x_i/n_i$. Denote the total number of successes over all groups by x, the total number of units over all groups by n, the overall proportion of successes by $\overline{p} = x/n$, and the overall proportion of failures by $\overline{q} = 1 - \overline{p}$.

(3) Construct a score variable S_i to correspond to the ith group. This variable will usually either be 1, 2, . . . , k for the k groups or be defined to correspond to some other numeric attribute of the group.

(4) More specifically, we wish to test the hypothesis H_0: There is no trend among the p_i versus H_1: The p_i are an increasing or decreasing function of the S_i, expressed in the form $p_i = \alpha + \beta S_i$ for some constants α, β. To relate p_i and S_i, compute the test statistic $X_1^2 = A^2/B$, where

$$A = \sum_{i=1}^{k} n_i\left(\hat{p}_i - \overline{p}\right)\left(S_i - \overline{S}\right)$$

$$= \left(\sum_{i=1}^{k} x_i S_i\right) - x\overline{S} = \left(\sum_{i=1}^{k} x_i S_i\right) - x\left(\sum_{i=1}^{k} n_i S_i\right)\bigg/ n$$

$$B = \overline{p}\,\overline{q}\left[\left(\sum_{i=1}^{k} n_i S_i^2\right) - \left(\sum_{i=1}^{k} n_i S_i\right)^2 \bigg/ n\right]$$

which under H_0 approximately follows a chi-square distribution with 1 *df*.

(5) For a two-sided level α test, if

$$X_1^2 > \chi^2_{1,1-\alpha}$$

then reject H_0. If $X_1^2 \le \chi^2_{1,1-\alpha}$ then accept H_0.

(6) The approximate p-value is given by the area to the right of X_1^2 under a χ_1^2 distribution.

(7) The direction of the trend in proportions is indicated by the sign of A. If $A > 0$, then the proportions increase with increasing score, if $A < 0$, then the proportions decrease with increasing score.

(8) Use this test only if $n\overline{p}\,\overline{q} \ge 5.0$.

The acceptance and rejection regions for this test are depicted in Figure 10.10. The computation of the p-value is illustrated in Figure 10.11.

The test statistic in Equation 10.24 is reasonable, since if $\hat{p}_i$ (or $\hat{p}_i - \overline{p}$) increases as S_i increases, then $A > 0$, whereas if $\hat{p}_i$ decreases as S_i increases, then $A < 0$. In either case A^2 and the test statistic X_1^2 will be large. However, if $\hat{p}_i$ shows no particular trend regarding S_i, then A will be close to 0 and the test statistic X_1^2 will be small. This test can be used even if some of the groups have small sample size, since the test is based on the overall trend in the proportions. This property is in

FIGURE 10.10 **Acceptance and rejection regions for the chi-square test for trend in binomial proportions**

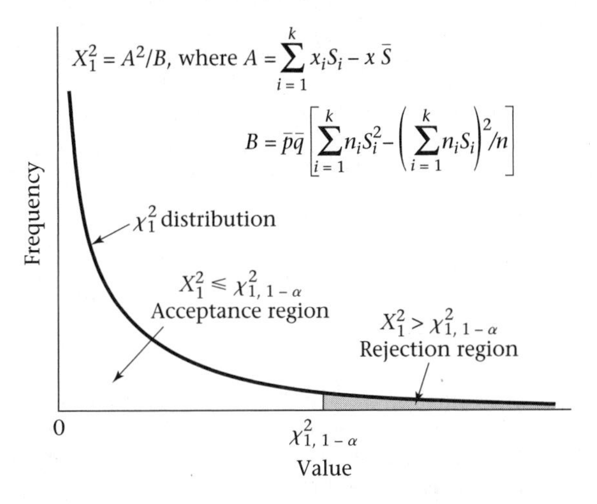

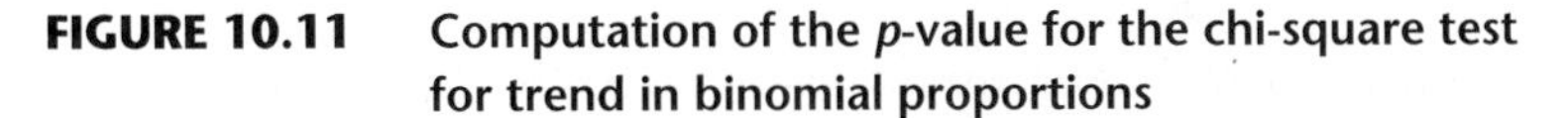

FIGURE 10.11 **Computation of the *p*-value for the chi-square test for trend in binomial proportions**

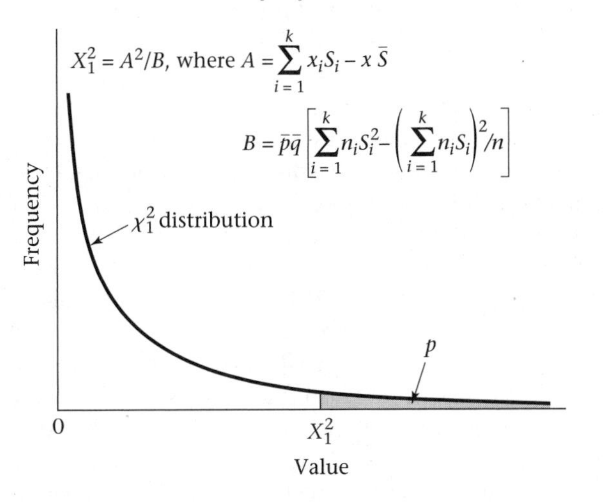

contrast to the overall chi-square test in Equation 10.23, which tests for heterogeneity among proportions and requires that the expected number of units in individual cells not be too small.

Example 10.37 **Cancer** Using the international study data in Table 10.18, assess whether or not there is an increasing trend in the proportion of breast-cancer cases as age at first birth increases.

Solution Note that $S_i = 1, 2, 3, 4, 5$ in the five groups, respectively. Furthermore, from Table 10.18, $x_i = 320, 1206, 1011, 463, 220$, and $n_i = 1742, 5638, 3904, 1555, 626$ in the five respective groups, whereas $x = 3220$, $n = 13{,}465$, $\bar{p} = x/n = .239$, $\bar{q} = 1 - \bar{p} = .761$. From Equation 10.24 it follows that

$$A = 320(1) + 1206(2) + \cdots + 220(5)$$
$$- (3220)[1742(1) + 5638(2) + \cdots + 626(5)]/13{,}465$$
$$= 8717 - (3220)(34{,}080)/13{,}465 = 8717 - 8149.84 = 567.16$$

$$B = (.239)(.761)\{1742(1^2) + 5638(2^2) + \cdots + 626(5^2)$$
$$- [1742(1) + (5638)(2) + \cdots + 626(5)]^2/13{,}465\}$$
$$= .239(.761)[99{,}960 - 34{,}080^2/13{,}465]$$
$$= .239(.761)(99{,}960 - 86{,}256.70) = 2493.33$$

Thus, $X_1^2 = A^2/B = \dfrac{567.16^2}{2493.33} = 129.01 \sim \chi_1^2$ under H_0

Since $\chi^2_{1,.999} = 10.83 < 129.01 = X_1^2$, H_0 can be rejected with $p < .001$ and we can conclude that there is a significant trend in the proportion of breast-cancer cases among age-at-first-birth groups. Because $A > 0$, it follows that as age at first birth increases, the proportion of breast-cancer cases rises.

With a $2 \times k$ table, the chi-square test for trend in Equation 10.24 is often more relevant to the hypotheses of interest than the chi-square test for heterogeneity in Equation 10.23, because the former procedure tests for specific trends in the proportions, whereas the latter tests for any differences in the proportions, where the proportions may follow any pattern. Other, more advanced methods for assessing $R \times C$ contingency tables are given in Maxwell's *Analyzing Qualitative Data* [4].

In this section, we have discussed tests for association between two categorical variables with R and C categories, respectively, where either $R > 2$ or $C > 2$. If both R and C are >2, then the chi-square test for $R \times C$ contingency tables is used. Referring to the flowchart at the end of this chapter (p. 412), we would answer no to (1) 2×2 contingency table? no to (2) $2 \times k$ contingency table? and yes to (3) $R \times C$ contingency table with $R > 2$ and $C > 2$, which leads to the box entitled "Use chi-square test for $R \times C$ tables." If either R or $C = 2$, then assume that we have rearranged the row and column variables so that the row variable has two categories. Let us designate the number of column categories by k (rather than C). If we are interested in assessing trend over the k binomial proportions formed by the proportions of units in the first row of each of the k columns, then we would use the chi-square test for trend in binomial proportions. Referring to the flowchart at the end of this chapter (p. 412), we would answer no to (1) 2×2 contingency table? yes to (2) $2 \times k$ contingency table? and yes to (3) interested in trend over k proportions? This leads us to the box entitled "Use chi-square test for trend if no confounding is present, or the Mantel Extension test if confounding is present."

10.6.3 Relationship Between the Wilcoxon Rank-Sum Test and the Chi-Square Test for Trend

The Wilcoxon rank-sum test given in Equation 9.7 is actually a special case of the chi-square test for trend.

EQUATION 10.25

Relationship Between the Wilcoxon Rank-Sum Test and the Chi-Square Test for Trend Suppose we have a $2 \times k$ table as shown in Table 10.20.

TABLE 10.20 **A hypothetical $2 \times k$ table relating a dichotomous disease variable D to a categorical exposure variable E with k ordered categories**

		E				
		1	2		k	
D	+	x_1	x_2	...	x_k	x
	−	$n_1 - x_1$	$n_2 - x_2$	...	$n_k - x_k$	$n - x$
		n_1	n_2		n_k	n
	Score	S_1	S_2	...	S_k	

The ith exposure category is assumed to have an associated score S_i, $i = 1, \ldots, k$. Let p_i = probability of disease in the ith exposure group. If $p_i = \alpha + \beta S_i$, and we wish to test the hypothesis H_0: $\beta = 0$ versus H_1: $\beta \neq 0$, then

(1) We can use the chi-square test for trend, whereby we can write the test statistic in the form

$$X^2 = \frac{\left(|O - E| - 0.5\right)^2}{V} \sim \chi_1^2, \text{ under } H_0$$

where

$$O = \text{observed total score among diseased subjects} = \sum_{i=1}^{k} x_i S_i$$

$$E = \text{expected total score among diseased subjects under } H_0 = \frac{x}{n} \sum_{i=1}^{k} n_i S_i$$

$$V = \frac{x(n-x)}{n(n-1)} \left[\sum_{i=1}^{k} n_i S_i^2 - \left(\sum_{i=1}^{k} n_i S_i \right)^2 \Big/ n \right]$$

and we reject H_0 if $X^2 > \chi_{1,1-\alpha}^2$

and accept H_0 otherwise.

(2) We can use the Wilcoxon rank-sum test as given in Equation 9.7 whereby we have the test statistic

$$T = \frac{\left| R_1 - \frac{x(n+1)}{2} \right| - \frac{1}{2}}{\sqrt{\left[\frac{x(n-x)}{12} \right] \left[n + 1 - \frac{\sum n_i (n_i^2 - 1)}{n(n-1)} \right]}}$$

and reject H_0 if $T > z_{1-\alpha/2}$

and accept H_0 if $T \le z_{1-\alpha/2}$

where $z_{1-\alpha/2}$ = upper $\alpha/2$ percentile of an $N(0, 1)$ distribution.

(3) If the scores S_i are set equal to the midrank for the ith group as defined in Equation 9.6, where the midrank for the ith exposure category = number of observations in the first $i-1$ groups $+\left(\frac{1+n_i}{2}\right)$

$$= \sum_{j=1}^{i-1} n_j + \frac{(1+n_i)}{2} \quad \text{if } i > 1$$

$$= \frac{1+n_1}{2} \quad \text{if } i = 1$$

then the test procedures in steps (1) and (2) yield the same p-values and are equivalent. In particular,

$$O = R_1 = \text{Rank sum in the first row, } E = \frac{x(n+1)}{2}$$

$$V = \left[\frac{x(n-x)}{12}\right]\left[n+1-\frac{\sum_{i=1}^{k} n_i(n_i^2-1)}{n(n-1)}\right], \text{ and } X_1^2 = T^2$$

Example 10.38 **Ophthalmology** Test the hypothesis that the average visual acuity is different for dominant and sex-linked individuals in Table 9.2 or, equivalently, that the proportion of dominant subjects changes in a consistent manner as visual acuity declines, using the chi-square test for trend.

Solution We have the following 2×8 table:

	20–20	20–25	20–30	20–40	20–50	20–60	20–70	20–80	
Dominant	5	9	6	3	2	0	0	0	25
Sex-linked	1	5	4	4	8	5	2	1	30
	6	14	10	7	10	5	2	1	55
Score	3.5	13.5	25.5	34.0	42.5	50.0	53.5	55.0	

If the scores are set equal to the average ranks given in Table 9.2, then we have

$$O = 5(3.5) + 9(13.5) + 6(25.5) + 3(34.0) + 2(42.5) = 479$$

$$E = \frac{25}{55}[6(3.5) + 14(13.5) + 10(25.5) + 7(34.0) + 10(42.5) + 5(50.0) + 2(53.5) + 1(55.0)]$$

$$= \frac{25}{55}(1540) = 700$$

$$V = \frac{25(30)}{55(54)}\left\{\left[6(3.5)^2 + 14(13.5)^2 + 10(25.5)^2 + 7(34.0)^2 + 10(42.5)^2 + 5(50.0)^2 + 2(53.5)^2 + 1(55.0)^2\right] - \frac{1540^2}{55}\right\}$$

$$= \frac{25(30)}{55(54)}(56{,}531.5 - 43{,}120)$$

$$= \frac{25(30)}{55(54)}(13{,}411.5) = 3386.74$$

$$X^2 = \frac{\left(|479 - 700| - 0.5\right)^2}{3386.74} = 14.36 \sim \chi_1^2 \text{ under } H_0$$

The p-value $= Pr(\chi_1^2 > 14.36) < .001$. Also, referring to Example 9.17, we see that $O = R_1 = 479$, $E = E(R_1) = 700$, $V = V(R_1)$ corrected for ties $= 3386.74$ and

$$X^2 = 14.36 = T^2 = 3.79^2$$

Thus the two test procedures are equivalent. However, if we had chosen different scores (e.g., 1, . . . ,8) for the 8 visual-acuity groups, then the test procedures would *not* be the same. The choice of scores is somewhat arbitrary. If each column corresponds to a specific quantitative exposure category, then it is reasonable to use the average exposure within the category as the score. If the exposure level is not easily quantified, then either midranks or consecutive integers are reasonable choices for scores. If the number of subjects in each exposure category is the same, then these two methods of scoring will yield identical test statistics and p-values using the chi-square test for trend.

The estimate of the variance (V) given in Equation 10.25 is derived from the hypergeometric distribution and differs slightly from the variance estimate for the chi-square test for trend in Equation 10.24 given by

$$V = \frac{x(n-x)}{n^2}\left[\sum_{i=1}^{k} n_i S_i^2 - \left(\sum_{i=1}^{k} n_i S_i\right)^2 \Big/ n\right]$$

which is based on the binomial distribution. The hypergeometric distribution is more appropriate, although the difference is usually slight, particularly for large n. Also, a continuity correction of 0.5 is used in the numerator of X^2 in Equation 10.25, but not in A in the numerator of X_1^2 in Equation 10.24. This difference is also usually slight.

SECTION 10.7 Chi-Square Goodness-of-Fit Test

In our previous work on estimation and hypothesis testing, we usually assumed that the data came from a specific underlying probability model and then proceeded either to estimate the parameters of the model or test hypotheses concerning different possible values of the parameters. This section presents a general method of testing for the *goodness of fit of a probability model*. Consider the problem in Example 10.39.

Example 10.39 **Hypertension** Diastolic blood-pressure measurements were collected at home in a community-wide screening program of 14,736 adults ages 30–69 in East Boston, Massachusetts, as part of a nationwide study to detect and treat hypertensive people [5]. The people in the study were each screened in the home with two measurements taken at one visit. A frequency distribution of the mean diastolic blood pressure is given in Table 10.21 in 10-mm Hg intervals.

We would like to assume that these measurements came from an underlying normal distribution, because standard methods of statistical inference could then be applied on these data as presented in this text. How can the validity of this assumption be tested?

TABLE 10.21 **Frequency distribution of mean diastolic blood pressure for adults 30–69 years old in a community-wide screening program in East Boston, Massachusetts**

Group (mm Hg)	Observed frequency	Expected frequency	Group	Observed frequency	Expected frequency
<50	57	78	≥80, <90	4604	4479
≥50, <60	330	547	≥90, <100	2119	2431
≥60, <70	2132	2127	≥100, <110	659	684
≥70, <80	4584	4283	≥110	251	107
			Total	14,736	14,736

This assumption can be tested by first computing what the expected frequencies would be in each group if the data did come from an underlying normal distribution and then comparing these expected frequencies with the corresponding observed frequencies.

Example 10.40 **Hypertension** Compute the expected frequencies for the data in Table 10.21 assuming an underlying normal distribution.

Solution Assume that the mean and standard deviation of this hypothetical normal distribution are given by the sample mean and standard deviation, respectively ($\overline{x} = 80.68$, $s = 12.00$). The expected frequency within a group interval from a to b would then be given by

$$14{,}736\left\{\Phi\left[(b-\mu)/\sigma\right]-\Phi\left[(a-\mu)/\sigma\right]\right\}$$

Thus, the expected frequency within the (≥50, <60) group would be

$$\begin{aligned}&14{,}736\times\left\{\Phi\left[(60-80.68)/12\right]-\Phi\left[(50-80.68)/12\right]\right\}\\&=14{,}736\times\left[\Phi(-1.72)-\Phi(-2.56)\right]\\&=14{,}736\times(.0424-.0053)=14{,}736(.0371)=547\end{aligned}$$

Also, the expected frequency less than a would be $\Phi[(a-\mu)/\sigma]$, and the expected frequency greater than or equal to b would be $1-\Phi[(b-\mu)/\sigma]$. The expected frequencies for all the groups are given in Table 10.21.

The same measure of agreement between the observed and expected frequencies in a group will be used as was used in our work on contingency tables, namely, $(O-E)^2/E$. Furthermore, the agreement between observed and expected frequencies can be summarized over the whole table by summing $(O-E)^2/E$ over all the groups. If we have the correct underlying model, then this sum will approximately follow a chi-square distribution with $g-1-k$ *df*, where g = the number of groups and k = the number of parameters estimated from the data to compute the expected frequencies. This approximation will again be valid only if the expected values in the groups are not too small. In particular, the requirement is that no expected value can be <1 and not more than $\frac{1}{5}$ of the expected values can be <5. If there are too many groups with small expected frequencies, then some of them should be combined with other adjacent groups so that the preceding rule is not violated. The test procedure can be summarized as follows:

EQUATION 10.26

Chi-Square Goodness-of-Fit Test To test for the goodness of fit of a probability model, use the following procedure:

(1) Divide the raw data into groups. The considerations for grouping data are similar to those given in Section 2.7. In particular, the groups must not be too small, so that step 7 is not violated.

(2) Estimate the k parameters of the probability model from the data using the methods of Chapter 6.

(3) Use the estimates in step 2 to compute the probability $\hat{p}$ of obtaining a value within a particular group and the corresponding expected frequency within that group $(n\hat{p})$, where n is the total number of data points.

(4) If O_i and E_i are, respectively, the observed and expected number of units within the ith group, then compute

$$X^2 = (O_1 - E_1)^2/E_1 + (O_2 - E_2)^2/E_2 + \cdots + (O_g - E_g)^2/E_g$$

where g = the number of groups.

(5) For a test with significance level α, if

$$X^2 > \chi^2_{g-k-1,1-\alpha}$$

then reject H_0; if

$$X^2 \le \chi^2_{g-k-1,1-\alpha}$$

then accept H_0.

(6) The approximate p-value for this test is given by

$$Pr\left(\chi^2_{g-k-1} > X^2\right)$$

(7) Use this test only if

(a) No more than $\frac{1}{5}$ of the expected values are < 5.

(b) No expected value is < 1.

The acceptance and rejection regions for this test are depicted in Figure 10.12. The computation of the p-value for this test is illustrated in Figure 10.13.

Example 10.41

Hypertension Test for the goodness of fit of the normal probability model using the data in Table 10.21.

Solution

Two parameters have been estimated from the data (μ, σ^2), and there are 8 groups. Therefore, $k = 2$, $g = 8$. Under H_0, X^2 follows a chi-square distribution with $8 - 2 - 1 = 5$ *df*.

$$\begin{aligned} X^2 &= (O_1 - E_1)^2/E_1 + \cdots + (O_8 - E_8)^2/E_8 \\ &= (57 - 78)^2/78 + \cdots + (251 - 107)^2/107 = 350.2 \sim \chi^2_5 \text{ under } H_0 \end{aligned}$$

FIGURE 10.12 **Acceptance and rejection regions for the chi-square goodness-of-fit test**

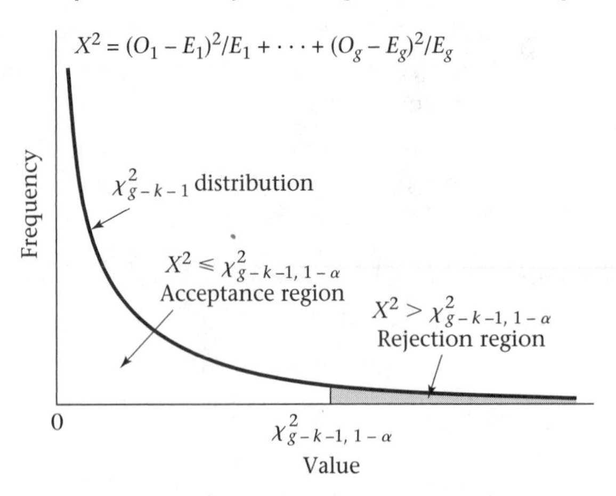

FIGURE 10.13 **Computation of the *p*-value for the chi-square goodness-of-fit test**

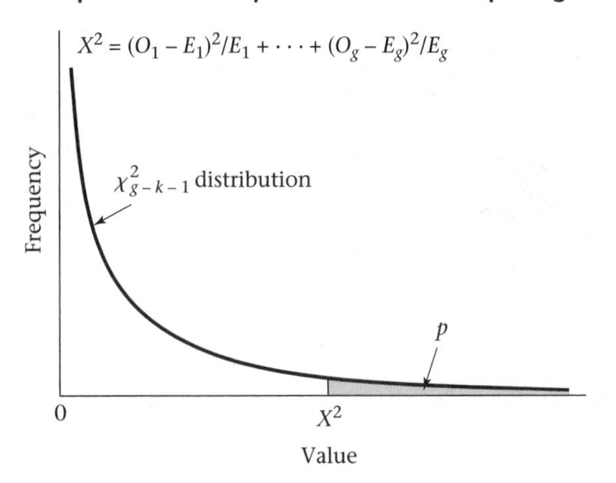

Since $\chi^2_{5,.999} = 20.52 < 350.2 = X^2$, the p-value $< 1 - .999 = .001$ and the results are very highly significant.

Thus, the normal model does not provide an adequate fit to the data. The normal model appears to fit fairly well in the middle of the distribution (between 60 and 110 mm Hg) but fails badly in the tails, predicting too many blood pressures below 60 mm Hg and too few over 110 mm Hg.

The test procedure in Equation 10.26 can be used to assess the goodness of fit of any probability model, not just the normal model. The expected frequencies would be computed from the probability distribution of the proposed model and then the same goodness-of-fit test statistic as given in Equation 10.26 would be used.

SECTION 10.8 The Kappa Statistic

Most of our previous work has been concerned with *tests of association* between two categorical variables (usually a disease and an exposure variable). In some instances, it is expected that there will be some association between the variables and the issue is quantifying the *degree of association*. This is particularly true in **reliability studies**, where one wants to quantify the reproducibility of the same variable (e.g., dietary intake of a particular food) measured more than once.

Example 10.42 **Nutrition** A diet questionnaire was administered by mail to 537 female American nurses on two separate occasions several months apart. The questions asked included the quantities eaten of more than 100 separate food items. The data obtained from the two surveys for the amount of beef consumption are presented in Table 10.22. Notice that the responses on the two surveys are the same only for 136 + 240 = 376 out of 537 (70.0%) women. How can the reproducibility of response for the beef consumption data be quantified?

A chi-square test for association between the survey 1 and survey 2 responses could be performed. However, this test would not give a quantitative measure of reproducibility between the responses at the two surveys. Instead, we will focus on the percentage of women with concordant responses in the two surveys. We noted in Example 10.42 that 70.0% of the women gave concordant responses. We would like to compare the observed concordance rate (p_o) with the expected concordance rate (p_e) if the responses of the women in the two surveys were statistically independent. The motivation behind this definition is that the questionnaire would be virtually worthless if the frequency of consumption reported at one survey had no relationship to the frequency of consumption reported at a second survey. Suppose there are c response categories and the probability of response in the ith category is a_i for the first survey and b_i for the second survey. These probabilities can be estimated from the row and column margins of the contingency table. The expected concordance rate (p_e) if the survey responses are independent is given by $\sum a_i b_i$.

TABLE 10.22 **Amount of beef consumption reported by 537 female American nurses at two different surveys**

	Survey 2		
Survey 1	≤1 serving/week	>1 serving/week	Total
≤1 serving/week	136	92	228
>1 serving/week	69	240	309
Total	205	332	537

Example 10.43 **Nutrition** Compute the expected concordance rate using the beef-consumption data in Table 10.22.

Solution From Table 10.22

$$a_1 = \frac{228}{537} = .425$$

$$a_2 = \frac{309}{537} = .575$$

$$b_1 = \frac{205}{537} = .382$$

$$b_2 = \frac{332}{537} = .618$$

Thus, $p_e = (.425 \times .382) + (.575 \times .618) = .518$

Therefore, 51.8% concordance would be expected if the subjects were responding independently regarding beef consumption at the two surveys.

We could use $p_o - p_e$ as the measure of reproducibility. However, it is preferable to use a measure that equals +1.0 in the case of perfect agreement and 0.0 if the responses on the two surveys are completely independent. Indeed, the maximum possible value for $p_o - p_e$ is $1 - p_e$, which is achieved with $p_o = 1$. Therefore, the Kappa statistic, which is defined as $(p_o - p_e)/(1 - p_e)$, is used as the measure of reproducibility:

EQUATION 10.27

The Kappa Statistic

(1) If a categorical variable is reported at two surveys by each of n subjects, then the Kappa statistic (κ) is used to measure reproducibility between surveys, where

$$\kappa = \frac{p_o - p_e}{1 - p_e}$$

and p_o = observed probability of concordance between the two surveys

p_e = expected probability of concordance between the two surveys

$= \sum a_i b_i$

where a_i, b_i are the marginal probabilities for the ith category in the $c \times c$ contingency table relating response at the two surveys.

(2) Furthermore,

$$se(\kappa) = \sqrt{\frac{1}{n(1-p_e)^2} \times \left\{p_e + p_e^2 - \sum_{i=1}^{c}\left[a_i b_i (a_i + b_i)\right]\right\}}$$

To test the one-sided hypothesis H_0: $\kappa = 0$ versus H_1: $\kappa > 0$, use the test statistic

$$z = \frac{\kappa}{se(\kappa)}$$

which follows an $N(0, 1)$ distribution under H_0.

(3) Reject H_0 at level α if $z > z_{1-\alpha}$ and accept H_0 otherwise.

(4) The exact p-value is given by $p = 1 - \Phi(z)$.

The acceptance and rejection regions for this test are depicted in Figure 10.14. The computation of the p-value is shown in Figure 10.15.

FIGURE 10.14 **Acceptance and rejection regions for the significance test for Kappa**

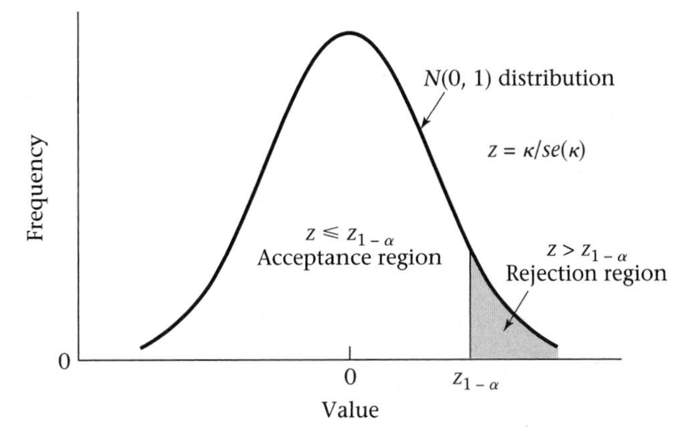

FIGURE 10.15 **Computation of the *p*-value for the significance test for Kappa**

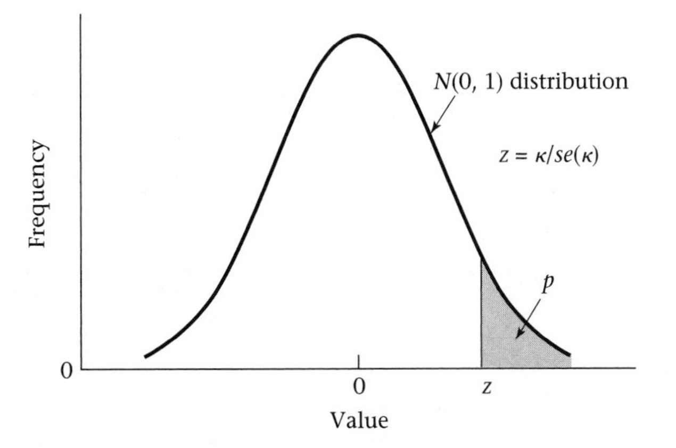

Note that we are customarily interested in one-tailed tests in Equation 10.27, because negative values for Kappa usually have no biological significance.

Example 10.44 **Nutrition** Compute the Kappa statistic and assess its statistical significance using the beef-consumption data in Table 10.22.

Solution From Examples 10.42 and 10.43,

$$p_o = .700$$

$$p_e = .518$$

Therefore, the Kappa statistic is given by

$$\kappa = \frac{.700 - .518}{1 - .518} = \frac{.182}{.482} = .378$$

Furthermore, from Equation 10.27 and the results of Example 10.43, the standard error of κ is given by

$$se(\kappa) = \sqrt{\frac{1}{537(1-.518)^2} \times \left\{.518 + .518^2 - \sum\left[a_i b_i (a_i + b_i)\right]\right\}}$$

where

$$\sum\left[a_i b_i (a_i + b_i)\right] = .425 \times .382 \times (.425 + .382) + .575 \times .618 \times (.575 + .618)$$
$$= .555$$

Thus, $se(\kappa) = \sqrt{\frac{1}{537 \times .232} \times (.518 + .268 - .555)} = \sqrt{\frac{1}{124.8} \times .231} = .0430$

The test statistic is given by

$$z = \frac{.378}{.0430} = 8.8 \sim N(0,1) \text{ under } H_0$$

The p-value is $p = 1 - \Phi(8.8) < .001$

Thus, the Kappa statistic indicates highly significant reproducibility between the first and second surveys for beef consumption.

Although the Kappa statistic was significant in Example 10.44, it still shows that the reproducibility was far from perfect. Indeed, Landis and Koch (1977) provide the following guidelines for the evaluation of Kappa [6]:

EQUATION 10.28

Guidelines for the Evaluation of Kappa

$\kappa > .75$ denotes *excellent* reproducibility.

$.4 \le \kappa \le .75$ denotes *good* reproducibility.

$0 \le \kappa < .4$ denotes *marginal* reproducibility.

In general, reproducibility is not good for many items in dietary surveys, indicating the need for multiple dietary assessments to reduce variability. See Fleiss [7] for further information about the Kappa statistic, including assessments of reproducibility for more than two surveys.

Kappa is usually used as a measure of reproducibility between repeated assessments of the same variable. If we are interested in the concordance between responses on two different variables, where one of the variables can be considered to be a gold standard, then sensitivity and specificity (see Chapter 3) are more appropriate indices of reliability than Kappa.

SECTION 10.9 Summary

In this chapter the most widely used techniques for analyzing qualitative (or categorical) data were discussed. First, the problem of how to compare binomial proportions from two independent samples was studied. For the large-sample case, this problem was solved in two different (but equivalent) ways: using either the two-sample test for binomial proportions or the chi-square test for 2×2 contingency tables. The former method is similar to the t test methodology introduced in Chapter 8, whereas the contingency-table approach can be easily generalized to more complex problems involving qualitative data. For the small-sample case, Fisher's exact test is used to compare binomial proportions in two independent samples. To compare binomial proportions in paired samples, such as when a person is used as his or her own control, McNemar's test for correlated proportions should be used.

The 2×2 contingency-table problem was extended to the investigation of the relationship between two qualitative variables, in which one or both variables have more than two possible categories of response. A chi-square test for $R \times C$ contingency tables was developed, which is a direct generalization of the 2×2 contingency-table test. Also, the problem of how to assess the goodness of fit of the probability models proposed in earlier chapters was studied. The chi-square goodness-of-fit test was used to address this problem. Furthermore, the Kappa statistic was introduced as an index of reproducibility for categorical data.

Finally, formulas to compute sample size and power for comparing two binomial proportions were provided in the independent-sample and paired-sample case. Special considerations in computing sample size and power were discussed in a clinical trial setting The methods in this chapter are illustrated in the shaded boxes of the flowchart in Figure 10.16.

In Chapters 8, 9, and 10, we considered the comparison between two groups for variables measured on a continuous, ordinal, and categorical scale, respectively. In Chapter 11, we discuss methods for studying the relationship between a continuous response variable and one or more predictor variables, which can be either continuous or categorical.

PROBLEMS

Cardiovascular Disease

Consider the Physicians' Health Study data presented in Example 10.32.

10.1 How many subjects need to be enrolled in each group to have a 90% chance of detecting a significant difference using a two-sided test with $\alpha = .05$ if compliance is perfect?

10.2 Answer Problem 10.1 if compliance is as given in Example 10.32.

10.3 Answer Problem 10.1 if a one-sided test with power = .8 is used and compliance is perfect.

10.4 Suppose 11,000 men are actually enrolled in each treatment group. What would be the power of such a study if a two-sided test with $\alpha = .05$ were used and compliance is perfect?

10.5 Answer Problem 10.4 if compliance is as given in Example 10.32.

Refer to Table 2.11.

10.6 What significance test can be used to assess whether there is a relationship between receiving an antibiotic and receiving a bacterial culture, while in the hospital?

10.7 Perform the test in Problem 10.6 and report a p-value.

FIGURE 10.16 Flowchart for appropriate methods of statistical inference for categorical data (shaded)

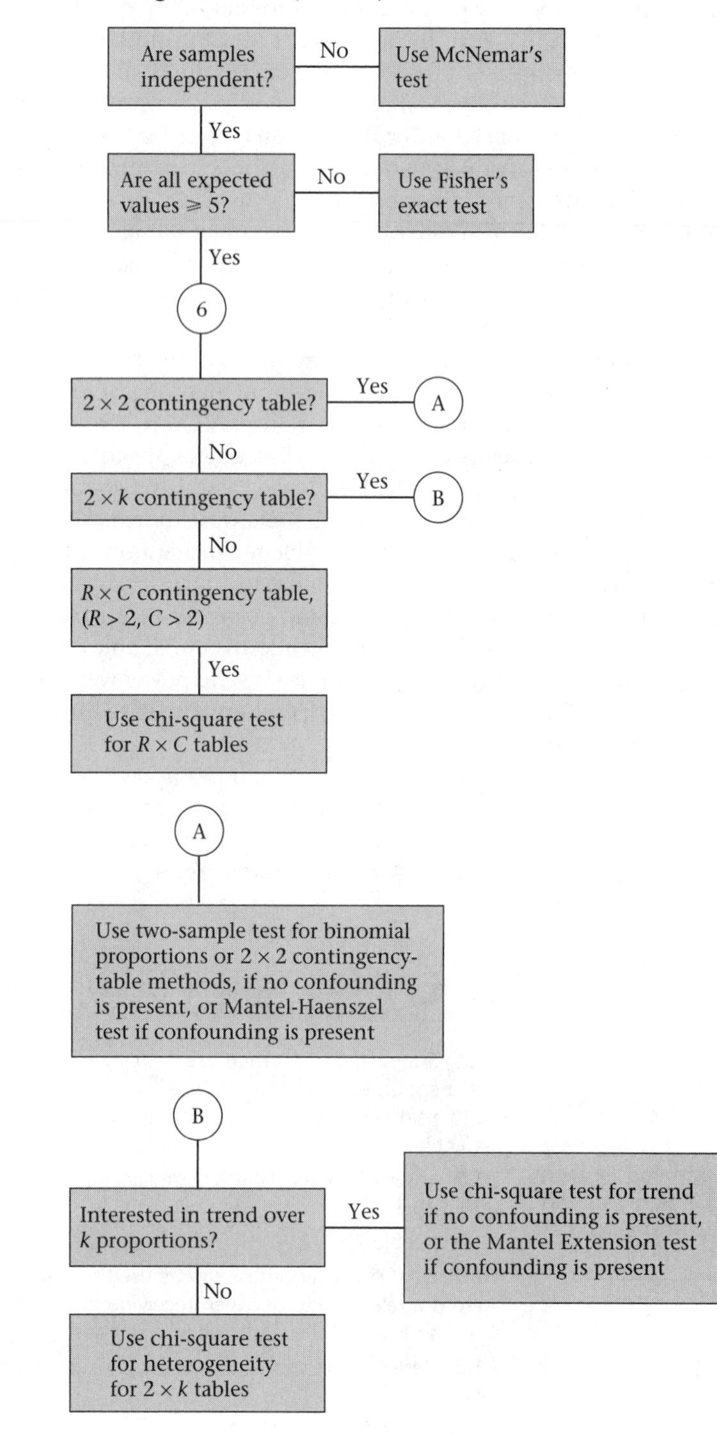

Gastroenterology

Two drugs (A, B) are compared for the medical treatment of duodenal ulcer. For this purpose, patients are carefully matched on age, sex, and clinical condition. The treatment results based on 200 matched pairs show that for 89 matched pairs both treatments are effective; for 90 matched pairs both treatments are ineffective; for 5 matched pairs drug A is effective, whereas drug B is ineffective; for 16 matched pairs drug B is effective, whereas drug A is ineffective.

***10.8** What test procedure can be used to assess the results?

***10.9** Perform the test in Problem 10.8 and report a *p*-value.

In the same study, if the focus is on the 100 matched pairs consisting of male patients, then the following results are obtained; for 52 matched pairs both drugs are effective; for 35 matched pairs both drugs are ineffective; for 4 matched pairs drug A is effective, whereas drug B is ineffective; for 9 matched pairs drug B is effective, whereas drug A is ineffective.

***10.10** How many concordant pairs are there among the male matched pairs?

***10.11** How many discordant pairs are there among the male matched pairs?

***10.12** Perform a significance test to assess any differences in effectiveness between the drugs among males. Report a *p*-value.

Gynecology

A study was performed in 1985 relating the duration of IUD use to infertility [8]. A group of 89 infertile IUD users and a group of 640 control IUD users were identified. The women were subdivided by the duration of IUD use with the data presented in Table 10.23.

TABLE 10.23 Relationship between duration of IUD use and infertility among IUD users

	Duration of IUD use (months)			
	<3	$\geq 3, <18$	$\geq 18, \leq 36$	>36
Cases	10	23	20	36
Controls	53	200	168	219

Source: Reprinted with permission of the *New England Journal of Medicine, 312*(15), 941–947, 1985.

10.13 Perform a test for heterogeneity of the proportions of cases in the four groups.

10.14 Suppose the score variable 1, 2, 3, 4 is assigned to the four duration groups. Perform a test for trend on these data; that is, does the proportion of cases increase or decrease as the duration of IUD use increases?

10.15 Interpret your results in Problems 10.13 and 10.14.

10.16 Test for the adequacy of the normal model for the distribution of duration of hospitalization given in Table 2.11.

10.17 Answer Problem 10.16 for the distribution of ln (duration of hospitalization).

Sexually Transmitted Disease

Suppose an epidemiologic investigation of people entering a STD clinic is performed. It is found that 160 of 200 patients who are diagnosed as having gonorrhea and 50 of 105 patients who are diagnosed as having nongonococcal urethritis (NGU) have had previous episodes of urethritis.

***10.18** Is there an association between the present diagnosis and prior episodes of urethritis?

Cancer

10.19 A 1980 study investigated the relationship between the use of oral contraceptives and the development of endometrial cancer [9]. It was found that of 117 endometrial-cancer patients, 6 had used the oral contraceptive Oracon at some time in their lives, whereas of 395 controls, 8 had used this agent. Test for an association between the use of Oracon and the incidence of endometrial cancer using a two-tailed test.

Ophthalmology

Retinitis pigmentosa is a disease that manifests itself via different genetic modes of inheritance. Cases have been documented with a dominant, recessive, and sex-linked mode of inheritance. It has been conjectured that the mode of inheritance is related to the ethnic origin of the individual. Cases of the disease have been surveyed in an English and a Swiss population with the following results: Out of 125 English cases, 46 had sex-linked disease, 25 had recessive disease, and 54 had dominant disease. Out of 110 Swiss cases, 1 had sex-linked disease, 99 had recessive disease, and 10 had dominant disease.

***10.20** Do these data show a significant association between ethnic origin and genetic type?

Sexually Transmitted Disease

Suppose we are interested in comparing the effectiveness of two different antibiotics, A and B, in treating gonorrhea. Each person receiving antibiotic A is matched with an equivalent person (age within 5 years, same sex), to whom antibiotic B is given. These people are asked to return to the clinic within 1 week to see if the gonorrhea has been eliminated. Suppose the results are as follows:

(1) For 40 pairs of people, both antibiotics are successful.

(2) For 20 pairs of people, antibiotic A is effective whereas antibiotic B is not.

(3) For 16 pairs of people, antibiotic B is effective whereas antibiotic A is not.

(4) For 3 pairs of people, neither antibiotic is effective.

10.21 Test for the relative effectiveness of the two antibiotics.

10.22 How many matched pairs should be enrolled in a future study if we wish to conduct a two-sided test with $\alpha = .05$ and power = .80? Use the results in (1), (2), (3), and (4) for planning purposes.

Pulmonary Disease

One important aspect of medical diagnosis is its reproducibility. Suppose that two different doctors examine 100 patients for dyspnea in a respiratory-disease clinic and that 15 patients are diagnosed as having dyspnea by Doctor A, 10 patients are diagnosed as having dyspnea by Doctor B, and 7 patients are diagnosed as having dyspnea by *both* Doctor A and Doctor B.

10.23 Compute the Kappa statistic and its standard error regarding the reproducibility of the diagnosis of dyspnea in this clinic.

Infectious Disease

Suppose there is a computerized data bank consisting of all charts of patients at nine hospitals in Cleveland, Ohio. One concern of the group conducting the study is the possibility that the attending physician under- or overreports various diagnoses that seem consistent with a patient's chart. An investigator notes that 50 out of the 10,000 people in the data bank are reported as having a particular viral infection by their attending physician. A computer using an automated method of diagnosis claims that 68 out of the 10,000 people have the infection, 48 of them from the attending physician's 50 positives and 20 from the attending physician's 9950 negatives.

10.24 Test the hypothesis that the computer's and the attending physician's diagnoses are comparable.

Cardiovascular Disease

An investigator wishes to study the effect of cigarette smoking on the development of myocardial infarction (MI) in women. In particular, there is some question in the literature as to the relationship of the timing of cigarette smoking to the development of the disease. One school of thought says that current smokers are at much higher risk than ex-smokers. Another school of thought says that a considerable latent period of nonsmoking is needed before the risk of ex-smokers becomes less than that of current smokers. A third school of thought is that ex-smokers may actually have a higher incidence of MI than current smokers because they may include more women with some prior cardiac symptoms (e.g., angina) than current smokers. To test this hypothesis, 2000 disease-free currently smoking women and 1000 disease-free ex-smoking women, aged 50–59, are identified in 1996, and the incidence of MI between 1996 and 1998 is noted at follow-up visits 2 years later. Investigators find that 40 currently smoking women and 10 ex-smoking women have developed the disease.

***10.25** Is a one-sample or a two-sample test needed here?

***10.26** Is a one-sided or a two-sided test needed here?

***10.27** Which of the following test procedures should be used to test this hypothesis? (More than one may be necessary.) (*Hint:* Use the flowchart in Figure 10.16.)

(1) χ^2 test for 2×2 contingency tables

(2) Fisher's exact test

(3) McNemar's test

(4) One-sample binomial test

(5) One-sample *t* test

(6) Two-sample *t* test with equal variances

***10.28** Carry out the test procedure(s) mentioned in Problem 10.27 and report a *p*-value.

Cardiovascular Disease

A longitudinal study in apparently normal men is organized to relate *changes* in cardiovascular risk parameters to subsequent mortality. The hypothesis being tested is that men whose cholesterol level rises have a different subsequent mortality than those whose cholesterol level drops. In particular, two groups of 50- to 59-year-old men with initially normal cholesterol levels are identified: (1) group A = 25 men whose cholesterol level rises by 50 mg/dL over a 5-year period, and (2) group B = 25 men whose cholesterol level drops by 50 mg/dL over a 5-year period. The groups are then followed for mortality over the next 5 years. The results are given in Table 10.24.

10.29 Is a one-sample or two-sample test needed here?

10.30 Is a one-sided or two-sided test needed here?

10.31 Which of the following test procedures should be used? (More than one may be necessary.) (Consult the flowchart in Figure 10.16.)

TABLE 10.24 Association between cardiovascular mortality and cholesterol change (+ = dead within the next 5 years; – = alive after 5 years)

Number	Mortality outcome, group A	Mortality outcome, group B
1	–	–
2	+	–
3	–	–
4	–	+
5	–	–
6	–	–
7	–	–
8	+	–
9	–	–
10	–	–
11	+	–
12	–	–
13	–	–
14	–	–
15	–	–
16	–	–
17	+	–
18	–	–
19	–	–
20	–	–
21	+	–
22	–	–
23	–	–
24	–	–
25	–	–

(1) Paired *t* test
(2) Two-sample *t* test with equal variances
(3) χ^2 test for 2 × 2 tables
(4) Fisher's exact test
(5) McNemar's test
(6) One-sample binomial test

10.32 Carry out the test procedure in Problem 10.31, and report a *p*-value.

10.33 How many people should be enrolled in a future study if 90% power is required, a two-sided test with $\alpha = .05$ is used, and there are an equal number of subjects in each group? (Use the results given in Table 10.24 for planning purposes.)

10.34 Suppose 200 people are enrolled in the study, 100 in group A and 100 in group B. How much power would such a study have? (Use the results given in Table 10.24 for planning purposes.)

Cancer

The following data are survival rates for cancer of the pancreas for the years 1955–1964, published by the U.S. Department of Health, Education and Welfare [10]: There were 256 reported cases of disease in the under-45 age group, of whom 8% survived for at least 3 years; 710 reported cases of disease in the 45–54 age group, of whom 2% survived for at least 3 years; 1348 reported cases of disease in the 55–64 age group, of whom 2% survived for at least 3 years; 1768 cases of disease in the 65–74 age group, of whom 1% survived for at least 3 years; 1292 cases of disease in the 75+ age group, of whom 1% survived for at least 3 years.

***10.35** What significance test can be used to test if there is an age trend in the 3-year survival rates?

***10.36** Perform the test in Problem 10.35, and report a *p*-value.

Sexually Transmitted Disease

Suppose a study to examine the relative efficacy of penicillin and spectinomycin in the treatment of gonorrhea is conducted. Three treatments are looked at (1) penicillin, (2) spectinomycin, low dose, and (3) spectinomycin, high dose. Three possible responses are recorded: (1) positive smear, (2) negative smear, positive culture, (3) negative smear, negative culture. The data in Table 10.25 are obtained.

TABLE 10.25 Efficacy of different treatments for gonorrhea

	Response			
Treatment	+ Smear	– Smear + culture	– Smear – culture	Total
Penicillin	40	30	130	200
Spectinomycin (low dose)	10	20	70	100
Spectinomycin (high dose)	15	40	45	100
Total	65	90	245	400

10.37 Is there any relationship between the type of treatment and the response? What form does the relationship take?

10.38 Suppose either a positive smear or a positive culture is regarded as a positive response and distinguished from the negative smear, negative culture response. Is there an association between the type of treatment and this measure of response?

Diabetes

Improvement in control of blood-glucose levels is an important motivation for the use of insulin pumps for diabetic patients. However, certain side effects have been reported with pump therapy. Table 10.26 provides data on the occurrence of diabetic ketoacidosis (DKA) in patients before and after the onset of pump therapy [11].

TABLE 10.26 Occurrence of DKA in patients before and after the onset of insulin-pump therapy

	Before pump therapy	
After pump therapy	No DKA	DKA
No DKA	128	7
DKA	19	7

Source: Reprinted with permission of *JAMA, 252*(23), 3265–3269, 1984.

***10.39** What is the appropriate procedure to test if the rate of DKA is different before and after the onset of pump therapy?

***10.40** Perform the significance test in Problem 10.39, and report a *p*-value.

Renal Disease

A study group of 576 working women 30–49 years old who took phenacetin-containing analgesics and a control group of 533 comparably aged women without such intake were identified in 1968 and followed for mortality and morbidity outcomes. One hypothesis to be tested was that phenacetin intake may influence renal (kidney) function and hence have an effect on specific indices of renal morbidity and mortality. The mortality status of these women was determined from 1968 to 1987. It was found that 16 of the women in the study group and 1 of the women in the control group died, where at least one of the causes of death was deemed to be renal [12].

10.41 To test for differences in renal mortality between the two groups in either direction, what statistical test should be used?

10.42 Implement the test in Problem 10.41, and report a *p*-value.

The cohort was also followed for total mortality. It was found that 74 women in the study group died, compared with 27 in the control group.

10.43 What statistical test should be used to compare the total mortality experience of the study group with that of the control group?

10.44 Implement the test in Problem 10.43, and report a *p*-value.

Mental Health

A study was performed in Lebanon looking at the effect of widowhood on mortality [13]. Each of 151 widowers and 544 widows were matched to a person married at the time of widowhood and of the same age (±2 years) and sex. The people in the matched pairs were followed until one member of the matched pair died. The results in Table 10.27 were obtained for those matched pairs in which at least one member had died by 1980.

***10.45** Suppose all the matched pairs in Table 10.27 are considered. What method of analysis can be used to test if there is an association between widowhood and mortality?

TABLE 10.27 Effect of widowhood on mortality

	Males		Females	
Age (years)	n_1[a]	n_2[b]	n_1	n_2
36–45	4	8	3	2
46–55	20	17	17	10
56–65	42	26	16	15
66–75	21	10	18	11
Unknown	0	2	3	2
Total	87	63	57	40

[a] n_1 = number of pairs in which the widowed subject is deceased and the married subject is alive.
[b] n_2 = number of pairs in which the widowed subject is alive and the married subject is deceased.
Source: Reprinted with the permission of the *American Journal of Epidemiology, 125*(1), 127–132, 1987.

***10.46** Implement the test in Problem 10.45 and report a *p*-value.

***10.47** Answer the same question as Problem 10.45 considering 36- to 45-year-old males only.

***10.48** Implement the test in Problem 10.47 and report a *p*-value.

***10.49** How much power did the study just mentioned based on all matched pairs have versus the alternative hypothesis that a widower is twice as likely to die before a married person of the same age and sex, assuming that all age groups are considered?

Hepatic Disease

Refer to Data Set HORMONE.DAT, on the data disk. (See p. 323 for a description of this Data Set.)

10.50 What test procedure can be used to compare the percentage of hens whose pancreatic secretions increased (post–pre) among the 5 treatment regimens?

10.51 Implement the test procedure in Problem 10.50, and report a *p*-value.

10.52 Answer Problem 10.51 for biliary secretions.

10.53 For all hormone groups except saline, different doses of hormones were administered to different groups of hens. Is there a dose–response relationship between the proportion of hens with increasing pancreatic secretions and the hormone dose? This should be assessed separately for each specific active hormone.

10.54 Answer Problem 10.53 for biliary secretions.

Cardiovascular Disease

A study was performed to look at the association between a parental history of myocardial infarction (MI) and coronary heart disease in women. In the subgroup of 50- to 55-year-old women, it was found that 4 of 900 women with a parental history of MI at age >60 and 1 of 1700 women without any parental history of MI had died from coronary heart disease over a 4-year period. All women were disease-free at the beginning of the period.

10.55 What test procedure can be used to test if there is an association between a parental history of MI at age >60 and the occurrence of fatal coronary heart disease?

10.56 Implement the procedure in Problem 10.55, and report a *p*-value.

Cardiovascular Disease

A secondary prevention trial of lipid lowering is planned in patients with previous myocardial infarction (MI). Patients are to be randomized to either a treatment group getting diet therapy and cholesterol-lowering drugs or a control group getting diet therapy and placebo pills. The study endpoint is to be a combined endpoint consisting of either definite fatal coronary heart disease or nonfatal MI (i.e., a new nonfatal MI distinct from previous events). Suppose it is projected that the incidence of combined events among controls is 7% per year.

***10.57** What proportion of controls will have events over 5 years?

Suppose the treatment benefit is projected to be a reduction in the event rate by 30%.

***10.58** What is the expected event rate in the treated group?

***10.59** How many subjects will be needed in each group if a one-sided test with $\alpha = .05$ is to be used and an 80% chance of finding a significant difference is desired if the rates in Problems 10.57 and 10.58 are the true rates?

An expectation of the investigators is that not all subjects will comply. In particular, it is projected that 5% of the treatment group will not comply with drug therapy, while 10% of the control group will start taking cholesterol-lowering drugs outside the study.

***10.60** What will be the expected rates in Problems 10.57 and 10.58 if this level of lack of compliance is realized?

***10.61** What will be the revised sample size estimate in Problem 10.59 if the lack of compliance is taken into account?

Pediatrics, Endocrinology

A study was performed among 40 boys in a school in Edinburgh to look at the presence of spermatozoa in urine samples according to age [14]. The boys entered the study at age 8–11 years and left the study at age 12–18 years. A 24-hour urine sample was supplied every 3 months by each boy. Table 10.28 gives the presence or absence of sperm cells in the urine samples for each boy together with the ages at entrance and exit of the study and the age at the first sperm-positive urine sample.

For all parts of this question, we will exclude boys who exited this study without 1 sperm-positive urine sample (i.e., boys 8, 9, 14, 25, 28, 29, 30).

10.62 Provide a stem-and-leaf plot of the age at first sperm-positive urine specimen.

***10.63** If we assume that all boys have no sperm cells at age 11 (11.0 years) and all have sperm cells at age 18, then estimate the probability of first developing sperm cells at age 12 (i.e., between 12.0 and 12.9 years), at age 13, at age 14, at age 15, at age 16, at age 17.

***10.64** Suppose the mean age at spermatogenesis = 13.67 years, with standard deviation = 0.89 years and we assume that the age at spermatogenesis follows a normal distribution. The pediatrician would like to know what is the earliest age (in months) before which 95% of boys experience spermatogenesis since he or she would like to refer boys who haven't experienced spermatogenesis by this age to a specialist for further follow-up. Can you estimate this age from the information provided in this part of the problem?

***10.65** Suppose we are uncertain whether a normal distribution provides a good fit to the distribution of the age at spermatogenesis. Answer this question using the results

TABLE 10.28 **Presence (+) or absence (–) of spermatazoa in consecutive urine samples for all boys; age at first collected urine sample, at first positive, and at last sample**

Boy	Age at Entrance	Age at First positive	Age at Exit	Observations
1	10.3	13.4	16.7	− − − − − − − − − − + + − − − − + + + − −
2	10.0	12.1	17.0	− − − − − − − − + − − + + − + − − + − + − − − − − + +
3	9.8	12.1	16.4	− − − − − − − − + − + + − + + + + + + − − + + − +
4	10.6	13.5	17.7	− − − − − − − − − − − + + − − − + − − − −
5	9.3	12.5	16.3	− − − − − − − − − − − − + + − − − − + − − − − − − − −
6	9.2	13.9	16.2	− − − − − − − − − − − − − − − − − + − − − − − − −
7	9.6	15.1	16.7	− − − − − − − − − − − − − − − − − − − + − − − +
8	9.2	—	12.2	− − − − − − − − − − − −
9	9.7	—	12.1	− − − − − − − − −
10	9.6	12.7	16.4	− − − − − − − − − − − − + − + + + + + − − + + − +
11	9.6	12.5	16.7	− − − − − − − − − − + − − + − + − − + + +
12	9.3	15.7	16.0	− + +
14	9.6	—	12.0	− − − − − − − − −
16	9.4	12.6	13.1	− − − − − − − − − − + + + +
17	10.5	12.6	17.5	− − − − − − − + − + + + + + + + + − − + − − + +
18	10.5	13.5	14.1	− − − − − − − − − − + − −
19	9.9	14.3	16.8	− − − − − − − − − − − − − − − + − − − − − + − +
20	9.3	15.3	16.2	− + + +
21	10.4	13.5	17.3	− − − − − − − − + + − + − + + − + − + + +
22	9.8	12.9	16.7	− − − − − − − − − − − + + + + − + + + + − + + − + − −
23	10.8	14.2	17.3	− − − − − − − − − − − − + − − + + + − +
24	10.9	13.3	17.8	− − − − − − − − + + + + − + + + + + − + + − −
25	10.6	—	13.8	− − − − − − − − − − −
26	10.6	14.3	16.3	− − − − − − − − − − − − − + − − − + − − −
27	10.5	12.9	17.4	− − − − − − − − + − + + + + − − − + + − − + + + +
28	11.0	—	12.4	− − − − − −
29	8.7	—	12.3	− − − − − − − − − − − − − −
30	10.9	—	14.5	− − − − − − − − − − − − −
31	11.0	14.6	17.5	− − − − − − − − − − − − + + + + + + + + + + − +
32	10.8	14.1	17.6	− − − − − − − − − − − + + − − + − − − − − −
33	11.3	14.4	18.2	− − − − − − − − − − − + + − + + − − + − − − − −
34	11.4	13.8	18.3	− − − − − − − + − − − + − − − + + + − − + − +
35	11.3	13.7	17.8	− − − − − − − + + + − + − − − + + + − + +
36	11.2	13.5	15.7	− − − − − − − − − + − − − − − − − −
37	11.3	14.5	16.3	− − − − − − − − − − − + − + + − − −
38	11.2	14.3	17.2	− − − − − − − − − − − + − − + − + + + + + + −
39	11.6	13.9	14.7	− − − − − + − − −
40	11.8	14.1	17.9	− − − − + − + − + − + + + + − − − −
41	11.4	13.3	18.2	− − − − + + + − + − − − − − + + + + + − −
42	11.5	14.0	17.9	− − − − − − − + + − − − − − − − + + − + −

from Problems 10.62–10.64. (Assume that the large-sample method discussed in this chapter is applicable to these data.)

Health Services Administration

In the Harvard Medical Practice Study [15], a sample of 31,429 medical records of hospital patients were reviewed

TABLE 10.29 Reproducibility of types of malpractice

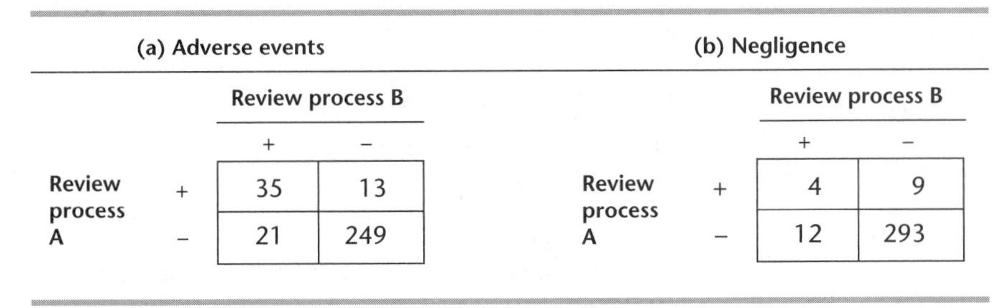

(a) Adverse events

		Review process B	
		+	–
Review process A	+	35	13
	–	21	249

(b) Negligence

		Review process B	
		+	–
Review process A	+	4	9
	–	12	293

to assess the frequency of medical malpractice. Two types of malpractice were identified:

(1) An *adverse event* was defined as an injury caused by medical management (rather than by the underlying disease).

(2) *Negligence* was defined as care that fell below the standard expected of physicians in the community.

An approximate 1% sample of records was reviewed on two different occasions by different review teams. The data in Table 10.29 were obtained.

10.66 Are the frequencies of reporting of adverse events or negligence comparable in review process A (the original review) and review process B (the re-review)?

10.67 Can you assess the reproducibility of adverse events and negligence designations? Which seems to be more reproducible?

Infectious Disease

A study was conducted looking at risk factors for HIV infection among intravenous drug users enrolled in a methadone program [16]. The data in Table 10.30 were presented regarding the presence of HIV antibody among 120 non-Hispanic white subjects by total family income as assessed by blood testing at one point in time.

TABLE 10.30 Presence of HIV antibody among 120 non-Hispanic whites, by income level

Group	Total family income ($)	Number of HIV-positive	Number tested
A	<10,000	13	72
B	10,000–20,000	5	26
C	>20,000	2	22

10.68 What test can be used to assess if total family income is related in a consistent manner to percentage of HIV-positive?

10.69 Implement the test and report a *p*-value.

Sports Medicine

Data from a study concerning tennis elbow (a painful condition experienced by many tennis players) are given in TENNIS1.DAT, and the format is given in TENNIS1.DOC (both on the data disk).

Members of several tennis clubs in the greater Boston area were surveyed. Each participant was asked whether they had ever had tennis elbow, and if so, how many episodes they had. An attempt was made to enroll roughly an equal number of subjects with at least one episode of tennis elbow (the cases) and subjects with no episodes of tennis elbow (the controls). The interviewer asked the participants about other possibly related factors, including demographic factors (e.g., age, sex) and characteristics of their tennis racquet (e.g., string type of racquet used, materials of racquet). This type of study is a case–control study and also can be considered as an **observational study.** It is distinctly different from a clinical trial, where treatments are assigned at random. Because of randomization, subjects receiving different treatments in a clinical trial will, on average, tend to be comparable. In an observational study, we are interested in relating risk factors to disease outcomes. However, it is difficult to make causal inferences (e.g., "wood racquets cause tennis elbow") because subjects are not assigned to a type of racquet at random. Indeed, if we find differences in the frequency of tennis elbow by type of racquet, there may be some other variable(s) that are related to both tennis elbow and to the type of racquet that are more direct "causes" of tennis elbow. Nevertheless, observational studies are useful in obtaining important clues as to disease etiology. One interesting aspect of observational studies is that there are often no prior leads as to which risk factors are even associated with disease. Therefore, investigators tend to ask many questions about possible risk factors without having a firm idea as to which risk factors are really important.

10.70 In this problem, act like a detective and look at each risk factor in the data set separately and relate this risk factor to tennis elbow. How you define tennis elbow is somewhat arbitrary. You may want to compare subjects with 1+ episodes of tennis elbow versus subjects with no episodes. Or you may want to focus specifically on subjects with multiple episodes of tennis elbow, or perhaps create a graded scale according to the number of episodes of tennis elbow (e.g., 0 episodes, 1 episode, 2+ episodes), etc. In this exercise, consider each risk factor separately. In Chapter 13, we will discuss logistic regression methods, where we will be able to study the effects of more than one risk factor simultaneously on disease.

Genetics

Consider Data Set SEXRAT.DAT, on the data disk. This data set consists of the sex of successive births and was first described on page 113. We would like to test the hypothesis that the sex of different children, in the same family, are independent random variables. We will focus on families who have had several successive children of the same sex.

10.71 Consider families with more than 2 children where the first 2 children are of the same sex. Compare the proportion of male third offspring for families where the first 2 children are boys versus families where the first 2 children are girls.

10.72 Consider families with more than 3 children where the first 3 children are of the same sex. Compare the proportion of male fourth offspring for families where the first 3 children are boys versus families where the first 3 offspring are girls.

10.73 Consider families with 5 children where the first 4 children are of the same sex. Compare the proportion of male fifth offspring for families where the first 4 children are boys versus families where the first 4 children are girls.

10.74 What is your overall opinion concerning the hypothesis of independence of gender of successive offspring?

Otolaryngology

Consider Data Set EAR.DAT, on the data disk.

10.75 For children with one affected ear at baseline, compare the efficacy of the two study medications.

10.76 In children with two affected ears at baseline, compare the efficacy of the two study medications, treating the response at 14 days as a graded scale (2 cleared ears, 1 cleared ear, 0 cleared ears).

10.77 For children with two affected ears, test the hypothesis that response to the study medications for the first and second ears are independent.

Hospital Epidemiology

The death of a patient in the hospital is a high-priority medical outcome. Some hospital deaths may be due to inadequate care and are potentially preventable. An *adverse event* (AE) during hospital stay is defined as a problem of any nature and seriousness experienced by a patient during the stay in the hospital that is potentially attributable to clinical or administrative management, rather than the underlying disease. A study was performed in a hospital in Granada, Spain, to assess whether there was a relationship between adverse events and deaths during hospital stay [17]. In this study, 524 cases (i.e., persons who died in the hospital) were identified between January 1, 1990, and January 1, 1991. For each case, a control patient was selected who matched the case on admission diagnosis and admission date. A retrospective chart review was performed to determine the occurrence of adverse events among all cases and controls. The following data were reported: There were 299 adverse events occurring among the cases and 225 among the controls. Among the 299 cases where an adverse event occurred, for 126 their corresponding matched control also had an adverse event.

10.78 What type of study design was used in this investigation?

10.79 What method of analysis can be used to compare the proportion of adverse events between cases and controls?

10.80 Implement the method suggested in Problem 10.79, and report a p-value (two-tail).

Cancer

A topic of current interest is whether abortion is a risk factor for breast cancer. On issue is whether women who have had abortions are comparable to women who have not had abortions in terms of other breast-cancer risk factors. One of the most well-known breast-cancer risk factors is parity (i.e., number of children) with parous women with many children having about a 30% lower risk of breast cancer than nulliparous women (i.e., women with no children). Hence, it is important to assess whether the parity distribution of women with and without previous abortions are comparable. The data in Table 10.31 were obtained from the Nurses' Health Study on this issue.

10.81 What test can be performed to compare the parity distribution of women with and without induced abortions?

10.82 Implement the test in Problem 10.81, and report a p-value (two-tail).

TABLE 10.31 Parity distribution of women with abortions and women without abortions

	Induced abortion	
Parity	Yes (n = 16,353)	No (n = 77,220)
0	34%	29%
1	23%	18%
2	30%	34%
3	10%	15%
4+	3%	4%

Suppose that with each additional child, breast-cancer risk is reduced by 10% (i.e., women with 1 child have a risk of breast cancer that is 90% of that of a nulliparous woman of the same age; women with 2 children have a risk that is $.9^2$ or 81% of that of a nulliparous woman, etc.). (Consider women with 4+ births as having exactly 4 births for the purposes of this problem.)

10.83 Suppose there is no causal effect of induced abortion on breast cancer. Based on the parity distribution in the two groups, would women with induced abortion be expected to have the same, a higher, or lower risk of breast cancer and if higher or lower, by how much? (Assume that the age distributions of women who have or have not had previous abortions are the same.)

Ophthalmology

A 5-year study was performed among 601 subjects with retinitis pigmentosa to assess the effects of high-dose vitamin A (15,000 IU per day) and vitamin E (400 IU per day) on the course of their disease. One issue is to what extent supplementation with vitamin A affected their serum-retinol levels. The following serum-retinol data were obtained over 3 years of follow-up among 73 males taking 15,000 IU/day of vitamin A (vitamin A group) and among 57 males taking 75 IU/day of vitamin A (the trace group) (a negligible amount compared to usual dietary intake of 3000 IU/day):

		Year 0	Year 3
Retinol (mmol/L)	N	Mean ± *sd*	Mean ± *sd*
Vitamin A group	73	1.89 ± 0.36	2.06 ± 0.53
Trace group	57	1.83 ± 0.31	1.78 ± 0.30

10.84 What test can be used to assess if mean serum retinol has increased over 3 years among subjects in the vitamin A group?

10.85 Can the test be implemented based on the data just presented? Why or why not? If so, implement the test and report a two-tailed p-value.

10.86 One assumption of the test in Problem 10.84 is that the distribution of serum retinol is approximately normal. To verify this assumption, the investigators obtained a frequency distribution of serum retinol at year 0 among males in the vitamin A group, with data shown as follows:

Serum-retinol group (μmol/L)	N
≤1.40	6
1.41–1.75	22
1.76–2.10	22
2.11–2.45	20
≥2.46	3
	73

Perform a statistical test to check on the normality assumption. Based on your results, do you feel the assumption of normality is warranted? Why or why not?

Ophthalmology

One interesting aspect of the study described in Problem 10.84 is to assess changes in other parameters as a result of supplementation with vitamin A. One quantity of interest is the level of serum triglycerides. it was found that among 133 subjects in the vitamin A group (males and females combined) who were in the normal range at baseline (<2.13 μmol/L), 15 were above the upper limit of normal at each of their last 2 consecutive study visits. Similarly, among 138 subjects in the trace group who were in the normal range at baseline (<2.13 μmol/L), 2 were above the upper limit of normal at each of their last 2 consecutive study visits [18].

10.87 What proportion of people in each group developed abnormal triglyceride levels over the course of the study? Are these proportions measures of prevalence, incidence, or neither?

10.88 What test can be performed to compare the percent of subjects who develop abnormal triglyceride levels between the vitamin A group and the trace group?

10.89 Implement the test in Problem 10.88, and report a p-value (two-tail).

Pulmonary Disease

A study was conducted relating morbidity in children with asthma to exposure to cockroach allergen (Rosenstreich et al. [19]). Household surveillance was performed, and assays were obtained of dust samples to detect cockroach, dust mite, and cat allergen. The investigators wanted to compare their results to a larger study (The National Cooperative Inner-City Asthma Study). The following data were presented comparing ethnic distributions in the two studies:

Ethnic group	Study sample ($n = 475$)	National Cooperative Inner-City Asthma Study ($n = 1512$)
Hispanic	78	295
Black	371	1111
Other	26	106

10.90 What test can be used to compare the ethnic distribution between the two studies?

10.91 Implement the test in Problem 10.90 and report a *p*-value (two-tail).

Zoology

A study was performed to look at the preference of different species of birds for different types of sunflower seeds. Two bird feeders were set up with different types of sunflower seeds, one with a black oil seed and one with a striped seed. The bird feeders were observed for a 1-hour period for each of 12 days over a 1-month period. The number of birds of different species who ate seeds from a specific bird feeder was counted for each bird feeder for each of a number of species of birds. (The data for this problem were supplied by David Rosner.)

On the first day of testing, there was 1 titmouse who ate the black oil seeds and 4 titmice who ate the striped seeds. There were a total of 19 goldfinches who ate the black oil seeds and 5 goldfinches who ate the striped seeds.

10.92 What test can be performed to assess if the feeding preferences of titmice and goldfinches are comparable on the first day of testing?

10.93 Implement the test in Problem 10.92 and report a *p*-value.

One assumption in the entire experiment is that the feeding preferences of the same species of bird remains the same over time. To test this assumption, the data for goldfinches were separated by the 6 different days on which they were observed (they were not observed at all for the other 6 days). For 2 of the 6 days there were small numbers of goldfinches observed (2 on one day and 1 on another day). Thus, data from these two days were not included. The results are as follows for the remaining 4 days:

Feeding preferences of goldfinches on different days

	Day				
Type of seed	1	2	3	4	Total
Black oil	19	14	9	45	87
Striped	5	10	6	39	60

10.94 What test can be used to assess if the feeding preference of goldfinches are the same on different days?

10.95 Implement the test in Problem 10.94 and report a *p*-value.

Cancer

The Physicians' Health Study was a randomized double-blind placebo-controlled trial of beta-carotene (50 mg every other day). 22,071 male physicians age 40–84 were enrolled in 1982. The subjects were followed until December 31, 1995, for the development of new cancers (malignant neoplasms). The following results were reported (Hennekens et al. [20]).

	Beta-carotene ($n = 11{,}036$)	Placebo ($n = 11{,}035$)
Malignant neoplasms	1273	1293

10.96 What test can be used to compare cancer incidence rates between the two treatment groups?

10.97 Implement the test in Problem 10.96, and report a *p*-value (2-tail).

10.98 The expectation before the study started was that beta-carotene might prevent 10% of incident cancers relative to placebo. How much power did the study have to detect an effect of this magnitude if a two-sided test is used with $\alpha = .05$ and we assume that the true incidence rate in the placebo group is the same as the observed incidence rate?

Demography

A common assumption is that the gender of successive offspring are independent. To test this assumption, birth records were collected from the first 5 births in 51,868 families. In families with exactly 5 children, the following is a frequency distribution of number of male offspring from the Data Set SEXRAT.DAT (see p. 113).

Frequency distribution of number of male offspring in families of size 5

Number of male offspring	n
0	518
1	2245
2	4621
3	4753
4	2476
5	549
Total	15,162

Suppose the investigators doubt that the probability of a male birth is exactly 50%, but are willing to assume that the gender of successive offspring are independent.

10.99 What is the best estimate of the probability of a male birth based on the observed data?

10.100 What is the probability of 0, 1, 2, 3, 4, and 5 male births out of 5 births based on the estimate in Problem 10.99?

10.101 Test the hypothesis that the gender of successive offspring are independent based on the model in Problem 10.100. What are your conclusions concerning the hypothesis?

REFERENCES

[1] MacMahon, B., Cole, P., Lin, T. M., Lowe, C. R., Mirra, A. P., Ravnihar, B., Salber, E. J., Valaoras, V. G., & Yuasa, S. (1970). Age at first birth and breast cancer risk. *Bulletin of the World Health Organization, 43,* 209–221.

[2] Doll, R., Muir, C., & Waterhouse, J. (Eds.). (1970). *Cancer in five continents* (Vol. II). Berlin: Springer-Verlag.

[3] Cochran, W. G. (1954). Some methods for strengthening the common χ^2 test. *Biometrics, 10,* 417–451.

[4] Maxwell, A. E. (1961). *Analyzing qualitative data.* London: Methuen.

[5] Hypertension Detection and Follow-up Program Cooperative Group. (1977). Blood pressure studies in 14 communities—A two-stage screen for hypertension. *JAMA, 237*(22), 2385–2391.

[6] Landis, J. R., & Koch, G. G. (1977). The measurement of observer agreement for categorical data. *Biometrics, 33,* 159–174.

[7] Fleiss, J. (1981). *Statistical methods for rates and proportions.* New York: Wiley.

[8] Cramer, D. W., Schiff, I., Schoenbaum, S. C., Gibson, M., Belisle, J., Albrecht, B., Stillman, R. J., Berger, M. J., Wilson, E., Stadel, B. V., & Seibel, M. (1985). Tubal infertility and the intrauterine device. *New England Journal of Medicine, 312*(15), 941–947.

[9] Weiss, N. S., & Sayetz, T. A. (1980). Incidence of endometrial cancer in relation to the use of oral contraceptives. *New England Journal of Medicine, 302*(10), 551–554.

[10] U.S. Department of Health, Education, and Welfare. (1972). *End results in cancer* (Report No. 4).

[11] Mecklenburg, R. S., Benson, E. A., Benson, J. W., Fredlung, P. N., Guinn, T., Metz, R. J., Nielsen, R. L., & Sannar, C. A. (1984). Acute complications associated with insulin pump therapy: Report of experience with 161 patients. *JAMA, 252*(23), 3265–3269.

[12] Dubach, U. C., Rosner, B., & Stürmer, T. (1991). An epidemiological study of abuse of analgesic drugs: Effects of phenacetin and salicylate on mortality and cardiovascular morbidity (1968–1987). *New England Journal of Medicine, 324,* 155–160.

[13] Armenian, H., Saadeh, F. M., & Armenian, S. L. (1987). Widowhood and mortality in an Armenian church parish in Lebanon. *American Journal of Epidemiology, 125*(1), 127–132.

[14] Jorgensen, M., Keiding, N., & Skakkebaek, N. E. (1991). Estimation of spermache from longitudinal spermaturia data. *Biometrics, 47,* 177–193.

[15] Brennan, T. A., Leake, L. L., Laird, N. M., Hebert, L., Localio, A. S., Lawthers, A. G., Newhouse, J. P., Weiler, P. G., & Hiatt, H. H. (1991). Incidence of adverse events and negligence in hospitalized patients. Results of the Harvard Medical Practice Study I. *New England Journal of Medicine, 324*(6), 370–376.

[16] Schoenbaum, E. E., Hartel, D., Selwyn, P. A., Klein, R. S., Davenny, K., Rogers, M., Feiner, C., & Friedland, G. (1989). Risk factors for human immunodeficiency virus infection in intravenous drug users. *New England Journal of Medicine, 321*(13), 874–879.

[17] García-Martín, M., Lardelli-Claret, P., Bueno-

Cavanillas, A., Luna-del-Castillo, J. D., Espigares-Garcia, M., & Galvez-Vargas, R. (1997). Proportion of hospital deaths associated with adverse events. *Journal of Clinical Epidemiology, 50*(12), 1319–1326.

[18] Sibulesky, L., Hayes, K. C., Pronczuk, A., Weigel-DiFranco, C., Rosner, B., & Berson, E. L. (1999). Safety of <7500 RE (<25000 IU) vitamin A daily in adults with retinitis pigmentosa. *American Journal of Clinical Nutrition 69*(4): 656–663.

[19] Rosenstreich, D. L., Eggleston, P., Kattan, M., Baker, D., Slavin, R. G., Gergen, P., Mitchell, H., McNiff-Mortimer, K., Lynn, H., Ownby, D., & Malveaux, F. (1997). The role of cockroach allergy and exposure to cockroach allergen in causing morbidity among inner-city children with asthma. *New England Journal of Medicine, 336*(19), 1356–1363.

[20] Hennekens, C. H., Buring, J. E., Manson, J. E., Stampfer, M., Rosner, B., Cook, N. R., Belanger, C., LaMotte, F., Gaziano, J. M., Ridker, P. M., Willett, W., & Peto, R. (1996). Lack of effect of long-term supplementation with beta carotene on the incidence of malignant neoplasms and cardiovascular disease. *New England Journal of Medicine, 334*(18), 1145–1149.

11 Regression and Correlation Methods

SECTION 11.1 Introduction

In Chapter 8, statistical methods for comparing the means of a normally distributed outcome variable between two populations were presented based on t tests. Suppose we call the outcome variable y and we call the group classification (or class) variable x. For t test applications, x takes on two values. Another way of looking at the methods in Chapter 8 is as methods for assessing the possible association between a normally distributed variable y and a categorical variable x. We will see that these methods are special cases of **linear-regression** techniques. In linear regression, we will study how to relate a normally distributed outcome variable y to one or more predictor variables $x_1, \ldots, x_k$ where the x's may be either continuous or categorical variables.

Example 11.1 **Obstetrics** Obstetricians sometimes order tests for estriol levels from 24-hour urine specimens taken from pregnant women who are near term, since the level of estriol has been found to be related to the birthweight of the infant. The test can provide indirect evidence of an abnormally small fetus. The relationship between estriol level and birthweight can be quantified by fitting a *regression line* that relates the two variables.

In Chapter 10, we also studied the Kappa statistic, which is a measure of association between two categorical variables. This index is useful when we are interested in how strong the association is between two categorical variables rather than in predicting one variable as a function of the other variable. To quantify the association between two continuous variables, we can use the correlation coefficient introduced in Section 5.6.1. In this chapter, we will consider hypothesis-testing methods for correlation coefficients and will extend the concept of a correlation coefficient to describe association among several continuous variables.

Example 11.2 **Hypertension** Much discussion has taken place in the literature concerning the familial aggregation of blood pressure. In general, children whose parents have high blood pressure tend to have higher blood pressure than their peers. One way of expressing this relationship is to compute a *correlation coefficient* relating the blood pressure of parents and children over a large collection of families.

In this chapter, we discuss methods of regression and correlation analysis in which *two* different variables in the same sample are related. The extension of these methods to the case of multiple-regression analysis, where the relationship between more than two variables at a time is considered, is also discussed.

SECTION 11.2 General Concepts

Example 11.3 **Obstetrics** Greene and Touchstone conducted a study to relate birthweight to the estriol level of pregnant women [1]. Figure 11.1 is a plot of the data from the study, and the actual data points are listed in Table 11.1. As can be seen from the figure, there appears to be a linear relationship between estriol level and birthweight, although this relationship is not consistent and considerable scatter exists throughout the plot. How can this relationship be quantified?

If x = estriol level and y = birthweight, then a relationship between y and x that is of the following form may be postulated:

EQUATION 11.1

$$E(y|x) = \alpha + \beta x$$

That is, for a given estriol level x, the average birthweight $E(y|x)$ is $\alpha + \beta x$.

DEFINITION 11.1 The line $y = \alpha + \beta x$ is the **regression line**, where α is the intercept and β is the slope of the line.

FIGURE 11.1 **Data from Greene-Touchstone study relating birthweight and estriol level in pregnant women near term**

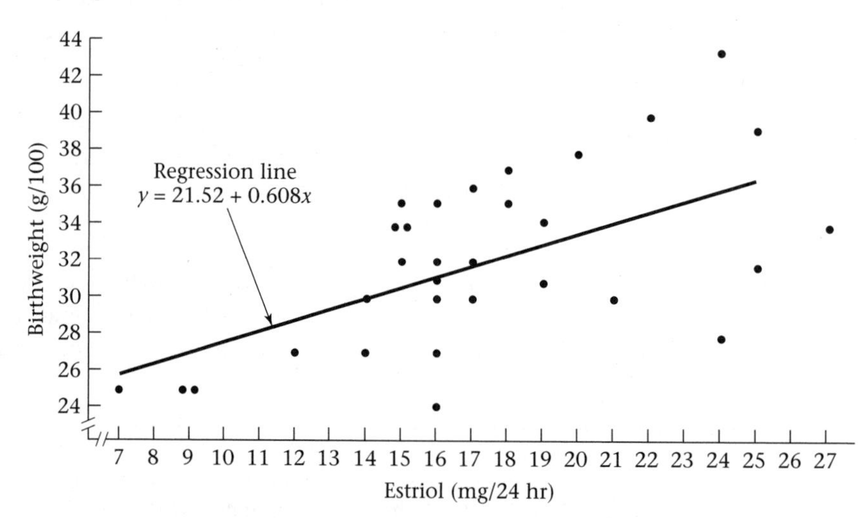

Source: Reprinted with permission of the *American Journal of Obstetrics and Gynecology, 85*(1), 1–9, 1963.

TABLE 11.1 **Sample data from the Greene-Touchstone study relating birthweight and estriol level in pregnant women near term**

i	Estriol (mg/24 hr) x_i	Birthweight (g/100) y_i	i	Estriol (mg/24 hr) x_i	Birthweight (g/100) y_i
1	7	25	17	17	32
2	9	25	18	25	32
3	9	25	19	27	34
4	12	27	20	15	34
5	14	27	21	15	34
6	16	27	22	15	35
7	16	24	23	16	35
8	14	30	24	19	34
9	16	30	25	18	35
10	16	31	26	17	36
11	17	30	27	18	37
12	19	31	28	20	38
13	21	30	29	22	40
14	24	28	30	25	39
15	15	32	31	24	43
16	16	32			

Source: Reprinted with permission of the *American Journal of Obstetrics and Gynecology, 85*(1), 1–9, 1963.

The relationship $y = \alpha + \beta x$ is not expected to hold exactly for every woman. For example, not all women with a given estriol level will have babies with identical birthweights. Thus, an error term e, which represents the variance of birthweight among all babies of women with a given estriol level x, is introduced into the model. We will assume that e follows a normal distribution with mean 0 and variance σ^2. The full linear-regression model then takes the following form:

EQUATION 11.2

$$y = \alpha + \beta x + e$$

where e is normally distributed with mean 0 and variance σ^2.

DEFINITION 11.2

For any linear-regression equation of the form $y = \alpha + \beta x + e$, y is referred to as the **dependent variable** and x as the **independent variable**, because we are trying to predict y from x.

Example 11.4 **Obstetrics** Birthweight is the dependent variable and estriol is the independent variable for the problem posed in Example 11.3, because estriol levels are being used to try to predict birthweight.

One interpretation of the regression line is that for a woman with estriol level x, the corresponding birthweight will be normally distributed with mean $\alpha + \beta x$ and variance σ^2. If σ^2 were 0, then every point would fall exactly on the regression

FIGURE 11.2 **The effect of σ^2 on the goodness of fit of a regression line**

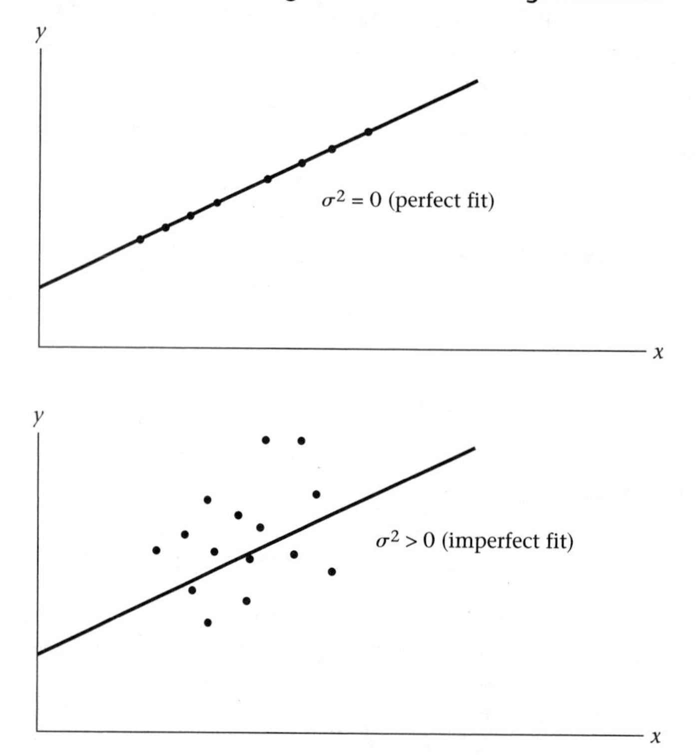

line, whereas the larger σ^2 is, the more scatter occurs about the regression line. This relationship is illustrated in Figure 11.2.

How can β be interpreted? If β is greater than 0, then as x increases, the expected value of $y = \alpha + \beta x$ will increase.

Example 11.5 **Obstetrics** This situation appears to be the case in Figure 11.3a for birthweight (y) and estriol (x), since as estriol increases, the average birthweight correspondingly increases.

If β is less than 0, then as x increases, the expected value of y will decrease.

Example 11.6 **Pediatrics** This situation might occur in a plot of pulse rate (y) versus age (x), as illustrated in Figure 11.3b, because infants are born with rapid pulse rates that gradually decline with age.

If β is equal to 0, then there is no linear relationship between x and y.

Example 11.7 This situation might occur in a plot of birthweight versus birthday, as shown in Figure 11.3c, because there is no relationship between birthweight and birthday.

FIGURE 11.3 The interpretation of the regression line for different values of β

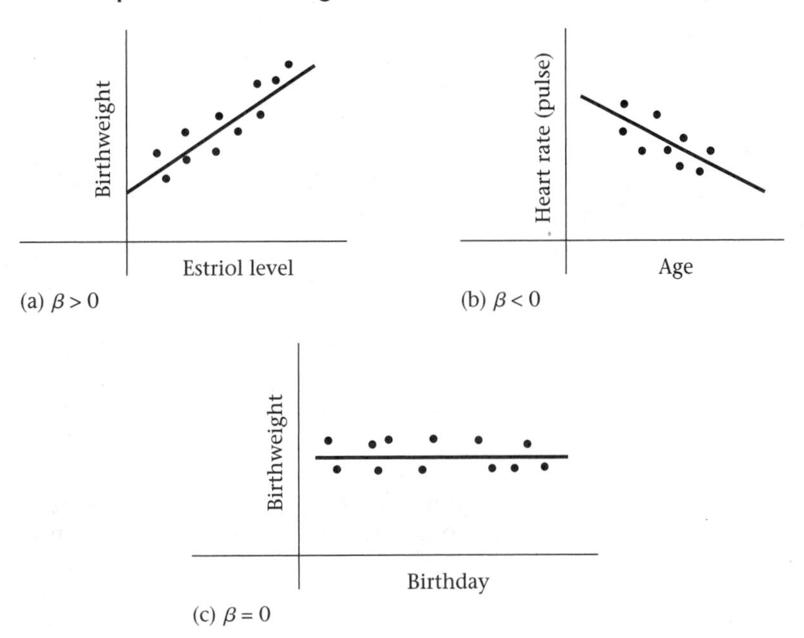

SECTION 11.3 Fitting Regression Lines— The Method of Least Squares

The question remains as to how to fit a regression line (or, equivalently, to obtain estimates of α and β, denoted by a and b, respectively) when data appear in the form of Figure 11.1. We could eyeball the data and draw a line that is not too distant from any of the points, but this approach is difficult in practice and can be quite imprecise with either a large number of points or a lot of scatter. A better method is to set up a specific criterion that defines the closeness of a line to a set of points and to find the line closest to the sample data according to this criterion.

Consider the data in Figure 11.4 and the estimated regression line $y = a + bx$. The distance d_i of a typical sample point (x_i, y_i) from the line could be measured along a direction parallel to the y-axis. If we let $(x_i, \hat{y}_i) = (x_i, a + bx_i)$ be the point on the estimated regression line at x_i, then this distance is given by $d_i = y_i - \hat{y}_i = y_i - a - bx_i$. A good-fitting line would make these distances as small as possible. Since the d_i cannot all be 0, the criterion S_1 = sum of the absolute deviations of the sample points from the line = $\sum_{i=1}^{n} |d_i|$ can be used and the line that minimizes S_1 can be found. This strategy has proven analytically difficult. Instead, for both theoretical reasons and ease of derivation, the following least-squares criterion is commonly used:

$$S = \text{sum of the squared distances of the points from the line}$$
$$= \sum_{i=1}^{n} d_i^2 = \sum_{i=1}^{n} (y_i - a - bx_i)^2$$

FIGURE 11.4 **Possible criteria for judging the fit of a regression line**

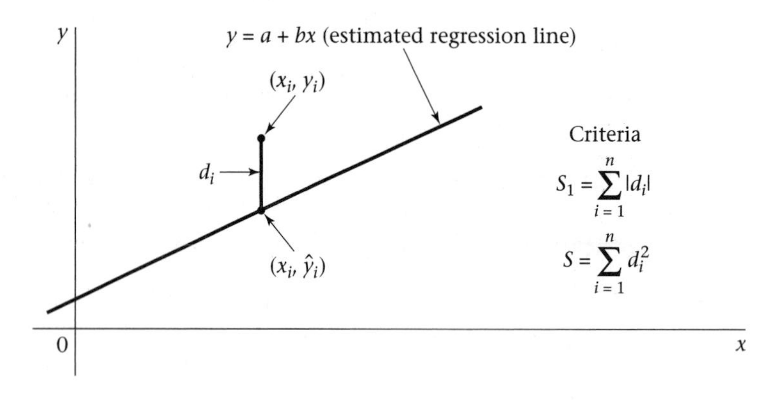

DEFINITION 11.3 The **least-squares line,** or **estimated regression line,** is the line $y = a + bx$ that minimizes the sum of squared distances of the sample points from the line given by

$$S = \sum_{i=1}^{n} d_i^2$$

This method of estimating the parameters of the regression line is known as the **method of least squares.**

The following notation is needed to define the slope and intercept of a regression line:

DEFINITION 11.4 The **raw sum of squares for x** is defined by

$$\sum_{i=1}^{n} x_i^2$$

The **corrected sum of squares for x** is denoted by L_{xx} and defined by

$$\sum_{i=1}^{n} (x_i - \bar{x})^2 = \sum_{i=1}^{n} x_i^2 - \left(\sum_{i=1}^{n} x_i\right)^2 \Big/ n$$

It represents the sum of squares of the deviations of the x_i from the mean. Similarly, the **raw sum of squares for y** is defined by

$$\sum_{i=1}^{n} y_i^2$$

The **corrected sum of squares for y** is denoted by L_{yy} and defined by

$$\sum_{i=1}^{n} (y_i - \bar{y})^2 = \sum_{i=1}^{n} y_i^2 - \left(\sum_{i=1}^{n} y_i\right)^2 \Big/ n$$

Notice that L_{xx} and L_{yy} are simply the numerators of the expressions for the sample variances of x (i.e., s_x^2) and y (i.e., s_y^2), respectively, since

$$s_x^2 = \sum_{i=1}^{n} (x_i - \bar{x})^2 \big/ (n-1) \quad \text{and} \quad s_y^2 = \sum_{i=1}^{n} (y_i - \bar{y})^2 \big/ (n-1)$$

DEFINITION 11.5

The **raw sum of cross products** is defined by

$$\sum_{i=1}^{n} x_i y_i$$

The **corrected sum of cross products** is defined by

$$\sum_{i=1}^{n} (x_i - \bar{x})(y_i - \bar{y})$$

which is denoted by L_{xy}.

It can be shown that a short form for the corrected sum of cross products is given by

$$\sum_{i=1}^{n} x_i y_i - \left(\sum_{i=1}^{n} x_i\right)\left(\sum_{i=1}^{n} y_i\right) \Big/ n$$

What does the corrected sum of cross products mean? Suppose that $\beta > 0$. From Figure 11.3a, we see that if $\beta > 0$, then as x increases, y will tend to increase as well. Another way of expressing this relationship is that if $(x_i - \bar{x})$ is greater than 0 (which will be true for large values of x_i), then y_i will tend to be large or $y_i - \bar{y}$ will be greater than 0 and $(x_i - \bar{x})(y_i - \bar{y})$ will be the product of two positive numbers and thus will be positive. Similarly, if $x_i - \bar{x}$ is < 0, then $y_i - \bar{y}$ will also tend to be < 0 and $(x_i - \bar{x})(y_i - \bar{y})$ will be the product of two negative numbers and thus will be positive. Thus, if $\beta > 0$, the sum of cross products will tend to be positive. Suppose that $\beta < 0$. From Figure 11.3b, when x is small, y will tend to be large and when x is large, y will tend to be small. In both cases, $(x_i - \bar{x})(y_i - \bar{y})$ will often be the product of 1 positive and 1 negative number and will be negative. Thus, if $\beta < 0$, the sum of cross products will tend to be negative. Finally, if $\beta = 0$, then x and y bear no linear relation to each other and the sum of cross products will be close to 0.

It can be shown that the best estimate of the underlying slope β is given by $b = L_{xy}/L_{xx}$. Because L_{xx} is always positive (except in the degenerate case when all x's in the sample are the same), the sign of b is the same as the sign of the sum of cross products L_{xy}. This makes good intuitive sense based on the preceding discussion. Furthermore, for a given estimate of the slope b, it can be shown that the value of the intercept for the line that satisfies the least-squares criterion is given by $a = \bar{y} - b\bar{x}$. We summarize these results as follows:

EQUATION 11.3

Estimation of the Least-Squares Line The coefficients of the least-squares line $y = a + bx$ are given by

$$b = L_{xy}/L_{xx} \quad \text{and} \quad a = \bar{y} - b\bar{x} = \left(\sum_{i=1}^{n} y_i - b\sum_{i=1}^{n} x_i\right) \Big/ n$$

Example 11.8 **Obstetrics** Derive the regression line for the data in Table 11.1.

Solution First

$$\sum_{i=1}^{31} x_i \quad \sum_{i=1}^{31} x_i^2 \quad \sum_{i=1}^{31} y_i \quad \sum_{i=1}^{31} x_i y_i$$

must be obtained so as to compute the corrected sum of squares (L_{xx}) and cross products (L_{xy}). These quantities are given as follows:

$$\sum_{i=1}^{31} x_i = 534 \quad \sum_{i=1}^{31} x_i^2 = 9876 \quad \sum_{i=1}^{31} y_i = 992 \quad \sum_{i=1}^{31} x_i y_i = 17{,}500$$

Then compute L_{xy} and L_{xx} as follows:

$$L_{xy} = \sum_{i=1}^{31} x_i y_i - \left(\sum_{i=1}^{31} x_i\right)\left(\sum_{i=1}^{31} y_i\right)\Bigg/31 = 17{,}500 - (534)(992)/31 = 412$$

$$L_{xx} = \sum_{i=1}^{31} x_i^2 - \left(\sum_{i=1}^{31} x_i\right)^2\Bigg/31 = 9876 - 534^2/31 = 677.42$$

Finally, compute the slope of the regression line as follows:

$$b = L_{xy}/L_{xx} = 412/677.42 = 0.608$$

The intercept of the regression line can also be computed. Note from Equation 11.3 that

$$a = \left(\sum_{i=1}^{31} y_i - 0.608\sum_{i=1}^{31} x_i\right)\Bigg/31 = [992 - 0.608(534)]/31 = 21.52$$

Thus the regression line is give by $y = 21.52 + 0.608x$. This regression line is shown in Figure 11.1.

How can the regression line be used? One of its uses is to *predict* values of y for given values of x.

DEFINITION 11.6 The **predicted,** or **average, value of** y for a given value of x, as estimated from the fitted regression line, is denoted by $\hat{y} = a + bx$. Thus the point $(x, a + bx)$ is always on the regression line.

Example 11.9 **Obstetrics** What is the estimated average birthweight if a pregnant woman has an estriol level of 15 mg/24 hr?

Solution If the estriol level were 15 mg/24 hr, then the best prediction of the average value of y would be

$$\hat{y} = 21.52 + 0.608(15) = 30.65$$

Since y is in the units of birthweight (g)/100, the estimated average birthweight $= 30.65 \times 100 = 3065$ g.

One possible use of estriol levels is to identify women who are carrying a low-birthweight fetus. If such women can be identified, then drugs might be used to prolong the pregnancy until the fetus grows larger, because low-birthweight infants are at greater risk than normal infants for infant mortality in the first year of life and for poor growth and development in childhood.

Example 11.10 **Obstetrics** Low birthweight is defined here as ≤2500 g. For what estriol level would the predicted birthweight be 2500 g?

Solution Note that the predicted value of y (birthweight/100) is

$$\hat{y} = 21.52 + 0.608x$$

If $\hat{y} = 2500/100 = 25$, then x can be obtained from the equation

$$25 = 21.52 + 0.608x \qquad \text{or} \qquad x = (25 - 21.52)/0.608 = 3.48/0.608 = 5.72$$

Thus if a woman has an estriol level of 5.72 mg/24 hr, then the predicted birthweight would be 2500 g. Furthermore, the predicted birthweight for all women with estriol levels of ≤ 5 mg/24 hr would be < 2500 g (assuming that estriol can only be measured in increments of 1 mg/24 hr). This level could serve as a critical value for identifying high-risk women and attempting to prolong their pregnancies.

How can the slope of the regression line be interpreted? The slope of the regression line tells us the amount that y increases per unit increase in x.

Example 11.11 **Obstetrics** Interpret the slope of the regression line for the birthweight–estriol data in Example 11.1.

Solution The slope of 0.608 tells us that the predicted y increases by about 0.61 units per 1 mg/24 hr. Thus, the predicted birthweight increases by 61 g per 1 mg/24 hr, increase in estriol.

In this section, we learned how to fit regression lines using the method of least squares based on the linear-regression model in Equation 11.1 Note that the method of least squares is appropriate whenever the average residual for each given value of x is 0—that is, when $E(e|X = x) = 0$ in Equation 11.1. Normality of the residuals is not strictly required. However, the normality assumption in Equation 11.1 is necessary to perform hypothesis tests concerning regression parameters, as discussed in the next section.

Referring to the flowchart at the end of this chapter (Figure 11.32, p. 502), we would answer yes to (1) interested in relationships between two variables? (2) both variables continuous? and (3) interested in predicting one variable from another? This would lead us to the box entitled "Simple linear regression."

SECTION 11.4 Inferences About Parameters from Regression Lines

In Section 11.3, the fitting of regression lines using the method of least squares was discussed. Since this method can be used with any set of points, criteria for distinguishing regression lines that fit the data well from those that do not must be established. Consider the regression line depicted in Figure 11.5.

FIGURE 11.5 **Goodness of fit of a regression line**

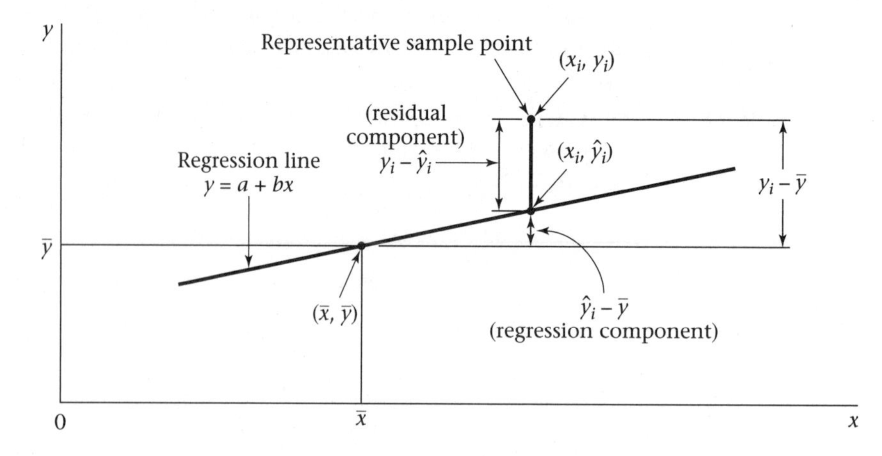

A hypothetical regression line and a representative sample point have been drawn. First, notice that the point $(\bar{x}, \bar{y})$ falls on the regression line. This feature is common to all regression lines, because a regression line can be represented as

$$y = a + bx = \bar{y} - b\bar{x} + bx = \bar{y} + b(x - \bar{x})$$

or, equivalently,

EQUATION 11.4

$$y - \bar{y} = b(x - \bar{x})$$

If $\bar{x}$ is substituted for x and $\bar{y}$ for y in Equation 11.4, then 0 is obtained on both sides of the equation, which shows that *the point* $(\bar{x}, \bar{y})$ *must always fall on the regression line.* If a typical sample point (x_i, y_i) is selected and a line is drawn through this point parallel to the y-axis, then the representation in Figure 11.5 is obtained.

DEFINITION 11.7

For any sample point (x_i, y_i), the **residual**, or **residual component**, of that point about the regression line is defined by $y_i - \hat{y}_i$.

DEFINITION 11.8

For any sample point (x_i, y_i), the **regression component** of that point about the regression line is defined by $(\hat{y}_i - \bar{y})$.

In Figure 11.5 the deviation $y_i - \bar{y}$ can be separated into residual $(y_i - \hat{y}_i)$ and regression $(\hat{y}_i - \bar{y})$ components. Note that if the point (x_i, y_i) fell exactly on the regression line, then $y_i = \hat{y}_i$ and the residual component $y_i - \hat{y}_i$ would be 0 and $y_i - \bar{y} = \hat{y}_i - \bar{y}$. Generally speaking, a good-fitting regression line will have regression components large in absolute value relative to the residual components, whereas the opposite is true for poor-fitting regression lines. Some typical situations are depicted in Figure 11.6.

FIGURE 11.6 **Regression lines with varying goodness-of-fit relationships**

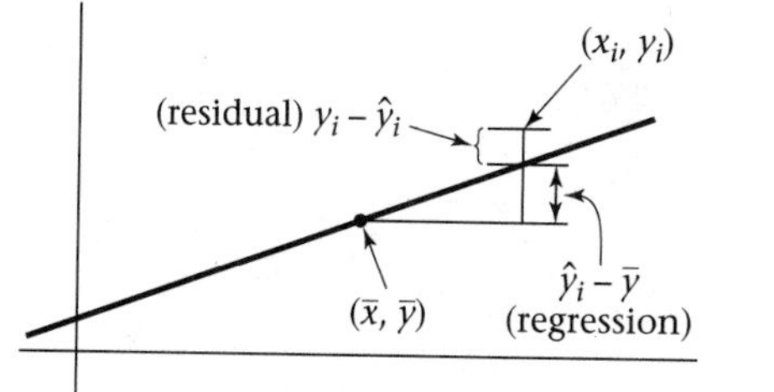

(a) Large regression, small residual components

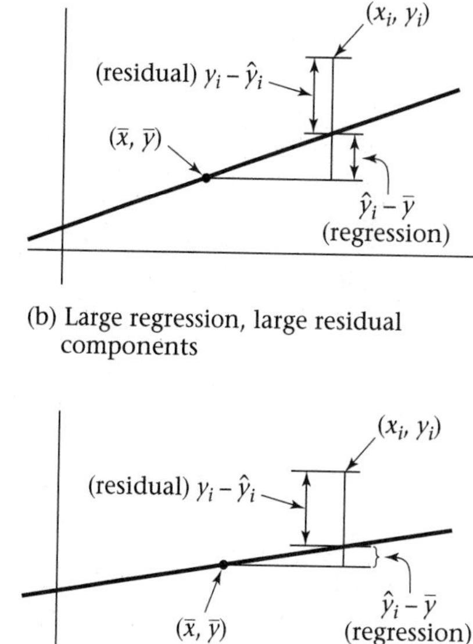

(b) Large regression, large residual components

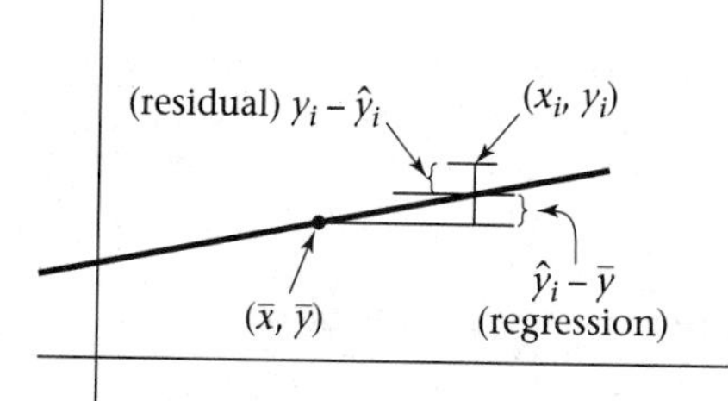

(c) Small regression, small residual components

(d) Small regression, large residual components

The best-fitting regression line is depicted in Figure 11.6a, with large regression components and small residual components. The worst-fitting regression line is depicted in Figure 11.6d, which has small regression components and large residual components. Intermediate situations for goodness of fit are depicted in Figures 11.6b and 11.6c.

How can the plots in Figure 11.6 be quantified? One strategy is to square the deviations about the mean $y_i - \bar{y}$, sum them up over all points, and decompose this sum of squares into regression and residual components.

DEFINITION 11.9 The **total sum of squares**, or **Total SS**, is the sum of squares of the deviations of the individual sample points from the sample mean:

$$\sum_{i=1}^{n} \left(y_i - \bar{y}\right)^2$$

DEFINITION 11.10 The **regression sum of squares**, or **Reg SS**, is the sum of squares of the regression components:

$$\sum_{i=1}^{n} \left(\hat{y}_i - \bar{y}\right)^2$$

DEFINITION 11.11 The **residual sum of squares**, or **Res SS**, is the sum of squares of the residual components:

$$\sum_{i=1}^{n}\left(y_i - \hat{y}_i\right)^2$$

It can be shown that the following relationship is true:

EQUATION 11.5 **Decomposition of the Total Sum of Squares into Regression and Residual Components**

$$\sum_{i=1}^{n}\left(y_i - \bar{y}\right)^2 = \sum_{i=1}^{n}\left(\hat{y}_i - \bar{y}\right)^2 + \sum_{i=1}^{n}\left(y_i - \hat{y}_i\right)^2$$

or Total SS = Reg SS + Res SS

11.4.1 *F* Test for Simple Linear Regression

The criterion for goodness of fit that will be used is the ratio of the regression sum of squares to the residual sum of squares. A large ratio indicates a good fit, whereas a small ratio indicates a poor fit. In hypothesis-testing terms we wish to test the hypothesis H_0: $\beta = 0$ versus H_1: $\beta \neq 0$, where β is the underlying slope of the regression line in Equation 11.2.

The following terms are introduced for ease of notation in describing the hypothesis test:

DEFINITION 11.12 The **regression mean square**, or **Reg MS**, is the Reg SS divided by the number of predictor variables (k) in the model. Thus Reg MS = Reg SS/k. For simple linear regression, which we have been discussing, $k = 1$ and thus Reg MS = Reg SS. For multiple regression in Section 11.9, k will be >1. We will refer to k as the degrees of freedom for the regression sum of squares, or Reg *df*.

DEFINITION 11.13 The **residual mean square**, or **Res MS**, is the ratio of the Res SS divided by $(n - k - 1)$, or Res MS = Res SS/$(n - k - 1)$. For simple linear regression, $k = 1$ and Res MS = Res SS/$(n - 2)$. We will refer to $n - k - 1$ as the degrees of freedom for the residual sum of squares, or Res *df*. Res MS is also sometimes denoted by $s^2_{y \cdot x}$ in the literature.

Under H_0, F = Reg MS/Res MS follows an F distribution with 1 and $n - 2$ *df*, respectively. H_0 should be rejected for large values of F. Thus, for a level α test, H_0 will be rejected if $F > F_{1,n-2,1-\alpha}$ and accepted otherwise.

The expressions for the regression and residual sums of squares in Equation 11.5 simplify for computational purposes as follows:

EQUATION 11.6

Short Computational Form for Regression and Residual SS

$$\text{Regression SS} = bL_{xy} = b^2 L_{xx} = L_{xy}^2 / L_{xx}$$

$$\text{Residual SS} = \text{Total SS} - \text{Regression SS} = L_{yy} - L_{xy}^2 / L_{xx}$$

Thus the test procedure can be summarized as follows:

EQUATION 11.7

***F* Test for Simple Linear Regression** To test H_0: $\beta = 0$ versus H_1: $\beta \neq 0$, use the following procedure:

(1) Compute the test statistic

$$F = \text{Reg MS/Res MS} = \left(L_{xy}^2 / L_{xx}\right) / \left[\left(L_{yy} - L_{xy}^2 / L_{xx}\right) / (n-2)\right]$$

that follows an $F_{1,n-2}$ distribution under H_0.

(2) For a two-sided test with significance level α, if

$F > F_{1,n-2,1-\alpha}$ then reject H_0; if

$F \le F_{1,n-2,1-\alpha}$ then accept H_0.

(3) The exact *p*-value is given by $Pr(F_{1,n-2} > F)$.

The acceptance and rejection regions for the regression *F* test are illustrated in Figure 11.7. The computation of the *p*-value for the regression *F* test is depicted in Figure 11.8. These results are typically summarized in an analysis-of-variance (ANOVA) table, as shown in Table 11.2.

FIGURE 11.7 Acceptance and rejection regions for the simple linear regression *F* test

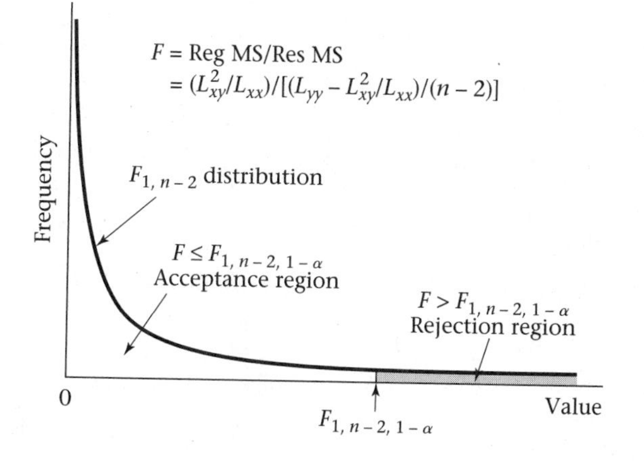

FIGURE 11.8 **Computation of the *p*-value for the simple linear regression *F* test**

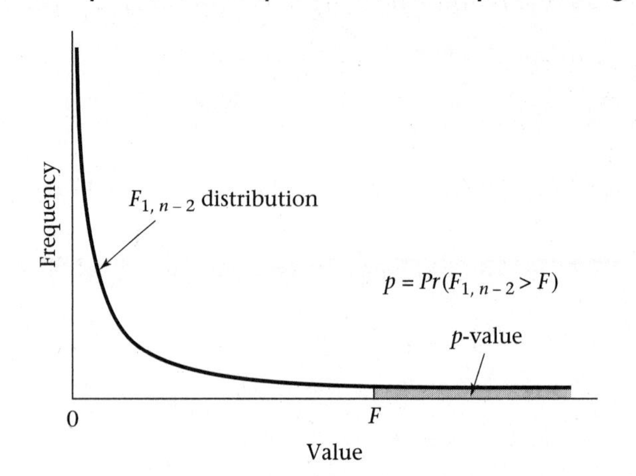

TABLE 11.2 **ANOVA table for displaying regression results**

	SS	df	MS	F statistic	p-value
Regression	$(a)^a$	1	$(a)/1$	$F = [(a)/1]/[(b)/(n-2)]$	$Pr(F_{1,n-2} > F)$
Residual	$(b)^b$	$n-2$	$(b)/(n-2)$		
Total	$(a)+(b)$				

[a] (a) = Regression SS.
[b] (b) = Residual SS.

Example 11.12 **Obstetrics** Test for the significance of the regression line derived for the birthweight–estriol data in Example 11.8.

Solution From Example 11.8,

$$L_{xy} = 412, \ L_{xx} = 677.42$$

Furthermore,

$$\sum_{i=1}^{31} y_i^2 = 32{,}418 \qquad L_{yy} = \sum_{i=1}^{31} y_i^2 - \left(\sum_{i=1}^{31} y_i\right)^2 \bigg/ 31 = 32{,}418 - 992^2/31 = 674$$

Therefore,

$$\begin{aligned}
\text{Reg SS} &= L_{xy}^2/L_{xx} = \text{Reg MS} = 412^2/677.42 = 250.57 \\
\text{Total SS} &= L_{yy} = 674 \\
\text{Res SS} &= \text{Total SS} - \text{Reg SS} = 674 - 250.57 = 423.43 \\
\text{Res MS} &= \text{Res SS}/(31-2) = \text{Res SS}/29 = 423.43/29 = 14.60 \\
F &= \text{Reg MS}/\text{Res MS} = 250.57/14.60 = 17.16 \sim F_{1,29} \text{ under } H_0
\end{aligned}$$

From Table 9 in the Appendix,

$$F_{1,29,.999} < F_{1,20,.999} = 14.82 < 17.16 = F$$

Therefore, $p < .001$

and H_0 is rejected and the alternative hypothesis, namely, that the slope of the regression line is significantly different from 0, is accepted, implying a *significant linear relationship* between birthweight and estriol level. These results are summarized in the ANOVA table (Table 11.3) using the MINITAB REGRESSION program.

TABLE 11.3 **ANOVA results for the birthweight–estriol data in Example 11.12**

Analysis of Variance

SOURCE	DF	SS	MS	F	p
Regression	1	250.57	250.57	17.16	0.000
Error	29	423.43	14.60		
Total	30	674.00			

A summary measure of goodness of fit frequently referred to in the literature is R^2.

DEFINITION 11.14 R^2 is defined as Reg SS/Total SS.

R^2 can be thought of as the proportion of the variance of y that can be explained by the variable x. If $R^2 = 1$, then all the variation in y can be explained by the variation in x, and all the data points fall on the regression line. In other words, once x is known, y can be predicted exactly, with no error or variability in the prediction. If $R^2 = 0$, then x gives no information about y, and the variance of y is the same with or without knowing x. If R^2 is between 0 and 1, then for a given value of x, the variance of y is lower than it would be if x were unknown but is still greater than 0. In particular, the best estimate of the variance of y given x (or σ^2 in the regression model in Equation 11.2) is given by Res MS (or $s_{y \cdot x}^2$). For large n, $s_{y \cdot x}^2 \approx s_y^2(1 - R^2)$. Thus R^2 represents the proportion of the variance of y that is explained by x.

Example 11.13 **Obstetrics** Compute and interpret R^2 and $s_{y \cdot x}^2$ for the birthweight–estriol data in Example 11.12.

Solution From Table 11.3, the R^2 for the birthweight–estriol regression line is given by $250.57/674 = .372$. Thus about 37% of the variance of birthweight can be explained by estriol level. Furthermore $s_{y \cdot x}^2 = 14.60$, as compared with

$$s_y^2 = \sum_{i=1}^{n} (y_i - \bar{y})^2 / (n-1) = 674/30 = 22.47$$

Thus, for the subgroup of women with a specific estriol level, such as 10 mg/24 hr, the variance of birthweight is 14.60, whereas for *all* women with any estriol level, the variance of birthweight is 22.47. Note that

$$s_{y \cdot x}^2 / s_y^2 = 14.60/22.47 = .650 \approx 1 - R^2 = 1 - .372 = .628$$

Example 11.14 **Pulmonary Function** Forced expiratory volume (FEV) is a standard measure of pulmonary function. To identify people with abnormal pulmonary function, standards of FEV for normal people must be established. One problem here is that FEV is related to both age and height. Let us focus on boys who are ages 10–15 and postulate a regression model of the form FEV = $\alpha + \beta$(height) + e. Data were collected on FEV and height for 655 boys in this age group residing in Tecumseh, Michigan [2]. The mean FEV in liters is presented for each of twelve 4-cm height groups in Table 11.4. Find the best-fitting regression line and test it for statistical significance. What proportion of the variance of FEV can be explained by height?

TABLE 11.4 **Mean FEV and height for boys ages 10–15 in Tecumseh, Michigan**

Height (cm)	Mean FEV (L)	Height (cm)	Mean FEV (L)
134[a]	1.7	158	2.7
138	1.9	162	3.0
142	2.0	166	3.1
146	2.1	170	3.4
150	2.2	174	3.8
154	2.5	178	3.9

[a]The middle value of each 4-cm height group is given here.
Source: Reprinted with permission of the *American Review of Respiratory Disease, 108*, 258–272, 1973.

Solution A linear-regression line is fitted to the points in Table 11.4:

$$\sum_{i=1}^{12} x_i = 1872 \qquad \sum_{i=1}^{12} x_i^2 = 294{,}320 \qquad \sum_{i=1}^{12} y_i = 32.3$$

$$\sum_{i=1}^{12} y_i^2 = 93.11 \qquad \sum_{i=1}^{12} x_i y_i = 5156.20$$

Therefore,

$$L_{xy} = 5156.20 - \frac{1872(32.3)}{12} = 117.4$$

$$L_{xx} = 294{,}320 - \frac{1872^2}{12} = 2288$$

$$b = L_{xy}/L_{xx} = 0.0513$$

$$a = \left(\sum_{i=1}^{12} y_i - b\sum_{i=1}^{12} x_i\right)\Big/12 = [32.3 - 0.0513(1872)]/12 = -5.313$$

Thus the fitted regression line is

$$\text{FEV} = -5.313 + 0.0513 \times \text{height}$$

Statistical significance is assessed by computing the F statistic in Equation 11.7 as follows:

$$\text{Reg SS} = L_{xy}^2/L_{xx} = 117.4^2/2288 = 6.024 = \text{Reg MS}$$

$$\text{Total SS} = L_{yy} = 93.11 - 32.3^2/12 = 6.169$$

$$\text{Res SS} = 6.169 - 6.024 = 0.145$$

$$\text{Res MS} = \text{Res SS}/(n-2) = 0.145/10 = 0.0145$$

$$F = \text{Reg MS/Res MS} = 414.8 \sim F_{1,10} \text{ under } H_0$$

Clearly, the fitted line is statistically significant, because from Table 9 in the Appendix, $F_{1,10,.999} = 21.04$, so $p < .001$. These results can be displayed in an ANOVA table (Table 11.5).

TABLE 11.5 **ANOVA table for the FEV–height regression results in Example 11.14**

Analysis of Variance

SOURCE	DF	SS	MS	F	p
Regression	1	6.0239	6.0239	414.78	0.000
Error	10	0.1452	0.0145		
Total	11	6.1692			

Finally, the proportion of the variance of FEV that is explained by height is given by $R^2 = 6.024/6.169 = .976$. Thus, differences in height explain almost all the variability in FEV among boys in this age group.

11.4.2 *t* Test for Simple Linear Regression

In this section an alternative method for testing the hypothesis H_0: $\beta = 0$ versus H_1: $\beta \neq 0$ is presented. This method is based on the *t* test and is equivalent to the *F* test presented in Equation 11.7. The procedure is widely used and also provides interval estimates for β.

The hypothesis test here is based on the sample regression coefficient b, or, more specifically, on $b/se(b)$, and H_0 will be rejected if $|b|/se(b) > c$ for some constant c and will be accepted otherwise.

The sample regression coefficient b is an **unbiased estimator** of the population regression coefficient β and, in particular, under H_0, $E(b) = 0$. Furthermore, the variance of b is given by

$$\sigma^2 \bigg/ \sum_{i=1}^{n} \left(x_i - \bar{x}\right)^2 = \sigma^2 / L_{xx}$$

In general, σ^2 is unknown. However, the best estimate of σ^2 is given by $s_{y \cdot x}^2$. Hence

$$se(b) \approx s_{y \cdot x} \big/ (L_{xx})^{1/2}$$

Finally, under H_0, $t = b/se(b)$ follows a t distribution with $n - 2$ *df*. Therefore, the following test procedure for a two-sided test with significance level α is used:

EQUATION 11.8

***t* Test for Simple Linear Regression** To test the hypothesis H_0: $\beta = 0$ versus H_1: $\beta \neq 0$, use the following procedure:

(1) Compute the test statistic

$$t = b/(s^2_{y \cdot x}/L_{xx})^{1/2}$$

(2) For a two-sided test with significance level α, if

$$t > t_{n-2,1-\alpha/2}$$

$$\text{or} \quad t < t_{n-2,\alpha/2} = -t_{n-2,1-\alpha/2}$$

then reject H_0; if $-t_{n-2,1-\alpha/2} \le t \le t_{n-2,1-\alpha/2}$

then accept H_0.

(3) The p-value is given by

$$p = 2 \times (\text{area to the left of } t \text{ under a } t_{n-2} \text{ distribution}) \text{ if } t < 0$$

$$p = 2 \times (\text{area to the right of } t \text{ under a } t_{n-2} \text{ distribution}) \text{ if } t \ge 0$$

The acceptance and rejection regions for this test are depicted in Figure 11.9. The computation of the p-value is illustrated in Figure 11.10.

The t test in this section and the F test in Equation 11.7 are equivalent in that they always provide the same p-values. Which test is used is a matter of personal preference; both appear in the literature.

Example 11.15 **Obstetrics** Assess the statistical significance for the birthweight–estriol data using the t test in Equation 11.8.

Solution From Example 11.8, $b = L_{xy}/L_{xx} = 0.608$. Furthermore, from Table 11.3 and Example 11.12,

$$se(b) = (s^2_{y \cdot x}/L_{xx})^{1/2} = (14.60/677.42)^{1/2} = 0.147$$

FIGURE 11.9 **Acceptance and rejection regions for the t test for simple linear regression**

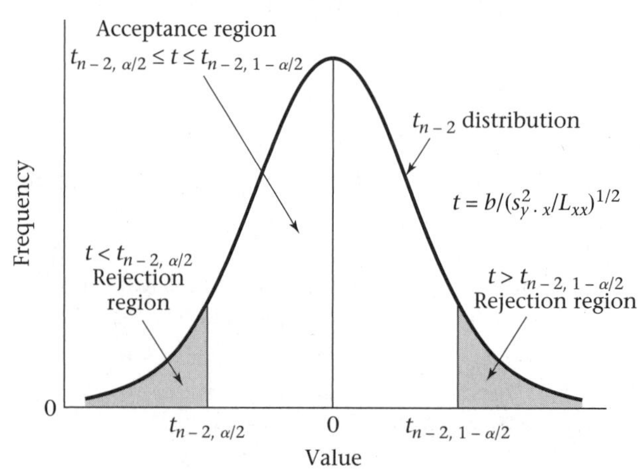

FIGURE 11.10 Computation of the *p*-value for the *t* test for simple linear regression

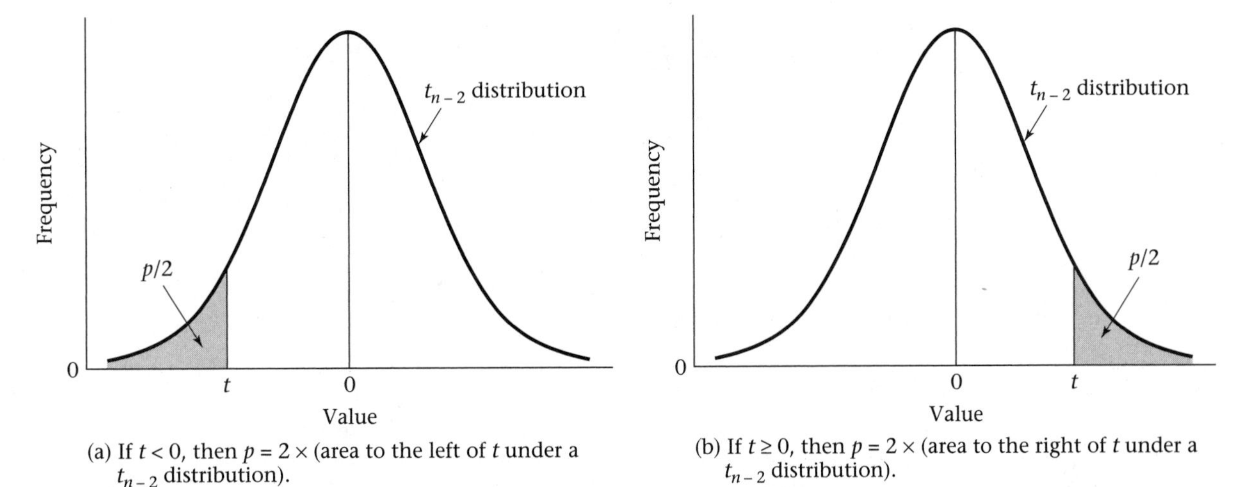

(a) If $t < 0$, then $p = 2 \times$ (area to the left of t under a t_{n-2} distribution).

(b) If $t \ge 0$, then $p = 2 \times$ (area to the right of t under a t_{n-2} distribution).

Thus, $t = b/se(b) = 0.608/0.147 = 4.14 \sim t_{29}$ under H_0

Since $\quad t_{29,.9995} = 3.659 < 4.14 = t$

we have $\quad p < 2 \times (1 - .9995) = .001$

This information is summarized in Table 11.6. Note that the *p*-values based on the *F* test in Table 11.3 and the *t* test in Table 11.6 are the same ($p = .000$).

TABLE 11.6 The *t* test approach for the birthweight–estriol example

```
The regression equation is
birthwt = 21.5 + 0.608 estriol
Predictor         Coef        Stdev      t-ratio         P
Constant        21.523        2.620         8.21     0.000
estriol         0.6082       0.1468         4.14     0.000
```

SECTION 11.5 Interval Estimation for Linear Regression

11.5.1 Interval Estimates for Regression Parameters

Standard errors and interval estimates for the parameters of a regression line are often computed to obtain some idea of the precision of the estimates. Furthermore, if we want to compare our regression coefficients with previously published regression coefficients β_0 and α_0, where these estimates are based on much larger samples than ours, then, based on our data, we can check whether β_0 and α_0 fall within the 95% confidence intervals for β and α, respectively, to decide whether the two sets of results are comparable.

The standard errors of the estimated regression parameters are given as follows:

EQUATION 11.9

Standard Errors of Estimated Parameters in Simple Linear Regression

$$se(b) = \sqrt{\frac{s_{y \cdot x}^2}{L_{xx}}}$$

$$se(a) = \sqrt{s_{y \cdot x}^2 \left(\frac{1}{n} + \frac{\bar{x}^2}{L_{xx}} \right)}$$

Furthermore, the two-sided 100% × (1 – α) confidence intervals for β and α are given by

EQUATION 11.10

Two-Sided 100% × (1 – α) Confidence Intervals for the Parameters of a Regression Line If b and a are, respectively, the estimated slope and intercept of a regression line as given in Equation 11.3 and $se(b)$, $se(a)$ are the estimated standard errors as given in Equation 11.9, then the two-sided 100% × (1 – α) confidence intervals for β and α are given by

$$b \pm t_{n-2,1-\alpha/2}se(b) \quad \text{and} \quad a \pm t_{n-2,1-\alpha/2}se(a) \quad \text{respectively.}$$

Example 11.16 **Obstetrics** Provide standard errors and 95% confidence intervals for the regression parameters of the birthweight–estriol data in Table 11.1.

Solution From Example 11.15, the standard error of b is given by

$$\sqrt{14.60/677.42} = 0.147$$

Thus, a 95% confidence interval for β is obtained from

$$0.608 \pm t_{29,.975}(0.147) = 0.608 \pm 2.045(0.147) = 0.608 \pm 0.300 = (0.308, 0.908)$$

Compute $\bar{x}$ to obtain the standard error of a. From Example 11.8,

$$\bar{x} = \frac{\sum_{i=1}^{31} x_i}{31} = \frac{534}{31} = 17.23$$

Thus, the standard error of a is given by

$$\sqrt{14.60\left(\frac{1}{31} + \frac{17.23^2}{677.42} \right)} = 2.62$$

It follows that a 95% confidence interval for α is provided by

$$21.52 \pm t_{29,.975}(2.62) = 21.52 \pm 2.045(2.62) = 21.52 \pm 5.36 = (16.16, 26.88)$$

These intervals are rather wide, which is not surprising due to the small sample size.

Suppose another data set based on 500 pregnancies, where the birthweight–estriol regression line is estimated as $y = 25.04 + 0.52x$ is found in the literature. Since 0.52 is within the 95% confidence interval for the slope and 25.04 is within the 95% confidence interval for the intercept, our results are comparable with the earlier study.

11.5.2 Interval Estimation for Predictions Made from Regression Lines

One important use for regression lines is in making predictions. Frequently, the accuracy of these predictions must be assessed.

Example 11.17 **Pulmonary Function** Suppose we wish to use the FEV–height regression line computed in Example 11.14 to develop normal ranges for 10- to 15-year-old boys of particular heights. In particular, consider John H., who is 12 years old and 160 cm tall and whose FEV is 2.5 L. Can his FEV be considered abnormal for his age and height?

In general, if all boys of height x are considered, then the average FEV for such boys can be best estimated from the regression equation by $\hat{y} = a + bx$. How accurate is this estimate? The answer to this question depends on whether we are making predictions for *one specific boy* or for the *mean value of all boys of a given height.* The first estimate would be useful to a pediatrician interested in assessing the lung function of a particular patient, whereas the second estimate would be useful to a researcher interested in relationships between pulmonary function and height over large populations of boys. The standard error (se_1) of the first type of estimate and the resulting interval estimate are given as follows:

EQUATION 11.11

Predictions Made from Regression Lines for Individual Observations Suppose we wish to make predictions from a regression line for an individual observation with independent variable x that was not used in constructing the regression line. The distribution of observed y values for the subset of individuals with independent variable x is normal with mean = $\hat{y} = a + bx$ and standard deviation given by

$$se_1(\hat{y}) = \sqrt{s_{y \cdot x}^2 \left[1 + \frac{1}{n} + \frac{(x - \bar{x})^2}{L_{xx}}\right]}$$

Furthermore, 100% × (1 – α) of the observed values will fall within the interval

$$\hat{y} \pm t_{n-2,1-\alpha/2}\, se_1(\hat{y})$$

This interval is sometimes called a 100% × (1 – α) prediction interval for y.

Example 11.18 **Pulmonary Function** Can the FEV of John H. in Example 11.17 be considered abnormal for his age and height?

Solution John's observed FEV is 2.5 L. The regression equation relating FEV and height was computed in Example 11.14 and is given by $y = -5.313 + 0.0513 \times \text{height}$. Thus, the estimated average FEV for 12-year-old boys of height 160 cm is

$$\hat{y} = -5.313 + 160 \times 0.0513 = 2.90 \text{ L}$$

We need to obtain $\bar{x}$ before computing the $se_1(\hat{y})$. From Example 11.14,

$$\bar{x} = \frac{\sum_{i=1}^{12} x_i}{12} = \frac{1872}{12} = 156.0$$

Thus, $se_1(\hat{y})$ is given by

$$se_1(\hat{y}) = \sqrt{0.0145\left[1 + \frac{1}{12} + \frac{(160-156)^2}{2288}\right]} = \sqrt{0.0145(1.090)} = 0.126$$

Furthermore, 95% of boys of this age and height will have an FEV between

$$2.90 \pm t_{10,.975}(0.126) = 2.90 \pm 2.228(0.126) = 2.90 \pm 0.28 = (2.62, 3.18)$$

How can this prediction interval be used? Since the observed FEV for John H. (2.5 L) does not fall within this interval, we can say that John's lung function is abnormally low for a boy of his age and height, and that, if possible, further exploration is needed to find a reason for this abnormality.

The magnitude of the standard error in Equation 11.11 depends on how far the observed value of x for the new sample point is from the mean value of x for the data points used in computing the regression line ($\bar{x}$). The standard error is smaller when x is close to $\bar{x}$ than when x is far from $\bar{x}$. In general, making predictions from a regression line for values of x that are very far from $\bar{x}$ is risky, because the predictions are likely to be very inaccurate.

Example 11.19 **Pulmonary Function** Suppose that Bill Y. has a height of 190 cm with an FEV of 3.5 L. Compare the standard error of his predicted value with that for John H. given in Example 11.18.

Solution From Equation 11.11,

$$se_1(\hat{y}) = \sqrt{0.0145\left[1 + \frac{1}{12} + \frac{(190-156)^2}{2288}\right]}$$

$$= \sqrt{0.0145(1.589)} = 0.152 > 0.126 = se_1 \quad \text{(computed in Example 11.18)}$$

This result is expected, because 190 cm is further than 160 cm from $\bar{x} = 156$ cm.

Suppose we want to assess the mean value of FEV for a large number of boys of a particular height rather than for one particular boy. This parameter might be of interest to a researcher interested in growth curves of pulmonary function in children. How can the estimated mean FEV and the standard error of the estimate be obtained? The procedure is given as follows:

EQUATION 11.12

Standard Error and Confidence Interval for Predictions Made from Regression Lines for the Average Value of y for a Given x The best estimate of the average value of y for a given x is $\hat{y} = a + bx$. Its standard error, denoted by $se_2(\hat{y})$, is given by

$$se_2(\hat{y}) = \sqrt{s_{y\cdot x}^2\left[\frac{1}{n} + \frac{(x-\bar{x})^2}{L_{xx}}\right]}$$

Furthermore, a two-sided 100% × (1 − α) confidence interval for the average value of y is

$$\hat{y} \pm t_{n-2,1-\alpha/2} se_2(\hat{y})$$

Example 11.20 **Pulmonary Function** Compute the standard error and 95% confidence interval for the average value of FEV over a large number of boys with height of 160 cm.

Solution Refer to the results of Example 11.18 for the necessary raw data to perform the computations. The best estimate of the mean value of FEV is the same as given in Example 11.18, which was 2.90 L. However, the standard error is computed differently. From Equation 11.12,

$$se_2(\hat{y}) = \sqrt{0.0145\left[\frac{1}{12} + \frac{(160-156)^2}{2288}\right]} = \sqrt{0.0145(0.090)} = 0.036$$

Therefore, a 95% confidence interval for the mean value of FEV over a large number of boys with height 160 cm is given by

$$2.90 \pm t_{10,.975}(0.036) = 2.90 \pm 2.228(0.036) = 2.90 \pm 0.08 = (2.82, 2.98)$$

Notice that this interval is much narrower than the interval computed in Example 11.18 (2.62, 3.18), which is a range encompassing approximately 95% of individual boys' FEVs. This disparity reflects the intuitive idea that there is much more precision in estimating the mean value of y for a large number of boys with the same height x than in estimating y for one particular boy with height x.

Note again that the standard error for the average value of y for a given value of x is not the same for all values of x, but gets larger the further x is from the mean value of x $(\bar{x})$ used to estimate the regression line.

Example 11.21 **Pulmonary Function** Compare the standard error of the average FEV for boys of height 190 cm with that for boys of 160 cm.

Solution From Equation 11.12,

$$se_2(\hat{y}) = \sqrt{0.0145\left[\frac{1}{12} + \frac{(190-156)^2}{2288}\right]} = \sqrt{0.0145(0.589)}$$

$$= 0.092 > 0.036 = se_2(\hat{y}) \text{ for } x = 160 \text{ cm}$$

This result is expected, because 190 cm is further than 160 cm from $\bar{x} = 156$ cm.

SECTION 11.6 Assessing the Goodness of Fit of Regression Lines

A number of assumptions were made in using the methods of simple linear regression in the previous sections of this chapter. What are some of these assumptions, and what possible situations could be encountered that would make these assumptions not viable?

EQUATION 11.13

Assumptions Made in Linear-Regression Models

(1) For any given value of x, the corresponding value of y has an average value $\alpha + \beta x$, which is a linear function of x.

(2) For any given value of x, the corresponding value of y is normally distributed about $\alpha + \beta x$ with the same variance σ^2 for any x.

(3) For any two data points (x_1, y_1), (x_2, y_2), the error terms e_1, e_2 are independent of each other.

Let us now reassess the birthweight–estriol data for possible violation of linear regression assumptions. To assess whether these assumptions are reasonable, we can use several different kinds of plots. The simplest plot is the $x - y$ scatter plot. Here we plot the dependent variable y versus the independent variable x and superimpose the regression line $y = a + bx$ on the same plot. We have constructed a scatter plot of this type for the birthweight–estriol data in Figure 11.1. The linearity assumption appears reasonable in that there is no obvious curvilinearity in the raw data. However, there is a hint that there is more variability about the regression line for higher estriol values than for lower estriol values. To focus more clearly on this issue, we can compute the residuals about the fitted regression line and then construct a scatter plot of the residuals versus either the estriol values (x) or the predicted birthweights $(\hat{y} = a + bx)$.

From Equation 11.2, we see that the errors (e) about the true regression line ($y = \alpha + \beta x$) have the same variance σ^2. However, it can be shown that the residuals about the fitted regression line ($y = a + bx$) have different variances depending on how far an individual x value is from the mean x value used to generate the regression line. Specifically, residuals for points (x_i, y_i) where x_i is close to the mean x value for all points used in constructing the regression line (i.e., $|x_i - \bar{x}|$ is small) will tend to be larger than residuals where $|x_i - \bar{x}|$ is large. Interestingly, if $|x_i - \bar{x}|$ is very large, then the regression line is forced to go through the point (x_i, y_i) (or nearly through it) with a small residual for this point. The standard deviation of the residuals is given in Equation 11.14.

EQUATION 11.14

Standard Deviation of Residuals About the Fitted Regression Line Let (x_i, y_i) be a sample point used in estimating the regression line, $y = \alpha + \beta x$.

If $y = a + bx$ is the estimated regression line, and

$\hat{e}_i$ = residual for the point (x_i, y_i) about the estimated regression line, then

$\hat{e}_i = y_i - (a + bx_i)$ and

$$sd(\hat{e}_i) = \sqrt{\hat{\sigma}^2\left[1 - \frac{1}{n} - \frac{(x_i - \bar{x})^2}{L_{xx}}\right]}$$

The **Studentized residual** corresponding to the point (x_i, y_i) is $\hat{e}_i / sd(\hat{e}_i)$.

FIGURE 11.11 **Plot of Studentized residuals versus the predicted value of birthweight for the birthweight–estriol data in Table 11.1**

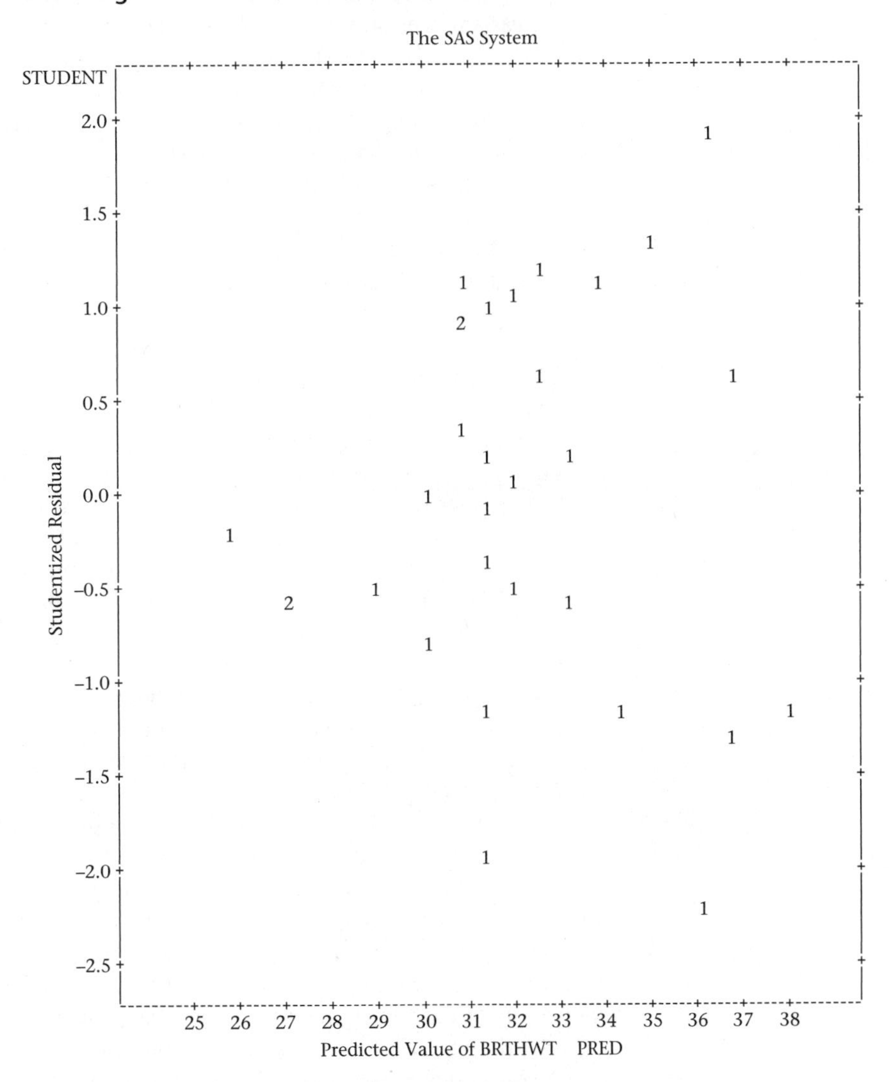

In Figure 11.11, we have plotted the Studentized residuals (i.e., the individual residuals divided by their standard deviations) versus the predicted birthweights ($\hat{y} = 21.52 + 0.608 \times \text{estriol}$).

A point labeled 2 indicates that there are two identical data points—that is, the second and third points in Table 11.1 are both (9, 25). There is still a hint that the spread increases slightly as the predicted birthweight increases. However, this impression is mainly due to the four data points with the lowest predicted values, all of which have residuals that are close to 0. One commonly used strategy that can be employed if unequal residual variances are present is to transform the dependent variable (y) to a different scale. This type of transformation is called a **variance-stabilizing transformation.** The goal of using such a transformation is to make the residual variances approximately the same for each level of x (or, equivalently, each level of the predicted value). The most common transformations when the residual variance is an increasing function of x are either the ln or square-root transformations. The square-root transformation is useful when the residual variance is proportional to the average value of y (e.g., if the average value goes up by a factor of 2, then the residual variance goes up by a factor of 2 also). The log transformation is useful when the residual variance is proportional to the square of the average values (e.g., if the average value goes up by a factor of 2, then the residual variance goes up by a factor of 4). For purposes of illustration, we have computed the regression using the ln transformation for birthweight (i.e., $y = \ln$ birthweight). The residual plot is shown in Figure 11.12.

The plots in Figures 11.11 and 11.12 look similar. The plot using the square-root transformation for birthweight is also similar. Therefore, we would probably choose to keep the data in the original scale for the sake of simplicity. However, in other data sets the use of the appropriate transformation is crucial and each of the linearity, equal-variance, and normality assumptions can be made more plausible using a transformed scale. However, occasionally a transformation may make the equal-variance assumption more plausible but the linearity assumption less plausible. Another possibility is to keep the data in the original scale, but employ a weighted regression where the weight is approximately inversely proportional to the residual variance. This may be reasonable if the data points consist of averages over varying numbers of individuals (e.g., people living in different cities, where the weight is proportional to the size of the city). Weighted regression is beyond the scope of this text (see [3] for a more complete discussion of this technique).

Other issues of concern in judging the goodness of fit of a regression line are **outliers** and **influential points.** In Section 8.9, we discussed methods for the detection of outliers in a sample, where only a single variable is of interest. However, it is more difficult to detect outliers in a regression setting than in univariate problems, particularly if multiple outliers are present in a data set. *Influential points* are defined heuristically as points that have an important influence on the coefficients of the fitted regression lines. Suppose we delete the ith sample point, and refit the regression line from the remaining $n - 1$ data points. If we denote the estimated slope and intercept for the reduced data set by $b^{(i)}$ and $a^{(i)}$, respectively, then the sample point will be influential if either $|b - b^{(i)}|$ or $|a - a^{(i)}|$ are large. Outliers and influential points are not necessarily the same. An outlier (x_i, y_i) may or may not be influential, depending on its location relative to the remaining sample points. For example, if $|x_i - \bar{x}|$ is small, then even a gross outlier will have a relatively small influence on the slope estimate, but will have an important influence on the intercept estimate. Conversely, if $|x_i - \bar{x}|$ is large, then even a data point that is not a gross outlier may be influential. The reader is referred to Draper and Smith [3] and Weisberg [4] for a more complete description of residual analysis, detection of outliers, and influential points in a regression setting.

FIGURE 11.12 Plot of Studentized residuals versus the predicted value of ln(birthweight) for the birthweight–estriol data in Table 11.1

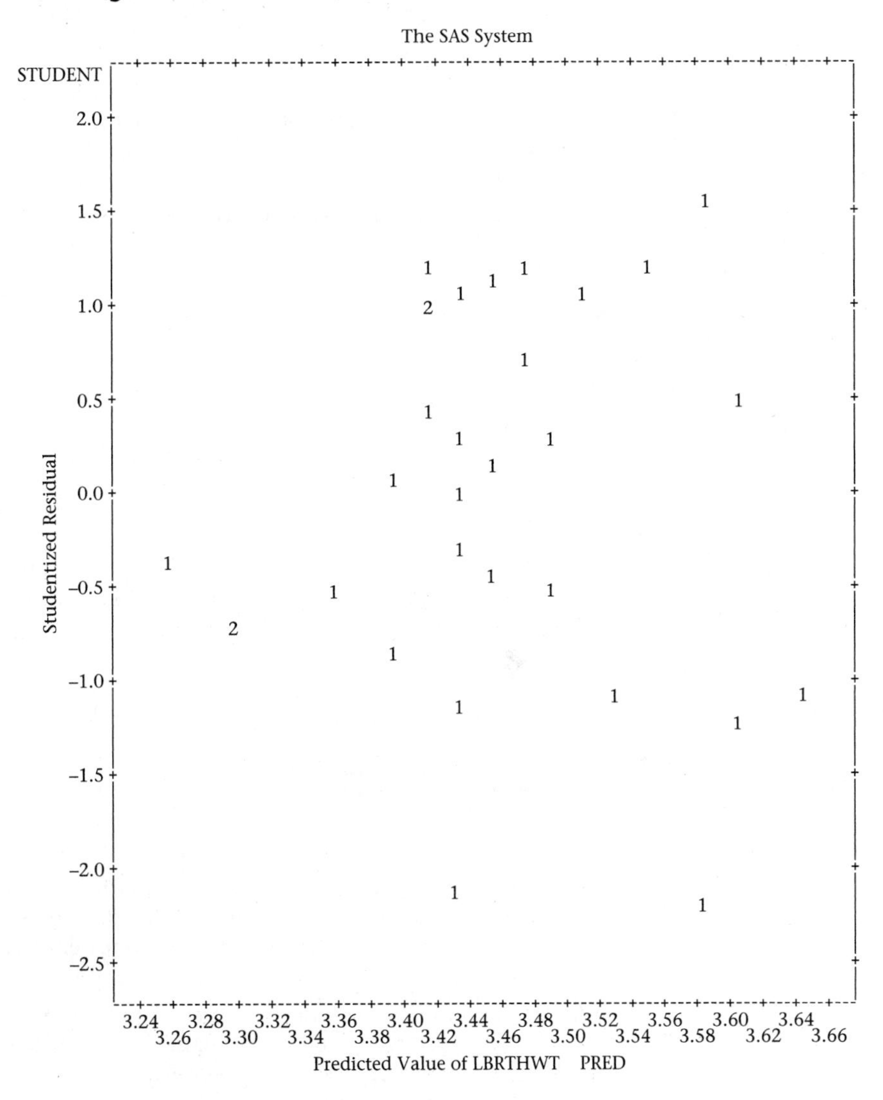

SECTION 11.7 The Correlation Coefficient

The primary focus of the discussion of linear-regression analysis in Sections 11.2–11.6 was on methods of predicting one dependent variable (y) from an independent variable (x). Often we are interested not in predicting one variable from another but rather in investigating whether or not there is a relationship between two variables. The **correlation coefficient**, introduced in Definition 5.13, is a useful tool for quantifying the relationship between variables and is better suited for this purpose than the regression coefficient.

Example 11.22 **Cardiovascular Disease** Serum cholesterol is an important risk factor in the etiology of cardiovascular disease. Much research has been devoted to understanding the environmental factors that cause elevated cholesterol levels. For this purpose, cholesterol levels were measured on 100 genetically unrelated spouse pairs. We are not interested in predicting the cholesterol level of a husband from that of his wife but rather would like some quantitative measure of the relationship between their levels. We will use the correlation coefficient for this purpose.

In Definition 5.13, we defined the population correlation coefficient ρ. In general, ρ is unknown and we have to estimate ρ by the sample correlation coefficient r.

DEFINITION 11.15 The **sample (Pearson) correlation coefficient (r)** is defined by

$$L_{xy} / \sqrt{L_{xx} L_{yy}}$$

The correlation is not affected by changes in location or scale in either variable and must lie between −1 and +1. The sample correlation coefficient can be interpreted in a similar manner to the population correlation coefficient ρ:

EQUATION 11.15

Interpretation of the Sample Correlation Coefficient

(1) If the correlation is greater than 0, such as for birthweight and estriol, then the variables are said to be **positively correlated.** Two variables (x, y) are positively correlated if as x increases, y tends to increase, whereas as x decreases, y tends to decrease.

(2) If the correlation is less than 0, such as for pulse rate and age, then the variables are said to be **negatively correlated.** Two variables (x, y) are negatively correlated if as x increases, y tends to decrease, whereas as x decreases, y tends to increase.

(3) If the correlation is exactly 0, such as for birthweight and birthday, then the variables are said to be **uncorrelated.** Two variables (x, y) are uncorrelated if there is no linear relationship between x and y.

Thus the correlation coefficient provides a *quantitative* measure of the dependence between two variables: the closer $|r|$ is to 1, the more closely related the variables are; if $|r| = 1$, then one variable can be predicted exactly from the other.

As was the case for the population correlation coefficient (ρ), the interpretation of the sample correlation coefficient (r) in terms of degree of dependence is only correct if the variables x and y are normally distributed and in certain other special cases. If the variables are not normally distributed, then the interpretation may not be correct (see Example 5.30 for an example of two random variables that have correlation coefficient = 0, but are completely dependent).

Example 11.23 Suppose the two variables under study are temperature in °F (y) and temperature in °C (x). The correlation between these two variables must be 1, since one variable can be predicted exactly from the other $\left(y = \frac{9}{5}x + 32\right)$.

Example 11.24 **Obstetrics** Compute the sample correlation coefficient for the birthweight–estriol data presented in Table 11.1.

Solution From Examples 11.8 and 11.12,

$$L_{xy} = 412 \qquad L_{xx} = 677.42 \qquad L_{yy} = 674$$

Therefore, $r = L_{xy}\big/\sqrt{L_{xx}L_{yy}} = 412\big/\sqrt{677.42(674)} = 412/675.71 = .61$

11.7.1 Relationship Between the Sample Correlation Coefficient (r) and the Population Correlation Coefficient (ρ)

We can relate the sample correlation coefficient r and the population correlation coefficient ρ more clearly by dividing the numerator and denominator of r by $(n - 1)$ in Definition 11.15, whereby

EQUATION 11.16
$$r = \frac{L_{xy}/(n-1)}{\sqrt{\left(\frac{L_{xx}}{n-1}\right)\left(\frac{L_{yy}}{n-1}\right)}}$$

We note that $s_x^2 = L_{xx}/(n-1)$ and $s_y^2 = L_{yy}/(n-1)$. Furthermore, if we define the *sample covariance* by $s_{xy} = L_{xy}/(n-1)$, then we can re-express Equation 11.16 in the following form:

EQUATION 11.17
$$r = \frac{s_{xy}}{s_x s_y} = \frac{\text{sample covariance between } x \text{ and } y}{(\text{sample standard deviation of } x)(\text{sample standard deviation of } y)}$$

This is completely analogous to the definition of the population correlation coefficient ρ given in Definition 5.13 with the population quantities, $Cov(X, Y)$, σ_x, and σ_y replaced by their sample estimates s_{xy}, s_x, and s_y.

11.7.2 Relationship Between the Sample Regression Coefficient (b) and the Sample Correlation Coefficient (r)

What is the relationship between the sample regression coefficient (b) and the sample correlation coefficient (r)? Note from Equation 11.3 that $b = L_{xy}/L_{xx}$ and from Definition 11.15 that $r = L_{xy}\big/\sqrt{L_{xx}L_{yy}}$. Therefore, if r is multiplied by $\sqrt{L_{yy}/L_{xx}}$,

EQUATION 11.18

$$r\sqrt{\frac{L_{yy}}{L_{xx}}} = \frac{L_{xy}}{\sqrt{L_{xx}L_{yy}}} \times \frac{\sqrt{L_{yy}}}{\sqrt{L_{xx}}} = \frac{L_{xy}}{L_{xx}} = b$$

Furthermore, from Definition 11.4,

$$s_y^2 = \frac{L_{yy}}{n-1}$$

$$s_x^2 = \frac{L_{xx}}{n-1}$$

$$s_y^2/s_x^2 = L_{yy}/L_{xx}$$

or $s_y/s_x = \sqrt{L_{yy}/L_{xx}}$

Substituting s_y/s_x for $\sqrt{L_{yy}/L_{xx}}$ on the left-hand side of Equation 11.18 yields the following relationship:

EQUATION 11.19

$$b = \frac{rs_y}{s_x}$$

How can Equation 11.19 be interpreted? The regression coefficient (b) can be interpreted as a rescaled version of the correlation coefficient (r), where the scale factor is the ratio of the standard deviation of y to that of x. Note that r will be unchanged by a change in the units of x or y (or even by which variable is designated as x and which is designated as y), whereas b is in the units of y/x.

Example 11.25 **Pulmonary Function** Compute the correlation coefficient between FEV and height for the pulmonary-function data in Example 11.14.

Solution From Example 11.14,

$$L_{xy} = 117.4 \qquad L_{xx} = 2288 \qquad L_{yy} = 6.169$$

Therefore, $r = \dfrac{117.4}{\sqrt{2288(6.169)}} = \dfrac{117.4}{118.81} = .988$

Thus, a very strong positive correlation exists between FEV and height. The sample regression coefficient b was calculated as 0.0513 in Example 11.14. Furthermore, the sample standard deviation of x and y can be computed as follows:

$$s_x = \sqrt{\frac{\sum_{i=1}^{n}(x_i - \bar{x})^2}{n-1}} = \sqrt{\frac{L_{xx}}{n-1}} = \sqrt{\frac{2288}{11}} = \sqrt{208} = 14.42$$

$$s_y = \sqrt{\frac{\sum_{i=1}^{n}(y_i - \bar{y})^2}{n-1}} = \sqrt{\frac{L_{yy}}{n-1}} = \sqrt{\frac{6.169}{11}} = \sqrt{0.561} = 0.749$$

and their ratio is thus given by

$$s_y/s_x = 0.749/14.42 = .0519$$

Finally, b can be expressed as a rescaled version of r as

$$b = r(s_y/s_x) \quad \text{or} \quad .0513 = .988(.0519)$$

Notice that if height is re-expressed in inches rather than centimeters (1 in. = 2.54 cm), then s_x is divided by 2.54, and b is multiplied by 2.54; that is,

$$b_{\text{in.}} = b_{\text{cm}} \times 2.54 = .0513 \times 2.54 = .130$$

where $b_{\text{in.}}$ is in the units of liters per inch and b_{cm} is in the units of liters per centimeter. However, the correlation coefficient remains the same at .988.

When should the regression coefficient be used and when should the correlation coefficient be used? The regression coefficient is used when we specifically want to predict one variable from another. The correlation coefficient is used when we simply want to describe the linear relationship between two variables but do not want to make predictions. In cases when it is not clear which of these two aims is primary, both a regression and a correlation coefficient can be reported.

Example 11.26 **Obstetrics, Pulmonary Disease, Cardiovascular Disease** For the birthweight–estriol data in Example 11.4, the obstetrician is interested in using a regression equation to predict birthweight from estriol levels. Thus the regression coefficient is more appropriate. Similarly, for the FEV–height data in Example 11.14, the pediatrician is interested in using a growth curve relating a child's pulmonary function to height, and again the regression coefficient is more appropriate. However, in collecting data on cholesterol levels in spouse pairs in Example 11.22, the geneticist is interested simply in describing the relationship between cholesterol levels of spouse pairs and is not interested in prediction. Thus the correlation coefficient is more appropriate here.

In this section, we have introduced the concept of a correlation coefficient. In the next section, we will discuss various hypothesis tests concerning correlation coefficients. Correlation coefficients are used when we are interested in studying the association between two variables but are not interested in predicting one variable from another. Referring to the flowchart at the end of this chapter (Figure 11.32, p. 502), we answer yes to (1) interested in relationships between two variables? and (2) both variables continuous? no to (3) interested in predicting one variable from another? and yes to (4) interested in studying the relationships between two variables? and to (5) both variables normal? This would lead us to the box entitled "Pearson correlation methods."

SECTION 11.8 Statistical Inference for Correlation Coefficients

In the previous section, the sample correlation coefficient was defined. Based on Equation 11.17, if every unit in the reference population could be sampled, then the sample correlation coefficient (r) would be the same as the population correlation coefficient, denoted by ρ, which was introduced in Definition 5.13.

In this section, we will use r, which is computed from finite samples to test various hypotheses concerning ρ.

11.8.1 One-Sample *t* Test for a Correlation Coefficient

Example 11.27 **Cardiovascular Disease** Suppose serum-cholesterol levels in spouse pairs are measured to determine whether or not there is a correlation between cholesterol levels in spouses. Specifically, we wish to test the hypothesis H_0: $\rho = 0$ versus H_1: $\rho \neq 0$. Suppose that $r = .25$ based on 100 spouse pairs. Is this evidence sufficient to warrant rejecting H_0?

In this instance, the hypothesis test would naturally be based on the sample correlation coefficient r and H_0 would be rejected if $|r|$ is sufficiently far from 0. Assuming that each of the random variables x = serum-cholesterol level for the husband and y = serum-cholesterol level for the wife is normally distributed, then the best procedure for testing the hypothesis is given as follows:

EQUATION 11.20

One-Sample *t* Test for a Correlation Coefficient To test the hypothesis H_0: $\rho = 0$ versus H_1: $\rho \neq 0$, use the following procedure:

(1) Compute the sample correlation coefficient r.

(2) Compute the test statistic

$$t = r(n-2)^{1/2} \big/ (1-r^2)^{1/2}$$

which under H_0 follows a t distribution with $n - 2$ *df*.

(3) For a two-sided level α test, if

$$t > t_{n-2,1-\alpha/2} \quad \text{or} \quad t < -t_{n-2,1-\alpha/2}$$

then reject H_0. If

$$-t_{n-2,1-\alpha/2} \le t \le t_{n-2,1-\alpha/2}$$

then accept H_0.

(4) The p-value is given by

$$p = 2 \times (\text{area to the left of } t \text{ under a } t_{n-2} \text{ distribution}) \qquad \text{if } t < 0$$

$$p = 2 \times (\text{area to the right of } t \text{ under a } t_{n-2} \text{ distribution}) \qquad \text{if } t \ge 0$$

(5) We assume an underlying normal distribution for each of the random variables used to compute r.

The acceptance and rejection regions for this test are depicted in Figure 11.13. The computation of the p-value is illustrated in Figure 11.14.

Example 11.28 Perform a test of significance for the data in Example 11.27.

Solution We have $n = 100$, $r = .25$. Thus in this case,

$$t = .25\sqrt{98} \big/ \sqrt{1-.25^2} = 2.475/.968 = 2.56$$

FIGURE 11.13 **Acceptance and rejection regions for the one-sample *t* test for a correlation coefficient**

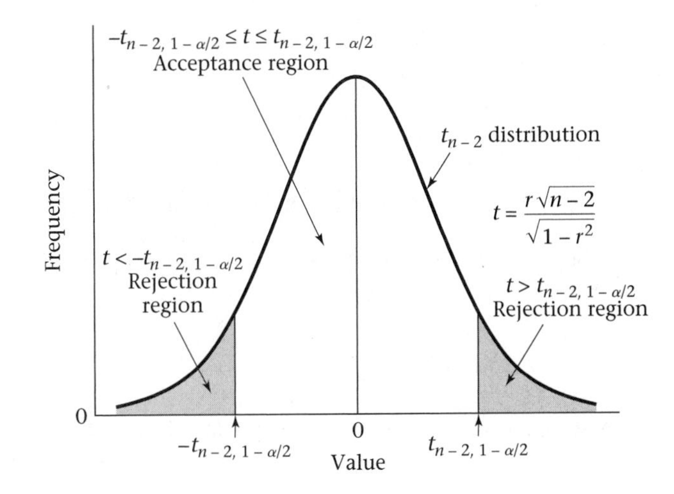

FIGURE 11.14 **Computation of the *p*-value for the one-sample *t* test for a correlation coefficient**

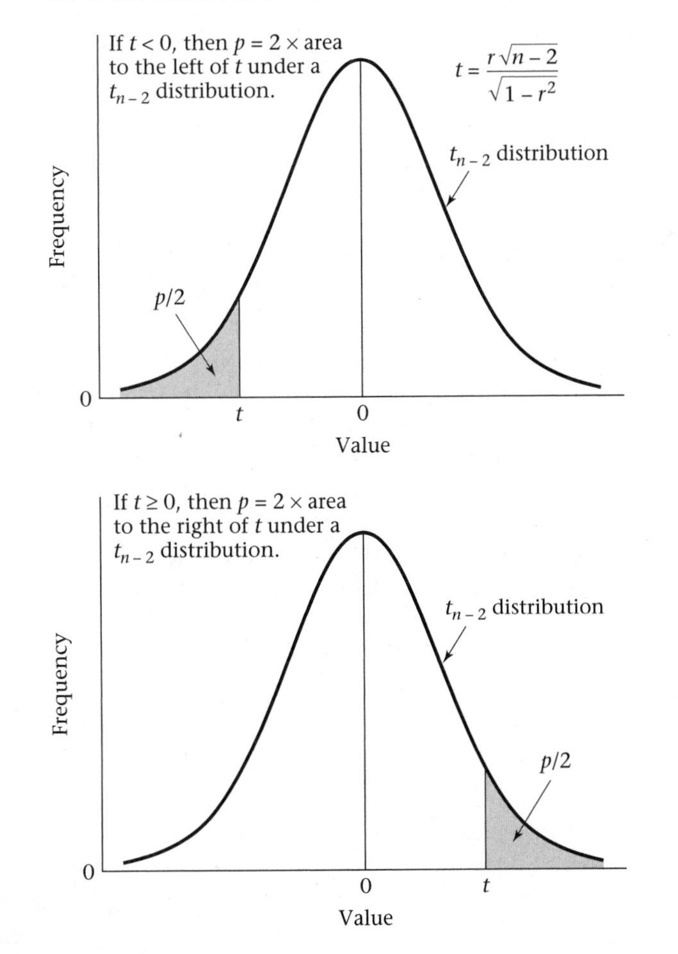

From Table 5 in the Appendix,

$$t_{60,\,.99} = 2.39 \qquad t_{60,\,.995} = 2.66 \qquad t_{120,\,.99} = 2.358 \qquad t_{120,\,.995} = 2.617$$

Therefore, since $60 < 98 < 120$,

$$.005 < p/2 < .01 \qquad \text{or} \qquad .01 < p < .02$$

and H_0 is rejected. We conclude that there is a significant aggregation of cholesterol levels between spouses. This result is possibly due to common environmental factors such as diet. But it could also be due to the tendency for people of similar body build to marry each other, and their cholesterol levels may have been correlated at the time of marriage.

Interestingly, the one-sample t test for correlation coefficients in Equation 11.20 is mathematically equivalent to the F test in Equation 11.7 and the t test in Equation 11.8 for simple linear regression, in that they always yield the same p-values. The question as to which test is more appropriate is best answered by whether a regression or a correlation coefficient is the parameter of primary interest.

11.8.2 One-Sample z Test for a Correlation Coefficient

In Section 11.8.1, a test of the hypothesis H_0: $\rho = 0$ versus H_1: $\rho \neq 0$ was considered. Sometimes the correlation between two random variables is expected to be some quantity ρ_0 other than 0 and we wish to test the hypothesis H_0: $\rho = \rho_0$ versus H_1: $\rho \neq \rho_0$.

Example 11.29 Suppose the body weights of 100 fathers (x) and first-born sons (y) are measured and a sample correlation coefficient r of .38 is found. We might ask whether or not this sample correlation is compatible with an underlying correlation of .5 that might be expected on genetic grounds. How can this hypothesis be tested?

In this case, we want to test the hypothesis H_0: $\rho = .5$ versus H_1: $\rho \neq .5$. The problem with using the t test formation in Equation 11.20 is that the sample correlation coefficient r has a skewed distribution for nonzero ρ that cannot be easily approximated by a normal distribution. Fisher considered this problem and proposed the following transformation to better approximate a normal distribution:

EQUATION 11.21

Fisher's z Transformation of the Sample Correlation Coefficient r The z transformation of r given by

$$z = \frac{1}{2}\ln\left(\frac{1+r}{1-r}\right)$$

is approximately normally distributed under H_0 with mean

$$z_0 = \frac{1}{2}\ln\left[(1+\rho_0)/(1-\rho_0)\right]$$

and variance $1/(n-3)$. The z transformation is very close to r for small values of r but tends to deviate substantially from r for larger values of r. A table of the z transformation is given in Table 13 in the Appendix.

Example 11.30 Compute the z transformation of $r = .38$.

Solution The z transformation can be computed from Equation 11.21 as follows:

$$z = \frac{1}{2}\ln\left(\frac{1+0.38}{1-0.38}\right) = \frac{1}{2}\ln\left(\frac{1.38}{0.62}\right) = \frac{1}{2}\ln(2.226) = \frac{1}{2}(0.800) = 0.400$$

Alternatively, we could refer to Table 13 in the Appendix with $r = .38$ to obtain $z = 0.400$.

Fisher's z transformation can be used to conduct the hypothesis test as follows: Under H_0, z is approximately normally distributed with mean z_0 and variance $1/(n-3)$ or, equivalently,

$$\lambda = (z - z_0)\sqrt{n-3} \sim N(0,1)$$

H_0 will be rejected if z is far from z_0. Thus, the following test procedure for a two-sided level α test is used:

EQUATION 11.22

One-Sample z Test for a Correlation Coefficient To test the hypothesis H_0: $\rho = \rho_0$ versus H_1: $\rho \neq \rho_0$, use the following procedure:

(1) Compute the sample correlation coefficient r and the z transformation of r.

(2) Compute the test statistic

$$\lambda = (z - z_0)\sqrt{n-3}$$

(3) If $\lambda > z_{1-\alpha/2}$ or $\lambda < -z_{1-\alpha/2}$

reject H_0. If $-z_{1-\alpha/2} \le \lambda \le z_{1-\alpha/2}$

accept H_0.

(4) The exact p-value is given by

$$p = 2 \times \Phi(\lambda) \qquad \text{if } \lambda \le 0$$

$$p = 2 \times [1 - \Phi(\lambda)] \qquad \text{if } \lambda > 0$$

(5) Assume an underlying normal distribution for each of the random variables used to compute r and z.

The acceptance and rejection regions for this test are depicted in Figure 11.15. The computation of the p-value is illustrated in Figure 11.16.

Example 11.31 Perform a test of significance for the data in Example 11.29.

Solution In this case $r = .38$, $n = 100$, $\rho_0 = .50$. From Table 13 in the Appendix,

$$z_0 = \frac{1}{2}\ln\left(\frac{1+.5}{1-.5}\right) = .549 \qquad z = \frac{1}{2}\ln\left(\frac{1+.38}{1-.38}\right) = .400$$

Hence,

$$\lambda = (0.400 - 0.549)\sqrt{97} = (-0.149)(9.849) = -1.47 \sim N(0,1)$$

Thus, the p-value is given by

$$2 \times [1 - \Phi(1.47)] = 2 \times (1 - .9292) = .142$$

FIGURE 11.15 **Acceptance and rejection regions for the one-sample z test for a correlation coefficient**

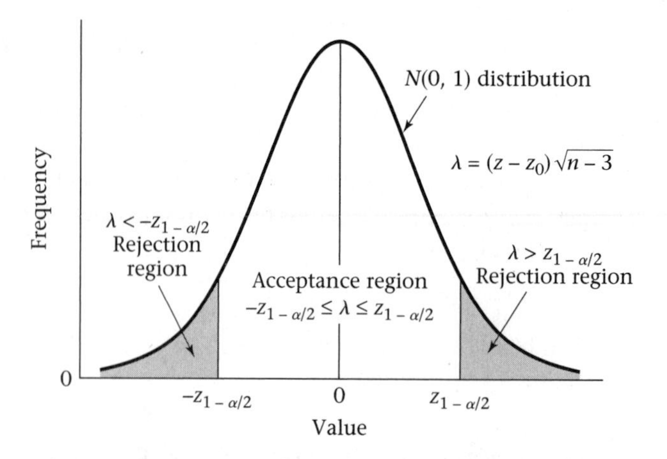

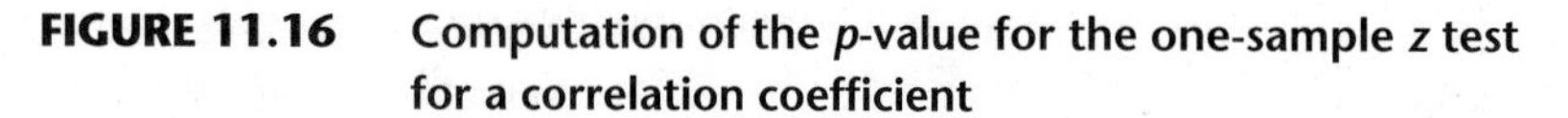

FIGURE 11.16 **Computation of the p-value for the one-sample z test for a correlation coefficient**

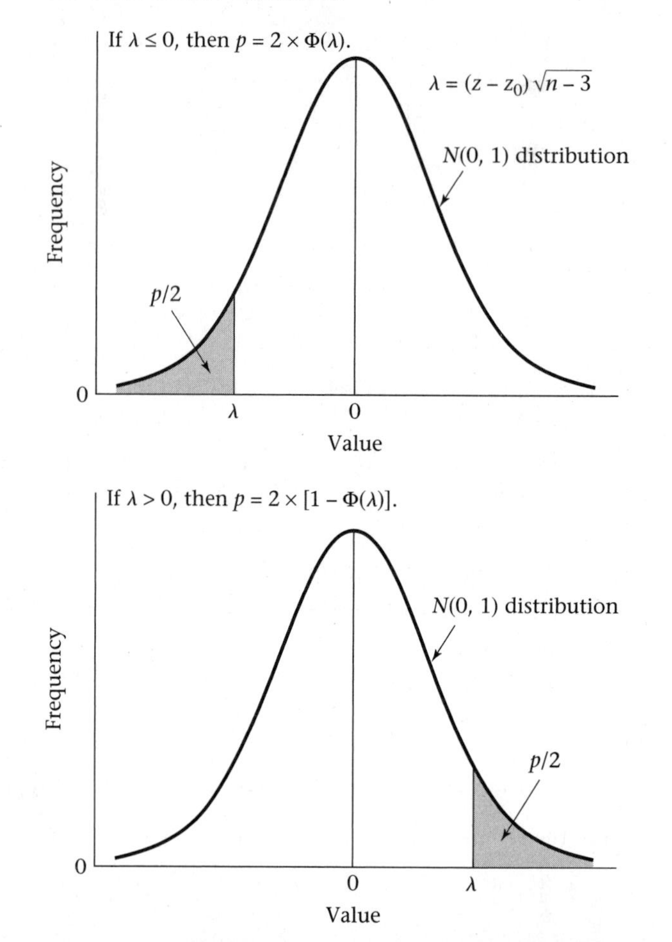

Therefore, we accept H_0 that the sample estimate of .38 is compatible with an underlying correlation of .50; this would be expected on purely genetic grounds.

To sum up, the z test in Equation 11.22 is used to test hypotheses about nonzero null correlations, whereas the t test in Equation 11.20 is used to test hypotheses about null correlations of zero. The z test can also be used to test correlations of zero under the null hypothesis, but the t test is slightly more powerful in this case and is preferred. However, if $\rho_0 \neq 0$, then the one-sample z test is very sensitive to nonnormality of either x or y. This is also true for the two-sample correlation test presented in Section 11.8.5.

11.8.3 Interval Estimation for Correlation Coefficients

In the previous two sections, we learned how to estimate a correlation coefficient ρ and how to perform appropriate hypothesis tests concerning ρ. It is also of interest to obtain confidence limits for ρ. An easy method for obtaining confidence limits for ρ can be derived based on the approximate normality of Fisher's z transformation of r. This method is given as follows:

EQUATION 11.23

Interval Estimation of a Correlation Coefficient (ρ) Suppose we have a sample correlation coefficient r based on a sample of n pairs of observations. To obtain a two-sided 100% × (1 − α) confidence interval for the population correlation coefficient (ρ):

(1) Compute Fisher's z transformation of $r = z = \frac{1}{2}\ln\left(\frac{1+r}{1-r}\right)$.

(2) Let z_0 = Fisher's z transformation of $\rho = \frac{1}{2}\ln\left(\frac{1+\rho}{1-\rho}\right)$.

A two-sided 100% × (1 − α) CI is given for $z_0 = (z_1, z_2)$ where

$$z_1 = z - z_{1-\alpha/2}\big/\sqrt{n-3}$$

$$z_2 = z + z_{1-\alpha/2}\big/\sqrt{n-3}$$

and $z_{1-\alpha/2} = 100\% \times (1-\alpha/2)$ percentile of an $N(0, 1)$ distribution

(3) A two-sided 100% × (1 − α) confidence int erval for ρ is then given by (ρ_1, ρ_2) where

$$\rho_1 = \frac{e^{2z_1}-1}{e^{2z_1}+1}$$

$$\rho_2 = \frac{e^{2z_2}-1}{e^{2z_2}+1}$$

The interval (z_1, z_2) in Equation 11.23 can be derived in a similar manner to the confidence interval for the mean of a normal distribution with known variance (see Equation 6.7), which is given by

EQUATION 11.24

$$(z_1, z_2) = z \pm z_{1-\alpha/2}/\sqrt{n-3}$$

We then solve Equation 11.24 for r in terms of z, whereby

EQUATION 11.25

$$r = \frac{e^{2z} - 1}{e^{2z} + 1}$$

We now substitute the confidence limits for z_0—that is, (z_1, z_2) in Equation 11.24—into Equation 11.25 to obtain the corresponding confidence limits for ρ given by (ρ_1, ρ_2) in Equation 11.23.

Example 11.32 In Example 11.29, a sample correlation coefficient of .38 was obtained between the body weights of 100 pairs of fathers (x) and first-born sons (y). Provide a 95% confidence interval for the underlying correlation coefficient ρ.

Solution From Example 11.31, the z transformation of $r = 0.400$. From step 2 of Equation 11.23, a 95% CI for z_0 is given by (z_1, z_2), where

$$z_1 = 0.400 - 1.96/\sqrt{97} = 0.400 - 0.199 = 0.201$$

$$z_1 = 0.400 + 1.96/\sqrt{97} = 0.400 + 0.199 = 0.599$$

From step 3 of Equation 11.23, a 95% CI for ρ is given by (ρ_1, ρ_2) where

$$\rho_1 = \frac{e^{2(0.201)} - 1}{e^{2(0.201)} + 1} = \frac{e^{.402} - 1}{e^{.402} + 1} = \frac{1.4950 - 1}{1.4950 + 1} = \frac{0.4950}{2.4950} = .198$$

$$\rho_2 = \frac{e^{2(.599)} - 1}{e^{2(.599)} + 1} = \frac{e^{1.198} - 1}{e^{1.198} + 1} = \frac{2.3139}{4.3139} = .536$$

Thus, a 95% CI for $\rho = (.198, .536)$.

Notice that the confidence interval for z_0, given by $(z_1, z_2) = (0.201, 0.599)$, is symmetric about z (0.400). However, when the confidence limits are transformed back to the original scale (i.e., the scale of ρ) the corresponding confidence limits for ρ are given by $(\rho_1, \rho_2) = (.198, .536)$, which are not symmetric about $r = .400$. The reason for this is that Fisher's z transformation is a nonlinear function of r, which only becomes approximately linear when r is small (i.e., $|r| \le .2$).

11.8.4 Sample-Size Estimation for Correlation Coefficients

Example 11.33 **Nutrition** Suppose a new dietary questionnaire is constructed to be administered over the Internet, based on dietary recall over the past 24 hours. To validate this questionnaire, participants are given 3 days' worth of food diaries, in which they fill out exactly what they eat for 3 days, spaced about 1 month apart. The average intake over 3 days will be considered a gold standard. The correlation between the 24-hour recall and the gold standard will be an index of validity. How large a sample is needed to have 80% power to detect a significant correlation between these measures, if it is expected that the true correlation is .5 and a one-sided test is used with $\alpha = .05$?

To address this question, we use the Fisher's z-transform approach. Specifically, we wish to test the hypothesis H_0: $\rho = 0$ versus H_1: $\rho = \rho_0 > 0$. Under H_0,

$$z \sim N\left[0, 1/(n-3)\right]$$

We will reject H_0 at level α if $z\sqrt{n-3} > z_{1-\alpha}$. Suppose z_0 is the Fisher's z transform of ρ_0. If we subtract $z_0\sqrt{n-3}$ from both sides of the equation, it follows that

$$\lambda = \sqrt{n-3}\left(z - z_0\right) > z_{1-\alpha} - \sqrt{n-3}\, z_0$$

Furthermore, under H_1, $\lambda \sim N(0, 1)$. Therefore,

$$\begin{aligned} Pr\left(\lambda > z_{1-\alpha} - \sqrt{n-3}\, z_0\right) &= 1 - \Phi\left(z_{1-\alpha} - \sqrt{n-3}\, z_0\right) \\ &= \Phi\left(\sqrt{n-3}\, z_0 - z_{1-\alpha}\right) \end{aligned}$$

If we require a power of $1-\beta$, then we set the right-hand side to $1-\beta$ or, equivalently,

$$\sqrt{n-3}\, z_0 - z_{1-\alpha} = z_{1-\beta}$$

If follows that

$$\text{Power} = 1 - \beta = \Phi\left(\sqrt{n-3}\, z_0 - z_{1-\alpha}\right)$$

The corresponding sample-size estimate is obtained by solving for n, whereby

$$n = \left[\left(z_{1-\alpha} + z_{1-\beta}\right)^2 / z_0^2\right] + 3$$

This procedure is summarized as follows:

EQUATION 11.26

Power and Sample-Size Estimation for Correlation Coefficients Suppose we wish to test the hypothesis H_0: $\rho = 0$ versus H_1: $\rho = \rho_0 > 0$. For the specific alternative $\rho = \rho_0$, to test the hypothesis with a one-sided significance level of α and specified sample size n, the power is given by

$$\text{Power} = \Phi\left(\sqrt{n-3}\, z_0 - z_{1-\alpha}\right)$$

For the specific alternative $\rho = \rho_0$, to test the hypothesis with a one-sided significance level of α and specified power of $1-\beta$, we require a sample size of

$$n = \left[\left(z_{1-\alpha} + z_{1-\beta}\right)^2 / z_0^2\right] + 3$$

Solution to Example 11.33 In this case, we have $\rho_0 = .5$. Therefore, from Table 13 in the Appendix, $z_0 = .549$. Also, $\alpha = .05$, $1 - \beta = .80$. Thus, from Equation 11.26, we have

$$\begin{aligned} n &= \left[\left(z_{.95} + z_{.80}\right)^2 / .549^2\right] + 3 \\ &= \left[\left(1.645 + 0.84\right)^2 / .549^2\right] + 3 \\ &= 23.5 \end{aligned}$$

Therefore, to have 80% power, we need 24 subjects in the validation study.

Example 11.34 **Nutrition** Suppose that 50 subjects are actually enrolled in the validation study. What power will the study have if the true correlation is .5 and a one-sided test is used with $\alpha = .05$?

Solution We have $\alpha = .05$, $\rho_0 = .50$, $z_0 = .549$, $n = 50$. Thus, from Equation 11.26,

$$\begin{aligned} \text{Power} &= \Phi\left(\sqrt{47}(.549) - z_{.95}\right) \\ &= \Phi(3.764 - 1.645) \\ &= \Phi(2.119) = .983 \end{aligned}$$

Therefore, the study will have 98.3% power.

11.8.5 Two-Sample Test for Correlations

The use of Fisher's z transformation can be extended to two-sample problems.

Example 11.35 **Hypertension** Suppose there are two groups of children. Children in one group live with their natural parents, whereas children in the other group live with adopted parents. One question that arises is whether or not the correlation between the blood pressure of a mother and a child is different in these two groups. A different correlation would suggest a genetic effect on blood pressure. Suppose there are 1000 mother–child pairs in the first group, with correlation .35, and 100 mother–child pairs in the second group, with correlation .06. How can this question be answered?

We wish to test the hypothesis H_0: $\rho_1 = \rho_2$ versus H_1: $\rho_1 \neq \rho_2$. It is reasonable to base the test on the difference between the z's in the two samples. If $|z_1 - z_2|$ is large, then H_0 will be rejected; otherwise, H_0 will be accepted. This principle suggests the following test procedure for a two-sided level α test:

EQUATION 11.27

Fisher's *z* Test for Comparing Two Correlation Coefficients To test the hypothesis H_0: $\rho_1 = \rho_2$ versus H_1: $\rho_1 \neq \rho_2$, use the following procedures:

(1) Compute the sample correlation coefficients (r_1, r_2) and Fisher's z transformation (z_1, z_2) for each of the two samples.

(2) Compute the test statistic

$$\lambda = \frac{z_1 - z_2}{\sqrt{\dfrac{1}{n_1 - 3} + \dfrac{1}{n_2 - 3}}} \sim N(0,1) \text{ under } H_0$$

(3) If $\lambda > z_{1-\alpha/2}$ or $\lambda < -z_{1-\alpha/2}$
reject H_0. If $-z_{1-\alpha/2} \le \lambda \le z_{1-\alpha/2}$
accept H_0.

(4) The exact p-value is given by

$$p = 2\,\Phi(\lambda) \qquad \text{if } \lambda \le 0$$
$$p = 2 \times [1 - \Phi(\lambda)] \qquad \text{if } \lambda > 0$$

(5) Assume an underlying normal distribution for each of the random variables used to compute r_1, r_2 and z_1, z_2.

The acceptance and rejection regions for this test are depicted in Figure 11.17. The computation of the p-value is illustrated in Figure 11.18.

Example 11.36 Perform a significance test for the data in Example 11.35.

Solution

$$r_1 = .35 \qquad n_1 = 1000 \qquad r_2 = .06 \qquad n_2 = 100$$

Thus, from Table 13 in the Appendix,

$$z_1 = 0.365 \qquad z_2 = 0.060$$

and

$$\lambda = \frac{0.365 - 0.060}{\sqrt{\frac{1}{997} + \frac{1}{97}}} = 9.402(0.305) = 2.87 \sim N(0,1) \text{ under } H_0$$

Hence the p-value is given by

$$2 \times [1 - \Phi(2.87)] = .004$$

Therefore, there is a significant difference between the mother–child correlations in the two groups, implying a significant genetic effect on blood pressure.

FIGURE 11.17 **Acceptance and rejection regions for Fisher's z test for comparing two correlation coefficients**

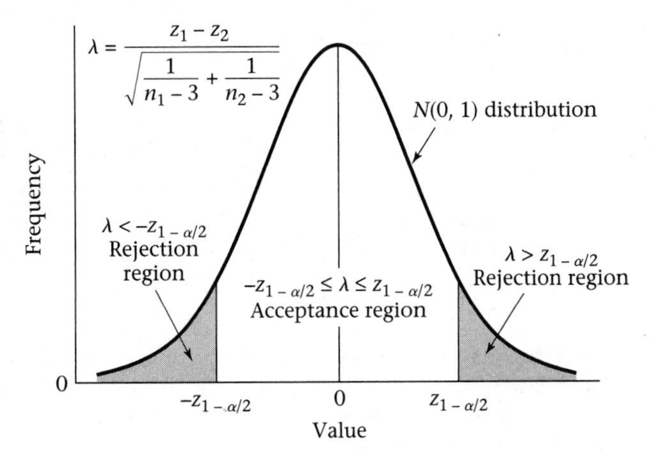

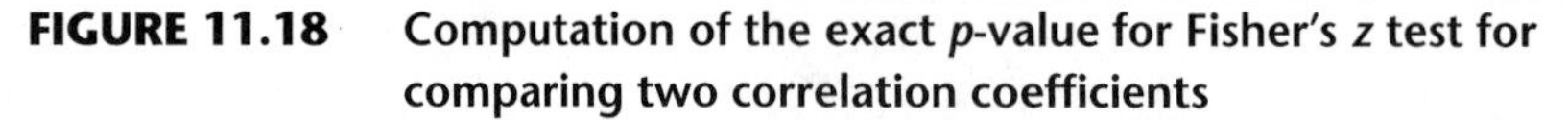

FIGURE 11.18 **Computation of the exact *p*-value for Fisher's *z* test for comparing two correlation coefficients**

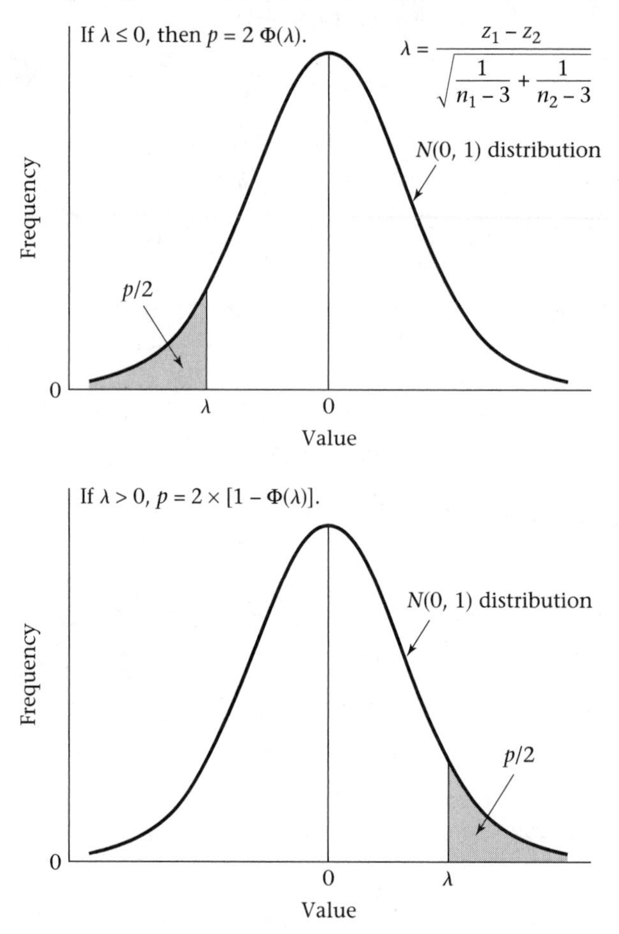

SECTION 11.9 Multiple Regression

In Sections 11.2 through 11.6 problems in linear-regression analysis in which there is one independent variable (x), one dependent variable (y), and a linear relationship between x and y were discussed. In practice, there is often more than one independent variable and we would like to look at the relationship between each of the independent variables ($x_1, \ldots, x_k$) and the dependent variable (y) after taking into account the remaining independent variables. This type of problem is the subject matter of **multiple-regression analysis.**

Example 11.37 **Hypertension, Pediatrics** A topic of interest in hypertension research is how the relationship between the blood-pressure levels of newborns and blood-pressure levels of infants relates to subsequent adult blood pressure. One problem that arises is that the blood pressure of a newborn is affected by several extraneous factors that make this relationship difficult to study. In particular, newborn blood pressures are affected by (1) birthweight and (2) the day of life on which blood pressure is measured. In this study, the infants were weighed at the time of the blood-pressure measurements. We refer to this weight as the "birthweight" although it differs somewhat from their actual weight at birth. Since the infants grow in the first few days of life, we would expect that infants seen at 5 days of life would on average have a greater weight than those seen at 2 days of life. We would like to be able to adjust the observed blood pressure for these two factors before looking at other factors that may influence newborn blood pressure.

11.9.1 Estimation of the Regression Equation

Suppose a relationship is postulated between systolic blood pressure (y), birthweight (x_1), and age in days (x_2) that is of the form

EQUATION 11.28

$$y = \alpha + \beta_1 x_1 + \beta_2 x_2 + e$$

where e is an error term that is normally distributed with mean 0 and variance σ^2. We would like to estimate the parameters of this model and test various hypotheses concerning it. The same method of least squares that was introduced in Section 11.3 for simple linear regression will be used to fit the parameters of this multiple-regression model. In particular, α, β_1, β_2 will be estimated by a, b_1, and b_2, where we choose a, b_1, and b_2 to minimize the sum of

$$\left[y-\left(a+b_1 x_1+b_2 x_2\right)\right]^2$$

over all the data points.

In general, if we have k independent variables $x_1, \ldots, x_k$, then a linear-regression model relating y to $x_1, \ldots, x_k$, is of the form

EQUATION 11.29

$$y = \alpha + \sum_{j=1}^{k} \beta_j x_j + e$$

where e is an error term that is normally distributed with mean 0 and variance σ^2.

We estimate $\alpha, \beta_1, \ldots, \beta_k$ by $a, b_1, \ldots, b_k$ using the method of least squares, where we minimize the sum of

$$\left[y-\left(a+\sum_{j=1}^{k} b_j x_j\right)\right]^2$$

Example 11.38 **Hypertension, Pediatrics** Suppose systolic blood pressure, birthweight (oz), and age (days) are measured for 16 infants and the data are as shown in Table 11.7. Estimate the parameters of the multiple-regression model in Equation 11.28.

Solution Use the SAS PROC REG program to obtain the least-squares estimates. The results are given in Table 11.8.

According to the parameter estimate column, the regression equation is given by

$$y = 53.45 + 0.126x_1 + 5.89x_2$$

TABLE 11.7 Sample data for infant blood pressure, age, and birthweight for 16 infants

i	Birthweight in oz (x_1)	Age in days (x_2)	Systolic blood pressure (mm Hg) (y)
1	135	3	89
2	120	4	90
3	100	3	83
4	105	2	77
5	130	4	92
6	125	5	98
7	125	2	82
8	105	3	85
9	120	5	96
10	90	4	95
11	120	2	80
12	95	3	79
13	120	3	86
14	150	4	97
15	160	3	92
16	125	3	88

TABLE 11.8 Least-squares estimates of the regression parameters for the newborn blood-pressure data in Table 11.7 using the SAS PROC REG program

```
                                The SAS System
Model: MODEL1
Dependent Variable: SYSBP
                              Analysis of Variance
                           Sum of            Mean
     Source      DF       Squares          Square         F Value        Prob>F
     Model        2     591.03564       295.51782          48.081        0.0001
     Error       13      79.90186         6.14630
     C Total     15     670.93750

         Root MSE       2.47917       R-square        0.8809
         Dep Mean      88.06250       Adj R-sq        0.8626
         C.V.           2.81524

                              Parameter Estimates
                                                                                Squared
                 Parameter    Standard    T for H0:                Standarized  Partial
Variable   DF    Estimate        Error  Parameter=0  Prob > |T|     Estimate  Corr Type II
INTERCEP    1   53.450194   4.53188859      11.794      0.0001   0.00000000        .
BRTHWGT     1    0.125583   0.03433620       3.657      0.0029   0.35207698   0.50714655
AGEDYS      1    5.887719   0.68020515       8.656      0.0001   0.83323075   0.85214312
```

The regression equation tells us that for a newborn the average blood pressure increases by an estimated 0.126 mm Hg per ounce of birthweight and 5.89 mm Hg per day of age.

Example 11.39 **Hypertension, Pediatrics** Calculate the predicted average systolic blood pressure of a baby with birthweight 8 lb (128 oz) measured at 3 days of life.

Solution The average systolic blood pressure is given by

$$53.45 + 0.126(128) + 5.89(3) = 87.2 \text{ mm Hg}$$

The regression coefficients in Table 11.8 are called *partial-regression coefficients.*

DEFINITION 11.16 Suppose we consider the multiple-regression model $y = \alpha + \sum_{j=1}^{k} \beta_j x_j + e$, where e follows a normal distribution with mean 0 and variance σ^2. The β_j, $j = 1, 2, \ldots, k$ are referred to as **partial-regression coefficients.** β_j represents the predicted increase in y per unit increase in x_j, with all other variables held constant (or stated another way, after adjusting for all other variables in the model), and is estimated by the parameter b_j.

Partial-regression coefficients differ from simple linear-regression coefficients as given in Equation 11.2. The latter represent the average increase in y per unit increase in x, without considering any other independent variables. If there are strong relationships among the independent variables in a multiple-regression model, then the partial-regression coefficients may differ considerably from the simple linear-regression coefficients obtained from considering each independent variable separately.

Example 11.40 **Hypertension** Interpret the regression coefficients in Table 11.8.

Solution The partial-regression coefficient for birthweight = b_1 = 0.126 mm Hg/oz represents the average increase in systolic blood pressure per 1 oz increase in birthweight for infants of the same age. The regression coefficient for age = b_2 = 5.89 mm Hg/day represents the average increase in systolic blood pressure per 1-day increase in age for infants of the same birthweight.

We are often interested in ranking the independent variables according to their predictive relationship with the dependent variable y. It is difficult to rank the variables based on the magnitude of the partial-regression coefficients, since the independent variables are often in different units. Specifically, from the multiple-regression model in Equation 11.29, we see that b estimates the increase in y per unit increase in x, while holding the values of all other variables in the model constant. If x is increased by 1 standard deviation unit (i.e., s_x) to $x + s_x$, then y would be expected to increase by $b \times s_x$ raw units or $(b \times s_x)/s_y$ standard deviation units of y (s_y).

DEFINITION 11.17

The **standardized regression coefficient (b_s)** is given by $b \times (s_x/s_y)$. It represents the average increase in y (expressed in standard deviation units of y) per standard deviation increase in x, after adjusting for all other variables in the model.

Thus, the standardized regression coefficient is a useful measure for comparing the predictive value of several independent variables, since it tells us the predicted increase in standard deviation units of y per standard deviation increase in x. By expressing change in standard deviation units of x, we can control for differences in the units of measurement for different independent variables.

Example 11.41 Compute the standardized regression coefficients for birthweight and age in days using the data in Tables 11.7 and 11.8.

Solution From Table 11.7, $s_y = 6.69$, $s_{x_1} = 18.75$, $s_{x_2} = 0.946$. Therefore, referring to the standardized estimate column in Table 11.8,

$$b_s(\text{birthweight}) = \frac{0.1256 \times 18.75}{6.69} = 0.352$$

$$b_s(\text{age in days}) = \frac{5.888 \times 0.946}{6.69} = 0.833$$

Thus, the average increase in systolic blood pressure is 0.352 standard deviation units of blood pressure per standard deviation increase in birthweight holding age constant and 0.833 standard deviation units of blood pressure per standard deviation increase in age, holding birthweight constant. Thus, age appears to be the more important variable after controlling for both variables simultaneously in the multiple-regression model.

11.9.2 Hypothesis Testing

Example 11.42 **Hypertension, Pediatrics** We would like to test various hypotheses concerning the data in Table 11.7. First, we would like to test the overall hypothesis that birthweight and age when considered together are significant predictors of blood pressure. How can this be done?

Specifically, we will test the hypothesis H_0: $\beta_1 = \beta_2 = \cdots = \beta_k = 0$ versus H_1: at least one of $\beta_1, \ldots, \beta_k \neq 0$. The test of significance is similar to the F test in Section 11.4. The test procedure for a level α test is given as follows:

EQUATION 11.30

F Test for Testing the Hypothesis H_0: $\beta_1 = \beta_2 = \cdots = \beta_k = 0$ versus H_1: At Least One of the $\beta_j \neq 0$ in Multiple Linear Regression

(1) Fit the regression parameters using the method of least squares, and compute Reg SS and Res SS,

where
$$\text{Res SS} = \sum_{i=1}^{n} (y_i - \hat{y}_i)^2$$

$$\text{Reg SS} = \text{Total SS} - \text{Res SS}$$

$$\text{Total SS} = \sum_{i=1}^{n} (y_i - \bar{y})^2$$

$$\hat{y}_i = a + \sum_{j=1}^{k} b_j x_{ij}$$

$x_{ij} = j$th independent variable for the ith subject, $j = 1, \ldots, k$; $i = 1, \ldots, n$

(2) Compute Reg MS = Reg SS/k, Res MS = Res SS/$(n - k - 1)$.

(3) Compute the test statistic

$$F = \text{Reg MS/Res MS}$$

which follows an $F_{k,n-k-1}$ distribution under H_0.

(4) For a level α test, if

$$F > F_{k,n-k-1,1-\alpha}$$

then reject H_0; if $F \le F_{k,n-k-1,1-\alpha}$

then accept H_0.

(5) The exact p-value is given by the area to the right of F under an $F_{k,n-k-1}$ distribution = $Pr(F_{k,n-k-1} > F)$.

The acceptance and rejection regions for this test procedure are depicted in Figure 11.19. The computation of the exact p-value is illustrated in Figure 11.20.

Example 11.43 **Hypertension, Pediatrics** Test the hypothesis H_0: $\beta_1 = \beta_2 = 0$ versus H_1: either $\beta_1 \ne 0$ or $\beta_2 \ne 0$ using the data in Tables 11.7 and 11.8.

Solution Refer to Table 11.8 and note that

$$\text{Reg SS} = 591.04 \text{ (referred to as Model SS)}$$

$$\text{Reg MS} = 591.04/2 = 295.52$$

$$\text{Res SS} = 79.90 \text{ (referred to as Error SS)}$$

FIGURE 11.19 **Acceptance and rejection regions for testing the hypothesis H_0: $\beta_1 = \beta_2 = \cdots = \beta_k = 0$ versus H_1: at least one of the $\beta_j \ne 0$ in multiple linear regression**

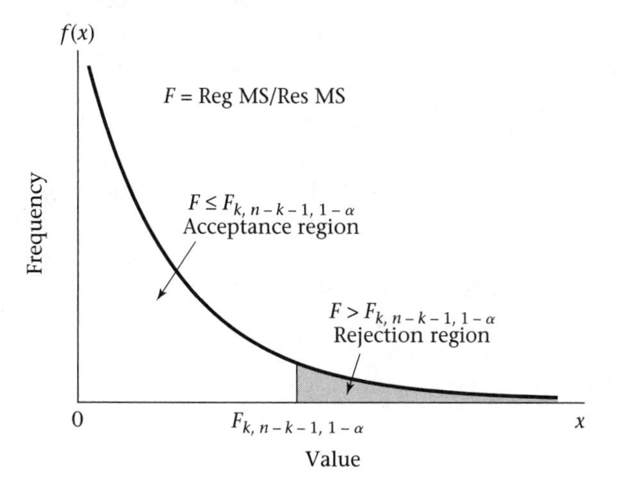

FIGURE 11.20 **Computation of the p-value for testing the hypothesis H_0: $\beta_1 = \beta_2 = \cdots = \beta_k = 0$ versus H_1: at least one of the $\beta_j \neq 0$ in multiple linear regression**

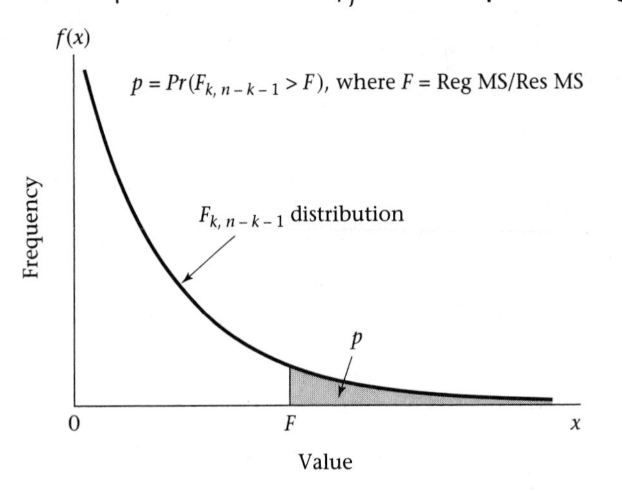

$$\text{Res MS} = 79.90/13 = 6.146$$

$$F = \text{Reg MS/Res MS} = 48.08 \sim F_{2,13} \text{ under } H_0$$

Since $F_{2,13,.999} < F_{2,12,.999} = 12.97 < 48.08 = F$

$p < .001$. Thus, we can conclude that the two variables when considered together are significant predictors of blood pressure.

The significant p-value for this test could be attributed to either variable. We would like to perform significance tests to identify the independent contributions of each variable. How can this be done?

In particular, to assess the independent contribution of birthweight, we will assume that age is making a contribution under either hypothesis, and we will test the hypothesis H_0: $\beta_1 = 0$, $\beta_2 \neq 0$ versus H_1: $\beta_1 \neq 0$, $\beta_2 \neq 0$. Similarly, to assess the independent contribution of age, we will assume that birthweight is making a contribution under either hypothesis and will test the hypothesis H_0: $\beta_2 = 0$, $\beta_1 \neq 0$ versus H_1: $\beta_2 \neq 0$, $\beta_1 \neq 0$. In general, if we have k independent variables, then to assess the specific effect of the lth independent variable (x_l), on y after controlling for the effects of all other variables, we wish to test the hypothesis H_0: $\beta_l = 0$, all other $\beta_j \neq 0$ versus H_1: all $\beta_j \neq 0$. We will focus on assessing the independent contribution of birthweight. Our approach will be to compute the standard error of the partial-regression coefficient for birthweight and base our test on $t = b_1/se(b_1)$, which will follow a t distribution with $n - k - 1$ df under H_0. Specifically, the following test procedure for a level α test is used:

EQUATION 11.31

t Test for Testing the Hypothesis H_0: $\beta_\ell = 0$, All Other $\beta_j \neq 0$ versus H_1: $\beta_\ell \neq 0$, All Other $\beta_j \neq 0$ in Multiple Linear Regression

(1) Compute

$$t = b_\ell / se(b_\ell)$$

which should follow a t distribution with $n-k-1$ df under H_0.

(2) If

$$t < t_{n-k-1,\alpha/2} \quad \text{or} \quad t > t_{n-k-1,1-\alpha/2}$$

then reject H_0; if $t_{n-k-1,\alpha/2} \le t \le t_{n-k-1,1-\alpha/2}$

then accept H_0.

(3) The exact p-value is given by

$$2 \times Pr(t_{n-k-1} > t) \quad \text{if } t \ge 0$$

$$2 \times Pr(t_{n-k-1} \le t) \quad \text{if } t < 0$$

The acceptance and rejection regions for this test are depicted in Figure 11.21. The computation of the exact p-value is illustrated in Figure 11.22.

Example 11.44 **Hypertension, Pediatrics** Test for the independent contributions of birthweight and age in predicting systolic blood pressure in infants, using the data in Table 11.8.

Solution From Table 11.8,

$$b_1 = 0.1256$$

$$se(b_1) = 0.0343$$

$$t(\text{birthweight}) = b_1 / se(b_1) = 3.66$$

FIGURE 11.21 **Acceptance and rejection regions for the t test for multiple linear regression**

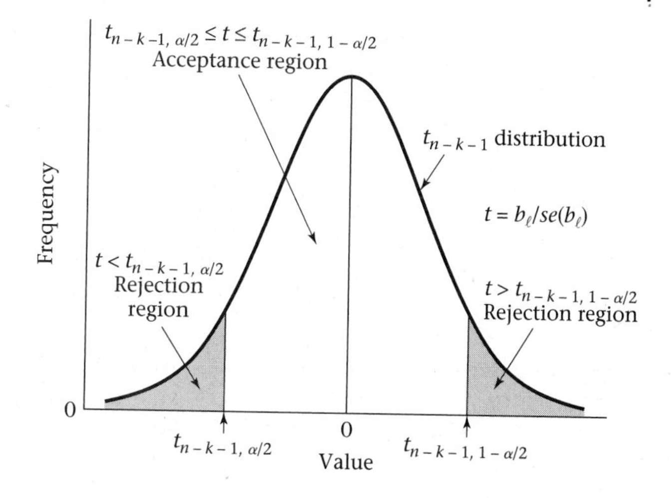

FIGURE 11.22 Computation of the exact *p*-value for the *t* test for multiple linear regression

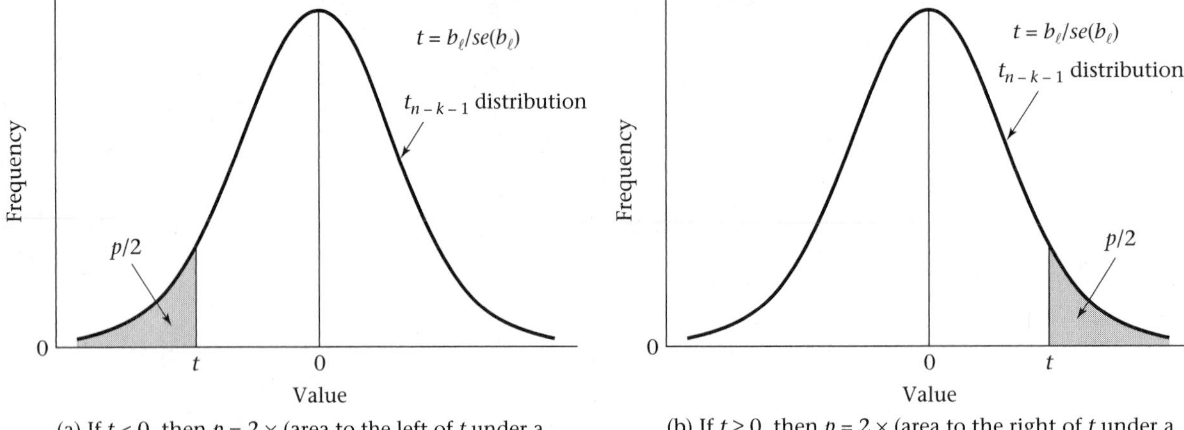

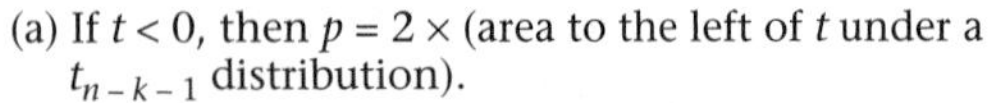

(a) If $t < 0$, then $p = 2 \times$ (area to the left of t under a t_{n-k-1} distribution).

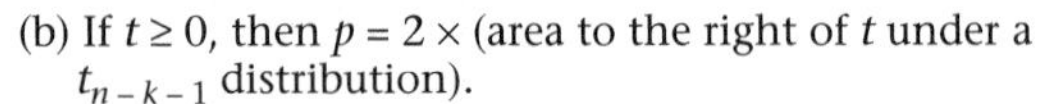

(b) If $t \ge 0$, then $p = 2 \times$ (area to the right of t under a t_{n-k-1} distribution).

$$p = 2 \times Pr(t_{13} > 3.66) = .003$$

$$b_2 = 5.888$$

$$se(b_2) = 0.6802$$

$$t(\text{age}) = b_2 / se(b_2) = 8.66$$

$$p = 2 \times Pr(t_{13} > 8.66) < .001$$

Therefore, both birthweight and age have highly significant associations with systolic blood pressure even after controlling for the other variable.

It is possible that an independent variable (x_1) will seem to have an important effect on a dependent variable (y) when considered by itself, but will not be significant after adjusting for another independent variable (x_2). This usually occurs when x_1 and x_2 are strongly related to each other and when x_2 is also related to y. We refer to x_2 as a confounder of the relationship between y and x_1. We will discuss confounding in more detail in Chapter 13. Indeed, one of the advantages of multiple-regression analysis is that it allows us to identify which few variables among a large set of independent variables have a significant relationship to the dependent variable *after adjusting for other important variables.*

Example 11.45 **Hypertension, Pediatrics** Suppose we consider the two independent variables, x_1 = birthweight, x_2 = body length and try to use these variables to predict systolic blood pressure in newborns (y). Perhaps both x_1 and x_2, *when considered separately* in a simple linear-regression model as given in Equation 11.2, have a significant relationship to blood pressure. However, since birthweight and body length are closely related to each other, after adjusting for birthweight, body length may not be significantly related to blood pressure based on the test procedure in Equation 11.31. One possible interpretation of this result is that the effect of body length on blood pressure can be explained by its strong relationship to birthweight.

In some instances, two strongly related variables are entered into the same multiple-regression model and neither variable is significant after controlling for the effect of the other variable. Such variables are referred to as **collinear.** It is best to avoid the use of highly collinear variables in the same multiple-regression model since their simultaneous presence can make it impossible to identify the specific effects of each variable.

Example 11.46 **Hypertension** A commonly used measure of obesity is **body-mass index (BMI)**, which is defined as weight/(height)^2. It is well known that both weight and BMI, when considered separately, are strongly related to level of blood pressure. However, if they are entered simultaneously in the same multiple-regression model, then it is possible that neither would be significant, since they are strongly related to each other. Thus, if we control for weight, there may not be any additional predictive power for BMI, and vice versa.

In Equation 11.31, we have considered the test of the hypothesis H_0: that a specific partial-regression coefficient ($\beta_\ell = 0$) versus the alternative hypothesis H_1: that $\beta_\ell \neq 0$. Under both H_0 and H_1, all other partial-regression coefficients are allowed to be different from 0. We used a t statistic to test these hypotheses. Another way to perform this test is in the form of a partial F test, which is given as follows:

EQUATION 11.32

Partial *F* Test for Partial-Regression Coefficients in Multiple Linear Regression To test the hypothesis H_0: $\beta_\ell = 0$, all other $\beta_j \neq 0$, versus H_1: $\beta_\ell \neq 0$, all other $\beta_j \neq 0$ in multiple linear regression, we

(1) Compute

$$F = \frac{\text{Regr SS}_{\text{full model}} - \text{Regr SS}_{\text{all variables except } \beta_\ell \text{ in the model}}}{\text{Res MS}_{\text{full model}}}$$

which should follow an $F_{1,n-k-1}$ distribution under H_0.

(2) The exact p-value is given by $Pr(F_{1,n-k-1} > F)$.

(3) It can be shown that the p-value from using the partial F test given in (2) is the same as the p-value obtained from using the t test in Equation 11.31.

Many statistical packages use variable selection strategies such as forward and backward selection based on a succession of partial F tests. A complete discussion of variable selection strategies is provided in [3] and [4].

11.9.3 Criteria for Goodness of Fit

In Section 11.6, we discussed criteria for goodness of fit in simple linear-regression models, based on residual analysis. Similar criteria can be used in a multiple-regression setting.

Example 11.47 **Hypertension** Assess the goodness of fit of the multiple-regression model in Table 11.8 fitted to the infant blood-pressure data in Table 11.7.

Solution We compute the residual for each of the 16 sample points in Table 11.7. The standard error of each of the fitted residuals is different and depends on the distance of the corresponding sample point from the average of the sample points used in fitting the regression line. Thus, we will usually be interested in the Studentized residuals = STUDENT(i) = $\hat{e}_i/\text{sd}(\hat{e}_i)$. (See [3] or [4] for the formulas used to compute $\text{sd}(\hat{e}_i)$ in a multiple-regression setting.) We have plotted the Studentized residuals versus the predicted blood pressure (Figure 11.23a) and each of the independent variables (Figures 11.23b and 11.23c). This will allow us to identify any outlying values as well as violations of the linearity and equal-variance assumptions in the multiple-regression model.

FIGURE 11.23a **Plot of Studentized residuals versus predicted values of systolic blood pressure for the multiple-regression model in Table 11.8**

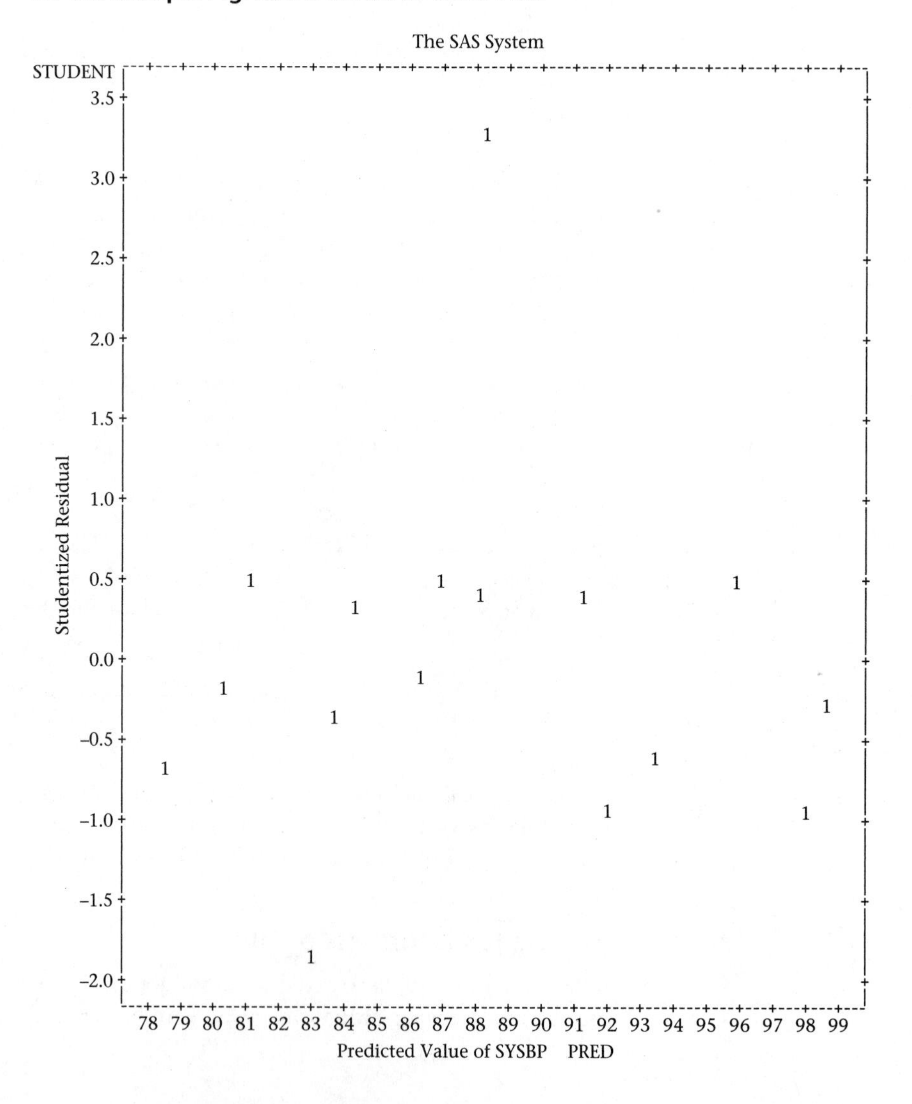

FIGURE 11.23b **Plot of Studentized residuals versus birthweight for the multiple-regression model in Table 11.8**

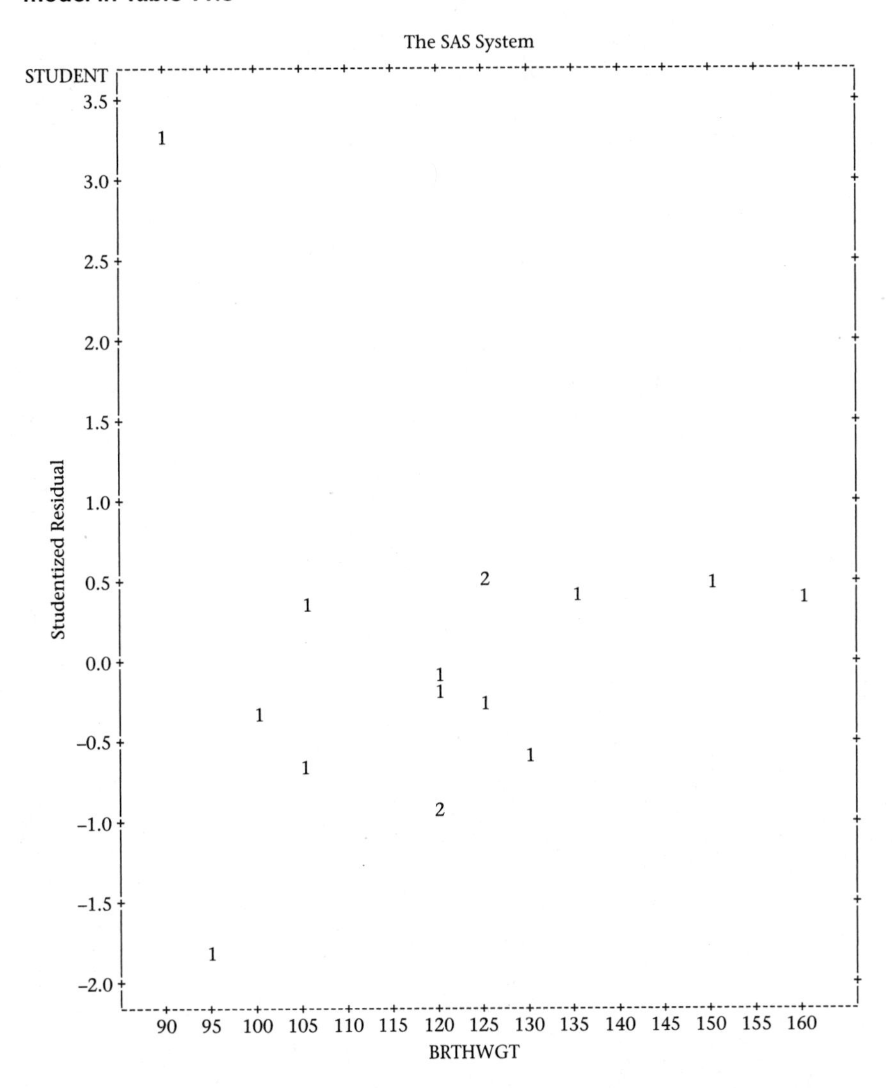

There appears to be a possible outlying value with a Studentized residual ≅ 3.0 corresponding to a birthweight of 90 oz and an age in days = 4.0 (observation 10). To focus more clearly on outlying values, some computer packages allow the user to delete an observation, recompute the regression model from the remaining data points, and compute the residual of the deleted observation based on the recomputed regression. The rationale for this procedure is that the outlying value may have affected the estimates of the regression parameters. Let

$$y = a^{(i)} + b_1^{(i)} x_1 + \cdots + b_k^{(i)} x_k$$

denote the estimated regression model with the ith sample point deleted. The residual of the deleted point from this regression line is

FIGURE 11.23c **Plot of Studentized residuals versus age for the multiple-regression model in Table 11.8**

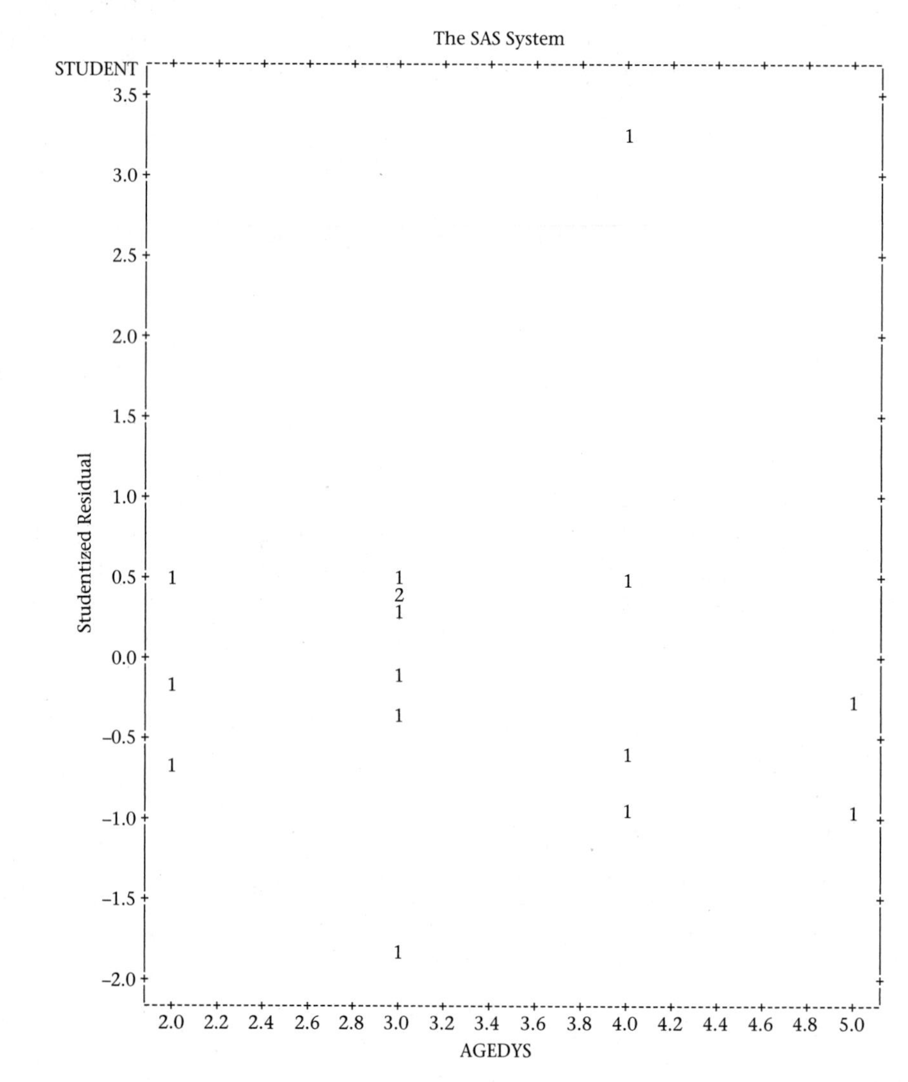

$$\hat{e}^{(i)} = y_i - \left[a^{(i)} + b_1^{(i)}x_{i1} + \cdots + b_k^{(i)}x_{ik}\right]$$

with standard error $\text{sd}(\hat{e}^{(i)})$. The corresponding Studentized residual is $\hat{e}^{(i)}/\text{sd}(\hat{e}^{(i)})$ and is denoted by RSTUDENT(i). It is sometimes referred to as an **externally Studentized residual**, because the ith data point was not used in estimating the regression parameters as opposed to STUDENT(i), which is sometimes referred to as an **internally Studentized residual,** because the ith data point was used in estimating the regression parameters. We have plotted the externally Studentized residuals [RSTUDENT(i)] versus the predicted blood pressure (Figure 11.24a) and each of the independent variables (Figures 11.24b and 11.24c). These plots really highlight the outlying value. Data point 10 has a value of RSTUDENT that is approximately 7 standard deviations above zero, which indicates a gross outlier.

FIGURE 11.24a **Plot of RSTUDENT versus the predicted systolic blood pressure for the multiple-regression model in Table 11.8**

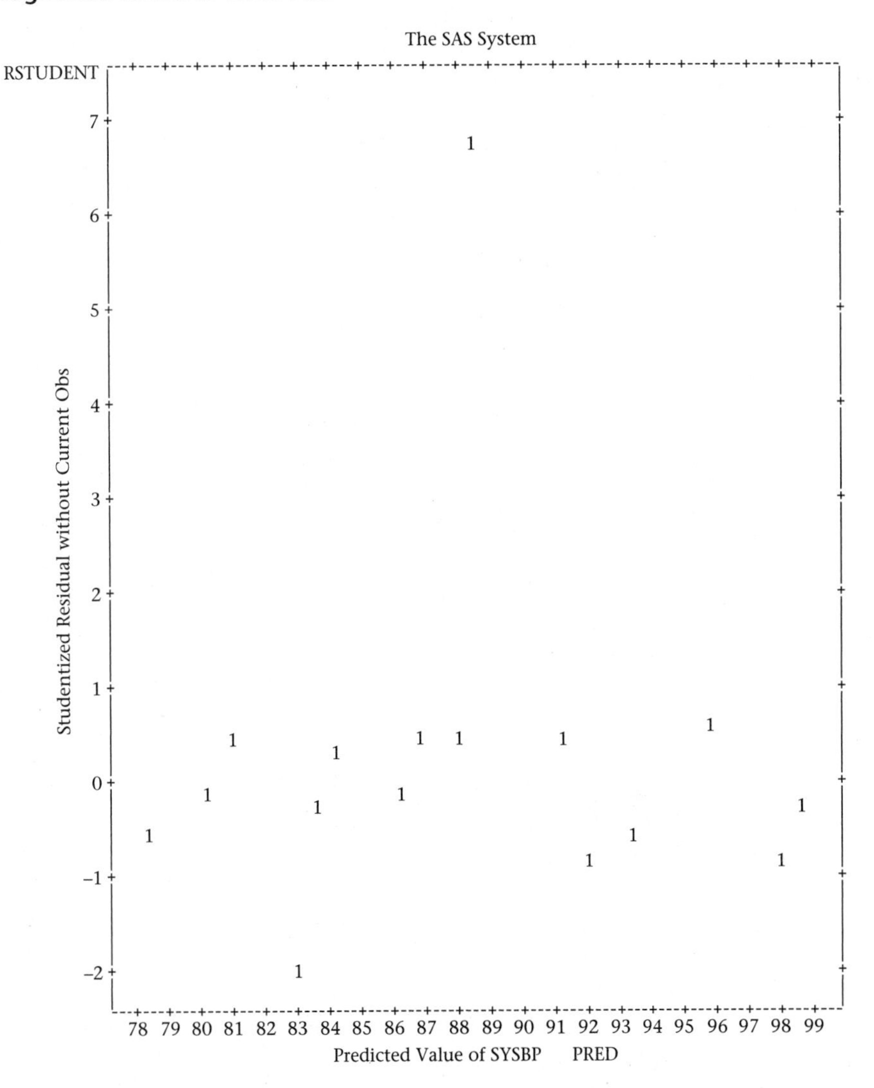

The plots in Figures 11.24a–11.24c do not really reflect the multivariate nature of the data. Specifically, under the multiple-regression model in Equation 11.29, the relationship between y and a specific independent variable x_ℓ is characterized as follows:

EQUATION 11.33

y is normally distributed with expected value = $\alpha_\ell + \beta_\ell x_\ell$ and variance s^2

where

$$\alpha_\ell = \alpha + \beta_1 x_1 + \cdots + \beta_{\ell-1} x_{\ell-1} + \beta_{\ell+1} x_{\ell+1} + \cdots + \beta_k x_k$$

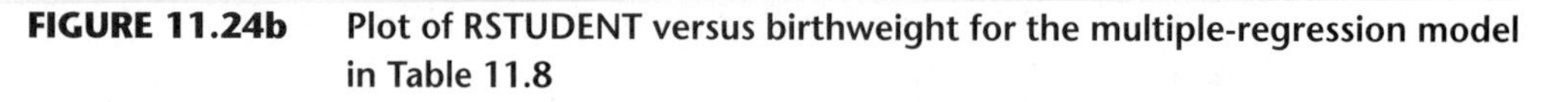

FIGURE 11.24b **Plot of RSTUDENT versus birthweight for the multiple-regression model in Table 11.8**

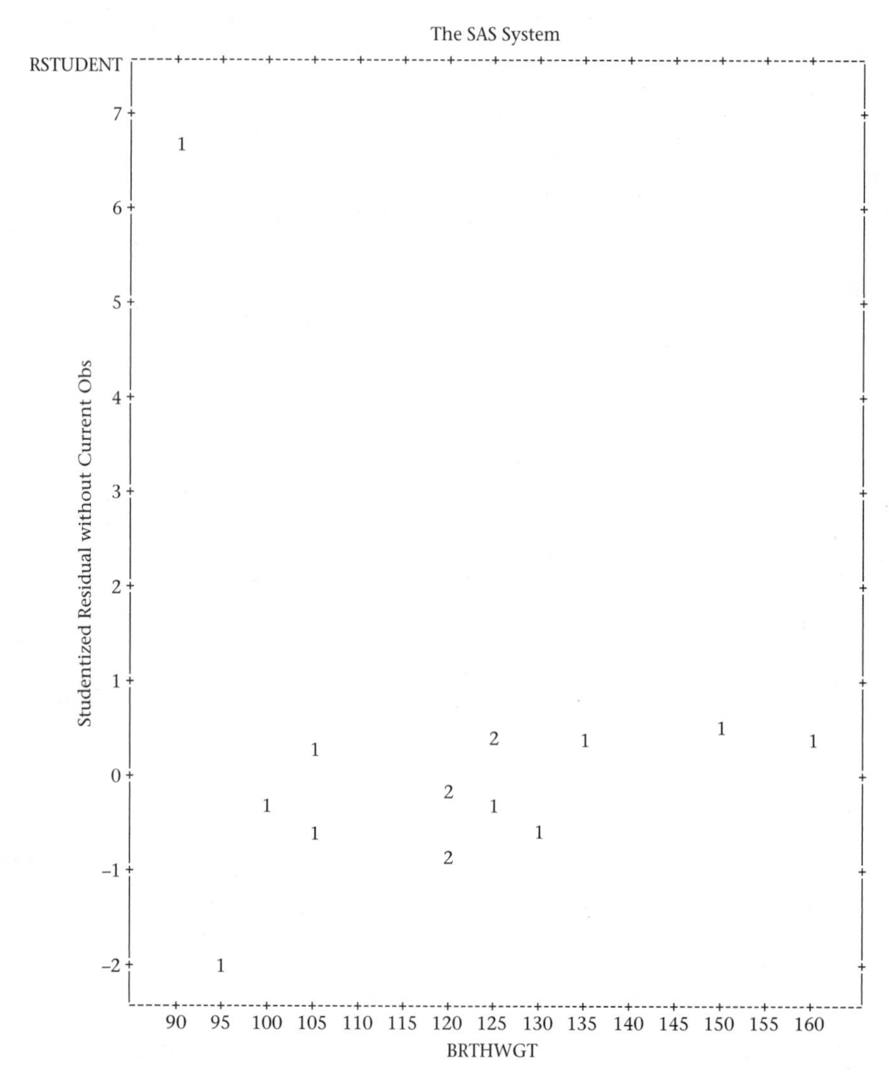

Thus, given the values of all other independent variables $(x_1, \ldots, x_{\ell-1}, x_{\ell+1}, \ldots, x_k)$,

(1) the average value of y is linearly related to x_ℓ

(2) the variance of y is constant (i.e., σ^2)

(3) y is normally distributed

A partial-residual plot is a good way to check the validity of the assumptions in Equation 11.33.

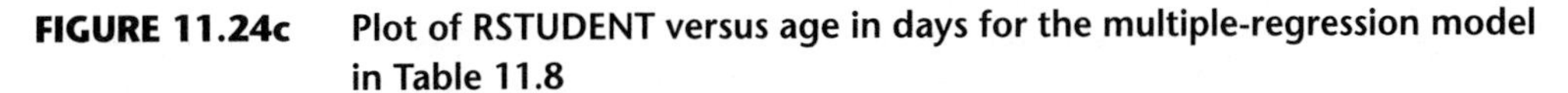

FIGURE 11.24c **Plot of RSTUDENT versus age in days for the multiple-regression model in Table 11.8**

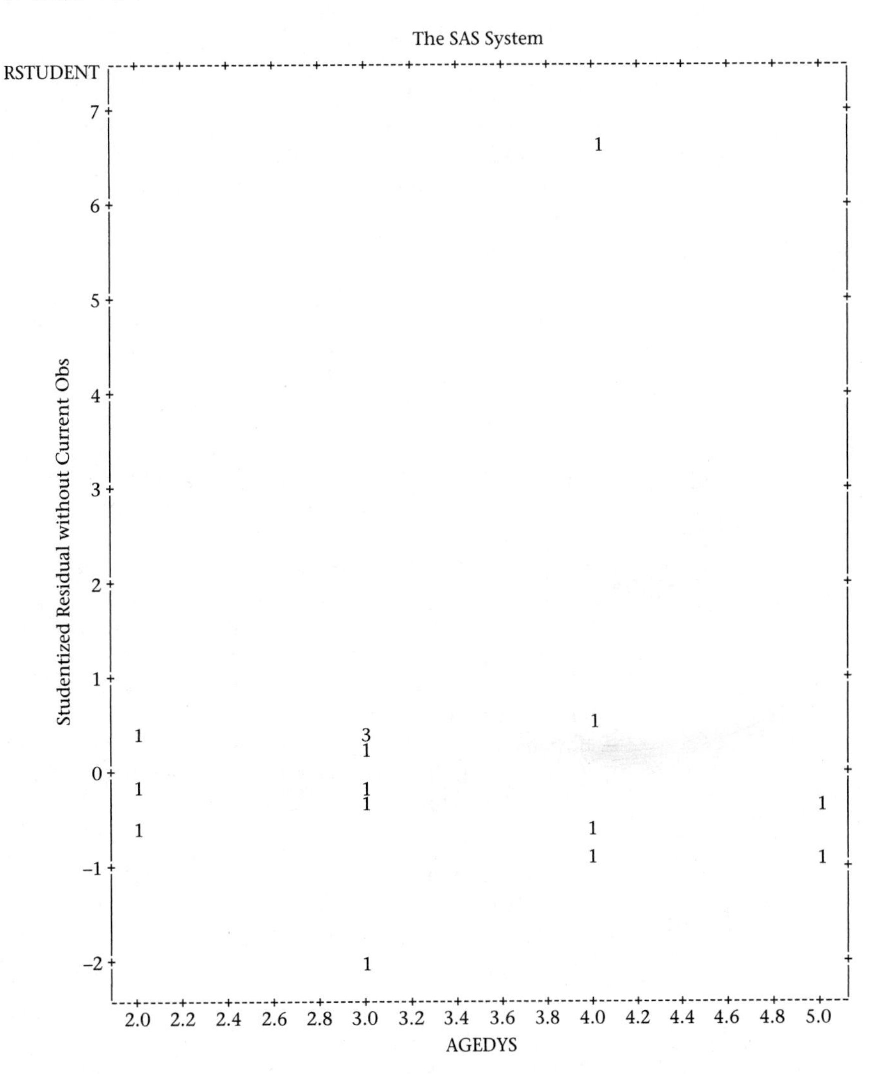

DEFINITION 11.18

A **partial-residual plot** characterizing the relationship between the dependent variable y and a specific independent variable x_ℓ in a multiple-regression setting is constructed as follows:

(1) A multiple regression is performed of y on all predictors other than x_ℓ (i.e., $x_1, \ldots, x_{\ell-1}, x_{\ell+1}, \ldots, x_k$) and the residuals are saved.

(2) A multiple regression is performed of x_ℓ on all other predictors (i.e., $x_1, \ldots, x_{\ell-1}, x_{\ell+1}, \ldots, x_k$) and the residuals are saved.

(3) The partial-residual plot is a scatter plot of the residuals in step 1 on the y-axis versus the residuals in step 2 on the x-axis.

Many computer packages compute partial-residual plots as an option in their multiple-regression routines, so that the user does not have to perform the individual steps 1 to 3. The partial-residual plot reflects the relationship between y and x_ℓ after each variable is adjusted for all other predictors in the multiple-regression model, which is a primary goal of performing a multiple-regression analysis. It can be shown that if the multiple-regression model in Equation 11.29 holds, then the residuals in step 1 should be linearly related to the residuals in step 2 with slope = β_ℓ (i.e., the partial-regression coefficient pertaining to x_ℓ in the multiple-regression model in Equation 11.29) and constant residual variance σ^2. A separate partial-residual plot can be constructed relating y to each predictor $x_1, \ldots, x_k$.

Example 11.48 Construct a separate partial-residual plot relating systolic blood pressure to birthweight and age for the data in Table 11.7.

Solution We refer to the SAS output in Figures 11.25a and 11.25b. The y-axis in Figure 11.25a corresponds to residuals of systolic blood pressure after adjusting for age in days. The x-axis corresponds to residuals of birthweight after adjusting for age in days. Figure 11.25b is defined similarly. Hence, the x- and y-axes correspond to residuals and are not in the familiar units of blood pressure and birthweight in Figure 11.25a, for example. In Figure 11.25a, we notice that the relationship between systolic blood pressure and birthweight is approximately linear (perhaps slightly curvilinear) with the exception of observation 10, which we previously identified as an outlier. In Figure 11.25b, the relationship between systolic blood pressure and age appears to be linear with the exception of observation 10. Notice that the three x values clustered in the lower left of the plot all correspond to age = 2 days. However, they have different abscissas in this plot, because they reflect the residual of age after correcting for birthweight and the three birthweights are all different (see observations 4, 7, and 11 with birthweights = 105, 125, and 120 oz, respectively). In these data, the fitted regression line of age on birthweight is given by age = 2.66 + 0.0054 × birthweight.

Since we identified observation 10 as an outlier, we deleted this observation and reran the regression analysis based on the reduced sample of size 15. The regression model is given in Table 11.9 and the partial-residual plots in Figures 11.26a and 11.26b. The estimated multiple-regression model is

$$y = 47.94 + 0.183 \times \text{birthweight} + 5.28 \times \text{age}$$

which differs considerably from the multiple-regression model in Table 11.8 ($y = 53.45 + 0.126 \times \text{birthweight} + 5.89 \times \text{age}$), particularly for the estimated regression coefficient for birthweight, which increased by about 50%. No outliers are evident in either of the partial-residual plots in Figure 11.26. There is only a slight hint of curvilinearity in Figure 11.26a, which relates systolic blood pressure to birthweight (after controlling for age).

In Section 11.9, we were introduced to multiple linear regression. This technique is used when we wish to relate a normally distributed outcome variable y (called the dependent variable) to several (more than one) independent variables $x_1, \ldots, x_k$. The independent variables do not have to be normally distributed. Indeed, the independent variables can even be categorical as discussed further in the

FIGURE 11.25a Partial-residual plot of systolic blood pressure versus birthweight for the data in Table 11.7

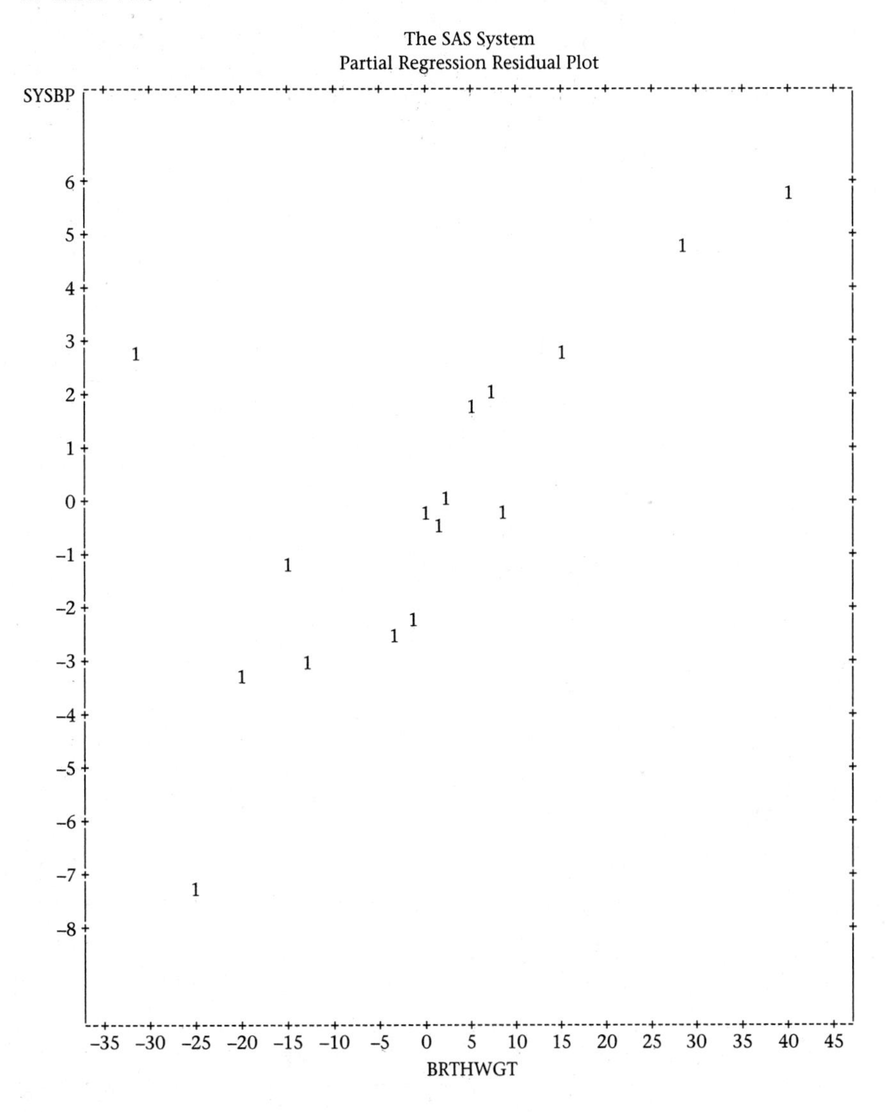

Case Study in Sections 11.10 and 12.5. Referring to the flowchart at the end of this chapter (Figure 11.32), we would answer no to (1) interested in relationships between two variables? and yes to (2) outcome variable continuous? leading us to the box entitled "Multiple-Regression Methods."

FIGURE 11.25b Partial-residual plot of systolic blood pressure versus age for the data in Table 11.7

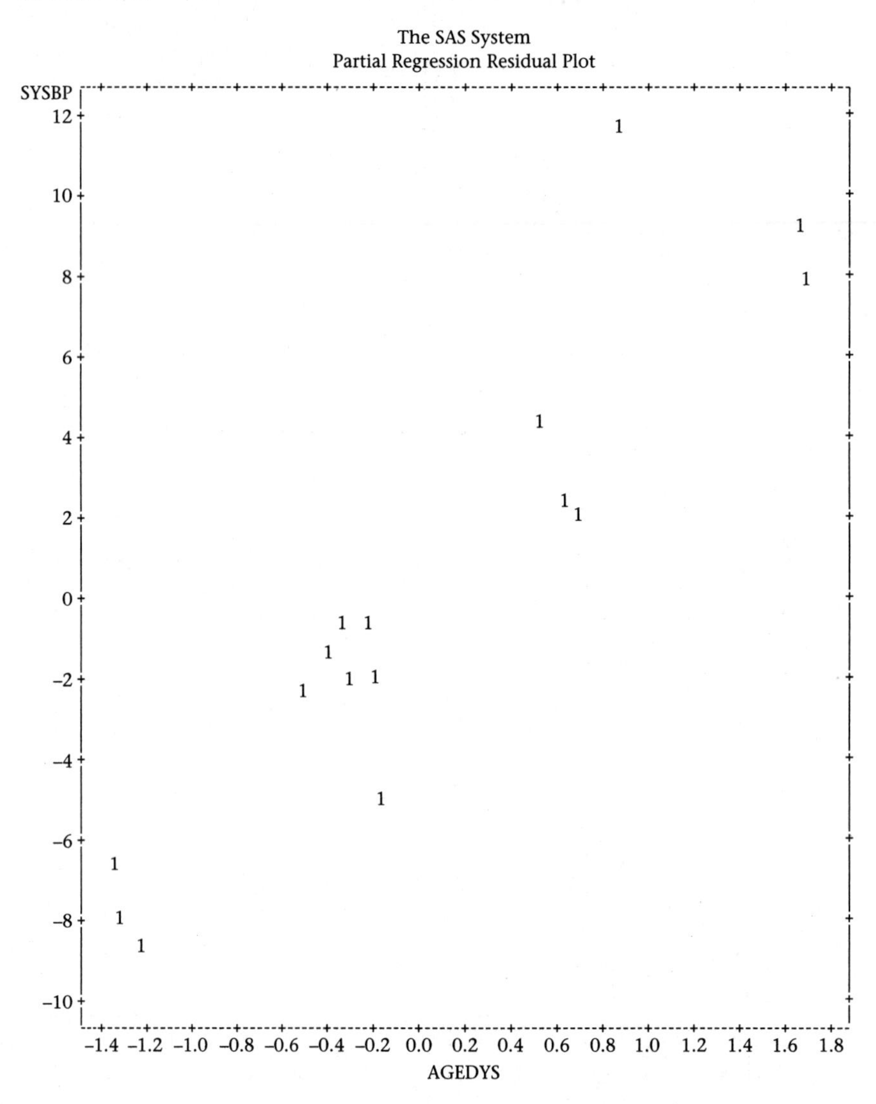

FIGURE 11.26a **Partial-residual plot of systolic blood pressure versus birthweight based on the data in Table 11.7 after deleting one outlier (observation 10) (n = 15)**

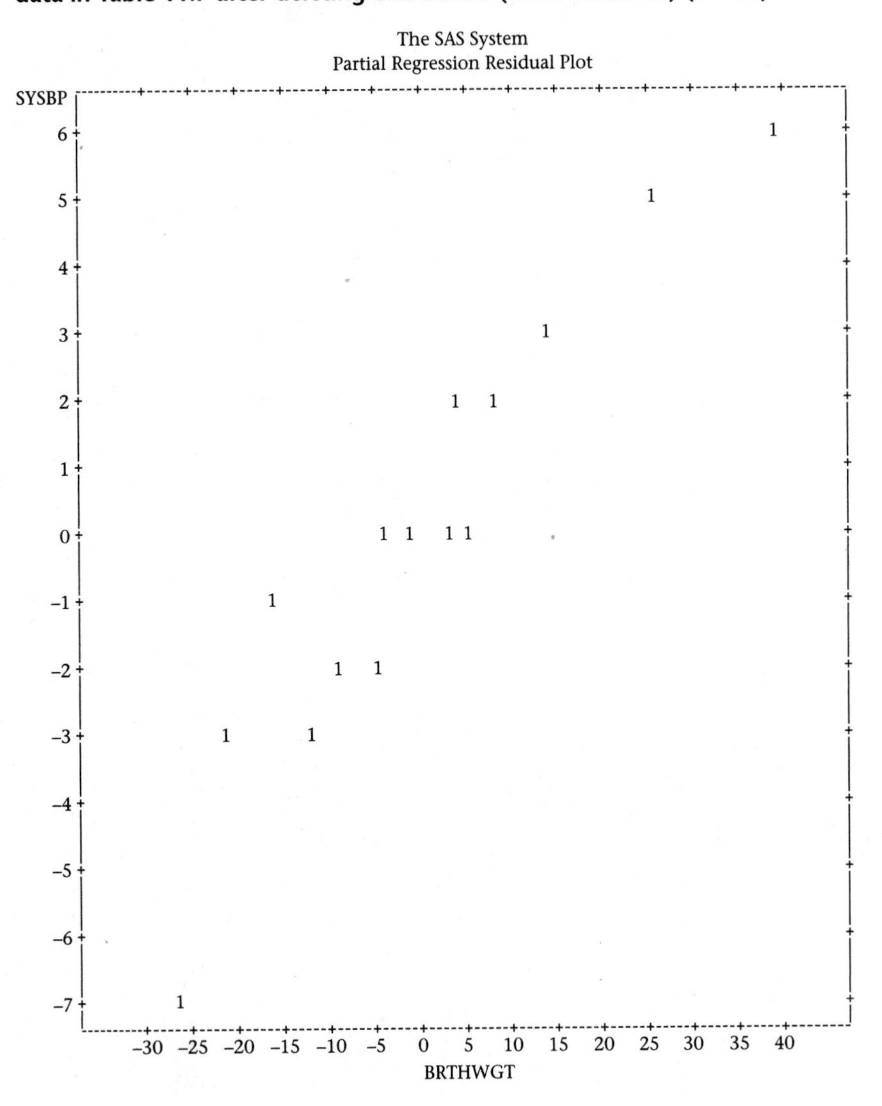

FIGURE 11.26b **Partial-residual plot of systolic blood pressure versus age based on the data in Table 11.7 after deleting one outlier (observation 10) (n = 15)**

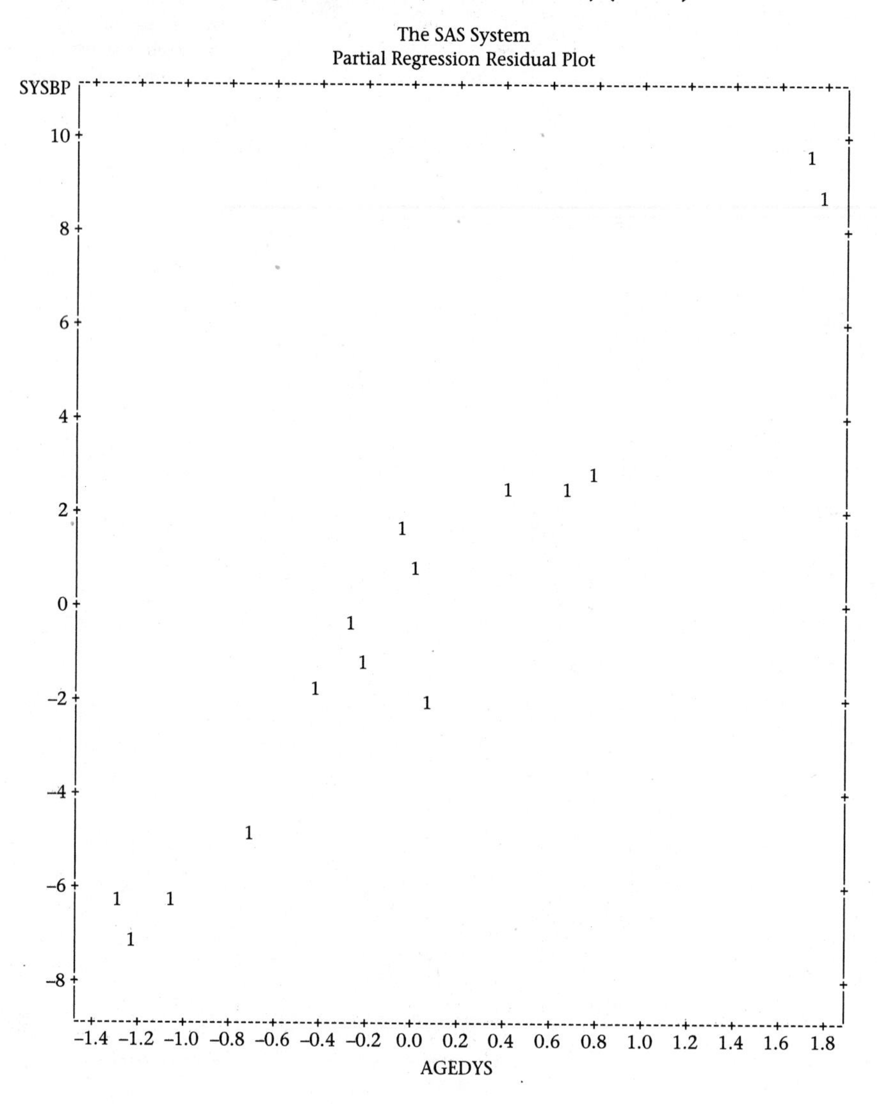

TABLE 11.9 Multiple-regression model of systolic blood pressure on birthweight and age based on the data in Table 11.7 after deleting one outlier (observation 10) ($n = 15$)

```
                                   The SAS System

Model: MODEL1
Dependent Variable: SYSBP
                                Analysis of Variance
                               Sum of           Mean
           Source     DF      Squares          Square      F Value      Prob>F
           Model       2    602.96782       301.48391      217.519      0.0001
           Error      12     16.63218         1.38601
           C Total    14    619.60000

              Root MSE      1.17729      R-square      0.9732
              Dep Mean     87.60000      Adj R-sq      0.9687
              C.V.          1.34394

                                Parameter Estimates
                                                                                  Squared
                 Parameter      Standard      T for H0:                Standardized    Partial
Variable   DF    Estimate          Error   Parameter = 0   Prob > |T|      Estimate  Corr Type II
INTERCEP    1   47.937689    2.30154115          20.829       0.0001    0.00000000        .
BRTHWGT     1    0.183157    0.01839777           9.955       0.0001    0.48213133    0.89199886
AGEDYS      1    5.282477    0.33520248          15.759       0.0001    0.76319791    0.95390786
```

SECTION 11.10 Case Study: Effects of Lead Exposure on Neurological and Psychological Function in Children

In Table 8.11, we compared the mean finger-wrist tapping score (MAXFWT) between exposed and control children using a two-sample t test. However, another approach would be to use regression methods to compare the two groups using dummy-variable coding to denote group membership.

DEFINITION 11.19 A **dummy variable** is a binary variable that is used to represent a categorical variable with two categories (say A and B). The dummy variable is set to the value c_1 if a subject is in category A and c_2 if a subject is in category B. The most common choices for the values c_1 and c_2 are 1 and 0, respectively.

Example 11.49 **Environmental Health, Pediatrics** Use regression methods to compare the mean MAXFWT between exposed and control children.

Solution We will represent group membership by the dummy variable CSCN2 defined by

$$\text{CSCN2} = \begin{cases} 1 \text{ if child is exposed} \\ 0 \text{ if child is control} \end{cases}$$

We then can run a simple linear-regression model of the following form:

EQUATION 11.34

$$\text{MAXFWT} = \alpha + \beta \times \text{CSCN2} + e$$

What do the parameters of this model mean? If a child is in the exposed group, then the average value of MAXFWT for that child is $\alpha + \beta$; if a child is in the control group, then the average value of MAXFWT for that child is α. Thus, β represents the difference between the average value of MAXFWT for children in the exposed versus the control group. Our best estimate of $\alpha + \beta$ is given by the sample mean of MAXFWT for children in the exposed group; our best estimate of α is given by the sample mean of MAXFWT for children in the control group. Thus, our best estimate of β is given by the mean difference in MAXFWT between the exposed and control groups. Another way to interpret β is as the average increase in MAXFWT per 1-unit increase in CSCN2. However, a 1-unit increase in CSCN2 corresponds to the difference between the exposed and control groups for CSCN2. We have run the regression model in Equation 11.34 using the SAS PROC REG procedure. The results are given in Table 11.10. We see that there is a significant difference between the mean MAXFWT for the two groups ($p = .003$). The estimated mean difference between the MAXFWT scores for the two groups is −6.66 taps/10 seconds with standard error = 2.22 taps per 10 seconds. The regression coefficient b corresponds exactly to the mean difference reported in Table 8.11 (mean for exposed group = 48.44 taps per 10 seconds; mean for control group = 55.10 taps per 10 seconds, mean difference = −6.66 taps per 10 seconds). The standard error for b corresponds exactly to the standard error of the mean difference given by a two-sample t test with *equal* variances. The p-values for the two procedures are also the same.

TABLE 11.10 Simple linear-regression model comparing exposed and control children for MAXFWT ($n = 95$)

The SAS System

Model: MODEL1
Dependent Variable: MAXFWT

Analysis of Variance

Source	DF	Sum of Squares	Mean Square	F Value	Prob>F
Model	1	940.63327	940.63327	9.021	0.0034
Error	93	9697.30357	104.27208		
C Total	94	10637.93684			

Root MSE	10.21137	R-square	0.0884
Dep Mean	52.85263	Adj R-sq	0.0786
C.V.	19.32046		

Parameter Estimates

Variable	DF	Parameter Estimate	Standard Error	T for H0: Parameter=0	Prob > |T|	Standardized Estimate
INTERCEP	1	55.095238	1.28651172	42.825	0.0001	0.00000000
CSCN2	1	-6.657738	2.21666753	-3.003	0.0034	-0.29735926

This leads to the following general principle:

EQUATION 11.35

Relationship Between Simple Linear Regression and *t* Test Approaches Suppose we wish to compare the underlying mean between two groups where the observations in group 1 are assumed to be normally distributed with mean μ_1 and variance σ^2 and the observations in group 2 are assumed to be normally distributed with mean μ_2 and variance σ^2. To test the hypothesis H_0: $\mu_1 = \mu_2$ versus H_1: $\mu_1 \neq \mu_2$, we can use one of two equivalent procedures:

(1) We can perform a two-sample t test with equal variances.

(2) We can set up a linear-regression model of the form

$$y = \alpha + \beta x + e$$

where y is the outcome variable and $x = 1$ if a subject is in group 1 and 0 if a subject is in group 2 and $e \sim N(0, \sigma^2)$.

The t statistic in procedure 1 = $(\bar{y}_1 - \bar{y}_2)/\sqrt{s^2(1/n_1 + 1/n_2)}$ is the same as the t statistic in procedure 2 = $b/se(b)$. The estimated mean difference between the groups in procedure 1 = $\bar{y}_1 - \bar{y}_2$ is the same as the estimated slope (b) in procedure 2. The standard error of the mean difference in procedure 1 = $\sqrt{s^2(1/n_1 + 1/n_2)}$ is the same as the standard error of b in procedure 2. The p-values are the same in procedure 1 and procedure 2.

One of the issues with neurological-function data in children is that they are often strongly related to age and in some cases to gender as well. Even slight age differences between the exposed and control groups could explain differences between the groups in neurological function. The first issue is to determine whether MAXFWT is related to age and/or to gender. For this purpose, we might construct a multiple-regression model of the form

EQUATION 11.36

$$\text{MAXFWT} = \alpha + \beta_1 \text{ age } + \beta_2 \text{ sex} + e$$

where age is in years and sex is coded as 1 if male and 2 if female. *Note:* A slightly more accurate measure of age could be obtained if we used age in years + (age in months)/12 rather than simply age in years. The results from fitting the model in Equation 11.36 are given in Table 11.11. We see that MAXFWT is strongly related to age ($p < 0.001$) and is slightly, but not significantly related to gender. Older children and males have higher values of MAXFWT (mean MAXFWT increases by 2.5 taps per 10 seconds for every 1 year increase in age for children of the same sex and is 2.4 taps per 10 seconds higher in boys than in girls of the same age). The partial-residual plots of MAXFWT versus age and sex are given in Figures 11.27a and 11.27b, respectively. From Figure 11.27a, we see that MAXFWT is strongly and approximately linearly related to age. No obvious outlying values are apparent. From Figure 11.27b, we see that males (corresponding to the left cloud of points) tend to have slightly higher values of MAXFWT (after correcting for age) than females. The Studentized residual plot versus the predicted MAXFWT is given in Figure 11.28. No outliers are apparent from this plot, either. Also, the variance of MAXFWT looks similar for different values of age, sex, and the predicted MAXFWT.

From Table 11.10, mean MAXFWT differs between the exposed and control groups by 6.66 taps per 10 seconds. Thus, even a 1-year difference in mean age between the two groups would account for approximately $(2.52/6.66) \times 100\%$ or 38%

TABLE 11.11 Multiple-regression model of MAXFWT on age and sex ($n = 95$)

```
                                The SAS System
Model: MODEL1
Dependent Variable: MAXFWT
                              Analysis of Variance
                             Sum of          Mean
             Source    DF    Squares         Square       F Value     Prob>F
             Model      2    5438.14592    2719.07296     48.109      0.0001
             Error     92    5199.79092      56.51947
             C Total   94   10637.93684

                Root MSE     7.51794     R-square     0.5112
                Dep Mean    52.85263     Adj R-sq     0.5006
                C.V.        14.22435
                              Parameter Estimates
                   Parameter      Standard       T for H0:                       Standardized
Variable    DF     Estimate          Error     Parameter=0     Prob > |T|           Estimate
INTERCEP     1    31.591389     3.16011063           9.997         0.0001         0.00000000
AGEYR        1     2.520683     0.25705630           9.806         0.0001         0.72618377
SEX          1    -2.365745     1.58721503          -1.491         0.1395        -0.11037958
```

of the observed mean difference in MAXFWT. Thus, it is essential to redo the crude analyses in Table 11.10, adjusting for possible age and sex differences between groups. We will use the multiple-regression model:

EQUATION 11.37

$$\text{MAXFWT} = \alpha + \beta_1 \times \text{CSCN2} + \beta_2 \times \text{age} + \beta_3 \times \text{sex} + e$$

where CSCN2 = 1 if in the exposed group and 0 if in the control group and sex = 1 if male and 2 if female. The fitted model is given in Table 11.12. We see that the estimated mean difference between the groups is -5.15 ± 1.56 taps/10 sec ($p = .001$), after controlling for age and sex. The effects of age and sex are similar to those seen in Table 11.11. The partial-residual plot of MAXFWT on group (CSCN2) is given in Figure 11.29. The left cloud of points corresponds to the control group and the right cloud to the exposed group. The control group does appear to have generally higher values than the exposed group, although there is considerable overlap between the two groups. Also, the range of residual values for the control group appears to be greater than for the exposed group. However, the control group ($n = 63$) is larger than the exposed group ($n = 32$), which would account, at least in part, for the difference in the range. In Table 8.11, when we performed a two-sample t test comparing the two groups we found that the within-group standard deviation was 10.9 for the control group and 8.6 for the exposed group ($p = .14$ using the F test). Another interesting finding is that the age- and sex-adjusted mean difference between the groups (–5.15 taps/10 sec) is smaller than the crude difference (–6.66 taps/10 seconds). This difference is explained in part by differences in the age–sex distribution between the two groups. The model in Equation 11.37 is referred to as an **analysis-of-covariance** model. This model is a general

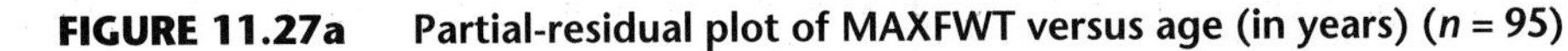

FIGURE 11.27a Partial-residual plot of MAXFWT versus age (in years) ($n = 95$)

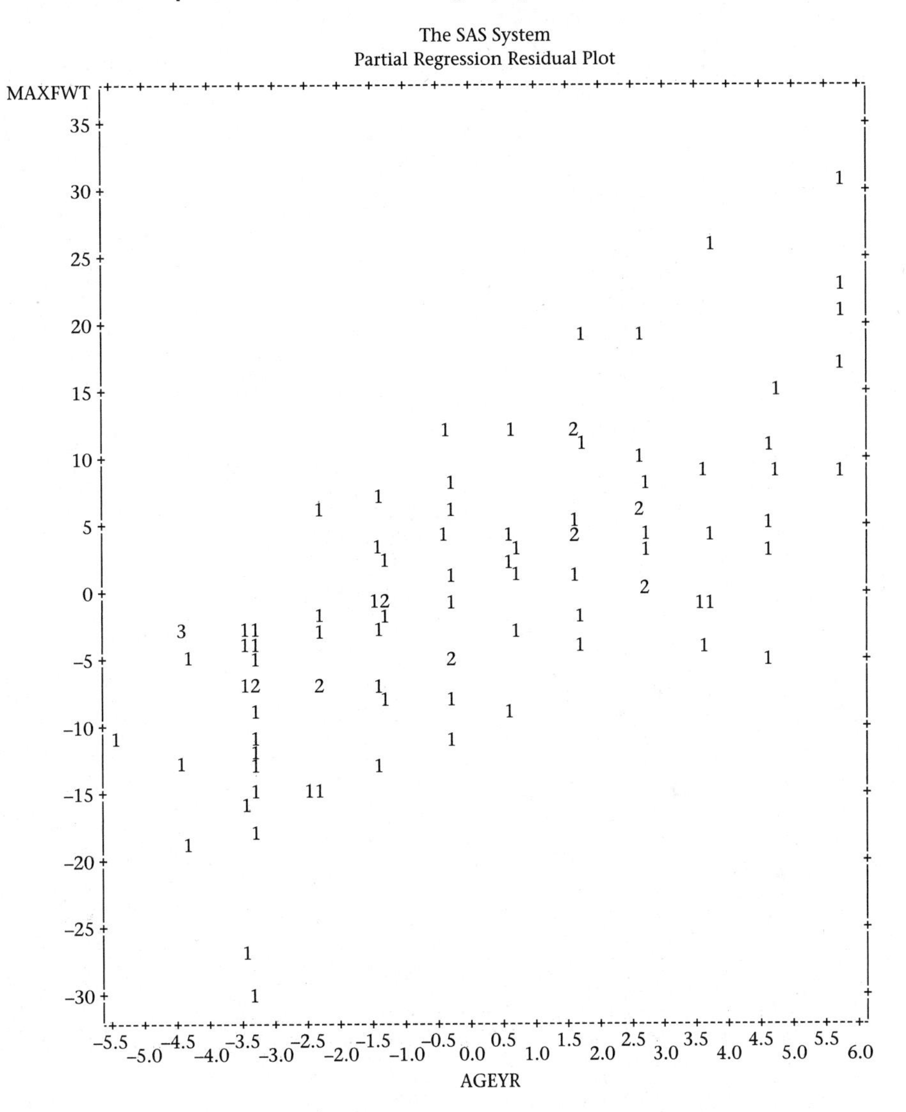

procedure for assessing the mean difference in a normally distributed outcome variable between groups after controlling for one or more confounding variables (sometimes referred to as *covariates*). Groups are defined by a categorical variable, which may have two or more categories. The covariates may be any combination of either continuous (e.g., age) or categorical (e.g., sex) variables. We will discuss analysis-of-covariance models in more detail in Chapter 12.

FIGURE 11.27b **Partial-residual plot of MAXFWT versus sex (n = 95)**

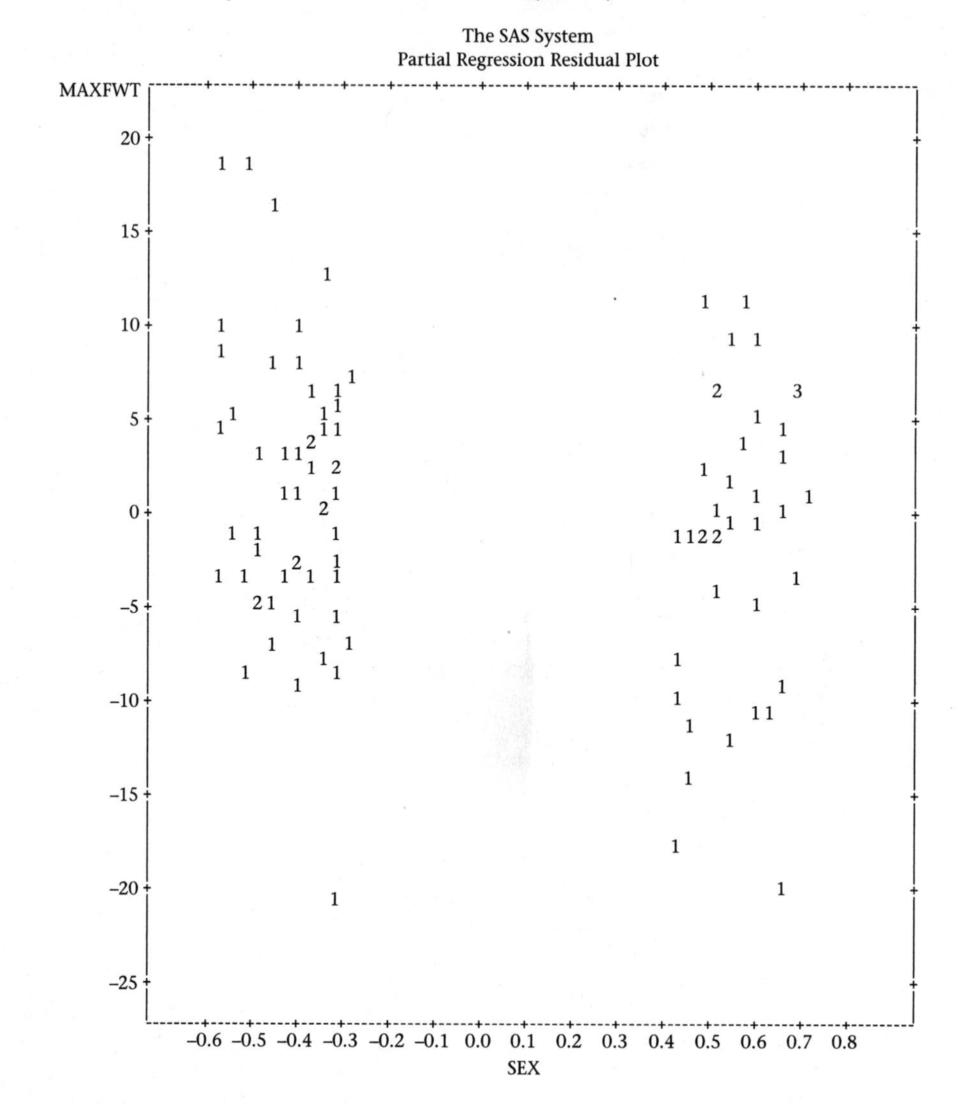

FIGURE 11.28 **Studentized residual versus predicted MAXFWT (n = 95)**

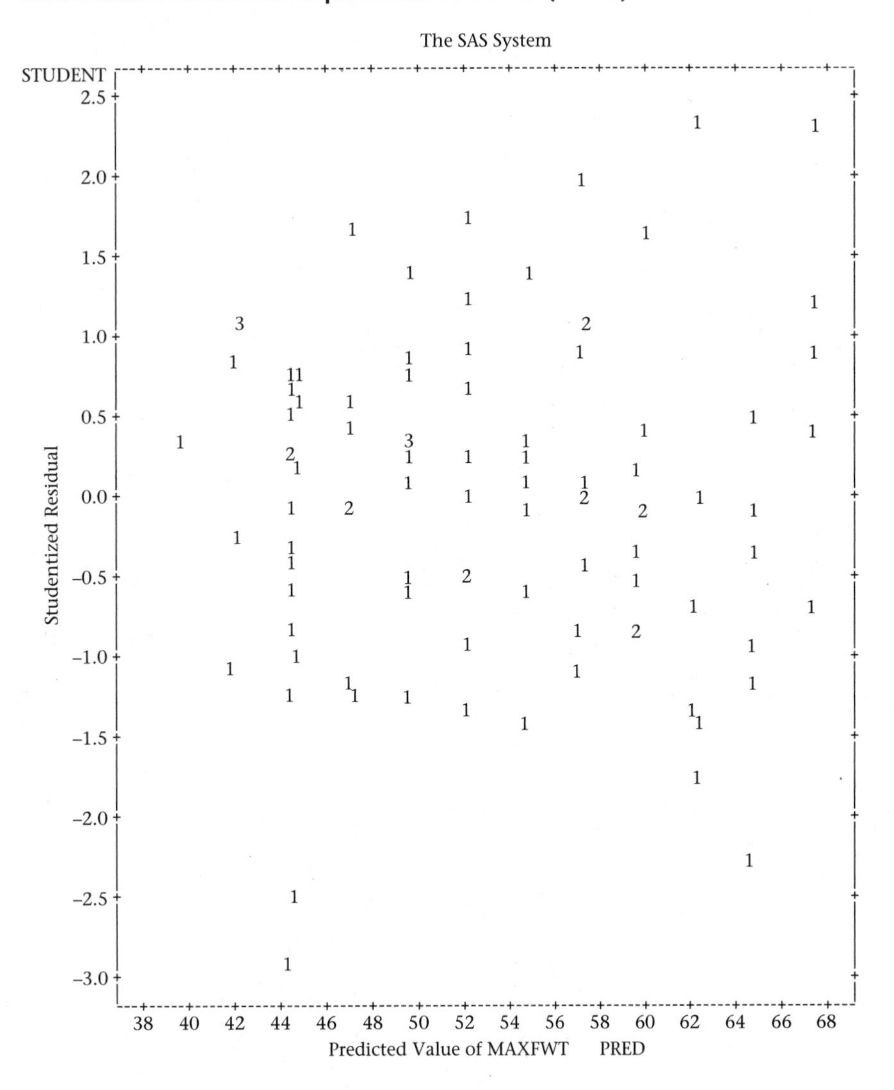

FIGURE 11.29 Partial-residual plot of MAXFWT on CSCN2 after correcting for age and sex ($n = 95$)

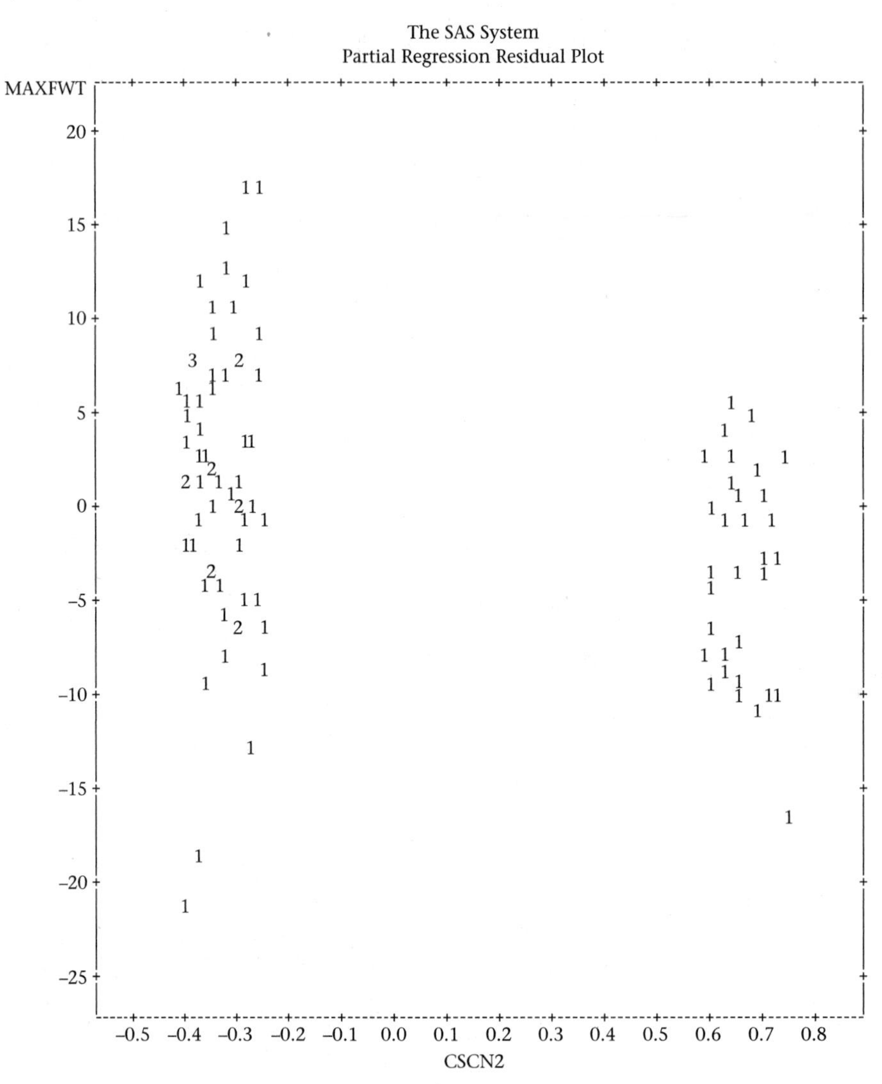

TABLE 11.12 Multiple-regression model comparing mean MAXFWT between exposed and control children after controlling for age and sex ($n = 95$)

```
                              The SAS System
Model: MODEL1
Dependent Variable: MAXFWT
                           Analysis of Variance
                         Sum of            Mean
         Source    DF     Squares          Square      F Value     Prob>F
         Model      3   5994.81260     1998.27087       39.164     0.0001
         Error     91   4643.12424       51.02334
         C Total   94  10637.93684

            Root MSE     7.14306     R-square      0.5635
            Dep Mean    52.85263     Adj R-sq      0.5491
            C.V.        13.51506
                           Parameter Estimates
                 Parameter      Standard      T for H0:                       Standardized
Variable   DF     Estimate         Error    Parameter=0     Prob > |T|          Estimate
INTERCEP    1    34.121285    3.09868543         11.012         0.0001        0.00000000
CSCN2       1    -5.147169    1.55831499         -3.303         0.0014       -0.22989163
AGEYR       1     2.442016    0.24539673          9.951         0.0001        0.70352057
SEX         1    -2.385209    1.50808045         -1.582         0.1172       -0.11128774
```

SECTION 11.11 Partial and Multiple Correlation

11.11.1 Partial Correlation

The correlation coefficient is a measure of linear association between two variables x and y. In some instances, it is important to assess the degree of association between two variables after controlling for other covariates. The partial correlation is a measure that accomplishes this goal.

DEFINITION 11.20 Suppose we are interested in the association between two variables x and y, but wish to control for other covariates $z_1, \ldots, z_k$. The **partial correlation** is defined to be the Pearson correlation between two derived variables e_x and e_y, where

e_x = the residual from the linear regression of x on $z_1, \ldots, z_k$

e_y = the residual from the linear regression of y on $z_1, \ldots, z_k$

Example 11.50 **Hypertension, Pediatrics** Consider the pediatric blood-pressure data given in Table 11.7, where we related systolic blood pressure to birthweight and age for 16 infants. Estimate the partial correlation between systolic blood pressure and each risk factor after controlling for the other risk factor.

Solution Refer to Table 11.8, under the last column, labeled Squared Partial Corr Type II. The partial correlation between systolic blood pressure and birthweight after correcting for age is $\sqrt{.5071} = .71$. The partial correlation between systolic blood pressure and age after controlling for birthweight is $\sqrt{.8521} = .92$. These relationships are displayed in the partial-residual plots of systolic blood pressure on birthweight (Figure 11.25a) and age (Figure 11.25b), respectively.

11.11.2 Multiple Correlation

The partial-correlation coefficient provides a measure of association between two variables while controlling for the effects of one or more other covariates. In a multiple-regression setting, where we have an outcome variable (y) and there are two or more predictor variables ($x_1, \ldots, x_k$), then the partial correlation between y and a single predictor x_j quantifies the specific association between y and x_j, while controlling for all other predictors $x_1, \ldots, x_{j-1}, x_{j+1}, \ldots, x_k$. However, we are also often interested in the association between y and *all* the predictors when considered as a group. This measure of association is given by the multiple correlation.

DEFINITION 11.21

Suppose we have an outcome variable y and a set of predictors $x_1, \ldots, x_k$. The maximum possible correlation between y and a linear combination of the predictors $c_1x_1 + \cdots + c_kx_k$ is given by the correlation between y and the regression function $\beta_1x_1 + \cdots + \beta_kx_k$ and is referred to as the **multiple correlation** between y and $\{x_1, \ldots, x_k\}$. It is estimated by the Pearson correlation between y and $b_1x_1 + \cdots + b_kx_k$, where $b_1, \ldots, b_k$ are the least-squares estimates of $\beta_1, \ldots, \beta_k$. The multiple correlation can also be shown to be equivalent to $\sqrt{\text{Reg SS}/\text{Total SS}} = \sqrt{R^2}$ from Equation 11.30.

Example 11.51 **Hypertension, Pediatrics** Compute the multiple correlation between systolic blood pressure and the predictors (birthweight, age) based on the data in Table 11.7.

Solution Refer to Table 11.8. The R^2 for the regression model is $591.04/670.94 = .8809$. The multiple correlation is $\sqrt{.8809} = .94$. This indicates a strong association between y and the set of predictors {birthweight, age}.

SECTION 11.12 Rank Correlation

Sometimes we may want to look at the relationship between two variables, but one or both of the variables are either ordinal or have a distribution that is far from normal. Significance tests based on the Pearson correlation coefficient will then no longer be valid, and nonparametric analogues to these tests are needed.

Example 11.52 **Obstetrics** The Apgar score was developed in 1952 as a measure of the physical condition of an infant at 1 and 5 minutes after birth [5]. The score is obtained by summing five components, each of which is rated as 0, 1, or 2 and represents different aspects of the condition of an infant at birth [6]. The method of scoring is indicated in Table 11.13. The score is routinely calculated for most newborn infants in U.S. hospitals. Suppose we are given the data in Table 11.14. We wish to relate the Apgar scores at 1 and 5 minutes and assess the significance of this relationship. How should this be done?

TABLE 11.13 Method of Apgar scoring

Sign	Score 0	1	2
Heart rate	Absent	Slow (<100)	≥100
Respiratory effort	Absent	Weak cry; hypoventilation	Good; strong cry
Muscle tone	Limp	Some flexion of extremities	Well flexed
Reflex irritability	No response	Some motion	Cry
Color	Blue; pale	Body pink; extremities blue	Completely pink

Source: Reprinted with permission of *JAMA*, *168*(15), 1985–1988, 1958.

TABLE 11.14 Apgar scores at 1 and 5 minutes for 24 newborns

Infant	Apgar score, 1 min	Apgar score, 5 min	Infant	Apgar score, 1 min	Apgar score, 5 min
1	10	10	13	6	9
2	3	6	14	8	10
3	8	9	15	9	10
4	9	10	16	9	10
5	8	9	17	9	10
6	9	10	18	9	9
7	8	9	19	8	10
8	8	9	20	9	9
9	8	9	21	3	3
10	8	9	22	9	9
11	7	9	23	7	10
12	8	9	24	10	10

The ordinary correlation coefficient developed in Section 11.7 should not be used, because the significance of this measure can be assessed only if the distribution of each Apgar score is assumed to be normally distributed. Instead, a nonparametric analogue to the correlation coefficient based on ranks is used.

DEFINITION 11.22 The **Spearman rank-correlation coefficient (r_s)** is an ordinary correlation coefficient based on ranks.

Thus $$r_s = \frac{L_{xy}}{\sqrt{L_{xx} \times L_{yy}}}$$

where the L's are computed from the ranks rather than from the actual scores.

The rationale for this estimator is that if there were a perfect correlation between the two variables, then the ranks for each person on each variable would be the same and $r_s = 1$. The less perfect the correlation, the closer to zero r_s would be.

Example 11.53 **Obstetrics** Compute the Spearman rank-correlation coefficient for the Apgar-score data in Table 11.14.

Solution We use MINITAB to rank the Apgar 1-minute and 5-minute scores as shown in Table 11.15 under APGAR_1R and APGAR_5R, respectively. We then compute the correlation coefficient between APGAR_1R and APGAR_5R and obtain $r_s = .593$.

TABLE 11.15 **Ranks of Apgar-score data from Table 11.14**

```
MTB > Rank C1 C3.
MTB > Rank C2 C4.
MTB > Print C1-C4
ROW      APGAR_1M      APGAR_5M      APGAR_1R      APGAR_5R
  1            10            10          23.5          19.5
  2             3             6           1.5           2.0
  3             8             9          10.0           8.5
  4             9            10          18.5          19.5
  5             8             9          10.0           8.5
  6             9            10          18.5          19.5
  7             8             9          10.0           8.5
  8             8             9          10.0           8.5
  9             8             9          10.0           8.5
 10             8             9          10.0           8.5
 11             7             9           4.5           8.5
 12             8             9          10.0           8.5
 13             6             9           3.0           8.5
 14             8            10          10.0          19.5
 15             9            10          18.5          19.5
 16             9            10          18.5          19.5
 17             9            10          18.5          19.5
 18             9             9          18.5           8.5
 19             8            10          10.0          19.5
 20             9             9          18.5           8.5
 21             3             3           1.5           1.0
 22             9             9          18.5           8.5
 23             7            10           4.5          19.5
 24            10            10          23.5          19.5
MTB > Correlation C3 C4.
Correlation of APGAR_1R and APGAR_5R = 0.593
```

We would now like to test the rank correlation for statistical significance. A similar test to that given in Equation 11.20 for the Pearson correlation coefficient can be performed, as follows:

EQUATION 11.38

***t* Test for Spearman Rank Correlation**

(1) Compute the test statistic

$$t_s = \frac{r_s\sqrt{n-2}}{\sqrt{1-r_s^2}}$$

which under the null hypothesis of no correlation follows a t distribution with $n-2$ degrees of freedom.

(2) For a two-sided level α test, if

$$t_s > t_{n-2,1-\alpha/2} \quad \text{or} \quad t_s < t_{n-2,\alpha/2} = -t_{n-2,1-\alpha/2}$$

then reject H_0; otherwise, accept H_0.

(3) The exact p-value is given by

$$p = 2 \times (\text{area to the left of } t_s \text{ under a } t_{n-2} \text{ distribution}) \quad \text{if } t_s < 0$$

$$p = 2 \times (\text{area to the right of } t_s \text{ under a } t_{n-2} \text{ distribution}) \quad \text{if } t_s \ge 0$$

(4) This test is valid only if $n \ge 10$.

The acceptance and rejection regions for this test are given in Figure 11.30. The computation of the exact p-value is illustrated in Figure 11.31.

Example 11.54 **Obstetrics** Perform a significance test for the Spearman rank-correlation coefficient based on the Apgar-score data in Table 11.14.

Solution Note that $r_s = .593$ from Example 11.53. The test statistic is given by

$$t_s = \frac{r_s\sqrt{n-2}}{\sqrt{1-r_s^2}} = \frac{.593\sqrt{22}}{\sqrt{1-.593^2}} = \frac{2.781}{.805} = 3.45$$

which follows a t_{22} distribution under H_0. Note that

$$t_{22,.995} = 2.819 \qquad t_{22,.9995} = 3.792$$

Thus, the two-tailed p-value is given by

$$2 \times (1 - .9995) < p < 2 \times (1 - .995) \quad \text{or} \quad .001 < p < .01$$

Thus, there is a significant rank correlation between the two scores.

FIGURE 11.30 **Acceptance and rejection regions for the *t* test for a Spearman rank-correlation coefficient**

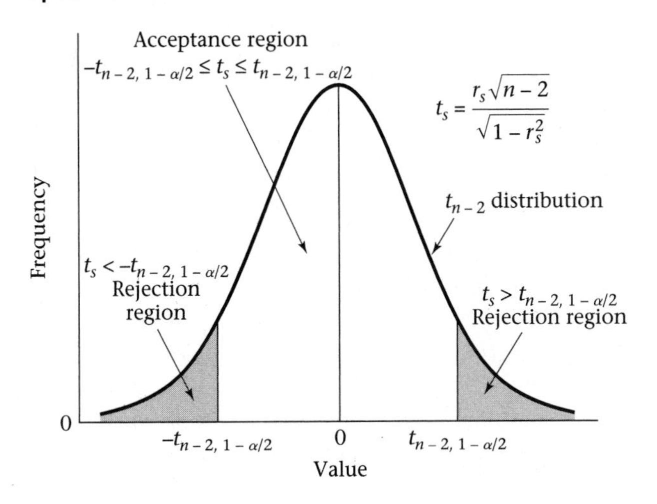

FIGURE 11.31 **Computation of the exact p-value for the t test for a Spearman rank-correlation coefficient**

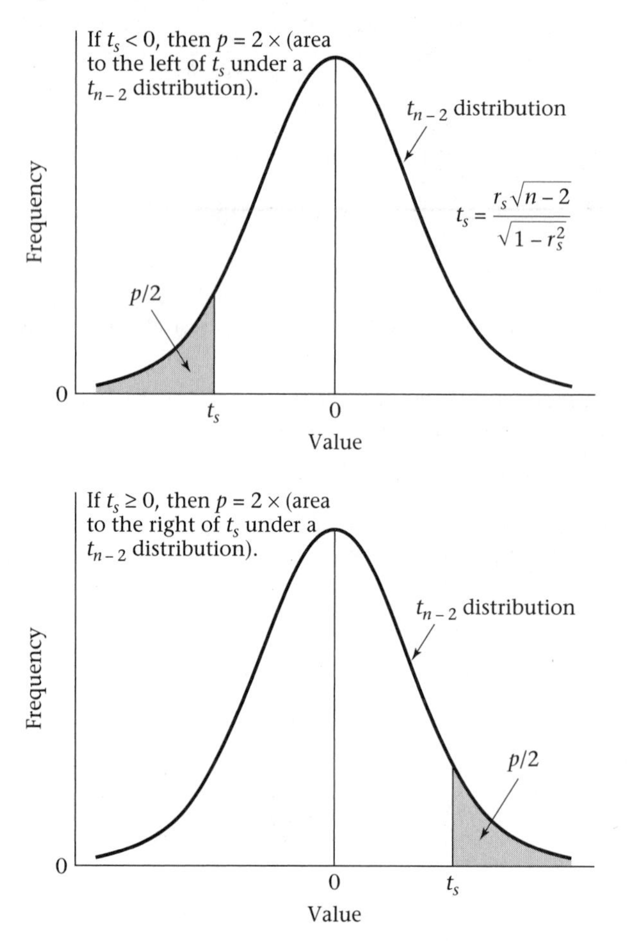
If $t_s < 0$, then $p = 2 \times$ (area to the left of t_s under a t_{n-2} distribution).
t_{n-2} distribution
$t_s = \frac{r_s\sqrt{n-2}}{\sqrt{1-r_s^2}}$
$p/2$
Frequency
0
t_s
0
Value
If $t_s \geq 0$, then $p = 2 \times$ (area to the right of t_s under a t_{n-2} distribution).
t_{n-2} distribution
$p/2$
Frequency
0
0
t_s
Value

Note that the test procedure given in Equation 11.38 is valid only for $n \geq 10$. If $n < 10$, then the t distribution is not a good approximation to the distribution of t_s, and a table giving exact significance levels must be used. For this purpose, exact two-sided critical values for r_s when $n \leq 9$ are presented in Table 14 in the Appendix. This table can be used in the following way:

(1) Suppose the critical value in the table for significance level α is c.

(2) Reject H_0 using a two-sided test with significance level α if $r_s \geq c$ or $r_s \leq -c$ and accept H_0 otherwise.

Example 11.55 Suppose that $r_s = .750$ based on a sample of size 9. Assess the statistical significance of the results.

Solution From Table 14 in the Appendix, the critical value for $\alpha = .05$, $n = 9$ is .683 and for $\alpha = .02$, $n = 9$ is .783. Since $.683 < .750 < .783$, it follows that the two-tailed p-value is given by $.02 \leq p < .05$.

In this section, we have discussed rank correlation methods. These methods are used when we are interested in studying the association between two continuous variables, where at least one of the variables is not normally distributed. Referring to the flowchart at the end of this chapter (Figure 11.32, p. 502), we would answer yes to (1) interested in relationships between two variables? and (2) both variables continuous? no to (3) interested in predicting one variable from another? yes to (4) interested in studying the relationship between two variables? and no to (5) both variables normal? This leads us to the box entitled "Rank correlation methods."

Rank correlation methods are also useful for studying the association between two ordinal variables. Again referring to the flowchart at the end of this chapter (Figure 11.32), we would answer yes to (1) interested in relationships between two variables? no to (2) both variables continuous? no to (3) one variable continuous and one categorical? and yes to (4) ordinal data? This also leads us to the box entitled "Rank correlation methods."

SECTION 11.13 Summary

In this chapter, methods of statistical inference that are appropriate for investigating the relationship between two or more variables were studied. If only two variables, both of which are continuous, are being studied, and we wish to predict one variable (the dependent variable) as a function of the other variable (the independent variable), then simple linear-regression analysis is used. If we simply wish to look at the association between two normally distributed variables without distinguishing between dependent and independent variables, then Pearson correlation methods are more appropriate. If both variables are continuous, but are not normally distributed, or are ordinal variables, then rank correlation can be used instead. If both variables of interest are categorical and we are interested in the association between the two variables, then the contingency-table methods of Chapter 10 can be used. If, on the other hand, we are almost certain that there will be some association between the two variables and we wish to quantify the degree of association, then the Kappa statistic can be used.

In many instances we are interested in more than two variables and we wish to predict the value of one variable (the dependent variable) as a function of several independent variables. If the dependent variable is normally distributed, then multiple-regression methods can be used. Multiple-regression methods can be very powerful, because the independent variables can be either continuous or categorical, or a combination of both.

In many situations we have a continuous outcome variable that we wish to relate to one or more categorical variables. In general, this situation can be handled with ANOVA (analysis of variance) methods. However, in many instances the formulation is easier if multiple-regression methods are used. We will discuss these alternative approaches in Chapter 12. The preceding methods are summarized in the flowchart in Figure 11.32 (p. 502) and again in the back of the book.

FIGURE 11.32 Flowchart for appropriate methods of statistical inference (regression and correlation methods are shaded)

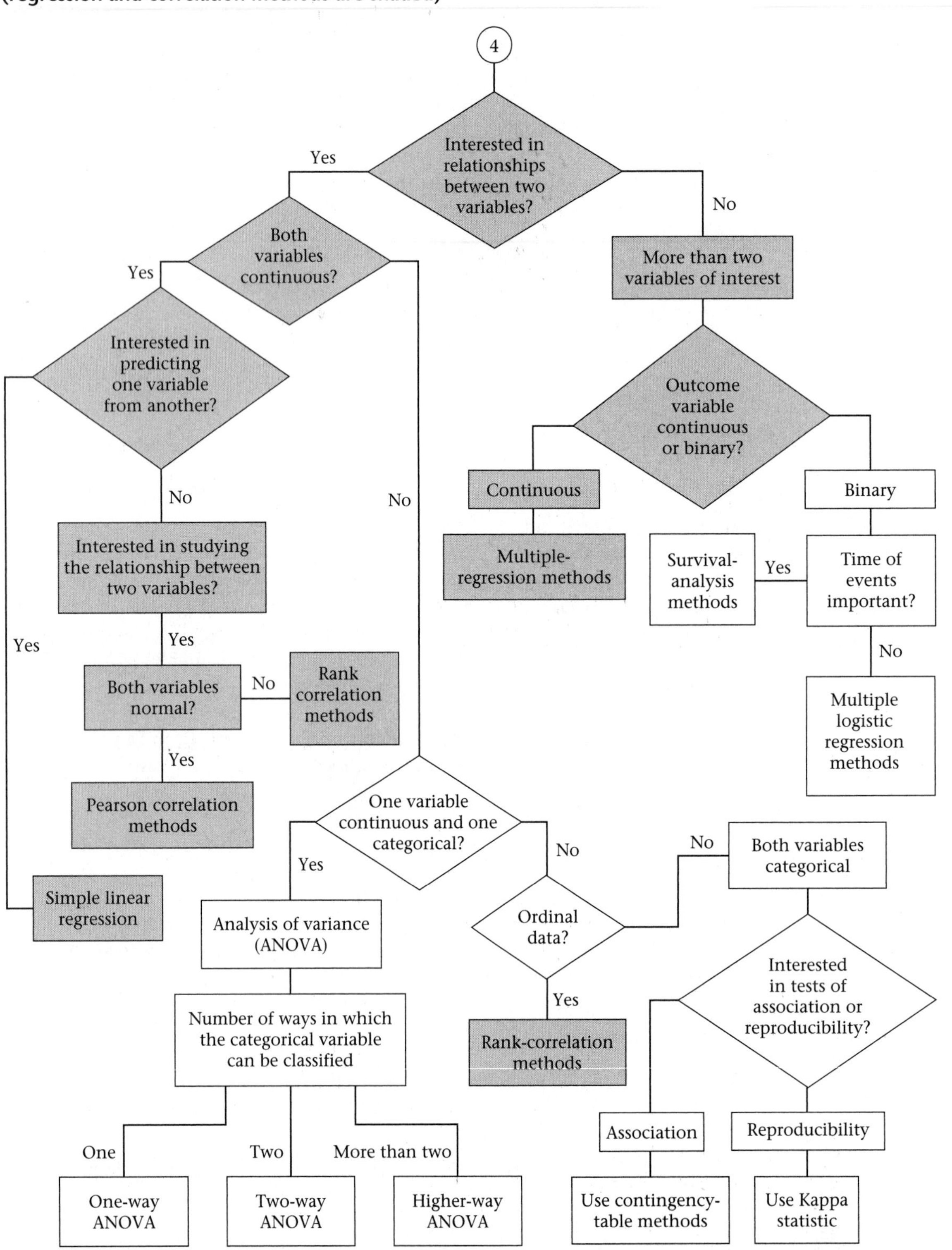

PROBLEMS

Hematology

The data in Table 11.16 are given for 9 patients with aplastic anemia [7].

TABLE 11.16 Hematologic data for patients with aplastic anemia

Patient number	% reticulytes	Lymphocytes (per mm^2)
1	3.6	1700
2	2.0	3078
3	0.3	1820
4	0.3	2706
5	0.2	2086
6	3.0	2299
7	0.0	676
8	1.0	2088
9	2.2	2013

Source: Reprinted with permission of the *New England Journal of Medicine, 312*(16), 1015–1022, 1985.

***11.1** Fit a regression line relating the percentage of reticulytes (x) to the number of lymphocytes (y).

***11.2** Test for the statistical significance of this regression line using the F test.

***11.3** What is R^2 for the regression line in Problem 11.1?

***11.4** What does R^2 mean is Problem 11.3?

***11.5** What is $s_{y \cdot x}^2$?

***11.6** Test for the statistical significance of the regression line using the t test.

***11.7** What are the standard errors of the slope and intercept for the regression line in Problem 11.1?

11.8 What is the z transformation of .34?

Pulmonary Function

Suppose the correlation coefficient between FEV for 100 sets of identical twins is .7, whereas the comparable correlation for 120 sets of fraternal twins is .38.

***11.9** What test procedure can be used to compare the two correlation coefficients?

***11.10** Perform the procedure in Problem 11.9 using the critical-value method.

***11.11** What is the p-value of the test?

Suppose the correlation coefficient between weight is .78 for the 100 sets of identical twins and .50 for the 120 sets of fraternal twins.

***11.12** Test for whether or not the true correlation coefficients are different between these groups. Report a p-value.

Infectious Disease

Refer to the data on hospital stays given in Table 2.11.

11.13 Find the best-fitting linear relationship between duration of hospitalization and age.

11.14 Test for the significance of this relationship. State any underlying assumptions you have used.

11.15 What is R^2 for this regression?

11.16 Assess the goodness of fit of the regression line.

Environmental Health

Suppose we are interested in the relation between carbon-monoxide concentration and the density of cars in a geographical area. The number of cars per hour (to the nearest 500 cars per hour) and the concentration of carbon monoxide (CO) in parts per million at a particular street corner are measured and the data are grouped by cars per hour. The data are given in Table 11.17.

TABLE 11.17 CO concentration and car density at a particular street corner

Cars/hour ($\times 10^3$)	CO concentrations				Number of samples
1.0	9.0	6.8	7.7		3
1.5	9.6	6.8	11.3		3
2.0	12.3	11.8			2
3.0	20.7	19.2	21.6	20.6	4

11.17 Is the CO concentration related to the number of cars per hour?

11.18 What is the average CO concentration if 2500 cars per hour are on the road?

11.19 What is the standard error for the average CO concentration over a large number of days when 2500 cars per hour are on the road?

11.20 Assess the goodness of fit of the regression line.

Obstetrics

The data in Table 11.18 give the infant-mortality rates per 1000 livebirths in the United States for the period 1960–1979 [8].

TABLE 11.18 U.S. infant-mortality rates per 1000 livebirths, 1960–1979

x^*	y^*	x	y
1960	26.0	1974	16.7
1965	24.7	1975	16.1
1970	20.0	1976	15.2
1971	19.1	1977	14.1
1972	18.5	1978	13.8
1973	17.7	1979	13.0

*x = year, y = infant mortality rate per 1000 live births.

Suppose the following information is given:

$$\sum_{i=1}^{12} x_i = 23{,}670$$

$$\sum_{i=1}^{12} x_i^2 = 46{,}689{,}410$$

$$\sum_{y=1}^{12} y_i = 214.9$$

$$\sum_{i=1}^{12} y_i^2 = 4033.83$$

$$\sum_{i=1}^{12} x_i y_i = 423{,}643.3$$

***11.21** Fit a linear-regression line relating infant-mortality rate to chronological year using these data.

***11.22** Test for the significance of the linear relationship developed in Problem 11.21.

***11.23** If the present trends continue for the next 10 years, then what would be the predicted infant-mortality rate in 1989?

***11.24** Provide a standard error for the estimate in Problem 11.23.

***11.25** Can the linear relationship developed in Problem 11.21 be expected to continue indefinitely? Why or why not?

Nutrition

The assessment of the relationship between dietary intake and disease is one of the more prominent areas of current medical research. One of the problems is the difficulty in accurately assessing a person's diet, because reported dietary intake varies over different surveys because of both (1) faulty memory of the actual diet and (2) true changes in diet over time. To assess reproducibility, a food-frequency questionnaire with over 100 food items was administered to 537 American nurses at two different points in time, 6 months apart. A number of nutrients, such as total protein and fat, were calculated on the basis of the individual items, and correlations were calculated over the responses to the two questionnaires. Suppose the correlation coefficient for total protein intake as assessed at two points in time is .362. We wish to test for a significant relationship between reported total protein intake at two points in time.

11.26 Is a one-sided or two-sided test appropriate here?

11.27 Perform a significance test for this relationship and report a *p*-value.

An alternative method for assessing dietary intake is the 7-day diet record, in which a person writes down each food item eaten over a 1-week period, and the total nutrient intake is computed from these data. Suppose that two 7-day diet records 6 months apart are completed for a group of 50 nurses (different from the first group of 537 nurses), and it is found that the correlation between total protein intake is .45 using this method. We wish to test if the two methods are equally reproducible.

11.28 Is a one-sided or two-sided test needed here?

11.29 Perform a significance test for this hypothesis and report a *p*-value.

Hypertension

The Second Task Force on Blood Pressure Control in Children [9] reported the observed 90th percentile of systolic blood pressure in single years of age from age 1 to 18 based on prior studies. The data for boys are given in Table 11.19.

TABLE 11.19 90th percentile of systolic blood pressure (SBP) in boys ages 1–18

Age (x)	SBP[a] (y)	Age (x)	SBP[a] (y)
1	105	10	117
2	106	11	119
3	107	12	121
4	108	13	124
5	109	14	126
6	111	15	129
7	112	16	131
8	114	17	134
9	115	18	136

[a]90th percentile for each 1-year age group.

Suppose we seek a more efficient way to display the data and choose linear regression to accomplish this task.

11.30 Fit a regression line relating age to SBP, using the data in Table 11.19.

Note:

$$\sum_{i=1}^{18} x_i = 171, \quad \sum_{i=1}^{18} x_i^2 = 2109, \quad \sum_{i=1}^{18} y_i = 2124,$$

$$\sum_{i=1}^{18} y_i^2 = 252{,}338, \quad \sum_{i=1}^{18} x_i y_i = 21{,}076$$

11.31 Provide a 95% CI for the parameters of the regression line.

11.32 Do you think that the linear regression provides a good fit to the data? Why or why not? Use residual analysis to justify your answer.

11.33 What is the predicted blood pressure for an average 13-year-old boy as estimated from the regression line?

11.34 What is the standard error of the estimate in Problem 11.33?

11.35 Answer Problems 11.33 and 11.34 for an 18-year-old boy.

Cancer

The following statistics are taken from an article by P. Burch relating cigarette smoking to lung cancer [10]. The article presents data relating mortality from lung cancer to average cigarette consumption (lb/person) for females in England and Wales over a 40-year period. The data are given in Table 11.20.

$$\left(\sum_{i=1}^{8} x_i = 2.38, \quad \sum_{i=1}^{8} x_i^2 = 1.310, \quad \sum_{i=1}^{8} y_i = -15.55, \quad \sum_{i=1}^{8} y_i^2 = 30.708, \quad \sum_{i=1}^{8} x_i y_i = -4.125 \right)$$

11.36 Compute the correlation between 5-year mortality and annual cigarette consumption when each is expressed in the $\log_{10}$ scale.

11.37 Test this correlation for statistical significance and report a *p*-value.

11.38 Fit a regression line relating 5-year mortality to annual cigarette consumption.

11.39 To test the significance of this regression line, is it necessary to perform any additional tests other than those in Problem 11.37? If so, perform them.

11.40 What is the expected mortality rate with an annual cigarette consumption of 1 lb/person?

TABLE 11.20 Cigarette consumption and lung-cancer mortality in England and Wales, 1930–1969

Period	$\log_{10}$ mortality (over 5 years), y	$\log_{10}$ annual cigarette consumption (lb/person), x
1930–1934	–2.35	–0.26
1935–1939	–2.20	–0.03
1940–1944	–2.12	0.30
1945–1949	–1.95	0.37
1950–1954	–1.85	0.40
1955–1959	–1.80	0.50
1960–1964	–1.70	0.55
1965–1969	–1.58	0.55

Source: Reprinted with permission of the *Journal of the Royal Statistical Society, A., 141*, 437–477, 1978.

11.41 Why are the variables mortality rate and annual cigarette consumption expressed in the log scale?

Hypertension

Refer to the data in Table 3.9. Another method for relating measures of reactivity for the automated and manual blood pressures is the correlation coefficient. Suppose the correlation coefficient relating these two measures of reactivity is .19, based on 79 individuals having reactivity measured by each type of blood-pressure monitor.

11.42 What is the appropriate procedure to test if there is a relationship between reactivity as measured by the automated and manual monitors?

11.43 Conduct the test procedure in Problem 11.42 and report a *p*-value. What do the results mean in words?

11.44 Provide a 95% confidence interval for the correlation coefficient between these two measures of reactivity.

Nutrition

Refer to Data Set VALID.DAT on the data disk.

11.45 Assess the agreement between the food-frequency questionnaire and the dietary record with regard to total fat intake, saturated fat intake, alcohol intake, and total caloric intake. Quantify the level of agreement by representing the dietary intake both in the original continuous scale and on a quintile scale.

Pulmonary Disease

Refer to Data Set FEV.DAT on the data disk.

11.46 Use regression methods to look at the relationship between level of pulmonary function and other factors (e.g., age, height, and personal smoking) when considered separately and simultaneously. Assess the goodness of fit of the models you develop. Perform both analyses for males and females combined (where gender is controlled for in the analysis), as well as gender-specific analyses.

Hepatic Disease

Refer to Data Set HORMONE.DAT on the data disk.

11.47 Use methods of linear-regression analysis to assess whether these are dose–response relationships with regard to biliary and pancreatic secretion levels. Perform separate analyses for each of the four active hormones tested.

11.48 Use methods of linear-regression analysis to assess whether there are dose–response relationships with regard to biliary and pancreatic pH levels. Perform separate analyses for each of the four active hormones tested.

Pediatrics, Cardiovascular Disease

An important area of investigation is the development of strategies for altering cardiovascular risk factors in children. LDL (low-density lipoprotein) cholesterol has consistently been shown to be related to cardiovascular disease in adults. A study was conducted in Bogalusa, Louisiana, and Brooks County, Texas, to identify modifiable variables that are related to LDL cholesterol in children [11]. It was found that the correlation coefficient between LDL cholesterol and ponderal index [weight (kg)/height^3 (cm^3)], which is a measure of obesity, was .28 among 903 white boys and .14 among 474 black boys.

***11.49** What test can be used to assess if there is a significant association between LDL cholesterol and ponderal index?

***11.50** Implement the test in Problem 11.49 for white boys and report a *p*-value (two-sided).

***11.51** What test can be used to compare correlations between white and black boys?

***11.52** Implement the test in Problem 11.51 and report a *p*-value.

***11.53** Provide a 95% confidence interval for the true correlation among white boys (ρ_1) and black boys (ρ_2), respectively.

Infectious Disease

Refer to the Data Set HOSPITAL.DAT with documentation in HOSPITAL.DOC (both on the data disk). The data are also displayed in Table 2.11.

11.54 Construct a multiple-regression model relating duration of hospitalization to the other variables listed in Table 2.11. Comment on the goodness of fit of the regression model. Use appropriate data transformations if necessary.

Hypertension, Pediatrics

In Problems 6.67–6.69, we described the Data Set INFANTBP.DAT (see also on the data disk). The data set concerns the possible relationship between infant blood pressure and responsiveness to salt and sugar, respectively. Indices were constructed summarizing responsiveness to salt and sugar.

11.55 Use linear-regression methods to relate each salt and sugar index proposed to systolic blood pressure. Assess the goodness of fit of the proposed models; use appropriate data transformations, if necessary.

11.56 Answer Problem 11.55 for diastolic blood pressure.

Environmental Health, Pediatrics

Refer to the Data Set LEAD.DAT on the data disk. In Section 11.10, we used regression methods to relate finger-wrist tapping score (MAXFWT) to lead exposure group.

11.57 Use regression methods to assess the relationship between full-scale IQ (IQF) and lead-exposure group, where lead exposure is quantified as a categorical variable with 2 categories (exposed and control). Control for the possible confounding effects of age and gender in your analysis. Assess the goodness of fit of the model(s) you propose.

11.58 The actual blood-lead levels are available in the data set. LD72 and LD73 (variables 22 and 23) are the actual blood-lead levels in 1972 and 1973, respectively. Use regression methods to assess the relationship between MAXFWT and the actual blood-lead level(s), while controlling for age and sex. Assess the goodness of fit of the model(s) you propose.

11.59 Answer the same questions posed in Problem 11.58 for IQF.

Pediatrics, Endocrinology

Transient hypothyroxinemia, a common finding in premature infants, is not thought to have long-term consequences, or to require treatment. A study was performed to investigate whether hypothyroxinemia in premature infants is a cause of subsequent motor and cognitive abnormalities [12]. Blood thyroxine values were obtained on routine screening in the first week of life from 536 infants who weighed 2000 g or less at birth and were born at 33 weeks gestation or earlier. The data in Table 11.21 were presented concerning the relationship between mean thyroxine level and gestational age.

11.60 What is the best-fitting regression line relating mean

TABLE 11.21 Relationship between mean thyroxine level and gestational age among 536 premature infants

(x) Gestational age	(y) Mean thyroxine level (μg/dL)
≤ 24[a]	6.5
25	7.1
26	7.0
27	7.1
28	7.2
29	7.1
30	8.1
31	8.7
32	9.5
33	10.1

[a]Treated as 24 in subsequent analyses.

thyroxine level to gestational age?

Hint:

$$\sum_{i=1}^{10} x_i = 285, \sum_{i=1}^{10} x_i^2 = 8205, \sum_{i=1}^{10} y_i = 78.4,$$

$$\sum_{i=1}^{10} y_i^2 = 627.88, \sum_{i=1}^{10} x_i y_i = 2264.7$$

11.61 Is there a significant association between mean thyroxine level and gestational age? Please report a p-value.

11.62 Assess the goodness of fit of the regression line fitted in Problem 11.60.

To test the primary hypothesis of the study, infants were categorized by gestational age and were designated as having severe hypothyroxinemia if their thyroxine concentration was 2.6 sd below the mean score for the assay that their specimen came from. Thyroxine assays typically were done on every group of 240 specimens; a mean and sd were calculated for each assay, based on a sample size of 240. Children in the study were given the Bayley Mental Development Index at <30 months of age. The Bayley test is a commonly used test of mental development in young children. The results were as follows for the subgroup of children with gestational age 30–31 weeks.

Bayley Mental Development Index

Severe hypothyroxinemia	Mean score ± sd	n
No	106 ± 21	138
Yes	88 ± 25	17

11.63 Perform a test to compare the mean Bayley score between children with and without severe hypothyroxinemia (please report a p-value).

11.64 Suppose we wanted to use data on children of all gestational ages in the study. Suggest a type of analysis that could be used to relate the Bayley score to severe hypothyroxinemia, while controlling for gestational age. (Do not actually carry out the analysis.)

Hypertension

Endothelin is a powerful vasoconstrictor peptide derived from the endothelium. The contribution of endothelin to blood-pressure regulation in patients with hypertension was assessed by studying the effect of an endothelin-receptor antagonist, bosentan. 293 patients with mild to moderate hypertension were randomized to receive one of four oral doses of bosentan (100, 500, 1000, or 2000 mg daily), a placebo, or the ACE (angiotensin-converting enzyme) inhibitor enalapril (an established antihypertensive drug) [13]. The reported mean changes in systolic blood pressure over a 24-hour period were as shown in Table 11.22:

TABLE 11.22 Mean change in systolic blood pressure (SBP) over 24 hours

Group	Mean change	Bosentan dose (mg)	ln(dose)
Placebo	−0.9	1	0
100 mg bosentan	−2.5	100	4.61
500 mg bosentan	−8.4	500	6.21
1000 mg bosentan	−7.4	1000	6.91
2000 mg bosentan	−10.3	2000	7.60

11.65 Fit a regression line relating the mean change in SBP to the ln(dose) of bosentan. (*Note:* For the placebo group, assume that the dose of bosentan = 1 mg; hence the ln(dose) = 0.)

11.66 What test can be used to assess if the mean change in SBP is significantly related to ln(dose) of bosentan?

11.67 Implement the method in Problem 11.66 and report a p-value (two-tail).

11.68 What is the estimated mean change in systolic blood pressure for an average patient taking 2000 mg of bosentan? Provide a 95% CI corresponding to this estimate.

Endocrinology

Refer to the Data Set BONEDEN.DAT on the data disk.

11.69 Use regression analysis to relate the number of

pack-years smoked to the bone density of the lumbar spine. Assess the goodness of fit of the regression line.

Hint: For a twinship, relate the difference in bone density between the heavier- and lighter-smoking twin to the difference in the number of pack-years of smoking.

11.70 Answer the question in Problem 11.69 for bone density at the femoral neck.

11.71 Answer the question in Problem 11.69 for bone density at the femoral shaft.

One of the issues in relating bone density to smoking is that smokers and nonsmokers differ in many other characteristics that may be related to bone density, referred to as confounders.

11.72 Compare the weight of the heavier- versus lighter-smoking twin using hypothesis-testing methods.

11.73 Repeat the analyses in Problem 11.69 controlling for weight differences between the heavier- and lighter-smoking twins.

11.74 Answer the question in Problem 11.73 for bone density at the femoral neck.

11.75 Answer the question in Problem 11.73 for bone density at the femoral shaft.

11.76 Consider other possible confounding variables in comparing the heavier- versus lighter-smoking twin. Repeat the analyses in Problems 11.72–11.75, adjusting for these confounding variables. What is your overall conclusion regarding the possible association between bone density and smoking?

Health Promotion

Refer to the Data Set SMOKE.DAT on the data disk.

11.77 Use rank-correlation methods to test whether the number of days abstinent from smoking is related to age. Compare your results with those obtained in Problem 9.29.

11.78 Use rank-correlation methods to test whether the amount previously smoked is related to the number of days abstinent from smoking. Compare your results with those obtained in Problem 9.30.

11.79 Use rank-correlation methods to test whether the adjusted CO level is related to the number of days abstinent from smoking. Compare your results with those obtained in Problem 9.31.

Nutrition

Refer to the Data Set VALID.DAT on the data disk.

11.80 Use rank-correlation methods to relate alcohol intake, as reported on the diet record, to alcohol intake as reported on the food-frequency questionnaire.

11.81 Answer the question in Problem 11.80 for total fat intake.

11.82 Answer the question in Problem 11.80 for saturated fat intake.

11.83 Answer the question in Problem 11.80 for total caloric intake.

11.84 Do you think that parametric or nonparametric methods are better suited to analyze the data in VALID.DAT and why?

11.85 Suppose we have an estimated rank correlation of .45 based on a sample of size 24. Assess the significance of the results.

11.86 Suppose we have an estimated rank correlation of .75 based on a sample of size 8. Assess the significance of the results.

REFERENCES

[1] Greene, J., & Touchstone, J. (1963). Urinary tract estriol: An index of placental function. *American Journal of Obstetrics and Gynecology, 85*(1), 1–9.

[2] Higgins, M., & Keller, J. (1973). Seven measures of ventilatory lung function. *American Review of Respiratory Disease, 108*, 258–272.

[3] Draper, N., & Smith, H. (1981). *Applied regression analysis* (2nd ed.). New York: Wiley.

[4] Weisberg, S. (1985). *Applied regression analysis* (2nd ed.). New York: Wiley.

[5] Apgar, V. (1953). A proposal for a new method of evaluation of the newborn infant. *Current Researches in Anesthesia and Analgesia*, 260–267.

[6] Apgar, V., et al. (1958). Evaluation of the newborn infant—second report. *JAMA, 168*(15), 1985–1988.

[7] Torok-Storb, B., Doney, K., Sale, G., Thomas, E. D., & Storb, R. (1985). Subsets of patients with aplastic anemia identified by flow microfluorometry. *New England Journal of Medicine, 312*(16), 1015–1022.

[8] National Center for Health Statistics. (1979). *Monthly vital statistics report, annual summary.*

[9] Report of the Second Task Force on Blood Pressure Control in Children. (1987). *Pediatrics, 79*(1), 1–25.

[10] Burch, P. R. B. (1978). Smoking and lung cancer: The problem of inferring cause. *Journal of the Royal Statistical Society, 141,* 437–477.

[11] Webber, L. S., Harsha, D. W., Phillips, G. T., Srinivasan, S. R., Simpson, J. W., & Berenson, G. S. (1991, April). Cardiovascular risk factors in Hispanic, white and black children: The Brooks County and Bogalusa Heart Studies. *American Journal of Epidemiology, 133*(7), 704–714.

[12] Reuss, M. L., Paneth, N., Pinto-Martin, J. A., Lorenz, J. M., & Susser, M. (1996). Relation of transient hypothyroxinemia in preterm infants to neurologic development at 2 years of age. *New England Journal of Medicine, 334*(13), 821–827.

[13] Krum, H., Viskoper, R. J., Lacourciere, V., Budde, M. & Charlon, V. for the Bosentan Hypertension Investigators. (1998). The effect of an endothelin-receptor antagonist, Bosentan, on blood pressure in patients with essential hypertension. *New England Journal of Medicine, 338*(12), 784–790.

12 Multisample Inference

SECTION 12.1 Introduction to the One-Way Analysis of Variance

In Chapter 8 we were concerned with comparing the means of two normal distributions using the two-sample t test for independent samples. Frequently, the means of more than two distributions need to be compared.

Example 12.1 **Pulmonary Disease** A topic of ongoing public health interest is whether or not *passive smoking* (i.e., exposure among nonsmokers to cigarette smoke in the atmosphere) has a measurable effect on pulmonary health. White and Froeb studied this question by measuring pulmonary function in several ways in the following six groups [1]:

(1) **Nonsmokers (NS)** People who themselves did not smoke and were not exposed to cigarette smoke either at home or on the job.

(2) **Passive smokers (PS)** People who themselves did not smoke and were not exposed to cigarette smoke in the home but were employed for 20 or more years in an enclosed working area that routinely contained tobacco smoke.

(3) **Noninhaling smokers (NI)** People who smoked pipes, cigars, or cigarettes, but who did not inhale.

(4) **Light smokers (LS)** People who smoked and inhaled 1–10 cigarettes per day for 20 or more years. (*Note:* There are 20 cigarettes in a pack.)

(5) **Moderate smokers (MS)** People who smoked and inhaled 11–39 cigarettes per day for 20 or more years.

(6) **Heavy smokers (HS)** People who smoked and inhaled 40 or more cigarettes per day for 20 or more years.

A principal measure used by the authors to assess pulmonary function was forced mid-expiratory flow (FEF). The authors were interested in comparing FEF in the six groups.

The t test methodology generalizes nicely in this case to a procedure referred to as the *one-way analysis of variance.*

SECTION 12.2 One-Way Analysis of Variance—Fixed-Effects Model

Example 12.2 **Pulmonary Disease** Refer to Example 12.1. The authors were able to identify 200 males and 200 females in each of the six groups except for the NI group, which, because of the small number of such people available, was limited to 50 males and 50 females. The mean and standard deviation of FEF for each of the six groups for males are presented in Table 12.1. How can the means of these six groups be compared?

TABLE 12.1 **FEF data for smoking and nonsmoking males**

Group number, i	Group name	Mean FEF (L/s)	sd FEF (L/s)	n_i
1	NS	3.78	0.79	200
2	PS	3.30	0.77	200
3	NI	3.32	0.86	50
4	LS	3.23	0.78	200
5	MS	2.73	0.81	200
6	HS	2.59	0.82	200

Source: Reprinted by permission of the *New England Journal of Medicine, 302*(13), 720–723, 1980.

Suppose that there are k groups with n_i observations in the ith group. The jth observation in the ith group will be denoted by y_{ij}. We will assume that the following model holds:

EQUATION 12.1

$$y_{ij} = \mu + \alpha_i + e_{ij}$$

where μ is a constant, α_i is a constant that is specific to the ith group, and e_{ij} is an error term, which is normally distributed with mean 0 and variance σ^2. Thus, a typical observation from the ith group is normally distributed with mean $\mu + \alpha_i$ and variance σ^2.

It is not possible to estimate both the overall constant μ as well as the k constants α_i, which are specific to each group. The reason is that we only have k observed mean values for the k groups, which are used to estimate $k + 1$ parameters. As a result, we need to put a constraint on the parameters so that only k parameters will be estimated. Some typical constraints are (1) the sum of the α_i's is set to 0, or (2) the α_i for the last group (α_k) is set to 0. We use the former approach in this text. However, the latter approach is the method that SAS uses.

DEFINITION 12.1

The model in Equation 12.1 is a **one-way analysis of variance**, or a **one-way ANOVA** model. With this model, the means of an arbitrary number of groups, each of which follows a normal distribution, can be compared. Whether the variability in the data comes mostly from variability within groups or can truly be attributed to variability between groups can also be determined.

The parameters in Equation 12.1 can be interpreted as follows:

EQUATION 12.2

Interpretation of the Parameters of a One-Way Analysis-of-Variance Fixed-Effects Model

(1) μ represents the underlying mean of all groups taken together.

(2) α_i represents the difference between the mean of the ith group and the overall mean.

(3) e_{ij} represents random error about the mean $\mu + \alpha_i$ for an individual observation from the ith group.

Intuitively, in Table 12.1 FEF will be predicted by an overall mean FEF plus the effect of each smoking group plus random variability within each smoking group. Group means will be compared within the context of this model.

SECTION 12.3 Hypothesis Testing in One-Way ANOVA—Fixed-Effects Model

The null hypothesis (H_0) in this case is that the underlying mean FEF of each of the six groups is the same. This hypothesis is equivalent to stating that each $\alpha_i = 0$, because the α_i sum up to 0. The alternative hypothesis (H_1) is that at least two of the group means are not the same. This hypothesis is equivalent to stating that at least one $\alpha_i \neq 0$. Thus, we wish to test the hypothesis H_0: all $\alpha_i = 0$ versus H_1: at least one $\alpha_i \neq 0$.

12.3.1 *F* Test for Overall Comparison of Group Means

The mean FEF for the ith group will be denoted by $\bar{y}_i$, and the mean FEF over all groups by $\bar{\bar{y}}$. The deviation of an individual observation from the overall mean can be represented as

EQUATION 12.3

$$y_{ij} - \bar{\bar{y}} = (y_{ij} - \bar{y}_i) + (\bar{y}_i - \bar{\bar{y}})$$

The first term on the right-hand side $(y_{ij} - \bar{y}_i)$ represents the deviation of an individual observation from the group mean for that observation and is an indication of *within-group variability*. The second term on the right-hand side $(\bar{y}_i - \bar{\bar{y}})$ represents the deviation of a group mean from the overall mean and is an indication of *between-group variability*. These terms are depicted in Figure 12.1.

Generally speaking, if the between-group variability is large and the within-group variability is small, as in Figure 12.1a, then H_0 will be rejected and the underlying group means will be declared significantly different. Conversely, if the between-group variability is small and the within-group variability is large, as in Figure 12.1b, then H_0, the hypothesis that the underlying group means are the same, will be accepted.

FIGURE 12.1 **Comparison of between-group and within-group variability**

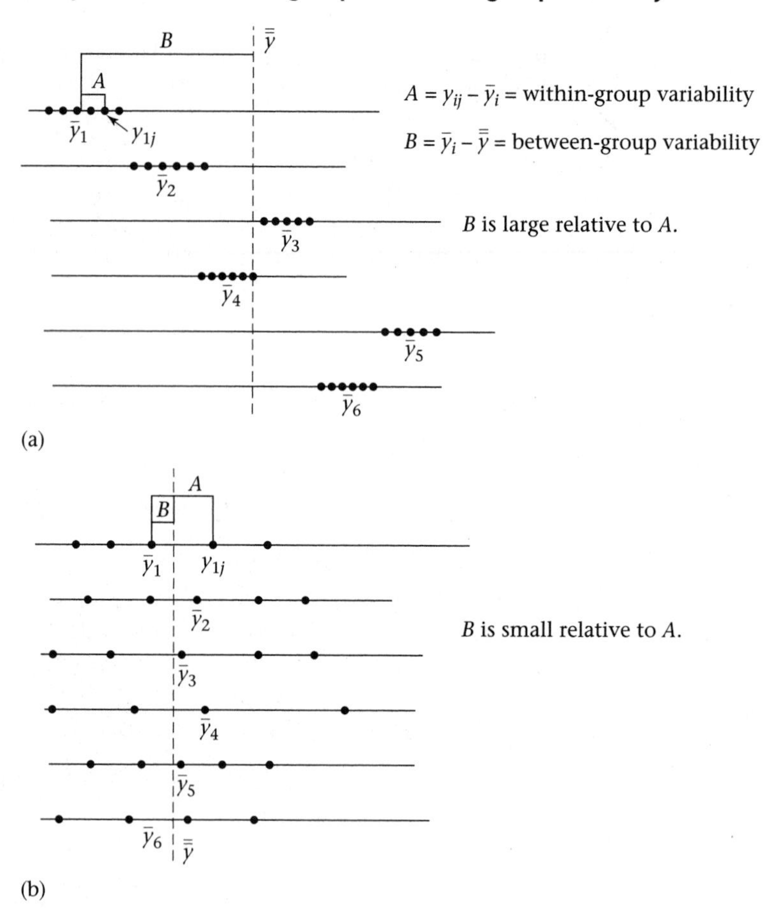

If both sides of Equation 12.3 are squared and the squared deviations are summed over all observations over all groups, then the following relationship is obtained

EQUATION 12.4

$$\sum_{i=1}^{k}\sum_{j=1}^{n_i}(y_{ij}-\bar{\bar{y}})^2 = \sum_{i=1}^{k}\sum_{j=1}^{n_i}(y_{ij}-\bar{y}_i)^2 + \sum_{i=1}^{k}\sum_{j=1}^{n_i}(\bar{y}_i-\bar{\bar{y}})^2$$

because the cross-product term can be shown to be zero.

DEFINITION 12.2 The term

$$\sum_{i=1}^{k}\sum_{j=1}^{n_i}(y_{ij}-\bar{\bar{y}})^2$$

is denoted as the **Total Sum of Squares** (Total SS).

DEFINITION 12.3

The term

$$\sum_{i=1}^{k}\sum_{j=1}^{n_i}(y_{ij}-\bar{y}_i)^2$$

is denoted as the **Within Sum of Squares** (Within SS).

DEFINITION 12.4

The term

$$\sum_{i=1}^{k}\sum_{j=1}^{n_i}(\bar{y}_i-\bar{\bar{y}})^2$$

is denoted as the **Between Sum of Squares** (Between SS).

Thus, the relationship in Equation 12.4 can be written as Total SS = Within SS + Between SS.

It will be easier to use the short computational form for the Within SS and Between SS in Equation 12.5 to perform the hypothesis test.

EQUATION 12.5

Short Computational Form for the Between SS and Within SS

$$\text{Between SS} = \sum_{i=1}^{k} n_i\bar{y}_i^2 - \frac{\left(\sum_{i=1}^{k} n_i\bar{y}_i\right)^2}{n} = \sum_{i=1}^{k} n_i\bar{y}_i^2 - \frac{y_{..}^2}{n}$$

$$\text{Within SS} = \sum_{i=1}^{k}(n_i-1)s_i^2$$

where $y_{..}$ = sum of the observations across all groups—i.e., the grand total of all observations over all groups—and n = total number of observations over all groups.

Example 12.3 **Pulmonary Disease** Compute the Within SS and Between SS for the FEF data in Table 12.1.

Solution We use Equation 12.5 as follows:

$$\text{Between SS} = \left[200(3.78)^2 + 200(3.30)^2 + \cdots + 200(2.59)^2\right] - \frac{\left[200(3.78) + 200(3.30) + \cdots + 200(2.59)\right]^2}{1050}$$

$$= 10{,}505.58 - 3292^2/1050 = 10{,}505.58 - 10{,}321.20 = 184.38$$

$$\text{Within SS} = 199(0.79)^2 + 199(0.77)^2 + 49(0.86)^2 + 199(0.78)^2 + 199(0.81)^2 + 199(0.82)^2$$

$$= 124.20 + 117.99 + 36.24 + 121.07 + 130.56 + 133.81 = 663.87$$

Finally, the following definitions are important:

DEFINITION 12.5

Between Mean Square = Between MS = Between SS/$(k-1)$

DEFINITION 12.6

Within Mean Square = Within MS = Within SS/$(n-k)$

The significance test will be based on the ratio of the Between MS to the Within MS. If this ratio is large, then we will reject H_0; if it is small, we will accept (or fail to reject) H_0. Furthermore, under H_0, the ratio of Between MS to Within MS follows an F distribution with $k-1$ and $n-k$ degrees of freedom. Thus, the following test procedure for a level α test is used:

EQUATION 12.6

Overall *F* Test for One-Way ANOVA To test the hypothesis H_0: $\alpha_i = 0$ for all i versus H_1: at least one $\alpha_i \neq 0$, use the following procedure:

(1) Compute the Between SS, Between MS, Within SS, and Within MS using Equation 12.5 and Definitions 12.5 and 12.6.

(2) Compute the test statistic F = Between MS/Within MS, which follows an F distribution with $k-1$ and $n-k$ *df* under H_0.

(3) If

$$F > F_{k-1,n-k,1-\alpha}$$

then reject H_0. If

$$F \le F_{k-1,n-k,1-\alpha}$$

then accept H_0.

(4) The exact p-value is given by the area to the right of F under an $F_{k-1,n-k}$ distribution $= Pr(F_{k-1,n-k} > F)$.

The acceptance and rejection regions for this test are depicted in Figure 12.2. The computation of the exact p-value is illustrated in Figure 12.3. The results from the analysis of variance are typically displayed in an ANOVA table, as in Table 12.2.

TABLE 12.2 **Display of one-way ANOVA results**

Source of variation	SS	*df*	MS	*F* statistic	*p*-value
Between	$\sum_{i=1}^{k} n_i \bar{y}_i^2 - \frac{y_{..}^2}{n} = A$	$k-1$	$\frac{A}{k-1}$	$\frac{A/(k-1)}{B/(n-k)} = F$	$Pr(F_{k-1,n-k} > F)$
Within	$\sum_{i=1}^{k} (n_i - 1)s_i^2 = B$	$n-k$	$\frac{B}{n-k}$		
Total	Between SS + Within SS				

FIGURE 12.2 Acceptance and rejection regions for the overall F test for one-way ANOVA

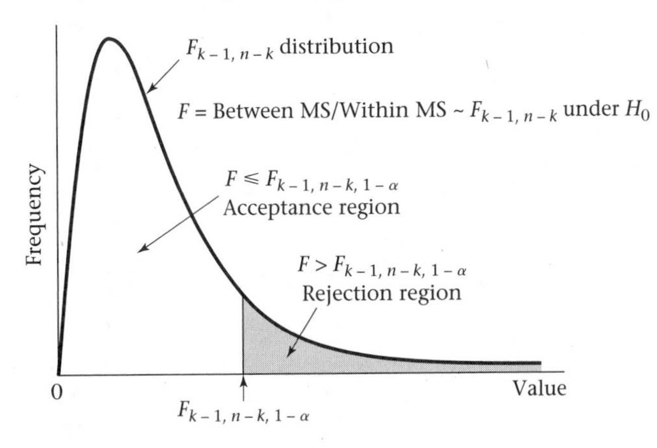

FIGURE 12.3 Computation of the exact p-value for the overall F test for one-way ANOVA

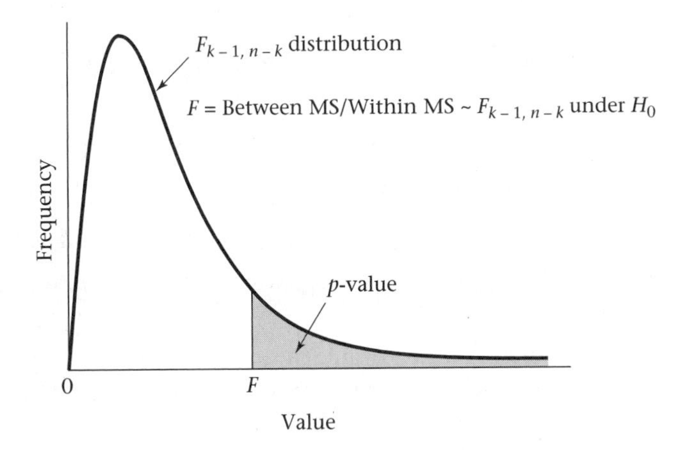

Example 12.4 **Pulmonary Disease** Test if the mean FEF scores are significantly different in the six groups in Table 12.1.

Solution From Example 12.3, Between SS = 184.38 and Within SS = 663.87. Therefore, since there are 1050 observations combined over all 6 groups, it follows that

$$\text{Between MS} = 184.38/5 = 36.875$$

$$\text{Within MS} = 663.87/(1050 - 6) = 663.87/1044 = 0.636$$

$$F = \text{Between MS}/\text{Within MS} = 36.875/0.636 = 58.0 \sim F_{5,1044} \text{ under } H_0$$

Refer to Table 9 in the Appendix and find that

$$F_{5,120,.999} = 4.42$$

Since $F_{5,1044,.999} < F_{5,120,.999} = 4.42 < 58.0 = F$

it follows that $p < .001$. Therefore, H_0, that all the means are equal, can be rejected and we can conclude that at least two of the means are not the same. These results are displayed in an ANOVA table (Table 12.3).

TABLE 12.3 ANOVA table for FEF data in Table 12.1.

	SS	df	MS	F statistic	p-value
Between	184.38	5	36.875	58.0	$p < .001$
Within	663.87	1044	0.636		
Total	848.25				

SECTION 12.4 Comparisons of Specific Groups in One-Way ANOVA

In the previous section a test of the hypothesis H_0: all group means are equal versus H_1: at least two group means are different was presented. This test enables us to detect when at least two groups have different underlying means, but it does not allow us to state which of the groups have means that are different from each other. The usual practice is to perform the overall F test just discussed. If H_0 is rejected, then specific groups are compared, as discussed in this section.

12.4.1 *t* Test for Comparison of Pairs of Groups

Suppose at this point we want to test if groups 1 and 2 have means that are significantly different from each other. From the underlying model in Equation 12.1, under either hypothesis,

EQUATION 12.7

$\bar{y}_1$ is normally distributed with mean $\mu + \alpha_1$ and variance σ^2/n_1
and $\bar{y}_2$ is normally distributed with mean $\mu + \alpha_2$ and variance σ^2/n_2

The difference of the sample means $(\bar{y}_1 - \bar{y}_2)$ will be used as a test criterion. Thus, from Equation 12.7, since the samples are independent, if follows that

EQUATION 12.8

$$\bar{y}_1 - \bar{y}_2 \sim N\left[\alpha_1 - \alpha_2, \sigma^2\left(\frac{1}{n_1} + \frac{1}{n_2}\right)\right]$$

However, under H_0, $\alpha_1 = \alpha_2$ and Equation 12.8 reduces to

EQUATION 12.9

$$\bar{y}_1 - \bar{y}_2 \sim N\left[0, \sigma^2\left(\frac{1}{n_1} + \frac{1}{n_2}\right)\right]$$

If σ^2 were known, then we could divide by the standard error

$$\sigma\sqrt{\frac{1}{n_1} + \frac{1}{n_2}}$$

and obtain the test statistic:

EQUATION 12.10

$$z = \frac{\bar{y}_1 - \bar{y}_2}{\sqrt{\sigma^2\left(\frac{1}{n_1} + \frac{1}{n_2}\right)}}$$

The test statistic z would follow an $N(0, 1)$ distribution under H_0. Because σ^2 is in general unknown, the best estimate of it, denoted by s^2, is substituted, and the test statistic is revised accordingly.

How should σ^2 be estimated? Recall from Equation 12.1 that $y_{ij} \sim N(\mu + \alpha_i, \sigma^2)$. Thus, the underlying variance of each group is the same. Therefore, a pooled estimate of group-specific variances is reasonable. Recall that when a pooled estimate of the variance from two independent samples was obtained in Chapter 8, a weighted average of the sample variances from the individual samples, where the weights were the number of degrees of freedom in each sample, was used. In particular, from Equation 8.10,

$$s^2 = \left[(n_1 - 1)s_1^2 + (n_2 - 1)s_2^2\right] / (n_1 + n_2 - 2)$$

For the one-way ANOVA, there are k sample variances and a similar approach is used to estimate σ^2 by computing a weighted average of k individual sample variances, where the weights are the number of degrees of freedom in each of the k samples. This formula is given as follows:

EQUATION 12.11

Pooled Estimate of the Variance for One-Way ANOVA

$$s^2 = \sum_{i=1}^{k}(n_i - 1)s_i^2 \Big/ \sum_{i=1}^{k}(n_i - 1) = \left[\sum_{i=1}^{k}(n_i - 1)s_i^2\right] \Big/ (n - k) = \text{Within MS}$$

However, note from Equations 12.5, 12.11, and Definition 12.6 that this weighted average is the same as the Within MS. Thus, the Within MS is used to estimate σ^2. Note that s^2 had $(n_1 - 1) + (n_2 - 1) = n_1 + n_2 - 2\,df$ in the two-sample case. Similarly, for the one-way ANOVA, s^2 has

$$(n_1 - 1) + (n_2 - 1) + \cdots + (n_k - 1)df = (n_1 + n_2 + \cdots + n_k) - k = n - k\,df$$

Example 12.5 **Pulmonary Disease** What is the best estimate of σ^2 for the FEF data in Table 12.1? How many *df* does it have?

Solution From Table 12.3, the best estimate of the variance is the Within MS = 0.636. It has $n - k\,df = 1044\,df$.

Hence, the test statistic z in Equation 12.10 will be revised, substituting s^2 for σ^2, with the new test statistic t distributed as t_{n-k} rather than $N(0, 1)$. The test procedure is given as follows:

EQUATION 12.12

***t* Test for the Comparison of Pairs of Groups in One-Way ANOVA (LSD Procedure)** Suppose we wish to compare two specific groups, arbitrarily labeled as group 1 and group 2, among k groups. To test the hypothesis H_0: $\alpha_1 = \alpha_2$ versus H_1: $\alpha_1 \neq \alpha_2$, use the following procedure:

(1) Compute the pooled estimate of the variance s^2 = Within MS from the one-way ANOVA.

(2) Compute the test statistic

$$t = \frac{\bar{y}_1 - \bar{y}_2}{\sqrt{s^2\left(\frac{1}{n_1} + \frac{1}{n_2}\right)}}$$

which follows a t_{n-k} distribution under H_0.

(3) For a two-sided level α test, if

$$t > t_{n-k,1-\alpha/2} \quad \text{or} \quad t < t_{n-k,\alpha/2}$$

then reject H_0; if

$$t_{n-k,\alpha/2} \le t \le t_{n-k,1-\alpha/2}$$

then accept H_0.

(4) The exact p-value is given by

$$p = 2 \times \text{the area to the left of } t \text{ under a } t_{n-k} \text{ distribution if } t < 0$$
$$= 2 \times Pr(t_{n-k} < t)$$

$$p = 2 \times \text{the area to the right of } t \text{ under a } t_{n-k} \text{ distribution if } t \ge 0$$
$$= 2 \times Pr(t_{n-k} > t)$$

The acceptance and rejection regions for this test are given in Figure 12.4. The computation of the exact p-value is illustrated in Figure 12.5. This test is often referred to as the *least significant difference (LSD) method.*

FIGURE 12.4 **Acceptance and rejection regions for the t test for the comparison of pairs of groups in one-way ANOVA (LSD approach)**

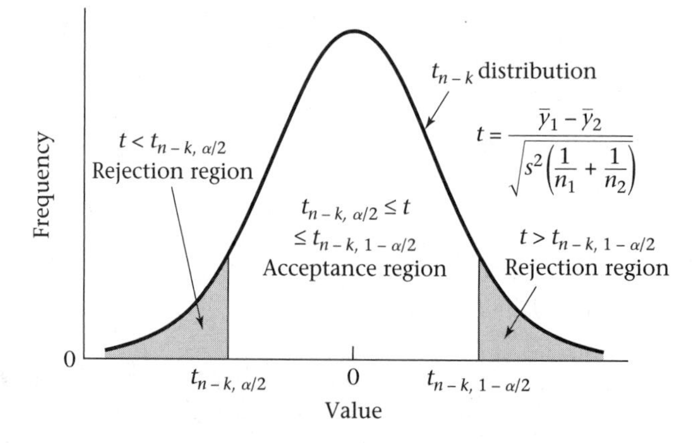

FIGURE 12.5 **Computation of the exact *p*-value for the *t* test for the comparison of pairs of groups in one-way ANOVA (LSD approach)**

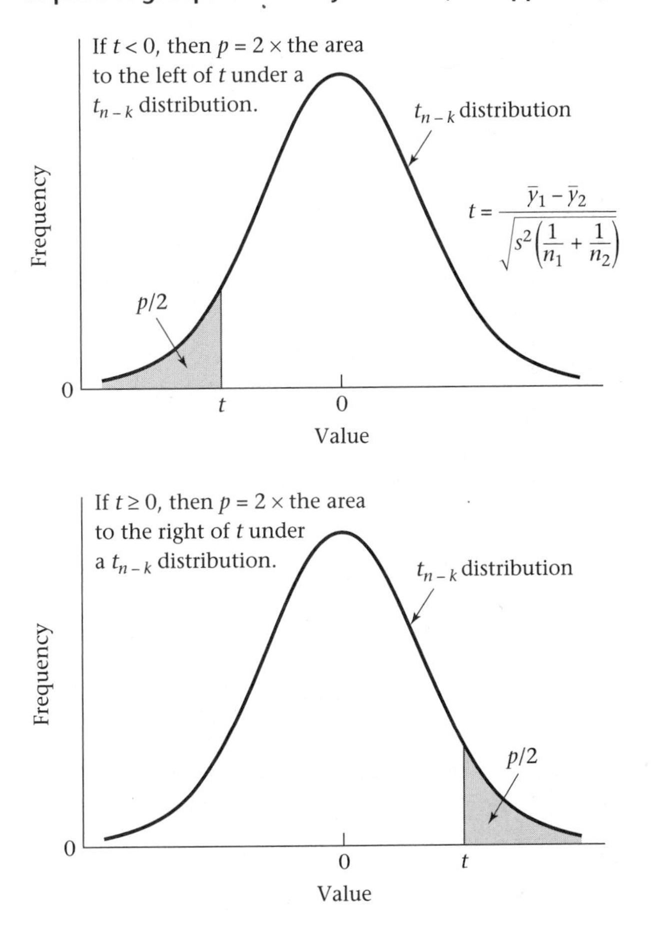
If $t < 0$, then $p = 2 \times$ the area to the left of t under a t_{n-k} distribution.
t_{n-k} distribution
$t = \frac{\bar{y}_1 - \bar{y}_2}{\sqrt{s^2\left(\frac{1}{n_1} + \frac{1}{n_2}\right)}}$
Frequency
$p/2$
0
t
0
Value
If $t \ge 0$, then $p = 2 \times$ the area to the right of t under a t_{n-k} distribution.
t_{n-k} distribution
Frequency
$p/2$
0
0
t
Value

Example 12.6 **Pulmonary Disease** Compare each pair of groups for the FEF data in Table 12.1 and report any significant differences.

Solution First plot the mean ± *se* of the FEF values for each of the six groups in Figure 12.6 to obtain some idea of the magnitude of the differences between groups. The standard error for an individual group mean is estimated by $s/\sqrt{n_i}$, where s^2 = Within MS. Notice that the nonsmokers have the best pulmonary function; the passive smokers, noninhaling smokers, and light smokers have about the same pulmonary function and are worse off than the nonsmokers; and the moderate and heavy smokers have the poorest pulmonary function. Note also that the standard error bars are wider for the noninhaling smokers than for the other groups, because this group has only 50 people compared with 200 for all other groups. Are the observed differences in the figure statistically significant as assessed by the *t* test procedure in Equation 12.12? The results are presented in Table 12.4.

FIGURE 12.6 **Mean ± *se* for FEF for each of six smoking groups**

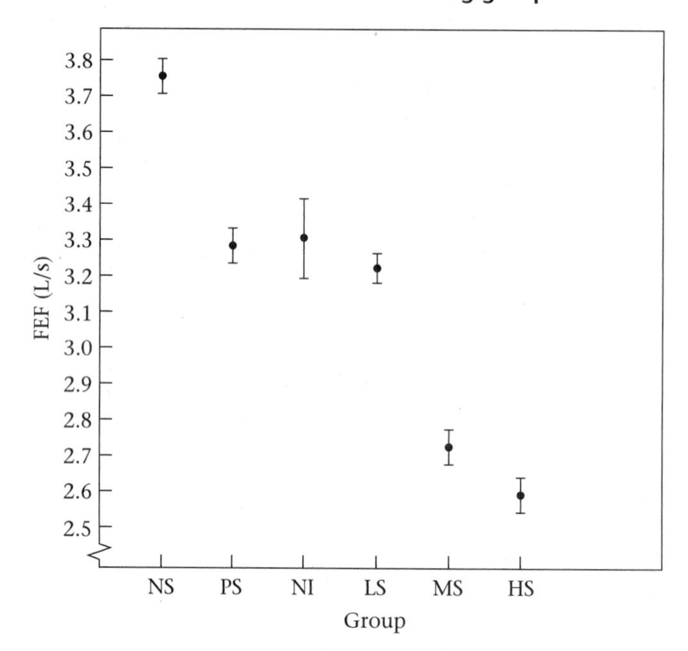

Source: Reprinted by permission of the *New England Journal of Medicine, 302*(13), 720–723, 1980.

There are very highly significant differences (1) between the nonsmokers and all other groups, (2) between the passive smokers and the moderate and heavy smokers, (3) between the noninhalers and the moderate and heavy smokers, and (4) between the light smokers and the moderate and heavy smokers. There are no significant differences between the passive smokers, noninhalers, and light smokers and no significant differences between the moderate and heavy smokers, although there is a trend toward significance with the latter comparison. Thus, these results tend to confirm what Figure 12.6 shows. They are very interesting because they show that the pulmonary function of passive smokers is significantly worse than that of nonsmokers and is essentially the same as that of noninhaling and light smokers (≤ 1/2 pack cigarettes per day).

A frequent error in performing the t test in Equation 12.12 when comparing groups 1 and 2 is to use only the sample variances from *these two groups* rather than from *all k groups* to estimate σ^2. If the sample variances from only two groups are used, then different estimates of σ^2 are obtained for each pair of groups considered, which is not reasonable because *all* the groups have the same underlying variance σ^2. Furthermore, the estimate of σ^2 obtained by using all k groups will be more accurate than that obtained from using any two groups, since the estimate of the variance will be based on more information. This is the principal advantage of performing the t tests in the framework of a one-way ANOVA rather than by considering each pair of groups separately and performing t tests for two independent samples as given in Equation 8.11 for each pair of samples. However, if there is

TABLE 12.4 **Comparisons of specific pairs of groups for the FEF data in Table 12.1 using the LSD *t* test approach**

Groups compared	Test statistic	*p*-value
NS, PS	$t = \frac{3.78 - 3.30}{\sqrt{0.636\left(\frac{1}{200} + \frac{1}{200}\right)}} = \frac{0.48}{0.08} = 6.02^{a}$	< .001
NS, NI	$t = \frac{3.78 - 3.32}{\sqrt{0.636\left(\frac{1}{200} + \frac{1}{50}\right)}} = \frac{0.46}{0.126} = 3.65$	< .001
NS, LS	$t = \frac{3.78 - 3.23}{\sqrt{0.636\left(\frac{1}{200} + \frac{1}{200}\right)}} = \frac{0.55}{0.08} = 6.90$	< .001
NS, MS	$t = \frac{3.78 - 2.73}{0.080} = \frac{1.05}{0.08} = 13.17$	< .001
NS, HS	$t = \frac{3.78 - 2.59}{0.080} = \frac{1.19}{0.08} = 14.92$	< .001
PS, NI	$t = \frac{3.30 - 3.32}{0.126} = \frac{-0.02}{0.126} = -0.16$	NS
PS, LS	$t = \frac{3.30 - 3.23}{0.080} = \frac{0.07}{0.08} = 0.88$	NS
PS, MS	$t = \frac{3.30 - 2.73}{0.080} = \frac{0.57}{0.08} = 7.15$	< .001
PS, HS	$t = \frac{3.30 - 2.59}{0.080} = \frac{0.71}{0.08} = 8.90$	< .001
NI, LS	$t = \frac{3.32 - 3.23}{0.126} = \frac{0.09}{0.126} = 0.71$	NS
NI, MS	$t = \frac{3.32 - 2.73}{0.126} = \frac{0.59}{0.126} = 4.68$	< .001
NI, HS	$t = \frac{3.32 - 2.59}{0.126} = \frac{0.73}{0.126} = 5.79$	< .001
LS, MS	$t = \frac{3.23 - 2.73}{0.08} = \frac{0.50}{0.08} = 6.27$	< .001
LS, HS	$t = \frac{3.23 - 2.59}{0.08} = \frac{0.64}{0.08} = 8.03$	< .001
MS, HS	$t = \frac{2.73 - 2.59}{0.08} = \frac{0.14}{0.08} = 1.76$	NS

[a]All test statistics follow a t_{1044} distribution under H_0.

reason to believe that not all groups have the same underlying variance (σ^2), then the one-way ANOVA should not be performed, and t tests based on pairs of groups should be used instead.

12.4.2 Linear Contrasts

In Section 12.4.1 methods for comparing specific groups within the context of the analysis of variance were developed. More general comparisons, such as the comparison of a collection of ℓ_1 groups with another collection of ℓ_2 groups, are frequently desired.

Example 12.7 **Pulmonary Disease** Suppose we want to compare the pulmonary function of the group of smokers who inhale cigarettes with the group of nonsmokers. The three groups of inhaling smokers in Table 12.1 could just be combined to form one group of 600 inhaling smokers. However, these three groups were selected so as to be of the same size, whereas in the general population the proportions of light, moderate, and heavy smokers are not likely to be the same. Suppose large population surveys report that 70% of inhaling smokers are moderate smokers, 20% are heavy smokers, and 10% are light smokers. How can inhaling smokers as a group be compared with nonsmokers?

It is for this type of question that the estimation and testing of hypotheses for linear contrasts is used.

DEFINITION 12.7 A **linear contrast** (L) is any linear combination of the individual group means such that the linear coefficients add up to 0. Specifically,

$$L = \sum_{i=1}^{k} c_i \bar{y}_i$$

where $\sum_{i=1}^{k} c_i = 0$

Notice that the comparison of two means that was considered in the previous section is a special case of a linear contrast.

Example 12.8 **Pulmonary Disease** Suppose we wish to compare the pulmonary function of the nonsmokers and the passive smokers. Represent this comparison as a linear contrast.

Solution Since the nonsmokers are the first group and the passive smokers are the second group, this comparison can be represented by the linear contrast

$$L = \bar{y}_1 - \bar{y}_2 \qquad \text{that is,} \qquad c_1 = +1 \qquad c_2 = -1$$

Example 12.9 **Pulmonary Disease** Suppose we wish to compare the pulmonary function of nonsmokers with that of inhaling smokers, assuming that 10% of inhaling smokers are light smokers, 70% are moderate smokers, and 20% are heavy smokers. Represent this comparison as a linear contrast.

Solution This comparison can be represented by the linear contrast

$$\bar{y}_1 - 0.1\bar{y}_4 - 0.7\bar{y}_5 - 0.2\bar{y}_6$$

since the nonsmokers are group 1, the light-smokers group 4, the moderate-smokers group 5, and the heavy-smokers group 6.

How can we test if the underlying mean of a linear contrast is different from 0? In general, for any linear contrast,

$$L = c_1\bar{y}_1 + c_2\bar{y}_2 + \cdots + c_k\bar{y}_k$$

we wish to test the hypothesis H_0: $\mu_L = 0$ versus H_1: $\mu_L \neq 0$, where μ_L is the mean of the linear contrast L:

$$c_1\alpha_1 + c_2\alpha_2 + \cdots + c_k\alpha_k$$

Since $Var(\bar{y}_i) = s^2/n_i$, we can derive $Var(L)$ using Equation 5.7, whereby

$$Var(L) = s^2 \sum_{i=1}^{k} c_i^2/n_i$$

Thus, the following test procedure, which is analogous to the LSD t test for pairs of groups in Equation 12.12, can be used:

EQUATION 12.13

***t* Test for Linear Contrasts in One-Way ANOVA** Suppose we wish to test the hypothesis H_0: $\mu_L = 0$ versus H_1: $\mu_L \neq 0$, using a two-sided test with significance level $= \alpha$, where $y_{ij} \sim N(\mu + \alpha_i, \sigma^2), \mu_L = \sum_{i=1}^{k} c_i\alpha_i$ and $\sum_{i=1}^{k} c_i = 0$.

(1) Compute the pooled estimate of the variance = s^2 = Within MS from the one-way ANOVA.

(2) Compute the linear contrast

$$L = \sum_{i=1}^{k} c_i\bar{y}_i$$

(3) Compute the test statistic

$$t = \frac{L}{\sqrt{s^2 \sum_{i=1}^{k} \frac{c_i^2}{n_i}}}$$

(4) If $t > t_{n-k,1-\alpha/2}$ or $t < t_{n-k,\alpha/2}$

then reject H_0. If $t_{n-k,\alpha/2} \le t \le t_{n-k,1-\alpha/2}$

then accept H_0.

(5) The exact p-value is given by

$$p = 2 \times \text{the area to the left of } t \text{ under a } t_{n-k} \text{ distribution}$$
$$= 2 \times Pr(t_{n-k} < t), \text{ if } t < 0$$

$$p = 2 \times \text{the area to the right of } t \text{ under a } t_{n-k} \text{ distribution}$$
$$= 2 \times Pr(t_{n-k} > t), \text{ if } t \ge 0$$

Example 12.10 **Pulmonary Disease** Test the hypothesis that the underlying mean of the linear contrast defined in Example 12.9 is significantly different from 0.

Solution From Table 12.3, $s^2 = 0.636$. Furthermore, the linear contrast L is given by

$$L = \bar{y}_1 - 0.1\bar{y}_4 - 0.7\bar{y}_5 - 0.2\bar{y}_6 = 3.78 - 0.1(3.23) - 0.7(2.73) - 0.2(2.59) = 1.03$$

The standard error of this linear contrast is given by

$$se(L) = \sqrt{s^2 \sum_{i=1}^{k} \frac{c_i^2}{n_i}} = \sqrt{0.636\left[\frac{(1)^2}{200} + \frac{(-0.1)^2}{200} + \frac{(-0.7)^2}{200} + \frac{(-0.2)^2}{200}\right]} = 0.070$$

Thus, $t = L/se(L) = 1.03/0.070 = 14.69 \sim t_{1044}$ under H_0

Clearly, this linear contrast is very highly significant ($p < .001$), and the inhaling smokers as a group have strikingly poorer pulmonary function than the nonsmokers.

Another useful application of linear contrasts is when the different groups correspond to different dose levels of a particular quantity, and the coefficients of the contrast are chosen to reflect a particular dose–response relationship. This application is particularly useful if the sample sizes of the individual groups are small and a comparison of any pair of groups does not show a significant difference, but the overall trend is consistent in one direction.

Example 12.11 **Pulmonary Disease** Suppose we wish to study whether or not the number of cigarettes smoked is related to the level of FEF among those smokers who inhale cigarettes. Perform a test of significance for this trend.

Solution Focus on the light smokers, moderate smokers, and heavy smokers in this analysis. We know the light smokers smoke from 1 to 10 cigarettes per day, and we will assume they smoke an average of $(1 + 10)/2 = 5.5$ cigarettes per day. The moderate smokers smoke from 11 to 39 cigarettes per day, and we will assume they smoke an average of $(11 + 39)/2 = 25$ cigarettes per day. The heavy smokers smoke at least 40 cigarettes per day. We will assume they smoke exactly 40 cigarettes per day, which will underestimate the trend but is the best we can do with the information presented. We wish to test the contrast

$$L = 5.5\bar{y}_4 + 25\bar{y}_5 + 40\bar{y}_6$$

for statistical significance. The problem is that the coefficients of this contrast do not add up to 0; indeed, they add up to $5.5 + 25 + 40 = 70.5$. However, if $70.5/3 = 23.5$ is subtracted from each coefficient, then they will add up to 0. Thus, we wish to test the contrast

$$L = (5.5 - 23.5)\bar{y}_4 + (25 - 23.5)\bar{y}_5 + (40 - 23.5)\bar{y}_6 = -18\bar{y}_4 + 1.5\bar{y}_5 + 16.5\bar{y}_6$$

for statistical significance. This contrast represents the increasing number of cigarettes smoked per day in the three groups. From Equation 12.13,

$$L = -18(3.23) + 1.5(2.73) + 16.5(2.59) = -58.14 + 4.10 + 42.74 = -11.31$$

$$se(L) = \sqrt{0.636\left[\frac{(-18)^2}{200} + \frac{1.5^2}{200} + \frac{16.5^2}{200}\right]} = \sqrt{0.636(2.99)} = \sqrt{1.903} = 1.38$$

Thus, $t = L/se(L) = -11.31/1.38 = -8.20 \sim t_{1044}$ under H_0

Clearly, this trend is very highly significant ($p < .001$), and we can say that among smokers who inhale, the greater the number of cigarettes smoked per day, the worse the pulmonary function.

12.4.3 Multiple Comparisons—Bonferroni Approach

In many studies, the comparisons of interest are specified before looking at the actual data, in which case the t test procedure in Equation 12.12 and the linear-contrast procedure in Equation 12.13 are appropriate. In other instances, the comparisons of interest will only be specified after looking at the data. In this case a large number of potential comparisons are often possible. Specifically, if there are a large number of groups and every pair of groups is compared using the t test procedure in Equation 12.12, then some significant differences are likely to be found just by chance.

Example 12.12 Suppose there are 10 groups. Thus, there are $\binom{10}{2} = 45$ possible pairs of groups to be compared. Using a 5% level of significance would imply that .05(45), or about two comparisons, are likely to be significant by chance alone. How can we protect ourselves against the detection of falsely significant differences resulting from making too many comparisons?

Several procedures, referred to as **multiple-comparisons procedures,** ensure that too many falsely significant differences are not declared. The basic idea of these procedures is to ensure that the *overall probability of declaring any significant differences between all possible pairs of groups* is maintained at some fixed significance level (say α). One of the simplest and most widely used such procedure is the method of *Bonferroni adjustment*. This method is summarized as follows.

EQUATION 12.14

Comparison of Pairs of Groups in One-Way ANOVA—Bonferroni Multiple-Comparisons Procedure Suppose we wish to compare two specific groups, arbitrarily labeled as group 1 and group 2, among k groups. To test the hypothesis H_0: $\alpha_1 = \alpha_2$ versus H_1: $\alpha_1 \neq \alpha_2$, use the following procedure:

(1) Compute the pooled estimate of the variance s^2 = Within MS from the one-way ANOVA.

(2) Compute the test statistic

$$t = \frac{\bar{y}_1 - \bar{y}_2}{\sqrt{s^2\left(\frac{1}{n_1} + \frac{1}{n_2}\right)}}$$

(3) For a two-sided level α test, let $\alpha^* = \alpha \Big/ \binom{k}{2}$.

If $t > t_{n-k,1-\alpha^*/2}$ or $t < t_{n-k,\alpha^*/2}$

then reject H_0; if

$$t_{n-k,\alpha^*/2} \le t \le t_{n-k,1-\alpha^*/2}$$

then accept H_0.

The acceptance and rejection regions for this test are given in Figure 12.7. This test is referred to as the *Bonferroni multiple-comparisons procedure.*

FIGURE 12.7 **Acceptance and rejection regions for the comparison of pairs of groups in one-way ANOVA (Bonferroni approach)**

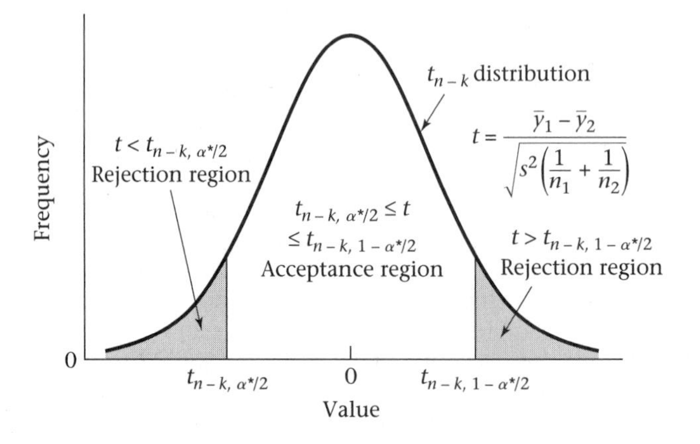

The rationale behind this procedure is that in a study with k groups, there are $\binom{k}{2}$ possible two-group comparisons. Suppose each two-group comparison is conducted at the α^* level of significance. Let E be the event that at least one of the two-group comparisons is statistically significant. $Pr(E)$ is sometimes referred to as the "experiment-wise type I error." We wish to determine the value α^* such that $Pr(E) = \alpha$. To find α^*, we note that

$$Pr(\bar{E}) = Pr(\text{none of the two-group comparisons is statistically significant}) = 1 - \alpha$$

If each of the two-group comparisons were independent, then from the multiplication law of probability, $Pr(\bar{E}) = (1 - \alpha^*)^c$, where $c = \binom{k}{2}$. Therefore,

EQUATION 12.15

$$1 - \alpha = \left(1 - \alpha^*\right)^c$$

If α^* is small, then it can be shown that the right-hand side of Equation 12.15 can be approximated by $1 - c\alpha^*$. Thus,

$$1 - \alpha \cong 1 - c\alpha^*$$

or

$$\alpha^* \cong \alpha/c = \alpha\bigg/\binom{k}{2} \text{ as given in Equation 12.14}$$

Usually all the two-group comparisons are *not* statistically independent, whereby the appropriate value α^* is greater than $\alpha \Big/ \binom{k}{2}$. Thus, the Bonferroni procedure will be conservative in the sense that $Pr(E) < \alpha$.

Example 12.13 Apply the Bonferroni multiple-comparisons procedure to the FEF data in Table 12.1.

Solution We wish to conduct a test with experiment-wise type I error = .05. We have a total of n = 1050 subjects and k = 6 groups. Thus, $n - k$ = 1044 and $c = \binom{6}{2} = 15$. Thus, $\alpha^* =$ .05/15 = .0033. Therefore, we conduct t tests between each pair of groups using the .0033 level of significance. From Equation 12.14, the critical value for each of these t tests is $t_{1044,1-.0033/2} = t_{1044,.99833}$. We will approximate a t distribution with 1044 df by an $N(0,1)$ distribution or, $t_{1044,.99833} \approx z_{.99833}$. From Table 3 in the Appendix, $z_{.99833} = 2.935$. We now refer to Table 12.4, which provides the t statistics for each two-group comparison. We notice that the absolute value of all t statistics for two-group comparisons that were statistically significant using the LSD approach are ≥3.65. Since 3.65 > 2.935, it follows that they will remain statistically significant under the Bonferroni procedure. Furthermore, the comparisons that were not statistically significant with the LSD procedure are also not significant under the Bonferroni procedure. This must be the case since the Bonferroni procedure is more conservative than the LSD procedure. In this example, the critical region using the LSD procedure with a two-sided test (α = .05) is $t < -1.96$ or $t > 1.96$, whereas the comparable critical region using the Bonferroni procedure is $t < -2.935$ or $t > 2.935$.

The results of the multiple-comparisons procedure are typically displayed as in Figure 12.8. A line is drawn between the names or numbers of each pair of means that are *not* significantly different. This plot enables us to visually summarize the results of many comparisons of pairs of means in one concise display.

FIGURE 12.8 **Display of results of Bonferroni multiple-comparisons procedure on FEF data in Table 12.1**

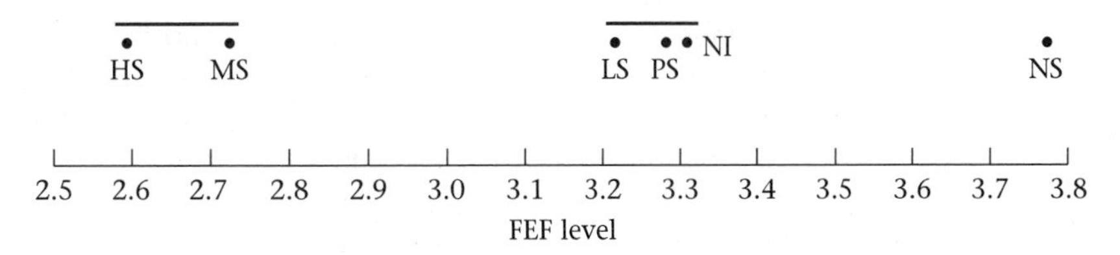

Note that the results of the LSD procedure in Table 12.4 and the Bonferroni procedure in Example 12.13 are the same. Namely, there are three distinct groups: heavy and moderate smokers; light smokers, passive smokers, and noninhaling smokers; and nonsmokers. In general, the multiple-comparisons procedures are more strict than the ordinary t tests (LSD procedure) if more than two means are being compared. That is, there are comparisons between pairs of groups for which the t test would declare a significant difference but the multiple-comparisons procedure would not. This is the price that is paid for trying to fix the α level of finding *any* significant difference among pairs of groups in using the multiple-comparisons procedure rather than for *particular* pairs of groups in using the t test. If only two means are being compared, then the p-values obtained from using the two procedures are identical.

Also note from Equation 12.14 that as the number of groups being compared (k) increases, the critical value for declaring statistical significance becomes larger. This is because as k increases, $c = \binom{k}{2}$ increases and therefore $\alpha^* = \alpha/c$ decreases. The critical value, $t_{n-k,1-\alpha^*/2}$, will therefore increase because as k increases, the degrees of freedom $(n - k)$ decreases and the percentile $1 - \alpha^*/2$ increases, both of which will result in a larger critical value. This is not true for the LSD procedure, where the critical value $t_{n-k,1-\alpha/2}$ remains roughly the same as k increases.

When should the more conservative multiple-comparisons procedure in Equation 12.14 rather than the LSD procedure in Equation 12.12 be used to identify specific differences between groups? This area is controversial. Some research workers routinely use multiple-comparisons procedures for all one-way ANOVA problems; others never use them. My opinion is that multiple-comparisons procedures should be used if there are many groups and not all comparisons between individual groups have been planned in advance. However, if there are relatively few groups and only specific comparisons of interest are intended, which have been planned in advance, preferably stated in a written set of procedures for a study (commonly called a *protocol*), then I prefer to use ordinary t tests (i.e., the LSD procedure) rather than multiple-comparisons procedures. In a sense, by first performing the overall F test for one-way ANOVA (Equation 12.6), and only comparing pairs of groups with the LSD procedure (Equation 12.12) if the overall F test is statistically significant, we have protected ourselves to some extent from the multiple comparisons problem, even with the LSD procedure.

12.4.4 Multiple-Comparisons Procedures for Linear Contrasts

The multiple-comparisons procedure in Section 12.4.3 is applicable for comparing pairs of means. In some situations, linear contrasts involving more complex comparisons than simple contrasts based on pairs of means are of interest. In this context, if linear contrasts, which have not been planned in advance, are suggested by looking at the data, then a multiple-comparisons procedure might be used to ensure that under H_0, the probability that any linear contrast will be significant is no larger than α. Scheffé's multiple-comparisons procedure is applicable in this situation and is summarized as follows:

EQUATION 12.16

Scheffé's Multiple-Comparisons Procedure Suppose we wish to test the hypothesis H_0: $\mu_L = 0$ versus H_1: $\mu_L \neq 0$, at significance level α, where

$$L = \sum_{i=1}^{k} c_i \bar{y}_i, \qquad \mu_L = \sum_{i=1}^{k} c_i \mu_i, \qquad \text{and} \qquad \sum_{i=1}^{k} c_i = 0$$

and we have k groups, with n_i subjects in the ith group and a total of $n = \sum_{i=1}^{k} n_i$ subjects overall. To use Scheffé's multiple-comparisons procedure in this situation, perform the following steps:

(1) Compute the test statistic

$$t = \frac{L}{\sqrt{s^2 \sum_{i=1}^{k} \frac{c_i^2}{n_i}}}$$

as given in Equation 12.13.

(2) If

$$t > a_2 = \sqrt{(k-1)F_{k-1,n-k,1-\alpha}} \quad \text{or} \quad t < a_1 = -\sqrt{(k-1)F_{k-1,n-k,1-\alpha}}$$

then reject H_0; if

$$a_1 \le t \le a_2$$

then accept H_0.

Example 12.14 **Pulmonary Disease** Test the hypothesis that the linear contrast defined in Example 12.11, representing the relationship between level of FEF and the number of cigarettes smoked among smokers who inhale cigarettes, is significantly different from 0 using Scheffé's multiple-comparisons procedure.

Solution From Example 12.11, $t = L/se(L) = -8.20$. There are six groups and 1050 subjects. Thus, since t is negative, the critical value is given by $a_1 = -\sqrt{(k-1)F_{k-1,n-k,1-\alpha}}$ $= -\sqrt{5F_{5,1044,0.95}}$. $F_{5,1044,.95}$ will be approximated by $F_{5,\infty,.95} = 2.21$. We have $a_1 =$ $-\sqrt{5(2.21)} = -3.32$. Since $t = -8.20 < c_1 = -3.32$, H_0 is rejected at the 5% level and a significant trend among inhaling smokers, with pulmonary function decreasing as the number of cigarettes smoked per day increases, is declared.

Scheffé's multiple-comparisons procedure could also have been used when pairs of means were being compared, since a difference between means is a special case of a linear contrast. However, the Bonferroni procedure introduced in Section 12.4.3 is preferable in this instance, because, if only pairs of means are being compared, then significant differences can appropriately be declared more often than with Scheffé's procedure (which is designed for a broader set of alternative hypotheses) when true differences exist in this situation. Indeed, from Example 12.13, the critical region using the Bonferroni procedure was $t < -2.935$ or $t > 2.935$, while the corresponding critical region using the Scheffé procedure is $t < -3.32$ or $t > 3.32$.

Once again, if a few linear contrasts, which have been specified in advance, are to be tested, then it may not be necessary to use a multiple-comparisons procedure, because if such procedures are used, there will be less power to detect differences for linear contrasts whose means are truly different from zero than the t tests introduced in Equation 12.13. Conversely, if many contrasts are to be tested, which have not been specified before looking at the data, then the multiple-comparisons procedure in this section may be useful in protecting against declaring too many significant differences.

Based on our work in Sections 12.1–12.4, Figure 12.9 summarizes the general procedure used to compare the means of k independent normally distributed samples.

FIGURE 12.9 General procedure for comparing the means of k independent normally distributed samples

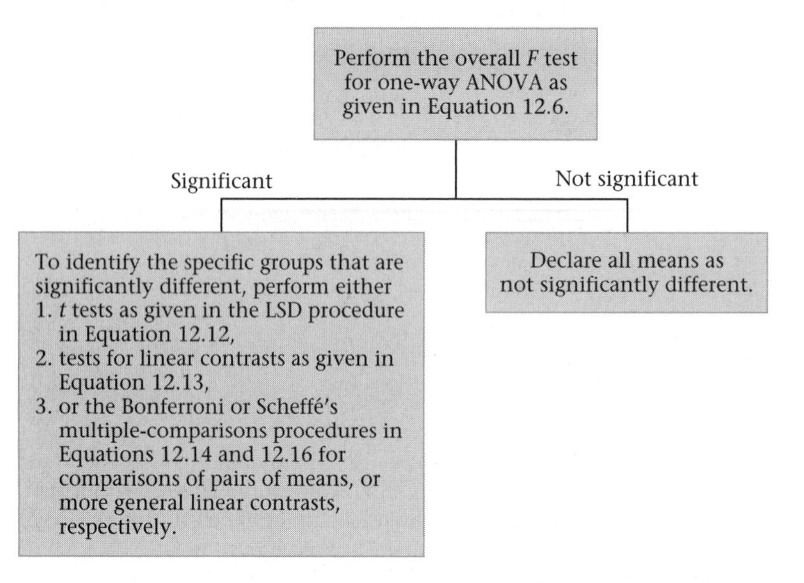

In this section, we have learned about one-way analysis-of-variance (ANOVA) methods. This technique is used to compare the means among several (>2) normally distributed samples. To place these methods in a broader perspective, refer to the master flowchart at the end of the book (p. 776). Starting at the Start box, we would answer no to (1) only one variable of interest? and proceed to 4. We would then answer yes to (2) interested in relationships between two variables? no to (3) both variables continuous? and yes to (4) one variable continuous and one categorical? This leads us to the box entitled "Analysis of Variance." We would then answer "one" to (5) number of ways in which the categorical variable can be classified, yes to (6) outcome variable normal, or can central-limit theorem be assumed to hold? and no to (7) other covariates to be controlled for? This leads us to the box entitled "One-Way ANOVA."

SECTION 12.5 Case Study: Effects of Lead Exposure on Neurological and Psychological Function in Children

12.5.1 Application of One-Way ANOVA

In Section 8.8 (Table 8.8), we analyzed the difference in mean finger-wrist tapping score (MAXFWT) by lead-exposure group. The children were subdivided into an exposed group who had elevated blood-lead levels (≥40 μg/100 mL) in either 1972 or 1973 and a control group who had normal blood-lead levels (<40 μg/100 mL) in both 1972 and 1973. We first removed outliers from each group using the ESD procedure and then used a two-sample t test to compare mean scores between the two groups (see Table 8.11). However, since the neurological and psychological tests were performed in 1973, it could be argued that it would be better to define an exposed group based on blood-lead levels in 1973 only. For this purpose, the variable LEAD_GRP in the data set allows us to subdivide the exposed group into two subgroups. Specifically, we will consider three lead-exposure groups according to the variable LEAD_GRP:

If LEAD_GRP = 1, then the child had normal blood-lead levels (<40 µg/100 mL) in both 1972 and 1973 (CONTROL group).
If LEAD_GRP = 2, then the child had elevated blood-lead levels (≥ 40 µg/100 mL) in 1973 (the currently exposed group).
If LEAD_GRP = 3, then the child had elevated blood-lead levels in 1972 and normal blood-lead levels in 1973 (the previously exposed group).

The mean and standard deviation of MAXFWT for each group are given in Table 12.5 and the corresponding box plots in Figure 12.10.

TABLE 12.5 **Descriptive statistics of MAXFWT by group**

Analysis Variable: MAXFWT

GRP	N Obs*	N*	Mean	Std Dev	Minimum	Maximum
1	77	63	55.0952381	10.9348721	23.0000000	84.0000000
2	22	17	47.5882353	7.0804204	34.0000000	58.0000000
3	21	15	49.4000000	10.1966381	35.0000000	70.0000000

*N Obs is the total number of subjects in each group. N is the number of subjects used in the analysis in each group (i.e., subjects who have a MAXFWT value that is not an outlier). *Note:* The test was given only to children ages ≥5.

It appears that the mean MAXFWT scores are similar in both the currently exposed and previously exposed groups (groups 2 and 3) and are lower than the corresponding mean score in the control group (group 1). To compare the mean scores in the three groups, we will use the one-way ANOVA. We begin by using the overall F test for one-way ANOVA given in Equation 12.6 to test the hypothesis H_0: $\alpha_1 = \alpha_2 = \alpha_3$ versus H_1: at least two of the α_i are different. The results are given in Table 12.6.

TABLE 12.6 **Overall *F* test for one-way ANOVA for MAXFWT**

The SAS System
General Linear Models Procedure

Dependent Variable: MAXFWT

Source	DF	Sum of Squares	Mean Square	F Value	Pr > F
Model	2	966.79062362	483.39531181	4.60	0.0125
Error	92	9671.14621849	105.12115455		
Corrected Total	94	10637.93684211			

We see that there is an overall significant difference among the mean MAXFWT scores in the three groups. The F statistic is given under F Value = 4.60. The p-value = $Pr(F_{2,92} > 4.60)$ is listed under $Pr > F$ and is .0125. Therefore, we will

FIGURE 12.10 Box plots of MAXFWT by group

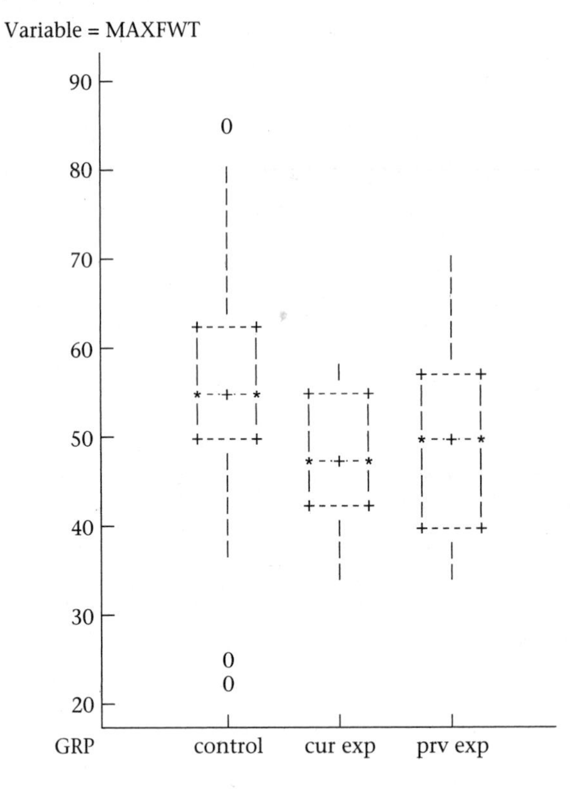

proceed to look at differences between each pair of groups. We will use the LSD procedure in Equation 12.12 because these comparisons are planned in advance. The results are given in Table 12.7.

TABLE 12.7 Comparison of group means for MAXFWT for pairs of specific groups (LSD procedure)

The SAS System
General Linear Models Procedure
Least Squares Means

GRP	MAXFWT LSMEAN	Std Err LSMEAN	Pr > |T| H0:LSMEAN=0	Pr > |T| H0: LSMEAN(i)=LSMEAN (j) i/j	1	2	3
1	55.0952381	1.2917390	0.0001	1	.	0.0087	0.0563
2	47.5882353	2.4866840	0.0001	2	0.0087	.	0.6191
3	49.4000000	2.6472773	0.0001	3	0.0563	0.6191	.

Note: To ensure overall protection level, only probabilities associated with preplanned comparisons should be used.

We see that there is a significant difference between the mean MAXFWT score for the currently exposed group and the control group ($p = .0087$, listed under $Pr > |T|$ in the first row corresponding to the control group and the second column corresponding to the currently exposed group). There is a strong trend toward a significant difference between the previously exposed group and the control group ($p = .0563$). There is clearly no significant difference between the mean MAXFWT scores for the currently and previously exposed groups (p-value = .6191). Thus, the currently exposed group is significantly different from the control group, and the previously exposed group shows a trend toward being significantly different from the control group. Other data given in Table 12.7 are the mean and standard error of the mean by group, listed under the first two LSMEAN columns. In this case, the first column contains the ordinary arithmetic mean (same as Table 12.5). The standard error is (Error Mean Square/n_i)$^{1/2}$ because our best estimate of the common within-group variance is the Error Mean Square (which we have previously referred to as the Within mean square). The third column provides a test of the hypothesis that the underlying mean in each specific group = 0 (specified as H_0: LSMEAN = 0). This is not relevant in this example, but would be if we were studying change scores over time. For uses of the general linear-model procedure other than for one-way ANOVA, the LSMEAN is different from the ordinary arithmetic mean. We discuss this issue in detail in Section 12.5.3.

Another approach for analyzing these data is to look at the 95% confidence interval for the difference in underlying means for pairs of specific groups. This is given by

$$\bar{y}_{i_1} - \bar{y}_{i_2} \pm t_{n-k,.975}\sqrt{\text{Within MS}(1/n_{i_1} + 1/n_{i_2})}$$

with results in Table 12.8. We note again that there are significant differences between the control group (Group 1) and the currently exposed group (Group 2) for MAXFWT—95% CI = (1.9, 13.1).

TABLE 12.8 **95% confidence intervals for mean difference in MAXFWT between pairs of groups**

```
                          The SAS System
                General Linear Models Procedure
              T tests (LSD) for variable: MAXFWT
NOTE: This test controls the type I comparisonwise error rate, not the
experimentwise error rate.
       Alpha = 0.05 Confidence = 0.95 df = 92 MSE = 105.1212
                  Critical Value of T = 1.98609
Comparisons significant at the 0.05 level are indicated by '***'.

                      Lower         Difference          Upper
         GRP        Confidence       Between          Confidence
      Comparison      Limit           Means             Limit
1         -3         -0.155           5.695            11.545
1         -2          1.942           7.507            13.072    ***

3         -1        -11.545          -5.695             0.155
3         -2         -5.402           1.812             9.025

2         -1        -13.072          -7.507            -1.942    ***
2         -3         -9.025          -1.812             5.402
```

12.5.2 Relationship Between One-Way ANOVA and Multiple Regression

In Section 12.5.1 we divided the exposed group into subgroups of currently exposed and previously exposed children. We then used a one-way ANOVA model to compare the mean MAXFWT among the currently exposed, previously exposed, and control groups. Another approach to this problem is to use a multiple-regression model with dummy variables. In Definition 11.19 we defined a single dummy variable to represent a categorical variable with two categories. This approach can be extended to represent a categorical variable with any number of categories.

EQUATION 12.17

Use of Dummy Variables to Represent a Categorical Variable with k Categories Suppose we have a categorical variable C with k categories. To represent that variable in a multiple-regression model, we construct $k - 1$ dummy variables of the form

$$x_1 = \begin{cases} 1 & \text{if subject is in category 2} \\ 0 & \text{otherwise} \end{cases}$$
$$x_2 = \begin{cases} 1 & \text{if subject is in category 3} \\ 0 & \text{otherwise} \end{cases}$$
$$\vdots$$
$$x_{k-1} = \begin{cases} 1 & \text{if subject is in category } k \\ 0 & \text{otherwise} \end{cases}$$

The category omitted (category 1) is referred to as the **reference group.**

It is arbitrary which group is assigned to be the reference group; the choice of a reference group is usually dictated by subject-matter considerations. In Table 12.9, we give the values of the dummy variables for subjects in different categories of C.

TABLE 12.9 **Representation of a categorical variable C by dummy variables**

	Dummy variables			
Category of C	x_1	x_2	...	x_{k-1}
1	0	0	...	0
2	1	0	...	0
3	0	1	...	0
⋮				
k	0	0	...	1

Notice that subjects in each category of C have a unique profile in terms of $x_1, \ldots, x_{k-1}$. To relate the categorical variable C to an outcome variable y, we use the multiple-regression model

EQUATION 12.18

$$y = \alpha + \beta_1 x_1 + \beta_2 x_2 + \cdots + \beta_{k-1} x_{k-1} + e$$

How can we use the multiple-regression model in Equation 12.18 to compare specific categories? From Equation 12.18, the average value of y for subjects in category 1 (the reference category) = α, the average value of y for subjects in category 2 = $\alpha + \beta_1$. Thus, β_1 represents the difference between the average value of y for subjects in category 2 and the average value of y for subjects in the reference category. Similarly, β_j represents the difference between the average value of y for subjects in category $(j + 1)$ versus the reference category, $j = 1, \ldots, k - 1$. In the fixed-effects one-way ANOVA model in Equation 12.1, we were interested in testing the hypothesis H_0: all underlying group means are the same versus H_1: at least two underlying group means are different. An equivalent way to specify these hypotheses in a multiple-regression setting is H_0: all $\beta_j = 0$ versus H_1: at least one of the $\beta_j \neq 0$. The latter specification of the hypotheses is the same as those given in Equation 11.30 where we used the overall F test for multiple linear regression. Thus, a fixed-effects one-way ANOVA model can be represented by a multiple linear-regression model based on a dummy-variable specification for the grouping variable. These results are summarized as follows:

EQUATION 12.19

Relationship Between Multiple Linear Regression and One-Way ANOVA Approaches Suppose we wish to compare the underlying mean among k groups where the observations in group j are assumed to be normally distributed with mean = $\mu_j = \mu + \alpha_j$ and variance = σ^2. To test the hypothesis: H_0: $\mu_j = 0$ for all $j = 1, \ldots, k$ versus H_1: at least two μ_j are different, we can use one of two equivalent procedures:

(1) We can perform the overall F test for one-way ANOVA.

(2) Or we can set up a multiple-regression model of the form

$$y = \alpha + \sum_{j=1}^{k-1} \beta_j x_j + e$$

where y is the outcome variable and $x_j = 1$ if subject is in group $(j + 1)$ and $= 0$ otherwise, $j = 1, \ldots, k - 1$.

The Between SS and Within SS for the one-way ANOVA model in procedure 1 are the same as the Regression SS and Residual SS for the multiple linear-regression model in procedure 2. The F statistics and p-values are the same as well.

To compare the underlying mean of the $(j + 1)$th group versus the reference group, we can use one of two equivalent procedures:

(3) We can use the LSD procedure based on one-way ANOVA, where we compute the t statistic

$$t = \frac{\bar{y}_{j+1} - \bar{y}_1}{s\sqrt{1/n_{j+1} + 1/n_1}} \sim t_{n-k}$$

and $\quad s^2 = \text{Within MS}, n = \sum_{j=1}^{k} n_j$

(4) Or we can compute the t statistic

$$t = \frac{b_j}{se(b_j)} \sim t_{n-k}$$

The test statistics and the p-values are the same under procedures 3 and 4.

To compare the underlying mean of the $(j + 1)$th and $(l + 1)$th groups, we can use one of two equivalent procedures:

(5) We can use the LSD procedure based on one-way ANOVA, whereby we compute the t statistic

$$t = \frac{\bar{y}_{j+1} - \bar{y}_{l+1}}{s\sqrt{1/n_{j+1} + 1/n_{l+1}}} \sim t_{n-k}$$

and $\quad s^2 = \text{Within MS}$

(6) Or $\quad t = \dfrac{b_j - b_l}{se(b_j - b_l)} \sim t_{n-k}$

The standard error of $b_j - b_l$ and test statistic t can usually be obtained as a linear-contrast option for multiple-regression programs available in most statistical packages.

Another way to compute $se(b_j - b_l)$ is to print out the **variance–covariance matrix** of the regression coefficients, which is an option in most statistical packages. If there are k regression coefficients, then the (j, j)th element of this matrix is the variance of $b_j = Var(b_j)$. The (j, l)th element of this matrix is the covariance between b_j and $b_l = Cov(b_j, b_l)$. Then using Equation 5.11, we have

$$Var(b_j - b_l) = Var(b_j) + Var(b_l) - 2Cov(b_j, b_l)$$

and

$$se(b_j - b_l) = \sqrt{Var(b_j - b_l)}$$

Example 12.15 **Environmental Health, Pediatrics** Compare the mean MAXFWT among control children, currently exposed children, and previously exposed children, respectively, using a multiple-regression approach.

Solution We set up dummy variables using the control group as the reference group and

$$\text{GRP2} = \begin{cases} 1 & \text{if currently exposed} \\ 0 & \text{otherwise} \end{cases}$$

$$\text{GRP3} = \begin{cases} 1 & \text{if previously exposed} \\ 0 & \text{otherwise} \end{cases}$$

The multiple-regression model is

$$y = \alpha + \beta_1 \times \text{GRP2} + \beta_2 \times \text{GRP3} + e$$

This model is fitted in Table 12.10 using SAS PROC REG.

The analysis-of-variance table reveals that there are significant differences among the three groups ($p = .0125$) and matches the p-value in Table 12.6 based on the F test for one-way ANOVA exactly. The parameter estimates reveal that currently exposed children (GRP2 = 1) have mean MAXFWT that is 7.51 taps/10 seconds slower than control children ($p = .009$), while previously exposed children (GRP3 = 1) have mean MAXFWT

TABLE 12.10 Multiple-regression model relating MAXFWT to group (control, currently exposed, previously exposed) ($n = 95$)

```
                                    The SAS System
Model: MODEL1
Dependent Variable: MAXFWT
                                 Analysis of Variance
                                   Sum of            Mean
           Source        DF       Squares          Square      F Value      Prob > F
           Model          2     966.79062       483.39531        4.598        0.0125
           Error         92    9671.14622       105.12115
           C Total       94   10637.93684

               Root MSE      10.25286      R-square      0.0909
               Dep Mean      52.85263      Adj R-sq      0.0711
               C.V.          19.39896

                                  Parameter Estimates
                      Parameter       Standard     T for H0:                     Standardized
Variable      DF       Estimate          Error   Parameter=0     Prob > |T|         Estimate
INTERCEP       1      55.095238     1.29173904        42.652         0.0001       0.00000000
GRP2           1      -7.507003     2.80217542        -2.679         0.0087      -0.27192418
GRP3           1      -5.695238     2.94561822        -1.933         0.0563      -0.19625106
```

that is 5.70 taps/10 seconds slower than control children, which is not quite statistically significant ($p = .056$).

If we wish to compare currently and previously exposed children, then we can use either a linear-contrast option for SAS PROC REG or a least squares means option for SAS PROC GLM. We have used the latter, as shown in Table 12.11. We see that the currently and previously exposed children are not significantly different ($p = .62$). In this case, the LSMEAN (least squares mean) is the same as the ordinary arithmetic mean. The standard error of a group mean $= s/\sqrt{n_j}$, $j = 1, \ldots, k$ where $s = \sqrt{\text{Residual MS}}$ $= \sqrt{\text{Error MS}}$. For example, for the control group, the $se = \sqrt{105.12115/63} = 1.292$.

TABLE 12.11 Comparison of specific group means for MAXFWT using the SAS PROC GLM (General Linear Model) procedure ($n = 95$)

```
                                    The SAS System
                            General Linear Models Procedure
                                 Least Squares Means
      GRP          MAXFWT       Std Err       Pr > |T|     Pr > |T| H0: LSMEAN(i)=LSMEAN(j)
                   LSMEAN        LSMEAN     H0:LSMEAN=0    i/j      1        2        3
   CONTROL     55.0952381     1.2917390          0.0001    1      .     0.0087   0.0563
   CUR EXP     47.5882353     2.4866840          0.0001    2  0.0087       .     0.6191
   PRV EXP     49.4000000     2.6472773          0.0001    3  0.0563   0.6191       .
```

Note: To ensure overall protection level, only probabilities associated with preplanned comparisons should be used.

12.5.3 One-Way Analysis of Covariance

In Equation 11.37, we compared the mean MAXFWT between exposed and control children after controlling for age and sex. We can also compare mean MAXFWT among the control group, the currently exposed group, and the previously exposed group, after controlling for age and gender by using the model:

EQUATION 12.20

$$y = \alpha + \beta_1 \times \text{GRP2} + \beta_2 \times \text{GRP3} + \beta_3 \times \text{age} + \beta_4 \times \text{sex} + e$$

The models in Equations 11.37 and 12.20 are referred to as **one-way analysis-of-covariance** (or **one-way ANCOVA**) **models.** In analysis of covariance, we wish to compare the mean of a continuous outcome variable among two or more groups defined by a single categorical variable, after controlling for other potential confounding variables (also called *covariates*). We have fitted this model using PROC GLM of SAS (see Table 12.12).

There are several hypotheses that can be tested using Equation 12.20. First, we can test the hypothesis H_0: $\beta_1 = \beta_2 = \beta_3 = \beta_4 = 0$ versus H_1: at least one $\beta_j \neq 0$. In words, this is a test of whether any of the variables in Equation 12.20 have any relationship to MAXFWT. The results are given at the top of Table 12.12 (F-value = 29.06, p-value = .0001) Thus, some of the variables are having a significant effect on MAXFWT. Second, to test for the effect of group after controlling for age and sex, we can test the hypothesis H_0: $\beta_1 = 0$, $\beta_2 = 0$, $\beta_3 \neq 0$, $\beta_4 \neq 0$ versus H_1: all $\beta_j \neq 0$, $j = 1, \ldots, 4$. This is given in the middle of Table 12.12, under the heading Type III SS. The Type III SS (sum of squares) provides an estimate of the effect of a specific risk factor, after controlling for the effects of all other variables in the model. In this case, the effect of group is significant after we control for age and sex (F statistic = 5.40, p = .0061). Third, we are interested in comparing specific categories of the group variable as shown in the middle of Table 12.12 under Least Squares Means. We see that both the currently exposed (p = .009) and the previously exposed (p = .018) groups have significantly lower mean MAXFWT scores than the control group, after adjusting for age and sex, while there is no significant difference between the currently exposed and previously exposed groups (p = .909). To estimate the mean difference between groups, we refer to the LSMEAN column in the middle of Table 12.12. In this case, the LSMEAN column is different from the ordinary arithmetic mean. Instead, it represents a mean value for each group that is, in a sense, adjusted for age and sex. Specifically, for each category of a categorical variable, the LSMEAN represents the average value of MAXFWT for a hypothetical sample of individuals who

(1) for each continuous variable in the model have a mean value equal to the overall sample mean (over all categories) for that variable,

(2) and for any other categorical variable in the model (with k categories) have a proportion of $1/k$ of individuals in each category.

Example 12.16 **Environmental Health, Pediatrics** Compute the LSMEAN for the control, currently exposed, and previously exposed groups in the multiple-regression model in Table 12.12.

TABLE 12.12 SAS PROC GLM output relating MAXFWT to group, age, and sex ($n = 95$)

```
                                  The SAS System
                         General Linear Models Procedure
Dependent Variable: MAXFWT
Source               DF       Sum of Squares        Mean Square      F Value        Pr > F
Model                 4        5995.49849815      1498.87462454        29.06        0.0001
Error                90        4642.43834395        51.58264827
Corrected Total      94       10637.93684211

Source               DF          Type III SS        Mean Square      F Value        Pr > F
GRP                   2         557.35257731       278.67628866         5.40        0.0061
AGEYR                 1        5027.97870763      5027.97870763        97.47        0.0001
SEX                   1         128.28154597       128.28154597         2.49        0.1183

                                          T for H0:        Pr > |T|      Std Error of
Parameter                  Estimate    Parameter=0                       Estimate
INTERCEPT              26.76514260 B          8.85          0.0001       3.02389880
GRP       CONTROL       4.99198543 B          2.42          0.0177       2.06543844
          CUR EXP      -0.29455702 B         -0.12          0.9085       2.55441917
          PRV EXP       0.00000000 B           .              .              .
AGEYR                   2.44032385            9.87          0.0001       0.24717388
SEX       FEMALE       -2.39491720 B         -1.58          0.1183       1.51865886
          MALE          0.00000000 B           .              .              .

                                  The SAS System
                         General Linear Models Procedure
                               Least Squares Means

     GRP          MAXFWT        Std Err      Pr > |T|     Pr > |T|   H0: LSMEAN(i)=LSMEAN(j)
                  LSMEAN         LSMEAN    H0:LSMEAN=0    i/j      1          2        3
   CONTROL    54.3977803      0.9139840        0.0001     1        .      0.0090   0.0177
   CUR EXP    49.1112378      1.7622906        0.0001     2   0.0090          .    0.9085
   PRV EXP    49.4057948      1.8550823        0.0001     3   0.0177     0.9085       .
```

Note: To ensure overall protection level, only probabilities associated with preplanned comparisons should be used.

Solution There is one continuous variable (age) and one other categorical variable (sex) in the model, besides group. The overall mean age for the entire sample ($n = 95$) is 9.768 years. Also, in the PROC GLM analyses in Table 12.12, the previously exposed group is the reference group (the SAS convention is to use the last group as the reference group) and females are coded as 1 and males as 0. Thus, the LSMEAN by group is

$$\text{Control: } 26.765 + 4.992 + 2.440(9.768) + \frac{1}{2}(-2.395)(1) = 54.4 \text{ taps/10 seconds}$$

$$\text{Currently exposed: } 26.765 - 0.295 + 2.440(9.768) + \frac{1}{2}(-2.395)(1) = 49.1 \text{ taps/10 seconds}$$

$$\text{Previously exposed: } 26.765 + 2.440(9.768) + \frac{1}{2}(-2.395)(1) = 49.4 \text{ taps/10 seconds}$$

Thus, both currently and previously exposed children have mean MAXFWT about 5 taps per 10 seconds slower than control children after adjusting for age and sex. Finally, we see that mean MAXFWT increases by 2.44 taps per 10 seconds for each year of age

($p < .001$) and that females have a lower mean MAXFWT score than males of the same age and group by 2.39 taps per 10 seconds, although the sex effect is not statistically significant ($p = .12$).

In this section, we have discussed the one-way analysis of covariance. The one-way analysis of covariance is used to assess mean differences among several groups after controlling for other covariates (which can be either continuous or categorical). Referring to the master flowchart (p. 776), we would answer (1) to (4) in the same way as for the analysis of variance (see p. 532). This leads us to the box entitled "Analysis of Variance." We could then answer "one" to (5) number of ways in which the categorical variable can be classified, yes to (6) outcome variable normal, or can central-limit theorem be expected to hold? and yes to (7) other covariates to be controlled for? This leads us to the box entitled "Analysis of Covariance."

SECTION 12.6 Two-Way Analysis of Variance

In Sections 12.1 through 12.5, the relationship between pulmonary function and cigarette smoking was used to illustrate the fixed-effects one-way analysis of variance. In this example, groups were defined by only one variable, cigarette smoking. In some instances, the groups being considered can be classified by two different variables and thus can be arranged in the form of an $R \times C$ contingency table. We would like to be able to look at the effects of each variable after controlling for the effects of the other variable. The latter type of data is usually analyzed using a technique known as the **two-way analysis of variance.**

Example 12.17 **Hypertension, Nutrition** A study was performed to look at the level of blood pressure in two different vegetarian groups, both compared with each other and with normals. A group of 226 strict vegetarians (SV), who eat no animal products of any kind, 63 lactovegetarians (LV), who eat dairy products but no other animal foods, and 460 normals (NOR), who eat a standard American diet, provided data for the study. Mean systolic blood pressure by dietary group and sex is given in Table 12.13.

TABLE 12.13 **Mean systolic blood pressure by dietary group and sex**

		Sex	
Dietary group		Male	Female
SV	mean	109.9	102.6
	n	138	88
LV	mean	115.5	105.2
	n	26	37
NOR	mean	128.3	119.6
	n	240	220

We are interested in the effects of sex and dietary group on the level of systolic blood pressure. The effects of sex and dietary group may by independent, or they may be related or "interact" with each other. One approach to this problem is to construct a two-way ANOVA model predicting mean systolic blood-pressure level as a function of sex and dietary group.

DEFINITION 12.8

An **interaction effect** between two variables is defined as one in which the effect of one variable depends on the level of the other variable.

Example 12.18

Hypertension, Nutrition Hypothesize that SV (strict vegetarian) males have mean systolic blood-pressure levels that are 10 mm Hg lower than those of normal males, whereas SV females have mean systolic blood-pressure levels identical to those of normal females. This relationship would be an example of an interaction effect between sex and dietary group, because the effect of diet on blood pressure would be different for males and females.

In general, if an interaction effect is present, then it becomes difficult to interpret the separate (or main) effects of each variable, because the effect of one factor (e.g., dietary group) depends on the level of the other factor (e.g., sex).

The general model for the two-way analysis of variance is given as follows:

EQUATION 12.21

Two-Way Analysis of Variance—General Model

$$y_{ijk} = \mu + \alpha_i + \beta_j + \gamma_{ij} + e_{ijk}$$

where

y_{ijk} is the systolic blood pressure of the kth person in the ith dietary group and the jth sex group

μ is a constant

α_i is a constant representing the effect of dietary group

β_j is a constant representing the effect of sex

γ_{ij} is a constant representing the interaction effect between dietary group and sex

e_{ijk} is an error term, which is assumed to be normally distributed with mean 0 and variance σ^2

By convention,

$$\sum_{i=1}^{r} \alpha_i = \sum_{j=1}^{c} \beta_j = 0, \quad \sum_{j=1}^{c} \gamma_{ij} = 0 \quad \text{for all } i$$

$$\sum_{i=1}^{r} \gamma_{ij} = 0 \quad \text{for all } j$$

Thus, from Equation 12.21, y_{ijk} is normally distributed with mean $\mu + \alpha_i + \beta_j + \gamma_{ij}$ and variance σ^2.

12.6.1 Hypothesis Testing in Two-Way ANOVA

Let us denote the mean systolic blood pressure for the ith row and jth column by $\bar{y}_{ij}$, the mean systolic blood pressure for the ith row by $\bar{y}_{i.}$, the mean systolic blood pressure for the jth column by $\bar{y}_{.j}$, and the overall mean by $\bar{y}_{..}$. The deviation of an individual observation from the overall mean can be represented as follows:

EQUATION 12.22

$$y_{ijk} - \bar{y}_{..} = \left(y_{ijk} - \bar{y}_{ij}\right) + \left(\bar{y}_{i.} - \bar{y}_{..}\right) + \left(\bar{y}_{.j} - \bar{y}_{..}\right) + \left(\bar{y}_{ij} - \bar{y}_{i.} - \bar{y}_{.j} + \bar{y}_{..}\right)$$

DEFINITION 12.9 The first term on the right-hand side $(y_{ijk} - \bar{y}_{ij})$ represents the deviation of an individual observation from the group mean for that observation. The expression is an indication of *within-group variability* and is called the **error term.**

DEFINITION 12.10 The second term on the right-hand side $(\bar{y}_{i.} - \bar{y}_{..})$ represents the deviation of the mean of the ith row from the overall mean and is called the **row effect.**

DEFINITION 12.11 The third term on the right-hand side $(\bar{y}_{.j} - \bar{y}_{..})$ represents the deviation of the mean of the jth column from the overall mean and is called the **column effect.**

DEFINITION 12.12 The fourth term on the right-hand side

$$\left(\bar{y}_{ij} - \bar{y}_{i.} - \bar{y}_{.j} + \bar{y}_{..}\right) = \left(\bar{y}_{ij} - \bar{y}_{i.}\right) - \left(\bar{y}_{.j} - \bar{y}_{..}\right)$$

represents the deviation of the column effect in the ith row $(\bar{y}_{ij} - \bar{y}_{i.})$ from the overall column effect $(\bar{y}_{.j} - \bar{y}_{..})$ and is called the **interaction effect.**

We would like to test the following hypotheses concerning these data:

(1) Test for the presence of row effects: H_0: all $\alpha_i = 0$ versus H_1: at least one $\alpha_i \neq 0$. This is a test for the effect of dietary group on systolic blood-pressure level after controlling for the effect of sex.

(2) Test for the presence of column effects: H_0: all $\beta_j = 0$ versus H_1: at least one $\beta_j \neq 0$. This is a test for the effect of sex on systolic blood-pressure level after controlling for the effect of dietary group.

(3) Test for the presence of interaction effects: H_0: all $\gamma_{ij} = 0$ versus H_1: at least one $\gamma_{ij} \neq 0$. This is a test of whether or not there is a differential effect of dietary group between males and females. For example, dietary group may have an effect on systolic blood pressure only among men.

For simplicity, we have ignored the interaction term in subsequent analyses. The SAS General Linear Model procedure (PROC GLM) has been used to analyze the data. In particular, two "indicator" or "dummy" variables were set up to represent study group (x_1, x_2), whereby

$x_1 = 1$ if a person is in the first (SV) group
$= 0$ otherwise

$x_2 = 1$ if a person is in the second (LV) group
$= 0$ otherwise

where the normal group is the reference group. A variable x_3 is also included to represent sex, whereby

$x_3 = 1$ if male
$= 0$ if female

The multiple-regression model can then be written as

EQUATION 12.23

$$y = \alpha + \beta_1 x_1 + \beta_2 x_2 + \beta_3 x_3 + e$$

The results from using the SAS GLM procedure are shown in Table 12.14. The program first provides a test of the overall hypothesis H_0: $\beta_1 = \beta_2 = \beta_3 = 0$ versus H_1: at least one of the $\beta_j \neq 0$, as given in Equation 11.30. The F statistic corresponding to this test is $105.85 \sim F_{3,\,745}$ under H_0, with p-value $< .001$. Thus, at least one of the effects (study group or sex) is significant. In the second part of the display, the program lists the type III SS and the corresponding F statistic (F-value) and p-value ($Pr > F$). The type III SS provides an estimate of the effects of specific risk factors after controlling for the effects of all other variables in the model. Thus, to test the effect of study group after controlling for sex, we wish to test the hypothesis H_0: $\beta_1 = \beta_2 = 0$, $\beta_3 \neq 0$, versus H_1: at least one of β_1, $\beta_2 \neq 0$, $\beta_3 \neq 0$. The F statistic for this comparison is obtained by dividing the study MS = (51,806.42/2) = 25,903.21 by the error MS = 195.89, yielding $132.24 \sim F_{2,\,745}$ under H_0, and a p-value ($Pr > F$) of $<.001$. Thus, there are highly significant effects of dietary group on systolic blood pressure even after controlling for the effect of sex. Similarly, to test for the effect of sex, we test the hypothesis H_0: $\beta_3 = 0$, at least one β_1, $\beta_2 \neq 0$, versus H_1: $\beta_3 \neq 0$, at least one β_1, $\beta_2 \neq 0$. The F statistic for the sex effect is given by (13,056/1)/195.89 = $66.65 \sim F_{1,745}$ under H_0, $p < .001$. Thus, there are highly significant effects of sex after controlling for the effect of dietary group, with males having higher blood pressure than females. SAS also displays a type I SS as well as an associated F statistic and p-value. The purpose of the type I SS is to enter and test the variables in the order specified by the user. In this case, study group was specified first, and sex was specified second. Thus, the effect of study group is assessed first (without controlling for sex), yielding an F statistic of $125.45 \sim F_{2,745}$ under H_0, $p < .001$. Second, the effect of sex is assessed after controlling for study group. This is the same hypothesis as was tested above using the type III SS. In general, except for the last user-specified risk factor, results from type I SS (where all variables above the current variable on the user-specified variable list are controlled for) and type III SS (where *all* other variables in the model are controlled for) will not necessarily be the same. Usually, unless we are interested in entering the variables in a prespecified order, hypothesis testing using the type III SS will be of greater interest.

TABLE 12.14 SAS GLM procedure output illustrating the effects of study group and sex on systolic blood pressure using the data set in Table 12.13

SAS
GENERAL LINEAR MODELS PROCEDURE

DEPENDENT VARIABLE: MSYS

SOURCE	DF	SUM OF SQUARES	MEAN SQUARE	F VALUE	PR > F	R-SQUARE	C.V.
MODEL	3	62202.79213079	20734.26404360	105.85	0.0001	0.298854	11.8858
ERROR	745	145934.76850283	195.88559531		ROOT MSE		MSYS MEAN
CORRECTED TOTAL	748	208137.56063362			13.99591352		117.75303516

SOURCE	DF	TYPE I SS	F VALUE	PR > F	DF	TYPE III SS	F VALUE	PR > F
STUDY	2	49146.49426085	125.45	0.0001	2	51806.42069945	132.24	0.0001
SEX	1	13056.29786994	66.65	0.0001	1	13056.29786994	66.65	0.0001

STUDY PROB > |T|	SV	LV	NOR
SV	.	0.0425	0.0001
LV	0.0425	.	0.0001
NOR	0.0001	0.0001	.

SEX PROB > |T|	MALE	FEMALE
MALE	.	0.0001
FEMALE	0.0001	.

PARAMETER		ESTIMATE	T FOR H0: PARAMETER=0	PR > |T|	STD ERROR OF ESTIMATE
INTERCEPT		119.75747587	141.53	0.0001	0.84614985
STUDY	SV	-17.86546724	-15.66	0.0001	1.14061756
	LV	-13.79147908	-7.32	0.0001	1.88356205
	NOR	0.00000000	.	.	.
SEX	MALE	8.42854624	8.16	0.0001	1.03239026
	FEMALE	0.00000000	.	.	.

Although there was a significant effect of study group after controlling for sex, this does not identify which specific dietary groups differ from one another on systolic blood pressure. For this purpose, t tests are provided comparing specific dietary groups (1 = SV, 2 = LV, 3 = NOR) after controlling for sex. Refer to the 3×3 table listed for STUDY under PROB > $|T|$. The (two-tailed) p-value comparing dietary group i versus dietary group j is given in the (i, j) cell of the table [as well as the (j, i) cell]. Thus referring to the (1, 2) cell, we see that the mean systolic blood pressure of people in group 1 (SV) is significantly different from people in group 2 (LV) after controlling for sex ($p = .0425$). Similarly, referring to the (1, 3) or (3, 1) cells, we see that the mean systolic blood pressure of people in group 1 (SV) is significantly different from people in group 3 (NORMAL) ($p = .0001$). Similar results are obtained from a comparison of people in groups 2 (LV) and 3 (NOR). Furthermore, a 2×2 table is listed for the sex effect, yielding a p-value for a comparison of the two sexes [refer to the (1, 2) or (2, 1) cell of the table] after controlling for the effect of study group ($p = .0001$). This test is actually superfluous in this instance, because there were only two groups under sex and the F test under type III SS for the sex effect is equivalent to the sex-effect t test.

Finally, note that at the bottom of the display there are estimates of the regression parameters as well as their standard errors and associated t statistics. These have a similar interpretation to that of the multiple-regression parameters in Definition 11.16. In particular, the regression coefficient $b_1 = -17.9$ mm Hg is an estimate of the difference in mean systolic blood pressure between the SV and NOR groups after controlling for the effect of sex. Similarly, the regression coefficient $b_2 = -13.8$ mm Hg is an estimate of the difference in mean systolic blood pressure between the LV and NOR groups after controlling for the effect of sex. Also, the estimated difference in mean systolic blood pressure between the SV and LV groups is given by $[-17.9 - (-13.8)] = -4.1$ mm Hg; thus, the strict vegetarians on average have systolic blood pressure 4.1 mm Hg lower than the lactovegetarians after controlling for the effect of sex. Since there was no explicit parameter entered for the third study group (the normal group), the program lists the default value of 0. Finally, the regression coefficient $b_3 = 8.4$ mm Hg tells us that males have mean systolic blood pressure 8.4 mm Hg higher than females, even after controlling for the effect of study group.

It is possible to assess interaction effects for two-way ANOVA models (e.g., using the SAS PROC GLM program), but these were not included in this example for the sake of simplicity. A more detailed discussion of two-way and higher-way ANOVA is given in Kleinbaum, Kupper, and Muller [2].

12.6.2 Two-Way Analysis of Covariance

We often will want to look at the relationship between one or more categorical variables and a continuous outcome variable. If there is one categorical variable, then one-way ANOVA can be used; if there are two (or more) categorical variables, then two-way (higher-way) ANOVA can be used. However, there may be other differences among the groups, which makes it difficult to interpret these analyses.

Example 12.19 **Hypertension, Nutrition** In Example 12.17, differences in mean systolic blood pressure (SBP) by dietary group and sex were presented using a two-way ANOVA model. Highly significant differences were found among dietary groups after controlling for sex, with mean SBP of SV < mean SBP of LV < mean SBP of NOR. However, there are other important differences between these groups, such as differences in weight, and possibly age, which might explain all or part of the apparent blood-pressure differences. How can we examine if these blood-pressure differences persist, after accounting for the confounding variables?

The multiple-regression model in Equation 12.23 can be extended to allow for the effects of other covariates using the **two-way** analysis of covariance. If weight is denoted by x_4 and age by x_5, then we have the multiple-regression model

EQUATION 12.24

$$y = \alpha + \beta_1 x_1 + \beta_2 x_2 + \beta_3 x_3 + \beta_4 x_4 + \beta_5 x_5 + e$$

where $e \sim N(0, \sigma^2)$. We have fit this model using the SAS PROC GLM program as shown in Table 12.15.

TABLE 12.15 SAS GLM procedure output illustrating the effects of study group, age, sex and weight on systolic blood pressure using the data set in Table 12.13

```
                                        SAS
                         GENERAL LINEAR MODELS PROCEDURE
DEPENDENT VARIABLE: MSYS
SOURCE            DF    SUM OF SQUARES      MEAN SQUARE  F VALUE        PR > F   R-SQUARE           C.V.
MODEL              5    85358.44910498   17071.68982100   103.16        0.0001   0.410402        10.9264
ERROR            741   122628.85342226     165.49103026                ROOT MSE                MSYS MEAN
CORRECTED TOTAL  746   207987.30252724                               12.86433171            117.73630968

SOURCE            DF         TYPE I SS   F VALUE     PR > F    DF       TYPE III SS    F VALUE     PR > F
STUDY GROUP        2    49068.28440076    148.25     0.0001     2     8257.21427825      24.95     0.0001
SEX                1    13092.51273176     79.11     0.0001     1     4250.57708379      25.68     0.0001
AGE                1    12978.84918739     78.43     0.0001     1    10524.41438768      63.60     0.0001
WGT                1    10218.80278507     61.75     0.0001     1    10218.80278507      61.75     0.0001

                                 STUDY PROB > |T|
                                               SV          LV        NOR
                                 SV             .      0.7012     0.0001
                                 LV        0.7012           .     0.0001
                                 NOR       0.0001      0.0001          .

                                 SEX    PROB > |T|
                                             MALE      FEMALE
                                 MALE           .      0.0001
                                 FEMALE    0.0001           .

                                                 T FOR H0:        PR > |T|        STD ERROR OF
         PARAMETER                    ESTIMATE   PARAMETER=0                          ESTIMATE
         INTERCEPT                 82.74987242         25.69          0.0001        3.22121552
         STUDY GROUP  SV           -8.22799340         -6.20          0.0001        1.32786689
                      LV           -8.95389632         -5.03          0.0001        1.78082376
                      NOR           0.00000000             .               .                 .
         SEX          MALE          5.50352855          5.07          0.0001        1.08593669
                      FEMALE        0.00000000             .               .                 .
         AGE                        0.47488301          7.97          0.0001        0.05954906
         WGT                        0.13011703          7.86          0.0001        0.01655851
```

Note from the top of Table 12.15 that the overall model is highly significant (*F*-value = 103.16, $p = .0001$), indicating that some of the variables are having a significant effect on SBP. To identify the effects on specific variables, refer to the type III SS. Note that each of the risk factors has a significant effect on SBP after controlling for the effects of all other variables in the model ($p = .0001$). Finally, of principal importance is whether there are differences in mean blood pressure by dietary group after controlling for the effects of age, sex, and weight. In this regard, different conclusions are reached from those reached in Table 12.14. Referring to the t statistics for STUDY, we see that there are no significant differences in mean SBP between the strict vegetarians (group 1) and the lactovegetarians (group 2) after controlling for the other variables ($p = .7012$). There are still highly significant differences between each of the vegetarian groups and normals ($p = .0001$). Thus, there must have been differences in either age and/or weight between the SV and LV groups, which accounted for the significant blood-pressure difference between these groups in Table 12.14. Finally, the estimates of specific regression parameters are given in the bottom of Table 12.15. Note that after controlling for age, sex, and weight, the estimated differences in mean SBP between the SV and NOR groups = $\beta_1 = -8.2$ mm Hg, between the LV and NOR groups = $\beta_2 = -9.0$ mm Hg, and between the SV and LV groups = $\beta_1 - \beta_2 = -8.23 - (-8.95) = 0.7$ mm Hg. These differences are all much smaller than the estimated differences in Table 12.14, of –17.9 mm Hg, –13.8 mm Hg, and –4.1 mm Hg, respectively, where age and weight were not controlled for. The difference in mean SBP between males and females is also much smaller in Table 12.15 after controlling for age and weight (5.5 mm Hg) than in Table 12.14 (8.4 mm Hg), where these factors were not controlled for. Also, we see from Table 12.15 that the estimated effects of age and weight on mean SBP are 0.47 mm Hg per year and 0.13 mm Hg per lb, respectively. Thus, it is important to control for the effects of possible explanatory variables in performing regression analyses.

In this section, we have learned about the two-way analysis of variance (ANOVA) and two-way analysis of covariance (ANCOVA). The two-way analysis of variance is used when we wish to simultaneously relate a normally distributed outcome variable to two categorical variables of primary interest. The two-way analysis of covariance is used when we wish to simultaneously relate a normally distributed outcome variable to two categorical variables of primary interest and, in addition, wish to control for one or more other covariates, which may be continuous or categorical. We saw that both two-way analysis of variance and two-way analysis of covariance models can be represented as special cases of multiple-regression models.

Referring to the master flowchart at the end of the book (p. 776), we would answer (1) to (4) in the same way as for the analysis of variance (see p. 532). This leads us to the box entitled "Analysis of Variance." We would then answer "two" to (5) number of ways in which the categorical variable can be classified, and yes to (6) outcome variable normal, or can central-limit theorem be expected to hold? If we have no other covariates to control for, then we answer no to (7) other covariates to be controlled for? and are led to the box entitled "Two-Way Analysis of Variance." If we have other covariates to be controlled for, then we answer yes to (7) and are led to the box entitled "Two-Way Analysis of Covariance."

If we wish to study the primary effect of more than two categorical variables as predictors of a continuous outcome variable, then two-way ANOVA and two-way ANCOVA generalize to multiway ANOVA and multiway ANCOVA, respectively. This is beyond the scope of this book; see [2].

SECTION 12.7 The Kruskal-Wallis Test

In some instances we wish to compare means among more than two samples, but either the underlying distribution is far from being normal or we have ordinal data. In these situations, a nonparametric alternative to the one-way ANOVA described in Sections 12.1–12.5 of this chapter must be used.

Example 12.20 **Ophthalmology** Arachidonic acid is well known to have an effect on ocular metabolism. In particular, topical application of arachidonic acid has caused lid closure, itching, and ocular discharge, among other effects. A study was conducted to compare the anti-inflammatory effects of four different drugs in albino rabbits after administration of arachidonic acid [3]. Six rabbits were studied in each group. Different rabbits were used in each of the four groups. For each animal in a group, one of the four drugs was administered to one eye and a saline solution was administered to the other eye. Ten minutes later arachidonic acid (sodium arachidonate) was administered to both eyes. Both eyes were evaluated every 15 minutes thereafter for lid closure. At each assessment the lids of both eyes were examined and a lid-closure score from 0 to 3 was determined, where 0 = eye completely open, 3 = eye completely closed, and 1, 2 = intermediate states. The measure of effectiveness (x) is the change in lid-closure score (from baseline to follow-up) in the treated eye minus the change in lid-closure score in the saline eye. A high value for x is indicative of an effective drug. The results, after 15 minutes of follow-up, are presented in Table 12.16. Since the scale of measurement was ordinal (0, 1, 2, 3), the use of a nonparametric technique to compare the four treatment groups is appropriate.

TABLE 12.16 **Ocular anti-inflammatory effects of four drugs on lid closure after administration of arachidonic acid**

	Indomethicin		Aspirin		Piroxicam		BW755C	
Rabbit Number	Score[a]	Rank	Score	Rank	Score	Rank	Score	Rank
1	+2	13.5	+1	9.0	+3	20.0	+1	9.0
2	+3	20.0	+3	20.0	+1	9.0	0	4.0
3	+3	20.0	+1	9.0	+2	13.5	0	4.0
4	+3	20.0	+2	13.5	+1	9.0	0	4.0
5	+3	20.0	+2	13.5	+3	20.0	0	4.0
6	0	4.0	+3	20.0	+3	20.0	–1	1.0

[a](Lid-closure score at baseline – lid-closure score at 15 minutes)$_{\text{drug eye}}$ – (lid-closure score at baseline – lid-closure score at 15 minutes)$_{\text{saline eye}}$

We would like to generalize the Wilcoxon rank-sum test to enable us to compare more than two samples. To accomplish this aim, the observations in all treatment groups are pooled and ranks are assigned to each observation in the combined sample. The average ranks ($\overline{R}_i$) in the individual treatment groups are then compared. If the average ranks are close to each other, then H_0, that the treatments are equally effective, will be accepted. If the average ranks are far apart, then H_0 will be rejected and we will conclude that at least some of the treatments are different. The test procedure for accomplishing this goal is known as the Kruskal-Wallis test.

EQUATION 12.25

The Kruskal-Wallis Test To compare the means of k samples ($k > 2$) using nonparametric methods, use the following procedure:

(1) Pool the observations over all samples, thus constructing a combined sample of size $N = \sum n_i$.

(2) Assign ranks to the individual observations, using the average rank in the case of tied observations.

(3) Compute the rank sum R_i for each of the k samples.

(4) If there are no ties, compute the test statistic

$$H = H^* = \frac{12}{N(N+1)} \times \sum_{i=1}^{k} \frac{R_i^2}{n_i} - 3(N+1)$$

If there are ties, compute the test statistic

$$H = \frac{H^*}{1 - \frac{\sum_{j=1}^{g}\left(t_j^3 - t_j\right)}{N^3 - N}}$$

where t_j refers to the number of observations (i.e., the frequency) with the same value in the jth cluster of tied observations and g is the number of tied groups.

(5) For a level α test, if

$$H > \chi^2_{k-1,1-\alpha}$$

then reject H_0; if

$$H \le \chi^2_{k-1,1-\alpha}$$

then accept H_0.

(6) To assess statistical significance, the p-value is given by

$$p = Pr\left(\chi^2_{k-1} > H\right)$$

(7) This test procedure should be used only if minimum $n_i \ge 5$ (i.e., if the smallest sample size for an individual group is at least 5).

The acceptance and rejection regions for this test are depicted in Figure 12.11. The computation of the exact p-value is given in Figure 12.12.

FIGURE 12.11 **Acceptance and rejection regions for the Kruskal-Wallis test**

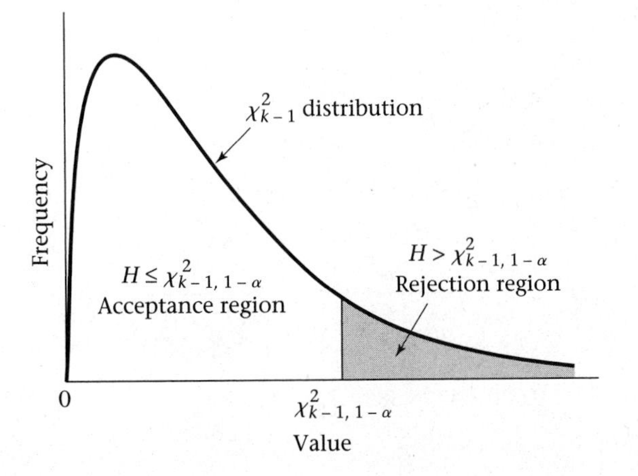

FIGURE 12.12 Computation of the exact p-value for the Kruskal-Wallis test

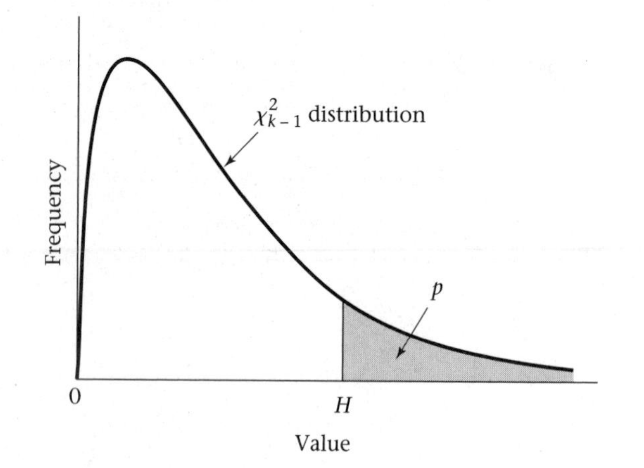

Example 12.21 **Ophthalmology** Apply the Kruskal-Wallis test procedure to the ocular data in Table 12.16 and assess the statistical significance of the results.

Solution First pool the samples together and assign ranks to the individual observations. This procedure is performed in Table 12.17 with ranks given in Table 12.16.

TABLE 12.17 Assignment of ranks to the individual observations in Table 12.16

Lid-closure score	Frequency	Range of ranks	Average rank
–1	1	1	1.0
0	5	2–6	4.0
+1	5	7–11	9.0
+2	4	12–15	13.5
+3	9	16–24	20.0

Then compute the rank sum in the four treatment groups:

$$R_1 = 13.5 + 20.0 + \cdots + 4.0 = 97.5$$

$$R_2 = 9.0 + 20.0 + \cdots + 20.0 = 85.0$$

$$R_3 = 20.0 + 9.0 + \cdots + 20.0 = 91.5$$

$$R_4 = 9.0 + 4.0 + \cdots + 1.0 = 26.0$$

Since there are ties, compute the Kruskal-Wallis test statistic H as follows:

$$H = \frac{\frac{12}{24 \times 25} \times \left(\frac{97.5^2}{6} + \frac{85.0^2}{6} + \frac{91.5^2}{6} + \frac{26.0^2}{6} \right) - 3(25)}{1 - \frac{(5^3 - 5) + (5^3 - 5) + (4^3 - 4) + (9^3 - 9)}{24^3 - 24}}$$

$$= \frac{0.020 \times 4296.583 - 75}{1 - \frac{1020}{13,800}} = \frac{10.932}{0.926} = 11.804$$

To assess statistical significance, compare H with a chi-square distribution with $k - 1 = 4 - 1 = 3$ *df*. Note from Table 6 in the Appendix that $\chi^2_{3,.99} = 11.34$, $\chi^2_{3,.995} = 12.84$. Since $11.34 < H < 12.84$, it follows that $.005 < p < .01$. Thus, there is a significant difference in the anti-inflammatory potency of the four drugs.

Note that although the sample sizes in the individual treatment groups were the same in Table 12.16, the Kruskal-Wallis test procedure can, in fact, be used for samples of unequal size. Also, if there are no ties, the Kruskal-Wallis test statistic H in Equation 12.25 can be written in the form

EQUATION 12.26

$$H = \frac{12}{N(N+1)} \sum_{i=1}^{k} n_i \left(\overline{R}_i - \overline{\overline{R}}\right)^2$$

where $\overline{R}_i$ = average rank in the ith sample and $\overline{\overline{R}}$ = average rank over all samples combined. Thus, if the average rank is about the same in all samples, then $\left|\overline{R}_i - \overline{\overline{R}}\right|$ will tend to be small and H_0 will be accepted. On the contrary, if the average rank is very different across samples, then $\left|\overline{R}_i - \overline{\overline{R}}\right|$ will tend to be large and H_0 will be rejected.

The test procedure in Equation 12.25 is only applicable if minimum $n_i \geq 5$. If one of the sample sizes is smaller than 5, then either the sample should be combined with another sample, or special small-sample tables should be used. Table 15 in the Appendix provides critical values for selected sample sizes for the case of three samples (i.e., $k = 3$). The procedure for using this table is as follows:

(1) Reorder the samples so that $n_1 \leq n_2 \leq n_3$, that is, so that the first sample has the smallest sample size and the third sample has the largest sample size.

(2) For a level α test, refer to the α column and the row corresponding to the sample sizes n_1, n_2, n_3 to find the critical value c.

(3) If $H \geq c$, then reject H_0 at level α (i.e., $p < \alpha$); if $H < c$, then accept H_0 at level α (i.e., $p \geq \alpha$).

Example 12.22 Suppose there are three samples of size 2, 4, and 5 and $H = 6.141$. Assess the statistical significance of the results.

Solution Refer to the $n_1 = 2$, $n_2 = 4$, $n_3 = 5$ row. The critical values for $\alpha = .05$ and $\alpha = .02$ are 5.273 and 6.541, respectively. Since $H \geq 5.273$, it follows that the results are statistically significant ($p < .05$). Since $H < 6.541$, it follows that $p \geq .02$. Thus, $.02 \leq p < .05$.

12.7.1 Comparison of Specific Groups Under the Kruskal-Wallis Test

In Example 12.21 we determined that the treatments in Table 12.16 were not all equally effective. To determine which pairs of treatment groups are different, use the following procedure:

EQUATION 12.27

Comparison of Specific Groups Under the Kruskal-Wallis Test (Dunn Procedure) To compare the ith and jth treatment groups under the Kruskal-Wallis test, use the following procedure:

(1) Compute

$$z = \frac{\overline{R}_i - \overline{R}_j}{\sqrt{\frac{N(N+1)}{12} \times \left(\frac{1}{n_i} + \frac{1}{n_j}\right)}}$$

(2) For a two-sided level α test, if

$$|z| > z_{1-\alpha^*}$$

then reject H_0; if $|z| \le z_{1-\alpha^*}$

then accept H_0, where $\alpha^* = \frac{\alpha}{k(k-1)}$

The acceptance and rejection regions for this test are depicted in Figure 12.13.

FIGURE 12.13 **Acceptance and rejection regions for the Dunn procedure**

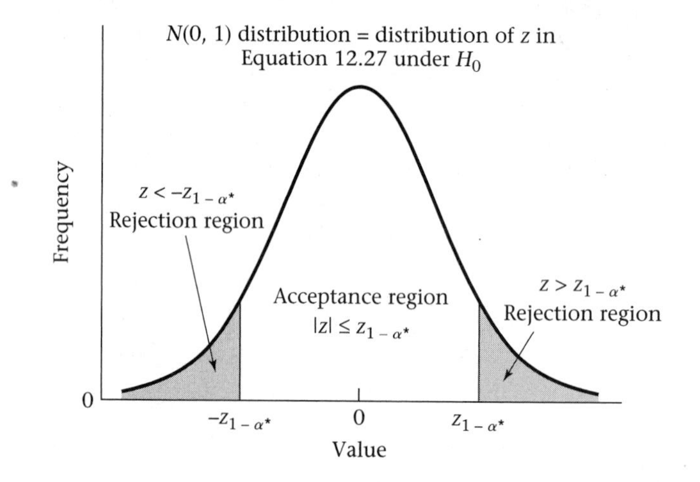

Example 12.23 **Ophthalmology** Determine which specific groups are different, using the ocular data in Table 12.16.

Solution From Example 12.21,

$$\overline{R}_1 = \frac{97.5}{6} = 16.25$$

$$\overline{R}_2 = \frac{85.0}{6} = 14.17$$

$$\overline{R}_3 = \frac{91.5}{6} = 15.25$$

$$\overline{R}_4 = \frac{26.0}{6} = 4.33$$

Therefore, the following test statistics are used to compare each pair of groups:

$$\text{Groups 1 and 2: } z_{12} = \frac{16.25 - 14.17}{\sqrt{\frac{24 \times 25}{12} \times \left(\frac{1}{6} + \frac{1}{6}\right)}} = \frac{2.08}{4.082} = 0.51$$

$$\text{Groups 1 and 3: } z_{13} = \frac{16.25 - 15.25}{4.082} = \frac{1.0}{4.082} = 0.24$$

$$\text{Groups 1 and 4: } z_{14} = \frac{16.25 - 4.33}{4.082} = \frac{11.92}{4.082} = 2.92$$

$$\text{Groups 2 and 3: } z_{23} = \frac{14.17 - 15.25}{4.082} = \frac{-1.08}{4.082} = -0.27$$

$$\text{Groups 2 and 4: } z_{24} = \frac{14.17 - 4.33}{4.082} = \frac{9.83}{4.082} = 2.41$$

$$\text{Groups 3 and 4: } z_{34} = \frac{15.25 - 4.33}{4.082} = \frac{10.92}{4.082} = 2.67$$

The critical value for $\alpha = .05$ is given by $z_{1-\alpha^*}$, where

$$\alpha^* = \frac{.05}{4 \times 3} = .0042$$

From Table 3 in the Appendix, $\Phi(2.635) = .9958 = 1 - .0042$. Thus $z_{1-.0042} = z_{.9958} = 2.635$ is the critical value. Since z_{14} and z_{34} are greater than the critical value, it follows that Indomethicin (group 1) and Piroxicam (group 3) have significantly better anti-inflammatory properties than BW755C (group 4), whereas the other treatment comparisons are not statistically significant.

In this section, we have discussed the Kruskal-Wallis test, which is a nonparametric test for the comparison of the median of several groups. It is used as an alternative to one-way ANOVA when the assumption of normality is questionable. Referring to the master flowchart at the end of the book (p. 776), we would answer (1) to (4) in the same way as for the analysis of variance (see p. 532). This leads to the box entitled "Analysis of Variance." We would then answer "one" to (5) number of ways in which the categorical variable can be classified, and no to (6) outcome variable normal, or can central-limit theorem be assumed to hold? This leads to the successive boxes entitled "Nonparametric ANOVA" and "Kruskal-Wallis Test."

SECTION 12.8 One-Way ANOVA—The Random-Effects Model

In Example 12.1, we studied the effect of both active and passive smoking on the level of pulmonary function. We were specifically interested in the difference in pulmonary function between the passive smoking (PS) and the nonsmoking (NS) groups. This is an example of the **fixed-effects** analysis-of-variance model, because the subgroups being compared have been fixed by the design of the study.

In other instances, we are interested in whether there are overall differences between groups and what percentage of the total variation is attributable to between-group versus within-group differences, but are not interested in comparing specific groups.

Example 12.24 **Endocrinology** The Nurses' Health Study is a large prospective study of approximately 100,000 American nurses who were sent a mail questionnaire every two years starting in 1976, inquiring about their health habits. One substudy is to obtain blood samples collected by mail from a subset of nurses and relate serum levels of various hormones to the development of disease. As a first step in this process, blood samples were obtained from 5 postmenopausal women. Each blood sample was split into two equal aliquots, which were sent in a blinded fashion to one laboratory for analysis. The same procedure was followed for each of four different laboratories. The goal of the study was to assess how much variation in the analyses was attributable to between-person versus within-person variation. Comparisons were made both between different hormones as well as between different laboratories [4].

Table 12.18 shows the reproducibility data for plasma estradiol from one laboratory. Can we estimate the degree of between-person and within-person variation from the data?

TABLE 12.18 **Reproducibility data for plasma estradiol (pg/mL), Nurses' Health Study**

Subject	Replicate 1	Replicate 2	Absolute value of difference between replicates	Mean value
1	25.5	30.4	4.9	27.95
2	11.1	15.0	3.9	13.05
3	8.0	8.1	0.1	8.05
4	20.7	16.9	3.8	18.80
5	5.8	8.4	2.6	7.10

It appears from Table 12.18 that the variation between replicates depends to some extent on the mean level. Subject 1, who has the largest absolute difference between replicates (4.9), also has the highest mean value (27.95). This is common with many laboratory measures. For this reason, we will analyze the reproducibility data on the ln scale. The justification for using this transformation is that it can be shown that if the within-person standard deviation is proportional to the mean level in the original scale, then the within-person standard deviation will be approximately independent of the mean level on the ln scale [5]. Using the ln scale will also enable us to estimate the coefficient of variation (a frequently used index in reproducibility studies).

To assess between- and within-person variability, consider the following model:

EQUATION 12.28

$$y_{ij} = \mu + \alpha_i + e_{ij},\ i = 1, \ldots, k, \text{ and } j = 1, \ldots, n_i$$

where

y_{ij} = jth replicate for ln (plasma estradiol) for the ith subject

α_i is a random variable representing between-subject variability, which is assumed to follow an $N(0, \sigma_A^2)$ distribution

e_{ij} is a random variable representing within-subject variability, which follows an $N(0, \sigma^2)$ distribution and is independent of α_i and any of the other e_{ij}

The model in Equation 12.28 is referred to as a **random-effects** one-way analysis-of-variance model. The underlying mean for the ith subject is given by $\mu + \alpha_i$, where α_i is drawn from a normal distribution with mean 0 and variance σ_A^2. Thus, two different individuals i_1, i_2 will have different underlying means $\mu + \alpha_{i_1}$ and $\mu + \alpha_{i_2}$, respectively. The extent of the between-subject variation is determined by σ_A^2. As σ_A^2 increases, the between-subject variation increases as well. The within-subject variation is determined by σ^2. Thus, if we have two replicates y_{i1}, y_{i2} from the same individual (i), they will be normally distributed with mean $\mu + \alpha_i$ and variance σ^2. An important goal in the random-effects analysis of variance is to test the hypothesis H_0: $\sigma_A^2 = 0$ versus H_1: $\sigma_A^2 > 0$. Under H_0, there is no underlying between-subject variation; all variation seen between individual subjects is attributable to within-person variation (or "noise"). Under H_1, there is a true underlying difference among means for individual subjects. How can we test these hypotheses? It can be shown under either hypothesis that

EQUATION 12.29

$$E(\text{Within MS}) = \sigma^2$$

In the balanced case, where there are an equal number of replicates per subject, it can be shown that under either hypothesis

EQUATION 12.30

$$E(\text{Between MS}) = \sigma^2 + n\sigma_A^2$$

where $n_1 = n_2 = \cdots = n_k = n$ = number of replicates per subject

In the unbalanced case, where there may be an unequal number of replicates per subject, it can be shown that under either hypothesis

EQUATION 12.31

$$E(\text{Between MS}) = \sigma^2 + n_0\sigma_A^2$$

where

$$n_0 = \left(\sum_{i=1}^{k} n_i - \sum_{i=1}^{k} n_i^2 \Big/ \sum_{i=1}^{k} n_i\right) \Big/ (k-1)$$

If the number of replicates are the same for each subject, then $n_1 = n_2 = \cdots = n_k = n$ and

$$\begin{aligned} n_0 &= \left(kn - kn^2/kn\right)/(k-1) \\ &= (kn - n)/(k-1) \\ &= n(k-1)/(k-1) = n \end{aligned}$$

Thus, the two formulas in Equations 12.30 and 12.31 agree in the balanced case. In general, in the unbalanced case, n_0 is always less than the average number of replicates per subject $\left(\bar{n} = \sum_{i=1}^{k} n_i / k\right)$, but the difference between n_0 and $\bar{n}$ is usually small.

We use the same test statistic (F = Between MS/Within MS) as was used for the fixed-effects model one-way ANOVA. Under the random-effects model, if H_1 is true $(\sigma_A^2 > 0)$, then F will be large, whereas if H_0 is true $(\sigma_A^2 = 0)$, then F will be small. It can be shown that under H_0, F will follow an F distribution with $k - 1$ and $N - k$ df, where $N = \sum_{i=1}^{k} n_i$. To estimate the variance components σ_A^2 and σ^2, we use Equations 12.29–12.31.

From Equation 12.29, an unbiased estimate of σ^2 is given by Within MS. From Equation 12.30, if we estimate σ^2 by Within MS, then in the balanced case

$$E\left(\frac{\text{Between MS} - \text{Within MS}}{n}\right)$$
$$= \frac{E(\text{Between MS} - \text{Within MS})}{n}$$
$$= \frac{\sigma^2 + n\sigma_A^2 - \sigma^2}{n} = \sigma_A^2$$

Thus, an unbiased estimate of σ_A^2 is given by $\hat{\sigma}_A^2$ = (Between MS – Within MS)/n. From Equation 12.31, an analogous result holds in the unbalanced case, where we replace n by n_0 in the estimation of σ_A^2. These results are summarized as follows:

EQUATION 12.32

One-Way Analysis of Variance—Random-Effects Model Suppose we have the model $y_{ij} = \mu + \alpha_i + e_{ij}$, $i = 1, \ldots, k$; $j = 1, \ldots, n_i$ where $\alpha_i \sim N(0, \sigma_A^2)$ and $e_{ij} \sim N(0, \sigma^2)$. To test the hypothesis H_0: $\sigma_A^2 = 0$ versus H_1: $\sigma_A^2 > 0$,

(1) Compute the test statistic $F = \dfrac{\text{Between MS}}{\text{Within MS}}$, which follows an $F_{k-1, N-k}$ distribution under H_0 where

$$\text{Between MS} = \sum_{i=1}^{k} n_i \left(\bar{y}_i - \bar{\bar{y}}\right)^2 \Big/ (k-1)$$

$$\text{Within MS} = \sum_{i=1}^{k} \sum_{j=1}^{n_i} \left(y_{ij} - \bar{y}_i\right)^2 \Big/ (N-k)$$

$$\bar{y}_i = \sum_{j=1}^{n_i} y_{ij} \Big/ n_i, \bar{\bar{y}} = \sum_{i=1}^{k} \sum_{j=1}^{n_i} y_{ij} \Big/ N = \sum_{i=1}^{k} n_i \bar{y}_i \Big/ N, \; N = \sum_{i=1}^{k} n_i$$

(2) If $F > F_{k-1, N-k, 1-\alpha}$, then reject H_0.
If $F \le F_{k-1, N-k, 1-\alpha}$, then accept H_0.

(3) The exact p-value is given by the area to the right of F under an $F_{k-1, N-k}$ distribution.

(4) The within-group variance component (σ^2) is estimated by the Within MS.

(5a) If we have a balanced design (i.e., $n_1 = n_2 = \cdots = n_k = n$), then the between-group variance component (σ_A^2) is estimated by

$$\hat{\sigma}_A^2 = \max\left[\left(\frac{\text{Between MS} - \text{Within MS}}{n}\right), 0\right]$$

(5b) If we have an unbalanced design (i.e., at least two of the n_i are unequal), then the between-group variance component (σ_A^2) is estimated by

$$\hat{\sigma}_A^2 = \max\left[\left(\frac{\text{Between MS} - \text{Within MS}}{n_0}\right), 0\right]$$

where

$$n_0 = \left(\sum_{i=1}^{k} n_i - \sum_{i=1}^{k} n_i^2 \Big/ \sum_{i=1}^{k} n_i\right) \Big/ (k-1)$$

Example 12.25 **Endocrinology** Test if the underlying mean plasma estradiol is the same for different subjects using the data in Table 12.18.

Solution We have used the SAS GLM (general linear model) procedure to perform the significance test in Equation 12.32 based on the ln estradiol values in Table 12.18. These results are given in Table 12.19.

TABLE 12.19 **Analysis of the plasma-estradiol data (ln scale) in Table 12.18 using the SAS GLM procedure**

```
                              The SAS System
                       General Linear Models Procedure
Dependent Variable: LESTRADL
Source             DF   Sum of Squares   Mean Square   F Value    Pr > F
Model               4       2.65774661    0.66443665     22.15    0.0022
Error               5       0.15001218    0.03000244
Corrected Total     9       2.80775879
         R-Square                 C.V.      Root MSE          LESTRADL Mean
         0.946572             6.744243    0.17321211             2.56829589
```

In Example 12.25 we have 5 subjects and 2 replicates per subject. The F statistic (given under F Value) = 22.15 ~ $F_{4,5}$ under H_0. The p-value = $Pr(F_{4,5} > 22.15) = .0022$ (given under $Pr > F$). Thus, there are significant differences among the underlying mean ln(plasma estradiol) values for different subjects.

Example 12.26 **Endocrinology** Estimate the between-person and within-person variance components for ln(plasma estradiol) using the data in Table 12.18.

Solution We have from Equation 12.32 that the within-person variance component is estimated by the Within Mean Square. In Table 12.19, this is referred to as the Mean Square for Error (or MSE) = 0.030. To estimate the between-person variance, we refer to Equation 12.32. Since this is a balanced design, we have

$$\hat{\sigma}_A^2 = \left[\max\left(\frac{\text{Between MS} - \text{Within MS}}{2}\right), 0\right]$$

$$= \frac{0.6644 - 0.0300}{2} = \frac{0.6344}{2} = 0.317$$

Thus, the between-person variance is about 10 times as large as the within-person variance, which indicates good reproducibility.

Alternatively, we could refer to the SAS GLM procedure output given in Table 12.20.

TABLE 12.20 **Representation of the Expected Mean Squares in terms of sources of variance using the data in Table 12.18**

```
The SAS System
General Linear Models Procedure
Source        Type III Expected Mean Square
PERSON        Var(Error) + 2 Var(PERSON)
```

We see that the PERSON (or Between Mean Square) gives an unbiased estimate of within-person variance + 2 between-person variance [which is denoted by Var(Error) + 2 Var(PERSON)] = $\sigma^2 + 2\sigma_A^2$. Thus, since we already have an estimate of Var(Error) we can compute Var(PERSON) by subtraction; that is, (PERSON Mean Square – Mean Square for Error)/2 = 0.317. The representation in Table 12.20 is most useful for unbalanced designs because it obviates the need for the user to compute n_0 in step 5b of Equation 12.32. In this case, the expected value of the PERSON Mean Square would be Var(Error) + n_0 Var(PERSON).

Another parameter that is often of interest in reproducibility studies is the coefficient of variation (CV). Generally speaking, coefficients of variation of <20% are desirable, whereas coefficients of variation of >30% are undesirable. The coefficient of variation in reproducibility studies is defined as

$$CV = 100\% \times \frac{\text{within-person standard deviation}}{\text{within-person mean}}$$

We could estimate the mean and standard deviation for each of the five subjects in Table 12.18 based on the raw plasma-estradiol values and average the individual CV estimates. However, if the standard deviation appears to increase as the mean increases, then a better estimate of the CV is given as follows [5]:

EQUATION 12.33

Estimation of the Coefficient of Variation in Reproducibility Studies Suppose we have k subjects enrolled in a reproducibility study where there are n_i replicates for the ith subject, $i = 1, \ldots, k$. To estimate the coefficient of variation,

(1) Apply the ln transformation to each of the values.

(2) Estimate the between- and within-subject variance components using a one-way random-effects model ANOVA as shown in Equation 12.32.

(3) The coefficient of variation in the *original scale* is estimated by

$$100\% \times \sqrt{\text{Within MS}} \quad \text{from step 2}$$

Example 12.27 **Endocrinology** Estimate the coefficient of variation for plasma estradiol given the data in Table 12.19.

Solution We have from Table 12.19 that the Within-person Mean Square based on ln-transformed plasma-estradiol values is given by 0.0300. Thus,

$$CV = 100\% \times \sqrt{0.0300} = 17.3\%$$

Alternatively, we could have used 100% × Root MSE = 100% × $\sqrt{\text{Mean Square for Error}}$ = 17.3%. Note that the *CV* given in Table 12.19 (6.74%) is *not* appropriate in this case, because it is simply based on 100% × $\sqrt{\text{MSE}}$/LESTRADL MEAN, which would give us the coefficient of variation for ln(plasma estradiol) rather than plasma estradiol itself. Similarly, if we had run the GLM procedure using the raw scale rather than the ln scale (see Table 12.21), the coefficient of variation given (16.4%) would also not be appropriate, because it is based on the assumption that the standard deviation is independent of the mean values, which is not true in the raw scale.

TABLE 12.21 **Analysis of the plasma-estradiol data (original scale) using the SAS GLM procedure**

```
                            The SAS System
                   General Linear Models Procedure
Dependent Variable: ESTRADIL
Source            DF   Sum of Squares      Mean Square   F Value    Pr > F
Model              4     593.31400000     148.32850000     24.55    0.0017
Error              5      30.21500000       6.04300000
Corrected Total    9     623.52900000
          R-Square               C.V.         Root MSE        ESTRADIL Mean
          0.951542           16.39928       2.45825141          14.99000000
```

In some examples, there are more than two sources of variation.

Example 12.28 **Hypertension** Suppose we obtain blood-pressure recordings from each of k subjects. We may ask each subject to return to the clinic at n_1 visits. At each of the n_1 visits, we may obtain n_2 blood-pressure readings. In this setting, we would be interested in three components of blood-pressure variation: (1) between persons, (2) between different visits for the same person, and (3) between different readings for the same person at the same visit.

This is referred to as a random-effects model analysis of variance with more than two levels of nesting. This type of problem is beyond the scope of this book. See Snedecor and Cochran [6] for a lucid description of this problem.

In this section, we have learned about the **one-way ANOVA random-effects model.** The random-effects model differs from the fixed-effects model in several important ways. First, with a random-effects model we are not interested in comparing mean levels of the outcome variable (e.g., estradiol) among specific levels of the grouping (or categorical variable). Thus, in Example 12.24, we were not interested in comparing mean estradiol levels among different specific women. Instead, the women in the study are considered a random sample of all women who could have participated in the study. It is usually a foregone conclusion that individual women will have different estradiol levels. Instead, what is of interest is estimating what proportion of the total variability of estradiol is attributable to between-person versus within-person variation. Conversely, in the fixed-effects ANOVA (e.g., Example 12.1) we were interested specifically in comparing mean FEF levels for nonsmokers versus passive smokers. In the fixed-effects case, the levels of the categorical variable have inherent meaning and the primary goal is to compare mean levels of the outcome variable (FEF) among different levels of the grouping variable.

SECTION 12.9 The Intraclass Correlation Coefficient

In Sections 11.7–11.8, we were concerned with Pearson correlation coefficients between two distinct variables denoted by x and y. For example, in Example 11.22 we were concerned with the correlation between cholesterol levels of a wife (x) and a husband (y). In Example 11.29 we were concerned with the correlation of body weight between a father (x) and first-born son (y). In some instances, we are interested in the correlation between variables that are not readily distinguishable from each other.

Example 12.29 **Endocrinology** In Example 12.24, we were concerned with the reproducibility of two replicate measures obtained from split samples of plasma estradiol from 5 women. In this instance, the replicate-sample determinations are indistinguishable from each other, and each plasma sample was split into two halves at random. Thus, it is impossible to specifically identify an x or y variable.

A more fundamental issue is that in Equation 11.17 the sample correlation coefficient was written in the form

$$s_{xy}/(s_x s_y)$$

Thus, from Equation 11.17 an alternative definition of the correlation coefficient r is the ratio of the sample covariance between x and y divided by the product of the standard deviation of x multiplied by the standard deviation of y. The implicit assumption in Equation 11.17 is that x and y are distinct variables and thus the sample mean and standard deviation are computed separately for x and y. In Example 12.24, we could, at random, assign one of the two replicates to be x and the other y for each woman and compute r in Equation 11.17 based on separate estimates of the sample mean and standard deviation for x and y. However, if x and y are indistinguishable from each other, a more efficient estimate of the mean and standard deviation can be obtained by using all available replicates for each woman. Thus, a special type of correlation is needed between repeated measures on the same subject that is referred to as an intraclass correlation coefficient.

DEFINITION 12.13

Suppose we have k subjects and obtain n_i replicates from the ith subject, $i = 1, \ldots, k$. Let y_{ij} represent the jth replicate from the ith subject. The correlation between two replicates from the same subject—i.e., between y_{ij} and y_{il} where $j \neq l$ and $1 \le j \le n_i$, $1 \le l \le n_i$—is referred to as the **intraclass correlation coefficient**, denoted by ρ_I.

If y_{ij} is distributed according to a one-way random-effects ANOVA model, where

$$y_{ij} = \mu + \alpha_i + e_{ij},\ \ \alpha_i \sim N(0, \sigma_A^2),\ \ e_{ij} \sim N(0, \sigma^2)$$

then it can be shown that $\rho_I = \sigma_A^2/(\sigma_A^2 + \sigma^2)$; i.e., ρ_I is the ratio of the between-person variance divided by the sum of the between-person and the within-person variance. The intraclass correlation coefficient is a measure of reproducibility of replicate measures from the same subject. It ranges between 0 and 1, with $\rho_I = 0$ indicating no reproducibility at all (i.e., large within-person variability and 0 between-person variability) and $\rho_I = 1$ indicating perfect reproducibility (i.e., 0 within-person variability and large between-person variability). According to Fleiss [7]:

EQUATION 12.34

Interpretation of Intraclass Correlation

$\rho_I < 0.4$ indicates poor reproducibility

$0.4 \le \rho_I < 0.75$ indicates fair to good reproducibility

$\rho_I \ge 0.75$ indicates excellent reproducibility

There are several methods of estimation for the intraclass correlation coefficient. The simplest and perhaps most widely used method is based on the one-way random-effects model ANOVA, which we discussed in Section 12.8.

EQUATION 12.35

Point and Interval Estimation of the Intraclass Correlation Coefficient Suppose that we have a one-way random-effects model ANOVA, where

$$y_{ij} = \mu + \alpha_i + e_{ij},\ \ e_{ij} \sim N(0, \sigma^2),\ \ \alpha_i \sim N(0, \sigma_A^2),\ i = 1, \ldots, k;\ j = 1, \ldots, n_i$$

The intraclass correlation coefficient $\rho_I = \sigma_A^2/(\sigma_A^2 + \sigma^2)$. A point estimate of ρ_I is given by

$$\hat{\rho}_I = \max\left[\hat{\sigma}_A^2/(\hat{\sigma}_A^2 + \hat{\sigma}^2), 0\right]$$

where $\hat{\sigma}_A^2$ and $\hat{\sigma}^2$ are the estimates of the between-subject and within-subject variance components from a one-way random-effects model ANOVA given in Equation 12.32. This estimator is sometimes referred to as the **analysis-of-variance estimator.**

An approximate two-sided $100\% \times (1-\alpha)$ CI for ρ_I is given by (c_1, c_2) where

$$c_1 = \max\left\{\frac{F/F_{k-1,N-k,1-\alpha/2}-1}{n_0+F/F_{k-1,N-k,1-\alpha/2}-1}, 0\right\}$$

$$c_2 = \max\left\{\frac{F/F_{k-1,N-k,\alpha/2}-1}{n_0+F/F_{k-1,N-k,\alpha/2}-1}, 0\right\}$$

F is the F statistic from the significance test of the hypothesis H_0: $\sigma_A^2 = 0$ versus H_1: $\sigma_A^2 > 0$ given in Equation 12.32, $N = \sum_{i=1}^{k} n_i$ and $n_0 = \left(\sum_{i=1}^{k} n_i - \sum_{i=1}^{k} n_i^2 \Big/ \sum_{i=1}^{k} n_i\right) \Big/ (k-1)$ as given in Equation 12.32.

If all subjects have the same number of replicates (n), then $n_0 = n$.

Example 12.30 **Endocrinology** Estimate the intraclass correlation coefficient between replicate plasma-estradiol samples based on the data in Table 12.18.

Solution We will use the ln(plasma estradiol) values as we did in Table 12.19, because we found that the within-person variance was relatively constant on the log scale, but depended on the mean value in the original (raw) scale. From Example 12.26, we have $\hat{\sigma}_A^2 = 0.317$, $\hat{\sigma}^2 = 0.030$. Therefore, our point estimate of the intraclass correlation is given by

$$\rho_I = \frac{0.317}{0.317 + 0.030} = \frac{0.317}{0.347} = 0.914$$

Thus, there is excellent reproducibility for ln(plasma estradiol). To obtain confidence limits about this estimate, we refer to Equation 12.35. From Table 12.19, we see that the F statistic from the SAS General Linear Model procedure based on a one-way random-effects model ANOVA is 22.15. Also, since this is a balanced design (i.e., we have the same number of replicates for each subject), we have $n_0 = n = 2$. We also need the critical values (which we obtain from Table 9 in the Appendix or from Excel) given by $F_{4,5,.975} = 7.39$ and $F_{4,5,.025} = 1/F_{5,4,.975} = 1/9.36 = 0.107$. Therefore, the 95% CI for ρ_I is given by (c_1, c_2), where

$$c_1 = \max\left\{\frac{\frac{22.15}{F_{4,5,.975}}-1}{2+\frac{22.15}{F_{4,5,.975}}-1}, 0\right\}$$

$$= \max\left\{\frac{\frac{22.15}{7.39}-1}{1+\frac{22.15}{7.39}}, 0\right\}$$

$$= \frac{1.997}{3.997} = .500$$

$$c_2 = \max\left\{\frac{\dfrac{22.15}{F_{4,5,.025}} - 1}{2 + \dfrac{22.15}{F_{4,5,.025}} - 1}, 0\right\}$$

$$= \max\left\{\frac{\dfrac{22.15}{0.107} - 1}{1 + \dfrac{22.15}{0.107}}, 0\right\}$$

$$= \frac{206.324}{208.324} = .990$$

Therefore, the 95% CI for $\rho_I = (.500, .990)$, which is quite wide.

Another interpretation for the intraclass correlation is based on *reliability* rather than *reproducibility*.

Example 12.31 **Hypertension** Suppose we wish to characterize blood pressure for an individual over a short time period (e.g., a 1-month period). Blood pressure is an imprecise measure with a lot of within-person variation. Therefore, the ideal way to characterize a subject's blood pressure is to take many replicate measures over a short period of time and use the average (T) as the true level of blood pressure. We then might ask to what extent does a single blood pressure (X) relate to T? The answer to this question is given by the intraclass correlation coefficient.

EQUATION 12.36

Alternative Interpretation of the Intraclass Correlation Coefficient as a Measure of Reliability Suppose that we have a one-way random-effects model ANOVA, where

$$y_{ij} = \mu + \alpha_i + e_{ij},\ e_{ij} \sim N(0, \sigma^2),\ \alpha_i \sim N(0, \sigma_A^2)$$

where y_{ij} denotes the jth replicate from the ith subject. The average of an infinite number of replicates for the ith subject is denoted by $Y_i = \mu + \alpha_i$. The square of the correlation between Y_i and a single replicate measure y_{ij} is given by the intraclass correlation coefficient. Therefore, the intraclass correlation coefficient can also be interpreted as a measure of reliability and is sometimes called the **reliability coefficient.**

Solution to Example 12.31 It has been shown that the intraclass correlation coefficient for diastolic blood-pressure measurements for 30- to 49-year-olds based on a single visit is .79 [8]. Thus, the correlation between diastolic blood pressure obtained at one visit and the "true" diastolic blood pressure is $\sqrt{.79} = .89$. To increase the reliability of blood-pressure measurements, clinicians will often take an average blood pressure over several visits. The rationale for this practice is that the reliability of a mean diastolic blood pressure over, say, three visits is higher (.92) than for a single visit (.79) in the sense that its correlation with the true diastolic blood pressure is higher. In each instance the average of three readings was used to characterize diastolic blood pressure at any one visit. This is the rationale for the screening design of TOHP (Trial of Hypertension Prevention) where approximately 20,000 persons were screened in order to identify approximately 2000 subjects with high normal diastolic blood pressure based on their having a mean diastolic blood pressure of 80–89 mm Hg based on an average of nine readings over three visits with three measurements per visit [9].

In this section, we have been introduced to the intraclass correlation coefficient. Suppose individuals (e.g., children) are categorized into groups by a grouping variable (e.g., families). The intraclass correlation coefficient is used to estimate the correlation between two separate members of the same group (e.g., two children in the same family). It is defined differently than an ordinary Pearson correlation coefficient (see Section 11.7). For a Pearson correlation, there are two distinct variables being compared (e.g., cholesterol levels for a husband versus cholesterol levels for a wife). The mean and variance of each variable is estimated separately. For an intraclass correlation coefficient, it is arbitrary which child is denoted as the x variable and which is denoted as the y variable. Thus, the estimated mean and variance of x and y are the same and are obtained by a pooled estimate over all children over all families. The intraclass correlation coefficient can also be interpreted as a measure of the percentage of total variation that is attributable to between-group (e.g., between-family) variation.

SECTION 12.10 Summary

One-way analysis-of-variance (ANOVA) methods enable us to relate a normally distributed outcome variable to the levels of a single categorical independent variable. Two different analysis-of-variance models based on a fixed- or random-effects model, respectively, were considered. In a fixed-effects model, the levels of the categorical variable are fixed in advance. A major objective of this design is to test the hypothesis that the mean level of the dependent variable is different for different groups defined by the categorical variable. Which specific groups are different can also be identified, using t tests based on the LSD (least significant difference) procedure if the comparisons have been planned in advance, or multiple-comparisons methods if they have not. More complex comparisons such as testing for dose–response relationships involving more than two levels of the categorical variable can also be accomplished using linear-contrast methods. Fixed-effects analysis-of-variance methods can be thought of as a generalization of two-sample inference based on t tests presented in Chapter 8.

In addition, nonparametric methods for fixed-effects ANOVA based on the Kruskal-Wallis test were also discussed for situations when the assumption of normality is questionable. This test can be thought of as a generalization of the Wilcoxon rank-sum test presented in Chapter 9, if more than two groups are being compared.

Under a random-effects model, the levels of the categorical variable are determined at random, with the levels being drawn from an underlying normal distribution whose variance characterizes between-subject variation in the study population.

For a given subject, there is, in addition, within-subject variation about the underlying mean for that subject. A major objective of a random-effects design is to estimate the between- and within-subject variance components. Random-effects models are also useful in computing coefficients of variation in reproducibility studies.

We also discussed the two-way analysis of variance (ANOVA). Under the two-way ANOVA, we are interested in jointly comparing the mean levels of an outcome variable according to the levels of two categorical variables (e.g., mean level of blood pressure by both sex and ethnic group). With two-way ANOVA, we can simultaneously estimate main effects of sex (e.g., the effect of sex on blood pressure after controlling for ethnic group), main effects of ethnic group (e.g., the effect of ethnic group on blood pressure after controlling for sex), and interaction effects between sex and ethnic group (e.g., estimates of differences of ethnic group effects on blood pressure between males and females). We also saw that both one-way and two-way fixed-effects ANOVA models can be considered as special cases of multiple-regression models by using dummy-variable coding for the effects of the categorical variable(s).

Finally, we learned about both one-way and two-way analysis of covariance (ANCOVA). In one-way ANCOVA, we are primarily interested in relating a continuous outcome variable to a categorical variable, but wish to control for other covariates. Similarly, in two-way ANCOVA we are primarily interested in relating a continuous outcome variable simultaneously to two categorical variables, but wish to control for other covariates. We also saw that both one-way and two-way ANCOVA models can be represented as special cases of multiple-regression models.

PROBLEMS

Nutrition

A comparison was made of protein intake among three groups of postmenopausal women: (1) women eating a standard American diet (STD), (2) women eating a lacto-ovo-vegetarian diet (LAC), and (3) women eating a strict vegetarian diet (VEG). The mean ±1 sd for protein intake (mg) is presented in Table 12.22.

TABLE 12.22 Protein intake (mg) among three dietary groups of postmenopausal women

Group	Mean	sd	*n*
STD	75	9	10
LAC	57	13	10
VEG	47	17	6

***12.1** Perform a statistical procedure to compare the means of the three groups using the critical-value method.

***12.2** What is the *p*-value from the test performed in Problem 12.1?

***12.3** Compare the means of each specific pair of groups using the *t* test methodology.

***12.4** Suppose that in the general population, 70% of vegetarians are lacto-ovo-vegetarians, whereas 30% are strict vegetarians. Perform a statistical procedure to test if the contrast $L = 0.7\bar{y}_2 + 0.3\bar{y}_3 - \bar{y}_1$ is significantly different from 0. What does the contrast mean?

12.5 Using the data in Table 12.22, perform a multiple-comparisons procedure to identify which specific underlying means are different.

Pulmonary Disease

Twenty-two young asthmatic volunteers were studied to assess the short-term effects of sulfur dioxide (SO_2) exposure under various conditions [10]. The baseline data in Table 12.23 were presented regarding bronchial reactivity to SO_2 stratified by lung function (as defined by FEV_1/FVC) at screening.

TABLE 12.23 Relationship of bronchial reactivity to SO_2 (cm H_2O/s) grouped by lung function at screening among 22 asthmatic volunteers

Lung function group		
Group A FEV_1/FVC ≤ 74%	Group B FEV_1/FVC 75–84%	Group C FEV_1/FVC ≥ 85%
20.8	7.5	9.2
4.1	7.5	2.0
30.0	11.9	2.5
24.7	4.5	6.1
13.8	3.1	7.5
	8.0	
	4.7	
	28.1	
	10.3	
	10.0	
	5.1	
	2.2	

Source: Reprinted with permission of the *American Review of Respiratory Disease, 131*(2), 221–225, 1985.

***12.6** Test the hypothesis that there is an overall mean difference in bronchial reactivity among the three lung-function groups.

***12.7** Compare the means of each pair of groups using the LSD method.

***12.8** Compare the means of each pair of groups using the Bonferroni method.

Hypertension

A recent phenomenon is the emergence of the automated blood-pressure device, which has appeared in many banks, drugstores, and other public places. A study was conducted to assess the comparability of the machine readings with those of the standard cuff [11]. Readings were taken using both the machine and the standard cuff at four separate locations. The results are given in Table 12.24. Suppose we wish to test if the mean difference between the machine and the standard cuff is consistent over the four locations (i.e., if the bias is comparable over all four locations).

12.9 Is a fixed-effects or a random-effects ANOVA appropriate here?

12.10 Test if the mean difference is consistent over all four locations.

12.11 Estimate the proportions of the variance attributable to between-machine versus within-machine variability.

Mental Health

For the purpose of identifying older nondemented people with early signs of senile dementia, a Mental Function Index was constructed based on three short tests of cognitive function. In Table 12.25, data relating the Mental Function Index at baseline to clinical status determined independently at baseline and follow-up, with a median follow-up period of 959 days, are presented [12].

12.12 What test procedure can be used to test for significant differences among the groups?

12.13 Perform the test mentioned in Problem 12.12 and report appropriate *p*-values identifying differences between specific groups.

TABLE 12.24 Mean systolic blood pressure (SBP) and difference between machine and human readings at four locations

	SBP machine (mm Hg)			SBP standard cuff (mm Hg)			SBP machine minus SBP standard cuff (mm Hg)		
Location	Mean	sd	*n*	Mean	sd	*n*	Mean	sd	*n*
A	142.5	21.0	98	142.0	18.1	98	0.5	11.2	98
B	134.1	22.5	84	133.6	23.2	84	0.5	12.1	84
C	147.9	20.3	98	133.9	18.3	98	14.0	11.7	98
D	135.4	16.7	62	128.5	19.0	62	6.9	13.6	62

Source: Reprinted with permission of the American Heart Association, *Hypertension, 2*(2), 221–227, 1980.

TABLE 12.25 Relationship between clinical status at baseline and follow-up (median follow-up period of 959 days) to mean Mental Function Index at baseline

Clinical Status				
Baseline	Follow-up	Mean	sd	*n*
Normal	Unchanged	0.04	0.11	27
Normal	Questionably or mildly affected	0.22	0.17	9
Questionably affected	Progressed	0.43	0.35	7
Definitely affected	Progressed	0.76	0.58	10

Source: Reprinted with permission of the *American Journal of Epidemiology, 120*(6), 922–935, 1984.

Obstetrics

The birthweight of an infant has been hypothesized to be associated with the smoking status of the mother during the first trimester of pregnancy. This hypothesis is tested by recording the birthweights of infants and the smoking status of the mother during pregnancy for all mothers who register at the prenatal clinic at a particular hospital within a 1-month period. The mothers are divided into four groups according to smoking habit, and the sample of birthweights in pounds within each group is given as follows:

Group 1: Mother is a nonsmoker (NON):

7.5 6.2 6.9 7.4 9.2 8.3 7.6

Group 2: Mother is an ex-smoker (smoked at some time prior to pregnancy but not during pregnancy) (EX):

5.8 7.3 8.2 7.1 7.8

Group 3: Mother is a current smoker and smokes less than 1 pack per day (CUR < 1):

5.9 6.2 5.8 4.7 8.3 7.2 6.2

Group 4: Mother is a current smoker and smokes greater than or equal to 1 pack per day (CUR ≥ 1):

6.2 6.8 5.7 4.9 6.2 7.1 5.8 5.4

12.14 Are the mean birthweights different overall in the four groups?

12.15 Test for all differences between each pair of groups and summarize your results using the LSD procedure.

12.16 Perform the same tests as in Problem 12.15 using the method of multiple comparisons.

12.17 Are there any differences between the results in Problems 12.15 and 12.16?

12.18 Suppose we assume that smokers of ≥ 1 pack per day smoke an average of 1.3 packs per day, whereas smokers of <1 pack per day smoke an average of 0.5 pack per day. Use the method of linear contrasts to test if the amount of current cigarette consumption among *ever* smokers is significantly related to birthweight.

12.19 Use a multiple-comparisons approach to answer the question posed in Problem 12.18.

Pharmacology

Suppose we wish to test the relative effects of three drugs (A, B, C) on the reduction of fever. Drug A is 100% aspirin; drug B is 50% aspirin and 50% other compounds; drug C is 25% aspirin and 75% other compounds. The drugs are prescribed to children aged 5–14 entering the outpatient ward of a hospital complaining of the "flu," with fever of 100.0°F to 100.9°F. The drugs are assigned in time-sequence order; that is, the first patient gets drug A, the second drug B, the third drug C, the fourth drug A, and so forth, until there is a set of 15 patients. The parents are then telephoned 4 hours after the administration of the drug and the reduction in fever is noted. The results are given in Table 12.26. We assume that the timing of prescribing the drugs (i.e., when it is prescribed during the day) is irrelevant to the reduction in fever.

TABLE 12.26 Reduction in fever for patients getting different doses of aspirin

Drug		Mean (°F)	*sd* (°F)	*n*
Drug A	2.0, 1.6, 2.1, 0.6, 1.3	1.52	0.61	5
Drug B	0.5, 1.2, 0.3, 0.2, –0.4	0.36	0.58	5
Drug C	1.1, –1.0, –0.2, +0.2, +0.3	0.08	0.77	5
Overall		0.65		15

12.20 What are the appropriate null and alternative hypotheses to test whether or not all three drugs are equally effective?

***12.21** Perform the significance test in Problem 12.20.

***12.22** Summarize the differences in treatment effects among the three drugs using the method of multiple comparisons.

***12.23** Provide a 95% CI for the mean difference in the reduction of fever between each pair of drugs tested.

Hypertension

Some common strategies for treating hypertensive patients by nonpharmacologic methods include (1) weight reduction and (2) trying to get the patient to relax more by meditational or other techniques. Suppose these strategies are evaluated by establishing four groups of hypertensive patients who receive the following types of nonpharmacologic therapy:

Group 1: Patients receive counseling for both weight reduction and meditation.

Group 2: Patients receive counseling for weight reduction but not for meditation.

Group 3: Patients receive counseling for meditation but not for weight reduction.

Group 4: Patients receive no counseling at all.

Suppose that 20 hypertensive patients are assigned at random to each of the four groups, and the change in diastolic blood pressure is noted in these patients after a 1-month period. The results are given in Table 12.27.

TABLE 12.27 Change in diastolic blood pressure (DBP) for four groups of hypertensive patients who receive different kinds of nonpharmacologic therapy

Group	Mean change in DBP (baseline – follow-up) (mm Hg)	sd change	n
1	8.6	6.2	20
2	5.3	5.4	20
3	4.9	7.0	20
4	1.1	6.5	20

12.24 Test the hypothesis that the mean change in DBP is the same among the four groups.

12.25 Analyze if counseling for weight reduction has a significant effect in reducing blood pressure.

12.26 Analyze if meditation instruction has a significant effect in reducing blood pressure.

12.27 Is there any relationship between the effects of weight-reduction counseling and meditation counseling on blood-pressure reduction? That is, does weight-reduction counseling work better for people who receive meditational counseling or for people who do not receive meditational counseling, or is there no difference in effect between these two subgroups?

Hypertension

An instructor in health education wants to familiarize her students with the measurement of blood pressure. Each student is given a portable blood-pressure machine to take home. Each student is told to obtain two readings on each of 10 consecutive days. The data for one student are given in Table 12.28.

TABLE 12.28 Systolic blood-pressure (SBP) recordings on 1 subject for 10 consecutive days with 2 readings per day

	Reading	
Day	1	2
1	98	99
2	102	93
3	100	98
4	99	100
5	96	100
6	95	100
7	90	98
8	102	93
9	91	92
10	90	94

***12.28** Estimate the between-day and within-day components of variance for this subject.

***12.29** Is there a difference in underlying mean blood pressure by day for this subject?

Bioavailability

The intake of high doses of beta-carotene in food substances has been associated in some observational studies with a decreased incidence of cancer. A clinical trial was planned comparing the incidence of cancer in a group getting beta-carotene in capsule form compared with a group getting beta-carotene placebo capsules. One issue in planning such a study is which preparation to use for the beta-carotene capsules. Four preparations were considered: (1) Solatene (30-mg capsules), (2) Roche (60-mg capsules), (3) BASF (30-mg capsules), and (4) BASF (60-mg capsules). To test the efficacy of the four agents in raising plasma-carotene levels, a small bioavailability study was conducted. After two consecutive-day fasting blood samples, 23 volunteers were randomized to one of the four preparations, taking 1 pill every other day for 12 weeks: (1) Solatene 30 mg,

TABLE 12.29 Format of BETACAR.DAT

Variable	Column	Code
Preparation	1	1 = SOL; 2 = ROCHE; 3 = BASF-30; 4 = BASF-60
Subject number	3–4	
First baseline level	6–8	
Second baseline level	10–12	
Week 6 level	14–16	
Week 8 level	18–20	
Week 10 level	22–24	
Week 12 level	26–28	

(2) Roche 60 mg, (3) BASF 30 mg, and (4) BASF 60 mg. The primary endpoint was the level of plasma carotene attained after a moderately prolonged steady ingestion. For this purpose, blood samples were drawn at 6, 8, 10, and 12 weeks, with results given in Data Set BETACAR.DAT, on the data disk. The format of the data is given in Table 12.29.

12.30 Use analysis-of-variance methods to estimate the coefficient of variation for plasma beta-carotene for the 23 subjects, based on the two baseline measurements.

12.31 Is there a significant difference in bioavailability of the 4 different preparations? Use analysis-of-variance methods to assess this based on the 6-week data in comparison with baseline.

12.32 Use the methods in Problem 12.31 to compare the bioavailability of the four preparations at 8 weeks versus baseline.

12.33 Use the methods in Problem 12.31 to compare the bioavailability of the four preparations at 10 weeks versus baseline.

12.34 Use the methods in Problem 12.31 to compare the bioavailability of the four preparations at 12 weeks versus baseline.

12.35 Use the methods in Problem 12.31 to compare the bioavailability of the four preparations based on the average plasma beta-carotene at (6, 8, 10, and 12 weeks) versus baseline.

12.36 Is there a difference in bioavailability at 6, 8, 10, or 12 weeks? If so, when is the bioavailability the greatest? (Provide a qualitative answer only; do not perform any significance tests.)

Hepatic Disease

Refer to Data Set HORMONE.DAT, on the data disk (see pp. 322–323 for a description of the data set).

12.37 Use analysis-of-variance methods to test if the change in biliary-secretion levels is comparable for the five hormone groups. Identify and test for any specific group differences.

12.38 Answer Problem 12.37 for changes in pancreatic-secretion levels.

12.39 Answer Problem 12.37 for changes in biliary-pH levels.

12.40 Answer Problem 12.37 for changes in pancreatic-pH levels.

Endocrinology

A study was conducted [13] concerning the effect of calcium supplementation on bone loss among postmenopausal women. Women were randomized to either (1) estrogen cream and calcium placebo ($n = 15$), (2) placebo estrogen cream and 2000 mg/day of calcium ($n = 15$), or (3) placebo estrogen cream and calcium placebo ($n = 13$). Subjects were seen every 3 months for a 2-year period. The rate of bone loss was computed for each woman and expressed as a percentage of the initial bone mass. The results are shown in Table 12.30.

TABLE 12.30 Mean (±1 sd) slope of total-body bone mass (percentage per year) in the three treatment groups

Treatment group		
(1) Estrogen ($n = 15$)	(2) Calcium ($n = 15$)	(3) Placebo ($n = 13$)
−0.43 ± 1.60	−2.62 ± 2.68	−3.98 ± 1.63

Source: Reprinted with permission from the *New England Journal of Medicine, 316*(4), 173–177, 1987.

12.41 What test can be used to compare the mean rate of bone loss in the three groups?

12.42 Implement the test in Problem 12.41 and report a *p*-value.

12.43 Identify which pairs of groups are different from each other, using both *t* tests and the method of multiple comparisons. Report a *p*-value for each comparison of treatment groups.

12.44 Which methodology do you think is more appropriate in Problem 12.43?

Endocrinology

Refer to Data Set ENDOCRIN.DAT, on the data disk. The data set consists of split-sample plasma determinations of 4 hormones for each of 5 subjects from one laboratory. The format of the data is given in Table 12.31.

TABLE 12.31 Format of ENDOCRIN.DAT

	Column	Units
Subject number	1	
Replicate number	3	
Plasma estrone	5–8	pg/mL
Plasma estradiol	10–14	pg/mL
Plasma androstenedione	16–19	ng/dL
Plasma testosterone	21–24	ng/dL

12.45 Estimate the between-subject and within-subject variation for plasma estrone, plasma androstenedione, and plasma testosterone.

12.46 Estimate the coefficient of variation for each of the hormones in Problem 12.45.

Environmental Health

A student wishes to determine if there are specific locations within a household that show heat loss. To assess this she records temperatures at 20 sites within a household for each of 30 days. In addition, she records the outside temperature. The data are given in Data Set TEMPERAT.DAT, on the data disk. The format of the data is given in Table 12.32.

12.47 Assume a random-effects model. Estimate the between-day versus within-day variation in temperature within this household.

12.48 Are there significant differences in temperature between different locations in the household?

12.49 Assume a fixed-effects model. Use the method of multiple comparisons to assess which specific locations in the household are different in mean temperature.

TABLE 12.32 Format of TEMPERAT.DAT

	Column	Comment
1. Date	1–6	(mo/da/yr)
2. Outside temperature	8–9	(°F)
3. Location within household	11–12	(1–20)
4. Inside temperature	14–17	(°F)

Note: The data were collected by Sarah Rosner.

Environmental Health

Refer to Data Set LEAD.DAT, on the data disk.

12.50 Use analysis-of-variance methods to assess if there are any overall differences between the control group, the currently exposed group, and the previously exposed group in mean full-scale IQ. Also, make a comparison between each pair of groups and report a *p*-value.

12.51 Determine a 95% CI for the underlying mean difference in full-scale IQ for each pair of groups considered.

Gastroenterology

In Table 12.33, we present data relating protein concentration to pancreatic function as measured by trypsin secretion among patients with cystic fibrosis [14].

12.52 If we do not wish to assume normality for these distributions, then what statistical procedure can be used to compare the three groups?

12.53 Perform the test mentioned in Problem 12.52, and report a *p*-value. How do your results compare to a parametric analysis of the data?

Environmental Health, Pediatrics

Refer to the Data Set LEAD.DAT, on the data disk.

12.54 Use nonparametric methods to compare MAXFWT among the three exposure groups defined by the variable LEAD_GRP.

12.55 Answer Problem 12.54 for IQF (full-scale IQ).

12.56 Compare your results in Problems 12.54 and 12.55 with the corresponding results in Table 12.6 for MAXFWT and Problem 12.50 for IQF where parametric methods were used.

Ophthalmology

A study was performed comparing the mean ERG (electroretinogram) amplitude of patients with different genetic types of retinitis pigmentosa (RP), a genetic eye disease that often results in blindness. The following results were ob-

TABLE 12.33 Relationship between protein concentration (mg/mL) of duodenal secretions to pancreatic function as measured by trypsin secretion [U/(kg/hr)]

Trypsin section [U/(kg/hr)]					
≤50		51–1000		>1000	
Subject number	Protein concentration	Subject number	Protein concentration	Subject number	Protein concentration
1	1.7	1	1.4	1	2.9
2	2.0	2	2.4	2	3.8
3	2.0	3	2.4	3	4.4
4	2.2	4	3.3	4	4.7
5	4.0	5	4.4	5	5.0
6	4.0	6	4.7	6	5.6
7	5.0	7	6.7	7	7.4
8	6.7	8	7.6	8	9.4
9	7.8	9	9.5	9	10.3
		10	11.7		

Source: Reprinted with permission of the *New England Journal of Medicine, 312*(6), 329–334, 1985.

tained for ln(ERG amplitude) among patients 18–29 years of age.

Genetic type	Mean ± sd[a]	*N*
Dominant	0.85 ± 0.18	62
Recessive	0.38 ± 0.21	35
X-linked	–0.09 ± 0.21	28

[a]Units are in ln(μ*V*).

12.57 Use analysis-of-variance methods to assess if there are overall group differences in mean ln(ERG amplitude) by genetic type.

12.58 Assess if there are differences between each pair of genetic types and report a two-sided *p*-value.

12.59 How do your results compare with those reported in Problem 8.124, where dominant and recessive individuals were compared with a two-sample *t* test?

Renal Disease

Refer to the Data Set SWISS.DAT on the data disk.

12.60 Use analysis-of-variance methods to compare the change in serum-creatinine values from the baseline visit to the 1978 visit among women in the high-NAPAP group, low-NAPAP group, and the control group.

One issue in Problem 12.60 is that only the first and last visits are used to assess change in serum creatinine over time.

12.61 Fit a regression line relating serum creatinine to time *for each person* in each of the high-NAPAP group, the low-NAPAP group, and the control group. How do you interpret the slope and intercept for each person?

12.62 Use regression analysis or analysis-of-variance methods to compare the slopes of the three groups.

12.63 Answer the question in Problem 12.62 for the intercepts in the three groups.

12.64 What are your overall conclusions regarding the comparison of serum creatinine among the three groups?

Note: One issue in the preceding analyses is that we have considered all persons as yielding identical information regardless of the number of visits available for analysis. A more precise method would use methods of *longitudinal data analysis* to weight subjects according to the number and spacing of their visits. This is beyond the scope of this text.

Bioavailability

Refer to Table 12.29.

12.65 Compute the intraclass correlation coefficient between replicate plasma beta-carotene blood samples at baseline. Provide a 95% CI about this estimate. Perform the analysis based on the entire data set.

12.66 Use linear-regression methods to assess whether plasma-carotene levels increase over the 12-week period. Use appropriate data transformations if necessary. Perform separate analyses for each of the four preparations.

12.67 Do you have any recommendations as to which preparation should be used in the main clinical trial?

Endocrinology

Refer to Table 12.31.

12.68 Estimate the intraclass correlation for each of plasma estrone, plasma androstenedione, and plasma testosterone, and provide associated 95% confidence limits. Is the reproducibility of these plasma-hormone levels excellent, good, or poor?

Environmental Health

Refer to Table 12.32.

12.69 Use regression methods to assess whether there is any relationship between the indoor and outdoor temperature readings.

Ophthalmology

The term *retinitis pigmentosa* refers to a group of hereditary, retinal pigmentary degenerations in which patients report night blindness and loss of visual field, usually between the ages of 10 and 40 years. Some patients lose all useful vision (i.e., become legally blind) by the age of 30 years, while others retain central vision even beyond the age of 60 years. A specific gene has been linked with some types of RP where the mode of genetic transmission is autosomal dominant. The most reliable methods of following the course of RP in humans is by using the electroretinogram (ERG), which is a measure of the electrical activity in the retina. As the disease progresses, the patient's ERG amplitude declines. The ERG amplitude has been strongly related to the patient's ability to perform routine activities, such as driving or walking unaided, especially at night.

One hypothesis is that direct exposure of the retina to sunlight is harmful to RP patients, so many patients wear sunglasses. To test the sunlight hypothesis, a group of mice had this gene introduced and were mated over many generations to produce a group of "RP mice." The mice were then randomly assigned to lighting conditions from birth that were either (1) light, (2) dim, or (3) dark. A control group of normal mice were also randomized to similar lighting conditions. The mice had their ERG amplitude (labeled BAMP and AAMP for B-wave amplitude and A-wave amplitude), which correspond to different frequencies of light, measured at 15, 20, and 35 days of life. In addition, the same protocol was used for a group of normal mice except that only B-wave amplitude was measured. The data for both RP mice and normal mice are available in the Data Set MICE.DAT and the documentation in MICE.DOC (both on the data disk).

12.70 Analyze the data regarding the sunlight hypothesis and summarize your findings. *Hint:* Estimate a slope for each ERG amplitude for each lighting-condition group, and compare the slopes among the light, dim, and dark groups using either ANOVA or regression methods. Do separate analyses for A-wave and B-wave amplitudes. Consider appropriate data transformations to ensure approximate normality for the outcome measure.

Hypertension

Refer to Table 12.14. A similar two-way ANOVA was run using PROC GLM of SAS comparing mean diastolic blood pressure by study group and sex. The results are given in Table 12.34.

12.71 Summarize the findings in a few sentences.

Hypertension

Refer to Table 12.15. An analysis of covariance was performed using PROC GLM of SAS comparing mean diastolic blood pressure by study group and sex after controlling for effects of age and weight. The results are given in Table 12.35.

12.72 Summarize the findings in a few sentences and compare them with the results in Problem 12.71.

REFERENCES

[1] White, J. R., & Froeb, H. F. (1980). Small-airways dysfunction in nonsmokers chronically exposed to tobacco smoke. *New England Journal of Medicine, 302*(13), 720–723.

[2] Kleinbaum, D. G., Kupper, L. L., & Muller, K. E. (1988). *Applied regression analysis and other multivariable methods* (2nd ed.). Boston: Duxbury.

[3] Abelson, M. B., Kliman, G. H., Butrus, S. I., & Weston, J. H. (1983). Modulation of arachidonic acid in the rabbit conjunctiva: Predominance of the cyclo-oxygenase pathway. Presented at the Annual Spring Meeting of the Association for Research in Vision and Ophthalmology, Sarasota, Florida, May 2–6, 1983.

TABLE 12.34 SAS GLM procedure output illustrating the effects of study group and sex on diastolic blood pressure using the data set in Example 12.17

```
                                          SAS
                             GENERAL LINEAR MODELS PROCEDURE
DEPENDENT VARIABLE: MDIAS
SOURCE             DF   SUM OF SQUARES      MEAN SQUARE   F VALUE        PR > F   R-SQUARE         C.V.
MODEL               3   48186.99270094   16062.33090031    134.15        0.0001   0.350741      15.0906
ERROR             745   89199.44205496     119.73079470                ROOT MSE              MDIAS MEAN
CORRECTED TOTAL   748  137386.43475590                              10.94215677             72.50972853

SOURCE             DF        TYPE I SS  F VALUE    PR > F       DF     TYPE III SS   F VALUE      PR > F
STUDY               2   45269.88509153   189.05    0.0001        2  46573.92818903    194.49      0.0001
SEX                 1    2917.10760942    24.36    0.0001        1   2917.10760942     24.36      0.0001

                              STUDY   PROB > |T|
                                             SV         LV        NOR
                              SV              .     0.0001     0.0001
                              LV         0.0001          .     0.0001
                              NOR        0.0001     0.0001          .

                              SEX          PROB > |T|
                                             MALE         FEMALE
                              MALE             .          0.0001
                              FEMALE      0.0001               .

                                                 T FOR H0:            PR > |T|         STD ERROR OF
    PARAMETER                   ESTIMATE      PARAMETER=0                                  ESTIMATE
    INTERCEPT                76.47708914           115.61               0.0001           0.66152912
    STUDY       SV          -17.30065001           -19.40               0.0001           0.89174716
                LV          -10.65302392            -7.23               0.0001           1.47258921
                NOR           0.00000000                .                    .                    .
    SEX         MALE          3.98399582             4.94               0.0001           0.80713389
                FEMALE        0.00000000                .                    .                    .
```


[4] Hankinson, S. E., Manson, J. E., Spiegelman, D., Willett, W. C., Longcope, C., & Speizer, F. E. (1995). Reproducibility of plasma hormone levels in post-menopausal women over a 2–3 year period. *Cancer Epidemiology, Biomarkers and Prevention, 4*(6), 649–654.

[5] Chinn, S. (1990). The assessment of methods of measurement. *Statistics in Medicine, 9*, 351–362.

[6] Snedecor, G., & Cochran, W. G. (1988). *Statistical methods.* Ames: Iowa State University Press.

[7] Fleiss, J. L. (1986). *The design and analysis of clinical experiments*. New York: Wiley.

[8] Cook, N. R., & Rosner, B. (1993). Screening rules for determining blood pressure status in clinical trials: Application to the Trials of Hypertension Prevention. *American Journal of Epidemiology, 137*(12), 1341–1352.

[9] Satterfield, S., Cutler, J. A., Langford, H. G., et al. (1991). Trials of Hypertension Prevention: Phase I design. *Annals of Epidemiology, 1*, 455–457.

[10] Linn, W. S., Shamoo, D. A., Anderson, K. R., Whynot, J. D., Avol, E. L., & Hackney, J. D. (1985). Effects of heat and humidity on the responses of exercising asthmatics to sulfur dioxide exposure. *American Review of Respiratory Disease, 131*(2), 221–225.

[11] Polk, B. F., Rosner, B., Feudo, R., & Van Denburgh, M. (1980). An evaluation of the Vita Stat automatic blood pressure measuring device. *Hypertension, 2*(2), 221–227.

[12] Pfeffer, R. I., Kurosaki, T. T., Chance, J. M., Filos, S., & Bates, D. (1984). Use of the mental function index in older adults: Reliability, validity and measurement of change over time. *American Journal of Epidemiology, 120*(6), 922–935.

TABLE 12.35 **SAS GLM procedure output illustrating the effects of study group and sex on diastolic blood pressure after controlling for age and weight using the data set in Example 12.17**

```
                                         SAS
                           GENERAL LINEAR MODELS PROCEDURE
DEPENDENT VARIABLE: MDIAS
SOURCE            DF   SUM OF SQUARES     MEAN SQUARE   F VALUE       PR > F   R-SQUARE          C.V.
MODEL              5   57521.19225928  11504.23845186    107.08       0.0001   0.419457       14.2973
ERROR            741   79611.39165542    107.43777551                ROOT MSE               MDIAS MEAN
CORRECTED TOTAL  746  137132.58391470                             10.36521951              72.49770638

SOURCE            DF       TYPE I SS   F VALUE    PR > F     DF      TYPE III SS   F VALUE      PR > F
STUDY              2  45234.31356527    210.51    0.0001      2   12349.74237359     57.47      0.0001
SEX                1   2912.79699312     27.11    0.0001      1     624.47946004      5.81      0.0162
AGE                1   5237.52247659     48.75    0.0001      1    4245.67574526     39.52      0.0001
WGT                1   4136.55922429     38.50    0.0001      1    4136.55922429     38.50      0.0001

                                STUDY    PROB > |T|
                                                  SV        LV       NOR
                                SV                 .    0.0186    0.0001
                                LV            0.0186         .    0.0001
                                NOR           0.0001    0.0001         .

                                SEX    PROB > |T|
                                            MALE      FEMALE
                                MALE           .      0.0162
                                FEMALE    0.0162           .

                                               T FOR H0:           PR > |T|          STD ERROR OF
    PARAMETER                  ESTIMATE      PARAMETER=0                                 ESTIMATE
    INTERCEPT               52.96724415            20.41             0.0001            2.59544038
    STUDY       SV         -11.18295628           -10.45             0.0001            1.06990647
                LV          -7.58825363            -5.29             0.0001            1.43486888
                NOR          0.00000000                .                  .                     .
    SEX         MALE         2.10948458             2.41             0.0162            0.87497527
                FEMALE       0.00000000                .                  .                     .
    AGE                      0.30162065             6.29             0.0001            0.04798065
    WGT                      0.08278540             6.20             0.0001            0.01334174
```

[13] Riis, B., Thomsen, K., & Christiansen, C. (1987). Does calcium supplementation prevent post-menopausal bone loss? A double-blind controlled study. *New England Journal of Medicine, 316*(4), 173–177.

[14] Kopelman, H., Durie, P., Gaskin, K., Weizman, Z., & Forstner, G. (1985). Pancreatic fluid secretion and protein hyperconcentration in cystic fibrosis. *New England Journal of Medicine, 312*(6), 329–334.

13 Design and Analysis Techniques for Epidemiologic Studies

SECTION 13.1 Introduction

In Chapter 10, we have discussed methods of analysis for categorical data. The data were displayed in a single 2×2 contingency table or more generally in a single $R \times C$ contingency table. In epidemiologic applications, the rows of the table refer to disease categories and the columns to exposure categories (or vice versa). It is natural in such applications to define a measure of effect based on the counts in the contingency table (such as the relative risk) and to obtain confidence limits for such measures. An important issue is whether a disease–exposure relationship is influenced by other variables (called *confounders*). In this chapter, we will

(1) learn about some common study designs used in epidemiologic work
(2) define several measures of effect that are commonly used for categorical data
(3) learn about techniques for assessing a primary disease–exposure relationship while controlling for confounding variable(s), including
    (a) Mantel-Haenszel methodology
    (b) logistic regression
(4) learn about meta-analysis, an increasingly popular methodology for combining results over more than one study
(5) consider several alternative study designs, including
    (a) active-control designs
    (b) cross-over designs
(6) learn about some newer data-analysis techniques in epidemiology when the assumptions of standard methods are not satisfied, including
    (a) methods for clustered binary data
    (b) methods for handling data with substantial measurement error

SECTION 13.2 Study Design

Suppose we consider Table 10.2. In this table, we looked at the association between the use of oral contraceptives (OC) at baseline and the development of MI over a

3-year follow-up period. In this setting, OC use is sometimes referred to as an *exposure variable,* and the occurrence of MI as a *disease variable.* We often are interested in exposure–disease relationships as shown in Table 13.1.

TABLE 13.1 Hypothetical table depicting exposure–disease relationship

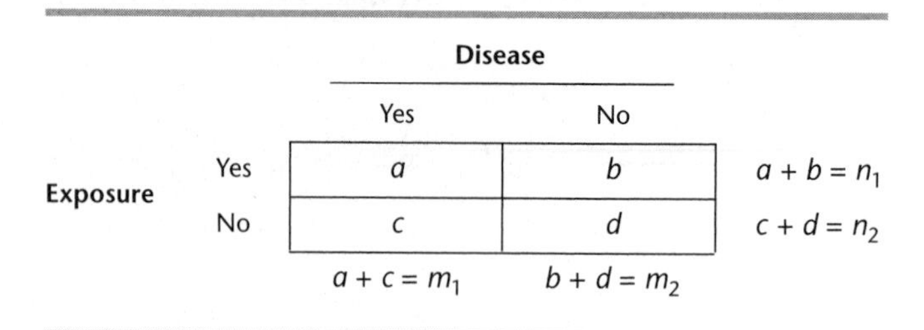

		Disease		
		Yes	No	
Exposure	Yes	a	b	$a + b = n_1$
	No	c	d	$c + d = n_2$
		$a + c = m_1$	$b + d = m_2$	

There are a total of $n_1 = a + b$ exposed subjects of whom a have disease and a total of $n_2 = c + d$ unexposed subjects, of whom c have disease. There are three principal study designs where such relationships are explored; that is, a prospective study design, a retrospective study design, and a cross-sectional study design.

DEFINITION 13.1 A **prospective study** is a study in which a group of disease-free individuals are identified at one point in time and are followed over a period of time until some of them develop the disease. The development of disease over time is then related to other variables measured at baseline, generally referred to as exposure variables. The study population in a prospective study is often referred to as a **cohort.** Thus, another name for this type of study is a **cohort study.**

DEFINITION 13.2 A **retrospective study** is a study in which two groups of individuals are initially identified: (1) a group that has the disease under study (the cases) and (2) a group that does not have the disease under study (the controls). An attempt is then made to relate their *prior* health habits to their current disease status. This type of study is also sometimes referred to as a **case–control study.**

DEFINITION 13.3 A **cross-sectional study** is a study in which a study population is ascertained at one point in time. All individuals in the study population are asked about their current disease status and their current or past exposure status. This type of study is sometimes called a **prevalence study,** because the prevalence of disease at one point in time is compared between exposed and unexposed individuals. This contrasts to a prospective study, where one is interested in the incidence rather than the prevalence of disease.

Example 13.1 **Cardiovascular Disease** What type of study design was used in Table 10.2?

Solution The study presented in Table 10.2 is an example of a prospective design. All subjects were disease free at baseline and had their exposure (OC use) measured at that time. They were followed for 3 years, during which some developed disease, whereas others remained disease free.

Example 13.2 **Cancer** What type of study design was used in the international breast-cancer study in Example 10.4?

Solution This is an example of a retrospective study. Breast-cancer cases were identified together with controls who were in the hospital at the same time as the cases, but did not have breast cancer and were of comparable age to the controls. Pregnancy history (age at first birth) was compared between cases and controls.

What are the advantages of the two types of studies? A prospective study is usually more definitive because the patients' knowledge of their current health habits will be more precise than their (or related individuals') recall of their past health habits. Second, there is a greater chance of bias with a retrospective study because (1) it is much more difficult to obtain a representative sample of people who already have the disease in question, because, for example, some of the diseased individuals may have already died and only the mildest cases (or if it is a study of deceased cases, the most severe cases) may be included, and (2) the diseased individuals, if still alive, or their surrogates will tend to give biased answers about prior health habits if they *believe* there is a relationship between these prior health habits and the disease. However, a retrospective study is much less expensive to perform and can be completed in much less time than a prospective study. For example, if the study in Example 10.4 were done as a prospective study, it would require a very large study population followed for many years before 3000 cases of breast cancer would occur. Thus an inexpensive retrospective study may initially be done as a justification for the ultimate, definitive prospective study.

Example 13.3 **Hypertension** Suppose a study is performed concerning infant blood pressure. All infants born in a specific hospital are ascertained within the first week of life while in the hospital and have their blood pressure measured in the newborn nursery. The infants are divided into two groups: a high-blood-pressure group, if their blood pressure is in the top 10% of infant blood pressure based on national norms; and a normal-blood-pressure group, otherwise. The infants' blood-pressure group is then related to their birthweight (low if ≤ 88 oz and normal otherwise). This is an example of a cross-sectional study since the blood pressures and birthweights are measured at approximately the same point in time.

Not all studies fit neatly into the characterizations given in Definitions 13.1–13.3. Indeed, some case–control studies are based on exposure variables that are collected prospectively.

Example 13.4 **Cardiovascular Disease** The Physicians' Health Study is a large, randomized clinical trial. Subjects were approximately 22,000 male physicians ages 40–84 who were initially (in 1982) free of coronary disease and cancer (except for nonmelanoma skin cancer). The principal aims of the study were to investigate the effect of aspirin use on coronary disease and the effect of beta-carotene use on cancer incidence. Accordingly, subjects were randomized to one of four treatment groups (group 1 received aspirin placebo and beta-carotene placebo capsules, group 2 received active aspirin and beta-carotene placebo capsules, group 3 received aspirin placebo and active beta-carotene capsules, and group 4 received active aspirin and active beta-carotene capsules). The aspirin arm of the study was stopped in 1990 when it became clear that aspirin had an important protective effect in preventing the development of coronary disease. The beta-carotene arm of the study was discontinued in 1997 when it became clear that beta-carotene had no effect on cancer incidence. As a secondary aim of the study, blood samples were collected from all physicians in the cohort at baseline. The goal of this part of the study was to relate lipid abnormalities identified in the blood samples to the occurrence of coronary disease. However, it would have been prohibitively expensive to analyze all the blood samples that were collected. Instead, all men who developed coronary disease (≈ 300) (the case group) and a random sample of physicians who did not develop coronary disease, but who had approximately the same age distribution as the case group (the control group)(≈ 600) were identified and their blood samples analyzed. The type of study is a case–control study nested within a prospective study that does not fit neatly into the characterizations in Definitions 13.1 and 13.2. Specifically, the issue of biased ascertainment of exposure in retrospective and case–control studies will not be an issue here, because blood samples were obtained at baseline. However, the methods of analysis to be described hereafter for case–control studies will also be applicable to this type of study.

In this section, we have discussed the principal designs used in observational epidemiologic studies. In a prospective study, a cohort of disease-free individuals are ascertained at baseline and are followed over time until some members of the cohort develop disease. It is generally considered the gold standard of designs for observational studies. However, it is relatively expensive because for a meaningful number of events to occur over time, a large number of subjects must often be followed. In a case–control design, a group of subjects with disease (the cases) and a group of subjects without disease (the controls) are recruited. Usually a retrospective history of health habits prior to getting disease is obtained. This design is relatively inexpensive, because we don't have to wait until subjects develop disease, which for rare diseases can often take a long time. However, the results from using this study design are sometimes problematic to interpret because of

(1) recall bias of previous exposures by people who already have disease

(2) potential selection bias of

 (a) the case group if, for example, a milder series of case subjects who are still alive is used

 (b) the control group if, for example, control selection is related, often unexpectedly, to the exposure

Thus, case–control studies are often used as preliminary steps to justify the ultimate, definitive prospective study. Cross-sectional studies are conducted at one point in time and have many of the same problems as case–control studies, except that the relative number of cases and controls is not fixed in advance.

SECTION 13.3 Measures of Effect for Categorical Data

We would like to compare the frequency of disease between exposed and unexposed subjects. This is most straightforward to do in the context of prospective studies where we compare incidence rates, or in cross-sectional studies where we compare prevalence rates between exposed and unexposed individuals. We will discuss these issues for prospective studies, in terms of comparing incidence rates, but the same measures of effect can be used for cross-sectional studies in terms of prevalence.

DEFINITION 13.4 Let

p_1 = probability of developing disease for exposed individuals

p_2 = probability of developing disease for unexposed individuals

The **risk difference** is defined as $p_1 - p_2$. The **risk ratio** or **relative risk** is defined as p_1/p_2.

13.3.1 The Risk Difference

Suppose that $\hat{p}_1$ and $\hat{p}_2$ are the sample proportions with disease for exposed and unexposed subjects, based on sample sizes of n_1 and n_2, respectively. An unbiased point estimate of $p_1 - p_2$ is given by $\hat{p}_1 - \hat{p}_2$. To obtain an interval estimate, we will assume that the normal approximation to the binomial distribution holds, whereby from Chapter 6, $\hat{p}_1 \sim N(p_1, p_1q_1/n_1), \hat{p}_2 \sim N(p_2, p_2q_2/n_2)$. Because there are two independent samples, from Equation 5.10,

$$\hat{p}_1 - \hat{p}_2 \sim N\left(p_1 - p_2, \frac{p_1q_1}{n_1} + \frac{p_2q_2}{n_2}\right)$$

Therefore, if $p_1q_1/n_1 + p_2q_2/n_2$ is approximated by $\hat{p}_1\hat{q}_1/n_1 + \hat{p}_2\hat{q}_2/n_2$, then

$$Pr\left(p_1 - p_2 - z_{1-\alpha/2}\sqrt{\frac{\hat{p}_1\hat{q}_1}{n_1} + \frac{\hat{p}_2\hat{q}_2}{n_2}} \le \hat{p}_1 - \hat{p}_2 \le p_1 - p_2 + z_{1-\alpha/2}\sqrt{\frac{\hat{p}_1\hat{q}_1}{n_1} + \frac{\hat{p}_2\hat{q}_2}{n_2}}\right) = 1 - \alpha$$

This can be rewritten as two inequalities:

$$p_1 - p_2 - z_{1-\alpha/2}\sqrt{\frac{\hat{p}_1\hat{q}_1}{n_1} + \frac{\hat{p}_2\hat{q}_2}{n_2}} \le \hat{p}_1 - \hat{p}_2$$

and $\hat{p}_1 - \hat{p}_2 \le p_1 - p_2 + z_{1-\alpha/2}\sqrt{\dfrac{\hat{p}_1\hat{q}_1}{n_1} + \dfrac{\hat{p}_2\hat{q}_2}{n_2}}$

If $z_{1-\alpha/2}\sqrt{\hat{p}_1\hat{q}_1/n_1 + \hat{p}_2\hat{q}_2/n_2}$ is added to both sides of the first inequality and subtracted from both sides of the second inequality, then we obtain

$$p_1 - p_2 \le \hat{p}_1 - \hat{p}_2 + z_{1-\alpha/2}\sqrt{\frac{\hat{p}_1\hat{q}_1}{n_1} + \frac{\hat{p}_2\hat{q}_2}{n_2}}$$

and $\hat{p}_1 - \hat{p}_2 - z_{1-\alpha/2}\sqrt{\dfrac{\hat{p}_1\hat{q}_1}{n_1} + \dfrac{\hat{p}_2\hat{q}_2}{n_2}} \le p_1 - p_2$

This leads to the following method for point and interval estimation of the risk difference:

EQUATION 13.1

Point and Interval Estimation for the Risk Difference Let $\hat{p}_1$, $\hat{p}_2$ represent the sample proportion who develop disease in a prospective study, based on sample sizes of n_1 and n_2, respectively. A point estimate of the risk difference is given by $\hat{p}_1 - \hat{p}_2$. A 100% × (1 − α) confidence interval for the risk difference is given by,

$$\hat{p}_1 - \hat{p}_2 - [1/(2n_1) + 1/(2n_2)] \pm z_{1-\alpha/2}\sqrt{\hat{p}_1\hat{q}_1/n_1 + \hat{p}_2\hat{q}_2/n_2} \text{ if } \hat{p}_1 > \hat{p}_2$$

$$\hat{p}_1 - \hat{p}_2 + [1/(2n_1) + 1/(2n_2)] \pm z_{1-\alpha/2}\sqrt{\hat{p}_1\hat{q}_1/n_1 + \hat{p}_2\hat{q}_2/n_2} \text{ if } \hat{p}_1 \le \hat{p}_2$$

Use these expressions for the confidence interval only if $n_1\hat{p}_1\hat{q}_1 \ge 5$ and $n_2\hat{p}_2\hat{q}_2 \ge 5$.

Example 13.5 **Cardiovascular Disease** Referring to the OC–MI data in Table 10.2, provide a point estimate and a 95% confidence interval for the difference between the proportion of women who develop MI among OC users and the comparable proportion among non-OC users.

Solution We have that $n_1 = 5000$, $\hat{p}_1 = 13/5000 = .0026$, $n_2 = 10{,}000$, $\hat{p}_2 = 7/10{,}000 = .0007$. Thus, a point estimate of the risk difference $(p_1 - p_2)$ is given by $\hat{p}_1 - \hat{p}_2 = .0019$. Since $n_1\hat{p}_1\hat{q}_1 = 5000(.0026)(.9974) = 13.0 \ge 5$, $n_2\hat{p}_2\hat{q}_2 = 10{,}000(.0007)(.9993) = 7.0 \ge 5$, the large-sample confidence interval in Equation 13.1 can be used. The 95% confidence interval is given by

$$.0026 - .0007 - \left[\frac{1}{2(5000)} + \frac{1}{2(10{,}000)}\right] \pm 1.96\sqrt{\frac{.0026(.9974)}{5000} + \frac{.0007(.9993)}{10{,}000}}$$

$$= .00175 \pm 1.96(.00077)$$

$$= .00175 \pm .00150 = (.0002, .0033)$$

13.3.2 The Risk Ratio

A point estimate of the risk ratio $(RR = p_1/p_2)$ is given by

EQUATION 13.2

$$\widehat{RR} = \hat{p}_1/\hat{p}_2$$

To obtain an interval estimate, we will assume that the normal approximation to the binomial distribution is valid. Under this assumption, it can be shown that the sampling distribution of $\ln(\widehat{RR})$ more closely follows a normal distribution than $\widehat{RR}$ itself.

We note that

$$\begin{aligned} Var[\ln(\widehat{RR})] &= Var[\ln(\hat{p}_1) - \ln(\hat{p}_2)] \\ &= Var[\ln(\hat{p}_1)] + Var[\ln(\hat{p}_2)] \end{aligned}$$

To obtain $Var[\ln(\hat{p}_1)]$, we employ a principle known as the *delta method.*

EQUATION 13.3

Delta Method The variance of a function of a random variable $f(X)$ is approximated by

$$Var[f(X)] \cong [f'(X)]^2 Var(X)$$

Example 13.6 Use the delta method to find the variance of $\ln(\hat{p}_1)$, $\ln(\hat{p}_2)$, and $\ln(\widehat{RR})$.

Solution In this case $f(X) = \ln(X)$. Since $f'(X) = \frac{1}{X}$, we obtain

$$Var[\ln(\hat{p}_1)] = \frac{1}{\hat{p}_1^2} Var(\hat{p}_1) = \frac{1}{\hat{p}_1^2}\left(\frac{\hat{p}_1\hat{q}_1}{n_1}\right) = \frac{\hat{q}_1}{\hat{p}_1 n_1}$$

However, from Table 13.1, $\hat{p}_1 = a/n_1$, $\hat{q}_1 = b/n_1$. Therefore,

$$Var[\ln(\hat{p}_1)] = \frac{b}{an_1}$$

Also, using similar methods,

$$Var[\ln(\hat{p}_2)] = \frac{\hat{q}_2}{\hat{p}_2 n_2} = \frac{d}{cn_2}$$

It follows that

$$Var[\ln(\widehat{RR})] = \frac{b}{an_1} + \frac{d}{cn_2}$$

or $se[\ln(\widehat{RR})] = \sqrt{\frac{b}{an_1} + \frac{d}{cn_2}}$

Therefore, a two-sided $100\% \times (1 - \alpha)$ CI for $\ln(RR)$ is given by

EQUATION 13.4

$$\left[\ln(\widehat{RR}) - z_{1-\alpha/2}\sqrt{\frac{b}{an_1} + \frac{d}{cn_2}},\quad \ln(\widehat{RR}) + z_{1-\alpha/2}\sqrt{\frac{b}{an_1} + \frac{d}{cn_2}}\right]$$

The antilog of each end of the interval in Equation 13.4 then provides a two-sided $100\% \times (1 - \alpha)$ CI for RR itself, given by

EQUATION 13.5

$$\left[e^{\ln(\widehat{RR}) - z_{1-\alpha/2}\sqrt{b/(an_1) + d/(cn_2)}},\ e^{\ln(\widehat{RR}) + z_{1-\alpha/2}\sqrt{b/(an_1) + d/(cn_2)}}\right]$$

The estimation procedures for the risk ratio are summarized as follows:

EQUATION 13.6

Point and Interval Estimation for the Risk Ratio (*RR*) Let $\hat{p}_1, \hat{p}_2$ represent the sample proportions of exposed and unexposed individuals who develop disease in a prospective study, based on samples of size n_1 and n_2, respectively. A point estimate of the risk ratio (or relative risk) is given by $\hat{p}_1/\hat{p}_2$. A 100% × (1 − α) confidence interval for the risk ratio is given by $[\exp(c_1), \exp(c_2)]$, where

$$c_1 = \ln(\widehat{RR}) - z_{1-\alpha/2}\sqrt{\frac{b}{an_1} + \frac{d}{cn_2}}$$

$$c_2 = \ln(\widehat{RR}) + z_{1-\alpha/2}\sqrt{\frac{b}{an_1} + \frac{d}{cn_2}}$$

where a, b = number of exposed subjects who do and do not develop disease, respectively, and c, d = number of unexposed subjects who do and do not develop disease, respectively. This method of interval estimation is only valid if $n_1\hat{p}_1\hat{q}_1 \geq 5$ and $n_2\hat{p}_2\hat{q}_2 \geq 5$.

Example 13.7 **Cardiovascular Disease** Referring to Table 10.2, provide a point estimate and a 95% CI for the relative risk of MI among OC users compared with non-OC users.

Solution We have from Example 13.5 that $\hat{p}_1 = 13/5000 = .0026$, $n_1 = 5000$, $\hat{p}_2 = 7/10{,}000 = .0007$, $n_2 = 10{,}000$. Thus our point estimate of RR is $\widehat{RR} = \hat{p}_1/\hat{p}_2 = .0026/.0007 = 3.71$. To compute a 95% CI, we obtain c_1, c_2 in Equation 13.6. We have $a = 13$, $b = 4987$, $c = 7$, and $d = 9993$. Thus,

$$\begin{aligned} c_1 &= \ln\left(\frac{.0026}{.0007}\right) - 1.96\sqrt{\frac{4987}{13(5000)} + \frac{9993}{7(10,000)}} \\ &= 1.312 - 1.96(0.4685) \\ &= 1.312 - 0.918 = 0.394 \\ c_2 &= 1.312 + 0.918 = 2.230 \end{aligned}$$

Therefore, our 95% CI for $RR = (e^{0.394}, e^{2.230}) = (1.5, 9.3)$.

13.3.3 The Odds Ratio

In Section 13.3.2, the risk ratio (or relative risk) was introduced. The relative risk can be expressed as the ratio of the probability of disease among exposed subjects (p_1) divided by the probability of disease among unexposed subjects (p_2). Although easily understood, the relative risk has the disadvantage of being constrained by the denominator probability (p_2). For example, if $p_2 = .5$, then the RR can be no larger than $1/.5 = 2$; if $p_2 = .8$, then the RR can be no larger than $1/.8 = 1.25$. To avoid this restriction, another comparative measure relating two proportions is sometimes used, called the *odds ratio*. The odds in favor of a success are defined as follows:

DEFINITION 13.5 If the probability of a success = p, then the **odds in favor of success** = $p/(1 - p)$.

If two proportions p_1, p_2 are considered and the odds in favor of success are computed for each proportion, then the ratio of odds, or odds ratio, becomes a useful measure for relating the two proportions.

DEFINITION 13.6 Let p_1, p_2 be the underlying probability of success for two groups. The **odds ratio (*OR*)** is defined as

$$OR = \frac{p_1/q_1}{p_2/q_2} = \frac{p_1 q_2}{p_2 q_1} \quad \text{and is estimated by} \quad \widehat{OR} = \frac{\hat{p}_1 \hat{q}_2}{\hat{p}_2 \hat{q}_1}$$

Equivalently, if the four cells of the 2 × 2 contingency table are labeled by *a*, *b*, *c*, *d*, as they are in Table 13.1, then

$$\widehat{OR} = \frac{[a/(a+b)] \times [d/(c+d)]}{[c/(c+d)] \times [b/(a+b)]} = \frac{ad}{bc}$$

In the context of a prospective study, the odds ratio can be interpreted as the odds in favor of disease for exposed subjects divided by the odds in favor of disease for the unexposed subjects. This is sometimes referred to as the *disease odds ratio.*

DEFINITION 13.7 The **disease odds ratio** is the odds in favor of disease for the exposed group divided by the odds in favor of disease for the unexposed group.

Example 13.8 **Cardiovascular Disease** Using the OC–MI data in Table 10.2, estimate the odds ratio in favor of MI for an OC user compared with a non-OC user (i.e., the disease odds ratio).

Solution We have $\hat{p}_1 = .0026$, $\hat{q}_1 = .9974$, $\hat{p}_2 = .0007$, $\hat{q}_2 = .9993$. Thus,

$$\widehat{OR} = \frac{.0026(.9993)}{.0007(.9974)} = 3.72$$

This means that the odds in favor of an MI for an OC user is 3.7 times the odds in favor of an MI for a non-OC user. The $\widehat{OR}$ could also have been computed from the contingency table in Table 10.2, whereby

$$\widehat{OR} = \frac{13 \times 9993}{7 \times 4987} = 3.72$$

If the probability of disease is the same for exposed and unexposed subjects, $OR = 1$. Conversely, odds ratios greater than 1 indicate a greater likelihood of disease among the exposed than among the unexposed, whereas odds ratios less than 1 indicate a greater likelihood of disease among the unexposed than among the exposed. Notice that there is no restriction on the odds ratio as there was for the risk ratio. Specifically, as the probability of disease among the exposed (p_1) approaches 0, *OR* approaches 0, whereas as p_1 approaches 1, *OR* approaches ∞, regardless of the value of the probability of disease among the unexposed (p_2). This property is particularly advantageous when combining results over several 2 × 2 tables, as discussed in Section 13.5. Finally, if the probabilities of success are low (i.e., p_1, p_2 are small), then $1 - p_1$ and $1 - p_2$ will each be close to 1, and the odds ratio will be approximately the same as the relative risk. Thus, the odds ratio is often used as an approximation to the relative risk for rare diseases.

In Example 13.8, we have computed the odds ratio as a disease odds ratio. However, another way to express the odds ratio is as an exposure odds ratio.

DEFINITION 13.8 The **exposure odds ratio** is the odds in favor of being exposed for diseased subjects divided by the odds in favor of being exposed for nondiseased subjects.

From Table 13.1, this is given by

EQUATION 13.7

$$\text{exposure odds ratio} = \frac{[a/(a+c)]/[c/(a+c)]}{[b/(b+d)]/[d/(b+d)]} = \frac{ad}{bc} = \text{disease odds ratio}$$

Therefore, the exposure odds ratio is the same as the disease odds ratio. This relationship is particularly useful for case–control studies. For prospective studies, we have seen that we can estimate either the risk difference, the risk ratio, or the odds ratio. For case–control studies, we cannot directly estimate either the risk difference or the risk ratio. To see why this is so, let A, B, C, and D represent the true number of subjects in the reference population, corresponding to the cells a, b, c, and d in our sample as shown in Table 13.2.

TABLE 13.2 **Hypothetical tables depicting exposure–disease relationships in a sample and a reference population**

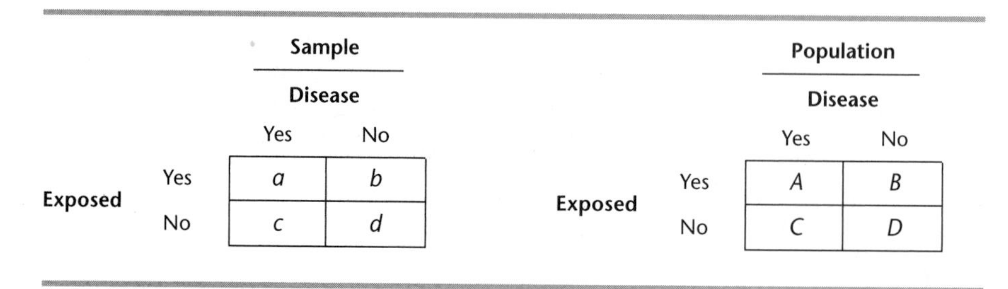

Sample

		Disease: Yes	Disease: No
Exposed	Yes	*a*	*b*
	No	*c*	*d*

Population

		Disease: Yes	Disease: No
Exposed	Yes	*A*	*B*
	No	*C*	*D*

In a case–control study, we assume that a random fraction f_1 of subjects with disease and a random fraction f_2 of subjects without disease in the reference population are included in our study sample. We also assume there is no sampling bias, so f_1 is the sampling fraction for both exposed and unexposed subjects with disease and f_2 is the sampling fraction for both exposed and unexposed subjects without disease. Therefore, $a = f_1A$, $c = f_1C$, $b = f_2B$, $d = f_2D$. If we attempt to estimate the risk ratio from our study sample, we obtain

EQUATION 13.8

$$\widehat{RR} = \frac{a/(a+b)}{c/(c+d)}$$

$$= \frac{f_1 A/(f_1 A + f_2 B)}{f_1 C/(f_1 C + f_2 D)}$$

$$= \frac{A/(f_1 A + f_2 B)}{C/(f_1 C + f_2 D)}$$

However, from Table 13.2, the true relative risk in the reference population is

EQUATION 13.9

$$RR = \frac{A/(A+B)}{C/(C+D)}$$

The expressions on the right-hand side of Equations 13.8 and 13.9 will only be the same if $f_1 = f_2$; that is, if the sampling fraction of subjects with disease and without disease are the same. However, this is very unlikely to be true in a case–control study, because the usual sampling strategy is to oversample subjects with disease.

Example 13.9 **Cancer** Consider a case–control study of the relationship of dietary factors to colon cancer. Suppose 100 colon-cancer cases are selected from a tumor registry, and 100 controls are chosen who live in the same census tract as the cases and have approximately the same age and sex distribution. Thus, there are an equal number of cases and controls in the sample even though the fraction of people with colon cancer in the census tract may be very low. Therefore, f_1 will be much larger than f_2. Thus, $\widehat{RR}$ will provide a biased estimate of RR in most case–control studies. This is also true for the risk difference. However, we can estimate the odds ratio from our sample in Table 13.2 given by

$$\widehat{OR} = \frac{ad}{bc}$$

$$= \frac{f_1 A(f_2 D)}{f_2 B(f_1 C)}$$

$$= \frac{AD}{BC} = OR$$

Thus, the odds ratio estimated from our sample will provide an unbiased estimate of the odds ratio from our reference population. We saw in Equation 13.7 that the exposure and disease odds ratios are the same for any 2×2 table relating exposure to disease, regardless of the sampling strategy. Therefore, the odds ratio from a case–control study provides an unbiased estimate of the true disease odds ratio. However, if the disease under study is rare, then the disease odds ratio is approximately the same as the risk ratio. This allows us to indirectly estimate the risk ratio for case–control studies.

The general method of estimation of the risk ratio in case–control studies is summarized as follows:

EQUATION 13.10

Estimation of the Risk Ratio for Case–Control Studies Suppose we have a 2×2 table relating exposure to disease as in Table 13.1. If the data are collected using a case–control study design, and the disease under study is rare (disease incidence < .10), then we can estimate the risk ratio by

$$\widehat{RR} \cong \widehat{OR} = \frac{ad}{bc}$$

Example 13.10 **Cancer** Estimate the risk ratio for breast cancer for women with a late age at first birth (≥30) compared to women with an early age at first birth (≤29) based on the data in Table 10.1.

Solution The estimated odds ratio is given by

$$\widehat{OR} = \frac{ad}{bc}$$

$$= \frac{683(8747)}{2537(1498)} = 1.57$$

This is an estimate of the risk ratio because, although breast cancer is one of the most common cancers in women, its incidence in the general population of women is relatively low, unless very old women are considered.

13.3.4 Interval Estimates for the Odds Ratio

In Section 13.3.3, we discussed how to estimate the odds ratio. We saw that, in a case–control study with a rare disease outcome, the odds ratio provides an approximate estimate of risk ratio. The issue remains as to how to obtain interval estimates for the odds ratio. Several methods exist for this purpose. One of the most popular approaches is the Woolf method [1]. Woolf showed that, approximately,

$$Var[\ln(\widehat{OR})] \cong \frac{1}{a} + \frac{1}{b} + \frac{1}{c} + \frac{1}{d}$$

where a, b, c, and d are the four cells of our 2×2 contingency table. To see this, suppose that we have a prospective design. From Definition 13.6 we can represent the estimated odds ratio as a disease odds ratio in the form $(\hat{p}_1/\hat{q}_1)/(\hat{p}_2/\hat{q}_2)$

where $\hat{p}_1 = a/(a+b)$, $\hat{p}_2 = c/(c+d)$, $\hat{q}_1 = 1-\hat{p}_1$, $\hat{q}_2 = 1-\hat{p}_2$

Furthermore,

$$\begin{aligned} Var[\ln(\widehat{OR})] &= Var[\ln(\hat{p}_1/\hat{q}_1)/(\hat{p}_2/\hat{q}_2)] \\ &= Var[\ln(\hat{p}_1/\hat{q}_1) - \ln(\hat{p}_2/\hat{q}_2)] \\ &= Var[\ln(\hat{p}_1/\hat{q}_1)] + Var[\ln(\hat{p}_2/\hat{q}_2)] \end{aligned}$$

To obtain $Var[\ln(\hat{p}_1/\hat{q}_1)]$, we use the delta method. We have that

$$\frac{d[\ln(\hat{p}_1/\hat{q}_1)]}{d\hat{p}_1} = \frac{1}{\hat{p}_1\hat{q}_1}$$

Furthermore, $Var(\hat{p}_1) = \hat{p}_1\hat{q}_1/(a+b)$

Hence,

$$Var[\ln(\hat{p}_1/\hat{q}_1)] = \left(\frac{1}{\hat{p}_1\hat{q}_1}\right)^2 \frac{\hat{p}_1\hat{q}_1}{a+b}$$

$$= \frac{1}{(a+b)\hat{p}_1\hat{q}_1}$$

$$= \frac{1}{(a+b)\left(\frac{a}{a+b}\right)\left(\frac{b}{a+b}\right)}$$

$$= \frac{a+b}{ab} = \frac{1}{a} + \frac{1}{b}$$

Similarly, $Var[\ln(\hat{p}_2/\hat{q}_2)] = \frac{1}{c} + \frac{1}{d}$. It follows that

$$Var[\ln(\widehat{OR})] = \frac{1}{a} + \frac{1}{b} + \frac{1}{c} + \frac{1}{d}$$

A similar result can be obtained if we have a case–control design instead of a prospective design.

If we assume approximate normality of $\ln(\widehat{OR})$, then a $100\% \times (1 - \alpha)$ CI for $\ln(OR)$ is given by

$$\ln(\widehat{OR}) \pm z_{1-\alpha/2}\sqrt{\frac{1}{a} + \frac{1}{b} + \frac{1}{c} + \frac{1}{d}}$$

If we take the antilog of each end of the CI, then if follows that a $100\% \times (1 - \alpha)$ CI for OR is given by

$$e^{\ln(\widehat{OR}) \pm z_{1-\alpha/2}\sqrt{1/a+1/b+1/c+1/d}} = \left(\widehat{OR}\, e^{-z_{1-\alpha/2}\sqrt{1/a+1/b+1/c+1/d}},\ \widehat{OR}\, e^{z_{1-\alpha/2}\sqrt{1/a+1/b+1/c+1/d}}\right)$$

This approach is summarized as follows.

EQUATION 13.11

Point and Interval Estimation for the Odds Ratio (Woolf Procedure) Suppose we have a 2×2 contingency table relating exposure to disease, with cell counts a, b, c, d as given in Table 13.1.

(1) A point estimate of the true odds ratio (OR) is given by $\widehat{OR} = ad/bc$.

(2) An approximate two-sided $100\% \times (1 - \alpha)$ CI for OR is given by (e^{c_1}, e^{c_2}), where

$$c_1 = \ln(\widehat{OR}) - z_{1-\alpha/2}\sqrt{\frac{1}{a} + \frac{1}{b} + \frac{1}{c} + \frac{1}{d}}$$

$$c_2 = \ln(\widehat{OR}) + z_{1-\alpha/2}\sqrt{\frac{1}{a} + \frac{1}{b} + \frac{1}{c} + \frac{1}{d}}$$

(3) In a prospective or a cross-sectional study, the CI in (2) should only be used if $n_1\hat{p}_1\hat{q}_1 \ge 5$ and $n_2\hat{p}_2\hat{q}_2 \ge 5$ where

n_1 = the number of exposed individuals

$\hat{p}_1$ = sample proportion with disease among exposed individuals and $\hat{q}_1 = 1 - \hat{p}_1$

n_2 = the number of unexposed individuals

$\hat{p}_2$ = the sample proportion with disease among unexposed individuals, and $\hat{q}_2 = 1 - \hat{p}_2$

(4) In a case–control study, the CI should only be used if $m_1\hat{p}_1^*\hat{q}_1^* \ge 5$ and $m_2\hat{p}_2^*\hat{q}_2^* \ge 5$ where

m_1 = the number of cases

$\hat{p}_1^*$ = the proportion of cases that are exposed, and $\hat{q}_1^* = 1 - \hat{p}_1^*$

m_2 = the number of controls

$\hat{p}_2^*$ = the proportion of controls that are exposed, and $\hat{q}_2^* = 1 - \hat{p}_2^*$

(5) If the disease under study is rare, then $\widehat{OR}$ and its associated $100\% \times (1 - \alpha)$ CI can be interpreted as approximate point and interval estimates of the risk ratio. This is particularly important in case–control studies where no direct estimate of the risk ratio is available.

Example 13.11 Compute a point estimate and a 95% CI for the odds ratio relating age at first birth to breast-cancer incidence based on the data in Table 10.1.

Solution From Example 13.10 we see that the point estimate of the odds ratio = 1.57. To obtain an interval estimate, we first compute a 95% CI for ln(*OR*) as follows:

$$= \ln(\widehat{OR}) \pm z_{.975}\sqrt{\frac{1}{a}+\frac{1}{b}+\frac{1}{c}+\frac{1}{d}}$$

$$= \ln(1.572) \pm 1.96\sqrt{\frac{1}{683}+\frac{1}{2537}+\frac{1}{1498}+\frac{1}{8747}}$$

$$= 0.452 \pm 1.96(0.0514)$$

$$= 0.452 \pm 0.101 = (0.352, 0.553)$$

A 95% CI for *OR* is thus given by

$$(e^{0.352},\ e^{0.553}) = (1.42,\ 1.74)$$

Because the 95% CI excludes 1, we can conclude that the true *OR* is significantly greater than 1. Also, this is a relatively rare disease, so we can also interpret this interval as an approximate 95% CI for the risk ratio.

In this section, we have learned about the risk difference, risk ratio, and odds ratio, the principal effect measures used in epidemiologic studies. The risk difference and risk ratio can be estimated directly from prospective studies, but not from case–control studies. The odds ratio is estimable from both prospective and case–control studies. In case–control studies with a rare disease outcome, the odds ratio provides an indirect estimate of the risk ratio. We also discussed large-sample methods for obtaining confidence limits for the preceding effect measures. To obtain confidence limits for the risk ratio and odds ratio, we introduced a general technique called the *delta method* to obtain the variance of a function of a random variable, such as ln(*X*), if the variance of the random variable (*X*) is already known.

SECTION 13.4 Confounding and Standardization

13.4.1 Confounding

When looking at the relationship between a disease and an exposure variable, it is often important to control for the effect of some other variable that is associated with both the disease and the exposure variable.

DEFINITION 13.9 A **confounding variable** is a variable that is associated with both the disease and the exposure variable. Such a variable must usually be controlled for before looking at the disease–exposure relationship.

Example 13.12 **Cancer** Suppose we are interested in the relationship between lung-cancer incidence and heavy drinking (defined as ≥2 drinks per day). We conduct a prospective study where drinking status is determined at baseline and the cohort is followed for 10 years to determine cancer endpoints. The following 2 × 2 table is constructed relating lung-cancer incidence to initial drinking status, where we compare heavy drinkers (≥2 drinks per day) versus nondrinkers. The results are given in Table 13.3.

TABLE 13.3 **Crude relationship between lung-cancer incidence and drinking status**

		Lung cancer		
		Yes	No	
Drinking status	Heavy drinker	33	1667	1700
	Nondrinker	27	2273	2300
		60	3940	4000

Because lung cancer is a relatively rare disease, we will estimate the risk ratio by the odds ratio = (33 × 2273)/(27 × 1667) = 1.67. Thus, it appears that heavy drinking is a risk factor for lung cancer.

We refer to Table 13.3 as expressing the "crude" relationship between lung-cancer incidence and drinking status. The adjective "crude" means that the relationship is being presented without any adjustment for possible confounding variables. One such confounding variable is smoking, because smoking is related to both drinking and lung-cancer incidence. Specifically, one explanation for the crude association found in Table 13.3 between lung-cancer incidence and drinking may be that smokers are more likely to both develop lung cancer and to be heavy drinkers than nonsmokers. To investigate this hypothesis, we can look at the relationship between lung-cancer incidence and drinking status after controlling for smoking (i.e., separately for smokers and nonsmokers at baseline). These data are given in Table 13.4.

We see that smoking is related to drinking status. Specifically, 800 of the 1000 smokers (80%) versus 900 of the 3000 nonsmokers (30%) are heavy drinkers. Also,

TABLE 13.4 Relationship between lung-cancer incidence and drinking status while controlling for smoking status at baseline

(a) Smokers at baseline

		Lung cancer		
		Yes	No	
Drinking status	Heavy drinker	24	776	800
	Nondrinker	6	194	200
		30	970	1000

(b) Nonsmokers at baseline

		Lung cancer		
		Yes	No	
Drinking status	Heavy drinker	9	891	900
	Nondrinker	21	2079	2100
		30	2970	3000

smoking is related to lung cancer. Specifically, 30 of the 1000 smokers (3%) versus 30 of the 3000 nonsmokers (1%) developed lung cancer.

Example 13.13 **Cancer** Investigate the relationship between lung-cancer incidence versus drinking status, while controlling for smoking.

Solution To investigate the relationship between lung-cancer incidence and drinking status, while controlling for smoking status, we can compute separate odds ratios for Table 13.4a and b. The odds ratio relating lung cancer to drinking status among smokers is $OR = (24 \times 194)/(6 \times 776) = 1.0$, while the comparable odds ratio among nonsmokers is $OR = (9 \times 2079)/(21 \times 891) = 1.0$. Thus, after controlling for the confounding variable smoking, there is *no* relationship between lung cancer and drinking status.

DEFINITION 13.10 The analysis of disease–exposure relationships in separate subgroups of the data, where the subgroups are defined by one or more potential confounders, is referred to as **stratification.** The subgroups themselves are referred to as **strata.**

We refer to smoking as a *positive confounder* because it is related in the same direction to both lung-cancer incidence and heavy drinking.

DEFINITION 13.11 A **positive confounder** is a confounder that either

(1) is positively related to both exposure and disease, or

(2) is negatively related to both exposure and disease

After adjusting for a positive confounder, an adjusted risk ratio or odds ratio is lower than the crude risk ratio or odds ratio.

DEFINITION 13.12 A **negative confounder** is a confounder that either

(1) is positively related to disease and negatively related to exposure, or

(2) is negatively related to disease and positively related to exposure

After adjusting for a negative confounder, an adjusted risk ratio or odds ratio is greater than the crude risk ratio or odds ratio.

Example 13.14 **Cancer** In Table 13.4, what type of confounder is smoking?

Solution Smoking is a positive confounder because it is positively related to both heavy drinking (the exposure) and lung cancer (the disease). Indeed, the crude positive association between lung cancer and drinking status ($OR = 1.67$) was reduced to no association at all ($OR = 1.0$) once smoking was controlled for.

Example 13.15 **Cardiovascular Disease** The relationship between oral-contraceptive (OC) use and myocardial infarction (MI) after stratification by age was considered by Shapiro et al. [2]. The data are given in Table 13.5.

We see that the age-specific odds ratios tend to be higher than the crude odds ratio (1.7) where age was not controlled for. Age is an example of a negative confounder here because it is negatively associated with OC use (older women use OC's less frequently than younger women) and positively associated with disease (older women are more likely to be cases than younger women). Thus, the age-specific odds ratios will tend to be higher than the crude odds ratio.

TABLE 13.5 **Association between myocardial infarction (MI) and oral-contraceptive (OC) use by age**

Age	Recent OC use	Cases (MI)	Controls	$\widehat{OR}$	Proportion OC user (%)	Proportion MI (%)
25–29	yes	4	62	7.2	23	2
	no	2	224			
30–34	yes	9	33	8.9	9	5
	no	12	390			
35–39	yes	4	26	1.5	8	9
	no	33	330			
40–44	yes	6	9	3.7	3	16
	no	65	362			
45–49	yes	6	5	3.9	3	24
	no	93	301			
Total	yes	29	135	1.7		
	no	205	1607			

An often asked question is, When is it reasonable to control for a confounder when exploring the relationship between an exposure and disease? This will depend on whether or not the confounder is in the "causal pathway" between exposure and disease.

DEFINITION 13.13 A confounder is said to be in the **causal pathway** between exposure and disease if (1) the exposure is causally related to the confounder and (2) the confounder is causally related to disease.

Example 13.16 **Cardiovascular Disease** Suppose we are interested in the possible association between obesity and the development of coronary disease. If we examine the crude relationship between obesity and coronary disease, we will usually find that obese people have higher rates of coronary disease than people of normal weight. However, obesity is positively related to both hypertension and diabetes. In some studies, once hypertension and/or diabetes are controlled for as confounders, there is a much weaker relationship or even no relationship between obesity and the development of coronary heart disease. Does this mean that there is no real association between obesity and coronary heart disease?

Solution No, it does not. If obesity is an important cause of both hypertension and diabetes and both are causally related to the development of coronary heart disease, then hypertension and diabetes are in the causal pathway between obesity and the development of coronary heart disease. It would be inappropriate to include hypertension or diabetes as confounders of the relationship between obesity and coronary heart disease, because they are in the causal pathway.

The decision as to which confounders are in the causal pathway should be made on the basis of biological rather than purely statistical considerations.

13.4.2 Standardization

Age is often an important confounder influencing both exposure and disease rates. For this reason, it is often a routine procedure to control for age when assessing disease–exposure relationships. A first step is sometimes to compute rates for the exposed and unexposed groups that have been "age standardized." The term "age standardized" means that the expected disease rate in the exposed and unexposed groups are each based on an age distribution from a standard reference population. If the same standard is used for both the exposed and unexposed groups, then a comparison can be made between the two standardized rates that is not confounded by possible age differences between the two populations.

Example 13.17 **Infectious Disease** The presence of bacteria in the urine (bacteriuria) has been associated with kidney disease. Conflicting results have been reported from several studies concerning the possible role of oral contraceptives (OC) on bacteriuria. The following data were collected in a population-based group of nonpregnant premenopausal women below the age of 50 [3]. The data are presented on an age-specific basis in Table 13.6.

TABLE 13.6 **Risk of bacteriuria among oral-contraceptive users and nonusers**

	% with bacteriuria			
	OC users		Non-OC users	
Age group	%	*n*	%	*n*
16–19	1.2	84	3.2	281
20–29	5.6	284	4.0	552
30–39	6.3	96	5.5	623
40–49	22.2	18	2.7	482

Source: Reprinted with permission of the *New England Journal of Medicine, 299*, 536–537, 1978.

The prevalence of bacteriuria generally increases with age. In addition, the age distribution of OC users and non-OC users differs considerably, with OC use more common among younger women. Thus, for descriptive purposes we would like to compute age-standardized rates of bacteriuria separately for OC users and non-OC users, and compare them using a risk ratio.

DEFINITION 13.14

Suppose people in a study population are stratified into k age groups. Let the risk of disease among the exposed in the ith age group = $\hat{p}_{i1} = x_{i1}/n_{i1}$ where x_{i1} = number of exposed subjects with disease in the ith age group and n_{i1} = total number of exposed subjects in the ith age group, $i = 1, \ldots, k$. Let the risk of disease among the unexposed in the ith age group = $\hat{p}_{i2} = x_{i2}/n_{i2}$, where x_{i2} = number of unexposed subjects with disease in the ith age group and n_{i2} = total number of unexposed subjects in the ith age group, $i = 1, \ldots, k$. Let n_i = number of subjects in the ith age group in a *standard* population, $i = 1, \ldots, k$.

The **age-standardized risk** of disease among the exposed = $\hat{p}_1^* = \sum_{i=1}^{k} n_i \hat{p}_{i1} \Big/ \sum_{i=1}^{k} n_i$.

The **age-standardized risk** of disease among the unexposed = $\hat{p}_2^* = \sum_{i=1}^{k} n_i \hat{p}_{i2} \Big/ \sum_{i=1}^{k} n_i$.

The **standardized risk ratio** = $\hat{p}_1^* / \hat{p}_2^*$.

Example 13.18 **Infectious Disease** Using the data in Table 13.6, compute the age-standardized risk of bacteriuria separately for OC users and non-OC users, and compute the standardized risk ratio for bacteriuria for OC users versus non-OC users.

Solution The age distribution of the total study population is

Age group	n
16–19	365
20–29	836
30–39	719
40–49	500
Total	2420

The age-standardized risk of bacteriuria for OC users (the exposed) is

$$\hat{p}_1^* = \frac{365(.012) + 836(.056) + 719(.063) + 500(.222)}{2420} = \frac{207.493}{2420} = .086$$

The age-standardized risk of bacteriuria for non-OC users (the unexposed) is

$$\hat{p}_2^* = \frac{365(.032) + 836(.040) + 719(.055) + 500(.027)}{2420} = \frac{98.165}{2420} = .041$$

The standardized risk ratio = .086/.041 = 2.1.

This method of standardization is sometimes referred to as **direct standardization.** The use of age-standardized risks is somewhat controversial because results may differ depending on which standard is used. However, space limitations often make it impossible to present age-specific results in a paper, and the reader can get a quick summary of the overall results from the age-standardized risks.

The use of standardized risks is a good descriptive tool for controlling for confounding. In the next section, we will discuss how to control for confounding in assessing disease–exposure relationships in a hypothesis-testing framework using the Mantel-Haenszel test. Finally, standardization can be performed based on stratification by factors other than age. For example, standardization by both age and sex is common. Similar methods can be used to obtain age–sex standardized risks, and standardized risk ratios as given in Definition 13.14.

In this section, we have introduced the concept of a confounding variable (C) which is a variable related to both the disease (D) and exposure (E) variables. Furthermore, we classified confounding variables as positive confounders if the directions of association between C versus D and C versus E, respectively, are in the same direction and as negative confounders if the associations between C versus D and C versus E are in opposite directions. We also discussed when it is or is not appropriate to control for a confounder, according to whether C is or is not in the causal pathway between E and D. Finally, because age is often an important confounding variable, it is reasonable to consider descriptive measures of proportions and relative risk that control for age. Age-standardized proportions and risk ratios are such measures.

SECTION 13.5 Methods of Inference for Stratified Categorical Data—The Mantel-Haenszel Test

Example 13.19 **Cancer** A 1985 study identified a group of 518 cancer cases ages 15–59 and a group of 518 age- and sex-matched controls by mail questionnaire [4]. The main purpose of the study was to look at the effect of passive smoking on cancer risk. In the study, passive smoking was defined as exposure to the cigarette smoke of a spouse who smoked at least one cigarette per day for at least 6 months. One potential confounding variable was smoking by the test subjects themselves (i.e., personal smoking), because personal smoking is related to both cancer risk and spouse smoking. Therefore, it was important to control for personal smoking before looking at the relationship between passive smoking and cancer risk.

To display the data, a 2×2 table relating case–control status to passive smoking can be constructed for both nonsmokers and smokers. The data are given in Table 13.7 for nonsmokers and Table 13.8 for smokers.

The passive-smoking effect can be assessed separately for nonsmokers and smokers. Indeed, we notice from Tables 13.7 and 13.8 that the odds ratio in favor of a case being exposed to cigarette smoke from a spouse who smokes versus a control is $(120 \times 155)/(80 \times 111) = 2.1$ for nonsmokers, whereas the corresponding odds ratio for smokers is $(161 \times 124)/(130 \times 117) = 1.3$. Thus for both subgroups the trend is in the direction of more passive smoking among cases than controls. The

TABLE 13.7 **Relationship of passive smoking to cancer risk among nonsmokers**

	Passive smoker		
Case–control status	Yes	No	Total
Case	120	111	231
Control	80	155	235
Total	200	266	466

Source: Reprinted with permission of the *American Journal of Epidemiology, 121*(1), 37–48, 1985.

TABLE 13.8 **Relationship of passive smoking to cancer risk among smokers**

	Passive smoker		
Case–control status	Yes	No	Total
Case	161	117	278
Control	130	124	254
Total	291	241	532

Source: Reprinted with permission of the *American Journal of Epidemiology, 121*(1), 37–48, 1985.

key question is how to combine the results of the two tables to obtain an overall test of significance for the passive-smoking effect.

In general, the data will be stratified into k subgroups according to one or more confounding variables to make the units within a stratum as homogeneous as possible. The data for each stratum consist of a 2×2 contingency table relating exposure to disease, as shown in Table 13.9 for the ith stratum.

TABLE 13.9 **Relationship of disease to exposure in the *i*th stratum**

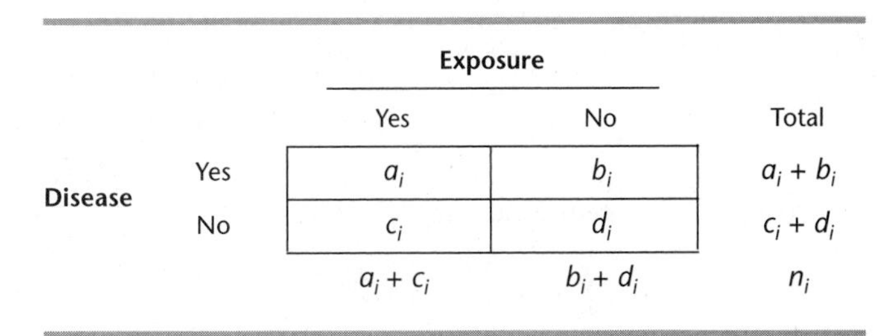

		Exposure		
		Yes	No	Total
Disease	Yes	a_i	b_i	$a_i + b_i$
	No	c_i	d_i	$c_i + d_i$
		$a_i + c_i$	$b_i + d_i$	n_i

Based on our work on Fisher's exact test, the distribution of a_i follows a **hypergeometric distribution.** The test procedure will be based on a comparison of the observed number of units in the (1, 1) cell of each stratum (denoted by $O_i = a_i$) with the expected number of units in that cell (denoted by E_i). The test procedure is the same regardless of the order of the rows and columns; that is, which row (or column) is designated as the first row (or column) is arbitrary. Based on the hypergeometric distribution (Equation 10.9), the expected number of units in the (1, 1) cell of the ith stratum is given by

EQUATION 13.12

$$E_i = \frac{(a_i + b_i)(a_i + c_i)}{n_i}$$

The observed and expected numbers of units in the (1, 1) cell are then summed over all strata, obtaining $O = \sum_{i=1}^{k} O_i$, $E = \sum_{i=1}^{k} E_i$, and the test is based on $O - E$. Based on the hypergeometric distribution (Equation 10.9), the variance of O_i is given by

EQUATION 13.13

$$V_i = \frac{(a_i + b_i)(c_i + d_i)(a_i + c_i)(b_i + d_i)}{n_i^2(n_i - 1)}$$

Furthermore, the variance of $O = V = \sum_{i=1}^{k} V_i$. The test statistic is given by $X_{MH}^2 = \left(|O - E| - .5\right)^2 / V$, which should follow a chi-square distribution with 1 df under the null hypothesis of no association between disease and exposure. H_0 is rejected if X_{MH}^2 is large. The abbreviation MH refers to Mantel-Haenszel; this procedure is known as the Mantel-Haenszel test and is summarized as follows:

EQUATION 13.14

Mantel-Haenszel Test To assess the association between a dichotomous disease and a dichotomous exposure variable after controlling for one or more confounding variables, use the following procedure:

(1) Form k strata, based on the level of the confounding variable(s), and construct a 2 × 2 table relating disease and exposure within each stratum, as shown in Table 13.9.

(2) Compute the total observed number of units (O) in the (1, 1) cell over all strata, where

$$O = \sum_{i=1}^{k} O_i = \sum_{i=1}^{k} a_i$$

(3) Compute the total expected number of units (E) in the (1, 1) cell over all strata, where

$$E = \sum_{i=1}^{k} E_i = \sum_{i=1}^{k} \frac{(a_i + b_i)(a_i + c_i)}{n_i}$$

(4) Compute the variance (V) of O, where

$$V = \sum_{i=1}^{k} V_i = \sum_{i=1}^{k} \frac{(a_i + b_i)(c_i + d_i)(a_i + c_i)(b_i + d_i)}{n_i^2(n_i - 1)}$$

(5) The test statistic is then given by

$$X_{MH}^2 = \frac{(|O - E| - .5)^2}{V}$$

which under H_0 follows a chi-square distribution with 1 *df*.

(6) For a two-sided test with significance level α, if

$$X_{MH}^2 > \chi_{1,1-\alpha}^2$$

then reject H_0. If $X_{MH}^2 \le \chi_{1,1-\alpha}^2$

then accept H_0.

(7) The exact *p*-value for this test is given by

$$p = Pr(\chi_1^2 > X_{MH}^2)$$

(8) Use this test only if the variance $V \ge 5$.

(9) Which row or column is designated as first is arbitrary. The test statistic X_{MH}^2 and the assessment of significance are the same regardless of the order of the rows and columns.

The acceptance and rejection regions for the Mantel-Haenszel test are depicted in Figure 13.1. The computation of the *p*-value for the Mantel-Haenszel test is illustrated in Figure 13.2.

Example 13.20 **Cancer** Assess the relationship between passive smoking and cancer risk using the data stratified by personal smoking status in Tables 13.7 and 13.8.

Solution Denote the nonsmokers as stratum 1 and the smokers as stratum 2.

O_1 = observed number of nonsmoking cases who are passive smokers = 120

O_2 = observed number of smoking cases who are passive smokers = 161

FIGURE 13.1 **Acceptance and rejection regions for the Mantel-Haenszel test**

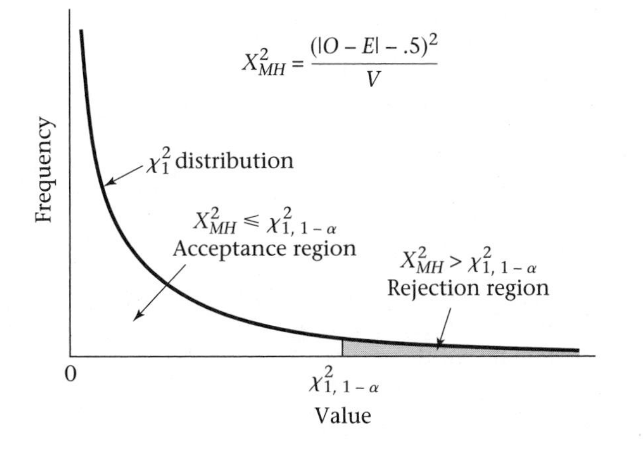

FIGURE 13.2 **Computation of the *p*-value for the Mantel-Haenszel test**

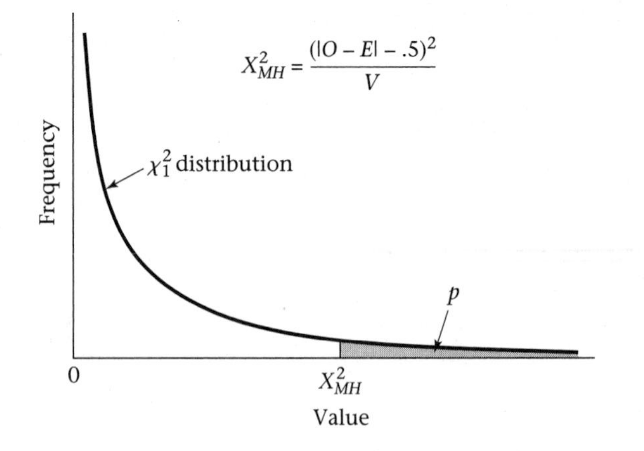

Furthermore,

$$E_1 = \frac{231 \times 200}{466} = 99.1$$

$$E_2 = \frac{278 \times 291}{532} = 152.1$$

Thus, the total observed and expected numbers of cases who are passive smokers are, respectively,

$$O = O_1 + O_2 = 120 + 161 = 281$$

$$E = E_1 + E_2 = 99.1 + 152.1 = 251.2$$

Therefore, there are more cases who are passive smokers than would be expected based on their personal smoking habits. Now compute the variance to assess if this difference is statistically significant.

$$V_1 = \frac{231 \times 235 \times 200 \times 266}{466^2 \times 465} = 28.60$$

$$V_2 = \frac{278 \times 254 \times 291 \times 241}{532^2 \times 531} = 32.95$$

Therefore, $V = V_1 + V_2 = 28.60 + 32.95 = 61.55$

Thus, the test statistic X^2_{MH} is given by

$$X^2_{MH} = \frac{\left(|281 - 251.2| - .5\right)^2}{61.55} = \frac{858.17}{61.55} = 13.94 \sim \chi^2_1 \text{ under } H_0$$

Since $\chi^2_{1,.999} = 10.83 < 13.94 = X^2_{MH}$, it follows that $p < .001$. Thus, there is a highly significant positive association between case–control status and passive-smoking exposure, even after controlling for personal cigarette-smoking habit.

13.5.1 Estimation of the Odds Ratio for Stratified Data

The Mantel-Haenszel test provides a test of significance of the relationship between disease and exposure. However, it does not give a measure of the strength of the association. Ideally, we would like a measure similar to the odds ratio presented for a single 2 × 2 contingency table in Definition 13.6. Assuming that the underlying odds ratio is the same for each stratum, an estimate of the common underlying odds ratio is provided by the Mantel-Haenszel estimator as follows:

EQUATION 13.15

Mantel-Haenszel Estimator of the Common Odds Ratio for Stratified Data
In a collection of k 2 × 2 contingency tables, where the ith table corresponding to the ith stratum is denoted as in Table 13.9, the Mantel-Haenszel estimator of the common odds ratio is given by

$$\widehat{OR}_{MH} = \frac{\sum_{i=1}^{k} a_i d_i / n_i}{\sum_{i=1}^{k} b_i c_i / n_i}$$

Example 13.21 **Cancer** Estimate the odds ratio in favor of being a passive smoker for cancer cases versus controls after controlling for personal smoking habit.

Solution From Equation 13.15, Table 13.7, and Table 13.8,

$$\widehat{OR}_{MH} = \frac{(120 \times 155/466) + (161 \times 124/532)}{(80 \times 111/466) + (130 \times 117/532)} = \frac{77.44}{47.65} = 1.63$$

Thus, the odds in favor of being a passive smoker for a cancer case is 1.6 times as large as that for a control. Since cancer is a relatively rare disease, we can also interpret these results as indicating that the risk of cancer for a passive smoker is 1.6 times as great as for a nonpassive smoker, even after controlling for personal smoking habit.

We are also interested in estimating confidence limits for the odds ratio in Equation 13.15. A variance estimate of $\ln(\widehat{OR}_{MH})$ has been provided by Robins et al. [5], which is accurate under a wide range of conditions, particularly if there are many strata with small numbers of subjects in each stratum. This variance estimate can be used to obtain confidence limits for ln(OR). We can then take the antilog of each of the confidence limits for ln(OR) to obtain confidence limits for OR. This procedure is summarized as follows:

EQUATION 13.16

Interval Estimate for the Common Odds Ratio from a Collection of *k* 2 × 2 Contingency Tables A two-sided 100% × (1 – α) CI for the common odds ratio from a collection of k 2 × 2 tables is given by

$$\exp\left[\ln \widehat{OR}_{MH} \pm z_{1-\alpha/2}\sqrt{Var(\ln \widehat{OR}_{MH})}\right]$$

where

$$Var(\ln \widehat{OR}_{MH}) = \frac{\sum_{i=1}^{k} P_i R_i}{2\left(\sum_{i=1}^{k} R_i\right)^2} + \frac{\sum_{i=1}^{k}(P_i S_i + Q_i R_i)}{2\left(\sum_{i=1}^{k} R_i\right)\left(\sum_{i=1}^{k} S_i\right)} + \frac{\sum_{i=1}^{k} Q_i S_i}{2\left(\sum_{i=1}^{k} S_i\right)^2} \equiv A + B + C$$

where A, B, and C correspond to the first, second, and third terms on the right-hand side of $Var(\ln \widehat{OR}_{MH})$, and

$$P_i = \frac{a_i + d_i}{n_i},\ Q_i = \frac{b_i + c_i}{n_i},\ R_i = \frac{a_i d_i}{n_i},\ S_i = \frac{b_i c_i}{n_i}$$

Example 13.22 **Cancer** Estimate 95% confidence limits for the common odds ratio using the data in Tables 13.7 and 13.8.

Solution Note from Example 13.21 that the point estimate of the odds ratio = $\widehat{OR}_{MH}$ = 1.63. To obtain confidence limits, we first compute P_i, Q_i, R_i, and S_i as follows

$$P_1 = \frac{120 + 155}{466} = .590, Q_1 = 1 - P_1 = .410$$

$$R_1 = \frac{120(155)}{466} = 39.91, S_1 = \frac{80(111)}{466} = 19.06$$

$$P_2 = \frac{161 + 124}{532} = .536, Q_2 = 1 - P_2 = .464$$

$$R_2 = \frac{161(124)}{532} = 37.53, S_2 = \frac{130(117)}{532} = 28.59$$

Thus,

$$Var(\ln OR_{MH}) = A + B + C$$

where

$$A = \frac{.590(39.91) + .536(37.53)}{2(39.91 + 37.53)^2} = 0.00364$$

$$B = \frac{.590(19.06) + .410(39.91) + .536(28.59) + .464(37.53)}{2(39.91 + 37.53)(19.06 + 28.59)} = 0.00818$$

$$C = \frac{.410(19.06) + .464(28.59)}{2(19.06 + 28.59)^2} = 0.00464$$

Thus, $Var(\ln \widehat{OR}_{MH})$ = 0.00364 + 0.00818 + 0.00464 = 0.01646. The 95% CI for ln(OR) is

$$\ln(1.63) \pm 1.96\sqrt{0.01646} = (0.234,\ 0.737)$$

The 95% CI for OR is

$$(e^{0.234},\ e^{0.737}) = (1.26,\ 2.09)$$

13.5.2 Effect Modification

One assumption made in the estimation of a common odds ratio in Equation 13.15 is that the strength of association is the same in each stratum. If the underlying odds ratio is different in the various strata, then it makes little sense to estimate a common odds ratio.

DEFINITION 13.15 Suppose we are interested in studying the association between a disease variable D and an exposure variable E, but are concerned about the possible confounding effect of another variable C. We stratify the study population into g strata according to the variable C and compute the odds ratio relating disease to exposure in each stratum. If the underlying (true) odds ratio is different across the g strata, then there is said to be **interaction** or **effect modification** between E and C, and the variable C is referred to as an **effect modifier.**

In other words, if C is an effect modifier, then the relationship between disease and exposure differs for different levels of C.

Example 13.23 **Cancer** Consider the data in Tables 13.7 and 13.8. We estimated that the odds ratio relating cancer and passive smoking is 2.1 for nonsmokers and 1.3 for smokers. If these were the underlying odds ratios in these strata, then personal smoking would be an effect modifier. Specifically, the relationship between passive smoking and cancer is much stronger for nonsmokers than for smokers. The rationale for this is that the home environment of active smokers already contains cigarette smoke and the extra degradation of the environment by spousal smoking may not be that meaningful.

The issue remains, how can we detect if another variable C is an effect modifier? We will use a generalization of the Woolf procedure for obtaining confidence limits for a single odds ratio given in Equation 13.11. Specifically, we wish to test the hypothesis H_0: $OR_1 = \cdots = OR_k$ versus H_1: at least two of the OR_i are different from each other. We will base our test on the test statistic $X^2 = \sum_{i=1}^{k} w_i(\ln \widehat{OR}_i - \overline{\ln OR})^2$ where $\ln \widehat{OR}_i$ = the estimated log odds ratio relating disease to exposure in the ith stratum of the potential effect modifier, C, $\overline{\ln OR}$ = the estimated "weighted average" log odds ratio over all strata, and w_i is a weight that is inversely proportional to the variance of $\ln \widehat{OR}_i$. The purpose of the weighting is to weight strata with lower variance (which usually correspond to strata with more subjects) more heavily. If H_0 is true, then X^2 will be small, because each of the stratum-specific log odds ratios will be relatively close to each other and to the "average" log odds ratio. Conversely, if H_1 is true, then X^2 will be large. Under H_0, it can be shown that X^2 follows a chi-square distribution with $k - 1$ df. Thus, we will reject H_0 if $X^2 > \chi^2_{k-1,1-\alpha}$ and accept H_0 otherwise. This procedure is summarized as follows:

EQUATION 13.17 **Chi-Square Test for Homogeneity of Odds Ratios over Different Strata (Woolf Method)** Suppose we have a dichotomous disease variable D and exposure variable E. We stratify our study population into k strata according to a confounding variable C. Let OR_i = underlying odds ratio in the ith stratum. To test the hypothesis H_0: $OR_1 = \cdots = OR_k$ versus H_1: at least two of the OR_i are different with a significance level α, use the following procedure.

(1) Compute the test statistic $X^2_{HOM} = \sum_{i=1}^{k} w_i(\ln \hat{OR}_i - \overline{\ln OR})^2 \sim \chi^2_{k-1}$ under H_0 where $\ln \hat{OR}_i$ = log odds ratio in the ith stratum = $\ln[a_i d_i/(b_i c_i)]$, and a_i, b_i, c_i, d_i are the cells of the 2 × 2 table relating disease to exposure in the ith stratum as shown in Table 13.9,

$$w_i = \left(\frac{1}{a_i} + \frac{1}{b_i} + \frac{1}{c_i} + \frac{1}{d_i}\right)^{-1}$$

$$\overline{\ln OR} = \sum_{i=1}^{k} w_i \ln \hat{OR}_i \Big/ \sum_{i=1}^{k} w_i$$

(1a) An alternative computational form of the test statistic is

$$X^2_{HOM} = \sum_{i=1}^{k} w_i (\ln \hat{OR}_i)^2 - \left(\sum_{i=1}^{k} w_i \ln \hat{OR}_i\right)^2 \Bigg/ \sum_{i=1}^{k} w_i$$

(2) If $X^2_{HOM} > \chi^2_{k-1,1-\alpha}$, then reject H_0.
If $X^2_{HOM} \le \chi^2_{k-1,1-\alpha}$, then accept H_0.

(3) The exact p-value = $Pr(\chi^2_{k-1} > X^2)$

Example 13.24 **Cancer** Assess whether the odds ratios relating passive smoking to cancer is different for smokers versus nonsmokers using the data in Tables 13.7 and 13.8.

Solution Let stratum 1 refer to nonsmokers and stratum 2 to smokers. Referring to Tables 13.7 and 13.8, we see that

$$\ln \hat{OR}_1 = \ln\left(\frac{120 \times 155}{80 \times 111}\right) = \ln(2.095) = 0.739$$

$$w_1 = \left(\frac{1}{120} + \frac{1}{111} + \frac{1}{80} + \frac{1}{155}\right)^{-1} = (0.036)^{-1} = 27.55$$

$$\ln \hat{OR}_2 = \ln\left(\frac{161 \times 124}{130 \times 117}\right) = \ln(1.313) = 0.272$$

$$w_2 = \left(\frac{1}{161} + \frac{1}{117} + \frac{1}{130} + \frac{1}{124}\right)^{-1} = (0.031)^{-1} = 32.77$$

Thus, based on step 1a in Equation 13.17, the test statistic is given by

$$\begin{aligned} X^2_{HOM} &= 27.55(0.739)^2 + 32.77(0.272)^2 - [27.55(0.739) + 32.77(0.272)]^2 / (27.55 + 32.77) \\ &= 17.486 - (29.284)^2 / 60.32 \\ &= 17.486 - 14.216 = 3.27 \sim \chi^2_1 \text{ under } H_0 \end{aligned}$$

Referring to Table 6 in the Appendix, we note that $\chi^2_{1,.90} = 2.71$, $\chi^2_{1,.95} = 3.84$. Since $2.71 < 3.27 < 3.84$, it follows that $.05 < p < .10$. Thus, there is no significant effect modification; i.e., the odds ratios in the two strata are not significantly different.

In general, it is important to test for homogeneity of the stratum-specific odds ratios. If the true odds ratios are different, then it makes no sense to obtain a pooled-odds ratio estimate such as given by the Mantel-Haenszel estimator in Equation 13.15. Instead, separate odds ratios should be reported.

13.5.3 Estimation of the Odds Ratio in Matched-Pair Studies

There is a close connection between McNemar's test for matched-pair data in Equation 10.12 and the Mantel-Haenszel test procedure for stratified categorical data in Equation 13.14. Matched pairs are a special case of stratification where each matched pair corresponds to a separate stratum of size 2. It can be shown that McNemar's test is a special case of the Mantel-Haenszel test for strata of size 2. Furthermore, the Mantel-Haenszel odds-ratio estimator in Equation 13.15 reduces to $\widehat{OR} = \frac{n_A}{n_B}$ for matched-pair data where n_A = number of discordant pairs of type A and n_B = number of discordant pairs of type B. Also, it can be shown that the variance of $\ln(\widehat{OR})$ for a matched-pair study is given by $Var[\ln(\widehat{OR})] = \frac{1}{n\hat{p}\hat{q}}$, where n = total number of discordant pairs = $n_A + n_B$, $\hat{p}$ = proportion of discordant pairs of type A = $n_A/(n_A + n_B)$, $\hat{q} = 1 - \hat{p}$. This leads to the following technique for estimating the disease–exposure odds ratio in matched-pair studies.

EQUATION 13.18

Estimation of the Odds Ratio in Matched-Pair Studies Suppose we wish to study the relationship between a dichotomous disease and exposure variable, in a case–control design. We control for confounding by forming matched pairs of subjects with disease (cases) and subjects without disease (controls), where the 2 subjects in a matched pair are the same or similar on one or more confounding variables.

(1) The odds ratio relating disease to exposure is estimated by

$$\widehat{OR} = n_A/n_B$$

where

n_A = number of matched pairs where the case is exposed and the control is not exposed

n_B = number of matched pairs where the case is not exposed and the control is exposed

(2) A two-sided $100\% \times (1 - \alpha)$ CI for OR is given by (e^{c_1}, e^{c_2}), where

$$c_1 = \ln(\widehat{OR}) - z_{1-\alpha/2}\sqrt{\frac{1}{n\hat{p}\hat{q}}}$$

$$c_2 = \ln(\widehat{OR}) + z_{1-\alpha/2}\sqrt{\frac{1}{n\hat{p}\hat{q}}}$$

$$n = n_A + n_B$$

$$\hat{p} = \frac{n_A}{n_A + n_B}, \quad \hat{q} = 1 - \hat{p}$$

(3) The same methodology can be used for prospective or cross-sectional studies where exposed and unexposed individuals are matched on one or more confounding variables and disease outcomes are compared between exposed and unexposed individuals. In this setting,

n_A = number of matched pairs where the exposed subject has disease and the unexposed subject does not

n_B = number of matched pairs where the exposed subject does not have disease and the unexposed subject does and steps 1 and 2 are as just indicated

(4) This method should only be used if n = number of discordant pairs ≥ 20.

Example 13.25 **Cancer** Estimate the odds ratio relating type of treatment to 5-year mortality using the matched-pair data in Table 10.14.

Solution We have from Table 10.14 that

n_A = number of matched pairs where treatment A patient dies within 5 years and treatment B patient survives for 5 years = 5

n_B = number of matched pairs where treatment B patient dies within 5 years and treatment A patient survives for 5 years = 16

Thus, $\widehat{OR} = 5/16 = 0.31$. To obtain 95% confidence limits we see that $n = 21$, $\hat{p} = 5/21 = .238$, $\hat{q} = .762$, and $n\hat{p}\hat{q} = 3.81$. Thus, $\ln(\widehat{OR}) = -1.163$, $Var[\ln(\widehat{OR})] = 1/3.81 = 0.263$ and a 95% CI for $\ln(OR)$ is $(-1.163 - 1.96\sqrt{0.263}, -1.163 + 1.96\sqrt{0.263}) = (-2.167, -0.159)$. The corresponding 95% CI for OR is $(e^{-2.167}, e^{-0.159}) = (0.11, 0.85)$.

13.5.4 Testing for Trend in the Presence of Confounding—Mantel-Extension Test

Example 13.26 **Sleep Disorders** Sleep-disordered breathing is very common among adults. To estimate the prevalence of this disorder, 3513 employees 30–60 years of age who worked for three large state agencies in Wisconsin were sent a mail questionnaire concerning their sleep habits [6]. Subjects were classified as habitual snorers if they reported either (1) snoring, snorting, or breathing pauses every night or almost every night or (2) extremely loud snoring. The results are given by age and sex group in Table 13.10.

TABLE 13.10 **Prevalence of habitual snoring by age and sex group**

	Women			Men		
Age	Yes	No	Total	Yes	No	Total
30–39	196	603	799	188	348	536
40–49	223	486	709	313	383	696
50–60	103	232	335	232	206	438
Total	522	1321	1843	733	937	1670

We would like to assess whether the prevalence of habitual snoring increases with age.

In this study, we would like to assess whether there is a trend in the prevalence rates with age after controlling for sex. To address this issue, we need to generalize the chi-square test for trend given in Equation 10.24 to allow for stratification of our study sample by relevant confounding variables. We can also describe this problem as a generalization of the Mantel-Haenszel test given in Equation 13.14 to the case where we are combining results from several $2 \times k$ tables (rather than just 2×2 tables). Suppose we have s strata and k ordered categories for the exposure variable. Consider the $2 \times k$ table relating the dichotomous disease variable D to the ordered categorical exposure variable E for subjects in the ith stratum (see Table 13.11). We assume that there is a score for the jth exposure category denoted by x_j, $j = 1, \ldots, k$.

TABLE 13.11 **Relationship of disease to exposure in the ith stratum, $i = 1, \ldots, s$**

		Exposure				
		1	2	...	k	
Disease	+	n_{i1}	n_{i2}	...	n_{ik}	n_i
	–	m_{i1}	m_{i2}	...	m_{ik}	m_i
		t_{i1}	t_{i2}	...	t_{ik}	N_i
Score		x_1	x_2	...	x_k	

The total observed score among subjects with disease in the ith stratum $= O_i = \sum_{j=1}^{k} n_{ij}x_j$. The expected score among diseased subjects in the ith stratum under the null hypothesis that the average score for subjects with and without disease in a stratum is the same $= E_i = \left(\sum_{j=1}^{k} t_{ij}x_j\right)\frac{n_i}{N_i}$. If diseased subjects tend to have higher exposure scores on average than nondiseased subjects, then O_i will be greater than E_i for most strata. If diseased subjects tend to have lower exposure scores than non-diseased subjects, then O_i will be less than E_i for most strata. Therefore, we will base our test on $O - E$ where $O = \sum_{i=1}^{s} O_i$, $E = \sum_{i=1}^{s} E_i$. The test procedure is given as follows:

EQUATION 13.19

Chi-Square Test for Trend-Multiple Strata (Mantel-Extension Test)

(1) Suppose we have s strata. In each stratum, we have a $2 \times k$ table relating disease (2 categories) to exposure (k ordered categories) with score for the jth category $= x_j$ as shown in Table 13.11.

(2) To test the hypothesis H_0: $\beta = 0$ versus H_1: $\beta \neq 0$, where

$$p_{ij} = \alpha_i + \beta x_j$$

p_{ij} = proportion of subjects with disease among subjects in the ith stratum and jth exposure category

We compute the test statistic

$$X_{TR}^2 = \left(|O - E| - 0.5\right)^2 / V \sim \chi_1^2 \text{ under } H_0$$

where

$$O = \sum_{i=1}^{s} O_i = \sum_{i=1}^{s} \sum_{j=1}^{k} n_{ij} x_j$$

$$E = \sum_{i=1}^{s} E_i = \sum_{i=1}^{s} \left[\left(\sum_{j=1}^{k} t_{ij} x_j \right) \frac{n_i}{N_i} \right]$$

$$V = \sum_{i=1}^{s} V_i = \sum_{i=1}^{s} \frac{n_i m_i \left(N_i s_{2i} - s_{1i}^2\right)}{N_i^2 (N_i - 1)}$$

$$s_{1i} = \sum_{j=1}^{k} t_{ij} x_j, i = 1, \ldots, s$$

$$s_{2i} = \sum_{j=1}^{k} t_{ij} x_j^2, i = 1, \ldots, s$$

(3) If $X_{TR}^2 > \chi_{1,1-\alpha}^2$, we reject H_0.
    If $X_{TR}^2 \le \chi_{1,1-\alpha}^2$, we accept H_0.

(4) The exact p-value $= Pr(\chi_1^2 > X_{TR}^2)$.

(5) This test should only be used if $V \ge 5$.

Example 13.27 Use the data in Table 13.10 to assess whether the prevalence of habitual snoring increases with age, after controlling for sex.

Solution In this example, we have two strata, corresponding to women ($i = 1$) and men ($i = 2$), respectively. We will use scores of 1, 2, 3 for the three age groups. We have

$$O_1 = 196(1) + 223(2) + 103(3) = 951$$

$$O_2 = 188(1) + 313(2) + 232(3) = 1510$$

$$O = 951 + 1510 = 2461$$

$$E_1 = [799(1) + 709(2) + 335(3)]522/1843 = 912.6$$

$$E_2 = [536(1) + 696(2) + 438(3)]733/1670 = 1423.0$$

$$E = 912.6 + 1423.0 = 2335.6$$

$$s_{11} = 799(1) + 709(2) + 335(3) = 3222$$

$$s_{21} = 799(1^2) + 709(2^2) + 335(3^2) = 6650$$

$$s_{12} = 536(1) + 696(2) + 438(3) = 3242$$

$$s_{22} = 536(1^2) + 696(2^2) + 438(3^2) = 7262$$

$$V_1 = \frac{522(1321)\left[1843(6650)-3222^2\right]}{1843^2(1842)} = 206.61$$

$$V_2 = \frac{733(937)\left[1670(7262)-3242^2\right]}{1670^2(1669)} = 238.59$$

$$V = 206.61 + 238.59 = 445.21$$

Thus the test statistic is given by

$$X_{TR}^2 = \frac{\left(|2461-2335.6|-.5\right)^2}{445.21} = \frac{124.9^2}{445.21} = 35.06 \sim \chi_1^2$$

Because $\chi_{1,.999}^2 = 10.83$ and $X_{TR}^2 = 35.06 > 10.83$, it follows that $p < .001$. Therefore, there is a significant trend relating the prevalence of habitual snoring with age, with older subjects snoring more frequently. This analysis was performed while controlling for the possible confounding effects of sex.

In this section, we have learned about analytic techniques for confounding in epidemiologic studies. If we have a dichotomous disease variable (D), a dichotomous exposure variable (E), and a categorical confounder (C), then we can use the Mantel-Haenszel test to assess the association between D and E while controlling for C. Referring to the master flowchart in the back of the book (p. 780), at box 6 we would answer yes to (1) 2×2 contingency table? and at A arrive at the box entitled "Use two-sample test for binomial proportions, or 2×2 contingency-table methods if no confounding is present, or the Mantel-Haenszel test if confounding is present."

If E is categorical but has more than two categories, then we can use the Mantel Extension test for this purpose. Referring to the master flowchart again, we would answer no to (1) 2×2 contingency table? yes to (2) $2 \times k$ contingency table? and yes to (3) interested in trend over k proportions? This would lead us to the box entitled "Use chi-square test for trend if no confounding is present, or the Mantel Extension test if confounding is present."

SECTION 13.6 Power and Sample-Size Estimation for Stratified Categorical Data

Example 13.28 **Cancer** A study was performed [7] based on a sample of 106,330 women enrolled in the Nurses' Health Study relating ever use of oral contraceptives (OC) at baseline (in 1976) to breast-cancer incidence from 1976 to 1980. Since both OC use and breast cancer are related to age, the data were stratified by 5-year age groups and the Mantel-Haenszel test was employed to test for this association. The results supported the null hypothesis. The estimated odds ratio ($\widehat{OR}_{MH}$) was 1.0 with 95% confidence interval = (0.8, 1.3). The issue is, What power did the study have to detect a significant difference if the underlying $OR = 1.3$?

The power formulas given in Section 10.5 are not applicable because a stratified analysis was used rather than a simple comparison of binomial proportions. However, an approximate power formula is available [8]. To use this formula, we need to know (1) the proportion of exposed subjects in each stratum, (2) the proportion of diseased subjects in each stratum, (3) the proportion of subjects in each stratum out of the total study population, and (4) the size of the total study population. The power formula is given as follows:

EQUATION 13.20

Power Estimation for a Collection of 2 × 2 Tables Based on the Mantel-Haenszel Test Suppose we wish to relate a dichotomous disease variable D to a dichotomous exposure variable E and want to control for a categorical confounding variable C. We subdivide the study population into k strata, where the 2×2 table in the ith stratum is given by

		Exposure +	Exposure −	
Disease	+	a_i	b_i	N_{1i}
	−	c_i	d_i	N_{2i}
		M_{1i}	M_{2i}	N_i

We wish to test the hypothesis H_0: $OR = 1$ versus H_1: $OR = \exp(\gamma)$ for $\gamma \neq 0$, where OR is the underlying stratum-specific odds ratio relating disease to exposure that is assumed to be the same in each stratum. Let

N = size of the total study population

r_i = proportion of exposed subjects in stratum i

s_i = proportion of diseased subjects in stratum i

t_i = proportion of total study population in stratum i

If we use the Mantel-Haenszel test with a significance level of α, then the power is given by

$$\text{Power} = \Phi\left[\frac{\sqrt{N}\left(\gamma B_1 + \frac{\gamma^2}{2} B_2\right) - z_{1-\alpha/2}\sqrt{B_1}}{\left(B_1 + \gamma B_2\right)^{1/2}}\right]$$

where

$$B_1 = \sum_{i=1}^{k} B_{1i}$$

$$B_{1i} = r_i s_i t_i (1 - r_i)(1 - s_i)$$

$$B_2 = \sum_{i=1}^{k} B_{2i}$$

$$B_{2i} = B_{1i}(1 - 2r_i)(1 - 2s_i)$$

Example 13.29 **Cancer** Estimate the power of the study described in Example 13.28 for the alternative hypothesis that $OR = 1.3$.

Solution The data were stratified by 5-year age groups. The age-specific proportion of ever OC users (r_i), the age-specific 4-year incidence of breast cancer (s_i), and the age distribution of the total study population (t_i) are given in Table 13.12 together with B_1 and B_2.

We see that the proportion of ever OC users goes down sharply with age and breast-cancer incidence rises sharply with age. Thus, there is evidence of strong negative confounding and a stratified analysis is essential. To compute power, we note that $N = 106,330$, $\gamma = \ln(1.3) = 0.262$, $z_{1-\alpha/2} = z_{.975} = 1.96$. Thus

$$\text{Power} = \Phi\left\{\frac{\sqrt{106,330}\left[0.262 \times 1.06 \times 10^{-3} + 0.262^2\left(2.32 \times 10^{-4}/2\right)\right] - 1.96\sqrt{1.06 \times 10^{-3}}}{\left[1.06 \times 10^{-3} + 0.262\left(2.32 \times 10^{-4}\right)\right]^{1/2}}\right\}$$

$$= \Phi\left(\frac{0.0296}{0.0335}\right) = \Phi(0.882) = .81$$

Thus, the study had 81% power to detect a true OR of 1.3.

An alternative (and simpler) method for computing power would be to pool data over all strata and compute crude power based on the overall 2×2 table relating disease to exposure. Generally, if there is positive confounding, then the true power (i.e., the power based on the Mantel-Haenszel test—Equation 13.14) is lower than the crude power; if there is negative confounding (as was the case in Example 13.29), then the true power is greater than the crude power.

TABLE 13.12 **Power calculation for studying the association between breast-cancer incidence and oral-contraceptive use based on Nurses' Health Study data**

Age group	Proportion ever OC use (r_i)	Incidence of breast cancer[a] (s_i)	Proportion of total study population (t_i)	B_{1i}	B_{2i}
30–34	.771	160	.188	5.30×10^{-5} [b]	-2.86×10^{-5} [b]
35–39	.629	350	.195	1.59×10^{-4}	-4.07×10^{-5}
40–44	.465	530	.209	2.74×10^{-4}	1.90×10^{-5}
45–49	.308	770	.199	3.24×10^{-4}	1.23×10^{-4}
50–55	.178	830	.209	2.52×10^{-4}	1.59×10^{-4}
Total				1.06×10^{-3} (B_1)	2.32×10^{-4} (B_2)

[a] $\times 10^{-5}$.
[b] e.g., $B_{11} = .771 \times 160 \times 10^{-5} \times .188 \times (1 - .771) \times (1 - 160 \times 10^{-5}) = 5.30 \times 10^{-5}$
$B_{21} = 5.30 \times 10^{-5} \times [1 - 2(.771)] \times [1 - 2(160 \times 10^{-5})] = -2.86 \times 10^{-5}$

Alternatively, before a study begins we may want to specify the power and compute the size of the total study population needed to achieve that power, given that we know the distribution of the study population by stratum and the overall exposure and disease rates within each stratum. The appropriate sample-size formula is given by

EQUATION 13.21

Sample-Size Estimation for a Collection of 2 × 2 Tables Based on the Mantel-Haenszel Test

N = total number of subjects in entire study needed for a stratified design using the Mantel-Haenszel test as the method of analysis

$$= \left(z_{1-\alpha/2}\sqrt{B_1} + z_{1-\beta}\sqrt{B_1 + \gamma B_2}\right)^2 \Big/ \left(\gamma B_1 + \frac{\gamma^2}{2} B_2\right)^2$$

where α = type I error, $1 - \beta$ = power, $\gamma = \ln(OR)$ under H_1, and B_1 and B_2 are defined in Equation 13.20.

SECTION 13.7 Multiple Logistic Regression

13.7.1 Introduction

In Section 13.5, we learned about the Mantel-Haenszel test and the Mantel Extension test, which are techniques for controlling for a single categorical covariate C while assessing the association between a dichotomous disease variable D and a categorical exposure variable E. If either

(1) E is continuous

(2) or C is continuous

(3) or there are several confounding variables $C_1, C_2, \ldots$, each of which may be either categorical or continuous

then it is either difficult or impossible to use the preceding methods to control for confounding. In this section, we will learn about the technique of multiple logistic regression, which can handle all the situations in Section 13.5 as well as those in (1), (2), and (3) above. Multiple logistic regression can be thought of as an analogue to multiple linear regression, discussed in Chapter 11, where the outcome (or dependent) variable is binary as opposed to normally distributed.

13.7.2 General Model

Example 13.30 **Infectious Disease** *Chlamydia trachomatis* is a microorganism that has been established as an important cause of nongonococcal urethritis, pelvic inflammatory disease, and other infectious diseases. A study of risk factors for *C. trachomatis* was conducted in a population of 431 female college students [9]. Since multiple risk factors may be involved, several risk factors must be controlled for simultaneously in the analysis of variables associated with *C. trachomatis*.

A model of the following form might be considered:

EQUATION 13.22

$$p = \alpha + \beta_1 x_1 + \cdots + \beta_k x_k$$

where p = probability of disease. However, since the right-hand side of Equation 13.22 could be less than 0 or greater than 1 for certain values of $x_1, \ldots, x_k$, predicted probabilities that are either less than 0 or greater than 1 could be obtained, which is impossible. Instead, the logit (logistic) transformation of p is often used as the dependent variable.

DEFINITION 13.16

The **logit transformation logit(p)** is defined as

$$\text{logit}(p) = \ln[p/(1-p)]$$

Unlike p, the logit transformation can take on any value from $-\infty$ to $+\infty$.

Example 13.31 Compute logit(.1), logit(.95).

Solution

$$\text{logit}(.1) = \ln(.1/.9) = \ln(1/9) = -\ln(9) = -2.20$$

$$\text{logit}(.95) = \ln(.95/.05) = \ln(19) = 2.94$$

If logit(p) is modeled as a linear function of the independent variables $x_1, \ldots, x_k$, then the following multiple logistic-regression model is obtained:

EQUATION 13.23

Multiple Logistic-Regression Model If $x_1, \ldots, x_k$ are a collection of independent variables and y is a binomial-outcome variable with probability of success = p, then the multiple logistic-regression model is given by

$$\text{logit}(p) = \ln\left(\frac{p}{1-p}\right) = \alpha + \beta_1 x_1 + \cdots + \beta_k x_k$$

or, equivalently, if we solve for p, then the model can be expressed in the form

$$p = \frac{e^{\alpha+\beta_1 x_1+\cdots+\beta_k x_k}}{1+e^{\alpha+\beta_1 x_1+\cdots+\beta_k x_k}}$$

In the second form of the model, we see that p must always lie between 0 and 1 regardless of the values of $x_1, \ldots, x_k$. Complex numeric algorithms are generally required to fit the parameters of the model in Equation 13.23. The best-fitting model relating the prevalence of *C. trachomatis* to the risk factors (1) race and (2) the lifetime number of sexual partners is presented in Table 13.13.

13.7.3 Interpretation of Regression Parameters

How can the regression coefficients in Table 13.13 be interpreted? The regression coefficients in Table 13.13 play a role similar to that played by partial-regression coefficients in multiple linear regression (See Definition 11.16). Specifically, suppose we consider two individuals with different values of the independent variables as shown in Table 13.14, where the jth independent variable is a binary variable.

If we refer to the independent variables as exposure variables, then individuals A and B are the same on all risk factors in the model except for the jth exposure variable, where individual A is exposed (coded as 1) and individual B is not exposed

TABLE 13.13 **Multiple logistic-regression model relating prevalence of *C. trachomatis* to race and number of lifetime sexual partners**

Risk factor	Regression coefficient ($\hat{\beta}_j$)	Standard error $se(\hat{\beta}_j)$	z $[\hat{\beta}_j/se(\hat{\beta}_j)]$
Constant	−1.637		
Black race	+2.242	0.529	+4.24
Lifetime number of sexual partners among users of nonbarrier[a] methods of contraception[b]	+0.102	0.040	+2.55

[a]Barrier methods of contraception include diaphragm, diaphragm and foam, and condom; nonbarrier methods include all other forms of contraception or no contraception.
[b]This variable is defined as 0 for users of barrier methods of contraception.
Source: Reprinted with permission of the *American Journal of Epidemiology, 121*(1), 107–115, 1985.

TABLE 13.14 **Two hypothetical subjects with different values for a binary independent variable (x_j) and the same values for all other variables in a multiple logistic-regression model**

	Independent variable					
Individual	1	2	. . .	$j-1$	j	$j+1 \ldots k$
A	x_1	x_2	. . .	x_{j-1}	1	$x_{j+1} \ldots x_k$
B	x_1	x_2	. . .	x_{j-1}	0	$x_{j+1} \ldots x_k$

(coded as 0). According to Equation 13.23, the logit of the probability of success for individuals A and B, denoted by logit(p_A), and logit(p_B), are given by

EQUATION 13.24

$$\text{logit}(p_A) = \alpha + \beta_1 x_1 + \cdots + \beta_{j-1}x_{j-1} + \beta_j(1) + \beta_{j+1}x_{j+1} + \cdots + \beta_k x_k$$
$$\text{logit}(p_B) = \alpha + \beta_1 x_1 + \cdots + \beta_{j-1}x_{j-1} + \beta_j(0) + \beta_{j+1}x_{j+1} + \cdots + \beta_k x_k$$

If we subtract logit(p_B) from logit(p_A) in Equation 13.24, we obtain

EQUATION 13.25

$$\text{logit}(p_A) - \text{logit}(p_B) = \beta_j$$

However, from Definition 13.16, logit(p_A) = $\ln[p_A/(1-p_A)]$, logit(p_B) = $\ln[p_B/(1-p_B)]$. Therefore, on substituting into Equation 13.25, we obtain

$$\ln[p_A/(1-p_A)] - \ln[p_B/(1-p_B)] = \beta_j$$

or

EQUATION 13.26

$$\ln\left[\frac{p_A/(1-p_A)}{p_B/(1-p_B)}\right] = \beta_j$$

If we take the antilog of each side of Equation 13.26, then we have

EQUATION 13.27

$$\frac{p_A/(1-p_A)}{p_B/(1-p_B)} = e^{\beta_j}$$

However, from the definition of an odds ratio (Definition 13.6), we know that the odds in favor of success for subject A (denoted by Odds_A) is given by $\text{Odds}_A = p_A/(1-p_A)$. Similarly, $\text{Odds}_B = p_B/(1-p_B)$. Therefore, we can rewrite Equation 13.27 as follows:

EQUATION 13.28

$$\frac{\text{Odds}_A}{\text{Odds}_B} = e^{\beta_j}$$

Thus, in words, the odds in favor of disease for subject A divided by the odds in favor of disease for subject $B = e^{\beta_j}$. However, we can also think of $\text{Odds}_A/\text{Odds}_B$ as the odds ratio relating disease to the jth exposure variable for two hypothetical individuals, one of whom is exposed for the jth exposure variable (subject A) and the other of whom is not exposed for the jth exposure variable (subject B), where the individuals are the same for all other risk factors considered in the model. Thus, this odds ratio is an odds ratio relating disease to the jth exposure variable, adjusted for the levels of all other risk factors in our model. This is summarized as follows:

EQUATION 13.29

Estimation of Odds Ratios in Multiple Logistic Regression for Dichotomous Independent Variables Suppose there is a dichotomous exposure variable (x_j), which is coded as 1 if present and 0 if absent. For the multiple logistic-regression model in Equation 13.23, the odds ratio (OR) relating this exposure variable to the dependent variable is estimated by

$$\widehat{OR} = e^{\hat{\beta}_j}$$

This relationship expresses the odds in favor of success if $x_j = 1$ divided by the odds in favor of success if $x_j = 0$ (i.e., the disease–exposure odds ratio) *after controlling for all other variables in the logistic-regression model*. Furthermore, a two-sided $100\% \times (1-\alpha)$ confidence interval for the true odds ratio is given by

$$\left[e^{\hat{\beta}_j - z_{1-\alpha/2}se(\hat{\beta}_j)}, e^{\hat{\beta}_j + z_{1-\alpha/2}se(\hat{\beta}_j)}\right]$$

Example 13.32 **Infectious Disease** Estimate the odds in favor of infection with *C. trachomatis* for black women compared with white women after controlling for previous sexual experience and provide a 95% confidence interval about this estimate.

Solution From Table 13.13,

$$\widehat{OR} = e^{2.242} = 9.4$$

Thus, the odds in favor of infection for black women is 9 times as great as that for white women after controlling for previous sexual experience. Furthermore, since $z_{1-\alpha/2} = z_{.975} = 1.96$ and $se(\hat{\beta}_j) = 0.529$, a 95% confidence interval for OR is given by

$$\left[e^{2.242-1.96(0.529)}, e^{2.242+1.96(0.529)}\right] = (e^{1.205}, e^{3.279}) = (3.3, 26.5)$$

We can also use Equation 13.29 to make a connection between logistic regression and contingency-table analysis for 2 × 2 tables given in Chapter 10. Specifically, suppose that there is only one risk factor in the model, which we denote by E and which takes on the value 1 if exposed and 0 if unexposed, and we have a dichotomous disease variable D. We can relate D to E using the logistic-regression model:

EQUATION 13.30

$$\log\left[p/(1-p)\right] = \alpha + \beta E$$

where p = probability of disease given a specific exposure status E. Therefore, the probability of disease among the unexposed = $e^{\alpha}/(1 + e^{\alpha})$ and among the exposed = $e^{\alpha+\beta}/(1 + e^{\alpha+\beta})$. Also, from Equation 13.29, e^{β} represents the odds ratio relating D to E and is the same odds ratio $[ad/(bc)]$ obtained from the 2 × 2 table in Table 10.7 relating D to E. We have formulated the models in Equations 13.23 and 13.30 under the assumption that we have conducted either a prospective study or a cross-sectional study (i.e., that our study population is representative of the general population and we have not oversampled cases in our study population, as would be true in a case–control study). However, logistic regression is applicable to data from case–control studies as well. Suppose we have a case–control study where we have a disease variable D and an exposure variable E and no other covariates. If we use the logistic-regression model in Equation 13.30—i.e., D as the outcome variable and E as the independent (or predictor) variable—then the probability of disease among the unexposed $[e^{\alpha}/(1 + e^{\alpha})]$ and the exposed $[e^{\alpha+\beta}/(1 + e^{\alpha+\beta})]$ will *not* be generalizable to the reference population because they are derived from a biased sample with a greater proportion of cases than in the reference population. However, the odds ratio e^{β} will be generalizable to the reference population. Thus, we can estimate odds ratios from case–control studies, but not probabilities of disease. This statement is also true if there are multiple exposure variables in a logistic-regression model derived from data from a case–control study. The relationships between logistic-regression analysis and contingency-table analysis are summarized as follows:

EQUATION 13.31

Relationship Between Logistic-Regression Analysis and Contingency-Table Analysis Suppose we have a dichotomous disease variable D and a single dichotomous exposure variable E, derived from either a prospective, cross-sectional, or case–control study design, and that the 2 × 2 table relating disease to exposure is given by

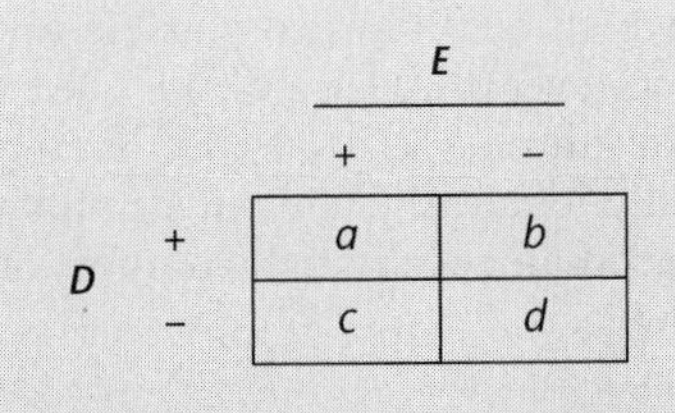

(1) We can estimate the odds ratio relating D to E in either of two equivalent ways:

(a) We can compute the odds ratio directly from the 2×2 table $= ad/bc$

(b) We can set up a logistic-regression model of the form

$$\ln[p/(1-p)] = \alpha + \beta E$$

where p = probability of disease D given exposure status E and where we estimate the odds ratio by e^{β}.

(2) For prospective or cross-sectional studies, we can estimate the probability of disease among exposed (p_E) subjects and unexposed $(p_{\bar{E}})$ subjects in either of two equivalent ways:

(a) From the 2×2 table, we have

$$p_E = a/(a+c),\ p_{\bar{E}} = b/(b+d)$$

(b) From the logistic-regression model,

$$p_E = e^{\hat{\alpha}+\hat{\beta}}/(1+e^{\hat{\alpha}+\hat{\beta}}),\ p_{\bar{E}} = e^{\hat{\alpha}}/(1+e^{\hat{\alpha}})$$

where $\hat{\alpha}$, $\hat{\beta}$ are the estimated parameters from the logistic-regression model.

(3) For case–control studies, it is impossible to estimate absolute probabilities of disease unless the sampling fraction of cases and controls from the reference population is known, which is almost always *not* the case.

Example 13.33 Assess the relationship between mother's age at first birth and breast-cancer incidence based on the data in Table 10.1 using logistic-regression analysis.

Solution We will use the logistic-regression model

$$\ln[p/(1-p)] = \alpha + \beta \times \text{AGEGE30}$$

where

$$p = \text{probability of breast cancer}$$

$$\text{AGEGE30} = 1 \text{ if age at first birth} \geq 30$$

$$= 0 \text{ otherwise}$$

The results using the SAS PROC LOGISIC program are given in Table 13.15. We see that the estimated odds ratio relating breast-cancer incidence to AGEGE30 = $e^{0.4523} = 1.57$. This is identical to the odds ratio estimated using contingency-table methods given in Example 13.10. Notice that although there are actually 13,465 subjects in the study, PROC LOGISTIC tells us that there are two observations. The reason is that the program allows us to enter data in grouped form with all observations with the same combination of independent variables entered as one record. In this case, there is only one covariate (AGEGE30), which only has two possible values (0 or 1); thus, there are two "observations." For each level of AGEGE30, we need to provide the number of cases (i.e., successes) and the number of trials (i.e., observations). Entering the data in grouped form will usually reduce the computer time needed to fit a logistic-regression model (in some data sets, dramatically if the number of covariate patterns is small relative to the number of subjects).

TABLE 13.15 Relationship of age at first birth to breast-cancer incidence based on the data in Table 10.1 using the SAS PROC LOGISTIC procedure

```
                    Case/Trials Model (Recommended Instead of Wts)
                                Logistic Regression
                               The LOGISTIC Procedure
Data Set: WORK.LGTFL
Response Variable (Events): CASES
Response Variable (Trials): TRIALS
Number of Observations: 2
Link Function: Logit

                                 Response Profile
                      Ordered      Binary
                        Value      Outcome           Count
                            1      EVENT              3220
                            2      NO EVENT          10245

                      Analysis of Maximum Likelihood Estimates
                   Parameter   Standard      Wald          Pr >      Standardized     Odds
Variable    DF      Estimate      Error   Chi-Square    Chi-Square     Estimate       Ratio
INTERCPT     1       -1.2377     0.0225    3012.7792        0.0001
AGEGE30      1        0.4523     0.0514      77.4982        0.0001      0.091884      1.572
```

We are also interested in expressing the strength of association between a continuous independent variable and the dependent variable in terms of an odds ratio after controlling for the other independent variables in the model.

Suppose we have two individuals A and B who are the same for all independent variables in the model except for a single continuous risk factor x_j, where they differ by an amount Δ (see Table 13.16).

TABLE 13.16 Two hypothetical subjects with different values for a continuous independent variable (x_j) in a multiple logistic-regression model and the same values for all other variables

	Independent variable			
Individual	1	$2 \ldots j-1$	j	$j+1 \ldots k$
A	x_1	$x_2 \cdots x_{j-1}$	$x_j + \Delta$	$x_{j+1} \cdots x_k$
B	x_1	$x_2 \cdots x_{j-1}$	x_j	$x_{j+1} \cdots x_k$

Following the same argument as in Equation 13.24, we have

EQUATION 13.32

$$\text{logit}(p_A) = \alpha + \beta_1 x_1 + \cdots + \beta_{j-1}x_{j-1} + \beta_j(x_j + \Delta) + \beta_{j+1}x_{j+1} + \cdots + \beta_k x_k$$
$$\text{logit}(p_B) = \alpha + \beta_1 x_1 + \cdots + \beta_{j-1}x_{j-1} + \beta_j x_j + \beta_{j+1}x_{j+1} + \cdots + \beta_k x_k$$

If we subtract logit(p_B) from logit(p_A) in Equation 13.32, we obtain

$$\text{logit}(p_A) - \text{logit}(p_B) = \beta_j \Delta$$

or

$$\ln\left(\frac{p_A}{1-p_A}\right) - \ln\left(\frac{p_B}{1-p_B}\right) = \beta_j \Delta$$

or

$$\ln\left[\frac{p_A/(1-p_A)}{p_B/(1-p_B)}\right] = \beta_j \Delta$$

or

$$OR = \frac{p_A/(1-p_A)}{p_B/(1-p_B)} = e^{\beta_j \Delta}$$

Thus, the odds in favor of disease for subject A versus subject $B = e^{\beta_j \Delta}$. This result is summarized as follows:

EQUATION 13.33

Estimation of Odds Ratios in Multiple Logistic Regression for Continuous Independent Variables Suppose there is a continuous independent variable (x_j). Consider two individuals who have values of $x + \Delta$ and x for x_j, respectively, and have the same values for all other independent variables in the model. The odds ratio in favor of success for the first individual versus the second individual is estimated by

$$\widehat{OR} = e^{\hat{\beta}_j \Delta}$$

Furthermore, a two-sided 100% × (1 – α) confidence interval for *OR* is given by

$$\left\{e^{[\hat{\beta}_j - z_{1-\alpha/2} se(\hat{\beta}_j)]\Delta}, e^{[\hat{\beta}_j + z_{1-\alpha/2} se(\hat{\beta}_j)]\Delta}\right\}$$

Thus, *OR* represents the odds in favor of success for an individual with level $x + \Delta$ for x_j versus an individual with level x for x_j, after controlling for all other variables in the model. Δ is usually chosen to represent a meaningful increment in the continuous variable x.

Example 13.34 **Infectious Disease** Based on the data in Table 13.13, what is the extra risk of infection for each additional sexual partner for women of a particular race who use nonbarrier methods of contraception? Provide a 95% confidence interval associated with this estimate.

Solution We have that Δ = 1. From Table 13.13, $\hat{\beta}_j = 0.102$, $se(\hat{\beta}_j) = 0.040$. Thus,

$$\widehat{OR} = e^{0.102 \times 1} = e^{0.102} = 1.11$$

Thus, the odds in favor of infection increase an estimated 11% for each additional sexual partner for women of a particular race who use nonbarrier methods of contraception. A 95% confidence interval for OR is given by

$$\left\{e^{[0.102-1.96(0.040)]}, e^{[0.102+1.96(0.040)]}\right\} = (e^{0.0236}, e^{0.1804}) = (1.02, 1.20)$$

13.7.4 Hypothesis Testing

How can the statistical significance of the risk factors in Table 13.13 be evaluated? The statistical significance of each of the independent variables after controlling for all other independent variables in the model should be assessed. This task can be accomplished by first computing the test statistic $z = \hat{\beta}_j/se(\hat{\beta}_j)$, which should follow an $N(0, 1)$ distribution under the null hypothesis that the jth independent variable has no association with the dependent variable after controlling for the other variables. H_0 will be rejected for either large positive or large negative values of z. This procedure is summarized as follows:

EQUATION 13.34

Hypothesis Testing in Multiple Logistic Regression To test the hypothesis H_0: $\beta_j = 0$, all other $\beta_l \neq 0$, versus H_1: all $\beta_j \neq 0$ for the multiple logistic-regression model in Equation 13.23, use the following procedure:

(1) Compute the test statistic $z = \hat{\beta}_j/se(\hat{\beta}_j) \sim N(0, 1)$ under H_0.

(2) To conduct a two-sided test with significance level α, if

$$z < z_{\alpha/2} \quad \text{or} \quad z > z_{1-\alpha/2}$$

then reject H_0; if

$$z_{\alpha/2} \leq z \leq z_{1-\alpha/2}$$

then accept H_0.

(3) The exact p-value is given by

$$2 \times [1 - \Phi(z)] \quad \text{if } z \geq 0$$

$$2 \times \Phi(z) \quad \text{if } z < 0$$

(4) This large-sample procedure should only be used if there are at least 20 successes and 20 failures, respectively, in the data set.

The acceptance and rejection regions for this test are depicted in Figure 13.3. The computation of the exact p-value is illustrated in Figure 13.4.

Example 13.35 **Infectious Disease** Assess the significance of the independent variables in the multiple logistic-regression model presented in Table 13.13.

Solution First compute the test statistic $z = \hat{\beta}_j/se(\hat{\beta}_j)$ for each of the independent variables, as shown in Table 13.13. For an α level of .05, compare $|z|$ with $z_{.975} = 1.96$ to assess statistical significance. Because both of the independent variables satisfy this criterion, they are both significant at the 5% level. The exact p-values are given by

$$p(\text{race}) = 2 \times [1 - \Phi(4.24)] < .001$$

$$p(\text{number of sexual partners}) = 2 \times [1 - \Phi(2.55)] = .011$$

FIGURE 13.3 **Acceptance and rejection regions for the test of the hypothesis H_0: $\beta_j = 0$, all other $\beta_l \neq 0$, versus H_1: all $\beta_j \neq 0$ in multiple logistic regression**

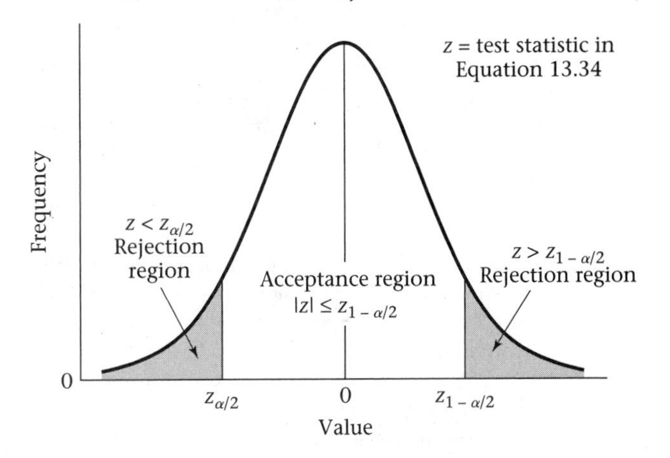

FIGURE 13.4 **Computation of the p-value for the test of the hypothesis H_0: $\beta_j = 0$, all other $\beta_l \neq 0$, versus H_1: all $\beta_j \neq 0$ in multiple logistic regression**

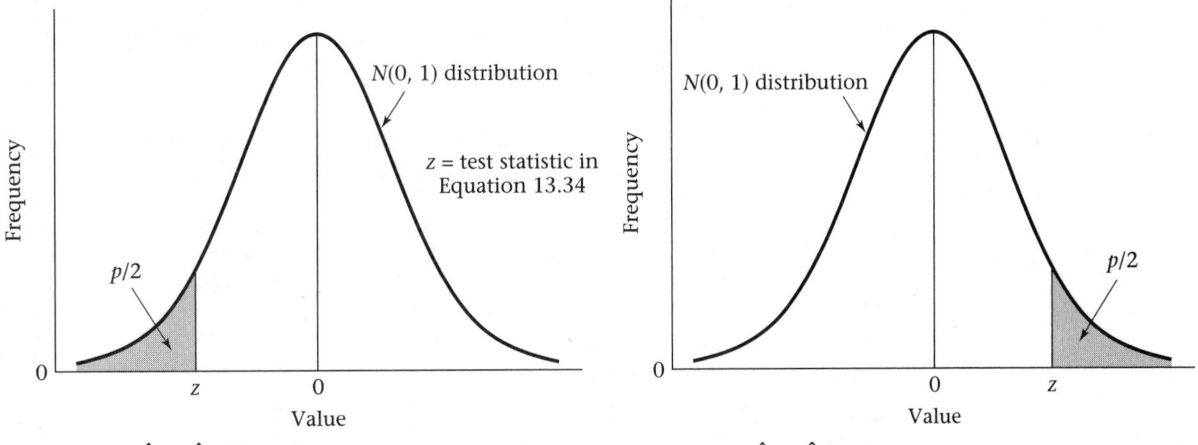

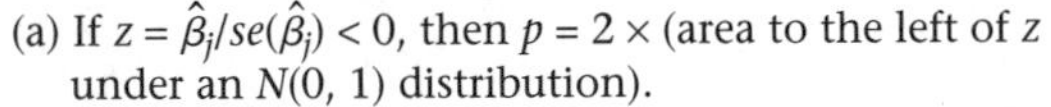

(a) If $z = \hat{\beta}_j/se(\hat{\beta}_j) < 0$, then $p = 2 \times$ (area to the left of z under an $N(0, 1)$ distribution).

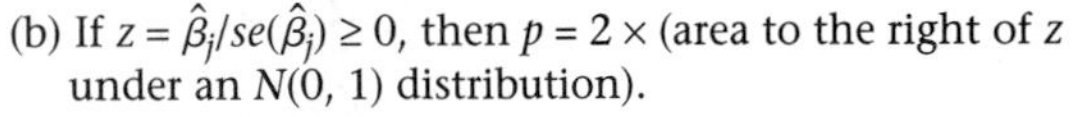

(b) If $z = \hat{\beta}_j/se(\hat{\beta}_j) \geq 0$, then $p = 2 \times$ (area to the right of z under an $N(0, 1)$ distribution).

Thus, both variables are significantly associated with *C. trachomatis*. Specifically, after controlling for the other variable in the model, there is an increased probability of infection for black women versus white women and for women with greater previous sexual experience versus women with lesser previous sexual experience.

We also can quantify the magnitude of the association. From Example 13.32, the odds in favor of *C. trachomatis* are 9.4 times as high for black women than for white women where both women either used barrier methods of contraception (in which case $x_2 = 0$) or used nonbarrier methods of contraception and had the same lifetime number of sexual partners ($x_2 > 0$). From Example 13.34, for women of the same race who used nonbarrier methods of contraception, the odds in favor of *C. trachomatis* increase by a factor of 1.11 for each extra sexual partner during her lifetime.

Finally, the 95% confidence intervals in Equations 13.29 and 13.33 will contain 1 only if there is a nonsignificant association between x_j and the dependent variable. Similarly, these intervals will not contain 1 only if there is a significant association between x_j and the dependent variable. Thus, since both independent variables in Table 13.13 are statistically significant, the confidence intervals in Examples 13.32 and 13.34 both exclude 1.

Example 13.36 **Cardiovascular Disease** The Framingham Heart Study began in 1950 by enrolling 2,282 men and 2,845 women aged 30–59 years, who have been followed up to the present [10]. Coronary risk-factor information was obtained at biannual examinations of cohort members. The relationship between the incidence of coronary heart disease and selected risk factors including age, sex, serum cholesterol, serum glucose, body-mass index, systolic blood pressure, and cigarette smoking was assessed [11]. For the analysis, men were chosen who were free of coronary heart disease (CHD) (either nonfatal myocardial infarction or fatal coronary heart disease) at examination 4 and for whom all risk-factor information from examination 4 was available. In this analysis, the cohort was assessed for the development of coronary heart disease over the next 10 years (examinations 5–9). There were 1,731 men who satisfied these criteria and constitute the study population. The baseline characteristics of the study population are presented in Table 13.17. Subjects were 50 years of age on average and exhibited a wide range for each of the coronary heart disease risk factors.

TABLE 13.17 **Baseline characteristics of study population, Framingham Heart Study**

Risk factor	Mean	Standard deviation	No.	%
Serum cholesterol (mg/dl)[a]	234.8	40.6		
Serum glucose (mg/dl)[b]	81.8	27.4		
Body-mass index (kg/m^2)	26.5	3.4		
Systolic blood pressure (mm Hg)[c]	132.1	20.1		
Age (years)	49.6	8.5		
35–39			228	13
40–49			670	39
50–59			542	31
60–69			291	17
Current smoking (cigarettes/day)[d]	13.1	13.5		
0			697	40
1–10			183	11
11–20			510	29
≥21			341	20

Note: The subjects were 1,731 men who were seen at examination 4 and were free of coronary heart disease at or before examination 4.
[a]Based on the Abell-Kendall method.
[b]Based on a casual specimen of the subject's whole blood, using the Nelson method.
[c]Average of two replicate measurements at examination 4, using a standard mercury sphygmomanometer.
[d]A current smoker is defined as a person who smoked within the last year.

In Equation 13.29, we studied how to assess logistic-regression coefficients for dichotomous exposure variables. In Equation 13.33, we studied how to assess logistic-regression coefficients for continuous exposure variables. In some instances, we wish to use categorical variables in a multiple logistic-regression model that have more than two categories. In this case, we can represent such variables by a collection of $k - 1$ dummy variables in a similar manner to that used for multiple regression in Equation 12.17.

In the analysis of the data in Example 13.36, age was treated as a categorical variable with categories (35–44, 45–54, 55–64, 65–69), while all other risk factors were treated as continuous variables. The reason for treating age as a categorical variable is that in other studies, the increase in incidence of CHD with age was reported to be nonlinear with age; e.g., the odds ratio relating incidence for 50- to 54-year-olds versus incidence for 45- to 49-year-olds is different from that for 60- to 64-year-olds versus 55- to 59-year-olds. If age is entered as a continuous variable, then the increase in risk for every 5-year increase in age (as measured by the odds ratio) is assumed to be the same. Therefore, we choose one category (35–44) to be the reference category and create three dummy variables to represent group membership in age groups 45–54, 55–64, and 65–69, respectively. It is arbitrary as to which category is assigned to be the reference category. In some instances, a particular category is a natural reference category based on scientific considerations. Also, all other risk factors except number of cigarettes currently smoked were converted to the ln scale to reduce the positive skewness. The resulting model is of the form

$$\ln\left(\frac{p}{1-p}\right) = \alpha + \beta_1 \text{age4554} + \beta_2 \text{age5564} + \beta_3 \text{age6569} + \beta_4 \text{LCHLD235} + \beta_5 \text{LSUGRD82} + \beta_6 \text{SMOKEM13} + \beta_7 \text{LBMID26} + \beta_8 \text{LMSYD132}$$

where

$$\begin{aligned}
p &= \text{probability of developing CHD over a 10-year period} \\
\text{age4554} &= 1 \text{ if a subject is age 45–54}, = 0 \text{ otherwise} \\
\text{age5564} &= 1 \text{ if a subject is age 55–64}, = 0 \text{ otherwise} \\
\text{age6569} &= 1 \text{ if a subject is age 65–69}, = 0 \text{ otherwise} \\
\text{LCHLD235} &= \ln(\text{serum cholesterol}/235) \\
\text{LSUGRD82} &= \ln(\text{serum glucose}/82) \\
\text{SMOKEM13} &= \text{number of cigarettes currently smoked} - 13 \\
\text{LBMID26} &= \ln(\text{body-mass index}/26) \\
\text{LMSYD132} &= \ln(\text{systolic blood pressure}/132)
\end{aligned}$$

Each of the risk factors (except for the age variables) have been **mean-centered;** i.e., the approximate mean has either been subtracted from each value (for cigarettes per day, which is in the original scale) or each value has been divided by the approximate mean (for all other risk factors, which are in the ln scale). The reason for doing this is to make the constant (α) more meaningful and in some instances, to reduce the amount of computer time needed to perform logistic-regression analyses. In this analysis, the constant α represents logit(p) for an "average" subject in the reference group (35–44 years of age); i.e., where all other risk factors are 0, which means that serum cholesterol = 235, serum glucose = 82, number of cigarettes per day = 13, body-mass index = 26, systolic blood pressure = 132. The model was fitted using SAS PROC LOGISTIC and the results are given in Table 13.18.

TABLE 13.18 **Multiple logistic-regression model for predicting the cumulative incidence of CHD over 10 years based on 1731 men in the Framingham Heart Study who were disease free at baseline using SAS PROC LOGISTIC**

```
                         Logistic Regression
                       The LOGISTIC Procedure

Data Set: WORK.FRMFL
Response Variable: CMBMICHD
Response Levels: 2
Number of Observations: 1731
Link Function: Logit
```

Response Profile

Ordered Value	CMBMICHD	Count
1	1	163
2	2	1568

Analysis of Maximum Likelihood Estimates

Variable	DF	Parameter Estimate	Standard Error	Wald Chi-Square	Pr > Chi-Square	Standardized Estimate	Odds Ratio
INTERCPT	1	-3.1232	0.2090	223.2961	0.0001	.	.
AGE4554	1	0.7199	0.2474	8.4686	0.0036	0.188857	2.054
AGE5564	1	1.1661	0.2551	20.9036	0.0001	0.283285	3.210
AGE6569	1	1.4583	0.3762	15.0251	0.0001	0.166826	4.299
LCHLD235	1	1.8303	0.5086	12.9537	0.0003	0.175202	6.236
LSUGRD82	1	0.5728	0.3261	3.0848	0.0790	0.070232	1.773
SMOKEM13	1	0.0177	0.00637	7.6910	0.0055	0.131608	1.018
LBMID26	1	1.4817	0.7012	4.4658	0.0346	0.106619	4.400
LMSYD132	1	2.7967	0.5737	23.7665	0.0001	0.222964	16.391

We see that each of the risk factors is significantly related to the incidence of CHD, with the exception of serum glucose. The odds ratio provides us with information as to the magnitude of the associations. The odds ratio for each of the age variables gives an estimate of the odds in favor of CHD for that age group as compared with the reference group (ages 35–44). The odds in favor of CHD for 45- to 54-year-olds are 2.1 ($e^{.7199}$) times as great as for the reference group holding all other variables constant. Similarly, the odds in favor of CHD are 3.2 and 4.3 times as great for 55- to 64-year-olds and 65- to 69-year-olds versus the reference group. The odds ratio for cholesterol ($e^{1.8303} = 6.2$) indicates that for two men who are 1 ln unit apart on ln serum cholesterol and are comparable on all other risk factors, the odds in favor of CHD for the man with the higher cholesterol is 6.2 times as great as for the man with the lower cholesterol. Recall that 1 ln unit apart is equivalent to a cholesterol ratio of $e^1 = 2.7$. If we want to compare men with a different cholesterol ratio (e.g., twice as great), we convert 2 to the ln scale and compute the odds ratio as $e^{1.8303 \ln 2} = e^{1.8303 \times 0.6931} = e^{1.27} = 3.6$. Thus, if man A has a cholesterol level twice as great as man B and is the same on all other risk factors, then the odds in favor of CHD over a 10-year period are 3.6 times as great for man A versus man B. The other continuous variables that were converted to the ln scale (glucose, body-mass index, and systolic blood pressure) are interpreted similarly. Cigarette smoking was left in the original scale, so the odds ratio of 1.018 provides a comparison of two men 1 cigarette per day apart. Since this is a trivial difference, a more meaningful comparison would be obtained if we compare a smoker of 1 pack per day (i.e., 20 cigarettes per day—man A) versus a nonsmoker (i.e., 0 cigarettes per day—man B). The odds in favor of CHD for man A versus man B = $e^{20(.0177)} = e^{0.354} = 1.42$. Thus, the smoker of 1 pack per day is 1.4 times as likely to develop CHD over a 10-year period as is the nonsmoker, given that they are the same for all other risk factors. Because we mean-centered all the risk factors except for age, the intercept allows us to estimate the 10-year cumulative incidence of CHD for an "average" man in the reference group (age 35–44). Specifically, the 10-year cumulative incidence of CHD for a 35- to 44-year-old man with cholesterol = 235, glucose = 82, number of cigarettes per day = 13, body-mass index = 26, and systolic blood pressure = 132 is $e^{-3.1232}/(1 + e^{-3.1232}) = .0440/1.0440 = .042$, or 4.2%.

13.7.5 Prediction with Multiple Logistic Regression

We can use a multiple logistic-regression model to predict the probability of disease for an individual subject with covariate values $x_1, \ldots, x_k$. If the regression parameters were known, then the probability of disease would be estimated using Equation 13.23 by

$$p = \frac{e^L}{1+e^L}$$

where $L = \alpha + \beta_1 x_1 + \cdots + \beta_k x_k$. L is sometimes called a **linear predictor.** Since the parameters are unknown, we substitute estimates of them to obtain the predicted probability

EQUATION 13.35

$$\hat{p} = \frac{e^{\hat{L}}}{1+e^{\hat{L}}}$$

where $\hat{L} = \hat{\alpha} + \hat{\beta}_1 x_1 + \cdots + \hat{\beta}_k x_k$

To obtain two-sided $100\% \times (1 - \alpha)$ confidence limits for the true probability p, we must first obtain confidence limits for the linear predictor L given by

$$\hat{L} \pm z_{1-\alpha/2} se(\hat{L}) = (L_1, L_2)$$

and then transform back to the probability scale to obtain the CI = (p_1, p_2) where

$$p_1 = e^{L_1}/(1+e^{L_1}),\ p_2 = e^{L_2}/(1+e^{L_2})$$

The actual expression for $se(\hat{L})$ is complex, requiring matrix algebra, and is beyond the scope of this text, although it can be easily evaluated on the computer. This approach is summarized as follows:

EQUATION 13.36

Point and Interval Estimation of Predicted Probabilities Using Logistic Regression Suppose we wish to estimate the predicted probability of disease (p) for a subject with covariate values $x_1, \ldots, x_k$ and obtain confidence limits about this prediction.

(1) We compute the linear predictor

$$\hat{L} = \hat{\alpha} + \hat{\beta}_1 x_1 + \cdots + \hat{\beta}_k x_k$$

where $\hat{\alpha}, \hat{\beta}_1, \ldots, \hat{\beta}_k$ are the estimated regression coefficients from the logistic-regression model.

(2) The point estimate of p is given by $e^{\hat{L}}/(1+e^{\hat{L}})$

(3) A two-sided 100% × (1 − α) CI for p is given by (p_1, p_2), where

$$p_1 = e^{L_1}/(1+e^{L_1}),\ p_2 = e^{L_2}/(1+e^{L_2})$$

and

$$L_1 = \hat{L} - z_{1-\alpha/2}se(\hat{L})$$

$$L_2 = \hat{L} + z_{1-\alpha/2}se(\hat{L})$$

To obtain $se(\hat{L})$ requires matrix algebra and is always obtained by computer. Also, $se(\hat{L})$ will vary, depending on the covariate values $x_1, \ldots, x_k$.

(4) These estimates are only valid for prospective or cross-sectional studies.

Example 13.37 **Cardiovascular Disease** Obtain a point estimate and 95% confidence limits for the predicted probability of CHD for subjects in the Framingham Heart Study data set in Example 13.36.

Solution We refer to Table 13.19, which provides the raw data, the predicted probability of CHD (labeled as PHAT), and the lower and upper 95% confidence limits about the point estimates (labeled as LCL and UCL) for subjects 951–1000. For example, for subject 969 the outcome variable is given in the second column (CMBMICHD) and = 2 (which indicates no event). The values of the independent variables are given in columns 3–10. For example, we know that the subject is 35–44 years old because all the age dummy variables (columns 3–5) are 0. Also, the subject is a nonsmoker, since SMOKEM13 = −13. His serum cholesterol = 235 × $e^{-.0085}$ = 233, etc. From Table 13.18, the linear predictor is

$$\hat{L} = -3.1232 + 0.7199(0) + 1.1661(0) + 1.4583(0) + 1.8303(-0.0085) + \cdots + 2.7967(0.0445) = -3.1192$$

The predicted probability is

$$\hat{p}_i = e^{-3.1192}/(1+e^{-3.1192}) = .0423 \text{ (see PHAT column)}$$

The lower and upper 95% confidence limits are LCL = .0269 and UCL = .0660.

13.7.6 Assessing Goodness of Fit of Logistic-Regression Models

We can use the predicted probabilities for individual subjects to define residuals and assess goodness of fit of logistic-regression models.

EQUATION 13.37

Residuals in Logistic Regression If our data are in ungrouped form—that is, each subject has a unique set of covariate values (as in Table 13.19)—then we can define the **Pearson residual** for the ith observation by

$$r_i = \frac{y_i - \hat{p}_i}{se(\hat{p}_i)}$$

where

y_i = 1 if the ith observation is a success and = 0 if it is a failure

$$\hat{p}_i = \frac{e^{\hat{L}_i}}{1+e^{\hat{L}_i}}$$

$$\hat{L}_i = \text{linear predictor for the } i\text{th subject} = \hat{\alpha} + \hat{\beta}_1 x_1 + \cdots + \hat{\beta}_k x_k$$

$$se(\hat{p}_i) = \sqrt{\hat{p}_i(1-\hat{p}_i)}$$

If our data are in grouped form—i.e., if the subjects with the same covariate values have been grouped together (as in Table 13.15)—then the Pearson residual for the ith group of observations is defined by

$$r_i = \frac{y_i - \hat{p}_i}{se(\hat{p}_i)}$$

where

y_i = proportion of successes among the ith group of observations

$$\hat{p}_i = \frac{e^{\hat{L}_i}}{1+e^{\hat{L}_i}} \text{ as defined for ungrouped data}$$

$$se(\hat{p}_i) = \sqrt{\frac{\hat{p}_i(1-\hat{p}_i)}{n_i}}$$

n_i = number of observations in the ith group

Thus the Pearson residual is similar to the Studentized residual in linear regression as defined in Equation 11.14. As was the case in linear regression, the residuals do not have the same standard error. The standard error is computed based on the binomial distribution, where the probability of success is estimated by $\hat{p}_i$. Thus, for ungrouped data, $se(\hat{p}_i) = \sqrt{\hat{p}_i(1-\hat{p}_i)}$ since each observation constitutes a sample of size 1. For grouped data, $se(\hat{p}_i) = \sqrt{\hat{p}_i(1-\hat{p}_i)/n_i}$ where n_i = number of observations in the ith group. The standard error decreases as $\hat{p}_i$ approaches either 0 or 1; for grouped data, the standard error decreases as n_i increases.

Example 13.38 **Cardiovascular Disease** Compute the Pearson residual for the 969th observation in the Framingham Heart Study data set for the logistic-regression model fitted in Table 13.18.

Solution From Example 13.37 we have the predicted probability $\hat{p}_i = .0423$. The standard error of $\hat{p}_i$ is $se(\hat{p}_i) = \sqrt{.0423(.9577)} = .2013$. Also, from Table 13.19, we saw that the subject did *not* have an event; thus, $y_i = 0$. Therefore, the Pearson residual is

$$r_i = \frac{0-.0423}{.2013} = -.2102$$

In Table 13.20, we display the Pearson residuals for a subset of the subjects in the Framingham Heart Study data set (listed under the column labeled Value). Thus, for observation 969, Value = –0.2102. The Pearson residuals are also displayed in graphic form at the right in Table 13.20.

We can use the Pearson residuals to identify outlying values. However, the utility of individual residuals is more limited for logistic regression than for linear regression,

TABLE 13.19 Raw data, predicted probabilities, and 95% confidence limits for a subset of the Framingham Heart Study data based on the logistic-regression model in Table 13.18

Logistic Regression
Predicted Probabilities and 95% Confidence Limits

OBS	CMBMICHD	AGE4554	AGE5564	AGE6569	LCHLD235	LSUGRD82	SMOKEM13	LBMID26	LMSYD132	PHAT	LCL	UCL
951	2	1	0	0	0.0579	-0.0892	-4	0.1716	0.0335	0.11192	0.08182	0.15127
952	2	0	0	1	-0.4162	-0.1726	-10	-0.1061	0.0225	0.05751	0.02768	0.11563
953	2	1	0	0	-0.1125	0.0241	17	-0.0506	-0.1643	0.05575	0.03713	0.08291
954	2	0	0	0	0.0294	0.0476	30	0.0117	-0.0153	0.07328	0.04603	0.11472
955	2	1	0	0	-0.0751	0.0931	-13	-0.0409	-0.1733	0.03688	0.02400	0.05628
956	1	0	1	0	0.2834	0.0121	17	0.2921	0.4353	0.62697	0.46703	0.76324
957	2	0	0	0	-0.1713	-0.1163	30	-0.0997	0.0939	0.05426	0.03103	0.09320
958	2	1	0	0	-0.1366	-0.0247	-13	0.2771	0.3429	0.17833	0.10975	0.27644
959	1	1	0	0	-0.0435	0.2567	17	0.1276	0.0113	0.14006	0.09710	0.19785
960	1	1	0	0	-0.2285	0.1041	7	0.0044	-0.0953	0.05224	0.03487	0.07757
961	2	0	0	0	0.1572	-0.1301	7	-0.0011	0.0075	0.05914	0.03975	0.08711
962	2	0	1	0	-0.1030	-0.1027	7	-0.0622	0.2530	0.18765	0.13240	0.25907
963	2	1	0	0	-0.1317	-0.1027	-13	-0.0997	-0.2053	0.02521	0.01556	0.04061
964	2	1	0	0	0.0738	0.0121	-13	0.0277	0.0588	0.09232	0.06695	0.12601
965	2	0	0	0	0.0376	-0.3463	-13	0.0903	-0.1037	0.02561	0.01533	0.04250
966	2	0	0	0	0.0579	-0.0760	17	-0.1202	-0.1643	0.03236	0.02017	0.05152
967	2	0	1	0	0.1894	-0.0123	-13	0.0385	0.0939	0.17834	0.13293	0.23507
968	2	0	1	0	-0.2126	0.2472	-13	0.0545	-0.0788	0.07083	0.04636	0.10675
969	2	0	0	0	-0.0085	0.0931	-13	0.0485	0.0445	0.04233	0.02691	0.06600
970	2	0	0	0	-0.0705	-0.0629	2	-0.0244	-0.2007	0.02083	0.01312	0.03292
971	2	0	0	0	0.0934	0.5631	7	0.1814	-0.0465	0.08570	0.04953	0.14428
972	2	1	0	0	-0.0843	-0.0500	-13	0.1702	0.0870	0.08943	0.06178	0.12778
973	2	1	0	0	0.0211	-0.0892	-13	0.2161	0.1924	0.14340	0.09817	0.20474
974	2	0	0	0	0.0085	-0.1726	-13	0.1803	0.1008	0.05279	0.03259	0.08442
975	2	0	1	0	-0.0983	0.3724	-13	-0.0210	0.0335	0.10998	0.07607	0.15646
976	1	1	0	0	0.5854	0.0931	7	-0.1561	0.0113	0.20513	0.11702	0.33447
977	2	0	0	0	-0.0085	-0.1027	17	0.1304	-0.0545	0.05435	0.03553	0.08231
978	2	1	0	0	-0.0524	0.8403	7	0.1029	0.1542	0.21241	0.12589	0.33555
979	2	1	0	0	-0.3416	0.2183	7	0.1333	0.0150	0.07308	0.04465	0.11738
980	2	0	0	0	-0.0705	-0.1163	-13	0.0297	-0.0666	0.02434	0.01512	0.03897
981	2	0	1	0	0.1859	-0.2020	-13	-0.1366	-0.1554	0.06918	0.04326	0.10887
982	2	0	0	0	-0.0660	-0.0373	7	-0.0588	-0.0076	0.03732	0.02454	0.05639
983	2	0	0	0	-0.0479	-0.1582	7	-0.0727	-0.2627	0.01763	0.01069	0.02896
984	2	0	0	0	-0.1866	-0.0247	-13	-0.0144	0.0660	0.02805	0.01660	0.04702
985	2	0	0	0	-0.4357	-0.2171	7	-0.1098	-0.1037	0.01244	0.00646	0.02382
986	2	1	0	0	0.1499	-0.1301	7	-0.1019	0.0730	0.11644	0.08325	0.16053
987	1	0	0	0	-0.0435	-0.0629	7	-0.1244	-0.1823	0.02168	0.01342	0.03485
988	1	1	0	0	0.2240	-0.2171	7	-0.0747	-0.2578	0.05595	0.03475	0.08890
989	2	0	0	0	-0.1317	0.0819	7	-0.1669	0.0113	0.03200	0.01912	0.05310
990	2	0	0	0	-0.2556	-0.1582	-13	0.1712	0.0939	0.03245	0.01840	0.05661
991	2	0	0	0	-0.1563	0.1366	7	0.1350	-0.0829	0.03772	0.02352	0.05996
992	2	0	0	0	0.2770	0.0706	17	-0.0022	0.0588	0.10773	0.06959	0.16311
993	2	0	0	0	-0.1125	-0.2796	7	0.0221	0.0225	0.03662	0.02355	0.05651
994	1	0	0	1	0.1929	-0.0629	2	0.1789	0.1606	0.35472	0.22029	0.51681
995	2	0	0	0	-0.0479	0.1680	-13	0.0999	-0.2007	0.02280	0.01337	0.03864
996	2	0	0	1	0.1011	0.2567	-13	0.1154	0.0225	0.20937	0.12173	0.33597
997	2	0	0	0	0.0000	-0.1872	7	-0.1808	-0.2627	0.01615	0.00933	0.02781
998	2	1	0	0	0.1425	-0.1301	-13	-0.0795	0.1638	0.10846	0.07181	0.16060
999	2	0	0	0	-0.2948	0.0000	-13	0.0221	-0.2384	0.01070	0.00578	0.01974
1000	2	0	0	0	-0.4162	0.0241	-13	0.0174	-0.0038	0.01653	0.00867	0.03130

TABLE 13.20 Display of Pearson residuals for a subset of the Framingham Heart Study data based on the logistic-regression model fitted in Table 13.18

```
Logistic Regression
The LOGISTIC Procedure
```

	Covariates								Regression Diagnostics Pearson Residual	
Case Number	AGE4554	AGE5564	AGE6569	LCHLD235	LSUGRD82	SMOKEM13	LBMID26	LMSYD132	Value	(1 unit = 0.9) -8 -4 0 2 4 6 8
969	0	0	0	-0.00850	0.0931	-13.0000	0.0485	0.0445	-0.2102	*
970	0	0	0	-0.0705	-0.0629	2.0000	-0.0244	-0.2007	-0.1458	*
971	0	0	0	0.0934	0.5631	7.0000	0.1814	-0.0465	-0.3062	*
972	1.0000	0	0	-0.0843	-0.0500	-13.0000	0.1702	0.0870	-0.3134	*
973	1.0000	0	0	0.0211	-0.0892	-13.0000	0.2161	0.1924	-0.4092	*
974	0	0	0	0.00850	-0.1726	-13.0000	0.1803	0.1008	-0.2361	*
975	0	1.0000	0	-0.0983	0.3724	-13.0000	-0.0210	0.0335	-0.3515	*
976	1.0000	0	0	0.5854	0.0931	7.0000	-0.1561	0.0113	1.9685	| *
977	0	0	0	-0.00850	-0.1027	17.0000	0.1304	-0.0545	-0.2397	*
978	1.0000	0	0	-0.0524	0.8403	7.0000	0.1029	0.1542	-0.5193	*|
979	1.0000	0	0	-0.3416	0.2183	7.0000	0.1333	0.0150	-0.2808	*
980	0	0	0	-0.0705	-0.1163	-13.0000	0.0297	-0.0666	-0.1580	*
981	0	1.0000	0	0.1859	-0.2020	-13.0000	-0.1366	-0.1554	-0.2726	*
982	0	0	0	-0.0660	-0.0373	7.0000	-0.0588	-0.00760	-0.1969	*
983	0	0	0	-0.0479	-0.1582	7.0000	-0.0727	-0.2627	-0.1340	*
984	0	0	0	-0.1866	-0.0247	-13.0000	-0.0144	0.0660	-0.1699	*
985	0	0	0	-0.4357	-0.2171	7.0000	-0.1098	-0.1037	-0.1122	*
986	1.0000	0	0	0.1499	-0.1301	7.0000	-0.1019	0.0730	-0.3630	*
987	0	0	0	-0.0435	-0.0629	7.0000	-0.1244	-0.1823	6.7172	| *
988	1.0000	0	0	0.2240	-0.2171	7.0000	-0.0747	-0.2578	4.1076	| *
989	0	0	0	-0.1317	0.0819	7.0000	-0.1669	0.0113	-0.1818	*
990	0	0	0	-0.2556	-0.1582	-13.0000	0.1712	0.0939	-0.1831	*
991	0	0	0	-0.1563	0.1366	7.0000	0.1350	-0.0829	-0.1980	*
992	0	0	0	0.2770	0.0706	17.0000	-0.00220	0.0588	-0.3475	*
993	0	0	0	-0.1125	-0.2796	7.0000	0.0221	0.0225	-0.1950	*
994	0	0	1.0000	0.1929	-0.0629	2.0000	0.1789	0.1606	1.3487	| *
995	0	0	0	-0.0479	0.1680	-13.0000	0.0999	-0.2007	-0.1528	*
996	0	0	1.0000	0.1011	0.2567	-13.0000	0.1154	0.0225	-0.5146	*|
997	0	0	0	0	-0.1872	7.0000	-0.1808	-0.2627	-0.1281	*
998	1.0000	0	0	0.1425	-0.1301	-13.0000	-0.0795	0.1638	-0.3488	*
999	0	0	0	-0.2948	0	-13.0000	0.0221	-0.2384	-0.1040	*
1000	0	0	0	-0.4162	0.0241	-13.0000	0.0174	-0.00380	-0.1296	*
1001	1.0000	0	0	0.4376	-0.1163	17.0000	0.0809	-0.1379	-0.4417	*
1002	0	1.0000	0	-0.1173	0.0931	-13.0000	0.0240	0.1008	-0.3623	*
1003	0	0	0	-0.2073	0.2567	30.0000	-0.1994	-0.2100	-0.1566	*
1004	0	0	0	-0.0705	-0.1301	7.0000	0.1291	-0.0788	-0.1987	*
1005	0	0	0	-0.1764	0.0706	17.0000	0.1471	-0.0625	-0.2163	*
1006	0	0	0	-0.0524	-0.1872	-10.0000	0.0762	0.0150	-0.1875	*
1007	0	0	0	0.0895	-0.0500	-13.0000	-0.1052	0.00750	-0.1871	*
1008	0	1.0000	0	-0.3179	-0.0760	7.0000	-0.0678	-0.0747	-0.2506	*
1009	0	0	0	-0.1030	-0.0500	2.0000	-0.1216	-0.0545	-0.1622	*
1010	1.0000	0	0	0.0252	-0.1163	-13.0000	0.0887	-0.1554	-0.2280	*
1011	0	0	0	-0.1464	-0.1027	7.0000	-0.0494	-0.2627	-0.1266	*
1012	0	0	0	-0.1077	0.0359	-13.0000	0.0347	0	-0.1757	*

particularly if the data are in ungrouped form. Nevertheless, Pearson residuals with large absolute values are worth further checking to be certain that the values of the dependent and independent variables are correctly entered and possibly to identify patterns in covariate values that consistently lead to outlying values. For ease of observation, the square of the Pearson residuals (referred to as DIFCHISQ) are displayed in Figure 13.5 for a subset of the data. The largest Pearson residual in this data set is for observation 646, which corresponds to a young smoker with no other risk factors who had a predicted probability of CHD of approximately 2% and developed CHD during the 10 years. He had a Pearson residual of 7.1, corresponding to a DIFCHISQ of $7.1^2 \approx 50$. If there are several other young smokers with no other risk factors who had events, we may want to modify our model to indicate interaction effects between smoking and age; i.e., to allow the effect of smoking to be different for younger versus older men.

FIGURE 13.5 **Display of DIFCHISQ (the square of the Pearson residual) for a subset of the Framingham Heart Study data**

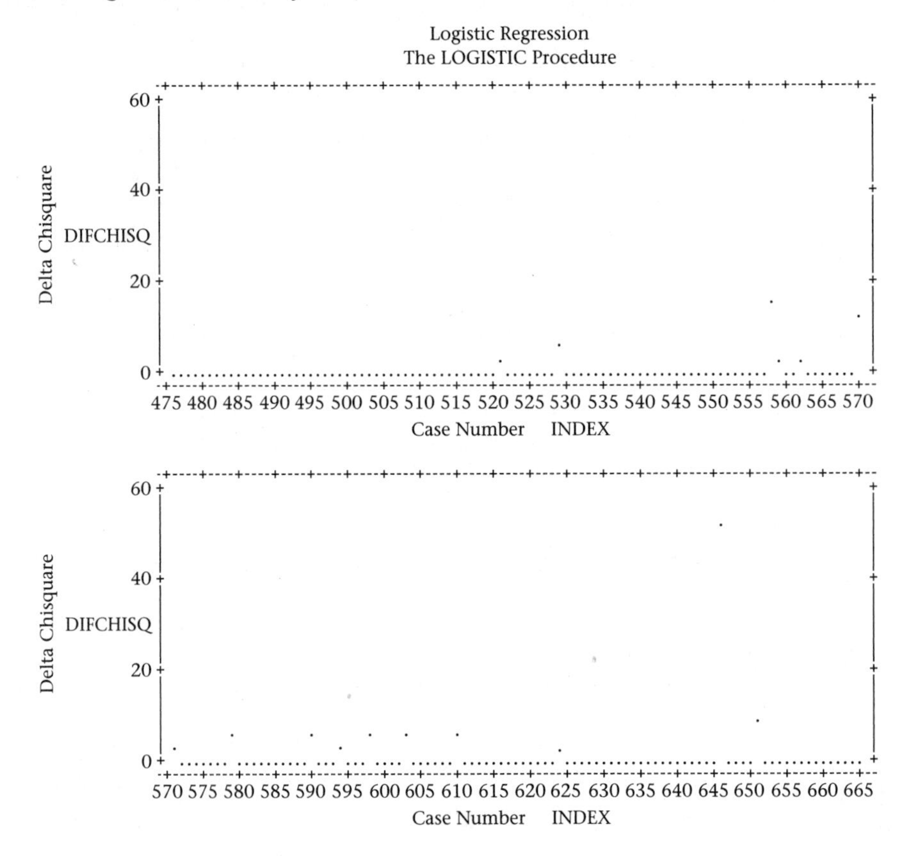

As in linear regression, another aspect of assessing goodness of fit is to determine how influential particular observations are in estimating the regression coefficients. Suppose the jth regression coefficient when estimated from the full data set is denoted by $\hat{\beta}_j$ and from the reduced data set obtained by deleting the ith individual by $\hat{\beta}_j^{(i)}$. A measure of influence of the ith observation on the estimation of β_j is given by

EQUATION 13.38

$$\Delta\beta_j^{(i)} = \frac{\hat{\beta}_j - \hat{\beta}_j^{(i)}}{se(\hat{\beta}_j)}$$

= influence of the ith observation on estimation of the jth regression coefficient

where $se(\hat{\beta}_j)$ is the standard error of $\hat{\beta}_j$ from the full data set. The ith observation will have a great influence on the estimation of β_j if $\left|\Delta\beta_j^{(i)}\right|$ is large.

Example 13.39 **Cardiovascular Disease** Assess the influence of individual observations on the estimation of the regression coefficient for systolic blood pressure for the model in Table 13.18.

Solution Systolic blood pressure is represented as ln(systolic blood pressure/132) and is denoted by LMSYD132 in Table 13.18. The influence measure in Equation 13.38 is displayed in Figure 13.6 for a subset of the data and is denoted by DFBETA8 (since this is the eighth regression coefficient in the model, other than the intercept). We see that none of the observations has a great influence in this subset of the data. The maximum value of $\left|\Delta\beta_j^{(i)}\right| \approx 0.2$, which is an actual difference of $0.2 \times se(\hat{\beta}_8) = 0.2 \times 0.5737 \approx 0.11$ (see Table 13.18). Because the actual value of $\hat{\beta}_j$ in the full model = 2.80, this is a relatively minor change (≈ 4%). This is also true for the remaining observations in the data set as well as for the other regression coefficients. However, influential observations are more likely to appear in smaller data sets, particularly for predictor variables for an individual subject that are far from the mean value.

In this section, we have learned about multiple logistic regression. This is an extremely important technique used to control for one or more continuous or categorical covariates (independent variables) when the outcome (dependent) variable is binary. It can be viewed as an extension of the Mantel-Haenszel type techniques and usually enables more complete control of confounding effects of several risk factors simultaneously. Also, it is analogous to multiple linear regression for normally distributed outcome variables.

Referring to the master flowchart (p. 778), we would start at ④ and generally answer no to (1) interested in relationships between two variables? This leads us to the box entitled "More than two variables of interest." We would then answer "binary" to (2) outcome variable continuous or binary? which leads us to the box entitled "Binary." We would then answer no to (3) time of events important? which leads us to the box entitled "Multiple logistic-regression methods."

It is of course also possible to use multiple logistic regression if only two variables are of interest, where the dependent variable is binary and the single independent variable is continuous. If the dependent variable is binary and we have a single independent variable that is categorical, then we can use either multiple logistic-regression or contingency-table methods, which should give identical results. The latter are probably preferable, for simplicity.

FIGURE 13.6 **Influence of individual observations on the estimation of the regression coefficient for systolic blood pressure for the Framingham Heart Study data fitted in Table 13.18**

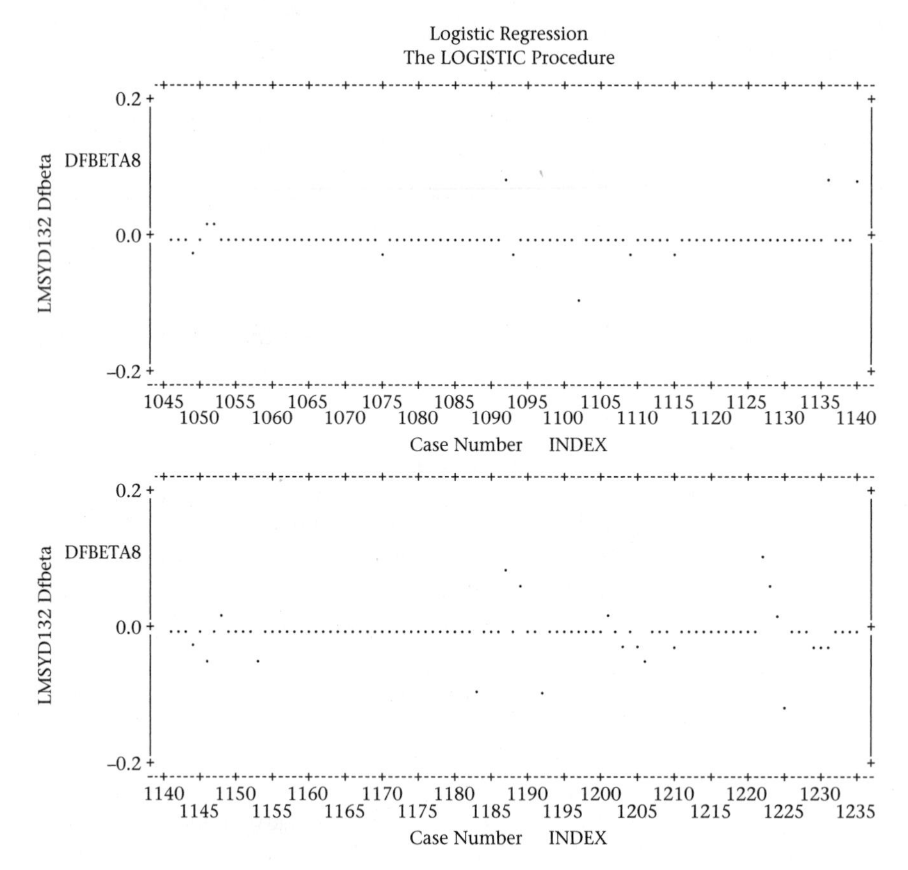

Section 13.8 Meta-Analysis

In the previous sections of this chapter, and in all previous chapters, we have learned about methods of analysis for a single study. However, it is often the case that more than one investigation is performed to study a particular research question, often by different research groups. In some instances, results are seemingly contradictory, with some research groups reporting significant differences for a particular finding while other research groups report no significant differences.

Example 13.40 **Renal Disease** In the Data Set NEPHRO.DAT we present data from a literature search comparing the nephrotoxicity (i.e., development of abnormal kidney function) of several different amminoglycosides (Buring et al. [12]). In Table 13.21, we focus on a subset of eight studies that compared two of the aminoglycosides—gentamicin and tobramycin. In seven out of eight studies, the odds ratio for tobramycin in comparison with gentamicin is less than 1, implying that there are fewer nephrotoxic side effects for tobramycin than for gentamicin. However, many of the studies are small and individually are likely to yield nonsignificant results. The issue is, What is the appropriate way to combine evidence across all the studies so as to reduce sampling error and increase the power of the investigation, and in some instances to resolve the inconsistencies among the study results? The technique for accomplishing this is called *meta-analysis*. In this section, we will present the methods of DerSimonian and Laird [13].

TABLE 13.21 **Comparison of nephrotoxicity of gentamicin versus tobramycin in NEPHRO.DAT**

	Gentamicin		Tobramycin					
Study	No. of subjects	No. of positives[a]	No. of subjects	No. of positives[a]	Odds ratio[b]	y_i[c]	w_i[d]	w_i^*[e]
1. Walker	40	7	40	2	0.25	−1.394	1.430	1.191
2. Wade	43	13	47	11	0.71	−0.349	4.367	2.709
3. Greene	11	2	15	2	0.69	−0.368	0.842	0.753
4. Smith	72	19	74	9	0.39	−0.951	5.051	2.957
5. Fong	102	18	103	15	0.80	−0.229	6.873	3.500
6. Brown	103	5	96	2	0.42	−0.875	1.387	1.161
7. Feig	25	10	29	8	0.57	−0.560	2.947	2.086
8. Matzke	99	9	97	17	2.13	+0.754	5.167	2.996

[a]Number who developed nephrotoxicity.
[b]Odds in favor of nephrotoxicity for tobramycin patients / odds in favor of nephrotoxicity for gentamicin patients.
[c]$y_i = \ln(\text{odds ratio}_i)$.
[d]$w_i = (1/a_i + 1/b_i + 1/c_i + 1/d_i)^{-1}$.
[e]$w_i^* = [(1/w_i) + \hat{\Delta}^2]^{-1}$.

Suppose there is an underlying log odds ratio θ_i for the ith study, which is estimated by $y_i = \ln(\hat{OR}_i)$, $i = 1, \ldots, 8$, where the estimated OR_i is given in Table 13.21 in the odds ratio column. We assume there is *within-study variation* of y_i about θ_i, where the variance of y_i is

$$s_i^2 = \frac{1}{a_i} + \frac{1}{b_i} + \frac{1}{c_i} + \frac{1}{d_i} = \frac{1}{w_i}$$

and a_i, b_i, c_i, and d_i are the cell counts in the 2×2 table for the ith study. We also assume that there is *between-study variation* of θ_i about an average true log odds ratio μ over all studies so that

$$\theta_i = \mu + \delta_i$$

and $Var(\delta_i) = \Delta^2$

This is similar to the random-effects ANOVA model presented in Section 12.8. To estimate μ, we calculate a weighted average of the study-specific log odds ratios given by

$$\hat{\mu} = \sum_{i=1}^{k} w_i^* y_i \Big/ \sum_{i=1}^{k} w_i^*$$

where

$$w_i^* = (s_i^{\ 2} + \hat{\Delta}^2)^{-1}$$

i.e., the weight for the ith study is inversely proportional to the total variance for that study (which equals $s_i^2 + \Delta^2$), and

$$se(\hat{\mu}) = 1\Big/\left(\sum_{i=1}^{k} w_i^*\right)^{1/2}$$

It can be shown that the best estimate of Δ^2 is given by

$$\hat{\Delta}^2 = \max\left\{0,\ [Q_w - (k-1)]\Big/\left[\sum_{i=1}^{k} w_i - \left(\sum_{i=1}^{k} w_i^2 \Big/ \sum_{i=1}^{k} w_i\right)\right]\right\}$$

where

$$Q_w = \sum_{i=1}^{k} w_i (y_i - \overline{y}_w)^2$$

and

$$\overline{y}_w = \sum_{i=1}^{k} w_i y_i \Big/ \sum_{i=1}^{k} w_i$$

This procedure is summarized as follows:

EQUATION 13.39

Meta-Analysis, Random-Effects Model Suppose we have k studies, each with the goal of estimating an odds ratio $\exp(\mu)$ defined as the odds of disease in a treated group compared with the odds of disease in a control group.

(1) The best estimate of the average study-specific log odds ratio from the k studies is given by

$$\hat{\mu} = \sum_{i=1}^{k} w_i^* y_i \Big/ \sum_{i=1}^{k} w_i^*$$

where y_i = log odds ratio for the ith study

$$w_i^* = \left(s_i^2 + \hat{\Delta}^2\right)^{-1}$$

$$1/w_i = s_i^2 = \frac{1}{a_i} + \frac{1}{b_i} + \frac{1}{c_i} + \frac{1}{d_i} = \text{within-study variance}$$

and a_i, b_i, c_i, and d_i are the cell counts for the 2×2 table for the ith study.

$$\hat{\Delta}^2 = \max\left\{0,\ [Q_w - (k-1)]\Big/\left[\sum_{i=1}^{k} w_i - \left(\sum_{i=1}^{k} w_i^2 \Big/ \sum_{i=1}^{k} w_i\right)\right]\right\}$$

$$Q_w = \sum_{i=1}^{k} w_i \left(y_i - \overline{y}_w\right)^2 = \sum_{i=1}^{k} w_i y_i^2 - \left(\sum_{i=1}^{k} w_i y_i\right)^2 \Big/ \sum_{i=1}^{k} w_i$$

and

$$\overline{y}_w = \sum_{i=1}^{k} w_i y_i \Big/ \sum_{i=1}^{k} w_i$$

The corresponding point estimate of the odds ratio = $\exp(\hat{\mu})$.

(2) The standard error of $\hat{\mu}$ = is given by $se(\hat{\mu}) = \left(1\Big/\sum_{i=1}^{k} w_i^*\right)^{1/2}$

(3) A $100\% \times (1 - \alpha)$ CI for μ is given by

$$\hat{\mu} \pm z_{1-\alpha/2} se(\hat{\mu}) = (\mu_1, \mu_2)$$

The corresponding $100\% \times (1 - \alpha)$ CI for OR is $[\exp(\mu_1), \exp(\mu_2)]$.

(4) To test the hypothesis H_0: $\mu = 0$ versus H_1: $\mu \neq 0$ (or equivalently to test H_0: $OR = 1$ versus H_1: $OR \neq 1$), we use the test statistic

$$z = \hat{\mu}/se(\hat{\mu})$$

which under H_0 follows an $N(0, 1)$ distribution. The two-sided p-value is given by $2 \times \left[1 - \Phi(|z|)\right]$.

Example 13.41 **Renal Disease** Estimate the combined nephrotoxicity odds ratio comparing tobramycin to gentamicin based on the data in Table 13.21. Obtain a 95% CI, and provide a two-sided p-value for the hypothesis that the 2 treatments have equal nephrotoxicity.

Solution We first compute the study-specific log odds ratio (y_i) and weight $w_i = (1/a_i + 1/b_i + 1/c_i + 1/d_i)^{-1}$ which are shown in Table 13.21.

Next we compute the estimated between-study variance $\hat{\Delta}^2$. We have that

$$\sum_{i=1}^{8} w_i = 1.430 + \cdots + 5.167 = 28.0646$$

$$\sum_{i=1}^{8} w_i y_i = 1.430(-1.394) + \cdots + 5.167(0.754) = -9.1740$$

$$\sum_{i=1}^{8} w_i y_i^2 = 1.430(-1.394)^2 + \cdots + 5.167(0.754)^2 = 13.2750$$

Hence,

$$Q_w = 13.2750 - (-9.1740)^2/28.0646 = 10.276$$

Furthermore,

$$\sum_{i=1}^{8} w_i^2 = 1.430^2 + \cdots + 5.167^2 = 131.889$$

$$\hat{\Delta}^2 = (10.276 - 7)/(28.0646 - 131.889/28.0646)$$

$$= 0.140 = \text{between-study variance}$$

Hence,

$$w_i^* = (1/w_i + \hat{\Delta}^2)^{-1}$$

as shown in Table 13.21. Finally,

$$\sum_{i=1}^{8} w_i^* y_i = 1.191(-1.394) + \cdots + 2.996(0.754) = -6.421$$

$$\sum_{i=1}^{8} w_i^* = 1.191 + \cdots + 2.996 = 17.3526$$

and

$$\hat{\mu} = -6.421/17.3526 = -0.370$$

with standard error given by

$$se(\hat{\mu}) = (1/17.3526)^{1/2} = 0.240$$

Hence, the point estimate of the average odds ratio $\exp(\mu)$ is given by $\exp(-0.370) = 0.69$. A 95% CI for $\exp(\mu)$ is given by $\exp[-0.370 \pm 1.96(0.240)] = (0.43, 1.11)$.

To test the hypothesis H_0: $\mu = 0$ versus H_1: $\mu \neq 0$, we use the test statistic

$$z = \hat{\mu}/se(\hat{\mu})$$
$$= -0.370/0.240 = -1.542$$

with corresponding two-sided p-value given by $p = 2 \times [1 - \Phi(1.542)] = 0.123$. Hence the odds ratio, although less than 1, is not significantly different from 1.

13.8.1 Test of Homogeneity of Odds Ratios

Some investigators feel that the procedure in Equation 13.39 should only be used if there is not significant heterogeneity among the k study-specific odds ratios. To test the hypothesis H_0: $\theta_1 = \cdots = \theta_k$ versus H_1: at least two θ_i's are different, we use Equation 13.40.

EQUATION 13.40

Test of Homogeneity of Study-Specific Odds Ratios in Meta-Analysis To test the hypothesis H_0: $\theta_1 = \cdots = \theta_k$ versus H_1: at least two θ_i's are different where θ_i = log odds ratio in the ith study, we use the test statistic

$$Q_w = \sum_{i=1}^{k} w_i (y_i - \bar{y}_w)^2$$

as defined in Equation 13.39. It can be shown under H_0 that $Q_w \sim \chi^2_{k-1}$. Hence, to compute the p-value, we have

$$p\text{-value} = Pr\left(\chi^2_{k-1} > Q_w\right)$$

Example 13.42 **Renal Disease** Test for the homogeneity of the odds ratios in Table 13.21.

Solution From the solution to Example 13.41, we have $Q_w = 10.276 \sim \chi^2_7$ under H_0. Since $\chi^2_{7,.75} = 9.04$ and $\chi^2_{7,.90} = 12.02$ and $9.04 < 10.276 < 12.02$, it follows that $1 - .90 < p < 1 - .75$ or $.10 < p < .25$. Hence, there is not significant heterogeneity among the study-specific odds ratios.

It is controversial among research workers whether a fixed- or random-effects model should be used when performing a meta-analysis. Under a fixed-effects model, the between-study variance (Δ^2) is ignored in computing the study weights and only the within-study variance is considered. Hence, one uses w_i for the weights in Equation 13.39 instead of w_i^*. It is argued that if there is substantial variation among the study-specific odds ratios, then one should investigate the source of the heterogeneity (e.g., different study designs, etc.) and not report an overall pooled estimate of the odds ratio as given in Equation 13.39. Others feel that between-study variation should always be considered in meta-analyses. Generally speaking, using a fixed-effects model results in tighter confidence limits and more significant results. However, note that the fixed-effects model gives different relative weights to the individual studies than the random-effects model. In a fixed-effects model, only within-study variation are considered. In a random-effects model, both between- and within-study variation are considered. If the between-study variation is substantial relative to the within-study variation (as is sometimes the case), then larger studies will get proportionally more weight under a fixed-effects model than under a random-effects model. Hence, the summary odds ratios under these two models may also be different. This is indeed the case in Table 13.21, where the relative weight of larger studies compared to smaller studies is greater for the fixed-effects model weights (w_i) than for the random-effects model weights (w_i^*). For example, for the data in Table 13.21, if one uses the w_i for weights instead of w_i^*, one obtains a point estimate for the overall odds ratio [$\exp(\hat{\mu})$] of 0.72 with 95% confidence limits of (0.49, 1.04) with p-value = .083 for testing H_0: $OR = 1$ versus H_1: $OR \neq 1$, compared with an odds ratio of 0.69 with 95% confidence limits of (0.43, 1.11) for the random-effects model.

One drawback to the random-effects approach is that one cannot use studies with no events in either treatment group. Under a fixed-effects model such studies get 0 weight. However, under a random-effects model such studies get nonzero weight if the between-study variance is greater than 0. This is problematic, because the log odds ratio is either $+\infty$ or $-\infty$ unless both groups have no events. We excluded one small study in our survey in Table 13.21 for this reason (Bodey et al.; see [12]), where there were 11 gentamicin patients who experienced 0 events and 11 tobramycin patients who experienced two events. A reasonable compromise might be to check for significant heterogeneity among the study-specific odds ratios using Equation 13.40 and use the decision rules in Table 13.22.

TABLE 13.22 **Models used for meta-analysis**

p-value for heterogeneity	Type of model used
≥0.5	Use fixed-effects model
$0.05 \le p < 0.5$	Use random-effects model
<0.05	Do not report pooled odds ratio; assess sources of heterogeneity

Meta-analysis methods can also be performed based on other effect measures (e.g., mean differences between treatment groups instead of odds ratios). However, this is beyond the scope of this text. For a complete description of meta-analysis, see Hedges and Olkin [14].

In this section, we have studied meta-analysis, a technique for formally combining results over more than one study to maximize precision in estimating parameters and to maximize power for testing hypotheses for particular research questions. We studied both a fixed-effects model, where the weight received by individual studies is determined only by within-study variation, and a random-effects model, where both between- and within-study variation determine the weight. We also discussed a possible strategy for determining which of the two models, if either, to use in specific situations.

SECTION 13.9 Equivalence Studies

13.9.1 Introduction

In Chapter 10, we have considered the estimation of sample size for studies where the null hypothesis is that two treatments are equally effective versus the alternative hypothesis that the effects of the two treatments are different from each other. These types of studies, which constitute the majority of clinical trials, are referred to as *superiority studies*. However, a newer type of study design has emerged in recent years where the major goal is to show that two treatments are equivalent rather than that one is superior to the other. Consider the following example, presented by Makuch and Simon [15].

Example 13.43 **Cancer** Suppose we want to design a clinical trial to compare two surgical treatments for early-stage breast cancer. The treatments are simple mastectomy and a more conservative tumor resection. In this setting, it would be unethical to compare the experimental treatment to a placebo. Instead, two active treatments are compared with each other. The former treatment is the standard and yields a 5-year survival rate of 80%. The latter is an experimental treatment that is less debilitating than the standard. However, it will only be considered acceptable if it can be shown in some statistical sense to be no more than 10% inferior to the standard treatment in terms of 5-year survival. How can we test whether the experimental treatment is acceptable, and how can we estimate sample size for such a study?

13.9.2 Interference Based on Confidence-Interval Estimation

DEFINITION 13.17 The type of study in Example 13.43 where the goal is to show approximate equivalence of two experimental treatments is called an **equivalence study.**

Suppose p_1 is the survival rate for the standard treatment and p_2 is the survival rate for the experimental treatment. The approach we will take is to determine a lower one-sided $100\% \times (1 - \alpha)$ confidence interval for $p_1 - p_2$. In Equation 13.1, we provide a two-sided CI for $p_1 - p_2$. The corresponding lower one-sided interval is given by

EQUATION 13.41

$$p_1 - p_2 < \hat{p}_1 - \hat{p}_2 + z_{1-\alpha}\sqrt{\hat{p}_1\hat{q}_1/n_1 + \hat{p}_2\hat{q}_2/n_2}$$

where we have ignored the generally small continuity correction.

We will consider the treatments as equivalent if the upper bound of this one-sided confidence interval does not exceed δ, where δ is a prespecified threshold.

Example 13.44 **Cancer** Suppose we have a clinical trial with 100 patients each on the standard treatment and on the experimental treatment. We find that 80% of patients on the standard treatment and 75% of the patients on the experimental treatment survive for 5 years. Can the treatments be considered equivalent if the threshold for equivalence is that the survival rate for the experimental treatment is no more than 10% worse than for the standard treatment based on a one-sided 95% CI approach?

Solution We construct a lower one-sided 95% CI for $p_1 - p_2$. From Equation 13.41, this is given by

$$p_1 - p_2 < .80 - .75 + z_{.95}\sqrt{.80(.20)/100 + .75(.25)/100} = .05 + 1.645(.0589)$$
$$= .05 + .097 = .147$$

The upper bound of the lower 95% CI exceeds 10%, so the treatments *cannot* be considered as equivalent. Thus, despite the fact that the observed survival rates are only 5% apart, the underlying rates may differ by as much as 15%, which implies the treatments cannot be considered equivalent.

13.9.3 Sample-Size Estimation for Equivalence Studies

It seems clear from Example 13.44 that large sample sizes are needed to demonstrate equivalence. In some cases, depending on the threshold δ specified for equivalence, the sample size needed may be considerably larger than for typical superiority studies. The approach we will take is to require a sample size large enough so that with high probability $(1 - \beta)$ the upper confidence limit in Equation 13.41 does not exceed δ. Hence, we want

EQUATION 13.42

$$Pr\left[\hat{p}_1 - \hat{p}_2 + z_{1-\alpha}\sqrt{\hat{p}_1\hat{q}_1/n_1 + \hat{p}_2\hat{q}_2/n_2} \le \delta\right] = 1 - \beta$$

However, if we subtract $p_1 - p_2$ from both sides of Equation 13.42, divide by $\sqrt{\hat{p}_1\hat{q}_1/n_1 + \hat{p}_2\hat{q}_2/n_2}$, and then subtract $z_{1-\alpha}$ from both sides, we obtain

$$Pr\left[\frac{\hat{p}_1 - \hat{p}_2 - (p_1 - p_2)}{\sqrt{\hat{p}_1\hat{q}_1/n_1 + \hat{p}_2\hat{q}_2/n_2}} \le \frac{\delta - (p_1 - p_2)}{\sqrt{\hat{p}_1\hat{q}_1/n_1 + \hat{p}_2\hat{q}_2/n_2}} - z_{1-\alpha}\right] = 1 - \beta$$

Under the hypothesis that the true difference in survival rates between treatment groups is $p_1 - p_2$, the random variable on the left side is a standard normal deviate. Therefore, to satisfy this equation we have

EQUATION 13.43

$$\frac{\delta - (p_1 - p_2)}{\sqrt{\hat{p}_1\hat{q}_1/n_1 + \hat{p}_2\hat{q}_2/n_2}} - z_{1-\alpha} = z_{1-\beta}$$

We now add $z_{1-\alpha}$ to both sides of Equation 13.43 and divide by $\delta - (p_1 - p_2)$ to obtain

$$\frac{1}{\sqrt{\hat{p}_1\hat{q}_1/n_1 + \hat{p}_2\hat{q}_2/n_2}} = \frac{z_{1-\alpha} + z_{1-\beta}}{\delta - (p_1 - p_2)}$$

If we assume that the experimental treatment sample size (n_2) is k times as large as the standard treatment sample size (n_1), we obtain

$$\frac{\sqrt{n_1}}{\sqrt{\hat{p}_1\hat{q}_1 + \hat{p}_2\hat{q}_2/k}} = \frac{z_{1-\alpha} + z_{1-\beta}}{\delta - (p_1 - p_2)}$$

Solving for n_1 yields

EQUATION 13.44

$$n_1 = \frac{\left(\hat{p}_1\hat{q}_1 + \hat{p}_2\hat{q}_2/k\right)\left(z_{1-\alpha} + z_{1-\beta}\right)^2}{\left[\delta - (p_1 - p_2)\right]^2}, \quad n_2 = kn_1$$

Under the null hypothesis, suppose $p_1 = p_2 = p$. Equation 13.44 can then be summarized as follows:

EQUATION 13.45

Sample-Size Estimation for Equivalence Studies Suppose we wish to establish equivalence between a standard treatment (treatment 1) and an experimental treatment (treatment 2), where p_1 and p_2 are treatment success rates in groups 1 and 2, respectively. The treatments will be considered equivalent (in the sense that the experimental treatment is not substantially worse than the standard treatment) if the upper bound of a lower $100\% \times (1 - \alpha)$ confidence interval for $p_1 - p_2$ is $\le \delta$. If $p_1 = p_2 = p$ and we wish to establish equivalence with a probability of $1 - \beta$, we require

$$n_1 = \frac{(pq)(1 + 1/k)(z_{1-\alpha} + z_{1-\beta})^2}{\delta^2} \quad \text{subjects in group 1}$$

$n_2 = kn_1$ subjects in group 2 where k is specified in advance

Example 13.45 **Cancer** Estimate the required sample size for the study described in Example 13.44 if (1) we wish to have a probability of 80% for establishing equivalence, (2) the sample sizes are the same in the two groups, (3) the underlying 5-year survival rate in both groups is 80%, (4) the threshold for equivalence is 10%, and (5) we establish equivalence based on an upper bound of a lower 95% CI.

Solution We have $p = .80$, $q = .20$, $k = 1$, $\alpha = .05$, $\beta = .20$, $\delta = .10$. Therefore,

$$n_1 = \frac{.80(.20)(2)(z_{.95} + z_{.80})^2}{(.10)^2}$$

$$= \frac{.32(1.645 + 0.84)^2}{.01} = 197.6 = n_2$$

Therefore, we require 198 subjects in each group to have an 80% probability of establishing equivalence under this design. This is larger than the sample size in Example 13.44, where we were unable to demonstrate equivalence.

In this section, we have considered methods of analysis and methods of sample-size estimation for equivalence studies (sometimes called active-control studies). An equivalence study is one where we want to establish with high probability that the difference in effect between two treatment groups does not exceed some prespecified threshold δ with high probability $(1 - \beta)$. The threshold δ and probability $1 - \beta$ need to be specified in advance.

When is it reasonable to consider an equivalence versus a superiority study? Some people feel that a superiority study based on placebo control is always the design of choice to establish the efficacy of a treatment [16]. Others feel that if a standard therapy has already proven its effectiveness, then it would be unethical to withhold treatment from patients (e.g., by using a placebo as one of the treatment groups in a clinical trial to establish the efficacy of a new treatment for schizophrenia). These issues are discussed in more detail by Rothman and Michels [17].

SECTION 13.10 The Cross-Over Design

Example 13.46 **Sports Medicine** In Problem 8.112 we were introduced to Data Set TENNIS2.DAT, on the data disk. This was a clinical trial comparing Motrin versus placebo for the treatment of tennis elbow. Each subject was randomized to receive either Motrin (group A) or placebo (group B) for a 3-week period. All subjects then had a 2-week washout period during which they received no study medication. All subjects were then "crossed-over" for a second 3-week period to receive the opposite study medication from that initially received. Subjects in group A received 3 weeks of placebo while subjects in group B received 3 weeks of Motrin. This type of design is called a cross-over design. How should we compare the efficacy of Motrin versus placebo using this design?

DEFINITION 13.18 A **cross-over design** is a type of randomized clinical trial. In this design, each subject is randomized to either group 1 or group 2. All subjects in group 1 receive drug A in the first treatment period and drug B in the second treatment period. All subjects in group 2 receive drug B in the first treatment period and drug A in the second treatment period. Often there is a *washout* period between the two active drug periods during which they receive no study medication. The purpose of the washout period is to reduce the likelihood that study medication taken in the first period will have an effect that carries over to the next period.

DEFINITION 13.19 A **washout period** in a cross-over design is a period between active drug periods, during which subjects receive no study medication.

DEFINITION 13.20 A **carry-over effect** in a cross-over design is when the effects of one or both study medications taken during the first active drug period have a residual biological effect during the second active drug period.

The cross-over design described in Definition 13.18 is actually a two-period cross-over design. There are also cross-over designs with more than two periods and/or more than two treatments being compared. These latter designs are beyond the scope of this text. See Fleiss [18] for a discussion of these designs.

13.10.1 Assessment of Treatment Effects

A 6-point scale for pain relief was used in the study. At the end of each active treatment period, the subjects were asked to rate their degree of pain relative to baseline—that is, the beginning of the study prior to either active treatment period. The rating scale was 1 if worse, 2 if unchanged, 3 if slightly improved (25%), 4 if moderately improved (50%), 5 if mostly improved (75%), and 6 if completely improved (100%). We wish to compare the degree of pain relief for subjects while on Motrin to the degree of pain relief while on placebo. Let x_{ijk} = the pain relief rating for the jth subject in the ith group during the kth period, where i = group (1 = group A, 2 = group B), j = subject ($j = 1, \ldots, n_1$ if subject is in group A, $j = 1, \ldots, n_2$ if subject is in group B), k = period (1 = 1st period, 2 = 2nd period). For the jth patient in group A, the measure of drug efficacy is $d_{1j} = x_{1j1} - x_{1j2}$, whereas for the jth patient in group B, the measure of drug efficacy is $d_{2j} = x_{2j2} - x_{2j1}$. In each case, a large number indicates that the patient experiences less pain while on Motrin than on placebo. The summary measure of efficacy for patients in group A is therefore

$$\bar{d}_1 = \sum_{j=1}^{n_1} d_{1j}/n_1$$

while for patients in group B it is

$$\bar{d}_2 = \sum_{j=1}^{n_2} d_{2j}/n_2$$

The overall measure of drug efficacy is

EQUATION 13.46

$$\bar{d} = \frac{1}{2}\left(\bar{d}_1 + \bar{d}_2\right)$$

To compute the standard error of $\bar{d}$, we assume that the underlying variance of the within-subject differences in group A and group B are the same, and estimate this variance (σ_d^2) by the pooled estimate

EQUATION 13.47

$$s_{d,\text{pooled}}^2 = \frac{\sum_{j=1}^{n_1}\left(d_{1j} - \bar{d}_1\right)^2 + \sum_{j=1}^{n_2}\left(d_{2j} - \bar{d}_2\right)^2}{n_1 + n_2 - 2} = \frac{(n_1 - 1)s_{d_1}^2 + (n_2 - 1)s_{d_2}^2}{n_1 + n_2 - 2}$$

Therefore,

$$\begin{aligned} Var(\bar{d}) &= \frac{1}{4}\left[Var(\bar{d}_1) + Var(\bar{d}_2)\right] \\ &= \frac{1}{4}\left[\frac{\sigma_d^2}{n_1} + \frac{\sigma_d^2}{n_2}\right] \\ &= \frac{\sigma_d^2}{4}\left(\frac{1}{n_1} + \frac{1}{n_2}\right) \end{aligned}$$

which is estimated by

$$\frac{s_{d,\text{pooled}}^2}{4}\left(\frac{1}{n_1} + \frac{1}{n_2}\right)$$

with $n_1 + n_2 - 2$ *df*. The standard error of $\bar{d}$ is thus

$$se(\bar{d}) = \sqrt{\frac{s_{d,\text{pooled}}^2}{4}\left(\frac{1}{n_1} + \frac{1}{n_2}\right)} = \frac{s_{d,\text{pooled}}}{2}\sqrt{\frac{1}{n_1} + \frac{1}{n_2}}$$

This leads to the following test procedure for assessing the overall treatment effect in cross-over designs.

EQUATION 13.48

The Cross-Over Design—Assessment of Overall Treatment Effects Let x_{ijk} represent the score for the jth patient in the ith group during the kth period for patients entered into a study using a cross-over design, $i = 1, 2; j = 1, \ldots, n_i; k = 1, 2$. Suppose patients in group 1 receive treatment 1 in period 1 and treatment 2 in period 2, while patients in group 2 receive treatment 2 in period 1 and treatment 1 in period 2. If we assume that no carry-over effect is present, then we use the following procedure to assess overall treatment efficacy:

(1) We compute

$$\bar{d} = \text{overall estimate of treatment efficacy} = \frac{1}{2}(\bar{d}_1 + \bar{d}_2), \text{ where}$$

$$\bar{d}_1 = \sum_{j=1}^{n_1} d_{1j}/n_1$$

$$\bar{d}_2 = \sum_{j=1}^{n_2} d_{2j}/n_2$$

$$d_{1j} = x_{1j1} - x_{1j2}, j = 1, \ldots, n_1$$

$$d_{2j} = x_{2j2} - x_{2j1}, j = 1, \ldots, n_2$$

(2) The standard error of $\bar{d}$ is estimated by

$$\sqrt{\frac{s^2_{d,\text{pooled}}}{4}\left(\frac{1}{n_1} + \frac{1}{n_2}\right)} = \frac{s_{d,\text{pooled}}}{2}\sqrt{\frac{1}{n_1} + \frac{1}{n_2}}$$

where

$$s^2_{d,\text{pooled}} = \frac{(n_1 - 1)s^2_{d_1} + (n_2 - 1)s^2_{d_2}}{n_1 + n_2 - 2}$$

$$s^2_{d_1} = \sum_{j=1}^{n_1} \left(d_{1j} - \bar{d}_1\right)^2 / (n_1 - 1)$$

$$s^2_{d_2} = \sum_{j=1}^{n_2} \left(d_{2j} - \bar{d}_2\right)^2 / (n_2 - 1)$$

(3) If Δ = underlying mean treatment efficacy, then to test the hypothesis H_0: $\Delta = 0$ versus H_1: $\Delta \neq 0$ using a two-sided level α significance test, compute the test statistic

$$t = \frac{\bar{d}}{\sqrt{\frac{s^2_{d,\text{pooled}}}{4}\left(\frac{1}{n_1} + \frac{1}{n_2}\right)}}$$

(4) If $t > t_{n_1+n_2-2,1-\alpha/2}$ or $t < t_{n_1+n_2-2,\alpha/2}$, then reject H_0. If $t_{n_1+n_2-2,\alpha/2} \le t \le t_{n_1+n_2-2,1-\alpha/2}$, then accept H_0.

(5) The exact p-value is given by

$$2 \times \text{area to the left of } t \text{ under a } t_{n_1+n_2-2} \text{ distribution if } t \le 0$$

or

$2 \times$ area to the right of t under a $t_{n_1+n_2-2}$ distribution if $t > 0$

(6) A $100\% \times (1 - \alpha)$ CI for the underlying treatment effect Δ is given by

$$\bar{d} \pm t_{n_1+n_2-2,1-\alpha/2}\sqrt{\frac{s^2_{d,\text{pooled}}}{4}\left(\frac{1}{n_1}+\frac{1}{n_2}\right)}$$

Example 13.47 **Sports Medicine** Test for whether the overall degree of pain as compared with baseline is different for patients while on Motrin than while on placebo. Estimate a 95% confidence interval for the improvement in the degree of pain while on Motrin versus placebo.

Solution There are 88 subjects in the data set, 44 in group 1 and 44 in group 2. However, 2 subjects in each group had missing pain scores in one or both periods. Hence, there are 42 subjects available for analysis in each group. We first present the mean pain score versus baseline for patients in each group during each period as well as the mean difference in pain scores (Motrin–placebo) and the average pain relief score over the two periods (see Table 13.23).

TABLE 13.23 Summary statistics of overall impression of drug efficacy compared with baseline ($n = 84$)

	Group							
	1				2			
	Motrin	*Placebo*	*Difference*[a]	*Average*[b]	*Motrin*	*Placebo*	*Difference*	*Average*
Mean	3.833	3.762	0.071	3.798	4.214	2.857	1.357	3.536
sd	1.188	1.574	1.813	1.060	1.353	1.160	1.376	1.056
n	42	42	42	42	42	42	42	42

Note: The data are obtained from the Data Set TENNIS2.DAT using variable 22 for overall impression of drug efficacy during period 1 and variable 43 for overall impression of drug efficacy during period 2.
[a]Pain score on Motrin – pain score on placebo
[b]Average of (pain score on Motrin, and pain score on placebo)

The overall measure of drug efficacy is

$$\bar{d} = \frac{0.071 + 1.357}{2} = 0.714$$

To compute the standard error of $\bar{d}$, we first compute the pooled variance estimate given by

$$s^2_{d,\text{pooled}} = \left[(n_1 - 1)s^2_{d_1} + (n_2 - 1)s^2_{d_2}\right] / (n_1 + n_2 - 2)$$

$$= \frac{41(1.813)^2 + 41(1.376)^2}{82} = 2.590$$

The standard error of $\bar{d}$ is

$$se(\bar{d}) = \sqrt{\frac{2.590}{4}\left(\frac{1}{42}+\frac{1}{42}\right)} = 0.176$$

The test statistic is

$$t = \frac{0.714}{0.176} = 4.07 \sim t_{82} \text{ under } H_0$$

The exact p-value $= Pr(t_{82} > 4.07)$. Because $4.07 > t_{60,\,.9995} = 3.460 > t_{82,\,.9995}$, it follows that $p < 2 \times (1 - .9995)$ or $p < .001$. Thus, there is a highly significant difference in the mean pain score on Motrin versus the mean pain score on placebo, with patients experiencing less pain when on Motrin.

A 95% CI for the treatment benefit Δ is

$$\begin{aligned}&\bar{d} \pm t_{n_1+n_2-2,.975}\,se(\bar{d})\\&= 0.714 \pm t_{82,.975}(0.176)\end{aligned}$$

Using Excel, we estimate $t_{82,.975} = 1.989$. Therefore, the 95% CI for $\Delta = 0.714 \pm 1.989(0.176) = (0.365, 1.063)$. Thus, the treatment benefit is likely to be between 1/3 of a unit and 1 unit on the pain-score scale.

13.10.2 Assessment of Carry-Over Effects

In the preceding section, when we computed the overall estimate of the treatment effect in Equation 13.48, we assumed that there was no carry-over effect. A carry-over effect is present when the true treatment effect is different for subjects in group A than for subjects in group B.

Example 13.48 **Sports Medicine** Suppose that Motrin is very effective in the relief of pain from tennis elbow and that the pain relief is long-lasting (i.e., pain relief continues even after the patients stop taking the medication, while placebo has no effect on pain). In this case, the difference between Motrin- and placebo-treated patients will be greater during the first treatment period than during the second treatment period. Another way of stating this is that the difference between Motrin and placebo will be smaller for subjects in group 1 than for subjects in group 2. This is because of the carry-over effect of Motrin taken in the first period into the second period. How can we identify such carry-over effects?

Notice that in Example 13.48, if there is a carry-over effect, then the average response for patients in group 1 over the two periods will be greater than for patients in group 2. This will form the basis for our test for identifying carry-over effects.

EQUATION 13.49

Assessment of Carry-Over Effects in Cross-Over Studies Let x_{ijk} represent the score for the jth patient in the ith group during the kth period. Define $\bar{x}_{ij} = (x_{ij1} + x_{ij2})/2$ = average response over both periods for the jth patient in the ith group and $\bar{x}_i = \sum_{j=1}^{n_i} \bar{x}_{ij}/n_i$ = average response over all patients in the ith group over both treatment periods. We assume that $\bar{x}_{ij} \sim N(\mu_i, \sigma^2),\ i = 1,\ 2;\ j = 1, \ldots, n_i$. To test the hypothesis H_0: $\mu_1 = \mu_2$ versus H_1: $\mu_1 \neq \mu_2$:

(1) We compute the test statistic

$$t = \frac{\bar{x}_1 - \bar{x}_2}{\sqrt{s^2\left(\frac{1}{n_1} + \frac{1}{n_2}\right)}}$$

where

$$s^2 = \frac{(n_1 - 1)s_1^2 + (n_2 - 1)s_2^2}{n_1 + n_2 - 2}$$

and

$$s_i^2 = \sum_{j=1}^{n_i} (\bar{x}_{ij} - \bar{x}_i)^2 / (n_i - 1),\ i = 1,\ 2$$

(2) If $t > t_{n_1+n_2-2,1-\alpha/2}$ or $t < t_{n_1+n_2-2,\alpha/2}$, we reject H_0. If $t_{n_1+n_2-2,\alpha/2} \le t \le t_{n_1+n_2-2,1-\alpha/2}$, we accept H_0.

(3) The exact p-value $= 2 \times Pr(t_{n_1+n_2-2} > t)$ if $t > 0$
$= 2 \times Pr(t_{n_1+n_2-2} < t)$ if $t \le 0$

Example 13.49 Assess if there are any carry-over effects, based on the tennis-elbow data in Example 13.48.

Solution We refer to Table 13.23 and note that

$$\bar{x}_1 = 3.798, s_1 = 1.060, n_1 = 42$$

$$\bar{x}_2 = 3.536, s_2 = 1.056, n_2 = 42$$

$$s^2 = \frac{41(1.060)^2 + 41(1.056)^2}{82} = 1.119$$

Therefore, the test statistic is given by

$$t = \frac{3.798 - 3.536}{\sqrt{1.119\left(\frac{1}{42} + \frac{1}{42}\right)}}$$

$$= \frac{0.262}{0.231} = 1.135 \sim t_{82} \text{ under } H_0$$

Since $t > 1.046 = t_{60,\,.85} > t_{82,\,.85}$, it follows that $p < 2 \times (1 - .85) = .30$. Since $t < 1.289 = t_{120,\,.90} < t_{82,\,.90}$, it follows that $p > 2 \times (1 - .90) = .20$. Therefore, we have $.20 < p < .30$ and there is no significant carry-over effect. We can also gain some insight into possible carry-over effects by referring to Table 13.23. We see that the treatment benefit during period 1 is 3.833 – 2.857 = 0.976, while the treatment benefit during period 2 is 4.214 – 3.762 = 0.452. Thus, there is some treatment benefit during each period. The degree of benefit is larger in period 1, but is not significantly larger. In general, the power of the test to detect carry-over effects is not great. Also, the effect of possible carry-over effects on identifying overall treatment benefit can be large. Therefore, some authors [19] have recommended that the p-value for declaring significant carry-over effects be set at .10 rather than the usual .05. Even with this more relaxed criterion for achieving statistical significance, we still don't declare a significant carry-over effect with the tennis-elbow data.

Another important insight into the data is revealed by looking for period effects. For example, in Table 13.23 the effect of period 2 versus period 1 is 4.214 – 3.833 = 0.381 while subjects were on Motrin and 3.762 – 2.857 = 0.905 while subjects were on placebo. Thus, subjects are experiencing less pain in period 2 compared to period 1 regardless of which medication they are taking.

What can we do if we identify a significant carry-over effect using Equation 13.49? In this case, the second-period data are not useful to us, because they provide a biased estimate of treatment effects, particularly for subjects who were on active drug in the first period and on placebo in the second period, and we must base our comparison of treatment efficacy on first-period data only. We can use an ordinary two-sample t test for independent samples based on the first-period data. This test usually has less power than the cross-over efficacy test in Equation 13.48, or alternatively requires a greater sample size to achieve a given level of power (see Example 13.50).

13.10.3 Sample-Size Estimation for Cross-Over Studies

A major advantage of cross-over studies is that they usually require many fewer subjects than the usual randomized clinical trials (which have only 1 period), if no carry-over effect is present. The sample-size formula is given as follows:

EQUATION 13.50

Sample-Size Estimation for Cross-Over Studies Suppose we wish to test the hypothesis H_0: $\Delta = 0$ versus H_1: $\Delta \neq 0$ using a two-sided test with significance level α, where Δ = underlying treatment benefit for treatment 1 versus treatment 2 using a cross-over study. If we require a power of $1 - \beta$, and we expect to randomize an equal number of subjects to each group (group 1 receives treatment 1 in period 1 and treatment 2 in period 2; group 2 receives treatment 2 in period 1 and treatment 1 in period 2), then the appropriate sample size per group = n, where

$$n = \frac{\sigma_d^2\left(z_{1-\alpha/2} + z_{1-\beta}\right)^2}{2\Delta^2}$$

where σ_d^2 = variance of difference scores = variance of (response on treatment 1 – response on treatment 2).

This sample-size formula is only applicable if no carry-over effects (as defined in Definition 13.20) are present.

Example 13.50 **Hypertension** Suppose we want to study the effect of postmenopausal hormones (PMH) on level of diastolic blood pressure (DBP). We intend to enroll n postmenopausal women per group. Women in group 1 will get PMH pills in period 1 (4 weeks) and placebo pills in period 2 (4 weeks). Women in group 2 will get placebo pills in period 1 and PMH pills in period 2. There will be a 2-week washout period between each 4-week active-treatment period. Women will have their blood pressure measured at the end of each active-treatment period based on a mean of 3 readings at a single visit. If we anticipate a 2-mm Hg treatment benefit and the within-subject variance of the difference in mean DBP between the two periods is estimated to be 31, based on pilot-study results, and we require 80% power, then how many subjects need to be enrolled in each group?

Solution We have $\sigma_d^2 = 31$, $\alpha = .05$, $\beta = .20$, $\Delta = 2$. Thus, from Equation 13.50, we have $z_{1-\alpha/2} = z_{.975} = 1.96$, $z_{1-\beta} = z_{.80} = 0.84$ and

$$n = \frac{31(1.96+0.84)^2}{2(4)} = \frac{31(7.84)}{8} = 30.4$$

Thus, we need to enroll 31 subjects per group, or 62 subjects overall, to achieve an 80% power using this design, if no carry-over effect is present.

An alternative design for such a study would be the so-called parallel-group design, where we randomize subjects to either PMH or placebo, measure their DBP at baseline and at the end of 4 weeks of follow-up, and base the measure of efficacy for an individual patient on (mean DBP follow-up − mean DBP at baseline). The sample size needed for such a study is given in Equation 8.27 by

$$n = \text{sample size per group} = \frac{(2\sigma_d^2)(z_{1-\alpha/2}+z_{1-\beta})^2}{\Delta^2}$$

$$= 4 \times \text{sample size per group for a cross-over study}$$

$$= 4(30.4) = 121.5 = 122 \text{ subjects per group or 244 subjects overall}$$

σ_d^2 is the within-subject variance of the difference in mean DBP (i.e., mean DBP follow-up − mean DBP baseline) = 31. Clearly, the cross-over design is much more efficient, *if the assumption of no carry-over effects is viable.* It is important in planning cross-over studies to include a baseline measurement prior to the active-treatment period. Although the baseline measurement is usually not useful in the analysis of cross-over studies, it can be very useful if it is subsequently found that a carry-over effect is present. In this case, one could use a parallel-group design based on the difference between period 1 scores and baseline as the outcome measure, rather than simply the period 1 scores. The difference score will generally have less variability than the period 1 score because it represents within-person variability, rather than both between-person and within-person variability as represented by the period 1 score.

In this section, we have learned about cross-over designs. Under a cross-over design, each subject receives both treatments, but at different times. Randomization is used to determine the ordering of the treatments for individual subjects. A cross-over design can be more efficient (i.e., require fewer subjects) than a traditional parallel-group design if no carry-over effects are present, but will be underpowered if unanticipated carry-over effects are present because the second-period data cannot be validly used. In the latter case, the power can be somewhat improved if a baseline score is obtained prior to receiving either treatment.

It is useful to consider types of studies where a cross-over design may be appropriate. In particular, studies based on objective endpoints such as blood pressure, where the anticipated period of drug efficacy occurs over a short time (i.e., weeks rather than years) and is not long-lasting after drug is withdrawn, are best suited for a cross-over design. On the other hand, most phase III clinical trials (i.e., definitive studies used by the FDA as a basis for establishing drug efficacy for new pharmaceutical products or existing products being tested for a new indication) are long-term studies that violate one or more of the preceding principles. Thus, in general, phase III clinical trials use the more traditional parallel-group design.

SECTION 13.11 Clustered Binary Data

13.11.1 Introduction

The two-sample test for the comparison of binomial proportions, discussed in Section 10.2, is one of the most frequently cited statistical procedures in applied research. An important assumption underlying this methodology is that the observations within the respective samples are statistically independent.

Example 13.51 **Infectious Disease, Dermatology** Rowe et al. [20] report on a clinical trial of topically applied 3% vidarbine versus placebo in the treatment of recurrent herpes labialis. During the medication phase of the trial, characteristics of 53 lesions observed on 31 patients receiving vidarbine were compared with the characteristics of 69 lesions observed on 39 patients receiving placebo. A question of interest is whether the proportion of lesions showing a significant reduction in size in the two groups is the same after 7 days. This requires the development of a test procedure that adjusts for dependencies in response among lesions observed on the same patient.

13.11.2 Hypothesis Testing

We assume that the sample data arise from two groups of individuals, n_1 individuals in group 1 and n_2 individuals in group 2. Suppose that individual j in group i contributes m_{ij} observations to the analysis, $j = 1, 2, \ldots, n_i$ and $M_i = \sum_{j=1}^{n_i} m_{ij}$ denotes the total number of observations in group i, each classified as either a success or a failure. Let a_{ij} denote the observed number of successes for individual j in group i, and define $A_i = \sum_{j=1}^{n_i} a_{ij}$. Then the overall proportion of successes in group i may be denoted by $\hat{p}_i = A_i/M_i = \sum_{j=1}^{n_i} m_{ij}\hat{p}_{ij}/M_i$, where $\hat{p}_{ij} = a_{ij}/m_{ij}$ denotes the observed success rate for individual j in group i. We further denote the total number of individuals as $N = n_1 + n_2$ and the total number of observations as $M = M_1 + M_2$.

Let p_i denote the underlying success rate among observations in group i, $i = 1, 2$. Then we wish to test the hypothesis H_0: $p_1 = p_2$ versus H_1: $p_1 \neq p_2$, assuming that the samples are large enough so that the normal approximation to the binomial distribution is valid.

An estimate of the degree of clustering within individuals is given by the intraclass correlation for clustered binary data. This is computed in a similar manner as for normally distributed data as given in Section 12.9. The mean square errors between and within individuals, respectively, are given in this case by

$$MSB = \sum_{i=1}^{2}\sum_{j=1}^{n_i} m_{ij}\left(\hat{p}_{ij} - \hat{p}_i\right)^2 \Big/ (N-2)$$

$$MSW = \sum_{i=1}^{2}\sum_{j=1}^{n_i} a_{ij}\left(1 - \hat{p}_{ij}\right) \Big/ (M-N)$$

The resulting estimate of intraclass correlation is given by

$$\hat{\rho} = (MSB - MSW)/[MSB + (m_A - 1)MSW]$$

where

$$m_A = \left[M - \sum_{i=1}^{2}\left(\sum_{j=1}^{n_i} m_{ij}^2 / M_i\right)\right] \Big/ (N-2)$$

The clustering correction factor in group i may now be defined as $C_i = \sum_{j=1}^{n_i} m_{ij} C_{ij} / M_i$, where $C_{ij} = 1 + (m_{ij} - 1)\hat{\rho}$.

The clustering correction factor is sometimes referred to as the *design effect*. Notice that if the intraclass correlation coefficient is 0, then no clustering is present and the design effects in the two samples are each 1. If the intraclass correlation coefficient is > 0, then the design effects will be greater than 1. The design effects in the two samples (C_1, C_2) are used as correction factors to modify the standard test statistic comparing two binomial proportions (Equation 10.3) for clustering effects. We have the following test procedure.

EQUATION 13.51

Two-Sample Test for Binomial Proportions (Clustered Data Case) Suppose we have two samples consisting of n_1 and n_2 individuals, respectively, where the jth individual in the ith group contributes m_{ij} observations to the analysis, of which a_{ij} are successes. To test the hypothesis H_0: $p_1 = p_2$ versus H_1: $p_1 \neq p_2$,

(1) We compute the test statistic

$$z = \left[|\hat{p}_1 - \hat{p}_2| - \left(\frac{C_1}{2M_1} + \frac{C_2}{2M_2}\right)\right] \Big/ \sqrt{\hat{p}\hat{q}(C_1/M_1 + C_2/M_2)}$$

where

$$\hat{p}_{ij} = a_{ij}/m_{ij}$$

$$\hat{p}_i = \sum_{j=1}^{n_i} a_{ij} \Big/ \sum_{j=1}^{n_i} m_{ij} = \sum_{j=1}^{n_i} m_{ij}\hat{p}_{ij} \Big/ \sum_{j=1}^{n_i} m_{ij}$$

$$\hat{p} = \sum_{i=1}^{2}\sum_{j=1}^{n_i} a_{ij} \Big/ \sum_{i=1}^{2}\sum_{j=1}^{n_i} m_{ij} = \sum_{i=1}^{2} M_i \hat{p}_i \Big/ \sum_{i=1}^{2} M_i, \ \hat{q} = 1 - \hat{p}$$

$$M_i = \sum_{j=1}^{n_i} m_{ij}$$

$$C_i = \sum_{j=1}^{n_i} m_{ij} C_{ij} \Big/ M_i$$

$$C_{ij} = 1 + (m_{ij} - 1)\hat{\rho}$$

$$\hat{\rho} = (MSB - MSW)/[MSB + (m_A - 1)MSW]$$

$$MSB = \sum_{i=1}^{2}\sum_{j=1}^{n_i} m_{ij}(\hat{p}_{ij} - \hat{p}_i)^2 \Big/ (N-2)$$

$$MSW = \sum_{i=1}^{2}\sum_{j=1}^{n_i} a_{ij}(1 - \hat{p}_{ij}) \Big/ (M - N)$$

$$m_A = \left[M - \sum_{i=1}^{2}\left(\sum_{j=1}^{n_i} m_{ij}^2 / M_i\right)\right] \Big/ (N-2)$$

$$N = \sum_{i=1}^{2} n_i$$

(2) To test for significance, we reject H_0 if $|z| > z_{1-\alpha/2}$, where $z_{1-\alpha/2}$ is the upper $\alpha/2$ percentile of a standard normal distribution.

(3) An approximate $100\% \times (1-\alpha)$ confidence interval for $p_1 - p_2$ is given by

$$\hat{p}_1 - \hat{p}_2 - [C_1/(2M_1) + C_2/(2M_2)] \pm z_{1-\alpha/2}\sqrt{\hat{p}_1\hat{q}_1 C_1/M_1 + \hat{p}_2\hat{q}_2 C_2/M_2} \text{ if } \hat{p}_1 > \hat{p}_2$$

$$\hat{p}_1 - \hat{p}_2 + [C_1/(2M_1) + C_2/(2M_2)] \pm z_{1-\alpha/2}\sqrt{\hat{p}_1\hat{q}_1 C_1/M_1 + \hat{p}_2\hat{q}_2 C_2/M_2} \text{ if } \hat{p}_1 \le \hat{p}_2$$

(4) This test should only be used if $M_1\hat{p}\hat{q}/C_1 \ge 5$ and $M_2\hat{p}\hat{q}/C_2 \ge 5$.

Example 13.52

Dentistry A longitudinal study of caries lesions on the exposed roots of teeth was reported in the literature (Banting et al. [21]). Forty chronically ill subjects were followed for the development of root lesions over a one-year period. The data are given in Table 13.24. Assess whether the male patients had a higher incidence of surfaces with root lesions than did female patients over this time period.

Solution

We note that 6 out of 27 (22.2%) surfaces among 11 male patients developed root lesions, compared with 6 out of 99 (6.1%) surfaces among 29 female patients. The standard normal deviate test statistic (Equation 10.3) for comparing these two proportions is given by

$$z = \left[|\hat{p}_1 - \hat{p}_2| - \left(\frac{1}{2M_1} + \frac{1}{2M_2}\right)\right] \Big/ \sqrt{\hat{p}\hat{q}(1/M_1 + 1/M_2)}$$

$$= \left\{|.2222 - .0606| - \left[\frac{1}{2(27)} + \frac{1}{2(99)}\right]\right\} \Big/ \sqrt{(12/126)(114/126)(1/27 + 1/99)}$$

$$= .1380/.0637 = 2.166 \sim N(0,1) \text{ under } H_0$$

which yields a p-value of $2 \times [1 - \Phi(2.166)] = .030$. However, application of this test procedure ignores the dependency of responses on different surfaces within the same patient. To incorporate this dependency, we use the test procedure in Equation 13.51. We must compute the intraclass correlation $\hat{\rho}$, which is given as follows

$$\hat{\rho} = (MSB - MSW)/[MSB + (m_A - 1)MSW]$$

where

$$MSB = \left[4(0/4 - .2222)^2 + \cdots + 2(0/2 - .2222)^2 + 2(1/2 - .0606)^2 + \cdots + 2(0/2 - .0606)^2\right] \Big/ 38$$

$$= 6.170/38 = 0.1624$$

$$MSW = [0(1 - 0/4) + \cdots + 0(1 - 0/2)]/(27 + 99 - 40)$$

$$= 5.133/86 = 0.0597$$

$$m_A = [126 - (77/27 + 403/99)]/(40 - 2)$$

$$= (126 - 6.923)/38 = 119.077/38 = 3.134$$

TABLE 13.24 **Longitudinal data of development of caries lesions over a one-year period**

	ID	Age	Sex	Lesions	Surfaces
	1	71	M	0	4
	5	70	M	1	1
	6	65	M	2	2
	7	53	M	0	2
	8	71	M	2	4
	11	74	M	0	3
	15	81	M	0	3
	18	64	M	0	3
	30	40	M	0	1
	32	78	M	1	2
	35	79	M	0	2
Total	11			6	27
	2	80	F	1	2
	3	83	F	1	6
	4	86	F	0	8
	9	69	F	1	5
	10	59	F	0	4
	12	88	F	0	4
	13	36	F	1	2
	14	60	F	0	4
	16	71	F	0	4
	17	80	F	0	4
	19	59	F	0	6
	20	65	F	0	2
	21	85	F	0	4
	22	72	F	0	4
	23	58	F	0	2
	24	65	F	0	3
	25	59	F	0	2
	26	45	F	0	2
	27	71	F	0	4
	28	82	F	2	2
	29	48	F	0	2
	31	67	F	0	2
	33	80	F	0	2
	34	69	F	0	4
	36	85	F	0	4
	37	77	F	0	4
	38	71	F	0	3
	39	85	F	0	2
	40	52	F	0	2
Total	29			6	99

$$\hat{\rho} = (0.1624 - 0.0597)/[0.1624 + (3.134 - 1)0.0597]$$
$$= 0.1027/0.2898 = .354$$

To compute the adjusted test statistic, we need to estimate C_1, C_2, where

$$C_1 = \frac{2.063(4) + \cdots + 1.354(2)}{4 + \cdots + 2}$$
$$= \frac{44.719}{27} = 1.656$$
$$C_2 = \frac{1.354(2) + \cdots + 1.354(2)}{2 + \cdots + 2}$$
$$= \frac{206.732}{99} = 2.088$$

Thus we have the adjusted test statistic

$$z = \frac{|.2222 - .0606| - \left[\frac{1.656}{2(27)} + \frac{2.088}{2(99)}\right]}{\sqrt{(12/126)(114/126)(1.656/27 + 2.088/99)}}$$
$$= \frac{.1204}{.08617(.08244)}$$
$$= \frac{.1204}{.0843} = 1.429$$

This yields a two-tailed p-value of $2 \times [1 - \Phi(1.429)] = 2 \times (.0766) = .153$, which is not statistically significant. Thus, the significance level attained from an analysis that ignores the dependency among surfaces within the same patient ($p = .030$) is considerably lower than the true significance level attained from the procedure in Equation 13.51, which accounts for the dependence. The existence of such dependence is biologically sensible, given the common factors that affect the surfaces within a mouth, such as nutrition, saliva production, and dietary habits (Imrey [22]).

Using Equation 13.51, we can also develop a 95% confidence interval for $p_1 - p_2$ = true difference in one-year incidence of root caries between males and females, which is given as follows:

$$.1204 \pm 1.96\sqrt{\frac{.2222(.7778)(1.656)}{27} + \frac{.0606(.9394)(2.088)}{99}}$$
$$= .1204 \pm 1.96(.1086)$$
$$= .1204 \pm .2129 = (-.092,\ .333)$$

Note that the inference procedure in Equation 13.51 reduces to the standard two-sample inference procedure when $\hat{\rho} = 0$ (i.e., Equation 10.3), or when $m_{ij} = 1$, $j = 1, 2, \ldots, n_i$; $i = 1, 2$. Finally, note that if all individuals in each group contribute exactly the same number of sites (m), then the test procedure in Equation 13.51 reduces as follows:

EQUATION 13.52

Two-Sample Test for Binomial Proportions (Equal Number of Sites per Individual) If each individual in each of two groups contributes m observations, then to test the hypothesis H_0: $p_1 = p_2$ versus H_1: $p_1 \neq p_2$, perform the following procedure:

(1) Let

$$z = \frac{|\hat{p}_1 - \hat{p}_2| - \frac{[1+(m-1)\hat{\rho}]}{2}\left(\frac{1}{M_1} + \frac{1}{M_2}\right)}{\sqrt{\hat{p}\hat{q}(1/M_1 + 1/M_2)}} \times \frac{1}{\sqrt{1+(m-1)\,\hat{\rho}}}$$

where $M_i = n_i m$, $i = 1, 2$, and $\hat{p}$ and $\hat{\rho}$ are defined in Equation 13.51.

(2) To test for significance, we reject H_0 if $|z| > z_{1-\alpha/2}$, where $z_{1-\alpha/2}$ is the upper $\alpha/2$ percentile of a standard normal distribution.

(3) An approximate $100\% \times (1-\alpha)$ confidence interval for $p_1 - p_2$ is given by

$$\left\{\hat{p}_1 - \hat{p}_2 - \frac{[1+(m-1)\hat{\rho}]}{2}(1/M_1 + 1/M_2) \pm z_{1-\alpha/2}\sqrt{[1+(m-1)\hat{\rho}](\hat{p}_1\hat{q}_1/M_1 + \hat{p}_2\hat{q}_2/M_2)}\right\}$$

if $\hat{p}_1 > \hat{p}_2$

$$\left\{\hat{p}_1 - \hat{p}_2 + \frac{[1+(m-1)\hat{\rho}]}{2}(1/M_1 + 1/M_2) \pm z_{1-\alpha/2}\sqrt{[1+(m-1)\hat{\rho}](\hat{p}_1\hat{q}_1/M_1 + \hat{p}_2\hat{q}_2/M_2)}\right\}$$

if $\hat{p}_1 \le \hat{p}_2$

(4) This test should only be used if $M_1\hat{p}\hat{q}/[1+(m-1)\hat{\rho}] \ge 5$ and $M_2\hat{p}\hat{q}/[1+(m-1)\hat{\rho}] \ge 5$.

13.11.3 Sample-Size Estimation for Clustered Binary Data

Suppose we wish to test the hypothesis H_0: $p_1 = p_2$ versus H_1: $p_1 \neq p_2$. If we assume that there is independence among the observations within an individual, then the required number of observations in each of the two samples needed to conduct a two-sided test with significance level α and power $1 - \beta$ is given by Equation 10.14, i.e.,

$$M = M_2 = \left(z_{1-\alpha/2}\sqrt{2\bar{p}\bar{q}} + z_{1-\beta}\sqrt{p_1 q_1 + p_2 q_2}\right)^2 \Big/ (p_1 - p_2)^2$$

where

$$\bar{p} = (p_1 + p_2)/2,\ \bar{q} = 1 - \bar{p}$$

To take the clustering into account, suppose each sample consists of n individuals supplying a variable number of observations to the analysis. The required total number of observations per group may be estimated as

EQUATION 13.53

$$M_s = M[1 + (\overline{m} - 1)\rho]$$

where $\bar{m}$ is the average number of observations per individual, and ρ is the intraclass correlation for outcome among observations within an individual. The number of individuals in each group is then given by $n = M_s/\bar{m}$.

In many investigations, the sample size is relatively fixed and it is important for the purpose of study design to determine the power of the study to detect specific alternatives. For this purpose, one can use Equation 13.53 to determine the power as a function of the total number of observations available per group (M_s), the average number of observations per individual ($\bar{m}$), the intraclass correlation among observations within an individual (ρ), and the two-sided type I error (α), as follows

EQUATION 13.54

$$\text{Power} = \Phi\left[\frac{\sqrt{M_s/[1+(\bar{m}-1)\rho]}\,|p_1 - p_2| - z_{1-\alpha/2}\sqrt{2\bar{p}\bar{q}}}{\sqrt{p_1 q_1 + p_2 q_2}}\right]$$

These results can be summarized as follows:

EQUATION 13.55

Sample-Size and Power Estimation for Comparing Two Binomial Proportions Suppose we wish to test the hypothesis H_0: $p_1 = p_2$ versus H_1: $p_1 \neq p_2$. If we intend to use a two-sided test with significance level α and require a power of $1 - \beta$, then the appropriate sample size (number of observations per group) is given by

$$M_s = M[1+(\bar{m}-1)\rho] = \text{number of observations per group}$$

where

$$M = \left(z_{1-\alpha/2}\sqrt{2\bar{p}\bar{q}} + z_{1-\beta}\sqrt{p_1 q_1 + p_2 q_2}\right)^2 \Big/ (p_1 - p_2)^2$$

and

$$\bar{p} = (p_1 + p_2)/2,\ \bar{q} = 1 - \bar{p}$$

The required number of individuals per group (n) is then given by

$$n = M_s/\bar{m}$$

In this setting, $\bar{m}$ is the average number of observations per individual, and ρ is the intraclass correlation for outcome between observations within the same individual.

If the total number of observations per group is fixed and one wishes to assess the power of a study versus particular alternatives, the power can be obtained as follows:

$$\text{Power} = \Phi(z_{1-\beta})$$

where

$$z_{1-\beta} = \frac{\sqrt{M_s/[1+(\bar{m}-1)\rho]}\,|p_1 - p_2| - z_{1-\alpha/2}\sqrt{2\bar{p}\bar{q}}}{\sqrt{p_1 q_1 + p_2 q_2}}$$

and M_s is the number of observations per group available for analysis, $\bar{m}$ is the average number of observations per individual, ρ is the intraclass correlation among observations within an individual, and α is the two-tailed type I error.

Example 13.53 **Dentistry** A clinical trial of a new therapeutic modality for the treatment of periodontal disease is being planned. The unit of observation is the surface within the patient's mouth. Two groups of patients, one randomly assigned to the new modality and the other randomly assigned to a standard treatment, will be monitored at six months after therapy to compare the percentage of surfaces over all patients that lose attachment. It is anticipated from previous studies that approximately 2/3 of the surfaces treated with the standard modality will lose attachment, and a reduction of this proportion to 1/2 would be considered clinically significant. Suppose each patient is required to contribute an average of 25 surfaces to the analysis. How many surfaces are required for each treatment group in order to have 80% power to detect this magnitude of effect if a two-sided test is to be used with significance level = .05?

Solution Because surfaces within patients cannot be regarded as independent, an estimate of the required sample size depends on the level of intrapatient correlation (ρ) with respect to the occurrence of attachment loss. Referring to Fleiss et al. [23], a reasonable estimate of ρ is given by .50. Also, $p_1 = .500$, $q_1 = .500$, $p_2 = .667$, $q_2 = .333$, $\bar{p} = (.500 + .667)/2 = .584$, $\bar{q} = .416$.The required number of surfaces per group is then obtained from Equation 13.55 as follows:

$$M_s = \frac{[1+(25-1).50]\left(z_{.975}\sqrt{2\bar{p}\bar{q}} + z_{.80}\sqrt{p_1q_1 + p_2q_2}\right)^2}{(p_1 - p_2)^2}$$

$$= \frac{13\left[1.96\sqrt{2(.584)(.416)} + 0.84\sqrt{.667(.333) + .50(.50)}\right]^2}{(.667 - .500)^2}$$

$$= \frac{13(1.3662 + 0.5772)^2}{.0279} = \frac{49.0983}{.0279} = 1760 \text{ sites per group}$$

Because there are 25 surfaces per subject, we need a sample size per group of $n = 1760/25 = 70.4$ or 71 subjects per group to have 80% power to detect a difference of the stated magnitude.

Example 13.54 **Dentistry** Suppose that the investigators feel they can recruit 100 subjects per group for the study mentioned in Example 13.53. How much power would such a study have with the parameters given in Example 13.53, if a two-sided test is to be used with $\alpha = .05$?

Solution We have from Equation 13.55 that

$$z_{1-\beta} = \frac{\sqrt{M_s/[1+(\bar{m}-1)\rho]}\,|p_1 - p_2| - z_{1-\alpha/2}\sqrt{2\bar{p}\bar{q}}}{\sqrt{p_1q_1 + p_2q_2}}$$

In this example, M_s = number of surfaces available for analysis per group = 100 subjects × 25 surfaces per subject = 2500 surfaces, $\bar{m} = 25$, $\rho = 0.50$, $p_1 = .500$, $q_1 = .500$, $p_2 = .667$, $q_2 = .333$, $\bar{p} = (p_1 + p_2)/2 = .584$, $\bar{q} = .416$, $z_{1-\alpha/2} = z_{.975} = 1.96$. Thus, we have

$$z_{1-\beta} = \frac{\sqrt{2500/[1+24(.50)]}\,|.667 - .500| - 1.96\sqrt{2(.584)(.416)}}{\sqrt{.500(.500) + .667(.333)}}$$

$$= \frac{2.3159 - 1.3662}{0.6871} = \frac{0.9497}{0.6871} = 1.382$$

$$\text{Power} = \Phi(1.382) = .917$$

Thus the study would have 91.7% power if 100 subjects per group were recruited.

In this section, we have learned about methods of analysis and sample-size estimation for clustered binary data (sometimes referred to as *correlated binary data*). Clustered binary data occur in clinical trials where the unit of randomization is different from the unit of analysis. For example, in dental clinical trials, randomization is usually performed at the person level, but the actual unit of analysis is usually either the tooth or the tooth surface. Similarly, in group randomized studies, a large group (such as an entire school) is the unit of randomization. For example, five schools may be randomized to an active nutritional intervention whose goal is to reduce dietary fat intake, and five other schools may be randomized to a control intervention. Suppose the outcome is reported dietary-fat intake <30% of calories after one year. The outcome is obtained on individual students within the school. In the former example, the outcomes on tooth surfaces represent correlated binary data because there is lack of independence of responses from different teeth or surfaces within the same mouth. In the latter example, outcomes on students represent correlated binary data because of the expected similarity of dietary habits of students from the same school due to similarity in, for example, socioeconomic status. Clustered binary data can also occur in observational studies, such as virtually any study in the field of ophthalmology where the eye is the unit of analysis.

It is possible to apply clustered binary data techniques to control for confounding based on the Mantel-Haenszel test [24]. It is also possible to extend these methods to continuous outcomes and to perform regression analyses, which takes into account the correlation of obervations within the same primary unit. This type of regression model is called either a correlated response model, a hierarchical model, a mixed-effects model, or a multilevel model. Special software is needed to fit these models (e.g., the PROC MIXED or GENMOD procedure of SAS or [25]). These methods are beyond the scope of this text.

SECTION 13.12 Measurement-Error Methods

13.12.1 Introduction

Exposure variables in epidemiology are often measured with error. An interesting question is, What effect does this measurement error have on the results obtained from standard analyses?

Example 13.55 **Cancer, Nutrition** A hypothesis has been proposed linking breast-cancer incidence to saturated-fat intake. To test this hypothesis, a group of 89,538 women, ages 34–59, who were free of breast cancer in 1980 were followed until 1984. During this period, 590 cases of breast cancer occurred. A logistic-regression model [26] of breast-cancer incidence from 1980 to 1984 on age (using dummy variables based on the age groups 34–39, 40–44, 45–49, 50–54, 55–59), saturated-fat intake as a continuous variable as reported on a 1980 food-frequency questionnaire (FFQ), and alcohol intake (using dummy variables based on the groups 0, 0.1–1.4, 1.5–4.9, 5.0–14.9, 15+ grams per day) was fit to the data. The results are given in Table 13.25.

The *OR* for a 10 g/day increase in calorie-adjusted saturated fat intake (henceforth, referred to as saturated fat) was 0.92 (95% CI = 0.80–1.05). The instrument used to assess diet for the analyses in Table 13.25 was the 1980 food-frequency questionnaire (FFQ). On the FFQ, subjects reported their average consumption of each of 61 foods over the past year. This instrument is known to have a large amount of measurement

TABLE 13.25 **Association between breast-cancer incidence and calorie-adjusted saturated-fat intake, Nurses' Health Study, 1980–1984, 590 events, 89,538 subjects**

Variable	β	*se*	*z*	*p*	*OR* (95% CI)
Saturated-fat intake (g) (X)	–0.0878	0.0712	–1.23	.22	0.92 (0.80–1.05)

Note: Based on a 10-g increase in saturated-fat intake.

error. It is sometimes referred to as a *surrogate* for the ideal instrument which could always measure dietary fat without error. What impact does the measurement error have on the results obtained?

13.12.2 Measurement-Error Correction with Gold-Standard Exposure

The model fit in Table 13.25 was of the form

EQUATION 13.56

$$\ln\left[p/(1-p)\right] = \alpha + \beta X + \sum_{j=1}^{m} \delta_j u_j$$

where X is saturated-fat intake from the FFQ that is measured with error and $u_1, \ldots, u_m$ are a set of variables always measured without error, which in this example represent dummy variables for age and alcohol intake. Alcohol intake is actually also measured with error, but the degree of measurement error is typically much smaller than for saturated fat [26]. For simplicity, we will assume that alcohol intake is measured without error.

To assess the impact of measurement error on the estimate of β in Equation 13.56, we would have to consider how the estimate of β would change if average daily fat intake could be ascertained with no error. The diet record (DR) is considered a gold standard by some nutritional epidemiologists. With a DR, a person records each food eaten and the corresponding portion size on a real-time basis. The foods and portion sizes are then entered onto a computer, and a computer program is used to calculate nutrients consumed during this period. Ideally, a DR would be filled out for each of 365 days in 1980 by each of the 89,538 nurses in the main study. However, it is very expensive to collect and process DR data. Instead, a *validation study* was performed among 173 nurses who filled out 4 weeks of DR with individual weeks spaced about 3 months apart. They then filled out an additional FFQ in 1981 to refer to the same time period as the DR. The data from the validation study were used to model the relationship between reported DR saturated-fat intake (x) and reported FFQ saturated-fat intake (X), using a linear-regression model of the form

EQUATION 13.57

$$x = \alpha' + \gamma X + e$$

TABLE 13.26 **Relationship between DR saturated fat-intake (x) and FFQ saturated-fat intake (X), Nurses' Health Study, 1981, 173 subjects**

Variable	γ	*se*	*t*	*p*-value
Saturated-fat intake FFQ (g) (X)	0.468	0.048	9.75	<.001

where e is assumed to be $N(0, \sigma^2)$. The results are given in Table 13.26.

We see that, as expected, there is a highly significant association between x and X. Our goal is to estimate the relationship between breast-cancer incidence and DR saturated-fat intake (x) after controlling for age and alcohol intake assuming a logistic model for this relationship of the form

EQUATION 13.58
$$\ln[p/(1-p)] = \alpha^* + \beta^* x + \sum_{j=1}^{m} \delta^*_j u_j$$

where $u_1, \ldots, u_m$ are a set of other covariates assumed to be measured without error that represent age and alcohol intake. The problem is that we only observe x directly on 173 out of 89,538 women. Therefore, instead of estimating the logistic regression directly from Equation 13.58, we use an indirect approach. Specifically, because we know X for each woman in the main study. For any particular woman with FFQ intake = X, we can estimate the average DR intake for that value of X, which we denote by $E(x|X)$, and use that as an estimate of x for that woman. From the linear regression in Equation 13.57, we have

EQUATION 13.59
$$E(x|X) = \alpha' + \gamma X$$

Substituting $E(x|X)$ from Equation 13.59 for x in Equation 13.58 yields

EQUATION 13.60
$$\ln[p/(1-p)] = (\alpha^* + \beta^*\alpha') + (\beta^*\gamma)X + \sum_{j=1}^{m} \delta^*_j u_j$$

If we compare Equation 13.60 with Equation 13.56, we see that the dependent and independent variables are the same. Thus, we can equate the regression coefficients corresponding to X (that is, FFQ saturated fat), yielding

EQUATION 13.61
$$\beta^*\gamma = \beta$$

If we divide both sides of Equation 13.61 by γ, we obtain

EQUATION 13.62
$$\beta^* = \beta/\gamma$$

Therefore, to estimate the logistic-regression coefficient of breast cancer on "true" saturated-fat intake, we divide the logistic-regression coefficient (β) of breast cancer on the surrogate saturated fat (X) from the main study by the linear-regression coefficient (γ) of true (x) on surrogate (X) saturated fat from the validation study. The equating of Equations 13.56 and 13.60 is an approximation since it ignores the impact of the error distribution from the validation study model in Equation 13.57 on the estimation of x for subjects in the main study. However, it is approximately valid if the disease under study is rare and the measurement error variance (σ^2 in Equation 13.57) is small.

To obtain the standard error of $\hat{\beta}^*$ and associated confidence limits for β^*, we use a multivariate extension of the delta method introduced in Equation 13.3. This approach for estimating β^* is called the regression-calibration method [26] and is summarized as follows:

EQUATION 13.63

Regression-Calibration Approach for Estimation of Measurement-Error-Corrected Odds Ratio Relating a Dichotomous Disease Variable (*D*) to a Single Exposure Variable (*X*) Measured with Error, when a Gold-Standard Exposure (*x*) Is Available Suppose we have

(1) A dichotomous disease variable (D), where $D = 1$ if disease is present, 0 if disease is absent

(2) A single exposure variable (X) measured with error (referred to as the *surrogate exposure*)

(3) A corresponding gold-standard exposure variable (x) that represents true exposure (or is at least an unbiased estimate of true exposure with errors that are uncorrelated with that of the surrogate)

(4) A set of other covariates $u_1, \ldots, u_m$, which are assumed to be measured without error

We wish to fit the logistic-regression model

$$\ln[p/(1-p)] = \alpha + \beta^* x + \sum_{j=1}^{m} \delta_j^* u_j$$

where $p = Pr(D = 1)$. We have available

(a) a main-study sample of size n (usually large), where D, X, and $u_1, \ldots, u_m$ are observed

(b) a validation-study sample of size n_1 (usually small), where x and X are observed. Ideally, the validation-study sample should be a representative sample from the main-study sample, or a comparable external sample.

Our goal is to estimate β^*. For this purpose,

(i) We use the main-study sample to fit the logistic-regression model of D on X and $u_1, \ldots, u_m$ of the form

$$\ln[p/(1-p)] = \alpha + \beta X + \sum_{j=1}^{m} \delta_j u_j$$

(ii) We use the validation-study sample to fit the linear-regression model of x on X of the form

$$x = \alpha' + \gamma X + e$$

where $e \sim N(0, \sigma^2)$

(iii) We use (i) and (ii) to obtain the point estimate of β^*, given by

$$\hat{\beta}^* = \hat{\beta}/\hat{\gamma}$$

The corresponding estimate of the odds ratio of D on x (OR) is given by

$$\widehat{OR} = \exp(\hat{\beta}^*)$$

(iv) We obtain the variance of $\hat{\beta}^*$ by computing

$$Var(\hat{\beta}^*) = (1/\hat{\gamma}^2)\ Var(\hat{\beta}) + (\hat{\beta}^2/\hat{\gamma}^4)\ Var(\hat{\gamma})$$

where $\hat{\beta}$ and $Var(\hat{\beta})$ are obtained from (i)

and $\hat{\gamma}$ and $Var(\hat{\gamma})$ are obtained from (ii)

(v) We obtain a $100\% \times (1-\alpha)$ CI for β^* by computing

$$\hat{\beta}^* \pm z_{1-\alpha/2} se(\hat{\beta}^*) = (\hat{\beta}_1^*, \hat{\beta}_2^*)$$

where $\hat{\beta}^*$ is obtained from (iii) and

$se(\hat{\beta}^*) = [Var(\hat{\beta}^*)]^{1/2}$ is obtained from (iv)

The corresponding $100\% \times (1-\alpha)$ CI for OR is given by

$$[\exp(\hat{\beta}_1^*),\ \exp(\hat{\beta}_2^*)]$$

This method should only be used if the disease under study is rare (incidence $< 10\%$) and the measurement error variance [σ^2 in (ii)] is small.

Example 13.56 **Cancer, Nutrition** Estimate the odds ratio relating breast-cancer incidence from 1980 to 1984 to DR intake of saturated fat in 1980 using the regression-calibration method based on the data in Tables 13.25 and 13.26.

Solution We have from Table 13.25 that $\hat{\beta} = -0.0878$, $se(\hat{\beta}) = 0.0712$. From Table 13.26, we have that $\hat{\gamma} = 0.468$, $se(\hat{\gamma}) = 0.048$. Thus, from step (iii) of Equation 13.63, the point estimate of β^* is $\hat{\beta}^* = -0.0878/0.468 = -0.1876$. The corresponding point estimate of the OR relating breast-cancer incidence to an increase of 10 g of "true" (DR) saturated-fat intake is exp(–0.1876) = 0.83. To obtain $Var(\hat{\beta}^*)$, we refer to step (iv) of Equation 13.63. We have

$$\begin{aligned} Var(\hat{\beta}^*) &= (1/0.468^2)(0.0712)^2 + [(-0.0878)^2/(0.468)^4](0.048)^2 \\ &= 0.02315 + 0.00037 = 0.02352 \end{aligned}$$

$$se(\hat{\beta}^*) = (0.02352)^{1/2} = 0.1533$$

Thus, from step (v) of Equation 13.63, a 95% CI for β^* is given by

$$-0.1876 \pm 1.96(0.1533) = -0.1876 \pm 0.3006 = (-0.488, 0.113) = (\hat{\beta}_1^*, \hat{\beta}_2^*)$$

The corresponding 95% CI for OR is given by

$$[\exp(-0.488), \exp(0.113)] = (0.61, 1.12)$$

Notice that the measurement-error-corrected estimate of OR (0.83) is farther away from 1 than the crude of uncorrected estimate (0.92) obtained in Table 13.25. The uncorrected estimate of 0.92 is attenuated (i.e., incorrectly moved closer to 1, the null value) due to the influence of measurement error. Therefore the corrected estimate (0.83) is sometimes referred to as a *deattenuated OR* estimate. Notice also that the CI for the corrected *OR* (0.61, 1.12) is much wider than the corresponding CI for the uncorrected *OR* (0.80, 1.05) in Table 13.25, which often occurs. Finally, the two terms in the expression for $Var(\hat{\beta}^*)$ (0.02315 and 0.00037) reflect error in the estimated main-study logistic-regression coefficient ($\hat{\beta}$) and the estimated validation-study linear-regression coefficient ($\hat{\gamma}$), respectively. Usually, the first term predominates, unless the validation-study sample size is very small.

13.12.3 Measurement-Error Correction Without a Gold-Standard Exposure

In Example 13.55, there existed a dietary instrument (the DR) that at least some nutritionists would regard as a gold standard. Technically, to use the regression-calibration method, the gold-standard instrument need only provide an unbiased estimate of "true" exposure rather than actually be "true" exposure with errors that are uncorrelated with that of the surrogate. Given that the DR in Example 13.55 consisted of average intake over 28 days spaced throughout the year, this seemingly would provide an unbiased estimate of intake over all 365 days, provided the DR is filled out accurately. However, for some exposures there doesn't exist even a potential gold-standard instrument.

Example 13.57 **Cancer, Endocrinology** Among postmenopausal women, a positive association has generally been observed between plasma-estrogen levels and breast-cancer risk. However, most studies have been small, and many have not evaluated specific estrogen fractions. A substudy of the Nurses' Health Study was conducted among 11,169 postmenopausal women who provided a blood sample during the period from 1989 to 1990 and were not using postmenopausal hormones at the time of the blood collection [27]. However, it was too expensive to analyze hormone levels for all 11,000 women. Instead, hormone levels were assayed from 156 women who developed breast cancer after blood collection but before June 1994. Two control women were selected for each breast-cancer case, who were matched with respect to age, menopausal status, and month and time of day of blood collection. In this example, we consider the relationship between ln(plasma estradiol) and the development of breast cancer. The ln transformation was used to better satisfy the linearity assumptions of logistic regression. The results indicated a *RR* of breast cancer of 1.77 (95% CI = 1.06–2.93) comparing women in the highest quartile of ln(estradiol) (median estradiol level = 14 pg/mL) with women in the lowest quartile of ln(estradiol) (median estradiol level = 4 pg/mL) based on the distribution of ln(estradiol) among the controls. However, it is known that plasma estradiol has some measurement error, and we would like to obtain a measurement-error-corrected estimate of the *RR*. How can we accomplish this?

Unlike the dietary study in Example 13.55, there is no gold-standard instrument for plasma estradiol similar to the DR for nutrient intake. However, it is reasonable to consider the average of a large number of ln(estradiol) measurements (x) as a gold-standard instrument that can be compared to the single ln(estradiol) measurement (X) obtained in the study. Although x is not directly measurable, we can consider a random-effects ANOVA model relating X to x, of the form

EQUATION 13.64

$$X_i = x_i + e_i$$

where X_i = a single ln(estradiol) measurement for the ith woman

x_i = underlying mean ln(estradiol) level for the ith woman

$x_i \sim N(\mu, \sigma_A^2),\ e_i \sim N(0, \sigma^2)$

σ_A^2 represents between-person variation and σ^2 represents within-person variation for ln(estradiol) levels.

To implement the regression-calibration method, we need to obtain an estimate of γ as in Equation 13.57—i.e., the regression coefficient of x [true ln(estradiol level)] on X [a single ln(estradiol value)]. We know from Equation 12.35 that

EQUATION 13.65

$$Corr(x, X) = \text{reliability coefficient} = (\rho_I)^{1/2}$$

where $\rho_I = \sigma_A^2/(\sigma_A^2 + \sigma^2)$ = intraclass correlation coefficient

Furthermore, from our work on the relationship between a regression coefficient and a correlation coefficient (Equation 11.19), we found that

EQUATION 13.66

$$b(x \text{ on } X) = Corr(x, X)\ \text{sd}(x)/\text{sd}(X)$$

Also, from Equation 13.64 we have

EQUATION 13.67

$$\text{sd}(x) = \sigma_A$$

EQUATION 13.68

$$\text{sd}(X) = (\sigma_A^2 + \sigma^2)^{1/2}$$

Therefore, on combining Equations 13.65–13.68, we obtain

EQUATION 13.69

$$\begin{aligned} b(x \text{ on } X) &= (\rho_I)^{1/2}\sigma_A/(\sigma_A^2 + \sigma^2)^{1/2} \\ &= (\rho_I)^{1/2}[\sigma_A^2/(\sigma_A^2 + \sigma^2)]^{1/2} \\ &= (\rho_I)^{1/2}(\rho_I)^{1/2} = \rho_I \end{aligned}$$

Thus, we can estimate the regression coefficient of x on X by the sample intraclass correlation coefficient r_I. To obtain r_I, we need to conduct a reproducibility study on a subsample of subjects with at least two replicates per subject. The reproducibility study plays the same role as a validity study, which is used when a gold standard is available. If we substitute r_I for γ in Equation 13.57, then we obtain the following regression-calibration procedure for measurement-error correction when no gold standard is available.

EQUATION 13.70

Regression-Calibration Approach for Estimation of Measurement-Error-Corrected Odds Ratio Relating a Dichotomous Disease Variable *D* to a Single Exposure Variable (*X*) Measured with Error When a Gold Standard Is Not Available Suppose we have

(1) A dichotomous disease variable D (= 1 if disease is present, = 0 if disease is absent)

(2) A single exposure variable X measured with error

(3) (Optionally) a set of other covariates $u_1, \ldots, u_m$ measured without error

We define x as the average of X over many replicates for an individual subject. We wish to fit the logistic-regression model

$$\ln[p/(1-p)] = \alpha^* + \beta^* x + \sum_{j=1}^{m} \delta_j^* u_j$$

where $p = Pr(D = 1)$.

We have available

(a) a main-study sample of size n (usually large), where D, X, and $u_1, \ldots, u_m$ are observed

(b) a reproducibility-study sample of size n_1 (usually small), where k_i replicate observations of X are obtained for the ith person, from which the estimated intraclass correlation coefficient r_I can be obtained (see Equation 12.35)

To estimate β^*, we

(i) Fit the main-study logistic-regression model of D on X and $u_1, \ldots, u_m$ of the form

$$\ln[p/(1-p)] = \alpha + \beta X + \sum_{j=1}^{m} \delta_j u_j$$

(ii) Use the reproducibility study to estimate ρ_I by r_I

(iii) Obtain the point estimate of β^*, given by

$$\hat{\beta}^* = \hat{\beta}/r_I$$

with corresponding odds ratio estimate $OR = \exp(\hat{\beta}^*)$

(iv) Obtain the variance of $\hat{\beta}^*$, given by

$$Var(\hat{\beta}^*) = (1/r_I^2)\ Var(\hat{\beta}) + (\hat{\beta}^2/r_I^4)\ Var(r_I) = A + B$$

where $Var(r_I)$ is obtained from [28] as follows:

$$Var(r_I) = 2(1-r_I)^2\left[1+(k_0-1)r_I\right]^2 \Big/ \left[k_0(k_0-1)(n_1-1)\right]$$

and

$$k_0 = \left(\sum_{i=1}^{n_1} k_i - \sum_{i=1}^{n_1} k_i^2 \Big/ \sum_{i=1}^{n_1} k_i\right) \Big/ (n_1 - 1)$$

(*Note:* If all subjects provide the same number of replicates (k), then $k_0 = k$.)

(v) Obtain a $100\% \times (1-\alpha)$ CI for β^* given by

$$\hat{\beta}^* \pm z_{1-\alpha/2}\, se(\hat{\beta}^*) = (\hat{\beta}_1^*, \hat{\beta}_2^*)$$

where $\hat{\beta}^*$ is obtained from (iii)

and $se(\hat{\beta}^*) = [Var(\hat{\beta}^*)]^{1/2}$ is obtained from (iv).

The corresponding $100\% \times (1-\alpha)$ CI for OR is given by

$$[\exp(\hat{\beta}_1^*), \exp(\hat{\beta}_2^*)]$$

Example 13.58 **Cancer, Endocrinology** Estimate the odds ratio relating breast-cancer incidence to plasma estradiol after correcting for measurement error for a woman with true plasma-estradiol level of 14 pg/mL compared with a woman with a true plasma-estradiol level of 4 pg/mL based on the data described in Example 13.57.

Solution We use the regression-calibration approach in Equation 13.70. We have from Example 13.57 that $\hat{\beta} = \ln(1.77) = 0.571$. Furthermore, the 95% CI for β = [ln(1.06), ln(2.93)] = (0.058, 1.075). The width of the 95% CI is

$$2(1.96)\ se(\hat{\beta}) = 3.92\ se(\hat{\beta}) = 1.075 - 0.058 = 1.017$$

and $se(\hat{\beta}) = 1.017/3.92 = 0.259$

Furthermore, a reproducibility study was conducted among a subset of 78 of the nurses [29]. The estimated intraclass correlation coefficient for ln(plasma estradiol) was 0.68. Sixty-five of the nurses provided 3 replicates, and 13 nurses provided 2 replicates. Therefore, from step (iii) of Equation 13.70, we have the point estimate, $\hat{\beta}^* = 0.571/0.68 = 0.840$, with corresponding $OR = \exp(0.840) = 2.32$. From step (iv) of Equation 13.70, we obtain $Var(\hat{\beta}^*)$. We have

$$A = (0.259)^2/(.68)^2 = 0.1451$$

To obtain B, we need to compute $Var(r_I)$. We have

$$Var(r_I) = 2(1 - .68)^2\ [1 + (k_0 - 1)(.68)]^2/[77k_0(k_0 - 1)]$$

To evaluate k_0, we have that there were 65(3) + 13(2) = 221 replicates over the entire sample. Thus,

$$\begin{aligned} k_0 &= \{221 - [65(3)^2 + 13(2)^2]/221\}/77 \\ &= 218.12/77 = 2.833 \end{aligned}$$

Therefore,

$$\begin{aligned} Var(r_I) &= 2(1 - .68)^2\ [1 + 1.833(.68)]^2/[2.833\,(1.833)77] \\ &= 1.033/399.74 = 0.0026 \end{aligned}$$

Thus,

$$\begin{aligned} B &= [(0.571)^2/(0.68)^4](0.0026) \\ &= 0.0040 \end{aligned}$$

It follows that

$$Var(\hat{\beta}^*) = 0.1451 + 0.0040 = 0.1490$$

$$se(\hat{\beta}^*) = (0.1490)^{1/2} = 0.386$$

From (v), a 95% CI for β^* is given by

$$\begin{aligned} 0.840 \pm 1.96\ (0.386) &= 0.840 \pm 0.757 \\ &= (0.083,\ 1.596) \end{aligned}$$

The corresponding 95% CI for OR is [exp(0.083), exp(1.596)] = (1.09, 4.94).

Thus, the measurement-error-corrected estimate of the odds ratio relating a woman with a true estradiol level of 14 pg/mL to a woman with a true estradiol level of 4 pg/mL is 2.32 (95% CI = 1.09, 4.94). The corresponding uncorrected estimate is 1.77 (95% CI = 1.06–2.93). The deattenuated (corrected) point estimate is substantially larger than the uncorrected estimate with much wider confidence limits.

Several comments are in order concerning this example. First, the design was a prospective case–control study nested within a cohort study. As discussed in Section 13.3, we cannot estimate the absolute risk of breast cancer because by design about 1/3 of the women were cases. However, it is possible to obtain valid estimates of relative risk for ln(estradiol). Second, there were several other variables in the model, all of which were assumed to be measured with little or no measurement error. Third, the point estimate and 95% CI for the relative risk differ slightly from those of [27] due to rounding error and slightly different approaches used to estimate the intraclass correlation coefficient.

In this section, we have discussed methods for obtaining point and interval estimates of relative risk from logistic-regression models that are corrected for measurement error in the covariate of interest using the regression-calibration approach. In Section 13.12.2, we assumed that a gold-standard exposure measure was available. To implement the methods in this setting, we need both a main study of size n in which the disease and surrogate exposure are measured and a validation study of size n_1 in which both the surrogate exposure and the gold-standard exposure are available on the same subjects. The validation-study sample may or may not be a subset of the main-study population. In Section 13.12.3 we assumed that a gold-standard exposure was not available. To implement the methodology in this setting, we need both a main-study sample where both the disease and the surrogate exposure are available, and a reproducibility study where replicate surrogate measurements are available. The reproducibility-study sample may or may not be a subset of the main-study sample. We have assumed in both Sections 13.12.2 and 13.12.3 that there is only a single covariate in the main study measured with error, although there may be several other covariates measured without error. The problem of multiple covariates measured with error is complex and beyond the scope of this text. An extension of the methods in this section when more than one exposure variable is measured with error is given in [30] when a gold standard is available and in [31] when a gold standard is not available. It is important to be aware that even if only a single exposure variable is measured with error, after correcting for measurement error the partial-regression coefficients of other covariates measured without error (e.g., age) may also be affected (see [31] for an example concerning this issue). In this section, we have only focused on the effects of measurement error on the regression coefficient for the covariate measured with error.

SECTION 13.13 Summary

In this chapter, we have learned about some of the principal design and analysis techniques used in epidemiologic studies. In Section 13.2, we learned about the principal study designs used in epidemiologic studies including cohort studies, case–control studies, and cross-sectional studies. In Section 13.3, we then learned about some common measures of effect used in these studies including the risk difference, the risk ratio, and the odds ratio. For each of the study designs, we discussed which of these parameters are estimable and which are not. In Section 13.4, we were introduced to the concept of a confounder and learned about standardization, which is a descriptive technique for obtaining measures of effect that are controlled for confounding variables. In Sections 13.5 and 13.6, we discussed Mantel-Haenszel–type methods, which are analytic procedures used to test hypotheses about effects of a primary exposure variable while controlling for other confounding variable(s). These techniques become cumbersome when there are many confounding variables to be controlled for. Thus, in Section 13.7 we learned about multiple logistic regression, which is a technique similar to multiple linear regression when the outcome variable is binary. Using this technique allows one to control for many confounding variables simultaneously.

The techniques in Sections 13.1–13.7 are standard methods of design and analysis used in epidemiologic studies. In recent years, there has been much interest in extensions of these techniques to nonstandard situations, some of which are discussed in Sections 13.8–13.12. In Section 13.8, we discuss the basic principles of meta-analysis. Meta-analysis is a popular methodology for combining results obtained from more than one study regarding a particular association of interest. In Section 13.9, we consider the emerging field of active-control (or equivalence) studies. In standard clinical trials, to demonstrate the efficacy of an agent, the active agent is usually compared to a placebo. In active-control studies, a new proposed active agent is compared with an existing active agent (which we refer to as *standard* therapy). The goal of the study is to show that the two treatments are roughly equivalent rather than that the new active treatment is superior to standard therapy. The rationale for active-control studies is that in some instances, it may be unethical to randomize a subject to placebo if a prior efficacious therapy already exists (e.g., in clinical trials of drugs used to treat schizophrenics). Another alternative design used in clinical studies is the cross-over design as discussed in Section 13.10. Under the usual parallel design for a clinical trial with two treatments, each subject is randomized to only one of two possible treatments. Under a cross-over design, each subject receives both treatments, but at different time periods. A washout period when no treatment is given is usually specified between the two active-treatment periods. The order of administration of the two treatments for an individual subject is randomized. The rationale for this design is that it usually requires fewer subjects than a parallel design, provided that the effect of treatment given in the first period does not carry over to the second active-treatment period. It is most appropriate for short-acting therapies with no carry-over effect. In Section 13.11, we considered the statistical treatment of clustered binary data. Clustered binary data occur in clinical trials when the unit of randomization is different from the unit of analysis. For example, in some lifestyle interventions (e.g., dietary interventions), the unit of randomization might be a school or school district, but the unit of analysis is the individual child. Modifications to ordinary techniques for analyzing 2×2 tables (discussed in Section 10.2) were introduced to account for the correlation of response from different children in the same school or school district. Finally, in Section 13.12 we considered the emerging field of measurement-error-correction methods. These techniques provide generalizations of standard techniques, such as logistic regression, that account for the common occurrence that a noisy convenient exposure is often used in epidemiologic studies, such as a single blood-pressure measurement, when what is really desired is a more accurate "gold standard" measurement (e.g., the "true" blood pressure) conceptualized as the average of a large number of blood-pressure measurements for an individual subject. Using these techniques, we can estimate the logistic regression that would have been obtained if the gold-standard exposure were available for all subjects instead of the surrogate exposure. We also introduced the concept of a validity study and a reproducibility study, which are ancillary studies that seek to estimate the relationship between the true and surrogate exposure, in the case when the gold standard is measurable and when it is not, respectively.

PROBLEMS

Gynecology

In a 1985 study of the relationship between contraceptive use and infertility, 89 out of 283 infertile women, compared with 640 out of 3833 control women, had used an IUD at some time in their lives [32].

***13.1** Use the normal-theory method to test for significant differences in contraceptive-use patterns between the two groups.

***13.2** Use the contingency-table method to perform the test in Problem 13.1.

***13.3** Compare your results in Problems 13.1 and 13.2.

***13.4** Compute a 95% confidence interval for the difference in the proportion of women who have ever used IUDs between the case and control groups in Problem 13.1.

***13.5** Compute the odds ratio in favor of ever using an IUD for infertile women versus control women.

***13.6** Provide a 95% confidence interval for the true odds ratio corresponding to your answer to Problem 13.5.

13.7 What is the relationship between your answers to Problems 13.2 and 13.6?

Renal Disease

13.8 Refer to Problem 10.41. Estimate the risk ratio for total mortality of the study group versus the control group. Provide 95% confidence limits for the risk ratio.

Cancer

Read "Smoking and Carcinoma of the Lung" by R. Doll and A. B. Hill in the *British Medical Journal,* September 30, 1950, pages 739–748. Refer to Table IV in this paper and answer the following questions based on it.

13.9 Test for the association between cigarette smoking and disease status among males only.

13.10 Compute the odds ratio in favor of cigarette smoking for male lung-cancer cases versus controls. How can the odds ratio be interpreted?

13.11 Compute a 95% confidence interval for the odds ratio computed in Problem 13.10.

13.12 Test for the association between cigarette smoking and disease status among females.

13.13 Answer Problem 13.10 for females.

13.14 Answer Problem 13.11 for females.

13.15 Assess if the odds ratio between cigarette smoking and disease status is the same for males and females.

The people in the study were also classified in Table V according to the number of cigarettes per day smoked regularly just prior to the onset of their present illness.

13.16 Is there a consistent trend between the number of cigarettes smoked and disease status? Perform the appropriate significance test. Combine available evidence for males and females.

Infectious Disease

Refer to Table 13.6.

13.17 Perform a significance test to examine the association between OC use and bacteriuria after controlling for age.

13.18 Estimate the odds ratio in favor of bacteriuria for OC users versus non-OC users after controlling for age.

13.19 Provide a 95% confidence interval for the odds ratio estimate in Problem 13.18.

13.20 Is the association between bacteriuria and OC use comparable among different age groups? Why or why not?

13.21 Suppose you did not control for age in the preceding analyses. Calculate the crude (unadjusted for age) odds ratio in favor of bacteriuria for OC users versus non-OC users.

13.22 How do your answers to Problems 13.18 and 13.21 relate to each other? Try to explain any differences found.

Endocrinology

A study was performed looking at the risk of fractures in three rural Iowa communities according to whether their

TABLE 13.27 Relationship of calcium content of drinking water to the rate of fractures in rural Iowa

Ages 20–35	Number of women with fractures	Total number of women	Ages 55–80	Number of women with fractures	Total number of women
Control	3	37	Control	11	121
Higher calcium	1	33	Higher calcium	21	148

drinking water was "higher calcium," "higher fluorides," or "control" as determined by water samples. Table 13.27 presents data comparing the rate of fractures (over 5 years) between the higher calcium versus the control communities for women ages 20–35 and 55–80, respectively [33].

***13.23** What test can be used to compare the fracture rates in these two communities while controlling for age?

***13.24** Implement the test in Problem 13.23 and report a *p*-value (two-sided).

***13.25** Estimate the odds ratio relating higher calcium and fractures while controlling for age.

***13.26** Provide a 95% CI for the estimate obtained in Problem 13.25.

Pulmonary Disease

Read "Influence of Passive Smoking and Parental Phlegm on Pneumonia and Bronchitis in Early Childhood" by J. R. T. Colley, W. W. Holland, and R. T. Corkhill in *Lancet*, November 2, 1974, pages 1031–1034, and answer the following questions based on it.

13.27 Perform a statistical test comparing the incidence rates of pneumonia and bronchitis for children in their first year of life in families in which both parents are nonsmokers versus families in which both parents are smokers.

13.28 Compute an odds ratio to compare the incidence rates of pneumonia and bronchitis in families in which both parents are smokers versus families in which both parents are nonsmokers.

13.29 Compute a 95% confidence interval corresponding to the odds ratio computed in Problem 13.28.

13.30 Compare the incidence rates of pneumonia and bronchitis for children in their first year of life in families in which both parents are nonsmokers versus families in which one parent is a smoker.

13.31 Answer Problem 13.28 comparing families in which one parent is a smoker with families in which both parents are nonsmokers.

13.32 Compute a 95% confidence interval corresponding to the odds ratio estimate computed in Problem 13.31.

13.33 Is there a significant trend in the percentage of children with pneumonia and bronchitis in the first year of life according to the number of smoking parents? Report a *p*-value.

13.34 Suppose we wish to compare the incidence rates of disease for children in their first and second years of life in families in which both parents are nonsmokers. Rates of 7.8% and 8.1% based on samples of size 372 and 358, respectively, are presented in Table II. Would it be reasonable to use a chi-square test to compare these rates?

13.35 Perform a statistical test comparing the incidence rates of pneumonia and bronchitis for children in the first year of life when stratified by number of cigarettes per day. (Use the groupings in Table IV.) Restrict your analysis to families in which one or both parents are current smokers. Specifically, is there a consistent trend in the incidence rates as the number of cigarettes smoked increases?

13.36 Does the number of siblings in the family affect the incidence rate of pneumonia and bronchitis for children in the first year of life? Use the data in Table VI to control for the confounding effects of (1) parental smoking habits and (2) parental respiratory-symptom history.

Mental Health

Refer to Problem 10.45.

***13.37** Estimate the odds ratio relating widowhood to mortality based on all the data in Table 10.27.

***13.38** Provide a 95% CI for the *OR*.

Hypertension

A study was conducted in Wales relating blood-pressure and blood-lead levels [34]. It was reported that 4 out of 455 men with blood-lead levels ≤ 11 µg/100 mL had elevated systolic blood pressure (SBP ≥ 160 mm Hg), whereas 16 out of 410 men with blood-lead levels ≥ 12 µg/100 mL also had elevated SBP. It was also reported that 6 out of 663 women with blood-lead levels ≤ 11 µg/100 mL had elevated SBP, whereas 1 out of 192 women with blood-lead levels ≥ 12 µg/100 mL had elevated SBP.

13.39 What is an appropriate procedure to test the hypothesis that there is an association between blood pressure and blood lead, while controlling for sex?

13.40 Implement the procedure in Problem 13.39 and report a p-value.

13.41 Estimate the odds ratio relating blood pressure to blood lead, and provide a 95% confidence interval about this estimate.

Infectious Disease

Aminoglycoside antibiotics are particularly useful clinically in the treatment of serious gram-negative bacterial infections among hospitalized patients. Despite their potential for toxicity, as well as the continued development of newer antimicrobial agents of other classes, it seems likely that the clinical use of aminoglycosides will continue to be widespread. The choice of a particular aminoglycoside antibiotic for a given patient depends on several factors, including the specific clinical situation, differences in antimicrobial spectrum and cost, and risks of side effects, particularly nephrotoxicity and auditory toxicity. Many randomized, controlled trials have been published that compare the various aminoglycoside antibiotics with respect to efficacy, nephrotoxicity, and, to a lesser extent, auditory toxicity. These individual trials have varied widely with respect to their design features and their conclusions. A major limitation to their interpretability is that the majority of the individual trials have lacked an adequate sample size to detect the small to moderate differences between treatment groups that are most plausible. As a result, the individual trials published to date have generally not permitted firm conclusions, especially concerning the relative potential for toxicity of aminoglycosides.

In these circumstances, one method to estimate the true effects of these agents more precisely is to conduct an overview, or *meta-analysis*, of the data from all randomized trials. In this way, a true increase in risk could emerge that otherwise would not be apparent in any single trial due to small sample size. Therefore, a quantitative overview of the results of all published randomized controlled trials that assessed the efficacy and toxicity of individual aminoglycoside antibiotics was undertaken.

Forty-five randomized clinical trials, published between 1975 and September 1985, were identified that compared two or more of five aminoglycoside antibiotics: amikacin, gentamicin, netilmicin, sisomicin, and tobramycin. Thirty-seven of these trials could provide data suitable for comparative purposes.

The specific endpoints of interest were efficacy, nephrotoxicity, and auditory toxicity (ototoxicity). Efficacy was defined as bacterial or clinical response to treatment as reported in each individual trial. Nephrotoxicity was defined as the percentage of toxic events to the kidney reported, whether or not the published paper suggested some explanation other than the use of the study drug, such as use of another potentially nephrotoxic agent, or the presence of an underlying disease affecting kidney function. Auditory toxicity was defined as reported differences between pre- and post-treatment audiograms.

The data are organized into three Data Sets: EFF.DAT, NEPHRO.DAT, and OTO.DAT, all on the data disk. A separate record is presented for each antibiotic studied for each endpoint. The format is given in the files EFF.DOC, NEPHRO.DOC, and OTO.DOC, on the data disk.

Columns	
1–8:	Study name
10–11:	Study number (number on reference list)
13:	Endpoint (1 = efficacy; 2 = nephrotoxicity; 3 = ototoxicity)
15:	antibiotic (1 = amikacin; 2 = gentamicin; 3 = netilmicin; 4 = sisomicin; 5 = tobramycin)
17–19:	Sample size
21–23:	Number cured (for efficacy) or number with side effect (for nephrotoxicity or ototoxicity)

Renal Disease

Refer to the Data Set NEPHRO.DAT on the data disk.

13.42 Use methods of meta-analysis to assess if there are differences in nephrotoxicity between each pair of antibiotics. Obtain point estimates and 95% confidence intervals for the odds ratio, and provide a two-sided p-value.

Refer to the Data Set OTO.DAT on the data set.

13.43 Answer the question in Problem 13.42 to assess if there are differences in ototoxicity between each pair of antibiotics.

Refer to the Data Set EFF.DAT on the data disk.

13.44 Answer the question in Problem 13.42 to assess if there are differences in efficacy between each pair of antibiotics.

Cardiology

A recent study compared the use of angioplasty (PTCA) with medical therapy in the treatment of single-vessel coronary-artery disease. A total of 105 patients were randomly assigned to PTCA and 107 to medical therapy. Over a period of six months, myocardial infarction (MI) occurred in 5 patients in the PTCA group and 3 patients in the medical-therapy group.

***13.45** Estimate the risk ratio of MI for patients assigned to PTCA versus patients assigned to medical therapy, and provide a 95% CI for this estimate.

At the six-month clinic visit, 61 of 96 patients seen in the PTCA group and 47 of 102 patients seen in the medical-therapy group were angina free.

***13.46** Answer Problem 13.45 for the endpoint of being angina free at 6 months.

Cardiovascular Disease

A study was performed to estimate the decline in ischemic heart disease mortality from 1965 to 1974 and to identify the causes behind it [35]. Representative samples of the residents of Alameda County, California, were obtained in 1965 ($n = 6928$) and 1974 ($n = 3119$). Each cohort was followed for 9 years for mortality outcomes. In Table 13.28, we present the age- and sex-specific 9-year ischemic heart disease mortality risk for white and black subjects age 40 and older ($n = 3742$ for the 1965 cohort; $n = 1549$ for the 1974 cohort) grouped by cohort (1965, 1974) and by whether the subject reported having heart trouble at baseline.

***13.47** Compute the age- and sex-adjusted 9-year mortality risk for people with heart trouble in the 1965 and 1974 cohorts, respectively. Compute the standardized risk ratio for people in the 1974 cohort compared with those in the 1965 cohort (use the total population in the combined 1965 and 1974 cohorts, including both persons with and without heart trouble, as the standard).

***13.48** Answer Problem 13.47 for those with no heart trouble at baseline.

***13.49** Compute the odds ratio for ischemic heart disease mortality in 1974 compared with 1965 for those with heart trouble at baseline. Compare your answer to that of Problem 13.47.

***13.50** Answer Problem 13.49 for those without heart trouble at baseline. Compare your answer to that for Problem 13.48.

***13.51** One of the hypotheses of the investigators was that the decline in heart-disease mortality is greater among those reporting heart trouble than among those not reporting heart trouble. Do the data support this hypothesis, given your answers to Problems 13.49 and 13.50? If so, what could be the explanation?

13.52 Use logistic-regression methods to assess the relationship between ischemic heart disease risk and age, sex, cohort (1965 versus 1974), and baseline heart-disease symptoms (heart trouble or no heart trouble). A key issue is to quantify whether ischemic heart disease mortality has declined between 1965 and 1974 and, if so, whether the rate of decline is different for different subgroups (e.g., age groups, sex groups, initial-symptom groups).

Sports Medicine

Refer to Problem 10.70. In this problem, we described the Data Set TENNIS1.DAT (on the data disk), which is an observational study relating episodes of tennis elbow to other risk factors.

13.53 Use logistic-regression methods to compare subjects with 1+ episodes of tennis elbow versus subjects with 0 episodes of tennis elbow, considering multiple risk factors in the same model.

13.54 Use linear-regression methods to predict the number of episodes of tennis elbow as a function of several risk factors in the same model.

Hypertension

A drug company proposes to introduce a new antihypertensive agent that is aimed at elderly hypertensive subjects

TABLE 13.28 Age- and sex-specific 9-year ischemic heart disease mortality risk by self-report of heart trouble at baseline for the 1965 and 1974 cohorts: Alameda County Study

	With heart trouble				No heart trouble			
	1965 cohort		1974 cohort		1965 cohort		1974 cohort	
Sex and age (years)	%	*No. at risk*	%	*No. at risk*	%	*No. at risk*	%	*No. at risk*
Males								
< 60	11.6	43	7.3	41	2.4	1,129	1.5	411
60–69	38.5	39	4.2	24	7.7	273	7.2	110
70+	47.1	34	25.0	12	25.8	178	16.9	89
Females								
< 60	0.0	32	0.0	26	0.5	1,304	0.8	497
60–69	23.4	47	12.9	31	5.6	324	2.9	137
70+	25.8	62	11.1	45	13.4	277	14.3	126

with prior heart disease. Because this is a high-risk group, the company is hesitant to withhold antihypertensive therapy from these patients and instead proposes an equivalence study comparing the new agent (drug A) to the current antihypertensive therapy used by those subjects. Hence, the subjects will be randomized to either maintenance of their current therapy or replacement of their current therapy with drug A. Suppose the endpoint is total cardiovascular disease (CVD) mortality and it is assumed that under their current therapy that 15% of subjects will die of CVD over the next 5 years.

13.55 Suppose drug A will be considered equivalent to the current therapy if the 5-year CVD mortality is not worse than 20%. How many subjects need to be enrolled in the study to ensure that there is at least an 80% chance of demonstrating equivalence if equivalence will be based on a one-sided 95% CI approach and an equal number of subjects are randomized to drug A and current therapy?

13.56 Suppose in the actual study that 200 subjects are randomized to each group. Forty-four subjects who receive drug A and 35 subjects who receive current therapy die of CVD in the next 5 years. Can the treatments be considered equivalent? Why or why not?

13.57 How much power did the study described in Problem 13.56 have of demonstrating equivalence under the assumptions in Problem 13.55?

TABLE 13.29 Multiple logistic-regression model relating 2-year incidence of sudden death in females without prior coronary heart disease (data taken from the Framingham Heart Study) to several risk factors

Risk factor	Regression coefficient, $\hat{\beta}_i$	$se(\hat{\beta}_i)$
Constant	–15.3	
Systolic blood pressure (mm Hg)	0.0019	0.0070
Framingham relative weight (%)	–0.0060	0.0100
Cholesterol (mg/100 ml)	0.0056	0.0029
Glucose (mg/100 ml)	0.0066	0.0038
Cigarette smoking (cigarettes/day)	0.0069	0.0199
Hematocrit (%)	0.111	0.049
Vital capacity (centiliters)	–0.0098	0.0036
Age (years)	0.0686	0.0225

Source: Reprinted with permission of the *American Journal of Epidemiology, 120*(6), 888–899, 1984.

Cardiovascular Disease

Sudden death is an important, lethal cardiovascular endpoint. Most previous studies of risk factors for sudden death have focused on men. Looking at this issue for women is important as well. For this purpose, data were used from the Framingham Heart Study [36]. Several potential risk factors, such as age, blood pressure, and cigarette smoking, are of interest and need to be controlled for simultaneously. Therefore, a multiple logistic-regression model was fitted to these data, as shown in Table 13.29.

13.58 Assess the statistical significance of the individual risk factors.

13.59 What do these statistical tests mean in this instance?

13.60 Compute the odds ratio relating the additional risk of sudden death per 100-centiliter decrease in vital capacity after adjustment for the other risk factors.

13.61 Provide a 95% confidence interval for the estimate in Problem 13.60.

Hepatic Disease

Refer to Data Set HORMONE.DAT on the data disk.

13.62 Use logistic-regression methods to assess whether the presence of biliary secretions during the second period (any or none) is related to the type of hormone used during the second period.

13.63 Answer the same question as in Problem 13.62 for the presence of pancreatic secretions.

13.64 Use logistic-regression methods to assess whether the presence of biliary secretions during the second period is related to the dose of hormone used during the second period (do separate analyses for each of the active hormones—hormones 2–5).

13.65 Answer the same question as in Problem 13.64 for the presence of pancreatic secretions.

Otolaryngology

Refer to the Data Set EAR.DAT (see Table 3.10, also on the data disk).

13.66 Consider a subject as "cured" if (1) the subject is a unilateral case and the ear is cleared by 14 days or (2) the subject is a bilateral case and both ears are cleared. Run a logistic regression with outcome variable = cured and independent variables (1) antibiotic, (2) age, and (3) type of case (unilateral or bilateral). Assess the goodness of fit of the model you obtain.

Sports Medicine

Refer to Data Set TENNIS2.DAT, on the data disk.

13.67 Assess whether there are significant treatment effects regarding pain during maximum activity.

13.68 Asses whether there are significant treatment effects regarding pain 12 hours after maximum activity

13.69 Assess whether there are significant treatment effects regarding pain on an average day.

13.70 Assess whether there are significant carry-over effects for the endpoint in Problem 13.67.

13.71 Assess whether there are significant carry-over effects for the endpoint in Problem 13.68.

13.72 Assess whether there are significant carry-over effects for the endpoint in Problem 13.69.

Hypertension

Refer to Data Set ESTROGEN.DAT, on the data disk. The format is in Table 13.30.

Three separate two-period cross-over studies were performed, based on different groups of subjects. In study 1, 0.625 mg estrogen was compared with placebo. In study 2, 1.25 mg estrogen was compared with placebo. In study 3, 1.25 mg estrogen was compared with 0.625 mg estrogen. Subjects received treatment for 4 weeks in each active-treatment period; a 2-week washout period separated the two active-treatment periods.

13.73 Assess if there are any significant treatment or carry-over effects of systolic (SBP) and diastolic blood pressure (DBP) in study 1.

13.74 Answer Problem 13.73 for study 2.

13.75 Answer Problem 13.73 for study 3.

13.76 Suppose we are planning a new study similar in design to study 1. How many subjects do we need to study to detect an underlying 3-mm Hg treatment effect for SBP with 80% power assuming that there is no carry-over effect and we perform a two-sided test with $\alpha = .05$?

Hint: Use the sample standard deviation of the difference scores from study 1 as an estimate of the true standard deviation of the difference scores in the proposed study.

13.77 Answer Problem 13.76 for an underlying 2-mm Hg treatment effect for DBP.

13.78 Answer Problem 13.76 for a new study similar in design to study 2.

13.79 Answer Problem 13.77 for a new study similar in design to study 2.

Otolaryngology

A longitudinal study was conducted among children in the Greater Boston Otitis Media Study [37]. Based on all doctor visits during the first year of life, children were classified as having 1+ episodes versus 0 episodes of otitis media (OTM). A separate classification was performed for the right and left ears. Several risk factors were studied as possible predictors of OTM. One such risk factor was a sibling history of ear infection, with relevant data displayed in Table 13.31.

13.80 Assess whether a sib history of ear infection is associated with OTM incidence in the first year of life.

13.81 Provide a 95% CI for the true difference in incidence rates for those children with sibs between those children with and without a sib history of ear infection.

Otolaryngology

Consider the Data Set EAR.DAT (see Table 3.10).

Suppose we use the ear as the unit of analysis, where the outcome is a success if an ear is cleared by 14 days and a failure otherwise.

13.82 Compare the percentage of cleared ears between the cefaclor-treated and the amoxicillin-treated groups. Please report a *p*-value (two-tail).

13.83 Compare the percentage of cleared ears among children 2–5 years of age versus the percentage of cleared ears among children <2 years of age. Please report a *p*-value (two-tail).

13.84 Compare the percentage of cleared ears among children 6+ years of age versus the percentage of cleared ears among children <2 years of age. Please report a *p*-value (two-tail).

TABLE 13.30 Format of Data Set ESTROGEN.DAT

Variable	Column	Comments
Subject	1–2	
Treatment	4	(1 = placebo, 2 = 0.625 mg estrogen, 3 = 1.25 mg estrogen)
Period	6	
Mean SBP	8–10	(mm Hg)
Mean DBP	12–14	(mm Hg)

TABLE 13.31 Association between sib history (Hx) of ear infection and number of episodes of OTM in the first year of life

Group 1			Group 2		
Sib Hx ear infection = yes			Sib Hx ear infection = no		
Right ear	*Left ear*	*n*	*Right ear*	*Left ear*	*n*
–	–	76	–	–	115
+	–	21	+	–	20
–	+	20	–	+	18
+	+	77	+	+	91
		194			244

Note: + = 1+ episodes of OTM in the first year of life in a specific ear; – = 0 episodes of OTM in the first year of life in a specific ear.

Cancer, Nutrition

A logistic-regression analysis similar to that presented in Example 13.55 was run relating breast-cancer incidence in 1980–1984 to calorie-adjusted total fat (heretofore referred to as total fat intake) as reported on a 1980 food-frequency questionnaire (FFQ). In addition, age in 5-year categories and alcohol in categories (0, 0.1–4.9, 5.0–14.9, 15+ g/day) were also controlled for. The regression coefficient for a 10 g/day increase in total fat intake was –0.163 with standard error = 0.135.

13.85 Obtain a point estimate and a 95% CI for the relative risk of breast cancer comparing women whose total fat intake differs by 10 g/day.

The validation-study data discussed in Section 13.12.2 are available in the Data Set VALID.DAT.

13.86 Use the data for total fat to fit the linear regression of diet record (DR) total fat intake on FFQ total fat intake. Obtain the regression coefficient, standard error, and *p*-value from this regression.

13.87 Using the results from Problems 13.85 and 13.86, obtain an estimate of the *RR* of breast cancer comparing women who differ by 10 g/day on total fat intake on the DR, assuming that age and alcohol intake have no measurement error.

13.88 Obtain a 95% CI for the point estimate in Problem 13.87.

13.89 Compare the measurement-error-corrected *RR* and CI in Problems 13.87 and 13.88 with the uncorrected *RR* and CI in Problem 13.85.

Cancer, Endocrinology

In the study presented in Section 13.12.3, other hormones were considered in addition to plasma estradiol. Table 13.32

TABLE 13.32 Relative-risk estimates and 95% CIs for breast-cancer incidence from 1989 to June 1, 1994, in a nested case–control study among 11,169 postmenopausal women in the Nurses' Health Study not taking hormone-replacement therapy in 1989, comparing women at the median value of the fourth quartile versus women at the median value of the first quartile of the hormone distributions

Hormone	Median value 1st quartile	Median value 4th quartile	*RR*	95% CI
Free estradiol (%)	1.33	1.82	1.69	1.03–2.80
Estrone (pg/mL)	17	45	1.91	1.15–3.16
Testosterone (ng/dL)	12	37	1.65	1.00–2.71

Note: Comparing women at the median value of the fourth quartile versus women at the median value of the first quartile, where the quartiles are determined from the distribution of hormones among controls.

TABLE 13.33 Intraclass correlation coefficients (ICCs) for selected hormones from the Nurses' Health Study reproducibility study, 1989

Hormone	ICC	Number of subjects with		Total number of measurements
		3 replicates	2 replicates	
Free estradiol (%)	0.80	79	0	237
Estrone (pg/mL)	0.74	72	6	228
Testosterone (ng/dL)	0.88	79	0	237

presents the uncorrected relative-risk estimates and 95% CIs for several other hormones [27].

13.90 Obtain the uncorrected logistic-regression coefficients and standard errors for each of the hormones in Table 13.32.

The hormones in Table 13.32 were also included in the reproducibility study mentioned in Section 13.12.3 [29]. The intraclass correlation coefficient and sample size used for each hormone are given in Table 13.33.

13.91 Obtain the measurement-error-corrected logistic-regression coefficient and standard error for each of the hormones in Table 13.32.

13.92 Using the results from Problem 13.91, obtain a measurement-error-corrected odds ratio and 95% CI for each of the hormones.

13.93 How do the results from Problem 13.92 compare with the results in Table 13.32?

REFERENCES

[1] Woolf, B. (1955). On estimating the relation between blood group and disease. *Annals of Human Genetics, 19,* 251–253.

[2] Shapiro, S., Slone, D., Rosenberg, L., et al. (1979). Oral contraceptive use in relation to myocardial infarction. *Lancet, 1,* 743–747.

[3] Evans, D. A., Hennekens, C. H., Miao, L., Laughlin, L. W., Chapman, W. G., Rosner, B., Taylor, J. O., & Kass, E. H. (1978). Oral contraceptives and bacteriuria in a community-based study. *New England Journal of Medicine, 299,* 536–537.

[4] Sandler, D. P., Everson, R. B., & Wilcox, A. J. (1985). Passive smoking in adulthood and cancer risk. *American Journal of Epidemiology, 121*(1), 37–48.

[5] Robins, J. M., Breslow, N., & Greenland, S. (1986). Estimators of the Mantel-Haenszel variance consistent in both sparse data and large strata limiting models. *Biometrics, 42,* 311–323.

[6] Young, T., Palta, M., Dempsey, J., Skatrud, J., Weber, S., & Badr, S. (1993). The occurrence of sleep-disordered breathing among middle-aged adults. *New England Journal of Medicine, 328,* 1230–1235.

[7] Lipnick, R. J., Buring, J. E., Hennekens, C. H., Rosner, B., Willett, W., Bain, C., Stampfer, M. J., Colditz, G. A., Peto, R., & Speizer, F. E. (1986). Oral contraceptives and breast cancer: A prospective cohort study. *JAMA, 255,* 58–61.

[8] Munoz, A., & Rosner, B. (1984). Power and sample size estimation for a collection of 2 × 2 tables. *Biometrics, 40,* 995–1004.

[9] McCormack, W. M., Rosner, B., McComb, D. E., Evrard, J. R., & Zinner, S. H. (1985). Infection with *Chlamydia trachomatis* in female college students. *American Journal of Epidemiology, 121*(1), 107–115.

[10] Dawber, T. R. (1980). *The Framingham Study.* Cambridge, MA: Harvard University Press.

[11] Rosner, B., Spiegelman, D., & Willett, W. C. (1992). Correction of logistic regression relative risk estimates and confidence intervals for random within-person measurement error. *American Journal of Epidemiology, 136,* 1400–1413.

[12] Buring, J. E., Evans, D. A., Mayrent, S. L., Rosner, B., Colton, T., & Hennekens, C. H. (1988). Randomized trials of aminoglycoside antibiotics. *Reviews of Infectious Disease, 10*(5), 951–957.

[13] DerSimonian, R., & Laird, N. M. (1986). Meta analysis in clinical trials. *Controlled Clinical Trials, 7,* 177–188.

[14] Hedges, L. V., & Olkin, I. (1985). *Statistical methods in meta analysis.* London: Academic Press.

[15] Makuch, R. & Simon, R. (1978). Sample size requirements for evaluating a conservative therapy. *Cancer Treatment Reports, 62,* 1037–1040.

[16] Temple, R. (1996). Problems in interpreting active control equivalence trials. *Accountability in Research, 4,* 267–275.

[17] Rothman, K. J., & Michels, K. B. (1994). The continuing unethical use of placebo controls. *New England Journal of Medicine, 331*(16), 394–398.

[18] Fleiss, J. L. (1986). *The design and analysis of clinical experiments.* New York: Wiley.

[19] Grizzle, J. E. (1965). The two-period change-over design and its use in clinical trials. *Biometics, 21,* 467–480.

[20] Rowe, N. H., Brooks, S. L., Young, S. K., Spencer, J., Petrick, T. J., Buchanan, R. A., Drach, J. C., & Shipman, C. (1979). A clinical trial of topically applied 3 percent vidarbine against recurrent herpes labialis. *Oral Pathology, 47,* 142–147.

[21] Banting, D. W., Ellen, R. P., & Fillery, E. D. (1985). A longitudinal study of root caries: Baseline and incidence data. *Journal of Dental Research, 64,* 1141–1144.

[22] Imrey, P. B. (1986). Considerations in the statistical analyses of clinical trials in periodontics. *Journal of Clinical Periodontology, 13,* 517–528.

[23] Fleiss, J. L., Park, M. H., & Chilton, N. W. (1987). Within-mouth correlations and reliabilities for probing depth and attachment level. *Journal of Periodontology, 58,* 460–463.

[24] Donald, A., & Donner, A. (1987). Adjustments to the Mantel-Haenszel chi-square statistic and odds ratio variance estimator when the data are clustered. *Statistics in Medicine, 6,* 491–500.

[25] Rosner, B. (1984). Multivariate methods in ophthalmology with application to other paired data situations. *Biometrics, 40,* 1025–1035.

[26] Rosner, B., Willett, W. C., & Spiegelman, D. (1989). Correction of logistic regression relative risk estimates and confidence intervals for systematic within-person measurement error. *Statistics in Medicine, 8,* 1051–1069.

[27] Hankinson, S. E., Willett, W. C., Manson, J. E., Colditz, G. A., Hunter, D. J., Spiegelman, D., Barbieri, R. L., & Speizer, F. E. (1998). Plasma sex steroid hormone levels and risk of breast cancer in postmenopausal women. *Journal of the National Cancer Institute, 90,* 1292–1299.

[28] Donner, A. (1986). A review of inference procedures for the intraclass correlation coefficient in the one-way random effects model. *International Statistical Review, 54,* 67–82.

[29] Hankinson, S. E., Manson, J. E., Spiegelman, D., Willett, W. C., Longcope, C., & Speizer, F. E. (1995). Reproducibility of plasma hormone levels in postmenopausal women over a 2–3 year period. *Cancer, Epidemiology, Biomarkers and Prevention, 4,* 649–654.

[30] Rosner, B., Spiegelman, D., & Willett, W. C. (1990). Correction of logistic regression relative risk estimates and confidence intervals for measurement error: The case of multiple covariates measured with error. *American Journal of Epidemiology, 132,* 734–745.

[31] Rosner, B., Spiegelman, D., & Willett, W. C. (1992). Correction of logistic regression relative risk estimates and confidence intervals for random within-person measurement error. *American Journal of Epidemiology, 136,* 1400–1413.

[32] Cramer, D. W., Schiff, I., Schoenbaum, S. C., Gibson, M., Belisle, J., Albrecht, B., Stillman, R. J., Berger, M. J., Wilson, E., Stadel, B. V., & Seibel, M. (1985). Tubal infertility and the intrauterine device. *New England Journal of Medicine, 312*(15), 941–947.

[33] Sowers, M. F. R., Clark, M. K., Jannausch, M. L., & Wallace, R. B. (1991). A prospective study of bone mineral content and fracture in communities with differential fluoride exposure. *American Journal of Epidemiology, 133*(7), 649–660.

[34] Elwood, P. C., Yarnell, J. W. G., Oldham, P. D., Catford, J. C., Nutbeam, D., Davey-Smith, G., & Toothill, C. (1988). Blood pressure and blood lead in surveys in Wales. *American Journal of Epidemiology, 127*(5), 942–945.

[35] Cohn, B. A., Kaplan, G. A., & Cohen, R. D. (1988). Did early detection and treatment contribute to the decline in ischemic heart disease mortality? Prospective evidence from the Alameda County Study. *American Journal of Epidemiology, 127*(6), 1143–1154.

[36] Schatzkin, A., Cupples, L. A., Heeren, T., Morelock, S., & Kannel, W. B. (1984). Sudden death in the Framingham Heart Study: Differences in incidence and risk factors by sex and coronary disease status. *American Journal of Epidemiology, 120*(6), 888–899.

[37] Teele, D. W., Klein, J. O., & Rosner, B. (1989). Epidemiology of otitis media during the first seven years of life in children in Greater Boston: A prospective, cohort study. *Journal of Infectious Disease, 160*(1), 83–94.

14 Hypothesis Testing: Person-Time Data

SECTION 14.1 Measure of Effect for Person-Time Data

In Chapter 10, we discussed the analysis of categorical data, where the person was the unit of analysis. In a prospective study design, we identified groups of exposed and unexposed individuals at baseline, and compared the proportion of subjects who developed disease over time between the two groups. We referred to these proportions as *incidence rates,* although a more technically appropriate term would be **cumulative incidence rates** (see Definition 3.17). Cumulative incidence (CI) rates are proportions where the person is the unit of analysis and must range between 0 and 1. In computing cumulative incidence rates, we implicitly assume that all subjects are followed for the same period of time T. This is not always the case, as is illustrated in the following example.

Example 14.1 **Cancer** A hypothesis of much recent interest is the possible association between the use of oral contraceptives (OC) and the development of breast cancer. To address this issue, data were collected in the Nurses' Health Study where disease-free women were classified in 1976 according to OC status (current user/past user/never user). A mail questionnaire was sent out every two years in which OC status was updated and breast-cancer status was ascertained over the next two years. For each woman, an amount of time that the woman is a current user or a never user of OC's (ignoring past use) can be calculated and this *person-time* can be accumulated over the entire cohort of nurses. Thus, each nurse contributes a different amount of person-time to the analysis. The data are presented in Table 14.1 for current and never users of OC's among women 45–49 years of age. How should these data be used to assess any differences in the incidence rate of breast cancer by OC-use group?

TABLE 14.1 Relationship between breast-cancer incidence and OC use among 45- to 49-year-old women in the Nurses' Health Study

OC-use group	Number of cases	Number of person-years
Current users	9	2,935
Never users	239	135,130

The first issue to consider is the appropriate unit of analysis for each group. If the woman is used as the unit of analysis, then the problem is that different women may contribute different amounts of person-time to the analysis, and the assumption of a constant probability of an event for each woman would then be violated. If a person-year is used as the unit of analysis (i.e., one person followed for one year), then since each woman can contribute more than one person-year to the analysis, the important assumption of independence for the binomial distribution would be violated.

For the purpose of allowing for varying follow-up time for each individual, we define the concept of incidence density:

DEFINITION 14.1

The **incidence density** in a group is defined by the number of events in that group divided by the total person-time accumulated during the study in that group.

The denominator used in computing incidence density is the person-year. Unlike cumulative incidence, incidence density may range from 0 to ∞.

Example 14.2 **Cancer** Compute the estimated incidence density among current and never OC users in Table 14.1.

Solution The incidence density among current users = 9/2935 = .00307 events per person-year = 307 events per 100,000 person-years. The incidence density among never users = 239/135,130 = .00177 events per person-year = 177 events per 100,000 person-years.

In following a subject, the incidence density may remain the same or may vary over time (e.g., as a subject ages over time, the incidence density generally increases). How can we relate cumulative incidence over time t to incidence density? Suppose for simplicity that incidence density remains the same over some time period t. If $CI(t)$ = cumulative incidence over time t and λ = incidence density, then it can be shown using calculus methods that

EQUATION 14.1

$$CI(t) = 1 - e^{-\lambda t}$$

If the cumulative incidence is low (<.1), then we can approximate $e^{-\lambda t}$ by $1 - \lambda t$ and $CI(t)$ by

EQUATION 14.2

$$CI(t) \cong 1 - (1 - \lambda t) = \lambda t$$

This relationship is summarized as follows:

EQUATION 14.3

Relationship Between Cumulative Incidence and Incidence Density Suppose we follow a group of individuals with constant incidence density λ = number of events per person-year. The exact cumulative incidence over time period t is

$$CI(t) = 1 - e^{-\lambda t}$$

If the cumulative incidence is low (<.1), then the cumulative incidence can be approximated by

$$CI(t) \cong \lambda t$$

Later in this chapter we refer to incidence density by the more commonly used term *incidence rate* (λ) and distinguish it from the cumulative incidence over some time period $t = CI(t)$. The former can range from 0 to ∞, while the latter is a proportion and must vary between 0 and 1.

Example 14.3 **Cancer** Suppose the incidence density of breast cancer in 40- to 44-year-old premenopausal women is 200 events per 100,000 person-years. What is the cumulative incidence of breast cancer over 5 years among 40-year-old initially disease-free women?

Solution From Equation 14.3, we have that $\lambda = 200/10^5$, $t = 5$ years. Thus, the exact cumulative incidence is given by

$$\begin{aligned} CI(5) &= 1 - e^{-(200/10^5)5} \\ &= 1 - e^{-1000/10^5} \\ &= 1 - e^{-10^{-2}} = 1 - e^{-.01} = .00995 = 995/10^5 \end{aligned}$$

The approximate cumulative incidence is given by

$$CI \cong (200/10^5) \times 5 = .01 = 1000/10^5$$

SECTION 14.2 One-Sample Inference for Incidence-Rate Data

14.2.1 Large-Sample Test

Example 14.4 **Cancer, Genetics** A registry is set up during the period 1990–1994 of women with a suspected genetic marker for breast cancer, but who have not yet had breast cancer. Five hundred 60- to 64-year-old women are identified and are followed until December 31, 2000. Thus, the length of follow-up is variable. The total length of follow-up is 4000 person-years, during which 28 new cases of breast cancer occurred. Is the incidence rate of breast cancer different in this group from the general population of 60- to 64-year-old women if the expected incidence rate is $400/10^5$ person-years in this age group?

We wish to test the hypothesis H_0: $ID = ID_0$ versus H_1: $ID \neq ID_0$, where ID = the unknown incidence density (rate) in the genetic-marker group and ID_0 = the known incidence density (rate) in the general population. We will base our test on the observed number of breast cancers, which we denote by a. We will assume that a approximately follows a Poisson distribution. Under H_0, a has mean = $t(ID_0)$ and variance = $t(ID_0)$, where t = total number of person-years. If we assume that the normal approximation to the Poisson distribution is valid, then this suggests the following test procedure:

EQUATION 14.4

One-Sample Inference for Incidence-Rate Data (Large-Sample Test) Suppose that a events are observed over t person-years of follow-up and that ID = underlying incidence density (rate). To test the hypothesis H_0: $ID = ID_0$ versus H_1: $ID \neq ID_0$,

(1) Compute the test statistic

$$X^2 = \frac{(a-\mu_0)^2}{\mu_0} \sim \chi_1^2 \text{ under } H_0$$

where

$$\mu_0 = t(ID_0)$$

(2) For a two-sided test at level α, H_0 is rejected if

$$X^2 > \chi^2_{1,\,1-\alpha}$$

and accepted if

$$X^2 \le \chi^2_{1,\,1-\alpha}$$

(3) The exact p-value = $Pr(\chi_1^2 > X^2)$.

(4) This test should only be used if $\mu_0 = t(ID_0) > 10$.

Example 14.5 **Cancer, Genetics** Perform a significance test based on the data in Example 14.4.

Solution We have that $a = 28$, $\mu_0 = (400/10^5)(4000) = 16$. Thus, the test statistic is given by

$$X^2 = \frac{(28-16)^2}{16} = \frac{144}{16} = 9.0 \sim \chi_1^2 \text{ under } H_0$$

Since $\chi^2_{1,.995} = 7.88$, $\chi^2_{1,.999} = 10.83$ and $7.88 < 9.0 < 10.83$, it follows that $.001 < p < .005$. Thus, there is a significant excess of breast cancers in the genetic-marker group.

14.2.2 Exact Test

Suppose that the number of events is too small to apply the large-sample test in Equation 14.4. In this case, an exact test based on the Poisson distribution must be used. If $\mu = t(ID)$, then we can restate the hypotheses in the form: H_0: $\mu = \mu_0$ versus H_1: $\mu \ne \mu_0$ and apply the one-sample Poisson test as follows:

EQUATION 14.5

One-Sample Inference for Incidence-Density (Rate) Data (Small-Sample Test) Suppose that a events are observed over t person-years of follow-up and that ID = underlying incidence density (rate). We wish to test the hypothesis H_0: $ID = ID_0$ versus H_1: $ID \ne ID_0$.

Under H_0, the observed number of events (a) will follow a Poisson distribution with parameter $\mu_0 = t(ID_0)$. Thus, the exact two-sided p-value is given by

$$\min\left(2 \times \sum_{k=0}^{a} \frac{e^{-\mu_0}\mu_0^k}{k!},\, 1\right) \text{ if } a < \mu_0$$

$$\min\left[2 \times \left(1 - \sum_{k=0}^{a-1} \frac{e^{-\mu_0}\mu_0^k}{k!}\right),\, 1\right] \text{ if } a \ge \mu_0$$

Example 14.6 **Cancer, Genetics** Suppose that 125 of the 500 women in Example 14.4 also have a family history of breast cancer in addition to having the genetic marker. Eight cases of breast cancer are observed in this subgroup over 1000 person-years. Does this subgroup have a significantly different breast-cancer incidence relative to the general population?

Solution The expected number of breast cancers in this subgroup = $1000(400/10^5) = 4$. Thus, the expected number of cases is too small to use the large-sample test in Equation 14.4. Instead, we use the small-sample test in Equation 14.5. We have $a = 8$, $\mu_0 = 4$. Since $8 \ge 4$, we have

$$p\text{-value} = 2 \times \left(1 - \sum_{k=0}^{7} \frac{e^{-4}4^k}{k!}\right)$$

From the Poisson tables (Table 2 in the Appendix), this is given by

$$\begin{aligned} p\text{-value} &= 2 \times [1 - (.0183 + .0733 + \cdots + .0595)] \\ &= 2 \times (1 - .9489) \\ &= .102 \end{aligned}$$

Hence, breast-cancer incidence in this subgroup is not significantly different from the general population. A larger number of person-years of follow-up are needed to obtain more power in this case.

14.2.3 Confidence Limits for Incidence Rates

To obtain confidence limits for *ID*, we obtain confidence limits for the expected number of events (μ) based on the Poisson distribution and then divide each confidence limit by t = number of person-years of follow-up. Specifically, we have $\hat{\mu} = a$, $Var(\hat{\mu}) = a$. Thus, if the normal approximation to the Poisson distribution holds, then a $100\% \times (1 - \alpha)$ confidence interval for μ is given by $a \pm z_{1-\alpha/2}\sqrt{a}$. The corresponding $100\% \times (1 - \alpha)$ confidence interval for $ID = (a \pm z_{1-\alpha/2}\sqrt{a})/t$. Otherwise, we obtain exact confidence limits for μ from Table 8, in the Appendix, and divide each confidence limit by t to obtain the corresponding confidence interval for *ID*. The procedure is summarized as follows:

EQUATION 14.6

Point and Interval Estimation for Incidence Rates Suppose that a events are observed over t person-years of follow-up.

(1) A point estimate of the incidence density rate = $\widehat{ID} = a/t$.

(2) To obtain a two-sided $100\% \times (1 - \alpha)$ confidence interval for μ,

  (a) if $a \ge 10$, then compute $a \pm z_{1-\alpha/2}\sqrt{a} = (c_1, c_2)$;

  (b) if $a < 10$, then obtain (c_1, c_2) from Table 8 by referring to the a row and the $1 - \alpha$ column.

(3) The corresponding two-sided $100\% \times (1 - \alpha)$ confidence interval for *ID* is given by

$$(c_1/t,\ c_2/t)$$

Example 14.7 **Cancer, Genetics** Obtain a point estimate and a two-sided 95% confidence interval for *ID* based on the data in Example 14.4.

Solution We have $a = 28$, $t = 4000$. Hence, the point estimate of $ID = 28/4000 = .007 = 700/10^5$

person-years = $\widehat{ID}$. Since $a \geq 10$, to obtain a 95% confidence interval for μ, we refer to 2(a) in Equation 14.6 and obtain the confidence limits

$$28 \pm 1.96\sqrt{28} = 28 \pm 10.4 = (17.6,\ 38.4) = (c_1,\ c_2)$$

The corresponding 95% confidence interval for ID = (17.6/4000, 38.4/4000) = (0.00440, 0.00960) or (440/10^5 person-years, 960/10^5 person-years). This interval excludes the null rate of 400/10^5 person-years given in Example 14.4.

Example 14.8 **Cancer, Genetics** Obtain a point estimate and a two-sided 95% confidence interval for ID based on the data in Example 14.6.

Solution From Example 14.6, the expected number of breast-cancer cases in this subgroup (i.e., μ_0) = 4. The point estimate of ID = 8/1000 = .008 = 800 per 10^5 person-years. Because $a = 8 < 10$, we use 2(b) of Equation 14.6 to obtain a 95% confidence interval for ID. Referring to the $a = 4$ row and 0.95 column of Table 8 in the Appendix, we have that a 95% confidence interval for μ = (1.09, 10.24). The corresponding 95% confidence interval for ID = (1.09/1000, 10.24/1000) = (109/10^5 person-years, 1024/10^5 person-years). This interval includes the general population rate of 400/10^5 person-years.

In this section, we have learned about incidence density (also called *incidence rate),* which is expressed as the number of events per unit time, and have distinguished it from cumulative incidence, which is the probability of an event occurring over time t. We have considered one-sample inference for incidence rates. The inference procedures are based on modeling the number of events over time t by a Poisson distribution. We have used a large-sample test based on the normal approximation to the Poisson distribution, when the expected number of events is ≥10, and a small-sample test based on the exact Poisson probabilities when the expected number of events is <10. Finally, we also discussed methods of point and interval estimation for incidence rates. In the next section, we extend this discussion to investigate methods for comparing incidence rates from two samples.

Referring to the flowchart at the end of this chapter (Figure 14.9), we would answer yes to (1) person-time data? and (2) one-sample problem? This leads us to the box entitled "Use one-sample test for incidence rates."

SECTION 14.3 Two-Sample Inference for Incidence-Rate Data

14.3.1 Hypothesis Testing—General Considerations

The question we wish to address in this section is, How can we compare the underlying incidence rates between two different exposure groups?

The approach we will take is to use a *conditional* test. Specifically, suppose we consider the case of two exposure groups and have the general table in Table 14.2. We wish to test the hypothesis H_0: $ID_1 = ID_2$ versus H_1: $ID_1 \neq ID_2$, where ID_1 = true incidence density in group 1 = the number of events per unit of person-time in group 1, and ID_2 is the comparable rate in group 2. Under the null hypothesis, the fraction $t_1/(t_1 + t_2)$ of the total number of events $(a_1 + a_2)$ would be expected to oc-

TABLE 14.2 **General observed table for comparing incidence rates between two groups**

Exposure group	Number of events	Person-time
1	a_1	t_1
2	a_2	t_2
Total	$a_1 + a_2$	$t_1 + t_2$

cur in group 1 and the fraction $t_2/(t_1 + t_2)$ of the total number of events to occur in group 2. Furthermore, under H_0, conditional on the observed total number of events = $a_1 + a_2$, the expected number of events in each group is given by

EQUATION 14.7

$$\text{Expected number of events in group 1} = E_1 = (a_1 + a_2)t_1/(t_1 + t_2)$$
$$\text{Expected number of events in group 2} = E_2 = (a_1 + a_2)t_2/(t_1 + t_2)$$

Example 14.9 **Cancer** Compute the expected number of events among current and never users for the OC–breast-cancer data in Table 14.1.

Solution We have that $a_1 = 9$, $a_2 = 239$, $t_1 = 2935$ person-years, $t_2 = 135{,}130$ person-years. Therefore, under H_0, from Equation 14.7, 2935/(2935 + 135,130) = .0213 of the cases would be expected to occur among current OC users and 135,130/(2935 + 135,130) = .9787 of the cases to occur among never OC users. Thus,

$$E_1 = .0213(248) = 5.27$$

$$E_2 = .9787(248) = 242.73$$

14.3.2 Normal-Theory Test

To assess statistical significance, the number of events in group 1 under H_0 is treated as a binomial random variable with parameters $n = a_1 + a_2$ and $p_0 = t_1/(t_1 + t_2)$. Under this assumption, the hypotheses can be stated as H_0: $p = p_0$ versus H_1: $p \neq p_0$, where p = the true proportion of events that are expected to occur in group 1. We will also assume that the normal approximation to the binomial distribution is valid. Using the normal approximation to the binomial distribution, the observed number of events in group 1 = a_1 is normally distributed with mean $= np_0 = (a_1 + a_2)t_1/(t_1 + t_2) = E_1$, and variance $= np_0q_0 = (a_1 + a_2)t_1t_2/(t_1 + t_2)^2 = V_1$. H_0 will be rejected if a_1 is much smaller or larger than E_1. This is an application of the large-sample one-sample binomial test, given by the following:

EQUATION 14.8

Comparison of Incidence Rates (Large-Sample Test) To test the hypothesis H_0: $ID_1 = ID_2$ versus H_1: $ID_1 \neq ID_2$, where ID_1 and ID_2 are the true incidence densities in groups 1 and 2, use the following procedure:

(1) Compute the test statistic

$$z = \frac{a_1 - E_1 - .5}{\sqrt{V_1}} \text{ if } a_1 > E_1$$

$$= \frac{a_1 - E_1 + .5}{\sqrt{V_1}} \text{ if } a_1 \le E_1$$

where

$$E_1 = (a_1 + a_2)t_1/(t_1 + t_2)$$

$$V_1 = (a_1 + a_2)t_1t_2/(t_1 + t_2)^2$$

a_1, a_2 = number of events in groups 1 and 2

t_1, t_2 = amount of person-time in groups 1 and 2

Under H_0, $z \sim N(0, 1)$

(2) For a two-sided level α test, if

$$z > z_{1-\alpha/2} \quad \text{or} \quad z < z_{\alpha/2}$$

then reject H_0. If

$$z_{\alpha/2} \le z \le z_{1-\alpha/2}$$

then accept H_0.

(3) Use this test only if $V_1 \ge 5$.

(4) The p-value for this test is given by

$$2 \times [1 - \Phi(z)] \text{ if } z \ge 0$$

$$2 \times \Phi(z) \text{ if } z < 0$$

The critical regions and p-value are illustrated in Figures 14.1 and 14.2, respectively.

FIGURE 14.1 **Acceptance and rejection regions for the two-sided test for incidence rates (normal-theory method)**

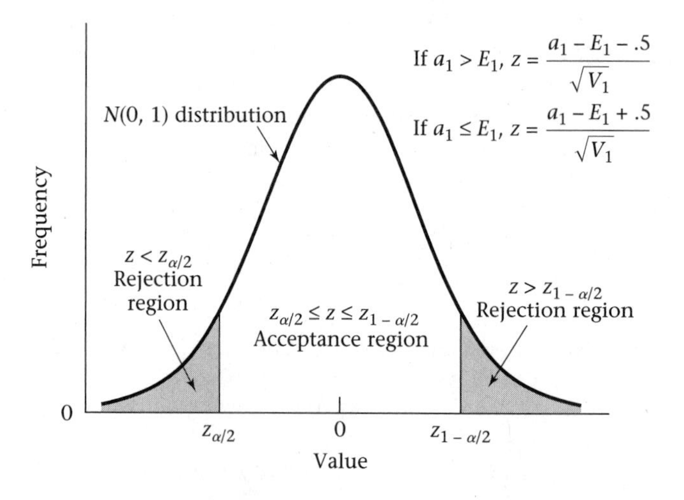

FIGURE 14.2 **Computation of the *p*-value for the two-sided test for incidence rates (normal-theory method)**

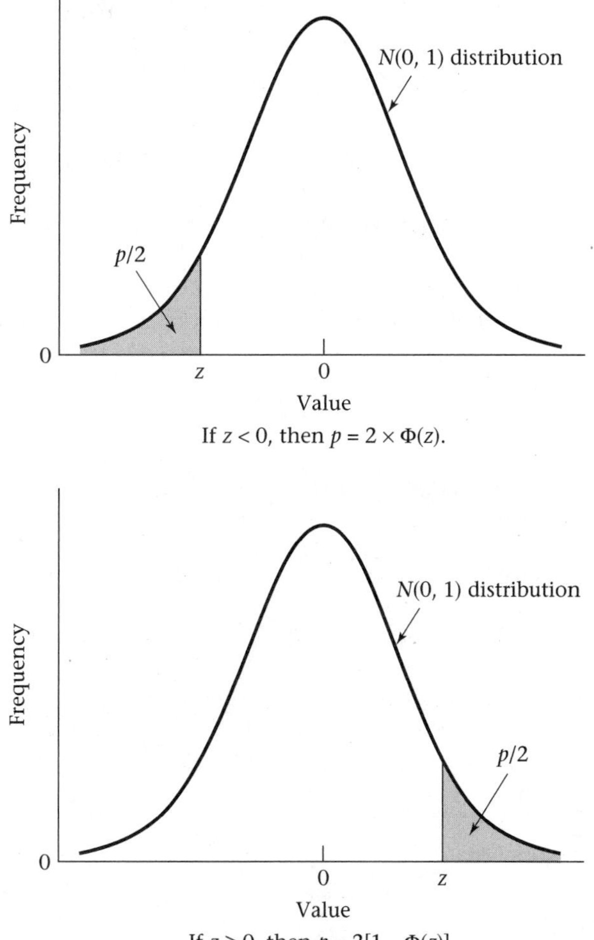

Example 14.10 **Cancer** Assess the statistical significance of the OC–breast-cancer data in Table 14.1.

Solution From Example 14.9, $a_1 = 9$, $a_2 = 239$, $t_1 = 2935$, $t_2 = 135{,}130$, $E_1 = 5.27$, $E_2 = 242.73$. Furthermore,

$$V_1 = \frac{(a_1 + a_2)t_1 t_2}{(t_1 + t_2)^2}$$

$$= \frac{(9 + 239)(2935)(135{,}130)}{(2935 + 135{,}130)^2} = \frac{9.8358 \times 10^{10}}{138{,}065^2} = 5.16$$

Since $V_1 \geq 5$, we can use the large-sample test in Equation 14.8.

Therefore, since $a_1 > E_1$, $z = \frac{9 - 5.27 - .5}{\sqrt{5.16}} = \frac{3.23}{2.27} = 1.42 \sim N(0, 1)$

The p-value $= 2 \times [1 - \Phi(1.42)] = 2 \times (1 - .9223) = .155$. Thus, the results are not statistically significant and there is no significant difference in the incidence rate of breast cancer between current OC users and never OC users in this age group.

14.3.3 Exact Test

Suppose that the number of events is too small to apply the normal-theory test (i.e., $V_1 < 5$). In this case, an exact test based on the binomial distribution must be used. From Section 14.3.1, under H_0, the number of events in group 1 (a_1) will follow a binomial distribution with parameters $n = a_1 + a_2$ and $p = p_0 = t_1/(t_1 + t_2)$. We wish to test the hypothesis H_0: $p = p_0$ versus H_1: $p \neq p_0$, where p is the underlying proportion of events that occur in group 1. This is an application of the exact one-sample binomial test. H_0 will be rejected if the observed number of events a_1 is much smaller or much larger than the expected number of events $= E_1 = np_0$. The following test procedure is used:

EQUATION 14.9

Comparison of Incidence Rates—Exact Test Let a_1, a_2 be the observed number of events and t_1, t_2 the amount of person-time in groups 1 and 2, respectively. Let p = true proportion of events in group 1. To test the hypothesis H_0: $ID_1 = ID_2$ (or equivalently, $p = p_0$) versus H_1: $ID_1 \neq ID_2$ (or equivalently, $p \neq p_0$), where

ID_1 = true incidence density in group 1

ID_2 = true incidence density in group 2

$p_0 = t_1/(t_1 + t_2)$, $q_0 = 1 - p_0$

using a two-sided test with significance level α, use the following procedure:

(1) If $a_1 < (a_1 + a_2)p_0$, then

$$p\text{-value} = 2 \times \sum_{k=0}^{a_1} \binom{a_1 + a_2}{k} p_0^k q_0^{a_1+a_2-k}$$

(2) If $a_1 \geq (a_1 + a_2)p_0$, then

$$p\text{-value} = 2 \times \sum_{k=a_1}^{a_1+a_2} \binom{a_1 + a_2}{k} p_0^k q_0^{a_1+a_2-k}$$

(3) This test is valid in general for comparing two incidence densities but is particularly useful when $V_1 < 5$, in which case the normal-theory test in Equation 14.8 should not be used.

The computation of the p-value is illustrated in Figure 14.3.

Example 14.11 **Cancer** Suppose we have the data in Table 14.3 relating OC use and incidence of breast cancer among women aged 30–34. Assess the statistical significance of the data.

Solution Note that $a_1 = 3$, $a_2 = 9$, $t_1 = 8250$, $t_2 = 17{,}430$. Thus,

$$V_1 = \frac{12(8250)(17,430)}{(8250 + 17,430)^2} = 2.62 < 5$$

FIGURE 14.3 **Illustration of the *p*-value for the two-sample test for incidence rates, exact method (two-sided alternative)**

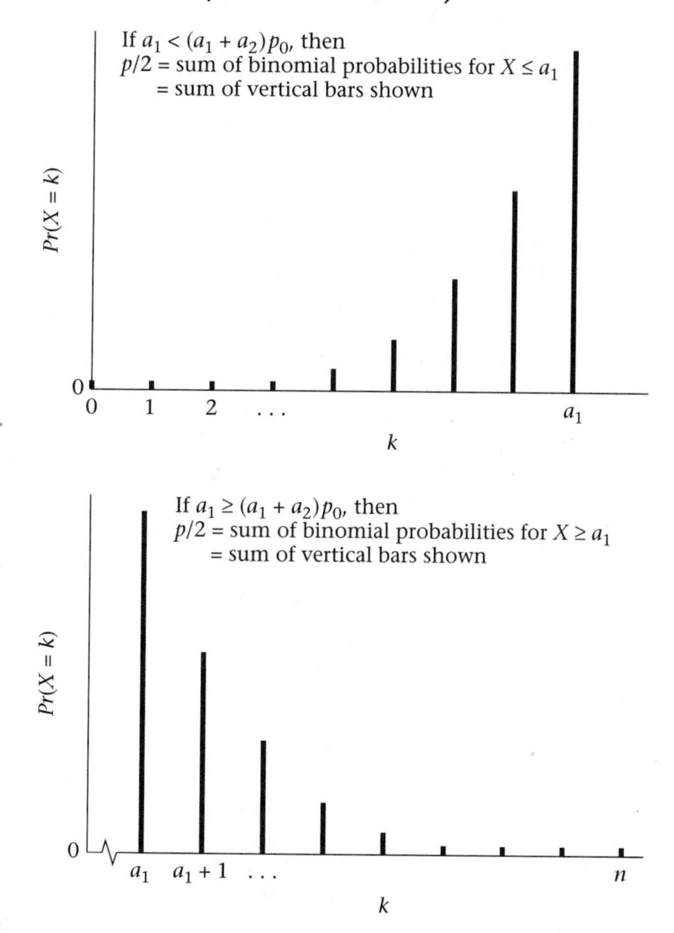
If $a_1 < (a_1 + a_2)p_0$, then
$p/2$ = sum of binomial probabilities for $X \le a_1$
= sum of vertical bars shown
$Pr(X = k)$
0
0 1 2 . . . a_1
k
If $a_1 \ge (a_1 + a_2)p_0$, then
$p/2$ = sum of binomial probabilities for $X \ge a_1$
= sum of vertical bars shown
$Pr(X = k)$
0
a_1 $a_1 + 1$ . . . n
k

TABLE 14.3 **Relationship between breast-cancer incidence and OC use among 30- to 34-year-old women in the Nurses' Health Study**

OC-use group	Number of cases	Number of person-years
Current users	3	8,250
Never users	9	17,430

Since $V_1 < 5$, the small-sample test must be used. From Equation 14.9, $p_0 = 8250/25{,}680 = .321$, $n = a_1 + a_2 = 12$. Since $a_1 = 3 < 12(.321) = 3.9$,

$$p\text{-value} = 2 \times \sum_{k=0}^{3} \binom{12}{k} (.321)^k (.679)^{12-k}$$

Let X be a random variable representing the number of events in group 1 and use Excel to evaluate the p-value:

n	12
p	0.321262
k	Pr(X=k)
0	0.009559
1	0.054296
2	0.141346
3	0.223008
Pr(X<=3)	0.428209
p-value	0.856418

Thus,

$$p\text{-value} = 2 \times (.0096 + .0543 + .1413 + .2230) = 2 \times .4282 = .856$$

Therefore, there is no significant effect of current OC use on breast-cancer incidence in the 30- to 34-year-old age group.

14.3.4 The Rate Ratio

In Section 13.3, we defined the risk ratio (RR) as a measure of effect for the comparison of two proportions. We applied this measure to compare cumulative incidence between two exposure groups in a prospective study, where the person was the unit of analysis. A similar concept can be employed to compare two incidence rates based on person-time data.

DEFINITION 14.2 Let λ_1, λ_2 be incidence rates for an exposed and an unexposed group, respectively. The **rate ratio** is defined as λ_1/λ_2.

Example 14.12 **Cancer** Suppose the incidence rate of breast cancer is 500 per 10^5 person-years among 40- to 49-year-old premenopausal women with a family history of breast cancer (either a mother or a sister history of breast cancer) (group 1) compared with 200 per 10^5 person-years among 40- to 49-year-old premenopausal women with no family history (group 2). What is the rate ratio of group 1 versus group 2?

Solution The rate ratio = $(500/10^5)/(200/10^5) = 2.5$.

What is the relationship between the rate ratio based on incidence rates and the risk ratio based on cumulative incidence? Suppose each person in a cohort is followed for T years, with incidence rate λ_1 in the exposed group and λ_2 in the unexposed group. If the cumulative incidence is low, then the cumulative incidence will be approximately $\lambda_1 T$ in the exposed group and $\lambda_2 T$ in the unexposed group. Thus, the risk ratio will be approximately $\lambda_1 T/(\lambda_2 T) = \lambda_1/\lambda_2$ = rate ratio.

How can we estimate the rate ratio from observed data? Suppose we have the number of events and person-years shown in Table 14.2. The estimated incidence rate in the exposed group = a_1/t_1 and in the unexposed group = a_2/t_2. A point estimate of the rate ratio is given by $\widehat{RR} = (a_1/t_1)/(a_2/t_2)$. To obtain an interval estimate, we assume approximate normality of $\ln(\widehat{RR})$. It can be shown that

EQUATION 14.10

$$Var[\ln(\widehat{RR})] \approx \frac{1}{a_1} + \frac{1}{a_2}$$

Therefore, a two-sided $100\% \times (1-\alpha)$ CI for $\ln(RR)$ is given by

$$(d_1, d_2) = \ln(\widehat{RR}) \pm z_{1-\alpha/2}\sqrt{\frac{1}{a_1} + \frac{1}{a_2}}$$

If we take the antilog of d_1 and d_2, we obtain a two-sided $100\% \times (1-\alpha)$ CI for RR. This is summarized as follows:

EQUATION 14.11

Point and Interval Estimation of the Rate Ratio Suppose we have observed a_1 events in t_1 person-years for an exposed group and a_2 events in t_2 person-years for an unexposed group. A point estimate of the rate ratio is given by

$$\widehat{RR} = (a_1/t_1)/(a_2/t_2)$$

A two-sided $100\% \times (1-\alpha)$ CI for RR is given by (c_1, c_2) where

$$c_1 = e^{d_1}, c_2 = e^{d_2} \text{ and}$$

$$d_1 = \ln(\widehat{RR}) - z_{1-\alpha/2}\sqrt{\frac{1}{a_1} + \frac{1}{a_2}}$$

$$d_2 = \ln(\widehat{RR}) + z_{1-\alpha/2}\sqrt{\frac{1}{a_1} + \frac{1}{a_2}}$$

This interval should only be used if $V_1 = (a_1 + a_2)t_1t_2/(t_1 + t_2)^2$ is greater than or equal to 5.

Example 14.13 **Cancer** Obtain a point estimate and associated 95% confidence limits for the rate ratio relating oral-contraceptive use to breast-cancer incidence based on the data in Table 14.1.

Solution From Table 14.1, the estimated rate ratio is

$$\widehat{RR} = \frac{9/2935}{239/135,130} = 1.73$$

To obtain an interval estimate, we refer to Equation 14.11. A 95% CI for $\ln(RR)$ is (d_1, d_2), where

$$d_1 = \ln(\widehat{RR}) - 1.96\sqrt{\frac{1}{9} + \frac{1}{239}}$$
$$= 0.550 - 0.666 = -0.115$$

$$d_2 = 0.550 + 0.666 = 1.216$$

Therefore, $c_1 = e^{-0.115} = 0.89$, $c_2 = e^{1.216} = 3.37$. Thus, the 95% confidence interval for RR is (0.89, 3.37).

In this section, we have introduced the rate ratio, which is a measure of effect for comparing two incidence rates. It is analogous to but not the same as the risk ratio. The latter was introduced in Chapter 13 as a measure of effect for comparing two cumulative incidence rates. Inference procedures for comparing two incidence rates are based on the one-sample binomial test, where the number of units of analysis = the number of events over the two samples combined, and the probability of success p = the probability that a subject is in group 1, given that he (she) has had an event. We considered both large-sample and small-sample inference procedures based on the normal approximation to the binomial distribution and exact binomial probabilities, respectively. In the next section, we consider power and sample-size estimation procedures for comparing two incidence rates.

Referring to the flowchart (Figure 14.9), we would answer yes to (1) person-time data? no to (2) one-sample problem? and yes to (3) incidence rates remain constant over time? and (4) two sample problem? This leads us to the box entitled "Use two-sample test for comparison of incidence rates, if no confounding is present, or methods for stratified person-time data, if confounding is present." In this section, we assume that no confounding is present. In Section 14.5, we consider two-sample inference for incidence rates when confounding is present.

SECTION 14.4 Power and Sample-Size Estimation for Person-Time Data

14.4.1 Estimation of Power

Example 14.14 **Cancer** Suppose that it is proposed to enroll 10,000 postmenopausal women who have not had any previous cancer for a clinical trial, where 5000 are to be randomized to received estrogen-replacement therapy (ERT) and 5000 are to be randomized to placebo. The endpoint is breast-cancer incidence. Subjects are enrolled from January 1, 1995, to December 31, 1996, and are followed until December 31, 2000, for an average of 5 years of follow-up for each subject (range from 4 to 6 years of follow-up). The expected incidence rate in the control group is $300/10^5$ person-years. If it is hypothesized that the effect of ERT is to increase the incidence rate of breast cancer by 25%, then how much power does the proposed study have?

We base our power calculations on the comparison of incidence rates as given in Equation 14.8. We wish to test the hypothesis H_0: rate ratio $(RR) = 1$ versus H_1: $RR \neq 1$, where $RR = ID_1/ID_2$.

As discussed in Section 14.3.2, another way to state the hypothesis is as follows: H_0: $p = p_0$ versus H_1: $p \neq p_0$, where $p_0 = t_1/(t_1 + t_2)$. This is a one-sample binomial test considered in Section 7.9, where n = total number of events over both groups and p = probability that an individual person with an event comes from group 1. The issue is, What specific value of p under H_1 (call it p_1) corresponds to a rate ratio of RR? To derive this, note that

$$\text{Expected number of events in group } 1 = 1 - \exp(-ID_1 t_1) \cong t_1 ID_1$$

$$\text{Expected number of events in group } 2 = 1 - \exp(-ID_2 t_2) \cong t_2 ID_2$$

Thus, the expected proportion of events in group 1 is

EQUATION 14.12

$$p = t_1ID_1 / (t_1ID_1 + t_2ID_2)$$

If we divide the numerator and denominator of Equation 14.12 by ID_2, we obtain

EQUATION 14.13

$$p = t_1RR/(t_1RR + t_2)$$

Under H_1, RR will be different from 1 and we denote p in Equation 14.13 by p_1. Under H_0, $p = t_1/(t_1 + t_2)$, which we denote by p_0. We now apply the power formula for the one-sample binomial test (Equation 7.37) and obtain

EQUATION 14.14

$$\text{Power} = \Phi\left[\sqrt{\frac{p_0q_0}{p_1q_1}}\left(z_{\alpha/2} + \frac{|p_0 - p_1|\sqrt{m}}{\sqrt{p_0q_0}}\right)\right]$$

where m = expected number of events over both groups combined = $m_1 + m_2$

We now wish to relate the expected number of events in groups 1 and 2 (m_1, m_2) to the number of subjects available in each group (n_1, n_2). Recall from Equation 14.1 that

EQUATION 14.15

$$CI(t) = 1 - \exp(-ID_1t^*)$$

where t^* = average number of person-years per subject

Applying Equation 14.15 to each group, we have

EQUATION 14.16

Cumulative incidence in group 1 = $m_1/n_1 = 1 - \exp(-ID_1t_1^*)$

Cumulative incidence in group 2 = $m_2/n_2 = 1 - \exp(-ID_2t_2^*)$

or

EQUATION 14.17

$$m_1 = n_1[1 - \exp(-ID_1t_1^*)]$$
$$m_2 = n_2[1 - \exp(-ID_2t_2^*)]$$

Substituting Equation 14.17 into Equation 14.14, we obtain the following power formula:

EQUATION 14.18

Power for the Comparison of Two Incidence Rates Suppose we wish to test the hypothesis H_0: $ID_1 = ID_2$ versus H_1: $ID_1 \neq ID_2$, where ID_1, ID_2 are incidence densities in groups 1 and 2. The power of the test for the specific alternative $ID_1/ID_2 = RR$ with two-sided significance level = α is given by

$$\text{Power} = \Phi\left[\sqrt{\frac{p_0q_0}{p_1q_1}}\left(z_{\alpha/2} + \frac{|p_0 - p_1|\sqrt{m}}{\sqrt{p_0q_0}}\right)\right]$$

where

$p_0 = t_1/(t_1 + t_2)$

$p_1 = t_1RR/(t_1RR + t_2)$

m = expected number of events in the two groups combined = $m_1 + m_2$

$m_1 = n_1\,[1 - \exp(-ID_1t_1^*)]$

$m_2 = n_2\,[1 - \exp(-ID_2t_2^*)]$

n_1, n_2 = number of subjects available in groups 1 and 2, respectively

t_1, t_2 = total number of person-years in groups 1 and 2, respectively

t_1^*, t_2^* = average number of person-years per subject in groups 1 and 2, respectively

ID_1, ID_2 = incidence density in groups 1 and 2, respectively, under H_1

Example 14.15 **Cancer** Answer the question posed in Example 14.14.

Solution We have $n_1 = n_2 = 5000$ subjects, $t_1^* = t_2^* = 5$, $ID_2 = 300/10^5$ person-years, $ID_1 = 1.25 \times 300/10^5 = 375/10^5$ person-years, $RR = 1.25$. Thus, from Equation 14.18,

$$\begin{aligned} m_1 &= 5000[1 - \exp[-(375/10^5)5] \\ &= 5000(1 - .98142) = 92.9 \end{aligned}$$

$$\begin{aligned} m_2 &= 5000[1 - \exp[-(300/10^5)5] \\ &= 5000(1 - .98511) = 74.4 \end{aligned}$$

$$m = 92.9 + 74.4 = 167.3$$

$$t_1 = t_2 = 5000 \times 5 = 25{,}000$$

$$p_0 = 1/2$$

$$\begin{aligned} p_1 &= 25{,}000(1.25)/[25{,}000(1.25) + 25{,}000] \\ &= 31{,}250/56{,}250 = .556 \end{aligned}$$

Thus,

$$\begin{aligned} \text{Power} &= \Phi\left[\sqrt{\frac{.5(.5)}{.556(.444)}}\left(-1.96 + \frac{|.50 - .556|\sqrt{167.3}}{\sqrt{.5(.5)}}\right)\right] \\ &= \Phi\left[1.0062\left(-1.96 + \frac{0.7186}{.50}\right)\right] \\ &= \Phi[1.0062(-0.5228)] \\ &= \Phi(-0.5261) = .299 \end{aligned}$$

Therefore, the study only has about 30% power to test the hypotheses.

14.4.2 Sample-Size Estimation

Clearly, the study proposed in Example 14.14 is too small to have sufficient power to test the hypotheses proposed. The issue is how large a study would be needed to have a prespecified (say, 80%) level of power. For this purpose, if we prespecify a power of $1 - \beta$, then we can solve for the required total number of events m from the one-sample binomial test. Specifically, from Equation 7.38 we have

EQUATION 14.19

$$m = \frac{\left(\sqrt{p_0 q_0}\, z_{1-\alpha/2} + \sqrt{p_1 q_1}\, z_{1-\beta}\right)^2}{|p_0 - p_1|^2}$$

To convert from the required number of events (m) to the required number of subjects (n), we refer to Equation 14.18, and get

EQUATION 14.20

$$m = m_1 + m_2 = n_1[1 - \exp(-ID_1 t_1^*)] + n_2[1 - \exp(-ID_2 t_2^*)]$$

If we prespecify the ratio of sample sizes in the two groups—i.e., $n_2/n_1 = k$; then it follows from Equation 14.20 that

EQUATION 14.21

$$n_1 = \frac{m}{\{1 - \exp(-ID_1 t_1^*) + k[1 - \exp(-ID_2 t_2^*)]\}}$$

$$n_2 = kn_1$$

Combining Equations 14.19–14.21 yields the following sample-size formula:

EQUATION 14.22

Sample-Size Estimation for the Comparison of Two Incidence Rates Suppose we wish to test the hypothesis H_0: $ID_1 = ID_2$ versus H_1: $ID_1 \neq ID_2$, where ID_1 and ID_2 are the incidence densities in groups 1 and 2, respectively. If we conduct a two-sided test with significance level α and power $1 - \beta$, then we require a total expected number of events over both groups of m, where

$$m = \frac{\left(\sqrt{p_0 q_0}\, z_{1-\alpha/2} + \sqrt{p_1 q_1}\, z_{1-\beta}\right)^2}{|p_0 - p_1|^2}$$

where

$$p_0 = t_1/(t_1 + t_2)$$

$$p_1 = t_1 RR/(t_1 RR + t_2)$$

$$RR = ID_1/ID_2$$

t_1, t_2 = total number of person-years in groups 1 and 2, respectively

ID_1, ID_2 = incidence densities in groups 1 and 2, respectively, under H_1

The corresponding number of subjects in each group is

$$n_1 = \frac{m}{(k+1) - \exp(-ID_1 t_1^*) - k\exp(-ID_2 t_2^*)}$$

$$n_2 = kn_1$$

Example 14.16 **Cancer** How many subjects need to be enrolled in the study proposed in Example 14.14 in order to have 80% power, if a two-sided test with significance level of .05 is used and an equal number of subjects are enrolled in each group?

Solution We have $\alpha = .05$, $1 - \beta = .80$. Also, from the solution to Example 14.15, we have $p_0 = .50$, $p_1 = .556$. Thus, from Equation 14.22, the required total number of events is

$$m = \frac{\left[\sqrt{.5(.5)}\,(1.96) + \sqrt{.556(.444)}\,(0.84)\right]^2}{(.50 - .556)^2}$$

$$= \frac{1.397^2}{.056^2}$$

$$= \frac{1.9527}{.00309} = 632.7 \text{ or } 633 \text{ events}$$

Thus, we need a total of 633 events to achieve 80% power. From Example 14.15, we have $t_1^* = t_2^* = 5$ years, $ID_1 = 375/10^5$ person-years, $ID_2 = 300/10^5$ person-years. Also, because the sample size in each group is the same, we have $k = 1$. Therefore, from Equation 14.22, the required number of subjects in each group is

$$n_1 = n_2 = \frac{633}{2 - \exp\left[(-375/10^5)5\right] - \exp\left[(-300/10^5)5\right]}$$

$$= \frac{633}{2 - .98142 - .98511}$$

$$= \frac{633}{.0335} = 18{,}916.2 \text{ or } 18{,}917 \text{ subjects}$$

Thus, we need to enroll 18,917 subjects in each group or a total of 37,834 subjects to have 80% power. This is about four times as large a study as the one originally contemplated in Example 14.14. This study would be expected to yield

$$m_1 = 18{,}917\{1 - \exp[(-375/10^5)5]\}$$
$$= 18{,}917(1 - .98142) = 351 \text{ events in the ERT group}$$

$$m_2 = 18{,}917\{1 - \exp[(-300/10^5)5]\}$$
$$= 18{,}917(1 - .98511) = 282 \text{ events in the control group}$$

for a total of 633 events. This is one study design used in the Women's Health Initiative, a large multicenter set of clinical trials, which has a similar sample size, number of events, and time frame as posed in Examples 14.14–14.16.

In this section, we have considered power and sample-size formulas for comparing two incidence rates. The formulas are special cases of similar formulas used for the one-sample binomial test in Equations 7.37 and 7.38, respectively. If the number of person-years of follow-up is the same for each subject, then these formulas should be approximately the same as the corresponding power and sample-size formulas for comparing two proportions, which are given in Equations 10.15 and 10.14, respectively. However, the advantage of the methods in this section is that they allow for a variable length of follow-up for individual subjects, which is more realistic in many clinical trial situations. In the next section, we consider inference procedures for comparing incidence rates between two groups, while controlling for confounding variables.

SECTION 14.5 Inference for Stratified Person-Time Data

14.5.1 Hypothesis Testing

It is very common in the analysis of person-time data to control for confounding variables before assessing the relationship between the main exposure of interest and disease. Confounding variables may include age and sex as well as other covariates that are related to exposure, disease, or both.

Example 14.17 **Cancer** An issue of continuing interest is the effect of postmenopausal hormone use on cardiovascular and cancer outcomes in postmenopausal women. Data were collected from postmenopausal women in the Nurses' Health Study to address this issue. Women were mailed an initial questionnaire in 1976 and follow-up questionnaires every 2 years thereafter. Data from 1976 to 1986, encompassing 352,871 person-years of follow-up and 707 incident cases of breast cancer, are given in Table 14.4 [1].

TABLE 14.4 **Current and past use of postmenopausal hormones and risk of breast cancer among postmenopausal participants in the Nurses' Health Study**

	Never users		Current users			Past users		
Age	No. of cases	Person-years	No. of cases	Person-years	*RR*	No. of cases	Person-years	*RR*
39–44	5	4,722	12	10,199	1.11	4	3,835	0.99
45–49	26	20,812	22	14,044	1.25	12	8,921	1.08
50–54	129	71,746	51	24,948	1.14	46	26,256	0.97
55–59	159	73,413	72	21,576	1.54	82	39,785	0.95
60–64	35	15,773	23	4,876	2.13	29	11,965	1.09

There were 23,607 women who were postmenopausal and did not have any type of cancer (except for nonmelanoma skin cancer) in 1976. Other women became postmenopausal during the follow-up period. Follow-up was terminated at the diagnosis of breast cancer, death, or the date of the last questionnaire return. Thus, each women had a variable duration of follow-up. Since breast-cancer incidence and possibly postmenopausal hormone use are related to age, it was important to control for age in the analysis.

We can use methods similar to the Mantel-Haenszel test used for cumulative incidence data (or generally for count data) as presented in Chapter 13.

Suppose we have k strata, where the number of events and the amount of person-time in the ith stratum are as shown in Table 14.5.

TABLE 14.5 General observed table for number of events and person-time in the ith stratum, $i = 1, \ldots, k$

Exposure group	Number of events	Person-time
Exposed	a_{1i}	t_{1i}
Unexposed	a_{2i}	t_{2i}
Total	$a_{1i} + a_{2i}$	$t_{1i} + t_{2i}$

Let us denote the incidence rate of disease among the exposed by p_{1i} and among the unexposed by p_{2i}. Therefore, the expected number of events among the exposed = $p_{1i}t_{1i}$ and among the unexposed = $p_{2i}t_{2i}$. Let p_i = the expected proportion of the total number of events over both groups that are among the exposed. We can relate p_i to p_{1i} and p_{2i} by

EQUATION 14.23

$$p_i = \frac{p_{1i}t_{1i}}{p_{1i}t_{1i} + p_{2i}t_{2i}}, \text{ which we denote by } p_i^{(0)}.$$

We will assume that the rate ratio relating disease to exposure is the same for each stratum and denote it by RR. Therefore, $RR = p_{1i}/p_{2i}$ and is the same for each $i = 1, \ldots, k$. If we divide numerator and denominator of Equation 14.23 by p_{2i}, and substitute RR for p_{1i}/p_{2i}, we obtain

EQUATION 14.24

$$\begin{aligned} p_i &= \frac{(p_{1i}/p_{2i})t_{1i}}{(p_{1i}/p_{2i})t_{1i} + t_{2i}} \\ &= \frac{RRt_{1i}}{RRt_{1i} + t_{2i}}, \text{ which we denote by } p_i^{(1)}. \end{aligned}$$

We wish to test the hypothesis H_0: $RR = 1$ versus H_1: $RR \neq 1$ or, equivalently, H_0: $p_i = p_i^{(0)}$ versus H_1: $p_i = p_i^{(1)}$, $i = 1, \ldots, k$. We will base our test on $A = \sum_{i=1}^{k} a_{1i}$ = total observed number of events for the exposed. Under H_0, we will assume that the total observed number of events for the ith stratum $(a_{1i} + a_{2i})$ is fixed. Therefore, under H_0,

EQUATION 14.25

$$\begin{aligned} E(a_{1i}) &= (a_{1i} + a_{2i})p_i^{(0)} = (a_{1i} + a_{2i})t_{1i}/(t_{1i} + t_{2i}) \\ Var(a_{1i}) &= (a_{1i} + a_{2i})p_i^{(0)}(1 - p_i^{(0)}) = (a_{1i} + a_{2i})t_{1i}t_{2i}/(t_{1i} + t_{2i})^2 \end{aligned}$$

and $E(A) = \sum_{i=1}^{k} E(a_{1i}), Var(A) = \sum_{i=1}^{k} Var(a_{1i})$. Under H_1, A will be larger than $E(A)$ if $RR > 1$ and will be smaller than $E(A)$ if $RR < 1$. We will use the test statistic

$X^2 = \left[|A - E(A)| - 0.5\right]^2 / Var(A)$, which follows a χ_1^2 distribution under H_0 and reject H_0 for large values of X^2. The test procedure is summarized as follows:

EQUATION 14.26

Hypothesis Testing for Stratified Person-Time Data Let p_{1i}, p_{2i} = incidence rate of disease for the exposed and unexposed groups in the ith stratum, respectively. Let a_{1i}, t_{1i} = the number of events and person-years for the exposed group in the ith stratum, a_{2i}, t_{2i} = the number of events and person-years for the unexposed group in the ith stratum, $i = 1, \ldots, k$.

We will assume that $RR = p_{1i}/p_{2i}$ is constant across all strata. To test the hypothesis H_0: $RR = 1$ versus H_1: $RR \neq 1$ using a two-sided test with significance level = α:

(1) We compute the total observed number of events among the exposed over all strata = $A = \sum_{i=1}^{k} a_{1i}$.

(2) We compute the total expected number of events under H_0 among the exposed over all strata = $E(A) = \sum_{i=1}^{k} E(a_{1i})$, where

$$E(a_{1i}) = (a_{1i} + a_{2i})t_{1i}/(t_{1i} + t_{2i}),\ i = 1, \ldots, k$$

(3) We compute $Var(A) = \sum_{i=1}^{k} Var(a_{1i})$ under H_0, where

$$Var(a_{1i}) = (a_{1i} + a_{2i})t_{1i}t_{2i}/(t_{1i} + t_{2i})^2,\ i = 1, \ldots, k$$

(4) We compute the test statistic

$$X^2 = \frac{\left(|A - E(A)| - 0.5\right)^2}{Var(A)}$$

which follows a chi-square distribution with 1 df under H_0.

(5) If $X^2 > \chi_{1,1-\alpha}^2$, then we reject H_0.

If $X^2 \le \chi_{1,1-\alpha}^2$, then we accept H_0.

(6) The p-value = $Pr(\chi_1^2 > X^2)$.

(7) This test should only be used if $Var(A) \ge 5$.

Example 14.18 **Cancer** Test the hypothesis that there is a significant association between breast-cancer incidence and current use of postmenopausal hormones based on the data in Table 14.4.

Solution We will compare current users of postmenopausal hormones (the exposed group) with never users of postmenopausal hormones (the unexposed group) using the method in Equation 14.26. For 39- to 44-year-old women, we have

$$a_{11} = 12$$

$$E(a_{11}) = \frac{(12+5)10{,}199}{10{,}199+4722} = 17 \times .684 = 11.62$$

$$Var(a_{11}) = 17 \times .684 \times .316 = 3.677$$

Similar computations are performed for each of the other four age groups, whereby

$$A = 12 + 22 + 51 + 72 + 23 = 180$$

$$E(A) = 11.62 + 19.34 + 46.44 + 52.47 + 13.70 = 143.57$$

$$Var(A) = 3.677 + 11.548 + 34.459 + 40.552 + 10.462 = 100.698$$

$$X^2 = \frac{\left(|180 - 143.57| - 0.5\right)^2}{100.698} = \frac{35.93^2}{100.698} = 12.82 \sim \chi_1^2 \text{ under } H_0$$

Since $X^2 > 10.83 = \chi^2_{1,\ .999}$, it follows that $p < .001$. Therefore, there is a highly significant association between breast-cancer incidence and postmenopausal hormone use.

14.5.2 Estimation of the Rate Ratio

We will use a similar approach to that considered in the estimation of a single rate ratio in Section 14.3.4. We obtain estimates of the ln(rate ratio) in each stratum and then compute a weighted average of the stratum-specific estimates to obtain an overall estimate of the ln(rate ratio). Specifically, let

EQUATION 14.27

$$\widehat{RR}_i = (a_{1i}/t_{1i})/(a_{2i}/t_{2i})$$

be the estimate of the rate ratio in the ith stratum. From Equation 14.10, we see that

EQUATION 14.28

$$Var[\ln(\widehat{RR}_i)] \doteq \frac{1}{a_{1i}} + \frac{1}{a_{2i}}$$

To obtain an overall estimate of $\ln(\widehat{RR})$ we now compute a weighted average of $\ln(\widehat{RR}_i)$ where the weights are the inverse of the variance of $\ln(\widehat{RR}_i)$ and then take the antilog of the weighted average:

EQUATION 14.29

$$\ln(\widehat{RR}) = \frac{\sum_{i=1}^{k} w_i \ln(\widehat{RR}_i)}{\sum_{i=1}^{k} w_i}$$

where $w_i = 1/Var[\ln(\widehat{RR}_i)]$. We then take the antilog of $\ln(\widehat{RR})$ to obtain an estimate of RR.

To obtain confidence limits for the rate ratio, we use Equation 14.29 to obtain the variance of $\ln(\widehat{RR})$ as follows:

EQUATION 14.30

$$Var[\ln(\widehat{RR})] = \frac{1}{\left(\sum_{i=1}^{k} w_i\right)^2} Var\left[\sum_{i=1}^{k} w_i \ln(\widehat{RR}_i)\right]$$

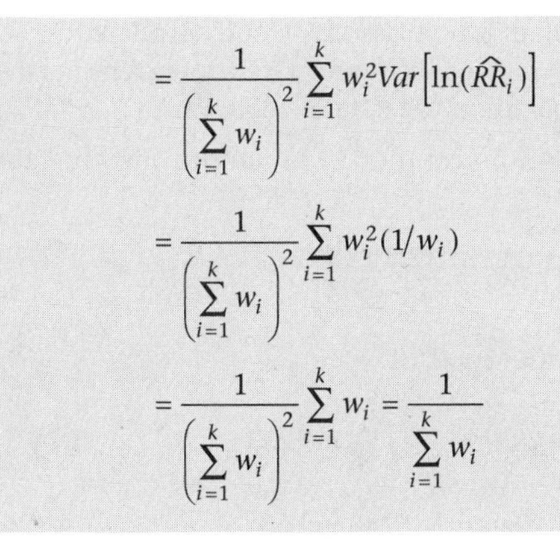

$$= \frac{1}{\left(\sum_{i=1}^{k} w_i\right)^2} \sum_{i=1}^{k} w_i^2 Var\left[\ln(\widehat{RR}_i)\right]$$

$$= \frac{1}{\left(\sum_{i=1}^{k} w_i\right)^2} \sum_{i=1}^{k} w_i^2 (1/w_i)$$

$$= \frac{1}{\left(\sum_{i=1}^{k} w_i\right)^2} \sum_{i=1}^{k} w_i = \frac{1}{\sum_{i=1}^{k} w_i}$$

Thus, a two-sided $100\% \times (1 - \alpha)$ CI for $\ln(RR)$ is given by $\ln(\widehat{RR}) \pm z_{1-\alpha/2} \times \sqrt{1/\sum_{i=1}^{k} w_i}$. We then take the antilog of each of the confidence limits for $\ln(RR)$ to obtain confidence limits for RR. This procedure is summarized as follows:

EQUATION 14.31

Point and Interval Estimation of the Rate Ratio (Stratified Data)

Let a_{1i}, t_{1i} = number of events and person-years for the exposed in the ith stratum

a_{2i}, t_{2i} = number of events and person-years for the unexposed in the ith stratum

A point estimate of the rate ratio (RR) is given by $\widehat{RR} = e^c$, where

$$c = \frac{\sum_{i=1}^{k} w_i \ln(\widehat{RR}_i)}{\sum_{i=1}^{k} w_i}$$

$$\widehat{RR}_i = \frac{a_{1i}/t_{1i}}{a_{2i}/t_{2i}}$$

$$w_i = \left(\frac{1}{a_{1i}} + \frac{1}{a_{2i}}\right)^{-1}$$

A two-sided $100\% \times (1 - \alpha)$ CI for RR is given by (e^{c_1}, e^{c_2}), where

$$c_1 = \ln(\widehat{RR}) - z_{1-\alpha/2}\sqrt{1/\sum_{i=1}^{k} w_i}$$

$$c_2 = \ln(\widehat{RR}) + z_{1-\alpha/2}\sqrt{1/\sum_{i=1}^{k} w_i}$$

This interval should only be used if $Var(A)$ as given in Equation 14.26 is ≥ 5.

Example 14.19 **Cancer** Obtain a point estimate and associated 95% confidence limits for the rate ratio for breast-cancer incidence rate for current versus never users of postmenopausal hormones, based on the data in Table 14.4.

Solution A summary of the computations are provided in Table 14.6. For example, for the age group 39–44,

$$\widehat{RR}_1 = \frac{12/10{,}199}{5/4722} = 1.11$$

$$\ln(\widehat{RR}_1) = 0.105$$

$$Var[\ln(\widehat{RR}_1)] = \frac{1}{12} + \frac{1}{5} = 0.283$$

$$w_1 = 1/0.283 = 3.53$$

Similar computations are performed for each of the other age strata. Thus, the overall estimate of the rate ratio = 1.41 with 95% confidence limits = (1.17, 1.69). This indicates that the incidence of breast cancer is estimated to be about 40% higher for current users of postmenopausal hormones than for never users even after controlling for age. Note that the crude $RR = (180/75{,}643)/(354/186{,}466) = 1.25 < 1.41$, which implies that age is a **negative confounder.** Age is a negative confounder because the percentage of postmenopausal women using postmenopausal hormones decreases with increasing age, whereas the incidence rate of breast cancer increases with increasing age.

TABLE 14.6 **Breast cancer versus estrogen-replacement therapy, estimation of the rate ratio after stratification by age**

Age group	$\widehat{RR}_i$	$\ln(\widehat{RR}_i)$	$Var[\ln(\widehat{RR}_i)]$	w_i	$w_i\ln(\widehat{RR}_i)$
39–44	1.11	0.105	0.283	3.53	0.372
45–49	1.25	0.226	0.084	11.92	2.697
50–54	1.14	0.128	0.027	36.55	4.691
55–59	1.54	0.432	0.020	49.56	21.423
60–64	2.13	0.754	0.072	13.88	10.467
Total				115.43	39.650

Note:

$\ln(\widehat{RR}) = \frac{39.650}{115.43} = 0.343$, $\widehat{RR} = \exp(0.343) = 1.41$

$Var[\ln(\widehat{RR})] = 1/115.43 = 0.0087$, $se[\ln(\widehat{RR})] = \sqrt{0.0087} = 0.0931$

95% CI for $\ln(RR) = 0.343 \pm 1.96(0.0931) = 0.343 \pm 0.182 = (0.161, 0.526)$

95% CI for $RR = [\exp(0.161), \exp(0.526)] = (1.17, 1.69)$

14.5.3 Testing the Assumption of Homogeneity of the Rate Ratio Across Strata

An important assumption made in the estimation methods in Section 14.5.2 is that the underlying rate ratio is the same in all strata. If this assumption is not true, then it makes little sense to estimate a common rate ratio. If the rate ratios in different strata are all in the same direction relative to the null hypothesis (i.e., all rate ratios > 1 or all rate ratios < 1), then the hypothesis-testing procedures in Section 14.5.1 will still be valid with only a slight loss of power. However, if the rate ratios are in different directions in different strata, or are null in some strata, then the power of the hypothesis-testing procedures will be greatly diminished.

To test this assumption, we will use similar methods to those used for testing the assumption of homogeneity of the odds ratio in different strata for count data given in Chapter 13. Specifically, we wish to test the hypothesis H_0: $RR_1 = \cdots = RR_k$ versus H_1: at least two of the RR_i are different. We will base our hypothesis test on the test statistic $X^2_{het} = \sum_{i=1}^{k} w_i\left[\ln(\widehat{RR}_i) - \ln(\widehat{RR})\right]^2 \sim \chi^2_{k-1}$ under H_0 and will reject H_0 for large values of X^2_{het}. The test procedure is summarized as follows:

EQUATION 14.32

Chi-Square Test for Homogeneity of Rate Ratios Across Strata Suppose we have incidence-rate data and wish to control for the confounding effect of another variable(s) that comprise k strata. To test the hypothesis H_0: $RR_1 = \cdots = RR_k$ versus H_1: at least two of the RR_i's are different, with significance level α, we use the following procedure:

(1) We compute the test statistic

$$X^2_{het} = \sum_{i=1}^{k} w_i\left[\ln(\widehat{RR}_i) - \ln(\widehat{RR})\right]^2 \sim \chi^2_{k-1} \text{ under } H_0$$

where $\widehat{RR}_i$ = estimated rate ratio in the ith stratum

$\widehat{RR}$ = estimate of the overall rate ratio as given in Equation 14.31

$w_i = 1/Var[\ln(\widehat{RR}_i)]$ as defined in Equation 14.31

(2) If $X^2_{het} > \chi^2_{k-1,1-\alpha}$, we reject H_0.

If $X^2_{het} \le \chi^2_{k-1,1-\alpha}$, we accept H_0.

(3) The p-value is given by $p = Pr\left(\chi^2_{k-1} > X^2_{het}\right)$.

(4) An alternative computational form for the test statistic in step 1 is

$$\sum_{i=1}^{k} w_i[\ln(\widehat{RR}_i)]^2 - \left(\sum_{i=1}^{k} w_i\right)[\ln(\widehat{RR})]^2$$

Example 14.20 **Cancer** Test for the assumption of the homogeneity of the rate ratio over the five age strata based on the data in Table 14.4.

Solution We have that

$$X^2_{het} = \sum_i w_i\left[\ln(\widehat{RR}_i) - \ln(\widehat{RR})\right]^2 = \sum_i w_i\left[\ln(\widehat{RR}_i)\right]^2 - \left(\sum_i w_i\right)\left[\ln(\widehat{RR})\right]^2 \sim \chi^2_{k-1}$$

$$= 18.405 - 115.43(0.343)^2 = 18.405 - 13.619 = 4.786 \sim \chi^2_4 \text{ under } H_0$$

$$p\text{-value} = Pr\left(\chi^2_4 > 4.786\right) = .31$$

Thus, there is no significant heterogeneity. However, it appears from Table 14.6 that the rate ratios are increasing with age. Thus, the test procedure in Equation 14.32 may not be sensitive to variation in the rate ratios in a specific direction with respect to the confounding variable(s). One possible explanation is that the average duration of use generally increases with increasing age. Thus the apparent increase in risk with increasing age may actually represent an increase in risk with increasing duration of use. To properly account for the effects of both age and duration of use, it is more appropriate to use Cox regression analyses, which are discussed later in this chapter.

In this section, we have considered a method for comparing incidence rates between two groups while controlling for a single categorical exposure variable. This method can also be used if there is more than one covariate to be controlled for, but it would be tedious to do so with many covariates. Instead, Poisson regression analysis can be used to accomplish this. This is a generalization of logistic regression for incidence-rate data, but is beyond the scope of this text.

Referring to the flowchart (Figure 14.9), we would answer yes to (1) person-time data? no to (2) one-sample problem? and yes to (3) incidence rates remain constant over time? and (4) two-sample problem? which leads us to the box entitled "Use two-sample test for comparison of incidence rates, if no confounding is present, or methods for stratified person-time data, if confounding is present." In this section, we have considered techniques for performing two-sample inference for incidence rates when confounding is present.

SECTION 14.6 Power and Sample-Size Estimation for Stratified Person-Time Data

14.6.1 Sample-Size Estimation

In Section 14.4, we studied how to obtain power and sample-size estimates for comparing two incidence rates. In this section, we extend this discussion to allow for power and sample-size estimation for comparison of incidence rates while controlling for confounding variables.

Example 14.21 **Cancer** Suppose we wish to study whether the positive association between breast-cancer incidence and postmenopausal hormone use is also present in another study population. We will assume that the age-specific incidence rates for unexposed women (i.e., never users of postmenopausal hormones) and the age distribution and percentage of women using postmenopausal hormones within specific age groups (as reflected by the percentage of total person-years realized by specific age–exposure groups) are the same as the Nurses' Health Study (Table 14.4). We will also assume that the true rate ratio within each age group = 1.5. How many subjects do we need to enroll if the average subject is followed for 5 years and we wish to achieve 80% power using a two-sided test with $\alpha = .05$?

The sample-size estimate will depend on

(1) the age-specific incidence rates of disease in the unexposed group

(2) the distribution of total person-years within specific age–exposure groups

(3) the true rate ratio under the alternative hypothesis

(4) the type I and type II errors

The sample-size estimate is given as follows:

EQUATION 14.33

Sample-Size Estimation for Incidence-Rate Data Suppose we have s strata. Let p_i = the probability that a case in the ith stratum comes from the exposed group. We wish to test the hypothesis

$$H_0: p_i = t_{1i}/(t_{1i} + t_{2i}) = p_i^{(0)} \text{ versus } H_1: p_i = RR\,t_{1i}/(RR\,t_{1i} + t_{2i}) = p_i^{(1)},\ i = 1, \ldots, s$$

where

t_{1i} = number of person-years of follow-up among exposed subjects in the ith stratum

t_{2i} = number of person-years of follow-up among unexposed subjects in the ith stratum

The equivalent hypotheses are H_0: $RR = 1$ versus H_1: $RR \neq 1$, where $RR = ID_{1i}/ID_{2i}$ = ratio of incidence densities of exposed compared to unexposed subjects in the ith stratum. RR is assumed to be the same for all strata. To test these hypotheses using a two-sided test with significance level α and power of $1 - \beta$ versus a true rate ratio of RR under H_1 requires a total expected number of cases over both groups = m, where

$$m = \frac{\left(z_{1-\alpha/2}\sqrt{C} + z_{1-\beta}\sqrt{D}\right)^2}{(A-B)^2}$$

where

$$A = \sum_{i=1}^{s} \lambda_i p_i^{(0)} = \sum_{i=1}^{s} A_i$$

$$B = \sum_{i=1}^{s} \lambda_i p_i^{(1)} = \sum_{i=1}^{s} B_i$$

$$C = \sum_{i=1}^{s} \lambda_i p_i^{(0)}\left[1 - p_i^{(0)}\right] = \sum_{i=1}^{s} C_i$$

$$D = \sum_{i=1}^{s} \lambda_i p_i^{(1)}\left[1 - p_i^{(1)}\right] = \sum_{i=1}^{s} D_i$$

$\lambda_i = G_i/G$ = proportion of cases in the ith stratum

where

$$G_i = \frac{\theta_i\left(k_i p_{2i} + p_{1i}\right)}{k_i + 1},\ i = 1, \ldots, s$$

p_{2i} = rate of disease among the unexposed subjects in stratum i
$= 1 - \exp(-ID_{2i}T),\ i = 1, \ldots, s$

p_{1i} = rate of disease among exposed subjects in stratum i
$= 1 - \exp[-RR(ID_{2i}T)],\ i = 1, \ldots, s$

T = average length of follow-up per subject

ID_{2i} = incidence density among unexposed subjects in the ith stratum, $i = 1, \ldots, s$

$k_i = t_{2i}/t_{1i}$, $i = 1, \ldots, s$

$\theta_i = n_i/n$ = overall proportion of subjects in the ith stratum, $i = 1, \ldots, s$

$$p_i^{(0)} = \frac{t_{1i}}{t_{1i} + t_{2i}}$$

= proportion of cases in the ith stratum that are exposed under H_0, $i = 1, \ldots, s$

$$p_i^{(1)} = \frac{t_{1i}RR}{t_{1i}RR + t_{2i}}$$

= proportion of cases in the ith stratum that are exposed under H_1, $i = 1, \ldots, s$

The required total sample size (n) is

$$n = \frac{m}{\sum_{i=1}^{s} \theta_i (k_i p_{2i} + p_{1i})/(k_i + 1)} = \frac{m}{\sum_{i=1}^{s} G_i}$$

Example 14.22 **Cancer** Compute the required sample size for the study proposed in Example 14.21 using a two-sided test with $\alpha = .05$, power = 80%, $RR = 1.5$, 5 years of follow-up for each subject, and incidence rate of disease among the unexposed is the same as in Table 14.4 for never users.

Solution We need to compute ID_{2i}, p_{2i}, p_{1i}, k_i, θ_i, $p_i^{(0)}$, and $p_i^{(1)}$ for each of the five age strata in Table 14.4. For example, in the first age group (age group = 39–44 years old), we have

$$ID_{2,1} = 105.9/10^5 \text{ person-years}$$

$$p_{2,1} = 1 - \exp[-5(105.9/10^5)] = .00528$$

$$p_{1,1} = 1 - \exp[-5(1.5)(105.9/10^5)] = .00791$$

$$k_1 = 4722/10{,}199 = 0.463$$

$$p_1^{(0)} = 10{,}199/(10{,}199 + 4722) = 10{,}199/14{,}921 = .684$$

$$p_1^{(1)} = 10{,}199(1.5)/[10{,}199(1.5) + 4722] = 15{,}299/20{,}021 = .764$$

The number of subjects in each age group is not given in Table 14.4. However, if we assume that the average length of follow-up is the same for each age group, then

$$\theta_i \cong t_i \Big/ \sum_{i=1}^{s} t_i$$

Thus,

$$\theta_1 = 14{,}921/262{,}109 = .0569$$

A summary of the computations for each age group is given in Table 14.7.

TABLE 14.7 Computations Needed for Sample-Size Estimate in Example 14.22

i	Age group	ID_{2i}[a]	p_{2i}	p_{1i}	t_{1i}	t_{2i}	k_i	θ_i	$p_i^{(0)}$	$p_i^{(1)}$	G_i	λ_i
1	39–44	105.9	.00528	.00791	10,199	4,722	0.463	0.0569	.684	.764	4.03×10^{-4}	.039
2	45–49	124.9	.00623	.00933	14,044	20,812	1.482	0.1330	.403	.503	9.94×10^{-4}	.095
3	50–54	179.8	.00895	.01339	24,948	71,746	2.876	0.3689	.258	.343	3.72×10^{-3}	.357
4	55–59	216.6	.01077	.01611	21,576	73,413	3.403	0.3624	.227	.306	4.34×10^{-3}	.416
5	60–64	221.9	.01103	.01650	4,876	15,773	3.235	0.0788	.236	.317	9.71×10^{-4}	.093
Total											1.04×10^{-2}	

i	Age group	A_i	B_i	C_i	D_i
1	39–44	.0264	.0295	.0084	.0070
2	45–49	.0384	.0479	.0229	.0238
3	50–54	.0921	.1223	.0683	.0804
4	55–59	.0945	.1273	.0731	.0884
5	60–64	.0220	.0295	.0168	.0201
Total		.2734	.3565	.1895	.2197

[a]Per 10^5 person-years.

Therefore, the required expected total number of events is

$$m = \frac{\left(z_{.975}\sqrt{.1895} + z_{.80}\sqrt{.2197}\right)^2}{(.2734 - .3565)^2}$$

$$= \frac{[1.96(.4353) + 0.84(.4687)]^2}{.0831^2}$$

$$= \frac{1.2469^2}{.0831^2} = 225.2 \text{ or } 226 \text{ events}$$

The corresponding total number of subjects is

$$n = \frac{226}{1.04 \times 10^{-2}} = 21{,}656.2 \text{ or } 21{,}657 \text{ subjects}$$

This constitutes approximately 108,285 person-years among current and never users combined. From Table 14.6, we see that 90,762 person-years (25.7%) are realized by past postmenopausal hormone (PMH) users out of a total of 352,871 person-years. Thus, we need to accrue 108,285/(1 – .257) = 145,781 person-years or enroll 29,156 postmenopausal women and follow them for 5 years to achieve 80% power in the comparison of current users to never users, if the underlying $RR = 1.5$ between these groups.

14.6.2 Estimation of Power

In some instances, the size of the study population and the average duration of follow-up are fixed by design and we wish to assess the power that can be obtained for a given rate ratio. In this instance, we can solve for the power as a function of the total number of person-years $\left(T = \sum_{j=1}^{2}\sum_{i=1}^{s} t_{ji}\right)$, stratum-specific incidence rates among the unexposed (ID_{2i}), distribution of person-years by stratum and exposure status (t_{1i}, t_{2i}), projected rate ratio (RR), and type I error (α). The power formula is given as follows:

EQUATION 14.34

Estimation of Power for Stratified Incidence-Rate Data Suppose we wish to compare the incidence rate of disease between exposed and unexposed subjects and wish to control for one (or more) covariates that can, as a group, be represented by a single categorical variable with k categories. Using the same notation as in Equation 14.33, if we wish to test the hypothesis H_0: RR (rate ratio) = 1 versus H_1: $RR \neq 1$, using a two-sided test with significance level α, then the power versus the specific rate ratio = RR under the alternative hypothesis is given by

$$\text{Power} = \Phi\left[\frac{\sqrt{m}\,|B-A| - z_{1-\alpha/2}\sqrt{C}}{\sqrt{D}}\right]$$

where m = the total expected number of events given by

$$m = \left(\sum_{i=1}^{s} G_i\right)n$$

n = total number of exposed and unexposed individuals over all strata, and A, B, C, D, and G_i are defined in Equation 14.33.

Example 14.23 **Cancer** Suppose we enroll 25,000 postmenopausal women and expect that 75% of the person-time is attributable to current or never PMH use with an average follow-up of 5 years per woman. If the same assumptions are made as in Example 14.22, then how much power will the study have if the true rate ratio = 1.5?

Solution Because the exposure-stratum-specific incidence rates (p_{1i}, p_{2i}) and person-year distribution are the same as in Example 14.22, we can use the same values for p_{1i}, p_{2i}, $p_i^{(0)}$, $p_i^{(1)}$, and λ_i. Thus, from Table 14.7, A = .2734, B = .3565, C = .1895, and D = .2197. To compute m, we note from Table 14.7 that $\sum_{i=1}^{5} G_i = 1.04 \times 10^{-2}$. Also, the number of women who are current or never users = 25,000(.75) = 18,750. Thus,

$$m = 1.04 \times 10^{-2}\,(18{,}750) = 195.7 \text{ or } 196 \text{ events}$$

To compute power, we refer to Equation 14.34 and obtain

$$\begin{aligned}\text{Power} &= \Phi\left[\frac{\sqrt{196}\,(.3565 - .2734) - z_{.975}\sqrt{.1895}}{\sqrt{.2197}}\right]\\ &= \Phi\left[\frac{1.1634 - 1.96(.4353)}{.4687}\right]\\ &= \Phi\left(\frac{0.3102}{0.4687}\right)\\ &= \Phi(0.662) = .746\end{aligned}$$

Thus, the study would have about 75% power. Note that from Example 14.22, to achieve 80% power, we needed to accrue 145,781 person-years among all postmenopausal women or 108,285 person-years among current or never PMH users to yield an expected 226 events. If we actually accrue 125,000 person-years among all postmenopausal women, as in this example, then this will result in 93,750 person-years among current or never PMH users, which yields an expected 196 events, and as a result obtains about 75% power.

Another approach to power estimation would be to ignore the effect of age and base the power computation on the comparison of overall incidence rates between current and never PMH users. However, from Example 14.19 we see that the age-adjusted $RR = 1.41$, whereas the crude $RR = 1.25$, between breast-cancer incidence and current PMH use. In this example, age is a negative confounder because it is positively related to breast-cancer incidence and is negatively related to PMH use. Thus, the power based on crude rates would be lower than the appropriate power based on rates stratified by age. In the case of a positive confounder, the power based on crude rates would be higher than the power based on rates stratified by age. In general, if confounding is present, then it is important to base power calculations on Equation 14.34, which takes confounding into account, rather than Equation 14.18, which does not.

SECTION 14.7 Testing for Trend: Incidence-Rate Data

In Sections 14.1–14.6, we were concerned with the comparison of incidence rates between an exposed and an unexposed group, possibly after controlling for other relevant covariates. In some instances, there are more than two exposure categories and it is desired to assess whether incidence rates are increasing or decreasing in a consistent manner as the level of exposure increases.

Example 14.24 **Cancer** The data in Table 14.8 display the relationship between breast-cancer incidence and parity (the number of children) by age, based on Nurses' Health Study data from 1976 to 1990. We see that within a given parity group, breast-cancer incidence rises sharply with age. Thus, it is important to control for age in the analysis. Also, within a given age group, breast-cancer incidence is somewhat higher for women with 1 birth than for nulliparous women (women with 0 births). However, for parous women (women with at least 1 child), breast-cancer incidence seems to decline with increasing parity. How should we assess the relationship between breast-cancer incidence and parity?

It is reasonable to study parous women as a group and to model ln(breast-cancer incidence) for the ith age group and the jth parity group as a linear function of parity:

EQUATION 14.35

$$\ln(p_{ij}) = \alpha_i + \beta(j-1)$$

where α_i represents ln(incidence) for women in the ith age group with 1 child and β represents the increase in ln(incidence) for each additional child. Notice that β is assumed to be the same for each age group (i). In general, if we have k exposure groups we might assign a score S_j for the jth exposure group, which might represent average exposure within that group, and consider a model of the form

TABLE 14.8 **Relationship of breast-cancer incidence to parity after controlling for age, Nurses' Health Study, 1976–1990**

	Parity			
	0	1	2	3+
Age	*Cases/person-years (incidence rate)*[a]	*Cases/person-years (incidence rate)*	*Cases/person-years (incidence rate)*	*Cases/person-years (incidence rate)*
30–39	13 / 15,265 (85)	18 / 20,098 (90)	72 / 87,436 (82)	60 / 86,452 (69)
40–49	44 / 30,922 (142)	73 / 31,953 (228)	245 / 140,285 (175)	416 / 262,068 (159)
50–59	102 / 35,206 (290)	94 / 31,636 (297)	271 / 103,399 (262)	608 / 262,162 (232)
60–69	32 / 11,594 (276)	50 / 10,264 (487)	86 / 29,502 (292)	176 / 64,448 (273)

[a]Per 100,000 person-years.

EQUATION 14.36

$$\ln(p_{ij}) = \alpha_i + \beta S_j$$

We would like to test the hypothesis H_0: $\beta = 0$ versus H_1: $\beta \neq 0$. The models in Equations 14.35 and 14.36 represent a more efficient use of the data than comparing individual pairs of groups, because we can use all the data to test for an overall trend. Comparing pairs of groups might yield contradictory results and would often have less power than the overall test for trend. We will use a "weighted regression approach" where incidence rates based on a larger number of cases are given more weight. The procedure is summarized as follows:

EQUATION 14.37

Test for Trend: Incidence-Rate Data Suppose we have an exposure variable E with k levels of exposure, where the jth exposure group is characterized by a score S_j, which may represent the average level of exposure within that group, if available. If no obvious scoring method is available, then integer scores $1, \ldots, k$ may be used instead. If p_{ij} = true incidence rate for the ith stratum and jth level of exposure, $i = 1, \ldots, s$; $j = 1, \ldots, k$; $\hat{p}_{ij}$ = the corresponding observed incidence rate and we assume that

$$\ln(p_{ij}) = \alpha_i + \beta S_j$$

then, to test the hypothesis H_0: $\beta = 0$ versus H_1: $\beta \neq 0$, using a two-sided test with significance level α:

(1) We compute a point estimate of β given by $\hat{\beta} = L_{xy}/L_{xx}$, where

$$L_{xy} = \sum_{i=1}^{s}\sum_{j=1}^{k} w_{ij} S_j \ln(\hat{p}_{ij}) - \left(\sum_{i=1}^{s}\sum_{j=1}^{k} w_{ij} S_j\right)\left[\sum_{i=1}^{s}\sum_{j=1}^{k} w_{ij} \ln(\hat{p}_{ij})\right] \Big/ \sum_{i=1}^{s}\sum_{j=1}^{k} w_{ij}$$

$$L_{xx} = \sum_{i=1}^{s}\sum_{j=1}^{k} w_{ij}S_j^2 - \left(\sum_{i=1}^{s}\sum_{j=1}^{k} w_{ij}S_j\right)^2 \Big/ \sum_{i=1}^{s}\sum_{j=1}^{k} w_{ij}$$

$w_{ij} = a_{ij}$ = number of cases in the ith stratum and jth level of exposure

(2) The standard error of $\hat{\beta}$ is given by

$$se(\hat{\beta}) = 1/\sqrt{L_{xx}}$$

(3) We compute the test statistic

$$z = \hat{\beta}/se(\hat{\beta}) \sim N(0,\ 1) \text{ under } H_0$$

(4) If $z > z_{1-\alpha/2}$ or $z < -z_{1-\alpha/2}$, then we reject H_0; if $-z_{1-\alpha/2} \le z \le z_{1-\alpha/2}$, then we accept H_0.

(5) The two-sided p-value $= 2\Phi(z)$ if $z < 0$

$= 2[1 - \Phi(z)]$ if $z \ge 0$

(6) A two-sided $100\% \times (1 - \alpha)$ CI for β is given by

$$\hat{\beta} \pm z_{1-\alpha/2}se(\hat{\beta})$$

Example 14.25 **Cancer** Assess if there is a significant trend between breast-cancer incidence and parity for parous women based on the data in Table 14.8.

Solution We have four age strata (30–39, 40–49, 50–59, 60–69) (s = 4) and three exposure (parity) groups (k = 3) to which we will assign scores of 1, 2, 3, respectively. The ln(incidence rate), score, and weight are given for each of the 12 cells in Table 14.9. We then proceed as in Equation 14.37, as follows:

$$\sum_{i=1}^{4}\sum_{j=1}^{3} w_{ij} = 2169$$

TABLE 14.9 **ln(incidence rate) of breast cancer and weight used in weighted regression analysis**

Age	i	Parity (j)	$\ln(\hat{p}_{ij})$	w_{ij}
30–39	1	1	–7.018	18
30–39	1	2	–7.102	72
30–39	1	3	–7.273	60
40–49	2	1	–6.082	73
40–49	2	2	–6.350	245
40–49	2	3	–6.446	416
50–59	3	1	–5.819	94
50–59	3	2	–5.944	271
50–59	3	3	–6.067	608
60–69	4	1	–5.324	50
60–69	4	2	–5.838	86
60–69	4	3	–5.903	176

$$\sum_{i=1}^{4}\sum_{j=1}^{3} w_{ij}\ln(\hat{p}_{ij}) = -13,408.7$$

$$\sum_{i=1}^{4}\sum_{j=1}^{3} w_{ij}S_j = 5363$$

$$\sum_{i=1}^{4}\sum_{j=1}^{3} w_{ij}S_j^2 = 14,271$$

$$\sum_{i=1}^{4}\sum_{j=1}^{3} w_{ij}S_j\ln(\hat{p}_{ij}) = -33,279.2$$

$$L_{xx} = 14,271 - 5363^2/2169 = 1010.6$$

$$L_{xy} = -33,279.2 - (-13,408.7)(5363)/2169 = -125.2$$

$$\hat{\beta} = -125.2/1010.6 = -0.124$$

$$se(\hat{\beta}) = \sqrt{1/1010.6} = 0.031$$

$$z = \hat{\beta}/se(\hat{\beta}) = -0.124/0.031 = -3.94 \sim N(0,1)$$

$$p\text{-value} = 2 \times \Phi(-3.94) < 0.001$$

Thus, there is a significant inverse association between ln(breast-cancer incidence) and parity among parous women. Breast-cancer incidence declines by $(1 - e^{-0.124}) = 11.7\%$ for each additional birth up to 3 births within a given age group.

In this section, we have considered the problem of relating the incidence density to a categorical exposure variable E, where E has more than two categories and the categories of E correspond to an ordinal scale with an associated score variable S_j for the jth category. The procedure is similar to the chi-square test for trend given in Chapter 10, except that here we are modeling trends in incidence rates based on person-time data, whereas in Chapter 10 we were modeling trends in proportions based on count data (which as a special case might correspond to cumulative incidence), where the person is the unit of analysis.

Referring to the flowchart (Figure 14.9), we would answer yes to (1) person-time data? no to (2) one-sample problem? yes to (3) incidence rates remain constant over time? no to (4) two-sample problem? and yes to (5) interested in test of trend over more than two exposure groups? This path leads us to the box entitled "Use test of trend for incidence rates."

SECTION 14.8 Introduction to Survival Analysis

In Sections 14.1–14.7, methods for comparing incidence rates between two groups, where the period of follow-up may be different for the two groups considered, were discussed. One assumption made in performing these analyses is that incidence rates remain *constant* over time. In many instances this assumption is not warranted and one wishes to compare the number of disease events between two groups where the incidence of disease varies over time.

Example 14.26 **Health Promotion** Consider Data Set SMOKE.DAT on the data disk. In this data set, 234 smokers who expressed a willingness to quit smoking were followed for 1 year to estimate the cumulative incidence of recidivism; that is, the proportion of smokers who quit for a time but who started smoking again. One hypothesis is that older smokers are less likely to be successful quitters (and more likely to be recidivists). How can this hypothesis be tested?

The data in Table 14.10 were obtained after subdividing the study population by age (>40/≤40).

TABLE 14.10 **Number of days quit smoking by age**

	Number of days quit smoking					
Age	≤90	91–180	181–270	271–364	365	Total
>40	92	4	4	1	19	120
≤40	88	7	3	2	14	114
Total	180	11	7	3	33	234
Percent	76.9	4.7	3.0	1.3	14.1	

We can compute the estimated incidence rate of disease (recidivism) within each 90-day period for the combined study population. We will assume that subjects who started smoking within a given period did so at the midpoint of the respective period. Thus, the number of person-days within days 1–90 = 180(45) + 54(90) = 12,960, and the incidence rate of recidivism = 180/12,960 = 0.014 events per person-day. For days 91–180, there were 11 recidivists and 43 successful quitters. Hence, the number of person-days = 11(45) + 43(90) = 4365 person-days and the incidence rate = 11/4365 = 0.0025 events per person-day. Similarly, the incidence rate over days 181–270 = 7/[7(45) + 36(90)] = 7/3555 = 0.0020 events per person-day. Finally, the incidence rate over days 271–365 = 3/[3(47) + 33(95)] = 3/3276 = 0.00092 events per person-day. Thus, the incidence rate of recidivism is much higher in the first 90 days and declines throughout the 365-day period. Incidence rates that vary substantially over time are more commonly referred to as **hazard rates.**

In Example 14.26, we have assumed, for simplicity, that the hazard remains constant during each 90-day period. One nice way of comparing incidence rates between two groups is to plot their hazard functions.

Example 14.27 **Health Promotion** Plot the hazard function for subjects age > 40 and age ≤ 40, respectively.

Solution The hazard functions are plotted in Figure 14.4. There is actually a slight tendency for younger smokers (≤40) to be more likely to start smoking (i.e., become recidivists) than older smokers (>40), particularly after the first 90 days.

Hazard functions are used extensively in biostatistical work to assess mortality risk.

FIGURE 14.4 **Hazard rates (per 1000 person-days) by age**

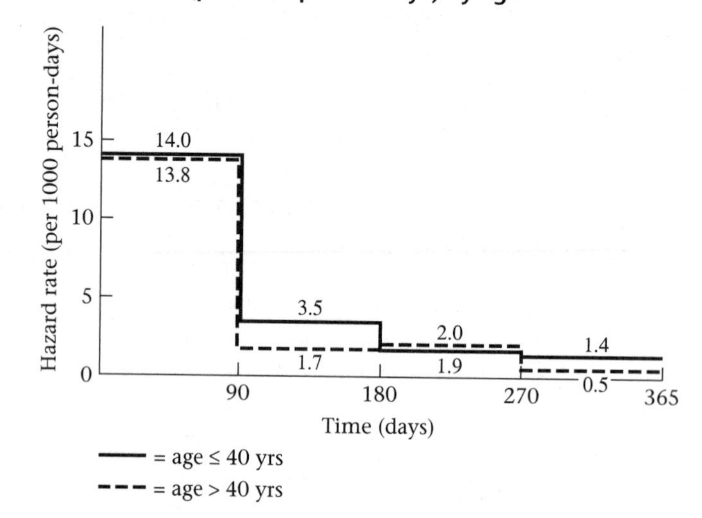

Another way of comparing disease incidence between two groups is by their cumulative incidence. If the incidence rate of disease over time is constant, as was assumed in Sections 14.1–14.7, then the cumulative incidence over time t is given exactly by $1 - e^{-\lambda t}$ and approximately by λt if the cumulative incidence is low (see Equation 14.3). We could also compute the probability of not developing disease = 1 – cumulative incidence = $e^{-\lambda t} \cong 1 - \lambda t$. The probability of not developing disease is commonly called the **survival probability.** We can plot the survival probability as a function of time. This function is referred to as the **survival function.**

DEFINITION 14.3 The **survival function** $S(t)$ gives the probability of survival up to time t for each $t \geq 0$.

The hazard at time t, denoted by $h(t)$, can be expressed in terms of the survival function $S(t)$ as follows:

DEFINITION 14.4 The **hazard function** $h(t)$ is the *instantaneous* probability of having an event at time t (per unit time), i.e, the instantaneous incidence rate, given that one has survived (i.e., has not had an event) up to time t. In particular,

$$h(t) = \left[\frac{S(t) - S(t + \Delta t)}{\Delta t}\right] \Big/ S(t) \quad \text{as } \Delta t \text{ approaches } 0$$

Example 14.28 **Demography** Use the life-table data in Table 4.13 to compute the approximate mortality hazard at age 60 and 80, respectively, for U.S. males in 1986.

Solution Table 4.13 shows that there were 80,908 men who survived up to age 60 and 79,539 men who survived up to age 61 among the original 100,000 men at time 0 (birth). Therefore, the hazard at age 60 is approximately

$$h(60) = \frac{80,908 - 79,539}{80,908} = \frac{1369}{80,908} = .017$$

Similarly, since there were 34,789 men who survived to age 80 and 31,739 men who survived to age 81, the hazard at age 80 is approximately given by

$$h(80) = \frac{34,789 - 31,739}{34,789} = \frac{3050}{34,789} = .088$$

Thus, in words, the probability of dying in the next year is 1.7% given that one has survived to age 60 and 8.8% given that one has survived to age 80. The percentages 1.7% and 8.8% represent the approximate hazard at ages 60 and 80, respectively. To improve the approximation, shorter time intervals than 1 year would need to be considered.

SECTION 14.9 Estimation of Survival Curves: The Kaplan-Meier Estimator

To estimate the survival probability when the incidence rate varies over time, we could use a more complex parametric survival model than the exponential model given in Equation 14.3 (see [2] for a good description of other parametric survival models). However, a more common approach is to use a nonparametric method referred to as the **product-limit** or **Kaplan-Meier estimator.**

Suppose individuals in the study population are assessed at times $t_1, \ldots, t_k$ where the times do not have to be equally spaced. If we wish to compute the probability of surviving up to time t_i, then we can write this probability in the form:

EQUATION 14.38

$$\begin{aligned} S(t_i) = \text{Prob}(\text{surviving to time } t_i) = {} & \text{Prob}(\text{surviving to time } t_1) \\ & \times \text{Prob}(\text{surviving to time } t_2 \mid \text{survived to time } t_1) \\ & \vdots \\ & \times \text{Prob}(\text{surviving to time } t_j \mid \text{survived to time } t_{j-1}) \\ & \vdots \\ & \times \text{Prob}(\text{surviving to time } t_i \mid \text{survived to time } t_{i-1}) \end{aligned}$$

Example 14.29 **Health Promotion** Estimate the survival curve for persons age >40 and age ≤40 for the subjects depicted in Table 14.10.

Solution We have for persons age >40,

$$S(90) = 1 - \frac{92}{120} = .233$$

$$S(180) = S(90) \times \left(1 - \frac{4}{28}\right) = .200$$

$$S(270) = S(180) \times \left(1 - \frac{4}{24}\right) = .167$$

$$S(365) = S(270) \times \left(1 - \frac{1}{20}\right) = .158$$

For persons age ≤40, we have

$$S(90) = 1 - \frac{88}{114} = .228$$

$$S(180) = S(90) \times \left(1 - \frac{7}{26}\right) = .167$$

$$S(270) = S(180) \times \left(1 - \frac{3}{19}\right) = .140$$

$$S(365) = S(270) \times \left(1 - \frac{2}{16}\right) = .123$$

These survival curves are plotted in Figure 14.5.

Subjects age >40 have a slightly higher estimated survival probability (i.e., probability of remaining a quitter) after the first 90 days.

FIGURE 14.5 **Survival probabilities by age**

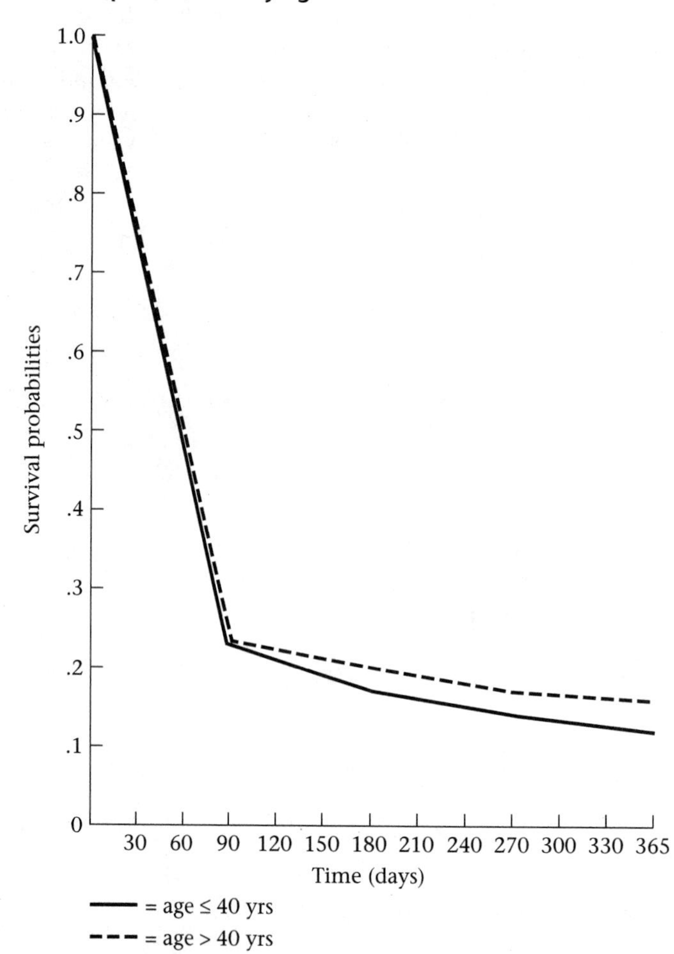

14.9.1 The Treatment of Censored Data

In Example 14.26, all members of the study population were followed until they started smoking again or for 1 year, whichever occurred first. In other instances, some subjects are not followed for the maximum period of follow-up, but have not yet had an event (i.e., have not yet failed).

Example 14.30 **Ophthalmology** A clinical trial was conducted to test the efficacy of different vitamin supplements in preventing visual loss in patients with retinitis pigmentosa [3]. Visual loss was measured by loss of retinal function as characterized by a 50% decline in the ERG (electroretinogram) 30 Hz amplitude, a measure of the electrical activity in the retina. In normal subjects, the normal range for ERG 30 Hz amplitude is >50 μV (microvolts). In patients with retinitis pigmentosa, ERG 30 Hz amplitude is usually <10 μV and is often <1 μV. Approximately 50% of patients with ERG 30 Hz amplitudes near 0.05 μV are legally blind as compared with <10% of patients whose ERG 30 Hz amplitudes are near 1.3 μV (the average ERG amplitude for patients in this clinical trial). Patients in the study were randomized to one of four treatment groups:

Group 1 received 15,000 IU of vitamin A and 3 IU (a trace amount) of vitamin E.

Group 2 received 75 IU (a trace amount) of vitamin A and 3 IU of vitamin E.

Group 3 received 15,000 IU of vitamin A and 400 IU of vitamin E.

Group 4 received 75 IU of vitamin A and 400 IU of vitamin E.

We will refer to these four groups as the A group, trace group, AE group, and E group, respectively. We wish to compare the proportion of patients who fail (i.e., lose 50% of ERG 30 Hz amplitude) in different treatment groups. Patients were enrolled in 1984–1987, and follow-up was terminated in September 1991. Since follow-up was terminated at the same point in *chronological time*, the period of follow-up was different for each patient. Patients who entered early in the study were followed for 6 years, while patients who enrolled later in the study were followed for 4 years. In addition, some patients dropped out of the study before September 1991 and had not failed. Dropouts were due to either death, other diseases, side effects possibly due to the study medications, or unwillingness to comply (i.e., to take study medications). How can we estimate the hazard and survival functions in each treatment group in the presence of variable follow-up for each patient?

DEFINITION 14.5 We will refer to patients who do not reach a disease endpoint during their period of follow-up as **censored observations.** A subject has been censored at time t if the subject has been followed up to time t and has not failed.

We will assume that censoring is noninformative, i.e., patients who are censored have the same underlying survival curve after their censoring time as patients who are not censored.

To estimate the survival function in the presence of censoring, suppose that S_{i-1} patients have survived through time t_{i-1} and are not censored at time t_{i-1}. Among these patients, S_i patients survive, d_i patients fail, and l_i patients are censored at time t_i. Thus, $S_{i-1} = S_i + d_i + l_i$. We can estimate the probability of surviving to time t_i given that one has survived up to time t_{i-1} by $(1 - d_i/S_{i-1}) = [1 - d_i/(S_i + d_i + l_i)]$. The

l_i patients who are censored at time t_i do not contribute to the estimation of the survival function at time $> t_i$. However, these patients do contribute to the estimation of the survival function at time $\le t_i$. We can summarize this procedure as follows:

EQUATION 14.39

Kaplan-Meier (Product-Limit) Estimator of Survival (Censored Data) Suppose that S_{i-1} subjects have survived up to time t_{i-1} and are not censored at time t_{i-1}, of whom S_i survive, d_i fail, and l_i are censored at time t_i, $i = 1, \ldots, k$. The **Kaplan-Meier estimator** of the survival probability at time t_i is

$$\hat{S}(t_i) = \left(1 - \frac{d_1}{S_0}\right) \times \left(1 - \frac{d_2}{S_1}\right) \times \cdots \times \left(1 - \frac{d_i}{S_{i-1}}\right),\ i = 1, \ldots, k$$

Example 14.31 **Ophthalmology** Estimate the survival probability at each of years 1–6 for subjects receiving 15,000 IU of vitamin A (i.e., groups A and AE combined) and subjects receiving 75 IU of vitamin A (i.e., groups E and trace combined), respectively, based on the data set mentioned in Example 14.30.

Solution The calculations are given in Table 14.11. For example, for the subjects receiving 15,000 IU of vitamin A the survival probability at year 1 = .9826. The probability of surviving to year 2 given that one survives to year 1 = 159/165 = .9636. Thus, the survival probability at year 2 = .9826 × .9636 = .9468, etc. The survival probabilities for the subjects receiving 15,000 IU of vitamin A tend to be higher than for the subjects receiving 75 IU of vitamin A, particularly at year 6.

TABLE 14.11 Survival probabilities for subjects receiving 15,000 IU of vitamin A and 75 IU of vitamin A, respectively

Time	Fail	Censored	Survive	Total	Prob(survive to time t_i | survived up to time t_{i-1})	$\hat{S}(t_i)$	$\hat{h}(t_i)$
15,000 IU of vitamin A daily							
1 yr = t_1	3	4	165	172	.9826	.9826	0.0174
2 yr = t_2	6	0	159	165	.9636	.9468	0.0364
3 yr = t_3	15	1	143	159	.9057	.8575	0.0943
4 yr = t_4	21	26	96	143	.8531	.7316	0.1469
5 yr = t_5	15	35	46	96	.8438	.6173	0.1563
6 yr = t_6	5	41	0	46	.8913	.5502	0.1087
75 IU of vitamin A daily							
1 yr = t_1	8	0	174	182	.9560	.9560	0.0440
2 yr = t_2	13	3	158	174	.9253	.8846	0.0747
3 yr = t_3	21	2	135	158	.8671	.7670	0.1329
4 yr = t_4	21	28	86	135	.8444	.6477	0.1556
5 yr = t_5	13	31	42	86	.8488	.5498	0.1512
6 yr = t_6	13	29	0	42	.6905	.3796	0.3095

Note: A person fails if his or her ERG 30 Hz amplitude declines by at least 50% from baseline to any follow-up visit, regardless of any subsequent ERG values obtained after the visit where the failure occurs.

There are other methods for calculating survival probabilities, such as assuming that censored subjects are only followed for half of the time interval during which they were measured. These methods are beyond the scope of this text.

14.9.2 Interval Estimation of Survival Probabilities

In Equation 14.39, we derived a point estimate of the survival probability at specific time points. We can derive an interval estimate as well. To obtain an interval estimate for $S(t)$, we consider $Var\{\ln[\hat{S}(t)]\}$, which is given by

EQUATION 14.40

$$Var\left\{\ln[\hat{S}(t_i)]\right\} = \sum_{j=1}^{i} \frac{d_j}{S_{j-1}(S_{j-1} - d_j)}$$

We then can obtain an approximate two-sided $100\% \times (1 - \alpha)$ confidence interval for $\ln[S(t_i)]$ given by

$$\ln[\hat{S}(t_i)] \pm z_{1-\alpha/2} \times \sqrt{Var\left\{\ln[\hat{S}(t_i)]\right\}} = (c_1, c_2)$$

The corresponding two-sided $100\% \times (1 - \alpha)$ CI for $S(t_i)$ is given by (e^{c_1}, e^{c_2}). This procedure is summarized as follows:

EQUATION 14.41

Interval Estimation of Survival Probabilities Suppose that S_{i-1} subjects have survived up to time t_{i-1} and are not censored at time t_{i-1}, of whom S_i survive, d_i fail and l_i are censored at time t_i, $i = 1, \ldots, k$. A two-sided $100\% \times (1 - \alpha)$ CI for the survival probability at time $t_i = S(t_i)$ is given by (e^{c_1}, e^{c_2}), where

$$c_1 = \ln[\hat{S}(t_i)] - z_{1-\alpha/2}se\left\{\ln[\hat{S}(t_i)]\right\}$$
$$c_2 = \ln[\hat{S}(t_i)] + z_{1-\alpha/2}se\left\{\ln[\hat{S}(t_i)]\right\}$$

where $\hat{S}(t_i)$ is obtained from the Kaplan-Meier estimator in Equation 14.39, and

$$se\left\{\ln[\hat{S}(t_i)]\right\} = \sqrt{\sum_{j=1}^{i} \frac{d_j}{S_{j-1}\left(S_{j-1} - d_j\right)}},\ i = 1, \ldots, k$$

Example 14.32 **Ophthalmology** Obtain a 95% CI for the survival probability at 6 years for the patients assigned to the 15,000 IU/day vitamin A group based on the data in Table 14.11.

Solution From Table 14.11, we have $\hat{S}(6) = .5502$. Thus, $\ln[\hat{S}(6)] = \ln(.5502) = -0.5975$

$$Var\left\{\ln[\hat{S}(6)]\right\} = \frac{3}{172(169)} + \frac{6}{165(159)} + \frac{15}{159(144)} + \frac{21}{143(122)} + \frac{15}{96(81)} + \frac{5}{46(41)}$$
$$= 6.771 \times 10^{-3}$$

Thus, a 95% CI for $\ln[S(6)]$ is given by

$$-0.5975 \pm 1.96\sqrt{6.771 \times 10^{-3}} = -0.5975 \pm 0.1613 = (-0.7588,\ -0.4362)$$

The corresponding 95% CI for $S(6)$ is $(e^{-0.7588},\ e^{-0.4362}) = (.4682,\ .6465)$.

14.9.3 Estimation of the Hazard Function: The Product-Limit Method

In Example 14.27, we considered the estimation of the hazard function in the context of the smoking-cessation data in Table 14.10. In this example, we estimated the hazard for each group, for each approximately 90-day period. We assumed that recidivists would resume smoking randomly throughout the 90-day period. Thus, to compute the hazard in the first 90 days for subjects with age >40, we assume that at day 45 half the recidivists (92/2 = 46) have resumed smoking. Thus, there remain 120 – 46 = 74 subjects who are still quitters. Among these subjects 92/90 = 1.022 subjects will resume smoking by day 46. Thus, the estimated hazard rate at day 45 = 1.022/74 = .0138 events per person-day. In a similar manner, the hazard for each 90-day period was approximated by the hazard at the midpoint of the period. This approach to hazard estimation is commonly referred to as the *actuarial method.*

In epidemiology, another approach is often used. A key assumption of the actuarial method is that events occur randomly throughout a defined follow-up period. Another approach is to assume that an event occurs at the precise time that it is either observed (e.g., if an abnormality such as a heart murmur is observed at a physical examination) or reported, if the patient reports a specific symptom (e.g., dizzy spells or breathlessness). This approach is referred to as the **product-limit method.**

Example 14.33 **Ophthalmology** Estimate the hazard at each year of follow-up for each vitamin A dose group based on the data in Table 14.11 using the product-limit method.

Solution For subjects taking 15,000 IU of vitamin A daily, the estimated hazard at year 1 = number of subjects with events at year 1/number of subjects available for examination at year 1 = 3/172 = .0174. At year 2, there were 165 subjects examined, of whom 6 failed. Thus, the estimated hazard at year 2 = 6/165 = .0364, . . . , etc. In general, the estimated hazard at year $t_i = h(t_i) = d_i/S_{i-1} = 1 - \text{Prob}(\text{survive to time } t_i | \text{survive to time } t_{i-1})$. The estimated hazard function for each group by year is given in the last column of Table 14.11.

We will use the product-limit approach to hazard estimation in the remainder of this chapter.

In this section, we introduced the basic concepts of survival analysis. The primary outcome measures in this type of analysis are the *survival function,* which provides the probability of surviving to time *t,* and the *hazard function,* which provides the instantaneous rate of disease per unit time given that one has survived to time *t.* A unique aspect of survival data is that usually not all subjects are followed for the same length of time. Thus, we introduced the concept of a *censored observation,* which is a subject who has not failed by time *t,* but is not followed any longer, so that the actual time of failure for censored observations is unknown. Finally, we learned about the *Kaplan-Meier product-limit method,* which is a technique for estimating the survival and hazard functions in the presence of censored data.

Referring to the flowchart (Figure 14.9), we would answer yes to (1) person-time data? and no to both (2) one-sample problem? and (3) incidence rates remain constant over time? This path leads us to the box entitled "Use survival-analysis methods."

In the next section, we will continue our discussion of survival analysis and learn about analytic techniques for comparing survival curves from two independent samples.

SECTION 14.10 The Log-Rank Test

In this section, we consider how to compare the two survival curves displayed in Figure 14.5 for the smoking-cessation data. We could compare survival at specific time points. However, we will usually gain power if we consider the entire survival curve. Suppose that we wish to compare the survival experience of two groups, which we will call the exposed and unexposed groups. Let $h_1(t)$ = hazard at time t for subjects in the exposed group, and $h_2(t)$ = hazard at time t for subjects in the unexposed group. We will assume that the hazard ratio is a constant $\exp(\beta)$.

EQUATION 14.42

$$h_1(t)/h_2(t) = \exp(\beta)$$

We wish to test the hypothesis H_0: $\beta = 0$ versus H_1: $\beta \neq 0$. If $\beta = 0$, then the survival curves of the two groups will be the same. If $\beta > 0$, then the exposed will consistently be at greater risk for disease than the unexposed, or equivalently the survival probability of the exposed will be less than the unexposed. If $\beta < 0$, then the exposed will be at lower risk than the unexposed and their survival probabilities will be greater than the unexposed. This is a similar hypothesis-testing situation to that given in Equation 14.8, where we were interested in comparing two incidence rates. The difference is that in Equation 14.8 we assumed that the incidence rate for a group was constant over time, while in Equation 14.42 we allow the hazard rate to vary over time.

Consider the data in Table 14.10. These data could be analyzed in terms of cumulative incidence over 1 year; that is, the percentage of older versus younger ex-smokers who were successful quitters for 1 year could be compared. However, if incidence changes greatly over the year, this will not be as powerful as the log-rank test described later in Equation 14.43. Using this procedure, *when* an event occurs (in this case the event is recidivism) rather than simply *whether* it occurs is taken into account.

To implement this procedure, the total period of follow-up is subdivided into shorter time periods over which incidence is relatively constant. In Example 14.26, the ideal situation would be to subdivide the 1-year interval into 365 daily time intervals. However, for the purpose of illustrating the method, time has been subdivided into 3-month intervals. For each time interval, the number of people who have been successful quitters up to the beginning of the interval are identified. These are the people who are at risk for recidivism during this time interval. This group is then categorized according to whether they remained successful quitters or became recidivists during the time interval. For each time interval, the data are displayed in the form of a 2×2 contingency table relating age to incidence of recidivism over the time interval.

Example 14.34 **Health Promotion** Display the smoking-cessation data in Table 14.10 in the form of incidence rates by age for each of the four time intervals, 0–90 days, 91–180 days, 181–270 days, and 271–365 days.

Solution For the first time interval, 0–90 days, there were 120 older smokers who were successful quitters at time 0, of whom 92 became recidivists during the 0- to 90-day period; similarly, there were 114 younger smokers, of whom 88 became recidivists during this time period. These data are shown in a 2×2 contingency table in Table 14.12. For the second time period, 91–180 days, 28 older smokers were successful quitters at day 90, of whom 4 became recidivists during the period from day 91 to day 180; similarly, 26 younger smokers were successful quitters at 90 days, of whom 7 became recidivists from day 91 to day 180. Thus, the second 2×2 contingency table would look like Table 14.13. Similarly, 2×2 contingency tables for the time periods 181–270 days and 271–365 days can be developed, as shown in Tables 14.14 and 14.15, respectively.

TABLE 14.12 **Incidence rates by age for the 0- to 90-day period**

	Outcome		
Age	Recidivist	Successful quitter	**Total**
>40	92	28	120
≤40	88	26	114
Total	180	54	234

TABLE 14.13 **Incidence rates by age for the 91- to 180-day period**

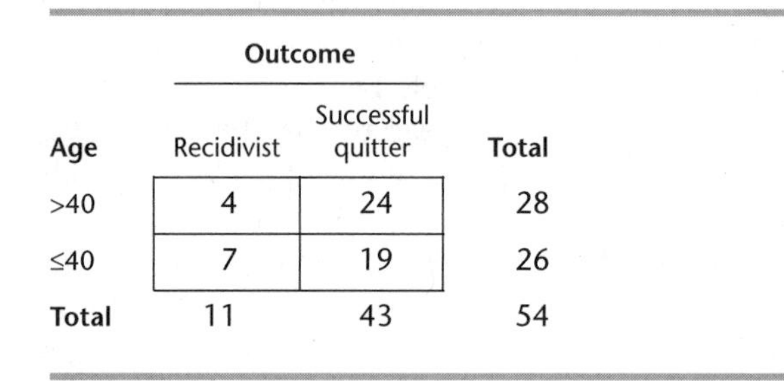

	Outcome		
Age	Recidivist	Successful quitter	**Total**
>40	4	24	28
≤40	7	19	26
Total	11	43	54

TABLE 14.14 **Incidence rates by age for the 181- to 270-day period**

	Outcome		
Age	Recidivist	Successful quitter	**Total**
>40	4	20	24
≤40	3	16	19
Total	7	36	43

TABLE 14.15 **Incidence rates by age for the 271- to 365-day period**

	Outcome		
Age	Recidivist	Successful quitter	**Total**
>40	1	19	20
≤40	2	14	16
Total	3	33	36

If age has no association with recidivism, then the incidence rate for recidivism for older and younger smokers within each of the four time intervals should be the same. Conversely, if it is harder for older smokers than younger smokers to remain quitters, then the incidence rate of recidivism should be consistently higher for older smokers within each of the four time intervals considered. Note that incidence is allowed to vary over different time intervals under either hypothesis. To accumulate evidence over the entire period of follow-up, the Mantel-Haenszel procedure in Equation 13.14, based on the 2×2 tables in Tables 14.12–14.15, is used. This procedure is referred to as the log-rank test and is summarized as follows:

EQUATION 14.43

The Log-Rank Test To compare incidence rates for an event between two exposure groups, where incidence varies over the period of follow-up (T), use the following procedure:

(1) Subdivide T into k smaller time intervals, where incidence is homogeneous over the shorter time intervals.

(2) Compute a 2×2 contingency table corresponding to each time interval relating incidence over the time interval to exposure status (+/–). The ith table is of the form of Table 14.16,

where n_{i1} = the number of exposed people who have not yet had the event at the beginning of the ith time interval and were not censored at the beginning of the interval

n_{i2} = the number of unexposed people who have not yet had the event at the beginning of the ith time interval and were not censored at the beginning of the interval

a_i = the number of exposed people who had an event during the ith time interval

b_i = the number of exposed people who did not have an event during the ith time interval

and c_i, d_i are defined similarly for unexposed people.

(3) Perform the Mantel-Haenszel test over the collection of 2×2 tables defined in step 2. Specifically, compute the test statistic

$$X_{LR}^2 = \frac{\left(|O - E| - .5\right)^2}{Var_{LR}}$$

where

$$O = \sum_{i=1}^{k} a_i$$

$$E = \sum_{i=1}^{k} E_i = \sum_{i=1}^{k} \frac{(a_i + b_i)(a_i + c_i)}{n_i}$$

$$Var_{LR} = \sum_{i=1}^{k} V_i = \sum_{i=1}^{k} \frac{(a_i + b_i)(c_i + d_i)(a_i + c_i)(b_i + d_i)}{n_i^2(n_i - 1)}$$

which under H_0 follows a chi-square distribution with one degree of freedom.

(4) For a two-sided test with significance level α, if $X_{LR}^2 > \chi_{1,1-\alpha}^2$, then reject H_0. If $X_{LR}^2 \le \chi_{1,1-\alpha}^2$, then accept H_0.

(5) The exact p-value for this test is given by

$$p\text{-value} = Pr\left(\chi_1^2 > X_{LR}^2\right)$$

(6) This test should be used only if $Var_{LR} \ge 5$.

The acceptance and rejection regions for the log-rank test are depicted in Figure 14.6. The computation of the exact p-value is depicted in Figure 14.7.

TABLE 14.16 **Relationship of disease incidence to exposure status over the *i*th time interval**

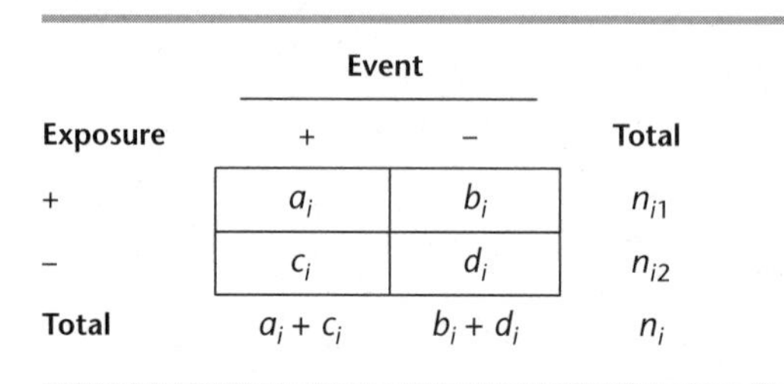

	Event		
Exposure	+	–	Total
+	a_i	b_i	n_{i1}
–	c_i	d_i	n_{i2}
Total	$a_i + c_i$	$b_i + d_i$	n_i

FIGURE 14.6 **Acceptance and rejection regions for the log-rank test**

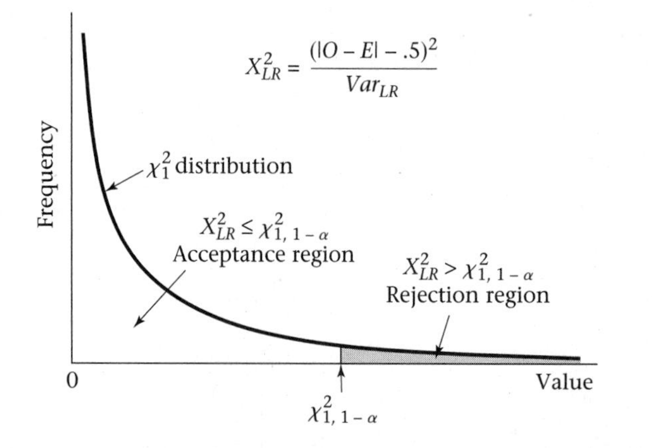

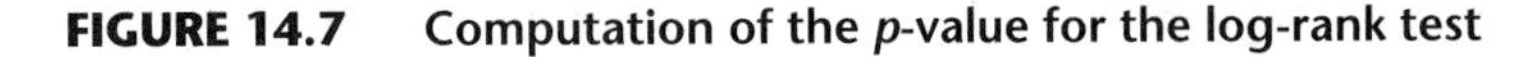

FIGURE 14.7 **Computation of the *p*-value for the log-rank test**

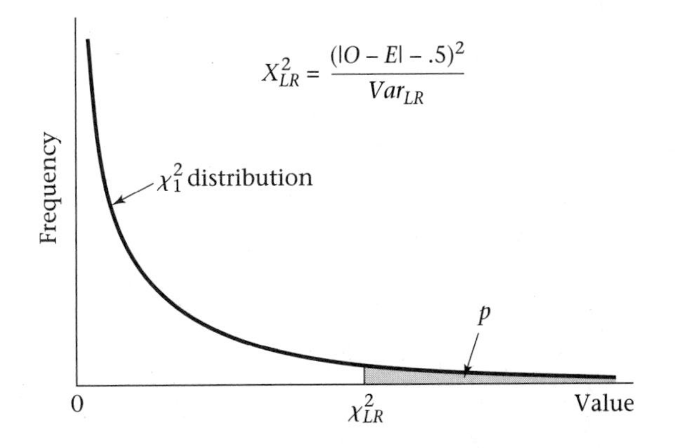

Example 14.35 **Health Promotion** Evaluate the statistical significance of the possible association between age and the incidence of recidivism based on the smoking-cessation data in Table 14.10.

Solution Refer to the four 2×2 tables (Tables 14.12–14.15) developed in Example 14.34. We have that

$$O = 92 + 4 + 4 + 1 = 101$$

$$E = \frac{120 \times 180}{234} + \frac{28 \times 11}{54} + \frac{24 \times 7}{43} + \frac{20 \times 3}{36}$$

$$= 92.308 + 5.704 + 3.907 + 1.667 = 103.585$$

$$Var_{LR} = \frac{120 \times 114 \times 180 \times 54}{234^2 \times 233} + \frac{28 \times 26 \times 11 \times 43}{54^2 \times 53}$$

$$+ \frac{24 \times 19 \times 7 \times 36}{43^2 \times 42} + \frac{20 \times 16 \times 3 \times 33}{36^2 \times 35}$$

$$= 10.422 + 2.228 + 1.480 + 0.698 = 14.829$$

Since $Var_{LR} \geq 5$, the log-rank test can be used. The test statistic is given by

$$X_{LR}^2 = \frac{\left(\left|101 - 103.585\right| - .5\right)^2}{14.829} = \frac{2.085^2}{14.829} = 0.29 \sim \chi_1^2 \text{ under } H_0$$

Since $\chi_{1,.95}^2 = 3.84 > 0.29$, it follows that the p-value $> .05$, and there is no significant difference in recidivism rates between younger and older smokers.

The data set in Table 14.10 did not have any censored data; that is, all subjects were followed until either 1 year had elapsed or they resumed smoking, whichever occurred first. However, the log-rank test can also be used if censored data are present. Here, for the year i table, S_i = number of subjects who survived to time t_i and l_i = number of subjects who were censored at time t_i and did not fail are combined into one group, because they all survived from time t_{i-1} to time t_i.

Example 14.36 **Ophthalmology** Compare the survival curves for subjects receiving 15,000 IU of vitamin A versus subjects receiving 75 IU of vitamin A, given the data in Table 14.11.

Solution We have 6 contingency tables corresponding to years 1, 2, 3, 4, 5, and 6. These are given as follows:

Year 1

	Fail	Survive	Total
15,000	3	169	172
75	8	174	182
	11	343	354

Year 2

	Fail	Survive	Total
15,000	6	159	165
75	13	161	174
	19	320	339

Year 3

	Fail	Survive	Total
15,000	15	144	159
75	21	137	158
	36	281	317

Year 4

	Fail	Survive	Total
15,000	21	122	143
75	21	114	135
	42	236	278

Year 5

	Fail	Survive	Total
15,000	15	81	96
75	13	73	86
	28	154	182

Year 6

	Fail	Survive	Total
15,000	5	41	46
75	13	29	42
	18	70	88

We use PROC LIFETEST of SAS to perform the log-rank test. Based on Equation 14.43, we have $O = 65$, $E = 78.432$, and $Var_{LR} = 33.656$. The chi-square statistic using PROC LIFETEST is 5.36 with 1 df, and yields a p-value of .021. Note that PROC LIFETEST does not use a continuity correction. Thus, the test statistic is

$$X^2_{LR,\text{ uncorrected}} = \frac{(O-E)^2}{Var_{LR}}$$

as opposed to

$$X^2_{LR} = \frac{\left(|O-E|-.5\right)^2}{Var_{LR}}$$

given in Equation 14.43. In this example, $X^2_{LR} = 4.97$, with p-value = .026. The Wilcoxon and likelihood ratio (*LR*) procedures are other approaches for comparing survival curves provided by PROC LIFETEST. These approaches are not discussed in this text, because the log-rank test is more widely used. PROC LIFETEST can also provide survival probabilities by treatment group similar to Table 14.11.

Therefore, there is a significant difference between the survival curves of the two groups. Since O = observed number of events in the 15,000 IU group = 65 < E = expected number of events in the 15,000 IU group = 78.432, it follows that the 15,000 IU group had a better survival experience than the 75 IU group. Stated another way, there were significantly fewer failures in the 15,000 IU group than in the 75 IU group.

In this section, we have presented the log-rank test, which is a procedure for comparing survival curves from two independent samples. It is similar to the Mantel-Haenszel test and can be used to compare survival curves with and without censored data. It allows one to compare the entire survival curve, which provides more power than focusing on survival at specific points in time.

Referring to the flowchart (Figure 14.9), we would answer yes to (1) person-time data? and no to (2) one-sample problem? and (3) incidence rates remain constant over time? This path leads to the box entitled "Use survival-analysis methods." We would then answer yes to (4) interested in comparison of survival curves between two groups with limited control of covariates? This leads to the box entitled "Use log-rank test."

SECTION 14.11 The Proportional-Hazards Model

The log-rank test is a very powerful method for analyzing data where the time to an event is important rather than simply whether or not the event occurs. The test can be used if variable periods of follow-up are available for each individual and/or if some of the data are censored. It can also be extended to allow one to look at the relationship between survival and a single primary exposure variable, while controlling for the effects of one or more other covariate(s). This can be accomplished by stratifying the data according to the levels of the other covariates; computing the observed number, expected number, and variance of the number of failures in each stratum; summing the respective values over all strata; and using the same test statistic as in Equation 14.43. However, if there are many strata and/or if there are several risk factors of interest, then a more convenient approach is to use a method of regression analysis for survival data.

Many different models can be used to relate survival to a collection of other risk factors. One of the most frequently used models was first proposed by D. R. Cox [4] and is called a proportional-hazards model.

EQUATION 14.44

Proportional-Hazards Model Under a **proportional-hazards model,** the hazard $h(t)$ is modeled as

$$h(t) = h_0(t)\exp(\beta_1 x_1 + \cdots + \beta_k x_k)$$

where $x_1, \ldots, x_k$ are a collection of independent variables, and $h_0(t)$ is the baseline hazard at time t, representing the hazard for a person with the value 0 for all the independent variables. The hypothesis H_0: $\beta_i = 0$ versus H_1: $\beta_i \neq 0$ can be tested as follows:

(1) Compute the test statistic $z = \hat{\beta}_i / se(\hat{\beta}_i)$.

(2) To conduct a two-sided level α significance test, if

$$z < z_{\alpha/2} \quad \text{or} \quad z > z_{1-\alpha/2}$$

reject H_0; if

$$z_{\alpha/2} \le z \le z_{1-\alpha/2}$$

accept H_0.

(3) The exact p-value is given by

$$2 \times [1 - \Phi(z)] \quad \text{if } z \ge 0$$

$$2 \times \Phi(z) \quad \text{if } z < 0$$

By dividing both sides of Equation 14.44 by $h_0(t)$ and taking logarithms, a proportional-hazards model can be written in the form

$$\ln\left[\frac{h(t)}{h_0(t)}\right] = \beta_1 x_1 + \cdots + \beta_k x_k$$

This representation allows us to interpret the coefficients of a proportional-hazards model in a similar manner to that of a multiple logistic-regression model. In particular, if x_i is a dichotomous independent variable, then the following principle applies:

EQUATION 14.45

Estimation of Hazard Ratio for Proportional-Hazards Models for Dichotomous Independent Variables Suppose there is a dichotomous independent variable (x_i) that is coded as 1 if present and 0 if absent. For the proportional-hazards model in Equation 14.44 the quantity $\exp(\beta_i)$ represents the ratio of hazards for two individuals, one with the risk factor present and the other with the risk factor absent, given that the individuals have the same values for all other covariates. The hazard ratio or relative hazard can be interpreted as the *instantaneous* relative risk of an event per unit time for an individual with the risk factor present compared with an individual with the risk factor absent, given that both individuals have survived to time t, and are the same on all other covariates.

A two-sided $100\% \times (1-\alpha)$ CI for β_i is given by (e^{c_1}, e^{c_2}), where

$$c_1 = \hat{\beta}_i - z_{1-\alpha/2} se(\hat{\beta}_i)$$

$$c_2 = \hat{\beta}_i + z_{1-\alpha/2} se(\hat{\beta}_i)$$

Similarly, if x_i is a continuous independent variable, then the following interpretation of the regression coefficient β_i is used:

EQUATION 14.46

Estimation of Hazard Ratio for Proportional-Hazards Models for Continuous Independent Variables Suppose there is a continuous independent variable (x_i). Consider two individuals who differ by the quantity Δ on the ith independent variable and are the same for all other independent variables. The quantity $\exp(\beta_i \Delta)$ represents the ratio of hazards between the two individuals. The hazard ratio can also be interpreted as the *instantaneous* relative risk of an event per unit time for an individual with risk-factor level $x_i + \Delta$ compared to an individual with risk-factor level x_i, given that both individuals have survived to time t, and are the same for all other covariates.

A two-sided $100\% \times (1-\alpha)$ CI for $\beta_i \Delta$ is given by (e^{c_1}, e^{c_2}) where

$$c_1 = \Delta\left[\hat{\beta}_i - z_{1-\alpha/2} se(\hat{\beta}_i)\right]$$

$$c_2 = \Delta\left[\hat{\beta}_i + z_{1-\alpha/2} se(\hat{\beta}_i)\right]$$

The Cox proportional-hazards model can also be thought of as an extension of multiple logistic regression where the time when an event occurs, rather than simply whether an event occurs, is taken into account.

Example 14.37 **Health Promotion** Fit a proportional-hazards model to the smoking-cessation data in Example 14.26 using the risk factors sex and adjusted log(CO concentration), which is an index of inhalation of smoking prior to quitting. Assess the statistical significance of the results, and interpret the regression coefficients.

Solution The SAS PHREG (Proportional Hazards Regression model) procedure has been used to fit the Cox model to the smoking-cessation data. For ease of interpretation, sex was recoded as (1 = male, 0 = female) from the original coding of (1 = male, 2 = female). For this example, the actual time of starting smoking was used based on the raw data in Data Set SMOKE.DAT, on the data disk, rather than on the grouped data given in Table 14.10. The results are given in Table 14.17.

TABLE 14.17 **Proportional-hazards model fitted to the smoking-cessation data in SMOKE.DAT**

Risk factor	Regression coefficient ($\hat{\beta}_i$)	Standard error $se(\hat{\beta}_i)$	z $[\hat{\beta}_i/se(\hat{\beta}_i)]$
$\log_{10}$ CO (adjusted)[a]	0.833	0.350	2.380
Sex (1 = M, 0 = F)	–0.117	0.135	–0.867

[a]This variable represents CO values adjusted for minutes elapsed since last cigarette smoked prior to quitting.

To assess significance of each regression coefficient, compute the test statistic given in Equation 14.44 as follows:

$$z(\log_{10}\text{CO}) = 0.833/0.350 = 2.380$$

$$p(\log_{10}\text{CO}) = 2\times\left[1-\Phi(2.380)\right] = 2\times(1-.9913) = .017$$

$$z(\text{sex}) = -0.117/0.135 = -0.867$$

$$p(\text{sex}) = 2\times\Phi(-0.867) = 2\times\left[1-\Phi(0.867)\right] = 2\times(1-.8069) = .386$$

Thus, there is a significant effect of CO concentration on the hazard or risk of recidivism (i.e., propensity to start smoking again), with the higher the CO concentration, the higher the hazard (risk). There is no significant effect of sex on risk of recidivism based on these data.

The effect of CO can be quantified in terms of relative risk. Specifically, if two individuals of the same sex who differ by one unit on adjusted $\log_{10}$CO are considered (i.e., who differ by 10-fold in CO concentration), then the instantaneous relative risk of recidivism for a person with adjusted $\log_{10}\text{CO} = x_i + 1$ (person A) compared to a person with adjusted $\log_{10}\text{CO} = x_i$ (person B) is given by

$$RR = \exp(0.833) = 2.30$$

Thus, given that person A and person B have not started smoking up to time t, person A is 2.3 times as likely to start smoking over a short period of time than person B.

The Cox proportional-hazards model can also be used with censored data.

Example 14.38 **Ophthalmology** Use the Cox proportional-hazards model to compare the survival curves for subjects receiving high dose (15,000 IU) versus low dose (75 IU) of vitamin A, based on the data in Table 14.11.

Solution We have used the SAS program PROC PHREG to compare the survival curves. In this case, there is only a single binary covariate x defined by

$$x = \begin{cases} 1 & \text{if high dose A} \\ 0 & \text{if low dose A} \end{cases}$$

The output from the program is given in Table 14.18. We see that subjects on 15,000 IU of vitamin A have a significantly lower hazard than subjects on 75 IU of vitamin A (p = .031, denoted by Pr > chi-square). The hazard ratio (denoted by Risk Ratio) is estimated by $e^{\hat{\beta}} = e^{-0.351729} = 0.703$. Thus, the failure rate at any point in time is approximately 30% lower for patients on 15,000 IU of vitamin A than for patients on 75 IU of vitamin A. We can obtain 95% confidence limits for the hazard ratio by (e^{c_1}, e^{c_2}), where

$$c_1 = \hat{\beta} - 1.96se(\hat{\beta}) = -0.352 - 1.96(0.163) = -0.672$$

$$c_2 = \hat{\beta} + 1.96se(\hat{\beta}) = -0.352 + 1.96(0.163) = -0.032$$

Thus, the $95\%\ CI = (e^{-0.672},\ e^{-0.032}) = (0.51,\ 0.97)$. The estimated survival curve(s) by year (LENFL30) are given at the bottom of Table 14.18 separately for the high-dose A group (HIGH_A = 1) and the low-dose A group (HIGH_A = 0), and are plotted in Figure 14.8. It is estimated that by year 6, 47% of subjects in the high-dose group (1 – .53) and 60% of subjects in the low-dose group (1 – .40) will have failed; that is, their ERG amplitude will decline by at least 50%.

If there are no ties—that is, if all subjects have a unique failure time—then the Cox proportional-hazards model with a single binary covariate and the log-rank test provide identical results. There are several different methods for handling ties with the Cox proportional-hazards model, and in general, in the presence of ties, the Cox proportional-hazards model and the log-rank test do not yield identical p-values, particularly in data sets with many tied observations, as in Table 14.18. Similarly, if there are many ties, then the survival curve estimated using the proportional-hazards model will not be exactly the same as that obtained from the Kaplan-Meier product-limit method.

Example 14.39 **Ophthalmology** The Cox proportional-hazards model can also be used to control for the effects of other covariates as well as the other treatment administered (denoted by HIGH_E). HIGH_E is defined by

$$\text{HIGH\_E} = \begin{cases} 1 & \text{if patient received 400 IU of vitamin E daily} \\ 0 & \text{if patient received 3 IU of vitamin E daily} \end{cases}$$

The other covariates considered were

AGEBAS = age at the baseline visit – 30 (in years)

$$\text{SEX} = \begin{cases} 1 & \text{if male} \\ 0 & \text{if female} \end{cases}$$

ER30OUCN = ln(ERG 30 Hz amplitude at baseline) – 0.215

BLRETLCN = baseline serum retinol – 50.0 (vitamin A) (μg/dL)

BLVITECN = baseline serum alpha-tocopherol – 0.92 (vitamin E) (mg/dL)

DRETINCN = dietary intake of retinol at baseline – 3624 (vitamin A) (IU)

DVTMNECN = dietary intake of alpha-tocopherol – 11.89 (vitamin E) (IU) at baseline

TABLE 14.18 Cox proportional-hazards model run on the RP data set in Table 14.11

```
                         The SAS System
                       The PHREG Procedure
     Data Set: WORK.TIMES2
     Dependent Variable: LENFL30
     Censoring Variable: FAIL30
     Censoring Value(s): 0
     Ties Handling: BRESLOW

                      Summary of the Number of
                     Event and Censored Values
                                                     Percent
             Total        Event        Censored     Censored
               354          154             200        56.50

               Analysis of Maximum Likelihood Estimates

                  Parameter    Standard     Wald          Pr >         Risk
Variable   DF      Estimate       Error  Chi-Square  Chi-Square       Ratio
HIGH_A      1     -0.351729     0.16322     4.64359      0.0312       0.703

                          The SAS System

               OBS      HIGH_A      LENFL30         S
                 1         1            0       1.00000
                 2         1            1       0.97436
                 3         1            2       0.92916
                 4         1            3       0.84090
                 5         1            4       0.73395
                 6         1            5       0.63853
                 7         1            6       0.52667
                 8         0            0       1.00000
                 9         0            1       0.96375
                10         0            2       0.90082
                11         0            3       0.78167
                12         0            4       0.64423
                13         0            5       0.52851
                14         0            6       0.40194
```

The mean value was subtracted from each covariate so as to minimize the computer time needed to obtain the parameter estimates. The results are given in Table 14.19.

We see that there are significant effects of both treatments administered, but in opposite directions. The estimated hazard ratio for subjects administered 15,000 IU of vitamin A versus subjects administered 75 IU of vitamin A was 0.69 ($p = .027$), while the estimated hazard ratio for subjects administered 400 IU of vitamin E versus subjects administered 3 IU of vitamin E was 1.47 ($p = .020$). Thus, vitamin A has a significant protective effect and vitamin E has a significant deleterious effect even after controlling for other baseline risk factors. Subjects administered high-dose vitamin A were about a third less likely to fail than subjects administered low-dose vitamin A, while subjects administered high-dose vitamin E were about 50% more likely to fail than subjects administered low-dose vitamin E. None of the other baseline covariates were statistically sig-

FIGURE 14.8 Survival curve for patients receiving 15,000 IU of vitamin A (HIGH_A = 1) and 75 IU of vitamin A (HIGH_A = 0)

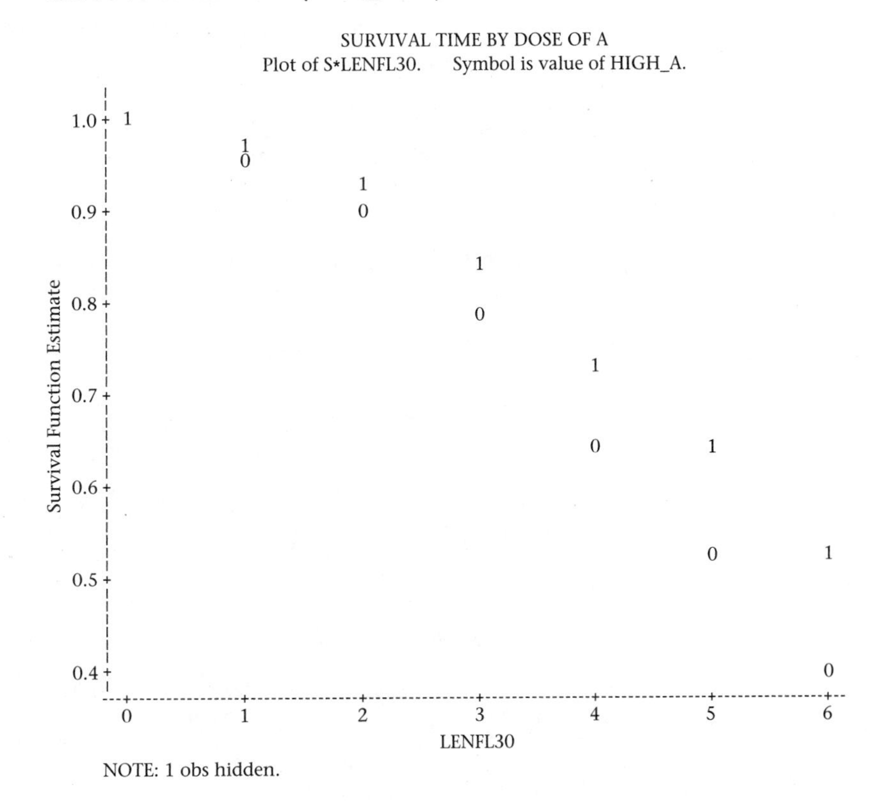

nificant, although baseline serum retinol was the closest to being significant (p = .11) with subjects with a high serum retinol being less likely to fail. One subject was lost in Table 14.19 versus Table 14.18 due to missing values on one of the other covariates.

Although the proportional-hazards model is perhaps the most frequently used model for survival data in the medical literature, many other models have been introduced for this type of data. An extensive survey of the field is given in Lee [5] and Miller [2].

In this section, we have been introduced to the Cox proportional-hazards model. This technique is analogous to multiple logistic regression and allows us to estimate the hazard ratio of a primary exposure variable while controlling for the effects of other covariates. An important assumption of this method is that the hazard ratio of the primary exposure (and also of any other covariates in the model) remains constant over time. This assumption can be tested by introducing a cross-product term of a specific variable of interest (x) by time (t) into the model and testing for statistical significance. Also, if this assumption is not satisfied, then we can still validly use the log-rank test, because this test does not require this assumption.

TABLE 14.19 **Effects of treatments administered in RP clinical trial, while controlling for the effects of other baseline covariates**

```
                         The SAS System
                      The PHREG Procedure
    Data Set: WORK.TIMES2
    Dependent Variable: LENFL30
    Censoring Variable: FAIL30
    Censoring Value(s): 0
    Ties Handling: BRESLOW

                     Summary of the Number of
                     Event and Censored Values
                                                  Percent
              Total       Event       Censored     Censored
                353         153            200        56.66

              Testing Global Null Hypothesis: BETA=0

                Without       With
  Criterion   Covariates   Covariates   Model Chi-Square
  -2 LOG L      1678.569     1660.859   17.710 with 9 DF (p=0.0387)
  Score              .            .     17.109 with 9 DF (p=0.0470)
  Wald               .            .     16.937 with 9 DF (p=0.0497)

              Analysis of Maximum Likelihood Estimates

                  Parameter     Standard       Wald         Pr >       Risk
Variable   DF     Estimate        Error     Chi-Square   Chi-Square   Ratio
AGEBAS      1     -0.004927      0.01111      0.19682      0.6573     0.995
SEX         1     -0.045872      0.17456      0.06905      0.7927     0.955
ER30OUCN    1     -0.087821      0.07337      1.43281      0.2313     0.916
BLRETLCN    1     -0.015270      0.00955      2.55691      0.1098     0.985
BLVITECN    1      0.410557      0.43412      0.89438      0.3443     1.508
DRETINCN    1  -0.000043831    0.0000480      0.83454      0.3610     1.000
DVTMNECN    1     -0.010197      0.01403      0.52812      0.4674     0.990
HIGH_A      1     -0.366556      0.16549      4.90616      0.0268     0.693
HIGH_E      1      0.386538      0.16604      5.41938      0.0199     1.472
```

Referring to the flowchart (Figure 14.9), we would answer yes to (1) person-time data? and no to each of (2) one-sample problem? and (3) incidence rates remain constant over time? which leads to the box entitled "Use survival-analysis methods." We would then answer no to (4) interested in comparison between survival curves of two groups with limited control of covariates? and yes to (5) interested in effects of several risk factors on survival? This leads us to the box entitled "Use Cox proportional-hazards model."

In the next section, we consider methods of power and sample-size estimation for proportional-hazards models.

SECTION 14.12 Power and Sample-Size Estimation under the Proportional-Hazards Model

14.12.1 Estimation of Power

Example 14.40 **Ophthalmology** Suppose the investigators consider repeating the study described in Example 14.30 to be sure that the protective effect of vitamin A was not a random occurrence. The study design of the new study would have only two treatment groups, the 15,000 IU per day vitamin A group and the 75 IU per day vitamin A group. The investigators feel that they can recruit 200 patients in each group who were not involved in the previous study. As in the previous study, the subjects would be enrolled over a 2-year period and followed for a maximum of 6 years. How much power would the study have to detect a *RR* of 0.7, where the endpoint is a 50% decline in ERG 30 Hz amplitude comparing the 15,000 IU per day group with the 75 IU per day group?

Several methods have been proposed for estimation of power and sample size for clinical trials based on survival curves that satisfy the proportional-hazards assumption. We present the method of Freedman [6] because it is relatively easy to implement and has fared relatively well in comparative simulation studies [7]. The method is as follows:

EQUATION 14.47

Estimation of Power for the Comparison of Survival Curves Between Two Groups under the Cox Proportional-Hazards Model Suppose we wish to compare the survival curves between an experimental group (E) and a control group (C) in a clinical trial with n_1 subjects in the E group and n_2 subjects in the C group, with a maximum follow-up of t years. We wish to test the hypothesis H_0: $RR = 1$ versus H_1: $RR \neq 1$, where RR = underlying hazard ratio for the E group versus the C group. We postulate a hazard ratio of RR under H_1, and wish to conduct a two-sided test with significance level α. If the ratio of subjects in group 1 compared to group 2 $= n_1/n_2 = k$, then the power of the test is

$$\text{Power} = \Phi\left(\frac{\sqrt{km}\,|RR-1|}{kRR+1} - z_{1-\alpha/2}\right)$$

where

m = expected total number of events over both groups

$= n_1 p_E + n_2 p_C$

n_1, n_2 = number of subjects in groups 1 and 2 (i.e., the E and C groups)

p_C = probability of failure in group C over the maximum time period of the study (t years)

p_E = probability of failure in group E over the maximum time period of the study (t years)

To calculate p_C and p_E, we let

(1) $\lambda_i = Pr$(failure at time i among subjects in the C group, given that a subject has survived to time $i-1$ and is not censored at time $i-1$) = approximate hazard at time i in the C group, $i = 1, \ldots, t$

(2) $RR\lambda_i = Pr$(failure at time i among subjects in the E group, given that a subject has survived to time $i-1$ and is not censored at time $i-1$) = approximate hazard at time i in the E group, $i = 1, \ldots, t$

(3) $\delta_i = Pr$(a subject is censored at time i given that he was followed up to time i and has not failed), $i = 0, \ldots, t$, which is assumed to be the same in each group

It follows from (1), (2), and (3) that

$$p_C = \sum_{i=1}^{t} \lambda_i A_i C_i = \sum_{i=1}^{t} D_i$$

$$p_E = \sum_{i=1}^{t} (RR\lambda_i) B_i C_i = \sum_{i=1}^{t} E_i$$

where

$$A_i = \prod_{j=0}^{i-1} (1-\lambda_j)$$

$$B_i = \prod_{j=0}^{i-1} (1-RR\lambda_j)$$

$$C_i = \prod_{k=0}^{i-1} (1-\delta_k)$$

Example 14.41 **Ophthalmology** Compute the power for the study proposed in Example 14.40.

Solution We have that $RR = 0.7$, $\alpha = .05$, $z_{1-\alpha/2} = z_{.975} = 1.96$, $k = 1$, $n_1 = n_2 = 200$, $t = 6$. To compute p_C and p_E, we must obtain λ_i, $RR\,\lambda_i$, and δ_i. We use the data in Table 14.11.

In this example, the 75 IU per day group is group C and the 15,000 IU per day group is group E. Also, no subjects are censored at year 0 (i.e., all subjects were followed for at least 1 year). We have $\lambda_1 = 8/182 = 0.0440$, $\lambda_2 = 13/174 = 0.0747, \ldots, \lambda_6 = 13/42 = 0.3095$. Also, $\delta_0 = 0$, $\delta_1 = 0$, $\delta_2 = 3/174 = .0172, \ldots, \delta_5 = 31/86 = .3605$, $\delta_6 = 29/42 = .6905$. The computations are displayed in Table 14.20.

TABLE 14.20 **Calculation of p_C and p_E for Example 14.41**

i	λ_i	$RR\lambda_i$	δ_i	A_i	B_i	C_i	D_i	E_i
0	0.0	0.0	0.0	—	—	1.0	—	—
1	0.0440	0.0308	0.0	1.0	1.0	1.0	0.0440	0.0308
2	0.0747	0.0523	0.0186	0.9560	0.9692	1.0	0.0714	0.0507
3	0.1329	0.0930	0.0146	0.8846	0.9185	0.9814	0.1154	0.0838
4	0.1556	0.1089	0.2456	0.7670	0.8331	0.9670	0.1154	0.0877
5	0.1512	0.1058	0.4247	0.6477	0.7424	0.7295	0.0714	0.0573
6	0.3095	0.2167	1.0	0.5498	0.6638	0.4197	0.0714	0.0604
Total							0.4890	0.3707

Thus, $p_C = .4890$, $p_E = .3707$, and $m = 200(.4890 + .3707) = 171.9$. Finally,

$$\text{Power} = \Phi\left(\frac{\sqrt{171.9}\,|0.7-1|}{0.7+1} - 1.96\right)$$

$$= \Phi\left[\frac{13.11(0.3)}{1.7} - 1.96\right]$$

$$= \Phi(2.314 - 1.96)$$

$$= \Phi(0.354) = .638$$

Thus the study would have about 64% power.

14.12.2 Estimation of Sample Size

Similarly, we can ask the following question: How many subjects are needed in each group to achieve a specified power of $1 - \beta$? The sample size can be obtained by solving for m and as a result n_1, n_2 in the power formula in Equation 14.47. The result is given as follows:

EQUATION 14.48

Sample-Size Estimation for the Comparison of Survival Curves Between Two Groups under the Cox Proportional-Hazards Model Suppose we wish to compare the survival curves between an experimental group (group E) and a control group (group C) in a clinical trial where the ratio of subjects in group E (n_1) to group C (n_2) is given by k and the maximum length of follow-up = t. We postulate a hazard ratio of RR for group E compared with group C and wish to conduct a two-sided test with significance level α. The number of subjects needed in each group to achieve a power of $1 - \beta$ is

$$n_1 = \frac{mk}{kp_E + p_C}, \quad n_2 = \frac{m}{kp_E + p_C}$$

where

$$m = \frac{1}{k}\left(\frac{kRR+1}{RR-1}\right)^2 \left(z_{1-\alpha/2} + z_{1-\beta}\right)^2$$

and p_E, p_C are the probabilities of failure over time t in groups E and C, respectively, given in Equation 14.47.

Example 14.42 **Ophthalmology** Estimate the required number of subjects needed in each group to achieve 80% power for the study proposed in Example 14.40.

Solution We have from Example 14.40 that $p_E = .3707$, $p_C = .4890$, and $k = 1$. Also, from Equation 14.48, we have

$$m = \left(\frac{1.7}{0.3}\right)^2 \left(z_{.975} + z_{.80}\right)^2$$
$$= 32.11(1.96 + 0.84)^2$$
$$= 32.11(7.84) = 251.8 \text{ events over both groups combined}$$

Thus

$$n_1 = n_2 = \frac{251.8}{.3707 + .4890}$$

$$= \frac{251.8}{.8597} = 293 \text{ subjects per group}$$

We therefore need to recruit 293 subjects in each group, or 586 subjects in total to achieve 80% power.

It may seem counterintuitive that we were able to achieve statistical significance based on the original study of 354 subjects in total over both groups, and yet the sample-size requirement for the new study is about 50% larger. The reason is that the results from the original study were only borderline significant ($p = .03$). If the p-value was exactly .05, and we used the same effect size in the proposed new study, then we would only achieve 50% power. Our power was slightly larger (64%) because the p-value was somewhat smaller than .05. To achieve 80% power, we need a larger sample size to allow for random fluctuations about the true effect size in finite-sample clinical trials.

The methods of power and sample-size estimation given in this section assume a proportional-hazards model relating the survival curves between the E and C groups. If the proportional-hazards assumption is not satisfied, then more complicated methods of power and sample-size estimation are needed. Other approaches for sample-size and power estimation are given in [7] and [8].

Example 14.43 **Cancer** Apply the method of sample-size estimation in Equation 14.48 to the study proposed in Example 14.14. Compare the results with those obtained in Example 14.16 using Equation 14.22.

Solution We have that $RR = 1.25$, $\alpha = .05$, $\beta = .20$, and $k = 1$. Thus,

$$m = \frac{(1.25+1)^2}{(1.25-1)^2}(1.96+0.84)^2$$

$$= 635.04$$

This is very similar to the required total number of events obtained in Example 14.16 (633). To obtain the corresponding sample-size estimate in each group, we use the Kaplan-Meier estimator to estimate p_C and p_E as follows:

$$p_C = 1 - \left[1 - (300 \times 10^{-5})\right]^5 = 1 - .98509 = .01491$$

$$p_E = 1 - \left[1 - (375 \times 10^{-5})\right]^5 = 1 - .98139 = .01861$$

Thus,

$$n_1 = n_2 = \frac{635.04}{.01861 + .01491} = 18{,}945 \text{ subjects per group}$$

or 37,890 subjects in total, to have 80% power.

This is also very similar to the total sample size projected in Example 14.16 (37,834 subjects). Thus, although somewhat different approaches were used to derive these sample-size formulas, the results in this example are very similar, which gives confidence in the validity of each of the approaches.

SECTION 14.13 Summary

In this chapter, we discussed how to analyze data where the unit of analysis is person-time. The incidence rate was defined as the number of events per unit of person-time and was compared with cumulative incidence, which is the proportion of people who develop disease over a specified period of time. The incidence rate is in units of events per unit time and has no upper bound, while cumulative incidence is a proportion that is bounded between 0 and 1. We discussed procedures for comparing a single estimated incidence rate with a known incidence rate and also for comparing two incidence rates both for crude data as well as for data stratified by other potential covariates. We then discussed methods of power and sample-size estimation for study designs based on comparisons of two incidence rates, both with and without adjusting for confounding variables. Finally, these methods were extended to exposure variables with more than two levels of exposure and a test of trend was introduced to analyze data of this form.

If incidence rates change greatly over time, then methods of survival analysis are appropriate. We introduced the concept of a hazard function, which is a function characterizing how incidence changes over time. We also introduced the concept of a survival curve, which is a function giving the cumulative probability of *not* having an event (i.e., surviving) as a function of time. The Kaplan-Meier estimator was introduced as a nonparametric method for estimating a survival curve. The log-rank test was then presented to enable us to statistically compare two survival curves (e.g., for an exposed versus an unexposed group). If we want to study the effects of several risk factors on survival, then the Cox proportional-hazards model can be used. This is a method analogous to multiple logistic regression where the time when an event occurs is considered rather than simply whether or not an event has occurred. Finally, we considered methods of power and sample-size estimation for studies where the primary method of analysis is the Cox proportional-hazards model. The methods in this chapter are outlined in the flowchart in Figure 14.9.

PROBLEMS

Cancer

The data relating oral-contraceptive use and the incidence of breast cancer in the age group 40–44 in the Nurses' Health Study are given in Table 14.21.

***14.1** Compare the incidence density of breast cancer in current users versus never users and report a p-value.

***14.2** Compare the incidence density of breast cancer in past users versus never users and report a p-value.

***14.3** Estimate the rate ratio comparing current users versus never users and provide a 95% CI about this estimate.

***14.4** Estimate the rate ratio comparing past users versus never users and provide a 95% CI about this estimate.

TABLE 14.21 Relationship between breast-cancer incidence and OC use among 40- to 44-year-old women in the Nurses' Health Study

OC-use group	Number of cases	Number of person-years
Current users	13	4,761
Past users	164	121,091
Never users	113	98,091

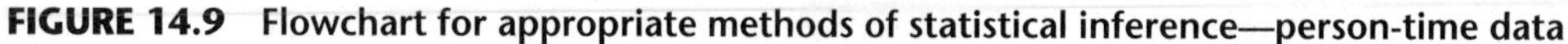

FIGURE 14.9 Flowchart for appropriate methods of statistical inference—person-time data

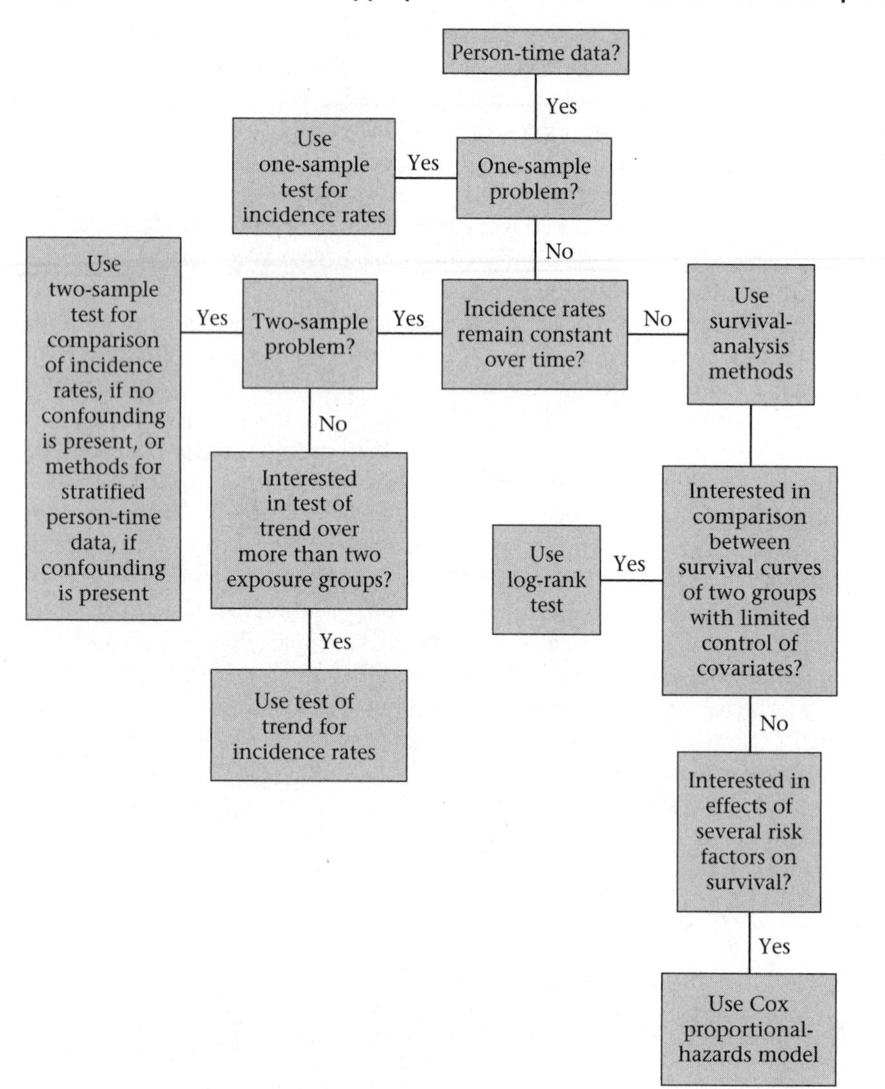

14.5 How much power did the study have of detecting a *RR* for breast cancer of 1.5, comparing current OC users versus never OC users among 40- to 44-year-old women if (a) the true incidence rate of breast cancer among never users and the amount of person-time for current and never users are the same as in Table 14.21, (b) the expected number of events for never OC users is the same as the observed number of events in Table 14.21, and (c) the average follow-up time per subject is the same for both current and never OC users?

14.6 What is the expected number of events that need to be realized in each group to achieve 80% power to detect a *RR* for breast cancer of 1.5 for current OC users versus never OC users under the same assumptions as in Problem 14.5?

Health Promotion

Refer to Data Set SMOKE.DAT on the data disk.

14.7 Divide subjects according to median $\log_{10}$CO (adjusted), and estimate survival curves for each subgroup.

14.8 Compare the survival curves of the two groups, using hypothesis-testing methods, and report a *p*-value.

A proportional-hazards model was fit to these data to assess the relationship of age, sex, number of cigarettes smoked, and log CO concentration, when considered simultaneously, on the ability to remain abstinent from smoking. The results are given in Table 14.22.

TABLE 14.22 Proportional-hazards model relating the hazard of recidivism to age, sex, number of cigarettes smoked prior to quitting, and log CO concentration

Risk factor	Regression coefficient ($\hat{\beta}_i$)	Standard error $se(\hat{\beta}_i)$
Age	0.0023	0.0058
Sex(1 = M/0 = F)	–0.127	0.143
Number of cigarettes smoked	–0.0038	0.0050
Log_{10}CO (adjusted)[a]	0.912	0.366

[a]This variable represents CO values adjusted for minutes elapsed since last cigarette smoked prior to quitting.

14.9 Assess the significance of each of the variables.

14.10 Estimate the effects of each variable in terms of hazard ratios and provide 95% confidence limits corresponding to each point estimate.

14.11 Compare the crude and adjusted analyses of the relationship of $\log_{10}$CO to recidivism in Problems 14.8 and 14.9.

Bioavailability

Refer to the Data Set BETACAR.DAT, on the data disk.

14.12 Suppose we regard a preparation as being bioavailable for a subject at the first week when the level of plasma carotene increased by 50% from the baseline level (based on an average of the first and second baseline determinations). Use survival-analysis methods to estimate the proportion of subjects for whom the preparation is not bioavailable at different points in time.

14.13 Assess whether there are significant differences among the survival curves obtained in Problem 14.12. (*Hint:* Use a dummy-variable approach with proportional-hazards models.)

14.14 Answer the same question as in Problem 14.12 if the criterion for bioavailability is a 100% increase in plasma-carotene level from baseline.

14.15 Answer the same question posed in Problem 14.13 if the criterion for bioavailability is a 100% increase in plasma-carotene level from baseline.

Ophthalmology

In Table 14.23, we present data from the RP clinical trial described in Example 14.30 concerning the effect of high-dose vitamin E (400 IU/day) versus low-dose vitamin E (3 IU/day) on survival (where failure is loss of at least 50% of initial ERG 30 Hz amplitude).

TABLE 14.23 Number of patients who failed, were censored, or survived by year in the 400 IU of vitamin E group and 3 IU of vitamin E group, respectively, RP clinical trial

	Fail	Censored	Survive	Total
400 IU of vitamin E daily				
0–1 yr	7	3	170	180
1–2 yr	9	2	159	170
2–3 yr	22	2	135	159
3–4 yr	24	27	84	135
4–5 yr	13	32	39	84
5–6 yr	11	28	0	39
3 IU of vitamin E daily				
0–1 yr	4	1	169	174
1–2 yr	10	3	156	169
2–3 yr	14	1	141	156
3–4 yr	16	27	98	141
4–5 yr	15	34	49	98
5–6 yr	7	42	0	49

Note: A person fails if his or her ERG amplitude declines by at least 50% from baseline to any follow-up visit.

14.16 Estimate the hazard function by year for each group.

14.17 Estimate the survival probability by year for each group.

14.18 Obtain a 95% CI for the survival probability at year 6 for each group.

14.19 Compare the overall survival curves of the two groups and obtain a *p*-value.

14.20 Suppose a new study is planned with 200 patients randomly assigned to each of a 400 IU per day of vitamin E group and a 3 IU per day of vitamin E group. If the survival experience in the 3 IU per day group is assumed to be the same as in Table 14.23, and the censoring experience of both groups are assumed to be the same as for the 3 IU per day group in Table 14.23, then how much power would a new study have if the maximum duration of follow-up is 4 years (rather than 6 years as in the original study) and a two-sided test is used with $\alpha = .05$?

14.21 How many subjects need to be enrolled in each group (assume an equal sample size in each group) to achieve 80% power if a two-sided test is used with $\alpha = .05$ and the same assumptions are made as in Problem 14.20?

Infectious Disease

Suppose that the rate of allergic reactions in a certain population is constant over time.

***14.22** An individual is selected randomly from the population and followed for 1.5 years. If the true rate of allergic reactions is 5 reactions per 100 person-years, what is the probability that the subject will have at least one allergic reaction during the follow-up period (i.e., cumulative incidence)?

***14.23** Two hundred subjects are selected randomly from the population and followed for various lengths of time. The average length of follow-up is 1.5 years. Suppose that at the end of the study, the estimated rate is 4 per 100 person-years. How many events must have been observed in order to yield the estimated rate of 4 per 100 person-years?

***14.24** Give a 95% confidence interval for the rate, based on the observed data in Problem 14.23. Express the answer in units of number of events per 100 person-years.

Cancer

The data in Table 14.24 provide the relationship between breast-cancer incidence rate and menopausal status by age, based on Nurses' Health Study data from 1976 to 1990.

TABLE 14.24 Relationship between breast-cancer incidence rate and menopausal status after controlling for age, Nurses' Health Study, 1976–1990

	Premenopausal	Postmenopausal
Age	Cases/person-years (incidence rate)[a]	Cases/person-years (incidence rate)
35–39	124 / 131,704 (94)	15 / 14,795 (101)
40–44	264 / 179,132 (147)	47 / 43,583 (108)
45–49	304 / 151,548 (201)	163 / 90,965 (179)
50–54	159 / 61,215 (260)	401 / 184,597 (217)
55–59	25 / 6,133 (408)	490 / 180,458 (272)

[a]Per 100,000 person-years.

14.25 Assess if there is a difference between the incidence rate of breast cancer for premenopausal versus postmenopausal women, while controlling for age. Please report a *p*-value.

14.26 Estimate the rate ratio for postmenopausal versus premenopausal women after controlling for age. Provide a 95% CI for the rate ratio.

Renal Disease

Refer to the Data Set SWISS.DAT on the data disk.

14.27 Suppose a serum-creatinine level of ≥1.5 mg/dL is considered a sign of possible kidney toxicity. Use survival-analysis methods to assess if there are differences in the incidence of kidney toxicity between the high-NAPAP group and the control group. In this analysis, exclude subjects who were ≥1.5 mg/dL at baseline.

14.28 Answer the question in Problem 14.27 comparing the low-NAPAP group to the control group.

One issue is that the groups in Problem 14.27 may not be exactly balanced by age and/or initial level of serum creatinine.

14.29 Answer the question in Problem 14.27, while controlling for possible age and initial-level differences between groups.

14.30 Answer the question in Problem 14.28, while controlling for possible age and initial-level differences between groups.

Infectious Disease

Suppose the incidence rate of influenza (flu) during the winter of 1998–1999 (i.e., from December 21, 1998, to March 20, 1999) was 50 events per 1000 person-months among students in high schools in a particular city.

14.31 A group of high-risk high school students were identified in the winter of 1998–1999 who had 3+ previous episodes of influenza before December 21, 1998. There were 20 students in this group, each followed for 90 days, of whom 8 developed influenza. Test the hypothesis that the high-risk students had a higher incidence rate of influenza than the average high school student during the winter of 1998–1999. Please report a *p*-value (one-tail).

14.32 Provide a 95% confidence interval for the incidence rate of flu among high-risk students during the winter of 1998–1999.

Among 1200 students in one high school in the city, 200 developed a new case of influenza over the 90 days from December 21, 1999, to March 20, 2000.

14.33 What is the estimated incidence rate of flu in the 1999–2000 winter season?

14.34 Provide a 95% confidence interval for the rate estimated in Problem 14.33.

14.35 Test the hypothesis that the rate of flu has changed from the 1998/1999 to 1999/2000 winter season. Please report a *p*-value (two-tail).

Orthopedics

A study was performed among patients with piriformis syndrome, a pelvic condition that involves malfunction of the piriformis muscle (a deep buttock muscle), which often causes lumbar and buttock pain with sciatica (pain radiating down the leg). A randomized double-blind clinical trial was performed whereby patients were injected with one of three substances.

Group 1 received an injection of a combination of triamcinolone and lidocaine (the TL group).

Group 2 received a placebo.

Group 3 received an injection of Botox.

The randomization schedule was set up such that, approximately, for every six patients, three were assigned to group 1, one to group 2, and two to group 3. All injections were directly into the piriformis muscle. Patients were asked to come back at 2 weeks post injection (0.5 month), 1 month post injection, and monthly thereafter up to 17 months, although there were many missed visits. At each visit the patients rated their percentage of improvement of pain versus baseline on a visual analog scale, with a maximum of 100% improvement (indicated by 100). Negative numbers indicate percentage of worsening. There were a total of 69 subjects and 70 legs (one patient, ID 23, had the condition in both legs). A priori it was of interest to compare the degree of efficacy between each pair of groups. There are three additional covariates that may influence the outcome, age (yrs), gender (1 = male, 0 = female), and the affected side (L = left, R = right). The data set is in BOTOX.DAT, with a description in BOTOX.DOC.

14.36 If the visual analog scale is treated as a continuous variable, assess whether there are any between-group differences in efficacy without considering the covariates. Try to do at least one analysis that uses the entire data set rather than focusing on specific time points.

14.37 Repeat the analysis in Problem 14.36, but account for covariate differences between groups.

14.38 Another way to score the visual analog scale is as a categorical variable where ≥50% improvement is considered a success and <50% improvement (either remaining the same or worsening) is considered a failure. Answer the question posed in Problem 14.36 using the success/failure scoring. Note that a patient may be a success at one visit but a failure at succeeding visits. (*Hint:* Either logistic regression methods or survival analysis methods may be applicable here.)

14.39 Repeat the analyses in Problem 14.38, but account for covariate differences between groups.

To reduce variability, the investigators also considered a criterion of at least 50% improvement on two successive visits as a definition of success.

14.40 Answer the question in Problem 14.38 under this definition of success.

14.41 Answer the question in Problem 14.39 under this definition of success.

REFERENCES

[1] Colditz, G. A., Stampfer, M. J., Willett, W. C., Hennekens, C. H., Rosner, B., & Speizer, F. E. (1990). Prospective study of estrogen replacement therapy and risk of breast cancer in post-menopausal women. *JAMA. 264*, 2648–2653.

[2] Miller, R. G. (1981). *Survival analysis*. New York: Wiley.

[3] Berson, E. L., Rosner, B., Sandberg, M. A., Hayes, K. C., Nicholson, B. W., Weigel-DiFranco, C., & Willett, W. C. (1993). A randomized trial of vitamin A and vitamin E supplementation for retinitis pigmentosa. *Archives of Ophthalmology, 111*, 761–772.

[4] Cox, D. R. (1972). Regression models and life tables (with discussion). *Journal of the Royal Statistical Society, Ser. B, 34*, 187–220.

[5] Lee, E. T. (1986). *Statistical methods for survival data analysis*. Belmont, CA: Wadsworth.

[6] Freedman, L. S. (1982). Tables of the number of patients required in clinical trials using the log-rank test. *Statistics in Medicine, 1*, 121–129.

[7] Lakatos, E., & Lan, K. K. G. (1992). A comparison of sample size methods for the log-rank statistic. *Statistics in Medicine, 11*, 179–191.

[8] Lachin, J. M., & Foulkes, M. A. (1986). Evaluation of sample size and power for analyses of survival with allowance for nonuniform patient entry, losses to follow-up, noncompliance, and stratification. *Biometrics, 42*, 507–519.

APPENDIX: Tables

TABLE 1 **Exact binomial probabilities $Pr(X = k) = \binom{n}{k} p^k q^{n-k}$**

n	k	.05	.10	.15	.20	.25	.30	.35	.40	.45	.50
2	0	.9025	.8100	.7225	.6400	.5625	.4900	.4225	.3600	.3025	.2500
	1	.0950	.1800	.2550	.3200	.3750	.4200	.4550	.4800	.4950	.5000
	2	.0025	.0100	.0225	.0400	.0625	.0900	.1225	.1600	.2025	.2500
3	0	.8574	.7290	.6141	.5120	.4219	.3430	.2746	.2160	.1664	.1250
	1	.1354	.2430	.3251	.3840	.4219	.4410	.4436	.4320	.4084	.3750
	2	.0071	.0270	.0574	.0960	.1406	.1890	.2389	.2880	.3341	.3750
	3	.0001	.0010	.0034	.0080	.0156	.0270	.0429	.0640	.0911	.1250
4	0	.8145	.6561	.5220	.4096	.3164	.2401	.1785	.1296	.0915	.0625
	1	.1715	.2916	.3685	.4096	.4219	.4116	.3845	.3456	.2995	.2500
	2	.0135	.0486	.0975	.1536	.2109	.2646	.3105	.3456	.3675	.3750
	3	.0005	.0036	.0115	.0256	.0469	.0756	.1115	.1536	.2005	.2500
	4	.0000	.0001	.0005	.0016	.0039	.0081	.0150	.0256	.0410	.0625
5	0	.7738	.5905	.4437	.3277	.2373	.1681	.1160	.0778	.0503	.0313
	1	.2036	.3280	.3915	.4096	.3955	.3602	.3124	.2592	.2059	.1563
	2	.0214	.0729	.1382	.2048	.2637	.3087	.3364	.3456	.3369	.3125
	3	.0011	.0081	.0244	.0512	.0879	.1323	.1811	.2304	.2757	.3125
	4	.0000	.0004	.0022	.0064	.0146	.0283	.0488	.0768	.1128	.1563
	5	.0000	.0000	.0001	.0003	.0010	.0024	.0053	.0102	.0185	.0313
6	0	.7351	.5314	.3771	.2621	.1780	.1176	.0754	.0467	.0277	.0156
	1	.2321	.3543	.3993	.3932	.3560	.3025	.2437	.1866	.1359	.0938
	2	.0305	.0984	.1762	.2458	.2966	.3241	.3280	.3110	.2780	.2344
	3	.0021	.0146	.0415	.0819	.1318	.1852	.2355	.2765	.3032	.3125
	4	.0001	.0012	.0055	.0154	.0330	.0595	.0951	.1382	.1861	.2344
	5	.0000	.0001	.0004	.0015	.0044	.0102	.0205	.0369	.0609	.0938
	6	.0000	.0000	.0000	.0001	.0002	.0007	.0018	.0041	.0083	.0156
7	0	.6983	.4783	.3206	.2097	.1335	.0824	.0490	.0280	.0152	.0078
	1	.2573	.3720	.3960	.3670	.3115	.2471	.1848	.1306	.0872	.0547
	2	.0406	.1240	.2097	.2753	.3115	.3177	.2985	.2613	.2140	.1641
	3	.0036	.0230	.0617	.1147	.1730	.2269	.2679	.2903	.2918	.2734
	4	.0002	.0026	.0109	.0287	.0577	.0972	.1442	.1935	.2388	.2734
	5	.0000	.0002	.0012	.0043	.0115	.0250	.0466	.0774	.1172	.1641
	6	.0000	.0000	.0001	.0004	.0013	.0036	.0084	.0172	.0320	.0547
	7	.0000	.0000	.0000	.0000	.0001	.0002	.0006	.0016	.0037	.0078

(continued on next page)

TABLE 1 Exact binomial probabilities $Pr(X = k) = \binom{n}{k} p^k q^{n-k}$ *(continued)*

n	k	.05	.10	.15	.20	.25	.30	.35	.40	.45	.50
8	0	.6634	.4305	.2725	.1678	.1001	.0576	.0319	.0168	.0084	.0039
	1	.2793	.3826	.3847	.3355	.2670	.1977	.1373	.0896	.0548	.0313
	2	.0515	.1488	.2376	.2936	.3115	.2965	.2587	.2090	.1569	.1094
	3	.0054	.0331	.0839	.1468	.2076	.2541	.2786	.2787	.2568	.2188
	4	.0004	.0046	.0185	.0459	.0865	.1361	.1875	.2322	.2627	.2734
	5	.0000	.0004	.0026	.0092	.0231	.0467	.0808	.1239	.1719	.2188
	6	.0000	.0000	.0002	.0011	.0038	.0100	.0217	.0413	.0703	.1094
	7	.0000	.0000	.0000	.0001	.0004	.0012	.0033	.0079	.0164	.0313
	8	.0000	.0000	.0000	.0000	.0000	.0001	.0002	.0007	.0017	.0039
9	0	.6302	.3874	.2316	.1342	.0751	.0404	.0207	.0101	.0046	.0020
	1	.2985	.3874	.3679	.3020	.2253	.1556	.1004	.0605	.0339	.0176
	2	.0629	.1722	.2597	.3020	.3003	.2668	.2162	.1612	.1110	.0703
	3	.0077	.0446	.1069	.1762	.2336	.2668	.2716	.2508	.2119	.1641
	4	.0006	.0074	.0283	.0661	.1168	.1715	.2194	.2508	.2600	.2461
	5	.0000	.0008	.0050	.0165	.0389	.0735	.1181	.1672	.2128	.2461
	6	.0000	.0001	.0006	.0028	.0087	.0210	.0424	.0743	.1160	.1641
	7	.0000	.0000	.0000	.0003	.0012	.0039	.0098	.0212	.0407	.0703
	8	.0000	.0000	.0000	.0000	.0001	.0004	.0013	.0035	.0083	.0176
	9	.0000	.0000	.0000	.0000	.0000	.0000	.0001	.0003	.0008	.0020
10	0	.5987	.3487	.1969	.1074	.0563	.0282	.0135	.0060	.0025	.0010
	1	.3151	.3874	.3474	.2684	.1877	.1211	.0725	.0403	.0207	.0098
	2	.0746	.1937	.2759	.3020	.2816	.2335	.1757	.1209	.0763	.0439
	3	.0105	.0574	.1298	.2013	.2503	.2668	.2522	.2150	.1665	.1172
	4	.0010	.0112	.0401	.0881	.1460	.2001	.2377	.2508	.2384	.2051
	5	.0001	.0015	.0085	.0264	.0584	.1029	.1536	.2007	.2340	.2461
	6	.0000	.0001	.0012	.0055	.0162	.0368	.0689	.1115	.1596	.2051
	7	.0000	.0000	.0001	.0008	.0031	.0090	.0212	.0425	.0746	.1172
	8	.0000	.0000	.0000	.0001	.0004	.0014	.0043	.0106	.0229	.0439
	9	.0000	.0000	.0000	.0000	.0000	.0001	.0005	.0016	.0042	.0098
	10	.0000	.0000	.0000	.0000	.0000	.0000	.0000	.0001	.0003	.0010
11	0	.5688	.3138	.1673	.0859	.0422	.0198	.0088	.0036	.0014	.0005
	1	.3293	.3835	.3248	.2362	.1549	.0932	.0518	.0266	.0125	.0054
	2	.0867	.2131	.2866	.2953	.2581	.1998	.1395	.0887	.0513	.0269
	3	.0137	.0710	.1517	.2215	.2581	.2568	.2254	.1774	.1259	.0806
	4	.0014	.0158	.0536	.1107	.1721	.2201	.2428	.2365	.2060	.1611
	5	.0001	.0025	.0132	.0388	.0803	.1321	.1830	.2207	.2360	.2256
	6	.0000	.0003	.0023	.0097	.0268	.0566	.0985	.1471	.1931	.2256
	7	.0000	.0000	.0003	.0017	.0064	.0173	.0379	.0701	.1128	.1611
	8	.0000	.0000	.0000	.0002	.0011	.0037	.0102	.0234	.0462	.0806
	9	.0000	.0000	.0000	.0000	.0001	.0005	.0018	.0052	.0126	.0269
	10	.0000	.0000	.0000	.0000	.0000	.0000	.0002	.0007	.0021	.0054
	11	.0000	.0000	.0000	.0000	.0000	.0000	.0000	.0000	.0002	.0005
12	0	.5404	.2824	.1422	.0687	.0317	.0138	.0057	.0022	.0008	.0002
	1	.3413	.3766	.3012	.2062	.1267	.0712	.0368	.0174	.0075	.0029
	2	.0988	.2301	.2924	.2835	.2323	.1678	.1088	.0639	.0339	.0161
	3	.0173	.0852	.1720	.2362	.2581	.2397	.1954	.1419	.0923	.0537
	4	.0021	.0213	.0683	.1329	.1936	.2311	.2367	.2128	.1700	.1208
	5	.0002	.0038	.0193	.0532	.1032	.1585	.2039	.2270	.2225	.1934
	6	.0000	.0005	.0040	.0155	.0401	.0792	.1281	.1766	.2124	.2256
	7	.0000	.0000	.0006	.0033	.0115	.0291	.0591	.1009	.1489	.1934

(continued on next page)

TABLE 1 Exact binomial probabilities $Pr(X = k) = \binom{n}{k} p^k q^{n-k}$ *(continued)*

n	k	.05	.10	.15	.20	.25	.30	.35	.40	.45	.50
	8	.0000	.0000	.0001	.0005	.0024	.0078	.0199	.0420	.0762	.1208
	9	.0000	.0000	.0000	.0001	.0004	.0015	.0048	.0125	.0277	.0537
	10	.0000	.0000	.0000	.0000	.0000	.0002	.0008	.0025	.0068	.0161
	11	.0000	.0000	.0000	.0000	.0000	.0000	.0001	.0003	.0010	.0029
	12	.0000	.0000	.0000	.0000	.0000	.0000	.0000	.0000	.0001	.0002
13	0	.5133	.2542	.1209	.0550	.0238	.0097	.0037	.0013	.0004	.0001
	1	.3512	.3672	.2774	.1787	.1029	.0540	.0259	.0113	.0045	.0016
	2	.1109	.2448	.2937	.2680	.2059	.1388	.0836	.0453	.0220	.0095
	3	.0214	.0997	.1900	.2457	.2517	.2181	.1651	.1107	.0660	.0349
	4	.0028	.0277	.0838	.1535	.2097	.2337	.2222	.1845	.1350	.0873
	5	.0003	.0055	.0266	.0691	.1258	.1803	.2154	.2214	.1989	.1571
	6	.0000	.0008	.0063	.0230	.0559	.1030	.1546	.1968	.2169	.2095
	7	.0000	.0001	.0011	.0058	.0186	.0442	.0833	.1312	.1775	.2095
	8	.0000	.0000	.0001	.0011	.0047	.0142	.0336	.0656	.1089	.1571
	9	.0000	.0000	.0000	.0001	.0009	.0034	.0101	.0243	.0495	.0873
	10	.0000	.0000	.0000	.0000	.0001	.0006	.0022	.0065	.0162	.0349
	11	.0000	.0000	.0000	.0000	.0000	.0001	.0003	.0012	.0036	.0095
	12	.0000	.0000	.0000	.0000	.0000	.0000	.0000	.0001	.0005	.0016
	13	.0000	.0000	.0000	.0000	.0000	.0000	.0000	.0000	.0000	.0001
14	0	.4877	.2288	.1028	.0440	.0178	.0068	.0024	.0008	.0002	.0001
	1	.3593	.3559	.2539	.1539	.0832	.0407	.0181	.0073	.0027	.0009
	2	.1229	.2570	.2912	.2501	.1802	.1134	.0634	.0317	.0141	.0056
	3	.0259	.1142	.2056	.2501	.2402	.1943	.1366	.0845	.0462	.0222
	4	.0037	.0349	.0998	.1720	.2202	.2290	.2022	.1549	.1040	.0611
	5	.0004	.0078	.0352	.0860	.1468	.1963	.2178	.2066	.1701	.1222
	6	.0000	.0013	.0093	.0322	.0734	.1262	.1759	.2066	.2088	.1833
	7	.0000	.0002	.0019	.0092	.0280	.0618	.1082	.1574	.1952	.2095
	8	.0000	.0000	.0003	.0020	.0082	.0232	.0510	.0918	.1398	.1833
	9	.0000	.0000	.0000	.0003	.0018	.0066	.0183	.0408	.0762	.1222
	10	.0000	.0000	.0000	.0000	.0003	.0014	.0049	.0136	.0312	.0611
	11	.0000	.0000	.0000	.0000	.0000	.0002	.0010	.0033	.0093	.0222
	12	.0000	.0000	.0000	.0000	.0000	.0000	.0001	.0005	.0019	.0056
	13	.0000	.0000	.0000	.0000	.0000	.0000	.0000	.0001	.0002	.0009
	14	.0000	.0000	.0000	.0000	.0000	.0000	.0000	.0000	.0000	.0001
15	0	.4633	.2059	.0874	.0352	.0134	.0047	.0016	.0005	.0001	.0000
	1	.3658	.3432	.2312	.1319	.0668	.0305	.0126	.0047	.0016	.0005
	2	.1348	.2669	.2856	.2309	.1559	.0916	.0476	.0219	.0090	.0032
	3	.0307	.1285	.2184	.2501	.2252	.1700	.1110	.0634	.0318	.0139
	4	.0049	.0428	.1156	.1876	.2252	.2186	.1792	.1268	.0780	.0417
	5	.0006	.0105	.0449	.1032	.1651	.2061	.2123	.1859	.1404	.0916
	6	.0000	.0019	.0132	.0430	.0917	.1472	.1906	.2066	.1914	.1527
	7	.0000	.0003	.0030	.0138	.0393	.0811	.1319	.1771	.2013	.1964
	8	.0000	.0000	.0005	.0035	.0131	.0348	.0710	.1181	.1647	.1964
	9	.0000	.0000	.0001	.0007	.0034	.0116	.0298	.0612	.1048	.1527
	10	.0000	.0000	.0000	.0001	.0007	.0030	.0096	.0245	.0515	.0916
	11	.0000	.0000	.0000	.0000	.0001	.0006	.0024	.0074	.0191	.0417
	12	.0000	.0000	.0000	.0000	.0000	.0001	.0004	.0016	.0052	.0139
	13	.0000	.0000	.0000	.0000	.0000	.0000	.0001	.0003	.0010	.0032
	14	.0000	.0000	.0000	.0000	.0000	.0000	.0000	.0000	.0001	.0005
	15	.0000	.0000	.0000	.0000	.0000	.0000	.0000	.0000	.0000	.0000

(continued on next page)

TABLE 1 **Exact binomial probabilities $Pr(X = k) = \binom{n}{k} p^k q^{n-k}$ *(continued)***

n	k	.05	.10	.15	.20	.25	.30	.35	.40	.45	.50
16	0	.4401	.1853	.0743	.0281	.0100	.0033	.0010	.0003	.0001	.0000
	1	.3706	.3294	.2097	.1126	.0535	.0228	.0087	.0030	.0009	.0002
	2	.1463	.2745	.2775	.2111	.1336	.0732	.0353	.0150	.0056	.0018
	3	.0359	.1423	.2285	.2463	.2079	.1465	.0888	.0468	.0215	.0085
	4	.0061	.0514	.1311	.2001	.2252	.2040	.1553	.1014	.0572	.0278
	5	.0008	.0137	.0555	.1201	.1802	.2099	.2008	.1623	.1123	.0667
	6	.0001	.0028	.0180	.0550	.1101	.1649	.1982	.1983	.1684	.1222
	7	.0000	.0004	.0045	.0197	.0524	.1010	.1524	.1889	.1969	.1746
	8	.0000	.0001	.0009	.0055	.0197	.0487	.0923	.1417	.1812	.1964
	9	.0000	.0000	.0001	.0012	.0058	.0185	.0442	.0840	.1318	.1746
	10	.0000	.0000	.0000	.0002	.0014	.0056	.0167	.0392	.0755	.1222
	11	.0000	.0000	.0000	.0000	.0002	.0013	.0049	.0142	.0337	.0667
	12	.0000	.0000	.0000	.0000	.0000	.0002	.0011	.0040	.0115	.0278
	13	.0000	.0000	.0000	.0000	.0000	.0000	.0002	.0008	.0029	.0085
	14	.0000	.0000	.0000	.0000	.0000	.0000	.0000	.0001	.0005	.0018
	15	.0000	.0000	.0000	.0000	.0000	.0000	.0000	.0000	.0001	.0002
	16	.0000	.0000	.0000	.0000	.0000	.0000	.0000	.0000	.0000	.0000
17	0	.4181	.1668	.0631	.0225	.0075	.0023	.0007	.0002	.0000	.0000
	1	.3741	.3150	.1893	.0957	.0426	.0169	.0060	.0019	.0005	.0001
	2	.1575	.2800	.2673	.1914	.1136	.0581	.0260	.0102	.0035	.0010
	3	.0415	.1556	.2359	.2393	.1893	.1245	.0701	.0341	.0144	.0052
	4	.0076	.0605	.1457	.2093	.2209	.1868	.1320	.0796	.0411	.0182
	5	.0010	.0175	.0668	.1361	.1914	.2081	.1849	.1379	.0875	.0472
	6	.0001	.0039	.0236	.0680	.1276	.1784	.1991	.1839	.1432	.0944
	7	.0000	.0007	.0065	.0267	.0668	.1201	.1685	.1927	.1841	.1484
	8	.0000	.0001	.0014	.0084	.0279	.0644	.1134	.1606	.1883	.1855
	9	.0000	.0000	.0003	.0021	.0093	.0276	.0611	.1070	.1540	.1855
	10	.0000	.0000	.0000	.0004	.0025	.0095	.0263	.0571	.1008	.1484
	11	.0000	.0000	.0000	.0001	.0005	.0026	.0090	.0242	.0525	.0944
	12	.0000	.0000	.0000	.0000	.0001	.0006	.0024	.0081	.0215	.0472
	13	.0000	.0000	.0000	.0000	.0000	.0001	.0005	.0021	.0068	.0182
	14	.0000	.0000	.0000	.0000	.0000	.0000	.0001	.0004	.0016	.0052
	15	.0000	.0000	.0000	.0000	.0000	.0000	.0000	.0001	.0003	.0010
	16	.0000	.0000	.0000	.0000	.0000	.0000	.0000	.0000	.0000	.0001
	17	.0000	.0000	.0000	.0000	.0000	.0000	.0000	.0000	.0000	.0000
18	0	.3972	.1501	.0536	.0180	.0056	.0016	.0004	.0001	.0000	.0000
	1	.3763	.3002	.1704	.0811	.0338	.0126	.0042	.0012	.0003	.0001
	2	.1683	.2835	.2556	.1723	.0958	.0458	.0190	.0069	.0022	.0006
	3	.0473	.1680	.2406	.2297	.1704	.1046	.0547	.0246	.0095	.0031
	4	.0093	.0700	.1592	.2153	.2130	.1681	.1104	.0614	.0291	.0117
	5	.0014	.0218	.0787	.1507	.1988	.2017	.1664	.1146	.0666	.0327
	6	.0002	.0052	.0301	.0816	.1436	.1873	.1941	.1655	.1181	.0708
	7	.0000	.0010	.0091	.0350	.0820	.1376	.1792	.1892	.1657	.1214
	8	.0000	.0002	.0022	.0120	.0376	.0811	.1327	.1734	.1864	.1669
	9	.0000	.0000	.0004	.0033	.0139	.0386	.0794	.1284	.1694	.1855
	10	.0000	.0000	.0001	.0008	.0042	.0149	.0385	.0771	.1248	.1669
	11	.0000	.0000	.0000	.0001	.0010	.0046	.0151	.0374	.0742	.1214
	12	.0000	.0000	.0000	.0000	.0002	.0012	.0047	.0145	.0354	.0708
	13	.0000	.0000	.0000	.0000	.0000	.0002	.0012	.0045	.0134	.0327
	14	.0000	.0000	.0000	.0000	.0000	.0000	.0002	.0011	.0039	.0117

(continued on next page)

TABLE 1 **Exact binomial probabilities** $Pr(X = k) = \binom{n}{k} p^k q^{n-k}$ ***(continued)***

n	k	.05	.10	.15	.20	.25	.30	.35	.40	.45	.50
	15	.0000	.0000	.0000	.0000	.0000	.0000	.0000	.0002	.0009	.0031
	16	.0000	.0000	.0000	.0000	.0000	.0000	.0000	.0000	.0001	.0006
	17	.0000	.0000	.0000	.0000	.0000	.0000	.0000	.0000	.0000	.0001
	18	.0000	.0000	.0000	.0000	.0000	.0000	.0000	.0000	.0000	.0000
19	0	.3774	.1351	.0456	.0144	.0042	.0011	.0003	.0001	.0000	.0000
	1	.3774	.2852	.1529	.0685	.0268	.0093	.0029	.0008	.0002	.0000
	2	.1787	.2852	.2428	.1540	.0803	.0358	.0138	.0046	.0013	.0003
	3	.0533	.1796	.2428	.2182	.1517	.0869	.0422	.0175	.0062	.0018
	4	.0112	.0798	.1714	.2182	.2023	.1491	.0909	.0467	.0203	.0074
	5	.0018	.0266	.0907	.1636	.2023	.1916	.1468	.0933	.0497	.0222
	6	.0002	.0069	.0374	.0955	.1574	.1916	.1844	.1451	.0949	.0518
	7	.0000	.0014	.0122	.0443	.0974	.1525	.1844	.1797	.1443	.0961
	8	.0000	.0002	.0032	.0166	.0487	.0981	.1489	.1797	.1771	.1442
	9	.0000	.0000	.0007	.0051	.0198	.0514	.0980	.1464	.1771	.1762
	10	.0000	.0000	.0001	.0013	.0066	.0220	.0528	.0976	.1449	.1762
	11	.0000	.0000	.0000	.0003	.0018	.0077	.0233	.0532	.0970	.1442
	12	.0000	.0000	.0000	.0000	.0004	.0022	.0083	.0237	.0529	.0961
	13	.0000	.0000	.0000	.0000	.0001	.0005	.0024	.0085	.0233	.0518
	14	.0000	.0000	.0000	.0000	.0000	.0001	.0006	.0024	.0082	.0222
	15	.0000	.0000	.0000	.0000	.0000	.0000	.0001	.0005	.0022	.0074
	16	.0000	.0000	.0000	.0000	.0000	.0000	.0000	.0001	.0005	.0018
	17	.0000	.0000	.0000	.0000	.0000	.0000	.0000	.0000	.0001	.0003
	18	.0000	.0000	.0000	.0000	.0000	.0000	.0000	.0000	.0000	.0000
	19	.0000	.0000	.0000	.0000	.0000	.0000	.0000	.0000	.0000	.0000
20	0	.3585	.1216	.0388	.0115	.0032	.0008	.0002	.0000	.0000	.0000
	1	.3774	.2702	.1368	.0576	.0211	.0068	.0020	.0005	.0001	.0000
	2	.1887	.2852	.2293	.1369	.0669	.0278	.0100	.0031	.0008	.0002
	3	.0596	.1901	.2428	.2054	.1339	.0716	.0323	.0123	.0040	.0011
	4	.0133	.0898	.1821	.2182	.1897	.1304	.0738	.0350	.0139	.0046
	5	.0022	.0319	.1028	.1746	.2023	.1789	.1272	.0746	.0365	.0148
	6	.0003	.0089	.0454	.1091	.1686	.1916	.1712	.1244	.0746	.0370
	7	.0000	.0020	.0160	.0546	.1124	.1643	.1844	.1659	.1221	.0739
	8	.0000	.0004	.0046	.0222	.0609	.1144	.1614	.1797	.1623	.1201
	9	.0000	.0001	.0011	.0074	.0271	.0654	.1158	.1597	.1771	.1602
	10	.0000	.0000	.0002	.0020	.0099	.0308	.0686	.1171	.1593	.1762
	11	.0000	.0000	.0000	.0005	.0030	.0120	.0336	.0710	.1185	.1602
	12	.0000	.0000	.0000	.0001	.0008	.0039	.0136	.0355	.0727	.1201
	13	.0000	.0000	.0000	.0000	.0002	.0010	.0045	.0146	.0366	.0739
	14	.0000	.0000	.0000	.0000	.0000	.0002	.0012	.0049	.0150	.0370
	15	.0000	.0000	.0000	.0000	.0000	.0000	.0003	.0013	.0049	.0148
	16	.0000	.0000	.0000	.0000	.0000	.0000	.0000	.0003	.0013	.0046
	17	.0000	.0000	.0000	.0000	.0000	.0000	.0000	.0000	.0002	.0011
	18	.0000	.0000	.0000	.0000	.0000	.0000	.0000	.0000	.0000	.0002
	19	.0000	.0000	.0000	.0000	.0000	.0000	.0000	.0000	.0000	.0000
	20	.0000	.0000	.0000	.0000	.0000	.0000	.0000	.0000	.0000	.0000

TABLE 2 **Exact Poisson probabilities $Pr(X = k) = \frac{e^{-\mu}\mu^k}{k!}$**

	μ									
k	0.5	1.0	1.5	2.0	2.5	3.0	3.5	4.0	4.5	5.0
0	.6065	.3679	.2231	.1353	.0821	.0498	.0302	.0183	.0111	.0067
1	.3033	.3679	.3347	.2707	.2052	.1494	.1057	.0733	.0500	.0337
2	.0758	.1839	.2510	.2707	.2565	.2240	.1850	.1465	.1125	.0842
3	.0126	.0613	.1255	.1804	.2138	.2240	.2158	.1954	.1687	.1404
4	.0016	.0153	.0471	.0902	.1336	.1680	.1888	.1954	.1898	.1755
5	.0002	.0031	.0141	.0361	.0668	.1008	.1322	.1563	.1708	.1755
6	.0000	.0005	.0035	.0120	.0278	.0504	.0771	.1042	.1281	.1462
7	.0000	.0001	.0008	.0034	.0099	.0216	.0385	.0595	.0824	.1044
8	.0000	.0000	.0001	.0009	.0031	.0081	.0169	.0298	.0463	.0653
9	.0000	.0000	.0000	.0002	.0009	.0027	.0066	.0132	.0232	.0363
10	.0000	.0000	.0000	.0000	.0002	.0008	.0023	.0053	.0104	.0181
11	.0000	.0000	.0000	.0000	.0000	.0002	.0007	.0019	.0043	.0082
12	.0000	.0000	.0000	.0000	.0000	.0001	.0002	.0006	.0016	.0034
13	.0000	.0000	.0000	.0000	.0000	.0000	.0001	.0002	.0006	.0013
14	.0000	.0000	.0000	.0000	.0000	.0000	.0000	.0001	.0002	.0005
15	.0000	.0000	.0000	.0000	.0000	.0000	.0000	.0000	.0001	.0002
16	.0000	.0000	.0000	.0000	.0000	.0000	.0000	.0000	.0000	.0000

	μ									
k	5.5	6.0	6.5	7.0	7.5	8.0	8.5	9.0	9.5	10.0
0	.0041	.0025	.0015	.0009	.0006	.0003	.0002	.0001	.0001	.0000
1	.0225	.0149	.0098	.0064	.0041	.0027	.0017	.0011	.0007	.0005
2	.0618	.0446	.0318	.0223	.0156	.0107	.0074	.0050	.0034	.0023
3	.1133	.0892	.0688	.0521	.0389	.0286	.0208	.0150	.0107	.0076
4	.1558	.1339	.1118	.0912	.0729	.0573	.0443	.0337	.0254	.0189
5	.1714	.1606	.1454	.1277	.1094	.0916	.0752	.0607	.0483	.0378
6	.1571	.1606	.1575	.1490	.1367	.1221	.1066	.0911	.0764	.0631
7	.1234	.1377	.1462	.1490	.1465	.1396	.1294	.1171	.1037	.0901
8	.0849	.1033	.1188	.1304	.1373	.1396	.1375	.1318	.1232	.1126
9	.0519	.0688	.0858	.1014	.1144	.1241	.1299	.1318	.1300	.1251
10	.0285	.0413	.0558	.0710	.0858	.0993	.1104	.1186	.1235	.1251
11	.0143	.0225	.0330	.0452	.0585	.0722	.0853	.0970	.1067	.1137
12	.0065	.0113	.0179	.0263	.0366	.0481	.0604	.0728	.0844	.0948
13	.0028	.0052	.0089	.0142	.0211	.0296	.0395	.0504	.0617	.0729
14	.0011	.0022	.0041	.0071	.0113	.0169	.0240	.0324	.0419	.0521
15	.0004	.0009	.0018	.0033	.0057	.0090	.0136	.0194	.0265	.0347
16	.0001	.0003	.0007	.0014	.0026	.0045	.0072	.0109	.0157	.0217
17	.0000	.0001	.0003	.0006	.0012	.0021	.0036	.0058	.0088	.0128
18	.0000	.0000	.0001	.0002	.0005	.0009	.0017	.0029	.0046	.0071
19	.0000	.0000	.0000	.0001	.0002	.0004	.0008	.0014	.0023	.0037
20	.0000	.0000	.0000	.0000	.0001	.0002	.0003	.0006	.0011	.0019
21	.0000	.0000	.0000	.0000	.0000	.0001	.0001	.0003	.0005	.0009
22	.0000	.0000	.0000	.0000	.0000	.0000	.0001	.0001	.0002	.0004
23	.0000	.0000	.0000	.0000	.0000	.0000	.0000	.0000	.0001	.0002
24	.0000	.0000	.0000	.0000	.0000	.0000	.0000	.0000	.0000	.0001
25	.0000	.0000	.0000	.0000	.0000	.0000	.0000	.0000	.0000	.0000

(continued on next page)

TABLE 2 **Exact Poisson probabilities** $Pr(X = k) = \frac{e^{-\mu}\mu^k}{k!}$ *(continued)*

	μ									
k	10.5	11.0	11.5	12.0	12.5	13.0	13.5	14.0	14.5	15.0
0	.0000	.0000	.0000	.0000	.0000	.0000	.0000	.0000	.0000	.0000
1	.0003	.0002	.0001	.0001	.0000	.0000	.0000	.0000	.0000	.0000
2	.0015	.0010	.0007	.0004	.0003	.0002	.0001	.0001	.0001	.0000
3	.0053	.0037	.0026	.0018	.0012	.0008	.0006	.0004	.0003	.0002
4	.0139	.0102	.0074	.0053	.0038	.0027	.0019	.0013	.0009	.0006
5	.0293	.0224	.0170	.0127	.0095	.0070	.0051	.0037	.0027	.0019
6	.0513	.0411	.0325	.0255	.0197	.0152	.0115	.0087	.0065	.0048
7	.0769	.0646	.0535	.0437	.0353	.0281	.0222	.0174	.0135	.0104
8	.1009	.0888	.0769	.0655	.0551	.0457	.0375	.0304	.0244	.0194
9	.1177	.1085	.0982	.0874	.0765	.0661	.0563	.0473	.0394	.0324
10	.1236	.1194	.1129	.1048	.0956	.0859	.0760	.0663	.0571	.0486
11	.1180	.1194	.1181	.1144	.1087	.1015	.0932	.0844	.0753	.0663
12	.1032	.1094	.1131	.1144	.1132	.1099	.1049	.0984	.0910	.0829
13	.0834	.0926	.1001	.1056	.1089	.1099	.1089	.1060	.1014	.0956
14	.0625	.0728	.0822	.0905	.0972	.1021	.1050	.1060	.1051	.1024
15	.0438	.0534	.0630	.0724	.0810	.0885	.0945	.0989	.1016	.1024
16	.0287	.0367	.0453	.0543	.0633	.0719	.0798	.0866	.0920	.0960
17	.0177	.0237	.0306	.0383	.0465	.0550	.0633	.0713	.0785	.0847
18	.0104	.0145	.0196	.0255	.0323	.0397	.0475	.0554	.0632	.0706
19	.0057	.0084	.0119	.0161	.0213	.0272	.0337	.0409	.0483	.0557
20	.0030	.0046	.0068	.0097	.0133	.0177	.0228	.0286	.0350	.0418
21	.0015	.0024	.0037	.0055	.0079	.0109	.0146	.0191	.0242	.0299
22	.0007	.0012	.0020	.0030	.0045	.0065	.0090	.0121	.0159	.0204
23	.0003	.0006	.0010	.0016	.0024	.0037	.0053	.0074	.0100	.0133
24	.0001	.0003	.0005	.0008	.0013	.0020	.0030	.0043	.0061	.0083
25	.0001	.0001	.0002	.0004	.0006	.0010	.0016	.0024	.0035	.0050
26	.0000	.0000	.0001	.0002	.0003	.0005	.0008	.0013	.0020	.0029
27	.0000	.0000	.0000	.0001	.0001	.0002	.0004	.0007	.0011	.0016
28	.0000	.0000	.0000	.0000	.0001	.0001	.0002	.0003	.0005	.0009
29	.0000	.0000	.0000	.0000	.0000	.0001	.0001	.0002	.0003	.0004
30	.0000	.0000	.0000	.0000	.0000	.0000	.0000	.0001	.0001	.0002
31	.0000	.0000	.0000	.0000	.0000	.0000	.0000	.0000	.0001	.0001
32	.0000	.0000	.0000	.0000	.0000	.0000	.0000	.0000	.0000	.0001
33	.0000	.0000	.0000	.0000	.0000	.0000	.0000	.0000	.0000	.0000

	μ									
k	15.5	16.0	16.5	17.0	17.5	18.0	18.5	19.0	19.5	20.0
0	.0000	.0000	.0000	.0000	.0000	.0000	.0000	.0000	.0000	.0000
1	.0000	.0000	.0000	.0000	.0000	.0000	.0000	.0000	.0000	.0000
2	.0000	.0000	.0000	.0000	.0000	.0000	.0000	.0000	.0000	.0000
3	.0001	.0001	.0001	.0000	.0000	.0000	.0000	.0000	.0000	.0000
4	.0004	.0003	.0002	.0001	.0001	.0001	.0000	.0000	.0000	.0000
5	.0014	.0010	.0007	.0005	.0003	.0002	.0002	.0001	.0001	.0001
6	.0036	.0026	.0019	.0014	.0010	.0007	.0005	.0004	.0003	.0002
7	.0079	.0060	.0045	.0034	.0025	.0019	.0014	.0010	.0007	.0005
8	.0153	.0120	.0093	.0072	.0055	.0042	.0031	.0024	.0018	.0013

(continued on next page)

TABLE 2 Exact Poisson probabilities $Pr(X = k) = \frac{e^{-\mu}\mu^k}{k!}$ *(continued)*

	μ									
k	15.5	16.0	16.5	17.0	17.5	18.0	18.5	19.0	19.5	20.0
9	.0264	.0213	.0171	.0135	.0107	.0083	.0065	.0050	.0038	.0029
10	.0409	.0341	.0281	.0230	.0186	.0150	.0120	.0095	.0074	.0058
11	.0577	.0496	.0422	.0355	.0297	.0245	.0201	.0164	.0132	.0106
12	.0745	.0661	.0580	.0504	.0432	.0368	.0310	.0259	.0214	.0176
13	.0888	.0814	.0736	.0658	.0582	.0509	.0441	.0378	.0322	.0271
14	.0983	.0930	.0868	.0800	.0728	.0655	.0583	.0514	.0448	.0387
15	.1016	.0992	.0955	.0906	.0849	.0786	.0719	.0650	.0582	.0516
16	.0984	.0992	.0985	.0963	.0929	.0884	.0831	.0772	.0710	.0646
17	.0897	.0934	.0956	.0963	.0956	.0936	.0904	.0863	.0814	.0760
18	.0773	.0830	.0876	.0909	.0929	.0936	.0930	.0911	.0882	.0844
19	.0630	.0699	.0761	.0814	.0856	.0887	.0905	.0911	.0905	.0888
20	.0489	.0559	.0628	.0692	.0749	.0798	.0837	.0866	.0883	.0888
21	.0361	.0426	.0493	.0560	.0624	.0684	.0738	.0783	.0820	.0846
22	.0254	.0310	.0370	.0433	.0496	.0560	.0620	.0676	.0727	.0769
23	.0171	.0216	.0265	.0320	.0378	.0438	.0499	.0559	.0616	.0669
24	.0111	.0144	.0182	.0226	.0275	.0328	.0385	.0442	.0500	.0557
25	.0069	.0092	.0120	.0154	.0193	.0237	.0285	.0336	.0390	.0446
26	.0041	.0057	.0076	.0101	.0130	.0164	.0202	.0246	.0293	.0343
27	.0023	.0034	.0047	.0063	.0084	.0109	.0139	.0173	.0211	.0254
28	.0013	.0019	.0028	.0038	.0053	.0070	.0092	.0117	.0147	.0181
29	.0007	.0011	.0016	.0023	.0032	.0044	.0058	.0077	.0099	.0125
30	.0004	.0006	.0009	.0013	.0019	.0026	.0036	.0049	.0064	.0083
31	.0002	.0003	.0005	.0007	.0010	.0015	.0022	.0030	.0040	.0054
32	.0001	.0001	.0002	.0004	.0006	.0009	.0012	.0018	.0025	.0034
33	.0000	.0001	.0001	.0002	.0003	.0005	.0007	.0010	.0015	.0020
34	.0000	.0000	.0001	.0001	.0002	.0002	.0004	.0006	.0008	.0012
35	.0000	.0000	.0000	.0000	.0001	.0001	.0002	.0003	.0005	.0007
36	.0000	.0000	.0000	.0000	.0000	.0001	.0001	.0002	.0003	.0004
37	.0000	.0000	.0000	.0000	.0000	.0000	.0001	.0001	.0001	.0002
38	.0000	.0000	.0000	.0000	.0000	.0000	.0000	.0000	.0001	.0001
39	.0000	.0000	.0000	.0000	.0000	.0000	.0000	.0000	.0000	.0001
40	.0000	.0000	.0000	.0000	.0000	.0000	.0000	.0000	.0000	.0000

TABLE 3 The normal distribution

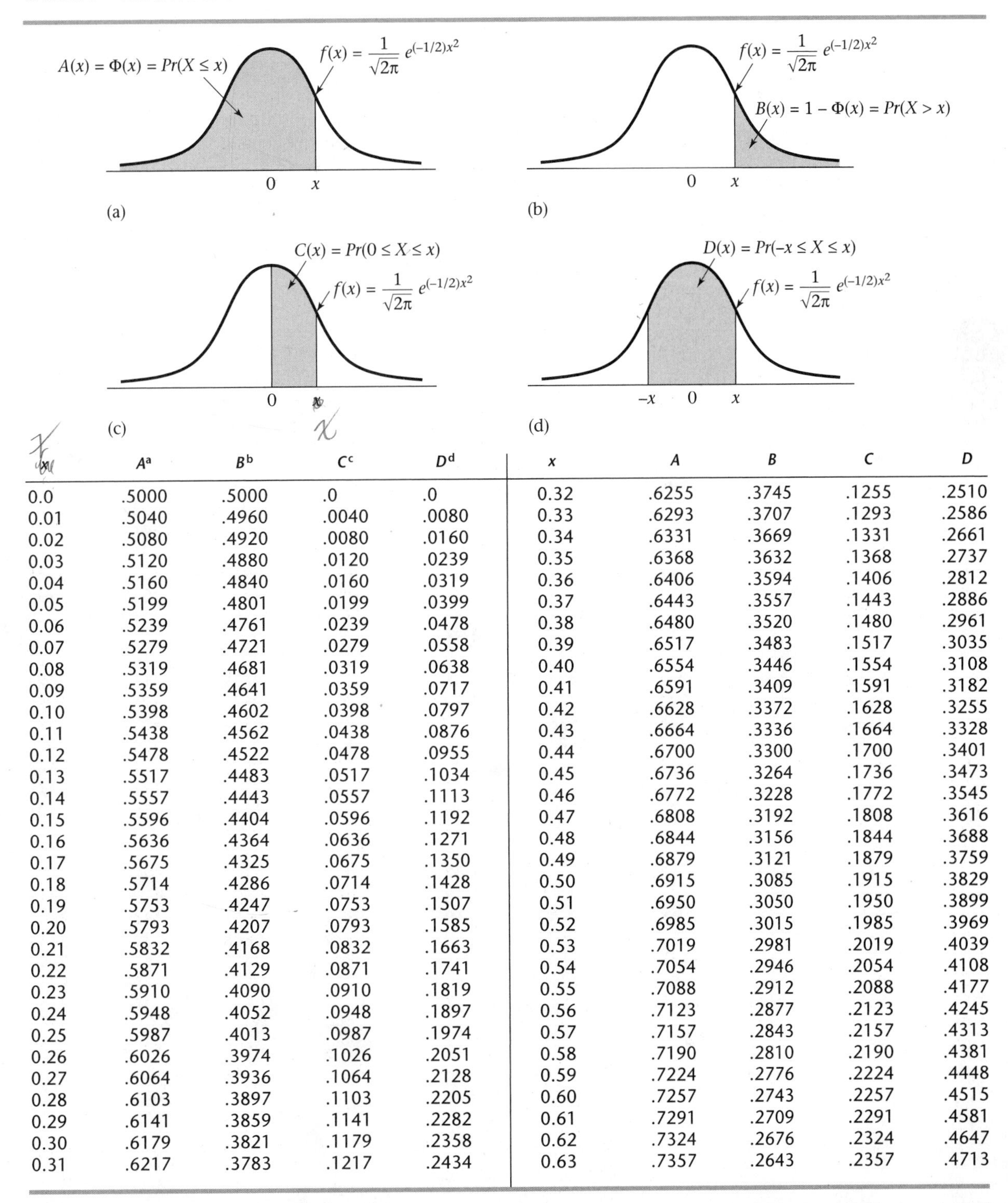

x	A[a]	B[b]	C[c]	D[d]
0.0	.5000	.5000	.0	.0
0.01	.5040	.4960	.0040	.0080
0.02	.5080	.4920	.0080	.0160
0.03	.5120	.4880	.0120	.0239
0.04	.5160	.4840	.0160	.0319
0.05	.5199	.4801	.0199	.0399
0.06	.5239	.4761	.0239	.0478
0.07	.5279	.4721	.0279	.0558
0.08	.5319	.4681	.0319	.0638
0.09	.5359	.4641	.0359	.0717
0.10	.5398	.4602	.0398	.0797
0.11	.5438	.4562	.0438	.0876
0.12	.5478	.4522	.0478	.0955
0.13	.5517	.4483	.0517	.1034
0.14	.5557	.4443	.0557	.1113
0.15	.5596	.4404	.0596	.1192
0.16	.5636	.4364	.0636	.1271
0.17	.5675	.4325	.0675	.1350
0.18	.5714	.4286	.0714	.1428
0.19	.5753	.4247	.0753	.1507
0.20	.5793	.4207	.0793	.1585
0.21	.5832	.4168	.0832	.1663
0.22	.5871	.4129	.0871	.1741
0.23	.5910	.4090	.0910	.1819
0.24	.5948	.4052	.0948	.1897
0.25	.5987	.4013	.0987	.1974
0.26	.6026	.3974	.1026	.2051
0.27	.6064	.3936	.1064	.2128
0.28	.6103	.3897	.1103	.2205
0.29	.6141	.3859	.1141	.2282
0.30	.6179	.3821	.1179	.2358
0.31	.6217	.3783	.1217	.2434
0.32	.6255	.3745	.1255	.2510
0.33	.6293	.3707	.1293	.2586
0.34	.6331	.3669	.1331	.2661
0.35	.6368	.3632	.1368	.2737
0.36	.6406	.3594	.1406	.2812
0.37	.6443	.3557	.1443	.2886
0.38	.6480	.3520	.1480	.2961
0.39	.6517	.3483	.1517	.3035
0.40	.6554	.3446	.1554	.3108
0.41	.6591	.3409	.1591	.3182
0.42	.6628	.3372	.1628	.3255
0.43	.6664	.3336	.1664	.3328
0.44	.6700	.3300	.1700	.3401
0.45	.6736	.3264	.1736	.3473
0.46	.6772	.3228	.1772	.3545
0.47	.6808	.3192	.1808	.3616
0.48	.6844	.3156	.1844	.3688
0.49	.6879	.3121	.1879	.3759
0.50	.6915	.3085	.1915	.3829
0.51	.6950	.3050	.1950	.3899
0.52	.6985	.3015	.1985	.3969
0.53	.7019	.2981	.2019	.4039
0.54	.7054	.2946	.2054	.4108
0.55	.7088	.2912	.2088	.4177
0.56	.7123	.2877	.2123	.4245
0.57	.7157	.2843	.2157	.4313
0.58	.7190	.2810	.2190	.4381
0.59	.7224	.2776	.2224	.4448
0.60	.7257	.2743	.2257	.4515
0.61	.7291	.2709	.2291	.4581
0.62	.7324	.2676	.2324	.4647
0.63	.7357	.2643	.2357	.4713

(continued on next page)

TABLE 3 The normal distribution *(continued)*

x	A[a]	B[b]	C[c]	D[d]	x	A	B	C	D
0.64	.7389	.2611	.2389	.4778	1.10	.8643	.1357	.3643	.7287
0.65	.7422	.2578	.2422	.4843	1.11	.8665	.1335	.3665	.7330
0.66	.7454	.2546	.2454	.4907	1.12	.8686	.1314	.3686	.7373
0.67	.7486	.2514	.2486	.4971	1.13	.8708	.1292	.3708	.7415
0.68	.7517	.2483	.2517	.5035	1.14	.8729	.1271	.3729	.7457
0.69	.7549	.2451	.2549	.5098	1.15	.8749	.1251	.3749	.7499
0.70	.7580	.2420	.2580	.5161	1.16	.8770	.1230	.3770	.7540
0.71	.7611	.2389	.2611	.5223	1.17	.8790	.1210	.3790	.7580
0.72	.7642	.2358	.2642	.5285	1.18	.8810	.1190	.3810	.7620
0.73	.7673	.2327	.2673	.5346	1.19	.8830	.1170	.3830	.7660
0.74	.7703	.2297	.2703	.5407	1.20	.8849	.1151	.3849	.7699
0.75	.7734	.2266	.2734	.5467	1.21	.8869	.1131	.3869	.7737
0.76	.7764	.2236	.2764	.5527	1.22	.8888	.1112	.3888	.7775
0.77	.7793	.2207	.2793	.5587	1.23	.8907	.1093	.3907	.7813
0.78	.7823	.2177	.2823	.5646	1.24	.8925	.1075	.3925	.7850
0.79	.7852	.2148	.2852	.5705	1.25	.8944	.1056	.3944	.7887
0.80	.7881	.2119	.2881	.5763	1.26	.8962	.1038	.3962	.7923
0.81	.7910	.2090	.2910	.5821	1.27	.8980	.1020	.3980	.7959
0.82	.7939	.2061	.2939	.5878	1.28	.8997	.1003	.3997	.7995
0.83	.7967	.2033	.2967	.5935	1.29	.9015	.0985	.4015	.8029
0.84	.7995	.2005	.2995	.5991	1.30	.9032	.0968	.4032	.8064
0.85	.8023	.1977	.3023	.6047	1.31	.9049	.0951	.4049	.8098
0.86	.8051	.1949	.3051	.6102	1.32	.9066	.0934	.4066	.8132
0.87	.8078	.1922	.3078	.6157	1.33	.9082	.0918	.4082	.8165
0.88	.8106	.1894	.3106	.6211	1.34	.9099	.0901	.4099	.8198
0.89	.8133	.1867	.3133	.6265	1.35	.9115	.0885	.4115	.8230
0.90	.8159	.1841	.3159	.6319	1.36	.9131	.0869	.4131	.8262
0.91	.8186	.1814	.3186	.6372	1.37	.9147	.0853	.4147	.8293
0.92	.8212	.1788	.3212	.6424	1.38	.9162	.0838	.4162	.8324
0.93	.8238	.1762	.3238	.6476	1.39	.9177	.0823	.4177	.8355
0.94	.8264	.1736	.3264	.6528	1.40	.9192	.0808	.4192	.8385
0.95	.8289	.1711	.3289	.6579	1.41	.9207	.0793	.4207	.8415
0.96	.8315	.1685	.3315	.6629	1.42	.9222	.0778	.4222	.8444
0.97	.8340	.1660	.3340	.6680	1.43	.9236	.0764	.4236	.8473
0.98	.8365	.1635	.3365	.6729	1.44	.9251	.0749	.4251	.8501
0.99	.8389	.1611	.3389	.6778	1.45	.9265	.0735	.4265	.8529
1.00	.8413	.1587	.3413	.6827	1.46	.9279	.0721	.4279	.8557
1.01	.8438	.1562	.3438	.6875	1.47	.9292	.0708	.4292	.8584
1.02	.8461	.1539	.3461	.6923	1.48	.9306	.0694	.4306	.8611
1.03	.8485	.1515	.3485	.6970	1.49	.9319	.0681	.4319	.8638
1.04	.8508	.1492	.3508	.7017	1.50	.9332	.0668	.4332	.8664
1.05	.8531	.1469	.3531	.7063	1.51	.9345	.0655	.4345	.8690
1.06	.8554	.1446	.3554	.7109	1.52	.9357	.0643	.4357	.8715
1.07	.8577	.1423	.3577	.7154	1.53	.9370	.0630	.4370	.8740
1.08	.8599	.1401	.3599	.7199	1.54	.9382	.0618	.4382	.8764
1.09	.8621	.1379	.3621	.7243	1.55	.9394	.0606	.4394	.8789

(continued on next page)

TABLE 3 The normal distribution *(continued)*

x	A[a]	B[b]	C[c]	D[d]	x	A	B	C	D
1.56	.9406	.0594	.4406	.8812	2.03	.9788	.0212	.4788	.9576
1.57	.9418	.0582	.4418	.8836	2.04	.9793	.0207	.4793	.9586
1.58	.9429	.0571	.4429	.8859	2.05	.9798	.0202	.4798	.9596
1.59	.9441	.0559	.4441	.8882	2.06	.9803	.0197	.4803	.9606
1.60	.9452	.0548	.4452	.8904	2.07	.9808	.0192	.4808	.9615
1.61	.9463	.0537	.4463	.8926	2.08	.9812	.0188	.4812	.9625
1.62	.9474	.0526	.4474	.8948	2.09	.9817	.0183	.4817	.9634
1.63	.9484	.0516	.4484	.8969	2.10	.9821	.0179	.4821	.9643
1.64	.9495	.0505	.4495	.8990	2.11	.9826	.0174	.4826	.9651
1.65	.9505	.0495	.4505	.9011	2.12	.9830	.0170	.4830	.9660
1.66	.9515	.0485	.4515	.9031	2.13	.9834	.0166	.4834	.9668
1.67	.9525	.0475	.4525	.9051	2.14	.9838	.0162	.4838	.9676
1.68	.9535	.0465	.4535	.9070	2.15	.9842	.0158	.4842	.9684
1.69	.9545	.0455	.4545	.9090	2.16	.9846	.0154	.4846	.9692
1.70	.9554	.0446	.4554	.9109	2.17	.9850	.0150	.4850	.9700
1.71	.9564	.0436	.4564	.9127	2.18	.9854	.0146	.4854	.9707
1.72	.9573	.0427	.4573	.9146	2.19	.9857	.0143	.4857	.9715
1.73	.9582	.0418	.4582	.9164	2.20	.9861	.0139	.4861	.9722
1.74	.9591	.0409	.4591	.9181	2.21	.9864	.0136	.4864	.9729
1.75	.9599	.0401	.4599	.9199	2.22	.9868	.0132	.4868	.9736
1.76	.9608	.0392	.4608	.9216	2.23	.9871	.0129	.4871	.9743
1.77	.9616	.0384	.4616	.9233	2.24	.9875	.0125	.4875	.9749
1.78	.9625	.0375	.4625	.9249	2.25	.9878	.0122	.4878	.9756
1.79	.9633	.0367	.4633	.9265	2.26	.9881	.0119	.4881	.9762
1.80	.9641	.0359	.4641	.9281	2.27	.9884	.0116	.4884	.9768
1.81	.9649	.0351	.4649	.9297	2.28	.9887	.0113	.4887	.9774
1.82	.9656	.0344	.4656	.9312	2.29	.9890	.0110	.4890	.9780
1.83	.9664	.0336	.4664	.9327	2.30	.9893	.0107	.4893	.9786
1.84	.9671	.0329	.4671	.9342	2.31	.9896	.0104	.4896	.9791
1.85	.9678	.0322	.4678	.9357	2.32	.9898	.0102	.4898	.9797
1.86	.9686	.0314	.4686	.9371	2.33	.9901	.0099	.4901	.9802
1.87	.9693	.0307	.4693	.9385	2.34	.9904	.0096	.4904	.9807
1.88	.9699	.0301	.4699	.9399	2.35	.9906	.0094	.4906	.9812
1.89	.9706	.0294	.4706	.9412	2.36	.9909	.0091	.4909	.9817
1.90	.9713	.0287	.4713	.9426	2.37	.9911	.0089	.4911	.9822
1.91	.9719	.0281	.4719	.9439	2.38	.9913	.0087	.4913	.9827
1.92	.9726	.0274	.4726	.9451	2.39	.9916	.0084	.4916	.9832
1.93	.9732	.0268	.4732	.9464	2.40	.9918	.0082	.4918	.9836
1.94	.9738	.0262	.4738	.9476	2.41	.9920	.0080	.4920	.9840
1.95	.9744	.0256	.4744	.9488	2.42	.9922	.0078	.4922	.9845
1.96	.9750	.0250	.4750	.9500	2.43	.9925	.0075	.4925	.9849
1.97	.9756	.0244	.4756	.9512	2.44	.9927	.0073	.4927	.9853
1.98	.9761	.0239	.4761	.9523	2.45	.9929	.0071	.4929	.9857
1.99	.9767	.0233	.4767	.9534	2.46	.9931	.0069	.4931	.9861
2.00	.9772	.0228	.4772	.9545	2.47	.9932	.0068	.4932	.9865
2.01	.9778	.0222	.4778	.9556	2.48	.9934	.0066	.4934	.9869
2.02	.9783	.0217	.4783	.9566	2.49	.9936	.0064	.4936	.9872

(continued on next page)

TABLE 3 The normal distribution *(continued)*

x	A[a]	B[b]	C[c]	D[d]	x	A	B	C	D
2.50	.9938	.0062	.4938	.9876	2.97	.9985	.0015	.4985	.9970
2.51	.9940	.0060	.4940	.9879	2.98	.9986	.0014	.4986	.9971
2.52	.9941	.0059	.4941	.9883	2.99	.9986	.0014	.4986	.9972
2.53	.9943	.0057	.4943	.9886	3.00	.9987	.0013	.4987	.9973
2.54	.9945	.0055	.4945	.9889	3.01	.9987	.0013	.4987	.9974
2.55	.9946	.0054	.4946	.9892	3.02	.9987	.0013	.4987	.9975
2.56	.9948	.0052	.4948	.9895	3.03	.9988	.0012	.4988	.9976
2.57	.9949	.0051	.4949	.9898	3.04	.9988	.0012	.4988	.9976
2.58	.9951	.0049	.4951	.9901	3.05	.9989	.0011	.4989	.9977
2.59	.9952	.0048	.4952	.9904	3.06	.9989	.0011	.4989	.9978
2.60	.9953	.0047	.4953	.9907	3.07	.9989	.0011	.4989	.9979
2.61	.9955	.0045	.4955	.9909	3.08	.9990	.0010	.4990	.9979
2.62	.9956	.0044	.4956	.9912	3.09	.9990	.0010	.4990	.9980
2.63	.9957	.0043	.4957	.9915	3.10	.9990	.0010	.4990	.9981
2.64	.9959	.0041	.4959	.9917	3.11	.9991	.0009	.4991	.9981
2.65	.9960	.0040	.4960	.9920	3.12	.9991	.0009	.4991	.9982
2.66	.9961	.0039	.4961	.9922	3.13	.9991	.0009	.4991	.9983
2.67	.9962	.0038	.4962	.9924	3.14	.9992	.0008	.4992	.9983
2.68	.9963	.0037	.4963	.9926	3.15	.9992	.0008	.4992	.9984
2.69	.9964	.0036	.4964	.9929	3.16	.9992	.0008	.4992	.9984
2.70	.9965	.0035	.4965	.9931	3.17	.9992	.0008	.4992	.9985
2.71	.9966	.0034	.4966	.9933	3.18	.9993	.0007	.4993	.9985
2.72	.9967	.0033	.4967	.9935	3.19	.9993	.0007	.4993	.9986
2.73	.9968	.0032	.4968	.9937	3.20	.9993	.0007	.4993	.9986
2.74	.9969	.0031	.4969	.9939	3.21	.9993	.0007	.4993	.9987
2.75	.9970	.0030	.4970	.9940	3.22	.9994	.0006	.4994	.9987
2.76	.9971	.0029	.4971	.9942	3.23	.9994	.0006	.4994	.9988
2.77	.9972	.0028	.4972	.9944	3.24	.9994	.0006	.4994	.9988
2.78	.9973	.0027	.4973	.9946	3.25	.9994	.0006	.4994	.9988
2.79	.9974	.0026	.4974	.9947	3.26	.9994	.0006	.4994	.9989
2.80	.9974	.0026	.4974	.9949	3.27	.9995	.0005	.4995	.9989
2.81	.9975	.0025	.4975	.9950	3.28	.9995	.0005	.4995	.9990
2.82	.9976	.0024	.4976	.9952	3.29	.9995	.0005	.4995	.9990
2.83	.9977	.0023	.4977	.9953	3.30	.9995	.0005	.4995	.9990
2.84	.9977	.0023	.4977	.9955	3.31	.9995	.0005	.4995	.9991
2.85	.9978	.0022	.4978	.9956	3.32	.9995	.0005	.4995	.9991
2.86	.9979	.0021	.4979	.9958	3.33	.9996	.0004	.4996	.9991
2.87	.9979	.0021	.4979	.9959	3.34	.9996	.0004	.4996	.9992
2.88	.9980	.0020	.4980	.9960	3.35	.9996	.0004	.4996	.9992
2.89	.9981	.0019	.4981	.9961	3.36	.9996	.0004	.4996	.9992
2.90	.9981	.0019	.4981	.9963	3.37	.9996	.0004	.4996	.9992
2.91	.9982	.0018	.4982	.9964	3.38	.9996	.0004	.4996	.9993
2.92	.9982	.0018	.4982	.9965	3.39	.9997	.0003	.4997	.9993
2.93	.9983	.0017	.4983	.9966	3.40	.9997	.0003	.4997	.9993
2.94	.9984	.0016	.4984	.9967	3.41	.9997	.0003	.4997	.9993
2.95	.9984	.0016	.4984	.9968	3.42	.9997	.0003	.4997	.9994
2.96	.9985	.0015	.4985	.9969	3.43	.9997	.0003	.4997	.9994

TABLE 3 The normal distribution ***(continued)***

x	A[a]	B[b]	C[c]	D[d]	x	A	B	C	D
3.44	.9997	.0003	.4997	.9994	3.72	.9999	.0001	.4999	.9998
3.45	.9997	.0003	.4997	.9994	3.73	.9999	.0001	.4999	.9998
3.46	.9997	.0003	.4997	.9995	3.74	.9999	.0001	.4999	.9998
3.47	.9997	.0003	.4997	.9995	3.75	.9999	.0001	.4999	.9998
3.48	.9997	.0003	.4997	.9995	3.76	.9999	.0001	.4999	.9998
3.49	.9998	.0002	.4998	.9995	3.77	.9999	.0001	.4999	.9998
3.50	.9998	.0002	.4998	.9995	3.78	.9999	.0001	.4999	.9998
3.51	.9998	.0002	.4998	.9996	3.79	.9999	.0001	.4999	.9998
3.52	.9998	.0002	.4998	.9996	3.80	.9999	.0001	.4999	.9999
3.53	.9998	.0002	.4998	.9996	3.81	.9999	.0001	.4999	.9999
3.54	.9998	.0002	.4998	.9996	3.82	.9999	.0001	.4999	.9999
3.55	.9998	.0002	.4998	.9996	3.83	.9999	.0001	.4999	.9999
3.56	.9998	.0002	.4998	.9996	3.84	.9999	.0001	.4999	.9999
3.57	.9998	.0002	.4998	.9996	3.85	.9999	.0001	.4999	.9999
3.58	.9998	.0002	.4998	.9997	3.86	.9999	.0001	.4999	.9999
3.59	.9998	.0002	.4998	.9997	3.87	.9999	.0001	.4999	.9999
3.60	.9998	.0002	.4998	.9997	3.88	.9999	.0001	.4999	.9999
3.61	.9998	.0002	.4998	.9997	3.89	.9999	.0001	.4999	.9999
3.62	.9999	.0001	.4999	.9997	3.90	1.0000	.0000	.5000	.9999
3.63	.9999	.0001	.4999	.9997	3.91	1.0000	.0000	.5000	.9999
3.64	.9999	.0001	.4999	.9997	3.92	1.0000	.0000	.5000	.9999
3.65	.9999	.0001	.4999	.9997	3.93	1.0000	.0000	.5000	.9999
3.66	.9999	.0001	.4999	.9997	3.94	1.0000	.0000	.5000	.9999
3.67	.9999	.0001	.4999	.9998	3.95	1.0000	.0000	.5000	.9999
3.68	.9999	.0001	.4999	.9998	3.96	1.0000	.0000	.5000	.9999
3.69	.9999	.0001	.4999	.9998	3.97	1.0000	.0000	.5000	.9999
3.70	.9999	.0001	.4999	.9998	3.98	1.0000	.0000	.5000	.9999
3.71	.9999	.0001	.4999	.9998	3.99	1.0000	.0000	.5000	.9999

[a] $A(x) = \Phi(x) = Pr(X \le x)$, where X is a standard normal distribution.
[b] $B(x) = 1 - \Phi(x) = Pr(X > x)$, where X is a standard normal distribution.
[c] $C(x) = Pr(0 \le X \le x)$, where X is a standard normal distribution.
[d] $D(x) = Pr(-x \le X \le x)$, where X is a standard normal distribution.

TABLE 4 Table of 1000 random digits

01	32924	22324	18125	09077	26	96772	16443	39877	04653
02	54632	90374	94143	49295	27	52167	21038	14338	01395
03	88720	43035	97081	83373	28	69644	37198	00028	98195
04	21727	11904	41513	31653	29	71011	62004	81712	87536
05	80985	70799	57975	69282	30	31217	75877	85366	55500
06	40412	58826	94868	52632	31	64990	98735	02999	35521
07	43918	56807	75218	46077	32	48417	23569	59307	46550
08	26513	47480	77410	47741	33	07900	65059	48592	44087
09	18164	35784	44255	30124	34	74526	32601	24482	16981
10	39446	01375	75264	51173	35	51056	04402	58353	37332
11	16638	04680	98617	90298	36	39005	93458	63143	21817
12	16872	94749	44012	48884	37	67883	76343	78155	67733
13	65419	87092	78596	91512	38	06014	60999	87226	36071
14	05207	36702	56804	10498	39	93147	88766	04148	42471
15	78807	79243	13729	81222	40	01099	95731	47622	13294
16	69341	79028	64253	80447	41	89252	01201	58138	13809
17	41871	17566	61200	15994	42	41766	57239	50251	64675
18	25758	04625	43226	32986	43	92736	77800	81996	45646
19	06604	94486	40174	10742	44	45118	36600	68977	68831
20	82259	56512	48945	18183	45	73457	01579	00378	70197
21	07895	37090	50627	71320	46	49465	85251	42914	17277
22	59836	71148	42320	67816	47	15745	37285	23768	39302
23	57133	76610	89104	30481	48	28760	81331	78265	60690
24	76964	57126	87174	61025	49	82193	32787	70451	91141
25	27694	17145	32439	68245	50	89664	50242	12382	39379

TABLE 5 Percentage points of the t distribution ($t_{d,u}$)[a]

Degrees of freedom, d	u (confidence level) .75	.80	.85	.90	.95	.975	.99	.995	.9995
1	1.000	1.376	1.963	3.078	6.314	12.706	31.821	63.657	636.619
2	0.816	1.061	1.386	1.886	2.920	4.303	6.965	9.925	31.598
3	0.765	0.978	1.250	1.638	2.353	3.182	4.541	5.841	12.924
4	0.741	0.941	1.190	1.533	2.132	2.776	3.747	4.604	8.610
5	0.727	0.920	1.156	1.476	2.015	2.571	3.365	4.032	6.869
6	0.718	0.906	1.134	1.440	1.943	2.447	3.143	3.707	5.959
7	0.711	0.896	1.119	1.415	1.895	2.365	2.998	3.499	5.408
8	0.706	0.889	1.108	1.397	1.860	2.306	2.896	3.355	5.041
9	0.703	0.883	1.100	1.383	1.833	2.262	2.821	3.250	4.781
10	0.700	0.879	1.093	1.372	1.812	2.228	2.764	3.169	4.587
11	0.697	0.876	1.088	1.363	1.796	2.201	2.718	3.106	4.437
12	0.695	0.873	1.083	1.356	1.782	2.179	2.681	3.055	4.318
13	0.694	0.870	1.079	1.350	1.771	2.160	2.650	3.012	4.221
14	0.692	0.868	1.076	1.345	1.761	2.145	2.624	2.977	4.140
15	0.691	0.866	1.074	1.341	1.753	2.131	2.602	2.947	4.073
16	0.690	0.865	1.071	1.337	1.746	2.120	2.583	2.921	4.015
17	0.689	0.863	1.069	1.333	1.740	2.110	2.567	2.898	3.965
18	0.688	0.862	1.067	1.330	1.734	2.101	2.552	2.878	3.922
19	0.688	0.861	1.066	1.328	1.729	2.093	2.539	2.861	3.883
20	0.687	0.860	1.064	1.325	1.725	2.086	2.528	2.845	3.850
21	0.686	0.859	1.063	1.323	1.721	2.080	2.518	2.831	3.819
22	0.686	0.858	1.061	1.321	1.717	2.074	2.508	2.819	3.792
23	0.685	0.858	1.060	1.319	1.714	2.069	2.500	2.807	3.767
24	0.685	0.857	1.059	1.318	1.711	2.064	2.492	2.797	3.745
25	0.684	0.856	1.058	1.316	1.708	2.060	2.485	2.787	3.725
26	0.684	0.856	1.058	1.315	1.706	2.056	2.479	2.779	3.707
27	0.684	0.855	1.057	1.314	1.703	2.052	2.473	2.771	3.690
28	0.683	0.855	1.056	1.313	1.701	2.048	2.467	2.763	3.674
29	0.683	0.854	1.055	1.311	1.699	2.045	2.462	2.756	3.659
30	0.683	0.854	1.055	1.310	1.697	2.042	2.457	2.750	3.646
40	0.681	0.851	1.050	1.303	1.684	2.021	2.423	2.704	3.551
60	0.679	0.848	1.046	1.296	1.671	2.000	2.390	2.660	3.460
120	0.677	0.845	1.041	1.289	1.658	1.980	2.358	2.617	3.373
∞	0.674	0.842	1.036	1.282	1.645	1.960	2.326	2.576	3.291

[a]The uth percentile of a t distribution with d degrees of freedom.

Source: Table 5 is taken from Table III of Fisher and Yates: "Statistical Tables for Biological, Agricultural and Medical Research," published by Longman Group Ltd., London (previously published by Oliver and Boyd Ltd., Edinburgh), and by permission of the authors and publishers.

.005 .99 .005
Normal

t table .995
99% confidence interval

TABLE 6 Percentage points of the chi-square distribution ($\chi^2_{d,u}$)[a]

	u													
d	.005	.01	.025	.05	.10	.25	.50	.75	.90	.95	.975	.99	.995	.999
1	0.0^4393[b]	0.0^3157[c]	0.0^3982[d]	0.00393	0.02	0.10	0.45	1.32	2.71	3.84	5.02	6.63	7.88	10.83
2	0.0100	0.0201	0.0506	0.103	0.21	0.58	1.39	2.77	4.61	5.99	7.38	9.21	10.60	13.81
3	0.0717	0.115	0.216	0.352	0.58	1.21	2.37	4.11	6.25	7.81	9.35	11.34	12.84	16.27
4	0.207	0.297	0.484	0.711	1.06	1.92	3.36	5.39	7.78	9.49	11.14	13.28	14.86	18.47
5	0.412	0.554	0.831	1.15	1.61	2.67	4.35	6.63	9.24	11.07	12.83	15.09	16.75	20.52
6	0.676	0.872	1.24	1.64	2.20	3.45	5.35	7.84	10.64	12.59	14.45	16.81	18.55	22.46
7	0.989	1.24	1.69	2.17	2.83	4.25	6.35	9.04	12.02	14.07	16.01	18.48	20.28	24.32
8	1.34	1.65	2.18	2.73	3.49	5.07	7.34	10.22	13.36	15.51	17.53	20.09	21.95	26.12
9	1.73	2.09	2.70	3.33	4.17	5.90	8.34	11.39	14.68	16.92	19.02	21.67	23.59	27.88
10	2.16	2.56	3.25	3.94	4.87	6.74	9.34	12.55	15.99	18.31	20.48	23.21	25.19	29.59
11	2.60	3.05	3.82	4.57	5.58	7.58	10.34	13.70	17.28	19.68	21.92	24.72	26.76	31.26
12	3.07	3.57	4.40	5.23	6.30	8.44	11.34	14.85	18.55	21.03	23.34	26.22	28.30	32.91
13	3.57	4.11	5.01	5.89	7.04	9.30	12.34	15.98	19.81	22.36	24.74	27.69	29.82	34.53
14	4.07	4.66	5.63	6.57	7.79	10.17	13.34	17.12	21.06	23.68	26.12	29.14	31.32	36.12
15	4.60	5.23	6.27	7.26	8.55	11.04	14.34	18.25	22.31	25.00	27.49	30.58	32.80	37.70
16	5.14	5.81	6.91	7.96	9.31	11.91	15.34	19.37	23.54	26.30	28.85	32.00	34.27	39.25
17	5.70	6.41	7.56	8.67	10.09	12.79	16.34	20.49	24.77	27.59	30.19	33.41	35.72	40.79
18	6.26	7.01	8.23	9.39	10.86	13.68	17.34	21.60	25.99	28.87	31.53	34.81	37.16	42.31
19	6.84	7.63	8.91	10.12	11.65	14.56	18.34	22.72	27.20	30.14	32.85	36.19	38.58	43.82
20	7.43	8.26	9.59	10.85	12.44	15.45	19.34	23.83	28.41	31.41	34.17	37.57	40.00	45.32
21	8.03	8.90	10.28	11.59	13.24	16.34	20.34	24.93	29.62	32.67	35.48	38.93	41.40	46.80
22	8.64	9.54	10.98	12.34	14.04	17.24	21.34	26.04	30.81	33.92	36.78	40.29	42.80	48.27
23	9.26	10.20	11.69	13.09	14.85	18.14	22.34	27.14	32.01	35.17	38.08	41.64	44.18	49.73
24	9.89	10.86	12.40	13.85	15.66	19.04	23.34	28.24	33.20	36.42	39.36	42.98	45.56	51.18
25	10.52	11.52	13.12	14.61	16.47	19.94	24.34	29.34	34.38	37.65	40.65	44.31	46.93	52.62
26	11.16	12.20	13.84	15.38	17.29	20.84	25.34	30.43	35.56	38.89	41.92	45.64	48.29	54.05
27	11.81	12.88	14.57	16.15	18.11	21.75	26.34	31.53	36.74	40.11	43.19	46.96	49.64	55.48
28	12.46	13.56	15.31	16.93	18.94	22.66	27.34	32.62	37.92	41.34	44.46	48.28	50.99	56.89
29	13.12	14.26	16.05	17.71	19.77	23.57	28.34	33.71	39.09	42.56	45.72	49.59	52.34	58.30
30	13.79	14.95	16.79	18.49	20.60	24.48	29.34	34.80	40.26	43.77	46.98	50.89	53.67	59.70
40	20.71	22.16	24.43	26.51	29.05	33.66	39.34	45.62	51.81	55.76	59.34	63.69	66.77	73.40
50	27.99	29.71	32.36	34.76	37.69	42.94	49.33	56.33	63.17	67.50	71.42	76.15	79.49	86.66
60	35.53	37.48	40.48	43.19	46.46	52.29	59.33	66.98	74.40	79.08	83.30	88.38	91.95	99.61
70	43.28	45.44	48.76	51.74	55.33	61.70	69.33	77.58	85.53	90.53	95.02	100.42	104.22	112.32
80	51.17	53.54	57.15	60.39	64.28	71.14	79.33	88.13	96.58	101.88	106.63	112.33	116.32	124.84
90	59.20	61.75	65.65	69.13	73.29	80.62	89.33	98.64	107.56	113.14	118.14	124.12	128.30	137.21
100	67.33	70.06	74.22	77.93	82.36	90.13	99.33	109.14	118.50	124.34	129.56	135.81	140.17	149.45

[a] $\chi^2_{d,u}$ = *u*th percentile of a χ^2 distribution with *d* degrees of freedom.
[b] = 0.0000393
[c] = 0.000157
[d] = 0.000982

Source: Reproduced in part with permission of the Biometrika Trustees, from Table 3 of *Biometrika Tables for Statisticians,* Volume 2, edited by E. S. Pearson and H. O. Hartley, published for the Biometrika Trustees, Cambridge University Press, Cambridge, England, 1972.

TABLE 7a Exact two-sided 100% × (1 − α) confidence limits for binomial proportions (α = .05)

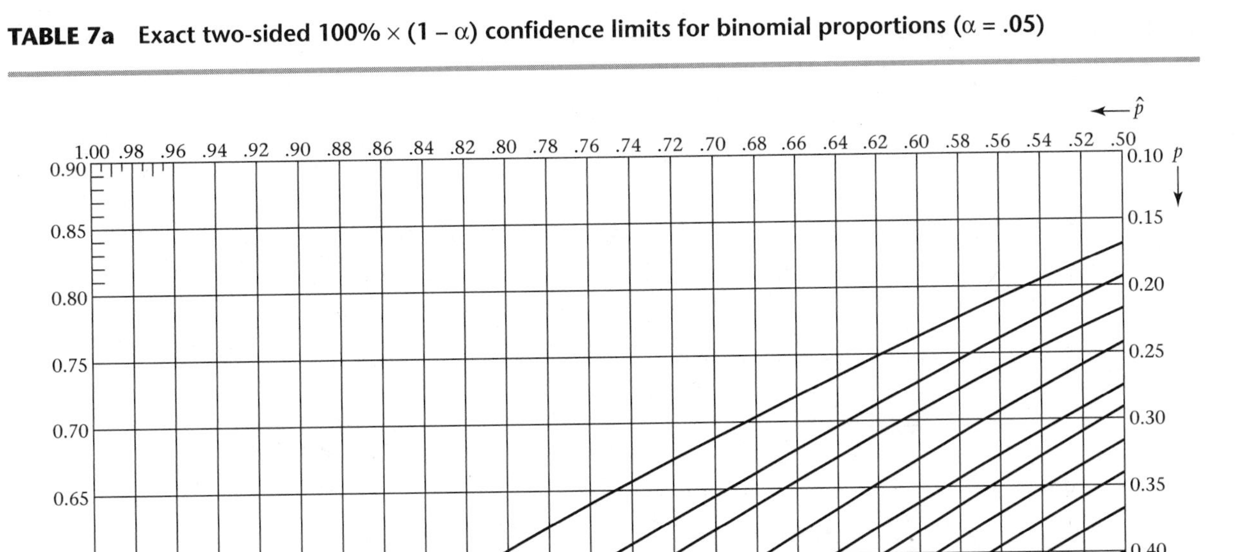

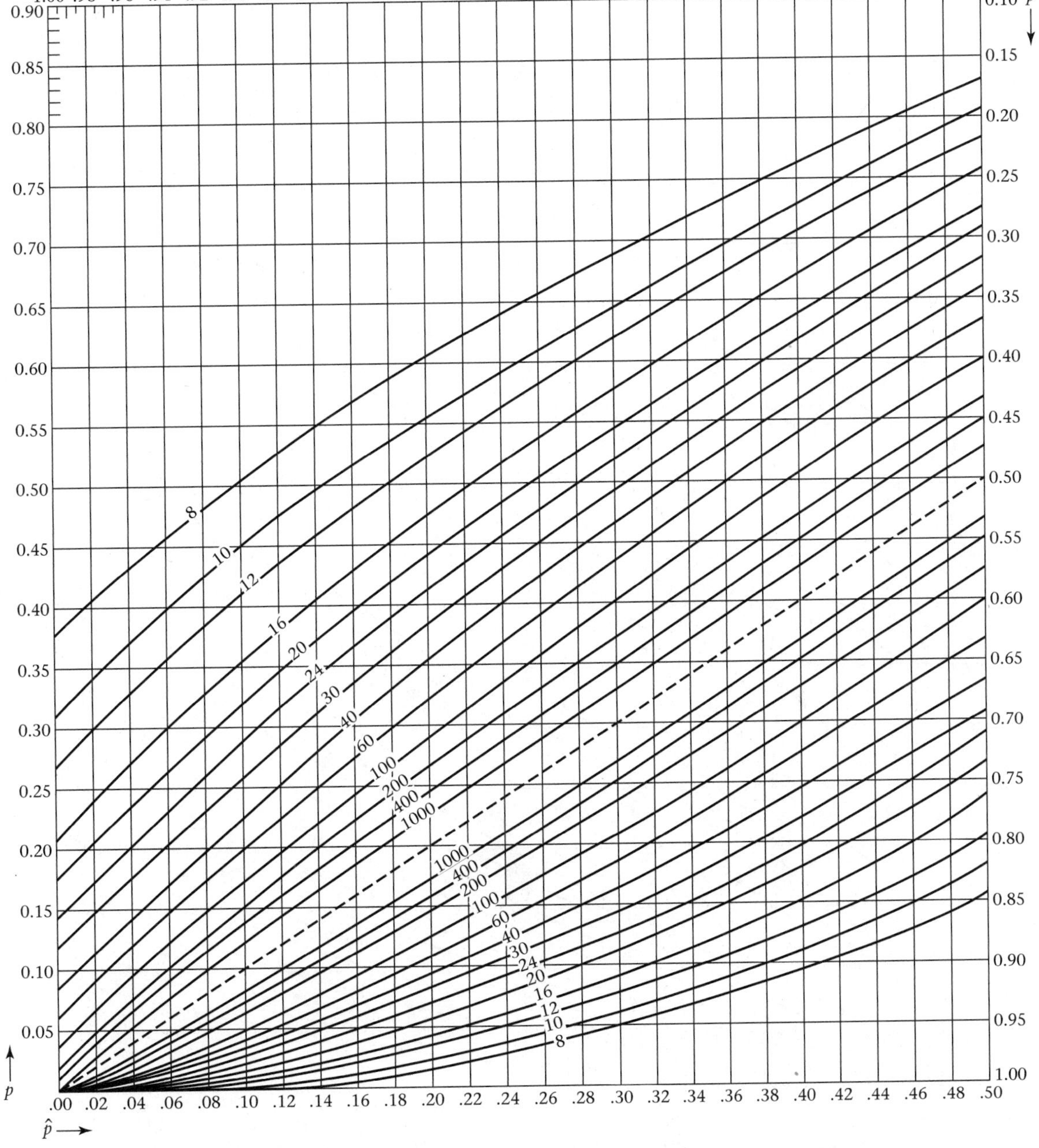

TABLE 7b **Exact two-sided 100% × (1 – α) confidence limits for binomial proportions (α = .01)**

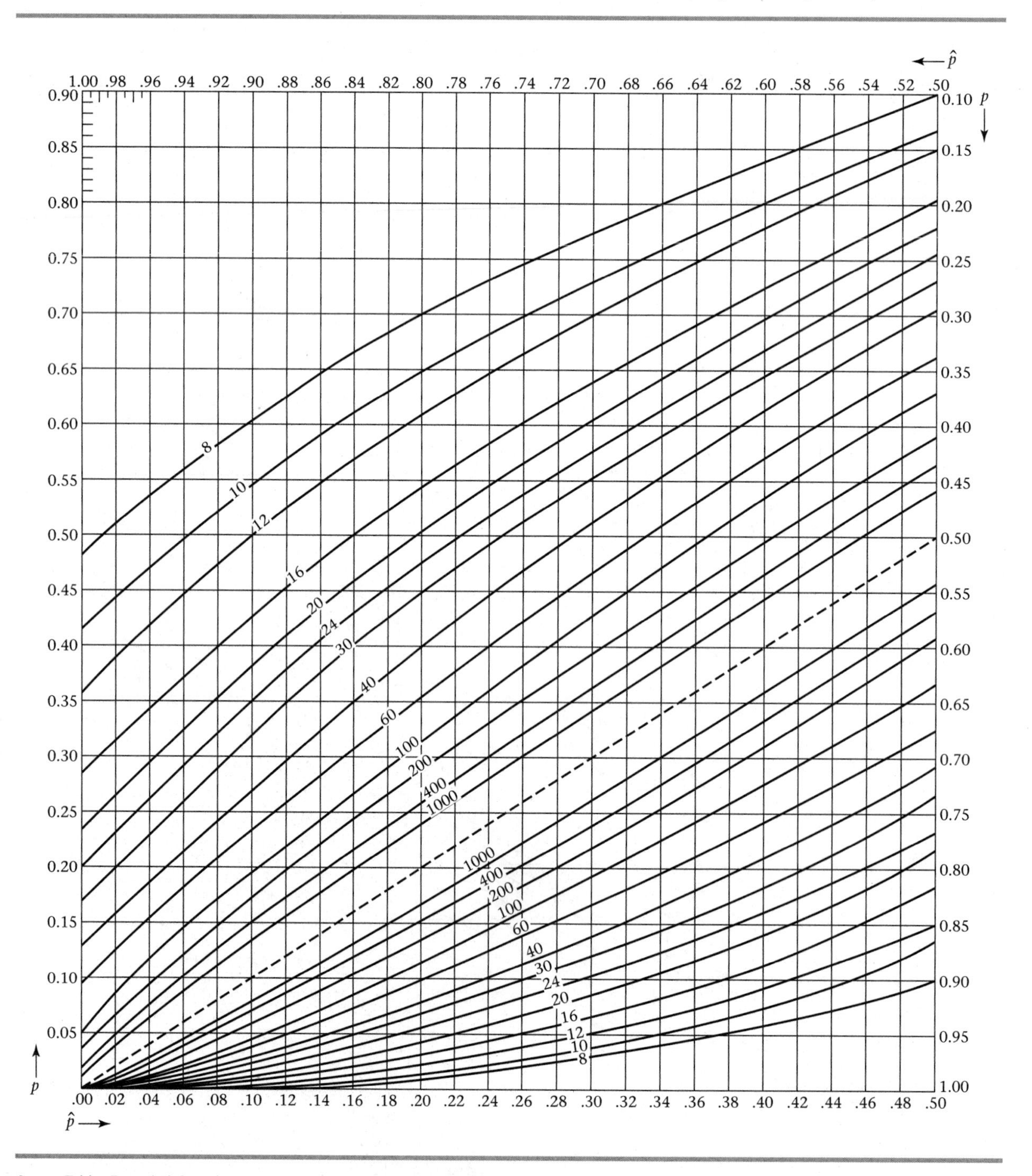

Source: Tables 7a and 7b have been reproduced with permission of Biometrika Trustees, from Table 41 of *Biometrika Tables for Statisticians,* 3rd edition, Volume 1, edited by E. S. Pearson and H. O. Hartley. Published for the Biometrika Trustees, Cambridge, England, 1966.

TABLE 8 Confidence limits for the expectation of a Poisson variable (μ)

	Confidence level $(1 - \alpha)$										
$(1 - \alpha)$	0.998		0.99		0.98		0.95		0.90		$(1 - 2\alpha)$
x	Lower	Upper	Lower	Upper	Lower	Upper	Lower	Upper	Lower	Upper	x
0	0.00000	6.91	0.00000	5.30	0.0000	4.61	0.0000	3.69	0.0000	3.00	**0**
1	.00100	9.23	.00501	7.43	.0101	6.64	.0253	5.57	.0513	4.74	**1**
2	.0454	11.23	.103	9.27	.149	8.41	.242	7.22	.355	6.30	**2**
3	.191	13.06	.338	10.98	.436	10.05	.619	8.77	.818	7.75	**3**
4	.429	14.79	.672	12.59	.823	11.60	1.09	10.24	1.37	9.15	**4**
5	0.739	16.45	1.08	14.15	1.28	13.11	1.62	11.67	1.97	10.51	**5**
6	1.11	18.06	1.54	15.66	1.79	14.57	2.20	13.06	2.61	11.84	**6**
7	1.52	19.63	2.04	17.13	2.33	16.00	2.81	14.42	3.29	13.15	**7**
8	1.97	21.16	2.57	18.58	2.91	17.40	3.45	15.76	3.98	14.43	**8**
9	2.45	22.66	3.13	20.00	3.51	18.78	4.12	17.08	4.70	15.71	**9**
10	2.96	24.13	3.72	21.40	4.13	20.14	4.80	18.39	5.43	16.96	**10**
11	3.49	25.59	4.32	22.78	4.77	21.49	5.49	19.68	6.17	18.21	**11**
12	4.04	27.03	4.94	24.14	5.43	22.82	6.20	20.96	6.92	19.44	**12**
13	4.61	28.45	5.58	25.50	6.10	24.14	6.92	22.23	7.69	20.67	**13**
14	5.20	29.85	6.23	26.84	6.78	25.45	7.65	23.49	8.46	21.89	**14**
15	5.79	31.24	6.89	28.16	7.48	26.74	8.40	24.74	9.25	23.10	**15**
16	6.41	32.62	7.57	29.48	8.18	28.03	9.15	25.98	10.04	24.30	**16**
17	7.03	33.99	8.25	30.79	8.89	29.31	9.90	27.22	10.83	25.50	**17**
18	7.66	35.35	8.94	32.09	9.62	30.58	10.67	28.45	11.63	26.69	**18**
19	8.31	36.70	9.64	33.38	10.35	31.85	11.44	29.67	12.44	27.88	**19**
20	8.96	38.04	10.35	34.67	11.08	33.10	12.22	30.89	13.25	29.06	**20**
21	9.62	39.38	11.07	35.95	11.82	34.36	13.00	32.10	14.07	30.24	**21**
22	10.29	40.70	11.79	37.22	12.57	35.60	13.79	33.31	14.89	31.42	**22**
23	10.96	42.02	12.52	38.48	13.33	36.84	14.58	34.51	15.72	32.59	**23**
24	11.65	43.33	13.25	39.74	14.09	38.08	15.38	35.71	16.55	33.75	**24**
25	12.34	44.64	14.00	41.00	14.85	39.31	16.18	36.90	17.38	34.92	**25**
26	13.03	45.94	14.74	42.25	15.62	40.53	16.98	38.10	18.22	36.08	**26**
27	13.73	47.23	15.49	43.50	16.40	41.76	17.79	39.28	19.06	37.23	**27**
28	14.44	48.52	16.24	44.74	17.17	42.98	18.61	40.47	19.90	38.39	**28**
29	15.15	49.80	17.00	45.98	17.96	44.19	19.42	41.65	20.75	39.54	**29**
30	15.87	51.08	17.77	47.21	18.74	45.40	20.24	42.83	21.59	40.69	**30**
35	19.52	57.42	21.64	53.32	22.72	51.41	24.38	48.68	25.87	46.40	**35**
40	23.26	63.66	25.59	59.36	26.77	57.35	28.58	54.47	30.20	52.07	**40**
45	27.08	69.83	29.60	65.34	30.88	63.23	32.82	60.21	34.56	57.69	**45**
50	30.96	75.94	33.66	71.27	35.03	69.07	37.11	65.92	38.96	63.29	**50**

Note: If X is the random variable denoting the observed number of events and μ_1, μ_2 are the lower and upper confidence limits for its expectation, μ, then $Pr(\mu_1 \leq \mu \leq \mu_2) = 1 - \alpha$.

Source: Biometrika Tables for Statisticians, 3rd edition, Volume 1, edited by E. S. Pearson and H. O. Hartley. Published for the Biometrika Trustees, Cambridge University Press, Cambridge, England, 1966.

TABLE 9 Percentage points of the F distribution ($F_{d_1,d_2,p}$)

df for denominator, d_2	p	df for numerator, d_1: 1	2	3	4	5	6	7	8	12	24	∞
1	.90	39.86	49.50	53.59	55.83	57.24	58.20	58.91	59.44	60.71	62.00	63.33
	.95	161.4	199.5	215.7	224.6	230.2	234.0	236.8	238.9	243.9	249.1	254.3
	.975	647.8	799.5	864.2	899.6	921.8	937.1	948.2	956.7	976.7	997.2	1018.
	.99	4052.	5000.	5403.	5625.	5764.	5859.	5928.	5981.	6106.	6235.	6366.
	.995	16211.	20000.	21615.	22500.	23056.	23437.	23715.	23925.	24426.	24940.	25464.
	.999	405280.	500000.	540380.	562500.	576400.	585940.	592870.	598140.	610670.	623500.	636620.
2	.90	8.53	9.00	9.16	9.24	9.29	9.33	9.35	9.37	9.41	9.45	9.49
	.95	18.51	19.00	19.16	19.25	19.30	19.33	19.35	19.37	19.41	19.45	19.50
	.975	38.51	39.00	39.17	39.25	39.30	39.33	39.36	39.37	39.42	39.46	39.50
	.99	98.50	99.00	99.17	99.25	99.30	99.33	99.36	99.37	99.42	99.46	99.50
	.995	198.5	199.0	199.2	199.2	199.3	199.3	199.4	199.4	199.4	199.5	199.5
	.999	998.5	999.0	999.2	999.2	999.3	999.3	999.4	999.4	999.4	999.5	999.5
3	.90	5.54	5.46	5.39	5.34	5.31	5.28	5.27	5.25	5.22	5.18	5.13
	.95	10.13	9.55	9.28	9.12	9.01	8.94	8.89	8.85	8.74	8.64	8.53
	.975	17.44	16.04	15.44	15.10	14.88	14.74	14.62	14.54	14.34	14.12	13.90
	.99	34.12	30.82	29.46	28.71	28.24	27.91	27.67	27.49	27.05	26.60	26.13
	.995	55.55	49.80	47.47	46.20	45.39	44.84	44.43	44.13	43.39	42.62	41.83
	.999	167.00	148.5	141.1	137.1	134.6	132.8	131.6	130.6	128.3	125.9	123.5
4	.90	4.54	4.32	4.19	4.11	4.05	4.01	3.98	3.95	3.90	3.83	3.76
	.95	7.71	6.94	6.59	6.39	6.26	6.16	6.09	6.04	5.91	5.77	5.63
	.975	12.22	10.65	9.98	9.60	9.36	9.20	9.07	8.98	8.75	8.51	8.26
	.99	21.20	18.00	16.69	15.98	15.52	15.21	14.98	14.80	14.37	13.93	13.46
	.995	31.33	26.28	24.26	23.16	22.46	21.98	21.62	21.35	20.70	20.03	19.32
	.999	74.14	61.25	56.18	53.44	51.71	50.53	49.66	49.00	47.41	45.77	44.05
5	.90	4.06	3.78	3.62	3.52	3.45	3.40	3.37	3.34	3.27	3.19	3.10
	.95	6.61	5.79	5.41	5.19	5.05	4.95	4.88	4.82	4.68	4.53	4.36
	.975	10.01	8.43	7.76	7.39	7.15	6.98	6.85	6.76	6.52	6.28	6.02
	.99	16.26	13.27	12.06	11.39	10.97	10.67	10.46	10.29	9.89	9.47	9.02
	.995	22.78	18.31	16.53	15.56	14.94	14.51	14.20	13.96	13.38	12.78	12.14
	.999	47.18	37.12	33.20	31.09	29.75	28.83	28.16	27.65	26.42	25.13	23.79
6	.90	3.78	3.46	3.29	3.18	3.11	3.05	3.01	2.98	2.90	2.82	2.72
	.95	5.99	5.14	4.76	4.53	4.39	4.28	4.21	4.15	4.00	3.84	3.67
	.975	8.81	7.26	6.60	6.23	5.99	5.82	5.70	5.60	5.37	5.12	4.85
	.99	13.75	10.92	9.78	9.15	8.75	8.47	8.26	8.10	7.72	7.31	6.88
	.995	18.64	14.54	12.92	12.03	11.46	11.07	10.79	10.57	10.03	9.47	8.88
	.999	35.51	27.00	23.70	21.92	20.80	20.03	19.46	19.03	17.99	16.90	15.75
7	.90	3.59	3.26	3.07	2.96	2.88	2.83	2.78	2.75	2.67	2.58	2.47
	.95	5.59	4.74	4.35	4.12	3.97	3.87	3.79	3.73	3.57	3.41	3.23
	.975	8.07	6.54	5.89	5.52	5.29	5.12	4.99	4.90	4.67	4.42	4.14
	.99	12.25	9.55	8.45	7.85	7.46	7.19	6.99	6.84	6.47	6.07	5.65
	.995	16.24	12.40	10.88	10.05	9.52	9.16	8.89	8.68	8.18	7.65	7.08
	.999	29.25	21.69	18.77	17.20	16.21	15.52	15.02	14.63	13.71	12.73	11.70
8	.90	3.46	3.11	2.92	2.81	2.73	2.67	2.62	2.59	2.50	2.40	2.29
	.95	5.32	4.46	4.07	3.84	3.69	3.58	3.50	3.44	3.28	3.12	2.93
	.975	7.57	6.06	5.42	5.05	4.82	4.65	4.53	4.43	4.20	3.95	3.67
	.99	11.26	8.65	7.59	7.01	6.63	6.37	6.18	6.03	5.67	5.28	4.86
	.995	14.69	11.04	9.60	8.81	8.30	7.95	7.69	7.50	7.01	6.50	5.95
	.999	25.42	18.49	15.83	14.39	13.49	12.86	12.40	12.04	11.19	10.30	9.33
9	.90	3.36	3.01	2.81	2.69	2.61	2.55	2.51	2.47	2.38	2.28	2.16
	.95	5.12	4.26	3.86	3.63	3.48	3.37	3.29	3.23	3.07	2.90	2.71
	.975	7.21	5.71	5.08	4.72	4.48	4.32	4.20	4.10	3.87	3.61	3.33
	.99	10.56	8.02	6.99	6.42	6.06	5.80	5.61	5.47	5.11	4.73	4.31
	.995	13.61	10.11	8.72	7.96	7.47	7.13	6.88	6.69	6.23	5.73	5.19
	.999	22.86	16.39	13.90	12.56	11.71	11.13	10.70	10.37	9.57	8.72	7.81

(continued on next page)

TABLE 9 Percentage points of the F distribution ($F_{d_1,d_2,p}$) *(continued)*

df for denominator, d_2	p	*df* for numerator, d_1: 1	2	3	4	5	6	7	8	12	24	∞
10	.90	3.29	2.92	2.73	2.61	2.52	2.46	2.41	2.38	2.28	2.18	2.06
	.95	4.96	4.10	3.71	3.48	3.33	3.22	3.14	3.07	2.91	2.74	2.54
	.975	6.94	5.46	4.83	4.47	4.24	4.07	3.95	3.85	3.62	3.37	3.08
	.99	10.04	7.56	6.55	5.99	5.64	5.39	5.20	5.06	4.71	4.33	3.91
	.995	12.83	9.43	8.08	7.34	6.87	6.54	6.30	6.12	5.66	5.17	4.64
	.999	21.04	14.91	12.55	11.28	10.48	9.93	9.52	9.20	8.45	7.64	6.76
12	.90	3.18	2.81	2.61	2.48	2.39	2.33	2.28	2.24	2.15	2.04	1.90
	.95	4.75	3.89	3.49	3.26	3.11	3.00	2.91	2.85	2.69	2.51	2.30
	.975	6.55	5.10	4.47	4.12	3.89	3.73	3.61	3.51	3.28	3.02	2.72
	.99	9.33	6.93	5.95	5.41	5.06	4.82	4.64	4.50	4.16	3.78	3.36
	.995	11.75	8.51	7.23	6.52	6.07	5.76	5.52	5.35	4.91	4.43	3.90
	.999	18.64	12.97	10.80	9.63	8.89	8.38	8.00	7.71	7.00	6.25	5.42
14	.90	3.10	2.73	2.52	2.39	2.31	2.24	2.19	2.15	2.05	1.94	1.80
	.95	4.60	3.74	3.34	3.11	2.96	2.85	2.76	2.70	2.53	2.35	2.13
	.975	6.30	4.86	4.24	3.89	3.66	3.50	3.38	3.29	3.05	2.79	2.49
	.99	8.86	6.51	5.56	5.04	4.69	4.46	4.28	4.14	3.80	3.43	3.00
	.995	11.06	7.92	6.68	6.00	5.56	5.26	5.03	4.86	4.43	3.96	3.44
	.999	17.14	11.78	9.73	8.62	7.92	7.44	7.08	6.80	6.13	5.41	4.60
16	.90	3.05	2.67	2.46	2.33	2.24	2.18	2.13	2.09	1.99	1.87	1.72
	.95	4.49	3.63	3.24	3.01	2.85	2.74	2.66	2.59	2.42	2.24	2.01
	.975	6.12	4.69	4.08	3.73	3.50	3.34	3.22	3.12	2.89	2.63	2.32
	.99	8.53	6.23	5.29	4.77	4.44	4.20	4.03	3.89	3.55	3.18	2.75
	.995	10.58	7.51	6.30	5.64	5.21	4.91	4.69	4.52	4.10	3.64	3.11
	.999	16.12	10.97	9.01	7.94	7.27	6.80	6.46	6.19	5.55	4.85	4.06
18	.90	3.01	2.62	2.42	2.29	2.20	2.13	2.08	2.04	1.93	1.81	1.66
	.95	4.41	3.55	3.16	2.93	2.77	2.66	2.58	2.51	2.34	2.15	1.92
	.975	5.98	4.56	3.95	3.61	3.38	3.22	3.10	3.01	2.77	2.50	2.19
	.99	8.29	6.01	5.09	4.58	4.25	4.01	3.84	3.71	3.37	3.00	2.57
	.995	10.22	7.21	6.03	5.37	4.96	4.66	4.44	4.28	3.86	3.40	2.87
	.999	15.38	10.39	8.49	7.46	6.81	6.35	6.02	5.76	5.13	4.45	3.67
20	.90	2.97	2.59	2.38	2.25	2.16	2.09	2.04	2.00	1.89	1.77	1.61
	.95	4.35	3.49	3.10	2.87	2.71	2.60	2.51	2.45	2.28	2.08	1.84
	.975	5.87	4.46	3.86	3.51	3.29	3.13	3.01	2.91	2.68	2.41	2.09
	.99	8.10	5.85	4.94	4.43	4.10	3.87	3.70	3.56	3.23	2.86	2.42
	.995	9.94	6.99	5.82	5.17	4.76	4.47	4.26	4.09	3.68	3.22	2.69
	.999	14.82	9.95	8.10	7.10	6.46	6.02	5.69	5.44	4.82	4.15	3.38
30	.90	2.88	2.49	2.28	2.14	2.05	1.98	1.93	1.88	1.77	1.64	1.46
	.95	4.17	3.32	2.92	2.69	2.53	2.42	2.33	2.27	2.09	1.89	1.62
	.975	5.57	4.18	3.59	3.25	3.03	2.87	2.75	2.65	2.41	2.14	1.79
	.99	7.56	5.39	4.51	4.02	3.70	3.47	3.30	3.17	2.84	2.47	2.01
	.995	9.18	6.35	5.24	4.62	4.23	3.95	3.74	3.58	3.18	2.73	2.18
	.999	13.29	8.77	7.05	6.12	5.53	5.12	4.82	4.58	4.00	3.36	2.59
40	.90	2.84	2.44	2.23	2.09	2.00	1.93	1.87	1.83	1.71	1.57	1.38
	.95	4.08	3.23	2.84	2.61	2.45	2.34	2.25	2.18	2.00	1.79	1.51
	.975	5.42	4.05	3.46	3.13	2.90	2.74	2.62	2.53	2.29	2.01	1.64
	.99	7.31	5.18	4.31	3.83	3.51	3.29	3.12	2.99	2.66	2.29	1.80
	.995	8.83	6.07	4.98	4.37	3.99	3.71	3.51	3.35	2.95	2.50	1.93
	.999	12.61	8.25	6.59	5.70	5.13	4.73	4.44	4.21	3.64	3.01	2.23
60	.90	2.79	2.39	2.18	2.04	1.95	1.87	1.82	1.77	1.66	1.51	1.29
	.95	4.00	3.15	2.76	2.53	2.37	2.25	2.17	2.10	1.92	1.70	1.39
	.975	5.29	3.93	3.34	3.01	2.79	2.63	2.51	2.41	2.17	1.88	1.48
	.99	7.08	4.98	4.13	3.65	3.34	3.12	2.95	2.82	2.50	2.12	1.60
	.995	8.49	5.80	4.73	4.14	3.76	3.49	3.29	3.13	2.74	2.29	1.69
	.999	11.97	7.77	6.17	5.31	4.76	4.37	4.09	3.86	3.32	2.69	1.89

(continued on next page)

TABLE 9 Percentage points of the *F* distribution ($F_{d_1,d_2,p}$) *(continued)*

df for denominator, d_2	p	*df* for numerator, d_1 1	2	3	4	5	6	7	8	12	24	∞
120	.90	2.75	2.35	2.13	1.99	1.90	1.82	1.77	1.72	1.60	1.45	1.19
	.95	3.92	3.07	2.68	2.45	2.29	2.17	2.09	2.02	1.83	1.61	1.25
	.975	5.15	3.80	3.23	2.89	2.67	2.52	2.39	2.30	2.05	1.76	1.31
	.99	6.85	4.79	3.95	3.48	3.17	2.96	2.79	2.66	2.34	1.95	1.38
	.995	8.18	5.54	4.50	3.92	3.55	3.28	3.09	2.93	2.54	2.09	1.43
	.999	11.38	7.32	5.78	4.95	4.42	4.04	3.77	3.55	3.02	2.40	1.54
∞	.90	2.71	2.30	2.08	1.94	1.85	1.77	1.72	1.67	1.55	1.38	1.00
	.95	3.84	3.00	2.60	2.37	2.21	2.10	2.01	1.94	1.75	1.52	1.00
	.975	5.02	3.69	3.12	2.79	2.57	2.41	2.29	2.19	1.94	1.64	1.00
	.99	6.63	4.61	3.78	3.32	3.02	2.80	2.64	2.51	2.18	1.79	1.00
	.995	7.88	5.30	4.28	3.72	3.35	3.09	2.90	2.74	2.36	1.90	1.00
	.999	10.83	6.91	5.42	4.62	4.10	3.74	3.47	3.27	2.74	2.13	1.00

Note: $F_{d_1,d_2,p}$ = pth percentile of an *F* distribution with d_1 and d_2 degrees of freedom.

Source: This table has been reproduced in part with the permission of the Biometrika Trustees, from *Biometrika Tables for Statisticians,* Volume 2, edited by E. S. Pearson and H. O. Hartley, published for the Biometrika Trustees, Cambridge University Press, Cambridge, England, 1972.

TABLE 10 Critical values for the ESD (Extreme Studentized Deviate) outlier statistic ($ESD_{n,1-\alpha}$, $\alpha = .05, .01$)

n	$1-\alpha$ .95	.99	n	$1-\alpha$ .95	.99
5	1.72	1.76	25	2.82	3.14
6	1.89	1.97	26	2.84	3.16
7	2.02	2.14	27	2.86	3.18
8	2.13	2.28	28	2.88	3.20
9	2.21	2.39	29	2.89	3.22
10	2.29	2.48	30	2.91	3.24
11	2.36	2.56	35	2.98	3.32
12	2.41	2.64	40	3.04	3.38
13	2.46	2.70	45	3.09	3.44
14	2.51	2.75	50	3.13	3.48
15	2.55	2.81	60	3.20	3.56
16	2.59	2.85	70	3.26	3.62
17	2.62	2.90	80	3.31	3.67
18	2.65	2.93	90	3.35	3.72
19	2.68	2.97	100	3.38	3.75
20	2.71	3.00	150	3.52	3.89
21	2.73	3.03	200	3.61	3.98
22	2.76	3.06	300	3.72	4.09
23	2.78	3.08	400	3.80	4.17
24	2.80	3.11	500	3.86	4.23

Note: For values of n not found in the table, the percentiles can be evaluated using the formula $ESD_{n,1-\alpha} = \dfrac{t_{n-2,p}(n-1)}{\sqrt{n(n-2+t^2_{n-2,p})}}$, where $p = 1 - [\alpha/(2n)]$.

TABLE 11 Two-tailed critical values for the Wilcoxon signed-rank test

	.10		.05		.02		.01	
n^a	Lower	Upper	Lower	Upper	Lower	Upper	Lower	Upper
1	—		—		—		—	
2	—		—		—		—	
3	—		—		—		—	
4	—		—		—		—	
5	0	15	—		—		—	
6	2	19	0	21	—		—	
7	3	25	2	26	0	28	—	
8	5	31	3	33	1	35	0	36
9	8	37	5	40	3	42	1	44
10	10	45	8	47	5	50	3	52
11	13	53	10	56	7	59	5	61
12	17	61	13	65	9	69	7	71
13	21	70	17	74	12	79	9	82
14	25	80	21	84	15	90	12	93
15	30	90	25	95	19	101	15	105

[a] n = number of untied pairs.
Source: Figures from "Documenta Geigy Scientific Tables," 6th edition. Reprinted with the kind permission of CIBA-GEIGY Limited, Basel, Switzerland.

TABLE 12 Two-tailed critical values for the Wilcoxon rank-sum test

	$\alpha = .10$ n_1[a]						$\alpha = .05$ n_1					
n_2[b]	4	5	6	7	8	9	4	5	6	7	8	9
	T_l[c] T_r[d]	T_l T_r	T_l T_r	T_l T_r	T_l T_r	T_l T_r	T_l T_r	T_l T_r	T_l T_r	T_l T_r	T_l T_r	T_l T_r
4	11–25	17–33	24–42	32–52	41–63	51–75	10–26	16–34	23–43	31–53	40–64	49–77
5	12–28	19–36	26–46	34–57	44–68	54–81	11–29	17–38	24–48	33–58	42–70	52–83
6	13–31	20–40	28–50	36–62	46–74	57–87	12–32	18–42	26–52	34–64	44–76	55–89
7	14–34	21–44	29–55	39–66	49–79	60–93	13–35	20–45	27–57	36–69	46–82	57–96
8	15–37	23–47	31–59	41–71	51–85	63–99	14–38	21–49	29–61	38–74	49–87	60–102
9	16–40	24–51	33–63	43–76	54–90	66–105	14–42	22–53	31–65	40–79	51–93	62–109
10	17–43	26–54	35–67	45–81	56–96	69–111	15–45	23–57	32–70	42–84	53–99	65–115
11	18–46	27–58	37–71	47–86	59–101	72–117	16–48	24–61	34–74	44–89	55–105	68–121
12	19–49	28–62	38–76	49–91	62–106	75–123	17–51	26–64	35–79	46–94	58–110	71–127
13	20–52	30–65	40–80	52–95	64–112	78–129	18–54	27–68	37–83	48–99	60–116	73–134
14	21–55	31–69	42–84	54–100	67–117	81–135	19–57	28–72	38–88	50–104	62–122	76–140
15	22–58	33–72	44–88	56–105	69–123	84–141	20–60	29–76	40–92	52–109	65–127	79–146
16	24–60	34–76	46–92	58–110	72–128	87–147	21–63	30–80	42–96	54–114	67–133	82–152
17	25–63	35–80	47–97	61–114	75–133	90–153	21–67	32–83	43–101	56–119	70–138	84–159
18	26–66	37–83	49–101	63–119	77–139	93–159	22–70	33–87	45–105	58–124	72–144	87–165
19	27–69	38–87	51–105	65–124	80–144	96–165	23–73	34–91	46–110	60–129	74–150	90–171
20	28–72	40–90	53–109	67–129	83–149	99–171	24–76	35–95	48–114	62–134	77–155	93–177
21	29–75	41–94	55–113	69–134	85–155	102–177	25–79	37–98	50–118	64–139	79–161	95–184
22	30–78	43–97	57–117	72–138	88–160	105–183	26–82	38–102	51–123	66–144	81–167	98–190
23	31–81	44–101	58–122	74–143	90–166	108–189	27–85	39–106	53–127	68–149	84–172	101–196
24	32–84	45–105	60–126	76–148	93–171	111–195	27–89	40–110	54–132	70–154	86–178	104–202
25	33–87	47–108	62–130	78–153	96–176	114–201	28–92	42–113	56–136	72–159	89–183	107–208
26	34–90	48–112	64–134	81–157	98–182	117–207	29–95	43–117	58–140	74–164	91–189	109–215
27	35–93	50–115	66–138	83–162	101–187	120–213	30–98	44–121	59–145	76–169	93–195	112–221
28	36–96	51–119	67–143	85–167	103–193	123–219	31–101	45–125	61–149	78–174	96–200	115–227
29	37–99	53–122	69–147	87–172	106–198	126–225	32–104	47–128	63–153	80–179	98–206	118–233
30	38–102	54–126	71–151	89–177	109–203	129–231	33–107	48–132	64–158	82–184	101–211	121–239
31	39–105	55–130	73–155	92–181	111–209	132–237	34–110	49–136	66–162	84–189	103–217	123–246
32	40–108	57–133	75–159	94–186	114–214	135–243	34–114	50–140	67–167	86–194	106–222	126–252
33	41–111	58–137	77–163	96–191	117–219	138–249	35–117	52–143	69–171	88–199	108–228	129–258
34	42–114	60–140	78–168	98–196	119–225	141–255	36–120	53–147	71–175	90–204	110–234	132–264
35	43–117	61–144	80–172	100–201	122–230	144–261	37–123	54–151	72–180	92–209	113–239	135–270
36	44–120	62–148	82–176	102–206	124–236	148–266	38–126	55–155	74–184	94–214	115–245	137–277
37	45–123	64–151	84–180	105–210	127–241	151–272	39–129	57–158	76–188	96–219	117–251	140–283
38	46–126	65–155	85–185	107–215	130–246	154–278	40–132	58–162	77–193	98–224	120–256	143–289
39	47–129	67–158	87–189	109–220	132–252	157–284	41–135	59–166	79–197	100–229	122–262	146–295
40	48–132	68–162	89–193	111–225	135–257	160–290	41–139	60–170	80–202	102–234	125–267	149–301
41	49–135	69–166	91–197	114–229	138–262	163–296	42–142	61–174	82–206	104–239	127–273	151–308
42	50–138	71–169	93–201	116–234	140–268	166–302	43–145	63–177	84–210	106–244	129–279	154–314
43	51–141	72–173	95–205	118–239	143–273	169–308	44–148	64–181	85–215	108–249	132–284	157–320
44	52–144	74–176	96–210	120–244	146–278	172–314	45–151	65–185	87–219	110–254	134–290	160–326
45	53–147	75–180	98–214	123–248	148–284	175–320	46–154	66–189	88–224	112–259	137–295	163–332
46	55–149	77–183	100–218	125–253	151–289	178–326	47–157	68–192	90–228	114–264	139–301	165–339
47	56–152	78–187	102–222	127–258	154–294	181–332	48–160	69–196	92–232	116–269	141–307	168–345
48	57–155	79–191	104–226	129–263	156–300	184–338	48–164	70–200	93–237	118–274	144–312	171–351
49	58–158	81–194	106–230	132–267	159–305	187–344	49–167	71–204	95–241	120–279	146–318	174–357
50	59–161	82–198	107–235	134–272	162–310	190–350	50–170	73–207	97–245	122–284	149–323	177–363

[a] n_1 = minimum of the two sample sizes.
[b] n_2 = maximum of the two sample sizes.
[c] T_l = lower critical value for the rank sum in the first sample.
[d] T_r = upper critical value for the rank sum in the first sample.

TABLE 12 Two-tailed critical values for the Wilcoxon rank-sum test *(continued)*

	$\alpha = .02$						$\alpha = .01$					
	n_1[a]						n_1					
n_2[b]	4	5	6	7	8	9	4	5	6	7	8	9
	T_l[c] T_r[d]	T_l T_r	T_l T_r	T_l T_r	T_l T_r	T_l T_r	T_l T_r	T_l T_r	T_l T_r	T_l T_r	T_l T_r	T_l T_r
4	— —	15– 35	22– 44	29– 55	38– 66	48– 78	— —	— —	21– 45	28– 56	37– 67	46– 80
5	10– 30	16– 39	23– 49	31– 60	40– 72	50– 85	— —	15– 40	22– 50	29– 62	38– 74	48– 87
6	11– 33	17– 43	24– 54	32– 66	42– 78	52– 92	10– 34	16– 44	23– 55	31– 67	40– 80	50– 94
7	11– 37	18– 47	25– 59	34– 71	43– 85	54– 99	10– 38	16– 49	24– 60	32– 73	42– 86	52–101
8	12– 40	19– 51	27– 63	35– 77	45– 91	56–106	11– 41	17– 53	25– 65	34– 78	43– 93	54–108
9	13– 43	20– 55	28– 68	37– 82	47– 97	59–112	11– 45	18– 57	26– 70	35– 84	45– 99	56–115
10	13– 47	21– 59	29– 73	39– 87	49–103	61–119	12– 48	19– 61	27– 75	37– 89	47–105	58–122
11	14– 50	22– 63	30– 78	40– 93	51–109	63–126	12– 52	20– 65	28– 80	38– 95	49–111	61–128
12	15– 53	23– 67	32– 82	42– 98	53–115	66–132	13– 55	21– 69	30– 84	40–100	51–117	63–135
13	15– 57	24– 71	33– 87	44–103	56–120	68–139	13– 59	22– 73	31– 89	41–106	53–123	65–142
14	16– 60	25– 75	34– 92	45–109	58–126	71–145	14– 62	22– 78	32– 94	43–111	54–130	67–149
15	17– 63	26– 79	36– 96	47–114	60–132	73–152	15– 65	23– 82	33– 99	44–117	56–136	69–156
16	17– 67	27– 83	37–101	49–119	62–138	76–158	15– 69	24– 86	34–104	46–122	58–142	72–162
17	18– 70	28– 87	39–105	51–124	64–144	78–165	16– 72	25– 90	36–108	47–128	60–148	74–169
18	19– 73	29– 91	40–110	52–130	66–150	81–171	16– 76	26– 94	37–113	49–133	62–154	76–176
19	19– 77	30– 95	41–115	54–135	68–156	83–178	17– 79	27– 98	38–118	50–139	64–160	78–183
20	20– 80	31– 99	43–119	56–140	70–162	85–185	18– 82	28–102	39–123	52–144	66–166	81–189
21	21– 83	32–103	44–124	58–145	72–168	88–191	18– 86	29–106	40–128	53–150	68–172	83 196
22	21– 87	33–107	45–129	59–151	74–174	90–198	19– 89	29–111	42–132	55–155	70–178	85–203
23	22– 90	34 111	47–133	61–156	76–180	93–204	19– 93	30–115	43–137	57–160	71–185	88–209
24	23– 93	35 115	48 138	63–161	78–186	95–211	20– 96	31–119	44–142	58–166	73–191	90–216
25	23– 97	36 119	50 142	64–167	81–191	98–217	20–100	32–123	45–147	60–171	75–197	92–223
26	24–100	37–123	51–147	66–172	83–197	100–224	21–103	33–127	46–152	61–177	77–203	94–230
27	25–103	38–127	52–152	68–177	85–203	103–230	22–106	34–131	48–156	63–182	79–209	97–236
28	26–106	39–131	54–156	70–182	87–209	105–237	22–110	35–135	49–161	64–188	81–215	99–243
29	26–110	40–135	55–161	71–188	89–215	108–243	23–113	36–139	50–166	66–193	83–221	101–250
30	27–113	41–139	56–166	73–193	91–221	110–250	23–117	37–143	51–171	68–198	85–227	103–257
31	28–116	42–143	58–170	75–198	93–227	112–257	24–120	37–148	53–175	68–204	87–233	106–263
32	28–120	43–147	59–175	77–203	95–233	115–263	24–124	38–152	54–180	71–209	89–239	108–270
33	29–123	44–151	61–179	78–209	97–239	117–270	25–127	39–156	55–185	72–215	90–246	110–277
34	30–126	45–155	62–184	79–215	99–245	120–276	26–130	40–160	56–190	73–221	92–252	112–284
35	30–130	46–159	63–189	81–220	101–251	122–283	26–134	41–164	57–195	75–226	94–258	114–291
36	31–133	47–163	65–193	83–225	103–257	125–289	27–137	42–168	58–200	76–232	96–264	117–297
37	32–136	48–167	66–198	84–231	105–263	127–296	28–140	43–172	60–204	78–237	98–270	119–304
38	32–140	49–171	67–203	86–236	107–269	129–303	28–144	44–176	61–209	79–243	100–276	121–311
39	33–143	50–175	69–207	88–241	109–275	132–309	29–147	45–180	62–214	81–248	102–282	123–318
40	34–146	51–179	70–212	90–246	111–281	134–316	29–151	46–184	63–219	82–254	103–289	126–324
41	34–150	52–183	72–216	91–252	113–287	137–322	30–154	46–189	65–223	84–259	105–295	128–331
42	35–153	53–187	73–221	93–257	116–292	139–329	31–157	47–193	66–228	85–265	107–301	130–338
43	35–157	54–191	74–226	95–262	118–298	142–335	31–161	48–197	67–233	87–270	109–307	133–344
44	36–160	55–195	76–230	97–267	120–304	144–342	32–164	49–201	68–238	88–276	111–313	135–351
45	37–163	56–199	77–235	98–273	122–310	147–348	32–168	50–205	69–243	90–281	113–319	137–358
46	37–167	57–203	78–240	100–278	124–316	149–355	33–171	51–209	71–247	91–287	115–325	139–365
47	38–170	58–207	80–244	102–283	126–322	152–361	34–174	52–213	72–252	93–292	117–331	142–371
48	39–173	59–211	81–249	103–289	128–328	154–368	34–178	53–217	73–257	95–297	118–338	144–378
49	39–177	60–215	82–254	105–294	130–334	157–374	35–181	54–221	74–262	96–303	120–344	146–385
50	40–180	61–219	84–258	107–299	132–340	159–381	36–184	55–225	76–266	98–308	122–350	148–392

Source: The data of this table are from *Documenta Geigy Scientific Tables,* 6th edition. Reprinted with the kind permission of CIBA–GEIGY Limited, Basel, Switzerland.

TABLE 13 Fisher's z transformation

r	z	r	z	r	z	r	z	r	z
.00	.000								
.01	.010	.21	.213	.41	.436	.61	.709	.81	1.127
.02	.020	.22	.224	.42	.448	.62	.725	.82	1.157
.03	.030	.23	.234	.43	.460	.63	.741	.83	1.188
.04	.040	.24	.245	.44	.472	.64	.758	.84	1.221
.05	.050	.25	.255	.45	.485	.65	.775	.85	1.256
.06	.060	.26	.266	.46	.497	.66	.793	.86	1.293
.07	.070	.27	.277	.47	.510	.67	.811	.87	1.333
.08	.080	.28	.288	.48	.523	.68	.829	.88	1.376
.09	.090	.29	.299	.49	.536	.69	.848	.89	1.422
.10	.100	.30	.310	.50	.549	.70	.867	.90	1.472
.11	.110	.31	.321	.51	.563	.71	.887	.91	1.528
.12	.121	.32	.332	.52	.576	.72	.908	.92	1.589
.13	.131	.33	.343	.53	.590	.73	.929	.93	1.658
.14	.141	.34	.354	.54	.604	.74	.950	.94	1.738
.15	.151	.35	.365	.55	.618	.75	.973	.95	1.832
.16	.161	.36	.377	.56	.633	.76	.996	.96	1.946
.17	.172	.37	.388	.57	.648	.77	1.020	.97	2.092
.18	.182	.38	.400	.58	.662	.78	1.045	.98	2.298
.19	.192	.39	.412	.59	.678	.79	1.071	.99	2.647
.20	.203	.40	.424	.60	.693	.80	1.099		

TABLE 14 **Two-tailed upper critical values for the Spearman rank-correlation coefficient (r_s)**

	α			
n	.10	.05	.02	.01
1	—	—	—	—
2	—	—	—	—
3	—	—	—	—
4	1.0	—	—	—
5	.900	1.0	1.0	—
6	.829	.886	.943	1.0
7	.714	.786	.893	.929
8	.643	.738	.833	.881
9	.600	.683	.783	.833

Source: The data for this table have been adapted with permission from E. G. Olds (1938), "Distributions of Sums of Squares of Rank Differences for Small Numbers of Individuals," *Annals of Mathematical Statistics, 9,* 133–148.

TABLE 15 **Critical values for the Kruskal-Wallis test statistic (H) for selected sample sizes for $k = 3$**

n_1	n_2	n_3	α .10	α .05	α .02	α .01
1	1	2	—	—	—	—
1	1	3	—	—	—	—
1	1	4	—	—	—	—
1	1	5	—	—	—	—
1	2	2	—	—	—	—
1	2	3	4.286	—	—	—
1	2	4	4.500	—	—	—
1	2	5	4.200	5.000	—	—
1	3	3	4.571	5.143	—	—
1	3	4	4.056	5.389	—	—
1	3	5	4.018	4.960	6.400	—
1	4	4	4.167	4.967	6.667	—
1	4	5	3.987	4.986	6.431	6.954
1	5	5	4.109	5.127	6.146	7.309
2	2	2	4.571	—	—	—
2	2	3	4.500	4.714	—	—
2	2	4	4.500	5.333	6.000	—
2	2	5	4.373	5.160	6.000	6.533
2	3	3	4.694	5.361	6.250	—
2	3	4	4.511	5.444	6.144	6.444
2	3	5	4.651	5.251	6.294	6.909
2	4	4	4.554	5.454	6.600	7.036
2	4	5	4.541	5.273	6.541	7.204
2	5	5	4.623	5.338	6.469	7.392
3	3	3	5.067	5.689	6.489	7.200
3	3	4	4.709	5.791	6.564	7.000
3	3	5	4.533	5.648	6.533	7.079
3	4	4	4.546	5.598	6.712	7.212
3	4	5	4.549	5.656	6.703	7.477
3	5	5	4.571	5.706	6.866	7.622
4	4	4	4.654	5.692	6.962	7.654
4	4	5	4.668	5.657	6.976	7.760
4	5	5	4.523	5.666	7.000	7.903
5	5	5	4.580	5.780	7.220	8.000

Source: The data for this table have been adapted from Table F of *A Nonparametric Introduction to Statistics* by C. H. Kraft and C. Van Eeden, Macmillan, New York, 1968, with the permission of the publisher and the authors.

Answers to Selected Problems

CHAPTER 2

2.4–2.7 The median, mode, geometric mean, and range are each multiplied by *c*. **2.11** $\bar{x}$ = 19.54 mg/dL **2.12** s = 16.81 mg/dL **2.14** Median = 19 mg/dL

CHAPTER 3

3.1 At least one parent has influenza. **3.2** Both parents have influenza. **3.3** No **3.4** At least one child has influenza. **3.5** The 1st child has influenza. **3.6** $C = A_1 \cup A_2$ **3.7** $D = B \cup C$ **3.8** The mother does not have influenza. **3.9** The father does not have influenza. **3.10** $\overline{A}_1 \cap \overline{A}_2$ **3.11** $\overline{B} \cap \overline{C}$ **3.29** .0167 **3.30** 180 **3.49** .069 **3.51** .541 **3.55** .20 **3.56** .5. This is a conditional probability. The probability in Problem 3.55 is a joint unconditional probability. **3.57** .20 **3.58** No, because $Pr(M|F) = .6 \neq Pr(M|\overline{F}) = .2$ **3.59** .084 **3.60** .655 **3.61** .690 **3.62** .486 **3.63** .373 **3.64** No. **3.65** No. **3.68** .05 **3.69** .326 **3.70** .652 **3.71** .967 **3.72** .479 **3.73** .893 **3.74** .975 **3.75** .630. It is lower than the predictive value negative based on self-report (i.e., .893). **3.94** .95 **3.95** .99 **3.96** .913 **3.97** The new test has a 13.6% lower cost.

CHAPTER 4

4.1 $Pr(0) = .72$, $Pr(1) = .26$, $Pr(2) = .02$ **4.2** .30 **4.3** .25 **4.4** $F(x) = 0$ if $x < 0$; $F(x) = .72$ if $0 \le x < 1$; $F(x) = .98$ if $1 \le x < 2$; $F(x) = 1.0$ if $x \ge 2$. **4.8** 362,880 **4.11** .1042 **4.12** .2148 **4.13** $E(X) = Var(X) = 4.0$ **4.23** $Pr(X \ge 6)$ = .010 **4.24** $Pr(X \ge 4) = .242$ **4.26** .62 **4.27** .202 **4.28** .385 **4.29** .471 **4.30** .144 **4.31** 1.24 **4.34** $Pr(X \ge 4) = .241 > .05$. Thus, there is no excess risk of malformations. **4.35** $Pr(X \ge 8) = .0006 < .05$. Thus, there is an excess risk of malformations. **4.39** .23 **4.40** .882 **4.41** $Pr(X = 0) = .91$, $Pr(X = 1) = .08$, $Pr(X = 2) = .01$ **4.42** 0.100 **4.43** 0.110 **4.53** 6 months, .25; 1 year, .52 **4.54** .435 **4.55** .104 **4.56** 10.4 **4.63** Based on the Poisson distribution, $Pr(X \ge 27) = .049 < .05$. Thus, there is a significant excess. **4.64** .0263 **4.65** If Y = number of cases of cleft palate, then based on the Poisson distribution, $Pr(Y \ge 12) = .0532 > .05$. This is a borderline result, since this probability is close to .05.

CHAPTER 5

5.1 .6915 **5.2** .3085 **5.3** .7745 **5.4** .0228 **5.5** .0441 **5.17** .079 **5.18** .0004 **5.20** .352 **5.21** .268 **5.22** .380 **5.27** .023 **5.28** .067 **5.29** .168 **5.30** .061 **5.35** .018 **5.36** .123 **5.37** .0005 **5.38** ≥43 **5.39** ≥69 **5.40** ≥72 **5.46** .851 **5.47** Sensitivity. **5.48** .941 **5.49** Specificity. **5.50** Δ = 0.2375 mg/dL, compliance = 88% in each group **5.58** .635 **5.59** .323 **5.60** No. The distributions are very skewed.

CHAPTER 6

6.5 0.079 for normal men, 0.071 for men with chronic airflow limitation. **6.18** .44 **6.19** .099 **6.20** (.25, .63)

6.24

	Point Estimate	95% CI
E. coli	25.53	(24.16, 26.90)
S. aureus	26.79	(24.88, 28.70)
P. aeruginosa	19.93	(18.60, 21.27)

6.25

	Point Estimate	95% CI
E. coli	25.06	(23.73, 26.38)
S. aureus	25.44	(24.60, 26.29)
P. aeruginosa	17.89	(17.09, 18.69)

6.26

	Point Estimate	95% CI
E. coli	1.78	(1.21, 3.42)
S. aureus	2.49	(1.68, 4.77)
P. aeruginosa	1.74	(1.17, 3.32)

6.27

	Point Estimate	95% CI
E. coli	1.73	(1.17, 3.31)
S. aureus	1.10	(0.75, 2.12)
P. aeruginosa	1.04	(0.70, 1.99)

6.35 6/46 = .130 **6.36** (.033, .228) **6.37** Since 10% is within the 95% CI, the two drugs are equally effective. **6.41** (6.17, 7.83) **6.42** (2.11, 9.71) **6.43** $n \doteq 246$ **6.47** 0.544 **6.48** (0.26, 1.81) **6.49** .958 **6.50** .999 **6.63** .615 **6.64** .918 **6.65** 0.5 lb, observed proportion = .615. 1 lb, observed proportion = .935. The observed and expected proportions are in good agreement. **6.66** Yes. **6.86** 95% CI = (2.20, 13.06). Because this interval does not include 1.8, there are an excess number of cases of bladder cancer among tire workers. **6.87** 95% CI = (1.09, 10.24). Because this interval includes 2.5, there are not an excess number of cases of stomach cancer among tire workers.

CHAPTER 7

7.1 $z = 1.732$, accept H_0 at the 5% level. **7.2** $p = .083$ **7.4** $t = 1.155 \sim t_{11}$, $.2 < p < .3$ **7.8** (0.82, 1.58) **7.9** The 95% CI contains 1.0, which is consistent with our decision to accept H_0 at the 5% level. **7.15** $z = 1.142$, accept H_0 at the 5% level. **7.16** $p = .25$ **7.17** Accept H_0 at the 5% level. **7.18** $p = .71$ **7.23** $z = 7.72$, $p < .001$ **7.32** 31 **7.33** .770 **7.39** H_0: $\mu = \mu_0$ vs. H_1: $\mu \neq \mu_0$. σ^2 unknown. μ = true mean daily iron intake for 9- to 11-year-old boys below the poverty level, μ_0 = true mean daily iron intake for 9- to 11-year-old boys in the general population. **7.40** $t = -2.917 \sim t_{50}$, reject H_0 at the 5% level. **7.41** $.001 < p < .01$ (exact p-value = .005) **7.42** H_0: $\sigma^2 = \sigma_0^2$ vs. H_1: $\sigma^2 \neq \sigma_0^2$. σ^2 = true variance in the low-income population, σ_0^2 = true variance in the general population. **7.43** $X^2 = 36.49 \sim \chi^2_{50}$, accept H_0 at the 5% level. **7.44** $.1 < p < .2$ (exact p-value = .15) **7.45** (15.80, 34.86). Since the interval contains $\sigma_0^2 = 5.56^2 = 30.91$, the underlying variances of the low-income and the general population are not significantly different. **7.57** One-sample binomial test, exact method. **7.58** $p = .28$ **7.59** One-sample binomial test, large-sample method. $z = 3.24$, $p = .0012$ **7.60** (.058, .142)

CHAPTER 8

8.15 135 girls in each group or a total of 270 overall. **8.16** 106 girls in each group or a total of 212 overall. **8.17** 96 girls in the below-poverty group, 192 girls in the above-poverty group. **8.18** Power = .401 **8.19** Power = .525 **8.20** Power = .300 **8.21** Power = .417 **8.25** Use the paired t test. $t = -3.37 \sim t_9$, $.001 < p < .01$ (exact p-value = .008) **8.26** Use the paired t test. $t = -1.83 \sim t_{29}$, $.05 < p < .10$ (exact p-value = .078)

8.27

Group	95% CI
Methazolamide and topical glaucoma medications	(–2.67, –0.53)
Topical drugs only	(–1.48, 0.08)

8.28 Use the two-sample t test with equal variances. $t = -1.25 \sim t_{38}$, $p > .05$ (exact p-value = .22). **8.31** H_0: $\mu_d = 0$ vs. H_1: $\mu_d \neq 0$. μ_d = mean difference in one-hour concentration (drug A – drug B) in a specific person. **8.32** Use a paired t test to test these hypotheses. **8.33** Use the paired t test. $t = 3.67 \sim t_9$, $.001 < p < .01$ (exact p-value = .005) **8.34** 3.60 mg % **8.35** (1.38, 5.82) mg % **8.59** H_0: $\mu_1 = \mu_2$ vs. H_1: $\mu_1 \neq \mu_2$, where μ_1 = true mean FEV of children both of whose parents smoke, μ_2 = true mean FEV of children neither of whose parents smoke. **8.60** First, perform F test for equality of two variances, $F = 3.06 \sim F_{22, 19}$, $p < .05$. Therefore, use the two-sample t test with unequal variances. **8.61** $t = -1.17 \sim t_{35}$, accept H_0 at the 5% level. **8.62** (–0.55, 0.15) **8.63** 212 children in each group. **8.64** 176 children in each group. **8.65** .363 **8.66** .486 **8.72** The paired t test. **8.73** Raw scale, $t = 3.49 \sim t_9$, $.001 < p < .01$ (exact p-value = .007), log scale, $t = 3.74 \sim t_9$, $.001 < p < .01$ (exact p-value = .005). The log scale is preferable since the change in the raw scale seems to be related to the initial level. **8.74** Urinary protein has declined by 56.7% over 8 weeks. **8.75** 95% CI for 8-week decline = (28.2%, 73.9%) **8.81** Two-sample t test with equal variances **8.82** H_0: $\mu_1 = \mu_2$ vs. H_1: $\mu_1 \neq \mu_2$; μ_1 = mean cholesterol level for men; μ_2 = mean cholesterol level for women; $t = -1.92 \sim t_{90}$, $.05 < p < .10$ (exact p-value = .058) **8.83** H_0: $\mu_1 = \mu_2$ vs. H_1: $\mu_1 > \mu_2$; $t = -1.92 \sim t_{90}$, $.95 < p < .975$ (exact p-value = .97) **8.84** No. The twin pairs are not independent observations. **8.88** F test for the equality of two variances, $F = 1.15 \sim F_{35, 29}$, $p > .05$. Therefore, use the two-sample t test with equal variances. **8.89** $t = 1.25 \sim t_{64}$, $.2 < p < .3$ (exact p-value = .22). **8.90** RA, .32; OA, .43 **8.91** 133 subjects in each of the RA and OA groups. **8.97** Paired t test. **8.98** $t = 2.27 \sim t_{99}$, $.02 < p < .05$ (exact p-value = .025) **8.99** F test for the equality of two variances, $F = 1.99 \sim F_{98, 99}$, $p < .05$. Use the two-sample t test with unequal variances. **8.100** $t = -4.20 \sim t_{176}$, $p < .001$

CHAPTER 9

9.1 Use the sign test. The critical values are $c_1 = 6.3$ and $c_2 = 16.7$. Since $c_1 < C < c_2$, where C = number of patients who improved = 15, we accept H_0 at the 5% level. **9.9** The distribution of length of stay is very skewed and far from being normal, which makes the t test not very useful here. **9.10** Use the Wilcoxon rank-sum test (large-sample test). $R_1 = 83.5$, $T = 3.10 \sim N(0,1)$, $p = .002$ **9.15** H_0: $\text{med}_1 = \text{med}_2$ vs. H_1: $\text{med}_1 < \text{med}_2$, where med_1 = median duration of effusion for breast-fed babies, med_2 = median duration of effusion for bottle-fed babies. **9.16** The distribution of duration of effusion is very skewed and far from being normal. **9.17** Wilcoxon signed-rank test (large-sample test). **9.18** $R_1 = 215$, $T = 2.33 \sim N(0, 1)$, $p = .010$ (one-tail). Breast-fed babies have a shorter duration of effusion than bottle-fed babies. **9.24** The Wilcoxon signed-rank test (large-sample test). **9.25** $R_1 = 33.5$, $T = 1.76 \sim N(0, 1)$, $p = .078$. The mean SBP is slightly, but not significantly, higher with the standard cuff. **9.26** The Wilcoxon signed-rank test (large-sample test). **9.27** $R_1 = 32.0$, $T = 1.86 \sim N(0, 1)$, $p = .062$. Variability with the standard cuff is slightly, but not significantly, lower than with the random zero.

CHAPTER 10

10.8 McNemar's test for correlated proportions. **10.9** $X^2 = 4.76 \sim \chi_1^2$, $.025 < p < .05$ **10.10** 87 **10.11** 13 **10.12** McNemar's test for correlated proportions, exact test; $p = .267$ **10.18** Use chi-square test for 2×2 tables. $X^2 = 32.17 \sim \chi_1^2$, $p < .001$ **10.20** Use chi-square test for $R \times C$ tables. $X^2 = 117.02 \sim \chi_2^2$, $p < .001$. There is a significant association between ethnic origin and genetic type. **10.25** Two-sample test. **10.26** Two-sided test. **10.27** Chi-square test for 2×2 tables. **10.28** $X^2 = 3.48 \sim \chi_1^2$, $.05 < p < .10$ **10.35** Chi-square test for trend in binomial proportions **10.36** Use scores of 1, 2, . . . , 5 for the age groups < 45, 45–54, . . . , 75+. $X^2 = 35.09 \sim \chi_1^2$, $p < .001$ **10.39** McNemar's test for correlated proportions. **10.40** $X^2 = 4.65 \sim \chi_1^2$, $.025 < p < .05$ **10.45** McNemar's test for correlated proportions (large-sample test). **10.46** $X^2 = 6.48 \sim \chi_1^2$, $.01 < p < .025$ **10.47** McNemar's test for correlated proportions (exact method). **10.48** $p = .387$ **10.49** .9997 **10.57** .304 **10.58** .213 **10.59** 284 subjects in each group. **10.60** Cholesterol-lowering drug patients, .218; placebo pill patients, .295 **10.61** 390 subjects in each group. **10.63** 12, .273; 13, .333; 14, .303; 15, .091; 16, 0; 17, 0 **10.64** 15.13 years ≐ 15 years, 2 months **10.65** We use the age groups, ≤12.9, 13.0–13.9, 14.0–14.9, ≥15.0, and perform the chi-square goodness-of-fit test. $X^2 = 1.27 \sim \chi_1^2$, $.25 < p < .50$. The goodness of fit of the normal model is adequate.

CHAPTER 11

11.1 $y = 1894.8 + 112.1x$ **11.2** $F = 180{,}750/490{,}818 = 0.37 \sim F_{1,7}$, $p > .05$ **11.3** .05 **11.4** R^2 = % variance of lymphocyte count that is explained by % reticulytes (≈ 5%). **11.5** 490,818 **11.6** $t = 0.61 \sim t_7$, $p > .05$ **11.7** $se(b) = 184.7$, $se(a) = 348.5$ **11.9** Two-sample z test to compare two correlation coefficients. **11.10** $\lambda = 3.40 \sim N(0, 1)$, reject H_0 at the 5% level. **11.11** $p < .001$ **11.12** Use the two-sample z test to compare two correlation coefficients. $\lambda = 3.61$, $p < .001$. The correlation coefficients are significantly different. **11.21** $y = 1472.0 - 0.737x$, where y = infant mortality rate, x = year. **11.22** Use F test for simple linear regression. $F = 182.04/0.329 = 553.9 \sim F_{1,10}$, $p < .001$ **11.23** 5.7 per 1000 livebirths. **11.24** 0.79 per 1000 livebirths. **11.25** No. If the linear relationship persisted, the expected mortality rate would eventually be projected as negative, which is impossible. **11.49** The one-sample t test for correlation **11.50** $t = 8.75 \sim t_{901}$, $p < .001$ **11.51** The two-sample z test for correlation **11.52** $\lambda = 2.581 \sim N(0, 1)$, $p = .010$ **11.53** White boys (.219, .339); black boys (.051, .227)

CHAPTER 12

12.1 $F = 1643.08/160.65 = 10.23 \sim F_{2,23}$, $p < .05$. The means of the three groups are significantly different. **12.2** $p < .001$

12.3

Groups	Test statistic	p-value
STD, LAC	$t = 3.18 \sim t_{23}$	$.001 < p < .01$
STD, VEG	$t = 4.28 \sim t_{23}$	$p < .001$
LAC, VEG	$t = 1.53 \sim t_{23}$	NS

12.4 $t = -4.09 \sim t_{23}$, $p < .001$. The contrast is an estimate of the difference in mean protein intake between the general vegetarian population and the general nonvegetarian population. **12.6** $F = 251.77/50.46 = 4.99 \sim F_{2,19}$, $p < .05$

12.7

Groups	Test statistic	p-value
A, B	$t = 2.67 \sim t_{19}$	$.01 < p < .02$
A, C	$t = 2.94 \sim t_{19}$	$.001 < p < .01$
B, C	$t = 0.82 \sim t_{19}$	NS

12.8 A,B $p < .05$. A,C $p < .05$. B,C NS. **12.21** $F = 2.915/0.429 = 6.79 \sim F_{2,12}$, $.01 < p < .025$ **12.22** Bonferroni critical value = 2.78. A vs. B, $t = 2.80$, $p < .05$; A vs. C, $t = 3.48$, $p < .05$; B vs. C, $t = 0.68$, p = NS. **12.23** $\mu_A - \mu_B$, (0.26, 2.06); $\mu_A - \mu_C$, (0.54, 2.34); $\mu_B - \mu_C$, (–0.62, 1.18) **12.28** Between-day variance = $\hat{\sigma}_A^2 = 1.19$, within-day variance = $\hat{\sigma}^2 = 14.50$ **12.29** $F = 16.89/14.50 = 1.16 \sim F_{9,10}$, $p > .05$. There is no significant between-day variance.

CHAPTER 13

13.1 $z = 6.27$, $p < .001$ **13.2** $X^2 = 38.34 \sim \chi_1^2$, p < .001 **13.3** The conclusions are the same. Also, $z^2 = X^2_{\text{uncorrected}} = 39.35$. **13.4** (.090, .201) **13.5** 2.29 **13.6** (1.76, 2.98) **13.23** The Mantel-Haenszel test **13.24** $X^2_{MH} = 0.51 \sim \chi_1^2$, $p > .05$ **13.25** 1.38 **13.26** (0.68, 2.82) **13.37** 1.40 **13.38** (1.09, 1.80) **13.45** $RR = 1.70$, 95% CI = (0.42, 6.93) **13.46** $RR = 1.38$, 95% CI = (1.06, 1.79) **13.47** 1965 cohort, .145; 1974 cohort, 0.65; Standardized risk ratio = 0.45 **13.48** 1965 cohort, .049; 1974 cohort, .040. Standardized risk ratio = 0.81 **13.49** 0.33 **13.50** 0.78 **13.51** Yes. The odds ratio is lower for those with heart trouble than for those without heart trouble. One possible explanation could be better surveillance of those with heart trouble in 1974 versus 1965 to prevent a fatal coronary event.

CHAPTER 14

14.1 Incidence density = 273.1 cases per 10^5 person-years for current users, 115.2 cases per 10^5 person-years for never users, $z = 6.67/2.359 = 2.827 \sim N(0, 1)$, $p = .005$. There is a significant excess of breast cancer among current OC users vs. never OC users. **14.2** Incidence density = 135.4 cases per 10^5 person-years for past users, 115.2 cases per 10^5

person-years for never users. $z = 10.47/8.276 = 1.265 \sim N(0, 1)$, $p = .21$. There is no significant excess (or deficit) of breast cancer among past OC users vs. never OC users. **14.3** $\widehat{RR} = 2.37$, 95% CI = (1.34, 4.21) **14.4** $\widehat{RR} = 1.18$, 95% CI = (0.93, 1.49) **14.22** .072 **14.23** 12 **14.24** (2.1 events per 100 person-years, 7.0 events per 100 person-years)

Flowchart: Methods of Statistical Inference

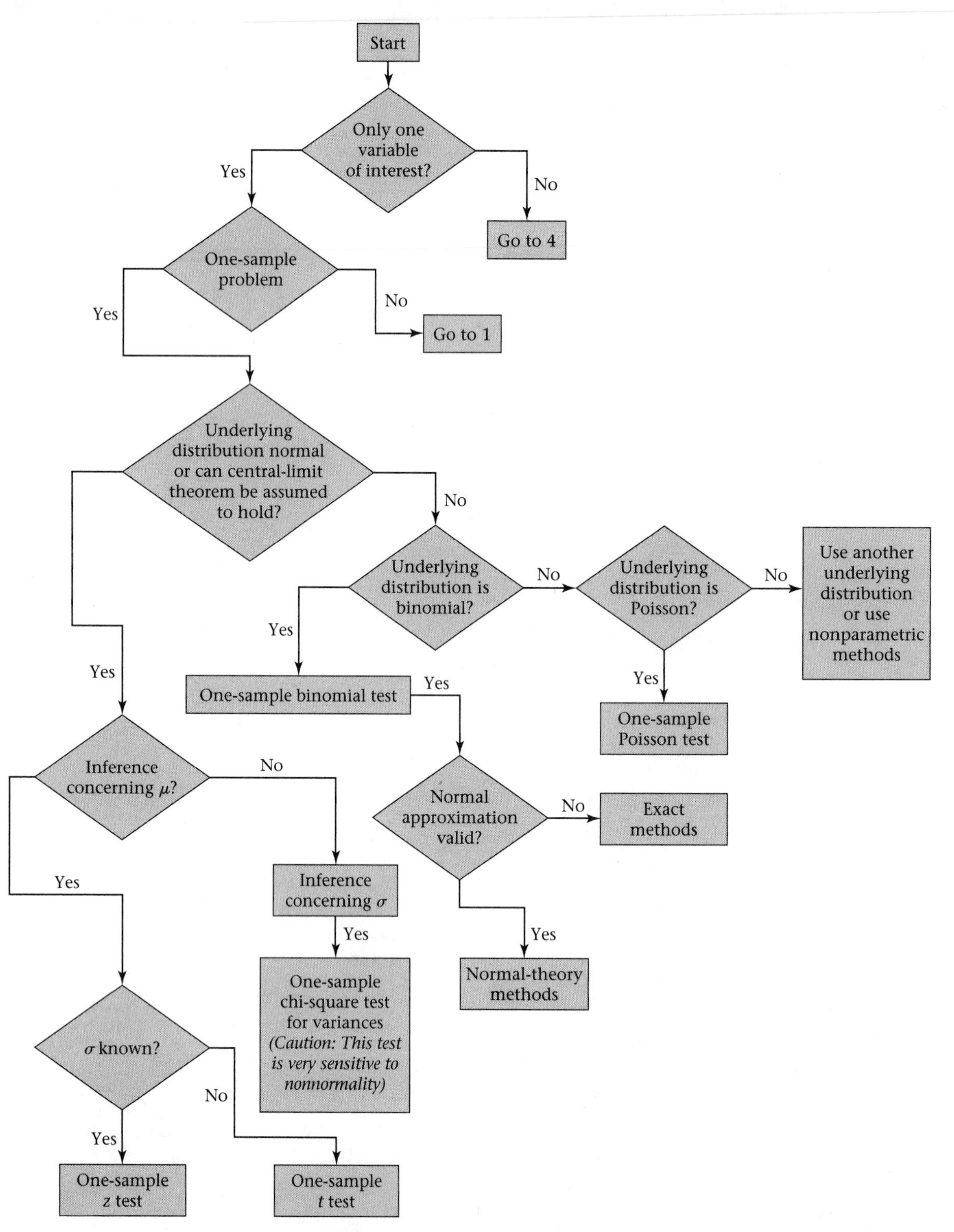

Start
Only one variable of interest?
Yes
No
Go to 4
One-sample problem
Yes
No
Go to 1
Underlying distribution normal or can central-limit theorem be assumed to hold?
No
Underlying distribution is binomial?
No
Underlying distribution is Poisson?
No
Use another underlying distribution or use nonparametric methods
Yes
Yes
Yes
One-sample binomial test
Yes
One-sample Poisson test
Inference concerning μ?
No
Normal approximation valid?
No
Exact methods
Yes
Inference concerning σ
Yes
Yes
One-sample chi-square test for variances (Caution: This test is very sensitive to nonnormality)
Normal-theory methods
σ known?
No
Yes
One-sample z test
One-sample t test

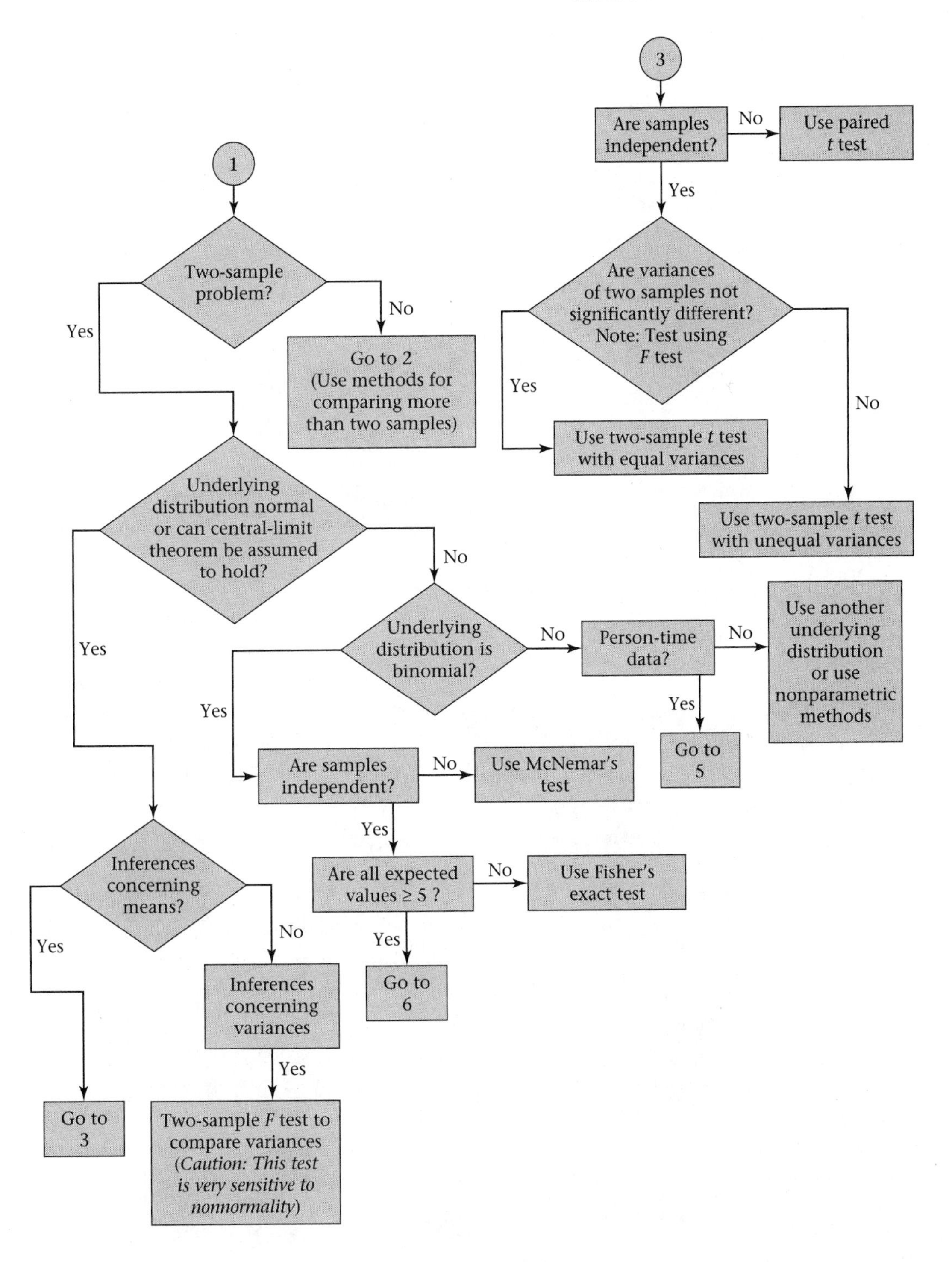

3
Are samples independent?
No
Use paired t test
Yes
Are variances of two samples not significantly different? Note: Test using F test
Yes
No
Use two-sample t test with equal variances
Use two-sample t test with unequal variances
1
Two-sample problem?
Yes
No
Go to 2 (Use methods for comparing more than two samples)
Underlying distribution normal or can central-limit theorem be assumed to hold?
Yes
No
Underlying distribution is binomial?
No
Person-time data?
No
Use another underlying distribution or use nonparametric methods
Yes
Go to 5
Yes
Are samples independent?
No
Use McNemar's test
Yes
Are all expected values ≥ 5 ?
No
Use Fisher's exact test
Yes
Go to 6
Inferences concerning means?
Yes
No
Inferences concerning variances
Yes
Go to 3
Two-sample F test to compare variances (Caution: This test is very sensitive to nonnormality)

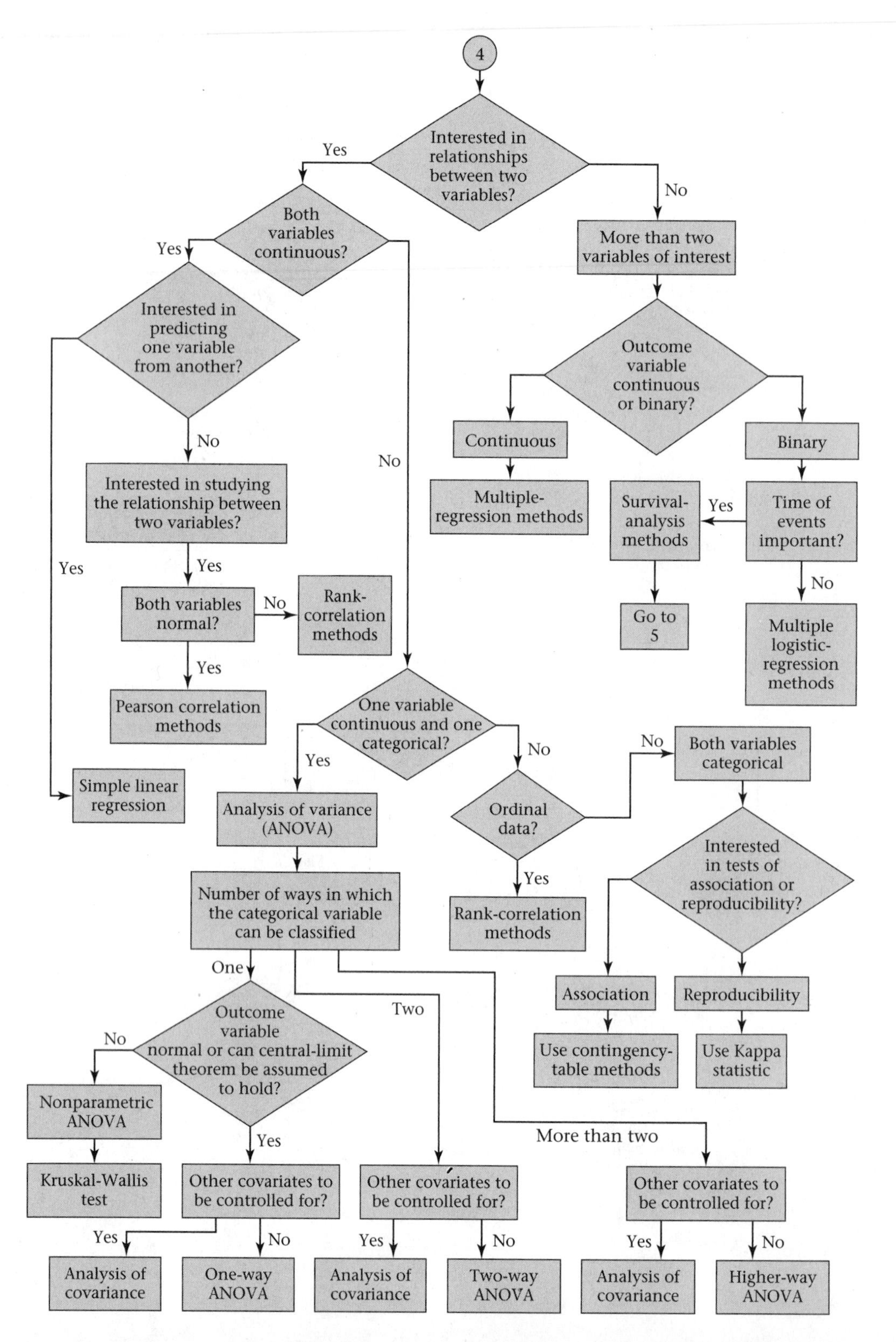
4
Interested in relationships between two variables?
Yes
No
Both variables continuous?
More than two variables of interest
Yes
Interested in predicting one variable from another?
Outcome variable continuous or binary?
Continuous
Binary
No
No
Interested in studying the relationship between two variables?
Multiple-regression methods
Survival-analysis methods
Yes
Time of events important?
Yes
Yes
No
Both variables normal?
No
Rank-correlation methods
Go to 5
Multiple logistic-regression methods
Yes
Pearson correlation methods
One variable continuous and one categorical?
No
No
Both variables categorical
Yes
Simple linear regression
Analysis of variance (ANOVA)
Ordinal data?
Interested in tests of association or reproducibility?
Yes
Number of ways in which the categorical variable can be classified
Rank-correlation methods
One
Association
Reproducibility
Two
Outcome variable normal or can central-limit theorem be assumed to hold?
No
Use contingency-table methods
Use Kappa statistic
Nonparametric ANOVA
More than two
Yes
Kruskal-Wallis test
Other covariates to be controlled for?
Other covariates to be controlled for?
Other covariates to be controlled for?
Yes
No
Yes
No
Yes
No
Analysis of covariance
One-way ANOVA
Analysis of covariance
Two-way ANOVA
Analysis of covariance
Higher-way ANOVA

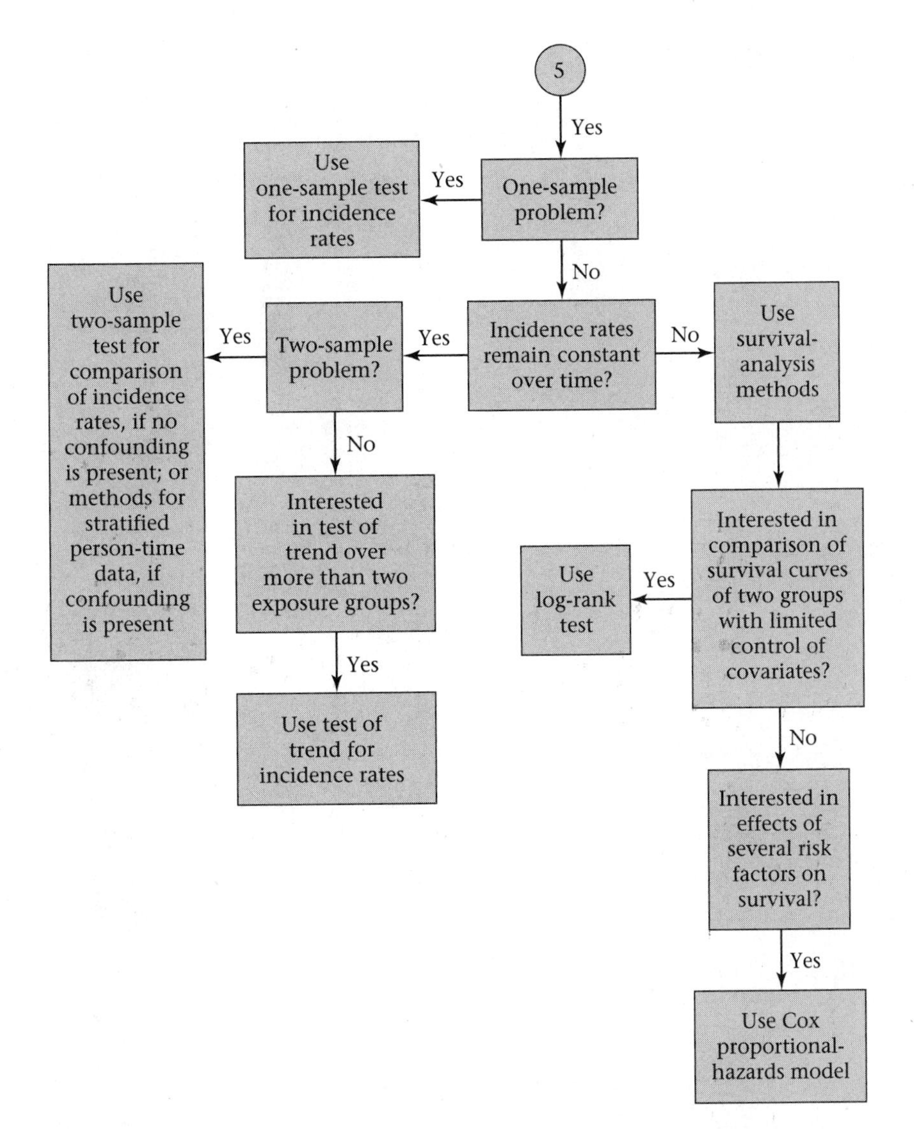
5
Yes
One-sample problem?
Yes
Use one-sample test for incidence rates
No
Incidence rates remain constant over time?
Yes
Two-sample problem?
Yes
Use two-sample test for comparison of incidence rates, if no confounding is present; or methods for stratified person-time data, if confounding is present
No
Interested in test of trend over more than two exposure groups?
Yes
Use test of trend for incidence rates
No
Use survival-analysis methods
Interested in comparison of survival curves of two groups with limited control of covariates?
Yes
Use log-rank test
No
Interested in effects of several risk factors on survival?
Yes
Use Cox proportional-hazards model

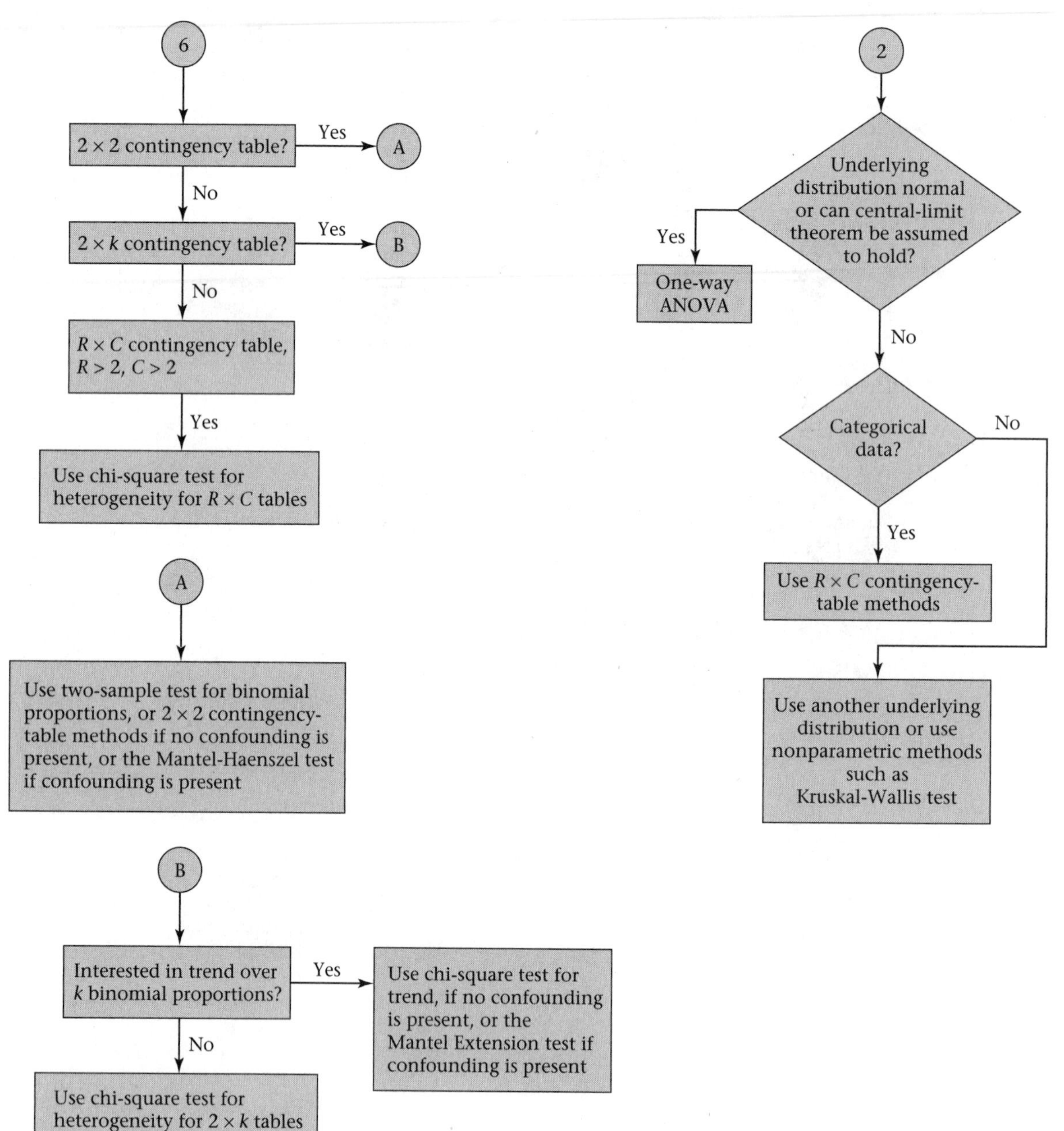
6
2 × 2 contingency table?
Yes
A
No
2 × k contingency table?
Yes
B
No
R × C contingency table, R > 2, C > 2
Yes
Use chi-square test for heterogeneity for R × C tables
A
Use two-sample test for binomial proportions, or 2 × 2 contingency-table methods if no confounding is present, or the Mantel-Haenszel test if confounding is present
B
Interested in trend over k binomial proportions?
Yes
Use chi-square test for trend, if no confounding is present, or the Mantel Extension test if confounding is present
No
Use chi-square test for heterogeneity for 2 × k tables
2
Underlying distribution normal or can central-limit theorem be assumed to hold?
Yes
One-way ANOVA
No
Categorical data?
No
Yes
Use R × C contingency-table methods
Use another underlying distribution or use nonparametric methods such as Kruskal-Wallis test

Index of Data Sets

Index

GYNECOLOGY *(continued)*

HEALTH PROMOTION

HEALTH SERVICES ADMINISTRATION

HEMATOLOGY

HEPATIC DISEASE

HOSPITAL EPIDEMIOLOGY

HYPERTENSION

INFECTIOUS DISEASE

MENTAL HEALTH

MICROBIOLOGY

NUTRITION

NUTRITION *(continued)*

OBSTETRICS

OCCUPATIONAL HEALTH

OPHTHALMOLOGY

INDEX OF APPLICATIONS (continued)